UNITS

Mass
$1 \text{ kg} = 1000 \text{ g} = 0.001 \text{ metric ton} = 2.20462 \text{ lb}_m$
$1 \text{ lb}_m = 16 \text{ oz} = 5\text{E-}4 \text{ ton} = 453.59 \text{ g} = 0.45359 \text{ kg}$

Length
$1 \text{ m} = 100 \text{ cm} = 1000 \text{ mm} = 1\text{E6 } \mu\text{m} = 1\text{E9 nm} = 39.370 \text{ in} = 3.2808 \text{ ft} = 1.0936 \text{ yd}$
$1 \text{ ft} = 12 \text{ in} = 0.30480 \text{ m} = 30.480 \text{ cm}$
$1 \text{ in} = 2.5400 \text{ cm}$

Volume
$1 \text{ m}^3 = 1000 \text{ L} = 1\text{E6 cm}^3 = 1\text{E6 ml} = 35.315 \text{ ft}^3 = 264.17 \text{ gal}$
$1 \text{ ft}^3 = 1728.0 \text{ in}^3 = 7.4805 \text{ gal} = 0.028317 \text{ m}^3 = 28.317 \text{ L} = 28317 \text{ ml}$

Force
$1 \text{ N} = 1 \text{ kg-m/s}^2 = 1\text{E5 dynes} = 1\text{E5 g-cm/s}^2 = 0.22481 \text{ lb}_f$
$1 \text{ lb}_f = 32.174 \text{ lb}_m\text{-ft/s}^2 = 4.4482 \text{ N}$

Pressure
$1 \text{ atm} = 1.01325\text{E5 N/m}^2\text{(Pa)} = 1.01325 \text{ bar} = 760 \text{ mm}_{\text{Hg at 0°C}} = 33.9 \text{ ft}_{\text{H2O at 4°C}}$
$1 \text{ atm} = 14.696 \text{ psia}$
$1 \text{ bar} = 0.1 \text{ MPa} = 0.98692 \text{ atm} = 14.504 \text{ psia} = 750.06 \text{ mm}_{\text{Hg at 0°C}} = 10.197 \text{ m}_{\text{H2O at 4°C}}$

Energy
$1 \text{ J} = 1 \text{ N-m} = 1 \text{ MPa-cm}^3 = 1 \text{ kgm}^2/\text{s}^2 = 0.23901 \text{ cal} = 0.73756 \text{ ft-lb}_f$
$1 \text{ J} = 1\text{E7 ergs} = 1\text{E7 g-cm}^2/\text{s}^2$
$1 \text{ kJ} = 0.94781 \text{ Btu}^{\text{(see note 1)}} = 2.7778\text{E-}4 \text{ kW-h} = 0.23901 \text{ food calorie}$

Power
$1 \text{ W} = 1 \text{ J/s} = 0.2390 \text{ cal/s}^{\text{(see note 2)}} = 0.73756 \text{ ft-lb}_f/\text{s} = 3.4121 \text{ Btu/h}^{\text{(see note 1)}}$
$1 \text{ hp} = 550 \text{ ft-lb}_f/\text{s} = 0.70726 \text{ Btu/s}^{\text{(see note 1)}} = 0.74570 \text{ kW}$

Gas Constant, R
$= 8.31447 \text{ J/mole-K} = 8.31447 \text{ cm}^3\text{-MPa/mole-K} = 8.31447 \text{ m}^3\text{-Pa/mole-K}$
$= 8,314.47 \text{ cm}^3\text{-kPa/mole-K} = 83.1447 \text{ cm}^3\text{-bar/mole-K} = 1.9859 \text{ Btu/lbmole-R}^{\text{(see note 1)}}$
$= 82.057 \text{ cm}^3\text{-atm/mole-K} = 1.9872 \text{ cal/mole-K}^{\text{(see note 2)}} = 10.731 \text{ ft}^3\text{-psia/lbmole-R}$

Gravitational Constants at sea level
$g = 9.8066 \text{ m/s}^2 \quad g/g_c = 9.8066 \text{ N/kg} \quad g_c = 1 \text{ (kg-m/s}^2)/\text{N}$
$g = 32.174 \text{ ft/s}^2 \quad g/g_c = 1 \text{ lb}_f/\text{lb}_m \quad g_c = 32.174 \text{ (lb}_m\text{-ft/s}^2)/\text{lb}_f$

Faraday's Constant
$F = 96,485 \text{ J/V}$

IUPAC Standard Conditions of Temperature and Pressure
$T = 0°C = 273.15 \text{ K}; P = 0.1 \text{ MPa}; V(\text{ideal gas}) = 22711 \text{ cm}^3/\text{mole};$
$\rho(\text{water}) = 0.99984 \text{ g/cm}^3 = 8.3441 \text{ lb}_m/\text{gal}$

---

1. The International Steam Table (IT) BTU.
2. The thermochemical calorie.

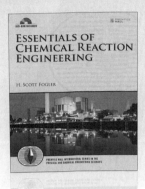

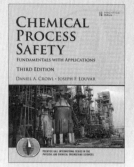

Introductory Chemical Engineering Thermodynamics, Second Edition

Introductory Chemical Engineering Thermodynamics, Second Edition

J. Richard Elliott

Carl T. Lira

PRENTICE
HALL

Upper Saddle River, NJ • Boston • Indianapolis • San Francisco
New York • Toronto • Montreal • London • Munich • Paris • Madrid
Capetown • Sydney • Tokyo • Singapore • Mexico City

Many of the designations used by manufacturers and sellers to distinguish their products are claimed as trademarks. Where those designations appear in this book, and the publisher was aware of a trademark claim, the designations have been printed with initial capital letters or in all capitals.

The authors and publisher have taken care in the preparation of this book, but make no expressed or implied warranty of any kind and assume no responsibility for errors or omissions. No liability is assumed for incidental or consequential damages in connection with or arising out of the use of the information or programs contained herein.

For information about buying this title in bulk quantities, or for special sales opportunities (which may include electronic versions; custom cover designs; and content particular to your business, training goals, marketing focus, or branding interests), please contact our corporate sales department at corpsales@pearsoned.com or (800) 382-3419.

For government sales inquiries, please contact governmentsales@pearsoned.com.

For questions about sales outside the U.S., please contact intlcs@pearson.com.

Visit us on the Web: informit.com/ph

Library of Congress Cataloging-in-Publication Data

Elliott, J. Richard.
  Introductory chemical engineering thermodynamics / J. Richard Elliott, Carl T. Lira.—2nd ed.
    p.   cm.
  Includes index.
  ISBN 978-0-13-606854-9 (hardcover : alk. paper)
  1. Thermodynamics. 2. Chemical engineering. I. Lira, Carl T. II. Title.
  TP149.E45 2012
  660'.2969—dc23

                                                                              2011050292

ISBN-13:   978-0-13-606854-9
ISBN-10:   0-13-606854-5

Seventh printing, October 2017

7   17

CONTENTS

PREFACE

"No happy phrase of ours is ever quite original with us; there is nothing of our own in it except some slight change born of our temperament, character, environment, teachings and associations."

Mark Twain

This textbook is designed for chemical engineering students from the sophomore level to the first year of graduate school. The approach blends molecular perspective with principles of thermodynamics to build intuitive reasoning regarding the behavior of species in chemical engineering processes and formulations. The molecular perspective is represented by descriptions encompassing: the relation of kinetic energy to temperature; the origin and consequences of intermolecular potentials; molecular acidity and basicity; methods used to incorporate molecular properties into molecular simulations; and the impact of molecular properties on macroscopic energy and entropy. This text is distinctive in making molecular perspectives accessible at the introductory level and connecting properties with practical implications.

This second edition offers enhanced coverage of biological, pharmaceutical, and electrolyte applications including osmotic pressure, solid solubility, and coupled reactions. Throughout the text, topics are organized to implement hierarchical instruction with increasing levels of detail. Content requiring deeper levels of theory is clearly delineated in separate sections and chapters. Less complex empirical model approaches have been moved forward to provide introductory practice with concepts and to provide motivation for understanding models more fully. The approach also provides more instructor flexibility in selecting topics to cover. Learning objectives have been clearly stated for each chapter along with chapter summaries including "important equations" to enhance student focus. Every chapter includes practice problems with complete solutions available online, as well as numerous homework problems. Online supplements include practice tests spanning many years, coursecasts describing difficult concepts or how to use computational tools, ConcepTests to quickly check comprehension, and objective lists that can be customized for greater detail. We also recommend the related resources available at the www.learncheme.com.

Unique features of the text include the level of pedagogical development of excess function models and electrolytes. For mixture models, the key assumptions and derivation steps are presented, stimulating readers to consider how the molecular phenomena are represented. For electrolytes and biological systems, the text makes connections between pH and speciation and provides tools for rapidly estimating concentrations of dissociated species. We emphasize speciation and problem solving in this introduction, instead of focusing on advanced theories of electrolyte activity. The material is written at an intermediate level to bridge students from the introductions in chemistry to the more complex models of electrolytes provided by process simulators.

We have created a number of homework problems with many variants, intending that different parts can be assigned to different classes or groups, not intending that each student work all parts.

NOTES TO STUDENTS

Thermodynamics is full of terminology and defined properties. Please note that the textbook provides a glossary and a summary of notation just before Unit I. Also consider the index a resource.

We consider the examples to be an integral part of the text, and we use them to illustrate important points. Examples are often cross-referenced and are therefore listed in the table of contents. We enclose important equations in boxes and we use special notation by equation numbers: (*) means that the equation assumes temperature-independent heat capacity; (ig) means the equation is limited to ideal gases. We include margin notes to highlight important concepts or supplemental information.

Computer programs facilitate the solutions to homework problems, but they should not be used to replace an understanding of the material. Computers are tools for calculating, not for thinking. To evaluate your understanding, we recommend that you know how to solve the problem by hand calculations. If you do not understand the formulas in the programs it is a good indication that you need to do more studying before using the program so that the structure of the program makes sense. It is also helpful to rework example problems from the text using the software.

ACKNOWLEDGMENTS

As the above quote from Mark Twain alludes, we are indebted to many others, from informal hallway conversations at meetings, to e-mail messages with suggestions and errata, to classroom testing. In many cases, we are merely the conveyors of others' suggestions. In particular, for the first edition, Dave Hart, Joan Brennecke, Mike Matthews, Bruce Poling, Ross Taylor, and Mark Thies worked with early versions of the text. We have benefited from classroom testing of the second edition by Margot Vigeant, Victor Vasquez, and Joan Brennecke. We have benefited from reviews by Keith Johnston, Ram Gupta, John O'Connell, Mike Greenfield (electrolytes), Andre Anderko (electrolytes), and Paul Mathias (electrolytes). We have adapted some example problems developed by John O'Connell at the NSF BioEMB Workshop, San Jose, CA, 2010. CTL would like to thank Ryoko Yamasaki for her work in typing many parts of the first edition manuscript and problem solutions. CTL also thanks family members Gail, Nicolas, and Adrienne for their patience, as many family sacrifices helped make this book possible. JRE thanks family members Guliz, Serra, and Eileen for their similar forbearance. We acknowledge Dan Friend and NIST, Boulder, for contributions to the steam tables and thermodynamic charts. Lastly, we acknowledge the influences of the many authors of previous thermodynamics texts. We hope we have done justice to this distinguished tradition, while simultaneously bringing deeper insight to a broader audience.

ABOUT THE AUTHORS

J. Richard Elliott is Professor of Chemical Engineering at the University of Akron in Ohio. He has taught courses ranging from freshman tools to senior process design as well as thermodynamics at every level. His research interests include: thermodynamics of polymer solutions and hydrogen bonding using molecular simulations and perturbation theory; thermodynamics of supercritical fluids and hydrocarbon processing; biorefining pretreatments; and experimental phase equilibrium measurements. He has worked with the NIST lab in Boulder and ChemStations in Houston. He holds a Ph.D. in chemical engineering from Pennsylvania State University.

jelliott@uakron.edu

Carl T. Lira is Associate Professor in the Department of Chemical Engineering and Materials Science at Michigan State University. He teaches thermodynamics at all levels, chemical kinetics, and material and energy balances. His research accomplishments include experimental measurements and modeling for liquid metals, supercritical fluids, adsorptive separations, and liquid-vapor, solid-liquid, and liquid-liquid phase equilibria. Currently, Professor Lira specializes in the study of thermodynamic properties of bio-derived fuels and chemicals via experiments and molecular simulations, and he collaborates in the MSU Reactive Distillation Facility. He has been recognized with the Amoco Excellence in Teaching Award, and multiple presentations of the MSU Withrow Teaching Excellence Award. He holds a B.S. from Kansas State University, and an M.S. and Ph.D. from the University of Illinois, Champaign-Urbana, all in chemical engineering.

lira@egr.msu.edu

GLOSSARY

Adiabatic—condition of zero heat interaction at system boundaries.

Association—description of complex formation where all molecules in the complex are of the same type.

Azeotrope—mixture which does not change composition upon vapor-liquid phase change.

Barotropy—the state of a fluid in which surfaces of constant density (or temperature) are coincident with surfaces of constant pressure.

Binodal—condition of binary phase equilibrium.

Dead state—a description of the state of the system when it is in equilibrium with the surroundings, and no work can be obtained by interactions with the surroundings.

Diathermal—heat conducting, and without thermal resistance, but impermeable to mass.

Efficiency—see *isentropic efficiency, thermal efficiency, thermodynamic efficiency.*

EOS—Equation of state.

Fugacity—characterizes the escaping tendency of a component, defined mathematically.

Heteroazeotrope—mixture that is not completely miscible in all proportions in the liquid phase and like an azeotrope cannot be separated by simple distillation. The heteroazeotropic vapor condenses to two liquid phases, each with a different composition than the vapor. Upon partial or total vaporization, the original vapor composition is reproduced.

Infinite dilution—description of a state where a component's composition approaches zero.

Irreversible—a process which generates entropy.

Isenthalpic—condition of constant enthalpy.

Isentropic—condition of constant entropy.

Isentropic efficiency—ratio characterizing actual work relative to ideal work for an isentropic process with the same inlet (or initial) state and the same outlet (or final) pressure. See also *thermodynamic efficiency, thermal efficiency.*

Isobaric—condition of constant pressure.

Isochore—condition of constant volume. See *isosteric*.

Isopiestic—constant or equal pressure.

Isopycnic—condition of equal or constant density.

Isolated—A system that has no interactions of any kind with the surroundings (e.g. mass, heat, and work interactions) is said to be isolated.

Isosteric—condition of constant density. See *isochore*.

Isothermal—condition of constant temperature.

LLE—liquid-liquid equilibria.

Master equation—$U(V,T)$.

Measurable properties—variables from the set $\{P, V, T, C_P, C_V\}$ and derivatives involving only $\{P, V, T\}$.

Metastable—signifies existence of a state which is non-equilibrium, but not unstable, e.g., superheated vapor, subcooled liquid, which may persist until a disturbance creates movement of the system towards equilibrium.

Nozzle—a specially designed device which nearly reversibly converts internal energy to kinetic energy. See *throttling*.

Polytropic exponent—The exponent n in the expression $PV^n = $ constant.

Quality—the mass fraction of a vapor/liquid mixture that is vapor.

rdf—radical distribution function.

Reference state—a state for a pure substance at a specified (T,P) and type of phase (S,L,V). The reference state is invariant to the system (P,T) throughout an entire thermodynamic problem. A problem may have various standard states, but only one reference state. See also *standard state*.

Sensible heat changes—heat effects accompanied by a temperature change.

Specific heat—another term for C_P or C_V with units per mass.

Specific property—an intensive property per unit mass.

SLE—solid-liquid equilibria.

Solvation—description of complex formation where the molecules involved are of a different type.

Spinodal—condition of instability, beyond which metastability is impossible.

Standard conditions—273.15 K and 0.1 MPa (IUPAC), *standard temperature and pressure*.

Standard state—a state for a pure substance at a specified (T,P) and type of phase (S,L,V). The standard state T is always at the T of interest for a given calculation within a problem. As the T of the system changes, the standard state T changes. The standard state P may be a fixed P or may be the P of the system. Gibbs energies and chemical potentials are commonly calculated relative to the standard state. For reacting systems, enthalpies and Gibbs energies of formation are commonly tabulated at a fixed pressure of 1 bar and 298.15 K. A temperature correction must be applied to calculate the standard state value at the temperature of interest. A problem may have various standard states, but only one reference state. See also *reference state*.

State of aggregation—solid, liquid, or gas.

Steady-state—open flow system with no accumulation of mass and where state variables do not change with time inside system boundaries.

STP—standard temperature and pressure, 273.15 K and 1 atm. Also referred to as *standard conditions*.

Subcooled—description of a state where the temperature is below the saturation temperature for the system pressure, e.g., subcooled vapor is metastable or unstable, subcooled liquid is stable relative to the bubble-point temperature; superheated vapor is stable, superheated liquid is metastable or unstable relative to the dew-point temperature; subcooled liquid is metastable or unstable relative to the fusion temperature.

Superheated—description of a state where the temperature is above the saturation temperature for the system pressure. See *subcooled*.

Thermal efficiency—the ratio or work obtained to the heat input to a heat engine. No engine may have a higher thermal efficiency than a Carnot engine.

Thermodynamic efficiency—ratio characterizing actual work relative to reversible work obtainable for exactly the same change in state variables for a process. The heat transfer for the reversible process will differ from the actual heat transfer. See also *isentropic efficiency, thermal efficiency*.

Throttling—a pressure drop without significant change in kinetic energy across a valve, orifice, porous plug, or restriction, which is generally irreversible. See *nozzle*.

Unstable—a state that violates thermodynamic stability, and cannot persist. See also *metastable, spinodal*.

VLE—vapor-liquid equilibrium.

Wet steam—a mixture of water vapor and liquid.

NOTATION

General Symbols

a Activity, or dimensional equation of state parameter or energetic parameter, or heat capacity or other constant

A Intensive Helmholtz energy, or dimensionless constant for equation of state, or Antoine, Margules, or other constant

b Dimensional equation of state parameter or heat capacity or other constant

B Virial coefficient, or dimensionless constant for other equation of state, or Antoine or other constant

$C, c,...$ Constants, c is a shape factor for the ESD equation of state

C_P Intensive constant pressure heat capacity

C_V Intensive constant volume heat capacity

F Feed

f Pure fluid fugacity

$\hat{f}_i$ Fugacity of component in mixture

G Intensive Gibbs energy

g Gravitational constant (9.8066 m/s^2) or radial distribution function

g_c Gravitational conversion factor (1 kg-m/N-s^2) (32.174[(1b$_m$-ft)/s^2]/1b$_f$)

H Intensive enthalpy

K_a Reaction equilibrium constant

k Boltzmann's constant $= R/N_A$

k_{ij} Binary interaction coefficient (Eqn. 1.8, 15.9)

K Distribution coefficient (vapor-liquid, liquid-liquid, etc.)

M Generic property or molarity when used as units

M_w Molecular weight

m Mass (energy balances), molality (electrolytes)

N Number of molecules

n Number of moles

N_A Avogadro's number $= 6.0221$ E 23 mol^{-1}

P Pressure

$\underline{Q}$	Extensive heat transfer across system boundary
Q	Intensive heat transfer across system boundary
q	Quality (mass% vapor)
R	Gas constant (8.31446 cm^3-MPa/mole-K)
S	Intensive entropy
T	Temperature
U	Intensive internal energy
u	Pair potential energy function
V	Intensive volume
v	Velocity
$\underline{W}$	Extensive work done at boundary
W	Intensive work done at boundary
Z	Compressibility factor
z	Height

Greek Symbols

α	Isobaric coefficient of thermal expansion, also Peng-Robinson equation of state parameter, and also an ESD equation of state variable
β	$1/kT$, where k is Boltzmann's constant and T is temperature.
ε	Potential energy parameter
η	Thermal θ, compressor or pump C, turbine or expander E, efficiency, or reduced density, $\eta_P = b/V$
γ	C_P/C_V
γ_i	Activity coefficient
κ_T	Isothermal compressibility
κ_S	Isentropic compressibility

κ	Parameter for Peng-Robinson equation of state
λ	Universal free volume fraction
μ_i	Chemical potential for a component in a mixture
ω	Acentric factor
φ	Pure fluid fugacity coefficient
$\hat{\varphi}_i$	Component fugacity coefficient in a mixture
Φ	Osmotic coefficient or electric potential
Φ_i	Volume fraction of component i
ρ	Intensive density
σ	Molecular diameter
ξ	Reaction coordinate

Operators

Δ	Denotes change: (1) for a closed system, denotes (final state − initial state); (2) for an open steady-state system, denotes (outlet state − inlet state)
ln	Natural logarithm (base e)
log	Common logarithm (base 10)
∂	Partial differential
Π	Cumulative product operator
Σ	Cumulative summation

Special Notation

ig	Equation applies to ideal gas only
*	Equation assumes heat capacity is temperature-independent.
$^{-}$ as in $\bar{H}_i$	Partial molar property
$\_$ as in $\underline{U}$	Extensive property
$\hat{}$	Mixture property

[≡]	Specifies units for variable
≡	Equivalence or definition
≈	Approximately equal to

Subscripts

b	Property at normal boiling point temperature
EC	Expansion/contraction work
f	Denotes property for formation of molecule from atoms in their naturally occurring pure molecular form
gen	Generated entropy
i	Component in a mixture
m	Property at melting point
mix	Used with Δ to denote property change on mixing
r	Property reduced by critical constant
R	Reference state
S	Shaft work

Superscripts

∞	Infinite dilution
□	Molality standard state
o	Standard state (usually pure)

*	Henry's law (rational) standard state
E	Excess property for a mixture
f	Property at final state
fus	Fusion (melting) process
i	Property at initial state
α, β	Denotes phase of interest
ig	Ideal gas property
in	Property at inlet (open system)
is	Ideal solution property
L	Liquid phase
out	Property at outlet (open system)
S	Solid phase
sat	Saturation property
V	Vapor phase
vap	Vaporization process

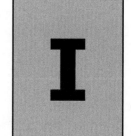
FIRST AND SECOND LAWS

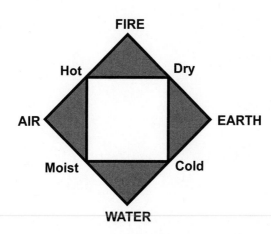

Aristotle, 384–322 BC

The ancient Greeks thought that there were only four elements: earth, air, fire, and water. As a matter of fact, you can explain a large number of natural phenomena with little more. The first and second laws of thermodynamics can be developed and illustrated quite completely with just solid blocks(earth), ideal gases (air), steam property tables (water), and heat (fire). Without significantly more effort, we can include a number of other "elements": methane, carbon dioxide, and several refrigerants. These additional species are quite common, and charts that are functionally equivalent to the steam property tables are readily available.

The first and second laws provide the foundation for all of thermodynamics, and their importance should not be underestimated. Many engineering disciplines typically devote an entire semester to the "earth, air, fire, and water" concepts. This knowledge is so fundamental and so universal that it is essential to any applied scientist. Nevertheless, chemical engineers must quickly lay this foundation and move on to other issues covered in Units II, III, and IV. The important thing for chemical engineers to anticipate as they move through Unit I is that the principles are at the core of the entire text and it will be necessary to integrate information from Unit I in the later units. The key is to follow the methods of applying systematically the first law (energy balance) and the second law (entropy balance). Watch carefully how the general equations are quickly reduced to the specific problem at hand. Especially watch how the systems of equations are developed to match the unknown variables in the problem. Learn to perform similar reductions quickly and accurately for yourself. It takes practice, but thorough knowledge of that much will help immensely when it comes to Unit II.

CHAPTER 1

BASIC CONCEPTS

"Aside from the logical and mathematical sciences, there are three great branches of natural science which stand apart by reason of the variety of far reaching deductions drawn from a small number of primary postulates. They are mechanics, electromagnetics, and thermodynamics.

These sciences are monuments to the power of the human mind; and their intensive study is amply repaid by the aesthetic and intellectual satisfaction derived from a recognition of order and simplicity which have been discovered among the most complex of natural phenomena... Yet the greatest development of applied thermodynamics is still to come. It has been predicted that the era into which we are passing will be known as the chemical age; but the fullest employment of chemical science in meeting the various needs of society can be made only through the constant use of the methods of thermodynamics."

Lewis and Randall (1923)

Lewis and Randall eloquently summarized the broad significance of thermodynamics as long ago as 1923. They went on to describe a number of the miraculous scientific developments of the time and the relevant roles of thermodynamics. Historically, thermodynamics has guided the development of steam engines, refrigerators, nuclear power plants, and rocket nozzles, to name just a few. The principles remain important today in the refinement of alternative refrigerants, heat pumps, and improved turbines, and also in technological advances including computer chips, superconductors, advanced materials, fermentations, biological cycles, and bioengineered pharmaceuticals. These latter-day "miracles" might appear to have little to do with power generation and refrigeration cycles at first thought. Nevertheless, as Lewis and Randall point out, the implications of the postulates of thermodynamics are far-reaching and will continue to be important in the development of even newer technologies. Much of modern thermodynamics focuses on characterization of the properties of mixtures, as their constituents partition into stable phases or inhomogeneous domains, and react. The capacity of thermodynamics to bring "quantitative precision in place of the old, vague ideas"[1] is as germane today as it was then.

---

1. Lewis, G.N., Randall, M. 1923. *Thermodynamics and the Free Energy of Chemical Substances,* New York: McGraw-Hill.

Before overwhelming you with the details that comprise thermodynamics, we outline a few "primary postulates" as clearly as possible and put them into the context of what we will refer to as classical equilibrium thermodynamics. In casual terms, our primary premises can be expressed as follows:

1. You can't get something for nothing. (Energy is conserved.)
2. Maintaining order requires work. (Entropy generation leads to lost work.)[2]

Occasionally, it may seem that we are discussing principles that are much more sophisticated. But the fact is that all of our discussions can be reduced to these fundamental principles. The first principle is a casual statement of the **first law of thermodynamics** (conservation of energy) which will be introduced in Chapters 2 and 3. The second principle is a casual statement of the **second law of thermodynamics** (entropy balance) which will be introduced in Chapter 4. When you find yourself in the midst of a difficult problem, it may be helpful to remember the underlying principles. We will see that coupling these two principles with some slightly sophisticated reasoning (mathematics included) leads to many clear and reliable insights about a wide range of subjects from energy crises to high-tech materials, from environmental remediation to biosynthesis. The bad news is that the level of sophistication required is not likely to be instantly assimilated by the average student. The good news is that many students have passed this way before, and the proper trail is about as well marked as one might hope.

There is less-than-universal agreement on what comprises "thermodynamics." If we simply take the word apart, "thermo" sounds like "thermal," which ought to have something to do with heat, temperature, or energy. "Dynamics" ought to have something to do with movement. And if we could just leave the identification of thermodynamics as the study of "energy movements," it would be sufficient for the purposes of this text. Unfortunately, such a definition would not clarify what distinguishes thermodynamics from, say, transport phenomena or kinetics, so we should spend some time clarifying the definition of thermodynamics in this way before moving on to the definitions of temperature, heat, energy, and so on.

The definition of thermodynamics as the study of energy movements has evolved considerably to include classical equilibrium thermodynamics, quantum thermodynamics, statistical thermodynamics, and irreversible thermodynamics as well as nonequilibrium thermodynamics. Classical thermodynamics has the general connotation of referring to the implications of constraints related to multivariable calculus as developed by J.W. Gibbs. We spend a significant effort applying these insights in developing generalized equations for the thermodynamic properties of pure substances. Statistical thermodynamics focuses on the idea that knowing the precise states of 10^{23} atoms is not practical and prescribes ways of computing the average properties of interest based on very limited measurements. We touch on this principle in our introduction to entropy, in our kinetic theory and molecular dynamics, and in the formulation of the internal energy relative to the intermolecular potential energy. We generally refrain from detailed formulation of all the statistical averages, however, maintaining the focus on simple concepts of molecular interactions. Irreversible thermodynamics and nonequilibrium thermodynamics emphasize the ways that local concentrations of atoms and energy evolve over periods of time. At this point, it becomes clear that such a broad characterization of thermodynamics would overlap with transport phenomena and kinetics in a way that would begin to be confusing at the introductory level. Nevertheless, these fields of study represent legitimate subtopics within the general realm of thermodynamics.

---

2. The term "lost work" refers to the loss of capability to perform useful work, and is discussed in more detail in Sections 2.4 on page 42, 4.2 on page 132, and 4.3 on page 142.

1.1   INTRODUCTION

These considerations should give you some idea of the potential range of applications possible within the general study of thermodynamics. This text will try to find a happy medium. One general unifying principle about the perspective offered by thermodynamics is that there are certain properties that are invariant with respect to time. For example, the process of diffusion may indicate some changes in the system with time, but the diffusion coefficient is a property which only depends on a temperature, density, and composition profile. A thermodynamicist would consider the diffusion process as something straightforward given the diffusion coefficient, and focus on understanding the diffusion coefficient. A transport specialist would just estimate the diffusion coefficient as best as he could and get on with it. A kineticist would want to know how fast the diffusion was relative to other processes involved. In more down-to-earth terms, if we were touring about the countryside, the thermodynamicists would want to know where we were going, the transport specialists would want to know how long it takes to get there, and the kineticists would want to know how fast the fuel was running out.

In thermodynamics we utilize a few basic concepts: energy, entropy, and equilibrium. The ways in which these are related to one another and to temperature, pressure, and density are best understood in terms of the connections provided by molecular mechanisms. These connections, in turn, can be summarized by the thermodynamic model (e.g., ideal gas), our quantitative description of the substance. Showing how energy and entropy couple with molecular characteristics to impact chemical process applications is the primary goal of this text. These insights should stick with you long after you have forgotten how to estimate any particular thermodynamic property, a heat capacity or activity coefficient, for example. We will see how assuming a thermodynamic model and applying the rules of thermodynamics leads to precise and extremely general insights relevant to many applications. A general theme throughout the text (and arguably throughout engineering) is: observe, predict, test, and evaluate. The prediction phase usually involves a model equation. Testing and evaluation expose limitations of the prospective model, which leads to a new cycle of observation, prediction... We terminate this hierarchy at an introductory level, but it never really ends. Extending this hierarchy is the source of innovation that must serve you for the next 50 years.

Chapter Objectives: You Should Be Able to...

1. Explain the definitions and relations between temperature, molecular kinetic energy, molecular potential energy and macroscopic internal energy, including the role of intermolecular potential energy and how it is modeled. Explain why the ideal gas internal energy depends only on temperature.

2. Explain the molecular origin of pressure.

3. Apply the vocabulary of thermodynamics with words such as the following: work, quality, interpolation, sink/reservoir, absolute temperature, open/closed system, intensive/extensive property, subcooled, saturated, superheated.

4. Explain the advantages and limitations of the ideal gas model.

5. Sketch and interpret paths on a P-V diagram.

6. Perform steam table computations like quality determination, double interpolation.

1.2   THE MOLECULAR NATURE OF ENERGY AND TEMPERATURE

Energy is a term that applies to many aspects of a system. Its formal definition is in terms of the capability to perform work. We will not quantify the potential for work until the next chapter, but you should have some concept of work from your course in introductory physics. Energy may take the form of kinetic energy or potential energy, and it may refer to energy of a macroscopic or a molecular scale.

Energy is the sum total of all capacity for doing work that is associated with matter: kinetic, potential, submolecular (i.e., molecular rearrangements by reaction), or subatomic (e.g., ionization, fission).

Kinetic energy is the energy associated with motion of a system. Motion can be classified as translational, rotational, or vibrational.

Temperature is related to the "hotness" of a substance, but is fundamentally related to the kinetic energy of the constituitive atoms.

Potential energy is the energy associated with a system due to its position in a force field.

In the study of "energy movements," we will continually ask, "How much energy is here now, and how much is there?" In the process, we need to establish a point for beginning our calculations. According to the definition above, we might intuitively represent zero internal energy by a perfect vacuum. But then, knowing the internal energy of a single proton inside the vacuum would require knowing how much energy it takes to make a proton from nothing. Since this is not entirely practical, this intuitive choice is not a good engineering choice usually. This is essentially the line of reasoning that gives rise to the convention of calculating energy changes relative to a **reference state.** Thus, there is no absolute reference point that is always the most convenient; there are only changes in energy from one state to another. We select reference conditions that are relevant throughout any particular process of interest. Depending on the complexity of the calculation, reference conditions may vary from, say, defining the enthalpy (to be defined later) of liquid water to be zero at 0.01°C (as in the steam tables) to setting it equal to zero for the molecular hydrogen and oxygen at 1 bar and 298.15 K (as in the heat of reaction), depending on the situation. Since this text focuses on changes in kinetic energy, potential energy, and energies of reaction, we need not specify reference states any more fundamental than the elements, and thus we do not consider subatomic particles.

> ❶ Energy will be tabulated relative to a convenient reference state.

Kinetic Energy and Temperature

Kinetic energy is commonly introduced in detail during introductory physics as $\frac{1}{2}mv^2$, where m is the mass of the object and v is the object velocity. Atomic species that make up solids are frozen in localized positions, but they are continuously vibrating with kinetic energy. Fluids such as liquids and gases are not frozen into fixed positions and move through space with kinetic energy and collide with one another.

The most reliable definition of temperature is that it is a numerical scale for uniquely ordering the "hotness" of a series of objects.[3] However, this "hotness" is coupled to the molecular kinetic energy of the constituent molecules in a fundamental way. The relation between kinetic energy and temperature is surprisingly direct. When we touch a hot object, the kinetic energy of the object is transferred to our hand via the atoms vibrating at the surface. Temperature is proportional to the

> ❶ Temperature primarily reflects the kinetic energy of the molecules.

---

3. Denbigh, K., 1971. *The Principles of Chemical Equilibrium,* London: Cambridge University Press, p. 9.

average molecular kinetic energy. The expression is easiest to use in engineering on a molar basis. For a monatomic substance

$$T = \frac{M_w}{3R}\langle v^2 \rangle \text{ (for 3D)} \quad T_{2D} = \frac{M_w}{2R}\langle v^2 \rangle \text{ (for 2D) monatomic fluid.} \qquad 1.1$$

❶ Check your units when using this equation.

where <> brackets denote an average, and M_w is the molecular weight. We use a subscript for the temperature of 2D motion to avoid confusion with the more common 3D motion. The differences between 2D and 3D temperature are explained on page 22. For a polyatomic molecule, the temperature is coupled to the average velocity of the individual atoms, but some of the motion of the bonded atoms results in vibrations and rotations rather than a direct translation of the center of mass and thus it is not directly related to the velocity of the center of mass. (See Section 7.10 on page 276.)

Eqn. 1.1 is applicable to any classical monatomic system, including liquids and solids. This means that for a pure system of a monatomic ideal gas in thermal equilibrium with a liquid, the average velocities of the molecules are independent of the phase in which they reside. We can infer this behavior by envisioning gas atoms exchanging energy with the solid container walls and then the solid exchanging energy with the liquid. At equilibrium, all exchanges of energy must reach the same kinetic energy distribution. The liquid molecular environment is different from the gas molecular environment because liquid molecules move within a crowded environment where the atoms collide with a higher frequency due to the smaller intermolcular distances. The potential energies are more significant for a liquid and a minimum kinetic energy is required to escape the molecular attractiveness due to potential energy (we discuss the potential energy next). What happens if the temperature is raised such that the liquid molecules can escape the potential energies of the neighbors? We call this phenomenon "boiling." Now you can begin to understand what temperature is and how it relates to other important thermodynamic properties.

We are guaranteed that a universal scale of temperature can be developed because of the **zeroth law of thermodynamics:** If two objects are in equilibrium with a third, then they are in equilibrium with one another as we discussed in the previous paragraph. The zeroth law is a law in the sense that it is a fact of experience that must be regarded as an empirical fact of nature. The significance of the zeroth law is that we can calibrate the temperature of any new object by equilibrating it with objects of known temperature. Temperature is therefore an empirical scale that requires calibration according to specific standards. The Celsius and Fahrenheit scales are in everyday use. The conversions are:

$$(T \text{ in } °C) = \frac{5}{9}((T \text{ in } °F) - 32)$$

When we perform thermodynamic calculations, we must usually use **absolute temperature** in Kelvin or Rankine. These scales are related by

$$(T \text{ in K}) = (T \text{ in } °C) + 273.15$$

$$(T \text{ in } °R) = (T \text{ in } °F) + 459.67$$

❶ Thermodynamic calculations use absolute temperature in °R or K.

$$(T \text{ in } °R) = 1.8 \cdot (T \text{ in K})$$

The absolute temperature scale has the advantage that the temperature can never be less than absolute zero. This observation is easily understood from the kinetic perspective. The kinetic energy cannot be less than zero; if the atoms are moving, their kinetic energy must be greater than zero.

Potential Energy

Solids and liquids exist due to the intermolecular **potential energy** (molecular "stickiness') of atoms. If molecules were not "sticky" all matter would be gases or solids. Thus, the principles of molecular potential energy are important for developing a molecular perspective on the nature of liquids, solids, and non-ideal gases. Potential energy is associated with the "work" of moving a system some distance through a force field. On the macroscopic scale, we are well aware of the effect of gravity. As an example, the Earth and the moon are two spherical bodies which are attracted by a force which varies as r^{-2}. The potential energy represents the work of moving the two bodies closer together or farther apart, which is simply the integral of the force over distance. (The force is the negative derivative of potential with respect to distance.) Thus, the potential function varies as r^{-1}. Potential energies are similar at the microscopic level except that the forces vary with position according to different laws. The gravitational attraction between two individual atoms is insignificant because the masses are so small. Rather, the important forces are due to the nature of the atomic orbitals. For a rigorous description, the origin of the intermolecular potential is traced back to the solution of Schrödinger's quantum mechanics for the motions of electrons around nuclei. However, we do not need to perform quantum mechanics to understand the principles.

Intermolecular Potential Energy

❗ Engineering model potentials permit representation of attractive and repulsive forces in a tractable form.

Atoms are composed of dense nuclei of positive charge with electron densities of negative charge built around the nucleus in shells. The outermost shell is referred to as the valence shell. Electron density often tends to concentrate in lobes in the valence orbitals for common elements like C, N, O, F, S, and Cl. These lobes may be occupied by bonded atoms that are coordinated in specific geometries, such as the tetrahedron in CH_4, or they may be occupied by unbonded electron pairs that fill out the valence as in NH_3 or H_2O, or they may be widely "shared" as in a resonance or aromatic structure. These elements (H, C, N, O, F, S, Cl) and some noble gases like He, Ne, and Ar provide virtually all of the building blocks for the molecules to be considered in this text.

By considering the implications of atomic structure and atomic collisions, it is possible to develop the following subclassifications of intermolecular forces:

1. Electrostatic forces between charged particles (ions) and between permanent dipoles, quadrupoles, and higher multipoles.

2. Induction forces between a permanent dipole (or quadrupole) and an induced dipole.

3. Forces of attraction (dispersion forces) due to the polarizability of electron clouds and repulsion due to prohibited overlap.

4. Specific (chemical) forces leading to association and complex formation, especially evident in the case of hydrogen bonding.

❗ Attractive forces and potential energies are negative. Repulsive forces and potential energies are positive.

Attractive forces are quantified by negative numerical values, and repulsive forces will be characterized by positive numerical values. To a first approximation, these forces can be characterized by a spherically averaged model of the intermolecular potential (aka. "potential" model). **The potential, $u(r)$, is the work (integral of force over distance) of bringing two molecules from infinite distance to a specific distance, r.** When atoms are far apart (as in a low-pressure gas), they do not sense one another and interaction energy approaches zero. When the atoms are within about two diameters, they attract, resulting in a negative energy of interaction. Because they have finite size, as they are brought closer, they resist overlap. Thus, at very close distances, the forces are repulsive and create very large positive potential energies. These intuitive features are illustrated graphically in Fig. 1.1. The discussion below provides a brief background on why these forces exist and how they vary with distance.

Electrostatic Forces

The force between two point charges described by Coulomb's Law is very similar to the law of gravitation and should be familiar from elementary courses in chemistry and physics,

$$F \propto \frac{q_i q_j}{r^2}$$

where q_i and q_j are the charges, and r is the separation of centers. Upon integration, $u = \int F dr$, the potential energy is proportional to inverse distance,

$$u \propto \frac{q_i q_j}{r} \hspace{4cm} 1.2$$

If all molecules were perfectly spherical and rigid, the only way that these electrostatic interactions could come into play is through the presence of ions. But a molecule like NH_3 is not perfectly spherical. NH_3 has three protons on one side and a lobe of electron density in the unbonded valence shell electron pair. This permanent asymmetric distribution of charge density is modeled mathematically with a dipole (+ and – charge separation) on the NH_3 molecule.[4] This means that ammonia molecules lined up with the electrons facing one another repel while molecules lined up with the electrons facing the protons will attract. Since electrostatic energy drops off as r^{-1}, one might expect that the impact of these forces would be long-range. Fortunately, with the close proximity of the positive charge to the negative charge in a molecule like NH_3, the charges tend to cancel one another as the molecule spins and tumbles about through a fluid. This spinning and tumbling makes it reasonable to consider a spherical average of the intermolecular energy as a function of distance that may be constructed by averaging over all orientations between the molecules at each distance. In a large collection of molecules randomly distributed relative to one another, this averaging approach gives rise to many cancellations, and the net impact is approximately

$$u_{dipole-dipole} \propto \frac{-\varepsilon_{dipole}}{r^6 kT} \hspace{4cm} 1.3$$

where $k = R/N_A$ is **Boltzmann's constant,** related to the **gas constant**, R, and **Avogadro's number,** N_A. This surprisingly simple result is responsible for a large part of the attractive energy between polar molecules. This energy is attractive because the molecules tend to spend somewhat more time lined up attractively than repulsively, and the r^{-6} power arises from the averaging that occurs as the molecules tumble and the attractive forces decrease with separation. A key feature of dipole-dipole forces is the temperature dependence.

---

4. The dipole is a model of the charge distribution on the molecule, and it is thus an approximate description of the charge distribution.

Induction Forces

When a molecule with a permanent dipole approaches a molecule with no dipole, the positive charge of the dipolar molecule will tend to pull electron density away from the nonpolar molecule and "induce" a dipole moment into the nonpolar molecule. The magnitude of this effect depends on the strength of the dipole and how tightly the orbitals of the nonpolar molecule constrain the electrons spatially in an electric field, characterized by the "polarizability."[5] For example, the pi bonding in benzene makes it fairly polarizable. A similar consideration of the spherical averaging described in relation to electrostatic forces results again in a dependence of r^{-6} as approximately

$$u_{in} \propto \frac{-\varepsilon_{in}}{r^6} \qquad 1.4$$

Disperse Attraction Forces (Dispersion Forces)

❗ The r^{-6} dependence of attractive forces has a theoretical basis.

When two nonpolar molecules approach, they may also induce dipoles into one another owing to fluctuating distributions of electrons. Their dependence on radial distance may be analyzed and gives the form for the attractive forces:

$$u_{disp}^{att} \propto \frac{-\varepsilon_{att}}{r^6} \qquad 1.5$$

Note that dipole-dipole, induction, and dispersion forces all vary as r^{-6}.

Repulsive Forces

The forces become repulsive rapidly as radial distance decreases, and quickly outweighs the attractive force as the atoms are forced into the same space. A common empirical equation is

$$u_{disp}^{rep} \propto \frac{\varepsilon_{rep}}{r^{12}} \qquad 1.6$$

Engineering Potential Models

Based on the forms of these electrostatic, induction, and dispersion forces, it should be easy to appreciate the form of the Lennard-Jones potential in Fig. 1.1. Other approximate models of the potential function are possible, such as the square-well potential or the Sutherland potential also shown in Fig. 1.1. These simplified potential models are accurate enough for many applications.

The key features of all of these potential models are the representation of the size of the molecule by the parameter σ and the attractive strength (i.e. "stickiness") by the parameter ε. We can gain considerable insight about the thermodynamics of fluids by intuitively reasoning about the relatively simple effects of size and stickiness. For example, if we represent molecules by lumping together all the atomic sites, a large molecule like buckminsterfullerene (solid at room temperature) would have a larger value for σ and ε than would methane (gas at room temperature). Water and methane are about the same size, but their difference in boiling temperature indicates a large difference in their stickiness. Considering the molecular perspective, it should become apparent that water

---

5. The polarizability is the linear proportionality constant in a model of how easily a dipole is "induced" when the molecule is placed in an electric field.

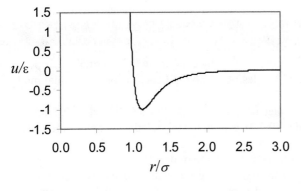

The Lennard-Jones potential.

$$u(r) = 4\varepsilon[(\sigma/r)^{12} - (\sigma/r)^6]$$

Note that $\varepsilon > 0$ is the depth of the well, and σ is the distance where $u(r) = 0$.

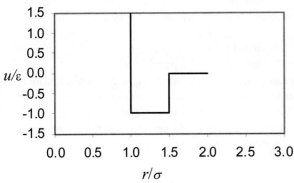

The square-well potential for $\lambda = 1.5$.

$$u(r) = \begin{cases} \infty & \text{if } r \le \sigma \\ -\varepsilon & \text{if } \sigma < r \le \lambda\sigma \\ 0 & \text{if } r > \lambda\sigma \end{cases}$$

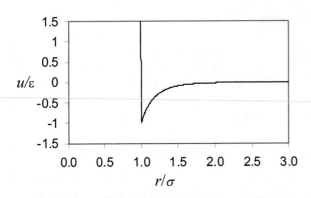

The Sutherland potential.

$$u(r) = \begin{cases} \infty & \text{if } r \le \sigma \\ -\varepsilon(\sigma/r)^6 & \text{if } r > \sigma \end{cases}$$

Figure 1.1 *Schematics of three engineering models for pair potentials on a dimensionless basis.*

has a higher boiling temperature because it sticks to itself more strongly than does methane. With these simple insights, you should be able to understand the molecular basis for many macroscopic phenomena. Example 1.1 illustrates several computations for intermolecular potential energy.

Example 1.1 The energy derived from intermolecular potentials

We can develop further appreciation for internal energy by computing the intermolecular potential energy for a well-defined system of molecules. Assume the Lennard-Jones potential model with $\sigma = 0.36$ nm and $\varepsilon = 1.38E-21$ J:

(a) Compute the potential energy for two molecules located at positions (0,0) and (0, 0.4 nm).

(b) Suppose a third molecule was located at (0.5,0). Compute the potential energy for the system.

(c) To develop a very crude insight on the methods of averaging, we can think of the average potential energy as defining an average distance between the molecules. As the volume expands, the average distance between molecules increases and the magnitude of the average potential energy decreases in accordance with the Lennard-Jones model. For the potential energy from (b), compute the average distance, $<r>$, that corresponds to the average potential energy for this system of molecules.

(d) Suppose the volume of the system in (c) expands by a factor of two. How would that affect the average distance, $<r>$, and what would you estimate as the new intermolecular energy?

(e) Assume approximately four molecules can fit around a central molecule in a liquid before it is too crowded and another layer starts to build up. Assuming the Lennard-Jones energy is practically zero beyond the first layer (i.e. ignore all but the first layer), and the average distance between the central molecule and its four neighbors is $<r> = 0.55$ nm, estimate the intermolecular energy around one single molecule and that for one mole of similar molecules.

Solution:

(a) The distance is $r_{12} = [\, (0-0)^2 + (0.4-0)^2 \,]^{1/2} = 0.4$ nm.
$u(r_{12}) = 4(1.38E-21)[(0.36/0.40)^{12} - (0.36/0.40)^6] = -1.375E-21$

(b) The distance $r_{13} = [\, (0.5-0)^2 + (0.5-0)^2 \,]^{1/2} = 0.5$. So $u(r_{13}) = -0.662E-21$.
But wait, there's more! $r_{23} = [\, (0.5-0)^2 + (0-0.4)^2 \,]^{1/2} = 0.6403$. So, $u(r_{23}) = -0.169E-21$.
The total intermolecular energy is: $-(1.375+0.662+0.169)(1E-21) = -2.206E-21$J.

(c) The average intermolecular energy for these three pairs is:
$<u> = -2.206E-21/3 = -0.735E-21$J.
Matching this value of $<u>$ by using a solver to adjust $<r>$ in the Lennard-Jones model gives
$<r> = 0.4895$ nm.

(d) Volume is related to the cubic of length. Expanding the volume by a factor of 2 changes the r-coordinates by a factor of $2^{1/3}$. So, $<r> = 0.4895(2^{1/3}) = 0.6167$, and $<u> = -0.210E-21$J.

(e) For $<r> = 0.55$, $<u(r)> = -0.400E-21$J and $u_1 = 4(-0.400E-21) = -1.600E-21$ per atom. For Avogadro's number of such molecules, the summed intermolecular energy becomes
$u_{NA} = N_A<u> = (602.22E21)(-1.600E-21) = -963$ J/mole.

When we sum the potential energy for a collection of molecules, we often call the sum **configurational energy** to differentiate quantity from the potential energy which is commonly used when discussing atoms or sites.

Configurational energy is the potential energy of a system of molecules in their "configuration."

Note that we would need a more complicated potential model to represent the shape of the molecule. Typically, molecules of different shapes are represented by binding together several potentials like those above with each potential site representing one molecular segment. For example, *n*-butane could be represented by four Lennard-Jones sites that have their relative centers located at distances corresponding to the bond-lengths in *n*-butane. The potential between two butane molecules would then be the sum of the potentials between each of the individual Lennard-Jones sites on the different molecules. In similar fashion, potential models for very complex molecules can be constructed.

Potentials in Mixtures

Our discussion of intermolecular potentials has focused on describing single molecules, but it is actually more interesting to contemplate the potential models for different molecules that are mixed together. Note that the square-well model provides a simple way for use to consider only the potential energy of the closest neighbors. We can use the square-well potential as the basis for this analysis and focus simply on the size (σ_{ij}) and stickiness (ε_{ij}) of each potential model, where the subscript ij indicates an interaction of molecule i with molecule j. Commonly, we assume that $\lambda = 1.5$ in discussions of the square-well potential, unless otherwise specified. For example, ε_{11} would be the stickiness of molecule 1 to itself, and ε_{12} would be its stickiness to a molecule of type 2 and $\varepsilon_{21} = \varepsilon_{12}$. We often calculate the interactions of dissimilar molecules by using **combining rules** that relate the interaction to the parameters of the sites. Commonly we use combining rules developed by **Lorentz and Bertholet.** The size parameter for interaction between different molecules is reasonably well represented by

$$\sigma_{12} = (\sigma_{11} + \sigma_{22})/2 \qquad\qquad 1.7$$

This rule simply states that the distance between two touching molecules is equal to the radius of the first one plus the radius of the second one. The estimation of the stickiness parameter for interaction between different molecules requires more empirical reasoning. It is conventional to estimate the stickiness by a geometric mean, but to permit flexibility to adjust the approximate rule by adding an adjustable constant that can be refined using experimental measurements, or sometimes using theories like quantum mechanical simulation. For historical reasons, this constant is typically referred to as "k_{ij}" or the **binary interaction parameter,** and defined through the following rule:

$$\varepsilon_{12} = (\varepsilon_{11}\,\varepsilon_{22})^{1/2}(1 - k_{12}) \qquad\qquad 1.8$$

The default value is $k_{12} = 0$.

Specific (Chemical) Forces Like Hydrogen Bonding

What happens when the strength of interaction between two molecules is so strong at certain orientations that it does not make sense to spherically average over it? Hydrogen bonding is an example of such an interaction, as you probably know from an introductory chemistry or biology course. For instance, it would not make sense to spherically average when two atoms preferentially interact in a specific orientation. But, if they were covalently bonded, we would call that a chemical reaction and handle it in a different way. An interesting problem arises when the strength of interaction is too strong to be treated entirely by spherically averaging and too weak to be treated as a normal chemical reaction which forms permanent stable chemical species. Clearly, this problem is difficult and it would be tempting to try to ignore it. In fact, most of this course will deal with theories that treat only what can be derived by spherically averaging. But it should be kept in mind that these specific forces are responsible for some rather important properties, especially in the form of hydrogen bonding. Since many systems are aqueous or contain amides or alcohols, ignoring hydrogen bonding entirely would substantially undermine the accuracy of our conceptual foundation. Furthermore, the concept of favorable energetic interactions between acids and bases can lend broad insights into the mysteries of chemical formulations. As an engineering approach, we can make large adjustments to the spherical nature of these forces such that we can often approximate them with a single characteristic constant to obtain a workable engineering model. Example 1.2 illustrates the concept of combining rules that pervades the entirety of mixture thermodynamics.

Example 1.2 Intermolecular potentials for mixtures

Methane (CH_4) has fewer atoms than benzene (C_6H_6), so it is smaller. Roughly, the diameter of methane is 0.36 nm and that of benzene is 0.52 nm. Similarly, methane's boiling temperature is lower so its stickiness must be smaller in magnitude. A crude approximation is $\varepsilon/k = T_c/1.25$, where T_c is the critical temperature listed on the back flap. Not knowing what to assume for k_{12}, we may consider three possibilities: $k_{12} = 0$, $k_{12} < 0$, $k_{12} > 0$. To illustrate, sketch on the same pair of axes the potential models for methane and benzene, assuming that the k_{12} parameter is given by: (a) $k_{12} = 0$ (b) $k_{12} = -0.2$ (c) $k_{12} = +0.2$. In each case, describe in words what is represented by each numerical value (e.g., favorable interactions, or unfavorable interactions...) Assume the square-well potential with $\lambda = 1.5$.

Solution: Following the suggested estimation formula, $\varepsilon/k = 190.6/1.25 = 152.5$ for methane and $562.2/1.25 = 449.8$ for benzene. Applying Eqns. 1.7 and Eqn. 1.8, we obtain Fig. 1.2 where $k_{12} = 0.0$ refers to case (a); -0.2 refers to the well location for case (b); and $+0.2$ refers to case (c).

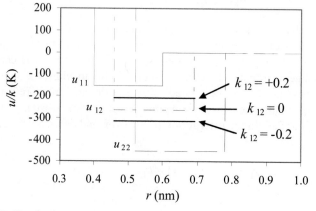

Figure 1.2 *Sketch of intermolecular square-well potential models for a mixture of methane and benzene for $\lambda = 1.5$ as explained in Example 1.1.*

Description of the interactions:

Case (a) corresponds to the molecular interactions being relatively neutral towards one another. This is the default assumption (i.e. $k_{12} = 0$). This would be the best description of methane + benzene (among these three choices) because both are nonpolar.

Case (b) corresponds to extremely favorable interactions, as indicated by the deep attractive well and strongly exothermic interaction. An Lewis acid might interact with a Lewis base in this way.

Case (c) corresponds to unfavorable interactions. The interactions are not zero exactly, so the molecules do attract one another, but the well for the 1-2 interactions is not as deep as expected from the 1-1 and 1-2 interactions. The molecules greatly prefer their own company. A mixture with this type of interaction may lead to liquid-liquid splitting, like oil and water.

Internal Energy

We have taken considerable time to develop the molecular aspects of kinetic and potential energy. These molecular properties are of great importance, but for large-scale macroscopic process system calculations, these microscopic energies are combined and we consider kinetic and configurational energy collectively as the **internal energy** of the system, which is given the symbol U. It may be somewhat confusing that kinetic and potential energy exist on the macroscopic level and the microscopic level. The potential energy (gravitational potential energy) and kinetic energy of the center of mass *of the system* are accounted for separately. The internal energy is a function of the temperature and density (the latter for non-ideal gases) of the system, and it does not usually change if the entire system is placed on, say, an airplane where the kinetic and potential energy of the center of mass differ considerably from a stationary position on the ground. This is the convention followed throughout the remainder of Unit I. In Units II and III, we reexamine the molecular potentials as to how they affect the bulk fluid properties. Thus, throughout the remainder of Unit I, when we refer to kinetic and potential energy of a body of fluid as a system, we are referring to the kinetic energy of the center of mass of the system and the gravitational potential energy of the center of mass of the system.

❶ The sum of microscopic random kinetic energy and intermolecular potential energies comprises the internal energy.

1.3 THE MOLECULAR NATURE OF ENTROPY

To be fair to both of the central concepts of thermodynamics, we must mention entropy at this point, in parallel with the mention of energy. Unfortunately, there is no simple analogy that can be drawn for entropy like that for potential energy using the gravitational forces between the Earth and moon. The study of entropy is fairly specific to the study of thermodynamics.

What we can say at this point is that entropy has been conceived to account for losses in the prospect of performing useful work. The energy can take on many forms and be completely accounted for without contemplating how much energy has been "wasted" by converting work into something like warm water. To picture how much work this would take, imagine yourself stirring water in a bath tub until the temperature rises by 5°C. Entropy accounts for this kind of wastefulness. It turns out that the generation of such waste can be directly related to the degree of disorder involved in conducting our process. Thus, generation of disorder results in **lost work**. Furthermore, work that is lost by not maintaining order cannot be converted into useful work. To see the difference between energy and entropy, consider the following example. Oxygen and nitrogen are mixed as ideal gases at room temperature and pressure. How much energy is involved in the mixing process? How much (energy) work must be exerted to separate them again? To completely answer the first question we must understand the ideal gas mixture more completely (Chapter 4). We note that ideal gases are point masses with no potential energy to affect the mixing. Thus, the answer to the first question is that no energy is involved. For the answer to the second question regarding work to separate them, however, we must acknowledge that a considerable effort would be involved. The minimum amount required can be calculated as we will show in Chapter 4. To avoid introducing too many concepts at one time, we defer the quantitative development of the topic until Chapter 4.

1.4 BASIC CONCEPTS

The System

A **system** is that portion of the universe which we have chosen to study.

A **closed system** is one in which no mass crosses the system boundaries.

An **open system** is one in which mass crosses the system boundaries. The system may gain or lose mass or simply have some mass pass through it.

An **isolated system** is one devoid of interactions of any kind with the surroundings (including mass exchange, heat, and work interactions).

System boundaries are established at the beginning of a problem, and simplification of balance equations depends on whether the system is open or closed. *Therefore, the system boundaries should be clearly identified. If the system boundaries are changed, the simplification of the mass and energy balance equations should be performed again, because different balance terms are likely to be necessary.* These guidelines become more apparent in Chapter 2. In many textbooks, especially those dealing with fluid mechanics, the system is called the **control volume.** The two terms are synonymous.

> ❶ The placement of system boundaries is a key step in problem solving.

Equilibrium

A system is in equilibrium when there is no driving force for a change of intensive variables within the system. The system is "relaxed" relative to all forces and potentials.[6]

An isolated system moves spontaneously to an equilibrium state. In the equilibrium state there are no longer any driving forces for spontaneous change of intensive variables.

The Mass Balance

Presumably, students in this course are familiar with mass balances from an introductory course in material and energy balances. The relevant relation is simply:

> ❶ The mass balance.

$$
\begin{bmatrix} \text{rate of mass} \\ \text{accumulation within} \\ \text{system boundaries} \end{bmatrix} = \begin{bmatrix} \text{rate of mass flow} \\ \text{into system} \end{bmatrix} - \begin{bmatrix} \text{rate of mass flow} \\ \text{out of system} \end{bmatrix}
$$

$$
\dot{m} = \sum_{inlets} \dot{m}^{in} - \sum_{outlets} \dot{m}^{out}
$$

1.9

where $\dot{m} = \dfrac{dm}{dt}$. $\dot{m}^{in}$ and $\dot{m}^{out}$ are the *absolute values* of mass flow rates entering and leaving, respectively.

We may also write

$$
dm = \sum_{inlets} dm^{in} - \sum_{outlets} dm^{out}
$$

1.10

---

6. We qualify this criterion for the purposes of chemical engineering that there is no driving force for "meaningful" change because most of our systems are technically metastable (in a state of local equilibrium). For example, considering air expansion in a piston/cylinder expansion, we neglect the potential corrosion of the piston/cylinder by air oxidation when we state the system has reached mechanical equilibrium.

where mass differentials dm^{in} and dm^{out} are *always positive*. When all the flows of mass are analyzed in detail for several subsystems coupled together, this simple equation may not seem to fully portray the complexity of the application. The general study of thermodynamics is similar in that regard. A number of simple relations like this one are coupled together in a way that requires some training to understand. *In the absence of chemical reactions,* we may also write a mole balance by replacing mass with moles in the balance.

Heat – Sinks and Reservoirs

Heat is energy in transit between the source from which the energy is coming and a destination toward which the energy is going. When developing thermodynamic concepts, we frequently assume that our system transfers heat to/from a **reservoir** or **sink.** A heat reservoir is an infinitely large source or destination of heat transfer. The reservoir is assumed to be so large that the heat transfer does not affect the temperature of the reservoir. A sink is a special name sometimes used for a reservoir which can accept heat without a change in temperature. The assumption of constant temperature makes it easier to concentrate on the thermodynamic behavior of the system while making a reasonable assumption about the part of the universe assigned to be the reservoir.

A reservoir is an infinitely large source or destination for heat transfer.

The mechanics of heat transfer are also easy to picture conceptually from the molecular kinetics perspective. In heat **conduction,** faster-moving molecules collide with slower ones, exchanging kinetic energy and equilibrating the temperatures. In this manner, we can imagine heat being transferred from the hot surface to the center of a pizza in an oven until the center of the pizza is cooked. In heat **convection,** packets of hot mass are circulated and mixed, accelerating the equilibration process. Heat convection is important in getting the heat from the oven flame to the surface of the pizza. Heat **radiation,** the remaining mode of heat transfer, occurs by an entirely different mechanism having to do with waves of electromagnetic energy emitted from a hot body that are absorbed by a cooler body. Radiative heat transfer is typically discussed in detail during courses devoted to heat transfer.

Work

Work is a familiar term from physics. We know that work is a force acting through a distance. There are several ways forces may interact with the system which all fit under this category, including pumps, turbines, agitators, and pistons/cylinders. We will discuss the details of how we calculate work and determine its impact on the system in the next chapter.

Density

Density is a measure of the quantity per unit volume and may be expressed on a molar basis (molar density) or a mass basis (mass density). In some situations, it is expressed as the number of particles per unit volume (number density).

Intensive Properties

Intensive properties are those properties which are independent of the size of the system. For example, in a system at equilibrium without internal rigid/insulating walls, the temperature and pressure are uniform throughout the system and are therefore intensive properties. **Specific properties** are the total property divided by the mass and are intensive. For example, the molar volume ($[\equiv]$ length^3/mole), mass density ($[\equiv]$ mass/length^3), and specific internal energy ($[\equiv]$ energy/mass) are intensive properties. In this text, intensive properties are not underlined.

The distinction between intensive and extensive properties is key in selecting and using variables for problem solving.

Extensive Properties

Extensive properties depend on the size of the system, for example the volume ($[\equiv]$ length^3) and energy ($[\equiv]$ energy). Extensive properties are underlined; for example, $\underline{U} = nU$, where n is the number of moles and U is molar internal energy.

States and State Properties – The Phase Rule

Two **state variables** are necessary to specify the state of a *single-phase* pure fluid, that is, two from the set P, V, T, U. Other state variables to be defined later in the text which also fit in this category are molar enthalpy, molar entropy, molar Helmholtz energy, and molar Gibbs energy. *State variables must be intensive properties.* As an example, specifying P and T permits you to find the specific internal energy and specific volume of steam. Note, however, that you need to specify only one variable, the temperature or the pressure, if you want to find the properties of saturated vapor or liquid. This reduction in the needed specifications is referred to as a reduction in the "**degrees of freedom.**" As another example in a ternary, two-phase system, the temperature and the mole fractions of two of the components of the lower phase are state variables (the third component is implicit in summing the mole fractions to unity), but the total number of moles of a certain component is not a state variable because it is extensive. In this example, the pressure and mole fractions of the upper phase may be calculated once the temperature and lower-phase mole fractions have been specified. The number of state variables needed to completely specify the state of a system is given by the Gibbs phase rule for a non-reactive system,[7]:

> The Gibbs phase rule provides the number of state variables (intensive properties) to specify the state of the system.

$$F = C - P + 2 \qquad 1.11$$

where F is the number of state variables that can be varied while P phases exist in a system where C is the number of components (F is also known as the number of degrees of freedom). More details on the Gibbs phase rule are given in Chapter 16.

Steady-State Open Systems

> Steady-state flow is very common in the process industry.

The term **steady state** is used to refer to open systems in which the inlet and outlet mass flow rates are invariant with time and there is no mass accumulation. In addition, steady state requires that state variables at all locations are invariant with respect to time. Note that state variables may vary with position. Steady state does not require the system to be at equilibrium. For example, in a heat exchanger operating at steady state with a hot and cold stream, each stream has a temperature gradient along its length, and there is always a driving force for heat transfer from the hotter stream to the colder stream. Section 2.13 describes this process in more detail.

The Ideal Gas Law

> The ideal gas law is a *model* that is not always valid, but gives an initial guess.

The **ideal gas** is a "model" fluid where the molecules have no attractive potential energy and no size (and thus, no repulsive potential energy). Properties of the ideal gas are calculated from the ideal gas model:

$$P\underline{V} = nRT \qquad \text{(ig) 1.12}$$

> An equation of state relates the P-V-T properties of a fluid.

Note that scientists who first developed this formula empirically termed it a "law" and the name has persisted, but it should be more appropriately considered a "model." In the terminology we develop, it is also an **equation of state,** relating the *P-V-T* properties of the ideal gas law to one

---

7. For a reactive system, C is replaced with the number of distinct species minus the number of independent reactions.

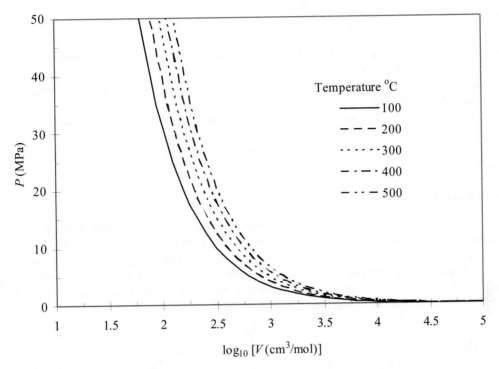

Figure 1.3 *Ideal gas behavior at five temperatures.*

another as shown in Eqn. 1.12 and Fig. 1.3. We know that real molecules have potential energy of attraction and repulsion. Due to the lack of repulsive forces, ideal gas particles can "pass through" one another. Ideal gas molecules are sometimes called "point masses" to communicate this behavior. While the assumptions may seem extreme, we know experimentally that the ideal gas model represents many compounds, such as air, nitrogen, oxygen, and even water vapor at temperatures and pressures near ambient conditions. Use of this model simplifies calculations while the concepts of the energy and entropy balances are developed throughout Unit I. This does not imply that the ideal gas model is applicable to all vapors at all conditions, even for air, oxygen, and nitrogen. Analysis using more complex fluid models is delayed until Unit II. We rely on thermodynamic charts and tables until Unit II to obtain properties for gases that may not be considered ideal gases.

Because kinetic energy is the only form of energy for an ideal gas, the internal energy of a monatomic ideal gas is given by summing the kinetic energy of the atoms and then relating this to temperature (*c.f.* Eqn. 1.1):

$$\underline{U}^{ig} = \frac{Nm\langle v^2 \rangle}{2} = \frac{nN_A m\langle v^2 \rangle}{2} = \frac{nM_w \langle v^2 \rangle}{2} = \frac{3}{2}nRT = \frac{3}{2}NkT \quad \text{(monatomic ideal gas) (ig) 1.13}$$

The proportionality constant between temperature and internal energy is known as the ideal gas **heat capacity** at constant volume, denoted C_V. Eqn 1.13 shows that $C_V = 1.5R$ for a monatomic ideal gas. If you refer to the tables of constant pressure heat capacities (C_P) on the back flap of the text and note that $C_P = C_V + R$ for an ideal gas, you may be surprised by how accurate this ultrasimplified theory actually is for species like helium, neon, and argon at 298 K.[8]

While the equality in Eqn. 1.13 is valid for monatomic fluids only, the functionality $U^{ig} = U^{ig}(T)$ is universal for all ideal gases. For more multi-atom molecules, the heat capacity depends on temperature because vibrations hold some energy in a manner that depends on temperature. However, the observation that $U^{ig} = U^{ig}(T)$ is true for any ideal gas, not only for ultrasimplified, monatomic ideal gases. We build on this relation in Chapters 6–8, where we show how to compute changes in energy for any fluid at any temperature and density by systematically correcting the relatively simple ideal gas result. Let us explore more completely the assumptions of the ideal gas law by investigating the molecular origins of pressure.

Pressure

Pressure is the force exerted per unit area. We will be concerned primarily with the pressure exerted by the molecules of fluids upon the walls of their containers. Briefly, when molecules collide with the container walls, they must change momentum. The change in momentum creates a force on the wall. As temperature increases, the particles have more kinetic energy (and momentum) when they collide, so the pressure increases. We can understand this more fully with an ultrasimplified analysis of kinetic theory as it relates to the ideal gas law.

Suppose we have two hard spherical molecules in a container that are bouncing back and forth with 1D velocity in the x-direction only and not contacting one another. We wish to quantify the forces acting on each wall. Since the particles are colliding only with the walls at A_1 and A_2 in our idealized motion, these are the only walls we need to consider. Let us assume that particles bounce off the wall with the same speed which they had before striking the wall, but in the opposite direction (a perfectly elastic collision where the wall is perfectly rigid and absorbs no momentum).

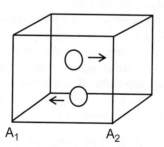

A_1 $\qquad$ A_2

Thus, the kinetic energy of the particles is fixed. If $\vec{v}$ is the initial velocity of the particle (recall that $\vec{v}$ is a vector quantity and notation v represents a scalar) before it strikes a wall, the change in velocity due to striking the wall is $-2\vec{v}$. The change in velocity of the particle indicates the presence of interacting forces between the wall and the particle. If we quantify the force necessary to change the velocity of the particle, we will also quantify the forces of the particle on the wall by Newton's third principle. To quantify the force, we may apply Newton's second principle stated in terms of momentum: *The time rate of change of the momentum of a particle is equal to the resultant force acting on the particle and is in the direction of the resultant force.*

---

8. We will formally define heat capacity and relations for C_P and C_V in Chapter 2.

$$\frac{d\vec{p}}{dt} = \vec{F} \qquad\qquad 1.14$$

The application of this formula directly is somewhat problematic since the change in direction is instantaneous, and it might seem that the time scale is important. This can be avoided by determining the time-averaged force,[9] $\vec{F}_{avg}$ exerted on the wall during time Δt,

$$\int_{t^i}^{t^f} \vec{F} dt = \vec{F}_{avg} \Delta t = \int_{t^i}^{t^f} \frac{d\vec{p}}{dt} dt = \Delta \vec{p} \qquad\qquad 1.15$$

where $\Delta \vec{p}$ is the total change in momentum during time Δt. The momentum change for each collision is $\Delta \vec{p} = -2m\vec{v}$ where m is the mass per particle. Collision frequency can be related easily to the velocity. Each particle will collide with the wall every Δt seconds, where $\Delta t = 2 L /v$, where L is the distance between A_1 and A_2. The average force is then

$$\vec{F}_{avg} = \frac{\Delta \vec{p}}{\Delta t} = -2m\vec{v}\frac{v}{2L} \qquad\qquad 1.16$$

where $\vec{v}$ is the velocity before the collision with the wall. Pressure is the force per unit area, and the area of a wall is L^2, thus

$$P = \frac{m}{L^3}(v_1^2 + v_2^2) \text{ (1D ideal gas motion)} \qquad\qquad \text{(ig) 1.17}$$

> *P* is proportional to the number of particles in a volume and to the kinetic energy of the particles.

where the subscripts denote the particles. If you are astute, you will recognize L^3 as the volume of the box and the kinetic energy which we have shown earlier to relate to the temperature.

If the particle motions are generalized to motion in arbitrary directions, collisions with additional walls in the analysis does not complicate the problem dramatically because each component of the velocity may be evaluated independently. To illustrate, consider a particle bouncing around the centers of four walls in a horizontal plane. From the top view, the trajectory would appear as below:

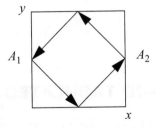

For the same velocity as the first case, the force of each collision would be reduced because the particle strikes merely a glancing blow. The time of collisions between walls is now dependent on the component of velocity perpendicular to the walls. We have chosen a special case to illustrate here, where the box is square and the particle impacts at a 45° angle in the center of each wall. The x-component of the force can be related to the magnitude of the velocity by noting that $v_x = v_y$, such that $v = (v_x^2 + v_y^2)^{1/2} =$

---

9. See an introductory physics text for further discussion of time-averaged force.

$v_x 2^{1/2}$. The time between collisions with wall A_1 would be $4L/(v2^{1/2})$. The formula for the average force in two dimensions then becomes:

$$F_{avg,A_1} = -2mv_x \frac{v\sqrt{2}}{4L} = -2m\frac{v}{\sqrt{2}}\frac{v\sqrt{2}}{4L} = -2mv\frac{v}{4L} = -\frac{mv^2}{2L} \tag{1.18}$$

and the pressure due to two particles that do not collide with one another in two dimensions becomes:

$$P = \frac{m}{2L^3}(v_1^2 + v_2^2) \quad \text{(2D motion)} \tag{ig) 1.19}$$

(More complicated impact angles and locations will provide the same results but require more tedious derivations.) The extension to three dimensions is more difficult to visualize, but comparing Eqn. 1.17 to Eqn. 1.19, you should not be surprised to learn that the pressure in three dimensions is:

$$P = \frac{m}{3L^3}(v_1^2 + (v_2)^2) = \frac{m}{3\underline{V}}(v_1^2 + v_2^2) \quad \text{(3D motion)} \tag{ig) 1.20}$$

The problem gets more complicated when collisions between particles occur. We ignored that possibility here because the ideal gases being considered are point masses that do not collide with one another. Including molecular collisions is a straightforward implementation of "billiard ball" physics. This subject is discussed further in Section 7.10 on page 276 and with great interactive graphics in the discontinuous molecular dynamics (DMD) module at Etomica.org.

We see a relation developing between P and kinetic energy. When we insert the relation between temperature and kinetic energy (Eqn. 1.1) into Eqn. 1.20 we find that the ideal gas law results for a spherical (monatomic) molecule in 3D,

⚠ Check your units when using this equation.

1J = 1kg-m²/s².

$$P\underline{V} = \left(\frac{m}{3\underline{V}}\sum_i^N v_i^2\right)\underline{V} = \frac{N}{3}m\langle v^2\rangle = \frac{nN_A}{3}m\langle v^2\rangle = \frac{nM_w}{3}\langle v^2\rangle = nRT \quad \text{(3D motion)} \tag{ig) 1.21}$$

where m is the mass per particle and M_w is the molecular weight. A similar derivation with Eqn. 1.19 gives the results for motions restricted to 2D,[10]

$$P\underline{V} = \left(\frac{m}{2\underline{V}}\sum_i v_i^2\right)\underline{V} = \frac{nN_A m}{2}\langle v^2\rangle = \frac{nM_w}{2}\langle v^2\rangle = nRT_{2D} \quad \text{(2D motion)} \tag{ig) 1.22}$$

1.5   REAL FLUIDS AND TABULATED PROPERTIES

The thermodynamic behavior of real fluids differs from the behavior of ideal gases in most cases. Real fluids condense, evaporate, freeze, and melt. Characterization of the volume changes and energy changes of these processes is an important skill for the chemical engineer. Many real fluids do behave *as if they are* ideal gases at typical process conditions. *P-V* behavior of a very common real fluid (i.e., water) and an ideal gas can be compared in Figs. 1.3 and 1.4. Application of the

⚠ A 3D steam diagram is available as a MATLAB file called PVT.m. The diagram can be rotated to view the 2-D projections.

---

10. This is a pressure [=] force/area where motion is in 2-D and forces are in only two dimensions. In an alternative perspective molecules would only exist in a 2D plane. Then the divisor should be $2L^2$ and we multiply by area L^2, and P_{2D} [=] MPa-m, $P_{2D}L^2 = nRT$.

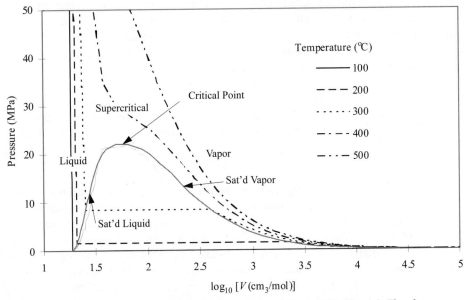

Figure 1.4 *P-V-T behavior of water at the same temperatures used in Fig. 1.3. The plot is prepared from the steam tables in Appendix E.*

ideal gas law simplifies many process calculations for common gases; for example, air at room temperature and pressures below 10 bars. However, you must always remember that the ideal gas law is an approximation (sometimes an excellent approximation) that must be applied carefully to any fluid. The behaviors are presented along **isotherms** (lines of constant temperature) and the deviations from the ideal gas law for water are obvious. Water is one of the most common substances that we work with, and water vapor behaves nearly as an ideal gas at 100°C ($P^{sat} = 0.1014$ MPa), where experimentally the vapor volume is 1.6718 m^3/kg (30,092 cm^3/mol) and by the ideal gas law we may calculate $V = RT/P = 8.314 \cdot 373.15 / 0.1014 = 30{,}595$ cm^3/mol. However, the state is the normal boiling point, and we are well aware that a liquid phase can co-exist at this state. This is because there is another density of water at these conditions that is also stable.[11]

We will frequently find it convenient to work mathematically in terms of molar density or mass density, which is inversely related to molar volume or mass volume, $\rho = 1/V$. Plotting the isotherms in terms of density yields a P-ρ diagram that qualitatively looks like the mirror image of the P-V diagram. Density is convenient to use because it always stays finite as $P \rightarrow 0$, whereas V diverges. Examples of P-ρ diagrams are shown in Fig. 7.1 on page 254.

The conditions where two phases coexist are called **saturation conditions.** The terms "saturation pressure" and "saturation temperature" are used to refer to the state. The volume (or density) is called the saturated volume (or saturated density). Saturation conditions are shown in Fig. 1.4 as the "hump" on the diagram. The hump is called the **phase envelope.** Two phases coexist when the system conditions result in a state *inside* or *on* the envelope. The horizontal lines *inside* the curves are called **tie lines** that show the two volumes (**saturated liquid** and **saturated vapor**) that can coexist. The curve labeled "Sat'd Liquid" is also called the **bubble line,** since it represents conditions where boiling (bubbles) can occur in the liquid. The curve labeled "Sat'd Vapor" is also called a

Real fluids have saturation conditions, bubble points, and dew points.

---

11. This stability is determined by the Gibbs energy and we will defer proof until Chapter 9.

dew line, since it is the condition where droplets (dew) can occur in the vapor. Therefore, saturation is a term that can refer to either bubble or dew conditions. When the total volume of a system results in a system state *on* the saturated vapor line, only an infinitesimal quantity of liquid exists, and the state is indicated by the term "saturated vapor. " Likewise, when a system state is *on* the saturated liquid line, only an infinitesimal quantity of vapor exists, and the state is indicated by the term "saturated liquid." When the total volume of the system results in a system *in between* the saturation vapor and saturation liquid volumes, the system will have vapor and liquid phases coexisting, each phase occupying a finite fraction of the overall system. Note that each isotherm has a unique saturation pressure. This pressure is known as the **saturation pressure** or **vapor pressure.** Although the vapor pressure is often used to characterize a pure liquid's bubble point, recognize that it also represents the dew point for the pure vapor.

Following an isotherm from the right side of the diagram along a path of decreasing volume, the isotherm starts in the vapor region, and the pressure rises as the vapor is isothermally compressed. As the volume reaches the saturation curve at the vapor pressure, a liquid phase begins to form. Notice that further volume decreases do not result in a pressure change until the system reaches the saturated liquid volume, after which further decreases in volume require extremely large pressure changes. Therefore, liquids are often treated as **incompressible** even though the isotherms really do have a finite rather than infinite slope. The accuracy of the incompressible assumption varies with the particular application.

❶ Liquids are quite incompressible.

As we work problems involving processes, we need to use properties such as the internal energy of a fluid.[12] Properties such as these are available for many common fluids in terms of a table or chart. For steam, both tables and charts are commonly used, and in this section we introduce the steam tables available in Appendix E. An online supplement is available to visualize the *P-V* and *P-T* representations in MATLAB permitting the user to interactively rotate the surface.

Steam Tables

When dealing with water, some conventions have developed for referring to the states which can be confusing if the terms are not clearly understood. **Steam** refers to a vapor state, and **saturated steam** is vapor at the dew point. For water, in the two-phase region, the term "wet steam" is used to indicate a vapor + liquid system.

Steam properties are divided into four tables. The first table presents saturation conditions indexed by temperature. This table is most convenient to use when the temperature is known. Each row lists the corresponding saturation values for pressure (vapor pressure), internal energy, volume, and two other properties we will use later in the text: enthalpy and entropy. Special columns represent the internal energy, enthalpy, and entropy of vaporization. These properties are tabulated for convenience, although they can be easily calculated by the difference between the saturated vapor value and the saturated liquid value. Notice that the **vaporization** values decrease as the saturation temperature and pressure increase. The vapor and liquid phases are becoming more similar as the saturation curve is followed to higher temperatures and pressures. At the **critical point,** the phases become identical. Notice in Fig. 1.4 that the two phases become identical at the highest temperature and pressure on the saturation curve, so this is the critical point. For a pure fluid, the **critical temperature** is the temperature at which vapor and liquid phases are identical on the saturation curve, and is given the notation T_c. The pressure at which this occurs is called the **critical pressure,** and is given the symbol P_c. A fluid above the critical temperature is often called **supercritical.**

❶ The critical temperature and critical pressure are key characteristic properties of a fluid.

---

12. Calculation of these properties requires mastery of several fundamental concepts as well as application of calculus and will be deferred. We calculate energies for ideal gas in Chapter 2 and for real fluids in Chapter 8.

The second steam table organizes saturation properties indexed by pressure, so it is easiest to use when the pressure is known. Like the temperature table, vaporization values are presented. The table duplicates the saturated temperature table, that is, plotting the saturated volumes from the two tables would result in the same curves. The third steam table is the largest portion of the steam tables, consisting of superheated steam values. **Superheated** steam is vapor above its saturation temperature at the given pressure. The adjective "superheated" specifies that the vapor is above the saturation temperature at the system pressure. The adjective is usually used only where necessary for clarity. The difference between the system temperature and the saturation temperature, $(T - T^{sat})$, is termed the **degrees of superheat.** The superheated steam tables are indexed by pressure and temperature. The saturation temperature is provided at the top of each pressure table so that the superheat may be quickly determined without referring to the saturation tables.

❶ Superheat.

The fourth steam table has liquid-phase property values at temperatures below the critical temperature and above each corresponding vapor pressure. Liquid at these states is sometimes called **subcooled liquid** to indicate that the temperature is below the saturation temperature for the specified pressure. Another common way to describe these states is to identify the system as **compressed liquid,** which indicates that the pressure is above the saturation pressure at the specified temperature. The adjectives "subcooled" and "compressed" are usually only used where necessary for clarity. Notice by scanning the table that pressure has a small effect on the volume and internal energy of liquid water. By looking at the saturation conditions together with the general behavior of Fig. 1.4 in our minds, we can determine the **state of aggregation** (vapor, liquid, or mixture) for a particular state.

❶ Subcooled, compressed.

Example 1.3   Introduction to steam tables

For the following states, specify if water exists as vapor, liquid, or a mixture: (a) 110°C and 0.12 MPa; (b) 200°C and 2 MPa; (c) 0.8926 MPa and 175°C.

Solution:
(a) Looking at the saturation temperature table, the saturation pressure at 110°C is 0.143 MPa. Below this pressure, water is vapor (steam).

(b) From the saturation temperature table, the saturation pressure is 1.5549 MPa; therefore, water is liquid.

(c) This is a saturation state listed in the saturation temperature table. The water exists as saturated liquid, saturated vapor, or a mixture.

Linear Interpolation

Since the information in the steam tables is tabular, we must interpolate to find values at states that are not listed. To interpolate, we assume the property we desire (e.g., volume, internal energy) varies linearly with the independent variables specified (e.g., pressure, temperature). The assumption of linearity is almost always an approximation, but is a close estimate if the interval of the calculation is small. Suppose we seek the value of volume, V, at pressure, P, and temperature, T, but the steam tables have only values of volume at P_1 and P_2 which straddle the desired pressure value as shown in Fig. 1.5. The two points represent values available in the tables and the solid line represents the true behavior. The dotted line represents a linear fit to the tabulated points.

❶ Linear interpolation is a necessary skill for problem solving using thermodynamic tables.

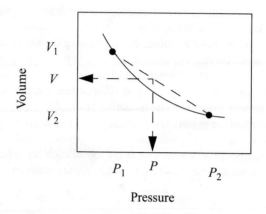

Figure 1.5 *Illustration of linear interpolation.*

If we fit a linear segment to the tabulated points, the equation form is $y = mx + b$, where y is the dependent variable (volume in this case), x is the independent variable (pressure in this case), m is the slope $m = \Delta y / \Delta x = (V_2 - V_1)/(P_2 - P_1)$, and b is the intercept. We can interpolate to find V without directly determining the intercept. Since the point we desire to calculate is also on the line with slope m, it also satisfies the equation $m = \Delta y / \Delta x = (V - V_1)/(P - P_1)$. We can equate the two expressions for m to find the interpolated value of V at P.

There are two quick ways to think about the interpolation. First, since the interpolation is linear, the fractional change in V relative to the volume interval is equal to the fractional change in P relative to the pressure interval. In terms of variables:

$$\frac{V - V_1}{V_2 - V_1} = \frac{P - P_1}{P_2 - P_1}$$

For example, $(V - V_1)$ is 10% of the volume interval $(V_2 - V_1)$, when $(P - P_1)$ is 10% of $(P_2 - P_1)$. We can rearrange this expression to find:

$$V = V_1 + \frac{P - P_1}{P_2 - P_1}(V_2 - V_1) \qquad\qquad 1.23$$

If we consider state "1" as the base state, we can think of this expression in words as

$$V = \text{base } V + (\text{fractional change in } P) \cdot (\text{volume interval size})$$

Another way to think of Eqn. 1.23 is by arranging it as:

$$V = V_1 + \frac{V_2 - V_1}{P_2 - P_1}(P - P_1) \qquad\qquad 1.24$$

which in words is

$$V = \text{base } V + \text{slope} \cdot (\text{change in } P \text{ from base state})$$

Note that subscripts for 1 and 2 can be interchanged in any of the formulas if desired, provided that *all* subscripts are interchanged. In general, interpolation can be performed for any generic property M such that (modifying Eqn. 1.23)

$$M = M_1 + \frac{x - x_1}{x_2 - x_1}(M_2 - M_1)$$ 1.25

where M represents the property of interest (e.g., V) and x is the property you know (e.g., P).

Example 1.4 Interpolation

Find the volume and internal energy for water at: (a) 5 MPa and 325°C and (b) 5 MPa and 269°C.

Solution:
(a) Looking at the superheated steam table at 5 MPa, we find the saturation temperature in the column heading as 263.9°C; therefore, the state is superheated. Values are available at 300°C and 350°C. Since we are halfway in the temperature interval, by interpolation the desired U and V will also be halfway in their respective intervals (which may be found by the average values):

$$U = (2699.0 + 2809.5)/2 = 2754.3 \text{ kJ/kg}$$

$$V = (0.0453 + 0.0520)/2 = 0.0487 \text{ m}^3/\text{kg}$$

(b) For this state, we are between the saturation temperature (263.9°C) and 300°C, and we apply the interpolation formula:

$$U = 2597.0 + \frac{269 - 263.9}{300 - 263.9}(2699.0 - 2597.0) = 2611.4 \text{ kJ/kg}$$

$$V = 0.0394 + \frac{269 - 263.9}{300 - 263.9}(0.0453 - 0.0394) = 0.0402 \text{ m}^3/\text{kg}$$

Double Interpolation

Occasionally, we must perform double or multiple interpolation to find values. The following example illustrates these techniques.

Example 1.5 Double interpolation

For water at 160°C and 0.12 MPa, find the internal energy.
Solution: By looking at the saturation tables at 160°C, water is below the saturation pressure, and will exist as superheated vapor, but superheated values at 0.12 MPa are not tabulated in the superheated table. If we tabulate the available values, we find

Example 1.5 Double interpolation (Continued)

	0.1 MPa	0.12 MPa	0.2 MPa
150°C	2582.9		2577.1
160°C			
200°C	2658.2		2654.6

We may either interpolate the first and third columns to find the values at 160°C, followed by an interpolation in the second row at 160°C, or interpolate the first and third rows, followed by the second column. The values found by the two techniques will not be identical because of the non-linearities of the properties we are interpolating. Generally, the more precise interpolation should be done first, which is over the smaller change in U, which is the pressure interpolation. The pressure increment is 20% of the pressure interval $[(0.12 - 0.1)/(0.2 - 0.1)]$; therefore, interpolating in the first row,

$$U = 2582.9 + 0.2 \cdot (2577.1 - 2582.9) = 2581.7 \text{ kJ/kg}$$

and in the third row,

$$U = 2658.2 + 0.2 \cdot (2654.6 - 2658.2) = 2657.5 \text{ kJ/kg}$$

and then interpolating between these values, using the value at 150°C as the base value,

$$U = 2581.7 + 0.2 \cdot (2657.5 - 2581.7) = 2596.9 \text{ kJ/kg}$$

The final results are tabulated in the boldface cells in the following table:

	0.1 MPa	0.12 MPa	0.2 MPa
150°C	2582.9	**2581.7**	2577.1
160°C		**2596.9**	
200°C	2658.2	**2657.5**	2654.5

We also may need to interpolate between values in different tables, like the saturated tables and superheated tables. This is also straightforward as shown in the following example.

Example 1.6 Double interpolation using different tables

Find the internal energy for water at 0.12 MPa and 110°C.

Solution: We found in Example 1.3 on page 25 that this is a superheated state. From the superheated table we can interpolate to find the internal energy at 110°C and 0.1 MPa:

$$U = 2506.2 + 0.2 \cdot (2582.9 - 2506.2) = 2521.5 \text{ kJ/kg}$$

Example 1.6 Double interpolation using different tables (Continued)

At 0.2 MPa, 110°C is not found in the superheated table because the saturation temperature is 120.3°C, so the values at this pressure cannot be used. Therefore, we can find the desired internal energy by interpolation using the value above and the saturation value at 110°C and 0.143 MPa from the saturation temperature table:

$$U = 2521.5 + \frac{0.12 - 0.1}{0.143 - 0.1}(2517.7 - 2521.5) = 2519.7 \text{ kJ/kg}$$

Computer-Aided Interpolation

Occasionally, interpolation must be performed when the T and P are both unknown. Computers or spreadsheets can be helpful as shown in the next example.

Example 1.7 Double interpolation using Excel

Steam undergoes a series of state changes and is at a final state where $U = 2650$ kJ/kg and $V = 0.185$ m³/kg. Find the T and P.

Solution: Scanning the steam tables, the final state is in the range 1.0 MPa < P < 1.2 MPa, 200°C < T < 250°C. The final state requires a double interpolation using U and V. One easy method is to set up the table in Excel. In each of the tables below, the pressure interpolation is performed first in the top and bottom rows, dependent on the pressure variable in the top of the center column, which can be set at any intermediate pressure to start. The temperature interpolation is then entered in the center cell of each table using the temperature variable. The formulas in both tables reference a common temperature variable cell and a common pressure variable cell. Solver is started and T and P are adjusted to make $U = 2650$ kJ/kg subject to the constraint $V = 0.185$ m³/kg. (See Appendix A for Solver instructions.) The converged result is shown at $T = 219.6$°C and $P = 1.17$ MPa.

Example use of Excel for double interpolation.

		P^f	
U(kJ/kg) table	P = 1 MPa	1.164752	P = 1.2 MPa
T = 200°C	2622.2	2614.539	2612.9
T^f 219.4486791		2650	
T = 250°C	2710.4	2705.705	2704.7

V(m³/kg) table	P = 1 MPa		P = 1.2 MPa
T = 200°C	0.2060	0.175768	0.1693
		0.185	
T = 250°C	0.2327	0.199502	0.1924

Extrapolation

Occasionally, the values we seek are not conveniently between points in the table and we can apply the "interpolation" formulas to extrapolate as shown in Fig. 1.6. In this case, T lies outside the interval. Extrapolation is much less certain than interpolation since we frequently do not know "what curve lies beyond" that we may miss by linear approximation. The formulas used for extrapolation are identical to those used for interpolation. With the steam tables, extrapolation is generally not necessary at normal process conditions and should be avoided if possible.

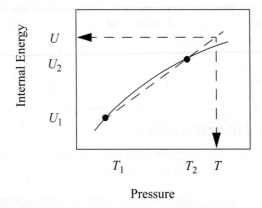

Figure 1.6 *Illustration of linear extrapolation.*

Phase Equilibrium and Quality

Along the saturation curve in Fig. 1.4 on page 23, there is just one **degree of freedom** ($F = C - P + 2$ $= 1 - 2 + 2 = 1$). If we seek saturation, we may choose either a T^{sat} or a P^{sat}, and the other is determined. The vapor pressure increases rapidly with temperature as shown in Fig. 1.7. A plot of $\ln P^{sat}$ versus $1/T^{sat}$ is nearly linear and over large intervals, so for accurate interpolations, vapor pressure data should be converted to this form before interpolation. However, the steam tables used

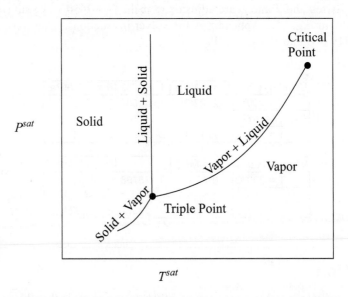

Figure 1.7 *P-T representation of real fluid behavior. Note that only vapor and liquid behavior is shown in Fig. 1.4 on page 23.*

with this text have small enough intervals that direct interpolation can be applied to P^{sat} and T^{sat} without appreciable error.

The saturation volume values of the steam tables were used to generate the phase diagram of Fig. 1.4 on page 23. Note that as the critical point is approached, the saturation vapor and liquid values approach one another. The same occurs for internal energy and two properties that will be used in upcoming chapters, enthalpy, H, and entropy, S. When a mixture of two phases exists, we must characterize the fraction that is vapor, since the vapor and liquid property values differ significantly.

The mass percentage that is vapor is called the **quality** and given the symbol q. The properties V, U, H, and S, may be represented with a generic variable M. The **overall value** of the state variable M is

Quality is the vapor mass percentage of a vapor/liquid mixture.

$$M = (1 - q) M^L + qM^V \qquad 1.26$$

which may be rearranged as

$$M = M^L + q(M^V - M^L)$$

but $(M^V - M^L)$ is just ΔM^{vap} and for internal energy, enthalpy, and entropy, it is tabulated in columns of the saturation tables. The value of overall M is

$$M = M^L + q\Delta M^{vap} \qquad 1.27$$

Look carefully at Eqn. 1.27 in comparison with Eqn. 1.25; it may be helpful to recognize a quality calculation as an interpolation between saturated liquid and saturated vapor. Two examples help demonstrate the importance of quality calculations.

Example 1.8 Quality calculations

Two kg of water coexists as vapor and liquid at 280°C in a 0.05 m³ rigid container. What is the pressure, quality, and overall internal energy of the mixture?

Solution: The overall mass volume is $V = 0.05$ m³/2 kg $= 0.025$ m³/kg. From the saturation temperature table, the pressure is 6.417 MPa. Using the saturation volumes at this condition to find q,

$$0.025 = 0.001333 + q \,(0.0302 - 0.0013) \text{ m}^3/\text{kg}$$

which leads to $q = 0.82$. The overall internal energy is

$$U = 1228.33 + 0.82 \cdot 1358.1 = 2342 \text{ kJ/kg}$$

Example 1.9 Constant volume cooling

Steam is initially contained in a rigid cylinder at $P = 30$ MPa and $V = 10^{2.498}$cm^3/mole. The cylinder is allowed to cool to 300°C. What is the pressure, quality, and overall internal energy of the final mixture?

Solution: The overall mass volume is $V = 10^{2.498}$cm^3-mole^{-1} · 10^{-6}(m^3/cm^3)/(18.02E-3kg/mole) = 0.01747 m^3/kg. From the superheated steam table at 30 MPa, the initial temperature is 900°C. When the cylinder is cooled to 300°C, the path is shown in Fig. 1.8 below. You should notice that there is no pressure in the superheated steam tables that provides a volume of $V =$ 0.01747 m^3/kg. Look hard, they are all too large. (Imagine yourself looking for this on a test when you are in a hurry.) Now look in the saturated steam tables at 300°C. Notice that the saturated vapor volume is 0.0217 m^3/kg. Since that is higher than the desired volume, but it is the lowest vapor volume at this temperature, we must conclude that our condition is somewhere between the saturated liquid and the saturated vapor at a pressure of 8.588 MPa. (When you are in a hurry, it is advisable to check the saturated tables first.) Using the saturation volumes at 300°C condition to find q,

$$0.01747 = 0.001404 + q \, (0.0217 - 0.001404) \text{ m}^3/\text{kg}$$

which leads to $q = (0.01747 - 0.001404)/(0.0217 - 0.001404) = 0.792$. The overall internal energy is

$$U = 1332.95 + 0.792 \cdot 1230.67 = 2308 \text{ kJ/kg}$$

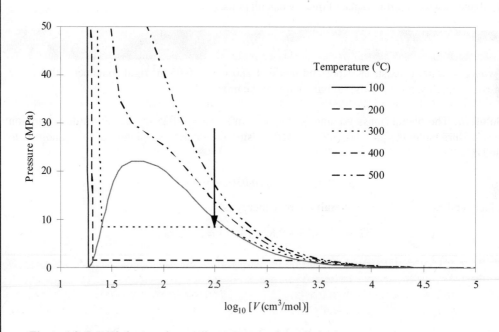

Figure 1.8 *P-V-T behavior of water illustrating a quality calculation.*

1.6   SUMMARY

Years from now you may have some difficulty recalling the details presented in this text. On the other hand, the two basic premises outlined in this introductory section are so fundamental to technically educated people that you really should commit them to long-term memory as soon as possible. Formally, we may state our two basic premises as the first and second "laws" of thermodynamics.[13]

> First Law: Overall energy is conserved (you can't get something for nothing).

> Second Law: Overall entropy changes are greater than or equal to zero (generation of disorder results in lost work).

The first law is further developed in Chapters 2 and 3. The concepts of entropy and the second law are developed in Chapter 4 and process applications in Chapter 5. The exact relationship between the two basic premises and these two laws may not become apparent until some time later in this text, but you should begin to absorb and contemplate these fundamentals now. There are times when the endeavor to apply these simple laws seems daunting, but the answer appears simple in retrospect, once obtained. By practicing adaptation of the basic principles to many specific problems, you slowly grasp the appropriate connection between the basic premises and finding the details. Try not to be distracted by the vocabulary or the tedious notation that goes into keeping all the coupled systems classified in textbook fashion. Keep in mind that other students have passed through this and found the detailed analysis to be worth the effort.

Important Equations

The content of this chapter is primarily about laying down the fundamental concepts more than deriving equations. Nevertheless, three concepts that we refer to repeatedly can be expressed by equations. Of course, the ideal gas law is important, but an implication of it that may be new is:

$$T = \frac{M_w}{3R}\langle v^2 \rangle \text{ (for monatomic molecules in 3D)} \qquad 1.28$$

This equation conveys that temperature is closely related to molecular kinetic energy. Although derived with the ideal gas assumption, it turns out to be true for real fluids as well. Every time you alter the temperature, you should think about the implications for molecular kinetic energy. Another important equation relates to deviations from the ideal gas law:

$$u(r) = \begin{cases} \infty \text{ if } r \leq \sigma \\ -\varepsilon \text{ if } \sigma < r \leq \lambda\sigma \\ 0 \text{ if } r > \lambda\sigma \end{cases} \qquad 1.29$$

This is the square well potential model, the simplest characterization of how real molecules attract and repel. As you add energy to the system, real fluids may absorb that energy by moving molecules from inside the square well to outside, converting potential energy into kinetic energy without altering the temperature as an ideal gas would. A simple example is boiling. Ideal gases cannot

---

13. There is also a **third law of thermodynamics,** as discussed by Denbigh, K., 1981. *The Principles of Chemical Equilibrium,* London: Cambridge University Press, p. 416. The third law is of less direct interest in this introductory text, however.

boil, but real fluids can. This interplay between kinetic energy, temperature, and potential energy pervades many discussions throughout the text.

Finally, we can write a generic equation that symbolizes the procedure for interpolation:

$$M = M_1 + \frac{x - x_1}{x_2 - x_1}(M_2 - M_1) \qquad \qquad 1.30$$

A similar equation is used for quality calculations which can be viewed as an interpolation between saturated liquid and saturated vapor. Throughout Unit I, we refer extensively to the steam tables and interpolation to account for deviations from the ideal gas law.

Test Yourself

1. Draw a sketch of the force model implied by the square-well potential, indicating the position(s) where the force between two atoms is zero and the positions where it is nonzero.

2. Explain in words how the pressure of a fluid against the walls of its container is related to the velocity of the molecules.

3. What is it about molecules that requires us to add heat to convert liquids to gases?

4. If the kinetic energy of pure liquid and vapor molecules at phase equilibrium must be the same, and the internal energy of a system is the sum of the kinetic and potential energies, what does this say about the intensive internal energy of a liquid phase compared with the intensive internal energy of the gas phase?

5. Explain the terms "energy," "potential energy," "kinetic energy," and "internal energy."

6. How is the internal energy of a substance related to the intermolecular pair potentials of the molecules?

7. Are T and P intensive properties? Name two intensive properties and two extensive properties.

8. How many degrees of freedom exist when a pure substance coexists as a liquid and gas?

9. Can an ideal gas condense? Can real fluids that follow the ideal gas law condense?

10. Give examples of bubble, dew, saturation, and superheated conditions. Explain what is meant when wet steam has a quality of 25%.

11. Create and solve a problem that requires double interpolation.

1.7 PRACTICE PROBLEMS

P1.1 Estimate the average speed (mph) of hydrogen molecules at 200 K and 3 bars. (ANS. 3532)

P1.2 Estimate the entropy (J/g-K) of steam at 27.5MPa and 425C. (ANS. 5.1847)

1.8   HOMEWORK PROBLEMS

Note: Some of the steam table homework problems involve enthalpy, H, which is defined for convenience using properties discussed in this chapter, $H \equiv U + PV$. The enthalpy calculations can be performed by reading the tabulated enthalpy values from the tables in an analogous manner used for internal energy. We expect that students will be introduced to this property in course lectures in parallel with the homework problems that utilize H.

1.1 In each of the following, sketch your estimates of the intermolecular potentials between the given molecules and their mixture on the same pair of axes.

   (a) Chloroform is about 20% larger than acetone and about 10% stickier, but chloroform and acetone stick to one another much more strongly than they stick to themselves.
   (b) You have probably heard that "oil and water don't mix." What does that mean in molecular terms? Let's assume that oil can be characterized as benzene and that benzene is four times larger than water, but water is 10% stickier than benzene. If the ε_{12} parameter is practically zero, that would represent that the benzene and water stick to themselves more strongly than to one another. Sketch this.

1.2 For each of the states below, calculate the number of moles of ideal gas held in a three liter container.

   (a) $T = 673$ K, $P = 2$ MPa
   (b) $T = 500$ K, $P = 0.7$ MPa
   (c) $T = 450$ K, $P = 1.5$ MPa

1.3 A 5 m^3 outdoor gas storage tank warms from 10°C to 40°C on a sunny day. If the initial pressure was 0.12 MPa at 10°C, what is the pressure at 40°C, and how many moles of gas are in the tank? Use the ideal gas law.

1.4 An automobile tire has a pressure of 255 kPa (gauge) in the summer when the tire temperature after driving is 50°C. What is the wintertime pressure of the same tire at 0°C if the volume of the tire is considered the same and there are no leaks in the tire?

1.5 A 5 m^3 gas storage tank contains methane. The initial temperature and pressure are $P = 1$ bar, $T = 18$°C. Using the ideal gas law, calculate the P following each of the successive steps.

   (a) 1 m^3 (at standard conditions) is withdrawn isothermally.
   (b) The sun warms the tank to 40°C.
   (c) 1.2 m^3 (at standard conditions) is added to the tank and the final temperature is 35°C.
   (d) The tank cools overnight to 18°C.

1.6 Calculate the mass density of the following gases at 298 K and 1 bar.

   (a) Nitrogen
   (b) Oxygen
   (c) Air (use average molecular weight)
   (d) CO_2
   (e) Argon

1.7 Calculate the mass of air (in kg) that is contained in a classroom that is 12m x 7m x 3m at 293 K and 0.1 MPa.

1.8 Five grams of the specified pure solvent is placed in a variable volume piston. What is the volume of the pure system when 50% and 75% have been evaporated at: (*i*) 30°C, (*ii*) 50°C? Use the Antoine equation (Appendix E) to relate the saturation temperature and saturation pressure. Use the ideal gas law to model the vapor phase. Show that the volume of the system occupied by liquid is negligible compared to the volume occupied by vapor.

(a) Hexane ($\rho^L = 0.66$ g/cm^3)
(b) Benzene ($\rho^L = 0.88$ g/cm^3)
(c) Ethanol ($\rho^L = 0.79$ g/cm^3)
(d) Water without using the steam tables ($\rho^L = 1$ g/cm^3)
(e) Water using the steam tables

1.9 A gasoline spill is approximately 4 liters of liquid. What volume of vapor is created at 1 bar and 293 K when the liquid evaporates? The density of regular gasoline can be estimated by treating it as pure isooctane (2,2,4-trimethylpentane $\rho^L = 0.692$ g/cm^3) at 298 K and 1 bar.

1.10 The gross lifting force of a balloon is given by $(\rho_{air} - \rho_{gas})\underline{V}_{balloon}$. What is the gross lifting force (in kg) of a hot air balloon of volume 1.5E6 L, if the balloon contains gas at 100°C and 1 atm? The hot gas is assumed to have an average molecular weight of 32 due to carbon dioxide from combustion. The surrounding air has an average molecular weight of 29 and is at 25°C and 1 atm.

1.11 LPG is a useful fuel in rural locations without natural gas pipelines. A leak during the filling of a tank can be extremely dangerous because the vapor is denser than air and drifts to low elevations before dispersing, creating an explosion hazard. What volume of vapor is created by a leak of 40L of LPG? Model the liquid before leaking as propane with $\rho^L = 0.24$ g/cm^3. What is the mass density of pure vapor propane after depressurization to 293 K and 1 bar? Compare with the mass density of air at the same conditions.

1.12 The gas phase reaction A → 2R is conducted in a 0.1 m^3 spherical tank. The initial temperature and pressure in the tank are 0.05 MPa and 400 K. After species A is 50% reacted, the temperature has fallen to 350 K. What is the pressure in the vessel?

1.13 A gas stream entering an absorber is 20 mol% CO_2 and 80 mol% air. The flowrate is 1 m^3/min at 1 bar and 360 K. When the gas stream exits the absorber, 98% of the incoming CO_2 has been absorbed into a flowing liquid amine stream.

(a) What are the gas stream mass flowrates on the inlet and outlets in g/min?
(b) What is the volumetric flowrate on the gas outlet of the absorber if the stream is at 320 K and 1 bar?

1.14 A permeation membrane separates an inlet air stream, *F*, (79 mol% N_2, 21 mol% O_2), into a permeate stream, *M*, and a reject stream, *J*. The inlet stream conditions are 293 K, 0.5 MPa, and 2 mol/min; the conditions for both outlet streams are 293 K and 0.1 MPa. If the permeate stream is 50 mol% O_2, and the reject stream is 13 mol% O_2, what are the volumetric flowrates (L/min) of the two outlet streams?

1.15 (a) What size vessel holds 2 kg water at 80°C such that 70% is vapor? What are the pressure and internal energy?

(b) A 1.6 m³ vessel holds 2 kg water at 0.2 MPa. What are the quality, temperature, and internal energy?

1.16 For water at each of the following states, determine the internal energy and enthalpy using the steam tables.

	$T(°C)$	$P(MPa)$
(a)	100	0.01
(b)	550	6.25
(c)	475	7.5
(d)	180	0.7

1.17 Determine the temperature, volume, and quality for one kg water under the following conditions:

(a) $U = 3000$ kJ/kg, $P = 0.3$ MPa
(b) $U = 2900$ kJ/kg, $P = 1.7$ MPa
(c) $U = 2500$ kJ/kg, $P = 0.3$ MPa
(d) $U = 350$ kJ/kg, $P = 0.03$ MPa

1.18 Two kg of water exist initially as a vapor and liquid at 90°C in a rigid container of volume 2.42 m³.

(a) At what pressure is the system?
(b) What is the quality of the system?
(c) The temperature of the container is raised to 100°C. What is the quality of the system, and what is the pressure? What are ΔH and ΔU at this point relative to the initial state?
(d) As the temperature is increased, at what temperature and pressure does the container contain only saturated vapor? What is ΔH and ΔU at this point relative to the initial state?
(e) Make a qualitative sketch of parts (a) through (d) on a P-V diagram, showing the phase envelope.

1.19 Three kg of saturated liquid water are to be evaporated at 60°C.

(a) At what pressure will this occur at equilibrium?
(b) What is the initial volume?
(c) What is the system volume when 2 kg have been evaporated? At this point, what is ΔU relative to the initial state?
(d) What are ΔH and ΔU relative to the initial state for the process when all three kg have been evaporated?
(e) Make a qualitative sketch of parts (b) through (d) on a P-V diagram, showing the phase envelope.

CHAPTER 2

THE ENERGY BALANCE

When you can measure what you are speaking about, and express it in numbers, you know something about it. When you cannot measure it, your knowledge is meager and unsatisfactory.

Lord Kelvin

The energy balance is based on the postulate of conservation of energy in the universe. This postulate is known as the **first law of thermodynamics**. It is a "law" in the same sense as Newton's laws. It is not refuted by experimental observations within a broadly defined range of conditions, but there is no mathematical proof of its validity. Derived from experimental observation, it quantitatively accounts for energy transformations (heat, work, kinetic, potential). We take the first law as a starting point, a postulate at the macroscopic level, although the conservation of energy in elastic collisions does suggest this inference in the absence of radiation. Facility with computation of energy transformations is a necessary step in developing an understanding of elementary thermodynamics. The first law relates work, heat, and flow to the internal energy, kinetic energy, and potential energy of the system. Therefore, we precede the introduction of the first law with discussion of work and heat.

 The energy balance is also known as the first law of thermodynamics.

Chapter Objectives: You Should Be Able to...

1. Explain why enthalpy is a convenient property to define and tabulate.

2. Explain the importance of assuming reversibility in making engineering calculations of work.

3. Calculate work and heat flow for an ideal gas along the following pathways: isothermal, isochoric, adiabatic.

4. Simplify the general energy balance for problems similar to the homework problems, textbook examples, and practice problems.

5. Properly use heat capacity polynomials and latent heats to calculate changes in *U, H* for ideal gases and condensed phases.

6. Calculate ideal gas or liquid properties relative to an ideal gas or liquid reference state, using the ideal gas law for the vapor phase properties and heats of vaporization.

2.1 EXPANSION/CONTRACTION WORK

There is a simple way that a force on a surface may interact with the system to cause expansion/contraction of the system in volume. This is the type of surface interaction that occurs if we release the latch of a piston, and move the piston in/out while holding the cylinder in a fixed location. Note that a moving boundary is not sufficient to distinguish this type of work—there must be movement of the system boundaries *relative to one another.* For expansion/contraction interactions, the size of the system *must change.* This distinction becomes significant when we contrast expansion/contraction work to flow work in Section 2.3.

How can we relate this amount of work to other quantities that are easily measured, like volume and pressure? For a force applied in the x direction, the work done on our system is

$$d\underline{W} = F_{applied}dx = -F_{system}dx$$

where we have used Newton's principle of equal and opposite forces acting on a boundary to relate the applied and system forces. Since it is more convenient to use the system force in calculations, we use the latter form, and drop the subscript with the understanding that we are calculating the work done on the system and basing the calculation on the system force. For a constant force, we may write

$$\underline{W} = -F\Delta x$$

If F is changing as a function of x then we must use an integral of F,

$$\underline{W} = -\int F dx \qquad \qquad 2.1$$

For a fluid acting on a surface of constant area A, the system force and pressure are related,

$$P = F/A \Rightarrow F = P \cdot A$$

$$\underline{W}_{EC} = -\int PA dx = -\int P d\underline{V} \qquad \qquad 2.2$$

where the subscript EC refers to expansion/contraction work.

In evaluating this expression, a nagging question of perspective comes up. It would be a trivial question except that it causes major headaches when we later try to keep track of positive and negative signs. The question is essentially this: In the discussion above, is positive work being done on the system, or is negative work being done by the system? When we add energy to the system, we consider it a positive input into the system; therefore, putting work into the system should also be considered as a positive input. On the other hand, when a system does work, the energy should go down, and it might be convenient to express work done *by* the system as positive. The problem is that both perspectives are equally valid—therefore, the choice is arbitrary. Since various textbooks choose differently, there is always confusion about sign conventions. The best we can hope for is to be consistent during our own discussions. We hereby consider **work to be positive when performed *on* the system.** Thus, energy put into the system is positive. Because volume decreases when performing work of compression, the sign on the integral for work is negative,

$$\boxed{\underline{W}_{EC} = -\int P\,d\underline{V}} \quad \text{(reversible change of system size)} \qquad 2.3$$

> ❶ Expansion/Contraction work is associated with a change in *system* size.

where P and $\underline{V}$ are of the *system*. Clarification of "reversible" is given in Section 2.4 on page 42. By comparing Eqn. 2.3 with the definitions of work given by Eqns. 2.1 and 2.2, it should be obvious that the $d\underline{V}$ term results from expansion/contraction of the boundary of the system. The P results from the force of the system acting at the boundary. Therefore, to use Eqn. 2.3, the pressure in the integral is the pressure *of the system* at the boundary, and the boundary must move. A system which does not have an expanding/contracting boundary does not have expansion/contraction work.[1]

2.2   SHAFT WORK

In a flowing system, we know that a propeller-type device can be used to push a fluid through pipes—this is the basis of a centrifugal pump. Also, a fluid flowing through a similar device could cause movement of a shaft—this is the basis for hydroelectric power generation and the water wheels that powered mills in the early twentieth century. These are the most commonly encountered forms of shaft work in thermodynamics, but there is another slight variation. Suppose an impeller was inserted into a cylinder containing cookie batter and stirred while holding the piston at a fixed volume. We would be putting work into the cylinder, but the system boundaries would neither expand nor contract. All of these cases exemplify shaft work. The essential feature of shaft work is that work is being added or removed without a change in volume of the system. We show in Section 2.8, page 54, that shaft work for a reversible flow process can be computed from

> ❶ Shaft work characterizes the work of a turbine or pump.

$$W_S = \int_{in}^{out} V\,dP \quad \text{reversible shaft work, flowing system} \qquad 2.4$$

Note that Eqns. 2.3 and 2.4 are distinct and should not be interchanged. Eqn. 2.4 is restricted to shaft work in an open system and Eqn. 2.3 is for expansion/contraction work in a closed system. We later show how selection of the system boundary in a flow system relates the two types of terms on page 54.

2.3   WORK ASSOCIATED WITH FLOW

In engineering applications, most problems involve flowing systems. This means that materials typically flow into a piece of equipment and then flow out of it, crossing well-defined system boundaries in the process. Thus, we need to introduce an additional characterization of work: the work interaction of the system and surroundings when mass crosses a boundary. For example, when a gas is released out of a tank through a valve, the exiting gas pushes the surrounding fluid, doing work on the surroundings. Likewise, when a tank valve is opened to allow gas from a higher pressure source to flow inward, the surroundings do work on the gas already in the system. We calculate the work in these situations most easily by first calculating the rate at which work is done.

---

1. Some texts refer to expansion/contraction work as *PV* work. This leads to confusion since Section 2.3 shows that work associated with flow is *PV,* and the types of work are distinctly different. We have chosen to use the term "expansion/contraction" for work involved in moving boundaries to help avoid this ambiguity.

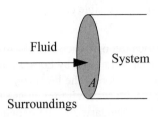

Figure 2.1 *Schematic illustration of flow work.*

Let us first consider a fluid entering a system as shown in Fig. 2.1. We have $d\underline{W} = F dx$, and the work interaction of the system is positive since we are pushing fluid into the system. The rate of work is $\dot{\underline{W}} = F\dot{x}$, but $\dot{x}$ is velocity, and $F = P \cdot A$. Further rearranging, recognizing $A\dot{x} = \dot{\underline{V}}$, and that the volumetric flow rate may be related to the mass specific volume and the mass flow rate, $\dot{\underline{V}} = V\dot{m}^{in}$,

<table><tr><td>Work associated with fluid flowing in/out of boundaries is called flow work.</td></tr></table>

$$\dot{\underline{W}}_{flow}^{in} = (PA\dot{x})^{in} = (P\dot{\underline{V}})^{in} = (PV)^{in}\dot{m}^{in} \qquad 2.5$$

where PV are the properties of the fluid at the point where it crosses the boundary, and $\dot{m}^{in}$ is the absolute value of the mass flow rate across the boundary. When fluid flows out of the system, work is done on the surroundings and the work interaction *of the system* is

$$\dot{\underline{W}}_{flow}^{out} = -(PA\dot{x})^{out} = -(P\dot{\underline{V}})^{out} = -(PV)^{out}\dot{m}^{out} \qquad 2.6$$

where $\dot{m}^{out}$ is the absolute value of the mass flow across the boundary, and since work is being done on the surroundings, the work interaction of the system is negative. When flow occurs both in and out, the net flow work is the difference:

$$\dot{\underline{W}}_{flow} = (PV)^{in}\dot{m}^{in} - (PV)^{out}\dot{m}^{out} \qquad 2.7$$

where $\dot{m}^{in}$ and $\dot{m}^{out}$ are absolute values of the mass flow rates. For more streams, we simply follow the conventions established, and add inlet streams and subtract outlet streams.

2.4   LOST WORK VERSUS REVERSIBILITY

In order to properly understand the various characteristic forms that work may assume, we must address an issue which primarily belongs to the upcoming chapter on entropy. The problem is that the generation of disorder reflected by entropy change results in conversion of potentially useful work energy into practically useless thermal energy. If "generation of disorder results in lost work," then operating in a disorderly fashion results in the lost capability to perform useful work, which we abbreviate by the term: "lost work." It turns out that the most orderly manner of operating is a hypothetical process known as a reversible process. Typically, this hypothetical, reversible process is applied as an initial approximation of the real process, and then a correction factor is applied to estimate the results for the actual process. It was not mentioned in the discussion of expansion/contraction work, but we implicitly assumed that the process was performed reversibly, so that all of the work on the system was stored in a potentially useful form. To see that this might not always be the case, and how this observation relates to the term "reversible," consider the problem of stirring

<table><tr><td>Real processes involve "lost work."</td></tr></table>

cookie batter. Does the cookie batter become unmixed if you stir in the reverse direction? Of course not. The shaft work of stirring has been degraded to effect the randomness of the ingredients. It is impossible to completely recover the work lost in the randomness of this irreversible process. Any real process involves some degree of stirring or mixing, so lost work cannot be eliminated, but we can hope to minimize unnecessary losses if we understand the issue properly.

Consider a process involving gas enclosed in a piston and cylinder. Let the piston be oriented upward so that an expansion of the gas causes the piston to move upward. Suppose that the pressure in the piston is great enough to cause the piston to move upward when the latch is released. How can the process be carried out so that the expansion process yields the maximum work? First, we know that we must eliminate friction to obtain the maximum movement of the piston.

> **Friction** decreases the work available from a process. Frequently we neglect friction to perform a calculation of maximum work.

❶ Friction results in "lost work."

If we neglect friction, what will happen when we release the latch? The forces are not balanced. Let us take z as our coordinate in the vertical direction, with increasing values in the upward direction. The forces downward on the piston are the force of atmospheric pressure ($-P_{atm} \cdot A$, where A is the cross-sectional area of the piston) and the force of gravity ($-m \cdot g$). These forces will be constant throughout movement of the piston. The upward force is the force exerted by the gas ($P \cdot A$). Since the forces are not balanced, the piston will begin to accelerate upward ($F = ma$). It will continue to accelerate until the forces become balanced.[2] However, when the forces are balanced, the piston will have a non-zero velocity. As it continues to move up, the pressure inside the piston continues to fall, making the upward force due to the inside pressure smaller than the downward force. This causes the piston to decelerate until it eventually stops. However, when it stops at the top of the travel, it is still not in equilibrium because the forces are again not balanced. It begins to move downward. In fact, *in the absence of dissipative mechanisms* we have set up a perpetual motion.[3] A reversible piston would oscillate continuously between the initial state and the state at the top of travel. This would not happen in a real system. One phenomenon which we have failed to consider is viscous dissipation (the effect of viscosity).

Let us consider how velocity gradients dissipate linear motion. Consider two diatomic molecules touching one another which both have exactly the same velocity and are traveling in exactly the same direction. Suppose that neither is rotating. They will continue to travel in this direction at the same velocity until they interact with an external body. Now consider the same two molecules in contact, again moving in exactly the same direction, but one moving slightly faster. Now there is a velocity gradient. Since they are touching one another, the fact that one is moving a little faster than the other causes one to begin to rotate clockwise and the other counter-clockwise because of friction as one tries to move faster than the other. Naturally, the kinetic energy of the molecules will stay constant, but the directional velocities are being converted to rotational (directionless) energies. This is an example of viscous dissipation in a shear situation. In the case of the oscillating piston, the viscous dissipation prevents complete transfer of the internal energy of the gas to the piston during expansion, resulting in a stroke that is shorter than a reversible stroke. During compression, viscous dissipation results in a fixed internal energy rise for a shorter stroke than a reversible process. In both expansion and compression, the temperature of the gas at the end of each stroke is higher than it would be for a reversible stroke, and each stroke becomes successively shorter.

---

2. Two other possibilities exist: 1) The piston may hit a stop before it has finished moving upward, a case that will be considered below, or; 2) The piston may fly out of the cylinder if the cylinder is too short, and there is no stop.
3. However, this is not a useful perpetual motion machine because the net effect on the surroundings and the piston is zero at the end of each cycle. If we tried to utilize the motion, we would damp it out.

Velocity gradients lead to dissipation of directional motion (kinetic energy) into random motion (internal energy) due to the viscosity of a fluid. Frequently, we neglect viscous dissipation to calculate maximum work. A fluid would need to have zero viscosity for this mechanism of dissipation to be non-existent. **Pressure gradients** within a viscous fluid lead to velocity gradients; thus, one type of gradient is associated with the other.

> ❶ Velocity gradients in viscous fluids lead to lost work.

We can see that friction and viscosity play an important role in the loss of capability to perform useful work in real systems. In our example, these forces cause the oscillations to decrease somewhat with each cycle until the piston comes to rest. Another possibility of motion that might occur with a piston is interaction with a stop, which limits the travel of the piston. As the piston travels upward, if it hits the stop, it will have kinetic energy which must be absorbed. In a real system, this kinetic energy is converted to internal energy of the piston, cylinder, and gas.

Kinetic energy is dissipated to internal energy when objects collide inelastically, such as when a moving piston strikes a stop. Frequently we imagine systems where the cylinder and piston can neither absorb nor transmit heat; therefore, the lost kinetic energy is returned to the gas as internal energy.

> ❶ Inelastic collisions result in lost work.

So far, we have identified three dissipative mechanisms. Additional mechanisms are **diffusion** along a **concentration gradient** and heat conduction along a **temperature gradient,** which will be discussed in Chapter 4. Velocity, temperature, and concentration gradients are always associated with losses of work. If we could eliminate them, we could perform maximum work (but it would require infinite time).

A process without dissipative losses is called **reversible.** A process is reversible if the system may be returned to a prior state by reversing the motion. We can usually determine that a system is not reversible by recognizing when dissipative mechanisms exist.

> ❶ A reversible process avoids lost work.

Approaching Reversibility

We can approach reversibility by eliminating gradients within our system. To do this, we can perform motion by differential changes in forces, concentrations, temperatures, and so on. Let us consider a piston with a weight on top, at equilibrium. If we slide the weight off to the side, the change in potential energy of the weight is zero, and the piston rises, so its potential energy increases. If the piston hits a stop, kinetic energy is dissipated. Now let us subdivide the weight into two portions. If we move off one-half of the weight, the piston strikes the stop with less kinetic energy than before, and in addition, we have now raised half of the weight. If we repeat the subdivision again we would find that we could move increasing amounts of weight by decreasing the weight we initially move off the piston. In the limit, our weight would become like a pile of sand, and we would remove one grain at a time. Since the changes in the system are so small, only infinitesimal gradients would ever develop, and we would approach reversibility. The piston would never develop kinetic energy which would need to be dissipated.

Reversibility by a Series of Equilibrium States

When we move a system differentially, as just discussed, the system is at equilibrium along each step of the process. To test whether the system is at equilibrium at a particular stage, we can imagine freezing the process at that stage. Then we can ask whether the system would change if we left it at those conditions. If the system would remain static (i.e., not changing) at those conditions, then it must be at equilibrium. Because it is static, we could just as easily go one way as another $\Rightarrow$ "reversible." Thus, reversible processes are the result of infinitesimal driving forces.

Reversibility by Neglecting Viscosity and Friction

Real processes are not done infinitely slowly. In the previous examples we have used idealized pistons and cylinders for discussion. Real systems can be far from ideal and may have much more complex geometry. For example, projectiles can be fired using gases to drive them, and we need a method to estimate the velocities with which they are projected into free flight. One application of this is the steam catapult used to assist airplanes in becoming airborne from the short flight decks of aircraft carriers. Another application would be determination of the exit velocity of a bullet fired from a gun. These are definitely not equilibrium processes, so how can we begin to calculate the exit velocities? Another case would be the centrifugal pump. The pump works by rapidly rotating a propeller-type device. The pump simply would not work at low speed without velocity gradients! So what do we do in these cases? The answer is that we perform a calculation ignoring viscosity and friction. Then we apply an efficiency factor to calculate the real work done. The efficiency factors are determined empirically from our experience with real systems of a similar nature to the problem at hand. Efficiencies are introduced in Chapter 4. In the remainder of this chapter, we concentrate on the first part of the problem, calculation of reversible work.

Viscosity and friction are frequently ignored for an estimation of optimum work, and an empirical efficiency factor is applied based on experience with similar systems.

Example 2.1   Isothermal reversible compression of an ideal gas

Calculate the work necessary to isothermally perform steady compression of two moles of an ideal gas from 1 to 10 bar and 311 K in a piston. An isothermal process is one at constant temperature. The steady compression of the gas should be performed such that the pressure of the system is always practically equal to the external pressure on the system. We refer to this type of compression as "reversible" compression.

Solution: System: closed; Basis: one mole

$$W_{EC} = -\int P\,dV$$

$$P = \frac{RT}{V} \Rightarrow W_{EC} = -\int_{V_1}^{V_2} \frac{RT}{V}\,dV = -RT\int_{V_1}^{V_2} \frac{dV}{V} = -RT\ln\left(\frac{V_2}{V_1}\right) \qquad \text{(ig) 2.8}$$

$$\frac{V_2}{V_1} = \frac{(RT)/P_2}{(RT)/P_1} = \frac{P_1}{P_2} = \frac{1}{10} \qquad \text{(ig) 2.9}$$

$$W_{EC} = -8.314 \text{ J/mol-K} \cdot 311 \text{ K } \ln(1/10) = 5954 \text{ J/mol}$$

$$\underline{W}_{EC} = 2(5954) = 11,908 \text{ J}$$

Note: Work is done on the gas since the sign is positive. This is the sign convention set forth in Eqn. 2.3. If the integral for Eqn. 2.3 is always written as shown with the initial state as the lower limit of integration and the P and V properties of the system, the work on the gas will always result with the correct sign.

2.5   HEAT FLOW

A very simple experiment shows us that heat transport is also related to energy. If two steel blocks of different temperature are placed in contact with one another, but otherwise are insulated from their surroundings, they will come to equilibrium at a common intermediate temperature. The warmer block will be cooled, and the colder block will be warmed.

$$\underline{Q}_{block\ 1} = -\underline{Q}_{block\ 2}$$

Heat is transferred at a boundary between the blocks. Therefore, heat is not a property of the system. It is a form of interaction at the boundary which transfers internal energy. If heat is added to a system for a finite period of time, then the energy of the system increases because the kinetic energy of the molecules is increased. When an object feels hot to our touch, it is because the kinetic energy of molecules is readily transferred to our hand.

Since the rate of heating may vary with time, we must recognize that the total heat flows must be summed (or integrated) over time. In general, we can represent a differential contribution by

$$d\underline{Q} = \dot{\underline{Q}} dt$$

We can also relate the internal energy change and heat transfer for either block in a differential form:

$$d\underline{U} = d\underline{Q} \quad \text{or} \quad \frac{d\underline{U}}{dt} = \dot{\underline{Q}} \qquad\qquad 2.10$$

❶ Diathermal.

An idealized system boundary that has no resistance to heat transfer but is impervious to mass is called a **diathermal** wall.

2.6   PATH PROPERTIES AND STATE PROPERTIES

❶ The terms "iso-thermal," "isobaric," "isochoric," and "adiabatic," describe pathways.

In the previous example, we have used an **isothermal** path. It is convenient to define other terms which describe pathways concisely. An **isobaric** path is one at constant pressure. An **isochoric** path is one at constant volume. An **adiabatic** path is one without heat transfer.

❶ The work and heat transfer neces-sary for a change in state are depen-dent on the pathway taken between the initial and final state.

The heat and work transfer necessary for a change in state are dependent on the pathway taken between the initial and final states. A state property is one that is independent of the pathway taken. For example, when the pressure and temperature of a gas are changed and the gas is returned to its initial state, the net change in temperature, pressure, and internal energy is zero, and these proper-ties are therefore state properties. However, the net work and net heat transfer will not necessarily be zero; their values will depend on the path taken. Also, it is helpful to recall that heat and work are not properties of the system; therefore, they are not state properties.

Example 2.2   Work as a path function

Consider 1.2 moles of an ideal gas in a piston at 298 K and 0.2 MPa and at volume $\underline{V}_1$. The gas is expanded isothermally to twice its original volume, then cooled isobarically to $\underline{V}_1$. It is then heated at constant volume back to T_1. Demonstrate that the net work is non-zero, and that the work depends on the path.

Example 2.2   Work as a path function (Continued)

Solution: First sketch the process on a diagram to visualize the process as shown in Fig. 2.2. Determine the initial volume:

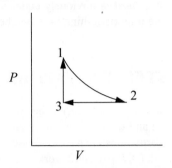

Figure 2.2 *Schematic for Example 2.2.*

$$\underline{V} = nR(T/P) = \frac{1.2 \text{ moles} \mid 8.314 \text{cm}^3 \text{MPa} \mid 298 \text{ K}}{\mid \text{mole K} \mid 0.2 \text{ MPa}} = 14{,}865 \text{ cm}^3 \qquad \text{(ig)}$$

1. Isothermally expand that gas:

$$\Rightarrow \underline{W}_{EC} = -\int Pd\underline{V} = -nRT_1 \ln(V_2/V_1) \qquad \text{(ig) 2.11}$$

$$= \frac{1.2 \text{ moles} \mid (8.314) \text{ cm}^3 MPa \mid 298 \text{ K}}{\mid \text{mole } K \mid} \ln(2) = -2060 \text{ J}$$

2. Isobarically cool down to V_1:

$$\underline{W}_{EC} = -\int_{V_2}^{V_1} P_2 d\underline{V} = -P_2(\underline{V}_1 - \underline{V}_2) = -0.1 \text{MPa}(-14{,}865 \text{ cm}^3) = 1487 \text{ J} \qquad 2.12$$

3. Heat at constant volume back to T_1:

$$\Rightarrow \underline{W}_{EC} = 0 \text{ (because } d\underline{V} = 0 \text{ over entire step)}$$

We have returned the system to its original state and all state properties have returned to their initial values. What is the total work done on the system?

$$\underline{W} = \underline{W}_{1 \rightarrow 2} + \underline{W}_{2 \rightarrow 3} + \underline{W}_{3 \rightarrow 1} = -nRT_1 \ln(V_2/V_1) + nP_2(V_2 - V_1) = -573 \text{ J} \quad \text{(ig) 2.13}$$

Therefore, we conclude that work is a path function, not a state function.

Exercise: If we reverse the path, the work will be different; in fact, it will be positive instead of negative (+573.6 J). If we change the path to isobarically expand the gas to double the volume

($\underline{W} = -2973$ J), cool to T_1 at constant volume ($\underline{W} = 0$ J), then isothermally compress to the original volume ($\underline{W} = -2060$ J), the work will be −913 J.

Note: Heat was added and removed during the process of Example 2.2 which has not been accounted for above. The above process transforms work into heat, and all we have done is computed the amount of work. The amount of heat is obviously equal in magnitude and opposite in sign, in accordance with the first law. The important thing to remember is that work is a path function, not a state function.

❶ Work and heat are path properties.

2.7   THE CLOSED-SYSTEM ENERGY BALANCE

A closed system is one in which no mass flows in or out of the system, as shown in Fig. 2.3. The introductory sections have discussed heat and work interactions, but we have not yet coupled these to the energy of the system. In the transformations we have discussed, energy can cross a boundary in the form of expansion/contraction work ($-\int PdV$), shaft work (W_S), and heat (Q)[4]. *There are only two ways a closed system can interact with the surroundings,* via heat and work interactions. If we put both of these possibilities into one balance equation, then developing the balance for a given application is simply a matter of analyzing a given situation and deleting the balance terms that do not apply. The equation terms can be thought of as a check list.

❶ A closed system interacts with the surroundings only through heat and work.

Experimentally, scientists discovered that if heat and work are measured for a cyclical process which returns to the initial state, the heat and work interactions together always sum to zero. *This is an important result!* This means that, in non-cyclical processes where the sum of heat and work is non-zero, the system has stored or released energy, depending on whether the sum is positive or negative. In fact, by performing enough experiments, scientists decided that the sum of heat and work interactions in a *closed* system *is* the change in energy of the system! To develop the closed-system energy balance, let us first express the balance in terms of words.

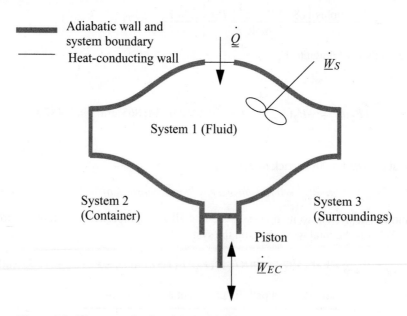

Figure 2.3 *Schematic of a closed system.*

---

4. Other possibilities include electric or magnetic fields, or mechanical springs, etc., which we do not address in this text.

$$\begin{bmatrix} \text{energy} \\ \text{accumulation within} \\ \text{system boundaries} \end{bmatrix} = \begin{bmatrix} \text{heat flow} \\ \text{into system} \end{bmatrix} + \begin{bmatrix} \text{work done} \\ \text{on system} \end{bmatrix} \qquad 2.14$$

Energy within the system is composed of the internal energy (e.g., $\underline{U}$), and the kinetic ($mu^2/2g_c$) and potential energy (mgz/g_c) of the center of mass. For closed systems, the "check list" equation is:

$$m\, d\left[U + \frac{v^2}{2g_c} + \frac{gz}{g_c}\right] = d\underline{Q} + d\underline{W}_S + d\underline{W}_{EC} \qquad 2.15$$

The left-hand side summarizes changes occurring *within* the system boundaries and the right-hand side summarizes changes due to interactions *at* the boundaries. It is a recommended practice to always write the balance in this convention when starting a problem. We will follow this convention throughout example problems in Chapters 2–4 and relax the practice subsequently. The kinetic and potential energy of interest in Eqn. 2.15 is for the center of mass, not the random kinetic and potential energy of molecules about the center of mass. The balance could also be expressed in terms of molar quantities, but if we do so, we need to introduce molecular weight in the potential and kinetic energy terms. Since the mass is constant in a closed system, we may divide the above equation by m,

$$\boxed{d\left[U + \frac{v^2}{2g_c} + \frac{gz}{g_c}\right] = dQ + dW_S + dW_{EC}} \qquad 2.16$$

❶ Closed-system balance. The left-hand side summarizes changes *inside* the boundaries, and the right-hand side summarizes interactions *at* the boundaries.

where heat and work interactions are summed for multiple interactions at the boundaries. We can integrate Eqn. 2.16 to obtain

$$\Delta\left(U + \frac{v^2}{2g_c} + \frac{gz}{g_c}\right) = Q + W_S + W_{EC} \qquad 2.17$$

We may also express the energy balance in terms of rates of change,

$$\frac{d}{dt}\left[U + \frac{v^2}{2g_c} + \frac{gz}{g_c}\right] = \dot{Q} + \dot{W}_S + \dot{W}_{EC} \qquad 2.18$$

where $\dot{Q} = dQ/dt$, $\dot{W}_S = dW_S/dt$, and $\dot{W}_{EC} = dW_{EC}/dt$. Frequently, the kinetic and potential energy changes are small (as we will show in Example 2.9), in a closed system shaft work is not common, and the balance simplifies to

$$\boxed{\frac{dU}{dt} = \dot{Q} + \dot{W}_{EC}} \quad \text{or} \quad \boxed{\Delta U = Q + W_{EC}} \qquad 2.19$$

Example 2.3 Internal energy and heat

In Section 2.5 on page 46 we discussed that heat flow is related to the energy of system, and now we have a relation to quantify changes in energy. If 2000 J of heat are passed from the hot block to the cold block, how much has the internal energy of each block changed?

Solution: First choose a system boundary. Let us initially place system boundaries around each of the blocks. Let the warm block be *block1* and the cold block be *block2*. Next, eliminate terms which are zero or are not important. The problem statement says nothing about changes in position or velocity of the blocks, so these terms can be eliminated from the balance. There is no shaft involved, so shaft work can be eliminated. The problem statement doesn't specify the pressure, so it is common to assume that the process is at a constant atmospheric pressure of 0.101 MPa. The cold block does expand slightly when it is warmed, and the warm block will contract; however, since we are dealing with solids, the work interaction is so small that it can be neglected. For example, the blocks together would have to change 10 cm^3 at 0.101 MPa to equal 1 J out of the 2000 J that are transferred.

Therefore, the energy balance for each block becomes:

$$d\left[U + \frac{\cancel{v^2}}{2g_c} + \frac{\cancel{gz}}{g_c}\right] = dQ + \cancel{dW_S} + \cancel{dW_{EC}}$$

We can integrate the energy balance for each block:

$$\Delta \underline{U}_{block1} = \underline{Q}_{block1} \qquad \Delta \underline{U}_{block2} = \underline{Q}_{block2}$$

The magnitude of the heat transfer between the blocks is the same since no heat is transferred to the surroundings, but how about the signs? Let's explore that further. Now, placing the system boundary around both blocks, the energy balance becomes:

$$d\left[U + \frac{\cancel{v^2}}{2g_c} + \frac{\cancel{gz}}{g_c}\right] = \cancel{dQ} + \cancel{dW_S} + \cancel{dW_{EC}}$$

Note that the composite system is an **isolated system** since all heat and work interactions across the boundary are negligible. Therefore, $\Delta \underline{U} = 0$ or by dividing in subsystems, $\Delta \underline{U}_{block1} + \Delta \underline{U}_{block2} = 0$ which becomes $\Delta \underline{U}_{block1} = -\Delta \underline{U}_{block2}$. Notice that the signs are important in keeping track of which system is giving up heat and which system is gaining heat. In this example, it would be easy to keep track, but other problems will be more complicated, and it is best to develop a good bookkeeping practice of watching the signs. In this example the heat transfer for the initially hot system will be negative, and the heat transfer for the other system will be positive. Therefore, the internal energy changes are $\Delta \underline{U}_{block1} = -2000$ J and $\Delta \underline{U}_{block2} = 2000$ J.

Although very simple, this example has illustrated several important points.

1. Before simplifying the energy balance, the boundary should be clearly described by a statement and/or a sketch.

2. A system can be subdivided into subsystems. The composite system above is isolated, but the subsystems are not. Many times, problems are more easily solved, or insight is gained by looking at the overall system. If the subsystem balances look difficult to solve, try an overall balance.

3. Positive and negative energy signs are important to use carefully.

4. Simplifications can be made when some terms are small relative to other terms. Calculation of the expansion contraction work for the solids is certainly possible above, but it has a negligible contribution. However, if the two subsystems had included gases, then this simplification would have not been reasonable.

Four important points about solving problems.

2.8 THE OPEN-SYSTEM, STEADY-STATE BALANCE

Having established the energy balance for a closed system, and, from Section 2.3, the work associated with flowing fluids, let us extend these concepts to develop the energy balance for a steady-state flow system. The term **steady-state** means the following:

1. All state properties throughout the system are invariant with respect to time. The properties may vary with respect to position within the system.

2. The system has constant mass, that is, the total inlet mass flow rate equals the total outlet mass flow rates, and all flow rates are invariant with respect to time.

3. The center of mass for the system is fixed in space. (This restriction is not strictly required, but will be used throughout this text.)

To begin, we write the balance in words, by adding flow to our previous closed-system balance. There are only three ways the surroundings can interact with the system: flow, heat, and work. A schematic of an open steady-state system is shown in Fig. 2.4. In consideration of the types of work encountered in steady-state flow, recognize that expansion/contraction work is rarely involved, so this term is omitted at this preliminary stage. This is because we typically apply the steady-state balance to systems of rigid mechanical equipment, and there is no change in the size of the system. Therefore, the expansion/contraction work term is set to 0.

Steady-state flow systems are usually fixed size, so $W_{EC} = 0$.

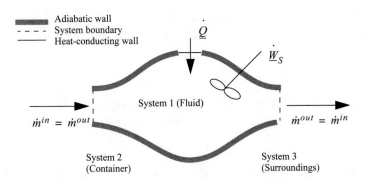

Figure 2.4 *Schematic of a steady-state flow system.*

The balance in words becomes time-dependent since we work with flow rates:

$$
\begin{bmatrix} \text{rate of energy} \\ \text{accumulation within} \\ \text{system boundaries} \end{bmatrix} = \begin{bmatrix} \text{energy per unit} \\ \text{mass of fluid at inlet} \end{bmatrix} \begin{bmatrix} \text{mass} \\ \text{flow rate} \\ \text{in} \end{bmatrix} \qquad 2.20
$$

$$
- \begin{bmatrix} \text{energy per unit} \\ \text{mass of fluid at outlet} \end{bmatrix} \begin{bmatrix} \text{mass} \\ \text{flow rate} \\ \text{out} \end{bmatrix} + \begin{bmatrix} \text{rate of heat flow} \\ \text{into system} \end{bmatrix}
$$

$$
+ \begin{bmatrix} \text{rate that work} \\ \text{is done on system} \end{bmatrix}
$$

Again, we follow the convention that the left-hand side quantifies changes *inside* our system. Consider the change of energy inside the system boundary given by the left-hand side of the equation. Due to the restrictions placed on the system by steady-state, there is no accumulation of energy within the system boundaries, so the left-hand side of Eqn. 2.20 becomes 0.

As a result,

$$
0 = \sum_{inlets} \left[U + \frac{v^2}{2g_c} + \frac{gz}{g_c} \right]^{in} \dot{m}^{in} - \sum_{outlets} \left[U + \frac{v^2}{2g_c} + \frac{gz}{g_c} \right]^{out} \dot{m}^{out} + \dot{Q} + \underline{\dot{W}}_S + \underline{\dot{W}}_{flow} \qquad 2.21
$$

where heat and work interactions are summed over all boundaries. The flow work from Eqn. 2.7 may be inserted and summed over all inlets and outlets,

$$
0 = \sum_{inlets} \left[U + \frac{v^2}{2g_c} + \frac{gz}{g_c} \right]^{in} \dot{m}^{in} - \sum_{outlets} \left[U + \frac{v^2}{2g_c} + \frac{gz}{g_c} \right]^{out} \dot{m}^{out} + \dot{Q} + \underline{\dot{W}}_S \qquad 2.22
$$

$$
+ \sum_{inlets} (PV)^{in} \dot{m}^{in} - \sum_{outlets} (PV)^{out} \dot{m}^{out}
$$

and combining flow terms:

$$
0 = \sum_{inlets} \left[U + PV + \frac{v^2}{2g_c} + \frac{gz}{g_c} \right]^{in} \dot{m}^{in} - \sum_{outlets} \left[U + PV + \frac{v^2}{2g_c} + \frac{gz}{g_c} \right]^{out} \dot{m}^{out} + \dot{Q} + \underline{\dot{W}}_S \qquad 2.23
$$

Enthalpy

Enthalpy is a mathematical property defined for convenience in problem solving.

Note that the quantity $(U + PV)$ arises quite naturally in the analysis of flow systems. Flow systems are very common, so it makes sense to define a single symbol that denotes this quantity:

$$
H \equiv U + PV
$$

Thus, we can tabulate precalculated values of H and save steps in calculations for flow systems. We call H the **enthalpy.**

The open-system, steady-state balance is then,

$$0 = \sum_{inlets} \left[H + \frac{v^2}{2g_c} + \frac{gz}{g_c} \right]^{in} \dot{m}^{in} - \sum_{outlets} \left[H + \frac{v^2}{2g_c} + \frac{gz}{g_c} \right]^{out} \dot{m}^{out} + \dot{\underline{Q}} + \dot{\underline{W}}_S \qquad 2.24$$

❶ Open-system, steady-state balance.

where the heat and work interactions are summations of the individual heat and work interactions over all boundaries.

> *Note: Q is positive when the system gains heat energy; W is positive when the system gains work energy; $\dot{m}^{in}$ and $\dot{m}^{out}$ are always positive; and $\dot{m}_{system} = \dot{m}^{in} - \dot{m}^{out}$ is positive when the systems gains mass and zero for steady-state flow. Mass may be replaced with moles in a non-reactive system with appropriate care for unit conversion.*

Note that the relevant potential and kinetic energies are for the fluid *entering* and *leaving* the boundaries, *not* for the fluid which is *inside* the system boundaries. When only one inlet and one outlet stream are involved, the steady-state flow rates must be equal, and

$$0 = \left[H + \frac{v^2}{2g_c} + \frac{gz}{g_c} \right]^{in} \dot{m}^{in} - \left[H + \frac{v^2}{2g_c} + \frac{gz}{g_c} \right]^{out} \dot{m}^{out} + \dot{\underline{Q}} + \dot{\underline{W}}_S \qquad \text{(one inlet/outlet)} \qquad 2.25$$

❶ Several common ways the steady-state balance can be written.

When kinetic and potential energy changes are negligible, we may write

$$0 = -\Delta H \dot{m} + \dot{\underline{Q}} + \dot{\underline{W}}_S \quad \text{(one inlet/outlet)} \qquad 2.26$$

where $\Delta H = H^{out} - H^{in}$. We could use molar flow rates for Eqns. 2.24 through 2.26 with the usual care for unit conversions of kinetic and potential energy. For an open steady-state system meeting the restrictions of Eqn. 2.26, we may divide through by the mass flow rate to find

$$0 = -\Delta H + Q + W_S \quad \text{(one inlet/outlet)} \qquad 2.27$$

In common usage, it is traditional to relax the convention of keeping only system properties on the left side of the equation. More simply we often write:

$$\Delta H = Q + W_S \quad \text{(one inlet/outlet)} \qquad 2.28$$

Compare Eqns. 2.19 and 2.28. Energy and enthalpy do not come from different energy balances, where the "closed system" balance uses U and W_{EC} and the "open system" balance uses H and W_s. Rather, the terms result from logically simplifying the generalized energy balance shown in the next section.

Comment on Δ Notation

In a closed system we use the Δ symbol to denote the change of a property from initial state to final state. In an open, steady-state system, the left-hand side of the energy balance is zero. Therefore, we frequently write Δ as a shorthand notation to combine the first two flow terms on the right-hand side of the balance, with the symbol meaning "outlet relative to inlet" as shown above. You need to learn to recognize which terms of the energy balance are zero or insignificant for a particular prob-

❶ Explanation of the use of Δ.

lem, whether a solution is for a closed or open system, and whether the Δ symbol denotes "outlet relative to inlet" or "final relative to initial."

Understanding Enthalpy and Shaft Work

Consider steady-state, adiabatic, horizontal operation of a pump, turbine, or compressor. It is possible to conceive of a closed packet of fluid as the system while it flows through the equipment. After analyzing the system from this perspective, we can switch to the open-system perspective to gain insight about the relation between open systems and closed systems, energy and enthalpy, and EC work and shaft work. As a bonus, we obtain a handy relation for estimating pump work and the enthalpy of compressed liquids.

In the conception of a closed-system fluid packet, no mass moves across the system boundary. The system, as we have chosen it, does not include a shaft even though it will move past the shaft. If you have trouble seeing this, remember that the system boundaries are defined by the conceived packet of mass. Since the system boundary does not contain the shaft before the packet enters, or after the packet exits, it cannot contain the shaft as it moves through the turbine. The system simply deforms to envelope the shaft. Therefore, all work for this closed system is technically expansion/contraction work; the closed-system expansion/contraction work is composed of the flow work and shaft work that we have seen from the open-system perspective. It is difficult to describe exactly what happens to the system at every point, but we can say something about how it begins and how it ends. This observation leads to what is called an integral method of analysis.

System: closed, adiabatic; Basis: packet of mass m. The kinetic and potential energy changes are negligible:

❶ Note that this derivation neglects kinetic and potential energy changes.

$$d\left[U + \frac{v^2}{2g_c} + \frac{g}{g_c}z\right] = dQ + dW_S + dW_{EC}$$

Integrating from the inlet (initial) state to the outlet (final) state:

$$U^{out} - U^{in} = W_{EC}$$

We may change the form of the integral representing work via integration by parts:

$$W_{EC} = -\int_{in}^{out} PdV = -[PV]_{in}^{out} + \int_{in}^{out} VdP$$

We recognize the term PV as representing the work done by the flowing fluid entering and leaving the system; it does not contribute to the work of the device. Therefore, the work interaction with the

turbine is the remaining integral, $W_S = \int_{in}^{out} VdP$. Substitution gives,

$$\Rightarrow [U + PV]^{out} - [U + PV]^{in} = \int_{in}^{out} VdP \qquad\qquad 2.29$$

Switching to the open-system perspective, Eqn. 2.28 gives

$$H^{out} - H^{in} = \cancel{Q} + W_S \qquad\qquad 2.30$$

Recalling that $H = U + PV$ and comparing the last two equations means $W_s = \int VdP$ is the work done using the pump, compressor, or turbine as the system. Furthermore, the appearance of the PV contribution in combination with U occurs naturally as part of the integration by parts. Physically, work is always "force times distance." Though this derivation has been restricted to an adiabatic device, the result is general to devices including heat transfer as we show later in Section 5.7.

$$\Delta H = W_S = \int_{in}^{out} VdP \qquad \text{reversible shaft work} \qquad 2.31$$

> ❗ Shaft work for a pump or turbine where kinetic and potential energy changes are small.

> **Note:** *The shaft work given by* $dW_S = VdP$ *is distinct from expansion/contraction work,* $dW_{EC} = PdV$. *Moreover, both are distinct from flow work,* $d\underline{W}_{flow} = PVdm$.

Several practical issues may be considered in light of Eqn. 2.31. First, the work done on the system is negative when the pressure change is negative, as in proceeding through a turbine or expander. This is consistent with our sign convention. Second, when considering gas flow, the integration may seem daunting if an ideal gas is not involved because of the complicated manner that V changes with T and P. Rather, for gases, we can frequently work with the enthalpy for a given state change. The enthalpy values for a state change read from a table or chart lead to W_s directly using Eqn. 2.30. For liquids, however, the integral can be evaluated quickly. Volume can often be approximated as constant, especially when $T_r < 0.75$. In that case, we obtain by integration an equation for estimating pump work:

> ❗ Shaft work for a *liquid* pump or turbine where kinetic and potential energy changes are small and $T_r < 0.75$ so that the fluid is incompressible.

$$\Delta H = W_S \approx V^L (P^{out} - P^{in}) = \frac{\Delta P}{\rho} \qquad \text{liquid} \qquad 2.32$$

Example 2.4   Pump work for compressing H₂O

Use Eqn. 2.32 to estimate the pump work of compressing 20°C H_2O from a saturated liquid to 5 and 50 MPa. Compare to the values obtained using the compressed liquid steam tables.

Solution: $\Delta H = W_S \approx V^L \Delta P$. Liquids are quite incompressible below $T_r = 0.75$. For water, that corresponds to 212°C, so we are safe on that count. We can calculate the pump work from Eqn. 2.32, reading $P^{sat} = 0.00234$ MPa and $V^L = 1.002$ cm³/g from the saturation tables at 20°C:

$$\Delta H = W_S \approx V^L \Delta P = 1.002 \text{ cm}^3/\text{g}(5 \text{ MPa} - 0.00234 \text{ MPa}) = 5.008 \text{ MPa-cm}^3/\text{g for 5 MPa}$$

$$\Delta H = W_S \approx V^L \Delta P = 1.002 \text{ cm}^3/\text{g}(50 \text{ MPa} - 0.00234 \text{ MPa}) = 50.1 \text{ MPa-cm}^3/\text{g for 50 MPa}$$

A convenient way of converting units for these calculations is to multiply and divide by the gas constant, noting its different units. This shortcut is especially convenient in this case, e.g.,

$$\Delta H = (5.008 \text{ MPa-cm}^3/\text{g}) \cdot (8.314 \text{ J/mole-K})/(8.314 \text{ MPa-cm}^3/\text{mole-K}) = 5.008 \text{ kJ/kg}$$

$$\Delta H = (50.1 \text{ MPa-cm}^3/\text{g}) \cdot (8.314 \text{ J/mole-K})/(8.314 \text{ MPa-cm}^3/\text{mole-K}) = 50.1 \text{ kJ/kg}$$

Example 2.4   Pump work for compressing H₂O   (Continued)

Note that, for water, the change in enthalpy in kJ/kg is roughly equal to the pressure rise in MPa because the specific volume is so close to one and $P^{sat} \ll P$. That is really handy.

The saturation enthalpy is read from the saturation tables as 83.95 kJ/kg. The values given in the compressed liquid table (at the end of the steam tables) are 88.6 kJ/kg at 5 MPa and 130 kJ/kg at 50 MPa, corresponding to estimated work values of 4.65 and 46.1 kJ/kg. The estimation error in the computed work is about 7 to 9%, and smaller for lower pressures. This degree of precision is generally satisfactory because the pump work itself is usually small relative to other work and terms (like the work produced by a turbine in a power cycle).

2.9   THE COMPLETE ENERGY BALANCE

An open-system that does not meet the requirements of a steady-state system is called an unsteady-state open-system as shown in Fig. 2.5. The mass-in may not equal the mass-out, or the system state variables (e.g., temperature) may change with time, so the system itself may gain in internal energy, kinetic energy, or potential energy. An example of this is the filling of a tank being heated with a steam jacket. Another example is the inflation of a balloon, where there is mass flow in and the system boundary expands. These considerations lead to a general equation which is applicable to open or closed systems,

❶ Complete
energy balance.

$$\frac{d}{dt}\left[m\left(U + \frac{v^2}{2g_c} + \frac{gz}{g_c}\right)\right] = \sum_{inlets}\left[H + \frac{v^2}{2g_c} + \frac{gz}{g_c}\right]^{in}\dot{m}^{in} - \sum_{outlets}\left[H + \frac{v^2}{2g_c} + \frac{gz}{g_c}\right]^{out}\dot{m}^{out} \qquad 2.33$$

$$+ \dot{Q} + \dot{W}_{EC} + \dot{W}_S$$

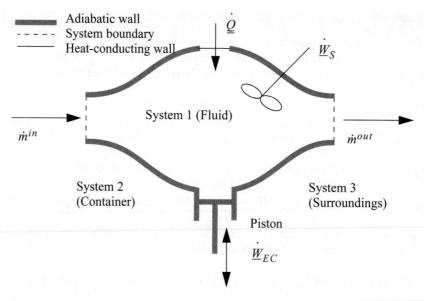

Figure 2.5 *Schematic of a general system.*

where the heat and work interactions are summations of the individual heat and work interactions over all boundaries. We also may write this with the time dependence implied:

$$d\left[m\left(U + \frac{v^2}{2g_c} + \frac{gz}{g_c}\right)\right] = \sum_{inlets}\left[H + \frac{v^2}{2g_c} + \frac{gz}{g_c}\right]^{in} dm^{in} - \sum_{outlets}\left[H + \frac{v^2}{2g_c} + \frac{gz}{g_c}\right]^{out} dm^{out} \qquad 2.34$$

$$+ d\underline{Q} + d\underline{W}_{EC} + d\underline{W}_S$$

Note: The signs and conventions are the same as presented following Eqn. 2.24.

Usually, the closed-system or the steady-state equations are sufficient by themselves. But for unsteady-state open systems, the entire equation must be considered. Fortunately, even when the entire energy balance is applied, some of the terms are usually not necessary for a given problem, so fewer terms are usually needed than shown in Eqn. 2.33. An objective of this text is to build your ability to recognize which terms apply to a given problem.

2.10   INTERNAL ENERGY, ENTHALPY, AND HEAT CAPACITIES

Before we proceed with more examples, we need to add another thermodynamic tool. Unfortunately, there are no "internal energy" or "enthalpy" meters. In fact, these state properties must be "measured" indirectly by other state properties. The Gibbs phase rule tells us that if two state variables are fixed in a pure single-phase system, then all other state variables will be fixed. Therefore, it makes sense to measure these properties in terms of P, V, and T. In addition, if this relation is developed, it will enable us to find P, V, and/or T changes for a given change in ΔU or ΔH. In Example 2.3, where a warm and cold steel blocks were contacted, we solved the problem without calculating the change in temperature for each block. However, if we had a relation between U and T, we could have calculated the temperature changes. The relations that we seek are the definitions of the heat capacities.

Constant Volume Heat Capacity

The constant of proportionality between the internal energy change *at constant volume* and the temperature change is known as the constant volume heat capacity. The constant volume heat capacity is defined by:

$$C_V \equiv \left(\frac{\partial U}{\partial T}\right)_V \qquad 2.35$$

❶ Definition of C_V.

Since temperature changes are easily measured, internal energy changes can be calculated once C_V is known. C_V is not commonly tabulated, but, as shown below, it can be easily determined from the constant pressure heat capacity, which is commonly available.

Constant Pressure Heat Capacity

In the last two sections, we have introduced enthalpy, and we can relate the change in enthalpy of a system to temperature in a manner analogous to the method used for internal energy. This relationship will involve a new heat capacity, the heat capacity at constant pressure defined by:

❗ Definition
of C_P.

$$C_P \equiv \left(\frac{\partial H}{\partial T}\right)_P$$ 2.36

where H is the enthalpy of the system.

The use of two heat capacities, C_V and C_P, forces us to think of constant volume or constant pressure as the important distinction between these two quantities. The important quantities are really internal energy versus enthalpy. You simply must convince yourself to remember that C_V refers to changes in U at constant volume, and C_P refers to changes in H at constant pressure.

Relations between Heat Capacities, *U* and *H*

We have said that C_V values are not readily available; therefore, how do we determine internal energy changes? Also, how do we determine enthalpy changes at constant volume or internal energy changes at constant pressure? We will return to the details of these questions in Chapters 6–8 and handle them rigorously, but the details have been rigorously followed by developers of thermodynamic charts and tables. Therefore, for relating the internal energy or enthalpy to temperature and pressure, a thermodynamic chart or table is preferred. If none is available, or properties are not tabulated in the state of interest, some exact relations and some approximate rules of thumb must be applied. The relations are also useful for introductory calculations while focus is on the energy balance rather than the property relations.

For an ideal gas,

$$\boxed{\Delta H \equiv \Delta U + \Delta(PV) \;=\; \Delta U + R(\Delta T)}$$ Exact for an ideal gas.   (ig) 2.37

❗ C_P, C_V and re-
lation between ΔU
and ΔH for an ideal
gas.

$$\boxed{C_P \;=\; C_V + R}$$ Exact for an ideal gas.   (ig) 2.38

Constant pressure heat capacities for ideal gases are tabulated in Appendix E. Constant volume heat capacities for ideal gases can readily be determined from Eqn. 2.38. For ideal gases, internal energy and enthalpy are independent of pressure as we implied with Eqn. 1.21. For real gases and for liquids, the relation between C_P and C_V is more complex, and derivatives of P-V-T properties must be used as shown rigorously in Examples 6.1, 6.6, and 6.9 and implemented thereafter. We will use thermodynamic tables and charts for real gases until these relations are developed.

For liquids or solids, we typically calculate ΔH and correct the calculation if necessary as explained below. For liquids, it has been experimentally determined that internal energy is only very weakly dependent on pressure below $T_r = 0.75$. In addition, the molar volume is insensitive to pressure below $T_r = 0.75$. We demonstrated in Example 2.4 that,

❗ Pressure de-
pendence of H for
condensed phases.

$$\boxed{\Delta H_T \approx V \Delta P_T}$$ Liquids below $T_r = 0.75$ and solids.   2.39

$T_r \equiv \dfrac{T}{T_c}$ is the **reduced temperature** calculated by dividing the absolute temperature by the critical temperature. (A rigorous evaluation is considered in Example 6.1 on page 233.) The relations for solids and liquids are important because frequently the properties have not been measured, or the measurements available in charts and tables are not available at the pressures of interest. We may then summarize the relations of internal energy and enthalpy with temperature.

$$\Delta U = \int_{T_1}^{T_2} C_V(T)dT$$

Ideal gas: exact.

Real gas: valid only if V = constant.

2.40

❶ Useful formula for relating T, P to U and H in the absence of phase changes.

$$\Delta H = \int_{T_1}^{T_2} C_P(T)dT$$

Ideal gas: exact.

Real gas: valid only if P = constant.

2.41

$$\Delta H \approx \int_{T_1}^{T_2} C_P(T)dT + V\Delta P$$

Liquid below $T_r = 0.75$ or solid: reasonable approximation.

2.42

$$\Delta U = \Delta H - \Delta(PV) \approx \Delta H - V\Delta P \approx \int_{T_1}^{T_2} C_P(T)dT$$

Liquid below $T_r = 0.75$ or solid: reasonable approximation when pressure change is below several MPa.

2.43

Note: These formulas do not account for phase changes which may occur.

Note that the heat capacity of a monatomic ideal gas can be obtained by differentiating the internal energy as given in Chapter 1, resulting in $C_V = 3/2\ R$ and $C_P = 5/2\ R$. Heat capacities for diatomics are larger, $C_P = 7/2\ R$, and $C_V = 5/2\ R$ near room temperature, and polyatomics are larger still. According to classical theory, each **degree of freedom**[5] contributes $1/2R$ to C_V. Kinetic and potential energy each contribute a degree of freedom in each dimension. A monatomic ideal gas has only three kinetic energy degrees of freedom, thus $C_V = 3/2\ R$. Diatomic molecules are linear so they have two additional degrees of freedom for the linear (one-dimensional) bond that has kinetic and potential energy both. In complicated molecules, the vibrations are characterized by *modes*. See the end flap to make a quick comparison. Monatomic solids have three degrees of freedom each for kinetic and vibrational energy, one for each principle direction, thus the law of Dulong and Petit, $C_V^S = 3R$ is a first approximation. Low-temperature heat capacities of monatomic solids are explored more in Example 6.8. If you become curious about the manner in which the heat capacities of polyatomic species differ from those of the spherical molecules discussed in Chapter 1, you will find introductions to statistical thermodynamics explain the contributions of translation, rotation, and vibration. For polyatomic molecules, the heat capacity increases with molecular weight due to the increased number of degrees of freedom for each bond. In this text, ideal gas heat capacity values at 298 K are summarized inside the back cover of the book, and may be assumed to be independent of temperature over small temperature ranges near room temperature. The increase in heat capacity with temperature for diatomics and polyatomics is dominated by the vibrational contribution. The treatment of heat capacity by statistical thermodynamics is particularly interesting because it is a theory[6] that often gives more accurate results than experimental calorimetric measurements. Commonly, engineers correlate ideal gas heat capacities with expressions like polynomials.

❶ Whenever we assume heat capacity to be temperature independent in this text, we mark the equation with a (*) symbol near the right margin.

---

5. The degrees of freedom discussed here are different from those discussed for the Gibbs phase rule.

6. This theory also requires experimental spectroscopic measurements, but those are quite different from the calorimetric measurement of enthalpy changes with respect to temperature.

We will frequently ignore the heat capacity dependence on T to make an approximate calculation. *Whenever we assume heat capacity to be temperature independent in this text, we mark the equation with a (*) symbol near the right margin.* Heat capacities represented as polynomials of temperature are available in Appendix E. The heat capacity depends on the **state of aggregation.** For example, water has a different heat capacity when solid (ice), liquid, or vapor (steam). The contribution of the heat capacity integral to the energy balance is frequently termed the **sensible heat** to communicate its contribution relative to **latent heat** (due to phase changes) or **heat of reaction** to be discussed later. Note that these are called "heats" even though they are enthalpy changes.

❶ The heat capacity of a substance depends on the state of aggregation.

Example 2.5 Enthalpy change of an ideal gas: Integrating $C_P^{ig}(T)$

Propane gas undergoes a change of state from an initial condition of 5 bar and 105°C to 25 bar and 190°C. Compute the change in enthalpy using the ideal gas law.

Solution: The ideal gas change is calculated via Eqn. 2.41 and is independent of pressure. The heat capacity constants are obtained from Appendix E.

$$H_2^{ig} - H_1^{ig} = \int_{T_1}^{T_2} C_P dT = \int_{T_1}^{T_2} (A + BT + CT^2 + DT^3)dT =$$

$$= A(T_2 - T_1) + \frac{B}{2}(T_2^2 - T_1^2) + \frac{C}{3}(T_2^3 - T_1^3) + \frac{D}{4}(T_2^4 - T_1^4)$$

$$= -4.224(463.15 - 378.15) + \frac{0.3063}{2}\ (463.15^2 - 378.15^2) +$$

$$\frac{-1.586 \times 10^{-4}}{3}\ (463.15^3 - 378.15^3) + \frac{3.215 \times 10^{-8}}{4}\ (463.15^4 - 378.15^4) = 8405 \text{ J/mol}$$

Example 2.6 Enthalpy of compressed liquid

The compressed liquid tables are awkward to use for compressed liquid enthalpies because the pressure intervals are large. Using saturated liquid enthalpy values for water and hand calculations, estimate the enthalpy of liquid water at 20°C H_2O and 5 and 50 MPa. Compare to the values obtained using the compressed liquid steam tables.

❶ This is a common calculation needed for working with power plant condensate streams at high pressure.

Solution: This is a common calculation needed for working with power plant condensate streams at high pressure. The relevant equation is Eqn. 2.42, but we can eliminate the temperature integral by selecting saturated water at the same temperature and then applying the pressure correction, i.e., applying Eqn. 2.39, $\Delta H \approx V\Delta P$ relative to the saturation condition, giving $H = H^{sat} + V\Delta P$. The numerical calculations have already been done in Example 2.4 on page 55. Both calculations use the same approximation, even though the paths are slightly different. A more rigorous analysis is shown later in Example 6.1.

Example 2.7 Adiabatic compression of an ideal gas in a piston/cylinder

Nitrogen is contained in a cylinder and is compressed adiabatically. The temperature rises from 25°C to 225°C. How much work is performed? Assume that the heat capacity is constant ($C_P/R = 7/2$), and that nitrogen follows the ideal gas law.

Solution: System is the gas. Closed system, system size changes, adiabatic.

$$d\left[U + \frac{v^2}{2g_c} + \frac{g}{g_c}z \right] = dQ + dW_S + dW_{EC}$$

$$\int dU = \int dW_{EC} = W_{EC}$$

$$dU = C_V dT \Rightarrow W_{EC} = \int C_V dT = C_V \Delta T = (C_P - R)\Delta T \qquad (\text{*ig})$$

$$= \left(\frac{5}{2}\right) 8.314(200) = 4157 \text{ J/mol}$$

Note that because the temperature rise is specified, we do not need to know if the process was reversible.

Relation to Property Tables/Charts

In Section 1.4, we used steam tables to find internal energies of water as liquid or vapor. Tables or charts usually contain enthalpy and internal energy information, which means that these properties can be read from the source for these compounds, eliminating the need to apply Eqns. 2.40–2.43. This is usually more accurate because the pressure dependence of the properties that Eqns. 2.40–2.43 neglect has been included in the table/chart, although the pressure correction method applied in the previous example for liquids is generally accurate enough for liquids. Energy and enthalpy changes spanning phase transitions can be determined directly from the tables since energies and enthalpies of phase transitions are implicitly included in tabulated values.

Estimation of Heat Capacities

If heat capacity information cannot be located from appendices in this text, from the NIST Chemistry WebBook[7], or from reference handbooks, it can be estimated by several techniques offered in the *Chemical Engineer's Handbook*[8] and *The Properties of Gases and Liquids.*[9]

Phase Transitions (Liquid-Vapor)

Enthalpies of vaporization are tabulated in Appendix E for selected substances at their **normal boiling temperatures** (their saturation temperatures at 1.01325 bar). In the case of the steam tables, Section 1.4 shows that the energies and enthalpies of vaporization of water are available

---

7. http://webbook.nist.gov/

8. Perry, R.M., Green, D.W., 2008. *Chemical Engineer's Handbook,* 8th ed., New York: McGraw-Hill.

9. Poling, B.E., Prausnitz, J.M., O'Connell, J.P., 2001. *The Properties of Gases and Liquids,* 5th ed., New York: McGraw-Hill.

along the entire saturation curve. Complete property tables for some other compounds are available in the literature or online, however, most textbooks present charts to conserve space, and we follow that trend. In the cases where tables or charts are available, their use is preferred for phase transitions away from the normal boiling point, although a hypothetical path that passes through the normal boiling point can usually be constructed easily.

The **energy of vaporization** is more difficult to find than the **enthalpy of vaporization.** It can be calculated from the enthalpy of vaporization and the P-V-T properties. Since $U = H - PV$,

$$\Delta U^{vap} = \Delta H^{vap} - \Delta(PV)^{vap} = \Delta H^{vap} - (P^{sat}V)^V - (P^{sat}V)^L = \Delta H^{vap} - P^{sat}(V^V - V^L)$$

Far from the critical point, the molar volume of the vapor is much larger than the molar volume of the liquid. Further, at the normal boiling point (the saturation temperature at 1.01325 bar), the ideal gas law is often a good approximation for the vapor volume,

> ⬤ Relation between ΔU^{vap} and ΔH^{vap} when the vapor follows the ideal gas law.

$$\boxed{\Delta U^{vap} \approx \Delta H^{vap} - P^{sat}V^V \approx \Delta H^{vap} - RT^{sat}}$$

(ig) 2.44

Estimation of Enthalpies of Vaporization

If the enthalpy of vaporization cannot be located in the appendices or a standard reference book, it may be estimated by several techniques offered and reviewed in the *Chemical Engineer's Handbook* and *The Properties of Gases and Liquids*. One particularly convenient correlation is[10]

> ⬤ Generalized correlation for ΔH^{vap}.

$$\frac{\Delta H^{vap}}{RT_c} \approx 7(1 - T_r)^{0.354} + 11\omega(1 - T_r)^{0.456}$$

2.45

where T_r is **reduced temperature** $T_r = T/T_c$, ω is the **acentric factor** (to be described in Chapter 7), also available on the back flap. If accurate vapor pressures are available, the enthalpy of vaporization can be estimated far from the critical point (i.e., $T_r < 0.75$) by the Clausius–Clapeyron equation:

$$\Delta H^{vap} \approx - R\frac{d\ln P^{sat}}{d(1/T)} \quad (T_r < 0.75)$$

(ig) 2.46

The background for this equation is developed in Section 9.2. Vapor pressure is often represented by the Antoine equation, $\log_{10}P^{sat} = A - B/(T(°C) + C)$. If Antoine parameters are available, they may be used to estimate the derivative term of Eqn. 2.46,

$$\frac{d\ln P^{sat}}{d(1/T)} = \frac{2.3026 d\log_{10}P^{sat}}{d(1/T)} = \frac{-2.3026B(T(°C) + 273.15)^2}{(T(°C) + C)^2}$$

(ig)

where T is in °C, and B and C are Antoine parameters for the common logarithm of pressure. For Antoine parameters intended for other temperature or pressure units, the equation must be carefully converted. The temperature limits for the Antoine parameters must be carefully followed because the Antoine equation does not extrapolate well outside the temperature range where the constants have been fit. If Antoine parameters are unavailable, they can be estimated to roughly 10% accuracy by the shortcut vapor pressure (SCVP) model, discussed in Section 9.3,

---

10. Pitzer, K.S., Lippmann, D.Z., Curl Jr., R.F., Huggins, C.M., Petersen, D.E. 1955. *J. Am. Chem. Soc.*, 77:3433.

$$A = \log_{10} P_c + 7(1 + \omega)/3; \quad B = 7(1 + \omega)T_c/3; \quad C = 273.15 \tag{ig) 2.47}$$

where the units of P_c match the units of P^{sat}, T_c is in K, and T is in °C.

Phase Transitions (Solid–Liquid)

Enthalpies of fusion (melting) are tabulated for many substances at the normal melting temperatures in the appendices as well as handbooks. Internal energies of fusion are not usually available, however the volume change on melting is usually very small, resulting in internal energy changes that are nearly equal to the enthalpy changes:

$$\Delta U^{fus} = \Delta H^{fus} - \Delta(PV)^{fus} = \Delta H^{fus} - P(V^L - V^S) \approx \Delta H^{fus} \tag{2.48}$$

❶ Relation between ΔU^{fus} and ΔH^{fus}.

Unlike the liquid-vapor transitions, where T^{sat} depends on pressure, the melting (solid-liquid) transition temperature is almost independent of pressure, as illustrated schematically in Fig. 1.7.

2.11   REFERENCE STATES

Notice that our heat capacities do not permit us to calculate absolute values of internal energy or enthalpy; they simply permit us to calculate *changes* in these properties. Therefore, when is internal energy or enthalpy equal to zero—at a temperature of absolute zero? Is absolute zero a reasonable place to assign a **reference state** from which to calculate internal energies and enthalpies? Actually, we don't usually solve this problem in engineering thermodynamics for the following two reasons:[11] 1) for a gas, there would almost always be at least two phase transitions between room temperature and absolute zero that would require knowledge of energy changes of phase transitions and heat capacities of each phase; and 2) even if phase transitions did not occur, the empirical fit of the heat capacity represented by the constants in the appendices are not valid down to absolute zero! Therefore, for engineering calculations, we arbitrarily set enthalpy *or* internal energy equal to zero at some *convenient* reference state where the heat capacity formula is valid. We calculate changes relative to this state. The *actual* enthalpy or internal energy is certainly not zero, it just makes our reference state location clear. If we choose to set the value of enthalpy to zero at the reference state, then $H_R = 0$, and $U_R = H_R - (PV)_R$ where we use subscript R to denote the reference state. Note that U_R and H_R cannot be precisely zero simultaneously at the reference state. The reference state for water (in the steam tables) is chosen to set enthalpy of water equal to zero at the triple point. Note that PV is negligible at the reference state so that it appears that U_R is also zero to the precision of the tabulated values, which is not rigorously correct. (Can you verify this in the steam tables? Which property is set to zero, and for which state of aggregation?). To clearly specify a reference state we must specify:

❶ Reference states permit the tabulation of *values* for *U, H*.

1. The composition which may or may not be pure.
2. The state of aggregation (*S, L,* or *V*).
3. The pressure.
4. The temperature.

---

11. In the most detailed calculations, absolute zero *is* used as a reference state to create some thermodynamic tables. This is based on a principle known as the **third law of thermodynamics,** that states that entropy goes to zero for a perfect crystal at absolute zero. The difficulties in the rigorous calculations are mentioned above, and although the principles are straightforward, the actual calculations are beyond the scope of this book.

As you will notice in the following problems, reference states are not necessary when working with a pure fluid in a closed system or in a steady-state flow system with a single stream. The numerical values of the changes in internal energy or enthalpy will be independent of the reference state.

When multiple components are involved, or many inlet/outlet streams are involved, definition of reference states is *recommended* since flow rates of the inlet and outlet streams will not necessarily match one-to-one. The reference state for each component may be different, so the reference temperature, pressure, and state of aggregation must be clearly designated.

For unsteady-state open systems that accumulate or lose mass, *reference states are imperative when values of ΔU or ΔH changes of the system or surroundings are calculated* as the numerical values depend on the reference state. It is only when the changes for the system and surroundings are summed together that the reference state drops out for unsteady-state open systems.

Ideal Gas Properties

For an ideal gas, we must specify only the reference T and P.[12] An ideal gas cannot exist as a liquid or solid, and this fact completely specifies the state of our system. In addition, we need to set H_R or U_R (but not both!) equal to zero.

$$U^{ig} = \int_{T_R}^{T} C_V dT + U_R^{ig} \qquad \text{(ig) 2.49}$$

$$H^{ig} = \int_{T_R}^{T} C_P dT + H_R^{ig} \qquad \text{(ig) 2.50}$$

Also at all states, including the reference state, $U^{ig} \equiv H^{ig} - PV = H^{ig} - RT$ so $U_R^{ig} = H_R^{ig} - RT_R$. The ideal gas approximation is reliable when contributions from intermolecular potential energy are relatively small. A convenient guideline is, in term of **reduced temperature** $T_r = T/T_c$, and **reduced pressure** $P_r = P/P_c$, where P_c is the critcal pressure.

$$\text{Assume ideal gas behavior if } P < P^{sat} \text{ and } T_r > 0.5 + 2P_r \qquad \text{(ig) 2.51}$$

State Properties Including Phase Changes

Problems will often involve phase changes. Throughout a problem, since the thermodynamic properties must always refer to the same reference state, phase changes must be incorporated into state properties relative to the state of aggregation of the reference state. To calculate a property for a fluid at T and P relative to a reference state in another phase, a sketch of the pathway from the reference state is helpful to be sure all steps are included. Several pathways are shown in Fig. 2.6 for different reference states. Note that the ideal gas reference state with the generalized correlation for the heat of vaporization (option (c)) is convenient because it does not require liquid heat capacities. The accuracy of the method depends on the accuracy of the generalized correlation or the technique used to estimate the heat of vaporization. Option (c) is frequently used in process simulators. In some cases the user may have flexibility in specifying the correlation used to estimate the heat of vaporization.

---

12. For calculation of ideal gas U and H, only a reference temperature is required; however, for the entropy introduced in the next chapter, a reference pressure is needed, so we establish the P requirement now.

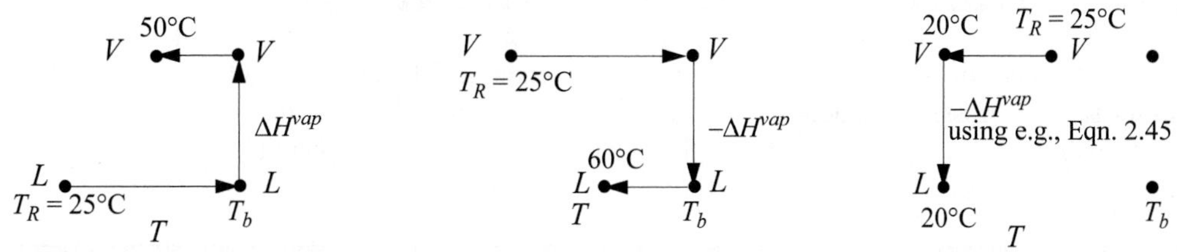

$$H^V = \int_{25}^{T^b} C_P^L dT + \Delta H^{vap}$$
$$+ \int_{T^b}^{50} C_P^V dT$$

$$H^L = \int_{25}^{T^b} C_P^V dT - \Delta H^{vap}$$
$$+ \int_{T^b}^{60} C_P^L dT$$

$$H^L = \int_{25}^{20} C_P^V dT - \Delta H^{vap}$$

(*a*) Enthalpy of vapor at 50 °C using liquid ref at 25°C and ΔH^{vap} at $T_b > 50$°C.

(*b*) Enthalpy of liquid at 60 °C using vapor ref at 25°C and ΔH^{vap} at $T_b > 60$°C.

(*c*) Enthalpy of liquid at 20 °C using vapor ref at 25°C and generalized correlation for ΔH^{vap}.

Figure 2.6 *Illustrations of state pathways to calculate properties involving liquid/vapor phase changes. The examples are representative, and modified paths would apply for states above the normal boiling point, T_b. Similar pathways apply for solid/liquid or solid/vapor transformations. Note that a generalized correlation is used for ΔH^{vap} which differs from the normal boiling point value. The method is intended to be used at subcritical conditions. Pressure corrections are not illustrated for any paths here.*

Example 2.8 Acetone enthalpy using various reference states

Calculate the enthalpy values for acetone as liquid at 20°C and vapor at 90°C and the difference in enthalpy using the following reference states: (a) liquid at 20°C; (b) ideal gas at 25°C and ΔH^{vap} at the normal boiling point; (c) ideal gas at 25°C and the generalized correlation for ΔH^{vap} at 20°C. Ignore pressure corrections and treat vapors as ideal gases.

Solution: Heat capacity constants are available in Appendix E. For all cases, 20°C is 293.15K, 90°C is 363.15K, and the normal boiling point is $T_b = 329.15$K.

(a) $H^L = 0$ because the liquid is at the reference state. The vapor enthalpy is calculated analogous to Fig. 2.6, pathway (a). The three terms of pathway (a) are $H^V = 4639 + 30200 + 2799 = 37,638$ J/mol. The difference in enthalpy is $\Delta H = 37,638$ J/mol.

(b) H^L will use a path analogous to Fig. 2.6, pathway (b). The three terms of pathway (b) are $H^L = 2366 - 30200 - 4638 = -32472$ J/mol. H^V is calculated using Eqn. 2.50, $H^V = 5166$ J/mol. The difference is $\Delta H = 5166 + 32472 = 37,638$ J/mol, same as part (a).

(c) H^L will use a path analogous to Fig. 2.6, pathway (c). The generalized correlation of Eqn. 2.45 predicts a heat of vaporization at T_b of 29,280 J/mol, about 3% low. At 20°C, the heat of vaporization is predicted to be 31,420 J/mol. The two steps in Fig. 2.6 (c) are $H^L = -365 - 31420 = -31785$ J/mol. The enthalpy of vapor is the same calculated in part (b), $H^V = 5166$ J/mol. The enthalpy difference is $\Delta H = 5166 + 31785 = 36,950$ J/mol, about 2% low relative to part (b).

2.12   KINETIC AND POTENTIAL ENERGY

The development of the energy balance includes potential and kinetic energy terms for the system and for streams crossing the boundary. When temperature changes occur, the magnitude of changes of U and H are typically so much larger than changes in kinetic and potential energy that the latter terms can be dropped. The next example demonstrates how this is justified.

Example 2.9 Comparing changes in kinetic energy, potential energy, internal energy, and enthalpy

For a system of 1 kg water, what are the internal energy and enthalpy changes for raising the temperature 1°C as a liquid and as a vapor from 24°C to 25°C? What are the internal energy enthalpy changes for evaporating from the liquid to the vapor state? How much would the kinetic and potential energy need to change to match the magnitudes of these changes?

Solution: The properties of water and steam can be found from the saturated steam tables, interpolating between 20°C and 25°C. For saturated water or steam being heated from 24°C to 25°C, and for vaporization at 25°C:

	ΔU(J)	ΔH(J)	ΔU^{vap}(kJ)	ΔH^{vap}(kJ)
water	4184	4184		
steam	1362	1816	2304.3	2441.7

Of these values, the values for ΔU of steam are lowest and can serve as the benchmark. How much would kinetic and potential energy of a system have to change to be comparable to 1000 J?

Kinetic energy: If $\Delta KE = 1000$ J, and if the kg is initially at rest, then the velocity change must be,

$$\Delta(v^2) = \frac{2(1000\text{J})}{1\text{kg}} \text{ or } \Delta v = 44.7 \text{ m/s}$$

This corresponds to a velocity change of 161 kph (100 mph). A velocity change of this order of magnitude is unlikely in most applications except nozzles (discussed below). Therefore, kinetic energy changes can be neglected in most calculations when temperature changes occur.

Potential energy: If $\Delta PE = 1000$ J, then the height change must be,

$$\Delta z = \frac{1000 \text{ J}}{1\left(9.8066\dfrac{\text{N}}{\text{kg}}\right)} = 102 \text{ m}$$

This is equivalent to about one football field in position change. Once again this is very unlikely in most process equipment, so it can usually be ignored relative to heat and work interactions. Further, when a phase change occurs, these changes are even less important relative to heat and work interactions.

⚠ Velocity and height changes must be large to be significant in the energy balance when temperature changes also occur.

Example 2.9 demonstrates that kinetic and potential energy changes of a fluid are usually negligible when temperature changes by a degree or more. Moreover, kinetic and potential energy changes are closely related to one another in the design of piping networks because the temperature changes *are* negligible. The next example helps illustrate the point.

Example 2.10 Transformation of kinetic energy into enthalpy

Water is flowing in a straight horizontal pipe of 2.5 cm ID with a velocity of 6.0 m/s. The water flows steadily into a section where the diameter is suddenly increased. There is no device present for adding or removing energy as work. What is the change in enthalpy of the water if the downstream diameter is 5 cm? If it is 10 cm? What is the maximum enthalpy change for a sudden enlargement in the pipe? How will these changes affect the temperature of the water?

Solution: A boundary will be placed around the expansion section of the piping. The system is fixed volume, ($\dot{W}_{EC} = 0$), adiabatic without shaft work. The open steady-state system is under steady-state flow, so the left side of the energy balance is zero.

$$0 = \sum_{inlets}\left[H + \frac{v^2}{2g_c} + \frac{gz}{g_c}\right]^{in} \dot{m}^{in} - \sum_{outlets}\left[H + \frac{v^2}{2g_c} + \frac{gz}{g_c}\right]^{out} \dot{m}^{out} + \cancel{\dot{Q}} + \cancel{\dot{W}_{EC}} + \cancel{\dot{W}_S}$$

Simplifying: $\Delta H = \dfrac{-\Delta(v^2)}{2g_c}$

Liquid water is incompressible, so the volume (density) does not change from the inlet to the outlet. Letting A represent the cross-sectional area, and letting D represent the pipe diameter, $\underline{V} = v_1 A_1 = v_2 A_2 \Rightarrow v_2 = v_1(A_1/A_2)$,

$$\Delta(v^2) = v_1^2\left[\left(\frac{D_1}{D_2}\right)^4 - 1\right]$$

$$\Delta H = \frac{-v_1^2}{2g_c}\left[\left(\frac{D_1}{D_2}\right)^4 - 1\right]$$

$D_2/D_1 = 2 \Rightarrow \Delta H = -6.0^2$ m²/s² (1J/1kg-m²/s²) ($\frac{1}{2}^4 - 1$)/2 = 17 J/kg
$D_2/D_1 = 4 \Rightarrow \Delta H = 18$ J/kg
$D_2/D_1 = \infty \Rightarrow \Delta H = 18$ J/kg

To calculate the temperature rise, we can relate the enthalpy change to temperature since they are both state properties. From Eqn. 2.42, neglecting the effect of pressure,

$$\Delta H = C_P \Delta T \qquad (*)$$

$$C_P = 4184\frac{J}{kgK} \Rightarrow \Delta T = \frac{18.00(J/kg)}{4184(J/(kgK))} = 0.004K \qquad (*)$$

Example 2.10 shows that the temperature rise due to velocity changes is very small. In a real system, the measured temperature rise will be slightly higher than our calculation presented here because irreversibilities are caused by the velocity gradients and swirling in the region of the sudden enlargement that we haven't considered. These losses increase the temperature rise. In fluid mechanics, irreversible losses due to flow are characterized by a quantity known as the **friction factor.** The losses of a valve, fitting, contraction, or enlargement can be characterized empirically by the *equivalent length* of straight pipe that would result in the same losses. We will introduce these topics in Section 5.7. However, we conclude that from the standpoint of the energy balance, the temperature rise is still small and can be neglected except in the most detailed analysis such as the design of the piping network. In Example 2.10 the velocity decreases, and enthalpy increases due to greater flow work on the inlet than the outlet. Note that the above result for a liquid does not depend on whether the enlargement is rapid or gradual. A gradual taper will give the same temperature change since the energy balance relates the enthalpy change to the initial and final velocities, but not on the manner in which the change occurs.

Applications where kinetic and potential energy changes are important include solids such as projectiles, where the temperature changes of the solids are negligible and the purpose of the work is to cause accelerate or elevate the system. One example of this application is a steam catapult used to assist in take-off from aircraft carriers. A steam-filled piston + cylinder device is expanded, and the piston drags the plane to a velocity sufficient for the jet engines to lift the plane. While the kinetic and potential energy changes for the *steam* are negligible, the work done by the steam causes important kinetic energy changes in the piston and plane because of their large masses.

2.13   ENERGY BALANCES FOR PROCESS EQUIPMENT

Several types of equipment are ubiquitous throughout industry, and facile abilities with the energy balance for these processes will permit more rapid analysis of composite systems where these units are combined. In this brief section we introduce valves and throttles used to regulate flow, nozzles, heat exchangers, adiabatic turbines and expanders, adiabatic compressors, and pumps.

Valves and Throttles

A throttling device is used to reduce the pressure of a flowing fluid *without extracting any shaft work and with negligible fluid acceleration.* Throttling is also known as **Joule-Thomson expansion** in honor of the scientists who originally studied the thermodynamics. An example of a throttle is the kitchen faucet. Industrial valves are modeled as throttles. Writing the balance for a boundary around the throttle valve, it is conventional to neglect any accumulation within the device since it is small relative to flow rates through the device, so the left-hand side is zero. At steady-state flow,

$$0 = \left[H + \frac{v^2}{2g_c} + \frac{gz}{g_c} \right]^{in} \dot{m}^{in} - \left[H + \frac{v^2}{2g_c} + \frac{gz}{g_c} \right]^{out} \dot{m}^{out} + \dot{Q} + \dot{W}_{EC} + \dot{W}_S$$

Changes in kinetic and potential energy are small relative to changes in enthalpy as we just discussed. When in doubt, the impact of changes in velocity can be evaluated as described in Example 2.9. The amount of heat transfer is negligible in a throttle. The boundaries are not expanding, and there is also no mechanical device for transfer of work, so the work terms vanish. Therefore, a throttle is isenthalpic:

$$\boxed{\Delta H = 0 \quad \text{for throttles}} \qquad 2.52$$

Nozzles

Nozzles are specially designed devices utilized to convert pressure drop into kinetic energy. Common engineering applications involve *gas flows*. An example of a nozzle is a booster rocket. Nozzles are also used on the inlets of impulse turbines to convert the enthalpy of the incoming stream to a high velocity before it encounters the turbine blades.[13] Δu is significant for nozzles. A nozzle is designed with a specially tapered neck on the inlet and sometimes the outlet as shown schematically in Fig. 2.7. Nozzles are optimally designed at particular velocities/pressures of operation to minimize viscous dissipation.

The energy balance is written for a boundary around the nozzle. Any accumulation of energy in the nozzle is neglected since it is small relative to flow through the device and zero at steady state. Velocity changes are significant by virtue of the design of the nozzle. However, potential energy changes are negligible. Heat transfer and work terms are dropped as justified in the discussion of throttles. Reducing the energy balance for a nozzle shows the following:

$$0 = \left[H + \frac{v^2}{2g_c} + \frac{gz}{g_c} \right]^{in} \dot{m}^{in} - \left[H + \frac{v^2}{2g_c} + \frac{gz}{g_c} \right]^{out} \dot{m}^{out} + \dot{Q} + \dot{W}_{EC} + \dot{W}_S$$

$$\boxed{\Delta H = (-\Delta(v^2)/(2g_c)) \quad \text{for nozzles}} \qquad 2.53$$

Properly designed nozzles cause an increase in the velocity of the vapor and a decrease in the enthalpy. A nozzle can be designed to operate nearly reversibly. Example 4.12 on page 162 describes a typical nozzle calculation.

Throttles are much more common in the problems we will address in this text. The meaning of "nozzle" in thermodynamics is much different from the common devices we term "nozzles" in everyday life. Most of the everyday devices we call nozzles are actually throttles.

Assessing when simplifications are justified requires testing the implications of eliminating assumptions. For example, to test whether a particular valve is acting more like a throttle or a nozzle, infer the velocities before and after the nozzle and compare to the enthalpy change. If the

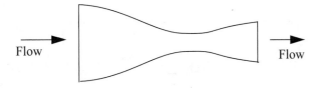

Flow   Flow

Figure 2.7 *Illustration of a converging-diverging nozzle showing the manner in which inlets and outlets are tapered.*

---

13. Turbine design is a specialized topic. Introductions to the actual operation are most readily available in mechanical engineering thermodynamics textbooks, such as Jones, J.B., Dugan, R.E. 1996. *Engineering Thermodynamics.* Upper Saddle River, NJ: Prentice-Hall, pp. 734–745.

kinetic energy change is negligible relative to the enthalpy change then call it a throttle. Take note of the magnitude of the terms in the calculation so that you can understand how to anticipate a similar conclusion. For example, the volume change of a liquid due to a pressure drop is much smaller than that of a gas. With less expansion, the liquids accelerate less, making the throttle approximation more reasonable. This kind of systematic analysis and reasoning is more important than memorizing, say, that throttles are for liquids.

Heat Exchangers

Heat exchangers are available in a number of flow configurations. For example, in an industrial heat exchanger, a hot stream flows over pipes that carry a cold stream (or vice versa), and the objective of operation is to cool one of the streams and heat the other. A generic tube-in-shell heat exchanger can be illustrated by a line diagram as shown in Fig. 2.8. Tube-in-shell heat exchangers consist of a shell (or outer sleeve) through which several pipes pass. (The figure just has one pipe for simplicity.) One of the process streams passes through the shell, and the other passes through the tubes. Stream *A* in our example passes through the shell, and Stream *B* passes through the tubes. *The streams are physically separated from one another by the tube walls and do not mix.* Let's suppose that Stream *A* is the hot stream and Stream *B* is the cold stream. In the figure, both streams flow from left to right. This type of flow pattern is called **concurrent.** The temperatures of the two streams will approach one another as they flow to the right. With this type of flow pattern, we must be careful that the hot stream temperature that we calculate is always higher than the cold stream temperature at every point in the heat exchanger.[14] If we reverse the flow direction of Stream *A,* a **countercurrent** flow pattern results. With a countercurrent flow pattern, the outlet temperature of the cold stream can be higher than the outlet temperature of the hot stream (but still must be lower than the inlet temperature of the hot stream). The hot stream temperature must always be above the cold stream temperature at all points along the tubes in this flow pattern also.

So far, our discussion has assumed that there are no phase transitions occurring in the heat exchanger. If Stream *A* is a hot stream, and Stream *B* is converted from liquid at the inlet to vapor at the outlet, we call the heat exchanger a **boiler** to bring attention to the phase transition occurring inside. The primary difference in the operation of a boiler to that of a generic heat exchanger is that the cold stream temperature change might be small or even zero. This is because the phase change will occur isothermally at the saturation temperature of the fluid corresponding to the boiler pressure, absorbing large amounts of heat. In a similar fashion, we could have Stream *A* be cooling water and Stream *B* be an incoming vapor which is condensed. We would call this heat exchanger a **condenser,** to clearly bring attention to the phase change occurring inside. In this case, the temper-

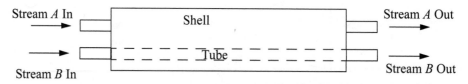

Figure 2.8 *Illustration of a generic heat exchanger with a concurrent flow pattern. The tube side usually consists of a set of parallel tubes which are illustrated as a single tube for convenience.*

---

14. This may seem like common sense, but sometimes when calculations are performed, it is surprisingly easy to overlook the fact that a valid mathematical result might be physically impossible to obtain.

ature change of the hot stream might be small. Another type of heat exchanger that we will use in Chapter 5 is the **superheater.** A superheater takes a vapor that is saturated and superheats it.

There are two more important points to keep in your mind as you perform thermodynamic calculations. For the purposes of this text we will neglect pressure drops in the heat exchangers; the outlet pressure will match the inlet pressure of Stream *A,* and a similar statement applies for Stream *B.* Note that this does not imply that streams *A* and *B* are at the *same* pressure. Also, we neglect heat transfer to or from the surroundings unless specified. Therefore, all heat transfer occurs inside the heat exchanger, not at the boundaries of the heat exchanger and the surroundings.

There are other configurations of heat exchangers such as kettle-type reboilers and plate-and-frame configurations; however, for thermodynamic purposes, only the flow pattern is important, not the details of material construction that lead to the flow pattern. Thus, the tube-in-shell concepts will be adequate for our needs.

The energy balance that we write depends on how we choose our system. Since the streams are physically separated from one another, we may write a balance for each of the streams independently, or we may place the system boundary around the entire heat exchanger and write a balance for both streams. Let us take the system to be Stream *B* and let us suppose that Stream *B* is boiled. In this case, there is just one inlet and outlet. There is no shaft work or expansion/contraction work. Even though the process fluid is expanding as it evaporates, the system boundaries are not expanding; expansion effects will be automatically included in the energy balance by the enthalpy terms which have the flow work embedded in them. If the system is operating at steady state, the left-hand side of the energy balance is zero,

$$0 = \left[H + \frac{v^2}{2g_c} + \frac{gz}{g_c} \right]^{in} \dot{m}^{in} - \left[H + \frac{v^2}{2g_c} + \frac{gz}{g_c} \right]^{out} \dot{m}^{out} + \underline{\dot{Q}} + \dot{W}_{EC} + \dot{W}_S$$

which simplifies to

⚠ The energy balance for a heat exchanger may be written in several ways.

$$0 = -\Delta H \dot{m} + \underline{\dot{Q}} \qquad\qquad 2.54$$

where $\underline{\dot{Q}}$ is the rate of heat transfer from the hot stream. On a molar (or mass) basis,

$$\boxed{0 = -\Delta H + Q \quad \text{half heat exchanger}} \qquad\qquad 2.55$$

If we take the system boundaries to be around the entire heat exchanger, then there are multiple streams, and all heat transfer occurs inside, resulting in

$$0 = \sum_{inlets} \left[H + \frac{v^2}{2g_c} + \frac{gz}{g_c} \right]^{in} \dot{m}^{in} - \sum_{outlets} \left[H + \frac{v^2}{2g_c} + \frac{gz}{g_c} \right]^{out} \dot{m}^{out} + \dot{Q} + \dot{W}_{EC} + \dot{W}_S$$

which simplifies to

$$0 = \sum_{inlets} H^{in} \dot{m}^{in} - \sum_{outlets} H^{out} \dot{m}^{out} \qquad\qquad 2.56$$

$$0 = -\Delta H_A \dot{m}_A - \Delta H_B \dot{m}_B \quad \text{overall heat exchanger} \qquad 2.57$$

Since Eqns. 2.55 and 2.57 look quite different for the same process, it is important that you understand the placement of boundaries and their implications on the balance expression.

Adiabatic Turbines or Adiabatic Expanders

A turbine or expander is basically a sophisticated windmill as shown in Fig. 2.9. The term "turbine" implies operation by steam and the term "expander" implies operation by a different process fluid, perhaps a hydrocarbon, although the term "turbine" is used sometimes for both. The objective of operation is to convert the kinetic energy from a gas stream to rotary motion of a shaft to produce work (shaft work). The enthalpy of the high-pressure inlet gas is converted to kinetic energy by special stators (stationary blades) or nozzles *inside the turbine shell*. The high-velocity gas drives the rotor. Turbines are designed to be adiabatic, although heat losses can occur. When heat losses are present, they decrease the output that would have otherwise been possible for the turbine. Therefore, when calculations are performed, we assume that turbines or expanders operate adiabatically, unless otherwise noted.

The energy balance for the turbine only involves the kinetic energy change for the *entering* and *exiting* fluid, not for the changes occurring inside the turbine. Since the nozzles which cause large kinetic energy changes are *inside* the turbine unit, these changes are irrelevant to the balance *around* the unit. Recall from the development of our energy balance that we are only interested in the values of enthalpy, kinetic, and potential energy for streams *as they cross the boundaries* of our system. The energy balance for a *steady-state* turbine involving one inlet and one outlet is:

$$0 = \left[H + \frac{v^2}{2g_c} + \frac{gz}{g_c} \right]^{in} \dot{m}^{in} - \left[H + \frac{v^2}{2g_c} + \frac{gz}{g_c} \right]^{out} \dot{m}^{out} + \dot{Q} + \dot{W}_{EC} + \dot{W}_S$$

which becomes

❶ Adiabatic turbine or expander.

$$0 = (H^{in} - H^{out})\dot{m} + \dot{W}_S = -\Delta H \dot{m} + \dot{W}_S \qquad 2.58$$

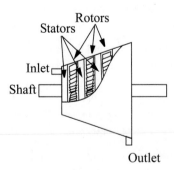

Figure 2.9 *Illustration of a turbine. The rotor (shaft) turns due to the flow of gas. The blades connected to the shell are stationary (stators), and are sometimes curved shapes to perform as nozzles. The stator blades are not shown to make the rotors more clear.*

and on a mass or molar basis becomes

$$0 = -\Delta H + W_S \quad \text{adiabatic turbine/expander/compressor/pump}$$

2.59

When we calculate values for the ΔH and work, they will be negative values.

Adiabatic Compressors

Adiabatic compressors can be constructed in a manner qualitatively similar to adiabatic turbines with stationary vanes (stators). This type of compressor is called an axial compressor. The main differences between turbines and axial compressors are: 1) the details of the construction of the vanes and rotors, which we won't be concerned with; 2) the direction of flow of the fluid; and 3) the fact that we must put work into the compressor rather than obtaining work from it. Thus, the energy balance is the same as the turbine (Eqns. 2.58 and 2.59). When we calculate values for the ΔH and work, they will be positive values, where they were negative values for a turbine. Compressors may also be constructed as reciprocating (piston/cylinder) devices. This modification has no impact on our energy balance, so it remains the same. Analogous to turbines, it is conventional to assume that compressors are adiabatic unless otherwise noted.

Pumps

Pumps are used to move liquids by creating the pressure necessary to overcome the resistance to flow. They are in principle just like compressors, except the liquid will not change density the way a gas does when it is compressed. Again, the energy balance will be the same as a turbine or compressor (Eqns. 2.58 and 2.59). The primary difference we will find in application of the energy balance is that tabulated enthalpies are difficult to find for compressed liquids. Therefore, if we want to calculate the work needed for a pump, we can find it from the energy balance after we have calculated or determined the enthalpy change.

❶ Compressors and pumps usually have the same energy balance as turbines.

Calculation of Shaft Work

Adiabatic steady-state turbines, compressors, and pumps all share a common energy balance, showing that the shaft work is related to the enthalpy change as shown on page 54. Often, it is helpful to calculate the shaft work directly and use the value of the shaft work to find the enthalpy change.

$$W_S = \int_{in}^{out} V dP \quad \text{reversible pump, compressor, turbine}$$

2.60

$$W_S \approx V^L (P^{out} - P^{in}) = \frac{\Delta P}{\rho} \quad \text{liquid pump}$$

2.61

When the work is to calculated, the adiabatic shaft work may, in principle, be analyzed using ΔH or Eqn 2.60. For gases, it is usually easier to use other constraints to find the enthalpy change and then calculate the work by equality, though in principle the integral can be evaluated. For the special case of liquids, Eqn 2.60 can be replaced by Eqn 2.61.

2.14   STRATEGIES FOR SOLVING PROCESS THERMODYNAMICS PROBLEMS

Before we start several more complicated example problems, it will be helpful to outline the strategies which will be applied. We provide these in a step form to make them easier to use. Many of these steps will seem obvious, but if you become stuck when working through a problem, it is usually because one of these steps was omitted or applied inconsistently with system boundaries.

1. Choose system boundaries; decide whether this boundary location will make the system **open or closed.**

2. Identify all given state properties of fluids in system and crossing boundaries. Identify which are invariant with time. **Identify your system as steady or unsteady state.** (For unsteady-state pumps, turbines, or compressors, the accumulation of energy within the device is usually neglected.) For open, steady-state systems, write the mass balance and **solve if possible.**

3. Identify how many state variables are unknown for the system. **Recall that only two state variables are required to specify the state of a pure, single-phase fluid.** The number of unknowns will equal the number of independent equations necessary for a solution. (Remember in a system of known total volume $\underline{V}$, that if n is known, the state variable V is known.)

> **The phase rule is important in determining the required number of equations.**

4. **Write the mass balance and the energy balance. These are the first equations to be used in the solution.** Specify reference states for all fluids if necessary. Simplify energy balance to **eliminate terms which are zero** for the system specified in step 1.[15] Combine the mass balance and the energy balance for open systems.

   For unsteady-state problems:

   (a) Identify whether the individual terms in the energy balance may be integrated directly without combining with other energy balance terms. Often the answer is obtained most easily this way. *This is almost always possible for closed-system problems.*

   (b) If term-by-term integration of the energy balance is not possible, rearrange the equation to simplify as much as possible before integration.

> **Always consider the overall balance.**

5. Look for any other information in the problem statement that will provide **additional equations** if unknowns remain. Look for key words such as **adiabatic, isolated, throttling, nozzle, reversible,** and **irreversible.** Using any applicable constraints of throttling devices, nozzles, and so on, relate stream properties for various streams to one another and to the system state properties. Constraints on flow rates, heat flow, and so on. provide additional equations. With practice, many of these constraints may be recognized immediately before writing the energy balance in steps 3 and 4.

> **Look for key words.**

6. Introduce the thermodynamic properties of the fluid (the equation of state). **This provides all equations relating $P, V, T, U, H, C_P,$ and C_V.** The information will consist of either 1) the ideal gas approximation; 2) a thermodynamic chart or table; or 3) a volumetric equation of state (which will be introduced in Chapter 7). Using more than one of these sources of information in the same problem may introduce inconsistencies in the properties used in the solution, depending on the accuracy of the methods used.

> **Use the same property method throughout the problem if possible.**

   Combine the thermodynamic information with the energy balance. Work to minimize the number of state variables which remain unknown. Many problems are solved at this point.

---

15. Section 2.17 on page 85 may be helpful for details on interpreting each term of the balance for new applications.

7. Do not hesitate to move your system boundary and try again if you are stuck. Do not forget to try an overall balance (frequently, two open systems can be combined to give an overall closed system, and strategy 4a can be applied). Make reasonable assumptions.

8. After an answer is obtained, verify assumptions that were made to obtain the solution.

❶ Try different system boundaries.

❷ Make sure answers seem reasonable.

2.15   CLOSED AND STEADY-STATE OPEN SYSTEMS

Several types of systems are quite common in chemical engineering practice. You need to be familiar with the results of their analysis and benefit if you memorize these results for rapid recall. You must simultaneously recall the assumptions underlying each simple model, however, to avoid incorrect applications.

Example 2.11   Adiabatic, reversible expansion of an ideal gas

Suppose an ideal gas in a piston + cylinder is adiabatically and reversibly expanded to twice its original volume. What will be the final temperature?

Solution: First consider the energy balance. The system will be the gas in the cylinder. The system will be closed. Since a basis is not specified, we can choose 1 mole. Since there is no mass flow, heat transfer, or shaft work, the energy balance becomes:

$$d\left[U + \frac{\cancel{v^2}}{2g_c} + \frac{\cancel{g}}{g_c}\cancel{z}\right] = \cancel{dQ} + \cancel{dW_S} + dW_{EC}$$

$$dU = -PdV$$

In this case, as we work down to step 4 in the strategy, we see that we cannot integrate the sides independently since P depends on T. Therefore, we need to combine terms before integrating.

$$C_V dT = -RT\frac{dV}{V} \quad \text{which becomes} \quad \frac{C_V}{T}dT = -\frac{R}{V}dV \qquad \text{(ig) 2.62}$$

The technique that we have performed is called separation of variables. All of the temperature dependence is on the left-hand side of the equation and all of the volume dependence is on the right-hand side. Now, if we assume a constant heat capacity for simplicity, we can see that this integrates to

$$\frac{C_V}{R}\ln\frac{T}{T^i} = \ln\frac{V^i}{V} \qquad (\text{*ig})$$

$$\boxed{\left(\frac{T}{T^i}\right)^{(C_V/R)} = \frac{V^i}{V}} \qquad (\text{*ig}) \ 2.63$$

❸ These boxed equations relate state variables for adiabatic reversible changes of an ideal gas in a closed system.

Example 2.11 Adiabatic, reversible expansion of an ideal gas (Continued)

Although not required, several rearrangements of this equation are useful for other problems. Note that we may insert the ideal gas law to convert to a formula relating T and P. Using $V = RT/P$,

$$\left(\frac{T}{T^i}\right)^{\frac{C_V}{R}} = \frac{T^i}{P^i}\frac{P}{T} \qquad (\text{*ig})$$

Rearranging,

$$\left(\frac{T}{T^i}\right)^{(C_V/R)}\frac{T}{T^i} = \left(\frac{T}{T^i}\right)^{(C_V/R)+1} = \frac{P}{P^i} \qquad (\text{*ig})$$

which becomes

$$\boxed{\left(\frac{T}{T^i}\right)^{(C_P/R)} = \frac{P}{P^i}} \qquad (\text{*ig}) \text{ 2.64}$$

We may also insert the ideal gas law into Eqn. 2.63 to convert to a formula relating P and V. Using $T = PV/R$,

$$\left(\frac{PV}{P^iV^i}\right)^{(C_V/R)} = \frac{V^i}{V} \qquad (\text{*ig})$$

$$\frac{P}{P^i} = \left(\frac{V^i}{V}\right)^{(R/C_V)}\frac{V^i}{V} = \left(\frac{V^i}{V}\right)^{(R/C_V)+1} = \left(\frac{V^i}{V}\right)^{(C_P/C_V)} \qquad (\text{*ig})$$

which may be written

$$\boxed{PV^{(C_P/C_V)} = \text{const}} \qquad (\text{*ig}) \text{ 2.65}$$

The analysis of a piston+cylinder implied the assumption of a closed system. This might be a reasonable approximation for a single stroke of a combustion engine, but most chemical engineering applications involve continuous operation. Nevertheless, we can apply the lessons learned from the analysis of the closed system when extending to steady-state systems, as exemplified below.

Example 2.12 Continuous adiabatic, reversible compression of an ideal gas

Suppose 1 kmol/h of air at 5 bars and 298 K is adiabatically and reversibly compressed in a continuous process to 25 bars. What will be the outlet temperature and power requirement for the compressor in hp?

Example 2.12   Continuous adiabatic, reversible compression of an ideal gas   (Continued)

Solution: Note that air is composed primarily of oxygen and nitrogen and these both satisfy the stipulations for diatomic gases with their reduced temperatures high and their reduced pressures low. In other words, the ideal gas approximation with $C_P/R = 7/2$ is clearly applicable. Next consider the energy balance. The system is the compressor. The system is open. Since it is a steady-state process with no heat transfer, the simplification of the energy balance has been discussed on page 73 and shown on page 72, and the energy balance becomes:

$$\Delta H = \cancel{Q} + W_S \qquad\qquad 2.66$$

We can adapt Eqn. 2.31 for an ideal gas as follows:

$$dW_S = dH = V dP$$

In this case, as we work down to step 4 in the strategy, we see that we cannot integrate the sides independently since P depends on T. Therefore, we need to combine terms before integrating.

$$C_P dT = RT \frac{dP}{P} \quad \text{which becomes} \quad \frac{C_P}{T} dT = \frac{R}{P} dP \qquad \text{(ig) 2.67}$$

Once again, we have performed separation of variables. The rest of the derivation is entirely analogous to Example 2.11, and, in fact, the resultant formula is identical.

$$\frac{T_2}{T_1} = \left(\frac{P_2}{P_1}\right)^{(R/C_P)} \qquad\qquad \text{(*ig) 2.68}$$

Steady-state adiabatic, reversible processing of an ideal gas results in the same relations as Example 2.11.

Note that this formula comes up quite often as an approximation for both open and closed systems. Making the appropriate substitutions,

$$T_2 = 298\left(\frac{25}{5}\right)^{\frac{2}{7}} = 472 \text{ K}$$

Adapting the adiabatic energy balance and assuming $C_P^{ig} = \text{constant}$,

$$W_S^{ig} = \Delta H^{ig} = C_P^{ig}\Delta T = C_P^{ig}T_1\left[\left(\frac{P_2}{P_1}\right)^{(R/C_P^{ig})} - 1\right] \qquad \text{(*ig) 2.69}$$

Steady-state adiabatic, reversible compression of an ideal gas.

Substituting, $W_S = 3.5 \cdot 8.314 \cdot (472 - 298) = 5063 \text{ J/mole}$

At the given flow rate, and reiterating that this problem statement specifies a reversible process:

$$W_S^{rev} = 5063 \text{ J/mole} \cdot [1000 \text{mole/hr}] \cdot [1\text{hr}/3600\text{sec}] \cdot [1\text{hp}/(745.7\text{J/s})] = 1.9\text{hp}$$

We have systematically extended our analysis from a single step of a closed system, to a continuous system with no heat loss. Let's consider isothermal operation.

Example 2.13 Continuous, isothermal, reversible compression of an ideal gas

Repeat the compression from the previous example, but consider steady-state isothermal compression. What will be the heat removal rate and power requirement for the compressor in hp?

Solution: Let's return to the perspective of the section 'Understanding Enthalpy and Shaft Work' on page 54 and analyze the EC work and flow work for an ideal gas packet of unit mass. The W_{EC} is,

$$W_{EC} = -\int P dV = -\int (RT/V) dV = -RT\ln(V_2/V_1) \qquad (*\text{ig})$$

For an isothermal, ideal gas, $V_2/V_1 = P_1/P_2$. Noting the reciprocal and negative logarithm,

$$W_{EC} = -RT\ln(V_2/V_1) = RT\ln(P_2/P_1) \qquad (*\text{ig})$$

This is the work to isothermally squeeze an ideal gas packet of unit mass to a given pressure. The flow work performed on an ideal gas packet of unit mass is,

$$W_{flow} = PV_{out} - PV_{in} = RT_{out} - RT_{in} = 0 \qquad (*\text{ig})$$

Therefore, the total requirement for isothermally compressing an ideal gas packet of unit mass is,

$$W_S = RT\ln(P_2/P_1) \qquad (*\text{ig})$$

$$W_S = RT\ln\left(\frac{P_2}{P_1}\right) = 8.314(298)\ln 5 = 3987\frac{J}{mol} \qquad (\text{ig}) \; 2.70$$

At the given flow rate, and reiterating that this problem statement specifies a reversible process,

$$W_S^{rev} = 3987 \text{ J/mole} \cdot [1000\text{mole/hr}] \cdot [1\text{hr}/3600\text{sec}] \cdot [1\text{hp}/745.7\text{J/s}] = 1.5 \text{ hp}$$

Compared to adiabatic compression, the isothermal compressor requires less work. This happens because cooling withdraws energy from the system. It is difficult to achieve perfectly adiabatic or isothermal operation in practice, but adiabatic operation is usually a better approximation because compression is so rapid that there is insufficient time for heat transfer. Usually fluids are cooled after compression as we discuss in later chapters.

Here is a brain teaser. Suppose the process fluid had been steam instead of an ideal gas. How would you have solved the problem then? Note, for steam $V \neq RT/P$ and $dH \neq C_P^{ig} dT$. We illustrate the answer in Example 4.16 on page 173.

Example 2.14 Heat loss from a turbine

High-pressure steam at a rate of 1100 kg/h initially at 3.5 MPa and 350°C is expanded in a turbine to obtain work. Two exit streams leave the turbine. Exiting stream (2) is at 1.5 MPa and 225°C and flows at 110 kg/h. Exiting stream (3) is at 0.79 MPa and is known to be a mixture of saturated vapor and liquid. A fraction of stream (3) is bled through a throttle valve to 0.10 MPa and is found to be 120°C. If the measured output of the turbine is 100 kW, estimate the heat loss of the turbine. Also, determine the quality of the steam in stream (3).

Solution: First draw a schematic. Designate boundaries. Both System A and System B are open steady-state systems.

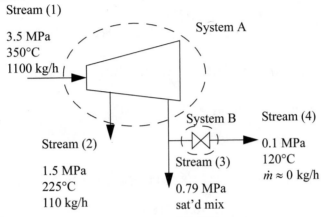

Stream (1)

3.5 MPa
350°C
1100 kg/h

System A

System B   Stream (4)

Stream (2)   0.1 MPa
Stream (3)   120°C
1.5 MPa $\dot{m} \approx 0$ kg/h
225°C
110 kg/h   0.79 MPa
            sat'd mix

The mass balance gives $\dot{m}_3$ = 990 kg/h. Next, determine which streams are completely specified: Streams (1), (2), and (4) are fully specified. Since Stream (3) is saturated, the temperature and pressure and specific enthalpies of the saturated vapor and liquid can be found, but the quality needs to be calculated to determine the overall molar enthalpy of the stream. From the steam tables we find H_1 directly. For H_2 we use linear interpolation. The value $H(1.5$ MPa, 225°C) is not available directly, so we need to first interpolate at 1.4 MPa between 200°C and 250°C to find $H(1.4$ MPa, 225°C) and then interpolate between this value and the value at 1.6 MPa:

$$H(1.4\text{MPa}, 225°C) = \frac{1}{2}(2803.0 + 2927.9) = 2865.5 \text{ kJ/kg}$$

$$H(1.6\text{MPa}, 225°C) = 2792.8 + \frac{23.6}{48.6}(2919.9 - 2792.8) = 2854.5 \text{ kJ/kg}$$

Then to find H_2: $H_2 = 0.5 \cdot (2865.5 + 2854.5) = 2860.0$ kJ/kg. For H_4 we can interpolate in the superheated steam tables:

$$H_4 = 2675.8 + \frac{20}{50}(2776.6 - 2675.8) = 2716.1 \text{ kJ/kg}$$

Example 2.14   Heat loss from a turbine (Continued)

Recognize System B as a throttle valve; therefore, $H_3 = H_4 = 2716.1$ kJ/kg. We make a table to summarize the results so that we can easily find values:

Stream	1	2	3, 4
H(kJ/kg)	3104.8	2860.0	2716.1

The energy balance for System A gives, using $\dot{W}_S = -100$ kW given in the problem statement,

$$0 = H_1 \dot{m}_1 - H_2 \dot{m}_2 - H_3 \dot{m}_3 + \dot{Q} + \dot{W}_S = 3104.8(1100) - 2860.0(110) - 2716.1(990) + \dot{Q} + \dot{W}_S$$

$$\dot{Q} = \frac{-411,741 \text{ kJ}}{\text{h}} \Bigg| \frac{\text{h}}{3600 \text{ s}} + \frac{100 \text{ kJ}}{s} = -14.4 \text{ kJ/s}$$

To find the quality of stream (3), $H_3 \dot{m}_3 = H^L \dot{m}^L - H^V \dot{m}^V$,

$$H_3 = H^L \frac{\dot{m}^L}{\dot{m}_3} - H^V \frac{\dot{m}^V}{\dot{m}_3} = H^L + q(\Delta H^{vap})$$

At 0.79 MPa from the sat'd P table, $H^L = 718.5$ kJ/kg and $\Delta H^V = 2049$ kJ/kg.

$$q = \frac{2716.1 - 718.5}{2049} = 0.975$$

The energy balance for a non-adiabatic turbine is identical to the balance for an isothermal compressor, but the conclusions are entirely different. In the compressor, we want to minimize work, so the heat loss works to our advantage. For the turbine, we want to maximize work, so any loss of energy should be avoided.

The examples in this section comprise several important common scenarios, but they also illustrate a procedure for analyzing systems with systematically increasing sophistication. In the context of certain simplifying assumptions, like the ideal gas model, we can derive final working equations applicable to process calculations. When those assumptions are invalid, however, we can still apply the energy balance, but we are more careful in the generality of the results we obtain. Processes involving steam, for example, require something more than the ideal gas model, and additional tools are required to develop a general analysis.

2.16   UNSTEADY-STATE OPEN SYSTEMS

In principle, most real systems are unsteady and open. A few systems couple the unsteady-state operation with flow across boundaries in a way that requires simultaneous analysis. We illustrate how to treat those systems with examples of leaking and filling tanks.

Example 2.15   Adiabatic expansion of an ideal gas from a leaky tank

An ideal gas is leaking from an insulated tank. Relate the change in temperature to the change in pressure for gas leaking from a tank. Derive an equation for $\Delta \underline{U}$ for the tank.

Solution: Let us choose our system as the gas in the tank at any time. This will be an open, unsteady-state system. There is no inlet stream and one outlet stream. The mass balance gives $dn = - dn^{out}$.

We can neglect kinetic and potential energy changes. Although the gas is expanding, the system size remains unchanged, and there is no expansion/contraction work. The energy balance becomes (on a molar basis):

$$d(nU) = H^{in}\cancel{dn^{in}} - H^{out}dn^{out} + \cancel{dQ} + \cancel{dW_{EC}} + \cancel{dW_S}$$

Since the enthalpy of the exit stream matches the enthalpy of the tank, $H^{out} = H$. $d(nU) = -H^{out}dn^{out} = Hdn$. Now H depends on temperature, which is changing, so we are not able to apply hint 4(a) from the problem-solving strategy. It will be necessary to combine terms before integrating. By the product rule of differentiation, the left-hand side expands to $d(nU) = ndU + Udn$. Collecting terms in the energy balance,

$$ndU = (H - U)dn$$

Performing some substitutions, the energy balance can be written in terms of T and n,

$$(H - U) = PV = RT; \quad dU = C_V dT; \quad \Rightarrow \frac{C_V}{R} \frac{dT}{T} = \frac{dn}{n} \tag{ig}$$

$$\boxed{\frac{C_V}{R}\ln\frac{T}{T^i} = \ln\frac{n}{n^i}} \tag{*ig}$$

The volume of the tank is constant, ($\underline{V}$ = constant); therefore,

$$\ln\frac{n}{n^i} = \ln\frac{PT^i}{TP^i} = -\ln\frac{T}{T^i} + \ln\frac{P}{P^i} = \frac{C_V}{R}\ln\frac{T}{T^i} \tag{ig}$$

substituting,

$$\left(\frac{C_V}{R} + 1\right)\ln\frac{T}{T^i} = \frac{C_P}{R}\ln\frac{T}{T^i} = \ln\frac{P}{P^i} \tag{*ig}$$

> ## Example 2.15   Adiabatic expansion of an ideal gas from a leaky tank (Continued)

For fluid *exiting* from an adiabatic tank, the results are the same as a closed system as in Example 2.11.

Recognizing the relation between C_V and C_P, defining $\gamma \equiv C_p/C_v \, (=1.4$ for an ideal diatomic gas), note $R/C_P = (C_P - C_V)/C_P = 1 - (1/\gamma) = (\gamma - 1)/\gamma$:

$$\frac{T}{T^i} = \left(\frac{P}{P^i}\right)^{R/C_p} = \left(\frac{P}{P^i}\right)^{(1 - (1/\gamma))} \qquad (*\text{ig}) \; 2.71$$

Through the ideal gas law $(PV = RT)$, we can obtain other arrangements of the same formula.

$$V/V^i = \left(\frac{P^i}{P}\right)^{1/\gamma}; \quad P^i/P = (V/V^i)^{\gamma} = (T^i/T)^{(1/\gamma) - 1}; \quad T^i/T = (V/V^i)^{\gamma/(\gamma - 1)} \quad (*\text{ig}) \; 2.72$$

The numerical value for the change in internal energy of the system depends on the reference state because the reference state temperature will appear in the result:

$$\Delta \underline{U} = n^f (C_V(T^f - T_R) + U_R) - n^i (C_V(T^i - T_R) + U_R) \qquad (*\text{ig})$$

At first glance, one might expect to use the same equation for a filling tank, but simply change the pressure ratio. Careful analysis shows that the energy balance is similar, but the final result is quite different.

> ## Example 2.16   Adiabatically filling a tank with an ideal gas

Helium at 300 K and 3000 bar is fed into an evacuated cylinder until the pressure in the tank is equal to 3000 bar. Calculate the final temperature of the helium in the cylinder $(C_P/R = 5/2)$.

Solution: The system will be the gas inside the tank at any time. The system will be an open, unsteady-state system. The mass balance is $dn = dn^{in}$. The energy balance becomes:

$$d(nU) = H^{in} dn^{in} - H^{out} \cancel{dn^{out}} + \cancel{dQ} + \cancel{dW_{EC}} + \cancel{dW_S}$$

We recognize that H^{in} will be constant throughout the tank filling. Therefore, by hint 4a from the problem-solving strategy, we can integrate terms individually. We need to be careful to keep the superscript since the incoming enthalpy is at a different state than the system. The right-hand side of the energy balance can be integrated to give

$$\int_i^f H^{in} dn = H^{in} \int_i^f dn = H^{in}(n^f - n^i) = H^{in} n^f$$

The left-hand side of the energy balance becomes

$$\Delta(Un) = U^f n^f - U^i n^i = U^f n^f$$

Example 2.16 Adiabatically filling a tank with an ideal gas (Continued)

Combining the result with the definition of enthalpy,

$$U^f = H^{in} = U^{in} + PV^{in} = U^{in} + RT^{in} \qquad \text{(ig) 2.73}$$

And with our definition of heat capacity, we can find temperatures:

$$\Delta U = C_V(T^f - T^{in}) = RT^{in} \Rightarrow T^f = T^{in}(R + C_v)/C_v = T^{in}C_p/C_v \qquad (*\text{ig})$$

Note that the final temperature is independent of pressure for the case considered here.

You should not get the impression that unsteady, open systems are limited to ideal gases. Energy balances are independent of the type of operating fluid.

Example 2.17 Adiabatic expansion of steam from a leaky tank

An insulated tank contains 500 kg of steam and water at 215°C. Half of the tank volume is occupied by vapor and half by liquid; 25 kg of dry vapor is vented slowly enough that temperature remains uniform throughout the tank. What is the final temperature and pressure?

Solution: There are some similarities with the solution to Example 2.15 on page 81; however, we can no longer apply the ideal gas law. The energy balance reduces in a similar way, but we note that the exiting stream consists of only vapor; therefore, it is not the overall average enthalpy of the tank:

$$d(mU) = -H^{out}dm^{out} = H^V dm$$

The sides of the equation can be integrated independently if the vapor enthalpy is constant. Looking at the steam table, the enthalpy changes only about 10 kJ/kg out of 2800 kJ/kg (0.3%) along the saturation curve down to 195°C. Let us assume it is constant at 2795 kJ/kg making the integral of the right-hand side simply $H^V \Delta m$. Note that this procedure is equivalent to a numerical integration by trapezoidal rule as given in Appendix B on page 822. Many students forget that analytical solutions are merely desirable, not absolutely necessary. The energy balance then can be integrated using hint 4a on page 74.

$$\Delta \underline{U} = m^f U^f - m^i U^i = 2795(m^f - m^i) = 2795(-25) = -69,875 \text{ kJ}$$

The quantity $m^f = 475$, and $m^i U^i$ will be easy to find, which will permit calculation of U^f. For each m^3 of the original saturated mixture at 215°C,

$$\frac{0.5 \text{ m}^3 \text{ vapor} \mid \text{kg vapor}}{\mid 0.0947 \text{ m}^3 \text{ vapor}} = 5.28 \text{ kg vapor}$$

$$\frac{0.5 \text{ m}^3 \text{ liquid} \mid \text{kg liquid}}{\mid 0.001181 \text{ m}^3 \text{ liquid}} = 423.4 \text{ kg liquid}$$

Example 2.17 Adiabatic expansion of steam from a leaky tank (Continued)

Therefore,

$$V^t = \frac{m^3}{423.4 + 5.28 \text{ kg}} = 0.00233 \text{ m}^3/\text{ kg}$$

So the tank volume, quality, and internal energy are:

$$\underline{V}_T = \frac{500 \text{ kg}}{} \left| \frac{m^3}{428.63} \right. = 1.166 \text{ m}^3$$

$$q^i = 5.28 \text{ kg vapor} / 428.63 \text{ kg} = 0.0123 \quad U^i = 918.04 + 0.0123(1681.9) = 938.7 \text{ kJ/kg}$$

$$\underline{U}^i = 938.7 \text{ kJ/kg} \cdot 500 \text{ kg} = 469{,}400 \text{ kJ}$$

Then, from the energy balance and mass balance,

$$U^f = (-69{,}875 + 469{,}400) \text{ kJ} / 475 \text{ kg} = 841.0 \text{ kJ/kg}$$

$$V^f = 1.166 \text{ m}^3 / 475 \text{ kg} = 0.00245 \text{ m}^3 /\text{kg}$$

We need to find P^f and T^f which correspond to these state variables. The answer will be along the saturation curve because the overall specific volume is intermediate between saturated vapor and liquid values at lower pressures. We will guess P^f (and the corresponding saturation T^f), find q from V^f, then calculate U^f_{calc} and compare to $U^f = 841.0$ kJ/kg. If U^f_{calc} is too high, the P^f (and T^f) guess will be lowered.

Since $V = V^L + q(V^V - V^L)$,

$$q = \frac{0.00245 - V^L}{V^V - V^L} \quad \text{and from this value of } q, \quad U^f_{calc} = U^L + q(U^V - U^L)$$

To guide our first guess, we need $U^L < U^f = 841.0$ kJ/kg. Our first guess is $T^f = 195°C$. Values for the properties from the steam tables are shown in the table below. This initial guess gives $U^f_{calc} = 845$ kJ/kg; no further iteration is necessary. The H^V at this state is 2789; therefore, our assumption of $H^{out} \approx$ constant is valid.

State	P(MPa)	T(°C)	V^L	V^V	U^L	ΔU^{vap}	q
Initial	2.106	215	0.001181	0.0947	918.04	1681.9	0.0123
Guess	1.3988	195	0.001149	0.1409	828.18	1763.6	0.0093

$$P^f = 1.4 \text{ MPa}, \ T^f = 195°C, \ \Delta P = 0.7 \text{ MPa}, \ \Delta T = 20°C$$

2.17   DETAILS OF TERMS IN THE ENERGY BALANCE

Generally, the strategies discussed in Section 2.14 are sufficient to simplify the energy balance. Occasionally, in applying the energy balance to a new type of system, simplification of the balances may require more detailed analysis of the background leading to the terms and/or details of interactions at boundaries. This section provides an overall summary of the details for the principles covered earlier in this chapter, and it is usually not necessary unless you are having difficulty simplifying the energy balance and need details regarding the meaning of each term.

The universe frequently consists of three subsystems, as illustrated in Fig. 2.10. The container (System 2) is frequently combined with System 1 (designated here as System (1 + 2)) or System 3 (designated here as System (2 + 3)). For every balance, all variables are of the *system* for which the balance is written,

$$\frac{d}{dt}\left[m\left(U + \frac{v^2}{2g_c} + \frac{gz}{g_c}\right)\right] = \sum_{inlets}\left(H + \frac{v^2}{2g_c} + \frac{gz}{g_c}\right)^{in}\dot{m}^{in}$$

$$-\sum_{outlets}\left(H + \frac{v^2}{2g_c} + \frac{gz}{g_c}\right)^{out}\dot{m}^{out} + \sum_{surfaces}\dot{Q} + \dot{\underline{W}}_{EC} + \dot{\underline{W}}_S$$

2.74

where superscripts "in" and "out" denote properties of the streams which cross the boundaries, which may or may not be equal to properties of the system.

1. Non-zero heat interactions of Systems 1 and 3 are not equal unless the heat capacity or mass of System 2 is negligible.

2. H^{out}, H^{in} account for internal energy changes and work done on the system   *due to flow across boundaries.*

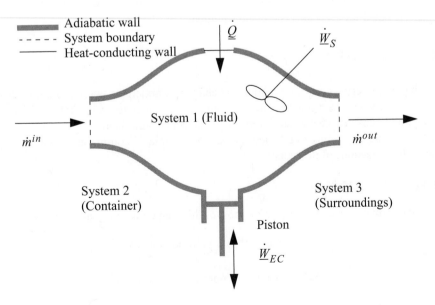

Figure 2.10 *Schematic of a general system.*

3. $\dot{\underline{W}}_{EC} = \left(-\dfrac{F_{boundary}}{A_{boundary}} \right) \dfrac{d\underline{V}_{system}}{dt}$ represents work done on the system due to *expansion or contraction of system size*. $F_{boundary}$ is the absolute value of the *system* force at the boundary. (By Newton's third principle, the forces are equal but act in opposite directions at any boundary.) Note that system 1 and system 3 forces are not required to be equal. Unequal forces create movement (acceleration) of any movable barrier (e.g., piston head).

4. $\dot{\underline{W}}_S$ represents the work done *on* the system resulting from *mechanical forces at the surface of the system* except work due to expansion/contraction or mass flow across boundaries. Turbines and compressors are part of system 2; thus they are involved with work interactions with the fluid in system 1. Note that piston movement is calculated as $\underline{W}_{EC}$ for systems 1, 3, (1+2), or (2+3), but the movement is calculated as $\underline{W}_S$ for system 2 alone. When a balance for system 2 is considered, the movement of the piston is technically shaft work, even though no shaft is involved. (The piston is a closed system, and it does not expand or contract when it moves.) As another example of the general definition of shaft work as it relates to forces at the surface of a system, consider the closed system of Fig. 2.3 on page 48 being raised 150 m or accelerated to 75 m/s. There is a work interaction at the surface of the system required for these energy changes even though there is not a "rotating" shaft.

5. Non-zero $\dot{\underline{W}}_{EC}$ or $\dot{\underline{W}}_S$ interactions of systems 1 and 3 are not equal unless changes in kinetic and potential energy of the moving portion of system 2 (e.g., piston head for $\underline{W}_{EC}$ or turbine for $\underline{W}_S$) are negligible and the movement is reversible.

6. Frictional forces, if present, *must* be attributed to one of the systems shown above. Irreversibility due to any cause does not require additional energy balance terms because energy is always conserved, even when processes are irreversible.

7. Electrical and magnetic fields have not been included.

8. On the left-hand side of the equation, kinetic and potential energy changes are calculated based on movement of the *center of mass*. In a composite system such as (1+2), they may be calculated for each subsystem and summed.

2.18   SUMMARY

We are trying to be very careful throughout this chapter to anticipate every possibility that might arise. As a result, the verbiage gets very dense. Think of the complete energy balance as a checklist, reminding you to consider whether each term may contribute significantly to a given problem, and learning to translate key terms like "frictionless," "reversible," "continuous," and "steady state" into meaningful reductions of the balance.

Important Equations

If we relax the formality, we can summarize most of this chapter casually as follows:

$$\Delta U = Q + W_{EC} \quad \text{closed systems}$$

$$\Delta H = Q + W_S \quad \text{open, steady systems}$$

$$d(nU) = H^{in} dn^{in} - H^{out} dn^{out} + dQ + d\underline{W}_{EC} + d\underline{W}_S \quad \text{open, unsteady-state systems}$$

Naturally, it is best to appreciate how these equations result from simplifications. Remember to check the general energy balance for terms that may be significant in exceptional situations.

A summary of expansion/contraction work relations for ideal gases is also presented here, however it is recommended that you become proficient in the manipulations leading to these formulas. Section 2.4 summarizes factors that may make a process irreversible. The following formulas represent reversible work done when the system pressure is inserted for the isothermal process. The isobaric formula is the only one that can be used to directly calculate work done on the surroundings in an irreversible process, and in that case the surrounding pressure is used instead of the system pressure.

Process Type	Work Formula (ig)
Isothermal	$W_{EC} = -\int PdV = -RT\int\dfrac{dV}{V} = -RT\ln\dfrac{V_2}{V_1}$ (ig)
Isobaric	$W_{EC} = -\int PdV = -P(V_2 - V_1)$
Adiabatic and reversible	$W_{EC} = -\int PdV = -\int \text{const}\dfrac{dV}{V^{(C_P/C_V)}}$ (*ig) or $\Delta U = C_V(T_2 - T_1) = W_{EC}$ (*ig) $\dfrac{T_2}{T_1} = \left(\dfrac{P_2}{P_1}\right)^{(R/C_P)} = \left(\dfrac{V_1}{V_2}\right)^{(R/C_V)}$ (*ig)

This last equation (for T_2/T_1) recurs frequently as we examine processes from various perspectives and simplify them to ideal gases for preliminary consideration. You should commit it to memory soon and learn to recognize when it is applicable.

Test Yourself

1. Write the energy balance without looking at the book. To help remember the terms, think about the properties the terms represent rather than memorizing the symbols.

2. In the presentation of the text, which side of the balance represents the system and which terms represent interactions at the boundaries?

3. Explain the terms "closed-system," "open-system," and "steady state" to a friend of the family member who is not an engineer.

4. Explain how a reference state helps to solve problems. Select a reference state for water that is different from the steam table reference state. Create a path starting from saturated liquid below the normal boiling point, through the normal boiling point, and cooling down to saturated vapor at the initial temperature. Use heat capacities and the latent heat at the normal boiling point to estimate the heat of vaporization and compare it with the steam table value.

5. Write a MATLAB, Excel, or calculator routine that will enable you to calculate heat capacity integrals easily.

6. Think of as many types of paths as you can from memory (isothermal, adiabatic, etc.) and try to derive the heat and work flow for a piston/cylinder system along each path.

2.19   PRACTICE PROBLEMS

A. General Reductions of the Energy Balance

The energy balance can be developed for just about any process. Since our goal is to learn how to develop model equations as well as to simply apply them, it is valuable practice to obtain the appropriate energy balance for a broad range of odd applications. If you can deduce these energy balances, you should be well prepared for the more common energy balances encountered in typical chemical engineering processes.

P2.1 A pot of water is boiling in a pressure cooker when suddenly the pressure relief valve becomes stuck, preventing any steam from escaping. System: the pot and its contents after the valve is stuck. (ANS. $d[mU]/dt = \dot{Q}$)

P2.2 The same pot of boiling water as above. System: the pot and its contents before the valve is stuck. (ANS. $d[mU] = inH^V + \dot{Q}$)

P2.3 An gas home furnace has been heating the house steadily for hours. System: the furnace. (ANS. $\Delta \underline{H} = \dot{Q}$) (gas furnace)

P2.4 An gas home furnace has been heating the house steadily for hours. System: the house and all contents. (ANS. $d[mU]/dt = \dot{Q}_{Heat} + \dot{Q}_{Loss}$) (gas furnace)

P2.5 A child is walking to school when he is hit by a snowball. He stops in his tracks. System: the child. (ANS. $\Delta[mU + mv^2/2g_c] = m_{snow}[v^2/(2g_c)]_{snow}$)

P2.6 A sealed glass bulb contains a small paddle-wheel (Crookes radiometer). The paddles are painted white on one side and black on the other. When placed in the sun, the paddle wheel begins to turn steadily. System: the bulb and its contents. (ANS. $\Delta U = 0$)

P2.7 A sunbather lays on a blanket. At 11:30 A.M., the sunbather begins to sweat. System: the sunbather at noon. (ANS. $(d[mU]/dt) = \dot{m}H^V + \dot{Q}$)

P2.8 An inflated balloon slips from your fingers and flies across the room. System: balloon and its contents. (ANS. $d[mU + mv^2{}_{balloon}/2g_c]/dt = [H + v^2/(2g_c)]^{out} dm/dt + \underline{W}_{EC}$)

B. Numerical Problems

P2.9 Consider a block of concrete weighing 1 kg.

(a) How far must it fall to change its potential energy by 1 kJ? (ANS. 100 m)
(b) What would be the value of its velocity at that stage? (ANS. 44.7 m/s)

P2.10 A block of copper weighing 0.2 kg with an initial temperature of 400 K is dropped into 4 kg of water initially at 300 K contained in a perfectly insulated tank. The tank is also made of copper and weighs 0.5 kg. Solve for the change in internal energy of both the water and the block given $C_V = 4.184$ J/g-K for water and 0.380 J/g-K for copper. (ANS. 7480 J, −7570 J)

P2.11 In the preceding problem, suppose that the copper block is dropped into the water from a height of 50 m. Assuming no loss of water from the tank, what is the change in internal energy of the block? (ANS. −7570 J)

P2.12 In the following take $C_V = 5$ and $C_P = 7$ cal/mol-K for nitrogen gas:

(a) Five moles of nitrogen at 100°C is contained in a rigid vessel. How much heat must be added to the system to raise its temperature to 300°C if the vessel has a negligible heat capacity? (ANS. 5000 cal) If the vessel weighs 80 g and has a heat capacity of 0.125 cal/g-K, how much heat is required? (ANS. 7000 cal)

(b) Five moles of nitrogen at 300°C is contained in a piston/cylinder arrangement. How much heat must be extracted from this system, which is kept at constant pressure, to cool it to 100°C if the heat capacity of the piston and cylinder is neglected? (ANS. 7000 cal)

P2.13 A rigid cylinder of gaseous hydrogen is heated from 300 K and 1 bar to 400 K. How much heat is added to the gas? (ANS. 2080 J/mole)

P2.14 Saturated steam at 660°F is adiabatically throttled through a valve to atmospheric pressure in a steady-state flow process. Estimate the outlet quality of the steam. (ANS. $q = 0.96$)

P2.15 Refer to Example 2.10 about transformation of kinetic energy to enthalpy. Instead of water, suppose N_2 at 1 bar and 298 K was flowing in the pipe. How would that change the answers? In particular, how would the temperature rise change? (ANS. max ~0.001K)

P2.16 Steam at 150 bars and 600°C passes through process equipment and emerges at 100 bars and 700°C. There is no flow of work into or out of the equipment, but heat is transferred.

(a) Using the steam tables, compute the flow of heat into the process equipment per kg of steam. (ANS. 288 kJ/kg)
(b) Compute the value of enthalpy at the inlet conditions, H^{in}, relative to an ideal gas at the same temperature, H^{ig}. Consider steam at 1 bar and 600°C as an ideal gas. Express your answer as $(H^{in} - H^{ig})/RT^{in}$. (ANS. −0.305)

P2.17 A 700 kg piston is initially held in place by a removable latch above a vertical cylinder. The cylinder has an area of 0.1 m^2; the volume of the gas within the cylinder initially is 0.1 m^3 at a pressure of 10 bar. The working fluid may be assumed to obey the ideal gas equation of state. The cylinder has a total volume of 0.25 m^3, and the top end is open to the surrounding atmosphere at 1 bar.

(a) Assume that the frictionless piston rises in the cylinder when the latches are removed and the gas within the cylinder is always kept at the same temperature. This may seem like an odd assumption, but it provides an approximate result that is relatively easy to obtain. What will be the velocity of the piston as it leaves the cylinder? (ANS. 13.8 m/s)
(b) What will be the maximum height to which the piston will rise? (ANS. 9.6 m)
(c) What is the pressure behind the piston just before it leaves the cylinder? (ANS. 4 bar)
(d) Now suppose the cylinder was increased in length such that its new total volume is 0.588 m^3. What is the new height reached by the piston? (ANS. ~13 m)
(e) What is the maximum height we could make the piston reach by making the cylinder longer? (ANS. ~13 m)

P2.18 A tennis ball machine fires tennis balls at 40 mph. The cylinder of the machine is 1 m long; the installed compressor can reach about 50 psig in a reasonable amount of time. The tennis ball is about 3 inches in diameter and weighs about 0.125 lb$_m$. Estimate the initial volume required in the pressurized firing chamber. [Hint: Note the tennis ball machine fires horizontally and the tennis ball can be treated as a frictionless piston. Don't be surprised if an iterative solution is necessary and ln (V_2/V_1) = ln$(1 + \Delta V/V_1)$]. (ANS. 390 cm^3)

P2.19 A 700 kg piston is initially held in place by a removable latch inside a horizontal cylinder. The totally frictionless cylinder (assume no viscous dissipation from the gas also) has an area of 0.1 m^2; the volume of the gas on the left of the piston is initially 0.1 m^3 at a pressure of 8 bars. The pressure on the right of the piston is initially 1 bar, and the total volume is 0.25 m^3. The working fluid may be assumed to follow the ideal gas equation of state. What would be the highest pressure reached on the right side of the piston and what would be the position of the piston at that pressure? (a) Assume isothermal; (b) What is the kinetic energy of the piston when the pressures are equal?[16] (ANS. (a) 12.5 MPa; (b) 17 kJ)

2.20   HOMEWORK PROBLEMS

2.1 Three moles of an ideal gas (with temperature-independent $C_P = (7/2)R$, $C_V = (5/2)R$) is contained in a horizontal piston/cylinder arrangement. The piston has an area of 0.1 m² and mass of 500 g. The initial pressure in the piston is 101 kPa. Determine the heat that must be extracted to cool the gas from 375°C to 275°C at: (a) constant pressure; (b) constant volume.

2.2 One mole of an ideal gas ($C_P = 7R/2$) in a closed piston/cylinder is compressed from $T^i = 100$ K, $P^i = 0.1$ MPa to $P^f = 0.7$ MPa by the following pathways. For each pathway, calculate ΔU, ΔH, Q, and W_{EC}: (a) isothermal; (b) constant volume; (c) adiabatic.

2.3 One mole of an ideal gas ($C_P = 5R/2$) in a closed piston/cylinder is compressed from $T^i = 298$ K, $P^i = 0.1$ MPa to $P^f = 0.25$ MPa by the following pathways. For each pathway, calculate ΔU, ΔH, Q, and W_{EC}: (a) isothermal; (b) constant volume; (c) adiabatic.

2.4 One mole of an ideal gas ($C_P = 7R/2$) in a closed piston/cylinder is expanded from $T^i = 700$ K, $P^i = 0.75$ MPa to $P^f = 0.1$ MPa by the following pathways. For each pathway, calculate ΔU, ΔH, Q, and W_{EC}: (a) isothermal; (b) constant volume; (c) adiabatic.

2.5 One mole of an ideal gas ($C_P = 5R/2$) in a closed piston/cylinder is expanded from $T^i = 500$ K, $P^i = 0.6$ MPa to $P^f = 0.1$ MPa by the following pathways. For each pathway, calculate ΔU, ΔH, Q, and W_{EC}: (a) isothermal; (b) constant volume; (c) adiabatic.

2.6 (a) What is the enthalpy change needed to change 3 kg of liquid water at 0°C to steam at 0.1 MPa and 150°C?

(b) What is the enthalpy change needed to heat 3 kg of water from 0.4 MPa and 0°C to steam at 0.1 MPa and 150°C?

(c) What is the enthalpy change needed to heat 1 kg of water at 0.4 MPa and 4°C to steam at 150°C and 0.4 MPa?

(d) What is the enthalpy change needed to change 1 kg of water of a water-steam mixture of 60% quality to one of 80% quality if the mixture is at 150°C?

(e) Calculate the ΔH value for an isobaric change of steam from 0.8 MPa and 250°C to saturated liquid.

(f) Repeat part (e) for an isothermal change to saturated liquid.

(g) Does a state change from saturated vapor at 230°C to the state 100°C and 0.05 MPa represent an enthalpy increase or decrease? A volume increase or decrease?

(h) In what state is water at 0.2 MPa and 120.21°C? At 0.5 MPa and 151.83°C? At 0.5 MPa and 153°C?

(i) A 0.15 m³ tank containing 1 kg of water at 1 MPa and 179.88°C has how many m³ of liquid water in it? Could it contain 5 kg of water under these conditions?

(j) What is the volume change when 2 kg of H_2O at 6.8 MPa and 93°C expands to 1.6 bar and 250°C?

(k) Ten kg of wet steam at 0.75 MPa has an enthalpy of 22,000 kJ. Find the quality of the wet steam.

---

16. This problem is reconsidered as an adiabatic process in problem P3.14.

2.7 Steam undergoes a state change from 450°C and 3.5 MPa to 150°C and 0.3 MPa. Determine ΔH and ΔU using the following:

(a) Steam table data.
(b) Ideal gas assumptions. (Be sure to use the ideal gas heat capacity for water.)

2.8 Five grams of the specified pure solvent is placed in a variable volume piston. What are the molar enthalpy and total enthalpy of the pure system when 50% and 75% have been evaporated at: (*i*) 30°C, (*ii*) 50°C? Use liquid at 25°C as a reference state.

(a) Benzene ($\rho^L = 0.88$ g/cm^3)
(b) Ethanol ($\rho^L = 0.79$ g/cm^3)
(c) Water without using the steam tables ($\rho^L = 1$ g/cm^3)
(d) Water using the steam tables

2.9 Create a table of *T, U, H* for the specified solvent using a reference state of $H = 0$ for liquid at 25°C and 1 bar. Calculate the table for: (*i*) liquid at 25°C and 1 bar; (*ii*) saturated liquid at 1 bar; saturated vapor at 1 bar; (*iii*) vapor at 110°C and 1 bar. Use the Antoine equation (Appendix E) to relate the saturation temperature and saturation pressure. Use the ideal gas law to model the vapor phase.

(a) Benzene
(b) Ethanol
(c) Water without using the steam tables
(d) Water using the steam tables

2.10 One kg of methane is contained in a piston/cylinder device at 0.8 MPa and 250°C. It undergoes a reversible isothermal expansion to 0.3 MPa. Methane can be considered an ideal gas under these conditions. How much heat is transferred?

2.11 One kg of steam in a piston/cylinder device undergoes the following changes of state. Calculate Q and W for each step.

(a) Initially at 350 kPa and 250°C, it is cooled at constant pressure to 150°C.
(b) Initially at 350 kPa and 250°C, it is cooled at constant volume to 150°C.

2.12 In one stroke of a reciprocating compressor, helium is isothermally and reversibly compressed in a piston + cylinder from 298 K and 20 bars to 200 bars. Compute the heat removal and work required.

2.13 Air at 30°C and 2MPa flows at steady state in a horizontal pipeline with a velocity of 25 m/s. It passes through a throttle valve where the pressure is reduced to 0.3 MPa. The pipe is the same diameter upstream and downstream of the valve. What is the outlet temperature and velocity of the gas? Assume air is an ideal gas with a temperature-independent $C_P = 7R/2$, and the average molecular weight of air is 28.8.

2.14 Argon at 400 K and 50 bar is adiabatically and reversibly expanded to 1 bar through a turbine in a steady process. Compute the outlet temperature and work derived per mole.

2.15 Steam at 500 bar and 500°C undergoes a throttling expansion to 1 bar. What will be the temperature of the steam after the expansion? What would be the downstream temperature if the steam were replaced by an ideal gas, $C_P/R = 7/2$?

2.16 An adiabatic turbine expands steam from 500°C and 3.5 MPa to 200°C and 0.3 MPa. If the turbine generates 750 kW, what is the flow rate of steam through the turbine?

2.17 A steam turbine operates between 500°C and 3.5 MPa to 200°C and 0.3 MPa. If the turbine generates 750 kW and the heat loss is 100 kW, what is the flow rate of steam through the turbine?

2.18 Valves on steam lines are commonly encountered and you should know how they work. For most valves, the change in velocity of the fluid flow is negligible. Apply this principle to solve the following problems.

   (a) A pressure gauge on a high-pressure steam line reads 80 bar absolute, but temperature measurement is unavailable inside the pipe. A small quantity of steam is bled out through a valve to atmospheric pressure at 1 bar. A thermocouple placed in the bleed stream reads 400°C. What is the temperature inside the high-pressure duct?

   (b) Steam traps are common process devices used on the lowest points of steam lines to remove condensate. By using a steam trap, a chemical process can be supplied with so-called *dry* steam, i.e., steam free of condensate. As condensate forms due to heat losses in the supply piping, the liquid runs downward to the trap. As liquid accumulates in the steam trap, it causes a float mechanism to move. The float mechanism is attached to a valve, and when the float reaches a control level, the valve opens to release accumulated liquid, then closes automatically as the float returns to the control level. Most steam traps are constructed in such a way that the inlet of the steam trap valve is always covered with saturated liquid when opened or closed. Consider such a steam trap on a 7 bar (absolute) line that vents to 1 bar (absolute). What is the quality of the stream that exits the steam trap at 1 bar?

2.19 An overall balance around part of a plant involves three inlets and two outlets which only contain water. All streams are flowing at steady state. The inlets are: 1) liquid at 1MPa, 25°C, $\dot{m}$ = 54 kg/min; 2) steam at 1 MPa, 250°C, $\dot{m}$ = 35 kg/min; 3) wet steam at 0.15 MPa, 90% quality, $\dot{m}$ = 30 kg/min. The outlets are: 1) saturated steam at 0.8 MPa, $\dot{m}$ = 65 kg/min; 2) superheated steam at 0.2 MPa and 300°C, $\dot{m}$ = 54 kg/min. Two kW of work are being added to the portion of the plant to run miscellaneous pumps and other process equipment, and no work is being obtained. What is the heat interaction for this portion of the plant in kW? Is heat being added or removed?

2.20 Steam at 550 kPa and 200°C is throttled through a valve at a flow rate of 15 kg/min to a pressure of 200 kPa. What is the temperature of the steam in the outlet state, and what is the change in specific internal energy across the value, $(U^{out} - U^{in})$?

2.21 A 0.1 m³ cylinder containing an ideal gas $(C_P/R = 3.5)$ is initially at a pressure of 10 bar and a temperature of 300 K. The cylinder is emptied by opening a valve and letting pressure drop to 1 bar. What will be the temperature and moles of gas in the cylinder if this is accomplished in the following ways:

   (a) Isothermally.
   (b) Adiabatically. (Neglect heat transfer between the cylinder walls and the gas.)

2.22 As part of a supercritical extraction of coal, an initially evacuated cylinder is fed with steam from a line available at 20 MPa and 400°C. What is the temperature in the cylinder immediately after filling?

2.23 A large air supply line at 350 K and 0.5 MPa is connected to the inlet of a well-insulated 0.002 m³ tank. The tank has mass flow controllers on the inlet and outlet. The tank is at 300 K and 0.1 MPa. Both valves are rapidly and simultaneously switched open to a flow of 0.1 mol/min. Model air as an ideal gas with $C_P = 29.3$ J/mol-K, and calculate the pressure and temperature as a function of time. How long does it take until the tank is within 5 K of the steady-state value?

2.24 An adiabatic tank of negligible heat capacity and 1 m³ volume is connected to a pipeline containing steam at 10 bar and 200°C, filled with steam until the pressure equilibrates, and disconnected from the pipeline. How much steam is in the tank at the end of the filling process, and what is its temperature if the following occurs:

(a) The tank is initially evacuated.
(b) The tank initially contains steam at 1 bar and 150°C.

CHAPTER

3

ENERGY BALANCES FOR COMPOSITE SYSTEMS

A theory is the more impressive the greater the simplicity of its premises is, the more different kinds of things it relates, and the more extended is its area of applicability. Therefore the deep impression which classical thermodynamics made upon me.

Albert Einstein

Having established the principle of the energy balance for individual systems, it is straightforward to extend the principle to a collection of several individual systems working together to form a composite system. One of the simplest and most enlightening examples is the Carnot cycle which is a benchmark system used to evaluate conversion of heat into work. The principle of the first law is perhaps most powerful when applied from an overall perspective. In other words, it is not necessary to deal with individual operations in order to draw conclusions about the overall system. Clever selections of composite systems or subsystems can then permit valuable insights about key behaviors and where to focus greater analysis. Implementing this overall perspective often requires dealing with multicomponent and reacting systems. For example, calorie counting for dietary needs must consider at least glucose, oxygen, CO_2, and water. It is not necessary at this stage to have a precise estimate of the mixture properties, but the reality of mixed systems must be acknowledged approximately. To illustrate practical implications, we consider distillation systems, reacting systems, and biological systems.

Chapter Objectives: You Should Be Able to...

1. Understand the steps of a Carnot engine and Carnot heat pump.

2. Analyze cycles to compute the work and heat input per cycle.

3. Apply the concepts of constant molar overflow in distillation systems.

4. Apply the concepts of ideal gas mixtures and ideal mixtures to energy balances.

5. Apply mole balances for reacting systems using reaction coordinates for a given feed properly using the stoichiometric numbers for single and multiple reactions.

6. Properly determine the standard heat of reaction at a specified temperature.

7. Use the energy balance properly for a reactive system.

3.1   HEAT ENGINES AND HEAT PUMPS – THE CARNOT CYCLE

In this section we introduce the Carnot cycle as a method to convert heat to work. Many power plants work on the same general principle of using heat to produce work. In a power plant, heat is generated by coal, natural gas, or nuclear energy. However, only a portion of this energy can be used to generate electricity, and the Carnot cycle analysis will be helpful in understanding those limitations. Before we start the analysis, let us define the ratio of net work produced to the heat input as the **thermal efficiency** using the symbol η_θ:

❶Thermal efficiency.

$$\eta_\theta \equiv \frac{-\dot{W}_{S,net}}{\dot{Q}_H} \equiv \frac{\text{work ouput}}{\text{heat input}} \qquad\qquad 3.1$$

We wish to make the thermal efficiency as large as possible. We prove in the next chapter that the Carnot engine matches the highest thermal efficiency for an engine operating between two isothermal reservoirs. Maximizing thermal efficiency is a design goal that pervades Units I and II. We reconsider the thermal efficiency each time we add a layer of sophistication in our analysis. The concept of entropy in Chapter 4 will help us to generalize from ideal gases to steam or other fluids with available tables and charts. The calculus of classical thermodynamics in Chapter 5 will help to generalize to any substance, making our own tables and charts in Chapters 6–9. Evaluating the thermal efficiency in many situations is a skill that any engineer should have. Chemical engineering embraces a broad scope of ... "chemicals."

The Carnot Engine

❶The Carnot cycle is *one* method of constructing a heat engine.

Nicolas Léonard Sadi Carnot (1796–1832) was a French scientist who developed the Carnot cycle to demonstrate the maximum conversion of heat into work.

The Carnot cycle was conceived by Sadi Carnot as a route to convert heat into work. In the previous chapter, we developed the energy balances and work calculations for reversible isothermal and adiabatic processes. The Carnot engine combines them in a cycle. Consider a piston in the vicinity of both a hot reservoir and a cold reservoir as illustrated in Fig. 3.1. The insulation on the piston may be removed to transfer heat from the hot reservoir during one step of the process, and also removed from the cold side to transfer heat to the cold reservoir during another step of the process. Carnot conceived of the cycle consisting of the steps shown schematically on the P-V diagram beginning from point a. Between points a and b, the gas undergoes an isothermal expansion, absorbing heat from the hot reservoir. From point b to c, the gas undergoes an adiabatic expansion. From point c to d, the gas undergoes an isothermal compression, rejecting heat to the cold reservoir. From point d to a, the gas undergoes an adiabatic compression to return to the initial state.

Let us consider the energy balance for the gas in the piston. Because the process is cyclic and returns to the initial state, the overall change in U is zero. The system is closed, so no flow work is involved. This work performed, $W_{EC} = \int PdV$ for the gas, is equal to the work done on the shaft plus the expansion/contraction work done on the atmosphere for each step. By summing the work terms for the entire cycle, the net work done on the atmosphere in a complete cycle is zero since the net atmosphere volume change is zero. Therefore, the work represented by the shaded portion of the P-V diagram is the useful work transferred to the shaft.

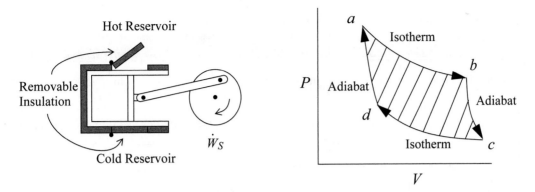

Figure 3.1 *Schematic of the Carnot engine, and the Carnot P-V cycle when using a gas as the process fluid.*

$$\Delta\left[m\left(\cancel{U}+\cancel{\frac{v^2}{2g_c}}+\cancel{\frac{gz}{g_c}}\right)\right] = \underline{Q}_H + \underline{Q}_C + \cancel{\underline{W}_{EC,net}} + \underline{W}_{S,net} \qquad 3.2$$

$$-\underline{W}_{S,net} = \underline{Q}_H + \underline{Q}_C \qquad 3.3$$

$$\eta_\theta = \frac{-\underline{W}_{S,net}}{\underline{Q}_H} = 1 + \frac{\underline{Q}_C}{\underline{Q}_H} \qquad 3.4$$

You can see from the shaded area in Fig. 3.1 that $-\underline{W}_{S,net} > 0$; therefore, since $\underline{Q}_H > 0$ and $\underline{Q}_C < 0$, $|\underline{Q}_H| > |\underline{Q}_C|$. The ratio $\underline{Q}_C/\underline{Q}_H$ is negative, and to maximize η we seek to make $\underline{Q}_C$ as small in magnitude as possible.

The heat transferred and work performed in the various steps of the process are summarized in Table 3.1. For this calculation we assume the gas within the piston follows the ideal gas law with temperature-independent heat capacities. We calculate reversible changes in the system; thus, we neglect temperature and velocity gradients within the gas (or we perform the process very slowly so that these gradients do not develop).

Comparing adiabats **b → c** and **d → a,** the work terms must be equal and opposite since the temperature changes are opposite. The temperature change in an adiabatic process is related to the volume change in Eqn. 2.63. In that equation, when the temperature ratio is inverted, the volume ratio is inverted. Therefore, we reason that for the two adiabatic steps, $V_b/V_a = V_c/V_d$. Using the ratio in the formulas for the isothermal steps, the ratio of heat flows becomes

$$\boxed{\frac{\underline{Q}_C}{\underline{Q}_H} = \frac{nRT_C\ln(V_d/V_c)}{nRT_H\ln(V_b/V_a)} = \frac{-T_C}{T_H}} \qquad 3.5$$

Inserting the ratio of heat flows into Eqn. 3.4 results in the **thermal efficiency.**

Table 3.1 *Illustration of Carnot Cycle Calculations for an Ideal Gas.*[a]

Step	Type	Q	W
a → b	Isotherm	$\underline{Q}_H = nRT_H\ln\dfrac{V_b}{V_a} > 0$ (ig)	$-\underline{Q}_H = -nRT_H\ln\dfrac{V_b}{V_a} < 0$ (ig)
b → c	Adiabat	0	$\Delta U = nC_V(T_C - T_H) < 0$ (*ig)
c → d	Isotherm	$\underline{Q}_C = nRT_C\ln\dfrac{V_d}{V_c} < 0$ (ig)	$-\underline{Q}_C = -nRT_C\ln\dfrac{V_d}{V_c} > 0$ (ig)
d → a	Adiabat	0	$\Delta U = nC_V(T_H - T_C) > 0$ (*ig)

a.   The Carnot cycle calculations are shown here for an ideal gas. There are no requirements that the working fluid is an ideal gas, but it simplifies the calculations.

> *Note: The temperatures T_H and T_C here refer to the hot and cold temperatures of the gas, which are not required to be equal to the temperatures of the reservoirs for the Carnot engine to be reversible. In Chapter 4 we will show that if these temperatures do equal the reservoir temperatures, the work is maximized.*

❶ The thermal efficiency of a heat engine is determined by the upper and lower operating temperatures *of the engine.*

$$\eta_\theta \equiv -\frac{\dot{W}_{S,net}}{\dot{\underline{Q}}_H} = 1 + \frac{\underline{Q}_C}{\underline{Q}_H} = 1 - \frac{T_C}{T_H} = \frac{T_H - T_C}{T_H} \qquad 3.6$$

You *must* use absolute temperature when applying Eqn. 3.6. We can skip the conversion to absolute temperature in the numerator of the last term because the subtraction means that the 273.15 (for units of K) in one term is canceled by the other. There is no such cancellation in the denominator.

Eqn. 3.6 indicates that we cannot achieve $\eta_\theta = 1$ unless the temperature of the hot reservoir becomes infinite or the temperature of the cold reservoir approaches 0 K. Such reservoir temperatures are not practical for real applications. For real processes, we typically operate between the temperature of a furnace and the temperature of cooling water. For a typical power-plant cycle based on steam as the working fluid, these temperatures might be 900 K for the hot reservoir and 300 K for the cold reservoir, so the maximum thermal efficiency for the process is near 67%, theoretically. Most real power plants operate with thermal efficiencies closer to 30% to 40% owing to inherent inefficiencies in real processes.

Perspective on the Heat Engine

The Carnot cycle provides a quick and convenient guideline for processes that seek to convert heat flow into work. The striking conclusion we will prove in Chapter 4 is that it is impossible to convert all of the heat flow from the hot reservoir into work at reasonable temperatures. From an overall perspective, the detailed steps of the Carnot engine can be ignored. The amount of work is given by Eqn. 3.6. We can simply state that heat comes in, heat goes out, and the difference is the net work. This situation is represented by Fig. 3.2(a). As an alternate perspective, any process with a finite temperature gradient should produce work. If it does not, then it must be irreversible, and

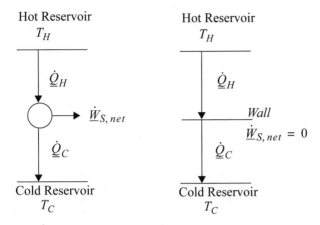

Figure 3.2 *The price of irreversibility. (a) Overall energy balance perspective for the reversible heat engine. (b) Zero work production in a temperature gradient without a heat engine, $\underline{Q}_H = \underline{Q}_C$.*

"wasteful." This situation is represented by Fig. 3.2(b). These observations are similar to previous statements about gradients and irreversibility, but Eqn. 3.6 establishes a quantitative connection between the temperature difference and the reversible work possible.

We will demonstrate in Chapter 4 that the Carnot thermal efficiency matches the upper limit attainable by any feasible process using heat flow as the source of work. On the other hand, it is not necessarily true that every engine is a heat engine. For instance, a fuel cell could conceivably convert the chemical energy of gasoline into electricity and then power a motor from the electricity. Fuel cells are not bound by the constraints of the Carnot cycle because they are not dependent on heat as the energy source. Also, not all engines operate between constant temperature reservoirs, so the direct comparison with the Carnot cycle is not possible.

Carnot Heat Pump

Suppose that we decided to reverse every step in the Carnot cycle. All the state variables would be the same, but all the heat and work flows would change signs. Instead of producing work, the process would require work, but it would absorb heat from a cold reservoir and reject heat to a warm reservoir. A Carnot heat pump may be used as a refrigerator/freezer or as a heater depending on whether the system of interest is on the cold or warm side of the heat pump. In a refrigeration system, the goal is to remove heat from the cold area and reject it to the warmer surroundings. Heat pumps can be purchased for home heating, and in that mode, they can extract heat from the colder outdoors and "pump" it to the warmer house.

❶ A Carnot heat pump results from running a heat engine backwards.

For a given transfer of heat from the cold reservoir, a Carnot heat pump requires the minimum amount of work for any conceivable process. The **coefficient of performance,** *COP*, is the ratio of heat transferred from the cold reservoir to the work required. COP is a mirror image of thermal efficiency, reappearing in Units I and II. We want to maximize COP when the objective is to cool a refrigerator with as little work as possible.

$$COP \equiv \frac{\dot{\underline{Q}}_C}{\dot{\underline{W}}_{S,net}} \qquad 3.7$$

Eqn. 3.5 has a subtle inversion of volumes that drops out:

$$\boxed{\frac{\underline{Q}_C}{\underline{Q}_H} = \frac{nRT_C\ln(V_c/V_d)}{nRT_H\ln(V_a/V_b)} = \frac{-T_C}{T_H}}$$ 3.8

Eqn. 3.3 still applies, and the *COP* is given by

$$\frac{\dot{W}_{S,net}}{\dot{Q}_C} = -\left(1 + \frac{\dot{Q}_H}{\dot{Q}_C}\right) = \left(\frac{T_H}{T_C} - 1\right),$$

❶ Coefficient of performance.

$$\boxed{COP \equiv \frac{\dot{Q}_C}{\dot{W}_{S,net}} = \left(\frac{T_H}{T_C} - 1\right)^{-1} = \frac{T_C}{T_H - T_C}}$$ 3.9

Example 3.1 Analyzing heat pumps for housing

Suppose your family is considering replacing your furnace with a heat pump. Work is necessary in order to "pump" the heat from the low outside temperature up to the inside temperature. The best you could hope for is if the heat pump acts as a reversible heat pump between a heat source (outdoors in this case) and the heat sink (indoors). The average winter temperature is 4°C, and the building is to be maintained at 21°C. The coils outside and inside for transferring heat are of such a size that the temperature difference between the fluid inside the coils and the air is 5°C. We generally refer to this as the **approach temperature.** What would be the maximum cost of electricity in ($/kW-h) for which the heat pump would be competitive with conventional heating where a fuel is directly burned for heat. Consider the cost of fuel as $7.00 per 10^9J, and electricity as $0.10 per kW-h. Consider only energy costs.

Solution: The Carnot heat pump COP, Eqn. 3.9

$$\dot{W}_{S,net} = \dot{Q}_C\left(\frac{T_H}{T_C} - 1\right) = \dot{Q}_H\left(\frac{-T_C}{T_H}\right)\left(\frac{T_H}{T_C} - 1\right) = \dot{Q}_H\left(\frac{T_C - T_H}{T_H}\right)$$

$$\dot{W}_{S,net} = \dot{Q}_H\frac{(-1-26)}{(26 + 273.15)} = -\frac{27}{299.15}\dot{Q}_H$$

where $\dot{Q}_H$ is the heating requirement in kW. Heat pump operating cost = $(0.09)\cdot\dot{Q}_H\cdot(\theta\text{ h})\cdot[x\ $/kW-h]$, where θ is an arbitrary time and x is the cost.

Direct heating operating cost = $\dot{Q}_H\cdot(3600\text{ s/h})\cdot(\theta\text{ h})\cdot($7/10^6\text{ kJ})$
For the maximum cost of electricity for competitive heat pump operation, let heat pump cost = direct heating cost at the breakeven point.

$$\Rightarrow (0.09)\cdot\dot{Q}_H\cdot(\theta\text{ h})\cdot[x\ $/kW-h] = \dot{Q}_H\cdot(3600\text{ s/h})\cdot(\theta\text{ h})\cdot($7/10^6\text{ kJ})\qquad x = $0.28/kW-h$$

Since the actual cost of electricity is given as $0.10/kW-h it might be worthwhile if the heat pump is reversible and does not break down. (Purchase, installation, and maintenance costs have been assumed equal in this analysis, although the heat pump is more complex.)

3.2 DISTILLATION COLUMNS

Roughly 80% of separations are conducted by distillation and a significant portion of the energy involved in the global chemical process industries is devoted to distillation. Why are more distillation columns needed than reactors? Reactors rarely run to 100% conversion and rarely produce a single product with 100% purity. There may be a by-product of the primary reaction or a solvent reaction medium. Additionally, there may be side reactions that can be minimized but not eliminated. Altogether, the reactor effluent almost always contains several components, and products need to be separated to high purity before they can be sent out of the process or recycled.

Analyzing distillation is important for a more philosophical reason as well. It is a common unit operation that involves mass and energy balances, heat and mass transfer, and phase partitioning. This kind of analysis pervades chemical engineering in general. The important consideration is the thought process that leads to the model equations. That thought process can be applied to any system, no matter how complex. As we proceed through the analysis distillation, try to imagine how you might develop similar approximations for other systems. Challenge yourself to anticipate the next step in every derivation. In the final analysis, your skill at developing simple models of complex phenomena is more valuable than memorizing model equations developed by somebody else.

We will address the phase equilibrium aspects of distillation in Section 10.6 on page 390, and you may wish to skim that section now. Briefly, the fraction with the lowest boiling point rises in the column and the fraction with the highest boiling point flows to the bottom. In this section, the focus is on mass and energy balances for distillation. A common model in distillation column screening is called **constant molar overflow**. In this model, the actual enthalpy of vaporization of a mixture is represented by the average enthalpy of vaporization, which can be assumed to be independent of composition for the purposes of this calculation, $\Delta H^{vap} = (\Delta H_1^{vap} + \Delta H_2^{vap})/2$. Also, all saturated liquid streams are considered to have the same enthalpy, and all saturated vapor streams are considered to have the same enthalpy. These assumptions may seem extreme, but the model provides an excellent overview of key operating conditions.

In the constant molar overflow model for a column with one feed, the column may be represented by five sections as shown in Fig. 3.3: a feed section where the feed enters; a **rectifying** section above the feed zone; a condenser above the rectifying section which condenses the vapors and returns a portion of the liquid condensate as **reflux** L_R to ensure that liquid remains on the trays to induce the liquid-vapor partitioning that enhances the compositions.; a **stripping** section below the feed section; and a **reboiler** that creates vapors from the liquid flowing down the column.

At the bottom of the column, heat is added in the **reboiler,** causing vapor to percolate up the column until it reaches the condenser. The heat requirement in the reboiler is called the heating **duty**. B is called the **bottoms** or **bottoms product**. The ratio V_S/B is called the **boilup ratio.** The energy requirement of the reboiler is known as the reboiler **duty** and is directly proportional to the moles of vapor produced as shown in the figure.

At the top of the column, Fig. 3.3(a) shows a **total condenser** where the vapor from the top of the column is totally condensed. The liquid flow rate leaving the condenser will be $V_R = (L_R + D)$. D is called the **overhead product** or **distillate. L_R** is called the **reflux.** The proportion of reflux is characterized by the **reflux ratio,** $R = L_R/D$. A **partial condenser** may also be used as shown in (b), and the overhead product leaves as a vapor and the condensed fraction is the reflux. The cooling

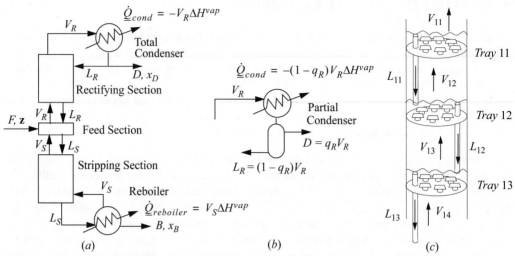

Figure 3.3 *(a) Overall schematic of a distillation column with a total condenser showing five sections of a distillation column. and conventional labels; (b) a partial condenser; (c) schematic of liquid levels on bubble cap trays with the downcomers used to maintain the liquid levels.*

requirement of the condenser is called the condenser **duty** and the duty depends on whether the condenser is total or partial as shown in the figure.

The rectifying and stripping sections of the column have either **packing** or **trays** to provide retention of the liquid and contact with vapors. Trays are easier to introduce as shown in Fig. 3.3(c). At steady state, each tray holds liquid and the vapor flows upwards through the liquid. The trays can be constructed with holes (sieve trays) or bubble caps (bubble cap trays) or valves (valve trays). The bubble caps sketched in the figure represent a short stub of pipe with a short inverted "cup" called the "cap" with slots in the sides (slots are not shown) supported with spacing so that vapor can flow upwards through the pipe and out through the slots in the cap. A **downcomer** controls the liquid level on the tray as represented by a simple pipe extending above the surface of the tray in the figure. and in an ideal column each tray creates a separation **stage.** Using multiple stages provides greater separation. By stacking the stages, rising vapor from a lower stage boils the liquid on the next stage. At steady state, streams V_S and V_R are assumed to be saturated vapor unless otherwise noted. Streams B, L_S, and L_R are assumed to be saturated liquid unless otherwise noted. D is either assumed to be either a saturated liquid or a vapor depending on whether the condenser is total or partial, respectively. According to the **constant molar overflow** model, at steady state the vapor and liquid flow rates are constant within the stripping and rectifying sections because all the internal streams in contact are saturated, and change only at the feed section as determined by mass and energy balances around the feed section.

One of the most challenging tasks in the undergraduate chemical engineering laboratory is to explain the dynamics of the distillation column during start-up. Students tend to assume that the entire distillation column starts working as soon as heat is applied to the reboiler. However, during column start-up, the internal flows are not constant in the column sections. Heat moves up the column one tray at a time as boiling vapor leaves one tray and enters another. Consider Fig. 3.3(c). When tray 12 starts to boil, rising vapors will be condensed by the cold liquid on tray 11 immediately above and by the cold column. In principle, the vapor flow leaving the upper tray 11, V_{11}, will be approximately zero until the column and liquid on the tray reaches the saturation temperature of the liquid on tray 11. Vapors reaching tray 11 are condensed and give up latent heat. Then tray 11 begins to warm. The condensed vapors create liquid overflow of subcooled liquid from tray 11 back down to tray 12. The lower tray 12 continues to boil, but the cool downcomer liquid causes the vapor rate V_{12} to be smaller than V_{13}. Tray 12 stays at a fairly constant temperature as it continues to boil because it stays saturated. Finally, when tray 11 begins to boil, the process repeats for tray 10 and so on.

This introduction is not intended to explain every detail of distillation and energy balances, but it should suffice for you to deduce the most important qualitative behaviors. An illustration of the kinds of possible inferences is given in Example 3.2. Homework problem 3.4 provides additional exercises related to steady-state distillation.

Example 3.2 Start-up for a distillation column

A particular bubble cap distillation column for methanol + water has 12 trays numbered from top to bottom. Each tray is composed of 4 kg of materials and holds 1 kg of liquid. The heat capacity of the tray materials is 6 J/g-K and the heat capacity of the liquid is 84 J/mol-K = 3.4 J/g-K. During start-up, the feed is turned off. Roughly 15 minutes after the reboiler is started, tray 12 has started to boil and the temperature on tray 11 begins to rise. The reboiler duty is 6 kW and the heat loss is negligible. Tray 11 starts at 25 °C and the temperature of the liquid and the tray materials is assumed the same during start-up. Assume the liquid inventory on Tray 11 is constant throughout start-up.

(a) Tabulate and plot the temperature versus time for tray 11 until it starts to boil at 70 °C.
(b) Plot the vapor flow V_{12} as a function of time.

Solution: (a) First, recognize that $V_{11} = 0$ since tray 11 is subcooled. No liquid is flowing down to tray 11. $L_{10} = 0$ since $F = 0$ and no vapors are being condensed yet ($L_R = 0$), even though the cooling water may be flowing.
Mass balance on tray 11 (all vapors from below are being condensed during start-up, $V_{11} = 0$, $L_{10} = 0$):

$$V_{12} = L_{11} \qquad\qquad 3.10$$

Mass balance on boundary around tray 11 + tray 12 ($V_{11} = 0$, $L_{10} = 0$, during start-up):

$$V_{13} = L_{12} \qquad\qquad 3.11$$

State 11 is subcooled during start-up and will warm until $T_{11} = 70$:

$$H_{11} = H^L + C_P(T_{11} - T_{11}^{sat}) = H^L + C_P^L(T_{11} - 70) \qquad\qquad 3.12$$

Energy balance on tray 11 during start-up (no work, no direct heat input, energy input by flow of hot vapors, negligible heat loss):

$$(m_{mat}C_P^{mat} + m_{11}C_p^L)\frac{dT_{11}}{dt} = V_{12}H^V - L_{11}H_{11} = V_{12}(\Delta H^{vap} - C_p^L(T_{11} - 70)) \qquad\qquad 3.13$$

where for the last equality we have inserted Eqn. 3.10 and then Eqn. 3.12.
An energy balance on tray 12 (which is at constant temperature) gives:

$$0 = V_{13}H^V + L_{11}H_{11} - V_{12}H^V - L_{12}H^L \qquad\qquad 3.14$$

Example 3.2 Start-up for a distillation column (Continued)

Using Eqn. 3.12 to eliminate H_{11}, Eqn. 3.10 to eliminate L_{11}, and Eqn. 3.11 to eliminate L_{12},

$$
\begin{aligned}
0 &= V_{13}H^V + V_{12}H^L + V_{12}C_p^L(T_{11} - 70) - V_{12}H^V - V_{13}H^L \\
&= V_{13}\Delta H^{vap} - V_{12}\Delta H^{vap} + V_{12}C_p^L(T_{11} - 70)
\end{aligned}
\tag{3.15}
$$

$$
V_{12} = \frac{V_{13}\Delta H^{vap}}{\Delta H^{vap} - C_p^L(T_{11} - 70)}
\tag{3.16}
$$

Inserting Eqn. 3.15 into Eqn. 3.13, and recognizing the constant vapor flow rate below tray 13,

$$
(m_{mat}C_P^{mat} + m_{11}C_p^L)\frac{dT_{11}}{dt} = V_{13}\Delta H^{vap} = V_S\Delta H^{vap} = Q_{reboiler}
\tag{3.17}
$$

where '*mat*' indicates column material. Inserting values from the problem statement gives,

$$
6000 \text{ J/s} = (4000 \text{ g } (6 \text{ J/g-K}) + 1000(3.4))dT_{11}/dt => dT_{11}/dt = 13.1 \text{ C/min}
\tag{3.18}
$$

The tray will require approximately $(70 - 25)/13.1 = 3.4$ min to reach saturation temperature. The plot is shown below.

(b) The vapor flow is given by Eqn. 3.13 using $T_{11} = 25 + 13.1(t - 15)$. The average heat of vaporization is $(40.7 + 35/3)/2 = 38$ kJ/mol. Between 15 and 18.4 min, the flow rate in mol/min

$$
V_{12} = \frac{6000(60)}{38000 - 84(13.1(t - 15) - 45)}
\tag{3.19}
$$

Note that we neglect details like imperfect mixing or bypass heating (vapor that does not get condensed) that would round the edges of the temperature profile.

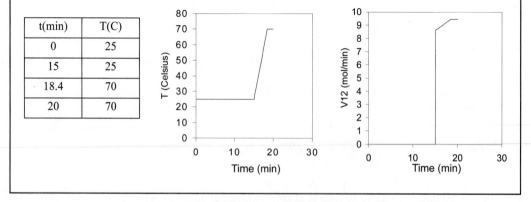

t(min)	T(C)
0	25
15	25
18.4	70
20	70

Note that as soon as vapors start to reach tray 11 in Example 3.2 the net energy input is constant even though the flow rate of hot vapors into the tray is increasing with time during start-up. Why? The answer is that energy is also being transported away from the tray by the flow down the downcomer. As the tray warms, this energy transport out is increasing at the same rate as the increased energy transport into the stage by the increasing flow of vapor. Analysis of subtleties such as this will deepen your understanding of physical phenomena and your appreciation for the utility of the energy balance.

3.3   INTRODUCTION TO MIXTURE PROPERTIES

The previous section used "average" properties for the streams. This level of approximation is too crude for most calculations. We therefore need to understand how to estimate mixture properties.

Property Changes of Mixing

Communication of the property changes is facilitated by defining the property change of mixing as the mixture property relative to the mole-fraction weighted sum of the component properties in the unmixed state. Using x to be a generic composition variable, the **energy of mixing** is

$$\Delta U_{mix} = U - \sum_i x_i U_i \qquad\qquad 3.20$$

The **enthalpy of mixing** is:

$$\Delta H_{mix} = H - \sum_i x_i H_i \qquad\qquad 3.21$$

The **volume of mixing** is the volume of the mixture relative to the volumes of the components before mixing.

$$\Delta V_{mix} \equiv V - \sum_i x_i V_i \qquad\qquad 3.22$$

Similar equations may be written for other properties that we will define later: **entropy of mixing, Gibbs energy of mixing,** and **Helmholtz energy of mixing** – but these must reflect the increased disorder inherent in creating mixtures from pure fluids. If the property change on mixing (e.g., ΔH_{mix}) has been measured and correlated in a reference book or database, we can use it to calculate the stream property at a later time:

$$U = \sum_i x_i U_i + \Delta U_{mix} \qquad\qquad 3.23$$

$$H = \sum_i x_i H_i + \Delta H_{mix} \qquad\qquad 3.24$$

$$V = \sum_i x_i V_i + \Delta V_{mix} \qquad\qquad 3.25$$

3.4   IDEAL GAS MIXTURE PROPERTIES

The ideal gas is a convenient starting point to introduce the calculation of mixture properties because the calculations are simple. Since ideal gas molecules do not have intermolecular potentials, the internal energy consists entirely of kinetic energy. When components are mixed at constant temperature and pressure, the internal energy is simply the sum of the component internal energies (kinetic energies), which can be written using y as a gas phase composition variable:

$$\boxed{U^{ig} = \sum_i y_i U_i^{ig}} \text{ or } \boxed{\underline{U}^{ig} = \sum_i n_i U_i^{ig}}$$

(ig) 3.26

The total volume of a mixture is related to the number of moles by **Amagat's law:**

$$\boxed{\underline{V}^{ig} = \left(RT\sum_i n_i\right)/P}$$

(ig) 3.27

We can thus see that the energy of mixing and the volume of mixing for an ideal gas are both zero. Combining U and V to obtain the definition of H, $H = U + PV$, and using Eqns. 3.26 and 3.27,

$$\boxed{\underline{H}^{ig} = \sum_i n_i U_i^{ig} + RT\sum_i n_i = \sum_i n_i(U_i^{ig} + RT) = \sum_i n_i H_i}$$

(ig) 3.28

Therefore the enthalpy of a mixture is given by the sum of the enthalpies of the components at the same temperature and pressure and the enthalpy of mixing is zero. On a molar basis,

$$\boxed{H^{ig} = \sum_i y_i H_i^{ig}}$$

(ig) 3.29

Furthermore, we can add the component enthalpies and internal energies for ideal gas mixtures using the mole fractions as the weighting factors. The energy of mixing, volume change of mixing, and enthalpy change of mixing are all zero.

$$\boxed{\Delta U_{mix}^{ig} = 0 \qquad \Delta V_{mix}^{ig} = 0 \qquad \Delta H_{mix}^{ig} = 0}$$

3.30

❶ The enthalpy of mixing is zero for an ideal gas.

3.5   MIXTURE PROPERTIES FOR IDEAL SOLUTIONS

Sometimes the simplest analysis deserves more consideration than it receives. Ideal solutions can be that way. For an ideal solution, there are no synergistic effects of the components being mixed together; each component operates independently. Thus, mixing will involve no energy change and no volume change. Using x as a generic composition variable,

❶ The energy of mixing and volume of mixing are zero for an ideal solution.

$$\boxed{U^{is} = \sum_i x_i U_i} \text{ or } \boxed{\underline{U}^{is} = \sum_i n_i U_i}$$

3.31

$$\boxed{V^{is} = \sum_i x_i V_i} \text{ or } \boxed{\underline{V}^{is} = \sum_i n_i V_i} \qquad 3.32$$

Though these restrictions were also followed by ideal gas solutions, the volumes for ideal solutions do not need to follow the ideal gas law, and can be liquids; thus, ideal gases are a subset of ideal solutions. Examples of ideal solutions are all ideal gases mixtures and liquid mixtures of family member pairs of similar size such as benzene + toluene, n-butanol + n-pentanol, and n-pentane + n-hexane.

Since $H \equiv U + PV$, and because the U and V are additive, the enthalpy of the mixture will simply be the sum of the pure component enthalpies times the number of moles of that component:

$$\boxed{H^{is} \equiv U^{is} + PV^{is} = \sum_i x_i(U_i + PV_i) = \sum_i x_i H_i} \boxed{\underline{H}^{is} = \sum_i n_i H_i} \qquad 3.33$$

Therefore, an ideal solution has a zero **energy of mixing, volume of mixing,** and **enthalpy of mixing** (commonly called the **heat of mixing**):

$$\boxed{\Delta U^{is}_{mix} = 0 \qquad \Delta V^{is}_{mix} = 0 \qquad \Delta H^{is}_{mix} = 0} \qquad 3.34$$

❶ The enthalpy of mixing is zero for an ideal solution.

The primary distinction between ideal gas mixtures and ideal solutions is the constraint of the ideal gas law for the volume of the former. Let us apply the principles of ideal solutions and ideal gas mixtures to an example that also integrates the principles of use of a reference state.

Example 3.3 Condensation of a vapor stream

A vapor stream of wt fractions 45% H_2O, 40% benzene, 15% acetone flows at 90°C and 1 bar into a condenser at 100 kg/h. The stream is condensed and forms two liquid phases. The water and benzene can be considered to be totally immiscible in one another. The acetone partitions between the benzene and water layer, such that the K-ratio, K = (wt. fraction in the benzene layer)/(wt. fraction in the water layer) = 0.9.[a] The liquid streams exit at 20°C and 1 bar. Determine the cooling duty, $\dot{Q}$ for the condenser. Assume the feed is an ideal gas and the liquid streams are ideal solutions once the immiscible component has been eliminated.

Solution: A schematic of the process is shown below. Using $\dot{m}_1$ as the flow rate of acetone in E and $\dot{m}_2$ as the flow rate in stream B, the K-ratio constraint is

$$\dot{m}_1/(\dot{m}_1 + 40) = 0.9(\dot{m}_2/(\dot{m}_2 + 45)) = 0.9((15 - \dot{m}_1)/(15 - \dot{m}_1 + 45))$$

where the acetone mass balance has been inserted in the second equality.
Using the first and third arguments, a quadratic equation results, which leads to $\dot{m}_1 = 6.6$ kg/h, and $\dot{m}_2 = 15 - 6.6 = 8.4$ kg/h.

Example 3.3 Condensation of a vapor stream (Continued)

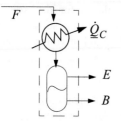

Basis: 100 kg/h Feed

kg/h	F	E	B
H₂O	45	–	45
Benzene	40	40	–
Acetone	15	6.6	8.4

Converted to kmol/h

kmol/h	F	E	B
H₂O	2.5	–	2.5
Benzene	0.5	0.5	–
Acetone	0.26	0.11	0.15

The energy balance for the process side of the dotted boundary is:

$$0 = \dot{n}_F H_F - \dot{n}_E H_E - \dot{n}_B H_B + \dot{Q}_C \qquad\qquad 3.35$$

We are free to choose a reference state for each component. Note that if the reference state is chosen as liquid at 20°C, then the enthalpies of E and B will both be zero since they are *at the* reference state temperature and pressure and the enthalpy of mixing is zero for the ideal solution assumption. This choice will greatly reduce the number of calculations. The energy balance with this reference state simplifies to the following:

$$\dot{n}_F H_F = -\dot{Q}_C \qquad\qquad 3.36$$

The enthalpy of F as an ideal gas is given by Eqn. 3.28:

$$\dot{n}_F H_F = \sum_i \dot{n}_i H_i = \dot{n}_{F,H2O} H_{F,H2O} + \dot{n}_{F,benz} H_{F,benz} + \dot{n}_{F,acet} H_{F,acet} \qquad 3.37$$

Refer back to Fig. 2.6 on page 65 to review the paths for calculation relative to a reference state. The path used here is similar to Fig. 2.6(a). To calculate the enthalpy for components in F, we can construct a path between the reference state and the feed state going through the normal boiling point, T_b, where the heat of vaporization is known.

$$\Delta H^L = \int_{20}^{T_b} C_p^L dT \qquad \Delta H^{vap} \qquad \Delta H^V = \int_{T_b}^{90} C_p^V dT$$

20°C   *liquid* T_b

The enthalpy of a component in the feed stream is a sum of the three steps, $H_{i,F} = \Delta H^L + \Delta H^{vap} + \Delta H^V$. Note that $T_b > 90°C$ for water. The ΔH^V term is calculated with the same formula, but results in a negative contribution as shown by the dotted line in the path calculation schematic. For benzene and acetone, $T_b < 90°C$, so the path shown by the solid line is used for ΔH^V. Note that, although the system is below the normal boiling point of water at 1 bar, the water can exist as a component in a mixture.

Example 3.3 Condensation of a vapor stream (Continued)

Using the heat capacity polynomials, and tabulating the three steps shown in the pathway schematic, programming the enthalpy integrals into Excel or MATLAB provides

$$H_{F, H_2O} = 6058 + 40656 - 342 = 46372 \text{ J/mol}$$
$$H_{F, benz} = 8624 + 30765 + 994 = 40383 \text{ J/mol}$$
$$H_{F, acet} = 4639 + 30200 + 2799 = 37638 \text{ J/mol}$$

Note that the last term in the sum is negative for water because the feed temperature of the mixture is below the normal boiling temperature. The cooling duty for the condenser is

$$\dot{Q}_C = -(2.5(46372) + 0.5(40383) + 0.26(37638)) = -146 \text{ kJ/h}$$

a. Throughout most of the text, we usually use K-ratios based on mole fraction ratio, not weight fraction. Nevertheless, many references use K-ratios based on weight fractions. You must read carefully and convert as needed.

3.6 ENERGY BALANCE FOR REACTING SYSTEMS

Chemical engineers must be proficient at including reacting systems into energy balances, and there are several key concepts that must be introduced. In reacting systems the number of moles is not conserved unless the number of moles of products is the same as the number of moles of reactants. Generally, the two best approaches for tracking species are to use an atom balance or to use the reaction coordinate. Here we will introduce the method of the reaction coordinate because it is much easier to incorporate into the energy balance. It is convenient to adopt the conventions of stoichiometry,

$$\nu_1 C_1 + \nu_2 C_2 + \nu_3 C_3 + \nu_4 C_4 = 0$$

where the C's represent the species, and reactants have negative ν's and products have positive ν's. The ν's are called the **stoichiometric numbers,** and the absolute values are called the **stoichiometric coefficients**. (e.g., $CH_4 + H_2O \rightleftarrows CO + 3H_2$, numbering from left to right),

$$\Rightarrow \nu_1 = -1; \quad \nu_2 = -1; \quad \nu_3 = +1; \quad \nu_4 = +3.$$

Consider what would happen if a certain amount of component 1 were to react with component 2 to form products 3 and 4. We see $dn_1 = dn_2 (\nu_1 / \nu_2)$ because ν_1 moles of component 1 requires ν_2 moles of component 2 in order to react. Rearranging:

$$\frac{dn_1}{\nu_1} = \frac{dn_2}{\nu_2} \text{ and similarly: } \frac{dn_1}{\nu_1} = \frac{dn_3}{\nu_3} \qquad 3.38$$

Since all these quantities are equal, it is convenient to define a variable which represents this quantity.

$$d\xi = dn_i / \nu_i \qquad 3.39$$

ξ is called the **reaction coordinate**[1] and is related to the **conversion.**[2] Integrating:

$$\int_0^\xi d\xi = \frac{1}{\nu} \int_{n^i}^{n^f} dn$$

$$\xi = \frac{1}{\nu}(n^f - n^i)$$

Or in a more useful form for any component i:

$$\boxed{n_i^f = n_i^i + \nu_i \xi} \qquad\qquad 3.40$$

In a flowing system, n_i^f represents the outlet, n_i^i represents the inlet, and thus for component i,

$$\boxed{\dot{n}_i^{out} = \dot{n}_i^{in} + \nu_i \dot{\xi}} \qquad\qquad 3.41$$

where $\dot{\xi}$ represents the overall rate of species interconversion. *Moles are not conserved in a chemical reaction,* which can be quantified by summing Eqn. 3.40 or 3.41 over all species—for a flowing system, $\dot{n}^{in} = \sum_i \dot{n}_i^{in}$ and $\dot{n}^{out} = \sum_i \dot{n}_i^{out}$, thus,

$$\boxed{\dot{n}^{out} = \dot{n}^{in} + \dot{\xi}\sum_i \nu_i} \qquad\qquad 3.42$$

In closed systems, the value of ξ is determined by chemical equilibria calculations; ξ may be positive or negative. The only limit on ξ is that $n_i^f \geq 0$ for all i. The boundary values of ξ may be determined in this manner before beginning an equilibrium calculation. Naturally, in a flowing system, the same arguments apply to $\dot{\xi}$ and $\dot{n}_i^{out} \geq 0$.

Example 3.4 Stoichiometry and the reaction coordinate

Five moles of hydrogen, two moles of CO, and 1.5 moles of CH_3OH vapor are combined in a closed system methanol synthesis reactor at 500 K and 1 MPa. Develop expressions for the mole fractions of the species in terms of the reaction coordinate. The components are known to react with the following stoichiometry:

$$2H_{2(g)} + CO_{(g)} \; \rightleftarrows \; CH_3OH_{(g)}$$

---

1. The reaction coordinate is in some texts called the **extent of reaction.** This is misleading because depending on conditions, it can be less than one at complete conversion, or it can be negative.

2. Another common measure of reaction progress is **conversion.** In reaction engineering, it is common to follow the conversion of a particular reactant species, say, species A. If X_A is the conversion of A, then $n_A = n_A^{in}(1 - X_A)$, and $X_A = a\,\xi/n_A^{in}$, where a is the stoichiometric coefficient for A as written in the reaction.

Example 3.4 Stoichiometry and the reaction coordinate (Continued)

Solution: Although the reaction is written as though it will proceed from left to right, the direction of the actual reaction does not need to be known. If the reverse reaction occurs, this will be obvious in the solution because a negative value of ξ will be found. The task at hand is to develop the mole balances that would be used toward determining the value of ξ. The table below presents a convenient format.

	n^i	n^f
H_2	5	$5 - 2\xi$
CO	2	$2 - \xi$
CH_3OH	1.5	$1.5 + \xi$
n_T		$8.5 - 2\xi$

The total number of moles at any time is $8.5 - 2\xi$. The mole fractions are

$$y_{H_2} = \frac{n^f_{H_2}}{n_T} = \frac{5 - 2\xi}{8.5 - 2\xi}$$

$$y_{CO} = \frac{2 - \xi}{8.5 - 2\xi}$$

$$y_{CH_3OH} = \frac{1.5 + \xi}{8.5 - 2\xi}$$

To ensure that all $n^f_i \geq 0$, the acceptable upper limit of ξ is determined by CO, and the acceptable lower limit is determined by CH_3OH,

$$-1.5 \leq \xi \leq 2$$

Example 3.5 Using the reaction coordinates for simultaneous reactions

For each independent reaction, a reaction coordinate variable is introduced. When a component is involved in two reactions, the moles are related to both reaction coordinates. This example is available as an online supplement.

Standard State Heat of Reaction

When a reaction proceeds, bonds are broken, and others are formed. Because the bond energies vary for each type of bond, there are energy and enthalpy changes on reaction. Bond changes have a significant effect on the energy balance because they are usually larger than the sensible heat effects. Because enthalpies are state properties, we can use **Hess's law** to calculate the enthalpy change. Hess's law states that the enthalpy change of a reaction can be calculated by summing any component reactions, or by calculating the reaction enthalpy along a convenient reaction pathway between the reactants and products. To organize calculations of the changes, enthalpies of components are usually available at a **standard state.** A standard state is slightly different from a **reference state** as discussed on page 63. A standard state requires all the specifications of a reference state, *except the T is the temperature of the system.* For reactions, the conventional standard state properties are at a specified composition, state of aggregation, and pressure, but they change with temperature. By combining Hess's law with the standard state concept, we may calculate the **standard state heat (or enthalpy) of reaction.** Suppose that we have the reaction of Fig. 3.4. For calculation of the heat of reaction, a convenient pathway is "decompose" the reactants into the constituent elements in their naturally occurring states at the standard state conditions, and then "reform" them into the products. The enthalpy of forming each product from the constituent elements is known as the **standard heat (or enthalpy) of formation.** The enthalpy change for "decomposing" the reactants is the negative of the heat of formation.

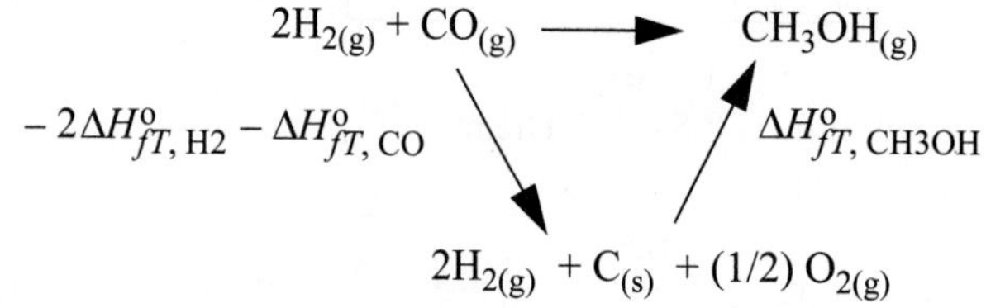

Figure 3.4 *Illustration of the calculation of the standard heat of reaction by use of standard heats of formation.*

We may write this mathematically using the stoichiometric numbers as:

$$\Delta H_T^o = \sum v_i H_{T,i}^o = \sum v_i \Delta H_{fT,i}^o = \sum_{products} |v_i| \Delta H_{fT,i}^o - \sum_{reactants} |v_i| \Delta H_{fT,i}^o \qquad 3.43$$

where every term in the equation varies with temperature. Frequently, the standard state pressure is 1 bar. When a reaction is not at 1 bar, the usual practice is to incorporate pressure effects into the energy balance as we will show later, rather than using a heat of reaction at the high pressure. If we specify a reference temperature in addition to the other properties used for the standard state, we can calculate the ΔH_T^o at any temperature by using the heat capacity of the reactants and products,

$$\Delta H_T^o = \Delta H_R^o + \int_{T_R}^{T} \Delta C_P dT \qquad 3.44$$

where $\Delta C_P = \sum_i v_i C_{P,i}$. A reaction with a negative value of ΔH_T^o is called an **exothermic** reaction. A reaction with a positive value of ΔH_T^o is called an **endothermic** reaction. In this equation, ΔH_T^o is easily determined if the standard heat of reaction ΔH_R^o is known at a single reference temperature and 1 bar.

$$\Delta H_R^o = \sum v_i H_{R,i}^o = \sum v_i \Delta H_{fR,i}^o = \underbrace{\sum |v_i| \Delta H_{fR,i}^o}_{products} - \underbrace{\sum |v_i| \Delta H_{fR,i}^o}_{reactants} \qquad 3.45$$

This is Eqn. 3.43, with an additional specification of temperature which creates a reference state. Almost always, the best reference state to use is $T_R = 298.15$ K and 1 bar, because this is the temperature where the standard state enthalpies (heats) of formation are commonly tabulated. The heat of formation is taken as zero for elements that naturally exist as molecules at 298.15 K and 1 bar. Then the zero value is set for the **state of aggregation** naturally occurring at 298.15 K and 1 bar. For example, H exists naturally as $H_{2(g)}$, so $\Delta H_{f\,298.15}^o$ is zero for $H_{2(g)}$. Carbon is a solid, so the value of $\Delta H_{f\,298.15}^o$ is zero for $C_{(s)}$. The zero values for elements in the naturally occurring state are often omitted in the tables in reference books. Enthalpies of formation are tabulated for many compounds in Appendix E at 298.15 K and 1 bar.

The state of aggregation must be specified in the reaction and care should be used to obtain the correct value from the tables. Some molecules, like water, commonly exist as vapor (g), or liquid (l). Note that for water, the difference between $\Delta H_{f\,298.15(l)}^o$ and $\Delta H_{f\,298.15(g)}^o$ is nearly the heat of vaporization at 298.15 K that can be obtained from the steam tables except that a minor pressure correction has been applied to correct the values from the vapor pressure to 1 bar.

The full form of the integral of Eqn. 3.44 is tedious to calculate manually, e.g., if $C_{P,i} = a_i + b_i T + c_i T^2 + d_i T^3$, Eqn. 3.44 becomes

$$\Delta H_T^o = \Delta H_R^o + \Delta a (T - T_R) + \frac{\Delta b}{2}(T^2 - T_R^2) + \frac{\Delta C}{3}(T^3 - T_R^3) + \frac{\Delta d}{4}(T^4 - T_R^4)$$

$$= J + \Delta a T + \frac{\Delta b}{2}T^2 + \frac{\Delta c}{3}T^3 + \frac{\Delta d}{4}T^4 \qquad 3.46$$

where $\Delta a = \sum v_i a_i$, and heat capacity constants b, c, and d are handled analogously. The value of the constant J is found by using a known numerical value of ΔH_R^o in the upper equation (e.g., 298.15K) and setting the temperature to T_R.

Energy Balances for Reactions

To calculate heat transfer to or from a reactor system, the energy balance used in earlier chapters requires further consideration. To simplify the derivation of the energy balance for reactive systems, consider a single inlet stream and single outlet stream flowing at steady state:

$$0 = H^{in} \dot{n}^{in} - H^{out} \dot{n}^{out} + \dot{\underline{Q}} + \dot{\underline{W}}_S \qquad 3.47$$

For either the inlet stream or the outlet stream, the total enthalpy can be calculated by summing the enthalpy of the components plus the enthalpy of mixing at the stream temperature and pressure. To properly use Eqn. 3.47, the enthalpies of the inlets and outlets need to be related to the reaction. Two methods are used for energy balances, and both are equally valid. An overview of the concept pathways is illustrated in Fig. 3.5. To make a thermodynamic connection with the reaction, the **Heat of Reaction method** replaces the first two terms in Eqn. 3.47 with the negative sum of the three steps shown by dashed lines in Fig. 3.5(a). In contrast, the **Heat of Formation method** uses an elemental reference state for every component, and the enthalpies of each component include the heat of formation as illustrated by each branch of Fig 3.5(b). The difference of enthalpies of the components then includes the generalized steps of Fig. 3.4, and the heat of reaction is included

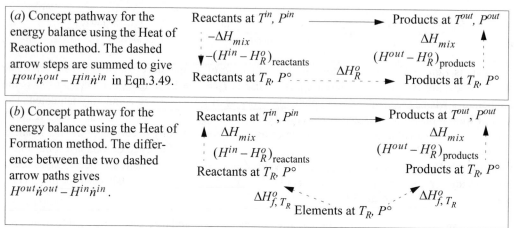

Figure 3.5 *Concept pathways for (a) the Heat of Reaction method and (b) the Heat of Formation method. Details for the steps are omitted as discussed in the text.*

implicitly when taking the difference in the two branches of Eqn. 3.5(b). The difference in the two branches represents the first two terms of Eqn. 3.47. The Heat of Reaction method is usually easiest for students to grasp, because of the explicit term for the heat of reaction. An advantage of the method is that an experimental heat of reaction can be readily used. Most process simulators use the Heat of Formation method. If you think about it, the Heat of Formation method does not require specification of exactly what reactions occur. Based on Hess's law, the overall results can be related to the differences in the heats of formation of the outlet and the inlet species. The notation and the manipulated energy balances for the two methods look different, and the choice of the method depends on data available. The numerical results are the same if the thermochemical data are reliable for each method. Differences will be due to accuracies in the properties used for the pathways.

Either method requires manipulation of stream enthalpies relative to the reference conditions. When discussing reference states in Section 2.11, convenient pathways were used. The reaction balance calculations require that the inlets and outlets be related to the standard state T_R and P^o using similar techniques, and often phase changes are necessary. Following our convention of hierarchical learning, we will use a simplified balance that ignores some of these terms (which are frequently small corrections anyway relative to the heat of reaction). In later chapters, we introduce methods to calculate them. By comparing the magnitude of terms for a particular application, you will then be able to evaluate their relative importance. Choices can be made for the route to calculate a stream enthalpy. One choice is to mix all the components at an ideal gas state and then correct for non-idealities of the mixture. Another route is to correct for non-ideal gas behavior of individual components, and then mix them together at the system T and P. For a system without phase transitions between T_R and T, when calculating the mixing process after the pressure correction, the stream enthalpy relative to species at the standard state looks like this,

$$\sum_i \dot{n}_i (H - H_R^o)_i + \dot{n}\Delta H_{mix} = \sum_i \dot{n}_i \left(\int_{T_R}^{T} \left(\frac{\partial H}{\partial T} \right)_{P^o, i} dT + \int_{P}^{P} \left(\frac{\partial H}{\partial P} \right)_{T, i} dP \right) + \dot{n}\Delta H_{mix}$$

$$\sim \sum_i \dot{n}_i \left(\int_{T_R}^{T} C_P dT \right)$$

small   usually small   small

3.48

❶ Enthalpy of a mixed stream where there are no phase changes between T_R and T.

where H_R^o is the enthalpy of the *species* at the reference state, the pressure dependence of the enthalpy and the heat of mixing have been assumed to be small relative to the heat of reaction. Details on the pressure effects are developed in Chapters 6–8 and are expressed as an enthalpy departure; they are usually small relative to heats of reaction for gases. When the standard state is an ideal gas and liquid streams are involved, the correction is very important and $-\Delta H^{vap}$ must be included in the path. The principle extends to solids as well. Heats of mixing are introduced beginning in Sections 11.4 and 11.10 and are typically small relative to reaction heats unless dissolving/neutralizing acids/bases or dissolving salts. Example 3.6 provides calculations including heats of vaporization for the components.

Heat of Reaction Method

It might not be immediately obvious that Eqn. 3.47 includes the heat of reaction. Considering *just the flow terms of the energy balance*, by plugging Eqn. 3.48 into Eqn. 3.47 the flow terms become

$$H^{in}\dot{n}^{in} - H^{out}\dot{n}^{out} = \sum_{\text{components}} (\dot{n}_i^{in} - \dot{n}_i^{out})H_{i,R}^o +$$
$$\sum_{\text{components}} \dot{n}_i^{in}(H^{in} - H_R^o)_i - \sum_{\text{components}} \dot{n}_i^{out}(H^{out} - H_R^o)_i \qquad 3.49$$

where the inlet temperature of all reactants is the same. The first term on the right of Eqn. 3.49 can be related to the heat of reaction using Eqn. 3.41 to introduce ξ and Eqn. 3.45 to insert the heat of reaction:

$$\sum_{\text{components}} (\dot{n}_i^{in} - \dot{n}_i^{out})H_{i,R}^o = -\dot{\xi} \sum_{\text{components}} \nu_i H_{i,R}^o = -\dot{\xi}(\Delta H_R^o) \qquad 3.50$$

Therefore, the steady-state energy balance can be calculated using Eqn 3.51 and the balance is known as the Heat of Reaction method:

$$0 = \sum_{\text{components}} \dot{n}_i^{in}(H^{in} - H_R^o)_i - \sum_{\text{components}} \dot{n}_i^{out}(H^{out} - H_R^o)_i + \dot{Q} + \dot{W}_S - \dot{\xi}\Delta H_R^o \qquad 3.51$$

If you consider the first two terms and the last term, you can see how they represent the negative of the sum of the steps in Fig. 3.5(a). When multiple reactions occur, a reaction term can be used for each reaction. To use this expression correctly, *the enthalpies of the inlet and outlet streams must be calculated relative to the same reference temperature where ΔH_R^o is known* and any phase transitions at temperatures between the reference state and the inlet or outlet states must be included in $(H - H_R^o)_i$. Also, the variable ξ must be determined for the same basis as the molar flows. The temperature of 298.15 K is almost always the reference temperature for balances involving chemical reactions. There is less flexibility in choosing the reference temperature than for non-reactive systems. This method is easiest to apply with one or two reactions where the stoichiometry is known.

An online supplement is available to relate the notation here to other common textbooks and includes other details.

Heat of Formation Method

The Heat of Formation method requires a reference state relative to the elements, and Eqn. 3.48 is modified by adding the heat of formation for each species. The stream enthalpy when there are no phase transitions between the reference state and the stream state looks like this:

Enthalpy of a mixed stream where there are no phase changes between T_R and T using the Heat of Formation method.

$$\sum_i \dot{n}_i[(H - H_R^o) + \Delta H_{f, T_R}^o]_i + \dot{n}\overbrace{\Delta H_{mix}}^{\text{small}} \sim \sum_i \dot{n}_i\left(\int_{T_R}^{T} C_P dT + \Delta H_{f, T_R}^o\right)_i \qquad 3.52$$

When phase changes are involved, the steps along the pathway are included as illustrated by several examples in Fig. Fig. 2.6 on page 65.

The energy balance is unmodified from Eqn. 3.47. The heat of reaction and the reaction coordinate are not needed explicitly, but the reaction coordinate is often helpful in determining the molar flows for the energy balance.

Work Effects

Usually shaft work and expansion/contraction work are negligible relative to other terms in the energy balance of a reactive system. They may usually be neglected without significant error.

Example 3.6 Reactor energy balances

Acetone (A) is reacted in the liquid phase over a heterogeneous acid catalyst to form mesityl oxide (MO) and water (W) at 80°C and 0.25 MPa. The reaction is $2A \rightleftarrows MO + W$. Conversion is to be 80%. The heat capacity of mesityl oxide has been estimated to be C_P^L (J/mol-K) = 131.16 + 0.2710T(K), C_P^V (J/mol-K) = 72.429 + 0.2645T(K), and the acentric factor is estimated to be 0.356. Other properties can be obtained from Appendix E or webbook.nist.gov. Ignore pressure corrections and assume ideal solutions.

mesityl oxide

(a) Estimate the heat duty for a steady-state reaction with liquid feed (100 mol/h) and liquid products. Use the Heat of Reaction method calculated using liquid heats of formation.
(b) Estimate the heat duty for the same conditions as (a), but use the Heat of Formation method incorporating heats of formation of ideal gases and Eqn 2.45 to estimate heat of vaporization. (This method is used by process simulators.)
(c) Repeat part (b) with the modification of using the experimental heat of vaporization.
(d) Estimate the heat duty for the same conditions as (a), but use the Heat of Formation method incorporating the heat of formation of liquids.

Example 3.6 Reactor energy balances (Continued)

Solution:

(a) For MO and A, $\Delta H^{oL}_{f, 298.15} = -221, -249.4$ kJ/mol respectively, from NIST. The liquid phase standard state heat of reaction is $-221 - 285.8 - 2(-249.4) = -8$ kJ/mol. Using a reference state of the liquid species at 298.15 K and 1 bar, the enthalpy of the each component is given by

$\int_{298.15}^{353.15} C_P^L dT$; the results are {A, 7.265 kJ/mol}, {MO, 12.068}, {W, 4.161}. The mass balance

for 100 mol/h A feed and 80% conversion gives an outlet of $100(1 - 0.8) = 20$ mol/h A, then,

$(\dot{n}_A^{out} = 20 = 100 - 2\dot{\xi}) \Rightarrow \dot{\xi} = 40$, $\dot{n}_{MO}^{out} = \dot{n}_W^{out} = \dot{\xi} = 40$.   The energy   balance   is

$$\dot{Q} = \sum_{\text{components}} \dot{n}_i^{out}(H^{out} - H_R^o)_i - \sum_{\text{components}} \dot{n}_i^{in}(H^{in} - H_R^o)_i + \dot{\xi}\Delta H_R^o, \text{ or}$$

$$\dot{Q} = (20(7.265) + 40(12.068) + 40(4.161)) - 100(7.265) + 40(-8) = -252\text{kJ/h}.$$

(b) The value of $\dot{\xi} = 40$ is the same. The path to calculate the component liquid enthalpies using the heat of formation for ideal gases is: form ideal gas at 298.15K → heat ideal gas to 353.15K → condense ideal gas at 353.15K (using Eqn. 2.45). For MO $\Delta H^{oV}_{f, 298.15} = -178.3$ kJ/mol, from NIST. The enthalpies of each component will be tabulated for each of the three steps: (A) $-215.7 + 4.320 - 27.71 = -239.1$ kJ/mol; (MO) $-178.3 + 8.72 - 39.0 = -208.6$ kJ/mol; (W) $-241.8 + 1.86 - 42.7 = -282.6$ kJ/mol. The energy balance is

$$\dot{Q} = \sum_{\text{components}} \dot{n}_i^{out} H_i^{out} - \sum_{\text{components}} \dot{n}_i^{in} H_i^{in}$$

$$= (20(-239.1) + 40(-208.6) + 40(-282.6)) - 100(-239.1) = -520 \text{ kJ/h.}$$

In principle, this method should have given the same result as (a), but the value differs significantly. The method is sensitive to the accuracy of the prediction for the heat of vaporization. When this method is used, the accuracy of the heat of vaporization needs to be carefully evaluated.

(c) To evaluate the effect of the prediction of the heat of vaporization, let us repeat with a modified path through the normal boiling point of the species, using the experimental heat of vaporization. The normal boiling point of MO from NIST is 403 K, and $\Delta H^{vap} = 42.7$ kJ/mol. The component enthalpy path is modified to: form ideal gas at 298.15 K → heat ideal gas to T_b → condense to liquid at T_b → change liquid to 353.15 K. The enthalpies of each step and totals are: (A) $-215.7 + 2.4 - 30.2 + 3.3 = -240.2$ kJ/mol; (MO) $-178.3 + 17.3 - 42.7 - 11.6 = -215.3$ kJ/mol; (W) $-241.9 + 2.5 - 40.7 - 1.5 = -281.6$ kJ/mol. The energy balance is

$$\dot{Q} = (20(-240.2) + 40(-215.3) + 40(-281.6)) - 100(-240.2) = -660 \text{ kJ/h.}$$

Parts (b) - (c) result in different heat transfer compared to (a). Note the difference in the heat of formation of vapor and liquid MO at 25°C matches the heat of vaporization at the normal boiling point and the difference would be expected to be larger. The original references for the thermochemical data should be consulted to decide which is most reliable.

> ## Example 3.6 Reactor energy balances (Continued)
>
> (d) This modification will not require heats of vaporization. The component enthalpy path is: form liquid at 298.15 K and heat liquid to 353.15 K. The sensible heat calculations are the same as tabulated in part (a). The enthalpies for the two steps and sum for each component are:
>
> (A) $-249.4 + 7.265 = -242.1$ kJ/mol; (MO) $-221 + 12.068 = -208.9$; (W) $-285.8 + 4.2 = -281.6$.
>
> The energy balance becomes:
> $$\dot{Q} = (20(-242.1) + 40(-208.9) + 40(-281.6)) - 100(-242.1) = -252 \text{ kJ/h}$$
>
> Comparing with (a), the Heat of Formation method and the Heat of Reaction method give the same results when the same properties are used.

Adiabatic Reactors

Suppose that a reactor is adiabatic ($\dot{Q} = 0$). For the Heat of Reaction method, the energy balance becomes (for a reaction without phase transformations between T_R and the inlets or outlets),

$$0 = \sum_{components} \dot{n}_i^{in} \int_{T_R}^{T^{in}} C_{P,i} dT - \sum_{components} \dot{n}_i^{out} \int_{T_R}^{T^{out}} C_{P,i} dT - \dot{\xi} \Delta H_R^o \qquad 3.53$$

and as before, any latent heats must be added to the flow terms. An exothermic heat of reaction will raise the outlet temperature above the inlet temperature. For an endothermic heat of reaction, the outlet temperature will be below the inlet temperature. At steady state, the system finds a temperature where the heat of reaction is just absorbed by the enthalpies of the process streams. This temperature is known as the adiabatic reaction temperature, and the maximum reactor temperature change is dependent on the kinetics and reaction time, or on equilibrium. For fixed quantities and temperature of feed, Eqn. 3.53 involves two unknowns, T^{out} and $\dot{\xi}$, and, *if the reaction is not limited by equilibrium,* the kinetic model and reaction time determine these variables. If a reaction time is sufficiently large, equilibrium may be approached. Equilibrium reactors will be considered in Chapter 17.

Graphical Visualization of the Energy Balance

The energy balance as presented by the Heat of Reaction Method (Eqn. 3.51) can be easily plotted for an adiabatic reaction. Let us replace the heat capacity polynomials with average heat capacities that are temperature independent. If we incorporate the material balance, Eqn. 3.41, the Heat of Reaction steady-state energy balance after rearranging becomes

$$T^{out} = T^{in} + \frac{\dot{Q} - \dot{\xi}\Delta H_R^o}{\left(\sum n_i^{in} C_{P,i}\right) + \dot{\xi}\Delta C_P} + c \qquad 3.54$$

where $c = (T_R - T^{in}) / \left(\left(\sum n_i^{in} C_{P,i}\right) / (\dot{\xi}\Delta C_P) + 1\right)$ and c is frequently small. Neglecting c and dropping heat for an adiabatic reactor,

$$T^{out} \approx T^{in} - \frac{\dot{\xi}\Delta H_R^o}{\left(\sum n_i^{in} C_{P,\,i}\right) + \dot{\xi}\Delta C_P} \quad \text{adiabatic reactor} \qquad 3.55$$

Consider the case of an exothermic reaction. A schematic of the energy balance is shown in Fig. 3.6 for an exothermic reaction. In the plot, we have neglected $\dot{\xi}\Delta C_P$ in the denominator which is often small relative to the summation and introduces a slight curvature if included. Note that an endothermic reaction will have an energy balance with a negative slope, making the plot for an endothermic reaction a mirror image of Fig. 3.6 reflected across a vertical line at T^{in}.

3.7 REACTIONS IN BIOLOGICAL SYSTEMS

Living systems constantly metabolize food. In a sophisticated system such as a human, the digestive system breaks down the carbohydrates (sugars and starches), protein (complex amino acids), and fats (glycerol esters of fatty acids) into the building block molecules. There are exothermic reactions as the chemical structure of the food is modified by breaking bonds. The small molecules that result can be transported through the body as sugars, amino acids, and fatty acids. The body transforms these basic molecules to create energy to constantly replace cells and also to provide the energy needed for physical activity. The process more closely resembles fuel cell operation than a Carnot cycle, but the concepts of the energy balance still apply.

A major reaction providing energy in the human body is the oxidation of sugars and starches. These compounds are known to provide "quick energy" because they are easily burned. As an example, consider the oxidation of glucose to CO_2 and H_2O. Oxygen taken in through the lungs is carried to the cells where the reaction takes place. CO_2 produced by the reaction is transported back to the lungs where it can be expelled. Water produced by the reaction largely is left as liquid, though respiration results in some transport. Although the actual process involves several intermediate steps, we know by **Hess's law** that the overall energy and enthalpy change depends on only the initial and final structures, not the intermediate paths. The process for oxidation of glucose is

$$C_6H_{12}O_{6(aq)} \;+\; 6O_{2(g)} \;\rightarrow\; 6CO_2(g) \;+ 6H_2O_{(l)} \qquad 3.56$$

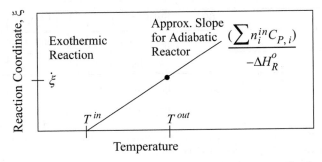

Figure 3.6 *Approximate energy balance for an adiabatic exothermic reaction. The dot represents the outlet reaction coordinate value and the adiabatic outlet temperature. The plot for an endothermic reaction will be a mirror image of this figure as explained in the text.*

Like other combustion reactions, this reaction is exothermic. When intense physical activity occurs, the body is not able to use oxidation quickly enough to produce energy. In this case the body can convert glucose (or other sugars and starches) anaerobically to lactic acid as follows:

$$C_6H_{12}O_{6(aq)} \quad \rightarrow \quad 2\ C_3H_6O_3 \qquad\qquad 3.57$$

This is also an exothermic reaction, and there is no gas produced. This mechanism is used by muscles to provide energy and a build-up of lactic acid causes the muscular aching during and after vigorous exercise. Fats have higher energy content per mass and may be oxidized in a reaction analogous to Eqn. 3.56. Fat is used by the body to store energy. To burn fat, the body usually needs to be starved of the more easily metabolized starches and sugars.

The nervous system regulates body temperature so that when energy is burned, there is little change in body temperature. Some energy is transported out via respiration, some through evaporation of moisture through the skin, and some by heat transfer at the skin surface. Blood vessels in extremities are flexible and change size to regulate the blood flow, which is used to modify the flow rates. On a cold winter day, when our hands start to feel cold, our body is sensing a need to preserve body heat, so the vessels contract to decrease blood flow, making our hands feel even colder! During exercise, the vessels expand to increase cooling, and perspiration starts to provide evaporative cooling. In any event, our core temperature is maintained at 37°C as much as possible. When the body temperature drops the condition is called hypothermia. When the body is unable to eliminate heat, the condition is called hyperthermia and results in heat exhaustion and heat stroke. Usually, the body is able to regulate temperature, and the temperature is stable, so the body does not hold or loose energy by temperature changes. In an adult, the mass is also constant except for the daily cycles of eating and excretion that are very small perturbations of the total body mass.

The body adjusts the metabolic rate as the level of physical activity changes. Data collected by performing energy balances on humans after 12 hours of fasting eliminate the heat effects due to digestion (about 30% higher) and provide an accurate measure of the metabolic rates. Examples of energy consumption are tabulated in Table 3.1.

Table 3.1 *Summary of Energy Expenditure by Various Physical Activities for a 70 kg Person[a]*

Activity	Energy Expenditure kcal h^{-1}	Activity	Energy Expenditure kcal h^{-1}
Lying still, awake	77	Bicycling on level (5.5 mi/h)	304
Sitting at rest	100	Walking on 3% grade, (2.6 mi/h)	357
Typing rapidly	140	Shoveling snow or sawing wood	480
Dressing/undressing	150	Jogging, (5.3 mi/h)	570
Walking on level surface (2.6 mi/h)	200	Rowing (20 strokes/min)	828
Sexual intercourse	280	Maximum activity (untrained)	1440

a. Vander, A.J., Serman, J.J., Luciano, D.S., 1985. *Human Physiology: The Mechanisms of Body Function*, 4th ed., ch. 15, New York: McGraw-Hill.

3.8   SUMMARY

This chapter started by introducing the concept of heat engines and heat pumps to interconverted heat and work, and the limitations in efficiency. As a cyclic process, the systems are simple, but practice builds confidence in working with multistep processes. In the distillation section we introduced quite a few terms because there are a lot of flow rates in the sections of a distillation column. This section provided practice in working with choice of boundaries for balances and working with many streams. Sorting out the streams that are relevant is a key step in the solution of problems. We introduced ideal gas mixtures and ideal solutions, stressing that the energy of mixing and volume of mixing are zero for both and so the enthalpy of a stream is the sum of the enthalpies of the constituents. We then used reference states to solve an energy balance on a mixed stream including a phase transition. In the section on reacting systems we set forth the procedures to properly account for energy flows in and out of the system. Finally we demonstrated that the energy balance was relevant for complex biochemical reactions. The energy balance for a reaction is independent of whether it occurs biologically or in an industrial reactor. The goal of this section was to demonstrate the breadth of applications and to build your confidence in solving problems. At this point most students still do not have a grasp on reversibility and irreversibility, which should be clarified in the next chapter as we build on this material.

Important Equations

Two equations that come up repeatedly are the Carnot efficiency (Eqn. 3.6) and Carnot COP (Eqn. 3.9). The Carnot efficiency teaches that the conversion of heat energy into mechanical energy cannot be 100% efficient, even if every operation in the process is 100% efficient. This has major implications throughout modern society, reflected in limitations on power plants and energy management. Much of Units I and II is devoted to understanding the details of these kinds of problems and how to solve them more precisely. The Carnot thermal efficiency and COP are benchmarks for real processes, though real processes are not always operating between reservoirs. Common errors applying the formulas are: (1) to interchange the formulas and use the COP formula when you want the efficiency; and (2) to use relative temperature (°C or °F) instead of absolute temperature.

This chapter introduced the concept of ideal solutions and methods of solving energy balances with mixtures, including phase transitions. Important equations are

$$\Delta U^{is}_{mix} = 0 \qquad \Delta V^{is}_{mix} = 0 \qquad \Delta H^{is}_{mix} = 0 \qquad\qquad 3.34$$

We also introduced the concept of the reaction coordinate and that all species in a single reaction can be related by

$$\dot{n}^{out} = \dot{n}^{in} + \dot{\xi}\sum_i \nu_i \qquad\qquad 3.42$$

We finished with the energy balance, which is most often expressed in the approximate form for the Heat of Reaction method:

$$0 = \sum_{components} \dot{n}^{in}_i (H^{in} - H^o_R)_i - \sum_{components} \dot{n}^{out}_i (H^{out} - H^o_R)_i + \dot{Q} + \underline{\dot{W}}_S - \dot{\xi}\Delta H^o_R \qquad 3.51$$

If the enthalpy of formation is incorporated into the enthalpy of the components, the Heat of Formation method looks unchanged from the energy balance of Chapter 2:

$$0 = H^{in}\dot{n}^{in} - H^{out}\dot{n}^{out} + \dot{Q} + \dot{W}_S \qquad \text{3.47}$$

3.9 PRACTICE PROBLEMS

P3.1 Dimethyl ether (DME) synthesis provides a simple prototype of many petrochemical processes. Ten tonnes (10,000 kg) per hour of methanol are fed at 25°C. The entire process operates at roughly 10 bar. It is 50% converted to DME and water at 250°C. The reactor effluent is cooled to 75°C and sent to a distillation column where 99% of the entering DME exits the top with 1% of the entering methanol and no water. This DME product stream is cooled to 45°C. The bottoms of the first column are sent to a second column where 99% of the entering methanol exits the top at 136°C, along with all the DME and 1% of the entering water, and is recycled. The bottoms of the second column exit at 179°C and are sent for wastewater treatment. Use the method of Example 3.6(b) to complete the following:

(a) Calculate the enthalpy in GJ/h of the feed stream of methanol.
(b) Calculate the enthalpy in GJ/h of the stream entering the first distillation column.
(c) Calculate the enthalpy in GJ/h of the DME product stream.
(d) Calculate the enthalpy in GJ/h of the methanol recycle stream.
(e) Calculate the enthalpy in GJ/h of the aqueous product stream.
(f) Calculate the energy balance in GJ/h of the entire process. Does the process involve a net energy need or surplus?
(ANS. -50, -106, -26, -51, -31, -7, i.e., energy surplus)

3.10 HOMEWORK PROBLEMS

3.1 Two moles of nitrogen are initially at 10 bar and 600 K (state 1) in a horizontal piston/cylinder device. They are expanded adiabatically to 1 bar (state 2). They are then heated at constant volume to 600 K (state 3). Finally, they are isothermally returned to state 1. Assume that N_2 is an ideal gas with a constant heat capacity as given on the back flap of the book. Neglect the heat capacity of the piston/cylinder device. Suppose that heat can be supplied or rejected as illustrated below. Assume each step of the process is reversible.

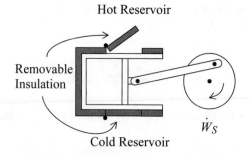

(a) Calculate the heat transfer and work done on the gas for each step and overall.
(b) Taking state 1 as the reference state, and setting $U_R^{ig} = 0$, calculate U and H for the nitrogen at each state, and ΔU and ΔH for each step and the overall $\underline{Q}$ and $\underline{W}_{EC}$.

(c) The atmosphere is at 1 bar and 298 K throughout the process. Calculate the work done on the atmosphere for each step and overall. (Hint: Take the atmosphere as the system.) How much work is transferred to the shaft in each step and overall?

3.2 One mole of methane gas held within a piston/cylinder, at an initial condition of 600 K and 5 MPa, undergoes the following reversible steps. Use a temperature-independent heat capacity of $C_P = 44$ J/mol-K.

(a) Step 1: The gas is expanded isothermally to 0.2 MPa, absorbing a quantity of heat $\underline{Q}_{H1}$.

Step 2: The gas is cooled at constant volume to 300 K. Step 3. The gas is compressed isothermally to the initial volume. Step 4: The gas is heated to the initial state at constant volume requiring heat transfer $\underline{Q}_{H2}$. Calculate ΔU, Q, and W_{EC} for each step and for the cycle. Also calculate the thermal efficiency that is the ratio of total work obtained to heat furnished, $\dfrac{-\underline{W}_{EC, cycle}}{(\underline{Q}_{H1} + \underline{Q}_{H2})}$.

(b) Step 1: The gas is expanded to 3.92 MPa isothermally, absorbing a quantity of heat $\underline{Q}_{H1}$. Step 2: The gas is expanded adiabatically to 0.1 MPa. Step 3: The gas is compressed isothermally to 0.128 MPa. Step 4: The gas is compressed adiabatically to the initial state. Calculate ΔU, Q, W_{EC} for each step and for the cycle. Also calculate the thermal efficiency for the cycle, $\dfrac{-\underline{W}_{EC, cycle}}{\underline{Q}_{H1}}$.

3.3 The Arrhenius model of global warming constitutes a very large composite system.[3] It assumes that a layer of gases in the atmosphere (A) absorbs infrared radiation from the Earth's surface and re-emits it, with equal amounts going off into space or back to the Earth's surface; 240 W/m² of the solar energy reaching the Earth's surface is reflected back as infrared radiation (S). The "emissivity value" (λ) of the ground (G) is set equal to that of A. λ characterizes the fraction of radiation that is not absorbed. For example, $(1-\lambda)$ would be the fraction of IR radiation that is absorbed by the ground, and the surface energy would be $S = A\lambda + G(1-\lambda)$. A similar balance for the atmosphere gives, $\lambda G = 2\lambda A$. This balance indicates that radiation received from G is balanced by that radiated from A. The factor of 2 on the right-hand side appears because radiation can be toward the ground or toward space. The equations for radiation are given by the relations: $G = \sigma T_G^4$ and $A = \sigma T_A^4$ where $\sigma =$ Stefan-Boltzmann constant 5.6704 x 10^{-8} (W/m²-K⁴).

(a) Noting that the average T_G is 300 K, solve for λ.
(b) Solve for T_G when $\lambda = 0$. This corresponds to zero global warming.
(c) Solve for T_G when $\lambda = 1$. This corresponds to perfect global warming.
(d) List at least three oversimplifications in the assumptions of this model. Discuss whether these lead to underestimating or overestimating T_G

3.4 A distillation column with a total condenser is shown in Fig. 3.3. The system to be studied in this problem has an average enthalpy of vaporization of 32 kJ/mol, an average C_P^L of 146 J/mol°-C, and an average C_P^V of 93 J/mol°-C. Variable names for the various stream flow rates and the heat flow rates are given in the diagram. The feed can be liquid, vapor, or

---

3. *cf.* RealClimate.org (… "simple model") and aip.org/history/climate/simple.htm, 8/2011

a mixture represented using subscripts to indicate the vapor and liquid flows, $F = F_V + F_L$. The enthalpy flow due to feed can be represented as: for saturated liquid, $F_L H^{satL}$; for saturated vapor, $F_V H^{satV}$; for subcooled liquid, $F_L H^{satL} + F_L C_P^L (T_F - T^{satL})$; for superheated vapor, $F_V H^{satV} + F_V C_P^V (T_F - T^{satV})$; and for a mix of vapor and liquid, $F_L H^{satL} + F_V H^{satV}$.

(a) Use a mass balance to show $F_V + V_S - V_R = L_S - L_R - F_L$.

[For parts (b)–(f), use the feed section mass and energy balances to show the desired result.]

(b) For saturated vapor feed, $F_L = 0$. Show $V_R = V_S + F_V$, $L_S = L_R$.
(c) For saturated liquid feed, $F_V = 0$. Show $V_S = V_R$, $L_S = L_R + F_L$.

(d) For subcooled liquid feed, $F_V = 0$. Show $V_R - V_S = F_L C_P (T_F - T^{sat}) / \Delta H^{vap}$.

(e) For superheated vapor feed, $F_L = 0$. Show $L_S - L_R = -F_V C_P (T_F - T^{sat}) / \Delta H^{vap}$.

(f) For a feed mixture of saturated liquid and saturated vapor. Show $V_R = V_S + F_V$, $L_S = L_R + F_L$.

(g) Use the mass and energy balances around the total condenser to relate the condenser duty to the enthalpy of vaporization, for the case of streams L_R and D being saturated liquid.

(h) Use the mass and energy balances around the reboiler to relate the reboiler duty to the enthalpy of vaporization.

(i) In the case of subcooled liquid streams L_R and D, the vapor flows out of the top of the column, and more variables are required. V_R' (into the condenser) will be smaller than the rectifying section flow rate V_R. Also, the liquid flow rate in the rectifying section, L_R, will be larger than the reflux back to the column, L_R'. Using the variables V_R', L_R' to represent the flow rate out of the top of the column and the reflux, respectively, relate V_R to V_R', L_R' and the degree of subcooling $T_{L'} - T^{satL}$.

[For parts (j)–(o), find all other flow rates and heat exchanger duties (Q values).]
(j) $F = 100$ mol/hr (saturated vapor), $B = 43$ mol/hr, $L_R/D = 2.23$.
(k) $F = 100$ mol/hr (saturated vapor), $D = 48$ mol/hr, $L_S/V_S = 2.5$.
(l) $F = 100$ mol/hr (saturated liquid), $D = 53$ mol/hr, $L_R/D = 2.5$.
(m) $F = 100$ mol/hr (half vapor, half liquid), $B = 45$ mol/hr, $L_S/V_S = 1.5$.
(n) $F = 100$ mol/hr (60°C subcooled liquid), $D = 53$ mol/hr, $L_R/D = 2.5$.
(o) $F = 100$ mol/hr (60°C superheated vapor), $D = 48$ mol/hr, $L_S/V_S = 1.5$.

3.5 Allyl chloride (AC) synthesis provides a simple prototype of many petrochemical processes. 869 kg per hour of propylene (C3) are fed with a 1% excess of chlorine (Cl2) at 25°C. The entire process operates at roughly 10 bar. Cl2 is recycled to achieve a 50% excess of Cl2 at the reactor inlet. The reactor conversion is 100% of the propylene to AC and hydrochloric acid (HCl) at 511°C.

The reactor effluent is cooled to 35°C and sent to a distillation column where 98% of the entering AC exits the bottom with 1% of the entering Cl2 and no HCl. This AC product stream exits at 57°C. The tops of the first column, at 36°C, are sent to a second column where 99% of the entering Cl2 exits the bottom at 36°C, along with 1% of the entering HCl, all of the AC, and is recycled. The tops of the second column exit at -31°C and are sent for waste treatment. Using the method of Heat of Formation method for the energy balance and ideal gas reference states with Eqn. 2.45 to estimate the heat of vaporization, complete the following.

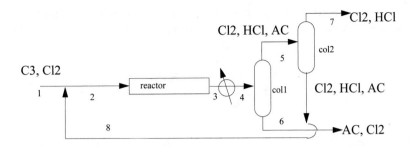

(a) Write a balanced stoichiometric equation for this reaction. (Hint: Check the NIST Web-Book for chemical names and formulas.)

(b) Perform a material balance to determine compositions and flow rates for all streams.

(c) Using only streams (1), (6), (7), calculate the energy balance in MJ/h of the entire process. Does the process involve a net energy need or surplus?

(d) Determine the heat load on the reactor in MJ/h.

(b) Calculate the enthalpies in MJ/h of the feed stream 1.

(c) Calculate the enthalpy in MJ/h of the stream 4 entering the first distillation column.

(d) Calculate the enthalpy in MJ/h of the AC product stream 6.

(e) Calculate the enthalpy in MJ/h of the Cl2 recycle stream 8.

(f) Calculate the enthalpy in MJ/h of the HCl waste stream 7.

3.6 Chlorobenzene$_{(l)}$ is produced by reacting benzene$_{(l)}$ initially at 30°C with Cl$_{2(g)}$ initially at 30°C in a batch reactor using AlCl$_3$ as a catalyst. HCl$_{(g)}$ is a by-product. During the course of the reaction, the temperature increases to 50°C. To avoid dichlorobenzenes, conversion of benzene is limited to 30%. On the basis of 1 mol of benzene, and 0.5 mol Cl$_{2(g)}$ feed, what heating or cooling is required using the specified method(s)? The NIST WebBook reports the heats of formation for liquid benzene and chlorobenzene at 25°C as 49 kJ/mol and 11.5 kJ/mol, respectively. The heat capacities of liquid benzene and chlorobenzene are 136 J/mol-K and 150 J/mol-K, respectively.

(a) Use liquid reference states for the benzenes and the heat of reaction method for the energy balance.

(b) Follow part (a), but instead, use the heat of formation method for the energy balance.

(c) Use an ideal gas-phase reference state for the benzenes, Eqn. 2.45 to estimate the heat of vaporization, and the Heat of Reaction method. Use path of Fig. 2.6(c) to avoid Cp^L.

3.7 Benzene and chlorobenzene produced from the reaction described in problem 3.6 are separated by distillation at 1 bar. The chlorine and HCl are removed easily and this problem concerns only a binary mixture. Suppose the liquid flow to the reboiler is 90 mol% chlorobenzene and 10 mol% benzene at 121.9°C. The boilup ratio is 0.7 at 127.8°C, and the vapor leaving the reboiler is 12.7 mol% benzene. The heat of vaporization of chlorobenzene is 41 kJ/mol. Heat capacities for liquids are in problem 3.6(a). Determine the heat duty for the reboiler (kJ/(mol of inlet flow)) assuming ideal solutions.

3.8 In the process of reactive distillation, a reaction occurs in a distillation column simultaneously with distillation, offering process intensification. Consider a reactive distillation including the reaction: cyclohexene$_{(l)}$ + acetic acid$_{(l)}$ $\underset{\leftarrow}{\rightarrow}$ cyclohexyl acetate$_{(l)}$. In a reactive distillation two feeds are used, one for each reactant. In a small-scale column, the flows, heat capacities, and temperatures are tabulated below in mol/min. Assume the heat capacities are temperature-independent. All streams are liquids. Use data from the NIST WebBook for heats of formation to estimate the net heat requirement for the column.

	C_P(J/mol-K)	bottom feed	top feed	distillate	bottoms
T(K)		304	298	328	386
cyclohexene	165	0.35	0	0.115	0.11
acetic acid	130	0	0.425	0.03	0.27
cyclohexyl acetate	290	0	0	0	0.125

3.9 This problem considers oxidation of glucose as a model carbohydrate.

(a) In the human body glucose is typically metabolized in aqueous solution. Ignoring the enthalpy change due to dissolving glucose in the aqueous solution, calculate the standard heat of reaction for Eqn. 3.56.

(b) Using the standard heat of reaction calculated in part (a) together with the steady-state energy expenditure for sitting at rest, determine the rate that glucose is consumed (g/day). You may ignore sensible heats because the respiration streams are very close to 298.15 K.

(c) Compare the energy consumption in part (b) with a light bulb by putting the energy consumption in terms of W, which is usually used to characterize light bulbs. How many 70 kg humans collectively produce energy equivalent to a 1500 W hair dryer?

(d) On a mass basis, all carbohydrates have about the same heat of reaction. A soft drink has approximately 73 g carbohydrates (sugar) per 20 fluid oz. bottle and the manufacturer's label reports that the contents provide 275 kcal. (Note: A dietary calorie is a thermodynamic kcal, so the conversion has already been made, and the bottle is labeled 275 calories.) Convert the answer from part (a) to kcal/g and compare with the manufacturer's ratio.

(e) What is the annual output of CO_2 (kg) for a 70 kg human using the energy expenditure for sitting as an average value?

3.10 Compressed air at room temperature (295 K) is contained in a 20-L tank at 2 bar. The valve is opened and the tank pressure falls slowly and isothermally to 1.5 bar. The frictionless piston-cylinder is isothermal and isobaric (P = 1.5 bar) during the movement. The surroundings are at 1 bar. The volumes of the piping and valve are negligible. During the expansion, the piston is pushing on external equipment and doing useful work such that the total resistance to the expansion is equivalent to 1.5 bar. The entire system is then cooled until all of the air is back in the original container. During the retraction of the piston, the piston must pull on the equipment, and the resistance of the external equipment is equivalent to 0.1 bar, so the total force on the piston is less than 1 bar, 1.0 − 0.1 = 0.9 bar. The valve is closed and the tank is heated back to room temperature. (Air can be considered an ideal gas with a T-independent C_P = 29 J/mol-K.)

Initial and Final States   Intermediate State

(a) The useful work done by the process is the total work done by the piston in the expansion step less the amount of work done on the atmosphere. Calculate the useful work done per expansion stroke in kJ.

(b) Calculate the amount of heat needed during the expansion in kJ. Neglect the heat capacity of the tank and cylinder.

(c) Calculate the amount of cooling needed during the retraction of the cylinder in kJ.

3.11 A well-insulated tank contains 1 mole of air at 2 MPa and 673 K. It is connected via a closed valve to an insulated piston/cylinder device that is initially empty. The piston may be assumed to be frictionless. The volumes of the piping and valve are negligible. The weight of the piston and atmospheric pressure are such that the total downward force can be balanced with gas pressure in the cylinder of 0.7 MPa. The valve between the tank and piston/cylinder is cracked open until the pressure is uniform throughout. The temperature in the tank is found to be 499.6 K. Air can be assumed to be an ideal gas with a temperature-independent heat capacity $C_P = 29.3$ J/mol-K.

(a) What is the number of moles left in the tank at the end of the process?

(b) Write and simplify the energy balance for the process. Determine the final temperature of the piston/cylinder gas.

3.12 A piston/cylinder has two chambers and includes a spring as illustrated below. The right-hand side contains air at 20°C and 0.2 MPa. The spring exerts a force to the right of $F = 5500x$ N where x is the distance indicated in the diagram, and $x^i = 0.3$ m. The piston has a cross-sectional area of 0.1 m^2. Assume that the piston/cylinder materials do not conduct heat, and that the piston/cylinder and spring do not change temperature. After oscillations cease, what is the temperature of the air in the right chamber and the final position of the piston, x^f, for the cases listed below? Use a temperature-independent heat capacity for air, $C_P = 7R/2$.

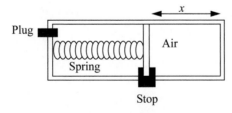

(a) The left chamber is evacuated and the plug remains in place.

(b) The plug is removed so that the left-side pressure stays at 0.1 MPa throughout the process.

3.13 We wish to determine the final state for the gas in an inflated balloon. Initially, the balloon has a volume of $\underline{V}^i$ at rest. The volume of the balloon is related to the internal pressure by $\underline{V} = a \cdot P + b$, where a and b are constants. The balloon is to be inflated by air from our lungs at P_{lung} and T_{lung} which are known and assumed to remain constant during inflation. Heat transfer through the walls of the balloon can be ignored. The system is defined to be the gas inside the balloon at any time. Starting with the general energy balance, simplify to write the balance in terms of $\{C_P, C_V, T_{lung}, P_{lung}, T^i, T^f, T_R, P_R\}$ and either $\{\underline{V}^i, \underline{V}^f\}$ or

$\{P^i, P^f\}$. This will demonstrate that measurement of V^f or P^f is sufficient to determine T^f. Assume air is an ideal gas.

ENTROPY

$$S = k \ln W$$

L. Boltzmann

We have discussed energy balances and the fact that friction and velocity gradients cause the loss of useful work. It would be desirable to determine maximum work output (or minimum work input) for a given process. Our concern for accomplishing useful work inevitably leads to a search for what might cause degradation of our capacity to convert any form of energy into useful work. As an example, isothermally expanding an ideal gas from V^i to $2V^i$ can produce a significant amount of useful work if carried out reversibly, or possibly zero work if carried out irreversibly. If we could understand the difference between these two operations, we would be well on our way to understanding how to minimize wasted energy in many processes. Inefficiencies are addressed by the concept of entropy.

Entropy provides a measure of the disorder of a system. As we will see, increased "disorder of the universe" leads to reduced capability for performing useful work. This is the second law of thermodynamics. Conceptually, it seems reasonable, but how can we define "disorder" mathematically? That is where Boltzmann showed the way:

$$\underline{S} = k \ln W$$

where $\underline{S}$ is the entropy, W is the number of ways of arranging the molecules given a specific set of independent variables, like T and V; k is known as Boltzmann's constant.

For example, there are more ways of arranging your socks around the entire room than in a drawer, basically because the volume of the room is larger than that of the drawer. We will see that $\Delta \underline{S} = Nk \ln(\underline{V}_2/\underline{V}_1)$ in this situation, where N is the number of socks and $Nk = nR$, where n is the number of moles, $\underline{V}$ is the volume, and R is the gas constant. In Chapter 1, we wrote $U^{ig} = 1.5NkT$ without thinking much about who Boltzmann was or how his constant became so fundamental to the molecular perspective. This connection between the molecular and macroscopic scales was Boltzmann's major contribution.

Chapter Objectives: You Should Be Able to...

1. Explain entropy changes in words and with numbers at the microscopic and macroscopic levels. Typical explanations involve turbines, pumps, heat exchangers, mixers, and power cycles.

2. Simplify the complete entropy balance to its most appropriate form for a given situation and solve for the productivity of a reversible process.

3. Sketch and interpret *T-S, T-V, H-S,* and *P-H* diagrams for typical processes.

4. Use inlet and outlet conditions and efficiency to determine work associated with turbines/compressors.

5. Determine optimum work interactions for reversible processes as benchmarks for real systems.

6. Sketch and interpret *T-S, T-V, H-S,* and *P-H* diagrams for typical processes.

4.1   THE CONCEPT OF ENTROPY

Chapters 2 and 3 showed the importance of irreversibility when it comes to efficient energy transformations. We noted that prospective work energy was generally dissipated into thermal energy (stirring) when processes were conducted irreversibly. If we only had an "irreversibility meter," we could measure the irreversibility of a particular process and design it accordingly. Alternatively, we could be given the efficiency of a process relative to a reversible process and infer the magnitude of the irreversibility from that. For example, experience might show that the efficiency of a typical 1000 kW turbine is 85%. Then, characterizing the actual turbine would be simple after solving for the reversible turbine (100% efficient).

In our initial encounters, entropy generation provides this measure of irreversibility. Upon studying entropy further, however, we begin to appreciate its broader implications. These broader implications are especially important in the study of multicomponent equilibrium processes, as discussed in Chapters 8–16. In Chapters 5–7, we learn to appreciate the benefits of entropy being a state property. Since its value is path independent, we can envision various ways of computing it, selecting the path that is most convenient in a particular situation.

> ❶ Entropy is a useful property for determining maximum/minimum work.
>
> Rudolf Julius Emanuel Clausius (1822–1888), was a German physicist and mathematician credited with formulating the macroscopic form of entropy to interpret the Carnot cycle and developed the second law of thermodynamics.

Entropy may be contemplated microscopically and macroscopically. The microscopic perspective favors the intuitive connection between entropy and "disorder." The macroscopic perspective favors the empirical approach of performing systematic experiments, searching for a unifying concept like entropy. Entropy was initially conceived macroscopically, in the context of steam engine design. Specifically, the term "entropy" was coined by Rudolf Clausius from the Greek for *transformation*.[1] To offer students connections with the effect of volume (for gases) and temperature, this text begins with the microscopic perspective, contemplating the detailed meaning of "disorder" and then demonstrating that the macroscopic definition is consistent.

To appreciate the distinction between the two perspectives on entropy, it is helpful to define the both perspectives first. The macroscopic definition is especially convenient for solving problems process problems, but the connection between this definition and disorder is not immediately apparent.

---

1. Denbigh, K. 1971. *The Principles of Chemical Equilibrium*. 3rd ed. New York: Cambridge University Press, p.33.

Macroscopic definition—Intensive entropy is a state property of the system. For a differential change in state *of a closed **simple system*** (no internal temperature gradients or composition gradients and no internal rigid, adiabatic, or impermeable walls),[2] the differential entropy change of the system is equal to the heat absorbed by the system *along a reversible path* divided by the absolute temperature of the system at the surface where heat is transferred.

$$dS \equiv \frac{dQ_{rev}}{T_{sys}}$$

4.1

where dS is the entropy change of the system. We will later show that this definition is consistent with the microscopic definition.

Microscopic definition—Entropy is a measure of the molecular disorder of the system. Its value is related to the number of microscopic states available at a particular macroscopic state. Specifically, for a system of fixed energy and number of particles, N,

$$S_i \equiv k \ln(p_i) \quad \text{or} \quad \Delta S \equiv k \ln\left(\frac{p_2}{p_1}\right)$$

4.2

where p_i is the number of microstates in the ith macrostate, $k = R/N_A$. We define microstates and macrostates in the next section.

The microscopic perspective is directly useful for understanding how entropy changes with volume (for a gas), temperature, and mixing. It simply states that disorder increases when the number of possible arrangements increases, like the socks and drawers mentioned in the introduction. Similarly, molecules redistribute themselves when a valve is opened until the pressures have equilibrated. From the microscopic approach, entropy is a specific mathematical relation related to the number of possible arrangements of the molecule. Boltzmann showed that this microscopic definition is entirely consistent with the macroscopic property inferred by Rudolf Clausius. We will demonstrate how the approaches are equivalent.

> *Entropy is a difficult concept to understand, mainly because its influence on physical situations is subtle, forcing us to rely heavily on the mathematical definition. We have ways to try to make some physical connection with entropy, and we will discuss these to give you every opportunity to develop a sense of how entropy changes. Ultimately, you must reassure yourself that entropy is defined mathematically, and like enthalpy, can be used to solve problems even though our physical connection with the property is occasionally less than satisfying.*

In Section 4.2, the microscopic definition of entropy is discussed. On the microscopic scale, S is influenced primarily by spatial arrangements (affected by volume and mixing), and energetic arrangements (occupation) of energy levels (affected by temperature). We clarify the meaning of the microscopic definition by analyzing spatial distributions of molecules. To make the connection between entropy and temperature, we outline how the principles of volumetric distributions extend to energetic distributions. In Section 4.3, we introduce the macroscopic definition of entropy and conclude with the second law of thermodynamics.

The microscopic approach to entropy is discussed first, then the macroscopic approach is discussed.

The second law is formulated mathematically as the entropy balance in Section 4.4. In this section we demonstrate how heat can be converted into work (as in an electrical power plant). How-

---

2. A simple system is not acted on by external force fields or inertial forces.

ever, the maximum thermal efficiency of the conversion of heat into work is less than 100%, as indicated by the Carnot efficiency. The thermal efficiency can be easily derived using entropy balances. This simple but fundamental limitation on the conversion of heat into work has a profound impact on energy engineering. Section 4.5 is a brief section, but makes the key point that pieces of an overall process can be reversible, even while the overall process is irreversible.

In Section 4.6 we simplify the entropy balance for common process equipment, and then use the remaining sections to demonstrate applications of system efficiency with the entropy balance. Overall, this chapter provides an understanding of entropy which is essential for Chapter 5 where entropy must be used routinely for process calculations.

4.2 THE MICROSCOPIC VIEW OF ENTROPY

Probability theory is nothing but common sense reduced to calculation.

LaPlace

❶ Configurational entropy is associated with spatial distribution. Thermal entropy is associated with kinetic energy distribution.

To begin, we must recognize that the disorder of a system can change in two ways. First, disorder occurs due to the physical arrangement (distribution) of atoms, and we represent this with the *configurational entropy*.[3] There is also a distribution of kinetic energies of the particles, and we represent this with the *thermal entropy*. For an example of kinetic energy distributions, consider that a system of two particles, one with a kinetic energy of 3 units and the other of 1 unit, is microscopically distinct from the same system when they both have 2 units of kinetic energy, even when the configurational arrangement of atoms is the same. This second type of entropy is more difficult to implement on the microscopic scale, so we focus on the configurational entropy in this section.[4]

Entropy and Spatial Distributions: Configurational Entropy

Given N molecules and M boxes, how can these molecules be distributed among the boxes? Is one distribution more likely than another? Consideration of these issues will clarify what is meant by microstates and macrostates and how entropy is related to disorder. Our consideration will focus on the case of distributing particles between two boxes.

❶ Distinguishability of particles is associated with microstates. Indistinguishability is associated with macrostates.

First, let us suppose that we distribute $N = 2$ ideal gas[5] molecules in $M = 2$ boxes, and let us suppose that the molecules are labeled so that we can identify which molecule is in a particular box. We can distribute the labeled molecules in four ways, as shown in Fig. 4.1. These arrangements are called *microstates* because the molecules are labeled. For two molecules and two boxes, there are four possible microstates. However, a macroscopic perspective makes no distinction between which molecule is in which box. The only macroscopic characteristic that is important is how many particles are in a box, rather than which particle is in a certain box. For *macrostates,* we just need to keep track of *how many* particles are in a given box, not *which particles* are in a given box. It might help to think about connecting pressure gauges to the boxes. The pressure gauge could distinguish

---

3. The term "configurational" is occasionally used in different contexts. We apply the term in the context of Denbigh, K. 1981. *The Principles of Chemical Equilibrium,* 4th ed. Cambridge University Press, pp. 54–55. Technically, the configurational entropy includes both the combinatorial contribution discussed here for ideal gases, and the entropy departure function discussed in Unit II. Note that configurational energy is equivalent to the energy departure function of Chapter 7 because the change in energy of spatially rearranging ideal gas particles is zero.
4. The distinctions between these types of entropy are discussed in more detail by Denbigh, K. 1981. *The Principles of Chemical Equilibrium,* 4th ed. Cambridge University Press, pg. 353.
5. Ideal gases are non-interacting. Non-interacting particles are oblivious to the presence of other particles and the energy is independent of the interparticle separations. In other words, potential energies are ignored.

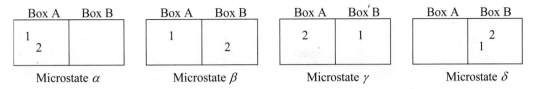

Figure 4.1 *Illustration of configurational arrangements of two molecules in two boxes, showing the microstates. Not that β and γ would have the same macroscopic value of pressure.*

between zero, one, and two particles in a box, but could not distinguish *which* particles are present. Therefore, microstates α and δ are different macrostates because the distribution of particles is different; however, microstates β and γ give the same macrostate. Thus, from our four microstates, we have only three macrostates.

To find out which arrangement of particles is most likely, we apply the "principle of equal *a priori* probabilities." This "principle" states that all microstates of a given energy are equally likely. Since all of the states we are considering for our non-interacting particles are at the same energy, they are all equally likely.[6] From a practical standpoint, we are interested in which macrostate is most likely. The probability of a macrostate is found by dividing the number of microstates in the given macrostate by the total number of microstates in all macrostates as shown in Table 4.1. For our example, the probability of the first macrostate is 1/4 = 0.25. The probability of the evenly distributed state is 2/4 = 0.5. That is, one-third of the macrostates possess 50% of the probability. The "most probable distribution" is the evenly distributed case.

Table 4.1 *Illustration of Macrostates for Two Particles and Two Boxes*

Macrostate		# of microstates	Probability of macrostate
# in box A	# in box B		
0	2	1	0.25
1	1	2	0.5
2	0	1	0.25

What happens when we consider more particles? It turns out that the total number of microstates for N particles in M boxes is M^N, so the counting gets tedious. For five particles in two boxes, the calculations are still manageable. There will be two microstates where all the particles are in one box or the other. Let us consider the case of one particle in box A and four particles in box B. Recall that the macrostates are identified by the number of particles in a given box, not by which particles are in which box. Therefore, the five microstates for this macrostate appear as given in Table 4.2(a).

---

6. Note that the number of particles and the energy are constant throughout the discussion presented here and the volume is specified at each stage. The constant energy for non-interacting particles means that the temperature will be constant; only the pressure will be reduced at larger volumes because it takes the molecules longer to get around the box and collide with a particular wall. We can think of this as an N, $\underline{V}$, U perspective, and we will demonstrate that entropy is maximized at equilibrium within this perspective, but some other quantity might characterize equilibrium if we held other quantities constant.

Table 4.2 *Microstates for the Second and Third Macrostates for Five Particles Distributed in Two Boxes*

(a) One particle in Box A		(b) Two particles in Box A			
Box A	Box B	Box A	Box B	Box A	Box B
1	2,3,4,5	1,2	3,4,5	2,4	1,3,5
2	1,3,4,5	1,3	2,4,5	2,5	1,3,4
3	1,2,4,5	1,4	2,3,5	3,4	1,2,5
4	1,2,3,5	1,5	2,3,4	3,5	1,2,4
5	1,2,3,4	2,3	1,4,5	4,5	1,2,3

The counting of microstates for putting two particles in box A and three in box B is slightly more tedious, and is shown in Table 4.2(b). It turns out that there are 10 microstates in this macrostate. The distributions for (three particles in A) + (two in B) and for (four in A) + (one in B) are like the distributions (two in A) + (three in B), and (one in A) + (four in B), respectively. These three cases are sufficient to determine the overall probabilities. There are $M^N = 2^5 = 32$ microstates total summarized in the table below.

Macrostate		# Microstates	Probability of Macrostate
Box A	Box B		
0	5	1	0.0313
1	4	5	0.1563
2	3	10	0.3125
3	2	10	0.3125
4	1	5	0.1563
5	0	1	0.0313

Note now that one-third of the macrostates (two out of six) possess 62.5% of the microstates. Thus, the distribution is now more peaked toward the most evenly distributed states than it was for two particles where one-third of the macrostates possessed 50% of the microstates. This is one of the most important aspects of the microscopic approach. As the number of particles increases, it won't be long before 99% of the microstates are in one-third of the macrostates. The trend will continue, and increasing the number of particles further will quickly yield 99% of the microstates in that one-tenth of the macrostates. In the limit as $N \rightarrow \infty$ (the "thermodynamic limit"), virtually all of the microstates are in just a few of the most evenly distributed macrostates, even though the system has a very slight finite possibility that it can be found in a less evenly distributed state. Based on the discussion, and considering the microscopic definition of entropy (Eqn. 4.2), entropy is maximized at equilibrium for a system of fixed energy and total volume.[7]

> With a large number of particles, the most evenly distributed configurational state is most probable, and the probability of any other state is small.

---

7. In an isolated system at constant (U, V), entropy will be generated as equilibrium is approached; S will increase and will be maximized at equilibrium. If the system is closed but not isolated, the property which is minimized is determined by the property which is a natural function of the controlled variables: H is minimized for constant (S,P); A for constant (T,V); G for constant (T,P). A and G will be introduced in future chapters.

Generalized Microstate Formulas

To extend the procedure for counting microstates to large values of N ($\sim 10^{23}$), we cannot imagine listing all the possibilities and counting them up. It would require 40 years to simply count to 10^9 if we did nothing but count night and day. We must systematically analyze the probabilities as we consider configurations and develop model equations describing the process.

How do we determine the number of microstates for a given macrostate for large N? For the first step in the process, it is fairly obvious that there are N ways of moving one particle to box B, i.e., 1 came first, or 2 came first, and so on, which is what we did to create Table 4.2(a). However, counting gets more complicated when we have two particles in a box. Since there are N ways of moving the first particle to box B, and there are $(N-1)$ particles left, we begin with the same logic for the $(N-1)$ remaining particles. For example, with five particles, there would then be five ways of placing the first particle, then four ways of placing the second particle for a total of 20 possible ways of putting two particles in box B. One way of writing this would be $5 \cdot 4$, which is equivalent to $(5 \cdot 4 \cdot 3 \cdot 2 \cdot 1)/(3 \cdot 2 \cdot 1)$, which can be generalized to $N!/(N-m)!$, where m is the number of particles we have placed in the first box.[8] ($N!$ is read "N factorial," and calculated as $N \cdot (N-1) \cdot (N-2) \ldots \ldots \cdot 2 \cdot 1$). Our formula gives 20 ways, but Table 4.2(b) shows only 10 ways. What are we missing? Answer: When we count this way, we are implicitly double counting some microstates. Note in Table 4.2(b) that although there are two ways that we could put the first particle in box B, the order in which we place them does not matter when we count microstates. Therefore, using $N!/(N-m)!$ implicitly distinguishes between the order in which particles are placed. For counting microstates, the history of how a particular microstate was achieved does not interest us. Therefore, we say there are only 10 *distinguishable* microstates.

It turns out that it is fairly simple to correct for this overcounting. For two particles in a box, they could have been placed in the order 1-2, or in the order 2-1, which gives two possibilities. For three particles, they could have been placed 1-2-3, 1-3-2, 2-1-3, 2-3-1, 3-1-2, 3-2-1, for six possibilities. For m particles in a box, without correction of the formula, we overcount by $m!$. Therefore, we modify the above formula by dividing by $m!$ to correct for overcounting. Finally, the number of microstates for arranging N particles in two boxes, with m particles in one of the boxes, is:[9]

$$p_j = \frac{N_j!}{m_j!(N_j - m_j)!} \qquad 4.3$$

The general formula for M boxes is:[10]

$$p_j = \frac{N_j!}{\displaystyle\prod_{i=1}^{M} m_{ij}!} \qquad 4.4$$

❶ Factorials are a quick tool for counting arrangements.

❶ General formula for number of microstates for N particles in M boxes.

8. In statistics, this is called the number of permutations.
9. The formula for the particles in boxes is an example of a *binomial coefficient*, fundamental in the study of probability and statistics. Detailed development of the binomial distribution and the issue of indistinguishability can be found in any textbook or handbook on the subject.
10. In statistics this is called the number of combinations. It is also known as the *multinomial coefficient*.

m_{ij} is the number of particles in the ith box at the jth macrostate. We will not derive this general formula, but it is a straightforward extension of the formula for two boxes which was derived above. Therefore, with 10 particles, and three in the first box, two in the second box and five in the third box, we have $10!/(3!2!5!) = 3,628,800/(6 \cdot 2 \cdot 120) = 2520$ microstates for this macrostate.

Recall the microscopic definition of entropy given by Eqn. 4.2. Let us use it to calculate the entropy change for an ideal gas due to an isothermal change in volume. The statistics we have just derived will apply since an ideal gas consists of non-interacting particles whose energy is independent of their nearest neighbors. During an expansion like that described, the energy is constant because the system is isolated. Therefore, the temperature is also constant because $dU = C_V\, dT$ for an ideal gas.

Entropy and Isothermal Volume/Pressure Change for Ideal Gases

Suppose an insulated container, partitioned into two equal volumes, contains N molecules of an ideal gas in one section and no molecules in the other. When the partition is withdrawn, the molecules quickly distribute themselves uniformly throughout the total volume. How is the entropy affected? Let subscript 1 denote the initial state and subscript 2 denote the final state. Here we take for granted that the final state will be evenly distributed.

We can develop an answer by applying Eqn. 4.4, and noting that $0! = 1$:

$$p_1 = \frac{N!}{N!0!} = 1; \qquad p_2 = \frac{N!}{(N/2)!(N/2)!}; \qquad \ln(p_2/p_1) = \ln\left(\frac{N!/[(N/2)!]^2}{1}\right)$$

Substituting into Eqn. 4.2, and recognizing $\ln\left[\left(\frac{N}{2}\right)!\right]^2 = 2\ln\left(\frac{N}{2}\right)!$,

$$\Delta \underline{S} = \underline{S}_2 - \underline{S}_1 = k\ln(p_2/p_1) = k\{\ln(N!) - 2\ln[(N/2)!]\}$$

Stirling's approximation may be used for $\ln(N!)$ when $N > 100$,

❶ Stirling's approximation.

$$\boxed{\ln(N!) \approx N\ln(N) - N} \tag{4.5}$$

The approximation is a mathematical simplification, and not, in itself, related to thermodynamics.

$$\Rightarrow \Delta \underline{S} = k[N\ln(N) - N - 2(N/2)\ln(N/2) + 2(N/2)]$$
$$= k[N\ln(N) - N - N\ln(N) + N\ln(2) + N]$$
$$= kN\ln(2) \Rightarrow \Delta \underline{S} = nR\ln(2)$$

❶ Entropy of a constant temperature system increases when volume increases.

Therefore, entropy of the system has increased by a factor of $\ln(2)$ when the volume has doubled at constant T. Suppose the box initially with particles is three times as large as the empty box. In this case the increase in volume will be 33%. Then what is the entropy change? The trick is to imagine four equal size boxes, with three equally filled at the beginning.

$$p_1 = \frac{N!}{[(N/3)!]^3 0!}; \qquad p_2 = \frac{N!}{[(N/4)!]^4}$$

A similar application of Stirling's approximation gives,

$$\Delta \underline{S} = k \ln \left\{ \frac{N!}{[(N/4)!]^4} \cdot \frac{[(N/3)!]^3 0! 11}{N!} \right\} = -k \left\{ N \ln \left(\frac{N}{4}\right) - N - N \ln \left(\frac{N}{3}\right) + N \right\} = nR \ln \left(\frac{4}{3}\right)$$

We may generalize the result by noting the pattern with this result and the previous result,

$$(\Delta S)_T = R \ln \left[\frac{V}{V^i} \right]$$

(ig) 4.6

where the subscript T indicates that this equation holds at constant T. For an isothermal ideal gas, we also may express this in terms of pressure by substituting $V = RT/P$ in Eqn. 4.6

$$(\Delta S)_T = -R \ln \left[\frac{P}{P^i} \right]$$

(ig) 4.7

> Formulas for *isothermal* entropy changes of an ideal gas.

Therefore, the entropy decreases when the pressure increases isothermally. Likewise, the entropy decreases when the volume decreases isothermally. These concepts are extremely important in developing an understanding of entropy, but by themselves, are not directly helpful in the initial objective of this chapter—that of determining inefficiencies and maximum work. The following example provides an introduction to how these conceptual points relate to our practical objectives.

Example 4.1 Entropy change and "lost work" in a gas expansion

An isothermal ideal gas expansion produces maximum work if carried out reversibly and less work if friction or other losses are present. One way of generating "other losses" is if the force of the gas on the piston is not balanced with the opposing force during the expansion, as shown in part (b) below. Consider a piston/cylinder containing one mole of nitrogen at 5 bars and 300 K is expanded isothermally to 1 bar.

(a) Suppose that the expansion is reversible. How much work could be obtained and how much heat is transferred? What is the entropy change of the gas?
(b) Suppose the isothermal expansion is carried out irreversibly by removing a piston stop and expanding against the atmosphere at 1 bar. Suppose that heat transfer is provided to permit this to occur isothermally. How much work is done by the gas and how much heat is transferred? What is the entropy change of the gas? How much work is lost compared to a reversible isothermal process and what percent of the reversible work is obtained (the efficiency)?

Solution:
Basis: 1 mole, closed unsteady-state system.
(a) The energy balance for the piston/cylinder is $\Delta \underline{U} = Q + \underline{W}_{EC} = 0$ because the gas is isothermal and ideal. $d\underline{W}_{EC} = -Pd\underline{V} = -(nRT/V)dV$; $\underline{W}_{EC} = -nRT \ln(V_2/V_1) = -nRT \ln(P_1/P_2) = -(1)8.314(300)\ln(5) = -4014 J$. By the energy balance $Q = 4014 J$.
The entropy change is by Eqn. 4.7, $\Delta \underline{S} = -nR \ln(P_2/P_1) = -(1)8.314 \ln(1/5) = 13.38$ J/K.

Example 4.1 Entropy change and "lost work" in a gas expansion (Continued)

(b) The energy balance does not depend on whether the work is reversible and is the same. Taking the atmosphere as the system, the work is $\underline{W}_{EC,atm} = -P_{atm}(\underline{V}_{2,atm}-\underline{V}_{1,atm}) = -\underline{W}_{EC} = -P_{atm}(\underline{V}_1-\underline{V}_2) = P_{atm}(nRT/P_2-nRT/P_1) = nRT(P_{atm}/P_2-P_{atm}/P_1) => \underline{W}_{EC} = nRT(P_{atm}/P_1-P_{atm}/P_2) = (1)8.314(300)(1/5-1) = -1995\text{J}, Q = 1995\text{J}$.
The entropy change depends on only the state change and this is the same as (a), 13.38 J/K. The amount of lost work is $\underline{W}_{lost} = 4014 - 1995 = 2019\text{J}$, the percent of reversible work obtained (efficiency) is $1995/4014 \cdot 100\% = 49.7\%$.

An important point is suggested by Example 4.1, even though the example is limited to ideal gas constraints. We saw that the isothermal entropy change for the gas was the same for the reversible and irreversible changes because the gas state change was the same. Though Eqn. 4.7 is limited to ideal gases, the relation between entropy changes and state changes is generalizable as we prove later. We will show later that case (b) always generates more entropy.

Entropy of Mixing for Ideal Gases

Mixing is another important process to which we may apply the statistics that we have developed. Suppose that one mole of pure oxygen vapor and three moles of pure nitrogen vapor at the same temperature and pressure are brought into intimate contact and held in this fashion until the nitrogen and oxygen have completely mixed. The resultant vapor is a uniform, random mixture of nitrogen and oxygen molecules. Let us determine the entropy change associated with this mixing process, assuming ideal-gas behavior.

Since the T^i and P^i of both ideal gases are the same, $V^i_{N2} = 3V^i_{O2}$ and $V^i_{tot} = 4V^i_{O2}$. Ideal gas molecules are point masses, so the presence of O_2 in the N_2 does not affect anything as long as the pressure is constant. The main effect is that the O_2 now has a larger volume to access and so does N_2. The component contributions of entropy change versus volume change can be simply added. Entropy change for O_2:

$$\Delta\underline{S} = n_{O2}R\ln(4) = n_{tot}R[-x_{O2}\ln(0.25)] = n_{tot}R[-x_{O2}\ln(x_{O2})]$$

Entropy change for N_2:

$$\Delta\underline{S} = n_{N2}R\ln\left(\frac{4}{3}\right) = n_{tot}R[-x_{N2}\ln(0.75)] = n_{tot}R[-x_{N2}\ln(x_{N2})]$$

Entropy change for total fluid:

$$\Delta\underline{S} = -n_{tot}R[x_{O2}\ln(x_{O2})+x_{N2}\ln(x_{N2})] = -4R(-0.562) = 18.7 \text{ J/K}$$

 Ideal entropy of mixing.

$$\boxed{\Delta S^{is}_{mix} = -R\sum_i x_i\ln x_i} \quad \text{In general, ideal mixing.} \qquad 4.8$$

This is an important result as it gives the entropy change of mixing for non-interacting particles. Remarkably, it is also a reasonable approximation for ideal solutions where energy and total volume do not change on mixing. This equation provides the underpinning for much of the discussion of mixtures and phase equilibrium in Unit III.

The entropy of a mixed ideal gas or an **ideal solution**, here both denoted with a superscript *"is:"*

$$\underline{S}^{is} = \sum_i n_i S_i + \Delta\underline{S}_{mix} = \sum_i n_i S_i - R \sum_i n_i \ln x_i \quad \text{or} \quad S^{is} = \sum_i x_i S_i - R \sum_i x_i \ln x_i \qquad 4.9$$

Note that these equations apply to ideal gases if we substitute y for x. In this section we have shown that a system of ideal gas molecules at equilibrium is most likely to be found in the most randomized (distributed) configuration because this is the macrostate with the largest fraction of microstates. In other words, the entropy of a state is maximized at equilibrium for a system of fixed $\underline{U}$, $\underline{V}$, and N.

But how can temperature be related to disorder? We consider this issue in the next subsection.

Entropy and Temperature Change: Thermal Entropy

One key to understanding the connection between thermal entropy and disorder is the appreciation that energy is quantized. Thus, there are discrete energy levels in which particles may be arranged. These energy levels are analogous to the boxes in the spatial distribution problem. The effect of increasing the temperature is to increase the energy of the molecules and make higher energy levels accessible.

To see how this affects the entropy, consider a system of three molecules and three energy levels ε_0, $\varepsilon_1 = 2\varepsilon_0$, $\varepsilon_3 = 3\varepsilon_0$. Suppose we are at a low temperature and the total energy is $\underline{U} = 3\varepsilon_0$. The only way this can be achieved is by putting all three particles in the lowest energy level. The other energy levels are not accessible, and $\underline{S} = \underline{S}_0$. Now consider raising the temperature to give the system $\underline{U} = 4\varepsilon_0$. One macrostate is possible (one molecule in ε_1 and two in ε_0), but there are now three microstates, $\Delta\underline{S} = k\ln(3)$. Can you show when $\underline{U} = 6\varepsilon_0$ that the macrostate with one particle in each level results in $\Delta\underline{S} = k\ln(6)$? Real systems are much larger and the molecules are more complex, but the same qualitative behavior is exhibited: increasing T increases the accessible energy levels which increases the microstates, increasing entropy.

We can advance our understanding of thermal effects on entropy by contemplating the Einstein solid.[11] Albert Einstein's (1907) proposal was that a solid could be treated as a large number of identical vibrating monatomic sites, modeling the potential energies as springs that follow Hooke's law. The quantum mechanical analog to the energy balance is known as Shrödinger's equation, which relates the momentum (kinetic energy) and potential energy. An exact solution is possible only for equally spaced energy levels, known as quantum levels. The equally spaced quantized states for each oscillator are separated by hf where h is Planck's constant and f is the frequency of the oscillator. Thus, the system is described as a system of harmonic oscillators. Assuming that each oscillator and each dimension (x,y,z) is independent, we can develop an expression for the internal energy and the heat capacity.

Albert Einstein (1879 – 1955) was a German-born physicist. He contributed to an understanding of quantum behavior and the general theory of relativity. He was awarded the 1921 Nobel Prize in physics.

The Einstein solid model was one of the earliest and most convincing demonstrations of the limitations of classical mechanics. It serves today as a simple illustration of the manner in which quantum mechanics and statistical mechanics lead to consistent and experimentally verifiable descriptions of thermodynamic properties. The assumptions of the Einstein solid model are as follows:

11. Moore, Schroeder, 1997. *Am. J. Phys.* 65:26–36.

- The total energy in a solid is the sum of M harmonic oscillator energies where $M/3$ is the number of atoms because the atoms oscillate independently in three dimensions. Since the energy of each oscillator is quantized, we can say that the total internal energy is

$$\underline{U} = q_M \varepsilon_q + M \varepsilon_q / 2 \qquad \text{for } M/3 \text{ atoms} \qquad \qquad 4.10$$

where ε_q is the (constant) energy step for each quantum level, and q_M gives the total quantum multiplier for all oscillator quantum energies added. The term $M\varepsilon_q/2$ represents the ground state energy that oscillators have in the lowest energy level. It is often convenient to relate the energy to the average quantum multiplier,

$$<q_M> = q_M/M = \underline{U}/M\varepsilon_q - 1/2 \qquad \qquad 4.11$$

- Each oscillator in each dimension is independent, so we can allocate integer multiples of ε_q to any oscillator as long as the total sum of multipliers is q_M. Each independent specification represents a microstate. For $M = 3$ oscillators, (3,1,1) specifies three units of energy in the first oscillator and one unit in each of the other two for a total of $q_M = 5$, $\underline{U} = 5\varepsilon_q + 3\varepsilon_q/2$.
- Raising the magnitude of q_M (by adding heat to raise T) makes more microstates accessible, increasing the entropy.

For $q_M=3$ units of energy distributed in an Einstein solid with $M=4$ oscillators, below is the detailed listing of the possible distributions of the energy, a total of 20 different distributions for three units of energy among four oscillators (a "multiplicity" of 20).

State	1	2	3	4	5	6	7	8	9	10	11	12	13	14	15	16	17	18	19	20
Osc1	3	0	0	0	2	2	2	1	1	1	0	0	0	0	0	0	0	1	1	1
Osc2	0	3	0	0	1	0	0	2	0	0	2	2	1	1	0	0	1	0	1	1
Osc3	0	0	3	0	0	1	0	0	2	0	1	0	2	0	2	1	1	1	0	1
Osc4	0	0	0	3	0	0	1	0	0	2	0	1	0	2	1	2	1	1	1	0

If we are trying to develop a description of a real solid with Avogadro's number of oscillators, enumeration is clearly impractical. Fortunately, mathematical expressions for the multiplicity make the task manageable. Callen[12] gives the general formula for the number of microstates as $p_i = (q_{Mi}+M_i-1)!/[q_{Mi}!(M_i-1)!]$. There is a clever way to understand this formula. Instead of distributing q_{Mi} quanta among M oscillator "boxes," consider that there are $M_i - 1$ "partitions" between oscillator "boxes." In the table above, there are four oscillators, but there are three row boundaries. Consider that the quanta can be redistributed by all the permutations of the particles *and* boundaries $(q_{Mi}+M_i-1)!$. However, the permutations overcount in that the q_{Mi} are indistinguishable, so we divide by $q_{Mi}!$, and that the (M_i-1) boundaries are indistinguishable, so we divide by $(M_i-1)!$. To apply the formula for $q_M=2$, $M=2$:

Instance	Osc1	Osc2
1	2	0
2	0	2
3	1	1

$$p_i = (2+2-1)!/[2!(1!)] = 3!/2! = 3. \text{ Check.}$$

12. Callen, H.B. 1985. *Thermodynamics and an Introduction to Thermostatistics*, 2ed, Indianapolis IN: Wiley, p.333.

For the case with $q_M = 3$ and $M = 4$, $p_i = (3+4-1)!/[3!(3!)] = 20$, as enumerated above.

Example 4.2   Stirling's approximation in the Einstein solid

(a) Show that Callen's formula is consistent with enumeration for:
   (1) $q_M = 3$, $M = 3$; (2) $q_M = 4$, $M = 3$
(b) Use the general formula to develop an expression for $\underline{S} = \underline{S}(q_M, M)$ when $M > 100$. Express the answer in terms of the average quantum multiplier, $<q_M>$.
(c) Plot $\underline{S}/Mk$ versus $<q_M>$. What does this indicate about entropy changes when heat is added?

Solution:

(a) (1): (3,0,0),(0,3,0),(0,0,3),(2,1,0),(2,0,1),(1,2,0),(1,0,2),(0,2,1),(0,1,2),(1,1,1) = 10. Check.

   (2): (4,0,0), (0,4,0), (0,0,4), (3,1,0), (3,0,1), (1,3,0), (1,0,3), (0,3,1), (0,1,3), (2,1,1),
   (1,2,1), (1,1,2), (2,0,2), (2,2,0), (0,2,2) = 15. Check.

(b) $\underline{S}_i = k \ln(p_i) = k \{ \ln[(q_{Mi}+M_i-1)!] - \ln[q_{Mi}!(M_i-1)!] \}$. Applying Stirling's approximation,

$$\underline{S}/k = [(q_M+M-1)\ln(q_M+M-1)-(q_M+M-1)] - q_M\ln q_M + q_M - (M-1)\ln(M-1) + M-1 =$$
$$\underline{S}/Mk = (q_M/M)\ln[(q_M+M-1)/q_M] + [(M-1)/M]\ln[(q_M+M-1)/(M-1)]$$

$$\underline{S}/Mk = <q_M>\ln(1+1/<q_M>) + \ln(<q_M>+1)$$

(c) Fig. 4.2 shows that $\underline{S}$ increases with $<q_M> = q_M/M$ ($=\underline{U}/M\varepsilon_q - \frac{1}{2}$). When T increases, $\underline{U}$ will increase, meaning that $<q_M>$ and $\underline{S}$ increase. It would be nice to relate the change in entropy quantitatively to the change in temperature, but a complete analysis of the entire temperature range requires advanced derivative manipulations that distracts from the main concepts at this stage. We return to this problem in Chapter 6.

You should notice that we represented the interactions of the monatomic sites as if they were connected by Hooke's law springs in the solid phase. They are not rigorously connected this way, but the simple model approximates the behavior of two interacting molecules in a potential energy

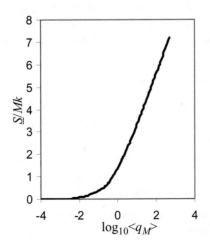

Figure 4.2 *Entropy of the Einstein solid with increasing energy and T as explained in Example 4.2.*

well. If the spring analogy were exact, the potential energy well would be a parabola (thus a harmonic oscillator). Look back at the Lennard-Jones potential in Chapter 1 and you can see that the shape is a good approximation if the atoms do not vibrate too far from the minimum in the well. The Einstein model gives qualitatively the right behavior, but the Debye model that followed in 1912 is more accurate because it represents collective waves moving through the solid. We omit discussion of the Debye model because our objectives are met with the Einstein model.[13] We briefly extend the concept of the Einstein model in Chapter 6 where we develop more powerful methods for manipulation of derivatives.

In the present day, the subtle relations between entropy and molecular distributions are complex but approachable. Imagine how difficult gaining this understanding must have been for Boltzmann in 1880, before the advent of quantum mechanics. Many scientists at the time refused even to accept the existence of molecules. Trying to explain to people the nature and significance of his discoveries must have been extremely frustrating. What we know for sure is that Boltzmann drowned himself in 1903. Try not to take your frustrations with entropy quite so seriously.

Test Yourself

1. Does molar entropy increase, decrease, or stay about the same for an ideal gas if: (a) volume increases isothermally?; (b) pressure increases isothermally?; (c) temperature increases isobarically?; (d) the gas at the vapor pressure condenses?; (e) two pure gas species are mixed?

2. Does molar entropy increase, decrease, or stay about the same for a liquid if: (a) temperature increases isobarically?; (b) pressure increases isothermally?; (c) the liquid evaporates at the vapor pressure?; (d) two pure liquid species are mixed?

4.3   THE MACROSCOPIC VIEW OF ENTROPY

In the introduction to this chapter, we alluded to the relation between entropy and maximum process efficiency. We have shown that entropy changes with volume (pressure) and temperature. How can we use entropy to help us determine maximum work output or minimum work input? The answer is best summarized by a series of statements. These statements refer to the entropy of the universe. This does not mean that we imagine measuring entropy all over the universe. It simply means that entropy may decrease in one part of a system but then it must increase at least as much in another part of a system. In assessing the reversibility of the overall system, we must sum all changes for each process in the overall system and its surroundings. This is the relevant part of the universe for our purposes.

Molar or specific entropy is a *state* property which will assist us in the following ways.

1. Irreversible processes will result in an increase in entropy of the *universe*. (Irreversible processes will result in entropy generation.) Irreversible processes result in loss of capability for performing work.

2. Reversible processes result in no increase in entropy of the *universe*. (Reversible processes result in zero entropy generation. This principle will be useful for calculation of maximum work output or minimum work input for a process.)

3. Proposed processes which would result in a decrease of entropy of the universe are impossible. (Impossible processes result in negative entropy generation.)

---

13. The Debye model is described by McQuarrie, D.A. 1976. *Statistical Mechanics*. Harper and Row.

These three principles are summarized in the second law of thermodynamics: Reversible processes and/or optimum work interactions occur without entropy generation, and irreversible processes result in entropy generation. The microscopic descriptions in the previous section teach us very effectively about the relation between entropy and disorder, but it is not fair to say that any increase in volume results in a loss of potentially useful work when the entropy of the system increases. (Note that it is the entropy change of the *universe* that determines irreversibility, not the entropy change of the *system*.) After all, the only way of obtaining any expansion/contraction work is by a change in volume. To understand the relation between lost work and volume change, we must appreciate the meaning of reversibility, and what types of phenomena are associated with entropy generation. We will explore these concepts in the next sections.

❶ The entropy balance is the second law of thermodynamics.

Entropy Definition (Macroscopic)

Let us define the differential change in entropy of a *closed* simple system by the following equation:

$$dS \equiv \frac{dQ_{rev}}{T_{sys}}$$

4.12

For a change in states, both sides of Eqn. 4.12 may be integrated,

$$\Delta S = \int_{state\ 1}^{state\ 2} \frac{dQ_{rev}}{T_{sys}}$$

4.13

where the following occurs:

1. The entropy change on the left-hand side of Eqn. 4.13 is dependent only on states 1 and 2 and not dependent on reversibility. However, to calculate the entropy change via the integral, the *integral* may be evaluated along any convenient *reversible* pathway between the actual states.

2. T_{sys} is the temperature of the system boundary where heat is transferred. Only if the system boundary temperature is constant along the pathway may this term be taken out of the integral sign.

A change in entropy is completely characterized for a pure single-phase fluid by any other two state variables. It may be surprising that the integral is independent of the path since Q is a path-dependent property. The key is to understand that the right-hand side integral is independent of the path, *as long as the path is reversible.* Thus, a process between two states does not need to be reversible to permit calculation of the entropy change, since we can evaluate it along *any* reversible path of choice. If the actual path is reversible, then the actual heat transfer and pathway may be used. If the process is *irreversible,* then any *reversible* path may be constructed for the calculation.

❶ Entropy is a state property. For a pure single-phase fluid, specific entropy is characterized by two state variables (e.g., T and P).

Look back at Example 4.1 and note that the entropy change for the isothermal process was calculated by the microscopic formula. However, look at the process again from the perspective of Eqn. 4.13 and subsequent statement 2. Because the process was isothermal, the entropy change can be calculated $\Delta S = Q_{rev}/T_{sys}$; try it! Note that the irreversible process in that example exhibits the same entropy change, calculated by the reversible pathway.

Now, look back at the integral of Eqn. 4.13 and consider an adiabatic, reversible process; the process will be **isentropic** (constant entropy, $\Delta S = 0$). Let us consider how the entropy can be used as a state property to identify the final state along a reversible adiabatic process. Further, the property does not depend on the limitations of the ideal gas law. The ideal gas law was convenient to introduce the property. Consider the adiabatic reversible expansion of steam, a non-ideal gas. We can read S values from the steam tables.

Example 4.3 Adiabatic, reversible expansion of steam

Steam is held at 450°C and 4.5 MPa in a piston/cylinder. The piston is adiabatically and reversibly expanded to 2.0 MPa. What is the final temperature? How much reversible work can be done?

Solution: The T, P are known in the initial state, and the value of S can be found in the steam tables. Steam is not an ideal gas, but by Eqn. 4.12, the process is isentropic because it is reversible and adiabatic. From the steam tables, the entropy at the initial state is 6.877 kJ/kgK. At 2 MPa, this entropy will be found between 300°C and 350°C. Interpolating,

$$T = 300 + \frac{6.877 - 6.7684}{6.9583 - 6.7684}(350 - 300) = 300 + 0.572(350 - 300) = 329°C$$

The P and $S^f = S^i$ are known in the final state and these two state properties can be used to find all the other final state properties. The work is determined by the energy balance: $\Delta U = Q + W_{EC}$. The initial value of U is 3005.8 kJ/kg. For the final state, interpolating U by using S^f at P^f, $U = 2773.2 + 0.572(2860.5 - 2773.2) = 2823.1$ kJ/kg, so

$$W_{EC} = (2823.1 - 3005.8) = -182.7 \text{ kJ/kg}$$

Let us revisit the Carnot cycle of Section 3.1 in light of this new state property, entropy. The Carnot cycle was developed with an ideal gas, but it is possible to prove that the cycle depends only on the combination of two isothermal steps and two adiabatic steps, not the ideal gas as the working fluid.[14] Because the process is cyclic, the final state and initial state are identical, so the overall entropy changes of the four steps must sum to zero, $\Delta S = 0$. Because the reversible, adiabatic steps are isentropic, $\Delta S = 0$, the entropy change for the two isothermal steps must sum to zero. As we discussed above, for an isothermal step Eqn. 4.13 becomes $\Delta S = Q_{rev}/T$. Therefore, an analysis of the Carnot cycle from the viewpoint of entropy is

$$\left(\Delta S = 0 = \frac{Q_H}{T_H} + \frac{Q_C}{T_C}\right) \Rightarrow \frac{Q_C}{Q_H} = \frac{-T_C}{T_H} \quad \text{Carnot cycle} \qquad 4.14$$

This can be inserted into the formula for Carnot efficiency, Eqn. 3.6. Note that this relation is not constrained to an ideal gas! In fact, there are only three constraints for this balance: The process is cyclic; all heat is absorbed at T_H; all heat is rejected at T_C. Example 4.4 shows how the Carnot cycle can be performed with steam including phase transformations.

---

14. Uffink, J. 2003. "Irreversibility and the Second Law of Thermodynamics," Chapter 7, in *Entropy*, Greven, A., Keller, G., Warnecke, G. eds., Princeton, NJ: Princeton University Press.

Example 4.4 A Carnot cycle based on steam

Fig. 4.3 shows the path of a Carnot cycle operating on steam in a continuous cycle that parallel the two isothermal steps and two adiabatic steps of Section 3.1. First, saturated liquid at 5 MPa is boiled isothermally to saturated vapor in step ($a{\to}b$). In step ($b{\to}c$), steam is adiabatically and reversibly expanded from saturated vapor at 5 MPa to 1 MPa. In ($c{\to}d$), heat is isothermally removed and the volume drops during condensation. Finally, in step ($d{\to}a$), the steam is adiabatically and reversibly compressed to 5 MPa and saturated liquid. (Hint: Challenge yourself to solve the cycle without looking at the solution.).

(a) Compute $W(a{\to}b)$ and Q_H.
(b) Compute $W(b{\to}c)$.
(c) Compute $W(d{\to}a)$. (The last step in the cycle).
(d) Compute $W(c{\to}d)$ and Q_C. (The third step in the cycle).
(e) For the cycle, compute the thermal efficiency by $\eta_\theta = -W_{net}/Q_H$ and compare to Carnot's efficiency, $\eta_\theta = (T_H - T_L)/T_H$.

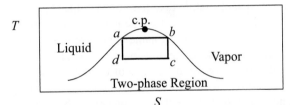

Figure 4.3 *A T-S diagram illustrating a Carnot cycle based on steam.*

Solution: The entropy change is zero for the expansion and compression steps because these steps are adiabatic and reversible, as indicated by the vertical line segments in Fig. 4.3.

(a) E-balance: fixed P,T vaporization, $Q_H = \Delta U - W_{EC} = (\Delta U + P\Delta V) = \Delta H^{vap} = 1639.57$ J/kg; $W_{EC}(a{\to}b) = -P\Delta V = -5(0.0394 - 0.001286){*}1000 = -191.1$ J/kg.

(b) E-balance: isentropic, $W_{EC}(b{\to}c) = \Delta U$; $U_b = U(\text{sat. vap., 5MPa}) = 2596.98$ kJ/kg; S-balance: $\Delta S = 0$; $S_c = S_b = 5.9737$ kJ/kg-K$= q_c(6.5850) + (1 - q_c)2.1381$; $q_c = 0.8625$; $U_c = 0.8625(2582.75) + (1 - 0.8625)761.39 = 2332.31$ kJ/kg; $W_{EC}(b{\to}c) = 2332.31 - 2596.98 = -264.67$ kJ/kg.

(c) This is the last step. E-balance: isentropic. $W_{EC}(d{\to}a) = \Delta U$; $U_a = U(\text{sat. liq., 5MPa}) = 1148.21$ kJ/kg; the quality at state d is not known, but we can use the entropy at state a to find it. S-balance: $\Delta S = 0$; $S_d = S_a = 2.9210$ kJ/kg-K$= q_d(6.5850) + (1 - q_d)2.1381$; $q_d = 0.1761$; $U_d = 0.1761(2582.75) + (1 - 0.1761)761.39 = 1082.13$ kJ/kg; $W_{EC}(d{\to}a) = 1148.21 - 1082.13 = 66.08$ kJ/kg.

Example 4.4   A Carnot cycle based on steam   (Continued)

(d) This is the third step using the quality for d calculated in part (c). This is a fixed T,P condensation. E-balance: $Q_C = \Delta U - W_{EC} = (\Delta U + P\Delta V) = \Delta H$; $H_d = 762.52 + 0.1761(2014.59) = 1117.29$ kJ/kg; $H_c = 762.52 + 0.8625(2014.59) = 2500.10$ kJ/kg; $Q_C = H_d - H_c = -1382.81$ kJ/kg

$W_{EC} = P\Delta V$; $V_c = 0.001127(1 - 0.8625) + 0.8625(0.1944) = 0.1678$ m^3/kg $= 167.8$ cm^3/g

$V_d = 0.001127(1 - 0.1761) + 0.1761(0.1944) = 0.0352$ m^3/kg $= 35.2$ cm^3/g

$W(c \rightarrow d) = -1.0(35.2 - 167.8) = 132.6$ MPa-cm^3/g $= 132.6$ kJ/kg

(e) $\eta_\theta = -W_{net}/Q_H$; $W_{net} = (-264.67 + 66.08 - 191.1 + 132.6) = -257.1$ kJ/kg;

$\eta_\theta = 257.1/1639.57 = 0.157$; η_θ(Carnot) $= (263.94 - 179.88)/(263.94 + 273.15) = 0.157$.
The actual cycle matches the Carnot formula. Note that the cyclic nature of this process means that we could have computed more quickly by

$W_{net} = -(Q_C + Q_H) = 1382.21 - 1639.57 = -257.4$ kJ/kg.

The macroscopic view of entropy can be bewildering when first studied because students strive to understand the physical connection. The microscopic view of entropy is helpful for some students but a significant disconnect often persists. In either case, if you review the definitions at the beginning of the chapter, both are mathematical, not physical. In Chapter 2 we discussed that Q was a path-dependent property, but now we are demonstrating that dividing the quantity by T and integrating along a path results in a quantity that is independent of the path![15] Some students find it helpful to accept this as a mathematical relationship with the name of "entropy." In fact, development of the macroscopic definition of entropy was not obvious to the scientists who eventually proved it to be a state property. The scientific literature at the time of Carnot can be confusing to read because the realization of the state nature of the integral was not obvious, but was developed by a significant amount of diligence and insight by the scientists of the day.

Note that entropy does not depend directly on the work done on a system. Therefore, it may be used to decouple heat and work in the energy balance for reversible processes. Also for reversible processes, entropy provides a second property that may be used to determine unknowns in a process as we have seen in the previous two examples. In fact, the power of this property is that it can be used to evaluate reversibility of processes, and such understanding is critical as we search sustainable energy management practices, such as CO_2 sequestration. Let us investigate some convenient pathways for the evaluation of entropy changes before we develop examples that utilize the pathways.

> ❶ Entropy can be used to decouple heat and work.

Calculation of Entropy Changes in Closed Systems

As with enthalpy and internal energy, tables and charts are useful sources for entropy information for common fluids. Note the tabulation of S in the tables of Appendix E. These tables and charts are calculated using the definition of entropy and procedures for non-ideal fluids that we will discuss in upcoming chapters.

For manual calculations of entropy, we develop some simple procedures here, and more rigorous procedures in Chapter 8. Since the integral of Eqn. 4.13 must be evaluated along a reversible path, let us consider some easy choices for paths. For a closed *reversible* system without shaft work, the energy balance in Eqn. 2.16 becomes,

---

15. Note, however, that Q is constrained to be reversible in the macroscopic definition, so it is not entirely arbitrary.

$$d\left[U + \frac{v^2}{2g_c} + \frac{g}{g_c}z \right] = dQ_{rev} + dW_S + dW_{EC} \qquad\qquad 4.15$$

Inserting Eqn. 2.3,

$$[dU + PdV] = dQ_{rev} \qquad\qquad 4.16$$

We now consider how this equation may be substituted in the integral of Eqn. 4.13 for calculating entropy changes in several situations.

Constant Pressure (Isobaric) Pathway

Many process calculations involve state changes at constant pressure. Recognizing $H = U + PV$, $dH = dU + PdV + VdP$. In the case at hand, dP happens to be zero; therefore, Eqn. 4.16 becomes

$$dH = dQ_{rev} \qquad\qquad 4.17$$

Since $dH = C_P dT$ at constant pressure, along a *constant-pressure pathway*, substituting for dQ_{rev} in Eqn. 4.13, the entropy change is

$$(dS)_P = \frac{C_P}{T}(dT)_P \qquad\qquad 4.18$$

$$\boxed{\Delta S = \int_{T_1}^{T_2} \frac{C_P}{T}dT} \qquad\qquad 4.19 \quad \textbf{❗} \text{ Constant pressure.}$$

Constant Volume Pathway

For a constant volume pathway, Eqn. 4.16 becomes

$$dU = dQ_{rev} \qquad\qquad 4.20$$

Since $dU = C_V dT$ along a *constant volume pathway*, substituting for dQ_{rev} in Eqn. 4.13, the entropy change is

$$\boxed{\Delta S = \int_{T_1}^{T_2} \frac{C_V}{T}dT} \qquad\qquad 4.21 \quad \textbf{❗} \text{ Constant volume.}$$

Constant Temperature (Isothermal) Pathway

The behavior of entropy at constant temperature is more difficult to generalize in the absence of charts and tables because dQ_{rev} depends on the state of aggregation. For the ideal gas, $dU = 0 = dQ - PdV$, $dQ = RTdV/V$, and plugging into Eqn. 4.13,

Isothermal.

$$\boxed{\Delta S^{ig} = R \ln\left[\frac{V_2}{V_1}\right]} \text{ or } \boxed{\Delta S^{ig} = -R \ln\left[\frac{P_2}{P_1}\right]}$$ (ig) 4.22

For a liquid or solid, the effect of isothermal pressure of volume change is small as a first approximation; the precise relations for detailed calculations will be developed in Chapters 6–8. Looking at the steam tables at constant temperature, entropy is very weakly dependent on pressure for *liquid* water. This result may be generalized to other liquids below $T_r = 0.75$ and also to solids. For *condensed* phases, to a first approximation, entropy can be assumed to be independent of pressure (or volume) at *fixed temperature*.

Adiabatic Pathway

A process that is adiabatic *and reversible* will result in an *isentropic path*. By Eqn. 4.13,

Adiabatic *and* reversible.

$$\boxed{\Delta S = \int_{state\ 1}^{state\ 2} \frac{dQ_{rev}}{T_{sys}} = 0} \text{ for reversible process only.}$$ 4.23

Note that a path that is adiabatic, but not reversible, will not be isentropic. This is because a reversible adiabatic process starting at the same state 1 will not follow the same path, so it will not end at state 2, and reversible heat transfer will be necessary to reach state 2.

Phase Transitions

In the absence of property charts or tables, entropy changes due to phase transitions can be easily calculated. Since equilibrium phase transitions for pure substances occur at constant temperature and pressure, for vaporization

$$\Delta S^{vap} = \int \frac{dQ_{rev}}{T} = \frac{1}{T^{sat}} \int dQ_{rev} = \frac{Q^{vap}}{T^{sat}}$$

where T^{sat} is the equilibrium saturation temperature. Likewise for a solid-liquid transition,

$$\Delta S^{fus} = \frac{Q^{fus}}{T_m}$$

where T_m is the equilibrium melting temperature. Since either transition occurs at constant pressure if along a reversible pathway, we may include Eqn. 4.17, giving

Phase transitions.

$$\boxed{\Delta S^{vap} = \frac{\Delta H^{vap}}{T^{sat}}} \text{ and } \boxed{\Delta S^{fus} = \frac{\Delta H^{fus}}{T_m}}$$ 4.24

Now let us examine a process from Chapter 2 that was reversible, and study the entropy change. We will show that the result is the same via two different paths, confirming that entropy is a state function.

Example 4.5   Ideal gas entropy changes in an adiabatic, reversible expansion

In Example 2.11 on page 75, we derived the temperature change for a closed-system adiabatic expansion of an ideal gas. How does the entropy change along this pathway, and what does this example show about changes in entropy with respect to temperature?

Solution: Reexamine the equation $(C_V/T)dT = -(R/V)dV$, which may also be written $(C_V/R)d\ln T = -d\ln V$. We can sketch this path as shown by the diagonal line in Fig. 4.4. Since our path is adiabatic $(dQ = 0)$ and reversible, and our definition of entropy is $dS = (dQ_{rev})/T$, we expect that this implies that the path is also isentropic (a constant-entropy path). Since entropy is a state property, we can verify this by calculating entropy along the other pathway of the figure consisting of a constant temperature (step A) and a constant volume (step B)

For the reversible isothermal step we have

$$dU = dQ_{rev} - PdV = 0 \quad \text{or} \quad dQ_{rev} = PdV \tag{ig}$$

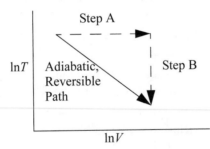

Figure 4.4 *Equivalence of an adiabatic and an alternate path on a T-V diagram.*

Thus,

$$(dS)_T = \frac{dQ_{rev}}{T} = \frac{PdV}{T} \tag{ig}$$

Substituting the ideal gas law,

$$(dS^{ig})_T = \frac{RTdV}{VT} = R\frac{dV}{V} \tag{ig} \quad 4.25$$

Example 4.5 Ideal gas entropy changes in an adiabatic, reversible expansion (Continued)

For the constant volume step, we have

$$dU = dQ_{rev} \quad \text{or} \quad C_V dT = dQ_{rev}$$

Thus,

$$(dS)_V = \frac{dQ_{rev}}{T} = C_V \frac{dT}{T} \tag{4.26}$$

We could replace a differential step along the adiabat (adiabatic pathway) with the equivalent differential steps along the alternate pathways; therefore, we can see that the change in entropy is zero,

> **!** The entropy change along the adiabatic, reversible path is the same as along (step A + step B) illustrating that S is a state property.

$$dS_{adiabat}^{ig} = (dS)_V + (dS^{ig})_T = \frac{C_V}{T} dT + \frac{R}{V} dV = 0 \tag{ig) 4.27}$$

which was shown by the energy balance in Eqn. 2.62, and we verify that the overall expansion is isentropic. Trials with additional pathways would show that ΔS is the same.

The method of subdividing state changes into individual temperature and volume changes can be generalized to any process, not just the adiabatic process of the previous example, giving

$$\boxed{dS^{ig} = \frac{C_V}{T} dT + \frac{R}{V} dV} \tag{ig) 4.28}$$

We may integrate steps A and B independently. We also could use temperature and pressure steps to calculate entropy changes, resulting in an alternate formula:

> **!** Formulas for an ideal gas.

$$\boxed{\Delta S^{ig} = C_V \ln \frac{T}{T^i} + R \ln \frac{V}{V^i}} \quad \text{or} \quad \boxed{\Delta S^{ig} = C_P \ln \frac{T}{T^i} - R \ln \frac{P}{P^i}} \tag{*ig) 4.29}$$

As an exercise, you may wish to choose two states and find the change in S along two different pathways: first with a step in T and then in P, and then by inverting the steps. The heat and work will be different along the two paths, but the change in entropy will be the same.

Looking back at Eqn. 4.26, we realize that it does not depend on the ideal gas assumption, and it is a general result,

> **!** Temperature derivatives of entropy are related to C_P and C_V.

$$\boxed{\left(\frac{\partial S}{\partial T}\right)_V = \frac{C_V}{T}} \tag{4.30}$$

which provides a relationship between C_V and entropy. Similarly, looking back at Eqn. 4.18,

$$\boxed{\left(\frac{\partial S}{\partial T}\right)_P = \frac{C_P}{T}} \tag{4.31}$$

Eqns. 4.30 and 4.31 are particularly easy to apply for ideal gases. In fact, the most common method for evaluating entropy changes with temperature applies Eqn. 4.31 in this way, as shown below.

Example 4.6 Ideal gas entropy change: Integrating $C_P^{ig}(T)$

Propane gas undergoes a change of state from an initial condition of 5 bar and 105°C to 25 bar and 190°C. Compute the change in entropy using the ideal gas law.

Solution: Because P and T are specified in each state, the ideal gas change is calculated most easily by combining an isobaric temperature step, Eqns. 4.19, and an isothermal pressure change, Eqn. 4.22. The heat capacity constants are obtained from Appendix E.

$$S_2^{ig} - S_1^{ig} = \int_{T_1}^{T_2} \frac{C_P^{ig}}{T} dT - R \ln\frac{P_2}{P_1} \; ; \int_{T_1}^{T_2} \frac{C_P^{ig}}{T} dT = \int_{T_1}^{T_2} \frac{(A + BT + CT^2 + DT^3)}{T} dT$$

$$S_2^{ig} - S_1^{ig} = A \ln\left(\frac{T_2}{T_1}\right) + B(T_2 - T_1) + \frac{C}{2}(T_2^2 - T_1^2) + \frac{D}{3}(T_2^3 - T_1^3) - R \ln\frac{P_2}{P_1}$$

$$= -4.224 \ln\frac{463.15}{378.15} + 0.3063(463.15 - 378.15) + \frac{-1.586 \times 10^{-4}}{2}(463.15^2$$

$$-378.15^2) + \frac{3.215 \times 10^{-8}}{3}(463.15^3 - 378.15^3) - 8.314\ln\frac{25}{5} = 6.613 \text{ J/mol-K}$$

Entropy Generation

At the beginning of Section 4.3 on page 142, statement number one declares that irreversible processes generate entropy. Now that some methods for calculating entropy have been presented, this principle can be explored.

Example 4.7 Entropy generation and "lost work"

In Example 4.1 consider the surroundings at 300 K: (a) Consider the entropy change in the surroundings and the universe for parts 4.1(a) and 4.1(b) and comment on the connection between entropy generation and lost work; (b) How would entropy generation be affected if the surroundings are at 310 K?

Example 4.7 Entropy generation and "lost work" (Continued)

Solution:

(a) For 4.1(a) the entropy change of the surroundings is

$$\Delta S_{surr} = \int \frac{dQ}{T_{surr}} = \frac{Q}{300} = \frac{-4014}{300} = -13.38 \text{J/K}.$$ This is equal and opposite to the entropy change of the piston/cylinder, so the overall entropy change is $\Delta S_{universe} = 0$.

For part 4.1(b), the entropy change of the universe is

$\Delta S_{surr} = Q/300 = -1995/300 = -6.65 \text{J/K}$. The total entropy change is

$\Delta S_{universe} = 13.38 - 6.65 = 6.73 \text{J/K} > 0$, thus entropy is generated when work is lost.

(b) If the temperature of the surroundings is raised to 310K, then for the reversible piston cylinder expansion for 4.1(a), $\Delta S_{surr} = -4014/310 = -12.948 \text{J/K}$, and

$\Delta S_{universe} = 13.38 - 12.95 = 0.43 \text{J/K} > 0$. This process now will have some 'lost work' due to the temperature difference at the boundary even though the piston/cylinder and work was frictionless without other losses. We will reexamine heat transfer in a gradient in a later example. For case 4.1(b), the entropy generation is still greater, indicating more lost work,

$\Delta S_{surr} = -1995/310 = -6.43 \text{J/K}$, $\Delta S_{universe} = 13.38 - 6.43 = 6.95 \text{J/K} > 0$.

---

❶ This is an irreversible process because entropy is generated.

Note: When the entropy change for the universe is positive the process is irreversible. Because entropy is a state property, the integrals that we calculate may be along any reversible pathway, and the time dependence along that pathway is unimportant.

We discussed in Chapter 2 that friction and velocity gradients result in irreversibilities and thus entropy generation occurs. Entropy generation can also occur during heat transfer, so let us consider that possibility.

Example 4.8 Entropy generation in a temperature gradient

A 500 mL glass of chilled water at 283 K is removed from a refrigerator. It slowly equilibrates to room temperature at 298 K. The process occurs at 1 bar. Calculate the entropy change of the water, ΔS_{water}, the entropy change of the surroundings, ΔS_{surr}, and the entropy change of the universe, ΔS_{univ}. Neglect the heat capacity of the container. For liquid water $C_P = 4.184$ J/g-K.

Solution:

Water: The system is closed at constant pressure with $T^i = 283$ K and $T^f = 298$ K. We choose any reversible pathway along which to evaluate Eqn. 4.13, a convenient path being constant-pressure heating. Thus,

$$dQ_{rev} = dH = mC_P dT$$

Example 4.8 Entropy generation in a temperature gradient (Continued)

Substituting this into our definition for a change in entropy, and assuming a T-independent C_P,

$$\Delta \underline{S}_{\text{water}} = \int_{T^i}^{T^f} \frac{mC_P dT}{T_{sys}} = mC_P \ln\left(\frac{T^f}{T^i}\right) = 500(4.184)\ln\left(\frac{298}{283}\right) = 108.0\frac{J}{K} \qquad (*)$$

Surroundings: The surroundings also undergo a constant pressure process as a closed system; however, the heat transfer from the glass causes no change in temperature—the surroundings act as a reservoir and the temperature is 298 K throughout the process. The heat transfer of the surroundings is the negative of the heat transfer of the water, so we have

$$\Delta \underline{S}_{\text{surr}} = \int_{T^i}^{T^f} \frac{dQ_{rev}}{T_{surr}} = -\int_{T^i}^{T^f} \frac{dQ_{rev,\,\text{water}}}{T_{surr}} = \frac{-mC_P \Delta T_{\text{water}}}{T_{surr}} = \frac{-31380J}{298K} = -105.3\frac{J}{K} \qquad (*)$$

Note that the temperature of the surroundings was constant, which simplified the integration.

Universe: For the universe we sum the entropy changes of the two subsystems that we have defined. Summing the entropy change for the water and the surroundings we have

$$\Delta \underline{S}_{\text{univ}} = 2.7\frac{J}{K}$$

Entropy has been generated. The process is irreversible.

4.4   THE ENTROPY BALANCE

In Chapter 2 we used the energy balance to track energy changes of the system by the three types of interactions with the surroundings—flow, heat, and work. This method was extremely helpful because we could use the balance as a checklist to account for all interactions. Therefore, we present a general entropy balance in the same manner. To solve a process problem we can use an analogous balance approach of starting with an equation including all the possible contributions that might occur and eliminating the balance terms that do not apply for the situation under consideration.

Entropy change within a system boundary will be given by the difference between entropy which is transported in and out, plus entropy changes due to the heat flow across the boundaries, and in addition, since entropy may be generated by an irreversible process, an additional term for entropy generation is added. A general entropy balance is

$$\frac{dS}{dt} = \sum_{inlets} S^{in} \dot{m}^{in} - \sum_{outlets} S^{out} \dot{m}^{out} + \sum_{surfaces} \frac{\dot{Q}}{T_{sys}} + \dot{S}_{gen} \qquad 4.32$$

General entropy balance.

Like the energy balance, the quantity to the left of the equals sign represents the entropy change of the system. The term representing heat transfer should be applied at each location where heat is transferred and the T_{sys} for each term is the system temperature at each boundary where the heat transfer occurs. The heat transfer represented in the general entropy balance is the heat transfer which occurs in the *actual* process. We may simplify the balance for steady-state or closed systems:

❶ Open, steady-state entropy balance.

$$0 = \dot{m}(S^{in} - S^{out}) + \frac{\dot{Q}}{T} + \dot{\underline{S}}_{gen} \qquad\qquad 4.33$$

❶ Closed system entropy balance.

$$d\underline{S} = \frac{d\underline{Q}}{T} + d\underline{S}_{gen} \qquad\qquad 4.34$$

The heat flow term(s) are always written in the balance with a "plus" from the perspective of the system for which the balance is written.

> **Caution:** The entropy balance provides us with an additional equation which may be used in solving thermodynamic problems; however, in irreversible processes, the entropy generation term usually cannot be calculated from first principles. Thus, it is an unknown in Eqns. 4.32–4.34. The balance equation is not useful for calculating any other unknowns when $\underline{S}_{gen}$ is unknown. In Example 4.8 the problem would have been difficult if we applied the entropy balance to the water or the surroundings independently, because we did not know how to calculate $\underline{S}_{gen}$ for each subsystem. However, we could calculate $\Delta\underline{S}$ for each subsystem along reversible pathways. Summing the entropy changes for the subsystems of the universe, we obtain the entropy change of the universe. Consider the right-hand side of the entropy balance when written for the universe in this example. There is no mass flow in and out of the universe—it all occurs between the subsystems of the universe. In addition, heat flow also occurs between subsystems of the universe, and the first three terms on the right-hand side of the entropy balance are zero. Therefore, the entropy change of the universe is equal to the entropy generated in the universe.

> **Caution:** The criterion for the feasibility of a process is that the entropy generation term must be greater than or equal to zero. The feasibility may not be determined unequivocally by ΔS for the system unless the system is the universe.

> *Note: As we work examples for* irreversible processes, *note that we do not apply the entropy balance to find entropy changes. We always calculate entropy changes by alternative reversible pathways that reach the same states, then we apply the entropy balance to find how much entropy was generated.*
>
> *Alternatively, for* reversible processes, *we do apply the entropy balance because we set the entropy generation term to zero.*

Let us now apply the entropy balance—first to another heat conduction problem. In Example 4.8 we studied an unsteady-state system. Now let us consider steady-state heat transfer to show which entropy balance terms are important in this application. In this example, we show that such heat conduction results in entropy generation because entropy generation occurs where the temperature gradient exists.

Example 4.9 Entropy balances for steady-state composite systems

Imagine heat transfer occurring between two reservoirs.

(a) A steady-state temperature profile for such a system is illustrated in Fig. (a) below. (Note that the process is an unsteady state with respect to the reservoirs, but the focus of the analysis here is on the wall.) The entire temperature gradient occurs within the wall. In this ideal case, there is no temperature gradient within either reservoir (therefore, the reservoirs are not a source of entropy generation). Note that the wall is at steady state. Derive the relevant energy and entropy balances, carefully analyzing three subsystems: the hot reservoir, the cold reservoir, and the wall. Note that a superficial view of the reservoirs and wall is shown in Fig. (b).

(b) Suppose the wall was replaced by a reversible Carnot engine across the same reservoirs, as illustrated in Fig. (c). Combine the energy and entropy balances to obtain the thermal efficiency.

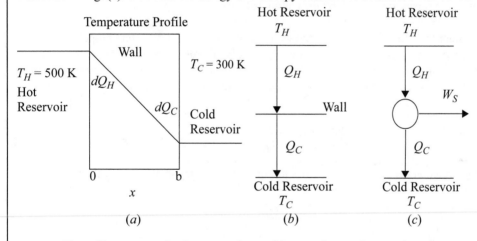

(a) $\qquad\qquad$ (b) $\qquad\qquad$ (c)

Note: Keeping track of signs and variables can be confusing when the universe is divided into multiple subsystems. Heat flow on the hot side of the wall will be negative for the hot reservoir, but positive for the wall. Since the focus of the problem is on the wall or the engine, we will write all symbols from the perspective of the wall or engine and relate to the reservoirs using negative signs and subscripts.

Solution:

(a) Since the wall is at steady state, the energy balance for the wall shows that the heat flows in and out are equal and opposite:

$$0 = \sum_{\text{boundaries}} \dot{Q} = \dot{Q}_H + \dot{Q}_C \Rightarrow \dot{Q}_H = -\dot{Q}_C \qquad\qquad 4.35$$

Example 4.9 Entropy balances for steady-state composite systems (Continued)

The entropy balance in each reservoir simplifies:

$$\frac{dS_{H \text{ or } C}}{dt} = \sum_{in} S^{in} \dot{m}^{in} - \sum_{out} S^{out} \dot{m}^{out} + \frac{\dot{Q}}{T_{sys}} + \dot{S}_{gen}$$

The entropy generation term drops out because there is no temperature gradient in the reservoirs. Taking the hot reservoir as the subsystem and noting that we have defined Q_H and Q_C to be based on the wall, we write:

$$\frac{d\underline{S}_H}{dt} = \frac{\dot{\underline{Q}}}{T_H} = \frac{-\dot{\underline{Q}}_{H,\text{wall}}}{T_H}; \quad \frac{d\underline{S}_C}{dt} = \frac{\dot{\underline{Q}}}{T_C} = \frac{-\dot{\underline{Q}}_{C,\text{wall}}}{T_C} = \frac{\dot{\underline{Q}}_{H,\text{wall}}}{T_C} \qquad 4.36$$

where the heat fluxes are equated by the energy balance.

Now consider the entropy balance for the wall subsystem. Entropy is a state property, and since no state properties throughout the wall are changing with time, entropy of the wall is constant, and the left-hand side of the entropy balance is equal to zero. Note that the entropy generation term is kept because we know there is a temperature gradient:

$$\frac{d\underline{S}_{\text{wall}}}{dt} = 0 = \frac{\dot{\underline{Q}}_H}{T_H} + \frac{\dot{\underline{Q}}_C}{T_C} + \dot{S}_{gen,\,\text{wall}}$$

Noting the relation between the heat flows in Eqn. 4.35, we may then write for the wall:

$$\dot{S}_{gen,\text{wall}} = \dot{\underline{Q}}_H\left(\frac{1}{T_C} - \frac{1}{T_H}\right) \qquad 4.37$$

Then the wall with the temperature gradient is a source of entropy generation. Summarizing,

$$\frac{d\underline{S}_{\text{univ}}}{dt} = \frac{d\underline{S}_H}{dt} + \frac{d\underline{S}_{\text{wall}}}{dt} + \frac{d\underline{S}_C}{dt} = \frac{-\dot{\underline{Q}}_{H,\text{wall}}}{T_H} + \frac{-\dot{\underline{Q}}_{C,\text{wall}}}{T_C} = \dot{S}_{gen,\text{wall}} \qquad 4.38$$

Hence we see that the wall is the source of entropy generation of the universe, which is positive. Notice that inclusion of the wall is important in accounting for the entropy generation by the entropy balance equations.

(b) The overall energy balance relative for the engine is:

$$0 = \sum_{\text{boundaries}} \dot{\underline{Q}} + \dot{\underline{W}} = \dot{\underline{Q}}_H + \dot{\underline{Q}}_C + \dot{\underline{W}}_S \Rightarrow \eta_\theta = \frac{-\dot{\underline{W}}_S}{\dot{\underline{Q}}_H} = \left(1 + \frac{\dot{\underline{Q}}_C}{\dot{\underline{Q}}_H}\right) \qquad 4.39$$

❗ Entropy is generated by a temperature gradient.

Example 4.9 Entropy balances for steady-state composite systems (Continued)

The engine operates a steady-state cycle, $d\underline{S}_{engine}/dt = 0$ (it is internally reversible):

$$0 = \sum_{in} S^{in}\dot{m}^{in} - \sum_{out} S^{out}\dot{m}^{out} + \sum_{boundaries} \frac{\dot{\underline{Q}}}{T} + \dot{\underline{S}}_{gen} = \frac{\dot{\underline{Q}}_H}{T_H} + \frac{\dot{\underline{Q}}_C}{T_C} \Rightarrow \frac{\dot{\underline{Q}}_C}{\dot{\underline{Q}}_H} = \frac{-T_C}{T_H} \qquad 4.40$$

As before, $\eta_\theta = -\underline{W}_S/\underline{Q}_H = (1 - T_C/T_H)$ and we have derived it using the entropy balance. Note that the heat flows are no longer equal and are such that the entropy changes of the reservoirs sum to zero.

We have concluded that heat transfer in a gradient results in entropy production. How can we transfer heat reversibly? If the size of the gradient is decreased, the right-hand side of Eqn. 4.37 decreases in magnitude; coincidentally, heat conduction slows, through the following relation:[16]

$$\dot{\underline{Q}} = Ak\frac{dT}{dx} \quad \text{where } A = \text{cross-sectional area}$$

A smaller temperature gradient decreases the rate of production of entropy, but from a practical standpoint, it requires a longer time to transfer a fixed amount of heat. If we wish to transfer heat reversibly from two reservoirs at finitely different temperatures, we must use a heat engine, as described in part (b). In addition to transferring the heat reversibly, use of a heat engine generates work.

> **Summary:** *This example has shown that boundaries (walls) between systems can generate entropy. In this example, entropy was not generated in either reservoir because no temperature profile existed. The entropy generation occurred within the wall.*

Entropy may be generated at system boundaries.

Note that we could have written the engine work of part (b) as follows:

$$\underline{W}_{S,net} = \underline{Q}_H\left(1 - \frac{T_C}{T_H}\right) = T_C \cdot \underline{Q}_H\left(\frac{1}{T_C} - \frac{1}{T_H}\right) \qquad 4.41$$

Compare this to Eqn. 4.37, the result of the steady-state entropy generation. If we run the heat transfer process without obtaining work, then the universe loses a quantity of work equal to the following:

$$\underline{\dot{W}}_{lost} = T_C\dot{\underline{S}}_{gen} \qquad 4.42$$

Note that T_C, the colder temperature of our engine, is important in relating the entropy generation to the lost work. T_C can be called the temperature at which the work is lost. Lost work is explored more completely in Section 4.12. Also note that we have chosen to operate the heat engine at temperatures which match the reservoir temperatures. This is arbitrary, but is required to obtain the maximum amount of work. The heat engine may be reversible without this constraint, but the entire process will not be reversible. These details are clarified in the next section.

---

16. This relation is known as Fourier's law and is studied in heat-transfer courses.

An entirely analogous analysis of heat transfer would apply if we ran the heat engine in reverse, as a heat pump. Only the signs would change on the direction the heat and work were flowing relative to the heat pump. Therefore, the use of entropy permits us to reiterate the Carnot formulas in the context of all fluids, not just ideal gases.

❶Carnot thermal efficiency.

$$\eta_\theta = \frac{-\dot{W}_{S,net}}{\dot{Q}_H} = \left(1 + \frac{\dot{Q}_C}{\dot{Q}_H}\right) = \left(1 - \frac{T_C}{T_H}\right), \text{ Carnot heat engine} \qquad 4.43$$

❶Carnot coefficient of performance.

$$COP \equiv \frac{\dot{Q}_C}{\dot{W}_{S,net}} = \left(\frac{T_H}{T_C} - 1\right)^{-1} = \frac{T_C}{T_H - T_C}, \text{ Carnot refrigerator} \qquad 4.44$$

4.5 INTERNAL REVERSIBILITY

A *process* may be irreversible due to interactions at the boundaries (such as discussed in Example 4.9 on page 155) even when each system in the process is reversible. Such a process is called **internally reversible.** Such a *system* has no entropy generation *inside* the system boundaries. We have derived equations for Carnot engines and heat pumps, assuming that the devices operate between temperatures that match the reservoir temperatures. While such restrictions are not necessary for internal reversibility, we show here that the work is maximized in a Carnot engine at these conditions and minimized in a Carnot heat pump. Note that in development of the Carnot devices, the only temperatures of concern are the operating temperatures at the hot and cold portions of the cycle. In the following illustrations, the internally reversible engine or pump operates between T_H and T_C, and the reservoir temperatures are T_2 and T_1.

Heat Engine

A schematic for a Carnot engine is shown in Fig 4.5(a). Heat is being transferred from the reservoir at T_2 to the reservoir at T_1, and work is being obtained as a result. In order for heat transfer to occur between the reservoirs and the heat engine in the desired direction, we must satisfy $T_2 \geq T_H > T_C \geq T_1$, and since the thermal efficiency is given by Eqn. 3.6, for maximum efficiency (maximum work), T_C should be as low as possible and T_H as high as possible, i.e., set $T_H = T_2$, $T_C = T_1$.

❶The operating temperatures of a reversible heat engine or heat pump are not necessarily equal to the surrounding's temperatures; however, the optimum work interactions occur if they match the surrounding's temperature because matching the temperatures eliminates the finite temperature-driving force that generates entropy.

(a) Heat engine   *(b) Heat pump*

Figure 4.5 *Schematic of a heat engine (a) and heat pump (b). The temperatures of the reservoirs are not required to match the reversible engine temperatures, but work is optimized if they do, as discussed in the text.*

Heat Pump

A schematic for a Carnot heat pump is shown in Fig. 4.5(b). Heat is being transferred from a reservoir at T_1 to the reservoir at T_2, and work is being supplied to achieve the transfer. In order for heat transfer to occur between the reservoirs and the heat engine in the desired direction, we must satisfy $T_2 \leq T_H > T_C \leq T_1$. Since the COP is given by Eqn. 4.44, for maximum COP (minimum work), T_C should be as high as possible and T_H as low as possible, i.e., set $T_C = T_1$, $T_H = T_2$. *Therefore, optimum work interactions occur when the Carnot device operating temperatures match the surrounding temperatures.* We use this feature in future calculations without special notice.

4.6   ENTROPY BALANCES FOR PROCESS EQUIPMENT

Before analysis involving multiple process units, it is helpful to consider the entropy balance for common steady-state process equipment. Familiarity with these common units will facilitate rapid analysis of situations with multiple units, because understanding these balances is a key step for the calculation of reversible heat and work interactions.

Simple Closed Systems

Changes in entropy affect all kinds of systems. We have previously worked with piston/cylinders and even a glass of water. You should be ready to adapt the entropy balance in creative ways to everyday occurrences as well as sophisticated equipment.

Example 4.10   Entropy generation by quenching

A carbon-steel engine casting [$C_P = 0.5$ kJ/kg°C] weighing 100 kg and having a temperature of 700 K is heat-treated to control hardness by quenching in 300 kg of oil [$C_P = 2.5$ kJ/kg°C] initially at 298 K. If there are no heat losses from the system, what is the change in entropy of: (a) the casting; (b) the oil; (c) both considered together; and (d) is this process reversible?

Solution: Unlike the previous examples, there are no reservoirs, and the casting and oil will both change temperature. The final temperature of the oil and the steel casting is found by an energy balance. Let T^f be the final temperature in K.

Energy balance: The total change in energy of the oil and steel is zero.
Heat lost by casting:
$\underline{Q} = mCp\Delta T = 100\,(0.5)\,(700 - T^f)$
Heat gained by oil:
$\underline{Q} = mCp\Delta T = 300\,(2.5)\,(T^f - 298) \Rightarrow$ balancing the heat flow, $T^f = 323.1$ K

Entropy balance: The entropy change of the universe will be the sum of the entropy changes of the oil and casting. We will not use the entropy balance directly except to note that $\Delta \underline{S}_{univ} = \underline{S}_{gen}$. We can calculate the change of entropy of the casting and oil by any reversible pathway which begins and ends at the same states. Consider an isobaric path:

$$\text{Using the macroscopic definition} \Rightarrow \quad \Delta \underline{S} = \int \frac{d\underline{Q}}{T} = m\int \frac{C_P}{T}\, dT = mC_P\, \ln\!\left(\frac{T_2}{T_1}\right) \qquad (*)$$

Example 4.10 Entropy generation by quenching (Continued)

(a) Change in entropy of the casting:

$$\Delta \underline{S} = 100 \ (0.5) \ \ln[323.1/700] = -38.7 \ \text{kJ/K} \qquad (*)$$

(b) Change in entropy of the oil (the oil bath is of finite size and will change temperature as heat is transferred to it):

$$\Delta \underline{S} = 300 \ (2.5) \ \ln[323.1/298] = 60.65 \ \text{kJ/K} \qquad (*)$$

(c) Total entropy change: $\underline{S}_{gen} = \Delta \underline{S}_{univ} = 60.65 - 38.7 = 21.9 \ \text{kJ/K}$

(d) $\underline{S}_{gen} > 0$; therefore irreversible; compare the principles with Example 4.8 on page 152 to note the similarities. The difference is that both subsystems changed temperature.

Compare with Example 4.8 on page 152.

Heat Exchangers

The entropy balance for a standard two-stream heat exchanger is given by Eqn. 4.45. Since the unit is at steady state, the left-hand side is zero. Applying the entropy balance around the *entire* heat exchanger, there is no heat transfer *across* the system boundaries (in the absence of heat loss), so the heat-transfer term is eliminated. Since heat exchangers operate by conducting heat across tubing walls with finite temperature driving forces, we would expect the devices to be irreversible. Indeed, if the inlet and outlet states are known, the flow terms may be evaluated, thus permitting calculation of entropy generation.

$$\frac{d\underline{S}}{dt} = \sum_{in} S^{in} \dot{m}^{in} - \sum_{out} S^{out} \dot{m}^{out} + \frac{\dot{Q}}{T_{sys}} + \dot{\underline{S}}_{gen} \qquad 4.45$$

We also may perform "paper" design of ideal heat transfer devices that operate reversibly. If we set the entropy generation term equal to zero, we find that the inlet and outlet states are constrained. Since there are multiple streams, the temperature changes of the streams are coupled to satisfy the entropy balance. In order to construct such a reversible heat transfer device, the unit would need to be impractically large to only have small temperature gradients.

Example 4.11 Entropy in a heat exchanger

A heat exchanger for cooling a hot hydrocarbon liquid uses 10 kg/min of cooling H_2O which enters the exchanger at 25°C. Five kg/min of hot oil enters at 300°C and leaves at 150°C and has an average specific heat of 2.51 kJ/kg-K.

(a) Demonstrate that the process is irreversible as it operates now.

(b) Assuming no heat losses from the unit, calculate the maximum work which could be obtained if we replaced the heat exchanger with a Carnot device which eliminates the water stream and transfers heat to the surroundings at 25°C

Example 4.11 Entropy in a heat exchanger (Continued)

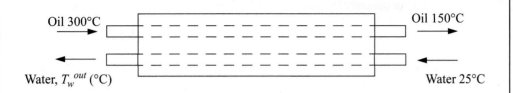

Oil 300°C → ⟶ → Oil 150°C

Water, T_w^{out} (°C) ← ← Water 25°C

Solution:

(a) System is heat exchanger (open system in steady-state flow)
Energy balance:

$$\Delta H_{oil}\dot{m}_{oil} + \Delta H_{water}\dot{m}_{water} = 0$$

$$10(4.184)(T_w^{out} - 25) + 5(2.51)(150 - 300) = 0; \quad T_w^{out} = 70°C$$

Entropy balance:

$$\Delta S_{oil}\dot{m}_{oil} + \Delta S_{water}\dot{m}_{water} = \dot{S}_{gen}$$

$$dS_i = dQ/T = C_P \, dT/T \Rightarrow \Delta S_i = C_{Pi} \ln (T_i^{\ out}/T_i^{in}) \tag{*}$$

$$\Delta S_{oil} = C_P \ln (T^{out}/T^{in}) = (2.51)\ln(423.15/573.15) = -0.7616 \text{ kJ/kgK} \tag{*}$$

$$\Delta S_{water} = C_P \ln (T^{out}/T^{in}) = (4.184) \ln(343.15/298.15) = 0.5881 \text{ kJ/kgK} \tag{*}$$

$$\dot{\underline{S}}_{gen} = \Delta S_{oil}\dot{m}_{oil} + \Delta S_{water}\dot{m}_{water} = 5(-0.7616) + 10(0.5881) = 2.073 \text{ kJ/K-min}$$

The process is irreversible because entropy is generated.

(b) The modified process is represented by the "device" shown below. Note that we avoid calling the device a "heat exchanger" to avoid confusion with the conventional heat exchanger. To simplify analysis, the overall system boundary is used.

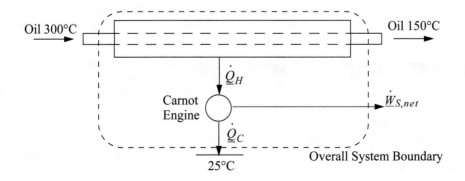

Oil 300°C → ⟶ → Oil 150°C

$\dot{Q}_H$

Carnot Engine → $\dot{W}_{S,net}$

$\dot{Q}_C$

Overall System Boundary

25°C

Example 4.11 Entropy in a heat exchanger (Continued)

By an energy balance around the overall system, $0 = \dot{n}(H^{in} - H^{out}) + \dot{\underline{Q}}_C + \dot{\underline{W}}_S$. We can only solve for the enthalpy term,

$$\dot{m}(H^{in} - H^{out}) = \dot{m}C_P(T^{in} - T^{out}) = 5(2.51)(300 - 150) = 1882.5 \text{ kJ/min}$$

Since heat and work are both unknown, we need another equation. Consider the entropy balance, which, since it is a reversible process, $\dot{\underline{S}}_{gen} = 0$, gives

$$0 = \dot{m}(S^{in} - S^{out}) + \frac{\dot{\underline{Q}}_C}{T}, \qquad \dot{\underline{Q}}_C = 298.15 \cdot (-0.7616) \cdot 5 = -1135 \text{ kJ/min}$$

Now inserting these results into the overall energy balance gives the work,

$$\dot{\underline{W}}_S^{net} = -1883. + 1135. = -748 \text{ kJ/min}$$

Throttle Valves

Steady-state throttle valves are typically assumed to be adiabatic, but a finite pressure drop with zero recovery of work or kinetic energy indicates that $\underline{S}_{gen} > 0$. Throttles are isenthalpic, and for an ideal gas, they are thus isothermal, $\Delta S^{ig} = C_P \ln(T/T^{in}) - R\ln(P^{out}/P^{in})$. For a real fluid, temperature changes can be significant. The entropy increase is large for gases, and small, but non-zero for liquids. It is important to recall that liquid streams near saturation may flash as they pass through throttle valves, which also produces large entropy changes and significant cooling of the process fluid even when the process is isenthalpic. Throttles involving flash are common in the liquefaction and refrigeration processes discussed in the next chapter. Throttles are always irreversible.

Nozzles

Steady-state nozzles can be designed to operate nearly reversibly; therefore, we may assume $\underline{S}_{gen} = 0$, and Eqn. 4.47 applies. Under these conditions, thrust is maximized as enthalpy is converted into kinetic energy. The distinction between a nozzle and a throttle is based on the reversibility of the expansion. Recall from Chapter 2 that a nozzle is specially designed with a special taper to avoid turbulence and irreversibilities. Naturally, any real nozzle will approximate a reversible one and a poorly designed nozzle may operate more like a throttle. Proper design of nozzles is a matter of fluid mechanics. We can illustrate the basic thermodynamic concepts of a properly designed nozzle with an example.

Example 4.12 Isentropic expansion in a nozzle

Steam at 1000°C and 1.1 bars passes through a horizontal adiabatic converging nozzle, dropping to 1 bar. Estimate the temperature, velocity, and kinetic energy of the steam at the outlet assuming the nozzle is reversible and the steam can be modeled with the ideal gas law under the conditions. Consider the initial velocity to be negligible. The highest exit velocity possible in a converging nozzle is the speed of sound. Use the NIST web site[a] as a resource for the speed of sound in steam at the exit conditions.

> **Example 4.12   Isentropic expansion in a nozzle   (Continued)**
>
> **Solution:**
>
> Energy balance: $\Delta H = -m \Delta v^2/2$ Entropy balance (reversible): $\Delta S = 0$.
>
> For an isentropic reversible expansion the temperature will drop. We will approximate the heat capacity with an average value. Let us initially use a C_P for 1273 K. Estimating the heat capacity from Appendix E at 1273 K, the polynomial gives $C_P = 44.37$ J/mol, $R/C_P \sim 8.314/44.37 = 0.1874$. The following relation satisfies the entropy balance for an adiabatic, reversible, ideal gas (Eqn. 4.29):
>
> $$\left(\frac{T_2}{T_1}\right) = \left(\frac{P_2}{P_1}\right)^{R/C_P} = (1.1)^{-0.1874} = 0.9823 \Rightarrow T_2 = 1273(0.9823) = 1250.5 \text{ K} \qquad (*)$$
>
> The temperature change is small, so the constant heat capacity assumption is fine. The enthalpy change is $-\Delta H = -C_P \Delta T = 44.6(1273 - 1250.5)(\text{J/mol}) = 1004$ J/mol.
>
> Assuming that the inlet velocity is low, $v_1 \sim 0$ and converting the enthalpy change to the change in velocity gives $v^2 = -2\Delta H/m = 2 \cdot 1004 \text{J/mol(mol/18.01g)(1000g/kg)(1kg-m}^2/\text{s}^2)/\text{J} = 111{,}500$ m^2/s^2, or $v = 334$ m/s. According to the NIST web site at 1250K and 0.1MPa, the speed of sound is 843 m/s. The design is reasonable.

a. Lemmon, E.W., McLinden, M.O., Friend, D.G. "Thermophysical Properties of Fluid Systems." in *NIST Chemistry WebBook*, NIST Standard Reference Database Number 69, P.J. Linstrom, W.G. Mallard (eds.) National Institute of Standards and Technology, Gaithersburg, MD. http://webbook.nist.gov, (retrieved November 12, 2011).

Adiabatic Turbine, Compressor, and Pump

The entropy balance for a steady-state *adiabatic* device is:

$$\frac{dS}{dt} = \sum_{in} S^{in} \dot{m}^{in} - \sum_{out} S^{out} \dot{m}^{out} + \frac{\dot{Q}}{T_{sys}} + \dot{S}_{gen} \qquad 4.46$$

The left-hand side drops out because the system is at steady state. If the device is reversible, $\dot{S}_{gen}$ is zero. Further, these devices typically have a single inlet or outlet,[17] and $\dot{m}^{in} = \dot{m}^{out}$, thus,

$$S^{out'} = S^{in} \qquad 4.47$$

❶ Adiabatic reversible turbine, compressor, and pump.

Therefore, if we know the inlet state, we can find S^{in}. The outlet pressure is generally given, so for a pure fluid, the reversible outlet state is completely specified by the two state variables $S^{out'}$ and P^{out}. We then use thermodynamic relations to find the other thermodynamic variables at this state, and use the energy balance at this state to find W_{rev}. Turbines, compressors, and pumps are very common equipment in chemical processes. Guidelines exist for estimating the degree of irreversibility in each piece of equipment based on experience. These guidelines take the form of an estimated efficiency. For example, a large expensive turbine might be 85% efficient, but a small cheap one might be 65% efficient. To apply these guidelines, we must formally define efficiency,

17. In multistage units, the stages may be considered individually.

then familiarize ourselves with variations on how to characterize the capacity and operating conditions of these operations.

4.7 TURBINE, COMPRESSOR, AND PUMP EFFICIENCY

Our analysis of the Carnot devices supports statement 2 at the beginning of Section 4.3. We have seen that work is maximized/minimized when the entropy generation is zero. Analysis of other processes would verify this useful conclusion. Work is lost by processes which generate entropy. If a device is not *internally reversible,* work will be lost within the device. Also, even if the device is internally reversible, work may be lost by irreversible interactions with the surroundings. Therefore, in setting up and solving problems to find maximum/minimum work, the objectives must be clear as to whether the system is internally reversible or whether the entire process is reversible. When we apply the entropy balance to a reversible process, the term representing entropy generation is zero.

In Chapter 2, both velocity gradients and friction were discussed as phenomena that lead to irreversibilities. Indeed, entropy is generated by both of these phenomena as well as by heat conduction along temperature gradients discussed in this chapter. Considering factors which affect reversibility (generation of entropy), you may have challenged yourself to consider a practical way to transfer heat with only infinitesimal temperature differences, or move fluid with only infinitesimal velocity gradients. You are probably convinced that such a process would not be practical. Indeed, the rate at which heat is transferred increases as the temperature driving forces increase, and we need finite temperature differences to transfer heat practically. Likewise, a pump will have large velocity gradients. Although we can measure changes in other properties by which we can calculate entropy changes arising from irreversibilities, an *a priori* prediction of lost work (W_{lost}) or entropy generated ($\underline{S}_{gen}$) is extremely difficult and generally impractical. Direct evaluation of lost work in process equipment, such as turbines and compressors, is far beyond routine calculation and determined by empirical experience. It may seem that all of the effort to characterize reversible processes will be difficult to relate to real processes.

However, we can use practical experience to relate real processes to the idealized reversible processes. Therefore, in analyzing or designing processes involving operations of this nature, it is often necessary to approximate the real situation with a reversible one in which (W_{lost}) = 0 (no entropy generation). Past experience with many devices, such as compressors and turbines, often permits the engineer to relate performance under hypothetical reversible conditions to actual operation under real conditions. The relation is usually expressed by means of an efficiency factor. Equipment manufacturers typically provide performance curves as a function of process conditions. For introductory purposes, in this text we use a fixed factor. For devices such as pumps and compressors which utilize work from the surroundings, efficiency is defined as

❶Primes are used to denote reversible processes.

$$\text{pump or compressor efficiency} = \boxed{\eta_C = \frac{W'}{\underline{W}} \times 100\%} \qquad 4.48$$

where the ' denotes the reversible work. This notation will be used throughout the text when irreversible and reversible calculations are performed in the same problem. For turbines and other expansion devices that supply work to the surroundings, the definition is inverted to give

$$\text{turbine or expander efficiency} = \boxed{\eta_E = \frac{W}{W'} \times 100\%} \qquad 4.49$$

For *adiabatic* pumps, compressors, turbines, or expanders, the work terms may be calculated from the reversible and irreversible enthalpy changes by application of the energy balance.

> *Note:* *The strategy is to first calculate the work involved in a reversible process, then apply an efficiency which is empirically derived from previous experience with similar equipment.* The outlet **pressure** of an irreversible adiabatic turbine or pump is always at the same pressure as a reversible device, but the enthalpy is always higher for the same inlet state. *This means that if the outlet of the reversible adiabatic device is a single phase, the outlet of the irreversible adiabatic device will be at a higher temperature. If the outlet of the reversible adiabatic device is a two-phase mixture, the quality for the irreversible adiabatic device will be higher or the outlet could potentially be a single phase.*

4.8   VISUALIZING ENERGY AND ENTROPY CHANGES

Turbines, compressors, and pumps occur so frequently that we need convenient tools to aid in process calculations. Visualization of the state change is possible by plotting entropy on charts. This technique also permits the charts to be used directly in the process calculations. One common representation is the *T-S* chart shown in Fig. 4.6. The phase envelope appears as a fairly symmetrical hump. A reversible turbine, compressor, or pump creates state changes along a vertical line on these coordinates. Lines of constant enthalpy and pressure are also shown on these diagrams, as sketched in the figure. Volumes are also usually plotted, but they lie so close to the pressure lines that they are not illustrated in the figure here to ensure clarity.

❶Visualizing state changes on charts will be helpful when using tables or computers for physical properties.

P-H diagrams shown in Fig. 4.7 are also useful; they are used frequently for refrigeration processes. The phase envelope tends to lean to the right because the enthalpies of vapor and liquid are both increasing along the saturation curve until the critical point is approached, where the vapor-phase enthalpy decreases due to significant non-idealities. Lines of constant entropy on these plots are slightly diagonal with a positive slope as shown in Fig. 4.7(a). For some hydrocarbons and

❶ A 3D diagram for steam is available in PHT.m. The diagram can be rotated.

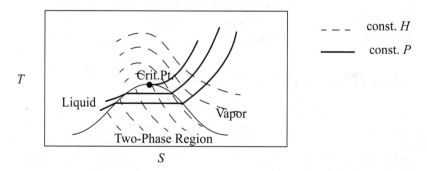

Figure 4.6 *Illustration of a T-S diagram showing lines of constant pressure and enthalpy.*

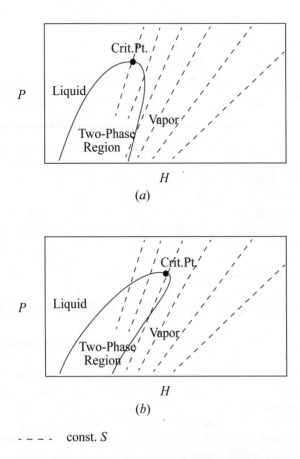

- - - - const. S

Figure 4.7 *Illustration of a P-H diagram showing (a) lines of constant entropy for a species where the saturation curve leans less than isentropes (e.g., water) and (b) illustration of a P-H diagram showing lines of constant entropy for a species where the saturation curve leans more than isentropes (e.g., hexane).*

halogenated compounds, the phase envelope can lean more sharply than the isentropic lines as shown in Fig. 4.7(b). A reversible compressor will operate along a line of constant entropy.

Another convenient representation of entropy is the *H-S* diagram (Mollier diagram). In this diagram, lines of constant pressure are diagonal, and isotherms have a downward curvature as in Fig. 4.8. The saturation curve is quite skewed.

4.9 TURBINE CALCULATIONS

The outlet entropy of an *irreversible* adiabatic turbine will be greater than the outlet entropy of a *reversible* adiabatic turbine with the same outlet pressure.

For a reversible adiabatic turbine, the entropy balance in Section 4.6 shows that the outlet entropy must equal the inlet entropy. For an irreversible turbine, the outlet entropy must be greater than the inlet entropy. We may now visualize the state change on the diagrams sketched in Section 4.8. For example, on a *T-S* diagram, the performance of a turbine can be visualized as shown in Fig. 4.9. Note that the isobars are important in sketching the behavior because the *outlet pressure must be the same* for the reversible and irreversible turbines, but the outlet enthalpies (not shown) and entropies must be different.

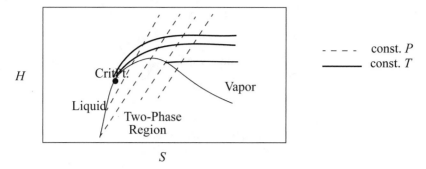

Figure 4.8 *Illustration of an H-S (Mollier) diagram showing lines of constant pressure and temperature.*

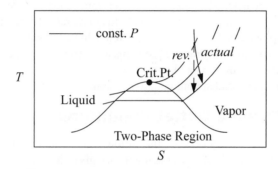

Figure 4.9 *Illustration of a reversible and actual (irreversible) turbine on a T-S diagram.*

Steam Quality Calculations

A common problem encountered when adiabatically reducing the pressure of real fluids like steam, methane, or refrigerants is the formation of a vapor-liquid mixture. Since the thermodynamic properties change dramatically depending on the mass fraction that is vapor (the quality), it is important to know how to calculate that fraction. The calculation procedure may differ from the case shown in Fig. 4.9 where the outlets for the reversible and irreversible cases are both one phase. Since the reversible adiabatic turbine is isentropic, the line representing the reversible process must be vertical. As shown in Fig. 4.10, if the upstream entropy is less than the saturated vapor entropy at the outlet pressure, the *reversible* outlet ends up *inside the liquid-vapor region,* to the left of the saturated vapor curve. In this case, we must perform a quality calculation to determine the vapor fraction. Since the actual turbine must have an outlet state of higher entropy, due to entropy generation, the outlet state can lie inside the phase envelope, on the saturation curve, or outside the phase envelope, depending on the proximity of the reversible outlet state to the saturation curve and also depending on the turbine efficiency. A frequent question is, "How do I know when I need a quality calculation?" The calculation is required if the *inlet* entropy is less than the saturation entropy at the *outlet* pressure as illustrated in the figure. A quality calculation may also be required for the actual state, if the actual enthalpy turns out to be less than the saturation enthalpy at the outlet pressure.

The actual outlet state might be in the one-phase region when the reversible outlet state is in the two-phase region.

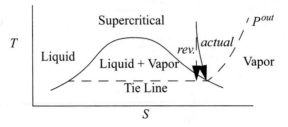

Figure 4.10 *Illustration of need for quality calculation on turbine outlet where the actual outlet is saturated steam.*

The best way to master turbine calculations is to practice; the examples in this section are designed to facilitate your effort. Example 4.13 explores the inference of outlet conditions and proper application of turbine efficiency. Example 4.14 illustrates calculation of turbine efficiency. Example 4.15 illustrates the determination of inlet conditions to match a desired outlet. The combinations of inlet and outlet specifications are too many to enumerate. Therefore, you need to practice inferring the necessary procedure for any given situation. Challenge yourself to repeat the examples without looking at the steps. Several practice problems are also given at the end of the chapter, with detailed solutions available at the textbook web site. As you practice, pay attention to the results and look for generalities that broaden your comprehension.

Determining Turbine Outlet Conditions

Let us work a series of examples illustrating the various situations that may arise in calculation of turbine outlets. Usually, the design of a turbine involves a given inlet state and outlet pressure. That outlet pressure may be specified explicitly, or it may be implicit in a statement giving the outlet temperature and the quality; it may be inferred then that the outlet pressure is the saturation pressure at the given temperature. An important skill is to quickly determine whether the reversible turbine follows Fig. 4.9 or whether it follows Fig. 4.10. Subsequently, for the cases that follow Fig. 4.10, the outlet state may lie inside or outside the phase envelope. The determination depends on the efficiency and the inlet entropy, with the following cases:

   (i) Reversible outlet one phase, actual outlet one phase;
   (ii) Reversible outlet two phase, actual outlet one phase;
   (iii) Reversible outlet two phase, actual outlet two phase.

   This example, though long, comprehensively covers the solution methods to determine turbine outlets for single-stage turbines from a known inlet state and specified pressure drop.

Example 4.13   Various cases of turbine outlet conditions

An adiabatic turbine inlet (state 1) is 500°C and 1.4 MPa. For each of the following outlet conditions (state 2), determine the specified quantities.

(a) $P_2 = 0.6$ MPa, $\eta_E = 0.85$. Find W_S, H_2, S_2, and T_2.
(b) $P_2 = 0.03$ MPa, $\eta_E = 0.85$. Find W_S, H_2, S_2, and T_2.
(c) $P_2 = 0.01$ MPa, $\eta_E = 0.9$. Find W_S, H_2, S_2, and T_2.

Example 4.13 Various cases of turbine outlet conditions (Continued)

Note that using a common inlet state for each of the cases will permit us to skip the steps to determine the inlet state as we work the different alternatives.

Solution: First, the inlet properties are determined: $H_1 = 3474.8$ kJ/kg, $S_1 = 7.6047$ kJ/kg-K. The reversible calculation is performed for each outlet condition, recognizing that a reversible turbine is isentropic.

(a) $S_2' = S_1 = 7.6047$ kJ/kg-K. Comparing with $S^{satV} = 6.7593$ kJ/kg-K at $P_2 = 0.6$ MPa, $S_2' > S^{satV}$, **so the reversible outlet state is superheated and any irreversibility must lead to greater entropy and greater superheat (case (i)).** This is the case of Fig. 4.9. Interpolating:

$T\,(°C)$	$H\,(kJ/kg)$	$S\,(kJ/kg\text{-}K)$
350	3166.1	7.5481
		7.6047
400	3270.8	7.7097

$$H_2' = 3166.1 + \frac{7.6047 - 7.5481}{7.7097 - 7.5481}(3270.8 - 3166.1) = 3202.8 \text{ kJ/kg}$$

By similar interpolation, $T_2' = 367.5°C$.

$$\Delta H' = W_S' = 3202.8 - 3474.8 = -272.0 \text{ kJ/kg}$$

Applying η_E calculation, $\Delta H = W_S = \eta_E \Delta H' = 0.85(-272) = -231.2$ kJ/kg,

$$H_2 = H_1 + \Delta H = 3474.8 - 231.2 = 3243.6 \text{ kJ/kg}$$

Preparing for interpolation:

$T\,(°C)$	$H\,(kJ/kg)$	$S\,(kJ/kg\text{-}K)$
350	3166.1	7.5481
	3243.6	
400	3270.8	7.7097

$$S_2 = 7.5481 + \frac{3243.6 - 3166.1}{3270.8 - 3166.1}(7.7097 - 7.5481) = 7.6677 \text{ kJ/kg-K}$$

By similar interpolation, $T_2 = 387.0°C$. We see that irreversibility has warmed the outlet, but not "heated" it, because it was adiabatic. With a one-phase outlet, $T_2 > T_2'$ if $\eta_E < 1$.

(b) The pressure is lower than part (a), and the saturated vapor S will be larger, and near the saturation boundary.
Recall that $S_2' = S_1 = 7.6047$ kJ/kg-K. Comparing with $S^{satV} = 7.7675$ kJ/kg-K at $P_2 = 0.03$ MPa, $S_2' < S^{satV}$, so the reversible outlet state is two-phase. This is the case of Fig. 4.10 and we need to proceed further to determine if the actual state is inside or outside the phase envelope. Interpolating using the saturation entropy values along with the S_2' at $T_2' = 69.1°C$,

$$q' = \frac{S - S^{satL}}{\Delta S^{vap}} = \frac{7.6047 - 0.9441}{6.8234} = 0.976$$

Using Eqn. 1.27:

Example 4.13 Various cases of turbine outlet conditions (Continued)

$$H_2' = 289.27 + 0.976(2335.28) = 2568.5 \text{ kJ/kg}$$

$$\Delta H' = W_S' = 2568.5 - 3474.8 = -906.3 \text{ kJ/kg}$$

Applying η_E calculation, $\Delta H = W_S = \eta_E \Delta H' = 0.85(-906.3) = -770.35 \text{ kJ/kg}$,

$$H_2 = H_1 + \Delta H = 3474.8 - 770.35 = 2704.4 \text{ kJ/kg}$$

Comparing H_2 with $H^{satV} = 2624.55 \text{ kJ/kg}$ at $P_2 = 0.03$ MPa, $H_2 > H^{satV}$, so the outlet state is superheated (outside the phase envelope). **This is an instance of case (ii).**

To conclude the calculations, a double interpolation is required. Performing the first interpolation between 0.01 and 0.05 MPa will bracket the outlet state. (Note: 0.03 MPa is halfway between 0.01 and 0.05 MPa, so tabulated values are obtained by averaging rather than by a slower interpolation.)

T (°C)	H (kJ/kg)	S (kJ/kg-K)
100	(2687.5 + 2682.4)/2 = 2684.95	(8.4489 + 7.6953)/2 = 8.0721
	2704.4	
150	(2783.0 + 2780.2)/2 = 2781.6	(8.6892 + 7.9413)/2 = 8.3153

Interpolating:

$$S_2 = 8.0721 + \frac{2704.4 - 2684.95}{2781.6 - 2684.95}(8.3153 - 8.0721) = 8.121 \text{ kJ/kg-K}$$

Similarly, by interpolation, $T_2 = 110.1°C$.

Note: The reversible state is two-phase, and the actual outlet is one-phase for part (b). Also, $S_2 > S_2' = S_1$ and $H_2 > H_2'$ which are always true for irreversible turbines. $T_2 > T_2'$, which is a general result for one-phase output.

(c) Very low-outlet pressures shifts the saturation value of S to even higher values, making it more likely that the outlet will be two phase, case (iii).
$S_2' = S_1 = 7.6047 \text{ kJ/kg-K}$. Comparing with $S^{satV} = 8.1488 \text{ kJ/kg-K}$ at $P_2 = 0.01$ MPa, $S_2' < S^{satV}$, so the reversible outlet state is two-phase. This is the case of Fig. 4.10 and we need to proceed further to determine if the actual state is inside or outside the phase envelope. Interpolating at $P = 0.01$ MPa ($T_2' = 45.81°C$),

$$q' = \frac{7.6047 - 0.6492}{7.4996} = 0.9274$$

Using Eqn. 1.27,

$$H_2' = 191.81 + 0.9274(2392.05) = 2410.2 \text{ kJ/kg}$$

$$\Delta H' = W_S' = 2410.3 - 3474.8 = -1064.6 \text{ kJ/kg}$$

Applying η_E calculation, $\Delta H = W_S = \eta_E \Delta H' = 0.90(-1064.6) = -958.1 \text{ kJ/kg}$,

$$H_2 = H_1 + \Delta H = 3474.8 - 958.1 = 2516.7 \text{ kJ/kg}.$$

Comparing H_2 with $H^{satV} = 2583.86 \text{ kJ/kg}$ at $P_2 = 0.01$ MPa, $H_2 < H^{satV}$, so **the actual outlet state is two-phase as well as the reversible outlet (case (iii)).** For the actual outlet, H_2 gives:

Example 4.13   Various cases of turbine outlet conditions (Continued)

$$q = \frac{2516.7 - 191.81}{2392.05} = 0.972$$

Using Eqn. 1.27,

$$S_2 = 0.6492 + 0.972(7.4996) = 7.9388 \text{ kJ/kg}$$

The actual outlet is wet steam at $T_2 = 45.81°C$. The reversible outlet and the actual outlet are both wet steam for part (c). Also, $S_2 > S_2' = S_1$ and $H_2 > H_2'$ which are always true for irreversible turbines. For case (c), $T_2 = T_2'$, however $q_2 > q_2'$, a general result for a two-phase outlet.

This example has exhaustively covered the possibilities that may occur when performing turbine analysis given a specified pressure drop and known inlet condition. The actual outlet enthalpy and entropy are always greater than the reversible values. The actual outlet T will be the same as the reversible T if both states are wet steam. Note that many variations could generate calculations that appear to be different from these cases, but are actually similar. For example, the quality could be specified at the outlet instead of the efficiency. Similarly, the steam outlet H or S, etc. could be specified rather than the quality. For any problem, the details of the interpolation may differ for a given application depending on the region of the steam tables. Nevertheless, the overall procedures of using entropy to identify the reversible state and then correcting for the actual state are always the same. The following example illustrates a typical turbine calculation that might be used to characterize the efficiency of a working turbine.

Example 4.14   Turbine efficiency calculation

An adiabatic turbine inlet is at 500°C and 1.4 MPa. Its outlet is at 0.01MPa and $q = 99\%$.

(a) Compute the work of the turbine.
(b) Compute the work of a reversible turbine.
(c) Compute the efficiency of the turbine and the entropy generation of the actual turbine.

Solution: The energy balance is $\Delta H = W_S$.

(a) The inlet is the same as Example 4.13: $H_1 = 3474.8$; $S_1 = 7.6047$. At the outlet,

$$H_2 = 191.81 + 0.99(2392.05) = 2559.9 \text{ kJ/kg}$$

$$\Delta H = W_S = 2559.9 - 3474.8 = -914.9 \text{ kJ/kg}$$

(b) Entropy balance: $\Delta S' = 0 \Rightarrow S_2' = S_1 = 7.6047 \text{ kJ/kg-K}$.
It is slightly ambiguous whether we should match the outlet pressure or the specification of quality. By convention, it is assumed that pressure is the desired criterion (or temperature in a similar situation) because this pertains to the physical constraints of the design. This means that the reversible work is the same as Example 4.13(c) and $W_S' = -1064.6 \text{ kJ/kg}$.

(c) The turbine efficiency is defined by $\eta_E = W_S/W_S' = 914.9/1064.6 = 85.9\%$.
The entropy generation is given by $S_{gen} = S_2 - S_2'$.

Example 4.14 Turbine efficiency calculation (Continued)

$$S_2 = 0.6492 + 0.99(7.4996) = 8.0738 \text{ kJ/kg-K}$$

Referring to the entropy balance, $S_2' = S_1 = 7.3046$ so $S_{gen} = 8.0738 - 7.6047 = 0.4691$ J/g-K.

Another type of calculation involves determining a turbine inlet that will result in a certain outlet. The procedure is to use the outlet state to estimate the inlet entropy as a crude guess, and then use trial and error inlet conditions until the desired outlet state is matched.

Example 4.15 Turbine inlet calculation given efficiency and outlet

An adiabatic turbine outlet (state 2) is 99% quality steam at 0.01 MPa, $\eta_E = 85\%$. The inlet pressure has been specified as 0.6 MPa. An absolute pressure of 0.6 MPa is conventionally defined as low pressure steam and is often applied in chemical processing. Find W_S, H_1, S_1, and T_1.

Solution: "Coincidentally," the outlet properties were determined in Example 4.14:
$H_2 = 2559.9$; $S_2 = 8.0738$. Referring to the superheated steam tables at 0.6 MPa, we seek an entropy value that is less than 8.0738 kJ/kg-K because $\eta_E < 100\%$ means entropy is generated. This occurs around 500°C. Trying 500°C, gives $H_1 = 3483.4$ kJ/kg and $S_1 = 8.0041$ kJ/kg-K. Then $W_S = -923.5$ kJ/kg; $q' = (8.0041 - 0.6492)/7.4996 = 0.9807$;
So $H_1' = 191.8 + 0.9807(2392) = 2537.6$ kJ/kg; $W_S' = 2537.6 - 3483.4 = -945.8$ kJ/kg;
$W_{\text{lost}} = 945.8 - 923.5 = 22.3$ kJ/kg; $\eta_E = -923.5/(-923.5 - 22.3) = 97.7\%$.

Further trials generate the values tabulated below. The last temperature is estimated by interpolation. (Hint: It would be great practice for you to compute these and check your answers.).

T	H_1	S_1	W_S	W_{lost}	η_E
450	3376.5	7.8611	−816.6	67.9	92.3%
400	3270.8	7.7097	−710.9	116.2	86.0%
350	3166.1	7.5481	−606.2	167.8	78.3%
393.8	3257.7	7.6895	−697.8	122.6	85.0%

Multistage Turbines

Commonly, turbines are staged for several reasons that we explore in Chapter 5. Generally, some steam is drawn off at intermediate pressures for other uses. The important point that needs to be stressed now is that the convention used for characterizing efficiency is important. Consider the three-stage turbine shown in Fig. 4.11 and the schematic that represents the overall reversible path and the actual path. The isobars on the H-S diagram for water curve slightly upward, and are spaced slightly closer together at the bottom of the diagram than at the top. The overall efficiency is given

by $\quad \eta_E^{overall} = \dfrac{H_4 - H_1}{\Delta H'_{overall}}$, and the efficiency of an individual stage is given by

$\eta_{E,i} = \dfrac{H_{i+1} - H_i}{\Delta H'_i}$. If we consider the reversible work as $\Delta H'_{overall}$, that quantity must be smaller

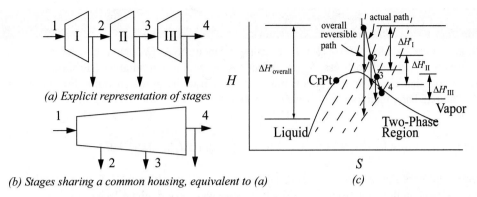

(a) Explicit representation of stages

(b) Stages sharing a common housing, equivalent to (a)

(c)

Figure 4.11 *Illustration that overall efficiency of an adiabatic turbine will be higher than the efficiency of the individual stages.*

than $\sum\limits_{stages} \Delta H'_i$. In fact, because the isobar spacing is increasing to the right of the diagram, the vertical drop between any isobars on the line marked as the overall reversible path must be smaller than the vertical drop between the same two isobars starting along the actual path (except for the very first turbine). Therefore, the efficiency calculated for the overall system must be higher than the efficiency for the individual stages. This comparison does not imply that staging turbines alters their performance. The difference in efficiencies is due to differences in what is considered to be the basis for the reversible calculation. The cautionary note to retain from this discussion is that the distinction between overall or individual efficiencies is important when communicating the performance of a staged turbine system.

Overall turbine efficiency will be greater than stage efficiencies for the same total work output.

4.10   PUMPS AND COMPRESSORS

An irreversible, adiabatic pump or compressor generates entropy. If these devices are reversible, they are isentropic. Examples of both are shown in Fig. 4.12. The calculations are generally straightforward. Consider the case where the inlet state and the outlet pressure is known. First, the reversible outlet state is determined based on the isentropic condition, and the enthalpy at the reversible state is known. The most common estimate for compressors is described in Example 2.12 on page 76. Even though it is intended for ideal gases, it is convenient for many applications and often provides a reasonable first approximation. The most common estimate for pumps is described in Eqn. 2.32 on page 55. These both pertain to reversible processes. The efficiency can then be used to determine the actual outlet enthalpy and work, using Eqn. 4.46.

Example 4.16 Isothermal reversible compression of steam

In Example 2.12 on page 76, we mentioned that computing the work for isothermal compression of steam was different from computing the work for an ideal gas. Now that you know about the entropy balance, use it to compute the work of continuously, isothermally, and reversibly compressing steam from 5 bars and 224°C to 25 bars. Compare to the result of the ideal gas formula.

Example 4.16 Isothermal reversible compression of steam (Continued)

Solution: Energy balance: $\Delta H = Q + W$. Entropy balance: $\Delta S = Q/T$. Note that $\Delta S \neq 0$, even though this is a reversible process. $S_{gen} = 0$, but the process is not adiabatic. From the steam tables, we note that 224°C and 25 bars is practically equal to the saturated vapor. For the vapor at 224°C and 25 bars, interpolation gives $H = 2910.5$, $S = 7.1709$. Noting $Q = T\Delta S$,
$Q = (224 + 273.15)(6.2558 - 7.1709) = -454.94$; $W = (2801.9 - 2910.5) + 454.94 = 346.3$ J/g.

By the ideal gas formula, $W = 8.314(4.04)(224 + 273.15)\ln(5)/18 = 1493.1$ J/g.

The work is less for the real vapor because of the intermolecular attractions. The difference was particularly large in this case because the final pressure was fairly high (> 10 bars).

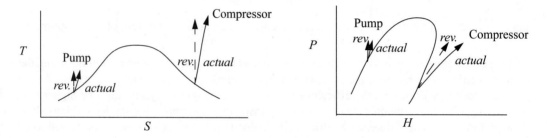

Figure 4.12 *Illustrations of pathways for reversible and irreversible pumps and compressors. The P-H diagram is for a system like Fig. 4.7(a).*

Multistage Compression

During adiabatic compression of vapors, the temperature rises. This can cause equipment problems if the temperature rise or pressure ratio (P^{out}/P^{in}) is too large. To address this problem, interstage cooling is used to lower the gas temperature between compression stages. Such operations are common when high pressures need to be reached. A schematic of a compressor with interstage cooling is shown in Fig. 4.13. The total work for multistage compression is generally given by summing the work of each stage using Eqn. 2.69 on page 77. However, the ideal gas law becomes less reliable as the stagewise inlet pressure increases. If the inlet pressure is above 10 bars and the reduced temperature is less than 1.5, nonideality effects should be evaluated. Methods to evaluate gas non-idealities and to calculate entropy for all manner of non-ideal gases are discussed in Unit II. For common refrigerants, it is convenient to apply charts that are functionally equivalent to the steam tables. The charts are difficult to read and precision is relatively low compared to using the steam tables. Example 4.17 illustrates the procedure using the refrigerant R134a.

Example 4.17 Compression of R134a using *P-H* chart

A compressor operates on R134a. The inlet to the compressor is saturated vapor at –20°C. The outlet of the compressor is at 7.7 bar and $\eta_C = 0.8$. Find the reversible and required work (kJ/kg) and the outlet temperature of the compressor.

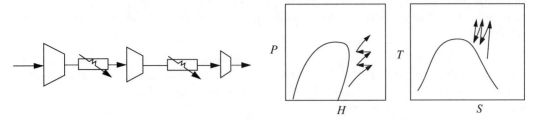

Figure 4.13 *Illustration of a multistage compression and the corresponding P-H diagram. On the P-H diagram, the compressors appear as the curves of increasing pressure and the heat exchangers are the horizontal lines at constant pressure.*

Example 4.17 Compression of R134a using *P-H* chart (Continued)

Solution: An inset of the *P-H* diagram from Appendix E is shown below. The axis labels and superheated temperature labels have been translated on the inset diagram.

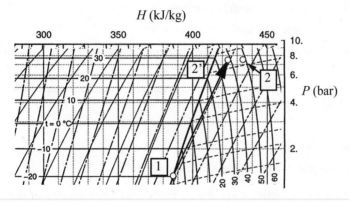

The inlet state is located at the intersection of the –20°C isotherm and the saturated vapor line. The enthalpy is found by following the vertical lines to the axis and $H = 386.5$ kJ/kg. (Note: This accurate value was found from the accompanying saturation table, but the schematic value is consistent, though less accurate.) The reversible outlet state is found by following an isentropic state up to 7.7 bars. One set of the diagonal lines are isentropes, and we visually interpolate to keep the same relative position between the isentropes at 7.7 bar at the state labeled 2'. By following the vertical lines to the axis, $H_2' = 424$ kJ/kg. The reversible work is $W_S' = 424 – 386.5 = 37.5$ kJ/kg. The actual work is $W_S = W_S' / 0.8 = 47$ kJ/kg. The actual outlet state is shifted to the right at 7.7 bar at an enthalpy value of $H_2 = 386.5 + 47 = 433.5$ kJ/kg. The reversible outlet is just near 38°C. The actual outlet is near 48°C.

4.11 STRATEGIES FOR APPLYING THE ENTROPY BALANCE

When solving thermodynamic problems, usually the best approach is to **begin by applying the mass and energy balances.** The entropy balance provides another balance, but it is not always necessary for every problem. In this chapter, we have introduced some new terms which can specify

New key words have been defined that specify constraints.

additional constraints when used in the problem statement, e.g., "isentropic," "reversible," "internally reversible," "irreversible," "thermal efficiency," and "turbine/expander or compressor/pump efficiency."

The entropy balance is useful to calculate maximum work available from a process or to evaluate reversibility. The entropy balance should be introduced with care because it is often redundant with the energy balance when simplified with information from step 5 from the strategies of Section 2.13. (For example, the entropy balance applied to Example 2.11 on page 75 results in the same simplified equation as the energy balance.) In general, if the pressures and temperatures of the process are already known, the entropies at each point, and the entropy changes, can be determined without direct use of the entropy balance. However, if either the pressure or the temperature is unknown for a process, the entropy balance may be the key to the solution.

Before beginning more examples, it is also helpful to keep in mind those processes which generate entropy. This is important because, in the event that such processes arise, the entropy-generation term cannot be set to zero unless we modify the process to eliminate the source of the generation. Entropy is generated by the following processes.

1. Heat conduction along a temperature gradient.

2. Diffusion along a concentration gradient.

3. Mixing of substances of different composition.

4. Adiabatic mixing at constant system volume of identical substances initially at different molar entropies due to (T, P) differences.

5. Velocity gradients within equipment. This is accounted for in pipe flow by the friction factor developed in textbooks on fluid flow.

6. Friction.

7. Electrical resistance.

8. Chemical reactions proceeding at measurable rates.

In an open system, irreversibilities are always introduced when streams of different temperatures are mixed at constant pressure (item 4 above) because we could have obtained work by operating a heat engine between the two streams to make them isothermal before mixing. If the streams are isothermal, but of different composition, mixing will still generate entropy (e.g., see Eqn. 4.8 on page 138), and we have not yet devised a general method to obtain work from motion on this molecular scale.

As chemical engineers, it is important to recognize that all chemical reactions proceeding at a finite rate generate entropy. The fundamental proof of this is provided in Section 17.16 and requires discussion of the chemical potential discussed much later in the text. Like the other processes listed here, reactions proceed spontaneously toward an equilibrium state due to finite driving forces. It is possible to calculate the rate of entropy generation if the chemical potentials are known at the reaction conditions. However, do not be deceived into thinking that a realistic reaction is thermodynamically reversible. Reaction engineering literature refers to reactions as reversible if the equilibrium constant (to be discussed later) is near 1 because the reaction can go "in either direction" (forward/backward) depending on the driving forces. This type of reversibility is not the same as thermodynamic reversibility. The reaction proceeding in either direction at measurable rates will generate entropy and be thermodynamically irreversible.

In some cases, T and P are known, so S can be determined without the entropy balance in a pure system.

Problem statements will rarely explicitly point out entropy generation, so you will need to look for causes.

When a situation requires the minimum work input, or the maximum work output, the system is designed to minimize entropy generation, or make it zero if possible considering the limitations discussed here. As we work examples, recall the comments from Section 3.4 which we repeat here:

> *Note: For* irreversible processes, *note that we* do not *apply the entropy balance to find entropy changes. We always calculate the entropy change by an alternate reversible pathway that reaches the final state, then we apply the entropy balance to find how much entropy was generated, or we find the reversible work, apply an efficiency factor, and identify the final state via the energy balance.*
>
> *Alternatively, for* reversible processes, *we do apply the entropy balance because we set the entropy generation term to zero.*

4.12   OPTIMUM WORK AND HEAT TRANSFER

Let us consider how to calculate the optimum work interactions for a general system. For an open system where kinetic energy and potential energy changes are negligible,

$$d\underline{U} = \sum_{\text{inlets}} H^{in}dn^{in} - \sum_{\text{outlets}} H^{out}dn^{out} + d\underline{Q} - Pd\underline{V} + d\underline{W}_S \qquad 4.50$$

$$d\underline{S} = \sum_{\text{inlets}} S^{in}dn^{in} - \sum_{\text{outlets}} S^{out}dn^{out} + \frac{d\underline{Q}}{T_{sys}} + d\underline{S}_{gen} \qquad 4.51$$

where $d\underline{S}_{gen} = 0$ for an internally reversible process. If all the heat is transferred at a single temperature T_{sys}, elimination of dQ in the first balance provides

$$d\underline{U} = \sum_{\text{inlets}} (H^{in} - T_{sys}S^{in})dn^{in} - \sum_{\text{outlets}} (H^{out} - T_{sys}S^{out})dn^{out} \qquad 4.52$$

$$+ T_{sys}d\underline{S} - Pd\underline{V} + d\underline{W}_S - T_{sys}d\underline{S}_{gen}$$

If we wish to apply this balance to a process that is conceptually reversible, we must use care to avoid any processes that are inherently irreversible (throttle valves, composition mixing processes, mixing of streams with identical composition but different temperatures, standard heat exchangers, chemical reactions at finite rates). Thus, if we consider a process that involves mixing compositions or reactions, we must include realistic estimates of these terms before determining the optimum work interaction. Once we recognize these limitations, we are ready to consider the general problem of finding optimum work interactions.

Availability (Exergy)

Section 4.5 considered optimum work interactions between a closed system and reservoir and found that optimum work interactions occur when the system temperature where heat transfer occurs is equal to the reservoir temperature. Therefore, for optimum work interactions with the surroundings, T_{sys} in Eqn. 4.52 should be replaced with the surrounding's temperature T_o. Though it leads to optimum work, it is an idealized condition because the *rate* of heat transfer is proportional to the temperature difference (as studied in heat transfer courses), and the heat transfer with infinitesimal temperature differences will be infinitesimally slow. The surrounding's temperature and pressure (T_o and P_o) are often considered the **dead state,** because when the system reaches this T and P, energy input of some type is necessary to obtain expansion/contraction work, shaft work, or heat transfer; without energy input the process is dead. However, departures from this dead state *do* provide opportunities for work and heat interactions. Further, it is desirable to give a name to the combination of variables that results. For the combination of variables in the summations of Eqn. 4.52 modified with T_o, we will use the term **availability,** or **exergy,** B,

$$B = H - T_o S \quad \text{availability or exergy} \tag{4.53}$$

where H and S are state properties of the system at T and P, but T_o is the temperature of the dead state. The terms "availability" and "exergy" are both used in literature for this property. At a given T and P, the availability changes with T_o, so B is somewhat different from other state properties used to this point. Inserting the availability into Eqn. 4.52, and collecting the state changes of the system on the left-hand side, results in a general balance (ignoring kinetic and potential energy like before),

$$d\underline{U} + Pd\underline{V} - T_o d\underline{S} = \sum_{\text{inlets}} (B^{in})dn^{in} - \sum_{\text{outlets}} (B^{out})dn^{out} + d\underline{W}_S - T_o d\underline{S}_{gen} \tag{4.54}$$

Steady-State Flow

For a system at steady-state flow, all terms on the left-hand side drop out, resulting in

$$\sum_{\text{outlets}} (B^{out})dn^{out} - \sum_{\text{inlets}} (B^{in})dn^{in} = d\underline{W}_S - T_o d\underline{S}_{gen} \quad \text{steady-state open system} \tag{4.55}$$

and we conclude that the difference in availability from the inlets to outlets is related to the optimum shaft work. Note that $T_o d\underline{S}_{gen} \geq 0$ and always subtracts from work input when $d\underline{W}_S \geq 0$ which means work input for an irreversible process is always greater than a reversible process for a given state change. Similarly, for a given state change producing work, an irreversible process will always produce less work compared to a reversible process. The quantity $T_o \underline{S}_{gen}$ is called the **lost work** and the reason for the term should now be obvious. T_o is ambiguous when no heat is transferred to the surroundings and drops out for an adiabatic process such as an adiabatic turbine without any loss in applicability of the equation,

Josiah Willard Gibbs (1839–1903) was an American chemist, mathematician, and physicist. Yale University awarded Gibbs the first American Ph.D. in Engineering in 1863 and Gibbs spent his career there. Gibbs is recognized for applying calculus to thermodynamics and combining the first and second laws. Gibbs studied the concept of chemical potential, the Gibbs phase rule, and many other relations.

$$(\Delta H - T_o \Delta S)dn^{out} = d\underline{W}_S - T_o d\underline{S}_{gen} \quad \text{but} \quad \Delta H = W_S \quad \text{and} \quad -T_o \Delta S = -T_o S_{gen} \tag{4.56}$$

Closed System

For a closed system, Eqn. 4.54 becomes

$$d\underline{U} + Pd\underline{V} - T_o d\underline{S} = d\underline{W}_S - T_o d\underline{S}_{gen} \quad \text{closed system} \tag{4.57}$$

For a constant-pressure closed system, $\underline{V}dP$ can be added to the left side (because it is zero in magnitude), which then results in $d\underline{U} + Pd\underline{V} + \underline{V}dP = d\underline{U} + d(P\underline{V}) = d\underline{H}$. Thus, the left-hand side can be replaced with $d\underline{B}$, though it is usually easier to calculate $d\underline{H}$ and $T_o d\underline{S}$ independently. We can define the **Gibbs energy** $G \equiv H - TS$. Then $dG = dU + PdV + VdP - TdS - SdT$. Thus, if the pressure is constant *and* the temperature is constant at T_o, then the change in Gibbs energy is related to the nonexpansion/contraction work. Another interesting analysis can be done if both work terms are on the right-hand side of the equation:

$$dU - T_o dS = dW_S - PdV - T_o dS_{gen} \quad \text{closed system} \qquad 4.58$$

We can define the **Helmholtz energy** $A \equiv U - TS$. Then $dA = dU - TdS - SdT$. When the system is isothermal at T_o, then the change in Helmholtz energy is related to the sum of all forms of work.

Hermann Ludwig Ferdinand von Helmholtz (1821–1894) was a German physician and physicist. Besides studying optics of the eye, he studied the concept of conservation of energy.

Availability Analysis

As the world population grows and energy use increases, energy conservation will become increasingly important. Not only is the energy balance important, but so is the wise use of existing resources. Availability analysis can be used to determine how much entropy is generated by a process. Availability analysis is sometimes used in process design analysis,[18] and may be used more widely in the future as we consider wise stewardship of energy resources.

Minimum Work for Separation

A key challenge in the development of a fermentation process is the **titer,** or yield from the fermentation. Fermentation is used in the pharmaceutical industry to develop natural molecules. For example, bacterial production of insulin is a feasible technology. Much of the cost of production depends on the dilution level. We can estimate the minimum energy cost for concentrating a product by combining the energy and entropy balances. Suppose that for a feed F we wish to obtain a pure product P and by-product water, B, in an isothermal process. For continuous separation of a mixed stream, the steady-state energy balance for a reversible separation is

$$0 = H_F \dot{n}_F - H_B \dot{n}_B - H_P \dot{n}_P + \dot{Q}_{rev} + \dot{W}_s \qquad 4.59$$

Assuming an ideal solution, we can recognize that because the heat of mixing is zero the energy balance becomes

$$H_F \dot{n}_F - H_B \dot{n}_B - H_P \dot{n}_P = \left(\sum x_i H_i\right)\dot{n}_f - H_B \dot{n}_B - H_P \dot{n}_P = 0 \qquad 4.60$$

The energy balance thus simplifies to

$$\dot{Q}_{rev} = -\dot{W}_s \qquad 4.61$$

The steady-state entropy balance for reversible separation exchanging heat at T is:

$$0 = T(S_F \dot{n}_F - S_B \dot{n}_B - S_P \dot{n}_P) + \dot{Q}_{rev} \qquad 4.62$$

The strategy will be to determine the heat transfer from the entropy balance and then use it in the energy balance to find the work.

The entropy of the feed stream will be (using Eqn. 4.9)

---

18. Bejan, A. 2006. *Advanced Engineering Thermodynamics,* 3ed, New York: John Wiley & Sons.

$$\dot{n}_F S_F = \dot{n}_F \sum_i x_i S_i - \dot{n}_F R \sum_i x_i \ln x_i \qquad 4.63$$

Inserting,

$$0 = T\left(\dot{n}_F \sum_i x_i S_i - \dot{n}_F R \sum_i x_i \ln x_i - S_B \dot{n}_B - S_P \dot{n}_P\right) + \dot{Q}_{rev} = -\dot{n}_F R T \sum_i x_i \ln x_i + \dot{Q}_{rev} \qquad 4.64$$

which becomes

$$RT \dot{n}_F (x_1 \ln x_1 + x_2 \ln x_2)_F = \dot{Q}_{rev} \qquad 4.65$$

Let us suppose that component 2 is the desired species. On the basis of heat transfer per mole of component 2,

$$\frac{RT \dot{n}_F (x_1 \ln x_1 + x_2 \ln x_2)_F}{\dot{n}_2} = \frac{RT(x_1 \ln x_1 + x_2 \ln x_2)_F}{x_2} = \frac{\dot{Q}_{rev}}{\dot{n}_2} \qquad 4.66$$

Combining with the energy balance, the minimum work per mole of desired species is

$$\frac{-RT(x_1 \ln x_1 + x_2 \ln x_2)_F}{x_2} = \frac{\dot{W}_s}{\dot{n}_2} \qquad 4.67$$

Example 4.18 Minimum heat and work of purification[a]

Products produced by biological systems can range over mole fractions from 10^{-1} to 10^{-9}.

(a) Estimate the minimum reversible heat and work requirement to purify one mole of product at 298.15 K over this range.
(b) To understand the concentrations in dilute mixtures, calculate the mole fraction of insulin in 0.1wt% aqueous solution.

Solution: The work is from Eqn. 4.67, and the heat will have the opposite sign.

$$\frac{-RT(x_1 \ln x_1 + x_2 \ln x_2)_F}{x_2} = \frac{\dot{W}_s}{\dot{n}_2}$$

For $x_2 = 0.10$, $x_1 = 0.9$. At 298.15K,
$\dot{W}_s / \dot{n}_2 = -8.314(298.15)(0.1\ln(0.1) + 0.9\ln(0.9)) / 0.1 = 8.1$kJ

Repeating the calculation for other values of x_2:.

x_2	10^{-1}	10^{-2}	10^{-3}	10^{-4}	10^{-5}	10^{-6}	10^{-7}	10^{-8}	10^{-9}
$\dot{W}_s / \dot{n}_2$ (kJ/mol)	8.1	13.9	19.6	25.3	31.0	36.7	42.4	48.1	53.8

> ### Example 4.18   Minimum heat and work of purification[a]   (Continued)
>
> Note that heat must be rejected. If a process is envisioned that requires heat, then the rejected heat must be increased by an equal amount. For example, if a solution of concentration $x_2 = 0.10$ is purified by adding 1 kJ of heat, then 9.1 kJ must be rejected.
>
> (b) Searching for the molecular weight of insulin reveals a value of 5808 g/mol. Therefore, $x_1 = 0.001/5808 = 1.72(10^{-7})$. The point is that biomolecules are often large, and therefore their concentrations can be quite small on a mole fraction basis.

a.   This problem was suggested by O'Connell, J.P. et al. July 2010. NSF BioEMB Workshop, San Jose, CA

4.13   THE IRREVERSIBILITY OF BIOLOGICAL LIFE

A fascinating feature of living systems is that they organize small molecules into large structures. Towering pines grow with energy from the sun, CO_2, water, and minerals extracted from the ground. Mammals grow into sophisticated thinking creatures by consuming small bits of food, consuming water, and breathing air. Small mindless flagella are known to swim "up" a concentration gradient toward a food source in a process known as chemotaxis. All of biological life builds molecules that are chiral rather than racemic. Don't these processes violate the principles developed thus far in this chapter where we indicated the tendency of a system to move toward randomness? A careful analysis shows that the answer is no.

The key is that the feasibility of a process is determined by the entropy change of the universe, not the system itself. If organisms build molecules with lower entropy than the reactants, then the surroundings must increase in entropy by a larger amount. These changes can occur by control of the flux of molecules in/out of the system or by heat transfer. Mammals in particular are warm-blooded, and are virtually isothermal. The body temperature of a healthy human being is universally 37°C (310 K). Rearrangement of the entropy balance for a human provides

$$(d(nS))/(dt) - T\dot{S}_{gen} = TS^{in}\dot{n}^{in} - TS^{out}\dot{n}^{out} + \dot{Q} \qquad 4.68$$

Technically, if we assume $T = 310K$, we also should recognize that entropy is generated at the boundary between the human and the surroundings (at a lower temperature) and this is not included.

Every biochemical reaction in the body continuously generates entropy. If the organism is fully grown, we can approximate an adult as a steady-state process, so the first term on the left is nearly zero. The second term on the left contributes a negative value because entropy is generated constantly by the biochemical reactions. Ignoring transport terms initially, we can see that the heat transfer is expected to be negative. The magnitude depends on the entropy flows entering and leaving the system. Considering the other limiting possibility of negligible heat transfer, the entropy flow "in" must be less than the entropy flow "out." In practice, these fluxes are not sufficient to sustain life. For humans, we know that our bodies reject heat at a rate of about 100 J/s. Thus, in addition to changing the entropy of our food to waste products, our existence depends on the ability to reject heat to the surroundings and thus contribute to increase the entropy of the surroundings via heat transfer. In the event that the surrounding temperature rises, humans experience heat stroke because the biological processes stop working when the heat transfer is not possible.

Trees and plants use photosynthesis to convert light energy (photons), to run the reaction

$$nCO_2 + nH_2O + \text{light} \quad \rightarrow \quad (CH_2O)_n + nO_2 \qquad 4.69$$

We have not explicitly included photons in our energy balance thus far, but it is a form of radiation similar to heat. A key point is that this reaction is not spontaneous as written. We discuss the driving forces for chemical reactions towards the end of the text, but a superficial discussion is relevant here. This reaction is an oversimplification of the actual process. In fact, more than 100 steps are needed for photosynthetic construction of glucose,[19] but these steps drive the overall reaction above. To create a forward reaction, biological systems have developed a complex series of steps, each spontaneous and coupled such that carbohydrate production is possible. Plants are able to maintain this reaction by increasing the entropy of the surroundings to a greater extent than this reaction decreases the entropy when turning small molecules into carbohydrates.

Energy usage is inherently less efficient as we move up the food chain. When a herbivore or omnivore eats plant material, the stomach and intestinal enzymes break down the carbohydrates to simple sugars in **catabolic reactions** (a step that increases entropy). These sugars are then "burned" to produce CO_2 (increasing entropy) as we showed in the previous chapter and the energy is used to maintain the life cycle and continuously produce new cells (**anabolic reactions,** decreasing entropy[20]) to replace dying cells and maintain tissue. Also, ingested proteins are broken into amino acids (increasing entropy) and then reassembled into new proteins (decreasing entropy[2]). As discussed above, only a portion of the energy provided by glucose can be utilized for biological maintenance; some must be rejected as heat. Each creature in the food chain repeats this "loss" of energy as heat is rejected. As engineers grapple with the challenge to use energy efficiently, it is helpful to keep in perspective that despite the complexity of living entities, the fundamentals of life are governed by the same principles as those developed in this chapter. Furthermore, life is made possible in humans by rejecting heat to the surroundings to drive the biological machinery. Despite their complexity, humans are not able to use all the energy generated by burning sugars.

4.14 UNSTEADY-STATE OPEN SYSTEMS

We end the chapter by providing examples of unsteady-state open systems. The first example shows that analysis of such systems can produce results quite consistent with expansion in a piston/cylinder.

Example 4.19 Entropy change in a leaky tank

Consider air (an ideal gas) leaking from a tank. How does the entropy of the gas in the tank change? Use this perspective to develop a relation between T^f and P^f and compare it to the expression we obtained previously by the energy balance.

---

19. Kondepudi, D. 2008. *Introduction to Modern Thermodynamics*. New York: Wiley, p. 386.
20. Though the structure building decreases entropy, the reactions are proceeding at a finite rate which generates entropy.

Example 4.19 Entropy change in a leaky tank (Continued)

Solution:

m-Balance: $dn = -dn^{out}$

S-Balance: $\dfrac{d(nS)}{dt} = -S^{out}\dfrac{dn^{out}}{dt} \Rightarrow ndS + Sdn = -S^{out}dn^{out}$

But physically, we know that the leaking fluid is at the same state as the fluid in the tank; therefore, the S-balance becomes $ndS + S\!\!\!\diagup\!\!dn = -S^{out}\!\!\!\diagup\!\!dn^{out}$, or $\Delta S = 0$.

For an ideal gas with a constant heat capacity:

$$\Delta S = C_V \ln(T_2/T_1) + R\ln(V_2/V_1) = 0 \qquad\qquad (\text{*ig})$$

$$\Delta S = C_V\ln(T_2/T_1) + R\ln((T_2 P_1)/(P_2 T_1)) = (C_V + R)\ln(T_2/T_1) - R\ln(P_2/P_1) \quad (\text{*ig})$$

$$(T_2/T_1) = (P_2/P_1)^{(R/C_P)} \qquad\qquad (\text{*ig})$$

Compare with Example 2.15 on page 81. The entropy balance and energy balance in this case are not independent. Either can be used to derive the same result. This also shows that our analysis in Example 2.15 was assumed to be reversible.

> ❶ Illustration that the energy and entropy balances may not be independent.

The next example builds on the first by adding a turbine to the tank. Note the method by which the system is subdivided to solve the problem.

Example 4.20 An ideal gas leaking through a turbine (unsteady state)

A portable power supply consists of a 28-liter bottle of compressed helium, charged to 13.8 MPa at 300 K, connected to a small turbine. During operation, the helium drives the turbine continuously until the pressure in the bottle drops to 0.69 MPa. The turbine exhausts at 0.1 MPa. Neglecting heat transfer, calculate the maximum possible work from the turbine. Assume helium to be an ideal gas with $C_P = 20.9$ J/mol-K.

Consider a balance on the *tank only*. The result of the balance will match the result of Example 4.19.

Writing an entropy balance for a reversible adiabatic *turbine only*,

$$(S^{out} - S^{in})\dot{n} = 0 \Rightarrow \Delta S = 0$$

Example 4.20 An ideal gas leaking through a turbine (unsteady state)

which shows that the turbine also does not change the molar entropy. Thus, the molar entropy of the exiting fluid is the same as the entropy in the tank, which is identical to the molar entropy at the start of the process. Therefore, the molar entropy and the pressure of the exiting gas are fixed. Since only two intensive properties fix all other intensive properties for a pure fluid, the exiting temperature is also fixed. The relation for an ideal gas along a reversible adiabat gives:

$$T^{out} = T^i \, (P^{out} / P^i)^{R/C_P} = 42.3 \text{ K} \qquad\qquad (\text{*ig})$$

$$\text{Likewise: } T^f = T^i \, (P^f / P^i)^{R/C_P} = 91.1 \text{ K} \qquad\qquad (\text{*ig})$$

Solution by overall energy balance:

$d(nU) = H^{out} \, dn + d\underline{W}_S$ and H^{out} is fixed since T^{out}, P^{out} are fixed; therefore, we may apply hint 4(a) from Section 2.14.

Integrating this expression:

$$n^f U^f - n^i U^i = H^{out}(n^f - n^i) + \underline{W}_S$$

Rearranging:

$$\underline{W}_S = n^f (U^f - H^{out}) - n^i (U^i - H^{out}) \qquad\qquad 4.70$$

Determining variables in the equation:

$$n^f = P^f \, \underline{V}/RT^f; \; n^f = 25.5 \text{gmol}; \; n^i = 154.9 \text{ gmol} \qquad\qquad (\text{ig})$$

Choose reference temperature, $T_R \equiv 300$ K, $\Rightarrow$ setting $U_R = 0$, then since $H_R = U_R + (PV)_R$, and since the fluid is an ideal gas, $C_V = C_P - R = 20.9 - 8.314 = 12.586$ J/mol-K:

$$H_R = (PV)_R = RT_R = R(300) \qquad\qquad (\text{ig}) \; 4.71$$

$$\text{Note:} \Rightarrow H(T) = C_P(T - T_R) + H_R = C_P(T - 300) + R(300) \qquad\qquad (\text{*ig})$$

$$H^{out} = -2892 \text{ J/mol} \qquad\qquad (\text{*ig})$$

$$U^f = C_V(T - T_R) + U_R = -2629 \text{ J/mol}; \; U^i = 0 \qquad\qquad (\text{*ig})$$

Now, plugging into Eqn. 4.70:

$$\underline{W}_S = 25.5(-2629 + 2892) - 154.9(0 + 2892)$$

$$\Rightarrow \underline{W}_S = -441{,}200 \text{ J}$$

❶ Illustration using a reference state.

4.15   THE ENTROPY BALANCE IN BRIEF

In this section, we refer to a division of the universe into the same three subsystems described in Section 2.14 on page 74.

$$\frac{dS}{dt} = \sum_{inlets} S^{in} \dot{m}^{in} - \sum_{outlets} S^{out} \dot{m}^{out} + \sum_{surfaces} \frac{\dot{Q}}{T} + \dot{S}_{gen}$$

1. T is the *system* temperature at the location where Q is transferred.
2. S^{in}, S^{out} are state variables, and *any* pathway may be used to calculate the change from inlet to outlet. The pathway for calculation does not need to be the pathway for the actual process.
3. $\dot{S}_{gen}$ represents entropy generation due to irreversibilities *within* the system, e.g., internal heat transfer or conduction, viscous dissipation or mixing of streams of differing composition, or degradation of mechanical energy to thermal energy. Entropy generation at system boundaries is not included in the balance.
4. Entropy generation may occur at container walls. The entropy generation of the universe must be calculated by summing $\dot{S}_{gen}$ for all three subsystems, not just system 1 and system 3.

Test Yourself

1. What are the constraints on the sign of $\dot{S}_{gen}$?
2. Consider two isothermal processes both rejecting heat at the same temperature. One process is reversible and the other is irreversible. Which has a larger absolute value of heat transfer?

4.16   SUMMARY

We began the chapter introducing microscopic methods to calculate entropy. We demonstrated that entropy increases when volume (for a gas) or temperature increases. Thermal energy is really a means of representing the randomness due to accessible microstates, such that the concept is best understood in terms of the microscopic definition of entropy. We showed that the macroscopic definition was consistent with the microscopic definition. We showed that entropy is essential for analysis of reversibility for processes, because irreversible processes generate entropy. We demonstrated that reversible Carnot cycle thermal efficiency was easily evaluated using entropy. The primary impact for pure-fluid applications is that compressors and turbines can be analyzed using empirical efficiencies relative to reversible devices. On a broader scale, however, you should appreciate the limitations of the conversion of heat into work.

This chapter is relatively long because mastering computations involving entropy can be challenging. Students may be familiar with energy balances and heats of reaction from previous courses, but entropy may seem new and abstract. Therefore, many examples have been provided. Students are encouraged to review these and the practice problems at the end of the chapter. The best way to develop a comfort level with entropy is to practice and learn by doing.

Important Equations

Entropy is a state change, and for an ideal gas,

$$\Delta S^{ig} = C_V \ln\frac{T}{T^i} + R\ln\frac{V}{V^i} \quad \text{or} \quad \Delta S^{ig} = C_P \ln\frac{T}{T^i} - R\ln\frac{P}{P^i} \qquad (\text{*ig})\ 4.29$$

For a condensed phase, the first term of the second equation should be used for a first approximation using the heat capacity for the appropriate phase.

Of course, the most important equation of this chapter is the complete entropy balance, but it may be convenient to remember some of its most common simplifications.

$\Delta S = 0$ for a fluid in an adiabatic reversible process, like across a steady-state **adiabatic reversible** compressor or turbine or within an unsteady-state adiabatic reversible piston/cylinder.

$\Delta S = Q_{rev}/T$ for a fluid in any process, like an **isothermal reversible** compressor. For an irreversible process, we design a reversible pathway to the actual final state.

$\Delta S = 0$ for a the entropy in a reversible adiabatic leaking tank.

Pay careful attention to the subtle distinctions between these equations. A common mistake is to write $\Delta S = 0$ whenever you see the word "reversible." Remember that reversibility is coupled to entropy generation, not directly to S or ΔS. For entropy changes of a fluid, a smart approach is to write $\Delta S = Q_{rev}/T$ always and then scratch a line through Q_{rev} after you deliberately determine that the process is adiabatic and reversible. A fluid in an irreversible adiabatic process will have an entropy change (and Q_{rev}) even though the actual heat transfer is zero. Note that the entropy balance uses the actual heat transfer, Q, not Q_{rev}. Only for a reversible process are they identical.

The distinction between $\Delta \underline{S} = 0$ and $\Delta S = 0$ is perhaps subtler. By writing $\Delta S = 0$, we emphasize that only the specific entropy of the fluid remains constant. When $\Delta S = 0$ for an open system, the quantity $\Delta \underline{S} = n^f S^f - n^i S^i = S\Delta n$ will be non-zero whenever S is not at the reference state.

Disorder must increase if two different gases are mixed slowly and adiabatically, but it is difficult to see how to compute the entropy change from the macroscopic definition of dS. For ideal solutions, the relation developed from the microscopic approach is:

$$\boxed{\Delta \underline{S}^{is}_{mix} = -R\sum_i x_i \ln x_i} \qquad 4.8$$

This equation is useful for estimating the effects of mixing in many situations, even beyond the assumption of ideal gases from which it derives. Eqn. 4.8 also conveys how disorder and mixing are not strictly related to heating, as one might infer from the macroscopic definition of entropy. The entropy of mixing will be of major importance in Unit III in the discussion of mixtures.

4.17 PRACTICE PROBLEMS

P4.1 Call placement of a particle in box A, "heads" and placement in box B, "tails." Given one particle, there are two ways of arranging it, H or T. For two particles, there are four ways of arranging them, {HH,HT,TH,TT}. We can treat the microstates by considering each particle in order. For example, {H T H H} means the first particle is in box A, the second in box B, the third in box A, and the fourth in box A.

(a) List and count the ways of arranging three particles. Now consider four particles. What is the general formula for the number of arrangements versus the number of particles? (ANS. 2^N)

(b) How many arrangements correspond to having two particles in box A and one in box B? What is the probability of {2H,1T}? (ANS. 3/8)

(c) How many arrangements correspond to {2H,2T}. {3H,2T}. {4H,2T}. {3H,3T}? (ANS. $N!/[(N-m)!m!]$)

(d) List the macrostates and corresponding number of microstates for an eight-particle, two-box system. What portion of all microstates are parts of either 5:3, 4:4, or 3:5 macrostates? (ANS. 71%)

(e) What is the change of entropy in going from a 5:3 macrostate to a 4:4 macrostate? (ANS. 3.08E-24 J/K)

(f) Use Stirling's approximation to estimate the change of entropy in going from a distribution of 50.1% of 6.022E23 in box A to a distribution of 50.001%, and from 50.001% to 50.000%. (ANS. 1.2E18 J/K)

P4.2 Twenty molecules are contained in a piston + cylinder at low pressure. The piston moves such that the volume is expanded by a factor of 4 with no work produced of any kind. Compute $\Delta \underline{S}/k$. (ANS. 23.19)

P4.3 Fifteen molecules are distributed as 9:4:2 between equal-sized boxes A:B:C, respectively. The partitions between the boxes are removed, and the molecules distribute themselves evenly between the boxes. Compute $\Delta \underline{S}/k$. (ANS. 11.23)

P4.4 Rolling two die (six-sided cubes with numbers between 1 and 6 on each side) is like putting two particles in six boxes. Compute $\Delta \underline{S}/k$ for going from double sixes to a four and three. (ANS. 0.693)

P4.5 Estimate the change in entropy when one mole of nitrogen is compressed by a piston in a cylinder from 300 K and 23 liters/mol to 400 K and 460 liters/mol. ($C_P = 7/2R$) (ANS. 1.07 kJ/kgK)

P4.6 Steam at 400°C and 10 bar is left in an insulated 10 m³ cylinder. The cylinder has a small leak, however. Compute the conditions of the steam after the pressure has dropped to 1 bar. What is the change in the specific entropy of the steam in the cylinder? Is this a reversible process? The mass of the cylinder is 600 kg, and its heat capacity is 0.1 cal/g-K. Solve the problem with and without considering the heat capacity of the cylinder. (ANS. (a)~120°C; (b) 360°C)

P4.7 A mixture of 1CO:2H₂ is adiabatically and continuously compressed from 5 atm and 100°F to 100 atm and 1100°F. Hint: For this mixture, $C_P = x_1 C_{P1} + x_2 C_{P2}$.

(a) Estimate the work of compressing 1 ton/h of the gas. ($C_P = 7/2R$) (ANS. 1.3E6 BTU/h)

(b) Determine the efficiency of the compressor. (ANS. 76%)

P4.8 An adiabatic compressor is used to continuously compress nitrogen ($C_P/R = 7/2$) from 2 bar and 300 K to 15 bar. The compressed nitrogen is found to have an outlet temperature of 625 K.

How much work is required (kJ/kg)? What is the efficiency of the compressor? (ANS. 9.46 kJ/mol, 72%)

P4.9 An adiabatic compressor is used to continuously compress low-pressure steam from 0.8 MPa and 200°C to 4.0 MPa and 500°C in a steady-state process. What is the work required per kg of steam through this compressor? Compute the efficiency of the compressor. (ANS. 606 J/g, 67%)

P4.10 An adiabatic turbine is supplied with steam at 2.0 MPa and 600°C and the steam exhausts at 98% quality and 24°C. Compute the work output per kg of steam. Compute the efficiency of the turbine. (ANS. 1.2E3 kJ, 85%)

P4.11 An adiabatic compressor has been designed to continuously compress 1 kg/s of saturated vapor steam from 1 bar to 100 bar and 1100°C. Estimate the power requirement of this compressor in horsepower. Determine the efficiency of the compressor. (ANS. 3000 hp, 60%)

P4.12 Ethylene gas is to be continuously compressed from an initial state of 1 bar and 20°C to a final pressure of 18 bar in an adiabatic compressor. If compression is 70% efficient compared with an isentropic process, what will be the work requirement and what will be the final temperature of the ethylene? Assume the ethylene behaves as an ideal gas with $C_P = 44$ J/mol-K. (ANS. 13.4 kJ/mol, 596 K)

P4.13 Operating a wind tunnel for aircraft experiments begins with adiabatically and reversibly compressing atmospheric air (300 K) into long cylinders comprising a total volume of 20 m^3 at 200 bars. The cylinders are initially at 1 bar. Estimate the minimal amount of work required (MJ) to perform the compression step. (ANS. online.)
(a) Write the most appropriate energy balance(s) for this process. Clearly identify the system(s) pertaining to your energy balance(s). Explain your reasoning *briefly*.
(b) Write the most appropriate entropy balance(s) for this process. Clearly identify the system(s) pertaining to your entropy balance(s). Explain your reasoning *briefly*.
(c) Solve for the minimal amount of work required (MJ) to perform the compression step.

P4.14 As part of a refrigeration cycle, Freon 134a is adiabatically compressed from the saturated vapor at –60°C (note the negative sign on temperature) to 1017 kPa and 100°C.
(a) How much work is required in kJ/kg?
(b) Estimate the efficiency of the compressor.
(ANS. 121, 75%)

P4.15 Steam is produced at 30 bar and some unknown temperature. A small amount of steam is bled off and goes through an adiabatic throttling valve to 1 bar. The temperature of the steam exiting the throttling valve is 110°C. What is the value of the specific entropy of the steam before entering the throttle? (ANS. 5.9736 J/g-K)

P4.16 Suppose the expansion in problem P2.19 was completely adiabatic instead of isothermal and $C_P = 7$ cal/(mol-K). How would the height of the piston be affected? Must we generate heat or consume heat to maintain isothermal operation? (ANS. decrease, generate)

P4.17 It is desired to determine the volume of an initially evacuated tank by filling it from an 80 liter cylinder of air at 300 bar and 300 K. The final pressure of both tanks is 5 bars. Estimate the volume in liters. (ANS. 4720 L)

P4.18 An insulated cylinder is fitted with a freely floating piston, and contains 0.5 kg of steam at 9 bar and 90% quality. The space above the piston, initially 0.05m^3, contains air at 300 K to maintain the pressure on the steam. Additional air is forced into the upper chamber, forcing the piston down and increasing the steam pressure until the steam has 100% quality. The final steam pressure is 30 bars, and the work done on the steam is 360 kJ, but the air above the steam has not had time to exchange heat with the piston, cylinder, or surroundings. The air supply line is at 50 bar and 300 K. What is the final temperature of the air in the upper chamber? (ANS. online)

P4.19 A well-insulated cylinder, fitted with a frictionless piston, initially contained 9 kg of liquid water and 0.4 kg of water vapor at a pressure of 1.4 MPa. 2 kg of steam at 1.6 MPa was admitted to the cylinder while the pressure was held constant by allowing the piston to expand. (ANS. online)
(a) Write the energy balance for this process.
(b) If the final volume of the contents of the cylinder was six times the initial volume, determine the temperature of the superheated steam that was admitted to the cylinder.

P4.20 Many action movies show gas cylinders that have their caps knocked off. The tanks go flying around wreaking havoc (only on the bad guys, of course). How much velocity could a tank like that really generate? For an upper bound, consider a tank traveling horizontally on a frictionless surface with an isentropic nozzle taking the place of the cap that has been knocked off. Suppose the cylinder weighs 70 kg and holds 50 L of He at 100bar, 300 K.
(a) Write the most appropriate energy balance(s) for this process. Clearly identify the system(s) pertaining to your energy balance(s). Explain your reasoning *briefly*.
(b) Write the most appropriate entropy balance(s) for this process. Clearly identify the system(s) pertaining to your entropy balance(s). Explain your reasoning *briefly*.
(c) Solve for the total kinetic energy (MJ) developed by the tank and its velocity.

4.18   HOMEWORK PROBLEMS

4.1 Extending Example 4.2 on page 141 from solids to gases is straightforward if you recall the development of Eqn. 1.13 on page 19. Consider N_2 for example. Being diatomic, we should expect that $U^{ig} = 2(3N_A kT/2) = 6RT/2$ in the limit of classical vibrations. Vibrational energy means that heat can be absorbed in the vibration of a bond. Since N_2 has only one bond, it can only absorb energy in one way, removing one degree of freedom. We show in Eqn. 6.49 on page 240 that $\Delta U^{vib}/RT = \beta\varepsilon/[\exp(-\beta\varepsilon) - 1]$ is the change in energy due to vibration. For now, without concern for the proof, assume ΔU^{vib} as given. Adapting Example 4.2 for N_2 then gives: $U^{ig}/RT = 2.5 + \beta\varepsilon/[\exp(-\beta\varepsilon) - 1]$.

   a. Use the NIST WebBook to plot data for C_V of N_2 at 0.1 MPa and $T = [150, 2000\ \text{K}]$.
   b. Derive an expression for C_V based on above disussion. Evaluate your expression at 1000K assuming $\varepsilon/k = 1000$K.
   c. Regress an optimal value for ε/k of N_2 and plot a comparison of the calculated results to experimental data. Show the calculated results as a curve with no points.

4.2 An ideal gas, with temperature-independent $C_P = (7/2)R$, at 15°C and having an initial volume of 60 m³, is heated at constant pressure ($P = 0.1013$ MPa) to 30°C by transfer of heat from a reservoir at 50°C. Calculate $\Delta \underline{S}_{gas}$, $\Delta \underline{S}_{heat\ reservoir}$, $\Delta \underline{S}_{universe}$. What is the irreversible feature of this process?

4.3 Steam undergoes a state change from 450°C and 3.5 MPa to 150°C and 0.3 MPa. Determine ΔH and ΔS using the following:

(a) Steam table data
(b) Ideal gas assumptions (be sure to use the ideal gas heat capacity for water)

4.4 The following problems involve one mole of an ideal monatomic gas, $C_P = 5R/2$, in a variable volume piston/cylinder with a stirring paddle, an electric heater, and a cooling coil through which refrigerant can flow (see figure). The piston is perfectly insulated. The piston contains 1 gmole of gas. Unless specified, the initial conditions are: $T^i = 25°C$, $P^i = 5$ bar.

Heater Coil

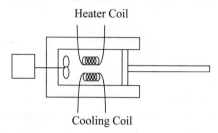

Cooling Coil

(a) Status: Heater on; cooler off; paddle off; piston fixed. Five kJ are added by the heater. Find ΔU, ΔS, ΔP, and ΔT.

(b) Status: Heater off; cooler off; paddle off; piston moveable. What reversible volume change will give the same temperature rise as in part (a)? Also find ΔU, ΔS, and ΔP.

(c) Status: Heater off; cooler off; paddle on; piston fixed. What shaft work will give the same ΔU, ΔS as part (a)?

(d) Status: Heater off; cooler off; paddle on; piston fixed. The stirring motor is consuming 55 watts and is 70% efficient. What rate is the temperature changing? At what initial rates are U and S changing?

(e) Status: Heater unknown; cooler unknown; paddle off; piston free. We wish to perform a reversible isothermal compression until the volume is half of the initial volume. If the volume is decreasing at 2.0 cm^3/s, at what rate should we heat or cool? Express your answer in terms of the instantaneous volume. What is the total heat transfer necessary?

4.5 When a compressed gas storage tank fails, the resultant explosion occurs so rapidly that the gas cloud can be considered adiabatic and assumed to not mix appreciably with the surrounding atmosphere. Consider the failure of a 2.5-m^3 air storage tank initially at 15 bar. Atmospheric pressure is 1 bar, $C_P = 7R/2$. Provide an estimate by assuming reversibility.

(a) Calculate the work done on the atmosphere. Does the reversibility approximation over-estimate or under-estimate the actual work?

(b) A detonation of 1 kg of TNT releases about 4.5 MJ of work. Calculate the equivalent mass of TNT that performs the same work as in part (a).

4.6 Work problem 4.5 but consider a steam boiler that fails. The boiler is 250 L in size, operating at 4 MPa, and half full of liquid.

4.7 An isolated chamber with rigid walls is divided into two equal compartments, one containing gas at 600 K and 1 MPa and the other evacuated. The partition between the two compartments ruptures. Compute the final T, P, and ΔS for the following:

(a) An ideal gas with $C_P/R = 7/2$
(b) Steam.

4.8 An isolated chamber is divided into two equal compartments, one containing gas and the other evacuated. The partition between the two compartments ruptures. At the end of the process, the temperature and pressure are uniform throughout the chamber.

(a) If the filled compartment initially contains an ideal gas at 25 MPa and 650 K, what is the final temperature and pressure in the chamber? What is ΔS for the process? Assume a constant heat capacity of $C_P/R = 4.041$.

(b) If the filled chamber initially contains steam at 25 MPa and 650 K, what is the final temperature and pressure in the chamber? What is ΔS for the process? (Use the steam tables.)

4.9 Airplanes are launched from aircraft carriers by means of a steam catapult. The catapult is a well-insulated cylinder that contains steam, and is fitted with a frictionless piston. The piston is connected to the airplane by a cable. As the steam expands, the movement of the piston causes movement of the plane. A catapult design calls for 270 kg of steam at 15 MPa and 450°C to be expanded to 0.4 MPa. How much work can this catapult generate during a single stroke? Compare this to the energy required to accelerate a 30,000 kg aircraft from rest to 350 km per hour.

4.10 We have considered heat and work to be path-dependent. However, if all heat transfer with surroundings is performed using a reversible heat transfer device (some type of reversible Carnot-type device), work can be performed by the heat transfer device during heat transfer to the surroundings. The net heat transferred to the surroundings and the net work done will be independent of the path. Demonstrate this by calculating the work and heat interactions for the system, the heat transfer device, and the sum for each of the following paths where the surroundings are at $T_{surr} = 273$ K. The state change is from state 1, $P_1 = 0.1$ MPa, $T_1 = 298$ K and state 2, $P_2 = 0.5$ MPa and T_2 which will be found in part (a). $C_P = 7R/2$.

(a) Consider a state change for an ideal gas in a piston/cylinder. Find T_2 by an adiabatic reversible path. Find the heat and work such that no entropy is generated in the universe. This is path a. Sketch path a qualitatively on a P-T diagram.

(b) Now consider a path consisting of step b, an isothermal step at T_1, and step c, an isobaric step at P_2. Sketch and label the step on the same P-T diagram created in (a). To avoid generation of entropy in the universe, use heat engines/pumps to transfer heat during the steps. Calculate the W_{EC} and W_S as well as the heat transfer with the surroundings for each of the steps and overall. Compare to part (a) the total heat and work interactions with the surroundings.

(c) Now consider a path consisting of step d, an isobaric step at P_1, and step e, an isothermal step at T_2. Calculate the W_{EC} and W_S as well as the heat transfer with the surroundings for each of the steps and overall. Compare to part (a) using this pathway and provide the same documentation as in (b).

4.11 Consider the wintertime heating of a house with a furnace compared to addition of Carnot heat engines/pumps. To compensate for heat losses to the surroundings, the house is maintained at a constant temperature T_{house} by a constant rate of heat transfer, Q_{house}. The furnace operates at a constant temperature T_F, and with direct heat transfer, the heat required from the furnace, Q_F is equal to Q_{house}.

(a) Instead of direct heat transfer, if we utilize the surroundings, at T_S, as an additional heat source and include heat pump technology, Q_F may be reduced by generating work from a heat engine operating between T_F and T_S, then applying that work energy to a heat pump operating between T_S and T_{house}. Given that $T_F = 800$ K, $T = 293$ K, $T_S = 265$ K, and $Q_{house} = 40$ kJ/h, determine Q_F utilizing heat pump technology. No other sources of energy may be used.

(b) Another option is to run a heat engine between T_F and T_{house} and the heat pump between T_S and T_{house}. Compare this method with part (a).

4.12 An ideal gas enters a valve at 500 K and 3 MPa at a steady-state rate of 3 mol/min. It is throttled to 0.5 MPa. What is the rate of entropy generation? Is the process irreversible?

4.13 SO_2 vapor enters a heat exchanger at 100°C and at a flowrate of 45 mole/h. If heat is transferred to the SO_2 at a rate of 1,300 kJ/h, what is the rate of entropy transport in the gas at the outlet relative to the inlet in kJ/K/h given by $(S^{out} - S^{in})\dot{n}$?

4.14 An ideal gas stream (Stream A), $C_P = 5R/2$, 50 mole/h, is heated by a steady-state heat exchanger from 20°C to 100°C by another stream (Stream B) of another ideal gas, $C_P = 7R/2$, 45 mole/h, which enters at 180°C. Heat losses from the exchanger are negligible.

(a) For concurrent flow in the heat exchanger, calculate the molar entropy changes $(S^{out} - S^{in})$ for each stream, and $\underline{S}_{gen}$ for the heat exchanger.
(b) For countercurrent flow in the heat exchanger, calculate the molar entropy changes $(S^{out} - S^{in})$ for each stream, and $\underline{S}_{gen}$ for the heat exchanger. Comment on the comparison of results from parts (a) and (b).

4.15 An inventor has applied for a patent on a device that is claimed to utilize 1 mole/min of air (assumed to be an ideal gas) with temperature independent $C_P = (7/2)R$ which enters at 500 K and 2 bar, and leaves at 350 K and 1 bar. The process is claimed to produce 2000 J/min of work and to require an undisclosed amount of heat transfer with a heat reservoir at 300 K. Should the inventor be issued a patent on this device?

4.16 Two streams of air are mixed in a steady-state process shown below. Assume air is an ideal gas with a constant heat capacity $C_P = 7R/2$.

(a) What is the temperature of the stream leaving the tank if the process is adiabatic?
(b) What is the rate of entropy generation within the tank if the process is adiabatic?
(c) If we duplicated the stream conditions (temperatures, pressures, and flowrates) with an internally reversible process, what is the maximum rate at which work could be obtained? If desirable, you are permitted to transfer heat to the surroundings at the surroundings' temperature of 295 K.

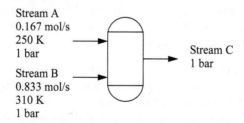

Stream A
0.167 mol/s
250 K
1 bar

Stream B
0.833 mol/s
310 K
1 bar

Stream C
1 bar

4.17 Air is flowing at steady state through a 5 cm diameter pipe at a flow rate of 0.35 mole/min at $P = 5$ bar and $T = 500$ K. It flows through a throttle valve and exits at 1 bar. Assume air is an ideal gas with $C_P = 29.1$ J/mol-K. If the throttle valve was replaced by a reversible steady-state flow device to permit exactly the same state change for the air in this steady-state process, at what rate could work could be obtained? Heat transfer, if desired, can occur with the surroundings at 298 K, which may be considered a reservoir.

4.18 A common problem in the design of chemical processes is the steady-state compression of gases from a low pressure P_1 to a much higher pressure P_2. We can gain some insight about optimal design of this process by considering adiabatic reversible compression of ideal gases with stage-wise intercooling. If the compression is to be done in two stages, first compressing the gas from P_1 to P^*, then cooling the gas at constant pressure down to the compressor inlet temperature T_1, and then compressing the gas to P_2, what should the value of the intermediate pressure be to accomplish the compression with minimum work?

4.19 Steam flowing at steady state enters a turbine at 400°C and 7 MPa. The exit is at 0.275 MPa. The turbine is 85% efficient. What is the quality of the exiting stream? How much work is generated per kg of steam?

4.20 An adiabatic steam turbine inlet is to be 4 MPa. The outlet of the turbine is to operate at 0.01 MPa, and provide saturated steam. The turbine has an efficiency of 85%. Determine the superheat which is required on the turbine inlet, and the work produced by the turbine.

4.21 Steam is fed to an adiabatic turbine at 4 MPa and 500°C. It exits at 0.1 MPa.

(a) If the turbine is reversible, how much work is produced per kg of steam?
(b) If the turbine is 80% efficient, how much work is produced per kg of steam?

4.22 Methane is compressed in a steady-state adiabatic compressor (87% efficient) to 0.4 MPa. What is the required work per mole of methane in kJ? If the flow is to be 17.5 kmol/h, how much work must be furnished by the compressor (in kW)? What is the rate of entropy generation (in kJ/K/h)? (a) the inlet is at 0.1013 MPa and −240°F; (b) the inlet is at 0.1013 MPa and 200 K.

4.23 Methane is to be compressed from 0.05 MPa and −120°F to 5 MPa in a two-stage compressor. In between adiabatic, reversible stages, a heat exchanger returns the temperature to −120°F. The intermediate pressure is 1.5 MPa.

(a) What is the work required (kJ/kg) in the first compressor of methane?
(b) What is the temperature at the exit of the first compressor (°C)?
(c) What is the cooling requirement (kJ/kg) in the interstage cooler?

4.24 A steady stream (1000 kg/hr) of air flows through a compressor, entering at (300 K, 0.1 MPa) and leaving at (425 K, 1 MPa). The compressor has a cooling jacket where water flows at 1500 kg/hr and undergoes a 20 K temperature rise. Assuming air is an ideal gas, calculate the work furnished by the compressor, and also determine the minimum work required for the same state change of air.

4.25 Propane is to be compressed from 0.4 MPa and 360 K to 4 MPa using a two-stage compressor. An interstage cooler returns the temperature of the propane to 360 K before it enters the second compressor. The intermediate pressure is 1.2 MPa. Both adiabatic compressors have a compressor efficiency of 80%.

(a) What is the work required in the first compressor per kg of propane?
(b) What is the temperature at the exit of the first compressor?
(c) What is the cooling requirement in the interstage cooler per kg of propane?

4.26 (a) A steam turbine in a small electric power plant is designed to accept 5000 kg/h of steam at 60 bar and 500°C and exhaust the steam at 1 bar. Assuming that the turbine is adiabatic and reversible, compute the exit temperature of the steam and the power generated by the turbine.

(b) If the turbine in part (a) is adiabatic but only 80% efficient, what would be the exit temperature of the steam? At what rate would entropy be generated within the turbine?

(c) One simple way to reduce the power output of the turbine in part (a) (100% efficient) is by adjusting a throttling valve that reduces the turbine inlet steam pressure to 30 bar. Compute the steam temperature to the turbine, the rate of entropy generation, and the power output of the turbine for this case. Is this a thermodynamically efficient way of reducing the power output? Can you think of a better way?

4.27 Steam is used in the following adiabatic turbine system to generate electricity; 15% of the mass flow from the first turbine is diverted for other use.

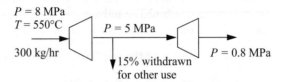

(a) How much work (in kJ/h) is generated by the first turbine which is 80% efficient?

(b) How much work (in kJ/h) is generated by the second turbine which is 80% efficient?

(c) Steam for the turbines is generated by a boiler. How much heat must be supplied to the boiler (not shown) which has 300 kg/h of flow? The stream entering the boiler is $T = 170°C, P = 8$ MPa. The stream exiting the boiler matches the inlet to the first turbine.

4.28 Liquid nitrogen is useful for medical purposes and for research laboratories. Determine the minimum shaft work needed to liquefy nitrogen initially at 298 K and 0.1013 MPa and ending with saturated liquid at the normal boiling point, 77.4 K and 0.1013 MPa. The heat of vaporization at the normal boiling point is 5.577 kJ/mol, and the surroundings are at 298 K. The constant pressure heat capacity of gaseous nitrogen can be assumed to be independent of temperature at $7/2R$ for the purpose of this calculation.

(a) Consider nitrogen entering a flow device at 1 mol/min. Give shaft work in kW.

(b) Consider nitrogen in a piston/cylinder device. Give the work in kJ per mole liquefied.

(c) Compare the minimum shaft work for the two processes. Is one of the processes more advantageous than the other on a molar basis?

4.29 Propane flows into a steady-state process at 0.2 MPa and 280 K. The final product is to be saturated liquid propane at 300 K. Liquid propane is to be produced at 1000 kg/h. The surroundings are at 295 K. Using a propane property chart, determine the rate of heat transfer and minimum work requirement if the process is to operate reversibly.

4.30 Propane (1000 kg/hr) is to be liquefied following a two-stage compression. The inlet gas is to be at 300 K and 0.1 MPa. The outlet of the adiabatic compressor I is 0.65 MPa, and the propane enters the interstage cooler where it exits at 320 K, then adiabatic compressor II raises the propane pressure to 4.5 MPa. The final cooler lowers the temperature to 320 K before it is throttled adiabatically to 0.1 MPa. The adiabatic compressors have an efficiency of 80%.

(a) Determine the work required by each compressor.

(b) If the drive motors and linkages are 80% efficient (taken together), what size motors (hp) are required?

(c) What cooling is required in the interstage cooler and the final cooler?

(d) What percentage of propane is liquefied, and what is the final temperature of the propane liquid?

4.31 A heat exchanger operates with the following streams: Water in at 20°C, 30 kg/hr; water out at 70°C; Organic in at 100°C, 41.8 kg/hr; organic out at 40°C.

(a) What is the maximum work that could be obtained if the flow rates and temperatures of the streams remain the same, but heat transfer is permitted with the surroundings at 298 K? ($C_{P\,water} = 4.184$ kJ/(kgK), $C_{P\,organic} = 2.5$ kJ/(kgK).)

(b) What is the maximum work that could be obtained by replacing the heat exchanger with a reversible heat transfer device, where the inlet flowrates and temperatures are to remain the same, the organic outlet temperature remains the same, and no heat transfer with the surroundings occurs?

4.32 Presently, benzene vapors are condensed in a heat exchanger using cooling water. The benzene (100 kmol/h) enters at 0.1013 MPa and 120°C, and exits at 0.1013 MPa and 50°C. Cooling water enters at 10°C and exits at 40°C.

(a) What is the current demand for water (kg/h)?

(b) To what flowrate could the water demand be lowered by introducing a reversible heat transfer device that is adiabatic with the surroundings? The temperature rise of water is to remain the same. What work could be obtained from the new heat transfer device?

4.33 A Hilsch vortex tube is an unusual device that takes an inlet gas stream and produces a hot stream and a cold stream without moving parts. A high-pressure inlet stream (A) enters towards one end of the tube. The cold gas exits at outlet B on the end of the tube near the inlet where the port is centered in the end cap. The hot stream exits at outlet C on the other end of the tube where the exit is a series of holes or openings around the outer edge of the end cap.

The tube works in the following way. The inlet stream A enters tangent to the edge of the tube, and swirls as it cools by expansion. Some of the cool fluid exits at port B. The remainder of the fluid has high kinetic energy produced by the volume change during expansion, and the swirling motion dissipates the kinetic energy back into internal energy, so the temperature rises before the gas exits at port C.

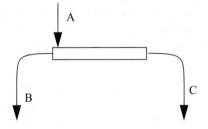

Inlet A is at 5 bar and 310 K and 3.2 mol/min. Outlet B is at 1 bar and 260 K. Outlet C is at 1 bar and 315 K. The tube is insulated and the fluid is air with $C_P = 7R/2$.

(a) Determine the flowrates of streams B and C.

(b) Determine S_{gen} for the Hilsch tube.

(c) Suppose a reversible heat engine is connected between the outlet streams B and C which is run to produce the maximum work possible. The proposed heat engine may only exchange heat between the streams and not with the surroundings as shown. The final temperature of streams B and C will be $T_{B'}$ as they exit the apparatus. What is $T_{B'}$?

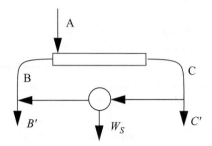

(d) What work output is possible in W? What is $\dot{S}_{gen}$ for the entire system including the tube plus the heat engine?

(e) Suppose that instead of using the heat engine, streams B and C were mixed directly with one another to form a single outlet stream. What would this temperature be, and how does it compare with $T_{B'}$ from part (c)? Calculate $\dot{S}_{gen}$ and compare it with $\dot{S}_{gen}$ from part (c). What do you conclude from the comparison?

4.34 Methane gas is contained in a 0.65-m³ gas cylinder at 6.9 MPa and 300 K. The cylinder is vented rapidly until the pressure falls to 0.5 MPa. The venting occurs rapidly enough that heat transfer between the cylinder walls and the gas can be neglected, as well as between the cylinder and the surroundings. What is the final temperature and the final number of moles of gas in the tank immediately after depressurization? Assume the expansion within the tank is reversible, and the following:

(a) Methane is considered to be an ideal gas with $C_P/R = 4.298$
(b) Methane is considered to be a real gas with properties given by a property chart.

4.35 A thermodynamically interesting problem is to analyze the fundamentals behind the product called "fix-a-flat." In reality, this product is a 500 mL can that contains a volatile compound under pressure, such that most of it is liquid. Nevertheless, we can make an initial approximation of this process by treating the contents of the can as an ideal gas. If the initial temperature of both the compressed air and the air in the tire is 300 K, estimate the initial pressure in the compressed air can necessary to reinflate one tire from 1 bar to 3 bar. Also, estimate the final air temperature in the tire and in the can. For the purposes of this calculation you may assume: air is an ideal gas with $C_P/R = 7/2$, the tire does not change its size or shape during the inflation process, and the inner tube of the tire has a volume of 40,000 cm³. We will reconsider this problem with liquid contents, after discussing phase equilibrium in a pure fluid.

4.36 Wouldn't it be great if a turbine could be put in place of the throttle in problem 4.35? Then you could light a small bulb during the inflation to see what you were doing at night. How much energy (J) could possibly be generated by such a turbine if the other conditions were the same as in problem 4.35?

4.37 A 1 m³ tank is to be filled using N_2 at 300 K and 20 MPa. Instead of throttling the N_2 into the tank, a reversible turbine is put in line to get some work out of the pressure drop. If the pressure in the tank is initially zero, and the final pressure is 20 MPa, what will be the final temperature in the tank? How much work will be accomplished over the course of the entire process? (Hint: Consider the entropy balance carefully.)

4.38 Two well-insulated tanks are attached as shown in the figure below. The tank volumes are given in the figure. There is a mass-flow controller between the two tanks. Initially, the flow controller is closed. At $t = 0$, the mass flow controller is opened to a flow of 0.1 mol/ s. After a time of 500 seconds, what are the temperatures of the two tanks? Neglect the heat capacity of the tanks and piping. No heat transfer occurs between the two tanks. (After 500 seconds, the pressure in the left tank is still higher than the pressure in the right tank.) The working fluid is nitrogen and the ideal gas law may be assumed. The ideal gas heat capacity $C_P = 7/2 \cdot R$ may be assumed to be independent of T.

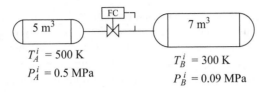

4.39 Two storage tanks (0.1 m³ each) contain air at 2 bar. They are connected across a small reversible compressor. The tanks, connecting lines, and compressor are immersed in a constant temperature bath at 280 K. The compressor will take suction from one tank, compress the gas, and discharge it to the other tank. The gas is at 280 K at all times. Assume that air is an ideal gas with $C_P = 29.3$ J/mol-K.

(a) What is the minimum work interaction required to compress the gas in one tank to 3 bar?
(b) What is the heat interaction with the constant temperature bath?

4.40 A constant pressure air supply is connected to a small tank (A) as shown in the figure below. With valves B and C, the tank can be pressurized or depressurized. The initial conditions are $T = 300$ K, $P = 1.013$ bar, $C_P = 29.3$ J/(mol-K). Consider the system adiabatic.

(a) The tank is pressurized with valve B open and valve C closed. What is the final temperature of the tank? Neglect the heat capacity of the tank and valves.
(b) Taking the system as the tank plus the valves, what is the entropy change of the system due to pressurization? What is the entropy change of the air supply reservoir? What is the entropy change of the universe? Use a reference state of 300 K and 1.013 bar.
(c) During depressurization with valve B closed and valve C open, how does the molar entropy entering valve C compare with the molar entropy leaving? What is the temperature of the tank following depressurization?

4.41 The pressurization of problem 4.40 is performed by replacing the inlet valve with a reversible device that permits pressurization that is internally reversible. The system is to remain adiabatic with respect to heat transfer of the surroundings.

(a) What is the final temperature of the tank?
(b) How much work could be obtained?

4.42 A 2m³ tank is at 292 K and 0.1 MPa and it is desired to pressurize the tank to 3 MPa. The gas is available from an infinite supply at 350 K and 5 MPa connected to the tank via a throttle valve. Assume that the gas follows the ideal gas law with a constant heat capacity of $C_P = 29$ J/(mol-K).

(a) Modeling the pressurization as adiabatic, what is the final temperature in the tank and the final number of moles when the pressure equals 3 MPa?

(b) Identify factors included in the idealized calculation of part (a) that contribute to irreversibilities.

(c) Identify factors neglected in the analysis of part (a) that would contribute to irreversibilities in a real process.

(d) If the pressurization could be performed reversibly, the final temperature might be different from that found in part (a). Clearly outline a procedure to calculate the temperature indicating that enough equations are provided for all unknowns. Also clearly state how you would use the equations. Additional equipment is permissible provided that the process remains adiabatic with regard to heat transfer to the surroundings.

(e) In part (d), would work be added, removed, or not involved in making the process reversible? Provide equations to calculate the work interaction.

4.43 Two gas storage tanks are interconnected through an isothermal expander. Tank 1 ($\underline{V} = 1$ m³) is initially at 298 K and 30 bar. Tank 2 ($\underline{V} = 1$ m³) is initially at 298 K and 1 bar. Reversible heat transfer is provided between the tanks, the expander, and the surroundings at 298 K. What is the maximum work that can be obtained from the expander when isothermal flow occurs from tank 1 to tank 2?

THERMODYNAMICS OF PROCESSES

5

There cannot be a greater mistake than that of looking superciliously upon practical applications of science. The life and soul of science is its practical application.

Lord Kelvin (William Thompson)

In the first four chapters, we have concentrated on applications of the first and second laws to simple systems (e.g., turbine, throttle). The constraints imposed by the second law should be clear. In this chapter, we show how the analyses we have developed for one or two operations at a time can be assembled into complex processes. In this way, we provide several specific examples of ways that operations can be connected to create power cycles, refrigeration cycles, and liquefaction cycles. We can consider these processes as paradigms for general observations about energy and entropy constraints.

Chapter Objectives: You Should Be Able to...

1. Write energy and entropy balances around multiple pieces of equipment using correct notation including mass flow rates.

2. Simplify energy balances by recognizing when streams have the same properties (e.g., splitter) or flow rates (heat exchanger inlet/outlet).

3. Apply the correct strategy for working through a power cycle with multiple reheaters and feedwater preheaters.

4. For ordinary vapor compression cycles, locate condenser/evaporator P or T given one or the other and plot the process outlet and P-H diagram.

5. Successfully approach complex processes by simplifying the E-balance and S-balance, solving for unknowns.

5.1   THE CARNOT STEAM CYCLE

We saw in Example 4.4 on page 145 how a Carnot cycle could be set up using steam as a working fluid. The addition of heat at constant temperature and the macroscopic definition of entropy establish a correspondence between temperature and heat addition/removal. Steam is especially well suited for isothermal heat exchange because boiling and condensation are naturally isothermal and exchange large amounts of heat. To review, we could plot this cycle in *T-S* coordinates and envision a flow process with a turbine to produce work during adiabatic expansion and some type of compressor for the adiabatic compression as shown in Fig. 5.1. The area inside the *P-V* cycle represents the work done by the gas in one cycle, and the area enclosed by the *T-S* path is equal to the net intake of energy as heat by the gas in one cycle.

The Carnot cycle has a major advantage over other cycles. It operates at the highest temperature available for as long as possible, reducing irreversibilities at the boundary because the system approaches the reservoir temperature during the entire heat transfer. In contrast, other cycles may only approach the hot reservoir temperature for a short segment of the heat transfer. A similar argument holds regarding the low temperature reservoir. Unfortunately, it turns out that it is impossible to make full use of the advantages of the Carnot cycle in practical applications. When steam is used as the working fluid, the Carnot cycle is impractical for three reasons: 1) It is impractical to stay inside the phase envelope because higher temperatures correlate with higher pressure. Higher pressures lead to smaller heat of vaporization to absorb heat. Since the critical point of water is only ~374°C, substantially below the temperatures from combustion, the temperature gradient between a fired heater and the steam would be large; 2) expanding saturated vapor to low-quality (very wet) steam damages turbine blades by rapid erosion due to water droplets; 3) compressing a partially condensed fluid is much more complex than compressing an entirely condensed liquid. Therefore, most power plants are based on modifications of the Rankine cycle, discussed below. Nevertheless, the Carnot cycle is so simple that it provides a useful estimate for checking results from calculations regarding other cycles.

5.2   THE RANKINE CYCLE

In a Rankine cycle, the vapor is superheated before entering the turbine. The superheat is adjusted to avoid the turbine blade erosion from low-quality steam. Similarly, the condenser completely reduces the steam to a liquid that is convenient for pumping.

In Fig. 5.2, state 4' is the outlet state for a reversible adiabatic turbine. *We use the prime (') to denote a reversible outlet state as in the previous chapter.* State 4 is the actual outlet state which is calculated by applying the efficiency to the enthalpy change.

The prime denotes a reversible outlet state.

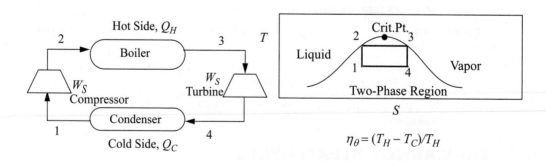

Figure 5.1 *Illustration of a Carnot cycle based on steam in T-S coordinates.*

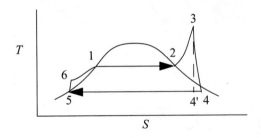
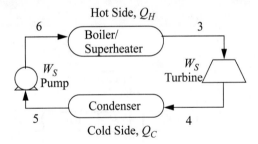

Figure 5.2 *Rankine cycle.*

$$\Delta H_{3 \to 4} = \eta_E \Delta H_{3 \to 4'} \quad H_4 = H_3 + \Delta H_{3 \to 4} \qquad 5.1$$

Because a real turbine always generates entropy, state 4 will always be to the right of 4' on a *T-S* diagram. States 4 and 4' can be inside or outside the phase envelope. Efficiencies are greater if state 4 is slightly inside the phase envelope because the enthalpy change will be larger for the same pressure drop due to the large enthalpy of vaporization; however, to avoid turbine blade damage, quality is kept above 90% in most cases.

Note in Fig. 5.2 that the superheater between the boiler and the turbine is not drawn, and only a single unit is shown. In actual power plants, separate superheaters are used; however, for the sake of simplicity in our discussions the boiler/superheater steam generator combination will be represented by a single unit in the schematic.

Most plants will have separate boilers and superheaters. We show just a boiler for simplicity.

Turbine calculation principles were covered in the last chapter. Now we recognize that the net work is the sum of the work for the turbine and pump and that some of the energy produced by the turbine is needed for the pump. In general, the thermal efficiency is given by:

$$\eta_\theta = \frac{-W_{S, net}}{Q_H} = \frac{-(\Delta H_{3 \to 4} + W_P)}{\Delta H_{6 \to 3}} \qquad 5.2$$

The boiler input can be calculated directly from the enthalpy out of the pump and the desired turbine inlet. The key steps are illustrated in Example 5.1.

Example 5.1 Rankine cycle

A power plant uses the Rankine cycle. The turbine inlet is 500°C and 1.4 MPa. The outlet is 0.01 MPa. The turbine has an efficiency of 90% and the pump has an efficiency of 80%. Determine:

(a) The work done by the turbine (kJ/kg)

(b) The work done by the pump, the heat required, and the thermal efficiency;

(c) The circulation rate to provide 1 MW net power output.

Solution: We will refer to Fig. 5.2 for stream numbers. The recommended method for solving process problems is to establish a table to record values as they are determined. *In this text we will show values in the tables with bold borders if they have been determined by balance calculations.* The turbine outlet can be read from the temperature table without interpolation. Cells with standard borders refer to properties determined directly from the problem statement

Example 5.1 Rankine cycle (Continued)

Boldfaced table cells show calculations that were determined by balances. We follow this convention in the following examples.

State	T (°C)	P (MPa)	q	H (kJ/kg)	S (kJ/kgK)	V (m³/kg)
3	500	1.4	supV	3474.8	7.6047	
4'	45.81	0.01	0.927	2410.2	7.6047	
4	45.81	0.01	0.972	2516.7	7.9388	
5, sat L	45.81	0.01	0.0	191.8		0.00101
6			compL	193.6		

Because the turbine inlet has two state variables specified, the remainder of the state properties are found from the steam tables and tabulated in the property table. We indicate a superheated vapor with "supV" compressed liquid with "compL."

(a) Stepping forward across the turbine involves the same specifications as part (c) of Example 4.13 on page 168. The properties from 4 and 4' are transferred from that example to the property table. The work done is –958.1 kJ/kg.

(b) The outlet of the condenser is taken as saturated liquid at the specified pressure, and those values are entered into the table. We must calculate $\dot{Q}_H$ and $\dot{W}_{net}$. So we need H_6 and $W_{S,pump}$ which are determined by calculating the adiabatic work input by the pump to increase the pressure from state 5. Although the reversible calculation for the pump is isentropic, we may apply Eqn. 2.61 without direct use of entropy, and then correct for efficiency. For the pump,

$$\Delta H'_{pump} = W'_{S,pump} = \int V dP \approx V \Delta P .$$

$$\Delta H'_{pump} = \frac{1010 \text{ cm}^3 (1.4 - 0.01)}{\text{kg}} \left| \frac{1\text{J}}{\text{cm}^3\text{MPa}} \right| \frac{\text{kJ}}{10^3 \text{J}} = 1.4 \text{ kJ/kg}$$

$$W_{S,pump} = W'_{S,pump} / \eta_C = 1.4/0.8 = 1.8 \text{ kJ/kg}$$

Thus, the work of the pump is small, resulting in $H_6 = 191.8 + 1.8 = 193.6$ kJ/kg. The net work is $W'_{S,net} = -958.1 + 1.8 = -956.3$ kJ/kg. The only source of heat for the cycle is the boiler/superheater. All of the heat input is at the boiler/superheater. The energy balance gives $Q_H = (H_3 - H_6) = 3281.2$ kJ/kg. The thermal efficiency is

$$\eta_\theta = \frac{-W_{S,net}}{Q_H} = \left(\frac{956.3}{3281.2} \right) = 0.2914$$

If we neglected the pump work, the efficiency would 29.2%. Note that the pump work has only a small effect on the thermal efficiency but is included for theoretical rigor.

Example 5.1 Rankine cycle (Continued)

(c) For 1 MW capacity, $\underline{\dot{W}}_S = \dot{m} W_{S,net}$, the circulation rate is

$$\dot{m} = \frac{-1000 \text{ kJ}}{\text{s}} \left| \frac{\text{kg}}{-956.3} \right| \frac{3600 \text{ s}}{\text{h}} = 3764 \text{ kg/h}$$

The cycle in Fig. 5.2 is idealized from a real process because the inlet to the pump is considered saturated. In a real process, it will be subcooled to avoid difficulties (e.g., cavitation[1]) in pumping. In fact, real processes will have temperature and pressure changes along the piping between individual components in the schematic, but these changes will be considered negligible in the Rankine cycle and all other processes discussed in the chapter, unless otherwise stated. These simplifications allow focus on the most important concepts, but the simplifications would be reconsidered in a detailed process design.

5.3 RANKINE MODIFICATIONS

Two modifications of the Rankine cycle are in common use to improve the efficiency. A Rankine cycle with reheat increases the boiler pressure but keeps the maximum temperature approximately the same. The maximum temperatures of the boilers are limited by corrosion concerns. This modification uses a two-stage turbine with reheat in-between. An illustration of the modified cycle is shown in Fig. 5.3. Crudely, adding multiple stages with reheat leads to the maximum temperature being applied as much as possible, while avoiding extremely wet steam during expansion. This moves the process efficiency in the direction of a Carnot cycle. The implication of this modification is shown in Example 5.2.

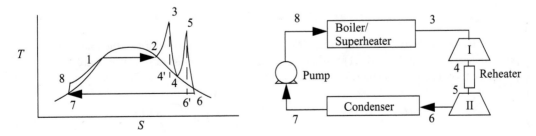

Figure 5.3 *Rankine cycle with reheat.*

---

1. Cavitation occurs when vapor bubbles form in the inlet line of a liquid pump, and the bubbles prevent the pump from drawing the liquid into the pump cavity.

Example 5.2 A Rankine cycle with reheat

Consider a modification of Example 5.1. Let us limit the process to a 500°C boiler/superheater and add reheat with an additional turbine to investigate improvement in efficiency and circulaion rate. For the additional turbine (as the first turbine), $\eta_E = 0.85$, while we keep the existing turbine (as the second turbine), $\eta_E = 0.9$ and pump $\eta_C = 0.80$. Let the feed to the first turbine be steam at 500°C and 6 MPa. Let the feed to the second turbine be 1.4 MPa and 500°C (the same as Example 5.1). Determine the improvement in efficiency and circulation rate relative to Example 5.1.

Solution: Refer to Fig. 5.3 for stream numbers. First, let us find state 3. The inlet state values are entered in the table in standard borders. $P_4 = P_5$ because we neglect the heat exchanger pressure drop. Upon expansion through the first turbine, we look at the S^{satV} at 1.4 MPa and find it is lower than $S_{4'}$. Therefore, the reversible state is superheated. Using $\{S', P\}$ to find H',

$$H_4' = 2927.9 + \frac{6.8826 - 6.7488}{6.9552 - 6.7488}(3040.9 - 2927.9) = 3001.2 \text{ kJ/kg}$$

Correcting for efficiency,

$$\Delta H_I = W_{S,I} = 0.85(3001.2 - 3423.1) = 0.85(-421.9) = -358.6 \text{ kJ/kg}$$

$$H_4 = 3423.1 - 358.6 = 3064.5 \text{ kJ/kg}$$

State	T (°C)	P (MPa)	H (kJ/kg)	S (kJ/kgK)	V (m³/kg)
3	500	6	3423.1	6.8826	
4'		1.4	3001.2	6.8826	
4			3064.5		
7, sat L	45.81	0.01	191.8		0.00101

State 5 was used in Example 5.1 (as state 3). Solving the energy balance for the reheater,

$$Q_{reheat} = (H_5 - H_4) = 3474.8 - 3064.5 = 410.3 \text{ kJ/kg}$$

Turbine II was analyzed in Example 5.1. We found $W_{S,II} = -958.1$ kJ/kg and the total work output is $W_{S,turbines} = (-358.6 - 958.1) = -1316.7$ kJ/kg. The pump must raise the pressure to 6 MPa. Using Eqn. 2.61, and correcting for efficiency,

$$\Delta H_{pump} = W_{S,pump} = 1010\frac{(6 - 0.01)}{1000}\frac{1}{0.8} = 7.6 \text{ kJ/kg}$$

State 7 is the same as state 5 in Example 5.1 and has been tabulated in the property table. $H_8 = H_7 + W_{S,pump} = 191.8 + 7.6 = 199.4$ kJ/kg. The net work is thus

$$W_{S,net} = -1316.7 + 7.6 = -1309.1 \text{ kJ/kg}$$

The heat for the boiler/superheater is given by $Q_{b/s} = H_3 - H_8 = 3423.1 - 199.4 = 3223.7$ kJ/kg.

Example 5.2 A Rankine cycle with reheat (Continued)

The thermal efficiency is

$$\eta = \frac{-W_{S,net}}{Q_H} = \frac{-W_{S,net}}{(Q_{reheat} + Q_{b/s})} = \frac{1309.1}{410.3 + 3223.7} = 0.36$$

$$\dot{m} = \frac{(-1000)(3600)}{(-1309.1)} = 2750 \text{ kg/h}$$

The efficiency has improved by $\frac{(0.36 - 0.29)}{0.29} \times 100 \% = 24\%$, and the circulation rate has been decreased by 27%.

Reheat improves thermal efficiency.

One variation of the Rankine cycle is for cogeneration as illustrated in Fig. 5.4. Most chemical plants need process steam for heating distillation columns or reactors as well as electrical energy. Therefore, they use steam at intermediate pressures, depending on the need, and circulate the condensate back to the boiler. Another very common modification of the Rankine cycle is a regenerative cycle using **feedwater preheaters.** A portion of high-pressure steam is used to preheat the water as it passes from the pump back to the boiler. A schematic of such a process is shown in Fig. 5.5 using **closed feedwater preheaters.** The economic favorability increases until about five preheaters are used, then the improvements are not worth the extra cost. Three preheaters are more

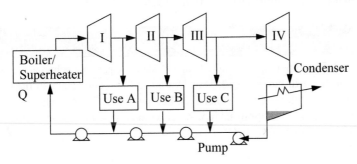

Figure 5.4 *Rankine cycle with side draws for process steam. Pumps and/or throttles may be used in returning process steam to the boiler.*

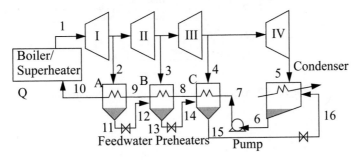

Figure 5.5 *Regenerative Rankine cycle using closed feedwater preheaters.*

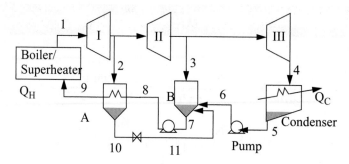

Figure 5.6 *Schematic for a system with a closed feedwater preheater, A, and an open feedwater preheater, B.*

common. As the condensate from each preheater enters the next preheater, it throttles through a valve to the next lower pressure and partially or totally vaporizes as it throttles. It is also common to withdraw some steam from turbine outlets for process use and heating. Often in actual processes, **open feedwater preheaters** are used. In an open feedwater preheater, all of the incoming streams mix. The advantage of this preheater is that dissolved oxygen in the returning condensate can be removed by heating, and if provision is made to vent the non-condensables from the open feedwater preheater, it may serve as a **deaerator** to remove dissolved air before feeding the boiler. A system with an open feedwater preheater is shown in Fig. 5.6. The vent on the open feedwater preheater is typically a small stream and omitted in the schematics and first order calculations. A regenerative Rankine cycle is illustrated in Example 5.3.

Example 5.3   Regenerative Rankine cycle

Steam (1) exits a boiler/superheater at 500°C and 5 MPa. A process schematic is shown in Fig. 5.7. The first stage of the turbine exits (2, 3) at 1MPa and the second stage of the turbine exits (4) at 0.1 MPa. A feed preheater is used to exchange heat with a 5°C approach temperature between streams 7 and 8. Given the conditions tabluated below in cells with standard borders, find the net power output per kg of flow in stream 1.

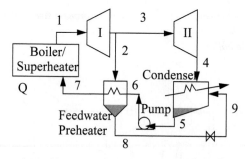

Figure 5.7 *Regenerative Rankine cycle for Example 5.3.*

Example 5.3 Regenerative Rankine cycle (Continued)

Solution: First, note that streams 2 and 8 are on the same side of the feedwater preheater and are thus at the same pressure. Streams 7 and 6 must be at the boiler pressure by similar arguments. And stream 5 must likewise be at the condenser pressure. Stream 8 leaves as saturated liquid at the 1 MPa, thus we find $H_8 = 762.5$ kJ/kg, and $T_8 = 180°C$. Stream 7 is thus at 175°C and 5 MPa. Following Example 2.6 for a compressed liquid, $H_7 = 741.02$ kJ/kg $+ (5–0.893$MPa$)(1.12$cm^3/g$)$ $= 745.6$ kJ/kg. Often, one of the key steps in working a problem involving a regenerative cycle is to solve for the fraction of each flow diverted rather than solving for the individual flow rates.

🛈 Solving for flowrate ratios in regenerative cycles can be helpful when the total flowrate is unknown.

Stream	T(°C)	P(MPa)	H(kJ/kg)	S(kJ/kg-K)
1	500	5	3434.7	6.9781
2	300	1	3051.6	7.1246
4		0.1	2615.7	7.2000
5		0.1	417.5	
6		5	422.7	
7	175	5	745.6	
8	180	1	762.5	

The flow rates of streams 7, 6, and 5 are equal to the flow rate of stream 1, and we may write the energy balance around the feedwater preheater using the mass flow rate of stream 1 together with the mass flow rate of stream 2, $\dot{m}_1(H_6 - H_7) + \dot{m}_2(H_2 - H_8) = 0$. Dividing by $\dot{m}_1$ and substituting values gives $(423 - 746) + (\dot{m}_2/\dot{m}_1)(3052 - 763) = 0$, and $\dot{m}_2/\dot{m}_1 = 0.14$.

The net work is given by

$$\dot{W}_{S,net} = \dot{m}_1(W_{S,I}) + \dot{m}_3(W_{S,II}) + \dot{m}_1(W_{S,pump})$$, and on the basis of one kg from the

boiler/superheater, $\dfrac{\dot{W}_{S,net}}{\dot{m}_1} = (W_{S,I}) + \left(1 - \dfrac{\dot{m}_2}{\dot{m}_1}\right)(W_{S,II}) + (W_{S,pump})$, and using enthalpies to

calculate the work of each turbine,

$$\frac{\dot{W}_{S,net}}{\dot{m}_1} = (3052 - 3435) + (1 - 0.14)(2616 - 3052) + (423 - 418) = -753\frac{kJ}{kg \text{ stream } 1}$$

Referring to the tabulated values,

$$\frac{Q_H}{\dot{m}_1} = H_1 - H_7 = 3435 - 746 = 2689\frac{kJ}{kg \text{ stream } 1} \Rightarrow \eta_\theta = \frac{753}{2689} = 0.280$$

When you consider all combinations of reheat and regeneration, it is clear that the number of configurations of turbine stages and heat exchangers is nearly endless. Practically speaking, one approaches a point of diminishing returns with each added complexity. The best alternative may depend on details of the specific application. Broadly speaking, Example 5.2 shows a clear 24%

gain in thermal efficiency by using reheat. Example 5.3 is not at the same conditions as Example 5.1, but we can quickly estimate the thermal efficiency at the same conditions as $\eta_\theta = (3435 - 2617)/(3435 - 422) = 0.271$, so regeneration alone offers just a 3% gain. A dedicated electric power facility would definitely want to make the most of every gain, but a small power generator for a chemical facility in an isolated rural area might be subject to other constraints. For example, the need for medium pressure steam to run distillation columns might dictate the pressure for the intermediate stage and a similar need for building heat could dictate a lower temperature requirement. In the final analysis, it is up to the engineer to devise the best solution by adapting these examples and general observations to any particular situation.

5.4   REFRIGERATION

Ordinary Vapor Compression Cycle

❶The ordinary vapor compression cycle is the most common refrigeration cycle.

The Carnot cycle is not practical for refrigeration for the same reasons as discussed for power production. Therefore, most refrigerators operate on the ordinary vapor-compression (OVC) cycle, shown in Fig. 5.8.

As with the Rankine cycle, we make some simplifications that would have to be reevaluated in a detailed engineering design. Again, we neglect pressure losses in piping. We assume that the vapor is saturated at the inlet to the compressor, and that the outlet of the condenser is saturated liquid. Thus, saturated vapor enters the compressor and exits heated above the condenser temperature, then cools in the condenser until it condenses to a saturated liquid. In the cyclic process, the saturated liquid is passed through a throttle valve at constant enthalpy and exits as a two-phase mixture. The evaporator is assumed to be isothermal, and accepts heat at the colder temperature to complete the vaporization. The OVC cycle is often characterized using a P-H diagram as shown in Fig. 5.9.

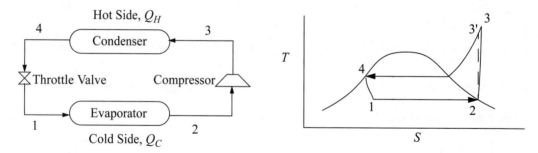

Figure 5.8 *OVC refrigeration cycle process schematic and T-S diagram.*

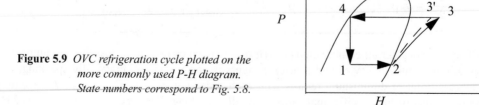

Figure 5.9 *OVC refrigeration cycle plotted on the more commonly used P-H diagram. State numbers correspond to Fig. 5.8.*

The COP can be related to the conditions of the process streams.

$$COP = Q_C/W_{S,net}; \quad Q_C = (H_2 - H_1)$$

$$Energy\ balance:\ W_{S,net} = \Delta H_{2\to3} = (H_3 - H_2)$$

$$\Rightarrow COP = (H_2 - H_1)/(H_3 - H_2) = (H_2 - H_4)/(H_3 - H_2) \qquad\qquad 5.3$$

❶ COP for ordinary vapor compression cycle.

Example 5.4 Refrigeration by vapor compression cycle

An industrial freezer room is to be maintained at –15°C by a cooling unit utilizing refrigerant R134a as the working fluid. The evaporator coils should be maintained at –20°C to ensure efficient heat transfer. Cooling water is available at 10°C from an on-site water well and can be warmed to 25°C in the condenser. The refrigerant temperature exiting the condenser is to be 30°C. The cooling capacity of the freezer unit is to be 120,000 BTU/h (126,500 kJ/h); 12,650 kJ/h is known as one *ton* of refrigeration because it is approximately the cooling rate (cooling duty) required to freeze one ton of water at 0°C to one ton of ice at 0°C in 24 h. So this refrigerator represents a 10 ton refrigerator. As a common frame of reference, typical home air conditioners are about 2–3 tons, but they typically weigh less than 100 kg. Calculate the COP and recirculation rate (except part (a)) for the industrial freezer in the following cases:

❶ 12,650 kJ/h is known as a *ton* of refrigeration capacity.

(a) Carnot cycle.
(b) Ordinary vapor compression cycle with a reversible compressor.
(c) Vapor compression cycle with the throttle valve replaced with an expander.
(d) Ordinary vapor compression cycle for which compressor is 80% efficient.

Solution: We will refer to Fig. 5.8 for identifying state by number. The operating temperatures of the refrigeration unit will be

$$T_H = T_4 = 303\ K \qquad\qquad T_C = T_2 = 253\ K$$

(a) Carnot cycle

Note that T_3 will be higher than T_4, but we use the condenser outlet T_4 as the benchmark temperature.

$$COP = \frac{T_C}{T_H - T_C} = \frac{253}{303 - 253} = 5.06$$

(b) Ordinary VC cycle with reversible compressor

We will create a table to summarize results. Values determined from balances are shown in bold-bordered table cells. Other valves are from the R-134a chart in Appendix E. State 2 is a convenient place to start since it is a saturated vapor and the temperature is known. $T_2 = -20°C$, from the chart, $H_2^{satV} = 386.5$ kJ/kg and $S_2^{satV} = 1.7414$ kJ/kg-K. The condenser outlet (state 4) is taken as saturated liquid at 30°C, so the pressure of the condenser will be $P_4^{sat}(30°C) = 0.77$ MPa, and $H_4 = 241.5$ kJ/kg, $S_4 = 1.1428$ kJ/kg-K. Because the throttle valve is isenthalpic (Section 2.13), $H_1 = H_4$.

Example 5.4 Refrigeration by vapor compression cycle (Continued)

State	T (K)	P (MPa)	H (kJ/kg)	S (kJ/kg-K)
1	253	0.13	241.5	
1' (for part (c))	253	0.13	235.0	1.1428
2, satV	253	0.13	386.5	1.7414
3'	311	0.77	424	1.7414
4	303	0.77	241.5	1.1428

The compressor calculation has already been performed in Example 4.17 on page 174. If the process is reversible, the entropy at state 3' will be the same as S_2. Finding H_3' from $S_3' = 1.7414$ kJ/kg-K and $P_3 = 0.77$ MPa, using the chart, $H_3' = 424$ kJ/kg. Note that the pressure in the condenser, not the condenser temperature, fixes the endpoint on the isentropic line from the saturated vapor.

$$COP = \frac{Q_C}{W_{S,net}} = \frac{(H_2 - H_1)}{(H_3' - H_2)} = \frac{(H_2 - H_4)}{(H_3' - H_2)} = \frac{386.5 - 241.5}{424 - 386.5} = 3.87$$

The required circulation rate is

$$\dot{Q} = 126,500 = \dot{m}Q_C = \dot{m}(H_2 - H_1) = \dot{m}(386.5 - 241.5), \text{ which gives } \dot{m} = 872 \text{ kg/h}.$$

(c) VC cycle with turbine expansion

The throttle valve will be replaced by a reversible expander. Therefore, $S_1' = S_4 = 1.1428$ kJ/kg-K. The saturation values at 253 K are $S^{satL} = 0.8994$ kJ/kg-K and $S^{satV} = 1.7414$ kJ/kg-K; therefore, $S_1' = 1.1428 = q' \cdot 1.7414 + (1 - q')0.8994$, which gives $q' = 0.289$.
Then, using the saturated enthalpy values and the quality, $H_1' = 235.0$ kJ/kg. In order to calculate the COP, we must recognize that we are able to recover some work from the expander, given by $H_1' - H_4$.

$$COP = \frac{Q_C}{W_{S,net}} = \frac{(H_2 - H_1')}{(H_3' - H_2) + (H_1' - H_4)} = \frac{386.5 - 235.0}{(424 - 386.5) + (235.0 - 241.5)} = 4.89$$

The increase in COP requires a significant increase in equipment complexity and cost, since a two-phase expander would probably have a short life due to erosion of turbine blades by droplets.

(d) Like (b) but with irreversible compressor

States 1, 2, and 4 are the same as in (b). The irreversibility simply changes state 3.

$$COP = \frac{Q_C}{W_{S,net}} = \frac{(H_2 - H_1)}{(H_3 - H_2)} = 0.8\frac{(H_2 - H_4)}{(H_3' - H_2)} = 0.8(3.87) = 3.10$$

Refrigerant choice is dictated by several factors:

1. Environmental impact (Freon R-12 depletes ozone and has been phased out; Freon R-22 is being phased out). HFO1234yf is beginning to supersede R134a.

2. Vapor pressure ~ atmospheric at T_{evap}. Consequently, the driving force for leakage will be small, but an evaporator pressure slightly above atmospheric pressure is desirable to avoid air leaking into the cycle.

3. Vapor pressure not too high at T_H so that the operating pressure is not too high; high pressure increases compressor and equipment costs.

4. High heat of vaporization per unit mass.

5. Small C_P/C_V of vapor to minimize temperature rise and work of compressor.

6. High heat transfer coefficient in vapor and liquid.

Flash Chamber (Economizer) Intercooling

When the temperature difference between the condenser and evaporator is increased, the compressor must span larger pressure ranges. If the compression ratio (P^{out}/P^{in}) becomes too large, interstage cooling can be used to increase efficiency. Because the process temperatures are usually below cooling water temperatures, a portion of the condensed refrigerant stream can be evaporated to provide the interstage cooling, as shown in Fig. 5.10. The interstage evaporative cooler is sometimes called an **economizer** or **flash chamber/drum**. The term *flash* characterizes the rapid vaporization. The economizer is considered adiabatic unless otherwise specified, and serves to disengage the liquid and vapor exiting the inlet valve. The quality out of the inlet valve is equal to $\dot{m}_7/\dot{m}_6$.

Flash chamber intercooling is a common method of increasing COP.

Cascade Refrigeration

In order to span extremely large temperature ranges, a single refrigerant becomes impractical, because the compression ratio (P^{out}/P^{in}) becomes too high and the COP decreases. A typical guideline is that the compression ratio should not be higher than about 8. Therefore, to span extremely large ranges, **binary** vapor cycles or **cascade** vapor cycles are used. In a binary cycle, a refrigerant with a normal boiling point below the coldest temperature is used on the cold cycle, and a refrigerant that condenses at a moderate pressure is used on the hot cycle. The two cycles are coupled at the

Cascade refrigeration is used to reach cryogenic temperatures.

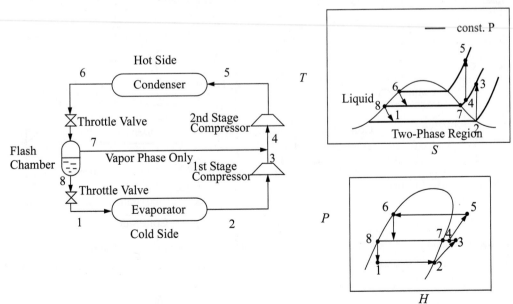

Figure 5.10 *Flash chamber intercooling.*

condenser of the cold cycle and the evaporator of the hot cycle as shown in Fig. 5.11. Because the heat of vaporization is coupled to the saturation temperature for any refrigerant, usually the operating temperatures are selected, and the circulation rates are determined for each cycle. Certainly, there are many variables to optimize in a process design of this type. For extremely large ranges, such as for cryogenic processing of liquefied gases, cascade refrigeration can be used with multiple cycles. For example, for the liquefaction of natural gas, the three cycles might be ammonia, ethylene, and methane. Note that the evaporator in each cycle must be colder than the condenser of the cycle below to ensure heat is transferred in the correct direction.

5.5   LIQUEFACTION

We have encountered liquefaction since our first quality calculation in dealing with turbines. In refrigeration, throttling or isentropic expansion results in a partially liquid stream. The point of a liquefaction process is simply to recover the liquid part as the primary product.

Linde Liquefaction

The Linde process works by throttling high-pressure vapor. The Joule-Thomson coefficient, $\left(\frac{\partial T}{\partial P}\right)_H$, must be such that the gas cools on expansion,[2] and the temperature must be low enough and the pressure high enough to ensure that the expansion will end in the two-phase region. Since less than 100% is liquefied, the vapor phase is returned to the compressor, and the liquid phase is withdrawn. Multistage compression is usually used in the Linde liquefaction process to achieve the required high

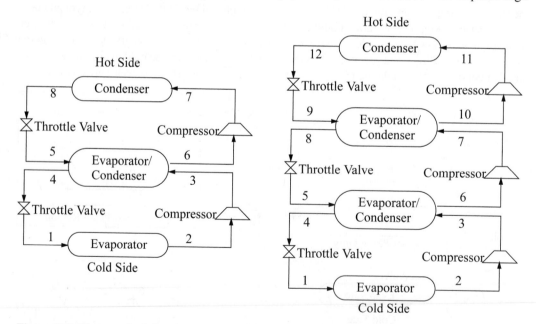

Figure 5.11 *Binary cycle (left) and three-cycle cascade (right) refrigeration cycles. The refrigerants do not mix in the evaporator/condensers.*

---

2. This criterion can be evaluated by looking at the dependence of temperature on pressure at constant enthalpy on a thermodynamic chart or table. In Chapter 6, we introduce principles for calculating this derivative from P, V, T properties.

pressures. An example of the process pathways on a T-S diagram is shown in Fig. 5.12. The actual state of the gas entering the multistage compressor depends on the state of the feed.

Example 5.5   Liquefaction of methane by the Linde process

Methane is to be liquefied in a simple Linde process. The feed and recycle are mixed, compressed to 60 bar, and precooled to 300 K. The vapor then passes through a heat exchanger for additional cooling before being throttled to 1 bar. The unliquefied fraction leaves the separator at the saturation temperature, and passes through the heat exchanger, then exits at 295 K. (a) What fraction of the gas is liquefied in the process; and (b) what is the temperature of the high-pressure gas entering the throttle valve?

Solution: The schematic is shown in Fig. 5.12. To solve this problem, first recognize that states 3, 6, 7, and 8 are known. State 3 is at 300 K and 60 bar; state 6 is saturated liquid at 1 bar; state 7 is saturated vapor at 1 bar; and state 8 is at 295 K and 1 bar. Use the furnished methane chart from Appendix E.

(a) The System I energy balance is: $H_3 - [qH_8 + (1 - q)H_6] = 0$

$$\Rightarrow q = \frac{H_3 - H_6}{H_8 - H_6} = \frac{H(60, 300) - H(1, satL)}{H(1, 295) - H(1, satL)} = \frac{1130 - 284}{1195 - 284} = 0.9286 \Rightarrow 7.14\% \text{ liquefied}$$

(b) The energy balance for System II is: $H_4 - H_3 = -q(H_8 - H_7) = -0.9286(1195 - 796.1)$
$$= -370.5 \Rightarrow H_4 = 780$$

$$\Rightarrow H_4 = 780 \text{ @ 60 bar} \quad \Rightarrow \text{chart gives } -95°F = 203 \text{ K}$$

Claude Liquefaction

The throttling process between states 4 and 5 in the Linde process is irreversible. To improve this, a reversible expansion is desirable; however, since the objective is to liquefy large fractions of the inlet stream, turbines are not practical because they cannot handle low-quality mixtures. One compromise, the Claude liquefaction, is to expand a portion of the high-pressure fluid in an expander

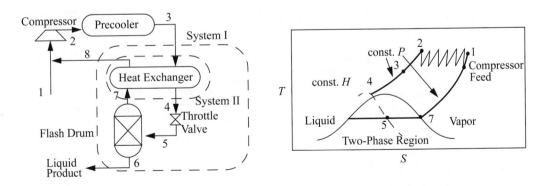

Figure 5.12 *Linde liquefaction process schematic. The system boundaries shown on the left are used in Example 5.5.*

under conditions that avoid the two-phase region, as shown in Fig. 5.13. Only a smaller fraction of the compressed gas enters the irreversible throttle valve, so the overall efficiency can be higher but more sophisticated equipment is required.

5.6   ENGINES

Steam is not the only working fluid that can be used in a power producing cycle. A common alternative is to use air, mixed with a small amount of fuel that is burned. The heat of combustion provides energy to heat the gas mixture before it does work in an expansion step. A major benefit of using air is that a physical loop is not necessary; we can imagine the atmosphere as the recycle loop. This approach forms the basis for internal combustion engines like lawn mowers, jet engines, diesels, and autos. An online supplement introduces the gas turbine, the turbofan jet engine, the internal combustion engine, and the diesel engine.

5.7   FLUID FLOW

This section is available as an on-line supplement and includes liquids and compressible gases. We discuss the energy balance, the Bernoulli equation, friction factor, and lost work. We also generalize that $W_s = \int_{Pin}^{Pout} V dP$ is a general result for open system compressors that are not adiabatic.

5.8   PROBLEM-SOLVING STRATEGIES

As you set up more complex problems, use the strategies in Section 2.14 on page 74, and incorporate the energy balances developed in Section 2.13 on page 68 for valves, nozzles, heat exchangers, turbines, and pumps and entropy balances developed in Section 4.6 on page 159 for turbines, compressors, and heat pumps/engines as you work through step 5 of the strategies. A stream that exits a

A review of common assumptions and hints.

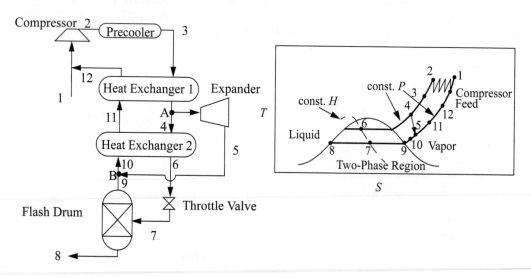

Figure 5.13 *The Claude liquefaction process.*

condenser is assumed to exit as saturated liquid unless otherwise specified. Likewise, a stream that is vaporized in a boiler is assumed to exit saturated unless otherwise specified; however, recall that in a Rankine cycle the steam is always superheated, and we omit the superheater unit in schematics for simplicity. Read problem statements carefully to identify the outlet states of turbines. Outlets of turbines *are not required to be saturated or in the two-phase region,* although operation in this manner is common. In a multistage turbine without reheat, only the last stages will be near saturation unless reheat is used. Unless specified, pressure drops are considered negligible in piping and heat exchangers as a first approximation. Throttle valves are assumed to be adiabatic unless otherwise stated, and they are always irreversible and *do* have an important pressure drop unless otherwise stated. Recognize that the entropy balances for throttle valves or heat exchangers are usually not helpful since, in practical applications, these devices are inherently irreversible and generate entropy.

Recognize that the energy balance must be used often to find mass flow rates. In order to do this for open steady-state flow systems, the enthalpies for all streams must be known in addition to the mass flow rates for all but one stream. You can try moving the system boundary as suggested in step 7 of the strategy in Section 2.14 on page 74 to search for balances that satisfy these conditions. Mass flow rates can be found using the entropy balance also, but this is not done very often, since the entropy balance is useful only if the process is internally reversible or if the rate of entropy generation is known (i.e., no irreversible heat exchangers or throttle valves or irreversible turbines/compressors inside the system boundary).

Basically, the energy and entropy balance and the *P-V-T* relation are the only equations that always apply. While we have shown common simplifications, there are always new applications that can arise, and it is wise to learn the principles involved in simplifying the balance to a given situation.

5.9   SUMMARY

Similar to energy balances in Chapter 3, entropy balances can be applied to composite systems. What is new in this chapter is the level of detail and the combination of the energy balance with the entropy balance. Instead of abstract processes like the Carnot cycle, the entropy balance enables us to compute the impacts of each individual step, whether isothermal, irreversible, or otherwise. This leads to composite processes with innumerable combinations of heat exchangers, multiple stages, and mixers, all striving to achieve the efficiency of the Carnot or Stirling cycles. For processes on the scale of commercial production, an efficiency improvement of 0.1% can mean millions of dollars, amply justifying an investment in clever engineering.

Important Equations

Once again the energy and entropy balances are the most important equations. We have incorporated most of the previous chapters, using state properties, interpolations, efficiency calculations, throttling, and so on. Carnot thermal efficiency and COP provide upper bounds on what can be achieved with any process and these are important.

5.10   PRACTICE PROBLEMS

P5.1 An ordinary vapor compression cycle is to operate a refrigerator on R134a between −40°C and 40°C (condenser temperatures). Compute the coefficient of performance and the heat removed from the refrigerator per day if the power used by the refrigerator is 9000 J per day. (ANS. 1.76)

P5.2 An ordinary vapor compression cycle is to be operated on methane to cool a chamber to –260°F. Heat will be rejected to liquid ethylene at –165°F. The temperatures in the condenser and evaporator are –160°F and –280°F. Compute the coefficient of performance. (ANS. 0.86)

P5.3 A simple Rankine cycle is to operate on steam between 200°C and 99.6°C, with saturated steam exhausting from the turbine. What is the maximum possible value for its thermodynamic efficiency? (ANS. 7.8%)

P5.4 An ordinary vapor compression cycle is to be operated on methane to cool a chamber to 112 K. Heat is rejected to liquid ethylene at 160 K. The temperatures in the coils are 168 K and 100 K. (ANS. online)
(a) Write the relevant energy and entropy balances for the compression step.
(b) Estimate the minimal work requirement (J/g) for the compressor assuming the ideal gas law.
(c) Estimate the coefficient of performance (COP) for this OVC cycle.
(d) Estimate the COP by the Carnot guideline.
(e) Estimate the minimal work requirement for the compressor using the chart in the text.

P5.5 A house has an effective heat loss of 100,000 Btu/hr. During the heating season of 160 days the average inside temperature should be 70°F while the outside is 45°F. Freon-134a is the working fluid and an ordinary vapor-compression cycle is used. A 10°F approach on each side may be assumed. Electricity costs $0.14/KW-hr.

(a) What is the cost in $/hr if the compressor is 100% efficient?
(b) What is the cost if the compressor is 80% efficient?
   (ANS. ~0.37, 0.45)

P5.6 An adiabatic turbine is supplied with steam at 300 psia and 550°F that exhausts at atmospheric pressure. The quality of the exhaust steam is 95%.

(a) What is the efficiency of the turbine? (ANS. 76%)
(b) What would be the thermodynamic efficiency of a Rankine cycle operated using this turbine at these conditions?   (ANS. 17%)

5.11   HOMEWORK PROBLEMS

5.1 A steam power plant operates on the Rankine cycle according to the specified conditions below. Using stream numbering from Fig. 5.2 on page 201, for each of the options below, determine:

(a) The work output of the turbine per kg of steam;
(b) The work input of the feedwater pump per kg of circulated water;
(c) The flowrate of steam required;
(d) The heat input required in the boiler/superheater;
(e) The thermal efficiency

Option	T_3(°C)	P_3(MPa)	P_4(MPa)	$\eta_{turbine}$	η_{pump}	Plant Capacity
i	600	10	0.01	0.8	0.75	80 MW
ii	600	8	0.01	0.8	0.75	80 MW

Option	$T_3(°C)$	$P_3(MPa)$	$P_4(MPa)$	$\eta_{turbine}$	η_{pump}	Plant Capacity
iii	400	4	0.01	0.83	0.95	75 MW
iv	500	4	0.02	0.85	0.8	80 MW
v	600	4	0.02	0.85	0.8	80 MW

5.2 A steam power plant operates on the Rankine cycle with reheat, using the specified conditions below. Using stream numbering from Fig. 5.3 on page 203, for each of the options below, determine

(a) The work output of each turbine per kg of steam;
(b) The work input of the feedwater pump per kg of circulated water;
(c) The flowrate of steam required;
(d) The heat input required in the boiler/superheater and reheater;
(e) The thermal efficiency.

Option	$T_3(°C)$	$P_3(MPa)$	$P_4(MPa)$	$T_5(°C)$	$P_6(MPa)$	$\eta_{turbines}$	η_{pump}	Plant Capacity
i	450	5	0.4	400	0.01	0.8	0.75	80 MW
ii	400	5	0.6	450	0.01	0.8	0.75	80 MW
iii	400	4	0.5	500	0.01	0.9	0.95	75 MW
iv	500	4	0.8	500	0.01	0.85	0.8	80 MW
v	600	4	1.2	600	0.01	0.85	0.8	80 MW

5.3 A modified Rankine cycle using a single feedwater preheater as shown in Fig. 5.7 on page 206 has the following characteristics.

(a) The inlet to the first turbine is at 500°C and 0.8 MPa.
(b) The feedwater preheater reheats the recirculated water so that stream 7 is 140°C, and steam at 0.4 MPa is withdrawn from the outlet of the first turbine to perform the heating.
(c) The efficiency of each turbine and pump is 79%.
(d) The output of the plant is to be 1 MW.
(e) The output of the second turbine is to be 0.025 MPa.

Determine the flow rates of streams 1 and 8 and the quality of stream 9 *entering* the condenser (after the throttle valve). Use the stream numbers from Fig. 5.7 on page 206 to label streams in your solution.

5.4 A modified Rankine cycle uses reheat and one closed feedwater preheater. The schematic is a modification of Fig. 5.7 on page 206 obtained by adding a reheater between the T-joint and turbine II. Letting stream 3 denote the inlet to the reheater, and stream 3a denote the inlet to the turbine, the conditions are given below. The plant capacity is to be 80 MW. Other constraints are as follows: The efficiency of each turbine stage is 85%; the pump efficiency is 80%; and the feedwater leaving the closed preheater is 5°C below the temperature of the condensate draining from the bottom of the closed preheater. For the options below, calculate:

(a) The flowrate of stream 1;
(b) The thermal efficiency of the plant;
(c) The size of the feedwater pump (kW);

Options:

(*i*) $T_1 = 500°C$, $P_1 = 4$ MPa, $P_2 = 0.8$ MPa, $T_{3a} = 500°C$, $P_4 = 0.01$ MPa.
(*ii*) $T_1 = 600°C$, $P_1 = 4$ MPa, $P_2 = 1.2$ MPa, $T_{3a} = 600°C$, $P_4 = 0.01$ MPa.

5.5 A regenerative Rankine cycle uses one open feedwater preheater and one closed feedwater preheater. Using the stream numbering from Fig. 5.6 on page 206, and the specified conditions below, the plant capacity is to be 75 MW. Other constraints are as follows: The efficiency of each turbine stage is 85%; the pump efficiencies are 80%; and the feedwater leaving the closed preheater is 5°C below the temperature of the condensate draining from the bottom of the closed preheater. For the options below, calculate

(a) The flowrate of stream 1.
(b) The thermal efficiency of the plant.
(c) The size of the feedwater pumps (kW).

Options:

(*i*) The conditions are $T_1 = 500°C$, $P_1 = 4$ MPa, $P_2 = 0.7$ MPa, $P_3 = 0.12$ MPa, and $P_4 = 0.02$ MPa.
(*ii*) The conditions are $T_1 = 600°C$, $P_1 = 4$ MPa, $P_2 = 1.6$ MPa, $P_3 = 0.8$ MPa, and $P_4 = 0.01$ MPa.

5.6 A regenerative Rankine cycle utilized the schematic of Fig. 5.6 on page 206. Conditions are as follows: stream 1, 450°C, 3 MPa; stream 2, 250°C, 0.4 MPa; stream 3, 150°C, 0.1 MPa; stream 4, 0.01 MPa; stream 9, 140°C, $H_9 = 592$ kJ/kg; stream 8, $H_8 = 421.1$ kJ/kg.

(a) Determine the pressures for streams 5, 6, 8, 9, and 10.
(b) Determine $\dot{m}_2 / \dot{m}_1$.
(c) Determine the enthalpies of streams 5 and 6 if the pump is 80% efficient.
(d) Determine the efficiency of turbine stage I.
(e) Determine the output of turbine stage III per kg of stream 4 if the turbine is 80% efficient.
(f) Determine $\dot{m}_3 / \dot{m}_1$.
(g) Determine the work output of the system per kg of stream 1 circulated.

5.7 A regenerative Rankine cycle uses three closed feedwater preheaters. Using the stream numbering from Fig. 5.5 on page 205, and the specified conditions below, the plant capacity is to be 80 MW. Other constraints are as follows: The efficiency of each turbine stage is 88%; the pump efficiency is 80%; and the feedwater leaving each preheater is 5°C below the temperature of the condensate draining from the bottom of each preheater. For the options below, calculate:

(a) The flowrate of stream 1
(b) The thermal efficiency of the plant
(c) The size of the feedwater pump (kW)

Options:

(*i*) The conditions are $T_1 = 700°C$, $P_1 = 4$ MPa, $P_2 = 1$ MPa, $P_3 = 0.3$ MPa, $P_4 = 0.075$ MPa, and $P_5 = 0.01$ MPa.
(*ii*) The conditions are $T_1 = 750°C$, $P_1 = 4.5$ MPa, $P_2 = 1.2$ MPa, $P_3 = 0.4$ MPa, $P_4 = 0.05$ MPa, and $P_5 = 0.01$ MPa.

5.8 An ordinary vapor compression refrigerator is to operate on refrigerant R134a with evaporator and condenser temperatures at −20°C and 35°C. Assume the compressor is reversible.

(a) Make a table summarizing the nature (e.g., saturated, superheated, temperature, pressure, and *H*) of each point in the process.
(b) Compute the coefficient of performance for this cycle and compare it to the Carnot cycle value.
(c) If the compressor in the cycle were driven by a 1 hp motor, what would be the tonnage rating of the refrigerator? Neglect losses in the motor.

5.9 An ordinary vapor compression refrigeration cycle using R134a is to operate with a condenser at 45°C and an evaporator at −10°C. The compressor is 80% efficient.

(a) Determine the amount of cooling per kg of R134a circulated.
(b) Determine the amount of heat rejected per kg of R134a circulated.
(c) Determine the work required per kg of R134a circulated, and the COP.

5.10 An ordinary vapor compression cycle using propane operates at temperatures of 240 K in the cold heat exchanger, and 280 K in the hot heat exchanger. How much work is required per kg of propane circulated if the compressor is 80% efficient? What cooling capacity is provided per kg of propane circulated? How is the cooling capacity per kg of propane affected by lowering the pressure of the hot heat exchanger, while keeping the cold heat exchanger pressure the same?

5.11 The low-temperature condenser of a distillation column is to be operated using a propane refrigeration unit. The evaporator is to operate at −20°C. The cooling duty is to be 10,000,000 kJ/hr. The compressor is to be a two-stage compressor with an adiabatic efficiency of 80% (each stage). The compression ratio (P^{out}/P^{in}) for each stage is to be the same. The condenser outlet is to be at 50°C. Refer to Fig. 5.8 on page 208 for stream numbers.

(a) Find the condenser, evaporator, and compressor interstage pressures.
(b) Find the refrigerant flowrate through each compressor.
(c) Find the work input required for each compressor.
(d) Find the cooling rate needed in the condenser.

5.12 Solve problem 5.11 using an economizer at the intermediate pressure and referring to Fig. 5.10 on page 211 for stream numbers.

5.13 A refrigeration process with interstage cooling uses refrigerant R134a. The outlet of the condenser is to be saturated liquid at 40°C. The evaporator is to operate at −20°C, and the outlet is saturated vapor. The economizer is to operate at 10°C. Refer to Fig. 5.10 on page 211 for stream numbers in your solution.

(a) Determine the required flowrate of stream 1 if the cooling capacity of the unit is to be 8250 kJ/h.
(b) Determine the pressure of stream 3, and the work required by the first compressor if it has an efficiency of 85%.
(c) What are the flowrates of streams 7 and 6?
(d) What is the enthalpy of stream 4?
(e) Determine the work required by the second compressor (85% efficient) and the COP.

5.14 A refrigeration process with interstage cooling uses refrigerant R134a, and the outlet of the condenser is to be saturated liquid at 40°C. Refer to Fig. 5.10 on page 211 for stream numbers in your solution. The pressure of the flash chamber and the intermediate pressure between compressors is to be 290 kPa. The evaporator is to operate at –20°C and the outlet is to be saturated vapor. The flow rate of stream 1 is 23 kg/h. The flash chamber may be considered adiabatic. The compressors may be considered to be 80% efficient. Attach the P-H chart with your solution.

(a) What is the work input required to the first compressor in kJ/h?
(b) What are the flow rates of streams 7 and 6?
(c) What is the enthalpy of stream 4?

5.15 The Claude liquefaction process is to be applied to methane. Using the schematic of Fig. 5.13 on page 214 for stream numbering, the key variables depend on the fraction of stream 3 that is liquefied, $\dot{m}_8/\dot{m}_3$, and the fraction of stream 3 that is fed through the expander, $\dot{m}_5/\dot{m}_3$. Create a table listing all streams from low to high stream numbers. Fill in the table as you complete the problem sections. Attach a P-H diagram with your solution.

(a) Write a mass balance for the system boundary encompassing all equipment except the compressor and precooler.
(b) Write an energy balance for the same boundary described in part (a), and show

$$\frac{\dot{m}_8}{\dot{m}_3} = \frac{(H_3 - H_{12}) + (\dot{m}_5/\dot{m}_3)W_{S\text{ expander}}}{(H_8 - H_{12})}$$

(c) Stream 3 is to be 300 K and 3 MPa, stream 4 is to be 280 K and 3 MPa, stream 12 is to be 290 K and 0.1 MPa, and the flash drum is to operate at 0.1 MPa. The expander has an efficiency of 91%. The fraction liquefied is to be $\dot{m}_8/\dot{m}_3 = 0.15$. Determine how much flow to direct through the expander, $\dot{m}_5/\dot{m}_3$.
(d) Find the enthalpies of streams 3–12, and the temperatures and pressures.

This problem references an online supplement at the URL given in the front flap.

5.16 A Brayton gas turbine typically operates with only a small amount of fuel added so that the inlet temperatures of the turbine are kept relatively low because of material degradation at higher temperatures, thus the flowing streams can be modeled as only air. Refer to the online supplement for stream labels. Consider a Brayton cycle modeled with air under the following conditions: $T_A = 298$ K, $P_A = P_D = 0.1$ MPa, $P_B = 0.6$ MPa, and $T_C = 973$ K. The efficiencies of the turbine and compressor are to be 85%. Consider air as an ideal gas stream with $C_P = 0.79 \cdot C_{P,N2} + 0.21 \cdot C_{P,O2}$. Determine the thermal efficiency, heat required, and net work output per mole of air assuming

(a) The heat capacities are temperature-independent at the values at 298 K.
(b) The heat capacities are given by the polynomials in Appendix E.

This problem references an online supplement at the URL given in the front flap.

5.17 The thermal efficiency of a Brayton cycle can be increased by adding a regenerator as shown in the schematic below. Consider a Brayton cycle using air under the following conditions: $T_A = 298$ K, $P_A = P_E = P_F = 0.1$ MPa, $P_B = 0.6$ MPa, $T_D = 973$ K, $T_F = 563$ K. The efficiency of the turbine and compressor are to be 85%. Consider air as an ideal gas stream with $C_P = 0.79 \cdot C_{P,N2} + 0.21 \cdot C_{P,O2}$, and assume the molar flows of B and E are equal. Determine the thermal efficiency, heat required, and net work output per mole of air, assuming

(a) The heat capacities are temperature-independent at the values at 298 K.

(b) The heat capacities are given by the polynomials in Appendix E.

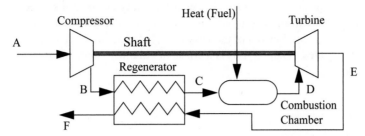

5.18 Consider the air-standard Otto cycle explained in the online supplement. At the beginning of the compression stroke, $P_1 = 95$ kPa, $T_1 = 298$ K. Consider air as an ideal gas stream with $C_P = 0.79 \cdot C_{P,N2} + 0.21 \cdot C_{P,O2}$. If the compression ratio is 6, determine T_2, T_4, and the thermal efficiency, if $T_3 = 1200$ K and the following are true

This problem references an online supplement at the URL given in the front flap.

(a) The heat capacities are temperature-independent at the values at 298 K.
(b) The heat capacities are given by the polynomials in Appendix E.

5.19 A hexane ($\rho \approx 0.66$ kg/L, $\mu = 3.2$ E-3 g/(cm-s)) storage tank in the chemical plant tank farm is 250 m from the 200 L solvent tank that is to be filled in 3 min. A pump is located at the base of the storage tank at ground level. The storage tank is large enough so that the liquid height doesn't change significantly when 200 L are removed. The bends and fittings in the pipe contribute lost work equivalent to 15 m of additional pipe. The pump and motor are to be sized based on a storage tank liquid level of 0.3 m above ground level to ensure adequate flow rate when the storage tank is nearly empty. Find the required power input to the pump and motor.

This problem references an online supplement at the URL given in the front flap.

(a) The pipe is to be 2.5 cm in diameter and the outlet is to be 10 m above ground level. The pump efficiency is 85%, the motor efficiency is 90%.
(b) The pipe is to be 3.0 cm in diameter and the outlet is to be 8.5 m above ground level. The pump efficiency is 87%, the motor efficiency is 92%.
(c) Determine the time required to fill the solvent tank using the pump and motor sized in part (a) if the storage tank liquid level is 6.5 m above ground.
(d) Answer part (c) except determine the filling time for part (b).

5.20 Consider problem 5.16(a). Determine the amount of fuel required per mole of air if the fuel is modeled as isooctane and combustion is complete.

5.21 Consider problem 5.18(a). Determine the amount of fuel required per mole of air if the fuel is modeled as isooctane and combustion is complete.

5.22 In the event of an explosive combustion of vapor at atmospheric pressure, the vapor cloud can be modeled as adiabatic because the combustion occurs so rapidly. The vapor cloud expands rapidly due to the increase in moles due to combustion, but also due to the adiabatic temperature rise. Estimate the volume increase of a 22°C, 1 m³ mixture of propane and a stoichiometric quantity of air that burns explosively to completion. Estimate the temperature rise.

5.23 (a) Derive the energy balance for a closed, constant-volume, adiabatic-system vapor phase chemical reaction, neglecting the energy of mixing for reactants and products, and assuming the ideal gas law.

(b) Suppose that a 200 L propane tank is at 0.09 MPa pressure and, due to an air leak, contains the propane with a stoichiometric quantity of air. If a source of spark is present, the system will burn so rapidly that it may be considered adiabatic, and there will not be time for any flow out of the vessel. If ignited at 20°C, what pressure and temperature are generated assuming this is a constant volume system and the reaction goes to completion?

GENERALIZED ANALYSIS OF FLUID PROPERTIES

Forming an intermediate state between liquids, in which we assume no external pressure, and gases, in which we omit molecular forces, we have the state in which both terms occur. As a matter of fact, we shall see further on, that this is the only state which occurs in nature.

van der Waals (1873, ch2)

In Unit I we focused predominantly on a relatively small number of pure fluids. But the number of chemical compounds encountered when considering all possible applications is vast, and new compounds are being invented and applied every day. Imagine how many charts and tables would be necessary to have properties available for all those compounds. Then imagine how many charts would be necessary to represent the properties of all the conceivable mixtures of those compounds. Clearly, we cannot address all problems by exactly the same techniques as applied in Unit I. We must still use the energy and entropy balance, but we need to be able to represent the physical properties of pure compounds and mixtures in some condensed form, and we desire to predict physical properties based on very limited data.

As one might expect, an excellent shorthand is offered by the language of mathematics. When we sought values in the steam tables, we noticed that specification of any two variables was sufficient to determine the variable of interest (e.g, S or H). This leads to an excellent application of the calculus of two variables. Changes in each value of interest may be expressed in terms of changes in whatever other two variables are most convenient. It turns out that the most convenient variables mathematically are temperature and density, and that the most convenient variables experimentally are temperature and pressure.

There is a limit to how condensed our mathematical analysis can be. That limit is dictated by how much physical insight is required to represent the properties of interest to the desired accuracy. With no physical insight, we can simply measure the desired values, but that is impractical. With maximum physical insight, we can represent all the properties purely in terms of their fundamental electronic structure as given by the periodic table and their known molecular structure. The current state-of-the-art lies between these limits, but somewhat closer to the fundamental side. By developing a sophisticated analysis of the interactions on the molecular scale, we can show that three carefully selected parameters characterizing physical properties are generally sufficient to characterize properties to the accuracy necessary in most engineering applications. This analysis leads to an equation of state, which is then combined with the necessary mathematics to provide methods for computing and predicting physical properties of interest. The development of van der Waals' equation of state provides an excellent case study in the development of engineering models based on insightful physics and moderately clever extrapolation. Note that before van der Waals the standard conception was that the vapor phase was represented by what we now refer to as an ideal gas, and the liquid was considered to be an entirely different species. Van der Waals' analysis led to a unification of these two conceptions of fluids that also set the stage for the analysis of interfaces and other inhomogeneous fluids. Van der Waals' approach lives on in modern research on inhomogeneous fluids.

CHAPTER

CLASSICAL THERMODYNAMICS — GENERALIZATIONS FOR ANY FLUID

When I first encountered the works of J.W. Gibbs, I felt like a small boy who had found a book describing the facts of life.

T. von Karmann

When people refer to "classical thermodynamics" with no context or qualifiers, they are generally referring to a subtopic of physical chemistry which deals with the mathematical transformations of energy and entropy in fluids. These transformations are subject to several constraints owing to the nature of state functions. This field was developed largely through the efforts of J.W. Gibbs during the late 1800's (Gibbs was granted the first engineering Ph.D. in the United States in 1863). Our study focuses on three aspects of the field:

1. The fundamental property relation: $dU(S,V) = TdS - PdV$;

2. Development of general formulas for property dependence of nonmeasurable properties in terms of on measurable variables, e.g., temperature and pressure dependence of U, H in terms of P, V, and T;

3. Phase equilibrium: e.g., quality and composition calculations.

The fundamental property relation provides a very general connection between the energy balance and the entropy balance. It relates state functions to one another mathematically such that no specific physical situation is necessary when considering how the energy and entropy change. That is, it tells how one variable changes with respect to some other variables that we may know something about. We implicitly applied this approach for solving several problems involving steam, determining the properties upstream of a throttle from the pressure and enthalpy, for instance. In this chapter we focus intensely on understanding how to transform from one set of variables to another as preparation for developing general formulas for property dependence on measured variables.

Through "classical thermodynamics," we can generalize our insights about steam to any fluid at any conditions. All engineering processes simply involve transitions from one set of conditions to another. To get from one state to another, we must learn to develop our own paths. It does not matter what path we take, only that we can compute the changes for each step and add them up. In Chapter 7, we present the insights that led van der Waals to formulate his equation of state, enabling the estimation of any fluid's pressure given the density and temperature. This analysis also illuminates the basis for development of current and future equations of state. In Chapter 8, we show the paths that are convenient for applying equations of state to estimate the thermodynamic properties for any fluid at any state based on a minimal number of experimental measurements.

Finally, a part of property estimation involves calculating changes of thermodynamic properties upon phase transitions so that they may be used in process calculations (e.g., formation of condensate during expansion through a turbine and characterization of the quality). The generalized analysis of phase changes requires the concept of phase equilibrium and an understanding of how the equilibrium is affected by changes in temperature and pressure. The skills developed in this chapter will be integrated in a slightly different form to analyze the thermodynamics of non-ideal fluid behavior in Chapter 8.

Chapter Objectives: You Should Be Able to...

1. Transform variables like dU into dH, dA, dG, dS, dP, dV, and dT given dU and the definition of H, A, and G. (e.g., Legendre transforms in Eqns. 6.5–6.7, Example 6.7).

2. Describe what is meant by a "measurable properties" and why they are useful.*

3. Apply the chain rule and other aspects of multivariable calculus to transform one derivative into another (e.g., Eqns. 6.11–6.17).

4. Use Maxwell relations to interchange derivatives.

5. Manipulate partial derivatives using two to three steps of manipulations to put in terms of measurable properties.

6. Substitute C_P and C_V for the appropriate temperature derivatives of entropy.

7. Recognize common characterizations of derivative properties like isothermal compressibility and the Joule-Thomson coefficient and express them in terms of measurable properties.

8. Devise step-wise paths to compute $\Delta U(T,V)$ and $\Delta S(T,P)$ or $\Delta S(T,V)$ from their ideal gas and density dependent terms.

6.1 THE FUNDAMENTAL PROPERTY RELATION

One equation underlies all the other equations to be discussed in this chapter. It is the combined energy and entropy balances for a closed system without shaft work. The only special feature that we add in this section is that we eliminate any references to specific physical situations. Transforming to a purely mathematical realm, we are free to apply multivariable calculus at will, transforming any problem into whatever variables seem most convenient at the time. Some of these relatively convenient forms appear frequently throughout the text, so we present them here as clear implications of the fundamental property relation changes in U.

The Fundamental Property Relation for *dU* in Simple Systems

We restrict our treatment here to systems without internal rigid, impermeable, or adiabatic walls, no internal temperature gradients, and no external fields. These restrictions comprise what we refer to as *simple* systems. This is not a strong restriction, however. Most systems can be treated as a sum of simple systems. Our goal is to transform the energy balance from extensive properties like heat and work to intensive (state) properties like density, temperature, and specific entropy. For this purpose, we may imagine any convenient physical path, recognizing that the final result will be independent of path as long as it simply relates state properties.

The energy balance for a closed simple system is

$$d(U + E_K + E_P) = dQ + dW_S + dW_{EC} \qquad 6.1$$

where E_K and E_P are the intensive kinetic and potential energies of the center of mass of the system. Eliminating all surface forces except those that cause expansion or contraction, because a simple system has no gradients or shaft work, and neglecting E_K and E_P changes by taking the system's center of mass as the frame of reference,

$$dU = dQ - PdV \qquad 6.2$$

Emphasizing the neglect of gradients, the reversible differential change between states is

$$dU_{rev} = dQ_{rev} - (PdV)_{rev} \qquad 6.3$$

but, by definition,

$$d\underline{S} = \frac{d\underline{Q}_{rev}}{T_{sys}} \qquad \Rightarrow T_{sys}d\underline{S} = d\underline{Q}_{rev} \quad (T_{sys}\text{-system temperature where } Q \text{ transferred})$$

Since the system is simple, for the process to be internally reversible, the temperature must be uniform throughout the system (no gradients). So the system temperature has a single value throughout. On a molar basis, the fundamental property relation for dU is

$$\boxed{dU = TdS - PdV} \qquad 6.4$$

🛑 dU for a closed simple system.

The significance of this relation is that changes in one state variable, dU, have been related to changes in two other state variables, dS and dV. Therefore, the physical problem of relating heat flow and volume changes to energy changes has been transformed into a purely mathematical problem of the calculus of two variables. This transformation liberates us from having to think of a physical means of attaining some conversion of energy. Instead, we can apply some relatively simple rules of calculus given changes in S and V.

Auxiliary Relations for Convenience Properties

Because dU is most simply written as a function of S and V it is termed a *natural function* of S and V. We can express changes of internal energy in terms of other state properties (such as $\{P, T\}$ or $\{T, V\}$), but when we do so, the expression always involves additional derivatives. We will show this in more detail in Example 6.11 on page 243. We also should explore the natural variables for the convenience properties.

We have defined enthalpy, $H \equiv U + PV$. Therefore, $dH = dU + PdV + VdP = TdS - PdV + PdV + VdP$,

$$\boxed{dH = TdS + VdP} \quad \text{which shows that } H \text{ is a natural function of } S \text{ and } P \qquad 6.5$$

🛑 dH for a reversible, closed simple system. Enthalpy is convenient when heat and pressure are manipulated.

The manipulation we have performed is known as a **Legendre transformation.** Note that $\{T, S\}$ and $\{P, V\}$ are paired in both U and the transform H. These pairs are known as **conjugate pairs** and will always stayed paired in all transforms.[1]

Enthalpy is termed a **convenience property** because we have specifically *defined* it to be useful in problems where reversible heat flow and pressure are manipulated. By now you have become so used to using it that you may not stop to think about what the enthalpy really is. If you look back to our introduction of enthalpy, you will see that we defined it in an arbitrary way when we needed a new tool. The fact that it relates to the heat transfer in a constant-pressure closed system, and relates to the heat transfer/shaft work in steady-state flow systems, is a result of our careful choice of its definition.

We may want to control T and V for some problems, particularly in statistical mechanics, where we create a system of particles and want to change the volume (intermolecular separation) at fixed temperature. Situations like this also arise quite often in our studies of pistons and cylinders. Since U is not a natural function of T and V, such a state property is convenient. Therefore, we define **Helmholtz energy** $A \equiv U - TS$. Therefore, $dA = dU - TdS - SdT = TdS - PdV - TdS - SdT,$

Helmholtz energy is convenient when T, V are manipulated.

$$\boxed{dA = -SdT - PdV}$$ which shows A is a natural function of T and V 6.6

Consider how the Helmholtz energy relates to expansion/contraction work for an isothermal system. Equilibrium occurs when the derivative of the Helmholtz energy is zero at constant T and V. The other frequently used convenience property is **Gibbs energy** $G \equiv U - TS + PV = A + PV = H - TS$. Therefore, $dG = dH - TdS - SdT = TdS + VdP - TdS - SdT.$

Gibbs energy is convenient when T, P are manipulated.

$$\boxed{dG = -SdT + VdP}$$ which shows G is a natural function of T and P 6.7

The Gibbs energy is used specifically in phase equilibria problems where temperature and pressure are controlled. We find that for systems constrained by constant T and P, the equilibrium occurs when the derivative of the Gibbs energy is zero ($\Rightarrow$ driving forces sum to zero and Gibbs energy is minimized). Note that $dG = 0$ when T and P are constant ($dT = 0$, $dP = 0$). The Helmholtz and Gibbs energies include the effects of entropic driving forces. The sign convention for Helmholtz and Gibbs energies are such that an increase in entropy detracts from our other energies, $A = U - TS$, $G = U - TS + PV$. In other words, increases in entropy detract from increases in energy. These are sometimes called the **free energies.** The relation between these free energies and maximum work is shown in Section 4.12.

In summary, the common Legendre transforms are summarized in Table 6.1. We will use other Legendre transforms in Chapter 18.

Often, students' first intuition is to expect that energy is minimized at equilibrium. But some deeper thought shows that equilibrium based purely on energy would eventually reach a state where all atoms are at the minimum of their potential wells with respect to one another. All the world would be a solid block. On the other hand, if entropy was always maximized, molecules would spread apart and everything would be a gas. Interesting phenomena are only possible over a narrow range of conditions (e.g., 298 K) where the spreading generated by entropic driving forces balances the compaction generated by energetic driving forces. A greater appreciation for how this balance occurs should be developed over the next several chapters.

---

1. Details are given by Tester, J.W., Modell, M. 1996. *Thermodynamics and Its Applications.* Upper Saddle River, NJ: Prentice-Hall.

Table 6.1 *Fundamental and Auxiliary Property Relations*

	Natural Variables	**Legendre Transformation**	**Transformed Variable Sets**
$dU = T\,dS - P\,dV$	$U(S,V)$		----------
$dH = T\,dS + V\,dP$	$H(S,P)$	$H = U + PV$	$\{V,P\}$
$dA = -S\,dT - P\,dV$	$A(T,V)$	$A = U - TS$	$\{S,T\}$
$dG = -S\,dT + V\,dP$	$G(T,P)$	$G = U - TS + PV$	$\{S,T\}, \{V,P\}$

6.2 DERIVATIVE RELATIONS

In Chapters 1–5, we analyzed processes using either the ideal gas law to describe the fluid or a thermodynamic chart or table. We have not yet addressed what to do in the event that a thermodynamic chart/table is not available for a compound of interest and the ideal gas law is not valid for our fluid. To meet this need, it would be ideal if we could express U or H in terms of other state variables such as P,T. In fact, we did this for the ideal gas in Eqns. 2.35 through 2.38. Unfortunately, such an expression is more difficult to derive for a real fluid. The required manipulations have been performed for us when we look at a thermodynamic chart or table. These charts and tables are created by utilizing the P-V-T properties of the fluid, together with their derivatives to calculate the values for H, U, S which you see compiled in the charts and tables. We explore the details of how this is done in Chapters 8–9 after discussing the equations of state used to represent the P-V-T properties of fluids in Chapter 7. The remainder of this chapter exploits primarily mathematical tools necessary for the manipulations of derivatives to express them in terms of measurable properties. By **measurable properties,** we mean

Generalized expression of U and H as functions of variables like T and P are desired. Further, the relations should use P-V-T properties and heat capacities.

1. P-V-T and partial derivatives involving only P-V-T.

2. C_P and C_V which are known functions of temperature at low pressure (in fact, C_P and C_V are special names for derivatives of entropy).

3. S is acceptable if it is not a derivative constraint or within a derivative term. S can be calculated once the state is specified.

Recall that the Gibbs phase rule specifies for a pure single-phase fluid that any state variable is a function of any two other state variables. For convenience, we could write internal energy in terms of $\{P,T\}$, $\{V,T\}$ or any other combination. In fact, we have already seen that the internal energy is a natural function of $\{S,V\}$:

$$dU = T\,dS - P\,dV$$

In real processes, this form is not the easiest to apply since $\{V,T\}$ and $\{P,T\}$ are more often manipulated than $\{S,V\}$. Therefore, what we seek is something of the form:

$$dU = f(P, V, T, C_P, C_V)\,dV + g(P, V, T, C_P, C_V)\,dT \qquad 6.8$$

The problem we face now is determining the functions $f(P, V, T, C_P, C_V)$ and $g(P, V, T, C_P, C_V)$. The only way to understand how to find the functions is to review multivariable calculus, then apply the results to the problem at hand. If you find that you need additional background to understand the steps applied here, try to understand whether you seek greater understanding of the mathematics or the thermodynamics. The mathematics generally involve variations of the chain rule or related derivatives. The thermodynamics pertain more to choices of preferred variables into which the final results should be transformed. Keep in mind that the development here is very mathematical, but the ultimate goal is to express U, H, A, and G in terms of measurable properties.

First, let us recognize that we have a set of state variables $\{T, S, P, V, U, H, A, G\}$ that we desire to interrelate. Further, we know from the phase rule that specification of any two of these variables will specify all others for a pure, simple system (i.e., we have two degrees of freedom). *The relations developed in this section are applicable to pure simple systems; the relations are entirely mathematical, and proofs do not lie strictly within the confines of "thermodynamics."* The first four of the state properties in our set $\{T, S, P, V\}$ are the most useful subset experimentally, so this is the subset we frequently choose to use as the controlled variables. Therefore, if we know the changes of any two of these variables, we will be able to determine changes in any of the others, including U, H, A, and G. Let's say we want to know how U changes with any two properties which we will denote symbolically as x and y. We express this mathematically as:

> The principles that we apply use multivariable calculus.

$$dU = (\partial U/\partial x)_y\, dx + (\partial U/\partial y)_x\, dy \qquad 6.9$$

where x and y are any two other variables from our set of properties. We also could write

$$dT = (\partial T/\partial x)_y\, dx + (\partial T/\partial y)_x\, dy \qquad 6.10$$

where x and y are any properties except T. The structure of the mathematics provides a method to determine how all of these properties are coupled. We could extend the analysis to all combinations of variables in our original set. As we will see in the remainder of the chapter, there are some combinations which are more useful than others. In the upcoming chapter on equations of state, some very specific combinations will be required. A peculiarity of thermodynamics that might not have been emphasized in calculus class is the significance of the quantity being held constant, e.g., the y in $(\partial U/\partial x)_y$. In mathematics, it may seem obvious that y is being held constant if there are only two variables and ∂x specifies the one that is changing. In thermodynamics, however, we have many more than two variables, although only two are varying at a time. For example, $(\partial U/\partial T)_V$ is something quite different from $(\partial U/\partial T)_P$, so the subscript should not be omitted or casually ignored.

Basic Identities

Frequently as we manipulate derivatives we obtain derivatives of the following forms which should be recognized.

$$\left(\frac{\partial x}{\partial y}\right)_z = \frac{1}{\left(\dfrac{\partial y}{\partial x}\right)_z} \qquad 6.11$$

> Basic identities.

$$\boxed{\left(\frac{\partial x}{\partial y}\right)_x = 0 \quad \text{and} \quad \left(\frac{\partial x}{\partial y}\right)_y = \infty} \qquad 6.12$$

$$\boxed{\left(\frac{\partial x}{\partial x}\right)_y = 1} \qquad 6.13$$

Triple Product Rule

Suppose $F = F(x,y)$, then

$$dF = (\partial F/\partial x)_y \, dx + (\partial F/\partial y)_x \, dy \qquad\qquad 6.14$$

Consider what happens when $dF = 0$ (i.e., at constant F). Then,

$$0 = \left(\frac{\partial F}{\partial x}\right)_y \left(\frac{\partial x}{\partial y}\right)_F + \left(\frac{\partial F}{\partial y}\right)_x \Rightarrow \left(\frac{\partial x}{\partial y}\right)_F = \frac{-\left(\frac{\partial F}{\partial y}\right)_x}{\left(\frac{\partial F}{\partial x}\right)_y} = \frac{-\left(\frac{\partial x}{\partial F}\right)_y}{\left(\frac{\partial y}{\partial F}\right)_x} \text{ or }$$

$$\boxed{\left(\frac{\partial x}{\partial y}\right)_F \left(\frac{\partial y}{\partial F}\right)_x \left(\frac{\partial F}{\partial x}\right)_y = -1} \qquad\qquad 6.15$$

❗Triple product rule.

Two Other Useful Relations

First, for any partial derivative involving three variables, say *x, y,* and *F,* we can interpose a fourth variable *z* using the **chain rule:**

$$\boxed{\left(\frac{\partial x}{\partial y}\right)_F = \left(\frac{\partial x}{\partial z}\right)_F \left(\frac{\partial z}{\partial y}\right)_F} \qquad\qquad 6.16$$

❗Chain rule interposing a variable.

Another useful relation is found by a procedure known as the **expansion rule.** The details of this expansion are usually not covered in introductory calculus texts:

$$\boxed{\left(\frac{\partial F}{\partial w}\right)_z = \left(\frac{\partial F}{\partial x}\right)_y \left(\frac{\partial x}{\partial w}\right)_z + \left(\frac{\partial F}{\partial y}\right)_x \left(\frac{\partial y}{\partial w}\right)_z} \qquad\qquad 6.17$$

❗The expansion rule.

Recall that we started with a function $F = F(x,y)$. If you look closely at the expansion rule, it provides a method to evaluate a partial derivative $(\partial F/\partial w)_z$ in terms of $(\partial x/\partial w)_z$ and $(\partial y/\partial w)_z$.

Thus, we have transformed the calculation of a partial derivative of F to partial derivatives of x and y. This relation is particularly useful in manipulation of the fundamental relations $S, U, H, A,$ and G when one of these properties is substituted for F, and the natural variables are substituted for x and y. We will demonstrate this in Examples 6.1 and 6.2. Look again at Eqn. 6.17. It *looks* like we have taken the differential expression of Eqn. 6.14 and divided through the differential terms by dw and constrained to constant z, but this procedure violates the rules of differential operators. *What we have actually done is not nearly this simple.*[2] However, looking at the equation this way provides a fast way to remember a complicated-looking expression.

2. Leithold, L., 1976. *The Calculus with Analytical Geometry.* 3rd ed. New York, NY: Harper & Row, p. 929.

Exact Differentials

In this section, we apply calculus to the fundamental properties. Our objective is to derive relations known as Maxwell's relations. We begin by reminding you that we can express any state property in terms of any other two state properties. For a function which is only dependent on two variables, we can obtain the following differential relation, called in mathematics an *exact differential*.

$$U = U(S,V) \Rightarrow dU = (\partial U/\partial S)_V \, dS + (\partial U/\partial V)_S \, dV \qquad 6.18$$

Developing the ability to express any state variable in terms of any other two variables from the set {P, T, V, S} as we have just done is very important. But the equation looks a little formidable. However, the fundamental property relationship says:

$$dU = TdS - PdV \qquad 6.19$$

Comparison of the above equations shows that:

$$T = (\partial U/\partial S)_V \quad \text{and} \quad -P = (\partial U/\partial V)_S \qquad 6.20$$

This means that the derivatives in Eqn. 6.18 are really properties that are familiar to us. Likewise, we can learn something about formidable-looking derivatives from enthalpy:

$$H = H(S,P) \Rightarrow dH = (\partial H/\partial S)_P dS + (\partial H/\partial P)_S \, dP \qquad 6.21$$

But the result of the fundamental property relationship is:

$$dH = TdS + VdP$$

Comparison shows that:

$$T = (\partial H/\partial S)_P \quad \text{and} \quad V = (\partial H/\partial P)_S$$

Now, we see that a definite pattern is emerging, and we could extend the analysis to Helmholtz and Gibbs energy. We can, in fact, derive relations between certain second derivatives of these relations. Since the properties $U, H, A,$ and G are state properties of only two other state variables, the differentials we have given in terms of two other state variables are known mathematically as exact differentials; we may apply properties of exact differentials to these properties. We show the features here; for details consult an introductory calculus textbook. Consider a general function of two variables: $F = F(x,y)$, and

$$dF = \left(\frac{\partial F}{\partial x}\right)_y dx + \left(\frac{\partial F}{\partial y}\right)_x dy$$

For an exact differential, differentiating with respect to x we can define some function M:

$$M \equiv (\partial F/\partial x)_y = M(x,y) \qquad 6.22$$

Exact differentials.

Similarly differentiating with respect to y:

$$N \equiv (\partial F/\partial y)_x = N(x,y) \qquad 6.23$$

Taking the second derivative and recalling from multivariable calculus that the order of differentiation should not matter,

$$\frac{\partial^2 F(x,y)}{\partial x \partial y} = \frac{\partial}{\partial x}\left(\left(\frac{\partial F(x,y)}{\partial y}\right)_x\right)_y = \frac{\partial}{\partial y}\left(\left(\frac{\partial F(x,y)}{\partial x}\right)_y\right)_x = \frac{\partial^2 F(x,y)}{\partial y \partial x} \qquad 6.24$$

$$\left(\frac{\partial N}{\partial x}\right)_y = \left(\frac{\partial M}{\partial y}\right)_x \qquad 6.25$$

Euler's reciprocity relation.

This simple observation is sometimes called Euler's reciprocity relation.[3] To apply the reciprocity relation, recall the total differential of enthalpy considering $H = H(S,P)$:

$$dH = (\partial H/\partial S)_P\, dS + (\partial H/\partial P)_S\, dP = TdS + VdP \qquad 6.26$$

Considering second derivatives:

$$\frac{\partial^2 H}{\partial S \partial P} = \frac{\partial^2 H}{\partial P \partial S} = \left[\frac{\partial}{\partial S}\left(\frac{\partial H}{\partial P}\right)_S\right]_P = \left[\frac{\partial}{\partial P}\left(\frac{\partial H}{\partial S}\right)_P\right]_S \qquad 6.27$$

$$\Rightarrow \left(\frac{\partial V}{\partial S}\right)_P = \left(\frac{\partial T}{\partial P}\right)_S \qquad 6.28$$

A similar derivation applied to each of the other thermodynamic functions yields the equations known as Maxwell's relations.

Maxwell's Relations

$$dU = TdS - PdV \;\Rightarrow\; -(\partial P/\partial S)_V = (\partial T/\partial V)_S \qquad 6.29$$

$$dH = TdS + VdP \;\Rightarrow\; (\partial V/\partial S)_P = (\partial T/\partial P)_S \qquad 6.30$$

Maxwell's relations.

$$dA = -SdT - PdV \;\Rightarrow\; (\partial P/\partial T)_V = (\partial S/\partial V)_T \qquad 6.31$$

$$dG = -SdT + VdP \;\Rightarrow\; -(\partial V/\partial T)_P = (\partial S/\partial P)_T \qquad 6.32$$

Example 6.1 Pressure dependence of H

Derive the relation for $\left(\dfrac{\partial H}{\partial P}\right)_T$ and evaluate the derivative for: a) water at 20°C where $\left(\dfrac{\partial V}{\partial T}\right)_P = 2.07\times10^{-4}$ cm^3/g-K and $\left(\dfrac{\partial V}{\partial P}\right)_T = -4.9\times10^{-5}$ cm^3/g-bar, $\rho = 0.998$ g/cm^3; b) an ideal gas.

Solution: First, consider the general relation $dH = TdS + VdP$. Applying the expansion rule,

$$\left(\frac{\partial H}{\partial P}\right)_T = T\left(\frac{\partial S}{\partial P}\right)_T + V\left(\frac{\partial P}{\partial P}\right)_T^{\;1} = T\left(\frac{\partial S}{\partial P}\right)_T + V$$

---

3. Other descriptions include "condition for exact differential." It was called by Rudolf Clausius the "condition of immediate integrability."

Example 6.1 Pressure dependence of H (Continued)

by a Maxwell relation, the entropy derivative may be replaced

$$\left(\frac{\partial H}{\partial P}\right)_T = -T\left(\frac{\partial V}{\partial T}\right)_P + V \qquad 6.33$$

which is valid for any fluid.

(a) Plugging in values for liquid water,

$$\left(\frac{\partial H}{\partial P}\right)_T = -293.15(2.07 \times 10^{-4}\text{cm}^3/\text{g-K}) + 1.002$$

$$= -0.061 + 1.002$$

Therefore, within 6% at room temperature, $\left(\frac{\partial H}{\partial P}\right)_T \approx V$ for liquid water as used in Eqn. 2.42 on page 59 and Example 2.6 on page 60.

(b) For an ideal gas, we need to evaluate $(\partial V/\partial T)_P$. Applying the relation to $V = RT/P$, $(\partial V/\partial T)_P = R/P$. Inserting into Eqn. 6.33, enthalpy is independent of pressure for an ideal gas.

$$\left(\frac{\partial H}{\partial P}\right)_T = -\frac{TR}{P} + V = -V + V = 0 \qquad \text{(ig) } 6.34$$

A non-ideal gas will have a different partial derivative, and the enthalpy will depend on pressure as we will show in Chapter 8.

Example 6.2 Entropy change with respect to T at constant P

Evaluate $(\partial S/\partial T)_P$ in terms of C_P, C_V, T, P, V, and their derivatives.

Solution: C_P is the temperature derivative of H at constant P. Let us start with the fundamental relation for enthalpy and then apply the expansion rule. Recall, $dH = TdS + VdP$.
Applying the expansion rule, Eqn. 6.17, we find,

$$\left(\frac{\partial H}{\partial T}\right)_P = T\left(\frac{\partial S}{\partial T}\right)_P + V\left(\frac{\partial P}{\partial T}\right)_P \qquad 6.35$$

Applying the basic identity of Eqn. 6.12 to the second term on the right-hand side, since P appears in the derivative and as a constraint the term is zero,

$$\left(\frac{\partial H}{\partial T}\right)_P = T\left(\frac{\partial S}{\partial T}\right)_P$$

Example 6.2 Entropy change with respect to T at constant P (Continued)

But the definition of the left-hand side is given by Eqn. 2.36: $C_P \equiv (\partial H / \partial T)_P$.
Therefore, $(\partial S / \partial T)_P = C_P / T$, which we have seen before as Eqn. 4.31, and we have found
that the constant-pressure heat capacity is related to the constant-pressure derivative of entropy
with respect to temperature. An analogous analysis of U at constant V results in a relation
between the constant-volume heat capacity and the derivative of entropy with respect to temperature at constant V. That is, Eqn. 4.30,

$$(\partial S / \partial T)_V = C_V / T$$

Example 6.3 Entropy as a function of T and P

Derive a general relation for entropy changes of any fluid with respect to temperature and pressure in terms of C_P, C_V, P, V, T, and their derivatives.

Solution: First, since we choose T, P to be the controlled variables, applying Eqn. 6.14

$$dS = (\partial S / \partial T)_P \, dT + (\partial S / \partial P)_T \, dP \tag{6.36}$$

but $(\partial S / \partial T)_P = C_P / T$ as derived above, and Maxwell's relations show that

$$(\partial S / \partial P)_T = -(\partial V / \partial T)_P$$

$$\Rightarrow dS \, (T, P) = C_P / T \ dT - (\partial V / \partial T)_P \ dP \tag{6.37}$$

This useful expression is ready for application, given an equation of state which describes
$V(T,P)$.

Note that expressions similar to Eqn. 6.37 can be derived for other thermodynamic variables in an
analogous fashion. These represent powerful short-hand relations that can be used to solve many
different process-related problems. In addition to Eqn. 6.37, some other useful expressions are
listed below. These are so frequently useful that they are also tabulated on the front flap of the text
for convenient reference.

$$dS \, (T, V) = C_V / T \ dT + (\partial P / \partial T)_V \ dV \tag{6.38}$$

$$dS \, (V, P) = C_P (\partial T / \partial V)_P / T \ dV + C_V (\partial T / \partial P)_V / T \ dP \tag{6.39}$$

$$dH \, (T, P) = C_P \ dT + [V - T(\partial V / \partial T)_P] \ dP \tag{6.40}$$

$$dU \, (T, V) = C_V \ dT + [T(\partial P / \partial T)_V - P] \ dV \tag{6.41}$$

A summary of useful relations.

The Importance of Derivative Manipulations

One may wonder, "What is so important about the variables C_P, C_V, P, V, T, and their derivatives?"
The answer is that these properties are **experimentally measureable**. Engineers have developed
equations of state written in terms of these fundamental properties. Briefly, an equation of state
provides the link between P, V, and T. So, we can solve for all the derivatives by knowing an equation for $P = P(V,T)$ and add up all the changes. Properties like H, U, and S are not considered mea-

An equation of state links the P, V, T properties of a fluid.

surable because we don't have gauges that measure them. Look back at the descriptions in Section 6.2. C_P and C_V are considered measurable though they are typically measured using the energy balance with temperature changes. A goal of derivative manipulations is to convert derivatives involving unmeasurable properties into derivatives involving measurable properties.

As for C_P and C_V, we should actually be very careful about specifying when we are referring to the C_P of a real fluid or the C_P^{ig} of an ideal gas, but the distinction is frequently not made clear in literature. You may need to recognize from the context of the source whether C_P is for the ideal gas or something else. In most cases, the intent is to apply C_P^{ig} at low density to account for temperature effects and then to apply a correction factor of dV or dP to account for non-ideal gas density or pressure effects. In this way, in Chapter 8 we compute ΔU, ΔH, ΔS, ΔA, or ΔG from any initial (or reference) condition to any final condition and we can imagine compiling the results in the format of the steam tables for any particular compound.

Test Yourself

Sketch two subcritical isotherms and two supercritical isotherms. Using the isotherms, describe how the following derivatives could be obtained numerically: $(\partial P/\partial T)_V$, $(\partial V/\partial T)_P$, $(\partial P/\partial V)_T$. Compare the relative sizes of the derivatives for liquids and gases.

Important Measurable Derivatives

Two measurable derivatives are commonly used to discuss fluid properties, the isothermal compressibility and the isobaric coefficient of thermal expansion. The **isothermal compressibility** is

❗ Isothermal compressibility.

$$\kappa_T \equiv \frac{-1}{V}\left(\frac{\partial V}{\partial P}\right)_T = \frac{1}{\rho}\left(\frac{\partial \rho}{\partial P}\right)_T \qquad 6.42$$

The **isobaric coefficient of thermal expansion** is

❗ Isobaric coefficient of thermal expansion.

$$\alpha_P \equiv \frac{1}{V}\left(\frac{\partial V}{\partial T}\right)_P = \frac{-1}{\rho}\left(\frac{\partial \rho}{\partial T}\right)_P \qquad 6.43$$

A similar commonly used property is the **Joule-Thomson coefficient** defined by

$$\mu_{JT} \equiv \left(\frac{\partial T}{\partial P}\right)_H \qquad 6.44$$

It is easy to see how this property relates to the physical situation of temperature changes as pressure drops through an isenthalpic throttle valve, though it is not considered measurable because of the constraint on H. The manipulation in terms of measurable properties is considered in Problem 6.8.

The next two examples illustrate the manner in which derivative manipulations are applied with a particularly simple equation of state to obtain an expression for ΔS. In this chapter, we establish this conceptual approach and the significant role of an assumed equation of state. The next chapter focuses on the physical basis of developing a reasonable equation of state. Since the derivative manipulations are entirely rigorous, we come to understand that all the approximations in all thermodynamic modeling are buried in the assumptions of an equation of state. Whenever an engineering thermodynamic model exhibits inaccuracy, the assumptions of the equation of state must be reconsidered and refined in the context of the particular application.

Example 6.4 Entropy change for an ideal gas

A gas is being compressed from ambient conditions to a high pressure. Devise a model equation for computing $\Delta S(T,P)$. Assume the ideal gas equation of state.

Solution: We begin with the temperature effect at (constant) low pressure. By Eqn. 6.37,

$$dS)_P^{ig} = C_P^{ig}\, dT/T \tag{ig}$$

Having accounted for the temperature effect at constant pressure, the next step is to account for the pressure effect at constant temperature. The derivative $(\partial V/\partial T)_P$ is required.

$$V = RT/P \;\Rightarrow (\partial V/\partial T)_P = R/P \tag{ig}$$

Putting it all together,

$$dS = \frac{C_P^{ig}}{T}dT - \frac{R}{P}dP = C_P^{ig}\,d\ln T - R\ln P \tag{ig}$$

Assuming C_P^{ig} is independent of T and integrating,

$$\Delta S = C_P^{ig}\ln(T_2/T_1) - R\ln(P_2/P_1) \tag{ig}$$

Once again, we arrive at Eqn. 4.29, but this time, it is easy to recognize the necessary changes for applications to non-ideal gases. That is, we must simply replace the P-V-T relation by a more realistic equation of state when we evaluate derivatives.

Example 6.5 Entropy change for a simple nonideal gas

A gas is compressed from ambient conditions to a high pressure. Devise a model equation for computing $\Delta S(T,P)$ with the equation of state: $V = RT/P + (a + bT)$, where a and b are constants.

Solution: Substituting the new equation of state and following the previous example,

$$(\partial V/\partial T)_P = R/P + b$$

We can still apply C_P^{ig} because we could be careful to calculate temperature effects at low P before calculating the pressure effect. Inserting into Eqn. 6.37,

$$dS = \frac{C_P^{ig}}{T}dT - \frac{R}{P}dP - bdP = C_P^{ig}\,d\ln T - R\ln P - bdP$$

Assuming C_P^{ig} is independent of T and integrating,

$$\Delta S = C_P^{ig}\ln(T_2/T_1) - R\ln(P_2/P_1) - b\Delta P \tag{*}$$

Equations of state can be much more complicated than this one, motivating us to carefully contemplate the most efficient manner to organize our derivative relations to transform complex equations of state into useful engineering models. The departure function formalism as described in Chapter 8 provides this kind of efficiency. A key manipulation in that chapter will be the volume dependence of properties at constant temperature.

Example 6.6 Accounting for T and V impacts on energy

Derive an expression for $\left(\dfrac{\partial U}{\partial V}\right)_T$ in terms of measurable properties. (a) Evaluate for the ideal gas.

(b) Evaluate for the van der Waals equation of state, $P = RT/(V-b) - a/V^2$.

Solution: Beginning with the fundamental relation for dU,

$$dU = TdS - PdV$$

Applying the expansion rule

$$\left(\frac{\partial U}{\partial V}\right)_T = T\left(\frac{\partial S}{\partial V}\right)_T - P\left(\frac{\partial V}{\partial V}\right)_T \tag{6.45}$$

Using a Maxwell relation and a basic identity

$$\left(\frac{\partial U}{\partial V}\right)_T = T\left(\frac{\partial P}{\partial T}\right)_V - P \tag{6.46}$$

(a) For an ideal gas, $P = RT/V$

$$\left(\frac{\partial P}{\partial T}\right)_V = \frac{R}{V}; \quad \left(\frac{\partial U}{\partial V}\right)_T^{ig} = \frac{RT}{V} - P = 0 \tag{ig}$$

Thus, internal energy of an ideal gas does not depend on volume (or pressure) at a given T.

(b) For the van der Waals equation,

$$\left(\frac{\partial P}{\partial T}\right)_V = \frac{R}{V-b}; \quad \left(\frac{\partial U}{\partial V}\right)_T = \frac{RT}{V-b} - \left(\frac{RT}{V-b} - \frac{a}{V^2}\right) = \frac{a}{V^2} \tag{ig}$$

Another important application of derivative manipulations is in deriving meaningful connections between U, H, A, G, and S. We illustrate this kind of application with two examples. The first is a fairly simple development of the relation between energy and Helmholtz energy; the relation arises frequently in applications in Unit III of the text. The second example pertains to the Einstein solid model that was used to demonstrate the connection between the microscopic and macroscopic definitions of entropy. This second example involves more tedious calculus, but reveals broad insights related to chemistry and spectroscopy as well as entropy and heat capacity.

Example 6.7 The relation between Helmholtz energy and internal energy

Express the following in terms of U, H, S, G, and their derivatives: $(\partial(A/(RT))/\partial T)_V$.

Solution: Applying the product rule,

$$\left(\frac{\partial(A/(RT))}{\partial T}\right)_V = \frac{1}{RT}\left(\frac{\partial A}{\partial T}\right)_V - \frac{A}{RT^2}$$

Applying Eqn. 6.6 and the definition of A,

$$\left(\frac{\partial(A/(RT))}{\partial T}\right)_V = \frac{-S}{RT} - \frac{(U-TS)}{RT^2} = \frac{-U}{RT^2}$$

Rearranging, and introducing a common definition $\beta \equiv 1/(kT)$,

$$\frac{U}{RT} = \left(\frac{-T\partial(A/(RT))}{\partial T}\right)_V = \left(\frac{\beta\,\partial(A/(RT))}{\partial\beta}\right)_V \qquad 6.47$$

The significance of Eqn. 6.47 is that one can easily transform from Helmholtz energy to internal energy and vice versa by integrating or differentiating. This is especially easy when the temperature dependence is expressed as a polynomial.

We are now in a position to return to the discussion of the relation between entropy and temperature that was begun in Chapter 4. Many experimental observations circa 1900 were challenging the conventional theories of atomic motions. It seemed that particles the size of atoms, and smaller, might be moving in discrete steps of energy, instead of continuous energies, and this altered the behavior that was being observed. For example, the covalent bond of nitrogen appears to vibrate classically at high temperature, but to become rigid at low temperature. The nature of this transition and its impact on heat capacity require explanation. The Einstein solid model was a major milestone in resolving many of these peculiar observations.

Albert Einstein wrote 45 articles between 1901 and 1907. Of these, 11 were about thermodynamics, 24 were reviews about thermodynamics, 7 were about relativity, 1 was about electromagnetism, and 2 were about the photoelectric effect. Of course, his revolutionary works were about relativity and the photoelectric effect because they opened new vistas of physical insight. But his work on heat capacity might arguably have had the broader immediate impact. Einstein showed that a very simple quantized theory predicted the experimental observations that classical theory could not explain. The following example applies derivative manipulations to obtain heat capacity from the entropy and demonstrate that the heat capacity approaches zero at zero Kelvin.

Example 6.8 A quantum explanation of low T heat capacity

The result of the Einstein solid model, from Example 4.2 on page 141, can be rearranged to:

$$\underline{S}/Mk = y = x\ln(1+x) - x\ln x + \ln(1+x) \qquad 6.48$$

where $y = \underline{S}/Mk$ and $x = <q_M> = \underline{U}/M\varepsilon_q - 1/2$. Use these results to derive the temperature dependence of $\underline{S}/Mk$, $\underline{U}/M\varepsilon_q - 1/2$, and C_V as instructed.

(a) Derive a formula for $\underline{C_V}/Mk$. Tabulate values of y, x, and $\underline{C_V}/Mk$ versus $\beta\varepsilon_q$ at $\beta\varepsilon_q = \{0.1, 0.2, 0.5, 1, 2, 5, 10\}$. Recall that $\beta = 1/kT$.

(b) The following data have been tabulated for silver by McQuarrie.[a] Iterate on ε_q/k to find the best fit and plot C_V/Mk versus kT/ε_q, comparing theory to experiment. Also indicate the result of classical mechanics with a dashed line.

T(K)	5	10	20	47.3	83.9	103.1	144.4	205.3
C_V(J/mol-K)	0.02	0.2	1.7	10.8	18.1	20.1	22.5	23.5

Solution:

(a) Noting that $(\partial \underline{S}/\partial T)_V = \underline{C_V}/T$, a logical step is to take $(\partial y/\partial T)_V = \underline{C_V}/MkT$.

$$\frac{dy}{dT} = \frac{dx}{dT}\left[\ln(1+x) - \ln x + \frac{x}{1+x} - \frac{x}{x} + \frac{1}{1+x}\right] = \frac{dx}{dT}\ln\left(1 + \frac{1}{x}\right) = \frac{\underline{C_V}}{MkT}$$

$$\frac{dx}{dT} = \left(\frac{1}{M\varepsilon_q}\right)\left(\frac{\partial \underline{U}}{\partial T}\right)_V = \frac{\underline{C_V}}{M\varepsilon_q}$$

Note that the $\underline{C_V}$ terms cancel, seeming to defeat the purpose, but substitute anyway and recognize internal energy is embedded.

$$\ln\left(1 + \frac{1}{x}\right) = \beta\varepsilon_q \Rightarrow x = \frac{\underline{U}}{M\varepsilon_q} - \frac{1}{2} = [\exp(\beta\varepsilon_q) - 1]^{-1} \qquad 6.49$$

We may not have obtained $\underline{C_V}$ as expected, but we obtained $\underline{U}(T)$, so it is straightforward to evaluate $\underline{C_V}$ from its definition: $\underline{C_V} = (\partial \underline{U}/\partial T)_V$.

Multiplying both sides by T for convenience,

$$\frac{\underline{C_V}}{M\varepsilon_q}T = \frac{T}{M\varepsilon_q}\left(\frac{d\underline{U}}{dT}\right) = \left(-\frac{\beta\varepsilon_q}{M\varepsilon_q}\right)\left(\frac{d\underline{U}}{d(\beta\varepsilon_q)}\right) = (-\beta\varepsilon_q)\left(\frac{dx}{d(\beta\varepsilon_q)}\right) = \frac{(\beta\varepsilon_q)\exp(\beta\varepsilon_q)}{[\exp(\beta\varepsilon_q) - 1]^2}$$

Multiplying and dividing by k on the left-hand side,

$$\left(\frac{\underline{C_V}}{Mk}\right)\left(\frac{kT}{\varepsilon_q}\right) = \frac{(\beta\varepsilon_q)\exp(\beta\varepsilon_q)}{[\exp(\beta\varepsilon_q) - 1]^2} \Rightarrow \left(\frac{\underline{C_V}}{Mk}\right) = \frac{(\beta\varepsilon_q)^2\exp(\beta\varepsilon_q)}{[\exp(\beta\varepsilon_q) - 1]^2} \qquad 6.50$$

Example 6.8 A quantum explanation of low T heat capacity (Continued)

Varying $\beta\varepsilon_q$ and tabulating energy and entropy, noting that the lowest temperatures are on the right, values of $\underline{U}$ and $x = <q_M>$ are calculated by Eqn. 6.49, $\underline{S}/Mk$ by 6.48, and $\underline{C_V}/Mk$ by 6.50. We can see that the heat capacity goes to zero as 0 K is approached.

$\beta\varepsilon_q = \varepsilon_q/kT$	0.1	0.2	0.5	1.0	2.0	5.0	10
$x = \underline{U}/M\varepsilon_q - 1/2$	9.508	4.517	1.541	0.582	0.157	0.007	4.5E-5
$y = \underline{S}/Mk$	3.303	2.611	1.703	1.041	0.458	0.041	0.0005
$\underline{C_V}/Mk$	0.9992	0.997	0.979	0.921	0.724	0.171	0.005

(b) As discussed in Example 4.2 on page 141, each atom has three possible directions of motion, so $M = 3N$. Therefore, the quantum mechanical formula is $\underline{C_V}/(Mk) = \underline{C_V}/(3Nk) = C_V/(3R)$.

The classical value of the solid heat capacity can be deduced by considering the degrees of freedom.[b] Kinetic and potential energy each contribute a degree of freedom for a bond. Recalling from Section 2.10 that each degree of freedom contributes $R/2$ to C_V, $\underline{C_V}^{cs} = 2(Nk/2)(M/N) = Mk$ $= 3nR$, or $C_V^{cs} = 3R$ where C_V^{cs} is the heat capacity of the classical solid. Note that the table in part (a) approaches the classical result at high temperature, $C_V^{cs}/3R = 1$.

$\underline{C_V}/Mk$ is fitted in Excel by naming a cell as ε_q/k and applying Eqn. 6.50. The function SUMXMY2 was used to compute $\Sigma(\text{calc} - \text{expt})^2$. Minimization with the Solver gives $\varepsilon_q/k =$ 159 K. Values of $C_V/3R$ and $\underline{C_V}/Mk$ at $\varepsilon_q/k = 159$ K are tabulated below.

T(K)	$C_V/3R$(expt)	$\underline{C_V}/Mk$(calc)
5.0	0.0009	1.6E-11
10.0	0.0080	3.1E-5
20.0	0.0670	0.0223
47.1	0.4331	0.4176
83.9	0.7257	0.7447
103.1	0.8047	0.8234
144.4	0.9013	0.9048
205	0.9402	0.9515

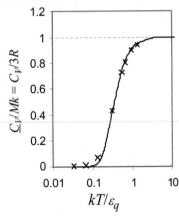

a. McQuarrie, D.A., 1976. *Statistical Mechanics*. New York, NY: Harper & Row, p. 205.
b. The degrees of freedom for kinetic and potential energy discussed here are different from the degrees of freedom for the Gibbs phase rule. See also, Section 2.10.

Einstein's expression approaches zero too quickly relative to the experimental data, but it explains the large qualitative difference between the classical theory and experiment. Several refinements have appeared over the years and these fit the data more closely, but the key point is that quantized energy leads to freezing out the vibrational degrees of freedom at temperatures substantially above 0 K. This phenomenology also applies to the vibrations of polyatomic molecules, as explored in the homework for a diatomic molecule. The weaker bond energy of bromine relative to nitrogen is also evident in its higher value of C_P/R as listed on the back flap. From a broader perspective, Einstein's profound insight was that quantized energy meant that thermal entropy could be conceived in a manner analogous to configurational entropy except by putting particles in energy levels rather than putting particles in boxes spatially. In this way, we can appreciate that the macroscopic and microscopic definitions of entropy are analogous applications of distributions.

Hints on Manipulating Partial Derivatives

Useful hints on manipulating derivatives.

1. Learn to recognize $\left(\frac{\partial S}{\partial T}\right)_P$ and $\left(\frac{\partial S}{\partial T}\right)_V$ as being related to C_P and C_V, respectively.

2. If a derivative involves entropy, enthalpy, or Helmholtz or Gibbs energy being held constant, e.g., $\left(\frac{\partial T}{\partial P}\right)_H$, bring it inside the parenthesis using the triple product relation (Eqn. 6.15). Then apply the expansion rule (Eqn. 6.17) to eliminate immeasurable quantities. The expansion rule is very useful when F of that equation is a fundamental property.

3. When a derivative involves $\{T, S, P, V\}$ only, look to apply a Maxwell relation.

4. When nothing else seems to work, apply the Jacobian method.[4] The Jacobian method will always result in derivatives with the desired independent variables.

Example 6.9 Volumetric dependence of C_V for ideal gas

Determine how C_V depends on volume (or pressure) by deriving an expression for $(\partial C_V/\partial V)_T$. Evaluate the expression for an ideal gas.

Solution: Following hint #1 and applying Eqn. 4.30:

$$C_V = T\left(\frac{\partial S}{\partial T}\right)_V$$

By the product rule:

$$\left(\frac{\partial C_V}{\partial V}\right)_T = \left(\frac{\partial S}{\partial T}\right)_V\left(\frac{\partial T}{\partial V}\right)_T + T\frac{\partial}{\partial V}\left[\left(\frac{\partial S}{\partial T}\right)_V\right]_T$$

Changing the order of differentiation:

$$\left(\frac{\partial C_V}{\partial V}\right)_T = T\frac{\partial}{\partial T}\left[\left(\frac{\partial S}{\partial V}\right)_T\right]_V = T\left(\frac{\partial^2 P}{\partial T^2}\right)_V$$

For an ideal gas, $P = RT/V$, we have $\left(\frac{\partial P}{\partial T}\right)_V$ in Example 6.6:

$$\frac{\partial}{\partial T}\left[\left(\frac{\partial P}{\partial T}\right)_V\right]_V = \frac{\partial}{\partial T}\left(\frac{R}{V}\right)_V = 0 \qquad \text{(ig) 6.51}$$

Thus, heat capacity of an ideal gas does not depend on volume (or pressure) at a fixed temperature. (We will reevaluate this derivative in Chapter 7 for a real fluid.)

---

4. This method is covered in optional Section 6.3.

Example 6.10 Application of the triple product relation

Evaluate $(\partial S/\partial V)_A$ in terms of C_P, C_V, T, P, and V. Your answer may include absolute values of S if it is not a derivative constraint or within a derivative term.

Solution: This problem illustrates a typical situation where the triple product rule is helpful because the Helmholtz energy is held constant (hint #2). It is easiest to express changes in the Helmholtz energies as changes in other variables. Applying the triple product rule:

$$(\partial S/\partial V)_A = -(\partial A/\partial V)_S \,/\,(\partial A/\partial S)_V$$

Applying the expansion rule twice, $dA = -PdV - SdT \Rightarrow (\partial A/\partial V)_S = -P - S(\partial T/\partial V)_S$ and $(\partial A/\partial S)_V = 0 - S(\partial T/\partial S)_V$. Recalling Eqn. 4.30 and converting to measurable derivatives:

$$\Rightarrow \left(\frac{\partial T}{\partial S}\right) = \frac{T}{C_V} \text{ and } (\partial T/\partial V)_S = \left(\frac{\partial T}{\partial S}\right)_V\left(\frac{\partial S}{\partial V}\right)_T = -\frac{T}{C_V}\left(\frac{\partial P}{\partial T}\right)_V$$

Substituting:

$$\Rightarrow \left(\frac{\partial S}{\partial V}\right)_A = \frac{-PC_V}{ST} + \left(\frac{\partial P}{\partial T}\right)_v$$

Example 6.11 Master equation for an ideal gas

Derive a master equation for calculating changes in U for an ideal gas in terms of $\{V, T\}$.

Solution:

$$dU = \left(\frac{\partial U}{\partial V}\right)_T dV + \left(\frac{\partial U}{\partial T}\right)_V dT$$

Applying results of the previous examples:

$$dU = C_V dT + \left[T\left(\frac{\partial P}{\partial T}\right)_V - P\right]dV \qquad\qquad 6.52$$

Notice that this expression is more complicated than the fundamental property relation in terms of $\{S, V\}$. As we noted earlier, this is why $\{S, V\}$ are the natural variables for dU, rather than $\{T,V\}$ or any other combination. For an ideal gas, we can use the results of Example 6.6 to find:

$$dU^{ig} = C_V^{ig} dT \qquad\qquad \text{(ig) 6.53}$$

Example 6.12 Relating C_P to C_V

Derive a general formula to relate C_P and C_V.

Solution: Start with an expression that already contains one of the desired derivatives (e.g., C_V) and introduce the variables necessary to create the second derivative (e.g., C_P). Beginning with Eqn. 6.38,

$$dS = \frac{C_V}{T}dT + \left(\frac{\partial P}{\partial T}\right)_V dV$$

and using the expansion rule with T at constant P,

$$\left(\frac{\partial S}{\partial T}\right)_P = \frac{C_V}{T}\left(\frac{\partial T}{\partial T}\right)_P^{\,1} + \left(\frac{\partial P}{\partial T}\right)_V\left(\frac{\partial V}{\partial T}\right)_P, \text{ where the left-hand side is } \frac{C_P}{T}.$$

$$C_P = C_V + T\left(\frac{\partial P}{\partial T}\right)_V\left(\frac{\partial V}{\partial T}\right)_P$$

Exercise:
Verify that the last term simplifies to R for an ideal gas.

Owing to all the interrelations between all the derivatives, there is usually more than one way to derive a useful result. This can be frustrating to the novice. Nevertheless, patience in attacking the problems, and attacking a problem from different angles, can help you to visualize the structure of the calculus in your mind. Each problem is like a puzzle that can be assembled in multiple ways. Patience in developing these tools is rewarded with a mastery of the relations that permit quick insight into the easiest way to solve problems.

6.3 ADVANCED TOPICS

Hints for Remembering the Auxiliary Relations

Auxiliary relations can be easily written by memorizing the fundamental relation for dU and the natural variables for the other properties. Note that $\{T,S\}$ and $\{P,V\}$ always appear in pairs, and each pair is a set of conjugate variables. A Legendre transformation performed on internal energy among conjugate variables changes the dependent variable and the sign of the term involving the conjugate variables. For example, to transform P and V, the product PV is added to U, resulting in Eqn. 6.5. To transform T and S, the product TS is subtracted: $A = U - TS$, $dA = dU - TdS - SdT = -SdT - PdV$. The pattern can be easily seen in the "Useful Derivatives" table on the front book end paper. Note that $\{T,S\}$ always appear together, and $\{P,V\}$ always appear together, and the sign changes upon transformation.

Jacobian Method of Derivative Manipulation

A partial derivative may be converted to derivatives of measurable properties with any two desired independent variables from the set $\{P,V,T\}$. Jacobian notation can be used to manipulate partial derivatives, and there are several useful rules for manipulating derivatives with the notation. The

Joule-Thomson coefficient is a derivative that indicates how temperature changes upon pressure change at fixed enthalpy, $\left(\dfrac{\partial T}{\partial P}\right)_H$, which is written in Jacobian notation as $\left(\dfrac{\partial T}{\partial P}\right)_H = \dfrac{\partial(T, H)}{\partial(P, H)}$. Note how the constraint of constant enthalpy is incorporated into the notation. The rules for manipulation of the Jacobian notation are,

1. Jacobian notation represents a determinant of partial derivatives,

$$\frac{\partial(K, L)}{\partial(X, Y)} = \left(\frac{\partial K}{\partial X}\right)_Y \left(\frac{\partial L}{\partial Y}\right)_X - \left(\frac{\partial K}{\partial Y}\right)_X \left(\frac{\partial L}{\partial X}\right)_Y = \begin{vmatrix} \left(\dfrac{\partial K}{\partial X}\right)_Y & \left(\dfrac{\partial K}{\partial Y}\right)_X \\ \left(\dfrac{\partial L}{\partial X}\right)_Y & \left(\dfrac{\partial L}{\partial Y}\right)_X \end{vmatrix} \qquad 6.54$$

The Jacobian is particularly simple when the numerator and denominator have a common variable,

$$\frac{\partial(K, L)}{\partial(X, L)} = \left(\frac{\partial K}{\partial X}\right)_L \qquad 6.55$$

which is a special case of Eqn. 6.54.

2. When the order of variables in the numerator or denominator is switched, the sign of the Jacobian changes. Switching the order of variables in both the numerator and denominator results in no sign change due to cancellation. Consider switching the order of variables in the numerator,

$$\frac{\partial(K, L)}{\partial(X, Y)} = -\frac{\partial(L, K)}{\partial(X, Y)} \qquad 6.56$$

3. The Jacobian may be inverted.

$$\frac{\partial(K, L)}{\partial(X, Y)} = \left[\frac{\partial(X, Y)}{\partial(K, L)}\right]^{-1} = \frac{1}{\dfrac{\partial(X, Y)}{\partial(K, L)}} \qquad 6.57$$

4. Additional variables may be interposed. When additional variables are interposed, it is usually convenient to invert one of the Jacobians.

$$\frac{\partial(K, L)}{\partial(X, Y)} = \frac{\partial(K, L)}{\partial(B, C)}\frac{\partial(B, C)}{\partial(X, Y)} = \frac{\dfrac{\partial(K, L)}{\partial(B, C)}}{\dfrac{\partial(X, Y)}{\partial(B, C)}} \qquad 6.58$$

Manipulation of Derivatives

Before manipulating derivatives, the desired independent variables are selected. The selected independent variables will be held constant outside the derivatives in the final formula. The general procedure is to interpose the desired independent variables, rearrange as much as possible to obtain Jacobians with common variables in the numerator and denominator, write the determinant for any

Jacobians without common variables; then use Maxwell relations, the expansion rule, and so on, to simplify the answer.

1. If the starting derivative already contains both the desired independent variables, the result of Jacobian manipulation is redundant with the triple product rule. The steps are: 1) write the Jacobian; 2) interpose the independent variables; 3) rearrange to convert to partial derivatives.

   Example: Convert $\left(\frac{\partial T}{\partial P}\right)_H$ to derivatives that use T and P as independent variables.

   $$\left(\frac{\partial T}{\partial P}\right)_H = \frac{\partial(T, H)}{\partial(P, H)} = \frac{\partial(T, H)}{\partial(T, P)}\frac{\partial(T, P)}{\partial(P, H)} = \frac{\frac{\partial(H, T)}{\partial(P, T)}}{\frac{\partial(H, P)}{\partial(T, P)}} = \frac{-\left(\frac{\partial H}{\partial P}\right)_T}{\left(\frac{\partial H}{\partial T}\right)_P} = \frac{-\left(\frac{\partial H}{\partial P}\right)_T}{C_P}$$

   and the numerator can be simplified using the expansion rule as presented in Example 6.1.

2. If the starting derivative has just one of the desired independent variables, the steps are: 1) write the Jacobian; 2) interpose the desired variables; 3) write the determinant for the Jacobian without a common variable; 4) rearrange to convert to partial derivatives.

   Example: Find a relation for the adiabatic compressibility, $\kappa_S = -\frac{1}{V}\left(\frac{\partial V}{\partial P}\right)_S$ in terms of derivatives using T, P as independent variables.

   $$\left(\frac{\partial V}{\partial P}\right)_S = \frac{\partial(V, S)}{\partial(P, S)} = \frac{\partial(V, S)}{\partial(P, T)}\frac{\partial(P, T)}{\partial(P, S)} = \left[\left(\frac{\partial V}{\partial P}\right)_T\left(\frac{\partial S}{\partial T}\right)_P - \left(\frac{\partial V}{\partial T}\right)_P\left(\frac{\partial S}{\partial P}\right)_T\right]\left(\frac{\partial T}{\partial S}\right)_P$$

   Now, including a Maxwell relation as we simplify the second term in square brackets, and then combining terms:

   $$\left(\frac{\partial V}{\partial P}\right)_S = \left(\frac{\partial V}{\partial P}\right)_T + \frac{T}{C_P}\left(\frac{\partial V}{\partial T}\right)_P^2$$

   $$\kappa_S = -\frac{1}{V}\left(\left(\frac{\partial V}{\partial P}\right)_T + \frac{T}{C_P}\left(\frac{\partial V}{\partial T}\right)_P^2\right) = \kappa_T - \frac{T}{VC_P}\left(\frac{\partial V}{\partial T}\right)_P^2$$

3. If the starting derivative has neither of the desired independent variables, the steps are: 1) write the Jacobian; 2) interpose the desired variables; 3) write the Jacobians as a quotient and write the determinants for both Jacobians; 4) rearrange to convert to partial derivatives.

   Example: Find $\left(\frac{\partial S}{\partial V}\right)_U$ in measurable properties using P and T as independent variables.

   $$\frac{\partial(S, U)}{\partial(V, U)} = \frac{\partial(S, U)}{\partial(P, T)}\frac{\partial(P, T)}{\partial(V, U)} = \frac{\frac{\partial(S, U)}{\partial(P, T)}}{\frac{\partial(V, U)}{\partial(P, T)}}$$

Writing the determinants for both Jacobians:

$$\frac{\left(\frac{\partial S}{\partial P}\right)_T \left(\frac{\partial U}{\partial T}\right)_P - \left(\frac{\partial U}{\partial P}\right)_T \left(\frac{\partial S}{\partial T}\right)_P}{\left(\frac{\partial V}{\partial P}\right)_T \left(\frac{\partial U}{\partial T}\right)_P - \left(\frac{\partial U}{\partial P}\right)_T \left(\frac{\partial V}{\partial T}\right)_P}$$

Now, using the expansion rule for the derivatives of U, and also introducing Maxwell relations,

$$\frac{-\left(\frac{\partial V}{\partial T}\right)_P \left[C_P - P\left(\frac{\partial V}{\partial T}\right)_P\right] + \frac{C_P}{T}\left[T\left(\frac{\partial V}{\partial T}\right)_P + P\left(\frac{\partial V}{\partial P}\right)_T\right]}{\left(\frac{\partial V}{\partial P}\right)_T \left[C_P - P\left(\frac{\partial V}{\partial T}\right)_P\right] + T\left(\frac{\partial V}{\partial T}\right)_P^2 + P\left(\frac{\partial V}{\partial P}\right)_T \left(\frac{\partial V}{\partial T}\right)_P} =$$

$$\frac{P\left(\frac{\partial V}{\partial T}\right)_P^2 + C_P \frac{P}{T}\left(\frac{\partial V}{\partial P}\right)_T}{T\left(\frac{\partial V}{\partial T}\right)_P^2 + C_P\left(\frac{\partial V}{\partial P}\right)_T} = \frac{P}{T}$$

The result is particularly simple. We could have derived this directly if we had recognized that S and V are the natural variables for U. Therefore, $dU = TdS - P\,dV = 0$, $TdS|_U = -PdV|_U$, $T\left(\frac{\partial S}{\partial V}\right)_U = P$, $\left(\frac{\partial S}{\partial V}\right)_U = \frac{P}{T}$. However, the exercise demonstrates the procedure and power of the Jacobian technique even though the result will usually not simplify to the extent of this example.

6.4 SUMMARY

We have seen in this chapter that calculus provides powerful tools permitting us to calculate changes in immeasurable properties in terms of other measurable properties. We started by defining additional convenience functions A, and G by performing Legendre transforms. We then reviewed basic calculus identities and extended throughout the remainder of the chapter. The ability to perform these manipulations lays the foundation for the development of general methods to calculate thermodynamic properties for any chemical from P-V-T relations. If we only had a general relation that perfectly described $P=P(V,T)$ for all the chemicals in the universe, it could be combined with the tools in this chapter to compute any property required by the energy and entropy balances. At present, no such perfect equation exists. This means that we need to understand what makes it so difficult to develop such an equation and how the various available equations can be applied in various situations to achieve reasonable and continuously improving estimates.

Important Equations

The procedures developed in this chapter are what is important. They provide a basis for transforming one set of derivatives into another and for thinking systematically about how variables relate to one another. The basic identities 6.11–6.17 combine with the fundamental properties for the remainder of the chapter. Nevertheless, several equations stand out as a summary of the results that can be rearranged to a desired form relatively quickly. These are the Maxwell Eqns. 6.29–6.32, and also some intermediate manipulations 6.37–6.41. They are included on the front flap of the textbook for your convenience. Eqn. 6.47 stands out as an equation for long term reference because it relates the Helmholtz energy to the internal energy. That is a key step when we turn to consideration of solution models. Also, when combined with a similar relation is for compressibility factor in Chapter 7, the central role of Helmholtz energy in thermodynamic properties becomes apparent.

Test Yourself

1. What are the restrictions necessary to calculate one state property in terms of only two other state variables?

2. When integrating Eqn. 6.53, under what circumstances may C_V be taken out of the integral?

3. May Eqn. 6.53 be applied to a condensed phase?

4. Is the heat capacity different for liquid acetone than for acetone vapor?

5. Can the tabulated heat capacities be used in Eqn. 6.53 for gases at high pressure?

6.5   PRACTICE PROBLEMS

P6.1 Express in terms of P, V, T, C_P, C_V, and their derivatives. Your answer may include absolute values of S if it is not a derivative constraint or within a derivative.

(a) $(\partial H/\partial S)_V$
(b) $(\partial H/\partial P)_V$
(c) $(\partial G/\partial H)_P$
(ANS. (a) $T[1+V/C_V(\partial P/\partial T)_v]$ (b) $C_v(\partial T/\partial P)_v+V$ (c) $-S/C_P$)

6.6   HOMEWORK PROBLEMS

6.1 CO_2 is given a lot of credit for global warming because it has vibrational frequencies in the infrared (IR) region that can absorb radiation reflected from the Earth and degrade it into thermal energy. Consider the vibration at $\varepsilon_q/k = 952K$ ($546cm^{-1}$).

(a) Plot C_v^{ig}/R versus T for CO_2 in the range 200–400 K using NIST Webbook data. Plot the polynomial expression in the back of the book on the same chart as a dashed line. Adapt the analysis of problem 4.1 to the vibration at 952K while noting that this vibration occurs twice. Plot your vibrational result as a solid line.

(b) Use your Internet search skills to learn the wavelength range of the IR spectrum. How many wavelengths are there? What fraction does the wavelength at $546cm^{-1}$ comprise? If the Earth's atmosphere was composed entirely of CO_2, what fraction of IR energy could be absorbed by CO_2?

(c) The Earth's atmosphere is really 380ppm CO_2. If the absorption efficiency is proportional to the concentration, then how much IR energy could be absorbed by CO_2?

6.2 Express in terms of P, V, T, C_P, C_V, and their derivatives. Your answer may include absolute values of S if it is not a derivative constraint or within a derivative.

(a) $(\partial G/\partial P)_T$	(d) $(\partial H/\partial T)_U$	(g) $(\partial T/\partial P)_H$
(b) $(\partial P/\partial A)_V$	(e) $(\partial T/\partial H)_S$	(h) $(\partial A/\partial S)_P$
(c) $(\partial T/\partial P)_S$	(f) $(\partial A/\partial V)_P$	(i) $(\partial S/\partial P)_G$

6.3 Express the following in terms of U, H, S, A, and their derivatives.

$$\left(\frac{\partial (G/(RT))}{\partial T}\right)_V$$

6.4 (a) Derive $\left(\frac{\partial H}{\partial P}\right)_T$ and $\left(\frac{\partial U}{\partial P}\right)_T$ in terms of measurable properties.

    (b) $dH = dU + d(PV)$ from the definition of H. Apply the expansion rule to show the difference between $\left(\frac{\partial H}{\partial P}\right)_T$ and $\left(\frac{\partial U}{\partial P}\right)_T$ is the same as the result from part (a).

6.5 In Chapter 2, internal energy of condensed phases was stated to be more weakly dependent on pressure than enthalpy. This problem evaluates that statement.

    (a) Derive $\left(\frac{\partial H}{\partial P}\right)_T$ and $\left(\frac{\partial U}{\partial P}\right)_T$ in terms of measurable properties.

    (b) Evaluate $\left(\frac{\partial H}{\partial P}\right)_T$ and compare the magnitude of the terms contributing to $\left(\frac{\partial H}{\partial P}\right)_T$ for the fluids listed in problem 6.10.

    (c) Evaluate $\left(\frac{\partial U}{\partial P}\right)_T$ for the fluids listed in problem 6.10 and compare with the values of

$$\left(\frac{\partial H}{\partial P}\right)_T.$$

6.6 Express $\left(\frac{\partial H}{\partial V}\right)_T$ in terms of α_P and/or κ_T.

6.7 Express the adiabatic compressibility, $\kappa_S = -\frac{1}{V}\left(\frac{\partial V}{\partial P}\right)_S$, in terms of measurable properties.

6.8 Express the Joule-Thomson coefficient in terms of measurable properties for the following:

    (a) Van der Waals equation given in Example 6.6
    (b) An ideal gas.

6.9 (a) Prove $\left(\frac{\partial P}{\partial T}\right)_S = \frac{C_P}{TV\alpha_P}$.

    (b) For an ideal gas reversible adiabat, the process is isentropic, and $(P/P^i) = (T/T^i)^{C_P/R}$. Demonstrate that this equation is consistent with the expression from part (a).

6.10 Determine the difference $C_P - C_V$ for the following liquids using the data provided near 20°C.

Liquid	MW	ρ (g/cm³)	$10^3 \alpha_P$ (K⁻¹)	$10^6 \kappa_T$ (bar⁻¹)
(a) Acetone	58.08	0.7899	1.487	111
(b) Ethanol	46.07	0.7893	1.12	100
(c) Benzene	78.12	0.87865	1.237	89
(d) Carbon disulfide	76.14	1.258	1.218	86
(e) Chloroform	119.38	1.4832	1.273	83

Liquid	MW	ρ (g/cm^3)	$10^3\alpha_P$ (K^{-1})	$10^6\kappa_T$ (bar^{-1})
(f) Ethyl ether	74.12	0.7138	1.656	188
(g) Mercury	200.6	13.5939	0.18186	3.95
(h) Water	18.02	0.998	0.207	49

6.11 A rigid container is filled with liquid acetone at 20°C and 1 bar. Through heat transfer at constant volume, a pressure of 100 bar is generated. $C_P = 125$ J/mol-K. (Other properties of acetone are given in problem 6.10.) Provide your best estimate of the following:

(a) The temperature rise
(b) ΔS, ΔU, and ΔH
(c) The heat transferred per mole

6.12 The fundamental internal energy relation for a rubber band is $dU = TdS - FdL$ where F is the system force, which is negative when the rubber band is in tension. The applied force is given by $F_{applied} = k(T)(L - L_o)$ where $k(T)$ is positive and increases with increasing temperature. The heat capacity at constant length is given by $C_L = \alpha(L) + \beta(L)\cdot T$. Stability arguments show that $\alpha(L)$ and $\beta(L)$ must provide for $C_L \geq 0$.

(a) Show that temperature should increase when the rubber band is stretched adiabatically and reversibly.
(b) Prof. Lira in his quest for scientific facts hung a weight on a rubber band and measured the length in the laboratory at room temperature. When he hung the rubber band with the same weight in the refrigerator, he noticed that the length of the rubber band had changed. Did the length increase or decrease?
(c) The heat capacity at constant force is given by

$$C_F = T\left(\frac{\partial S}{\partial T}\right)_F$$

Derive a relation for $C_F - C_L$ and show whether this difference is positive, negative, or zero.
(d) The same amount of heat flows into two rubber bands, but one is held at constant tension and the other at constant length. Which has the largest increase in temperature?
(e) Show that the dependence of $k(T)$ on temperature at constant length is related to the dependence of entropy on length at constant temperature. Offer a physical description for the signs of the derivatives.

CHAPTER

7

ENGINEERING EQUATIONS OF STATE FOR *PVT* PROPERTIES

I am more than ever an admirer of van der Waals.

Lord Rayleigh (1891)

From Chapter 6, it is obvious that we can calculate changes in *U, S, H, A,* and *G* by knowing changes in any two variables from the set $\{P\text{-}V\text{-}T\}$ plus C_P or C_V. This chapter introduces the various ways available for quantitative prediction of the *P-V-T* properties we desire in a general case. The method of calculation of thermodynamic properties like *U, H,* and so on. is facilitated by the use of **departure functions,** which will be the topic of the next chapter. The development of the departure functions is a relatively straightforward application of derivative manipulations. What is less straightforward is the logical development of a connection between *P, V,* and *T.* We introduced the concept in Chapter 1 that the pressure, temperature, and density (i.e., V^{-1}) are connected through intermolecular interactions. We must now apply that concept to derive quantitative relationships that are applicable to any fluid at any conditions, not simply to ideal gases. You will see that making the connection between *P, V,* and *T* hinges on the transition from the molecular-scale forces and potential energy to the macroscopic pressure and internal energy. Understanding the approximations inherent in a particular equation of state is important because effectively all of the approximations in a thermodynamic model can be traced to the assumed equation of state. Whenever deficiencies are found in a process model, the first place to look for improvement is in revisiting the assumptions of the equation of state.

Understanding the transition from the molecular scale to the macroscopic is a major contribution in our conceptual puzzle of calculating energy, entropy, and equilibrium. We made qualitative connections between the microscopic and macroscopic scales for entropy during our introduction to entropy. For energy, however, we have left a gap that you may not have noticed. We discussed the molecular energy in Chapter 1, but we did not quantify the macroscopic implications. We discussed the macroscopic implications of energy in Chapter 2, but we did not discuss the molecular basis. It is time to fill that gap, and in doing so, link the conceptual framework of the entire text.

From one perspective, the purpose of the examples in Chapter 6 was to explain the need for making the transition from the molecular scale to the macroscopic scale. The purpose of the material following this chapter is to demonstrate the reduction to practice of this conceptual framework in several different contexts. So in many ways, this chapter represents the conceptual kernel for all molecular thermodynamics.

Chapter Objectives: You Should Be Able to...

1. Explain and apply two- and three- parameter corresponding states.

2. Apply an equation of state to solve for the density given T and P, including liquid and vapor roots.

3. Evaluate partial derivatives like those in Chapter 6 using an equation of state for PVT properties.

4. Identify the repulsive and attractive contributions to an equation of state and critically evaluate their accuracy relative to molecular simulations and experimental data.

7.1   EXPERIMENTAL MEASUREMENTS

The preferred method of obtaining P-V-T properties is from experimental measurements of the desired fluid or fluid mixture. We spend most of the text discussing theories, but you should never forget the precious value of experimental data. Experimental measurements beat theories every time. The problem with experimental measurements is that they are expensive, especially relative to pushing a few buttons on a computer.

❶ The basic procedure for calculating properties involves using derivatives of P-V-T data.

To illustrate the difficulty of measuring all properties experimentally, consider the following case. One method to determine the P-V-T properties is to control the temperature of a container of fluid, change the volume of the container in carefully controlled increments, and carefully measure the pressure. The required derivatives are then calculated by numerical differentiation of the data obtained in this manner. It is also possible to make separate measurements of the heat capacity by carefully adding measured quantities of heat and determining changes in P, V, and T. These measurements can be cross-referenced for consistency with the estimated changes as determined by applying Maxwell's relations to the P-V-T measurements. Imagine what a daunting task this approach would represent when considering all fluids and mixtures of interest. It should be understandable that detailed measurements of this type have been made for relatively few compounds. Water is the most completely studied fluid, and the steam tables are a result of this study. Ammonia, carbon dioxide, refrigerants, and light hydrocarbons have also been quite thoroughly studied. The charts which have been used in earlier chapters are results of these careful measurements. Equations of state permit correlation and extrapolation of experimental data that can be much more convenient and more broadly applicable than the available charts.

An experimental approach is naturally impractical for all substances due to the large number of fluids needing to be characterized. The development of equations of state is the engineering approach to describing fluid behavior for prediction, interpolation, and extrapolation of data using the fewest number of adjustable parameters possible for the desired accuracy. Typically, when data are analyzed today, they are fitted with elaborate equations (embellishments of the equations of state discussed in this chapter) before determination of interpolated values or derivatives. The charts are generated from the fitted results of the equation of state.

As a summary of the experimental approach to equations of state, a brief review of the historical development of P-V-T measurements may be beneficial. First, it should be recalled that early measurements of P-V-T relations laid the foundation for modern physical chemistry. Knowing the densities of gases in bell jars led to the early characterizations of molecular weights, molecular formulas, and even the primary evidence for the existence of molecules themselves. At first, it seemed that gases like nitrogen, hydrogen, and oxygen were non-condensable and something quite different from liquids like water or wood alcohol (methanol). As technology advanced, however, experiments were performed at higher temperatures and pressures. Carbon dioxide was a very common compound in the early days (known as "carbonic acid" to van der Waals), and it soon became apparent that it showed a high degree of compressibility. Experimental data were carefully measured in 1871 for carbon dioxide ranging to 110 bars, and these data were referenced extensively by van der Waals. Carbon dioxide is especially interesting because it has some very "peculiar" properties that are exhibited near room temperature and at high pressure. At 31°C and about 70 bars, a very small change in pressure can convert the fluid from a gas-like density to a liquid density. Van der Waals showed that the cause of this behavior is the balance between the attractive forces from the intermolecular potential being accentuated at this density range and the repulsive forces being accentuated by the high-velocity collisions at this temperature. This "peculiar" range of conditions is known as the critical region. The precise temperature, pressure, and density where the vapor and the liquid become indistinguishable is called the critical point. Above the critical point, there is no longer an abrupt change in the density with respect to pressure while holding temperature constant. Instead, the balance between forces leads to a single-phase region spanning vapor-like densities and liquid-like densities. With the work of van der Waals, researchers began to recognize that the behavior was not "peculiar," and that all substances have critical points.[1]

Fortunately, P,V,T behavior of fluids follows the same trends for all fluids. All fluids have a critical point.

7.2   THREE-PARAMETER CORRESPONDING STATES

If we plot P versus ρ for several different fluids, we find some remarkably similar trends. As shown in Fig. 7.1 below, both methane and pentane show the saturated vapor density approaching the saturated liquid density as the temperature increases. Compare these figures to Fig. 1.4 on page 23, and note that the P versus ρ figure is qualitatively a mirror image of the P versus V figure. The isotherms are shown in terms of the **reduced temperature,** $T_r \equiv T/T_c$. Saturation densities are the values obtained by intersection of the phase envelope with horizontal lines drawn at the saturation pressures. The **isothermal compressibility** $\equiv -\dfrac{1}{V}\left(\dfrac{\partial V}{\partial P}\right)_T = \dfrac{1}{\rho}\left(\dfrac{\partial \rho}{\partial P}\right)_T$ is infinite, and its reciprocal is zero, at the critical point (e.g., 191 K and 4.6 MPa for methane). It is also worth noting that the critical temperature isotherm exhibits an inflection point at the **critical point.** This means that $(\partial^2 P/\partial \rho^2)_T = 0$ at the critical point as well as $(\partial P/\partial \rho)_T = 0$. The principle of **corresponding states** asserts that all fluid properties are similar if expressed properly in reduced variables.

The isothermal compressibility is infinite at the critical point.

Although the behaviors in Fig. 7.1 are globally similar, when researchers superposed the P-V-T behaviors based on only **critical temperature,** T_c and **critical pressure,** P_c, they found the superposition was not sufficiently accurate. For example, one way of comparing the behavior of fluids is to plot the **compressibility factor** Z. The compressibility factor is defined as

$$Z \equiv \frac{PV}{RT}$$

7.1

The compressibility factor.

---

1.   Naturally, some compounds decompose before their critical point is reached, or like carbon or tungsten they have such a high melting temperature that such a measurement is impossible even at the present time.

*Note: The compressibility factor **is not the same as** the isothermal compressibility. The similarity in names can frequently result in confusion as you first learn the concepts.*

The compressibility factor has a value of one when a fluid behaves as an ideal gas, but will be non-unity when the pressure increases. By plotting the data and calculations from Fig. 7.1 as a function of **reduced temperature** $T_r = T/T_c$, and **reduced pressure,** $P_r = P/P_c$, the plot of Fig. 7.2 results. Clearly, another parameter is needed to accurately correlate the data. Note that the vapor pressure for methane and pentane differs on the compressibility factor chart as indicated by the vertical lines on the subcritical isotherms. The same behavior is followed by other fluids. For example, the vapor pressures for six compounds are shown in Fig. 7.3, and although they are all nearly linear, the slopes are different. In fact, we may characterize this slope with a third empirical parameter, known as the acentric factor, ω. The **acentric factor** is a parameter which helps to specify the vapor pressure curve which, in turn, correlates the rest of the thermodynamic variables.[2]

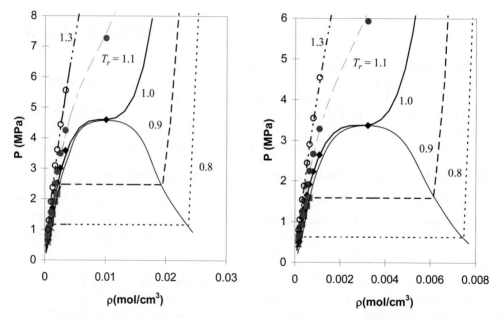

Figure 7.1 *Comparison of the PρT behavior of methane (left) and pentane (right) demonstrating the qualitative similarity which led to corresponding states' treatment of fluids. The lines are calculated with the Peng-Robinson equation to be discussed later. The phase envelope is an approximation sketched through the points available in the plots. The smoothed experimental data are from Brown, G.G., Sounders Jr., M., and Smith, R.L., 1932. Ind. Eng. Chem., 24:513. Although not shown, the Peng-Robinson equation is not particularly accurate for modeling liquid densities.*

---

2. The significance of the vapor pressure curve in determining the thermodynamic properties can be readily appreciated if you consider the difference between a vapor enthalpy and a liquid enthalpy. The detailed consideration of vapor pressure behavior is treated in Chapter 9.

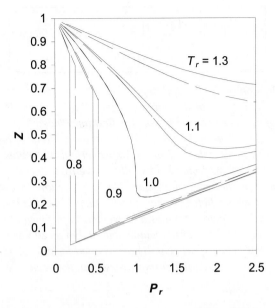

Figure 7.2 *The Peng-Robinson lines from Fig. 7.1 plotted in terms of the reduced pressure at $T_r = 0.8, 0.9$, 1.0, 1.1, and 1.3, demonstrating that critical temperature and pressure alone are insufficient to accurately represent the P-V-T behavior. Dashed lines are for methane, solid lines for pentane. The figure is intended to make an illustrative point. Accurate calculations should use the compressibility factor charts developed in the next section.*

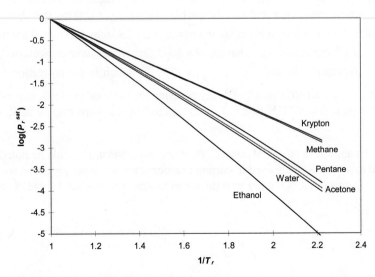

Figure 7.3 *Reduced vapor pressures plotted as a function of reduced temperature for six fluids demonstrating that the shape of the curve is not highly dependent on structure, but that the primary difference is the slope as given by the acentric factor.*

Critical temperature and pressure are insufficient characteristic parameters by themselves. The acentric factor serves as a third important parameter.

$$\omega \equiv -1 - \log_{10}\left(\frac{P^{sat}}{P_c}\right)\Bigg|_{T_r = 0.7} \equiv \text{acentric factor} \qquad 7.2$$

Note: *The specification of T_c, P_c, and ω provides two points on the vapor pressure curve. T_c and P_c specify the terminal point of the vapor pressure curve. ω specifies a vapor pressure at a reduced temperature of 0.7. The acentric factor was first introduced by Pitzer et al.[3] Its definition is arbitrary in that, for example, another reduced temperature could have been chosen for the definition. The definition above gives values of $\omega \sim 0$ for spherical molecules like argon, xenon, neon, krypton, and methane. Deviations from zero usually derive from deviations in spherical symmetry. Nonspherical molecules are "not centrally symmetric," so they are "acentric." In general, there is no direct theoretical connection between the acentric factor and the shape of the intermolecular potential. Rather, the acentric factor provides a convenient experimental vapor pressure which can be correlated with the shape of the intermolecular potential in an ad hoc manner. It is convenient in the sense that its value has been experimentally determined for a large number of compounds and that knowing its value permits a significant improvement in the accuracy of our engineering equations of state.*

The acentric factor is a measure of the slope of the vapor pressure curve plotted as ln P^{sat} versus $1/T$.

7.3   GENERALIZED COMPRESSIBILITY FACTOR CHARTS

P-V-T behavior can be generalized in terms of T_c, P_c, and ω. The original correlation was presented by Pitzer, and is given in the form

$$Z = Z^0 + \omega Z^1 \qquad 7.3$$

Pitzer correlation.

where tables or charts summarized the values of Z^0 and Z^1 at reduced temperature and pressure. The broad availability of computers and programmable calculators is making this approach somewhat obsolete, but it is worthwhile to visualize the trends. Fig. 7.4[4] may be applied for most hydrocarbons. The plot of Z^0 represents the behavior of a fluid that would have an acentric factor of 0, and the plot of Z^1 represents the quantity $Z|_{\omega = 1} - Z|_{\omega = 0}$, which is the correction factor for a hypothetical fluid with an acentric factor of 1. By perusing the table on the back flap of this book, you will notice that most fluids fall between these ranges so that the charts may be used for interpolation.

Eqn. 7.3 can be applied to any fluid once T_r, P_r, and ω are known. It should be noted, however, that this graphical approach is rarely used in current practice since computer programs are more conveniently written in terms of the equations of state as demonstrated in Section 7.5 and the homework.

3. Pitzer, K.S., Lippmann, D.Z., Curl Jr., R.F., Huggins, C.M., Petersen, D.E. 1955. *J. Am. Chem. Soc.* 77:3427–3433.
4. This is the Lee-Kesler equation. Lee, B. I., Kesler, M.G. 1975. *AIChE J.* 21:510.

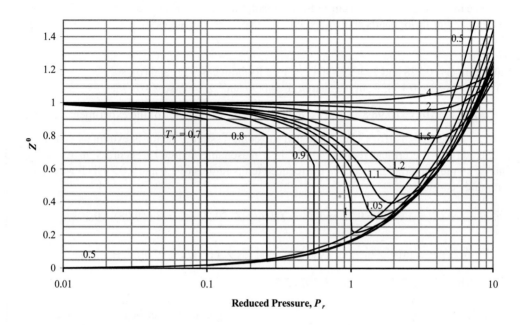

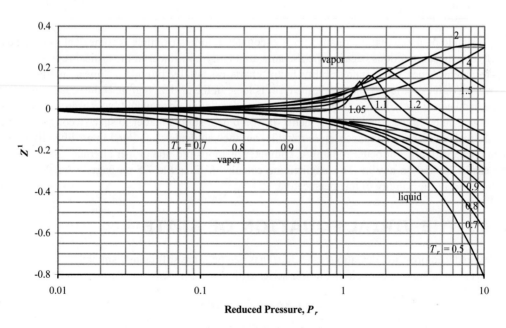

Figure 7.4 *Generalized charts for estimating the compressibility factor. (Z^0) applies the Lee-Kesler equation using $\omega = 0.0$, and (Z^1) is the correction factor for a hypothetical compound with $\omega = 1.0$. Note the semilog scale.*

Example 7.1 Application of the generalized charts

Estimate the specific volume in cm^3/g for carbon dioxide at 310 K and (a) 8 bar (b) 75 bar by the compressibility factor charts and compare to the experimental values[2] of 70.58 and 3.90, respectively.

Solution: $\omega = 0.228$ and $T_r = 310/304.2 = 1.02$ for both cases (a) and (b), so,
(a) $P_r = 8/73.82 = 0.108$; from the charts, $Z^0 = 0.96$ and $Z^1 = 0$, so $Z = 0.96$.

$V = ZRT/(P \cdot MW) = (0.96 \cdot 83.14 \cdot 310)/(8 \cdot 44) = 70.29$, within 0.4% of the experimental value.

(b) $P_r = 75/73.82 = 1.016 \approx 1.02$; Note that the compressibility factor is extremely sensitive to temperature in the critical region. To obtain a reasonable degree of accuracy in reading the charts, we must interpolate between the reduced temperatures of 1.0 and 1.05 which we can read with more confidence.

At $T_r = 1.0$, $Z^0 = 0.22$ and $Z^1 = -0.08$ so $Z = 0.22 + 0.228 \cdot (-0.08) = 0.202$

At $T_r = 1.05$, $Z^0 = 0.58$ and $Z^1 = 0.03$, so $Z = 0.58 + 0.228 \cdot (0.03) = 0.587$

Interpolating, $Z = 0.202 + (0.587 - 0.202) \cdot 2/5 = 0.356$

$V = ZRT/(P \cdot MW) = (0.356 \cdot 8.314 \cdot 310)/(7.5 \cdot 44) = 2.78$, giving 29% error relative to the experimental value.

It should be noted that the relative error encountered in this example is somewhat exaggerated relative to most conditions because the Z-charts are highly non-linear in the critical region used in this problem. Since the compressibility factor charts essentially provide a "linear interpolation" between Z values for $\omega = 0$ and $\omega = 1$, the error is large in the critical region. If the reduced temperature had been slightly higher, (e.g., $T_r = 1.1$), then the relative error would have been roughly 1%, as demonstrated in the homework problems. It would be a simple matter to specify conditions that would make the chart look much more reliable, but then students might tend to err liberally rather than conservatively. For better reliability, computer methods provide proper alternatives, and these are easily applied on any modern engineering calculator. Example 7.5 on page 268 will demonstrate in detail the validity of this perspective.

7.4   THE VIRIAL EQUATION OF STATE

At low reduced pressure, deviations from ideal gas behavior are sufficiently small that we can write our equation of state as explicit in a power series with respect to density. That is,

$$Z = 1 + B\rho + C\rho^2 + D\rho^3 + \ ... \tag{7.4}$$

where B, C, and D are the second, third, and fourth virial coefficients. This can be considered an expansion in powers of ρ. Coefficients C and D are rarely applied because this power series is not very accurate over a broad range of conditions. The most common engineering application of the virial equation of state is to truncate it after the second virial coefficient and to limit the range of application appropriately. It provides a simple equation which still has a reasonable number of viable applications. It has become common usage to refer to the equation truncated after the second virial coefficient as *the* virial equation, even though we know that it is really a specialized form. We, too, will follow this common usage. Furthermore, the truncated form may alternatively be expressed as $Z = Z(P,T)$. Hence, we often refer to *the* virial equation as:

$$Z = 1 + BP/(RT) \qquad\qquad 7.5$$

where B is a function of T. Note that Eqn. 7.5 indicates that Z varies linearly with pressure along an isotherm. Look back at Fig. 7.4 and notice that the region in which linear behavior occurs is limited, but in general, the approximation can be used at higher reduced pressures when the reduced temperature is higher. The virial equation can be generalized in reduced coordinates as given by Eqns. 7.6–7.9.[5] Eqn. 7.10 checks for restriction of the calculation to the linear Z region.

> Virial equation. B is known as the second virial coefficient, and it is a measure of the slope of the Z-chart isotherms in the linear region.

$$Z = 1 + (B^0 + \omega B^1)P_r/T_r \quad \text{or} \quad Z = 1 + BP/(RT) \qquad\qquad 7.6$$

$$\text{where } B(T) = (B^0 + \omega B^1)RT_c/P_c \qquad\qquad 7.7$$

$$B^0 = 0.083 - 0.422/T_r^{1.6} \qquad\qquad 7.8$$

$$B^1 = 0.139 - 0.172/T_r^{4.2} \qquad\qquad 7.9$$

$$\text{Subject to } T_r > 0.686 + 0.439P_r \text{ or } V_r > 2.0 \qquad\qquad 7.10$$

The temperature dependence of the slope of the Z lines is not sufficiently represented by $1/T$, so the temperature dependence of B in Eqns. 7.8 and 7.9 is required. The virial equation is limited in its range of applicability, but it has the advantage of simplicity. Its simplicity is especially advantageous when illustrating derivations of real-fluid behavior for the first time and extending thermodynamic relations to vapor mixtures. Unfortunately, the virial equation does *not* apply to liquids, and many interesting results in thermodynamics appear in the study of liquids. To develop a global perspective applicable to gases and liquids, we must consider the physics of fluids in a more sophisticated manner. The simplest form which still permits this level of sophistication is the cubic equation, discussed in the following section.

Example 7.2   Application of the virial equation

Estimate the specific volume in cm^3/g for carbon dioxide at 310 K and (a) 8 bar (b) 75 bar by the virial equation and compare to the experimental values of 70.58 and 3.90, respectively.

Solution: $\omega = 0.228$ and $T_r = 310/304.2 = 1.02$ for both cases (a) and (b), so,
$B^0 = 0.083 - 0.422/1.02^{1.6} = -0.326$
$B^1 = 0.139 - 0.172/1.02^{4.2} = -0.0193$
$B(T)P_c/(RT_c) = (B^0 + \omega B^1) = (-0.326 + 0.228 \cdot (-0.0193)) = -0.3304$

5. Smith, J.M., Van Ness, H.C. 1975. *Introduction to Chemical Engineering Thermodynamics,* 3rd ed, New York: McGraw-Hill, p. 87.

> **Example 7.2 Application of the virial equation (Continued)**
>
> (a) $P_r = 8/73.82 = 0.108$; so $Z = 1 + (B^0 + \omega B^1)P_r/T_r = 1 - 0.3304 \cdot 0.108/1.02 = 0.965$
>
> $V = ZRT/(P \cdot MW) = (0.965 \cdot 83.14 \cdot 310)/(8 \cdot 44) = 70.66$, within 0.1% of the experimental value.
>
> (b) $P_r = 75/73.82 = 1.016$; applying Eqn. 7.10, $0.686 + 0.439 \cdot 1.016 = 1.13 > T_r = 1.02$. Therefore, the virial equation may be inaccurate using only the second virial coefficient.

There is an adaptation of the form of the virial series which should be mentioned before concluding this discussion. It should not seem surprising that the inclusion of extra adjustable parameters in the form of the virial series is an extremely straightforward task—just add higher order terms to the series. In many cases, exponential terms are also included as in Eqn. 7.11. In this way, it is possible to fit the *P-V-T* behavior of the liquid as well as the vapor to a reasonable degree of accuracy. It turns out that the theoretical foundation for the series expansion in this way is tenuous, however. Reading "the fine print" in discussions of series expansions like the Taylor series shows that such an approach is only applicable to "analytic" functions. At present, there is a general acceptance that the behavior of real fluids is "non-analytic" at the critical point. This means that application of such a series expansion above the critical density and below the critical temperature is without a rigorous mathematical basis. Nevertheless, engineers occasionally invoke the motto that "we can fit the shape of an elephant with enough adjustable parameters." It is in this spirit that empirical equations like the Benedict-Webb-Rubin equation are best appreciated. One particular modification of the Benedict-Webb-Rubin form is given below. It is the form that Lee and Kesler[6] developed to render the Pitzer correlation in terms of computer-friendly equations. The Lee-Kesler equation was used to generate Fig. 7.4.

$$Z = 1 + \frac{B}{V_r} + \frac{C}{V_r^2} + \frac{D}{V_r^5} + \frac{E_0}{T_r^3 V_r^2}\left(E_1 + \frac{E_2}{V_r^2}\right)\exp\left(-\frac{E_2}{V_r^2}\right) \qquad 7.11$$

Twelve parameters are used to specify the temperature dependence of B, C, D, E_0, E_1, and E_2 for each compound. Readers are directed to the original article for the exact values of the parameters as part of the homework.

7.5 CUBIC EQUATIONS OF STATE

The acronym EOS will be used to mean equation of state.

To apply the relationships that we can develop for relating changes in properties to C_P, C_V, P, T, V, and their derivatives, we need really general relationships between P, V, and T. These relationships are dictated by the equation of state (**EOS**). Constructing an equation of state with a firm physical and mathematical foundation requires considering how the intermolecular forces affect the energy and pressure in a fluid. In a dense fluid, we know that the molecules are close together on the average, and such closeness gives rise to an attractive potential energy. A common practical manifestation of this attractive energy is the heat of vaporization of a boiling liquid. But how can we make a quantitative connection between molecular forces and macroscopic properties? A firm understanding of this physical and mathematical foundation is helpful to understand the extensions to multicomponent mixtures and multiphase equilibria. A proper derivation would provide a mathematical

6. Lee, B.I., Kesler, M.G. 1975. *AIChE J.* 21:510.

connection between the microscopic potential and the macroscopic properties. We will lay the groundwork for such a rigorous derivation later in the chapter. For introductory purposes, however, we would like to see how some typical equations look and how to use them in conjunction with the theorem of corresponding states.

The van der Waals Equation of State

One of the most influential equations of state has been the van der Waals (1873) equation. Even the most successful engineering equations currently used are only minor variations on the theme originated by van der Waals. The beauty of his model is that detailed knowledge of the molecular interactions is not necessary. Simply by noting that there are two characteristic molecular quantities (ε and σ) and two characteristic macroscopic quantities (T_c and P_c), he was able to infer a simple equation that captured the key features of each fluid through the principle of corresponding states. His final equation expressed the attractive energy in terms of a parameter which he referred to as a, and the size parameter b, but the choice of symbols was arbitrary. The key feature to recognize at this stage is that there are at least two parameters in all the equations and that these can be determined by matching experimental data.

Johannes Diderik van der Waals (1837–1923) was a Dutch physicist. He was awarded the 1910 Nobel Prize in physics.

The resulting equation of state is:

$$P = \frac{RT}{V-b} - \frac{a}{V^2} = \frac{\rho RT}{1-b\rho} - a\rho^2 \quad \text{or} \quad Z = \frac{1}{(1-b\rho)} - \frac{a\rho}{RT} \qquad 7.12$$

Van der Waals EOS.

where ρ = molar density = $n/\underline{V}$.

> *Note: Common engineering practice is to use ρ to denote intensive density. We follow that convention here, using ρ as molar density. Advanced chemistry and physics books and research publications frequently use ρ as number density $N/\underline{V} = nN_A/\underline{V}$, so the definitions must be carefully determined.*

ρ will be used to denote molar density.

The exact manner of determining the values for the parameters a and b is discussed in Section 7.8 on page 270.

$$a \equiv \frac{27}{64}\frac{R^2 T_c^2}{P_c} \qquad ; \qquad b \equiv \frac{RT_c}{8P_c} \qquad 7.13$$

We may write the equation of state as $Z = 1 + Z^{rep} + Z^{att}$, where 1 denotes the ideal gas behavior, Z^{rep} represents the deviations from the ideal gas law due to repulsive interactions, Z^{att} represents the deviations due to attractive interactions. For the van der Waals equation,

$$\boxed{Z = 1 + Z^{rep} + Z^{att} = 1 + \frac{b\rho}{1-b\rho} - \frac{a\rho}{RT}} \qquad 7.14$$

The van der Waals equation written in the form $Z = 1 + Z^{rep} + Z^{att}$.

where the second and third terms on the right-hand side are Z^{rep} and Z^{att}, respectively. Eqn. 7.12 is compact, but Eqn. 7.14 more clearly represents the origin of the contributions to Z. In many later applications we will need to use the departure of Z from ideal gas behavior, $Z–1$, and Eqn. 7.14 will fulfill this need. There are two key features of the van der Waals equation. First, note that the repulsive term accounts for the asymptotic divergence of the compressibility factor as the packing factor $b\rho$ increases. The divergence occurs because rapidly increasing large pressures are required to increase the density as close packing is approached. Second, note that the attractive term increases in magnitude as the temperature decreases, and contributes to smaller values of Z at low T. The contributions of the attractive forces increase at low temperature because the kinetic energy can no longer overwhelm the potential attractions at low temperature. As we have discussed, this eventually leads to condensation. The discussion in Section 7.11 provides a better understanding of the molecular basis of the van der Waals equation. Note that the van der Waals EOS does not incorporate the acentric factor and is incapable of representing different vapor pressure slopes. It is thus primarily a pedagogical tool to introduce the forms of cubic EOSs, not a practical tool.

The Peng-Robinson Equation of State

Since the time of van der Waals (1873), many approximate equations of state have been proposed. For the most part, these have been semi-empirical corrections to van der Waals' characterization of "a = constant," and most have taken the form $a = a(T,\omega)$. One of the most successful examples of this approach is that of Peng and Robinson (1976). We refer to this equation many times throughout this text and use it to demonstrate many central themes in thermodynamics theory as well as useful applications.

The Peng-Robinson equation of state (EOS) is given by:

> The Peng-Robinson EOS. Note that a is a temperature-dependent parameter, not a constant. Note the dependence on the acentric factor.

$$P = \frac{RT\rho}{(1-b\rho)} - \frac{a\rho^2}{1+2b\rho-b^2\rho^2} \quad \text{or} \quad Z = \frac{1}{(1-b\rho)} - \frac{a}{bRT}\cdot\frac{b\rho}{1+2b\rho-b^2\rho^2} \qquad 7.15$$

where ρ = molar density = $n/\underline{V}$, b is a constant, and a depends on temperature and acentric factor,[7]

$$a \equiv a_c\alpha; \quad a_c \equiv 0.45723553\frac{R^2T_c^2}{P_c} \qquad\qquad b \equiv 0.07779607R\frac{T_c}{P_c} \qquad 7.16$$

$$\alpha \equiv [1+\kappa(1-\sqrt{T_r})]^2 \qquad \kappa \equiv 0.37464+1.54226\,\omega-0.26992\,\omega^2 \qquad 7.17$$

Note that α is *not* the isobaric coefficient of thermal expansion and κ is *not* the isothermal compressibility – they are simply variables introduced by Peng and Robinson for notational convenience. The temperature derivative of a is useful when temperature derivatives of Z are needed:

$$\frac{da}{dT} = \frac{-a_c\kappa\sqrt{\alpha T_r}}{T} \qquad 7.18$$

T_c, P_c, and ω are reducing constants according to the principle of corresponding states. Expressing the contributions to Z in the manner we followed for the van der Waals equation,

$$\boxed{Z = 1 + Z^{rep} + Z^{att} = 1 + \frac{b\rho}{1-b\rho} - \frac{a}{bRT}\cdot\frac{b\rho}{1+2b\rho-b^2\rho^2}} \qquad 7.19$$

Comparison of the Peng-Robinson equation to the van der Waals equation shows one obvious similarity; the repulsive term is the same. There are some differences: the temperature dependence of the attractive parameter a is incorporated; dependence of a on the acentric factor is introduced; and the density dependence of Z^{att} is altered by the denominator of the attractive term. The manner

---

7. The number of significant figures presented in Eqn. 7.16 is important in reproducing the universal value of $Z_c = 0.307$ predicted by the Peng-Robinson equation of state.

in which these extra details were added was almost entirely empirical; different equations were tried until one was found which seemed to fit the data most accurately while retaining cubic behavior. (Many equations can be tried in 103 years.) There is not much to say about this empirical approach beyond the importance of including the acentric factor. The main reason for the success of the Peng-Robinson equation is that it is primarily applied to vapor-liquid equilibria and that the representation of vapor-liquid equilibria is strongly influenced by the more accurate representation of vapor pressure implicit in the inclusion of the acentric factor. Since the critical point and the acentric factor characterize the vapor pressure fairly accurately, it should not be surprising that the Peng-Robinson equation accurately represents vapor pressure.

One caution is given. The differences in accuracy between various equations of state are subtle enough that one equation may be most accurate for one narrow range of applications, while another equation of state is most accurate over another range. This may require a practicing engineer to adopt an equation of state other than the Peng-Robinson equation for his specific application. Nevertheless, the treatment of the Peng-Robinson equation presented here is entirely analogous to the treatment required for any equation of state. If you understand this treatment, you should have no problem adapting. A brief review of several thermodynamic models commonly encountered in chemical process simulations is given in Appendix D.

> *Note: The variables a and b are used throughout equation of state literature (and this text) to denote equation of state parameters. The formulas or values of these parameters for a given equation of state cannot be used directly with any other equation of state.*

There are a lot of EOSs. We focus on the Peng-Robinson to illustrate the concepts.

The variables a and b are commonly used in EOSs. Do not interchange the formulas.

7.6 SOLVING THE CUBIC EQUATION OF STATE FOR Z

In most applications we are given a pressure and temperature and asked to determine the density and other properties of the fluid. This becomes slightly difficult because the equation involves terms of density that are to the third power ("cubic") even when simplified as much as possible. Standard methods for solutions to cubic equations can be applied. The equation can be made dimensionless prior to application of the solution method. Note:

$$Z = PV/(RT) = P/(\rho RT) \qquad 7.20$$

Dimensional analysis is an important engineering tool. Here we make the EOS dimensionless so that it can be solved in a generalized way.

Defining dimensionless forms of the parameters

$$A \equiv aP/(R^2T^2) \qquad 7.21$$

$$B \equiv bP/(RT) \qquad 7.22$$

results in the lumped variables

$$b\rho = B/Z; \qquad a\rho/(RT) = A/Z \qquad 7.23$$

The Peng-Robinson equation of state becomes

$$\boxed{Z = \frac{1}{(1 - B/Z)} - \frac{A}{B} \cdot \frac{B/Z}{1 + 2B/Z - (B/Z)^2}} \qquad 7.24$$

Do not confuse the EOS parameters A and B with other uses of the variables. The intended use is almost always clear.

> *Note: The variable A is also used elsewhere in the text to denote Helmholtz energy. The variable B is used elsewhere in the text to represent the second virial coefficient and availability. The context of the variable usage should make the*

meaning of the variable clear. We choose to use A and B as reduced equation of state parameters for consistency with equation of state discussions in the literature.

Rearranging the dimensionless Peng-Robinson equation yields a cubic function in Z that must be solved for vapor, liquid, or fluid roots:

$$Z^3 - (1-B)Z^2 + (A - 3B^2 - 2B)Z - (AB - B^2 - B^3) = 0 \qquad 7.25$$

Isotherm Shape and the Real Roots

Fig. 7.5 shows several P-V isotherms for CO_2 as generated with Eqn. 7.15. Comparing to Fig. 1.4, note that the cubic EOS predicts "humps" when $T < T_c$. The humps are larger at lower temperature, and the pressure can be negative as shown by the 275 K isotherm. The cubic equation always has three roots, but at some conditions two are imaginary roots. For engineering, we are interested in the real roots. For the isotherm at 290 K, three real roots exist at pressures between approximately 39 bar and 58 bar. We will refer to this pressure region as the **three-root region** meaning that there are three real roots in this region. Above 58 bar, only a liquid root will result, and below 39 bar only a vapor root will result. These two regions are called **one-root regions** to communicate that only one real root exists in this region. The three-root region size depends on temperature. At 275 K the three-root region extends down to $P = 0$, but the region is very narrow near 300 K.

The cubic equation for Z in Eqn. 7.25 also has similar behavior. Naming this function $F(Z)$, we can plot $F(Z)$ versus Z to gain some understanding about its roots as shown in Fig. 7.6. Considering the case when $P = P^{sat}$, we see that three real roots exist; the larger root of $F(Z)$ will be the vapor root and will be the value of Z for saturated vapor. The smallest root will be the liquid root and will be the value of Z for saturated liquid. At all other pressures at this temperature, where three real

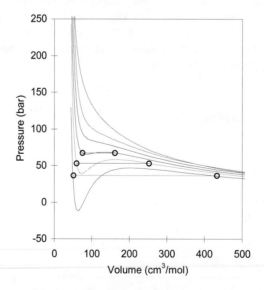

Figure 7.5 *Illustration of the prediction of isotherms by the Peng-Robinson equation of state for CO_2 ($T_c = 304.2$ K) at 275 K, 290 K, 300 K, 310 K, 320 K, and 350 K. Higher temperatures result in a high pressure for a given volume. The "humps" are explained in the text. The calculated vapor pressures are 36.42 bar at 275 K, 53.2 bar at 290 K, and 67.21 bar at 300 K.*

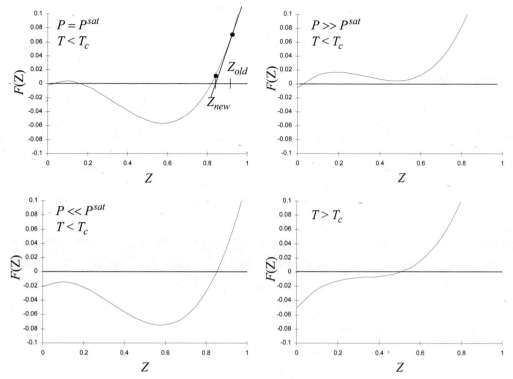

Figure 7.6 *Comparison of behavior of cubic in Z for the Peng-Robinson equation of state at several conditions. The labels Z_{new} and Z_{old} in the upper left are described in the iterative description in Appendix B.*

roots are found, one of the roots is always more stable. Below the critical temperature, when $P > P^{sat}$, the fluid will be a compressed liquid, and the liquid root is more stable. Below the critical temperature, when $P < P^{sat}$, the fluid will be a superheated vapor and the vapor root is more stable. When $T > T_c$, we have a supercritical fluid which can only have a single root but it may vary continuously between a "vapor-like" or "liquid-like" densities and compressibility factors. We will discuss stability and how to choose the stable root without generating an entire isotherm in an upcoming section.

Methods of Solving the Cubic Equation

Engineering applications typically specify P and T, and require information about V. Solution of the equation of state in terms of Z is preferred over solution for V, and we can subsequently find V using

$$V = ZRT/P \qquad\qquad 7.26$$

The value of Z often falls between 0 and 1. (See Fig. 7.4 on page 257.) V often varies from 50–100 cm^3/mole for liquids to near infinity for gases as P approaches zero. It is much easier to solve for roots over the smaller variable range using the compressibility factor Z. There are two basic approaches to solving cubic equations of state. First, we may use an iterative method. One such method is the Newton-Raphson method. Another method is to solve analytically. A computer or calculator program is helpful in either solution method.

Iterative Method

The Newton-Raphson method is described in Appendix B. The Newton-Raphson method uses an initial guess along with the derivative value to rapidly converge on the solution.

Analytical Solution

The other choice we have for solution of the cubic is to analytically obtain the roots as detailed in Appendix B. The method varies depending on whether one or three roots exist at the pressure of interest. Solutions are implemented in a spreadsheet (Preos.xlsx) or a MATLAB script (Preos.m). MATLAB includes a polynomial root finder, so the statement `Zvals=roots([1 a2 a1 a0])` results in both real and imaginary roots. The argument in the "roots" function is the vector of coefficients for the polynomial in Z. In MATLAB, the indexes of the real roots can by found with `index=find(imag(Zvals)==0)` followed by selecting the real parts of the roots using `Zreal=real(Zvals(index))`.

Preos.xlsx uses the procedures from Appendix B and shows the intermediate calculations.

Example 7.3 Peng-Robinson solution by hand calculation

Perform a hand calculation of the real roots for argon at 105.6 K and 0.498 MPa.

Solution: This example is available online and provides an example of hand calculation at the same conditions as the next example.

Example 7.4 The Peng-Robinson equation for molar volume

Preos.xlsx, Preos.m.

Find the molar volume predicted by the Peng-Robinson equation of state for argon at 105.6 K and 4.96 bar.

Solution: The critical data are entered from the table on the back flap of the text.

Preos.xlsx output is shown below. The state is in the three-root region, because the cells for the one-root region are labeled #NUM! by Excel. Many of the intermediate calculations are also shown. The volumes are 27.8, 134, and 1581 cm³/mole. The lower value corresponds to the liquid volume and the upper value corresponds to the vapor. Note that Z is close to zero for the liquid and close to one for the vapor.

The output from the Preos.m MATLAB script is also shown below. Though the default output does not include intermediate values, they may be obtained by removing the ";" at the end of any code line and rerunning the script.

Example 7.4 The Peng-Robinson equation for molar volume (Continued)

Output from Preos.m:

```
argon  Tc(K)= 150.9 Pc(MPa)= 4.898 w = -0.004
T(K)= 105.600000 P(MPa)= 0.496000
Zvals =
    0.8971
    0.0759
    0.0157
Z= 0.897123  0.015681
V(cm^3/mol)= 1588.066740  27.758560
fugacity (MPa)= 0.449384  0.449903
Hdep (J/mol)= -222.933032  -6002.507074
```

Peng-Robinson Equation of State (Pure Fluid)

Properties			
Gas	T_c (K)	P_c (MPa)	☐
ARGON	150.9	4.898	-0.004

Spreadsheet protected, but no password used.

Current State		Roots		
T (K)	105.6	Z	V	fugacity
P (MPa)	0.496		cm³/gmol	MPa
answers for three		0.897123	1588.067	0.449384
root region		0.075938	134.4247	
		0.015681	27.75856	0.449909
& for 1 root region		#NUM!	#NUM!	#NUM!

Stable Root has a lower fugacity

Intermediate Calculations		
R (cm³MPa/molK)	8.314472	
T_r	0.699801	a (MPa cm⁶/gmol²)
P_r	0.101266	165184.2
☐	0.368467	b (cm³/gmol)
☐	1.124086	19.92796
fugacity ratio		A 0.10628
	0.998832	B 0.011258

To find vapor pressure, or saturation temperature,
see cell A28 for instructions

Solution to Cubic	$Z^3 + a_2 Z^2 + a_1 Z + a_0 = 0$				$R = q^2/4 + p^3/27 =$ -1.76E-05
a_2	a_1	a_0	p	q	If Negative, three unequal real roots,
-0.988742	0.083385	-0.001068	-0.242486	-0.045187	If Positive, one real root

Method 1 - For region with one real root

P	Q	Root to equation in x	
#NUM!	#NUM!	#NUM!	

Solution methods are summarized
in the appendix of the text.

Method 2 - For region with three real roots

m	3q/pm	3*☐₁	☐₁	Roots to equation in x		
0.568607	0.983181	0.183666	0.061222	0.567542	-0.253642	-0.3139

Preos.xlsx,
Preos.m.

Example 7.5   Application of the Peng-Robinson equation

Estimate the specific volume in cm^3/g for carbon dioxide at 310 K and (a) 8 bars (b) 75 bars by the Peng-Robinson equation and compare to the experimental values of 70.58 and 3.90, respectively.[1]

Solution: $\omega = 0.228$, $T_c = 304.2$, $P_c = 73.82$, MW = 44 g/gmol,

(a) $Z = 0.961$

$V = ZRT/(P \cdot MW) = (0.961 \cdot 83.14 \cdot 310)/(8 \cdot 44) = 70.37$, within 0.3% of the experimental value.

(b) $Z = 0.492$

$V = ZRT/(P \cdot MW) = (0.492 \cdot 83.14 \cdot 310)/(75 \cdot 44) = 3.84$ giving 1.5% error relative to the experimental value.

Example 7.5 shows the equation of state is much more reliable than reading the compressibility factor charts in Example 7.1 on page 258.

Determining Stable Roots

When three real roots are found, which is most stable can be quickly determined. As mentioned in Example 7.4, the smallest root usually corresponds to a liquid state, and the largest root usually corresponds to the vapor state.[9] When three real roots are found over a range of pressures for a given temperature, these roots do not indicate vapor + liquid coexistence at all these conditions. Vapor and liquid phases coexist only at the vapor pressure—above the vapor pressure, the liquid root is most **stable**—below the vapor pressure, the vapor root is most stable. The most stable root represents the phase that will exist at equilibrium. When three roots are found, the most stable root has the lower **Gibbs energy** or **fugacity.** At phase equilibrium, the Gibbs energy and fugacity of the roots will be equal. Fugacity is closely related to the Gibbs energy and will be described in Chapter 9, but we will begin to use the calculated values before we explain the calculation procedures completely. When three roots exist, the center root is thermodynamically **unstable** because the derivative of pressure with respect to volume is positive, which violates our common sense, and is shown to be thermodynamically unstable in advanced thermodynamics texts. Physically, we can understand the meaning of the unstable root as follows. Imagine placing Lennard-Jones spheres all in a row such that the force pulling on each molecule from one direction is exactly balanced by the force in the other direction. Next, imagine placing similar rows perpendicular to that one until you obtain the desired density at the desired temperature. Physically and mathematically, this configuration is conceivable, but what will happen when one of these atoms moves? The perfect balance will be destroyed, and a large number of atoms will cluster together to form a liquid. The atoms that do not form a liquid will remain in the form of a low-density vapor. Although this discussion has discussed only the energy of interactions, the entropy is also important. It is actually a balance of enthalpy and entropy that results in phase equilibrium, as we will discuss in Chapter 9.

A simple way to visualize the conditions that lead to vapor-liquid equilibrium along an isotherm is to consider the P-V diagram illustrated in Fig. 7.5. We will show in Chapter 9 that the con-

The *stable* root represents the phase that will exist *at equilibrium*. The stable root has the lower Gibbs energy or fugacity.

---

8. Vargaftik, N.B. 1975. *Handbook of Physical Properties of Liquids and Gases.* New York: Hemisphere Publishers.
9. However, the Peng-Robinson equation does give small real roots at high pressures, and the smallest real root is not always the liquid root. See problem 7.11.

dition for equilibrium between vapor and liquid roots occurs when the horizontal line on the *P-V* diagram is positioned such that the area enclosed above the line is exactly equal to the area enclosed below the line. Even though the enclosed areas have different shapes, imagine moving this line up and down until it looks like the areas are equal. The dots in the figure are the predictions of the saturated liquid and vapor volumes, and form the phase envelope. The parts of the isotherms that are between the saturated vapor and saturated liquid roots are either **metastable** or **unstable.** To determine whether a given point is metastable or unstable, look for the point where the isotherm reaches a maximum or a minimum. Points between the liquid root and the minimum are considered metastable liquids; points between the maximum and the vapor root are considered metastable vapors. The metastable state can be experimentally obtained in careful experiments. Under clean conditions, it is possible to experimentally heat a liquid above its boiling point to obtain super-heated liquid. Likewise, under clean conditions, it is possible (though challenging) to experimentally obtain subcooled vapor. However, a metastable state is easily disrupted by vibrations or nucleation sites, e.g., provided by a boiling chip or dust, and once disrupted, the state decays rapidly to the equilibrium state. The boundary between the metastable and unstable states is known as the **spinodal** condition, predicted by the EOS by the maximum and minimum in the humps in sub-critical isotherms. We will discuss more details about characterizing proper fluid roots when we treat phase equilibrium in a pure fluid.

7.7 IMPLICATIONS OF REAL FLUID BEHAVIOR

There is one implication of non-ideal fluid behavior that should be clear from the equations presented above: Real fluids behave differently from ideal gases. How differently? An example provides the most straightforward answer to that question. Here we adapt some of the derivatives from Chapter 6.

Example 7.6 Derivatives of the Peng-Robinson equation

Determine $\left(\dfrac{\partial P}{\partial T}\right)_V$, $\left(\dfrac{\partial C_V}{\partial V}\right)_T$, and $\left(\dfrac{\partial U}{\partial V}\right)_T$ for the Peng-Robinson equation.

Solution: The derivatives $(\partial U/\partial V)_T$ and $(\partial C_V/\partial V)_T$ have been written in terms of measurable properties in Examples 6.6 and 6.9, respectively, and have been evaluated for an ideal gas. The analysis with the Peng-Robinson model provides more realistic representation of the properties of real substances. Beginning with the same analytical expressions set forth in the referenced examples, a key derivative is obtained for the Peng-Robinson equation,

$$\left(\frac{\partial P}{\partial T}\right)_V = \frac{R\rho}{1-b\rho} - \frac{\rho^2}{1+2b\rho-b^2\rho^2}\frac{da}{dT}$$

which approaches the ideal gas limit: $\lim_{\rho \to 0}\left(\dfrac{\partial P}{\partial T}\right)_V = R\rho = \dfrac{R}{V}$. The volume dependence of C_V is obtained by the second derivative:

$$\left(\frac{\partial C_V}{\partial V}\right)_T = T\left(\frac{\partial^2 P}{\partial T^2}\right)_V = \frac{-T\rho^2}{1+2b\rho-b^2\rho^2}\frac{d^2a}{dT^2} = \frac{-\rho^2}{1+2b\rho-b^2\rho^2}\frac{a_c\kappa}{2}\left(\frac{\kappa}{T_C}+\frac{\sqrt{\alpha T_r}}{T}\right)$$

Example 7.6 Derivatives of the Peng-Robinson equation (Continued)

which approaches the ideal gas limit of zero at low density,

$$\left(\frac{\partial U}{\partial V}\right)_T = T\left(\frac{\partial P}{\partial T}\right)_V - P = \frac{\rho^2}{1 + 2b\rho - b^2\rho^2}\left[a - \frac{da}{dT}\right] = \frac{\rho^2 a_C}{1 + 2b\rho - b^2\rho^2}[\alpha + \kappa\sqrt{\alpha T_r}],$$

which also approaches the ideal gas limit of zero at low density. We have thus shown that C_V depends on volume. To calculate a value of C_V, first we determine $C_V^{ig} = C_P^{ig} - R$, where C_P^{ig} is the heat capacity tabulated in Appendix E. Then, at a given $\{P,T\}$, the equation of state is solved for ρ. The resultant density is used as the limit in the following integrals, noting as $V \to \infty$, $\rho \to 0$, and $dV = -d\rho/\rho^2$: This method is used for departures from ideal gas properties in Chapter 8.

$$C_v - C_V^{ig} = \int_\infty^V\left(\frac{\partial C_V}{\partial V}\right)_T dV = \left(\frac{d^2a}{dT^2}\right)\int_0^\rho \frac{T\rho^2}{1 + 2b\rho - b^2\rho^2}\frac{d\rho}{\rho^2} = \frac{T}{2\sqrt{2}b}\left(\frac{d^2a}{dT^2}\right)\ln\left[\frac{1 + (1 + \sqrt{2})b\rho}{1 + (1 - \sqrt{2})b\rho}\right]$$

where $\left(\frac{d^2a}{dT^2}\right) = \frac{a_c\kappa}{2T_c^2 T_r}\left[\kappa + \sqrt{\frac{\alpha}{T_r}}\right]$

7.8   MATCHING THE CRITICAL POINT

The capability of a relatively simple equation to represent the complex physical phenomena illustrated in Figs. 7.5–7.6, and as shown later in Figs. 7.7 and 7.9, is a tribute to the genius of van der Waals. His method for characterizing the difference between subcritical and supercritical fluids was equally clever. He recognized that, at the critical point,

$$\left(\frac{\partial P}{\partial \rho}\right)_T = 0 \quad \text{and} \quad \left(\frac{\partial^2 P}{\partial \rho^2}\right)_T = 0 \quad \text{at} \quad T_c, P_c \qquad\qquad 7.27$$

You can convince yourself that this is true by looking at the P versus ρ plots of Fig. 7.1 on page 254. From this observation, we obtain two equations that characterize the equation of state parameters a and b in terms of the critical constants T_c and P_c. In principle, this is all we need to say about this problem. In practice, however, it is much simpler to obtain results by recognizing another key feature of the critical point: The vapor and liquid roots are exactly equal at the critical point (and the spurious middle root is also equal). We can apply this latter insight by specifying that $(Z - Z_c)^3 = 0 = Z^3 - 3Z_cZ^2 + 3Z_c^2 Z - Z_c^3 = Z^3 - a_2Z^2 + a_1Z - a_0$ (Appendix B). Equating the coefficients of these polynomials gives three equations in three unknowns: Z_c, A_c, and B_c.

Example 7.7 Critical parameters for the van der Waals equation

Apply the above method to determine the values of Z_c, A_c, and B_c for the van der Waals equation.

Solution: Rearranging the equation in terms of A_c and B_c we have:
$$0 = Z^3 - (1 + B_c)Z^2 + A_c Z - A_c B_c = 0 = Z^3 - 3Z_c Z^2 + 3Z_c^2 Z - Z_c^3$$

By comparing coefficients of Z^n: (1) $Z_c = (1 + B_c)/3$; (2) $A_c = 3Z_c^2$; (3) $A_c B_c = Z_c^3$.

Substituting A_c into the last equation, we have: $3Z_c^2 B_c = Z_c^2 (1 + B_c)/3$.

Cancelling the Z_c^2 and solving we have $B_c = 1/8 = 0.125$. The other equations then give $Z_c = 0.375$ and $A_c = 27/64$.

The solution is especially simple for the van der Waals equation, but the following procedure can be adapted for any cubic equation of state:
1. Rearrange the equation of state into its cubic form: $Z^3 - a_2 Z^2 + a_1 Z - a_0$.
2. Guess a value of Z_c (e.g., $Z_c \sim 1/3$).
3. Solve the equivalent of expression (1) for B_c.
4. Solve the equivalent of expression (2) for A_c.
5. Solve the equivalent of expression (3) for Z_c.
6. If Z_c = guess, then stop. Otherwise, repeat.

7.9 THE MOLECULAR BASIS OF EQUATIONS OF STATE: CONCEPTS AND NOTATION

In the previous sections we alluded to equations of state as empirical equations that may have appeared by magic. In this section and the next two, we attempt to de-mystify the origins behind equations of state by systematically describing the current outlook on equation of state development. It may seem like overkill to develop so much theory to justify such simple equations. As empirical equations go, equations of state are not much more difficult to accept than, say, Newton's laws of motion. Nevertheless, our general purpose is for readers to learn to develop their own engineering model equations and to refute models that are not sensible. By establishing the connection between the nanoscopic potential function and macroscopic properties, molecular modeling lays the foundation for design at the nanoscale.

It is feasible to develop equations of state based solely on fitting experimental data. If the fit is insufficiently precise for a given application, simply add more parameters. We see evidence of this approach in the Peng-Robinson equation, where temperature and density dependencies are added to the parameter "a" in order to fit vapor pressure and density better. A more extensive example of this approach is evident in the 32 parameter Benedict-Webb-Rubin equation that forms the basis of the Lee-Kesler model. The IAPWS model of H_2O is representative of the current state of this approach. It is the basis of the steam tables in Appendix E.

The shortcoming of this approach is that we lose the connection between the parameters in the equation of state and their physical meaning. For example, the Peng-Robinson "a" parameter must be related to attractive interactions, like the square-well parameter ε. But ε cannot be a function of temperature and density, so what part of the Peng-Robinson model is due to ε and what part is due to something else? If we could recover that physical connection, then all our efforts to fit data would result in systematically refined characterizations of the molecular interactions. With reliable characterizations of the molecular interactions, we could design molecules to assemble into a myriad of nanostructures: membranes for water purification, nanocomposites, polymer wrappers that block oxygen, artificial kidneys small enough to implant. The possibilities are infinite.

Since about 1960, computers have made it feasible to simulate macroscopic properties based on a specified intermolecular potential. With this tool, the procedure is clear: (1) Specify a potential model for a given molecule, (2) simulate the macroscopic properties, (3) evaluate the deviations between the simulated and experimental properties, (4) repeat until the deviations are minimized. This procedure is straightforward but tedious. Each simulation of $Z(T,\rho)$ can take an hour or so.

Corresponding States in Molecular Dimensions

As engineers, we would like to get results faster. One idea is to leverage the principle of corresponding states. We know that ε has dimensions of J/molecule, so $N_A\varepsilon$ has dimensions of J/mol. Therefore, $RT/(N_A\varepsilon)$ would be dimensionless and serve in similar fashion to the usual reduced temperature, T/T_c. Similarly, the molecular volume, v_{mol}, has dimensions of cm^3/molecule and $N_A v_{mol}$ has dimensions of cm^3/mol. Therefore, $N_A v_{mol}\rho$ would be dimensionless and serve in very similar fashion to the usual reduced density, ρ/ρ_c. Another idea would be to tabulate the dimensionless properties from the simulation at many state points, then interpolate, similar to the steam tables. The interpolating equations might even resemble traditional equations of state in form and speed. The difference would be that they retain the connection between the nanoscopic potential model and macroscopic properties. In other words, we can engineer our equations of state to be consistent with specific potential models by expressing our "reduced" temperature and density using molecular dimensions. Then the principle of corresponding states can be applied to match the ε and σ for a particular molecule in the same way that we match a and b parameters in the van der Waals model.

Perhaps the most difficult part of understanding the molecular perspective is making the transformations from the macroscopic scale to the nanoscopic. For example, the "b" parameter has dimensions of cm^3/mol. What does that imply about the diameter of the molecule in nm? As another example, the "a" parameter has dimensions of J-cm^3/mol^2. How does that relate to the molecular properties? Answering these questions leads to the introduction of a few "conversion shortcuts" to facilitate the scale transformations. One valuable conversion shortcut is to note that the transformation from cm^3 to nm^3 involves a factor of $(10^7)^3$ or 10^{21}. This transformation from cm^3 to nm^3 usually goes hand-in-hand with a transformation from moles to molecules, involving a factor of N_A. If we write $N_A = 602(10^{21})$ instead of $6.02(10^{23})$, then factors of 10^{21} cancel conveniently. This convenience motivates us to work in cm^3 at the macroscopic level. Another shortcut is to write the volume of a sphere in terms of diameter instead of radius. Finally, we note that the molecular volume on a molar basis is equivalent to the "b" parameter in cm^3/mol. Altogether,

$$v_{mol} = 4\pi r^3/3 = 4\pi(\sigma/2)^3/3 = \pi\sigma^3/6; \quad N_A v_{mol} = b \qquad 7.28$$

Example 7.8   Estimating molecular size

Example 7.4 shows that b = 19.9 cm^3/mol for argon. Estimate the diameter (nm) of argon according to the Peng-Robinson model.

Solution: $N_A \pi \sigma^3 / 6$ = 19.9 cm3/mol; σ^3 = 6(19.9cm3/mol)(1mol/602(10^{21})molecules)(10^{21} nm^3/cm^3)/π. Thus, σ^3 = 6(19.9)/(602π) = 0.06313 nm^3; σ = (0.06313)$^{1/3}$ = 0.398 nm.

Speaking of the "b" parameter, it is useful to note that the combination of $b\rho$ appears in the equations quite often. This combined variable is very important. In addition to being dimensionless, and a convenient reduced density, its meaning is quite significant. It represents the volume occupied by molecules divided by the total volume. It makes sense intuitively that the density cannot be higher than when the total volume is completely filled. So this explains why the van der Waals equation includes $(1-b\rho)$ in the denominator, forcing divergence as this limit is approached. The prevalence of this combined variable suggests that we give it a special symbol and name, $\eta_P = b\rho = b/V$, the packing efficiency (aka. **packing fraction**).[10]

Finally, we should consider the square-well energy parameter, ε, and the van der Waals parameter, a. Applying Eqn. 7.13 indicates that the dimensions of the "a" parameter are J-cm^3/mol^2. We can rewrite the van der Waals equation as $Z = 1/(1-\eta_P) - (a/b)\eta_P/(RT)$. In this format, it is clear that the combination of variables "a/b" represents an attractive energy in J/mol. In other words, $a/b \sim N_A \varepsilon$. Another shortcut for quickly transforming from the macro scale to the nano scale is to recognize that $\varepsilon/k = N_A\varepsilon/R$ and both have dimensions of absolute temperature, K. In this context, the combination of variables $\varepsilon/(kT) = \beta\varepsilon$ is an especially convenient characterization of dimensionless reciprocal temperature, where $\beta = 1/(kT)$.

Distinguishing Repulsive and Attractive Effects

One of the advantages of molecular modeling is that the potential model can be dissected into various parts: the repulsive core, attractive wells, dipole moments, hydrogen bonding, and so forth. The total potential function is the sum of all of these interactions, but simulations can be done separately with one, two, or all interactions. Then we can understand which parts of the equation of state come from each part of the potential model. Fig. 7.7 illustrates Z versus reciprocal temperature. This shows a specific y-intercept at infinite temperature. Analyzing the van der Waals equation shows that this y-intercept corresponds to $Z_0 = 1/(1-b\rho)$, where the subscript "0" designates the point where reciprocal temperature reaches zero. This contribution represents positive deviations from ideality, and therefore we can call it repulsive. The temperature-dependent part of the van der Waals equation is negative and represents attractive contributions. The reason that the attractive

---

10. To be more precise, however, it is impossible to pack spheres such that the space is completely filled. One example of highly efficient packing would be the body-centered-cubic (bcc) unit cell (Fig. 7.10). We can determine η_P for the bcc unit cell by noting that there are two atoms in the unit cell. The obvious atom is the one in the center. The second atom is actually the combination of pieces of atoms at the corners. There are eight corners and each one contributes one-eighth of an atom. To compute the packing fraction, we need to relate the box length, L, to the diameter, σ. Note that all the corner atoms are touching the atom in the center. Therefore, a diagonal line from the lower left to the upper right corner cuts through 2σ. In Cartesian coordinates, this same distance represents $(L^2+L^2+L^2)^{1/2} = L(3)^{1/2}$. Therefore, $L = 2\sigma/3^{1/2}$ and the packing fraction is $2\pi\sigma^3/(6L^3) = 0.68 = \eta_P^{bcc}$. Liquids at typical conditions cannot pack this efficiently, so typical packing fractions for liquids are 0.25–0.45.

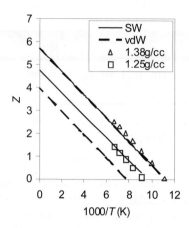

Figure 7.7 *Compressed liquid argon. Experimental data from NIST WebBook. Dashed lines characterize the van der Waals model and solid lines correspond to molecular simulation of the square-well model with $\lambda = 1.7$. The manner of fitting the molecular parameters (a and b or ε and σ) is described in Example 7.9.*

contribution becomes negligible at high temperature is that such a large molecular kinetic energy overwhelms the relatively small "stickiness" of the molecular attractions. Only the repulsive interaction is large enough to contribute at high temperature.

If the *y*-intercept is so important then what does the *x*-intercept mean? We have seen $Z \sim 0$ before, in Example 7.4, where $Z = 0.016$ for the liquid root. On the scale of Fig. 7.7, $Z = 0.016$ is practically zero. The pressure of the saturated liquid is low despite having a high density (and high repulsion) because the attractions are comparable to the repulsions when the temperature is low enough. Slow-moving molecules show a greater tendency to "stick together."

Ultimately, it is necessary to characterize the parameters that relate the intermolecular potential to experimental data. With sufficient data for the compressed liquid density, the problem of characterizing ε and σ becomes a simple matter of matching the slope and intercept of a plot like Fig. 7.7. The procedure is illustrated in Example 7.9(b). This is the most straightforward approach because it relates experimental *PVT* data directly to *PVT* data from a molecular simulation.

Fig. 7.7 also compares to experimental compressed liquid data for argon. These data transition quickly to supercritical temperatures and pressures, but the trend is smooth (almost linear) because the density is constant (i.e., isochoric). We can fit the van der Waals equation at one density by tuning the *a* and *b* parameters, as shown in Example 7.9(a). Deviations are large, however, when we apply the van der Waals model to a different density using the same *a* and *b*. This reflects deficiencies in the physics of the van der Waals model.

We can improve the characterization of argon by using the square-well potential with $\lambda = 1.7$. Once again, the parameters (ε and σ this time) are tuned to the Z versus $1/T$ data at the high density as shown in Example 7.9(b). The predictions (using the same ε and σ) are much better, as shown by the solid lines in Fig. 7.7, reflecting the improved physics underlying the square-well model coupled with molecular simulation. Systematically studying the square-well model, and dissecting the repulsive and attractive contributions, leads to a better understanding of the molecular interactions, and this leads to better predictions.

Other approaches exist to infer potential parameters from experimental data, but they are too complicated for our introductory treatment. One alternative is to apply experimental data for saturated vapor pressure and density. This approach accounts better for vapor pressure being a very important property in chemical engineering, and more data are available. As another alternative, you may be wondering about fitting the critical point, as done by van der Waals. Unfortunately, the behavior at the critical point does not conform to the rules of normal calculus and even simulations are challenging. If you read the fine print on the theorems of calculus, you find the stipulation that functions must be analytic for the theorems to apply. Phenomena in the critical region are non-analytic. The non-analytic behavior is universal for all compounds, so the principle of corresponding states is still valid. On the other hand, fitting a simple analytic function to data outside the critical region leads to inconsistencies inside the critical region, and vice versa. Cubic equations exhibit this inconsistency by predicting a *PV* phase envelope that is not flat enough on the top. Dealing further with these inconsistencies is a topic of current research. Methods that avoid the critical region are gaining favor at present.

We have barely scratched the surface of what is necessary to characterize the intermolecular forces between all the molecules that we can imagine. For example, we have only considered the square-well potential, but the Lennard-Jones model would be more realistic, and those are just two of the possibilities. As another example, the presentation here is limited to spherical molecules. The molecular perspective is not extended to non-spherical molecules until Chapter 19, and then only briefly.[11] You might say that the Peng-Robinson equation can be applied to non-spherical molecules, but only because of clever fitting. The physics behind the Peng-Robinson is simply the same as that of van der Waals: spherical. Simple physics and educated empirical fitting are cornerstones of engineering models. The Peng-Robinson model is a prime example of what can be accomplished with that approach. But recognize that we are always learning more about physics and those new insights are the cornerstones of new technology. Accurately characterizing intermolecular forces involves characterizing many small molecules that share common fragments. When those fragments are characterized, they can be assembled to predict the properties of large molecules. Ultimately, we can imagine a time when nanostructures can be designed and constructed the way civil engineers build bridges today. These structures occur naturally in everything from sea shells to proteins. Learning how to do it is a basis for modern research.

Example 7.9   Characterizing molecular interactions

Based on Fig. 7.7, trend lines indicate *y*-intercept values of, roughly, 5.7 and 4.7 when fit to the isochoric *PVT* data for argon at 1.38g/cm³ and 1.25 g/cm³, respectively. Similarly, the *x*-intercepts are roughly 11.2 and 9.5, respectively. Use these values to estimate the EOS parameters.
(a) Estimate the values of *a* and *b* at 1.38 g/cm³ according to the van der Waals model.
(b) Predict the values of *x*- and *y*- intercepts at 1.25 g/cm³ using the *a* and *b* from part (a).
(c) Suppose the square-well simulation data can be represented by:

$$Z = 1 + 4\,\eta_P/(1 - 1.9\,\eta_P) - 15.7\,\eta_P\,\beta\varepsilon/(1 - 0.16\,\eta_P)$$

Estimate the values of σ and ε/k at 1.38g/cm³ and predict the *x*- and *y*-intercepts at 1.25 g/cm³.

---

11. The principles are simple for extending molecular simulation to non-spherical molecules. These are described along with a number of implementation resources in an online supplement.

Example 7.9   Characterizing molecular interactions   (Continued)

Solution:

(a) At 1.38g/cm^3, y-intercept, $Z_0 = 1/(1 - \eta_P) = 5.7 \Rightarrow \eta_P = 1 - 1/5.7 = 0.825 = b\rho$.
$b = 0.825 \cdot 39.9(\text{g/mol})/1.38(\text{g/cm}^3) = 23.9 \text{ cm}^3/\text{mol}$.

At the x-intercept, $0 = 5.7 - (a/bRT) \cdot \eta_P = 5.7 - (a/bRT) \cdot 0.825 \Rightarrow a/bRT = 5.7/0.825 = 6.91$.
Using the x-intercept to determine temperature, $1000/T = 11.2 \Rightarrow T = 1000/11.2 = 89.3\text{K} \Rightarrow$
$a = 23.9(8.314)89.3(6.91) = 123 \text{ kJ-cm}^3/\text{mol}^2$.

(b) At 1.25 g/cm^3, $\eta_P = 23.9(1.25)/39.9 = 0.7487 \Rightarrow Z_0 = 1/(1 - \eta_P) = 4.0 = y$-intercept.
At the x-intercept, $0 = 4.0 - 123000/(23.9RT) \cdot 0.7487 = 4.0 - 463/T \Rightarrow T = 463/4 = 116$.
Therefore, the x-intercept is $1000/T = 1000/116 = 8.6$. These x- and y- intercepts form the basis
for the dashed line in Fig. 7.7 at 1.25 g/cm^3. The prediction of the van der Waals model is poor.

(c) The procedure for finding σ and ε/k is similar. At the 1.38g/cm^3,
$Z_0 = 1 + 4\eta_P/(1 - 1.9\,\eta_P) = 5.7 \Rightarrow \eta_P(4 + 4.7 \cdot 1.9) = 4.7 \Rightarrow \eta_P = 0.363 = b\rho$
$b = 0.363 \cdot 39.9/1.38 = 10.5 \text{ cm}^3/\text{mol} = N_A \pi \sigma^3/6 \Rightarrow \sigma = 0.322 \text{ nm}$

At the x-intercept, $0 = 5.7 - 15.7(0.363)\beta\varepsilon/(1 - 0.16 \cdot 0.363) \Rightarrow \beta\varepsilon = 0.942$;
$1000/T = 11.2 \Rightarrow T = 1000/11.2 = 89.3\text{K} \Rightarrow \varepsilon/k = (0.942)89.3 = 84.1 \text{ K}$

At 1.25 g/cm^3, following the same procedure:
$\eta_P = 10.5(1.25)/39.9 = 0.329 \Rightarrow Z_0 = 1 + 4\eta_P/(1 - \eta_P) = 4.5 = y$-intercept.
At the x-intercept, $0 = 4.5 - 15.7(0.329)\beta\varepsilon/(1 - 0.16 \cdot 0.329) = 4.5 - (5.452)\beta\varepsilon \Rightarrow \beta\varepsilon = 0.827$
$\Rightarrow T = 84.1/0.827 = 102$. Therefore, the x-intercept is $1000/T = 1000/102 = 9.8$. These x- and y-
intercepts form the basis for the solid line in Fig. 7.7 at 1.25 g/cm^3, and the prediction is quite
good.

To put the significance of this analysis in perspective, imagine you were designing a material
with pores just the right size to capture argon from air using the van der Waals model. The diameters would indicate the appropriate pore size. For the van der Waals model $\sigma = (6 \cdot 23.9/602\pi)^{1/3}$
$= 0.423\text{nm}$ compared to 0.322nm by the square-well estimate. This means that your pores could be
over sized by more than 30%. Improved physical insight can suggest more successful experiments.

7.10   THE MOLECULAR BASIS OF EQUATIONS OF STATE: MOLECULAR SIMULATION

In Chapter 1 we developed an ultrasimplified kinetic theory based on ideal gas interactions. Constructing a sophisticated kinetic theory simply means accounting for molecular interactions like
attractions and collisions. Unfortunately, accounting for these detailed interactions requires a
sequential calculation of collisions that does not lend itself to explicit solution. Fortunately, modern
computers make it easy to characterize these collisions over time.

Computation involving finite systems of molecules colliding with one another and with walls
is called molecular dynamics simulation. In this section, we apply a simple computation from elementary physics: the collision of two particles, as illustrated in Fig. 7.8. If we can compute a single
collision, a computer can be programmed to compute the next trillion collisions. Thus, we see how

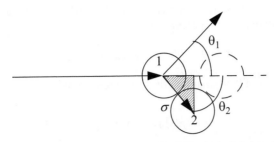

Figure 7.8 *Molecular collision in 2D. The dashed disk is a disk image that will be discussed in the text.*

the average properties over all collisions can be computed from the intermolecular potential. Comparing the computer model to experimental data leads to a connection between the molecular properties (e.g., ε, σ) and the macroscopic properties (e.g., a, b). Seeing the molecular motions and relating the molecular properties to changes in density, temperature, energy, and pressure sheds light on both the nanoscopic and macroscopic levels.

Elastic Collision of Two Particles in Two Dimensions

Tools for molecular simulation are freely available, highly visual, and conveniently interactive.[12] On the other hand, using a simulator as a mysterious "black box" is not satisfying. We should feel confident that a molecular dynamics simulation is nothing more than application of Newton's laws. In this way, we can leverage our confidence in Newton's laws to infer the behavior of large systems of molecules. Then the connection between elementary physics and molecular thermodynamics should not seem mysterious at all.

For simplicity consider two disks moving in two dimensions (2D). Typical physics courses describe 2D particle collision.[13] Consider the special case of a disk1 with an x-component of velocity and a stationary disk2 of identical size as shown in Fig. 7.8. Conservation of kinetic energy and momenta provides:

$$m_1 (v_1^o)^2 = m_1 v_1^2 + m_2 v_2^2 \qquad \text{Kinetic energy conserved (elastic)} \qquad 7.29$$

$$m_1 v_1^o = m_1 v_1 \cos \theta_1 + m_2 v_2 \cos \theta_2 \qquad \text{x-momentum conserved} \qquad 7.30$$

$$0 = m_1 v_1 \sin \theta_1 + m_2 v_2 \sin \theta_2. \qquad \text{y-momentum conserved} \qquad 7.31$$

This provides three equations and four unknowns (v_1, θ_1, v_2, θ_2). For hard disks, the fourth equation derives from noting that the force of impact changes the momentum of each particle. The collision force acts on disk2 in the direction of the vector between the two centers at contact and the force on disk1 is equal and opposite. The impact force changes the momentum in the corresponding direction. Referring to Fig. 7.8. disk2's angle of recoil θ_2 can be found by the shaded right triangle. For the purpose of this presentation, we consider the case of identical disk diameters, σ. Note that the shaded triangle hypotenuse is σ and the vertical side is $y_2^c - y_1^c$ where the "c" denotes collision location. The angle is given by

$$\sin \theta_2 = (y_2^c - y_1^c)/\sigma \qquad 7.32$$

---

12. For example, see the discontinuous molecular dynamics (DMD) module at Etomica.org.
13. For example, Giancoli, D.C. 2000. *Physics for Scientists and Engineers*, 3rd ed., Englewood Cliffs, NJ: Prentice-Hall, Example 7.6.

For purposes of this presentation, we solve the special case where both masses are equal, in which case Eqn. 7.29 and the Pythagorean theorem show that

$$\theta_1 - \theta_2 = 90° \text{ (Note that } \theta_2 < 0.)$$

7.33

The simultaneous solution of the remaining three equations provides:

$$v_2 = v_1°\cos\theta_2; \quad v_1 = \sqrt{(v_1^o)^2 - v_2^2}; \quad v_2\sin\theta_2 = -v_1\sin\theta_1$$

7.34

> An online supplement provides a generalized derivation for different sizes and masses.

Details of the simultaneous solution are provided as an online supplement. The solution in 2D is sufficient in principle because other orientations can be rotated into this reference frame.[14]

The final element of the computation is to calculate the collision time. The strategy is to calculate the collision time for all possible collisions, then execute the one that occurs first. Collisions with walls are the simplest. If the molecule has a positive x-velocity, then you know it collides with the east wall before it collides with the west wall. Similar reasoning applies for the y-velocity and north or south walls. A wall collision occurs when the molecule travels from its current position until its outer edge touches the wall. To be specific, we can define the "simulation box" to have its southwest corner at the origin and walls of length "L." For example, the collision time with the east wall (E) would be:

$$t_i^E = (L - x_i - \sigma/2)/v_{i,x}$$

7.35

The collision time between two molecules follows a similar procedure. In this case, they touch when their distance is $(\sigma/2 + \sigma/2) = \sigma$, but only if they approach at the proper angle. Approach angle is not a concern for a wall collision because the wall extends forever, but a molecule is finite in size. The strategy is to translate the molecules to the reference frame of Fig. 7.8, compute the time until crossing the x- position of molecule 2, then check if the molecules are close enough at that point to collide. The first step of the translation simply involves subtracting the position of the first molecule to put it at the origin, and subtracting the velocity of molecule 2 to make it stationary.

$$x_i' = x_i - x_1, y_i' = y_i - y_1; \quad v_{i,x}' = v_{i,x} - v_{2,x}, v_{i,y}' = v_{i,y} - v_{2,y};$$

7.36

The position of the second molecule must be rotated in accordance with the velocity of the first molecule being on the x-axis, such that $v_{1,x}'' = ((v_{1,x}')^2 + (v_{1,y}')^2)^{1/2}$. In polar coordinates,

$$\tan\phi_1 = v_{1,y}/v_{1,x}; \quad \tan\phi_2 = y_2'/x_2'; \quad r_2' = ((x_2')^2 + (y_2')^2)^{1/2}$$

7.37

Transforming the coordinates by rotation,[15]

$$x_2'' = r_2'\cos(\phi_2 - \phi_1)$$

7.38

$$y_2'' = r_2'\sin(\phi_2 - \phi_1)$$

7.39

A collision occurs if $|y_2''| < \sigma$. If there is a collision,

$$x_1^{c''} = x_2'' - \sigma\cos\theta_2, \text{ where } \sin\theta_2 = y_2''/\sigma,$$

7.40

---

14. The 2D perspective is not convenient for a general program. Therefore, the online supplement includes formulas for a general methodology in three dimensions, handling attractive collisions, extensions to multi-site molecules, and resources for implementation, all in vector notation.

15. Spiegel, M.R. 1968. *Schaum's Mathematical Handbook of Formulas and Tables*, New York:McGraw-Hill, p. 36.

$$t_{12}{}^c = x_1{}^{c''}/v_{i,x.}{}''$$

7.41

Immediately after the collision, the positions become

$$x_i^f = x_i + v_{i,x} \cdot t_{12}{}^c; \, y_i^f = y_i + v_{i,y} \cdot t_{12}{}^c;$$

7.42

The velocities can be reported in the original reference frame by reversing the ϕ_1 rotation,

$$v_{1,x}^f = v_1 \cos(\theta_1 + \phi_1); \, v_{1,y}^f = v_1 \sin(\theta_1 + \phi_1);$$

7.43

For the second molecule, it is most convenient to apply Eqns. 7.30 and 7.31 in their resolved forms:

$$v_{2,x}^f = v_{2,x}{}^o + v_{1,x}{}^o - v_{1,x}^f; \, v_{2,y}^f = v_{2,y}{}^o + v_{1,y}{}^o - v_{1,y}^f$$

7.44

Example 7.10 Computing molecular collisions in 2D

Let the diameters of two disks, σ, be 0.4 nm, the masses be 16 g/mole, and the length of the square box, L, be 5nm. Start the disks at [1.67 1.67], [3.33 3.33] and initial velocities (nm/ps): [0.167 0.222], [−0.167 −0.222] where 1 nm = 10^{-9} m and 1ps = 10^{-12} s. Note that the gas constant 8.314 J/mol-K = 8.314 kg-nm^2/(ns^2-mol-K) = 8.314(10^{-6}) kg-nm^2/(ps^2-mol-K).

(a) Compute the temperature (K).
(b) Compute the collision times with the walls.
(c) Compute the collision times with the disks. Which event occurs first?
(d) Compute the velocity vectors (m/s) after the first collision event.

Solution:

(a) $T_{2D} = M_w \langle v^2 \rangle/(2R)$; $\langle v^2 \rangle = (0.167^2 + 0.222^2 + 0.167^2 + 0.222^2)/2 = 0.07717$
   $T_{2D} = (0.016 \text{kg/mol})(0.07717 \text{ nm}^2/\text{ps}^2)/(2 \cdot 8.314(10^{-6}) \text{ kg-nm}^2/(\text{ps}^2\text{-mol-K})) = 74\text{K}.$

(b) The collision time with the walls depends on the wall being approached. Note that the molecular coordinate will be within $0.5\sigma = 0.2$ nm of the wall coordinate when a wall collision occurs. Disk1 is approaching the north wall and east wall (using superscripts to denote geographic directions), the collision times are $t_1{}^N = (4.8 - y_1{}^o)/v_{1,y} = (4.8 - 1.67)/0.222 = 14.10\text{ps}$, $t_1{}^E = (4.8 - x_1{}^o)/v_{1,x} = (4.8 - 1.67)/0.167 = 18.74\text{ps}$. Similarly, $t_2{}^S = (0.2 - y_2{}^o)/v_{2,y} = (0.2 - 3.33)/(-0.222) = 14.10\text{ps}$; $t_2{}^W = (0.2 - x_2{}^o)/v_{2,x} = (0.2 - 3.33)/(-0.167) = 18.74\text{ps}$. Molecule 1 collides with the north wall, and molecule 2 collides with the south wall at 14.10ps. The wall collisions corresponding to $t_1{}^E$ and $t_2{}^W$ will not occur.

(c) Translating by Eqn.7.36, $x_2' = y_2' = 3.33 - 1.67 = 1.66$. Translating the velocities to make molecule 2 stationary: $v_{1,x}' = 0.167 - (-0.167) = 0.334$. $v_{1,y}' = 0.444$. Using Eqn. 7.37, $\phi_1 = \tan^{-1}(v_{1,y}'/v_{1,x}') = \tan^{-1}(0.444/0.334) = 53.13°$. $\phi_2 = \tan^{-1}(1.66/1.66) = 45°$. $r_2' = 1.66(2)^{\frac{1}{2}} = 2.35$nm. $x_2'' = 2.35 \cos(45 - 53.13) = 2.33$; $y_2'' = 2.35 \sin(45 - 53.13) = -0.33$. Since $|y_2''| < \sigma$, these molecules do collide. By Eqn. 7.40, $\theta_2 = \sin^{-1}(-0.33/0.4) = -56.10°$. Then, $x_1{}^{c''} = 2.33 - 0.4 \cos(-56.10) = 2.103$; noting $v_{1,x}'' = (0.334^2 + 0.444^2)^{\frac{1}{2}} = 0.5556$. $t_{12}{}^c = 2.103/0.5556 = 3.78$ ps. The intermolecular collision occurs before the wall collisions calculated in part (b).

> ## Example 7.10 Computing molecular collisions in 2D (Continued)
>
> (d) Computing the velocities after collision requires Eqn. 7.34, noting by Eqn. 7.33 that $\theta_1 = 90 - 56.10 = 33.9$. $v_2 = v_1''\cos\theta_2 = 0.5556\cos(-56.10) = 0.3099$, $v_1 = (0.5556^2 - 0.3099^2)^{1/2} = 0.4611$. Also note that Eqn. 7.34 gives only the magnitude of the velocity. $v_{2,x} = v_2\cos\theta_2 = 0.3099\cos(-56.10) = 0.173$; $v_{2,y} = -0.257$; $v_{1,x} = 0.257$; $v_{1,y} = 0.173$; Returning to the original reference frame: $v_{1,x}{}^f = v_1\cos(\phi_1 + \theta_1) + v_{2,x}{}^o = 0.4611\cos(33.9 + 53.13) - 0.167 = -0.143$. $v_{1,y}{}^f = 0.4611\sin(33.9 + 53.13) - 0.222 = 0.238$; $v_{2,x}{}^f = 0.143$. $v_{2,y}{}^f = -0.238$. Finally, we update all the positions to the time of the collision. $x_1{}^f = [1.67 + 0.167 \cdot 3.78 \quad 1.67 + 0.222 \cdot 3.78] = [2.301 \quad 2.514]$; $x_2{}^f = [3.33 - 0.167 \cdot 3.78 \quad 3.33 - 0.222 \cdot 3.78] = [2.695 \quad 2.486]$. From this point, the procedure for the next collision is exactly the same.
>
> In retrospect, a major oversimplification of this problem deserves comment. By restricting the system to two particles, it is necessary that the components of velocity be equal and opposite in sign. Otherwise, the system itself would have a net velocity. You should not mistake this equality as a general result. If there were three particles, for example, the velocities would sum to zero, but the individual magnitudes could vary quite substantially.

Analyzing MD Results

For our purposes, we can assume that you have sufficiently grasped the principles of molecular simulation if you can compute a single collision. A second collision is much like the first and computers are made for these kinds of repetitive calculations. At that point, the challenge becomes analyzing the results of the simulations. We can illustrate this kind of analysis with simulations of the hard-sphere fluid to infer the repulsive contribution of the square-well fluid's equation of state. As shown in Fig. 7.9(a), the hard-sphere (HS) potential can be considered as a special case of the square-well potential when the depth of the well approaches zero. Thus, there are two ways that $\beta\varepsilon$ can approach zero: the temperature can approach infinity, or the well depth, ε, can approach zero. Both results lead to the hard-sphere repulsive term.

The results of hard-sphere simulations by Erpenbeck and Wood[16] are presented in Fig. 7.9(b). Three equations of state are compared to the simulation results: the van der Waals model, the Carnahan-Starling model, and the ESD model. These models are listed below, along with another called the Scott model.

$$Z^{HS} = 1/(1 - \eta_P); \text{ the van der Waals model} \qquad 7.45$$

$$Z^{HS} = (1 + 2\eta_P)/(1 - 2\eta_P); \text{ the Scott model} \qquad 7.46$$

$$Z^{HS} = 1 + 4\eta_P/(1 - 1.9\eta_P); \text{ the ESD model} \qquad 7.47$$

$$Z^{HS} = 1 + 4\eta_P(1 - \eta_P/2)/(1 - \eta_P)^3; \text{ the Carnahan-Starling model} \qquad 7.48$$

It is immediately apparent that the van der Waals model is quite inaccurate while the Carnahan-Starling model is practically quantitative. The ESD model is imprecise when the packing fraction exceeds $\eta_P > 0.40$, but it does preserve the prospect of forming the basis for a cubic equation of state. The Scott equation is not shown, but it is slightly less precise than the ESD model and slightly

16. Erpenbeck, J.J., Wood, W.W. 1984. *J. Stat. Phys.* 35:321.

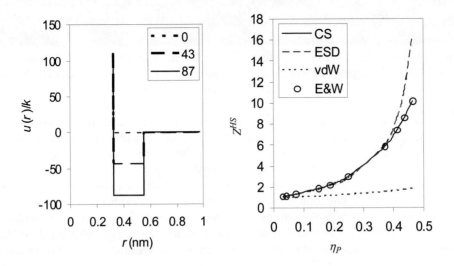

Figure 7.9 *(a) The hard-sphere potential as a special case of the square-well model; (b) results of DMD simulations for the hard-sphere potential compared to simulation data of Erpenbeck and Wood cited in the text.*

simpler. The precision of the Carnahan-Starling model makes it a popular choice for many of the equations of state discussed in the text. Nevertheless, it is feasible to mix and match various characterizations of the repulsive and attractive contributions to construct an equation of state that is applicable to any particular situation. Constructing your own equation of state is the best way to appreciate the advantages and disadvantages underlying models like the Peng-Robinson equation. We should probably warn you that it is hard to stop once you start down this path of "observe, predict, test, and evaluate." It is a very good sign, however, if you feel yourself being drawn that way.

Molecular dynamics simulation was first accomplished in 1959.[17] Until that time, it was impossible to resolve arguments about whose characterization of the hard-sphere reference system was best. In the final analysis, only a molecular simulation can resolve this debate conclusively. Today, several such programs can be accessed online and some are open source. In particular, the discontinuous molecular dynamics (DMD) module at Etomica.org has been designed to simplify visual and interactive exploration of the relations between temperature, pressure, density, internal energy, and the choice of potential model.

Online exercises may help your understanding of molecular simulations.

Example 7.11 Equations of state from trends in molecular simulations

Use the 3D DMD module at Etomica.org to characterize the trends of the attractive contributions for argon with $\lambda = 1.7$ at densities of 1.25 and 1.38 g/cm^3 assuming a diameter of 0.323 nm and $\varepsilon/k = 87$ K. Use the results to obtain a cubic equation of state.

---

17. Alder, B.J., Wainwright, T.E. 1959. *J. Chem. Phys.* 31:459.

> ### Example 7.11 Equations of state from trends in molecular simulations (Continued)
>
> **Solution:** It is straightforward to set a diameter of 0.323nm, $N_A \varepsilon = 87 \cdot 8.314 = 723$ J/mol, MW = 40, and $\lambda = 1.7$. For purposes of this problem, we assume the ESD form suffices over the density range of interest with the objective of obtaining a cubic equation.
>
> The next step is to simulate the full potential and solve for the attractive contribution by subtraction. Fig. 7.7 suggests that a linear function in $\beta\varepsilon$ should suffice, and we know that the attractive contribution increases with density. These observations suggest an equation of state of the form
> $$Z = 1 + 4\,\eta_P/(1 - 1.9\,\eta_P) - z_{11}\,\eta_P\,\beta\varepsilon,$$
> where z_{11} designates a constant corresponding to first order in both η_P and $\beta\varepsilon$. By regressing the slope of the attractive contribution at the two given densities, we can characterize z_{11} as a function of density. We can also infer the zero density limit of z_{11} from the second virial coefficient as $z_{11}(0) = 4(\lambda^3 - 1) = 15.7$. The results of these characterizations give $z_{11} = 16.3$ at $\eta_P = 0.333$ and 17.0 at $\eta_P = 0.367$. In order to obtain a cubic equation, we must restrict our attention to equations of the form,
> $$z_{11} = z_{11}(0)/(1 - z_{12}\,\eta_P).$$
> Plotting $z_{11}(0)/z_{11}$ and fitting a trendline gives $z_{12} = 0.16$ and the final model is,
> $$Z = 1 + 4\,\eta_P/(1 - 1.9\,\eta_P) - 15.7\,\eta_P\,\beta\varepsilon/(1 - 0.16\,\eta_P)$$
>
> This fit of the attractive trend is crude, but it would be difficult to improve given the constraints imposed by the cubic form. This leaves the door open to future improvements beyond the cubic form. The approach would be similar, however.

7.11 THE MOLECULAR BASIS OF EQUATIONS OF STATE: ANALYTICAL THEORIES

Molecular simulation provides a numerical connection between the intermolecular potential model and the macroscopic properties, but it does so one state point at a time. For an equation of state, we need an equation that makes this connection over all state points. The key to making this kind of connection is to consider the average number of neighbors for each molecule within range of the potential model. We alluded to this in Example 1.1(e), and simply called it "four," but this number must vary with density and strength of attraction and with the precise distance between molecules. Therefore, we must define a quantity representing the average number of molecules at each distance from the center of an average molecule, and study its dependence on density and temperature. To get the configurational internal energy,[18] multiply this average number of molecules by the amount of potential energy at that distance and integrate over all distances. To get the pressure, multiply this average number of molecules by the amount of force per unit area at that distance and integrate over all distances. The average number of molecules at a particular distance from an average molecule is characterized by the "radial distribution function," which is discussed in detail below. If you have ever seen a parking lot, you already know more about radial distribution functions than you may realize.

---

18. The configurational energy is that energy due solely to the intermolecular interactions at given distances, hence the adjective *configurational*.

The Energy Equation

The ideal gas continues to be an important concept, because it is a convenient reference fluid. To calculate the internal energy of a real gas, we simply need to compute the departure from the ideal gas. In this way, the kinetic energy of the gas is included in the ideal gas internal energy, and we calculate the contribution to internal energy due to the intermolecular potentials of the real gas,

$$U - U^{ig} = \frac{N_A \rho}{2} \int_0^\infty N_A u \ g(r) \ 4\pi r^2 dr \qquad\qquad 7.49$$

where u is the pair potential and $g(r) \equiv$ the radial distribution function defined by Eqn. 7.55. This is often called the **configurational energy** to denote that it relates to summed potential energy of the configuration. This equation can be written in dimensionless form as

$$\frac{U - U^{ig}}{RT} = \frac{N_A \rho}{2} \int_0^\infty \frac{N_A u}{RT} g(r) 4\pi r^2 dr \qquad\qquad 7.50$$

The Pressure Equation

We also may choose to solve for the pressure of our real fluid. Once again it is convenient to use the ideal gas as our reference fluid and calculate the pressure of the real fluid relative to the ideal gas law. Since intermolecular force is the derivative of the intermolecular potential, we note the derivative of the intermolecular potential in the following equation.

$$P = \rho RT - \frac{\rho^2 N_A^2}{6} \int_0^\infty r\left(\frac{du}{dr}\right) g(r) \ 4\pi r^2 \ dr \qquad\qquad 7.51$$

This equation is typically derived by determining the product PV, but we have multiplied by density to show the pressure.[19] This equation can also be written in dimensionless form, recalling the definition of the compressibility factor:

$$\frac{P}{\rho RT} = 1 - \frac{\rho N_A}{6} \int_0^\infty \frac{N_A r}{RT}\left(\frac{du}{dr}\right) g(r) \ 4\pi r^2 dr \qquad\qquad 7.52$$

Note in both the energy equation and the pressure equation, that our integral extends from 0 to infinity. Naturally, we never have a container of infinite size. How can we represent a real fluid this way? Look again at the form of the intermolecular potentials in Chapter 1. At long molecular distances, the pair potential and the derivative of the pair potential both go to zero. Long distances on the molecular scale are 4 to 5 molecular diameters (on the order of nanometers), and the integrand is practically zero outside this distance. Therefore, we may replace the infinity with dimensions of our container, and obtain the same numerical result in most situations. This substitution makes a single equation valid for all containers of any size greater than a few molecular diameters.[20]

---

19. Hansen, J-P., McDonald, I.R. 2006. *Theory of Simple Liquids,* 3rd ed. New York:Academic Press, p. 32.
20. The exception to this discussion occurs very near the critical point, but addressing this problem is beyond the scope of this text.

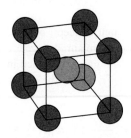

Figure 7.10 *The body centered cubic unit cell.*

An Introduction to the Radial Distribution Function

As a prelude to a general description of atomic distributions, it may be helpful to review the structure of crystal lattices like those in body-centered cubic (bcc) metals, as shown in Fig. 7.10. Such a lattice possesses long-range order due to repetitive arrangements of the unit cell in three dimensions. This close-packed arrangement of atoms gives a single value for the density, and the density correlates with many of the macroscopic properties of the material (e.g., strength, ductility). These are some of the key considerations fundamental to materials science, and more details are given in common texts on the subject. One goal of introducing the radial distribution function is to generalize the concept of atomic arrangements so that non-lattice fluids can be included.

The distribution of atoms in a bcc crystal is fairly easy to understand, but how can we address the distribution of atoms in a fluid? For a fluid, the positions of the atoms around a central atom are less well defined than in a crystal. To get started, think about the simplest fluid, an ideal gas.

The Fluid Structure of an Ideal Gas

Consider a fluid of point particles surrounding a central particle. What is the number of particles in a given volume element surrounding the central particle? Since they are point particles, they do not influence one another. This means that the number of particles is simply related to the density,

$$dN_V = N\frac{dV}{V} \qquad \text{(ig) 7.53}$$

where dN_V is the number of particles in the volume element, N is the total number of particles in the total volume, $\underline{V}$ is the total volume, $d\underline{V}$ is the size of the volume element, and $dN_V = N_A \rho \, d\underline{V}$

If we would like to know the number of particles within some spherical neighborhood of our central particle, then,

$$d\underline{V} = 4\pi r^2 \, dr$$

where r is the radial distance from our central particle,

$$N_c = \int_0^{N_C} dN_V = N_A \int_0^{R_o} \rho 4\pi r^2 dr \qquad \text{(ig) 7.54}$$

where R_o defines the range of our spherical neighborhood, N_c is the number of particles in the neighborhood (coordination number).

The Fluid Structure of a Low-Density Hard-Sphere Fluid

Now consider the case of atoms which have a finite size. In this case, the number of particles within a given neighborhood is strongly influenced by the range of the neighborhood. If the range of the neighborhood is less than two atomic radii, or one atomic diameter, then the number of particles in the neighborhood is zero (not counting the central particle). Outside the range of one atomic diameter, the exact variation in the number of particles is difficult to anticipate *a priori*. You can anticipate it, however, if you think about the way cars pack themselves into a parking lot. We can express these insights mathematically by defining a "weighting factor" which is a function of the radial distance. The weighting factor takes on a value of zero for ranges less than two atomic radii, and for larger ranges, we can consider its behavior undetermined as yet.

The hard-sphere fluid has been studied extensively to represent spherical fluids.

Then we may write

$$N_c = N_A \rho \int_0^{R_o} g(r) 4\pi r^2 dr \qquad\qquad 7.55$$

where $g(r)$ is our average "weighting function," called the radial distribution function. The radial distribution function is the number of atomic centers located in a spherical shell from r to $r + dr$ from one another, divided by the volume of the shell and the bulk number density.

This is a lot like algebra. It helps us to organize what we do know and what we do not know. The next task is to develop some insights about the behavior of this weighting factor so that we can make some engineering approximations.

As a first approximation, we might assume that atoms outside the range of two atomic radii do not influence one another. Then the number of particles in a given volume element goes back to being proportional to the size of the volume element, and the radial distribution function has a value of one for all r greater than one diameter. The approximation that atoms outside the atomic diameter do not influence one another is reasonable at low density. An analogy can be drawn between the problem of molecular distributions and the problem of parking cars. When the parking lot is empty, cars can be parked randomly at any position, as long as they are not parked on top of one another. Recalling the relation between a random distribution and a flat radial distribution function, Fig. 7.11 should seem fairly obvious at this point.

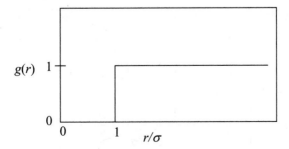

Figure 7.11 *The radial distribution function for the low-density hard-sphere fluid.*

The Structure of a bcc Lattice

Far from the low-density limit, the system is close-packed. The ultimate in close-packing is a crystal lattice. Let's clarify what is meant by the radial distribution function of a lattice. The radial distribution function of a bcc lattice can be deduced from knowledge of N_c and the defining relation for $g(r)$.

$$N_c = \rho N_A \int_0^{R_o} g(r) 4\pi r^2 dr \qquad 7.56$$

If we assume that the atoms in a crystal are located in specific sites, and no atoms are out of their sites, then $g(r)$ must be zero everywhere except at a site. For a body-centered cubic crystal these sites are at $r = \{\sigma, 1.15\sigma, 1.6\sigma,...\}$ $g(r)$ looks like a series of spikes. In the parking lot analogy, the best way of parking the most cars is to assign specific regular spaces with regular space between, as shown in Fig. 7.12.

The Fluid Structure of High-Density Hard-Sphere Fluid

The distributions of atoms in a fluid are most conveniently referred to as the fluid's **structure.** The structures of these simple cases clarify what is meant by structure in the context that we will be using, but the behavior of a dense liquid illustrates why this concept of structure is necessary. Dense-liquid behavior is something of a hybrid between the low-density fluid and the solid lattice. At large distances, atoms are too far away to influence one another and the radial distribution function approaches unity because the increase in neighbors becomes proportional to the size of the neighborhood. Near the atomic diameter, however, the central atom influences its neighbors to surround it in "layers" in an effort to approach the close packing of a lattice. Thus, the value of the radial distribution function is large, very close to one atomic diameter. Because liquids lack the long-range order of crystals, the influence of the central atom on its neighbors is not as well defined as in a crystal, and we get smeared peaks and valleys instead of spikes. Returning to the parking lot analogy once again, the picture of liquid structure is considerably more realistic than the assumption of a regular lattice structure. There are no "lines" marking the proper "parking spaces" in a real fluid. If a few individuals park out of line, the regularity of the lattice structure is disrupted, and it becomes impossible to say what the precise structure is at 10 or 20 molecular diameters. It is true, however, that the average parking around any particular object will be fairly regular for a somewhat shorter range, and the fluid structure in Fig. 7.13 reflects this by showing sharp peaks and valleys at short range and an approach to a random distribution at long range.

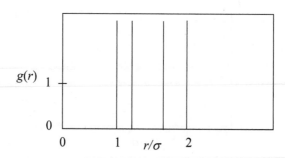

Figure 7.12 *The radial distribution function for the bcc hard-sphere fluid.*

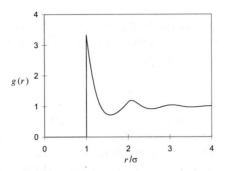

Figure 7.13 *The radial distribution function for the hard-sphere fluid at a packing fraction of $b\rho = 0.4$.*

The Structure of Fluids in the Presence of Attractions and Repulsions

As a final case, consider the influence of a square-well potential (presented in Section 1.2) on its neighbors. The range $r < \sigma$ is off-limits, and the value of the radial distribution function there is still zero. But what about the radial distribution function at low density for the range where the attractive potential is influential? We would expect some favoritism for atoms inside the attractive range, $\sigma < r < \lambda\sigma$, since that would release energy. How much favoritism? It turns out to be simply related to the energy inherent in the potential function.

$$\lim_{\rho \to 0} g(r) = \exp\left[-\frac{u(r)}{kT}\right] \qquad 7.57$$

The low-density limit of the radial distribution function is related to the pair potential.

This exponential function, known as a Boltzmann distribution, accounts for the off-limits range and the attractive range as well as the no-influence ($r > \lambda\sigma$) range. Referring to the parking lot analogy again, imagine the distribution around a coffee and doughnut vending truck early in the morning when the parking lot is nearly empty. Many drivers would be attracted by such a prospect and would naturally park nearby, if the density was low enough to permit it.

As for the radial distribution function at high density, we expect packing effects to dominate and attractive effects to subordinate because attaining a high density is primarily affected by efficient packing. At intermediate densities, the radial distribution function will be some hybrid of the high and low density limits, as shown in Fig. 7.14.

A mathematical formalization of these intuitive concepts is presented in several texts, but the difficulty of such a rigorous treatment is beyond the scope of our introductory presentation. For our purposes, we would simply like to understand that the number of particles around a central particle has some character to it that depends on the temperature and density, and that representing this temperature and density dependence in some way will be necessary in analyzing the energetics of how molecules interact. In other words, we would like to appreciate that something called "fluid structure" exists, and that it is described in detail by the "radial distribution function." This appreciation will be of use again when we extend these considerations to the energetics of mixing. Then we will develop expressions that can be used to predict partitioning of components between various phases (e.g., vapor-liquid equilibria).

The Virial Equation

The second virial coefficient can be easily derived using the concepts presented in this section, together with a little more mathematics. Advanced chemistry and physics texts customarily derive the virial equation as an expansion in density:

$$Z = 1 + B\rho + C\rho^2 + D\rho^3 + \ldots \tag{7.58}$$

The result of the advanced derivation is that each virial coefficient can be expressed exactly as an integral over the intermolecular interactions characterized by the potential function. Even at the introductory level we can illustrate this approach for the second virial coefficient. Comparing the virial equation at low density, $Z = 1 + B\rho$, with Eqn. 7.52, we can see that the second virial coefficient is related to the radial distribution function at low density. Inserting the low-density form of the radial distribution function as given by Eqn. 7.57, and subsequently integrating by parts (the topic of homework problem 7.29), we find

$$B = 2\pi N_A \int_0^\infty \left(1 - \exp\left(-\frac{u}{kT}\right)\right) r^2 dr \tag{7.59}$$

This relationship is particularly valuable, because experimental virial coefficient data may be used to obtain parameter values for pair potentials.

Example 7.12 Deriving your own equation of state

Appendix B shows how the following equation can be derived to relate the macroscopic equation of state to the microscopic properties in terms of the square-well potential for $\lambda = 1.5$.

$$Z = 1 + \frac{4\pi N_A \rho \sigma^3}{6}\{g(\sigma^+) - 1.5^3[1 - \exp(-\varepsilon/kT)g(1.5\sigma^-)]\} \tag{7.60}$$

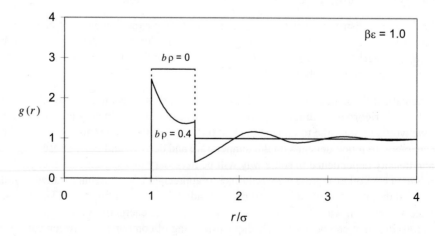

Figure 7.14 *The square-well fluid (R = 1.5) at zero density and at a packing fraction of bρ = 0.4.*
The variable $\beta \equiv 1/kT$.

> ### Example 7.12 Deriving your own equation of state (Continued)
>
> Apply this result to develop your own equation of state with a radial distribution function of the form:
>
> $$g(x) = \frac{\exp(-u/kT)}{(1 - 2b\rho/x^6)(1 + 2Sb\rho/x^6)} \qquad 7.61$$
>
> where $x = r/\sigma$, $b = \pi N_A \sigma^3/6$, and S is the "Student" parameter. You pick a number for S, and this will be your equation of state. Evaluate your equation of state at $\varepsilon/kT = 0.5$ and $b\rho = 0.4$.
>
> **Solution:** At first glance, this problem may look outrageously complicated, but it is actually quite simple. We only need to evaluate the radial distribution function at $x = 1$ and $x = 1.5$ and insert these two results into Eqn. 7.60.
>
> $$g(\sigma^+) = \frac{\exp(\varepsilon/kT)}{(1 - 2b\rho)\{1 + 2Sb\rho\}} \qquad 7.62$$
>
> $$g(1.5\sigma) = \frac{\exp(\varepsilon/kT)}{(1 - 0.176Sb\rho)\{1 + 0.176Sb\rho\}} \qquad 7.63$$
>
> $$Z = 1 + \frac{4b\rho}{1 - 2b\rho}\frac{\exp(\varepsilon/kT)}{1 + 2Sb\rho} - \frac{13.5b\rho[\exp(\varepsilon/kT) - 1]}{(1 - 0.176b\rho)(1 + 0.176Sb\rho)} \qquad 7.64$$
>
> Supposing $S = 3$, $Z(0.5, 0.4) = 1 + 4 \cdot 0.4 \cdot 1.649/(0.2 \cdot 2.4) - 13.5 \cdot 0.4 \cdot 0.648/(0.789 \cdot 1.211) = 2.83$.
>
> Congratulations! You have just developed your own equation of state. Have fun with it and feel free to experiment with different approximations for the radial distribution function. Hansen and McDonald[a] describe several systematic approaches to developing such approximations if you would like to know more.

a. Hansen, J.P., McDonald, I.R. 1986. *Theory of Simple Liquids*. New York:Academic Press.

We conclude these theoretical developments with the comment that a similar analysis appears in the treatment of mixtures. At that time, it should become apparent that the extension to mixtures is primarily one of accounting; the conceptual framework is identical. The sooner you master the concepts of separate contributions from repulsive forces and attractive forces, the sooner you will master your understanding of fluid behavior from the molecular scale to the macro scale.

7.12 SUMMARY

The simple physical observations and succinct mathematical models set forth in this chapter provide powerful tools for current chemical applications and excellent examples of model development that we would all do well to emulate. This chapter has illustrated applications of physical reasoning, dimensional analysis, asymptotic analysis, and parameter estimation that have set the standard for many modern engineering developments.

Furthermore, the final connection has been drawn between the molecular level and the macroscopic scale. In retrospect, the microscopic definition of entropy is relatively simple. It follows naturally from the elementary statistics of the binomial distribution. The qualitative description of molecular interaction energy is also simple; it was discussed in the introductory chapter. Last, but not least, the macroscopic description of energy is easy to understand; it gives the macroscopic energy balance. What is not so simple is the connection of the qualitative description of molecular energies with the macroscopic energy balance. This is the significant development of this chapter. Having complete descriptions of the molecular and macroscopic energy and entropy, all the "pieces to the puzzle" are now in our hands. What remains is to put the pieces together. This final step requires a fair amount of mathematics, but it is largely a straightforward application of tools that are readily available from elementary courses in calculus and the background of Chapter 6.

Important Equations

Several equations stand out in this chapter because we apply them repeatedly going forward. These are Eqns. 7.2, 7.12, and 7.15–7.19. Eqn. 7.2 is the definition of acentric factor (ω), which provides a convenient standard vapor pressure (*cf.* back flap) and a crude characterization of the molecular shape. Eqn. 7.12 is the van der Waals equation of state, one of the greatest model equations of all time. Eqns. 7.15–7.19 describe the Peng-Robinson equation of state, a remarkably small evolution from the van der Waals model considering the 100 years of intervening research. We refer to these often as the basis for further derivations and applications to energy and entropy balances for chemicals other than steam and refrigerants.

Finally, Eqns. 7.51 and 7.52 convey the foundation for understanding the connection between the molecular scale and the macroscopic scale. There is no simpler way to see the connection between u and U than Eqn. 7.51. Understanding the molecular interactions becomes essential when we consider why some mixtures behave ideally while others do not. This becomes apparent when we extend Eqn. 7.51 to mixtures.

7.13   PRACTICE PROBLEMS

P7.1   For $T_r < 1$ and $P_r \approx P_r^{sat}$, the Peng-Robinson equation of state has three roots corresponding to compressibility factors between zero and 10. The smallest root is the compressibility factor of the liquid. The largest root is the compressibility factor of the vapor and the middle root has no physical significance. This gives us a general method for finding the compressibility factor of any fluid obeying the Peng-Robinson equation. For the iterative method, use an initial guess of $Z = 0$ to find the liquid roots and $Z = 1$ to find the vapor roots of methane at the following conditions:

T_r	0.9	0.8	0.7	0.6
P_r	0.55	0.26	0.10	0.03

Compare to experimental data from N.B. Vargaftik. 1975. *Handbook of Physical Properties of Liquids and Gases,* 2nd ed., New York: Hemisphere.

T_r	0.9	0.8	0.7	0.6
Z^L	0.0908	0.0413	0.0166	0.0050
Z^V	0.6746	0.8124	0.9029	0.9608

ANS. The liquid roots are very close. The vapor roots are accurate for $T_r < 0.9$.

Z^L	0.0909	0.0397	0.0154	0.0049
Z^V	0.6374	0.7942	0.8977	0.9573

P7.2   (a) Estimate the value of the compressibility factor, Z, for neon at $P_r = 30$ and $T_r = 15$.
(b) Estimate the density of neon at $P_r = 30$ and $T_r = 15$. (ANS. 1.14, 0.25 g/cm^3)

P7.3 Above the critical point or far from the saturation curve,[21] only one real root to the cubic equation exists. If we are using Newton's method, we can check how many phases exist by trying the two different initial guesses and seeing if they both converge to the same root. If they do, then we can assume that only one real root exists. Find the compressibility factors for methane at the following conditions, and identify whether they are vapor, liquid, or supercritical fluid roots. Complete the table. Compare your results to Z-charts.

T_r	P_r	Z	Phase
0.9	0.77	ANS. online	
0.9	0.05		
1.1	2.00		

When Newton's method is applied with an initial guess of zero, erratic results are obtained at these conditions. Explain what is happening, and why, by plotting $F(Z)$ versus Z for each iteration.

P7.4 A rigid vessel is filled to one-half its volume with liquid methane at its normal boiling point (111 K). The vessel is then closed and allowed to warm to 77°F. Calculate the final pressure using the Peng-Robinson equation. (ANS. 33.8 MPa)

P7.5 4 m^3 of methane at 20°C and 1 bar is roughly equivalent to 1 gal of gasoline in an automotive engine of ordinary design. If methane were compressed to 200 bar and 20°C, what would be the required volume of a vessel to hold the equivalent of 10 gal of gasoline? (ANS. 16 L)

P7.6 A carbon dioxide cylinder has a volume of 0.15 m^3 and is filled to 100 bar at 38°C. The cylinder cools to 0°C. What is the final pressure in the cylinder and how much more CO_2 can be added before the pressure exceeds 100 bar? If you add that much CO_2 to the cylinder at 0 °C, what will the pressure be in the cylinder on a hot, 38°C day? What will happen if the cylinder can stand only 200 bar? [Hint: $\log (P_r^{sat}) \approx (7(1 + \omega)/3) (1 - 1/T_r)$] (ANS. 3.5 MPa, 38 MPa, boom!)

7.14   HOMEWORK PROBLEMS

7.1 The compressibility factor chart provides a quick way to assess when the ideal gas law is valid. For the following fluids, what is the minimum temperature in K where the fluid has a gas phase compressibility factor greater than 0.95 at 30 bar?

(a) Nitrogen
(b) Carbon dioxide
(c) Ethanol

7.2 A container having a volume of 40 L contains one of the following fluids at the given initial conditions. After a leak, the temperature and pressure are remeasured. For each option, determine the kilograms of fluid lost due to the leak, using:

(a) Compressibility factor charts
(b) The Peng-Robinson equation

Options:

---

21. Fig. 7.5 on page 264 shows that three roots will exist at all pressures below P^{sat} when the reduced temp is low, but over a limited range near $T_r = 1$.

(i)   Methane $T^i = 300$ K, $P^i = 100$ bar, $T^f = 300$ K, $P^f = 50$ bar
(ii) Propane $T^i = 300$ K, $P^i = 50$ bar, $T^f = 300$ K, $P^f = 0.9$ bar
(iii) n-butane $T^i = 300$ K, $P^i = 50$ bar, $T^f = 300$ K, $P^f = 10$ bar

7.3 Estimate the liquid density (g/cm³) of propane at 298 K and 10 bar. Compare the price per kilogram of propane to the price per kilogram of regular gasoline assuming the cost of 5 gal of propane for typical gas grills is roughly $20. The density of regular gasoline can be estimated by treating it as pure isooctane (2,2,4-trimethylpentane $\rho = 0.692$ g/cm³) at 298 K and 1 bar.

7.4 From experimental data it is known that at moderate pressures the volumetric equation of state may be written as $PV = RT + B \cdot P$, where the second virial coefficient B is a function of temperature only. Data for methane are given by Dymond and Smith (1969) as,[22]

T(K)	120	140	160	180	200	250	300	350	400	500	600
B(cm³/mole)	−284	−217	−169	−133	−107	−67	−42	−27.0	−15.5	−0.5	8.5

(a) Identify the Boyle temperature (the temperature at which $B = 0$) and the inversion temperature (the temperature at which $(\partial T/\partial P)_H = 0$) for gaseous methane. [Hint: Plot B versus T^{-1} and regress a trendline, then differentiate analytically.]
(b) Plot these data versus T^{-1} and compare to the curve generated from Eqn. 7.7. Use points without lines for the experimental data and lines without points for the theoretical curve.

7.5 Data for hydrogen are given by Dymond and Smith (1969) as,

T(K)	19	25	30	40	50	75	100	150	200	300	400
B(cm³/mole)	−164	−111	−85	−54	−35	−12	−1.9	7.1	11.3	14.8	15.2

(a) Plot these data versus T^{-1} and compare to the results from the generalized virial equation (Eqn. 7.7). Suggest a reason that this specific compound does not fit the generalized equation very accurately. Use points without lines for the experimental data and lines without points for the theoretical curve.
(b) Use the generalized virial equation to speculate whether a small leak in an H_2 line at 300 bar and 298 K might raise the temperature of H_2 high enough to cause it to spontaneously ignite.

7.6 N.B. Vargaftik (1975)[23] lists the following experimental values for the specific volume of isobutane at 175°C. Compute theoretical values and their percent deviations from experiment by the following:

(a) The generalized charts
(b) The Peng-Robinson equation

P (atm)	10	20	35	70
V (cm³/g)	60.7	27.79	13.36	3.818

7.7 Evaluate $(\partial P/\partial V)_T$ for the equation of state where b is a constant:

$$P = RT/(V - b)$$

22. Dymond, J.H., Smith, E.B. 1969. *The Virial Coefficients of Pure Gases and Mixtures,* New York: Oxford University Press.
23. Vargaftik, N.B. 1975. *Handbook of Physical Properties of Liquids and Gases,* 2nd ed. New York: Hemisphere.

7.8 Evaluate $(\partial P/\partial T)_V$ for the equation of state where a and b are constants:

$$P = RT/(V - b) + a/T^{3/2}$$

7.9 Evaluate $\left(\dfrac{\partial P}{\partial T}\right)_V$ for the Redlich-Kwong equation of state

$$P = \frac{RT}{V - b} - \frac{a}{T^{1/2}V(V + b)}\text{, where a and b are temperature-independent parameters.}$$

7.10 (a) The derivative $(\partial V/\partial T)_P$ is tedious to calculate by implicit differentiation of an equation of state such as the Peng-Robinson equation. Show that calculus permits us to find the derivative in terms of derivatives of pressure, which are easy to find, and provide the formula for this equation of state.

(b) Using the Peng-Robinson equation, calculate the isothermal compressibility of ethylene for saturated vapor and liquid at the following conditions: $\{T_r = 0.7, P = 0.414$ MPa$\}$; $\{T_r = 0.8, P = 1.16$ MPa$\}$; $\{T_r = 0.9, P = 2.60$ MPa$\}$.

7.11 When cubic equations of state give three real roots for Z, usually the smallest root is the liquid root and the largest is the vapor root. However, the Peng-Robinson equation can give real roots at high pressure that differ from this pattern. To study this behavior, tabulate all the roots found for the specified gas and pressures. As the highest pressures are approached at this temperature, is the fluid a liquid or gas? Which real root (smallest, middle, or largest) represents this phase at the highest pressure, and what are the Z values at the specified pressures?

(a) Ethylene at 250 K and 1, 3, 10, 100, 150, 170, 175, and 200 MPa
(b) *n*-Hexane at 400 K and 0.2, 0.5, 1, 10, 100, 130 and 150 MPa
(c) Argon at 420 K and 0.1, 1, 5, and 10 MPa

7.12 Plot P_r versus ρ_r for the Peng-Robinson equation with $T_r = [0.7, 0.9, 1.0]$, showing both vapor and liquid roots in the two-phase region. Assume $\omega = 0.040$ as for N_2. Include the entire curve for each isotherm, as illustrated in Fig. 7.1 on page 254. Also show the horizontal line that connects the vapor and liquid densities at the saturation pressure. Use lines without points for the theoretical curves. Estimate T_r^{sat} by $\log(P_r^{sat}) = 2.333(1 + \omega)$ $(1 - 1/T_r^{sat})$.

7.13 Within the two-phase envelope, one can draw another envelope representing the limits of supercooling of the vapor and superheating of liquid that can be observed in the laboratory; along each isotherm these are the points for which $(\partial P/\partial \rho)_T = 0$. Obtain this envelope for the Peng-Robinson equation, and plot it on the same figure as generated in problem 7.12. This is the spinodal curve. The region between the saturation curve and the curve just obtained is called the metastable region of the fluid. Inside the spinodal curve, the fluid is unconditionally unstable. The saturation curve is called the binodal curve. Outside, the fluid is entirely stable. It is possible to enter the metastable region with hot water by heating at atmospheric pressure in a very clean flask. Sooner or later, the superheated liquid becomes unstable, however. Describe what would happen to your flask of hot water under these conditions and a simple precaution that you might take to avoid these consequences.

7.14 Develop a spreadsheet that computes the values of the compressibility factor as a function of reduced pressure for several isotherms of reduced temperature using the Lee-Kesler (1975) equation of state (*AIChE J.*, 21:510). A tedious but straightforward way to do this is to tabulate reduced densities from 0.01 to 10 in the top row and reduced temperatures in the first column. Then, enter the Lee-Kesler equation for the compressibility factor of the simple fluid in one of the central cells and copy the contents of that cell to all other cells in the table. Next, copy that entire table to a location several rows lower. Replace the contents of the new cells by the relation $P_r = Z \cdot \rho_r \cdot T_r$. You now have a set of reduced pressures corresponding to a set of compressibility factors for each isotherm, and these can be plotted to reproduce the chart in the chapter, if you like. Copy this spreadsheet to a new one, and change the values of the B, C, D, and E parameters to correspond to the reference fluid. Finally, copy the simple fluid worksheet to a new worksheet, and replace the contents of the compressibility factor cells by the formula: $Z = Z_0 + \omega(Z_{ref} - Z_0)/\omega_{ref}$ where the Z_{ref} and Z_0 refer to numbers in the cells of the other worksheets.

7.15 The Soave-Redlich-Kwong equation[24] is given by:

$$P = \frac{RT\rho}{(1 - b\rho)} - \frac{a\rho^2}{1 + b\rho} \quad \text{or} \quad Z = \frac{1}{(1 - b\rho)} - \frac{a}{bRT} \cdot \frac{b\rho}{1 + b\rho} \qquad 7.65$$

where ρ = molar density = $n/\underline{V}$

$$a \equiv a_c \alpha; \quad a_c \equiv 0.42748 \frac{R^2 T_c^2}{P_c} \qquad\qquad b \equiv 0.08664 \frac{RT_c}{P_c} \qquad 7.66$$

$$\alpha \equiv [1 + \kappa(1 - \sqrt{T_r})]^2 \qquad \kappa \equiv 0.480 + 1.574\omega - 0.176\omega^2 \qquad 7.67$$

T_c, P_c, and ω are reducing constants according to the principle of corresponding states. Solve for the parameters at the critical point for this equation of state (a_c, b_c, and Z_c) and list the next five significant figures in the sequence 0.08664.......

7.16 Show that $B_c = bP_c/RT_c = 0.07780$ for the Peng-Robinson equation by setting up the cubic equation for B_c analogous to the van der Waals equation and solving analytically as described in Appendix B.

7.17 Determine the values of ε/kT_c, Z_c, and b_c in terms of T_c and P_c for the equation of state given by

$$Z = \frac{1 + 2b\rho}{1 - 2b\rho} - Fb\rho$$

where $F = \exp(\varepsilon/kT) - 1$. The first term on the right-hand side is known as the Scott equation for the hard-sphere compressibility factor.

7.18 Consider the equation of state

$$Z = \frac{1 - \eta_P + \eta_P^2 - \eta_P^3}{(1 - \eta_P)^3} - \frac{a\eta_P}{bRT}$$

where $\eta_P = b/V$. The first term on the right-hand side is known as the Carnahan-Starling equation for the hard-sphere compressibility factor.

24. Soave, G. 1972. *Chem. Eng. Sci.* 27:1197.

(a) Determine the relationships between a, b, and T_c, P_c, Z_c.

(b) What practical restrictions are there on the values of Z_c that can be modeled with this equation?

7.19 The ESD equation of state[25] is given by

$$\frac{PV}{RT} = 1 + \frac{4\langle c\eta_P \rangle}{1 - 1.9\eta_P} - \frac{9.5\langle qY\eta_P \rangle}{1 + 1.7745\langle Y\eta_P \rangle}$$

$\eta_P = b\rho$, c is a "shape parameter" which represents the effect of non-sphericity on the repulsive term, and $q = 1 + 1.90476(c - 1)$. A value of $c = 1$ corresponds to a spherical molecule. Y is a temperature-dependent function whose role is similar to the temperature dependence of the a parameter in the Peng-Robinson equation. Use the methods of Example 7.7 to fit b and Y to the critical point for ethylene using $c = 1.3$.

7.20 A molecular simulation sounds like an advanced subject, but it is really quite simple for hard spheres.[26] Furthermore, modern software is readily available to facilitate performing simulations, after an understanding of the basis for the simulations has been demonstrated. This problem provides an opportunity to demonstrate that understanding. Suppose that four hard disks are bouncing in two dimensions around a square box. Let the diameters of the disks, σ, be 0.4 nm, masses be 40 g/mole, and length of the square box, L, be 5 nm. Start the four disks at $(0.25L, 0.25L)$, $(0.75L, 0.25L)$, $(0.25L, 0.75L)$, $(0.75L, 0.75L)$ and with initial velocities of $(v, v/(1 + 2^{\frac{1}{2}}))$, $(-v, v)$, $(v/2^{\frac{1}{2}}, -v/2^{\frac{1}{2}})$, $(-v/2^{\frac{1}{2}}, -v/2^{\frac{1}{2}})$, where v designates an arbitrary velocity. (Hint: you may find useful information in the DMD module at Etomica.org.)

(a) Compute v initially assuming a temperature of 298 K.

(b) Sum the velocities of all four particles (x and y separately). Explain the significance of these sums.

(c) Sketch the disks using arrows to show their directions. Make the sizes of the arrows proportional to the magnitudes of their velocities.

(d) Solve for the time of the first collision. Is it with a wall or between particles? Compute the velocities of all disks after the first collision.

7.21 Suppose you had a program to simulate the motions of four molecules moving in 2D slowly enough that you could clearly see the velocities of all disks. (Hint: The Piston-Cylinder applet in the DMD module at Etomica.org is an example of such a program when kept in "adiabatic" mode.)

(a) Let the disk interactions be characterized by the ideal gas potential. Describe how the disks would move about. Note that the slow particles would always stay slow, and the fast particles stay fast. Why is that?

(b) Change the potential to "repulsion only" as modeled by a hard disk model. Compare the motions of the "repulsion only" particles to the ideal gas particles. Explain the differences. Which seems more realistic?

(c) Set the potential to "repulsion and attraction," as modeled by the square-well model with $\lambda = 2.0$. Compare the motions of these disks to the "repulsion only" particles and ideal gas particles. Explain the differences.

25. Elliott, J.R., Suresh, S.J., Donohue, M.D. 1990. *Ind. Eng. Chem. Res.* 29:1476.
26. Alder, B.J., Wainwright, T.E. 1959. *J. Chem. Phys.* 31:459.

7.22 Suppose you had a program to simulate the motions of N molecules moving in 2D. (Hint: The 2D applet in the DMD module at Etomica.org is an example of such a program when kept in "adiabatic" mode.)

(a) Simulate the motions of the disks using each potential model (ideal gas, hard disk, square well) for 1000 ps (1 picosecond=10^{-12} second) at a density of 2.86E-6mol/m2 with an initial temperature of 300K. Which would have the higher pressure, ideal gas or hard disks? Explain. Which would have the higher pressure, ideal gas or square well disks? Explain.

(b) Simulate the motions of the disks using each potential model for 1000 ps each at a density of 2.86E-6mol/m2 with an initial temperature of 300 K. Sketch the temperature versus time in each case. Explain your observations.

(f) Suppose you simulated the motions of the disks using each potential model for 1000 ps each at a density of 2.86E-6mol/m2 with an initial temperature of 300 K. How would the internal energy compare in each case? Explain.

7.23 Sphere and disk collisions can be expressed more compactly and computed more conveniently in vector notation. Primarily, this involves converting the procedures of Example 7.10 to use the dot product of the relative position and relative velocity. (Hint: You may find useful information in the DMD module at Etomica.org.)

(a) Write a vector formula for computing the center to center distance between two disks given their velocities, u, and their positions, r_0, at a given time, t_0.

(b) Write a vector formula for computing the distance of each disk from each wall. (Hint: Use unit vectors $x=(1, 0)$ and $y=(0, 1)$ to isolate vector components.)

(c) Noting that energy and momentum must be conserved during a collision, write a vector formula for the changes in velocity of two disks after collision.
Hints: (1) $ab=abcos\theta$. (2) A unit vector with direction of a is: a/a.

(d) Write a vector formula for the change in velocity of a disk colliding with a wall.

7.24 Molecular simulation can be used to explore the accuracy and significance of individual contributions to an equation of state. Here we explore how the σ parameter relates to experimental data.

(a) Erpenbeck and Wood have reported precise simulation results for hard spheres as listed below. Plot these data and compare the ESD and Carnahan-Starling (CS) equations for hard spheres.

η_P	0.0296	0.0411	0.0740	0.1481	0.1851	0.2468	0.3702	0.4114	0.4356	0.4628
Z^{HS}	1.128	1.183	1.359	1.888	2.244	3.031	5.850	7.430	8.600	10.194

(b) According to the CS equation, what value do you obtain for Z^{HS} at η_P=0.392?

(c) What value of b corresponds to $\eta_P = 0.392$ for Xenon at 22.14 mol/L? What value of σ corresponds to that value of b?

(d) The simulation results below have been tabulated at $\eta_P = 0.392$. Plot Z versus $\beta\varepsilon$ for these data. Estimate the value of $\beta\varepsilon$ that corresponds to the saturation temperature.

(e) Referring to Xenon on the NIST WebBook, estimate the saturation temperature at 22.14mol/L. Referring to part (d) for the value of $\beta\varepsilon$, estimate the value of ε(J/mol).

(f) Plot Z versus $1000/T$ for the simulation data using your best ε and σ at η_P=0.392. Referring to the "fluid properties" link, plot the isochoric data for Xenon from the NIST WebBook at 22.14mol/L on the same axes.

(g) What values of a and b of the vdW EOS match the simulation data of this plot? Compute Z^{vdw} versus $1000/T$ and show the vdW results as a dashed line on the plot.

(h) Using the values of ε and σ from parts (c) and (e), simulate the system "isothermally" at 225 K and 20.0 mol/L for ~400 ps (got pizza?). Use the CS equation to estimate the y-intercept for Z. Plot these points including a trendline with equation. Plot the NIST data for this isochore on the same axes. This represents a prediction of the data at 20.0 mol/L since the parameters were determined at other conditions.

(i) Using the values of a and b from part (g), plot the vdW results at 20.0 mol/L as a dashed line on the plot. This represents the vdW prediction.

(j) Which model (SW or vdW) matches the experimental trend best? Why?

(k) Neither prediction is perfect. Suggest ways that we may proceed to improve the predictions further.

SW results at $\eta_P = 0.392$, $\lambda = 2.0$.

η_P	$\beta\varepsilon$	t (ps)	t (min)	Z	$-(U-U^{ig})/(N_A\varepsilon)$	$g(\sigma)$	$g(2\sigma)$
0.392	0.493±0.002	2479	540	0.430±0.03	12.41±0.004	3.65	1.5
0.392	0.407±0.002	996	33	1.56 ±0.04	12.27±0.005	3.5	1.45
0.392	0.245±0.002	611	31	3.60 ±0.03	12.02±0.004	3.5	1.4
0.392	0.166±0.002	1411	44	4.63 ±0.08	11.91±0.004	3.3	1.3

7.25 Molecular simulation can be used to explore the accuracy and significance of individual contributions to an equation of state. Use the DMD module at Etomica.org to tune Xe's ε and σ parameters.

(a) According to the Carnahan-Starling (CS) model, what value do you obtain for Z^{HS} at $\eta_P = 0.375$?

(b) What value of σ corresponds to $\eta_P = 0.375$ for Xe at 22.14 mol/L?

(c) The simulation results below have been tabulated at $\eta_P = 0.375$, $\lambda = 1.7$. Plot Z versus $\beta\varepsilon$ for these data. Referring to the NIST WebBook for Xe, estimate the saturation T and Z at 22.14 mol/L. Estimate the value of $\beta\varepsilon$ that corresponds to the saturation Z. Estimate the value of ε(J/mol).

(d) Plot Z versus $1000/T$, using your best ε and σ at $\eta_P = 0.375$ and showing the fluid properties (isochoric) data from WebBook.nist.gov at 22.14 mol/L on the same axes.

(e) What values of a and b of the vdW EOS will match the simulation data of this plot? Show the vdW results as a dashed line on the plot.

(f) Using the values of ε and σ from parts (b) and (c), simulate the system at 225 K and 20.0 mol/L for ~400 ps (got pizza?). Use the CS equation to estimate the y-intercept value for Z and connect the dots on a new plot with a straight line extrapolating through the x-axis. Plot the NIST data for this isochore on the same axes. This represents a prediction of the data at 20.0 mol/L since the parameters were determined at other conditions.

(g) Using the values of a and b from part (e), plot the vdW results at 20.0 mol/L as a dashed line on the plot. This represents the vdW prediction. Comment critically.

(h) Compare to Problem 7.24. Summarize your observations.

SW results at $\eta_P = 0.375$, $\lambda = 1.7$.

η_P	$\beta\varepsilon$	t (ps)	t (min)	Z	$-(U-U^{ig})/(N_A\varepsilon)$	$g(\sigma)$	$g(\lambda\sigma)$
0.375	0.431±0.002	923	40	3.536 ±0.05	7.113±0.004	2.9	1.02
0.375	0.611±0.003	418	17	2.437 ±0.08	7.244±0.004	2.85	1.1
0.375	0.965±0.009	455	22	0.290 ±0.13	7.499±0.004	2.9	1.3

7.26 The discussion in the chapter focuses on the square-well fluid, but the same reasoning is equally applicable for any model potential function. Illustrate your grasp of this reasoning with some sketches analogous to those in the chapter.

(a) Sketch the radial distribution function versus radial distance for a low-density Lennard-Jones (LJ) fluid. Describe in words why it looks like that.
(b) Repeat the exercise for the high-density LJ fluid. Also sketch on the same plot the radial distribution function of a hard-sphere fluid at the same density. Compare and contrast the hard-sphere fluid to the LJ fluid at high density.

7.27 Suppose that a reasonable approximation to the radial distribution function is

$$g(x) = \begin{cases} 0 & r < \sigma \\ \dfrac{1}{1 - 2b\rho/x^6} + \dfrac{2.5F}{(1 + x^6 Fb\rho)} & r \geq \sigma \end{cases}$$

where $x = r/\sigma$, $F = \exp(\varepsilon/(kT)) - 1$ and $b = \pi N_A \sigma^3/6$. Derive an expression for the equation of state of the square-well fluid based on this approximation. Evaluate the equation of state at $\rho N_A \sigma^3 = 0.6$ and $\varepsilon/(kT) = 1$.

7.28 Suppose that a reasonable approximation for the radial distribution function is $g(r) = 0$ for $r < \sigma$, and

$$g(r) = \frac{\exp(-u(r)/(kT))}{1 - b\rho}$$

for $r \geq \sigma$, where u is the square-well potential and $b = \pi N_A \sigma^3/6$. Derive an equation of state for the square-well fluid based on this approximation.

7.29 The truncated virial equation (density form) is $Z = 1 + B\rho$. According to Eqn. 7.52, the virial coefficient is given by

$$B = -\frac{2}{3}\frac{\pi N_A}{kT}\int_0^\infty \left(\frac{du}{dr}\right)g(r)r^3\,dr$$

where the low pressure limit of $g(r)$ given by Eqn. 7.57 is to be used. Another commonly cited equation for the virial coefficient is Eqn. 7.59. Show that the two equations are equivalent by the following steps:

(a) Beginning with $B = -\frac{2}{3}\frac{\pi N_A}{kT}\int\limits_0^\infty \left(\frac{du}{dr}\right)g(r)r^3 dr$, insert the low-pressure limit for $g(r)$,

and simplify as much as possible.

(b) Integrate by parts to obtain

$$\frac{1}{3}\int\limits_0^\infty d(r^3\exp(-u/(kT))) = \int\limits_0^\infty \exp(-u/(kT))r^2 dr - \frac{1}{3kT}\int\limits_0^\infty \left(\frac{du}{dr}\right)g(r)r^3 dr$$

(c) Show that the left-hand side of the answer to part (b) may be written as $\int\limits_0^\infty r^2 dr$ for a physi-

cally realistic pair potential. Then combine integrals to complete the derivation of Eqn. 7.59.

7.30 The virial coefficient can be related to the pair potential by Eqn. 7.59.

(a) Derive the integrated expression for the second virial coefficient in terms of the square-well potential parameters ε/k, σ, and R.

(b) Fit the parameters to the experimental data for argon.[27]

T(K)	B (cm³/mole)	T(K)	B (cm³/mole)
85	−251	400	−1
100	−183.5	500	7
150	−86.2	600	12
200	−47.4	700	15
250	−27.9	800	17.7
300	−15.5	900	20

(c) Fit the parameters to the experimental data for propane.[1]

T(K)	B (cm³/mole)	T(K)	B (cm³/mole)
250	−584	350	−276
260	−526	375	−238
270	−478	400	−208
285	−424	430	−177
300	−382	470	−143
315	−344	500	−124
330	−313	550	−97

27. Dymond, J.H., Smith, E. B. 1980. *The Virial Coefficients of Pure Gases and Mixtures.* New York: Oxford University Press.

7.31 One suggestion for a simple pair potential is the triangular potential

$$u(r) = \infty \ \text{ for } r < \sigma \text{ and } u(r) = 0 \text{ for } r > R\sigma$$

$$u(r) = -\varepsilon \left[\frac{(r/\sigma) - R}{1 - R} \right] \text{ for } \sigma < r < R\sigma$$

Derive the second virial coefficient and fit the parameters σ, ε, and R to the virial coefficient data for argon tabulated in problem 7.30.

CHAPTER

DEPARTURE FUNCTIONS

All the effects of nature are only the mathematical consequences of a small number of immutable laws.

P.-S. LaPlace

Maxwell's relations make it clear that changes in any one variable can be represented as changes in some other pair of variables. In chemical processes, we are often concerned with the changes of enthalpy and entropy as functions of temperature and pressure. As an example, recall the operation of a reversible turbine between some specified inlet conditions of T and P and some specified outlet pressure. Using the techniques of Unit I, we typically determine the outlet T and q which match the upstream entropy, then solve for the change in enthalpy. Applying this approach to steam should seem quite straightforward at this stage. But what if our process fluid is a new refrigerant or a multicomponent natural gas, for which no thermodynamic charts or tables exist? How would we analyze this process? In such cases, we need to have a general approach that is applicable to any fluid. A central component of developing this approach is the ability to express changes in variables of interest in terms of variables which are convenient using derivative manipulations. The other important consideration is the choice of "convenient" variables. Experimentally, P and T are preferred; however, V and T are easier to use with cubic equations of state.

These observations combine with the observation that the approximations in equations of state themselves exhibit a certain degree of "fluidity." In other words, the "best" approximations for one application may not be the best for another application. Responding to this fluidity requires engineers to revisit the approximations *and* quickly reformulate the model equations for U, H, and S. Fortunately, the specific derivative manipulations required are similar regardless of the equation of state since equations of state are either in the $\{T,P\}$ or $\{T,V\}$ form. The formalism of departure functions streamlines the each formulation.

An equation of state describes the effects of pressure on our system properties, including the low pressure limit of the ideal gas law. However, integration of properties over pressure ranges is relatively complicated because most equations of state express changes in thermodynamic variables as functions of density instead of pressure, whereas we manipulate pressure as engineers. Recall that our engineering equations of state are typically of the pressure-explicit form:

Experimentally, P and T are usually specified. However, equations of state are typically density (volume) dependent.

$$P = P(T, V) \qquad\qquad 8.1$$

and general equations of state (e.g., cubic) typically cannot be rearranged to a volume explicit form:

$$V = V(T, P) \qquad\qquad 8.2$$

Therefore, development of thermodynamic properties based on $\{V, T\}$ is consistent with the most widely used equations of state, and deviations from ideal gas behavior will be expressed with the density-dependent formulas for departure functions in Sections 8.1–8.5. In Section 8.6, we present the pressure-dependent form useful for the virial equation. In Section 8.8, we show how reference states are used in tabulating thermodynamic properties.

Chapter Objectives: You Should Be Able to...

1. Choose between using the integrals in Section 8.5 or 8.6 for a given equation of state.

2. Evaluate the integrals of Section 8.5 or 8.6 for simple equations of state.

3. Combine departure functions with ideal gas calculations to determine numerical values of changes in state properties, and use a reference state.

4. Solve process thermodynamics problems using a tool like Preos.xlsx or PreosProps.m rather than a chart or table. This skill requires integration of several concepts covered by other topical objectives including selection of the correct root, and reading/interpreting the output file.

8.1   THE DEPARTURE FUNCTION PATHWAY

Suppose we desire to calculate the change in U in a process which changes state from (V_L, T_L) to (V_H, T_H). Now, it may seem unusual to pose the problem in terms of T and V, since we stated above that our objective was to use T and P. The choice of T and V as variables is because we must work often with equations of state that are functions of volume. The volume corresponding to any pressure is rapidly found by the methods of Chapter 7. We have two obvious pathways for calculating a change in U using $\{V, T\}$ as state variables as shown in Fig. 8.1. Path A consists of an isochoric step followed by an isothermal step. Path B consists of an isothermal step followed by an isochoric step. Naturally, since U is a state function, ΔU for the process is the same by either path. Recalling the relation for $dU(T, V)$, ΔU may be calculated by either.

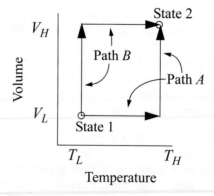

Figure 8.1 *Comparison of two alternate paths for calculation of a change of state.*

Path A:

$$\Delta U = \int C_V\Big|_{V_L} dT + \int \left[T\left(\frac{\partial P}{\partial T}\right)_V - P \right]\Bigg|_{T_H} dV \qquad 8.3$$

or Path B:

$$\Delta U = \int C_V\Big|_{V_H} dT + \int \left[T\left(\frac{\partial P}{\partial T}\right)_V - P \right]\Bigg|_{T_L} dV \qquad 8.4$$

We have previously shown, in Example 7.6 on page 269, that C_V depends on volume for a real fluid. Therefore, even though we could insert the equation of state for the integrand of the second integral, we must also estimate C_V by the equation of state for at least one of the volumes, using the results of Example 7.6. Not only is this tedious, but estimates of C_V by equations of state tend to be less reliable than estimates of other properties.

To avoid this calculation, we devise an equivalent pathway of three stages. First, imagine if we had a magic wand to turn our fluid into an ideal gas. Second, the ideal gas state change calculations would be pretty easy. Third, at the final state we could turn our fluid back into a real fluid. **Departure functions** represent the effect of the magic wand to exchange the real fluid with an ideal gas. Being careful with signs of the terms, we may combine the calculations for the desired result:

$$\Delta U = U_2 - U_1 = (U_2 - U_2^{ig}) + (U_2^{ig} - U_1^{ig}) - (U_1 - U_1^{ig}) \qquad 8.5$$

The calculation can be generalized to any fundamental property from the set $\{U, H, A, G, S\}$, using the variable M to denote the property

$$\Delta M = M_2 - M_1 = (M_2 - M_2^{ig}) + (M_2^{ig} - M_1^{ig}) - (M_1 - M_1^{ig}) \qquad 8.6$$

The steps can be seen graphically in Fig. 8.2. Note the dashed lines in the figure represent the calculations from our "magic wand" effect of turning on/off the nonidealities.

❶ Departure functions permit us to use the ideal gas calculations that are easy, and incorporate a departure property value for the initial and final states.

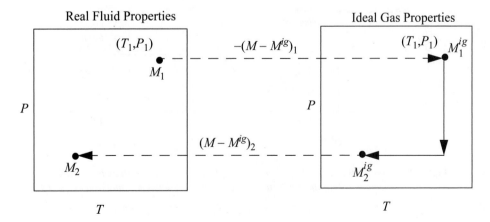

Figure 8.2 *Illustration of calculation of state changes for a generic property M using departure functions where M is U, H, S, G, or A.*

Note how all the ideal gas terms in Eqns. 8.5 and 8.6 cancel to yield the desired property difference. A common mistake is to get the sign wrong on one of the terms in these equations. Make sure that you have the terms in the right order by checking for cancellation of the ideal gas terms. The advantage of this pathway is that all temperature calculations are done in the ideal gas state where:

$$C_P^{ig} = C_V^{ig} + R$$

$$dU^{ig} = C_V^{ig} dT \qquad dH^{ig} = C_P^{ig} dT \qquad dS^{ig} = (C_P^{ig}/T)dT - (R/P)dP$$

8.7

and the ideal gas heat capacities are pressure- (and volume-) independent (see Example 6.9 on page 242).

To derive the formulas to be used in calculating the values of enthalpy, internal energy, and entropy for real fluids, we must apply our fundamental property relations once and our Maxwell's relations once.

8.2   INTERNAL ENERGY DEPARTURE FUNCTION

Fig. 8.3 schematically compares a real gas isotherm and an ideal gas isotherm at identical temperatures. At a given $\{T,P\}$ the volume of the real fluid is V, and the ideal gas volume is $V^{ig} = RT/P$. Similarly, the ideal gas pressure is not equal to the true pressure when we specify $\{T,V\}$. Note that we may characterize the departure from ideal gas behavior in two ways: 1) at the same $\{T,V\}$; or 2) at the same $\{T,P\}$. We will find it convenient to use both concepts, but we need nomenclature to distinguish between the two departure characterizations. When we refer to the departure of the real fluid property and the same ideal gas property at the same $\{T,P\}$, we call it simply the **departure function,** *and use the notation* $U - U^{ig}$. When we compare the departure at the same $\{T,V\}$ we call it the **departure function at fixed** $T,V,$ *and designate it as* $(U - U^{ig})_{TV}.$[1]

> The departure for property M is at fixed T and, P and is given by $(M{-}M^{ig})$. The departure at fixed T,V is also useful (particularly in Chapter 15) and is denoted by $(M{-}M^{ig})_{TV}$.

To calculate the change in internal energy along an isotherm for the real fluid, we write:

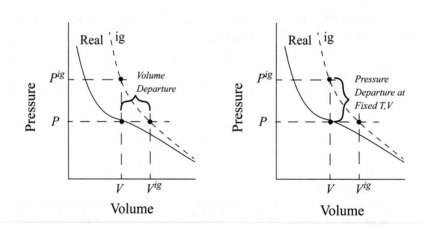

Figure 8.3 *Comparison of real fluid and ideal gas isotherms at the same temperature, demonstrating the departure function, and the departure function at fixed T,V.*

---

1.   Generalization of the ideal gas state is also possible beyond the two choices discussed here. For a discussion, see Reid, R.C., Prausnitz, J.M., Poling, B.E. 1987. *The Properties of Gases and Liquids.* 4th ed. New York: McGraw-Hill.

$$U(T, V) - U(T, \infty) = \int_{\infty}^{V} dU = \int_{\infty}^{V} \left(\frac{\partial U}{\partial V}\right)_T dV \qquad 8.8$$

For an ideal gas:

$$U^{ig}(T, V) - U^{ig}(T, \infty) = \int_{\infty}^{V} dU = \int_{\infty}^{V} \left(\frac{\partial U}{\partial V}\right)_T^{ig} dV \qquad 8.9$$

Since the real fluid approaches the ideal gas at infinite volume, we may take the difference in these two equations to find the departure function at fixed T, V,

$$(U - U^{ig})_{TV} - \cancel{(U - U^{ig})_{T, V = \infty}} = \int_{\infty}^{V} \left[\left(\frac{\partial U}{\partial V}\right)_T - \left(\frac{\partial U}{\partial V}\right)_T^{ig}\right] dV \qquad 8.10$$

where $(U - U^{ig})_{T, V = \infty}$ drops out because the real fluid energy approaches the ideal gas at infinite volume (low pressure). We have obtained a calculation with the real fluid in our desired state (T, P, V); however, we are referencing an ideal gas at the same volume rather than the same pressure. To see the difference, consider methane at 250 K, 10 MPa, and 139 cm^3/mole. The volume of the ideal gas should be $V^{ig} = 8.314 \cdot 250/10 = 208$ cm^3/mole. To obtain the departure function denoted by $(U - U^{ig})$ (which is referenced to an ideal gas at the same pressure), we must add a correction to change the ideal gas state to match the pressure rather than the volume. Note in Fig. 8.3 that the real state is the same for both departure functions—the difference between the two departure functions has to do with the volume used for the ideal gas part of the calculation. The result is

$$U - U^{ig} = (U - U^{ig})_{TP} = (U - U^{ig})_{TV} - (U_{TP}^{ig} - U_{TV}^{ig}) = (U - U^{ig})_{TV} - \int_{V}^{V^{ig}} \left(\frac{\partial U}{\partial V}\right)_T^{ig} dV \qquad 8.11$$

We have already solved for $(\partial U^{ig}/\partial V)_T$ (see Example 6.6 on page 238), and found that it is equal to zero. We are fortunate in this case because the internal energy of an ideal gas does not depend on the volume. When it comes to properties involving entropy, however, the dependency on volume requires careful analysis. Then the systematic treatment developed above is quite valuable.

$$\left(\frac{\partial U}{\partial V}\right)_T = T\left(\frac{\partial P}{\partial T}\right)_V - P \quad \text{and} \quad \left(\frac{\partial U}{\partial V}\right)_T^{ig} = 0 \qquad 8.12$$

Making these substitutions, we have

$$\boxed{U - U^{ig} = \int_{\infty}^{V} \left[T\left(\frac{\partial P}{\partial T}\right)_V - P\right] dV} \qquad 8.13$$

If we transform the integral to density, the resultant expression is easier to integrate for a cubic equation of state. Recognizing $dV = -d\rho/\rho^2$, and as $V \to \infty$, $\rho \to 0$, thus,

$$\frac{U - U^{ig}}{RT} = \int_0^\rho \left[\frac{P}{\rho RT} - \frac{1}{\rho R}\left(\frac{\partial P}{\partial T}\right)_\rho \right] \frac{d\rho}{\rho} = -\int_0^\rho T\left(\frac{\partial Z}{\partial T}\right)_\rho \frac{d\rho}{\rho} \qquad 8.14$$

The above equation applies the chain rule in a way that may not be obvious at first:

$$T\left(\frac{\partial Z}{\partial T}\right)_V = T\left(\frac{\partial(P/\rho RT)}{\partial T}\right)_V = \frac{T}{\rho RT}\left(\frac{\partial P}{\partial T}\right)_V - \frac{PT}{R\rho T^2}\left(\frac{\partial T}{\partial T}\right)_V = \frac{1}{\rho R}\left(\frac{\partial P}{\partial T}\right)_V - Z \qquad 8.15$$

We now have a compact equation to apply to any equation of state. Knowing $Z = Z(T, \rho)$, (e.g., Eqn. 7.15, the Peng-Robinson model), we simply differentiate once, cancel some terms, and integrate. This a perfect sample application of the multivariable calculus that should be familiar at this stage in the curriculum. More importantly, we have developed a systematic approach to solving for any departure function. The steps for a system where $Z = Z(T, \rho)$ are as follows.

1. Write the derivative of the property with respect to volume at constant T. Convert to derivatives of measurable properties using methods from Chapter 6.

2. Write the difference between the derivative real fluid and the derivative ideal gas.

3. Insert integral over dV and limits from infinite volume (where the real fluid and the ideal gas are the same) to the system volume V.

4. Add the necessary correction integral for the ideal gas from V to V^{ig}. (This will be more obvious for entropy.)

5. Transform derivatives to derivatives of Z. Evaluate the derivatives symbolically using the equation of state and integrate analytically.

6. Rearrange in terms of density and compressibility factor to make it more compact.

Some of these steps could have been omitted for the internal energy, because $(\partial U^{ig}/\partial V)_T = 0$. Steps 1 through 4 are slightly different when $Z = Z(T, P)$ such as with the truncated virial EOS. To see the importance of all the steps, consider the entropy departure function.

Example 8.1 Internal energy departure from the van der Waals equation

Derive the internal energy departure function for the van der Waals equation. Suppose methane is compressed from 200 K and 0.1 MPa to 220 K and 60 MPa. Which is the larger contribution in magnitude to ΔU, the ideal gas contribution or the departure function? Use C_P from the back flap and ignore temperature dependence.

Solution: For methane, $a = 230030$ J-cm³/mol² and $b = 43.07$ cm³/mol were calculated by the critical point criteria in Example 7.7 on page 271. Deriving the departure function, $-T(dZ/dT)\rho = -a\rho/RT$, because the repulsive part is constant with respect to T. Substituting,

$$\frac{U - U^{ig}}{RT} = -\int_0^\rho T\left(\frac{\partial Z}{\partial T}\right)_\rho \frac{d\rho}{\rho} = -\int_0^\rho \frac{a\rho d\rho}{RT \rho} = -\frac{a}{RT}\int_0^\rho d\rho = -\frac{a\rho}{RT}\bigg|_0^\rho = -\frac{a\rho}{RT}$$

> ### Example 8.1 Internal energy departure from the van der Waals equation (Continued)
>
> Because $T_r > 1$ there is only one real root. A quick but crude computation of ρ is to rearrange as $Zb\rho = bP/RT = b\rho/(1 - b\rho) - (a/bRT)(b\rho)^2$.
>
> At state 2, 220 K and 60 MPa,
> $60 \cdot 43.07/(8.314 \cdot 220) = b\rho/(1 - b\rho) - 230030/(43.07 \cdot 8.314 \cdot 220)(b\rho)^2$.
> Taking an initial guess of $b\rho = 0.99$ and solving iteratively gives $b\rho = 0.7546$, so
> $(U_2 - U_2^{ig})/RT = -230030 \cdot 0.7546/(43.07 \cdot 8.314 \cdot 220) = -2.203$.
>
> At state 1, 200 K and 0.1 MPa,
> $0.1 \cdot 43.07/(8.314 \cdot 200) = b\rho/(1 - b\rho) - 230030/(43.07 \cdot 8.314 \cdot 200)(b\rho)^2$.
> Taking an initial guess of $b\rho = 0.99$ and solving iteratively gives $b\rho = 0.00290$, so
> $(U_1 - U_1^{ig})/RT = -230030 \cdot 0.00290/(43.07 \cdot 8.314 \cdot 200) = -0.00931$.
>
> $$\Delta U = U_2 - U_1 = (U_2 - U_2^{ig}) + (U_2^{ig} - U_1^{ig}) - (U_1 - U_1^{ig})$$
>
> $\Delta U = -2.203(8.314)220 + (4.3 - 1) \cdot 8.314(220 - 200) + 0.00931(8.314)200 = -4030 + 549 + 15$
> $= -3466$ J/mol. The ideal gas part (549) is 14% as large in magnitude as the State 2 departure function (−4030) for this calculation. Clearly, State 2 is not an ideal gas.

Note that we do not need to repeat the integral for every new problem. For the van der Waals equation, the formula $(U - U^{ig})/(RT) = -(a\rho)/(RT)$ may readily be used whenever the van der Waals fluid density is known for a given temperature.

8.3 ENTROPY DEPARTURE FUNCTION

To calculate the entropy departure, adapt Eqn. 8.11,

$$S - S^{ig} = (S - S^{ig})_{TV} - \int_V^{V^{ig}} \left(\frac{\partial S}{\partial V}\right)_T^{ig} dV \qquad 8.16$$

Inserting the integral for the departure at fixed $\{T, V\}$, we have (using a Maxwell relation),

$$S - S^{ig} = \int_\infty^V \left[\left(\frac{\partial S}{\partial V}\right)_T - \left(\frac{\partial S}{\partial V}\right)_T^{ig}\right] dV - \int_V^{V^{ig}} \left(\frac{\partial S}{\partial V}\right)_T^{ig} dV = \int_\infty^V \left[\left(\frac{\partial P}{\partial T}\right)_V - \left(\frac{\partial P}{\partial T}\right)_V^{ig}\right] dV - \int_V^{V^{ig}} \left(\frac{\partial P}{\partial T}\right)_V^{ig} dV$$

$$8.17$$

Since $\left(\frac{\partial P}{\partial T}\right)_V^{ig} = \frac{R}{V}$, we may readily integrate the ideal gas integral (note that this is not zero whereas the analogous equation for energy was zero):

$$S - S^{ig} = \int_{\infty}^{V} \left[\left(\frac{\partial P}{\partial T} \right)_V - \frac{R}{V} \right] dV + R \ln \frac{V}{V^{ig}}$$ 8.18

Recognizing $V^{ig} = RT/P$, $V/V^{ig} = PV/RT = Z$,

$$\frac{S - S^{ig}}{R} = \int_{\infty}^{V} \left[\frac{1}{R} \left(\frac{\partial P}{\partial T} \right)_V - \frac{1}{V} \right] dV + \ln Z = \int_{0}^{\rho} \left[-T \left(\frac{\partial Z}{\partial T} \right)_\rho - (Z - 1) \right] \frac{d\rho}{\rho} + \ln Z$$ 8.19

where Eqn. 8.15 has been applied to the relation for the partial derivative of P.

Note the $\ln(Z)$ term on the end of this equation. It arises from the change in ideal gas $S_{TP}^{ig} - S_{TV}^{ig}$ represented by the integral in Eqn. 8.16. Changes in states like this may seem pedantic and arcane, but they turn out to be subtle details that often make a big difference numerically. In Example 7.4 on page 266, we determined vapor and liquid roots for Z. The vapor root was close to unity, so $\ln(Z)$ would make little difference in that case. For the liquid root, however, $Z = 0.016$, and $\ln(Z)$ makes a substantial difference. These arcane details surrounding the subject of state specification are the thermodynamicist's curse.

❶ The departures for U and S are the building blocks from which the other departures can be written by combining the relations derived in the previous sections.

8.4 OTHER DEPARTURE FUNCTIONS

The remainder of the departure functions may be derived from the first two and the definitions,

$$H = U + PV \Rightarrow \frac{H - H^{ig}}{RT} = \frac{U - U^{ig}}{RT} + \frac{PV - RT}{RT} = \frac{U - U^{ig}}{RT} + Z - 1$$

$$A = U - TS \Rightarrow \frac{A - A^{ig}}{RT} = \frac{U - U^{ig}}{RT} - \frac{S - S^{ig}}{R}$$

8.20

where we have used $PV^{ig} = RT$ for the ideal gas in the enthalpy departure. Using $H - H^{ig}$ just derived,

$$\frac{G - G^{ig}}{RT} = \frac{H - H^{ig}}{RT} - \frac{S - S^{ig}}{R}$$ 8.21

8.5 SUMMARY OF DENSITY-DEPENDENT FORMULAS

Formulas for departures at fixed T,P are listed below. These formulas are useful for an equation of state written most simply as $Z = f(T,\rho)$ such as cubic EOSs. For treating cases where an equation of state is written most simply as $Z = f(T,P)$ such as the truncated virial EOS, see Section 8.6.

$$\frac{(U - U^{ig})}{RT} = \int_{0}^{\rho} -T \left[\frac{\partial Z}{\partial T} \right]_\rho \frac{d\rho}{\rho}$$ 8.22

$$\frac{(S - S^{ig})}{R} = \int_0^\rho \left[-T \left[\frac{\partial Z}{\partial T} \right]_\rho - (Z - 1) \right] \frac{d\rho}{\rho} + \ln Z \qquad 8.23$$

$$\frac{(H - H^{ig})}{RT} = \int_0^\rho -T \left[\frac{\partial Z}{\partial T} \right]_\rho \frac{d\rho}{\rho} + Z - 1 \qquad 8.24$$

$$\frac{(A - A^{ig})}{RT} = \int_0^\rho \frac{(Z - 1)}{\rho} d\rho - \ln Z \qquad 8.25$$

$$\frac{(G - G^{ig})}{RT} = \int_0^\rho \frac{(Z - 1)}{\rho} d\rho + (Z - 1) - \ln Z \qquad 8.26$$

Useful formulas at fixed T, V include:

$$\frac{(A - A^{ig})_{TV}}{RT} = \int_0^\rho \frac{(Z - 1)}{\rho} d\rho \qquad 8.27$$

$$\frac{(S - S^{ig})_{TV}}{R} = \int_0^\rho \left[-T \left[\frac{\partial Z}{\partial T} \right]_\rho - (Z - 1) \right] \frac{d\rho}{\rho} \qquad 8.28$$

8.6   PRESSURE-DEPENDENT FORMULAS

Occasionally, our equation of state is difficult to integrate to obtain departure functions using the formulas from Section 8.5. This is because the equation of state is more easily arranged and integrated in the form $Z = f(T, P)$, such as the truncated virial EOS. For treating cases where an equation of state is written most simply as $Z = f(T, \rho)$ such as a cubic EOS, see Section 8.5. We adapt the procedures given earlier in Section 8.2.

1. Write the derivative of the property with respect to pressure at constant T. Convert to derivatives of measurable properties using methods from Chapter 6.

2. Write the difference between the derivative real fluid and the derivative ideal gas.

3. Insert integral over dP and limits from $P = 0$ (where the real fluid and the ideal gas are the same) to the system pressure P.

4. Transform derivatives to derivatives of Z. Evaluate the derivatives symbolically using the equation of state and integrate analytically.

5. Rearrange in terms of density and compressibility factor to make it more compact.

We omit derivations and leave them as a homework problem. The two most important departure functions at fixed T, P are

$$\left(\frac{H - H^{ig}}{RT}\right) = -\int_0^P T\left(\frac{\partial Z}{\partial T}\right)_P \frac{dP}{P} \qquad \text{8.29}$$

$$\left(\frac{S - S^{ig}}{R}\right) = -\int_0^P \left[(Z-1) + T\left(\frac{\partial Z}{\partial T}\right)_P\right] \frac{dP}{P} \qquad \text{8.30}$$

The other departure functions can be derived from these using Eqns. 8.20 and 8.21. Note the mathematical similarity between P in the pressure-dependent formulas and ρ in the density-dependent formulas.

8.7   IMPLEMENTATION OF DEPARTURE FORMULAS

The tasks that remain are to select a particular equation of state, take the appropriate derivatives, make the substitutions, develop compact expressions, and add up the change in properties. The good news is that many years of engineering research have yielded several preferred equations of state (see Appendix D) which can be applied generally to any application with a reasonable degree of accuracy. For the purposes of the text, we use the Peng-Robinson equation or virial equation to illustrate the principles of calculating properties. However, many applications require higher accuracy; new equations of state are being developed all the time. This means that it is necessary for each student to know how to adapt the departure function method to new situations as they come along.

The following example illustrates the procedure with an equation of state that is sufficiently simple that it can be applied with either the density-dependent formulas or the pressure-dependent formulas. Although the intermediate steps are a little different, the final answer is the same, of course.

Example 8.2   Real entropy in a combustion engine

A properly operating internal combustion engine requires a spark plug. The cycle involves adiabatically compressing the fuel-air mixture and then introducing the spark. Assume that the fuel-air mixture in an engine enters the cylinder at 0.08 MPa and 20°C and is adiabatically and reversibly compressed in the closed cylinder until its volume is 1/7 the initial volume. Assuming that no ignition has occurred at this point, determine the final T and P, as well as the work needed to compress each mole of air-fuel mixture. You may assume that C_V^{ig} for the mixture is 32 J/mole-K (independent of T), and that the gas obeys the equation of state,

$$PV = RT + aP$$

where a is a constant with value $a = 187$ cm^3/mole. Do not assume that C_V is independent of ρ. Solve using density integrals.

Example 8.2 Real entropy in a combustion engine (Continued)

Solution: The system is taken as a closed system of the gas within the piston/cylinder. Because there is no flow, the system is adiabatic and reversible, the entropy balance becomes

$$\frac{dS}{dt} = \sum_{in} S^{in} \dot{m}^{in} - \sum_{out} S^{out} \dot{m}^{out} + \frac{\dot{Q}}{T_{sys}} + \dot{S}_{gen} = 0 \qquad 8.31$$

showing that the process is isentropic. To find the final T and P, we use the initial state to find the initial entropy and molar volume. Then at the final state, the entropy and molar volume are used to determine the final T and P.

This example helps us to understand the difference between departure functions at fixed T and V and departure functions at fixed T and P. The equation of state in this case is simple enough that it can be applied either way. It is valuable to note how the $\ln(Z)$ term works out. Fixed T and V is convenient since the volume change is specified in this example, and we cover this as Method I, and then use fixed T and P as Method II.

This EOS is easy to evaluate with either the pressure integrals of Section 8.6 or the density integrals of Section 8.5. The problem statement asks us to use density integrals.[a] First, we need to rearrange our equation of state in terms of $Z = f(T, \rho)$. This rearrangement may not be immediately obvious. Note that dividing all terms by RT gives $PV/RT = 1 + aP/RT$. Note that $V\rho = 1$. Multiplying the last term by $V\rho$, $Z = 1 + aZ\rho$ which rearranges to

$$Z = \frac{1}{1 - a\rho}$$

Also, we find the density at the two states using the equation of state,

$$\rho = \frac{P}{RT + aP} \Rightarrow \rho_1 = 3.257\text{E-}5 \text{ gmole/cm}^3 \Rightarrow \rho_2 = 2.280\text{E-}4 \text{ gmole/cm}^3$$

Method I. In terms of fixed T and V, $\left(\frac{\partial Z}{\partial T}\right)_\rho = 0$; $\quad Z - 1 = \dfrac{1}{1 - a\rho} - \dfrac{1 - a\rho}{1 - a\rho} = \dfrac{a\rho}{1 - a\rho}$

$$\frac{(S - S^{ig})_{TV}}{R} = \int_0^\rho \left[-T\left[\frac{\partial Z}{\partial T}\right]_\rho - (Z - 1) \right] \frac{d\rho}{\rho} = -a \int_0^\rho \frac{d\rho}{1 - a\rho} = \ln(1 - a\rho)$$

$$S_2 - S_1 = (S - S^{ig})_{TV,2} + (S_2^{ig} - S_1^{ig}) - (S - S^{ig})_{TV,1}$$

$$= R \cdot [\ln(1 - 187 \cdot 2.28\text{E-}4) + \{(C_V/R)\ln(T_2/T_1) + \ln(V_2/V_1)\} - \ln(1 - 187 \cdot 3.257\text{E-}5)]$$

$$\Delta S/R = 0 = -0.04357 + 32/8.314 \cdot \ln(T_2/293.15) - \ln(7) + 0.00611 = 0 \Rightarrow T_2 = 490.8 \text{ K}$$

Example 8.2 Real entropy in a combustion engine (Continued)

Method II. In terms of T and P,

$$\frac{(S - S^{ig})}{R} = \int_0^\rho \left[-T\left[\frac{\partial Z}{\partial T}\right]_\rho - (Z - 1) \right]\frac{d\rho}{\rho} + \ln Z$$

$$= -a\int_0^\rho \frac{d\rho}{1 - a\rho} + \ln Z = \ln(1 - a\rho) + \ln\left[\frac{1}{1 - a\rho}\right] = 0$$

Since the departure is zero, it drops out of the calculations.

$S_2 - S_1 = S_2^{ig} - S_1^{ig} = C_p\ln(T_2/T_1) - R\ln(P_2/P_1)$. However, since we are given the final volume, we need to calculate the final pressure. Note that we cannot insert the ideal gas law into the pressure ratio in the last term even though we are performing an ideal gas calculation; we must use the pressure ratio for the real gas.

$$\Delta S = C_p\ln(T_2/T_1) - R\ln\left[\frac{RT_2}{V_2 - a}\Big/\frac{RT_1}{V_1 - a}\right] = (C_p - R)\ln(T_2/T_1) - R\ln\left(\frac{V_1 - a}{V_2 - a}\right)$$

Now, if we rearrange, we can show that the result is the same as Method I:

$$\Delta S = C_V\ln(T_2/T_1) + R\ln(V_2/V_1) + R\ln\left(\frac{1 - a\rho_2}{1 - a\rho_1}\right)$$

$$= R\ln(1 - a\rho_2) + C_V\ln(T_2/T_1) + R\ln(V_2/V_1) - R\ln(1 - a\rho_1)$$

This is equivalent to the equation obtained by Method I and $T_2 = 490.8$ K.

Finally, $P_2 = \dfrac{RT_2}{V_2 - a} = \dfrac{8.314(490.8)}{1/2.28\times10^{-4} - 187} = 0.972$ MPa

$W = \Delta U = (U - U^{ig})_2 + C_V\Delta T - (U - U^{ig})_1 = 0 + C_V\Delta T - 0 = 6325$ J/mole

a.   The solution to the problem using pressure integrals is left as homework problem 8.7.

Examples 8.1 and 8.2 illustrate the procedures for deriving and computing the impacts of departure functions, but the equations are too simple to merit broad application. The generalized virial equation helps to broaden the coverage of compound types while retaining a simple functional form.

Example 8.3 Compression of methane using the virial equation

Methane gas undergoes a continuous throttling process from upstream conditions of 40°C and 20 bars to a downstream pressure of 1 bar. What is the gas temperature on the downstream side of the throttling device? An expression for the molar ideal gas heat capacity of methane is
$C_P = 19.25 + 0.0523\,T + 1.197\text{E-}5\,T^2 - 1.132\text{E-}8\,T^3$; $T\,[\equiv]$ K; $C_P\,[\equiv]$ J/mol–K

Example 8.3 Compression of methane using the virial equation (Continued)

The virial equation of state (Eqns. 7.6–7.10) may be used at these conditions for methane:

$$Z = 1 + BP/(RT) = 1 + (B^0 + \omega B^1)P_r/T_r$$

where $B^0 = 0.083 - 0.422/T_r^{1.6}$ and $B^1 = 0.139 - 0.172/T_r^{4.2}$

Solution: Since a throttling process is isenthalpic, the enthalpy departure will be used to calculate the outlet temperature.

$$\Delta H = 0 = H_2 - H_1 = (H_2 - H_2^{ig}) + (H_2^{ig} - H_1^{ig}) - (H_1 - H_1^{ig})$$

The enthalpy departure for the first and third terms in parentheses on the right-hand side can be calculated using Eqn. 8.29. Because $Z(P,T)$, we use Eqn. 8.29. For the integrand, the temperature derivative of Z is required. Recognizing B is a function of temperature only and differentiating,

$$\left(\frac{\partial Z}{\partial T}\right)_P = 1 + \frac{P}{R}\left[\frac{\partial(B \cdot (1/T))}{\partial T}\right]_P = \frac{P}{R}\left[\left(\frac{1}{T}\right)\frac{dB}{dT} - \left(\frac{B}{T^2}\right)\right]$$

Inserting the derivative into Eqn 8.29,

$$\left(\frac{H - H^{ig}}{RT}\right) = \int_0^P \frac{1}{R}\left[\left(\frac{B}{T}\right) - \frac{dB}{dT}\right]dP$$

$$\left(\frac{H - H^{ig}}{RT}\right) = \frac{P}{R}\left(\frac{B}{T} - \frac{dB}{dT}\right)$$ 8.32

$$\left(\frac{S - S^{ig}}{R}\right) = -\frac{P}{R}\frac{dB}{dT}$$

We can easily show by differentiating Eqns. 7.8 and 7.9,

$$\frac{dB^0}{dT_r} = \frac{0.6752}{T_r^{2.6}} \qquad \frac{dB^1}{dT_r} = \frac{0.7224}{T_r^{5.2}}$$

Substituting the relations for B^0, B^1, dB^0/dT_r and dB^1dT_r into Eqn. 8.32 for the departure functions for a pure fluid, we get

$$\left(\frac{H - H^{ig}}{RT}\right) = -P_r\left[\frac{1.0972}{T_r^{2.6}} - \frac{0.083}{T_r} + \omega\left(\frac{0.8944}{T_r^{5.2}} - \frac{0.139}{T_r}\right)\right]$$ 8.33

$$\left(\frac{S - S^{ig}}{R}\right) = -P_r\left[\frac{0.675}{T_r^{2.6}} + \omega\frac{0.722}{T_r^{5.2}}\right]$$ 8.34

Example 8.3 Compression of methane using the virial equation (Continued)

For the initial state, 1,

$$\left(\frac{H - H^{ig}}{RT}\right) = -0.110 \qquad (H - H^{ig})_1 = -287 \text{ J/mole}$$

Assuming a small temperature drop, the heat capacity will be approximately constant over the interval, $C_P \approx 36$ J/mole-K.

For a throttle, $\Delta H = 0 \Rightarrow (H - H^{ig})_2 + 36(T_2 - 40) + 287 = 0$.

Trial and error at state 2 where $P = 1$ bar, $T_2 = 35°C \Rightarrow -13 + 36(35 - 40) + 287 = 94$.

$$T_2 = 30°C \Rightarrow -13 + 36(30 - 40) + 287 = -87$$

Interpolating, $T_2 = 35 + (35 - 30)/(94 + 87)(-94) = 32.4°C$, another trial would show this is close.

Finally, the Peng-Robinson model is sufficiently sophisticated to permit broad application, but the derivations are quite tedious. It may be helpful to see how simple the eventual computations are before getting overwhelmed with the mathematics. Hence, the first example below simply shows how to obtain results based on the computer programs furnished with the text. The subsequent examples confront the derivation of the departure function formulas that appear in the computer programs for the Peng-Robinson equation.

Example 8.4 Computing enthalpy and entropy departures from the Peng-Robinson equation

Preos.xlsx
or
PreosPropsMenu.m.

Propane gas undergoes a change of state from an initial condition of 5 bar and 105°C to 25 bar and 190°C. Compute the change in enthalpy and entropy.

Solution: For propane, $T_c = 369.8$ K; $P_c = 4.249$ MPa; $\omega = 0.152$. The heat capacity coefficients are given by $A = -4.224$, $B = 0.3063$, $C = -1.586\text{E-4}$, $D = 3.215\text{E-8}$. We may use the spreadsheet Preos.xlsx or PreosPropsMenu.m. If we select the spreadsheet, we can use the PROPS page to calculate thermodynamic properties. Using the m-file, we specify the species ID number in the function call to PreosPropsMenu.m and find the departures in the command window. We extract the following results:

For State 2:
$Z = 0.889058$ $V(\text{cm}^3/\text{mol}) = 1369.45$ $(H - H^{ig})$ (J/mol) = -1489.87
$(U - U^{ig})$ (J/mol) = -1062.65 $(S - S^{ig})$ (J/mol-K) = -2.29246

For State 1:
$Z = 0.957388$ $V(\text{cm}^3/\text{mol}) = 6020.28$ $(H - H^{ig})$ (J/mol) = -400.512
$(U - U^{ig})$ (J/mol) = -266.538 $(S - S^{ig})$ (J/mol-K) = -0.708254

Ignoring the specification of the reference state for now (refer to Example 8.8 on page 320 to see how to apply the reference state approach), divide the solution into the three stages described in Section 8.1: I. departure Function; II. ideal gas; III. departure function.

Example 8.4 Computing enthalpy and entropy departures from the Peng-Robinson equation (Continued)

The overall solution path for $H_2 - H_1$ is

$$\Delta H = H_2 - H_1 = (H_2 - H_2^{ig}) + (H_2^{ig} - H_1^{ig}) - (H_1 - H_1^{ig})$$

Similarly, for $S_2 - S_1 =$

$$\Delta S = S_2 - S_1 = (S_2 - S_2^{ig}) + (S_2^{ig} - S_1^{ig}) - (S_1 - S_1^{ig})$$

The three steps that make up the overall solution are covered individually.

Step I. Departures at state 2 from the spreadsheet:

$$(H_2 - H_2^{ig}) = -1490 \text{ J/mol}$$

$$(S_2 - S_2^{ig}) = -2.292 \text{ J/mol-K}$$

Step II. State change for ideal gas: The ideal gas enthalpy change has been calculated in Example 2.5 on page 60.

$$H_2^{ig} - H_1^{ig} = 8405 \text{ J/mol}$$

The ideal gas entropy has been calculated in Example 4.6 on page 151:

$$S_2^{ig} - S_1^{ig} = 6.613 \text{ J/mol-K}$$

Step III. Departures at state 1 from the spreadsheet:

$$(H_1 - H_1^{ig}) = -401 \text{ J/mole}$$

$$(S_1 - S_1^{ig}) = -0.708 \text{ J/mole-K}$$

The total changes may be obtained by summing the steps of the calculation.

$$\Delta H = -1490 + 8405 + 401 = 7316 \text{ J/mole}$$

$$\Delta S = -2.292 + 6.613 + 0.708 = 5.029 \text{ J/mole-K}$$

We have laid the foundation for deriving and applying departure functions given an equation of state. What remains is to organize our efforts to enable convenient and accurate calculations for the myriad of applications that we may encounter. Since these types of computations must be repeated many times, it is worthwhile to implement a broadly applicable ("one size fits all") thermodynamic model and construct a computer program that facilitates input and output. The derivations and programming are more tedious, but they repay the invested effort through multiple applications. The examples below illustrate the derivations for the Peng-Robinson model, but several alternative models could have formed the basis for broad application. You should be able to demonstrate your mastery of the model equations underlying the programming by performing equivalent derivations with alternative models like those illustrated in the homework problems

Example 8.5 Enthalpy departure for the Peng-Robinson equation

The final result of this example is incorporated into Preos.xlsx and PreosPropsMenu.m

Obtain a general expression for the enthalpy departure function of the Peng-Robinson equation.

Solution: Since the Peng-Robinson equation is of the form $Z(T,\rho)$, we can only solve with density integrals.

$$Z = \frac{1}{(1-b\rho)} - \frac{(a\rho)/(RT)}{(1 + 2b\rho + (-b^2\rho^2))}$$

$$-T\left(\frac{\partial Z}{\partial T}\right)_\rho = \frac{(\rho T)/R}{(1 + 2b\rho - b^2\rho^2)}\left[\frac{-a}{T^2} + \frac{1}{T}\left[\frac{da}{dT}\right]\right]$$

where da/dT is given in Eqn. 7.18. Inserting,

$$-T\left(\frac{\partial Z}{\partial T}\right)_\rho = \frac{b\rho}{(1 + 2b\rho - b^2\rho^2)}\left[\frac{-a}{bRT} - \frac{a_C\kappa\sqrt{\alpha T_r}}{bRT}\right]$$

We introduce $F(T_r)$ as a shorthand.

$$-T\left(\frac{\partial Z}{\partial T}\right)_\rho \equiv \frac{b\rho}{(1 + 2b\rho - b^2\rho^2)}F(T_r)$$

Also, $B \equiv bP/RT \Rightarrow b\rho = B/Z$ and $A \equiv aP/R^2T^2 \Rightarrow a/bRT = A/B$. Note that the integration is simplified by integration over $b\rho$ (see Eqn. B.34).

$$\int_0^{b\rho}\left(-T\left(\frac{\partial Z}{\partial T}\right)_\rho\right)\frac{d(b\rho)}{b\rho} = \int_0^{b\rho}\frac{b\rho}{(1 + 2b\rho - b^2\rho^2)}F(T_r)\frac{d(b\rho)}{\rho} =$$

$$\frac{F(T_r)}{\sqrt{8}}\left[\ln\left(\frac{1-\sqrt{2}}{1+\sqrt{2}}\right)\left(\frac{b\rho(1+\sqrt{2})+1}{b\rho(1-\sqrt{2})+1}\right)\right]_0^{b\rho} = \frac{F(T_r)}{\sqrt{8}}\ln\left[\frac{1 + (1+\sqrt{2})b\rho}{1 + (1-\sqrt{2})b\rho}\right]$$

$$\frac{B}{Z} = \frac{\dfrac{bP}{RT}}{\dfrac{P}{\rho RT}} = b\rho \Rightarrow \int_0^{b\rho}-T\left(\frac{\partial Z}{\partial T}\right)_\rho\frac{d(b\rho)}{b\rho} = \frac{F(T_r)}{\sqrt{8}}\ln\left[\frac{Z + (1+\sqrt{2})B}{Z + (1-\sqrt{2})B}\right]$$

$$\frac{(H - H^{ig})}{RT} = Z - 1 + \frac{1}{\sqrt{8}}\left(\frac{-a}{bRT} - \frac{a_C\kappa\sqrt{\alpha T_r}}{bRT}\right)\ln\left[\frac{Z + (1+\sqrt{2})B}{Z + (1-\sqrt{2})B}\right]$$

$$\boxed{\frac{(H - H^{ig})}{RT} = Z - 1 - \frac{A}{B\sqrt{8}}\left(1 + \frac{\kappa\sqrt{T_r}}{\sqrt{\alpha}}\right)\ln\left[\frac{Z + (1+\sqrt{2})B}{Z + (1-\sqrt{2})B}\right]}$$

8.35

Example 8.6 Gibbs departure for the Peng-Robinson equation

Obtain a general expression for the Gibbs energy departure function of the Peng-Robinson equation.

$$Z = \frac{1}{(1 - b\rho)} - \frac{a\rho/RT}{(1 + 2b\rho - b^2\rho^2)}$$

Solution: The answer is obtained by evaluating Eqn. 8.26. The argument for the integrand is

$$Z - 1 = \frac{1}{1 - b\rho} - \frac{1 - b\rho}{1 - b\rho} - \frac{(a\rho)/RT}{(1 + 2b\rho - b^2\rho^2)} = \frac{b\rho}{(1 - b\rho)} + \frac{a\rho/RT}{(1 + 2b\rho - b^2\rho^2)}$$

Evaluating the integral (similar to the integral in Example 8.5), noting again the change in integration variables,

$$\int_0^{b\rho} (Z - 1) \frac{d(b\rho)}{b\rho} = \int_0^{b\rho} \frac{d(b\rho)}{(1 - b\rho)} + \frac{a}{bRT} \int_0^{b\rho} \frac{d(b\rho)}{(1 + 2b\rho - b^2\rho^2)}$$

$$\frac{(A - A^{ig})_{T,V}}{RT} = -\ln(1 - b\rho) - \frac{a}{bRT\sqrt{8}} \ln\left[\frac{1 + (1 + \sqrt{2})b\rho}{1 + (1 - \sqrt{2})b\rho}\right]$$

Making the result dimensionless,

$$\boxed{\frac{(G - G^{ig})}{RT} = Z - 1 - \ln(Z - B) - \frac{A}{B\sqrt{8}} \ln\left[\frac{Z + (1 + \sqrt{2})B}{Z + (1 - \sqrt{2})B}\right]} \qquad 8.36$$

The final result of this example is incorporated into Preos.xlsx and PreosPropsMenu.m

It is often valuable to recognize simplifications that may circumvent the tedium. If two models are similar, you can reuse the part of the derivation that is equivalent. If thermodynamic identities can be used to substitute major portions of a derivation, so much the better. Productive engineers should be aware of opportunities to leverage their time efficiently, as illustrated below.

Example 8.7 *U* and *S* departure for the Peng-Robinson equation

Derive the departure functions for internal energy and entropy of the Peng-Robinson equation. Hint: You could start with Eqns. 8.22 and 8.23, or you could use the results of Examples 8.5 and 8.6 without further integration as suggested by Eqn. 8.20 and Eqn. 8.21.

Solution: By Eqn. 8.20, the *U* departure can be obtained by dropping the "*Z* – 1" term from Eqn. 8.35. We may immediately write:

$$\boxed{\frac{U - U^{ig}}{RT} = -\frac{A}{B\sqrt{8}}\left(1 + \frac{\kappa\sqrt{T_r}}{\sqrt{\alpha}}\right) \ln\left[\frac{Z + (1 + \sqrt{2})B}{Z + (1 - \sqrt{2})B}\right]} \qquad 8.37$$

Example 8.7 *U* and *S* departure for the Peng-Robinson equation (Continued)

By Eqn. 8.21, the entropy departure can be obtained by the difference between the enthalpy departure and Gibbs energy departure, available in Eqns. 8.35 and 8.36. Then, we may immediately write

$$\frac{S - S^{ig}}{R} = \ln(Z - B) - \frac{A}{B\sqrt{8}} \frac{\kappa\sqrt{T_r}}{\sqrt{\alpha}} \ln\left[\frac{Z + (1 + \sqrt{2})B}{Z + (1 - \sqrt{2})B}\right] \qquad 8.38$$

8.8   REFERENCE STATES

If we wish to calculate state changes in a property, then the reference state is not important, and all reference state information drops out of the calculation. However, if we wish to generate a chart or table of thermodynamic properties, or compare our calculations to a thermodynamic table/chart, then designation of a reference state becomes essential. Also, if we need to solve unsteady-state problems, the reference state is important because the answer may depend on the reference state as shown in Example 2.15 on page 81. The quantity $H_R - U_R = (PV)_R$ is non-zero, and although we may substitute $(PV)_R = RT_R$ for an ideal gas, for a real fluid we must use $(PV)_R = Z_R RT_R$, where Z_R has been determined at the reference state. We also may use a real fluid reference state or an ideal gas reference state. Whenever we compare our calculations with a thermodynamic chart/table, we must take into consideration any differences between our reference state and that of the chart/table. Therefore, to specify a reference state for a real fluid, we need to specify:

Pressure

Temperature

In addition we must specify the state of aggregation at the reference state from one of the following:

1. Ideal gas
2. Real gas
3. Liquid
4. Solid

Further, we set $S_R = 0$, and either (*but not both*) of U_R and H_R to zero. The principle of using a reference state is shown in Fig. 8.4 and is similar to the calculation outlined in Fig. 8.2 on page 303.

Ideal Gas Reference States

For an ideal gas reference state, to calculate a value for enthalpy, we have

$$H = (H - H^{ig})_{T, P} + \int_{T_R}^{T} C_P dT + H_R^{ig} \qquad 8.39$$

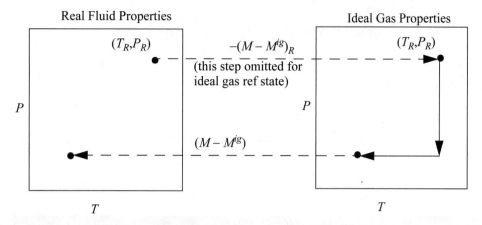

Figure 8.4 *Illustration of calculation of state changes for a generic property M using departure functions where M is U, H, S, G, or A. The calculations are an extension of the principles used in Fig. 8.2 where the initial state is designated as the reference state.*

where the quantity in parentheses is the departure function from Section 8.5 or 8.6 and H_R^{ig} may be set to zero. An analogous equation may be written for the internal energy. Because entropy of the ideal gas depends on pressure, we must include a pressure integral for the ideal gas,

$$S = (S - S^{ig})_{T,P} + \int_{T_R}^{T} \frac{C_P}{T} dT - R \ln \frac{P}{P_R} + S_R^{ig} \qquad 8.40$$

where the reference state value, S_R^{ig}, may be set to zero. From these results we may calculate other properties using relations from Section 6.1: $G = H - TS$, $A = U - TS$, and $U = H - PV$.

Real Fluid Reference State

For a real fluid reference state, to calculate a value for enthalpy, we adapt the procedure of Eqn. 8.5:

$$H = (H - H^{ig})_{T,P} + \int_{T_R}^{T} C_P dT - (H - H^{ig})_R + H_R \qquad 8.41$$

For entropy:

$$S = (S - S^{ig})_{T,P} + \int_{T_R}^{T} \frac{C_P}{T} dT - R \ln \frac{P}{P_R} - (S - S^{ig})_R + S_R \qquad 8.42$$

Changes in State Properties

Changes in state properties are independent of the reference state, or reference state method. To calculate changes in enthalpy, we have the analogy of Eqn. 8.5:

$$\Delta H = (H - H^{ig})_{T_2, P_2} + \int_{T_1}^{T_2} C_P dT - (H - H^{ig})_{T_1, P_1} \qquad 8.43$$

To calculate entropy changes:

$$\Delta S = (S - S^{ig})_{T_2, P_2} + \int_{T_1}^{T_2} \frac{C_P}{T} dT - R \ln \frac{P_2}{P_1} - (S - S^i)^{ig}_{T_1, P_1} \qquad 8.44$$

Example 8.8 Enthalpy and entropy from the Peng-Robinson equation

Preos.xlsx,
PreosPropsMenu.m

Propane gas undergoes a change of state from an initial condition of 5 bar and 105°C to 25 bar and 190°C. Compute the change in enthalpy and entropy. What fraction of the total change is due to the departure functions at 190°C? The departures have been used in Example 8.4, but now we can use the property values directly.

Solution: For propane, $T_c = 369.8$ K; $P_c = 4.249$ MPa; and $\omega = 0.152$. The heat capacity coefficients are given by $A = -4.224$, $B = 0.3063$, $C = -1.586\text{E-}4$, and $D = 3.215\text{E-}8$. For Preos.xlsx, we can use the "Props" page to specify the critical constants and heat capacity constants. The reference state is specified on the companion spreadsheet "Ref State." An arbitrary choice for the reference state is the liquid at 230 K and 0.1 MPa. Returning to the PROPS worksheet and specifying the desired temperature and pressure gives the thermodynamic properties for $V,U,H,$ and S.

State	T(K)	P(MPa)	V(cm^3/mole)	U(J/mole)	H(J/mole)	S(J/mole-K)
2	463.15	2.5	1369	33478	36902	109.15
1	378.15	0.5	6020	26576	29587	104.13

The changes in the thermodynamic properties are $\Delta H = 7315$J/mole and $\Delta S = 5.024$J/mole-K, identical to the more tediously determined values of Example 8.4 on page 314. The purpose of computing the fractional change due to departure functions is to show that we understand the roles of the departure functions and how they fit into the overall calculation. For the enthalpy, the appropriate fraction of the total change is 20%, for the entropy, 46%.

Example 8.9 Liquefaction revisited

Preos.xlsx,
PreosPropsMenu.m.

Reevaluate the liquefaction of methane considered in Example 5.5 on page 213 utilizing the Peng-Robinson equation. Previously the methane chart was used. Natural gas, assumed here to be pure methane, is liquefied in a simple Linde process. The process is summarized in Fig. 8.5. Compression is to 60 bar, and precooling is to 300 K. The separator is maintained at a pressure of 1.013 bar and unliquefied gas at this pressure leaves the heat exchanger at 295 K. What fraction of the methane entering the heat exchanger is liquefied in the process?

Example 8.9 Liquefaction revisited (Continued)

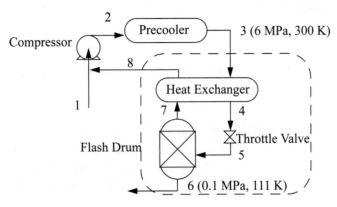

Figure 8.5 *Linde liquefaction schematic.*

State 8

Current State		Roots		Stable Root has a lower fugacity			
T (K)	295	Z	V	fugacity	H	U	S
P (MPa)	0.1013		cm³/gmol	MPa	J/mol	U/mol	J/molK
& for 1 root region		0.9976741	24156.607	0.101064	883.5888	-1563.476	35.86881

State 6

Current State		Roots		Stable Root has a lower fugacity			
T (K)	111	Z	V	fugacity	H	U	S
P (MPa)	0.1013		cm³/gmol	MPa	J/mol	U/mol	J/molK
answers for three		0.9666276	8806.5826	0.09802	-4736.595	-5628.701	6.759068
root region		0.0267407	243.62412		-6972.974	-6997.653	-26.6613
		0.0036925	33.640918	0.093712	-12954.48	-12957.88	-66.90221

State 8

```
┌─ Results ──────────────────────────────┐
│  methane                               │
│                                        │
│  T(K)    295    P(MPa)    0.1013        │
│  Z                       0.997674      │
│  V (cm^3/mol)            24156.1       │
│  U (J/mol)               -1563.48      │
│  H (J/mol)               883.589       │
│  S (J/mol-K)             35.8688       │
│  fugacity (MPa)          0.101064      │
└────────────────────────────────────────┘
```

State 6

```
┌─ Results ──────────────────────────────┐
│  methane                               │
│                                        │
│  T(K)    111    P(MPa)    0.1013        │
│  Z              0.966628  0.00369249   │
│  V (cm^3/mol)   8806.4    33.6402      │
│  U (J/mol)      -5628.7   -12957.9     │
│  H (J/mol)      -4736.59  -12954.5     │
│  S (J/mol-K)    6.75907   -66.9023     │
│  fugacity (MPa) 0.0980197 0.0937095    │
└────────────────────────────────────────┘
```

Figure 8.6 *Summary of enthalpy calculations for methane as taken from the files Preos.xlsx (above) and PreosPropsMenu.m below.*

Solution: Before we calculate the enthalpies of the streams, a reference state must be chosen. The reference state is arbitrary. Occasionally, an energy balance is easier to solve by setting one of the enthalpies to zero by selecting a stream condition as the reference state. To illustrate the results let us select a reference state of $H = 0$ at the real fluid at the state of Stream 3 (6 MPa and 300 K). Because state 3 is the reference state, the $H_3 = 0$. The results of the calculations from the Peng-Robinson equation are summarized in Fig. 8.6.

Example 8.9 Liquefaction revisited (Continued)

The fraction liquefied is calculated by the energy balance: $m_3H_3 = m_8H_8 + m_6H_6$; then incorporating the mass balance: $H_3 = (1 - m_6/m_3)H_8 + (m_6/m_3)H_6$.

The throttle valve is isenthalpic (see Section 2.13). The flash drum serves to disengage the liquid and vapor exiting the throttle valve. The fraction liquefied is $(1 - q) = m_6/m_3 = (H_3 - H_8)/(H_6 - H_8) = (0 - 883)/(-12,954 - 883) = 0.064$, or 6.4% liquefied. This is in good agreement with the value obtained in Example 5.5 on page 213.

Example 8.10 Adiabatically filling a tank with propane

Preos.xlsx,
PreosPropsMenu.m.

Propane is available from a reservoir at 350 K and 1 MPa. An evacuated cylinder is attached to the reservoir manifold, and the cylinder is filled adiabatically until the pressure is 1 MPa. What is the final temperature in the cylinder?

Solution: The critical properties, acentric factor and heat capacity constants, are entered on the "Props" page of Preos.xlsx. On the "Ref State" page, the reference state is arbitrarily selected as the real vapor at 298 K and 0.1 MPa, and $H_R = 0$. At the reservoir condition, propane is in the one-root region with $Z = 0.888$, $H = 3290$ J/mol, $U = 705$ J/mol, and $S = -7.9766$ J/mol-K. The same type of problem has been solved for an ideal gas in Example 2.16 on page 82; however, in this example the ideal gas law cannot be used. The energy balance reduces to $U^f = H^{in}$, where $H^{in} = 3290$ J/mol. In Excel, the answer is easily found by using Solver to adjust the temperature on the "Props" page until $U = 3290$ J/mol. The converged answer is 381 K. In MATLAB, the dialog boxes can be used to match $U = 3290$ J/mol by adjusting T. In the MATLAB window, note that the final T is shown in the "Results" box. The initial guess is preserved in the upper left.

Current State		Roots		Stable Root has a lower fugacity			
T (K)	381.365167	Z	V	fugacity	H	U	S
P (MPa)	1		cm³/gmol	MPa	J/mol	J/mol	J/molK
& for 1 root region		0.9153077	2902.3034	0.920298	6192.303	3290	-0.039259

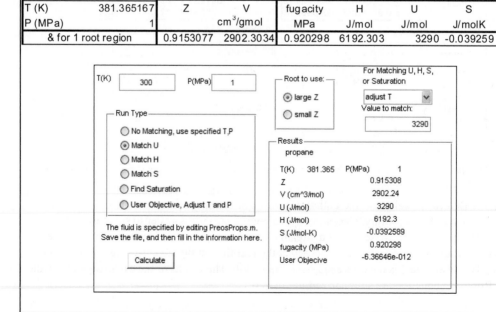

8.9 GENERALIZED CHARTS FOR THE ENTHALPY DEPARTURE

As in the case of the compressibility factor, it is often useful to have a visual idea of how generalized properties behave. Fig. 8.7 on page 324 is analogous to the compressibility factor charts from the previous chapter except that the formula for enthalpy is $(H - H^{ig}) = (H - H^{ig})^0 + \omega(H - H^{ig})^1$. Note that one primary influence in determining the liquid enthalpy departure is the heat of vaporization. Also, the subcritical isotherms shift to liquid behavior at lower pressures when the saturation pressures are lower. The enthalpy departure function is somewhat simpler than the compressibility factor in that the isotherms do not cross one another. Note that the temperature used to make the departure dimensionless is T_c. A sample calculation for propane at 463.15 K and 2.5 MPa gives $H^{ig} - H = [0.45 + 0.152(0.2)] (8.314) 369.8 = 1480$ J/mole compared to 1489.2 from the Peng-Robinson equation.

8.10 SUMMARY

The study of departure functions often causes students great difficulty. That is understandable since it involves simultaneous application of physics and multivariable calculus. This may be the first instance in which students have applied these subjects in combination to such an extent. On the other hand, it is impressive to see what can be accomplished with these tools: the functional equivalent of "steam tables" for any compound in the universe (given a reliable equation of state).

When you get beyond the technical details, however, it seems obvious that there is a difference between an ideal gas and a real fluid. As the accountants for energy movements, we need to be able to account for such contributions. Our method is to first add up all the contributions as if everything behaved like an ideal gas, then to compute and add up all the departures from ideal gas behavior. We apply this over and over again. The calculations are greatly facilitated by computers such that the minimum requirement is the knowledge of what calculation is required and which buttons to push. The purpose of this chapter was to turn your attention to developing a better understanding of the subtleties underlying the equations inside the computer programs.

Your understanding of departure functions is reflected in your ability to develop expressions for various equations of state, as well as the mechanics of adding up the numerical quantities. We covered several derivations, especially for the Peng-Robinson model, and you should be able to reproduce that procedure for other models. Obtaining numerical results occasionally requires iteration and careful consideration of the key constraints. For example, an isentropic compression may transition from the three-root region to the one-root region and your awareness of issues like this corresponds directly with your understanding of how the calculations are performed. Try to rewrite the Excel files yourself, to ensure that you fully comprehend them.

Important Equations

Eqns. 8.22–8.30 stand out in this chapter as the starting point for deriving the necessary departure function expressions for any equation of state. It is tempting to use spreadsheets or programs to add up the contributions from departure functions, reference states, and ideal gas temperature effects mindlessly, like a human computer. But keep in mind that a major goal is to teach the development of model equations, as well as their application. Your skill in developing model equations for novel applications can transcend the study of thermodynamics. Master the derivations behind the programs as well as the mechanics of implementing them.

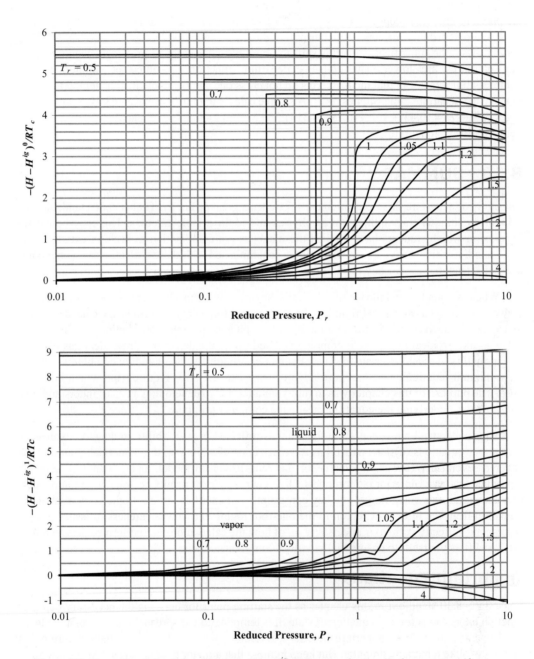

Figure 8.7 *Generalized charts for estimating $(H - H^{ig})/RT_c$ using the Lee-Kesler equation of state.*
$(H - H^{ig})^0/RT_c$ uses $\omega = 0.0$, and $(H - H^{ig})^1/RT_c$ is the correction factor for a hypothetical
compound with $\omega = 1.0$. Divide by reduced temperature to obtain the enthalpy departure
function.

8.11   PRACTICE PROBLEMS

P8.1 Develop an expression for the Gibbs energy departure function based on the Redlich-Kwong (1958) equation of state:

$$Z = 1 + \frac{b\rho}{1 - b\rho} - \frac{a\rho}{RT^{3/2}(1 + b\rho)}$$

(ANS. $(G - G^{ig})/RT = -\ln(1 - b\rho) - a\ln(1 + b\rho)/(bRT^{3/2}) + Z - 1 - \ln Z$)

P8.2 For certain fluids, the equation of state is given by $Z = 1 - b\rho/T_r$

Develop an expression for the enthalpy departure function for fluids of this type. (ANS. $-2b\rho/T_r$)

P8.3 In our discussion of departure functions we derived Eqn. 8.14 for the internal energy departure for any equation of state.

(a) Derive the analogous expression for $(C_V - C_V^{ig})/R$.
(b) Derive an expression for $(C_V - C_V^{ig})/R$ in terms of a, b, ρ and T for the equation of state:

$$Z \equiv 1 + \frac{b\rho}{1 + b\rho} - \rho[\exp(a/T) - 1]$$

(ANS. (a) $\displaystyle\int_0^\rho \left(-2T\left(\frac{\partial Z}{\partial T}\right) - T^2\left(\frac{\partial^2 Z}{\partial T^2}\right)\right)\frac{d\rho}{\rho}$; (b) $a^2\rho T^{-2}\exp(a/T)$)

P8.4 Even in the days of van der Waals, the second virial coefficient for square-well fluids ($\lambda = 1.5$) was known to be $B/b = 4 + 9.5\,[\exp(N_A\varepsilon/RT) - 1]$. Noting that $e^x \sim 1 + x + x^2/2 + \ldots$, this observation suggests the following equation of state:

$$Z = 1 + \frac{4b\rho}{1 - b\rho} - \frac{9.5 N_A \varepsilon}{RT} b\rho$$

Derive an expression for the Helmholtz energy departure function for this equation of state. (ANS. $-4\ln(1 - b\rho) - 9.5N_A\varepsilon b\rho/(RT)$)

P8.5 Making use of the Peng-Robinson equation, calculate ΔH, ΔS, ΔU, and ΔV for 1 gmol of 1,3-butadiene when it is compressed from 25 bar and 400 K to 125 bar and 550 K. (ANS. $\Delta H = 12,570$ J/mol; $\Delta S = 17.998$ J/mol-K; $\Delta U = 11,690$ J/mol; $\Delta V = -640.8$ cm^3/mol)

P8.6 Ethane at 425 K and 100 bar initially is contained in a 1 m^3 cylinder. An adiabatic, reversible turbine is connected to the outlet of the tank and exhausted to atmosphere at 1 bar absolute.

(a) Estimate the temperature of the first gas to flow out of the turbine. (ANS. 185 K)
(b) Estimate the rate of work per mole at the beginning of this operation. (ANS. 8880 J/mol)

P8.7 Ethylene at 350°C and 50 bar is passed through an adiabatic expander to obtain work and exits at 2 bar. If the expander has an efficiency of 80%, how much work is obtained per mole of ethylene, and what is the final temperature of the ethylene? How does the final temperature compare with what would be expected from a reversible expander? (ANS. 11 kJ/mole, 450 K versus 404 K)

P8.8 A Rankine cycle is to operate on methanol. The boiler operates at 200°C ($P^{sat} = 4.087$ MPa), and a superheater further heats the vapor. The turbine outlet is saturated vapor at 0.1027 MPa, and the condenser outlet is saturated liquid at 65°C ($P^{sat} = 0.1027$ MPa). What is the maximum possible value for the cycle thermal efficiency ($\eta_\theta = -W/Q_H$)? (ANS. 26%)

P8.9 An ordinary vapor-compression cycle is to be designed for superconductor application using N2 as refrigerant. The expansion is to 1 bar. A heat sink is available at 105 K. A 5 K approach should be sufficient. Roughly 100 Btu/hr must be removed. Compute the coefficient of performance (COP) and compare to the Carnot COP. Also, estimate the compressor's power requirement (hp) assuming it is adiabatic and reversible. (ANS. 1.33, 0.3)

P8.10 Suppose ethane was compressed adiabatically in a 70% efficient continuous compressor. The downstream pressure is specified to be 1500 psia at a temperature not to exceed 350°F. What is the highest that the upstream temperature could be if the upstream pressure is 200 psia? (Hint: Neglect the upstream departure function.) (ANS. 269 K)

P8.11 As part of a liquefaction process, ammonia is throttled to 80% quality at 1 bar. If the upstream pressure is 100 bar, what must be the upstream temperature?

P8.12 An alternative to the pressure equation route from the molecular scale to the macroscopic scale is through the energy equation (Eqn. 7.51). The treatment is similar to the analysis for the pressure equation, but the expression for the radial distribution function must now be integrated over the range of the potential function.

   (a) Suppose that $u(r)$ is given by the square-well potential ($R = 1.5$) and $g(r) = 10 - 5(r/\sigma)$ for $r > \sigma$. Evaluate the internal energy departure function where $\rho\sigma^3 = 1$ and $\varepsilon/kT = 1$. (ANS. -5.7π)

   (b) Suppose that the radial distribution function at intermediate densities can be reasonably represented by: $g \sim (1 + 2(\sigma/r)^2)$ at all temperatures. Derive an expression for the attractive contribution to the compressibility factor for fluids that can be accurately represented by the Sutherland potential. (ANS. $3\pi\rho N_A\sigma^3 N_A\varepsilon/RT$)

8.12   HOMEWORK PROBLEMS

8.1 What forms does the derivative $(\partial C_V/\partial V)_T$ have for a van der Waals gas and a Redlich-Kwong gas? (The Redlich-Kwong equation is given in Problem P8.1.) Comment on the results.

8.2 Estimate C_P, C_V, and the difference $C_P - C_V$ in (J/mol-K) for liquid n-butane from the following data.[2]

T(°F)	P(psia)	V(ft^3/lb)	H (BTU/lb)	U(BTU/lb)
20	14.7	0.02661	−780.22	−780.2924302
40	1400	0.02662	−765.05	−771.9507097
0	14.7	0.02618	−791.24	−791.3112598

8.3 Estimate C_P, C_V, and the difference $C_P - C_V$ in (J/mol-K) for saturated n-butane liquid at 298 K n-butane as predicted by the Peng-Robinson equation of state. Repeat for saturated vapor.

---

2. Starling, K.E. 1973. *Fluid Thermodynamic Properties for Light Petroleum Substances,* Houston, TX: Gulf Publishing.

8.4 Derive the integrals necessary for departure functions for U, G, and A for an equation of state written in terms of $Z = f(T,P)$ using the integrals provided for H and S in Section 8.6.

8.5 (a) Derive the enthalpy and entropy departure functions for a van der Waals fluid.
   (b) Derive the formula for the Gibbs energy departure.

8.6 The Soave-Redlich-Kwong equation is presented in problem 7.15. Derive expressions for the enthalpy and entropy departure functions in terms of this equation of state.

8.7 In Example 8.2 we wrote the equation of state in terms of $Z = f(T,\rho)$. The equation of state is also easy to rearrange in the form $Z = f(T,P)$. Rearrange the equation in this form, and apply the formulas from Section 8.6 to resolve the problem using departures at fixed T and P.

8.8 The ESD equation is presented in problem 7.19. Derive expressions for the enthalpy and entropy departure functions in terms of this equation of state.

8.9 A gas has a constant-pressure ideal-gas heat capacity of $15R$. The gas follows the equation of state,

$$Z = 1 + \frac{aP}{RT}$$

over the range of interest, where $a = -1000 \text{ cm}^3/\text{mole}$.

(a) Show that the enthalpy departure is of the following form:

$$\frac{H - H^{ig}}{RT} = \frac{aP}{RT}$$

(b) Evaluate the enthalpy change for the gas as it undergoes the state change:

$$T_1 = 300 \text{ K}, P_1 = 0.1 \text{ MPa}, T_2 = 400 \text{ K}, P_2 = 2 \text{ MPa}$$

8.10 Derive the integrated formula for the Helmholtz energy departure for the virial equation (Eqn. 7.7), where B is dependent on temperature only. Express your answer in terms of B and its temperature derivative.

8.11 Recent research suggests the following equation of state, known as the PC-SAFT model.

(a) Derive an expression for Z.
(b) Derive the departure function for $(U-U^{ig})$.
   Note: $\eta_P = b\rho$; m = constant proportional to molecular weight; a_i, b_i are constants.

$$\frac{(A - A^{ig})_{T,V}}{RT} = \frac{m\eta_P(4 - 3\eta_P)}{(1 - \eta_P)^2} - (1 - m)\ln\left(\frac{5 - 2\eta_P}{(1 - \eta_P)^3}\right) - A_1\beta\varepsilon - A_2(\beta\varepsilon)^2$$

$$A_1 = 12m\eta_P\sum_{i=0}^{6} a_i\eta_P^i ; A_2 = \frac{m^2 6\eta_P\sum_{i=0}^{6} b_i\eta_P^i}{1 + \frac{4m\eta_P(2 - \eta_P/2)}{(1 - \eta_P)^4} + \frac{(1 - m)\eta_P(20 - 27\eta_P + 12\eta_P^2 - 2\eta_P^3)}{[(1 - \eta_P)(2 - \eta_P)]^2}}$$

8.12 Recent research in thermodynamic perturbation theory suggests the following equation of state.

$$Z = 1 + \frac{4\eta_P}{(1 - 1.9\eta_P)} - \frac{9.5\eta_P}{T + 0.7\exp(-10\eta_P)}$$

(a) Derive the departure function for $(A - A^{ig})_{T,V}$.
(b) Derive the departure function for $(U - U^{ig})$.
   Hint: substitute $u = 0.7 + T\exp(10\eta_P)$; $\eta_P = b\rho$.

8.13 A gas is to be compressed in a steady-state flow reversible isothermal compressor. The inlet is to be 300 K and 1 MPa and the gas is compressed to 20 MPa. Assume that the gas can be modeled with equation of state

$$PV = RT - \frac{a}{T}P + bP$$

where $a = 385.2$ cm^3-K/mol and $b = 15.23$ cm^3/mol. Calculate the required work per mole of gas.

8.14 A 1 m^3 isolated chamber with rigid walls is divided into two compartments of equal volume. The partition permits transfer of heat. One side contains a nonideal gas at 5 MPa and 300 K and the other side contains a perfect vacuum. The partition is ruptured, and after sufficient time for the system to reach equilibrium, the temperature and pressure are uniform throughout the system. The objective of the problem statements below is to find the final T and P.

The gas follows the equation of state

$$\frac{PV}{RT} = 1 + \left(b - \frac{a}{T}\right)\frac{P}{RT}$$

where $b = 20$ cm^3/ mole; $a = 40{,}000$ cm^{3}K/mole; and $C_P = 41.84 + 0.084T$(K) J/mol-K.

(a) Set up and simplify the energy balance and entropy balance for this problem.
(b) Derive formulas for the departure functions required to solve the problem.
(c) Determine the final P and T.

8.15 P-V-T behavior of a simple fluid is found to obey the equation of state given in problem 8.14.

(a) Derive a formula for the enthalpy departure for the fluid.
(b) Determine the enthalpy departure at 20 bar and 300 K.
(c) What value does the entropy departure have at 20 bar and 300 K?

8.16 Using the Peng-Robinson equation, estimate the change in entropy (J/mole-K) for raising butane from a saturated liquid at 271 K and 1 bar to a vapor at 352 K and 10 bar. What fraction of this total change is given by the departure function at 271 K? What fraction of this change is given by the departure function at 352 K?

8.17 Suppose we would like to establish limits for the rule $T_2 = T_1(P_2/P_1)^{R/C_P}$ by asserting that the estimated T_2 should be within 5% of the one calculated using the departure functions.

For $\omega = 0$ and $T_r = [1, 10]$ at state 1, determine the values of P_r where this assertion holds valid by using the Peng-Robinson equation as the benchmark.

8.18 A piston contains 2 moles of propane vapor at 425 K and 8.5 MPa. The piston is taken through several state changes along a pathway where the total work done by the gas is 2 kJ. The final state of the gas is 444 K and 3.4 MPa. What is the change, $\Delta \underline{H}$, for the gas predicted by the Peng-Robinson equation and how much heat is transferred? Note: A reference state is optional; if one is desired, use vapor at 400 K and 0.1 MPa.

8.19 N.B. Vargaftik[3] (1975) lists the experimental values in the following table for the enthalpy departure of isobutane at 175°C. Compute theoretical values and their percent deviations from experiment by the following

P (atm)	10	20	35	70
$H-H^{ig}$(J/g)	−15.4	−32.8	−64.72	−177.5

(a) The generalized charts
(b) The Peng-Robinson equation

8.20 n-pentane is to be heated from liquid at 298 K and 0.01013 MPa to vapor at 360 K and 0.3 MPa. Compute the change in enthalpy using the Peng-Robinson equation of state. If a reference state is desired, use vapor at 310 K, 0.103 MPa, and provide the enthalpy departure at the reference state.

8.21 For each of the fluid state changes below, perform the following calculations using the Peng-Robinson equation: (a) Prepare a table and summarize the molar volume, enthalpy, and entropy for the initial and final states; (b) calculate ΔH and ΔS for the process; and (c) compare with ΔH and ΔS for the fluid modeled as an ideal gas. Specify your reference states.

(a) Propane vapor at 1 bar and 60°C is compressed to a state of 125 bar and 250°C.
(b) Methane vapor at −40°C and 0.1013 MPa is compressed to a state of 10°C and 7 MPa.

8.22 1 m³ of CO_2 initially at 150°C and 50 bar is to be isothermally compressed in a frictionless piston/cylinder device to a final pressure of 300 bar. Calculate the volume of the compressed gas, $\Delta \underline{U}$, the work done to compress the gas, and the heat flow on compression assuming

(a) CO_2 is an ideal gas.
(b) CO_2 obeys the Peng-Robinson equation of state.

8.23 Solve problem 8.22 for an adiabatic compression.

8.24 Consider problem 3.11 using benzene as the fluid rather than air and eliminating the ideal gas assumption. Use the Peng-Robinson equation. For the same initial state,

(a) The final tank temperature will not be 499.6 K. What will the temperature be?
(b) What is the number of moles left in the tank at the end of the process?
(c) Write and simplify the energy balance for the process. Determine the final temperature of the piston/cylinder gas.

8.25 Solve problem 8.24 using n-pentane.

---

3. See Vargaftik reference homework problem 7.6.

8.26 A tank is divided into two equal halves by an internal diaphragm. One half contains argon at a pressure of 700 bar and a temperature of 298 K, and the other chamber is completely evacuated. Suddenly, the diaphragm bursts. Compute the final temperature and pressure of the gas in the tank after sufficient time has passed for equilibrium to be attained. Assume that there is no heat transfer between the tank and the gas, and that argon:

(a) Is an ideal gas
(b) Obeys the Peng-Robinson equation

8.27 The diaphragm of the preceding problem develops a small leak instead of bursting. If there is no heat transfer between the gas and tank, what is the temperature and pressure of the gas in each tank after the flow stops? Assume that argon obeys the Peng-Robinson equation.

8.28 A practical application closely related to the above problem is the use of a compressed fluid in a small can to reinflate a flat tire. Let's refer to this product as "Fix-a-flat." Suppose we wanted to design a fix-a-flat system based on propane. Let the can be 500 cm^3 and the tire be 40,000 cm^3. Assume the tire remains isothermal and at low enough pressure for the ideal gas approximation to be applicable. The can contains 250 g of saturated liquid propane at 298 K and 10 bar. If the pressure in the can drops to 0.85 MPa, what is the pressure in the tire and the amount of propane remaining in the can? Assuming that 20 psig is enough to drive the car for a while, is the pressure in the tire sufficient? Could another tire be filled with the same can?

8.29 Ethylene at 30 bar and 100°C passes through a throttling valve and heat exchanger and emerges at 20 bar and 150°C. Assuming that ethylene obeys the Peng-Robinson equation, compute the flow of heat into the heat exchanger per mole of ethylene.

8.30 In the final stage of a multistage, adiabatic compression, methane is to be compressed from −75°C and 2 MPa to 6 MPa. If the compressor is 76% efficient, how much work must be furnished per mole of methane, and what is the exit temperature? How does the exit temperature compare with that which would result from a reversible compressor? Use the Peng-Robinson equation.

8.31 (a) Ethane at 280 K and 1 bar is continuously compressed to 310 K and 75 bar. Compute the change in enthalpy per mole of ethane using the Peng-Robinson equation.
(b) Ethane is expanded through an adiabatic, reversible expander from 75 bar and 310 K to 1 bar. Estimate the temperature of the stream exiting the expander and the work per mole of ethane using the Peng-Robinson equation. (Hint: Is the exiting ethane vapor, liquid, or a little of each? The saturation temperature for ethane at 1 bar is 184.3 K.)

8.32 Our space program requires a portable engine to generate electricity for a space station. It is proposed to use sodium (T_c = 2300 K; P_c = 195 bar; ω = 0; C_P/R = 2.5) as the working fluid in a customized form of a "Rankine" cycle. The high-temperature stream is not superheated before running through the turbine. Instead, the saturated vapor (T = 1444 K, P^{sat} = 0.828 MPa) is run directly through the (100% efficient, adiabatic) turbine. The rest of the Rankine cycle is the usual. That is, the outlet stream from the turbine passes through a condenser where it is cooled to saturated liquid at 1155 K (this is the normal boiling temperature of sodium), which is pumped (neglect the pump work) back into the boiler.

(a) Estimate the quality coming out of the turbine.
(b) Compute the work output per unit of heat input to the cycle, and compare it to the value for a Carnot cycle operating between the same T_H and T_C.

8.33 Find the minimum shaft work (in kW) necessary to liquefy n-butane in a steady-state flow process at 0.1 MPa pressure. The saturation temperature at 0.1 MPa is 271.7 K. Butane is to enter at 12 mol/min and 0.1 MPa and 290 K and to leave at 0.1 MPa and 265 K. The surroundings are at 298 K and 0.1 MPa.

8.34 The enthalpy of normal liquids changes nearly linearly with temperature. Therefore, in a single-pass countercurrent heat exchanger for two normal liquids, the temperature profiles of both fluids are nearly linear. However, the enthalpy of a high-pressure gas can be nonlinearly related to temperature because the constant pressure heat capacity becomes very large in the vicinity of the critical point. For example, consider a countercurrent heat exchanger to cool a CO_2 stream entering at 8.6 MPa and 115°C. The outlet is to be 8.6 MPa and 22°C. The cooling is to be performed using a countercurrent stream of water that enters at 10°C. Use a basis of 1 mol/min of CO_2.

(a) Plot the CO_2 temperature (°C) on the ordinate versus $\underline{H}$ on the abscissa, using $H = 0$ for the outlet state as the reference state.

(b) Since $d\underline{H}_{water}/dx = d\underline{H}_{CO2}/dx$ along a differential length, dx, of countercurrent of heat exchanger, the corresponding plot of T versus $\underline{H}$ for water (using the inlet state as the reference state) will show the water temperature profile for the stream that contacts the CO_2. The water profile must remain below the CO_2 profile for the water stream to be cooler than the CO_2. If the water profile touches the CO_2 profile, the location is known as a pinch point and the heat exchanger would need to be infinitely big. What is the maximum water outlet temperature that can be feasibly obtained for an infinitely sized heat exchanger?

(c) Approximately what water outlet temperature should be used to ensure a minimum approach temperature for the two streams of approximately 10°C?

8.35 An alternative to the pressure equation route from the molecular scale to the macroscopic scale is through the energy equation (Eqn. 7.50). The treatment is similar to the analysis for the pressure equation, but the expression for the radial distribution function must now be integrated over the range of the potential function. Suppose that the radial distribution function can be reasonably represented by:

$$g = 0 \text{ for } r < \sigma$$

$$g \sim 1 + \rho N_A \sigma^6 \varepsilon/(r^3 kT) \text{ for } r > \sigma$$

at all temperatures and densities. Use Eqn. 7.50 to derive an expression for the internal energy departure function of fluids that can be accurately represented by the following:

(a) The square-well potential with $\lambda_{sw} = 1.5$
(b) The Sutherland potential

Evaluate each of the above expressions at $\rho N_A \sigma^3 = 0.6$ and $\varepsilon/kT = 1$.

8.36 Starting with the pressure equation as shown in Chapter 6, evaluate the internal energy departure function at $\rho N_A \sigma^3 = 0.6$ and $\varepsilon/kT = 1$ by performing the appropriate derivatives and integrations of the equation of state obtained by applying

$$g = 0 \text{ for } r < \sigma$$

$$g \sim 1 + \rho N_A \sigma^6 \varepsilon/(r^3 kT) \text{ for } r > \sigma$$

at all temperatures and densities:

(a) The square-well potential with $\lambda_{sw} = 1.5$
(b) The Sutherland potential
(c) Compare these results to those obtained in problem 8.35 and explain why the numbers are not identical.

8.37 Molecular simulation can be used to explore the accuracy and significance of individual contributions to an equation of state. Use the DMD module at Etomica.org to explore Xe's energy departure.

(a) The simulation results below have been tabulated at $\eta_P = 0.167$, $\lambda = 1.7$. Plot $U/N_A\varepsilon$ versus $\beta\varepsilon$ for these data along with those at $\eta_P = 0.375$ from homework problem 7.25.
(b) Prepare a plot of Xe's simulated $U - U^{ig}$ versus $1000/T$ using your best ε and σ at $\eta_P = 0.375$ and showing the isochoric data for Xe from Webbook.nist.gov at 22.14 mol/L on the same axes.
(c) The data for $U - U^{ig}$ exhibit a linear trend with $\beta\varepsilon$. The data for Z also exhibit a linear trend with $\beta\varepsilon$. What trends do these two data sets indicate for $(A - A^{ig})_{TV}/RT$? Are they consistent? Explain.
(d) Use the trapezoidal rule and the energy equation (Eqn. 7.49) to estimate $A - A^{ig}$ and plot as a dashed line. How accurate are your estimates (AAD%) and how could you improve them?

SW results at $\eta_P = 0.167$, $\lambda = 1.7$.

η_P	$\beta\varepsilon$	Z	$-(U-U^{ig})/(N_A\varepsilon)$	$g(\sigma)$	$g(\lambda\sigma)$
0.167	0.335	0.988	2.99	1.6	1.2
0.167	0.384	0.820	3.04	1.6	1.2
0.167	0.512	0.456	3.23	1.7	1.3
0.167	0.624	0.156	3.47	2	1.5
0.167	0.709	-0.038	3.76	2.3	1.55

8.38 Suppose two molecules had similar potential functions, but they were mirror images of one another as shown in the figure below. Which one (A or B) would have the larger internal energy departure? You may assume that the radial distribution function is the same for both potential models.

(a) Reason qualitatively but refer to appropriate equations to explain your answer.
(b) Compute the values of $(U - U^{ig})/RT$ at a packing fraction of 0.4 and a temperature of 50 K. Assume values of the radial distribution function as follows:

r(nm)	0.3	0.45	0.6
$g(r)$	6.93	0.91	1.05

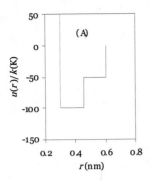

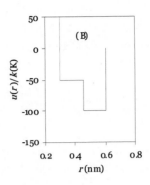

CHAPTER

PHASE EQUILIBRIUM IN A PURE FLUID

One of the principal objects of theoretical research is to find the point of view from which it can be expressed with greatest simplicity.

J.W. Gibbs (1881)

The problem of phase equilibrium is distinctly different from "(In – Out) = Accumulation." The fundamental balances were useful in describing many common operations like throttling, pumping, and compressing, and fundamentally, they provide the basis for understanding all processes. But the balances make a relatively simple contribution in solving problems of phase equilibrium—so much so that they are largely ignored, while simple questions like "How many phases are present?" take the primary role.

The general problem of phase equilibrium has a broad significance that begins to distinguish chemical thermodynamics from more generic thermodynamics. If we only care about steam, then it makes sense to concentrate on the various things we can do with steam and to use the steam tables for any properties we need. But, if our interest is in a virtually infinite number of chemicals and mixtures, then we need some unifying principles. Since our interest is chemical thermodynamics, we must deal extensively with property estimations. The determination of phase equilibrium is one of the most important and difficult estimations to make. The ability to understand, model, and predict phase equilibria is particularly important for designing separation processes. Typically, these operations comprise the most significant capital costs of plant facilities, and require knowledgeable engineers to design, maintain, and troubleshoot them.

In most separation processes, the controlled variables are the temperature and pressure. Thus, when we approach the modeling of phase behavior, we should seek thermodynamic properties that are natural functions of these two properties. In our earlier discussions of convenience properties, the Gibbs energy was shown to be such a function:

$$G \equiv U + PV - TS \qquad dG = -SdT + VdP \qquad \text{9.1}$$

As a defined mathematical property, the Gibbs energy remains abstract, in the same way that enthalpy and entropy are difficult to conceptualize. However, our need for a natural function of P and T requires the use of this property.

❶ Phase equilibrium at fixed T and P is most easily understood using G, which is a natural function of P, T.

Chapter Objectives: You Should Be Able to...

1. Use the Clapeyron and Clausius-Clapeyron equations to calculate thermodynamic properties from limited data.

2. Explain the origin of the shortcut vapor pressure equation and its limitations.

3. Use the Antoine equation to calculate saturation temperature or saturation pressure.

4. Describe in words the relationship between the Gibbs departure and the fugacity.

5. Given the pressure and temperature, estimate the value of fugacity (in appropriate units) using an ideal gas model, virial correlation, or cubic equation of state.

6. Calculate the fugacity coefficient of a vapor or liquid given an expression for a cubic equation of state and the parameter values Z, A, B, and decide which root among multiple roots is most stable.

7. Estimate the fugacity of a liquid or solid if given the vapor pressure.

8. Interpret equation of state results at saturation and apply the lever rule to properties like enthalpy, internal energy, and entropy for a two-phase mixture.

9. Solve throttling, compressor, and turbine expander problems using a cubic equation of state for thermodynamic properties rather than a chart or table.

9.1 CRITERIA FOR PHASE EQUILIBRIUM

As an introduction to the constraint of phase equilibrium, let us consider an example. A piston/cylinder contains both propane liquid and vapor at $-12°C$. The piston is forced down a specified distance. Heat transfer is provided to maintain isothermal conditions. Both phases still remain. How much does the pressure increase?

This is a trick question. As long as two phases are present for a single component and the temperature remains constant, then the system pressure remains fixed at the vapor pressure, so the answer is zero increase. The molar volumes of vapor and liquid phases also stay constant since they are state properties. However, as the total volume changes, the quantity of liquid increases, and the quantity of vapor decreases. We are working with a closed system where $n = n^L + n^V$. For the whole system: $\underline{V} = n^L V^{satL} + n^V V^{satV} = n \cdot V^{satL} + q \cdot n \cdot (V^{satV} - V^{satL})$ and since V^{satL} and V^{satV} are fixed and $V^{satL} < V^{satV}$, a decrease in $\underline{V}$ causes a decrease in q.[1]

Since the temperature and pressure from beginning to end are constant as long as two phases exist, applying Eqn. 9.1 shows that the change in Gibbs energy of each phase of the system from beginning to end must be zero, $dG^L = dG^V = 0$.

For the whole system:

$$\underline{G} = n^L G^L + n^V G^V, \text{ by the product rule} \Rightarrow d\underline{G} = n^L dG^L + n^V dG^V + G^L dn^L + G^V dn^V \qquad 9.2$$

But by the mass balance, $dn^L = -dn^V$ which reduces Eqn. 9.2 to $0 = G^L - G^V$ or

---

1. Once the system volume is decreased below a volume where $\underline{V} < nV^{satL}$, we are compressing a liquid, and the pressure could become quite high. We would need to compute how high using an equation of state. An analogous discussion could be developed for expansion of system volume showing that only vapor will exist for $\underline{V} > nV^{satV}$. The key to notice is that values of q are only physically meaningful in the range $0 < q < 1$.

$$\boxed{G^L = G^V}\ \text{pure fluid phase equilibria}$$

9.3 ❶ Gibbs energy is the key property for characterizing phase equilibria.

This is a very significant result. In other words, $G^L = G^V$ is a constraint for phase equilibrium. None of our other thermodynamic properties, $U, H, S,$ and A is equivalent in both phases. If we specify phase equilibrium must exist and one additional constraint (e.g., T), then all of our other state properties of each phase are fixed and can be determined by the equation of state and heat capacities.

Only needing to specify one variable at saturation to compute all state properties should not come as a surprise, based on our experience with the steam tables. The constraint of $G^L = G^V$ is simply a mathematical way of saying "saturated." As an exercise, select from the steam tables an arbitrary saturation condition and calculate $G = H - TS$ for each phase. The advantage of the mathematical expression is that it yields a specific equality applicable to many chemicals. The powerful insight of $G^L = G^V$ leads us to the answers of many more difficult and significant questions concerning phase equilibrium.

9.2 THE CLAUSIUS-CLAPEYRON EQUATION

We can apply these concepts of equilibrium to obtain a remarkably simple equation for the vapor-pressure dependence on temperature at low pressures. As a "point of view of greatest simplicity," the Clausius-Clapeyron equation is an extremely important example. Suppose we would like to find the slope of the vapor pressure curve, dP^{sat}/dT. Since we are talking about vapor pressure, we are constrained by the requirement that the Gibbs energies of the two phases remain equal as the temperature is changed. If the Gibbs energy in the vapor phase changes, the Gibbs energy in the liquid phase must change by the same amount. Thus,

$$dG^L = dG^V$$

Rewriting the fundamental property relation $\Rightarrow dG = V^V\, dP^{sat} - S^V\, dT = V^L\, dP^{sat} - S^L\, dT$ and rearranging,

$$\Rightarrow (V^V - V^L)dP^{sat} = (S^V - S^L)dT \qquad 9.4$$

Entropy is a difficult property to measure. Let us use a fundamental property to substitute for entropy. By definition of G: $G^V = H^V - TS^V = H^L - TS^L = G^L$

$$\boxed{S^V - S^L = \Delta S^{vap} = \frac{(H^V - H^L)}{T} = \frac{\Delta H^{vap}}{T}} \qquad 9.5$$

Substituting Eqn. 9.5 in for $S^V - S^L$ in Eqn. 9.4, we have the *Clapeyron equation* which is valid for pure fluids along the saturation line:

$$\boxed{\frac{dP^{sat}}{dT} = \frac{\Delta H^{vap}}{T(V^V - V^L)}} \qquad 9.6$$ ❶ Clapeyron equation.

> *Note: This general form of Clapeyron equation can be applied to any kind of phase equilibrium including solid-vapor and solid-liquid equilibria by substituting the alternative sublimation or fusion properties into Eqn. 9.6; we derived the current equation based on vapor-liquid equilibria.*

Several simplifications can be made in the application to vapor pressure (i.e., vapor-liquid equilibium). To write the equation in terms of Z^V and Z^L, we multiply both sides by T^2 and divide both sides by P^{sat}:

$$\frac{T^2}{P^{sat}}\frac{dP^{sat}}{dT} = \frac{\Delta H^{vap}}{R(Z^V - Z^L)}$$

We then use calculus to change the way we write the Clapeyron equation:

$$\frac{dP^{sat}}{P^{sat}} = d\ln P^{sat} \quad \text{and} \quad d\left(\frac{1}{T}\right) = -\frac{dT}{T^2}$$

Combining the results, we have an alternative form of the Clapeyron equation:

! Clapeyron equation.

$$\boxed{d\ln P^{sat} = \frac{-\Delta H^{vap}}{R(Z^V - Z^L)}d\left(\frac{1}{T}\right)}$$ 9.7

For a gas far from the critical point at "low" reduced temperatures, $Z^V - Z^L \approx Z^V$. In addition, for vapor pressures near 1 bar, where ideal gas behavior is approximated, $Z^V \approx 1$, resulting in the *Clausius-Clapeyron equation*:

! Clausius-Clapeyron equation.

$$\boxed{d\ln P^{sat} = \frac{-\Delta H^{vap}}{R}d\left(\frac{1}{T}\right)}$$ (ig) 9.8

Example 9.1 Clausius-Clapeyron equation near or below the boiling point

Derive an expression based on the Clausius-Clapeyron equation to predict vapor-pressure dependence on temperature.

Solution: If we assume that ΔH^{vap} is fairly constant in some range near the boiling point, integration of each side of the Clausius-Clapeyron equation can be performed from the boiling point to another state on the saturation curve, which yields

$$\ln\left[\frac{P^{sat}}{P_R^{sat}}\right] = \frac{-\Delta H^{vap}}{R}\left[\frac{1}{T} - \frac{1}{T_R}\right]$$ 9.9

where P_R^{sat} is 0.1013 MPa and T_R is the normal boiling temperature. This result may be used in a couple of different ways: (1) We may look up ΔH^{vap} so that we can calculate P^{sat} at a new temperature T; or (2) we may use two vapor pressure points to calculate ΔH^{vap} and subsequently apply method (1) to determine other P^{sat} values.

One vapor pressure point is commonly available through the acentric factor, which is the reduced vapor pressure at a reduced temperature of 0.7. (Frequently the boiling point is near this temperature.) That means, we can apply the definition of the acentric factor to obtain a value of the vapor pressure relative to the critical point.

9.3 SHORTCUT ESTIMATION OF SATURATION PROPERTIES

We found that the Clausius-Clapeyron equation leads to a simple, two-constant equation for the vapor pressure at low temperatures. What about higher temperatures? Certainly, the assumption of ideal gases used to derive the Clausius-Clapeyron equation is not valid as the vapor pressure becomes large at high temperature; therefore, we need to return to the Clapeyron equation. If $\Delta H^{vap}/\Delta Z^{vap}$ was constant over a wide range of temperature, then we could recover this simple form. Obviously, ΔZ^{vap} is not constant; as we approach the critical point, the vapor and liquid volumes get closer together until they eventually become equal and $\Delta Z^{vap} \to 0$. However, the enthalpies of the vapor and liquid approach one another at the critical point, so it is possible that $\Delta H^{vap}/\Delta Z^{vap}$ may be approximately constant. To analyze this hypothesis, let us plot the experimental data in the form of Eqn. 9.7 *assuming* that $\Delta H^{vap}/\Delta Z^{vap}$ is constant. A constant slope would confirm a constant value of $\Delta H^{vap}/\Delta Z^{vap}$. A plot is shown for two fluids in Fig. 9.1.

The conclusion is that setting $\Delta H/\Delta Z$ equal to a constant is a reasonable approximation, especially over the range of $0.5 < T_r < 1.0$. The plot for ethane shows another nearly linear region for $1/T_r > 2$ (temperatures below the normal boiling temperature), with a different slope and intercept. The approach of the previous section should be applied at $T_r < 0.5$. Integrating the Clapeyron equation for vapor pressure, we obtain,

$$\ln\left(\frac{P^{sat}}{P_R}\right) = -\frac{\Delta H^{vap}}{R\Delta Z^{vap}}\left(\frac{1}{T} - \frac{1}{T_R}\right)$$

9.10

🛈 The plot of $\ln P^{sat}$ versus $1/T$ is nearly linear.

Example 9.2 Vapor pressure interpolation

What is the value of the pressure in a piston/cylinder at $-12°C$ (261.2 K) with vapor and liquid propane present? Use only the boiling temperature (available from a handbook), critical properties, and acentric factor to determine the answer.

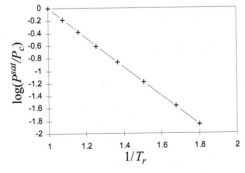
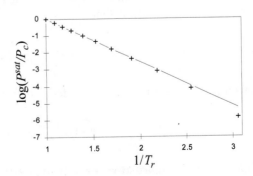

Figure 9.1 *Plot to evaluate Clausius-Clapeyron for calculation of vapor pressures at high pressures, argon (left) and ethane (right).*

Example 9.2 Vapor pressure interpolation (Continued)

Solution: We will use the boiling point and the vapor pressure given by the acentric factor to determine $(-\Delta H^{vap})/(R\Delta Z^{vap})$ for Eqn. 9.10, and then use the boiling temperature with $(-\Delta H^{vap})/(R\Delta Z^{vap})$ to determine the desired vapor pressure. First, let us use the acentric factor to determine the vapor pressure value at $T_r = 0.7$. For propane, $T_c = 369.8$ K, $P_c = 4.249$ MPa, and $\omega = 0.152$. Solving for the vapor pressure in terms of MPa by rearranging the definition of the acentric factor, $(P^{sat}|_{Tr=0.7}) = P_c \cdot 10^{(-(1+0.152))} = 0.2994$ MPa.[a] The temperature corresponding to this pressure is $T = T_r \cdot T_c = 0.7 \cdot 369.8 = 258.9$ K. The CRC handbook lists the normal boiling temperature of propane as $-42°C = 231.2$ K. Using these two vapor pressures in Eqn. 9.10:

$$\ln(0.2994/0.1013) = -\Delta H^{vap}/(R\Delta Z^{vap})(1/258.9 - 1/231.2) \Rightarrow -\Delta H^{vap}/(R\Delta Z) = -2342 \text{ K}$$

Therefore, using the boiling point and the value of $-\Delta H^{vap}/(R\Delta Z^{vap})$,

$$P^{sat}(261.2 \text{ K}) = 0.1013 \text{ MPa} \cdot \exp[-2342(1/261.2 - 1/231.2)] = 0.324 \text{ MPa}$$

The calculation is in excellent agreement with the experimental value of 0.324 MPa.

a. Could we use the Clausius-Clapeyron equation at this condition? Since the Clausius-Clapeyron equation requires the ideal gas law, the P^{sat} value must be low enough for the ideal gas law to be followed. The deviations at this state can be quickly checked with the virial equation, $P_r = 0.07$, $T_r = 0.7$, $B^0 = -0.664$, $B^1 = -0.630$, $Z = 0.924$; therefore, the Clausius-Clapeyron equation should probably not be used. Although you would get the same answer for vapor pressure over this narrow range, your inaccurate estimate of ΔH^{vap} might mislead you in a later calculation.

Since the linear relationship of Eqn. 9.10 applies over a broad range of temperatures, we can derive an approximate general estimate of the saturation pressure based on the critical point as the reference and acentric factor as a second point on the vapor pressure curve.

Setting $P_R = P_c$ and $T_R = T_c$,

$$\ln(P_r^{sat}) \approx \frac{-\Delta H^{vap}}{R\Delta Z^{vap} T_c}\left(\frac{T_c}{T} - \frac{T_c}{T_c}\right) = \frac{\Delta H^{vap}}{R\Delta Z^{vap} T_c}\left(1 - \frac{1}{T_r}\right)$$

Common logarithms are conventional for shortcut estimation, possibly because they are more convenient to visualize orders of magnitude.

$$\log_{10} P_r^{sat} = \frac{1}{2.303}\frac{\Delta H}{R\Delta Z T_c}\left(1 - \frac{1}{T_r}\right) \equiv A\left(1 - \frac{1}{T_r}\right)$$

Relating this equation to the acentric factor defined by Eqn. 7.2,

$$\log_{10} P_r^{sat}\Big|_{T_r=0.7} \equiv -(\omega + 1) = A\left(1 - \frac{1}{0.7}\right) = -\frac{3}{7}A \Rightarrow A = \frac{7}{3}(1 + \omega)$$

which results in a **shortcut vapor pressure** equation,

$$\log_{10} P_r^{sat} = \frac{7}{3}(1 + \omega)\left(1 - \frac{1}{T_r}\right)$$

9.11   🛑 Shortcut vapor pressure equation. Use care with the shortcut equation below $T_r = 0.5$.

> *Note: The shortcut vapor pressure equation must be regarded as an approximation for rapid estimates. The approximation is generally good above P = 0.5 bar; the percent error can become significant at lower pressures (and temperatures). Keep in mind that its estimates are based on the critical pressure which is generally 40–50 bar and acentric factor (at $T_r = 0.7$).*

Example 9.3 Application of the shortcut vapor pressure equation

Use the shortcut vapor pressure equation to calculate the vapor pressure of propane at −12°C, and compare the calculation with the results from Example 9.2.

Solution: For propane at −12°C, $T_r = 261.2/369.8 = 0.7063$,

$$P^{sat}(-12°C) \approx P_c \cdot 10^{\frac{7(1 + \omega)}{3}\left(1 - \frac{1}{0.7063}\right)} = 0.324 \text{ MPa}$$

This is in excellent agreement with the result of Example 9.2, with considerably less effort.

Example 9.4 General application of the Clapeyron equation

Liquid butane is pumped to a vaporizer as a saturated liquid under a pressure of 1.88 MPa. The butane leaves the exchanger as a wet vapor of 90 percent quality and at essentially the same pressure as it entered. From the following information, estimate the heat load on the vaporizer per gram of butane entering.

For butane, $T_c = 425.2$ K; $P_c = 3.797$ MPa; and $\omega = 0.193$. Use the shortcut method to estimate the temperature of the vaporizer, and the Peng-Robinson equation to determine the enthalpy of vaporization.

Solution: To find the T at which the process occurs:[a]

$$\log_{10}(P_r^{sat}) \approx \frac{7}{3}(1 + \omega)\left(1 - \frac{1}{T_r}\right) \Rightarrow T_r^{sat} = 0.90117, T = 383.2 \text{ K}$$

First, we use the Peng-Robinson equation to find departure functions for each phase, and subsequently determine the heat of vaporization at 383.2 K and 1.88 MPa:

$$\frac{H^V - H^{ig}}{RT} = -0.9949; \frac{H^L - H^{ig}}{RT} = -5.256;$$

Therefore, $\Delta H^{vap} = (-0.9949 + 5.256)8.314 \cdot 0.90117 \cdot 425.2 = 13,575$ J/mol

Since the butane enters as saturated liquid and exits at 90% quality, an energy balance gives
$Q = 0.9 \cdot 13,575 = 12,217$ J/mol $\cdot 1$mol/58g $= 210.6$ J/g

---

Example 9.4   General application of the Clapeyron equation   (Continued)

Alternatively, we could have used the shortcut equation another way by comparing the Clapeyron and shortcut equations:

Clapeyron: $\ln(P^{sat}) = -\Delta H^{vap}/RT(Z^V - Z^L) + \Delta H^{vap}/RT_c(Z^V - Z^L) + \ln P_c$

Shortcut: $\ln(P^{sat}) = 2.3025\dfrac{7}{3}(1 + \omega)\left(1 - \dfrac{1}{T_r}\right) + \ln P_c$

Comparing, we find: $\dfrac{\Delta H^{vap}}{R\Delta Z^{vap}} = 2.3025\dfrac{7}{3}(1 + \omega)T_c = 2725$ K

Therefore, using the Peng-Robinson equation at 383.3 K and 1.88 MPa to determine compressibility factor values,

$$\Delta H^{vap} = 2725R(Z^V - Z^L) = 2725(8.314)(0.6744 - 0.07854) = 13,500 \text{ J/mol}$$

which would give a result in good agreement with the first approach.

a.   In principle, since we are asked to use the Peng-Robinson equation for the rest of the problem, we could have used it to determine the saturation temperature also, but we were asked to use the shortcut method. The use of equations of state to calculate vapor pressure is discussed in Section 8.10.

The Antoine Equation

The simple form of the shortcut vapor-pressure equation is extremely appealing, but there are times when we desire greater precision than such a simple equation can provide. One obvious alternative would be to use the same form over a shorter range of temperatures. By fitting the local slope and intercept, an excellent fit could be obtained. To extend the range of applicability slightly, one modification is to introduce an additional adjustable parameter in the denominator of the equation. The resultant equation is referred to as the **Antoine equation**:

🛑 Antoine equation. Use with care outside the stated parameter temperature limits, and watch use of log, ln, and units carefully.

$$\log_{10} P^{sat} = A - B/(T + C) \tag{9.12}$$

where T is conventionally in Celsius.[2] Values of coefficients for the Antoine equation are widely available, notably in the compilations of vapor-liquid equilibrium data by Gmehling and coworkers.[3] The Antoine equation provides accurate correlation of vapor pressures over a narrow range of temperatures, *but a strong caution must be issued about applying the Antoine equation outside the stated temperature limits; it does not extrapolate well.* If you use the Antoine equation, you should be sure to report the temperature limits as well as the values of coefficients with every application. Antoine coefficients for some compounds are summarized in Appendix E and within the Excel workbook Antoine.xlsx.

💻 Antoine.xlsx and Matlab/Props have coefficients for many common substances.

9.4   CHANGES IN GIBBS ENERGY WITH PRESSURE

🛑 *Values* for Gibbs energy departures are needed for further generalization of phase equilibria.

We have seen that the Gibbs energy is the key property that must be used to characterize phase equilibria. In the previous section, we have used Gibbs energy in the derivation of useful relations

---

2.   What value of C would be common if the Clausius-Clapeyron equation was exact? Compare that value to tabulated C values.

3.   Gmehling, J. 1977-. *Vapor-liquid Equilibrium Data Collection*. Frankfurt, Germany: DECHEMA.

for vapor pressure. For our discussions here, we have been able to relate the two phases of a pure fluid to one another, and the actual calculation of *values* of the Gibbs energy were not needed. However, extension to general phase equilibria in the next chapters will require a capability to calculate departures of Gibbs energies of individual phases, sometimes using different techniques of calculation for each phase.

By observing the mathematical behavior of Gibbs energy for fluids derived from the above equations, some sense may be developed for how pressure affects Gibbs energy, and the property becomes somewhat more tangible. Beginning from our fundamental relation, $dG = -SdT + VdP$, the effect of pressure is most easily seen at constant temperature.

$$dG = VdP \text{(const. } T)$$ 9.13

❶ Starting point for many derivations.

Eqn. 9.13 is the basic equation used as a starting point for derivations used in phase equilibrium. In actual applications the appearance of the equation may differ, but it is useful to recall that most derivations originate with the variation of Eqn. 9.13. To evaluate the change in Gibbs energy, we simply need the P-V-T properties of the fluid. These P-V-T properties may be in the form of tabulated data from measurements, or predictions from a generalized correlation or an equation of state. For a change in pressure, Eqn. 9.13 may be integrated:

$$G_2 - G_1 = \int_{P_1}^{P_2} VdP \text{(const. } T)$$ 9.14

Methods for calculating Gibbs energies and related properties differ for gases, liquids, and solids. Each type of phase will be covered in a separate section to make the distinctions of the calculation methods clearer. Before proceeding to those analyses, however, we consider a problem which arises in the treatment of Gibbs energy at low pressure. This problem motivates the introduction of the term "fugacity" which takes the place of the Gibbs energy in the presentation in the following sections.

Gibbs Energy in the Low-Pressure Limit

The calculation of ΔG is illustrated in Fig. 9.2, where the shaded area represents the integral. The slope of a G versus P plot at constant temperature is equal to the molar volume. For a real fluid, the ideal-gas approximation is valid only at low pressures. The volume is given by $V = ZRT/P$; thus,

$$dG = RTZ \frac{dP}{P}$$ 9.15

which permits use of generalized correlations or volume-explicit equations of state to represent Z at T and P.[4] Of course, we may also use Eqn. 9.13 directly, using an equation of state to calculate V. Both techniques are shown later, but first the qualitative aspects of the calculations are illustrated.

For an ideal gas, we may substitute $Z = 1$ into Eqn. 9.15 to obtain

$$dG^{ig} = RT \frac{dP}{P} = RTd\ln P$$ (ig) 9.16

---

4. Naturally, the accuracy of our calculation is dependent on the accuracy of predicting Z, so we must use an accurate equation of state or correlation.

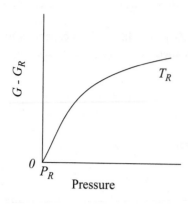
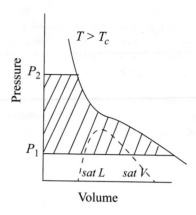

Figure 9.2 *Schematic of dependence of G on pressure for a real fluid at T_R, and an isothermal change on a P-V diagram for a change from P_1 to P_2.*

$$\Delta G^{ig} = \int_{P_1}^{P_2} \frac{RT}{P}\, dP = RT \ln \frac{P_2}{P_1} \qquad\text{(ig) 9.17}$$

Both dG and dG^{ig} become infinite as pressure approaches zero. This means that both Eqns. 9.15 and 9.16 are difficult to use directly at low pressure. However, as a real fluid state approaches zero pressure, Z will approach the ideal gas limit and dG approaches dG^{ig}. Thus, the difference $dG - dG^{ig}$ will remain finite, and goes to zero as P goes to zero. Thus,

$$dG - dG^{ig} = d(G - G^{ig})$$

which is simply the change in departure function. Therefore, we combine Eqns. 9.15 and 9.16 and write:

⓵ Differential form of the Gibbs departure.

$$d(G - G^{ig})/RT = (Z - 1)/P \; dP \qquad\qquad 9.18$$

This relates the departure function to the *P-V-T* properties in a way that we have seen before. If you look back to Eqn. 8.26, that equation looks different because we are integrating over volume rather than pressure, but they are really related. We use this departure to define a new state property, *fugacity,* to describe phase behavior. We reserve further discussion of pressure effects in gases for the following sections, where fugacity and Gibbs energy can be considered simultaneously. The generalized treatment by departure functions is also discussed there.

9.5   FUGACITY AND FUGACITY COEFFICIENT

⓵ Fugacity can be directly related to measurable properties under the correct conditions.

In principle, all pure-component, phase-equilibrium problems could be solved using Gibbs energy. Historically, however, an alternative property has been applied in chemical engineering calculations, the *fugacity.* The fugacity has one advantage over the Gibbs energy in that its application to mixtures is a straightforward extension of its application to pure fluids. It also has some empirical appeal because the fugacity of an ideal gas equals the pressure and the fugacity of a liquid equals the vapor pressure under common conditions, as we will show in Section 9.8. The vapor pressure was the original property used for characterization of phase equilibrium by experimentalists.

The forms of Eqns. 9.16 and 9.15 are similar, and the simplicity of Eqn. 9.16 is appealing. G.N. Lewis *defined* fugacity by

$$dG = VdP \equiv RTd\ln f \qquad\qquad 9.19$$

and comparing to Eqn. 9.16, we see that

$$d(G - G^{ig})/RT = d\ln(f/P) \qquad\qquad 9.20$$

Integrating from low pressure, at constant temperature, we have for the left-hand side,

$$\frac{1}{RT}\int_0^P d(G - G^{ig}) = \frac{1}{RT}\left[(G - G^{ig})\Big|_P - (G - G^{ig})\Big|_{P=0}\right] = \frac{(G - G^{ig})}{RT}$$

because $(G - G^{ig})$ approaches zero at low pressure. Integrating the right-hand side of Eqn. 9.20, we have

$$\ln\left(\frac{f}{P}\right)\Bigg|_P - \ln\left(\frac{f}{P}\right)\Bigg|_{P=0}$$

To complete the definition of fugacity, we define the low-pressure limit,

$$\lim_{P \to 0}\left(\frac{f}{P}\right) = 1 \qquad\qquad 9.21$$

and we define the ratio f/P to be the *fugacity coefficient*, φ.

$$\boxed{\frac{(G - G^{ig})}{RT} = \ln\left(\frac{f}{P}\right) = \ln\varphi} \qquad\qquad 9.22$$

The fugacity coefficient is simply another way of characterizing the Gibbs departure function at a fixed T,P. For an ideal gas, the fugacity will equal the pressure, and the fugacity coefficient will be unity. For representations of the P-V-T data in the form $Z = f(T,P)$ (like the virial equation of state), the fugacity coefficient is evaluated from an equation of the form:

$$\boxed{\frac{(G - G^{ig})}{RT} = \ln\left(\frac{f}{P}\right) = \ln\varphi = \frac{1}{RT}\int_0^P (V - V^{ig})dP = \int_0^P \frac{(Z-1)}{P}dP} \qquad\qquad 9.23$$

or the equivalent form for P-V-T data in the form $Z = f(T,V)$, which is essentially Eqn. 8.26,

$$\boxed{\frac{(G - G^{ig})}{RT} = \ln\left(\frac{f}{P}\right) = \int_0^\rho \frac{(Z-1)}{\rho}d\rho + (Z-1) - \ln Z} \qquad\qquad 9.24$$

which is the form used for cubic equations of state.

> ❗ Fugacity and fugacity coefficient are convenient ways to quantify the Gibbs departure.

> ❗ Fugacity has units of pressure, and the fugacity coefficient is dimensionless.

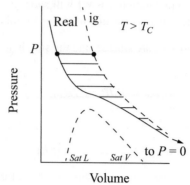

Figure 9.3 *Illustration of RT ln φ as a departure function.*

A graphical interpretation of the fugacity coefficient can be seen in Fig. 9.3. The integral of Eqn. 9.23 is represented by the negative value of the shaded region between the real gas isotherm and the ideal gas isotherm. The fugacity coefficient is a measure of non-ideality. *Under most common conditions, the fugacity coefficient is less than one.* At very high pressures, the fugacity coefficient can become greater than one.

> *Note: In practice, we do not evaluate the fugacity of a substance directly. Instead, we evaluate the fugacity coefficient, and then calculate the fugacity by*

$$\boxed{f = \varphi P}$$
9.25

9.6   FUGACITY CRITERIA FOR PHASE EQUILIBRIA

We began the chapter by showing that Gibbs energy was equivalent in phases at equilibrium. Here we show that equilibrium may also be described by equivalence of fugacities. Since

$$G^L = G^V$$
9.3

we may subtract G^{ig} from both sides and divide by RT, giving

$$\frac{(G^L - G^{ig})}{RT} = \frac{(G^V - G^{ig})}{RT}$$
9.26

Substituting Eqn. 9.22,

$$\ln\left(\frac{f^L}{P}\right) = \ln\left(\frac{f^V}{P}\right)$$

which becomes

$$\boxed{f^L = f^V}$$
9.27

Therefore, calculation of fugacity and equating in each phase becomes the preferred method of calculating phase equilibria. In the next few sections, we discuss the methods for calculation of fugacity of gases, liquids, and solids.

9.7 CALCULATION OF FUGACITY (GASES)

The principle of calculation of the fugacity coefficient is the same by all methods—Eqn. 9.23 or 9.24 is evaluated. The methods *look* considerably different, usually because the *P-V-T* properties are summarized differently. All methods use the formula below and differ only in the manner the fugacity coefficient is evaluated.

$$f = \varphi P \qquad\qquad 9.28$$

Equations of State

Equations of state are the dominant method used in process simulators because the EOS can be solved rapidly by computer. We consider two equations of state, the virial equation and the Peng-Robinson equation. We also consider the generalized compressibility factor charts as calculated with the Lee-Kesler equation.

Ideal Gas

$$\varphi^{ig} = 1 \text{ and } f^{ig} = P \qquad\qquad \text{(ig) } 9.29$$

The Virial Equation

The virial equation may be used to represent the compressibility factor in the *low-to-moderate pressure region* where Z is linear with pressure at constant temperature. Eqn. 7.10 should be used to evaluate the appropriateness of the virial coefficient method. Substituting $Z = 1 + BP/RT$, or $Z - 1 = BP/RT$ into Eqn. 9.23,

$$\ln \varphi = \int_0^P \frac{B}{RT} dP = \frac{BP}{RT} \qquad\qquad 9.30$$

Thus,

$$\ln \varphi = \frac{BP}{RT} \qquad\qquad 9.31$$

❶ The virial equation for gases.

Writing the virial coefficient in reduced temperature and pressure,

$$\ln \varphi = \frac{P_r}{T_r}(B^0 + \omega B^1) \qquad\qquad 9.32$$

where B^0 and B^1 are the virial coefficient correlations given in Eqns. 7.8 and 7.9 on page 259.

The Peng-Robinson Equation

Cubic equations of state are particularly useful in the petroleum and hydrocarbon-processing industries because they may be used to represent both vapor and liquid phases. Chapter 7 discussed how equations of state may be used to represent the volumetric properties of gases. The integral of Eqn. 9.23 is difficult to use for pressure-explicit equations of state; therefore, it is solved in the format of Eqn. 9.24. The integral is evaluated analytically by methods of Chapter 8. In fact, the result of Example 8.6 on page 317 is $\ln\varphi$ according to the Peng-Robinson equation.

❶ Peng-Robinson equation.

$$\ln\varphi = -\ln(Z-B) - \frac{A}{B\sqrt{8}}\ln\left[\frac{Z+(1+\sqrt{2})B}{Z+(1-\sqrt{2})B}\right] + Z - 1 \qquad 9.33$$

To apply, the technique is analogous to the calculation of departure functions. At a given P, T, the cubic equation is solved for Z, and the result is used to calculate φ and then fugacity is calculated, $f = \varphi P$. This method has been programmed into Preos.xlsx and Preos.m.

Below the critical temperature, equations of state may also be used to predict vapor pressure, saturated vapor volume, and saturated liquid volume, as well as liquid volumetric properties. While Eqn. 9.33 can be used to calculate fugacity coefficients for liquids, the details of the calculation will be discussed in the next section. Note again that Eqn. 9.24 is closely related to Eqn. 8.26 as used in Example 8.6 on page 317.

Generalized Charts

Properties represented by generalized charts may help to visualize the magnitudes of the fugacity coefficient in various regions of temperature and pressure. To use the generalized chart, we write

❶ Generalized charts.

$$\ln\varphi = \int_0^P (Z-1)\frac{dP}{P} = \frac{G-G^{ig}}{RT} = \int_0^\rho (Z-1)\frac{d\rho}{\rho} + (Z-1) - \ln Z \qquad 9.34$$

The Gibbs energy departure chart can be generated from the Lee-Kesler equation by specifying a particular value for the acentric factor. The charts are for the correlation $\ln\varphi = \ln\varphi^0 + \omega\ln\varphi^1$. The entropy departure can also be estimated by combining Fig. 9.4 with Fig. 8.7:

$$(S-S^{ig})/R = [(H-H^{ig})/RT]/T_r - (G-G^{ig})/(RT) \qquad 9.35$$

Fig. 9.4 can be useful for hand calculation, if you do not have a computer. A sample calculation for propane at 463.15 and 2.5 MPa gives

$$\ln(f/P) = \frac{G-G^{ig}}{RT} = -0.1 + 0.152(0.05) = -0.09$$

compared to the value of −0.112 from the Peng-Robinson equation.

9.8 CALCULATION OF FUGACITY (LIQUIDS)

To introduce the calculation of fugacity for liquids, consider Fig. 9.5. The shape of an isotherm below the critical temperature differs significantly from an ideal-gas isotherm. Such an isotherm is illustrated which begins in the vapor region at low pressure, intersects the phase boundary where vapor and liquid coexist, and then extends to higher pressure in the liquid region. Point A represents a vapor state, point B represents saturated vapor, point C represents saturated liquid, and point D represents a liquid.

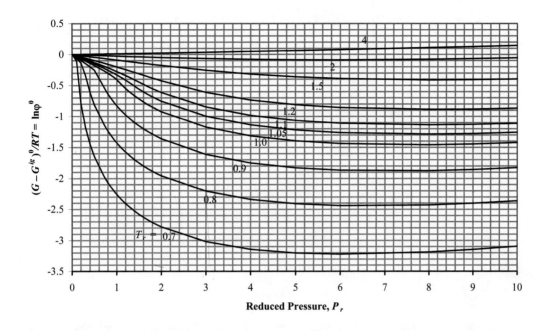

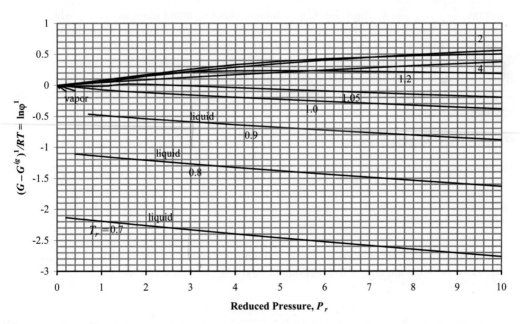

Figure 9.4 *Generalized charts for estimating the Gibbs departure function using the Lee-Kesler equation of state. $(G - G^{ig})^0/RT$ uses $\omega = 0.0$, and $(G - G^{ig})^1/RT$ is the correction factor for a hypothetical compound with $\omega = 1.0$.*

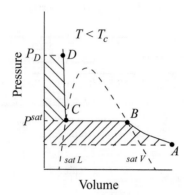

Figure 9.5 *Schematic for calculation of Gibbs energy
and fugacity changes at constant temperature
for a pure liquid.*

We showed in Section 9.6 on page 346 that

$$f_C = f_B = f^{sat}$$

9.36

Note that we have designated the fugacity at points C and B equal to f^{sat}. This notation signifies a saturation condition, and as such, it does not require a distinction between liquid or vapor. Therefore, we may refer to point B or C as saturated vapor or liquid interchangeably when we discuss fugacity. The calculation of the fugacity at point B (saturated vapor) is also adequate for calculation of the fugacity at point C, the fugacity of saturated liquid. Calculation of the saturation fugacity may be carried out by any of the methods for calculation of vapor fugacities from the above section. Methods differ slightly on how the fugacity is calculated between points C and D. There are two primary methods for calculating this fugacity change. They are the Poynting method and the equation of state method.

Poynting Method

The Poynting method applies Eqn. 9.19 between saturation (points B, C) and point D. The integral is

$$RT \ln \frac{f_D}{f^{sat}} = \int_{P^{sat}}^{P_D} V dP$$

9.37

Since liquids are fairly incompressible for $T_r < 0.9$, the volume is approximately constant, and may be removed from the integral, with the resultant Poynting correction becoming

⚠ Poynting
correction.

$$\frac{f}{f^{sat}} = \exp\left(\frac{V^L(P - P^{sat})}{RT}\right)$$

9.38

The fugacity is then calculated by

⚠ Poynting
method for liquids.

$$f = \varphi^{sat} P^{sat} \exp\left(\frac{V^L(P - P^{sat})}{RT}\right)$$

9.39

Saturated liquid volume can be estimated within a slight percent error using the **Rackett** equation

$$V^{satL} = V_c Z_c^{(1 - T_r)^{0.2857}}$$

9.40

The Poynting correction, Eqn. 9.38, is essentially unity for many compounds near room T and P; thus, it is frequently ignored.

$$\boxed{f^L \approx \varphi^{sat} P^{sat}} \quad \text{or commonly} \quad \boxed{f^L \approx P^{sat}}$$

9.41   ❗Frequent approximation.

Equation of State Method

Calculation of liquid fugacity by the equation of state method uses Eqn. 9.24 just as for vapor. To apply the Peng-Robinson equation of state, we can use Eqn. 9.33. *The only significant consideration is that the liquid compressibility factor must be used.* To understand the mathematics of the calculation, consider the isotherm shown in Fig. 9.6(a). When $T_r < 1$, the equation of state predicts an isotherm with "humps" in the vapor/liquid region. Surprisingly, these swings can encompass a range of negative values of the pressure near C' (although not shown in our example). The exact values of these negative pressures are not generally taken too seriously, however, because they occur in a region of the P-V diagram that is unimportant for routine calculations. Since the Gibbs energy from an equation of state is given by an integral of the volume with respect to pressure, the quantity of interest is represented by an integral of the humps. The downward and upward humps cancel one another in generating that integral. This observation gives rise to the equal area rule for computing saturation conditions to be discussed in Section 9.10 where we show that the shaded area above line $\overline{BC}$ is equal to the shaded area below, and that the pressure where the line is located represents the saturation condition (vapor pressure). With regard to fugacity calculations, it is sufficient simply to note that these humps are in fact integrable, and easily computed by the same formula derived for the vapor fugacity by an equation of state.

Fig. 9.6(b) shows that the molar Gibbs energy is indeed continuous as the fluid transforms from the vapor to the liquid. The Gibbs energy first increases according to Eqn. 9.14 based on the

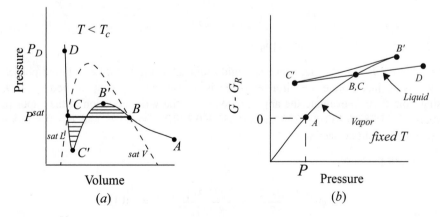

Figure 9.6 *Schematic illustration of the prediction of an isotherm by a cubic equation of state. Compare with Fig. 9.5 on page 350. The figure on the right shows the calculation of Gibbs energy relative to a reference state. The fugacity will have the same qualitative shape.*

vapor volume. Note that the volume and pressure changes are both positive, so the Gibbs energy relative to the reference value is monotonically increasing. During the transition from vapor to liquid, the "humps" lead to the triangular region associated with the name of **van der Waals loop.** Then the liquid behavior takes over and Eqn. 9.14 comes back into play, this time using the liquid volume. Note that the isothermal pressure derivative of the Gibbs energy is not continuous. Can you develop a simple expression for this derivative in terms of P, V, T, C_P, C_V and their derivatives? Based on your answer to the preceding question, would you expect the change in the derivative to be a big change or a small change?

Example 9.5 Vapor and liquid fugacities using the virial equation

Determine the fugacity (MPa) for acetylene at (a) 250 K and 10 bar, and (b) 250 K and 20 bar. Use the virial equation and the shortcut vapor pressure equation.

Solution: From the property files Props.xlsx or props.mat of the software, for acetylene: $T_c = 308.3$ K, $P_c = 6.139$, $\omega = 0.187$, $Z_c = 0.271$. For each part of the problem, the fluid state of aggregation is determined before the method of solution is specified. At 250 K, using the shortcut vapor pressure equation, Eqn 9.11, the vapor pressure is $P^{sat} = 1.387$ MPa.

We will calculate the virial coefficient at 250 K using Eqns. 7.7–7.9:
$T_r = 250/308.3 = 0.810$, $B^0 = -0.5071$, $B^1 = -0.2758$, $B = -233.3$ cm^3/mol.

(a) $P = 1$ MPa $< P^{sat}$ so the acetylene is vapor (between points A and B in Fig. 9.5). Using Eqn 7.10 to evaluate the appropriateness of the virial equation at 1 MPa, $P_r = 1/6.139 = 0.163$, and $0.686 + 0.439P_r = 0.76$ and $T_r = 0.810$, so the correlation should be accurate.

Using Eqn. 9.31,

$$\ln \varphi = \frac{BP}{RT} = \frac{-233.3(1)}{8.314(250)} = -0.11224$$

$(f = \varphi P = 0.894 (1) = 0.894$ MPa

(b) $P = 2$ MPa $> P^{sat}$, so the acetylene is liquid (point D of Fig. 9.5). For a liquid phase, the only way to incorporate the virial equation is to use the Poynting method, Eqn. 9.39. Using Eqn. 7.10 to evaluate the appropriateness of the virial equation at the vapor pressure, $P_r^{sat} = 1.387/6.139 = 0.2259$, and $0.686 + 0.439P_r^{sat} = 0.785$, and $T_r = 0.810$, so the correlation should be accurate.

At the vapor pressure,

$$\ln \varphi^{sat} = \frac{BP^{sat}}{RT} = \frac{-233.3(1.387)}{8.314(250)} = -0.156$$

$f^{sat} = \varphi^{sat} P^{sat} = 0.8558(1.387) = 1.187$ MPa

> ## Example 9.5 Vapor and liquid fugacities using the virial equation (Continued)
>
> Using the Poynting method to correct for pressure beyond the vapor pressure will require the liquid volume, estimated with the Rackett equation, Eqn. 9.40, using $V_c = Z_c RTc/P_c =$ 0.271(8.314)(308.3)/6.139 = 113.2 cm³/mol.
>
> $$V^{satL} = 113.2(0.271)^{(1-0.8109)^{0.2857}} = 50.3 \ cm^3/mol$$
>
> The Poynting correction is given by Eqn. 9.38,
>
> $$\frac{f}{f^{sat}} = \exp\left(\frac{50.3(2-1.387)}{8.314(250)}\right) = 1.015$$
>
> Thus, f = 1.187(1.015) = 1.20 MPa. The fugacity is close to the value of vapor pressure for liquid acetylene, even though the pressure is 2 MPa.

9.9   CALCULATION OF FUGACITY (SOLIDS)

Fugacities of solids are calculated using the Poynting method, with the exception that the volume in the Poynting correction is the volume of the solid phase.

$$f^S = \varphi^{sat} P^{sat} \exp\left(\frac{V^S(P-P^{sat})}{RT}\right)$$

9.42   ❶ Poynting method for solids.

Any of the methods for vapors may be used for calculation of φ^{sat}. P^{sat} is obtained from thermodynamic tables. Equations of state are generally not applicable for calculation of solid phases because they are used only to represent liquid and vapor phases. However, they may be used to calculate the fugacity of a vapor phase in equilibrium with a solid, given by $\varphi^{sat} P^{sat}$. As for liquids, the Poynting correction may be frequently set to unity with negligible error.

$$f^S \approx \varphi^{sat} P^{sat} \quad \text{or commonly} \quad f^S \approx P^{sat}$$

9.43   ❶ Frequent approximation.

9.10   SATURATION CONDITIONS FROM AN EQUATION OF STATE

The only thermodynamic specification that is required for determining the saturation temperature or pressure is that the Gibbs energies (or fugacities) of the vapor and liquid be equal. *This involves finding the pressure or temperature where the vapor and liquid fugacities are equal.* The interesting part of the problem comes in computing the saturation condition by iterating on the temperature or pressure.

Example 9.6 Vapor pressure from the Peng-Robinson equation

Preos.xlsx,
PreosPropsMenu.m.

Use the Peng-Robinson equation to calculate the normal boiling point of methane.

Solution: Vapor pressure calculations are available in Preos.xlsx and PreosPropsmenu.m. Let us discuss Preos.xlsx first. The spreadsheet is more illustrative in showing the steps to the calculation. Computing the saturation temperature or pressure in Excel is rapid using the Solver tool in Excel.

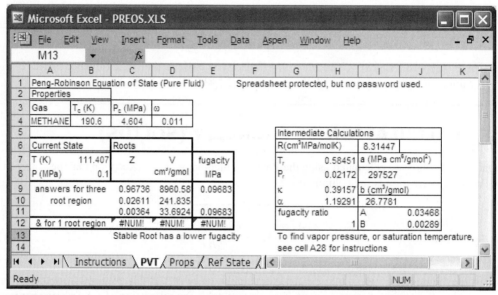

Figure 9.7 *Example of Preos.xlsx used to calculate vapor pressure.*

On the spreadsheet shown in Fig. 9.7, cell H12 is included with the fugacity ratio of the two phases; the cell can be used to locate a saturation condition. Initialize Excel by entering the desired P in cell B8, in this case 0.1 MPa. Then, adjust the temperature to provide a guess in the two-phase (three-root) region. Then, instruct Solver to set the cell for the fugacity ratio (H12) to a value of one by adjusting temperature (B7), subject to the constraint that the temperature (B7) is less than the critical temperature.

Phase equilib-
ria involves finding
the state where
$f^L = f^V$.

In MATLAB, the initial guess is entered in the upper left. The "Run Type" is set as a saturation calculation. The "Root to use" and "Value to match" are not used for a saturation calculation. The drop-down box "For Matching..." is set to adjust the temperature. The results are shown in Fig. 9.8.

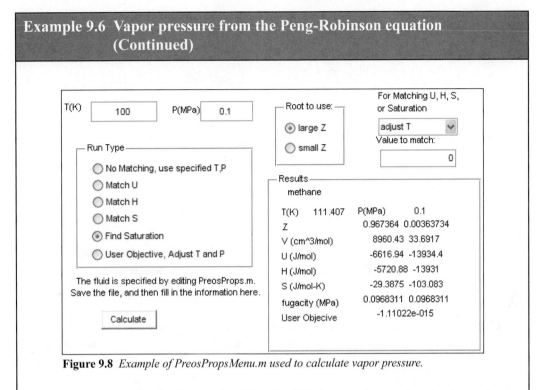

Example 9.6 Vapor pressure from the Peng-Robinson equation (Continued)

Figure 9.8 *Example of PreosPropsMenu.m used to calculate vapor pressure.*

For methane the solution is found to be 111.4 K which is very close to the experimental value used in Example 8.9 on page 320. Saturation pressures can also be found by adjusting pressure at fixed temperature.

Fugacity and P-V isotherms for CO_2 as calculated by the Peng-Robinson equation are shown in Fig. 9.9 and Fig. 7.5 on page 264. Fig. 9.9 shows more clearly how the shape of the isotherm is related to the fugacity calculation. Note that the fugacity of the liquid root at pressures between B and B' of Fig. 9.6 is lower than the fugacity of the vapor root in the same range, and thus is more stable because the Gibbs energy is lower. Analogous comparisons of vapor and liquid roots at pressures between C and C' show that vapor is more stable. In Chapter 7, we empirically instructed readers to use the lower fugacity. Now, in light of Fig. 9.9, readers can understand the reasons for the use of fugacity.

The term "fugacity" was defined by G. N. Lewis based on the Latin for "fleetness," which is often interpreted as "the tendency to flee or escape," or simply "escaping tendency." It is sometimes hard to understand the reasons for this term when calculating the property for a single root. However, when multiple roots exist as shown in Fig. 9.9, you may be able to understand how the system tries to "escape: from the higher fugacity values to the lower values. This perspective is especially helpful in mixtures, indicating the direction of driving forces to lower fugacities.

Just as the vapor pressure estimated by the shortcut vapor pressure equation is less than 100% accurate, the vapor pressure estimated by an equation of state is less than 100% accurate. For example, the Peng-Robinson equation tends to yield about 5% average error over the range $0.4 < T_r < 1.0$. This represents a significant improvement over the shortcut equation. The van der Waals equation, on the other hand, yields much larger errors in vapor pressure. One problem is that the van der Waals equation offers no means of incorporating the acentric factor to fine-tune the characterization

A stable system minimizes its Gibbs energy and its fugacity.

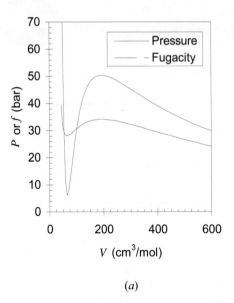
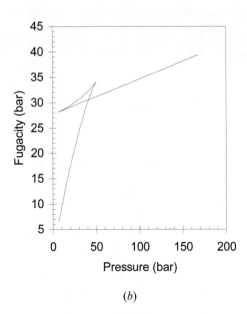

(a)   (b)

Figure 9.9 *Predictions of the Peng-Robinson equation of state for CO_2: (a) prediction of the P-V isotherm and fugacity at 280 K; (b) plot of data from part (a) as fugacity versus pressure, showing the crossover of fugacity at the vapor pressure. Several isotherms for CO_2 are shown in Fig. 7.5 on page 264.*

of vapor pressure. But the problems with the van der Waals equation go much deeper, as illustrated in the example below.

Example 9.7 Acentric factor for the van der Waals equation

Adapting Preos.xlsx to a different equation of state.

To clarify the problem with the van der Waals equation in regard to phase-equilibrium calculations, it is enlightening to compute the reduced vapor pressure at a reduced temperature of 0.7. Then we can apply the definition of the acentric factor to characterize the vapor pressure behavior of the van der Waals equation. If the acentric factor computed by the van der Waals equation deviates significantly from the acentric factor of typical fluids of interest, then we can quickly assess the magnitude of the error by applying the shortcut vapor-pressure equation. Perform this calculation and compare the resulting acentric factor to those on the inside covers of the book.

Solution: The computations for the van der Waals equation are very similar to those for the Peng-Robinson equation. We simply need to derive the appropriate expressions for a_0, a_1, and a_2, that go into the analytical solution of the cubic equation: $Z^3 + a_2 Z^2 + a_1 Z + a_0 = 0$.
Adapting the procedure for the Peng-Robinson equation given in Section 7.6 on page 263, we can make Eqn. 7.13 dimensionless:

$$Z = \frac{1}{1 - b\rho} - \frac{a\rho}{RT} = \frac{1}{1 - B/Z} - \frac{A}{Z} \qquad\qquad 9.44$$

where the dimensionless parameters are given by Eqns. 7.21–7.24; $A = (27/64)\, P_r/T_r^2$; $B = 0.125\, P_r/T_r$.

Example 9.7 Acentric factor for the van der Waals equation (Continued)

After writing the cubic in Z, the coefficients can be identified: $a_0 = -AB$; $a_1 = A$; and $a_2 = -(1 + B)$. For the calculation of vapor pressure, the fugacity coefficient for the van der Waals equation is quickly derived as the following:

$$\ln\left(\frac{f}{P}\right) = \int_0^\rho \frac{Z-1}{\rho}d\rho + Z - 1 - \ln Z = -\ln(Z - B) - \frac{A}{Z} + Z - 1 \qquad 9.45$$

Substituting these relations in place of their equivalents in Preos.xlsx, the problem is ready to be solved. Since we are not interested in any specific compound, we can set $T_c = 1$ and $P_c = 1$, $T_r = 0.7$. Setting an initial guess of $P_r = 0.1$, Solver gives the result that $P_r = 0.20046$.

Modification of PreosPropsMenu.m is accomplished by editing the routine PreosProps.m. Search for the text "global constants." Change the statements to match the a and b for the van der Waals equation. Search for "PRsolveZ" Two cases will be calls and you may wish to change the function name to "vdwsolveZ." The third case of "PRsolveZ" will be the function that solves the cubic. Change the function name. Edit the formulas used for the cubic coefficients. Finally, specify a fluid and find the vapor pressure at the temperature corresponding to $T_r = 0.7$.

The definition of the acentric factor gives

$$\omega = -\log(P_r) - 1 = -\log(0.20046) - 1 = -0.302$$

Comparing this value to the acentric factors listed in the table on the back flap, the only compound that even comes close is hydrogen, for which we rarely calculate fugacities at $T_r < 1$. This is the most significant shortcoming of the van der Waals equation. This shortcoming becomes most apparent when attempting to correlate phase-equilibria data for mixtures. Then it becomes very clear that accurate correlation of the mixture phase equilibria is impossible without accurate characterization of the pure component phase equilibria, and thus the van der Waals equation by itself is not useful for *quantitative* calculations. Correcting the repulsive contribution of the van der Waals equation using the Carnahan-Starling or ESD form gives significant improvement in the acentric factor. Another approach is to correct the attractive contribution in a way that cancels the error of the repulsive contribution. Cancellation is the approach that historically prevailed in the Redlich-Kwong, Soave, and Peng-Robinson equations.

The Equal Area Rule

As noted above, the swings in the P-V curve give rise to a cancellation in the area under the curve that becomes the free energy/fugacity. A brief discussion is helpful to develop an understanding of how the saturation pressure and liquid and vapor volumes are determined from such an isotherm.

To make this analysis quantitative, it is helpful to recall the formulas for the Gibbs departure functions, noting that the Gibbs departure is equal for the vapor and liquid phases (Eqn. 9.26).

$$\frac{G^L - G^V}{RT} = \frac{G^L - G^{ig}}{RT} - \frac{G^V - G^{ig}}{RT} = \int_{\rho^V}^{\rho^L} \frac{Z-1}{\rho}d\rho + Z^L - Z^V - \ln(Z^L/Z^V) \qquad 9.46$$

$$\frac{G^L - G^V}{RT} = -\int_{V^V}^{V^L} \left(\frac{P}{RT} - \frac{1}{V}\right) dV + \frac{1}{RT}(PV^L - PV^v) - \ln\left(\frac{PV^L/RT}{PV^V/RT}\right) \qquad 9.47$$

$$\frac{G^L - G^V}{RT} = \frac{1}{RT}(PV^L - PV^V) - \int_{V^V}^{V^L} \left(\frac{P}{RT}\right) dV + \int_{V^V}^{V^L} \left(\frac{1}{V}\right) dV - \ln\left(\frac{V^L}{V^V}\right) \qquad 9.48$$

$$\frac{G^L - G^V}{RT} = \frac{-1}{RT}\left\{ -[PV]_{V^V}^{V^L} + \int_{V^V}^{V^L} P \, dV \right\} \qquad 9.49$$

In the final equation, the second term in the right-hand side braces represents the area under the isotherm, and the first term on the right-hand side represents the rectangular area described by drawing a horizontal line at the saturation pressure from the liquid volume to the vapor volume in Fig. 9.6(a). Since this area is subtracted from the total inside the braces, the shaded area above a vapor pressure is equal to the shaded area below the vapor pressure for each isotherm. This method of computing the saturation condition is very sensitive to the shape of the P-V curve in the vicinity of the critical point and can be quite useful in estimating saturation properties at near-critical conditions.

Although Eqn. 9.49 illustrates the concept of the equal area rule most clearly, it is not in the form that is most useful for practical application. Noting that $G^L = G^V$ at equilibrium and rearranging Eqn. 9.47 gives

$$P = \left(\int_{\rho^V}^{\rho^L} \frac{Z-1}{\rho} d\rho - \ln\left(\frac{V^L}{V^V}\right)\right)\left[\frac{RT}{(V^V - V^L)}\right] \qquad 9.50$$

You should recognize the first term on the right-hand side as $(A^L - A^{ig})_{T,V} - (A^V - A^{ig})_{T,V}$. You probably have an expression for $(A - A^{ig})_{T,V}$ already derived. Solving for P is iterative in nature because we must first guess a value for P to solve for V^V and V^L. Five or six iterations normally suffice to converge to reasonable precision.[5] The method is guaranteed to converge as long as a maximum and minimum exist in the P-V isotherm. Therefore, initiation begins with finding the extrema, a form of "stability check" (see below) in the sense that the absence of extrema indicates a single phase. The search for extrema is facilitated by noting that the vapor maximum must appear at $\rho < \rho_c$ while the liquid minimum must appear at $\rho < \rho_c$. If $P_{min} > 0$, then initialize to $P_0 = (P_{min} + P_{max})/2$. Otherwise, initialize by finding V^V and V^L at $P = 10^{-12}$. The procedure is applied in Example 9.8 below.

5. Eubank, P.T., Elhassan, A.E., Barrufet, M.A. 1992. Whiting, W.B. *Ind. Eng. Chem Res.* 31:942. Tang, Y., Stephenson, G., Zhao, Z., Agrawal, M., Saha, S. 2011. *AIChE J.* 57:3333.

Example 9.8 Vapor pressure using equal area rule

Convergence can be tricky near the critical point or at very low temperatures when using the equality of fugacity, as in Example 9.6. The equal area rule can be helpful in those situations. As an example, try calling the solver for CO_2 at 30°C. Even though the initial guess from the shortcut equation is very good, the solver diverges. Alternatively, apply the equal area rule to solve as described above. Conditions in this range may be "critical" to designing CO_2 refrigeration systems, so reliable convergence is important.

Solution: The first step is to construct an isotherm and find the spinodal densities and pressures. Fig. 9.10 shows that $P_{min} = 7.1917$, $P_{max} = 7.2291$, $V_{max} = 117.98$, and $V_{min} = 94.509$. Following the procedure above, $P_0 = 7.2104$. Solving for the vapor and liquid roots at P_0 in the usual way gives $V_{vap} = 129.842$, and $V_{liq} = 88.160$. Similarly, $(A^L - A^{ig})_{T,V} = -1.0652$ and $(A^V - A^{ig})_{T,V} = -0.7973$, referring to the formula given in Example 8.6 on page 317:

$$\frac{(A - A^{ig})_{T,V}}{RT} = -\ln(1 - b\rho) - \frac{a}{bRT\sqrt{8}}\ln\left[\frac{1 + (1 + \sqrt{2})b\rho}{1 + (1 - \sqrt{2})b\rho}\right]$$

This leads to the next estimate of P^{sat} as,

$$P^{sat} = [-1.0652 + 0.7973 - \ln(88.160/129.842)][8.314(303.15)/(129.842 - 88.160)] = 7.2129$$

Solving for the vapor and liquid roots and repeating twice more gives: $P^{sat} = 7.21288$, shown below. Note the narrow range of pressures.

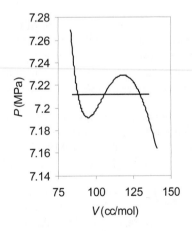

Figure 9.10 *Illustration of use of the equal area rule for a small van der Waals loop.*

9.11 STABLE ROOTS AND SATURATION CONDITIONS

When multiple real roots exist, the fugacity is used to determine which root is stable as explained in Section 9.10. However, often we are seeking a value of a state property and we are unable to find a stable root with the target value. This section explains how we handle that situation. We use entropy for the discussion, but calculations with other state properties would be similar.

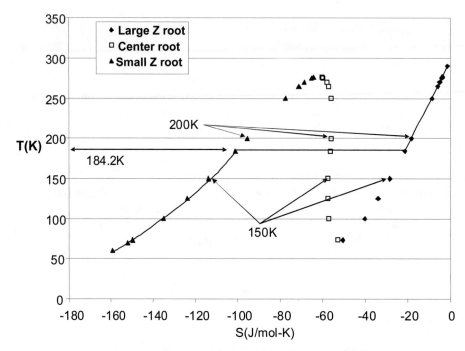

Figure 9.11 *Entropy values for ethane calculated from the Peng-Robinson equation along an isobar at 0.1 MPa. The largest Z root is shown as diamonds, the smallest Z root is shown as triangles, and the center root is shown as open squares. The stable behavior is indicated by the solid line.*

Fig. 9.11 shows the behavior of the entropy values for ethane at 0.1 MPa as calculated by the Peng-Robinson equation of state using a real gas reference state of $T = 298.15$K and $P = 0.1$ MPa. The figure was generated from the spreadsheet Preos.xlsx by adding the formula for entropy departure for the center root and then changing the temperature at a fixed pressure of 0.1 MPa. The corresponding values of S were recorded and plotted.

Suppose a process problem requires a state with $S = -18.185$ J/mol-K. At 200 K, the largest Z root has this value. The corresponding values of the fugacities from largest to smallest Z are 0.0976 MPa, 0.652 MPa, and 0.206 MPa, indicating that the largest root is most stable, so the largest root will give the remainder of the state variables.

Suppose a process problem requires a state with $S = -28.35$ J/mol-K. At 150 K, the largest Z root has this value. The corresponding values of the fugacities from largest to smallest Z are 0.0951 MPa, 0.313 MPa, and 0.0099 MPa, indicating that the smallest root is most stable. Even though the largest Z root has the correct value of S, the root is not the most stable root, and must be discarded.

Further exploration of roots would show that the desired value of S cannot be obtained by the middle or smallest roots, or any most stable root. Usually if this behavior is suspected, it is quickest to determine the saturation conditions for the given pressure and compare the saturation values to the specified value. (Think about how you handled saturated steam calculations from a turbine using the steam tables and used the saturation values as a guide.) The saturation conditions at 0.1 MPa can be found by adjusting the T until the fugacities become equal for the large Z and small Z roots, which is found to occur at 184.2 K. At this condition, the corresponding values of the fugacities from largest to smallest Z are 0.0971 MPa, 0.524 MPa, and 0.0971 MPa, indicating that largest and smallest Z roots are in phase equilibrium, and the center root is discarded as before. The corresponding values for saturated entropy are $S = -21.3021$ J/mol-K for the vapor phase and -100.955 J/mol-K for the liquid phase. For any condition at 0.1 MPa, any value of S between these two values will fall in the two-phase region. Therefore, the desired state of $S = -28.35$ J/mol-K is two-phase, with a quality calculated using the saturation values,

$$S = S^{satL} + q \, (S^{satV} - S^{satL})$$

-28.35 J/mol-K $= -100.955 + q(-21.3021 + 100.955)$. Solving, $q = 0.912$

The cautions highlighted in the example also apply when searching for specific values of other state properties by adjusting P and/or T. For example, it is also common to search for a state with specified values of $\{H,P\}$ by adjusting T. The user must make sure that the root selected is a stable root, or if the system is two-phase, then a quality calculation must be performed.[6]

9.12 TEMPERATURE EFFECTS ON *G* AND *f*

The effect of temperature at fixed pressure is

$$\left(\frac{\partial G}{\partial T}\right)_P = -S \qquad\qquad 9.51$$

The Gibbs energy change with temperature is then dependent on entropy. Gibbs energy will decrease with increasing temperature. Since the entropy of a vapor is higher than the entropy of a liquid, the Gibbs energy will change more rapidly with temperature for vapor. Since the Gibbs energy is proportional to the log of fugacity, the fugacity dependence will follow the same trends. Similar statements are valid comparing liquids and solids.[7]

9.13 SUMMARY

We began this chapter by introducing the need for Gibbs energy to calculate phase equilibria in pure fluids because it is a natural function of temperature and pressure. We also introduced fugacity, which is a convenient property to use instead of Gibbs energy because it resembles the vapor pressure more closely. We also showed that the fugacity coefficient is directly related to the deviation of a fluid from ideal gas behavior, much like a departure function (Eqns. 9.23 and 9.24). This principle of characterization of non-ideality extends into the next chapter where we consider non-idealities of mixtures. In fact, much of the pedagogy presented in this chapter finds its significance in the following chapters, where the phase equilibria of mixtures become much more complex.

---

6. It is possible to program conditionals to avoid unstable roots, but due to the importance of chemical engineers understanding the conditions, we require the users to make the determination. Can you see how to program the conditionals?

7. Frequently, we arbitrarily set $S_R = 0$ and either H_R or $U_R = 0$ at our reference states. For consistency in our calculations, $G_R = H_R - T_R S_R$. As a result, the calculated value of S at a given state depends on our current state relative to the reference sate. Calculated S values may be positive or negative due to our choice of $S_R = 0$, and Gibbs energy thus calculated may increase or decrease with temperature. Entropy does not actually go to zero except for a perfect crystal at absolute zero, and entropy of all substances at practical conditions is positive. The fact that our calculations result in negative numbers for S is purely a result of our choice of setting $S_R = 0$ at our reference state (to avoid more difficult calculation of the actual value relative to a perfect crystal at absolute zero). See third law of thermodynamics in Subject Index.

Methods for calculating fugacities were introduced using charts, equations of state, and derivative manipulations. (In the homework problems, we offer illustration of how tables may also be used.) Liquids and solids were considered in addition to gases, and the Poynting correction was introduced for calculating the effect of pressure on condensed phases. These pure component methods are summarized in Table 9.1, and they are applied often in the analysis of mixtures. Skills in their application must be kept ready throughout the following chapters.

Table 9.1 *Techniques for Calculation Pure Component Fugacities*

Gases	Liquids	Solids

1. Ideal gas law
2. Equation of state
  a. Virial equation ($V_r \geq 2$)
  b. Cubic equation
	1. Poynting method[a]
	2. Equation of state (cubic unless combined with Poynting).
		1. Poynting method[a]

a.   The saturation fugacity may be determined by any of the methods for gases, and the Poynting correction is omitted near the vapor pressure.

A critical concept in this chapter is that when multiple EOS roots exist from a process calculation, the stable root has a lower Gibbs energy or lower fugacity. We also provided methods to find saturation conditions for pure fluids.

Important Equations

Eqns. 9.3 and 9.27 basically state that equilibrium occurs when fugacity in each phase is equal. This is a general principle that can be applied to components in mixtures and forms the basis for phase equilibrium computations in mixtures. Eqn. 9.11 is a special form of Eqn. 9.6 that is particularly convenient for estimating the vapor pressure over wide ranges of temperature. It may not be as precise as the Antoine equation over a narrow temperature range, but it is less likely to lead to a drastic error when extrapolation is necessary. Nevertheless, Eqn. 9.11 is not a panacea. When you are faced with phase equilibrium problems other than vapor pressure, like solid-liquid (e.g., melting ice) or solid-vapor (e.g., dry ice), you must start with Eqn. 9.6 and re-derive the final equations subject to relevant assumptions for the problem of interest.

Fugacity is commonly calculated by Eqns. 9.28(Gases), 9.29(Ideal gases), 9.41(Liquids), and 9.43(Solids), and is dependent on the calculation of the fugacity coefficient. Fugacity is particularly helpful in identifying the stable root.

9.14   PRACTICE PROBLEMS

P9.1 Carbon dioxide ($C_P = 38$ J/mol-K) at 1.5 MPa and 25°C is expanded to 0.1 MPa through a throttle valve. Determine the temperature of the expanded gas. Work the problem as follows:

  (a) Assuming the ideal gas law (ANS. 298 K)
  (b) Using the Peng-Robinson equation (ANS. 278 K, sat L + V)
  (c) Using a CO_2 chart, noting that the triple point of CO_2 is at −56.6°C and 5.2 bar, and has a heat of fusion, ΔH^{fus}, of 43.2 cal/g. (ANS. 194 K, sat S + V)

P9.2 Consider a stream of pure carbon monoxide at 300 bar and 150 K. We would like to liquefy as great a fraction as possible at 1 bar. One suggestion has been to expand this high-pressure fluid across a Joule-Thompson valve and take what liquid is formed. What would be the fraction liquefied for this method of operation? What entropy is generated per mole processed? Use the Peng-Robinson equation. Provide numerical answers. Be sure to specify your reference state. (Assume $C_P = 29$ J/mol-K for a quick calculation.) (ANS. 32% liquefied)

P9.3 An alternative suggestion for the liquefaction of CO discussed above is to use a 90% efficient adiabatic turbine in place of the Joule-Thomson valve. What would be the fraction liquefied in that case? (ANS. 60%)

P9.4 At the head of a methane gas well in western Pennsylvania, the pressure is 250 bar, and the temperature is roughly 300 K. This gas stream is similar to the high-pressure stream exiting the precooler in the Linde process. A perfect heat exchanger (approach temperature of zero) is available for contacting the returning low-pressure vapor stream with the incoming high-pressure stream (similar to streams 3–8 of Example 8.9 on page 320). Compute the fraction liquefied using a throttle if the returning low-pressure vapor stream is 30 bar. (ANS. 30%)

9.15   HOMEWORK PROBLEMS

9.1 The heat of fusion for the ice-water phase transition is 335 kJ/kg at 0°C and 1 bar. The density of water is 1g/cm^3 at these conditions and that of ice is 0.915 g/cm^3. Develop an expression for the change of the melting temperature of ice as a function of pressure. Quantitatively explain why ice skates slide along the surface of ice for a 100 kg hockey player wearing 10 cm x 0.01 cm blades. Can it get too cold to ice skate? Would it be possible to ice skate on other materials such as solid CO_2?

9.2 Thermodynamics tables and charts may be used when both H and S are tabulated. Since $G = H - TS$, at constant temperature, $\Delta G = RT \ln(f_2/f_1) = \Delta H - T\Delta S$. If state 1 is at low pressure where the gas is ideal, then $f_1 = P_1$, $RT \ln(f_2/P_1) = \Delta H - T\Delta S$, where the subscripts indicate states. Use this method to determine the fugacity of steam at 400°C and 15 MPa. What value does the fugacity coefficient have at this pressure?

9.3 This problem reinforces the concepts of phase equilibria for pure substances.

(a) Use steam table data to calculate the Gibbs energy of 1 kg saturated steam at 150°C, relative to steam at 150°C and 50 kPa (the reference state). Perform the calculation by plotting the volume data and graphically integrating. Express your answer in kJ. (Note: Each square on your graph paper will represent [pressure·volume] corresponding to the area, and can be converted to energy units.)

(b) Repeat the calculations using the tabulated enthalpies and entropies. Compare your answer to part (a).

(c) The saturated vapor from part (a) is compressed at constant T and 1/2 kg condenses. What is the total Gibbs energy of the vapor liquid mixture relative to the reference state of part (a)? What is the total Gibbs energy relative to the same reference state when the mixture is completely condensed to form saturated liquid?

(d) What is the Gibbs energy of liquid water at 600 kPa and 150°C relative to the reference state from part (a)? You may assume that the liquid is incompressible.

(e) Calculate the fugacities of water at the states given in parts (a) and (d). You may assume that $f = P$ at 50 kPa.

9.4 Derive the formula for fugacity according to the van der Waals equation.

9.5 Use the result of problem 9.4 to calculate the fugacity of ethane at 320 K and at a molar volume of 150 cm^3/mole. Also calculate the pressure in bar.

9.6 Calculate the fugacity of ethane at 320 K and 70 bar using:

(a) Generalized charts
(b) The Peng-Robinson equation

9.7 CO_2 is compressed at 35°C to a molar volume of 200 cm^3/gmole. Use the Peng-Robinson equation to obtain the fugacity in MPa.

9.8 Use the generalized charts to obtain the fugacity of CO_2 at 125°C and 220 bar.

9.9 Calculate the fugacity of pure n-octane vapor as given by the virial equation at 580 K and 0.8 MPa.

9.10 Estimate the fugacity of pure n-pentane (C_5H_{12}) at 97°C and 7 bar by utilizing the virial equation.

9.11 Develop tables for H, S, and Z for N_2 over the range $P_r = [0.5, 1.5]$ and $T_r = [T_r^{sat}, 300$ K] according to the Peng-Robinson equation. Use the saturated liquid at 1 bar as your reference condition for $H = 0$ and $S = 0$.

9.12 Develop a P-H chart for saturated liquid and vapor butane in the range $T = [260, 340]$ using the Peng-Robinson equation. Show constant S lines emanating from saturated vapor at 260 K, 300 K, and 340 K. For an ordinary vapor compression cycle, what would be the temperature and state leaving an adiabatic, reversible compressor if the inlet was saturated vapor at 260 K? (Hint: This is a tricky question.)

9.13 Compare the Antoine and shortcut vapor-pressure equations for temperatures from 298 K to 500 K. (Note in your solution where the equations are extrapolated.) For the comparison, use a plot of $\log_{10} P^{sat}$ versus $1/T$ except provide a separate plot of P^{sat} versus T for vapor pressures less than 0.1 MPa.

(a) n-Hexane
(b) Acetone
(c) Methanol
(d) 2-propanol
(e) Water

9.14 For the compound(s) specified by your instructor in problem 9.13, use the virial equation to predict the virial coefficient for saturated vapor and the fugacity of saturated liquid at 80 °C. Compare the values of fugacity to the vapor pressure.

9.15 Compare the Peng-Robinson vapor pressures to the experimental vapor pressures (represented by the Antoine constants) for the species listed in problem 9.13.

9.16 Carbon dioxide can be separated from other gases by preferential absorption into an amine solution. The carbon dioxide can be recovered by heating at atmospheric pressure. Suppose pure CO_2 vapor is available from such a process at 80°C and 1 bar. Suppose the CO_2 is liquefied and marketed in 43-L laboratory gas cylinders that are filled with 90% (by mass) liquid at 295 K. Explore the options for liquefaction, storage, and marketing via the following questions. Use the Peng-Robinson for calculating fluid properties. Submit a copy of the H-U-S table for each state used in the solution.

(a) Select and document the reference state used throughout your solution.

(b) What is the pressure and quantity (kg) of CO_2 in each cylinder?

(c) A cylinder marketed as specified needs to withstand warm temperatures in storage/ transport conditions. What is the minimum pressure that a full gas cylinder must withstand if it reaches 373 K?

(d) Consider the liquefaction process via compression of the CO_2 vapor from 80 °C, 1 bar to 6.5 MPa in a single adiabatic compressor ($\eta_C = 0.8$). The compressor is followed by cooling in a heat exchanger to 295 K and 6.5 MPa. Determine the process temperatures and pressures, the work and heat transfer requirement for each step, and the overall heat and work.

(e) Consider the liquefaction via compression of the CO_2 vapor from 80°C, 1 bar to 6.5 MPa in a two-stage compressor with interstage cooling. Each stage ($\eta_C = 0.8$) operates adiabatically. The interstage pressure is 2.5 MPa, and the interstage cooler returns the CO_2 temperature to 295 K. The two-stage compressor is followed by cooling in a heat exchanger to 295 K and 6.5 MPa. Determine all process temperatures and pressures, the work and heat transfer requirement for each step, and the overall heat and work.

(f) Calculate the minimum work required for the state change from 80°C, 1 bar to 295 K, 6.5 MPa with heat transfer to the surroundings at 295 K. What is the heat transfer required with the surroundings?

9.17 A three-cycle cascade refrigeration unit is to use methane (cycle 1), ethylene (cycle 2), and ammonia (cycle 3). The evaporators are to operate at: cycle 1, 115.6 K; cycle 2, 180 K; cycle 3, 250 K. The outlets of the compressors are to be: cycle 1, 4 MPa; cycle 2, 2.6 MPa; cycle 3, 1.4 MPa. Use the Peng-Robinson equation to estimate fluid properties. Use stream numbers from Fig. 5.11 on page 212. The compressors have efficiencies of 80%.

(a) Determine the flow rate for cycle 2 and cycle 3 relative to the flow rate in cycle 1.

(b) Determine the work required in each compressor per kg of fluid in the cycle.

(c) Determine the condenser duty in cycle 3 per kg of flow in cycle 1.

(d) Suggest two ways that the cycle could be improved.

9.18 Consider the equation of state

$$Z = 1 + \frac{4c\,\eta_p}{1 - 1.9\,\eta_p} - \frac{a\,\eta_p}{bRT}$$

where $\eta_p = b/V$.

(a) Determine the relationships between a, b, c and T_c, P_c, Z_c.

(b) What practical restrictions are there on the values of Z_c that can be modeled with this equation?

(c) Derive an expression for the fugacity.

(d) Modify Preos.xlsx or Preos.m for this equation of state. Determine the value of c (+/– 0.5) that best represents the vapor pressure of the specified compound below. Use the shortcut vapor pressure equation to estimate the experimental vapor pressure for the purposes of this problem for the option(s) specifed by your instructor.

(i) CO_2

(ii) Ethane

(iii) Ethylene

(iv) Propane

(v) n-Hexane

UNIT

FLUID PHASE EQUILIBRIA IN MIXTURES

We have already encountered the phase equilibrium problem in our discussions of pure fluids. In Unit I, we were concerned with the quality of the steam. In Unit II, we developed generalized relations for the vapor pressure. These analyses enable us to estimate both the conditions when a liquid/vapor phase transition occurs and the ratio of vapor to liquid. In Unit III, we are not only concerned with the ratio of vapor to liquid, but also with the ratio of each component in the liquid to that in the vapor. These ratios may not be the same because all components are not equally soluble in all phases. These issues arise in a number of applications (e.g., distillation or extraction) that are extremely common. Unfortunately, prediction of the desired properties to the required accuracy is challenging. In fact, *no currently available method is entirely satisfactory,* even for the limited types of phase equilibria commonly encountered in the chemical processing industries. Nevertheless, the available methods do provide an adequate basis for correlating the available data and for making modest extrapolations, and the methods can be successfully applied to process design. Understanding of the difference between modest extrapolations and radical predictions is facilitated by a careful appreciation of the underlying theory as developed from the molecular level. Developing this understanding is strongly encouraged as a means of avoiding extrapolations that are unreasonable. This unit relies on straightforward extensions of the concepts of energy, entropy, and equilibrium to provide a solid background in the molecular thermodynamics of non-reactive mixtures. The final unit of the text, Unit IV, treats reactive systems.

CHAPTER

10

INTRODUCTION TO MULTICOMPONENT SYSTEMS

What we obtain too cheap we esteem too lightly.

Thomas Paine

Superficially, the extension of pure component concepts to mixtures may seem simple. In fact, this is a significant problem of modern science that impacts phase transitions in semi- and superconductors, polymer solutions and blends, alloy materials, composites, and biochemistry. Specialists in each of these areas devote considerable effort to the basic problem of segmental interactions between molecules. From a thermodynamic perspective, these various research efforts are very similar. The specific types of molecules differ, and the pair potential models may be different, but the means of connecting the molecular scale to the macro scale remains the same.

Our coverage of multicomponent systems consists of (1) a very brief extension to mixtures of the mathematical and physical principles; and (2) an introductory description of several common methods for reducing these principles to practice. This description is merely introductory because learning the specific details is partially what distinguishes the polymer scientist from the ceramicist. Such specialized study is greatly facilitated by having an appreciation of the types of molecular interactions that are most influential in each situation. Of similar importance is the ability to analyze thermodynamic data such that key aspects of processes are easily ascertained.

Chapter Objectives: You Should Be Able to...

1. Interpret phase diagrams to locate dew, bubble, and flash conditions. Use the lever rule in two-phase regions.

2. Identify when a bubble, dew, or flash computation is required and perform the computation subject to the assumptions of an ideal solution.

3. Know the assumptions of an ideal solution and where to start with the fundamental equations to develop models of nonideal solutions.

4. Recognize when the ideal solution model is reasonable and when it is not, including comparisons of theoretical results to experimental data, graphically and statistically, coupled with conceptual reasoning about the nature of the molecular interactions.

5. Understand how VLE relates conceptually to the process of distillation.

10.1   INTRODUCTION TO PHASE DIAGRAMS

Before we delve into the details of calculating phase equilibria, let us introduce elementary concepts of common vapor-liquid phase diagrams. For a pure fluid, the **Gibbs phase rule** shows vapor-liquid equilibrium occurs with only one **degree of freedom**, $F = C - P + 2 = 1 - 2 + 2 = 1$. At one atmosphere pressure, vapor-liquid equilibria will occur at only one temperature—the normal boiling point temperature. However, with a binary mixture, there are two degrees of freedom, $F = 2 - 2 + 2 = 2$. For a system with fixed pressure, phase compositions and temperature can both be varied over a finite range when two phases coexist. Experimental data for experiments of this type are usually presented as a function of T and composition on a plot known as a **T-x-y diagram,** such as that shown qualitatively in Fig. 10.1. At fixed temperature, pressure and composition may vary in a binary mixture and obtain data to create a **P-x-y diagram** as shown also in Fig. 10.1. The region where two phases coexist is shown by the area enclosed by the curved lines on either plot and is known as the **phase envelope.**[1] On the T-x-y diagram, the vapor region is at the top (raising T at fixed P causes vaporization of liquid). On the P-x-y diagram, the vapor region is at the bottom (lowering P at fixed T causes vaporization of liquid). Note that the intersections of the phase envelope with the ordinate scales at the pure component compositions give the pure component saturation temperatures on the T-x-y diagram, and the pure component vapor (saturation) pressures on the P-x-y diagram. Therefore, significant information about the shape of the diagram can often be deduced with a single mixture data point when combined with the pure component end points. Qualitatively, the shape of the P-x-y diagram can be found by inverting the T-x-y, and vice versa.[2] Customarily, for binary sys-

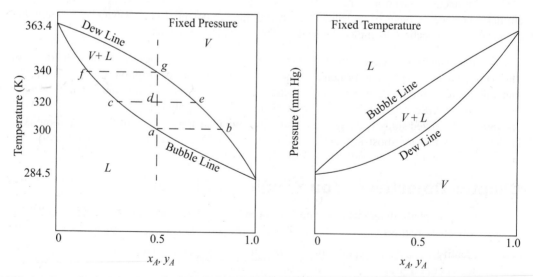

Figure 10.1 *Illustration of T-x-y (left) and P-x-y (right) diagrams.*

---

1. There are many variations of the diagrams, and this discussion is meant to introduce only the most commonly encountered types of diagrams. More complex diagrams are introduced gradually, and are discussed in depth in Chapter 16. These diagrams are cross sections of three-dimensional diagrams which are discussed in Chapter 16.
2. This is a convenient manipulation to visualize both diagrams if only one diagram is available.

tems in the separations literature, the more volatile component composition is plotted along the abscissa in mole fraction or percent.

The lower curve on the *T-x-y* diagram is next to the liquid region, and it is known as the **bubble line.** The bubble-temperature line gives the boiling temperature of the mixture as a function of composition at the specified pressure. The upper curve is next to the vapor region, and is known as the **dew line.** The two lines meet at the axes if the conditions are below the critical pressure of both components. At a given composition, the temperature along the bubble line is the temperature where an infinitesimal bubble of vapor coexists with liquid. Thus, at an over-all composition of 50 mole% *A,* the system of Fig. 10.1 at fixed pressure is 100% liquid below 300 K at the pressure of the diagram. As the temperature is raised, the overall composition is constrained to follow the vertical dashed line constructed on the diagram, and the first vapor bubble forms at the intersection of the bubble line at 300 K at point *a,* which is known as the **bubble temperature** for a 50 mole% mixture at the system pressure. Phase compositions at a given *P* and *T* may be found by reading the compositions from intersections of the bubble and dew lines with horizontal lines constructed on the diagram, such as the dashed line at 300 K. For our example at the bubble temperature, the liquid phase will be 50 mole% *A* because the first bubble of vapor has not yet caused a measurable change in the liquid composition. The vapor phase composition coexisting at the bubble-point temperature will be 80 mole% *A* (point *b*). As the temperature is increased to 320 K, the overall mixture is at point *d,* the liquid phase will be 30 mole% *A* (point *c*), and the vapor phase will be 70 mole% *A* (point *e*). Suppose we start an experiment with a 50 mole% mixture at 350 K, where the mixture is 100% vapor. As the temperature is lowered, the **dew temperature** is encountered at 340 K for the 50 mole% mixture at system pressure (point *g*), and the first drop of liquid is formed which is about 20 mole% *A* (point *f*). Note that the bubble and dew temperatures are composition-dependent. For example, the bubble temperature of a 30 mole% mixture is 320 K (point *c*), and the bubble temperature of a 20 mole% mixture is 340 K (point *f*). Similar discussion could be presented for the dew temperatures. The bubble and dew-point discussions could also be presented on the pressure diagram, but in this case we would refer to the **bubble and dew pressures.** Note that the relative vertical locations of the bubble- and dew lines are flipped on the two diagrams. The horizontal dotted lines connecting coexisting compositions are **tie lines.**

> The ability to quickly read phase diagrams is an essential skill.

When we speak of composition in a two-phase mixture, we must be clear about which phase we are discussing. We use *x* to denote a liquid phase mole fraction, *y* to denote a vapor phase mole fraction, and *z* to denote an overall mole fraction.[3] For the example, we have been discussing using a 50 mole% mixture: At the bubble point of 300 K we have $z_A = 0.5$, $x_A = 0.5$, $y_A = 0.8$; at 320 K we have $z_A = 0.5$, $x_A = 0.3$, $y_A = 0.7$; At 340 K we have $z_A = 0.5$, $x_A = 0.20$, $y_A = 0.5$. At 320 K, the system is in the two-phase region, and we may use the compositions of the vapor and liquid phases, together with an overall mass balance, to calculate the fraction of the overall mixture that is vapor or liquid. This is known as a **flash calculation.** If the initial number of moles is denoted by *F,* and it separates into *L* moles of liquid and *V* moles of vapor, the overall mole balance is $F = L + V$, which can be written $1 = L/F + V/F$. The *A* component balance is $z_A F = y_A V + x_A L$, which can be written $z_A = y_A \cdot V/F + x_A \cdot L/F$. Combining the two balances to eliminate *V/F*, the percentage that is liquid will be

$$\frac{L}{F} = \frac{z_A - y_A}{x_A - y_A} \qquad\qquad 10.1$$

---

3. Note that when at 100% liquid, $x = z$ and at 100% vapor, $y = z$. In some cases, such as formulas for an equation of state, we discuss a generic phase which may be liquid *or* vapor, and thus use either *x* or *y*.

which is simply given by line segment lengths, $\overline{de}/\overline{ce}$. Likewise the fraction that is vapor may be calculated

$$\frac{V}{F} = \frac{x_A - z_A}{x_A - y_A}$$

10.2

The lever rule.

which is given by line segment lengths, $\overline{cd}/\overline{ce}$. These balance equations are frequently called the **lever rule.** Note that the two fractions sum to one, $\frac{L}{F} + \frac{V}{F} = 1$.

10.2   VAPOR-LIQUID EQUILIBRIUM (VLE) CALCULATIONS

Classes of VLE Calculations

The classes of VLE calculations presented here will be used through the remainder of the text, so the concepts are extremely important.

Depending on the information provided, one may perform one of several types of vapor-liquid equilibrium **(VLE)** calculations to model the vapor-liquid partitioning. These are: **bubble-point pressure (BP)**, **dew-point pressure (DP)**, **bubble-point temperature (BT)**, **dew-point temperature (DT)**, and **isothermal flash (FL)** and **adiabatic flash (FA)**. The specifications of the information required and the information to be computed are tabulated below in Table 10.1. Also shown in the table are indications of the relative difficulty of each calculation. The best approach to understanding the calculations is to gain experience by plotting phase envelopes in various situations.

Principles of Calculations

Standard approaches to solving VLE problems utilize the ratio of vapor mole fraction to liquid mole fraction, known as the **VLE K-ratio:**

VLE K-ratio.

$$\boxed{K_i \equiv y_i/x_i}$$

10.3

The information available from a physical situation is combined with the K-ratio using one of the procedures shown in Table 10.1. The information available is in the second column of the table. The procedure involves combination of the known information together with a model-dependent K-ratio to calculate an objective function based on the estimated unknown compositions. For a bubble calculation, all the x_i are known, and we find the y_i by solving for the condition where $\sum_i y_i = 1$

written in terms of x_i, namely $\sum_i K_i x_i = 1$. For dew calculations, all y_i are known, and we solve

for the condition where $\sum_i x_i = 1$ written in terms of y_i and K_i. For a flash calculation, we solve

for the condition where $\sum_i x_i - \sum_i y_i = 0$ written in terms of the overall mole fraction z_i and K_i.

The information in Table 10.1 is rigorous. The method used to calculate K_i is model-dependent and will be the focus of the next few chapters of the text. The K_i ratios generally vary with composition, pressure, and temperature, though we focus in this chapter on the use of Raoult's law in situations where K_i ratios are dependent on only T and P.

Table 10.1 *Summary of the Types of Phase Equilibria Calculations (This Table is Independent of the VLE Model)*

Type	Information known	Information computed	Criteria	Effort
BP	$T, x_i = z_i$	P, y_i	$\sum_i y_i = \sum_i K_i x_i = 1$	Easiest
DP	$T, y_i = z_i$	P, x_i	$\sum_i x_i = \sum_i (y_i / K_i) = 1$	Not bad
BT	$P, x_i = z_i$	T, y_i	$\sum_i y_i = \sum_i K_i x_i = 1$	Difficult
DT	$P, y_i = z_i$	T, x_i	$\sum_i x_i = \sum_i (y_i / K_i) = 1$	Difficult
FL	P, T^{in}, T^{out}, z_i	$x_i, y_i, V/F, \underline{Q}$	$\sum_i \dfrac{z_i (1 - K_i)}{1 + (V/F)(K_i - 1)} = 0$, E-bal for $\underline{Q}$	Quite difficult
FA	$P, T^{in}, z_i, \underline{Q} = 0$	$x_i, y_i, V/F, T^{out}$	$\sum_i \dfrac{z_i (1 - K_i)}{1 + (V/F)(K_i - 1)} = 0$, E-bal	Most difficult

Strategies for Solving VLE Problems

Note that there are only six different types of calculations for VLE summarized in Table 10.1. Usually the solution of the VLE problem will be relatively straightforward after deciding which row of the table to use. A crucial skill in solving VLE problems is interpreting the physical situation to decide which of the five methods should be used, and how to calculate the K-ratio.

1. Decide if the liquid, vapor, or overall composition is known from the problem statement.
2. Identify if the fluid is at a bubble or dew point. If the fluid is at a bubble point, the overall composition will be the same as the liquid composition. At the dew point, the overall composition will be the same as the vapor.
3. Identify if the P, T, or both P and T are fixed. Decide if the system is adiabatic.
4. The information collected in the first three steps can be used with the second column in Table 10.1 to identify the method.
5. Select a method to calculate the VLE K-ratio.
6. Decide if the method will be iterative, and if so, generate an initial guess for the solution. Approaches for handling iterative calculations are introduced in the following chapters and examples.

Iterative Calculations

When the *K*-ratios vary, VLE calculations require iterative solutions from an initial guess. For performing iterative calculations, useful aids include the Solver tool of Excel or the `fzero()` or `fsolve()` function of MATLAB. Many of the following examples summarize detailed calculations to illustrate fully the iterative procedure. In practice, the detailed calculations can be performed rapidly using a solver. Online supplements summarize the use of the iterative aids and the methods for successive substitution.

10.3   BINARY VLE USING RAOULT'S LAW

For a small class of mixtures where the components have very similar molecular functionality and molecular size, the bubble-pressure line is found to be a linear function of composition as shown in Figs. 10.2 and 10.3. As was noted in Section 10.1, the *T-x-y* and *P-x-y* diagram shapes are related qualitatively by inverting one of the diagrams. Because the bubble pressure is a linear function of composition, we may write for a binary system,

ⓘ Raoult's Law for Bubble Pressure.

$$P = P_2^{sat} + x_1(P_1^{sat} - P_2^{sat}) = x_1 P_1^{sat} + (1 - x_1)P_2^{sat} = x_1 P_1^{sat} + x_2 P_2^{sat} \qquad 10.4$$

Dividing by *P*, we find the form of the bubble objective function summarized in Table 10.1,

$$1 = \frac{P_1^{sat}}{P} x_1 + \frac{P_2^{sat}}{P} x_2 = K_1 x_1 + K_2 x_2 \qquad 10.5$$

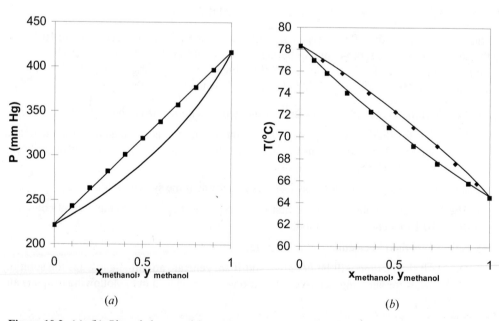

(a)   (b)

Figure 10.2 *(a), (b). Phase behavior of the methanol-ethanol system. Left figure at 50°C. Right figure at 760 mm Hg. (P-x-y from Schmidt, G.C. 1926. Z. Phys.Chem. 121:221, T-x-y from Wilson, A., Simons, E.L. 1952. Ind. Eng. Chem. 44:2214).*

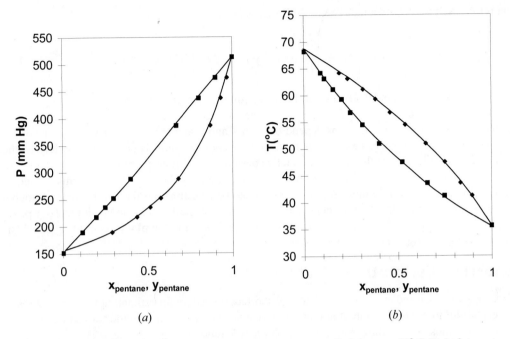

(a)   (b)

Figure 10.3 *(a), (b) Phase behavior of the pentane-hexane system. Left figure at 25°C. Right figure at 750 mm Hg. (P-x-y from Chen, S.S., Zwolinski, B.J. 1974. J.Chem. Soc. Faraday Trans. 70: 1133, T-x-y from Tenn, F.G., Missen, R.W. 1963. Can. J. Chem. Eng. 41:12).*

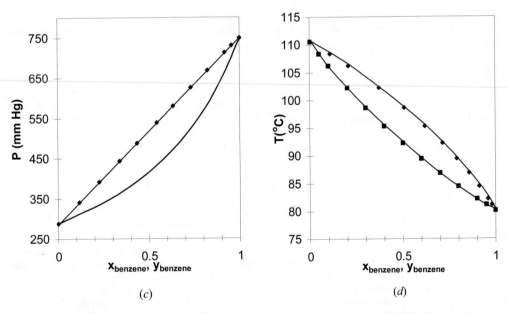

(c)   (d)

Figure 10.3 *(c); (d) Phase behavior of the benzene-toluene system. Left figure at 79.7°C. Right figure at 760 mm Hg. (P-x-y from Rosanoff, M.A., et al. 1914. J. Am.Chem.Soc. 36:1993, T-x-y from Delzenne, A.O. 1958. Ind. Eng. Chem., Chem. Eng. Data Series. 3:224).*

Therefore, we conclude that the K-ratio for Raoult's law[4] is

Raoult's Law K-ratio.

$$K_i = \frac{P_i^{sat}}{P} \quad \text{or} \quad y_i P = x_i P_i^{sat} \qquad \qquad 10.6$$

K-values are used for all types of separations.

François-Marie Raoult (1830–1901) was a French chemist who studied freezing point depression and the effect of mixing on bubble pressure. His first work on solution behavior in 1878 dealt with freezing point depression. His paper on "Raoult's Law" was published in 1892.

Throughout Chapters 10–16 and 18–19 we continually improve and refer to K-value models. The motivation for this focus on K-values is that they are used for all types of separations. For this select class of mixtures, the K-ratios depend only on T and P and are independent of composition. To calculate a K-ratio, we can use any method for the vapor pressure. Though we have derived the K-ratio empirically, we use this model to practice performing VLE calculations and return in future sections to a more fundamental derivation to explore more fully the restrictions for this model. In future chapters, we explore more sophisticated methods for estimating the K-ratio when Raoult's law does not apply and the K-ratios depend on composition. In addition to the P-x-y and T-x-y plots, another common presentation of data for distillation studies is the **x-y plot** of coexisting compositions as shown in Fig. 10.4.

Shortcut Estimation of VLE K-ratios

Given Raoult's law and recalling the Chapter 9 shortcut procedure for estimating vapor pressures, it is very useful to consider combining these to give a simple and quick method for estimating ideal solution K-ratios. Substituting, we obtain the shortcut K-ratio,

Shortcut K-ratio for Raoult's law. Not only is this expression restricted to Raoult's law, but it is also subject to the restrictions of Eqn. 9.11.

$$K_i = \frac{P_i^{sat}}{P} \approx \frac{P_{c,i} 10^{\frac{7}{3}(1+\omega)\left(1 - \frac{1}{T_{r,i}}\right)}}{P} \qquad \text{Shortcut } K\text{-ratio} \qquad 10.7$$

Since the ideal solution model is somewhat crude anyway, it is not unreasonable to apply the above equation as a first approximation for any ideal solution when rapid approximations are needed.[5]

Bubble Pressure

Note that as we begin applications, some of the equations are model-dependent.

For a **bubble-pressure** calculation, writing $\sum_i y_i = 1$ as $\sum_i K_i x_i = 1$, which is $\sum_i \frac{P_i^{sat}}{P} x_i = 1$. Multiplying by P, we may write

$$P = x_1 P_1^{sat} + x_2 P_2^{sat} \qquad \qquad 10.8$$

---

4. An interesting perspective on the contributions of Raoult is available in Wisniak, J. 2001. "François-Marie Raoult: Past and Modern Look". *The Chemical Educator* 6 (1): 41–49. doi:10.1007/s00897000432a.

5. If a supercritical component is present in significant quantity, the user must beware, because the shortcut K-ratio may falsely predict a liquid phase due to extrapolation of the vapor pressure. An interesting problem arises when we must calculate the VLE K-ratio for a component in the liquid phase but above its critical temperature. Carbon dioxide ($T_c = 31°C$) in soda pop on a 32°C day would be a common example. Since the saturation pressure of CO_2 does not exist above the critical temperature, and pure CO_2 cannot condense, we might consider that CO_2 would not exist in the liquid phase and that Raoult's law might indicate an infinite value for the K-ratio. Experience tells us that this component does exist in the liquid phase over a portion of the composition range. Remarkably, the extrapolated vapor pressures in the above formula give reasonably accurate results at small liquid phase concentrations of noncondensable components. Of course, it is more accurate for components that are only slightly above their critical temperature, because then the extrapolation is slight. Calculations with supercritical components are best done using a **Henry's law,** or a hypothetical liquid fugacity (Section 11.12 on page 443) or an equation of state (Example 15.9 on page 599).

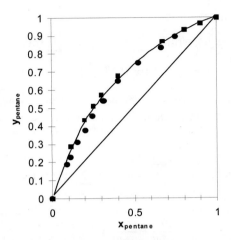

Figure 10.4 *Data from Fig. 10.3 plotted with coexisting liquid and vapor values for each experimental tie line, resulting in the x-y plot. Note that the data do not superimpose exactly because one data set is isobaric and the other set is isothermal. Squares are T-x-y data. Circles are P-x-y data. The diagonal is traditionally drawn in an x-y figure, and the data never cross the diagonal for systems that follow Raoult's law.*

where no iterations are required because temperature, and therefore vapor pressures, are known. Once the bubble pressure is found, the value can be reinserted into Eqn. 10.6 to find the vapor mole fractions:

$$y_i = x_i P_i^{sat}/P = K_i x_i \qquad\qquad 10.9$$

Dew Pressure

For a **dew-pressure** calculation, writing $\sum_i x_i = 1$ as $\sum_i (y_i/K_i) = 1$ and rearranging:

$$\frac{y_1 P}{P_1^{sat}} + \frac{y_2 P}{P_2^{sat}} = 1 \qquad\qquad 10.10$$

which may be rearranged and solved without iteration, because the vapor pressures are fixed at the specified temperature:

$$P = \frac{1}{\dfrac{y_1}{P_1^{sat}} + \dfrac{y_2}{P_2^{sat}}} \qquad\qquad 10.11$$

Once the dew pressure is calculated, the value can be reinserted into Eqn. 10.6 to find the liquid mole fractions:

$$x_i = y_i P/P_i^{sat} = y_i/K_i \qquad\qquad 10.12$$

Bubble Temperature

For a **bubble-temperature** calculation, writing $\sum_i y_i = 1$ as $\sum_i K_i x_i = 1$, and rearranging:

$$P = x_1 P_1^{sat} + x_2 P_2^{sat} \qquad\qquad 10.13$$

To solve this equation, it is necessary to iterate on temperature (which changes P_i^{sat}) until P equals the specified pressure. Then the vapor phase mole fractions are calculated using Eqn. 10.9.

Dew Temperature

For a **dew-temperature** calculation, writing $\sum_i x_i = 1$ as $\sum_i (y_i / K_i) = 1$, and rearranging:

$$P = \cfrac{1}{\cfrac{y_1}{P_1^{sat}} + \cfrac{y_2}{P_2^{sat}}} \qquad\qquad 10.14$$

To solve this equation, it is necessary to iterate on temperature (which changes P_i^{sat}) until P equals the specified pressure. Then the liquid phase mole fractions are calculated using Eqn. 10.12.

General Flash

Flash drums are frequently used in chemical processes. For an isothermal drum, the temperature and pressure of the drum are known. Consider that a feed stream is liquid which becomes partially vaporized after entering the drum as illustrated in Fig. 10.5. The name of the flash procedure implies that it is applicable only for flashing liquids, but in fact, the procedure is also valid for vapor entering a **partial condenser** or for any number of incoming vapor and/or liquid streams with overall compositions specified by $\{z_i\}$ with overall flow rate F. To apply the procedure, the overall composition of the components, z_i, total feed flow rate, F, and outlet T and P just need to be known before the procedure is started. Though the method is introduced using flowing streams, flow is not required for a flash calculation; the calculation can be performed for any overall composition constrained at a particular temperature and pressure even within a closed system. Also, the inlet stream temperature does not need to match the outlet temperature; the two outlet streams are at the specified temperature.

The flash equations are easily derived by modification of the overall and component balances used in the development of the lever rule in Section 10.1. If z is the feed composition and V/F is the vapor-to-feed ratio, then $L/F = 1 - V/F$, and the component balance is $z_i = x_i(L/F) + y_i(V/F)$. Substituting for L/F from the overall balance, and using $y_i = x_i K_i$, the component balance becomes $z_i = x_i[(1 - V/F) + K_i(V/F)]$, which can be solved for x_i:

$$x_i = \frac{z_i}{1 + (V/F)(K_i - 1)} \qquad\qquad 10.15$$

Figure 10.5 *Illustration of a flash drum and variable definitions for streams. Note that F need not be a liquid; F may be all vapor or partial vapor. The principles can be applied to a nonflowing system as described for a binary on page 371.*

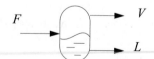

using $y_i = x_i K_i$, we may multiply Eqn. 10.15 by K_i to obtain y_i.

$$y_i = \frac{z_i K_i}{1 + (V/F)(K_i - 1)}$$

$\qquad$ 10.16

One obvious thing to do at this point is to iterate to find the V/F ratio which satisfies $\sum_i x_i = 1$. But

the flash problem is different from the dew- and bubble-point problems because we must also

solve $\sum_i y_i = 1$. Fortunately, a reliable successive substitution method has been developed by Rach-

ford and Rice[6] to solve this problem using the objective function $\sum_i x_i - \sum_i y_i = 0$. Introducing the

variable $D_i \equiv x_i - y_i$ to denote the difference between x_i and y_i for each component,

$\sum_i x_i - \sum_i y_i = \sum_i (x_i - y_i) = \sum_i D_i = 0$, we iterate on V/F until the sum approaches zero. For a

binary system, using the K_i ratios, the objective function becomes

$$\sum_i x_i - \sum_i y_i = (x_1 - y_1) + (x_2 - y_2) = D_1 + D_2$$

$\qquad$ 10.17

$$= \frac{z_1(1 - K_1)}{1 + (V/F)(K_1 - 1)} + \frac{z_2(1 - K_2)}{1 + (V/F)(K_2 - 1)} = 0$$

For Raoult's law, K_i is a function of temperature and pressure only, both of which are fixed for an general flash calculation. The outlets are assumed to exit at phase equilibrium, and the exit conditions are used to calculate K_i. Therefore, in Eqn. 10.17, the only unknown is V/F. Eqn. 10.17 is monotonic in V/F; the sum always increases as V/F increases. Therefore, we search for the value of V/F which satisfies the equation. Note that $0 < V/F < 1$ for a physically realistic answer. Outside this range, the system is either below the dew pressure (mathematically $V/F > 1$) or above the bubble pressure (mathematically $V/F < 0$). After finding V/F from our basis, then, $L/F = 1 - V/F$ and stream compositions can be found from Eqns. 10.15 and 10.16.

Before executing a flash calculation by hand, bubble and dew calculations at the overall composition are recommended to ensures that the flash drum is between the bubble and dew pressures at the given temperature. (These calculations are easier than the flash calculation and may save you from doing it if the system is outside the phase envelope.) When using a computer, the value of V/F can be used to ascertain if the system is outside the two-phase region. If a computer flash calculation does not converge, then the bubble or dew should be performed to troubleshoot.

For a binary system, a flash calculation may be avoided by plotting the overall composition on the P-x-y diagram between the dew and bubble pressures and reading the vapor and liquid compositions from the graph. Application of the lever rule permits calculation of the total fraction that is vapor; however, this method requires a lot of calculations to generate the diagram if it is not already available and is limited to binary mixtures. Also, plotting the curves is slower than a direct calculation.

---

6. Rachford Jr., H.H., Rice, J.D. 1952. *J. Petrol. Technol.* 4(sec. 1):19, 4(sec. 2):3.

The steady-state energy balance for an general flash is given by

$$0 = FH^F - LH^L - VH^V + \dot{Q}$$

$$= F\left(\sum_i z_i H_i^F + \Delta H_{mix}^F\right) - L\left(\sum_i x_i H_i^L + \Delta H_{mix}^L\right) - V\left(\sum_i y_i H_i^V + \Delta H_{mix}^V\right) + \dot{Q} \qquad 10.18$$

where we indicate a common method of calculation of the enthalpy of a mixture as the sum of the component enthalpies and the heat of mixing (Eqn. 3.24 on page 105). Writing in terms of V/F:

$$0 = \left(\sum_i z_i H_i^F + \Delta H_{mix}^F\right) - \left(1 - \frac{V}{F}\right)\left(\sum_i x_i H_i^L + \Delta H_{mix}^L\right) - \frac{V}{F}\left(\sum_i y_i H_i^V + \Delta H_{mix}^V\right) + \frac{\dot{Q}}{F} \qquad 10.19$$

A reference state must be specified for each component, and a method for calculation of the enthalpy must be selected. A simple method of calculating enthalpies of vapors and liquids relative to reference states has been illustrated in Example 3.3 on page 107. For an isothermal flash, the VLE constraint, Eqn. 10.17, can be solved independently of the energy balance, and the energy balance can then be solved for the required heat transfer.

Adiabatic Flash

An adiabatic flash differs from an isothermal flash because $\dot{Q} = 0$ in Eqn. 10.19. The adiabatic conditions will result in a temperature change from the feed conditions that is often significant. A typical scenario involves an outlet pressure less than the inlet pressure, resulting in an evaporation of a fraction of the feed. Because evaporation is endothermic, this type of flash results in a temperature drop (often significant). We have seen this type of calculation for pure fluids using throttles in Chapter 5. The additional complication with a mixture is that the components will distribute based on their different volatilities. The objective of an adiabatic flash calculation is to determine the outlet temperature in addition to the L and V compositions. An adiabatic flash requires that Eqn. 10.19, ($\dot{Q}=0$), must be solved simultaneously with Eqn. 10.17. The vapor and liquid mole fractions for Eqn. 10.19 are determined from Eqns. 10.15 and 10.16. The method is complex enough that even simple assumptions, such as ideal mixing ($\Delta H_{mix} = 0$), benefit from a computer algorithm.

The algorithm to solve for the adiabatic flash depends on the differences in boiling points of the components. Usually the boiling points are different enough that an initial guess of T is used in VLE Eqn. 10.17 (where V and L are assumed to exit at the same T) to find an initial V/F where $0 < V/F < 1$. With that initial V/F, then the V and L compositions (Eqns. 10.15 and 10.16) and then the energy balance (Eqn. 10.19) are evaluated, often scaling Eqn. 10.19 by dividing by 1000 when the enthalpy values are large. If the energy balance is not satisfied, then a new T trial is inserted into Eqn. 10.17 and the loop continues. Eqn. 10.19 is monotonic in T and increases when T decreases. When the boiling points are very close for all components, such as with isomers, the calculation converges better with an initial guess of V/F in Eqn. 10.17, which is solved by trial and error for the T which satisfies the equation. The L and V compositions and T are then used for the outlet enthalpies in Eqn. 10.19 to generate a new value for V/F and the iteration continues until convergence.

10.4 MULTICOMPONENT VLE RAOULT'S LAW CALCULATIONS

Extending our equations to multicomponent systems is straightforward. For a bubble calculation we have

$$1 = \left(\sum_i x_i P_i^{sat}\right) / P = \sum_i x_i K_i \qquad 10.20$$

For a dew calculation we have

$$1 = P\sum_i \frac{y_i}{P_i^{sat}} = \sum_i \frac{y_i}{K_i} \qquad 10.21$$

These equations may be used for bubble- or dew-pressure calculations without iterations. For bubble- or dew-point temperatures, iteration is required. A first guess may be obtained from one of the following formulas:

$$T_{\text{bubble guess}} = \sum_i x_i T_i^{sat}$$

$$T_{\text{dew guess}} = \frac{\sum_i y_i T_{r,i} T_i^{sat}}{\sum_i y_i T_{c,i}} \qquad \text{or} \qquad T_{\text{dew guess}} = \sum_i y_i T_i^{sat} \qquad 10.22$$

But these are somewhat inaccurate guesses which require subsequent iteration.[7]

For flash calculations, the general formula is:

$$\sum_i x_i - \sum_i y_i = \sum_i D_i = \sum_i \frac{z_i(1 - K_i)}{1 + (V/F)(K_i - 1)} = 0 \qquad 10.23$$

to find L/F and x_i and y_i are then found using Eqns. 10.15 and 10.16.

Use the next example to understand how to apply the strategies described in Section 10.2. Note how the column conditions are used to decide which routine to use. Note also that developing skill to determine *which* routine to run is as important as proficiency with the routines.

---

7. Note that Eqns. 10.20–10.22 can be used to estimate the bubble and dew points regardless of whether the components are supercritical or whether vapor and liquid phases are indeed possible. We will see in the discussion of equations of state that mixtures can have critical points, too, and this leads to a number of subtle complexities.

Example 10.1 Bubble and dew temperatures and isothermal flash of ideal solutions

The overhead from a distillation column is to have the following composition:

	z(Overhead)
Propane	0.23
Isobutane	0.67
n-Butane	0.10
Total	1.00

A schematic of the top of a distillation column is shown below. The overhead stream in relation to the column and condenser is shown where V_{prod} represents vapor flow and D_{prod} represents liquid flow. *In an ideal column, the vapor leaving each tray (going up) is in phase equilibrium with the liquid leaving the same tray (going down).* If the cooling water to the condenser is turned off, then only vapor product will be obtained, but this is not typical because the column works better with some liquid L returning to the column. To obtain liquid product only, cooling water is provided and the vapor product stream is turned off, and the condenser is known as a **total condenser.** If cooling water is provided to partially condense the vapor stream, the liquid product stream is typically turned off. Then the condenser provides additional separation, operating as a **partial condenser.** *In an ideal partial condenser, the exiting vapor and liquid leave in phase equilibrium with each other.*

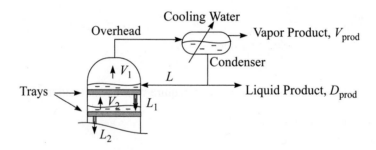

(a) Using the shortcut K-ratio, calculate the temperature at which the condenser must operate in order to condense the overhead product completely at 8 bar.

(b) Using the shortcut K-ratio, and assuming the overhead product vapors are in equilibrium with the liquid on the top plate of the column, calculate the temperature of the overhead vapors and the composition of the liquid on the top plate when operating at the pressure of part a.

(c) The vapors are condensed by a partial condenser operating at 8 bar and 320 K. Using the shortcut K-ratio, what fraction of liquid is condensed?

Example 10.1 Bubble and dew temperatures and isothermal flash of ideal solutions (Continued)

Solution: Use the shortcut estimates of the K-ratios. Use of a solver tool is recommended after developing an understanding of the manual iterations summarized below.

(a) To totally condense the overhead product, the mixture must be at the bubble-point temperature or lower. The maximum temperature is the bubble-point temperature. To find the bubble-point temperature for the ternary system, we apply Eqn. 10.20 extended to three components. The calculation requires trial-and-error iteration on temperature as summarized in Table 10.1.

It is easy to create an Excel spreadsheet to do these calculations quickly (e.g., Flshr.xlsx can be modified).

Tabulated below, the shortcut K-ratio is calculated using Eqn. 10.7 at each temperature guess at 8 bar, and the y values from Eqn. 10.9 are summed to check for convergence following the procedure set forth in Table 10.1. Iterations are repeated until the y's sum to 1.

	Guess $T = 310$ K		Guess $T = 320$ K	
	K_i	y_i	K_i	y_i
C_3	1.61	0.370	2.03	0.466
iC_4	0.616	0.413	0.80	0.536
nC_4	0.438	0.044	0.58	0.058
	$\Sigma y_i =$	0.827	$\Sigma y_i =$	1.061

The temperature has been bracketed, interpolating,

$$\Rightarrow T = 310 + \left(\frac{1.000 - 0.827}{1.061 - 0.827}\right)(320 - 310) = 317 \text{ K}$$

(b) The saturated vapor leaving the tray is in equilibrium with the liquid and is at its dew-point temperature at 8 bar. Eqn. 10.21 is used. The calculation requires iteration on temperature. Calculating the K_i ratios as in part (a), the liquid phase compositions are calculated at each iteration using Eqn. 10.12 until the values sum to 1.

	Guess $T = 325$ K		Guess $T = 320$ K	
	K_i	x_i	K_i	x_i
C_3	2.26	0.102	2.03	0.113
iC_4	0.905	0.740	0.80	0.838
nC_4	0.658	0.152	0.58	0.172
	$\Sigma x_i =$	0.994	$\Sigma x_i =$	1.123

The temperature has been bracketed, interpolating,

$$T = 325 + \left(\frac{1.00 - 0.994}{1.123 - 0.994}\right)(320 - 325) = 324.8 \text{ K} \qquad 10.24$$

Example 10.1 Bubble and dew temperatures and isothermal flash of ideal solutions (Continued)

Repeating the procedure at this temperature a final time results in the liquid compositions,

	Guess $T = 324.8$ K	
	K_i	x_i
C_3	2.25	0.102
iC_4	0.900	0.744
nC_4	0.654	0.153
	$\Sigma x_i =$	0.999

(c) Recognize that the solution involves an isothermal flash calculation because P and T are both specified. Begin by noting that the specified condenser temperature is between the bubble temperature, 317 K, and the dew temperature, 324 K, so vapor-liquid equilibrium is indeed possible. Because T and P are already set, Eqn. 10.7 is used to calculate the K-ratio for each component. Then, we seek a solution for Eqn 10.23 at 320 K and 8 bar. In the general flash routine, F_{flash}, V_{flash}, and L_{flash} are used to denote the flow rates for the flash drum and we must adapt the flash variables to the column stream names. We add "flash" and "column" subscript descriptors for increased clarity. In the solution, F_{flash}, V_{flash}, and L_{flash} represent the flow rates for the partial condenser, $F_{flash} = V_{1,column}$, $V_{flash} = V_{prod}$, and $L_{flash} = D_{prod} + L_{column}$. In the summarized calculations below, $V_{flash}/F_{flash} = V_{prod}/V_{1,column}$ and D_i in the table is the objective variable for the flash calculation as used in Eqn. 10.23, not the column liquid product flow rate D_{prod}. Each table column summarizes a guess for flash ratio V_{flash}/F_{flash} and the resultant flash objective variable D_i. The composition of the feed is given by z_i as conventional for a flash calculation, which is the composition of $V_{1,column}$.

Flshr.xlsx or
Chap10/Flshr.m
can be modified to
work this example.

A summary of the isothermal flash calculation is given below.:

	F_{flash}		Guess $V_{flash}/F_{flash} = 0.5$	Guess $V_{flash}/F_{flash} = 0.4$	Guess $V_{flash}/F_{flash} = 0.23$
	z_i	K_i	$D_i = x_i - y_i$	D_i	D_i
C_3	0.23	2.03	−0.1564	−0.1678	−0.1915
iC_4	0.67	0.80	0.1489	0.1457	0.1405
nC_4	0.10	0.58	0.0532	0.0505	0.0465
			sum = 0.0457	sum = 0.0284	sum = −0.0045

Example 10.1 Bubble and dew temperatures and isothermal flash of ideal solutions (Continued)

Interpolating between the last two results that bracket the answer,

$$\frac{V_{\text{flash}}}{F_{\text{flash}}} = 0.23 + \left(\frac{0.4 - 0.23}{0.0284 + 0.0045}\right)(0. + 0.0045) = 0.2533$$

$V_{\text{flash}}/F_{\text{flash}} = 0.25$, applying Eqns. 10.15 and 10.16,

$$\Rightarrow \{x_i\} = \{0.1829, 0.7053, 0.1117\} \text{ and } \{y_i\} = \{0.3713, 0.5642, 0.0648\}$$

The compositions can be confirmed to be converged. The outlet composition of the vapor, V_{prod}, is given by $\{y_i\}$ and it is clearly more enriched in the volatile components than the inlet from the top of the column $V_{1,\text{column}}$.

Note: The flash problem converges more slowly than the bubble- and dew-point calculations, so the third iteration is necessary.

Example 10.2 Adiabatic flash

Ethanol + methanol form a nearly ideal solution as shown in Fig. 10.2. An equimolar feed at 760 mmHg is subjected to an adiabatic flash operating at 200 mmHg. Feed enters at 70°C and 43 mol/min. Find the exiting stream temperatures, flow rates, and compositions. Assume ideal solutions and use the Antoine equation for vapor pressures.

Solution: This is a direct application of a procedure, so it is clear which VLE routine to use: the FA row of Table 10.1. We must combine the VLE procedure with energy balance. A bubble-pressure calculation at 70°C (not shown) shows that the feed is all liquid. Two solutions are provided using different pathways for the enthalpy calculations. Both solutions will use the same flash calculation procedures and the Antoine equation is used with {methanol, ethanol}:

Chap10/
Ex10_02.m.

A = {8.081, 8.1122}, B = {1582.3, 1592.9}, C = {239.73, 226.18}

Solution 1. This solution method calculates component enthalpies using a reference state of liquid at 25°C for all species where we set $H_R = 0$. The enthalpy calculations use the pathway of Fig. 2.6(a). The pathway is taken through the boiling point of each component, as in Example 3.3. To compute stream enthalpies, we use ideal solutions as shown in Example 3.3, ignoring heat of mixing. Heat capacity constants and heats of vaporization are taken from Appendix E.

The solution requires a guess of T resulting in $0 \leq V/F \leq 1$ that provides two phases, and then a check of the energy balance. Due to the complexity of the calculation, we iterate on the T guess manually, and automate the tedious parts of solving for V/F and checking the energy balance. The solution is provided in MATLAB file Ex10_02.m. Some intermediate results are tabulated.

Example 10.2 Adiabatic flash (Continued)

Because a computer is used, we skip preliminary bubble and dew calculations. Note that we do not tabulate all values until $0 \leq V/F \leq 1$. The first guess of 45°C is above the dew temperature. The second guess of 35°C is below the bubble temperature. The next guess happens to give a condition close to the bubble temperature, so we raise the temperature guess slightly. The column OBJEB = (Eqn. 10.19)/1000. The compositions and enthalpies are shown below and the last row is converged. The exiting flow rates are $V = V/F(43) = 0.09(43) = 3.87$ mol/min, and $L = 43 - 3.87 = 39.13$ mol/min. About 9% (molar basis) of the inlet is flashed, and the outlet temperature is 40.2°C compared to an inlet of 70°C.

T^{out}(C)	V/F	OBJEB	x_{EtOH}	y_{EtOH}	H^F(J/mol)	H^L(J/mol)	H^V(J/mol)
45	2.98						
35	−9.32						
40	0.001	3.137	0.50	0.33	4672	1481	39,149
41	0.43	−13.30	0.57	0.40	4672	1618	39,539
40.2	0.09	−0.22	0.515	0.349	4672	1508	39,225

Solution 2. This solution method calculates component enthalpies using the pathway of Fig. 2.6(c), the reference state of the elements at 25°C, the heats of formation of ideal gases, the generalized correlation for heat of vaporization in Eqn. 2.45, and the $Cp^{ig}(25°C)$ from the back flap. The results are slightly different from Solution 1, owing to the imprecision of Eqn. 2.45 and differences between the heat capacities. Process simulation software typically uses this enthalpy path and reference state.

We begin by finding the enthalpy of the feed relative to the elements at 25°C, noting that it is a liquid ideal solution. $H^F = H^L(70°C) = \Sigma x_i(\Delta H°_{f,i} + Cp^{ig}(T - T_R) - \Delta H_i^{vap}) = 0.5(-200,940 + 5.28(8.314)(70 - 25) - 35,976) + 0.5(-234,950 + 7.88(8.314)(70 - 25) - 38,595) = -252,769$ J/mol. This takes care of the first term in Eqn. 10.19. Noting that the feed is liquid, we might suspect the flash outlet to be mostly liquid. Performing a bubble-temperature calculation at 200 mmHg gives $T = 40.00°C$ and $H^L(40°C) = -256,901$ and with no vapor stream results in $Q = -4132$ J/mol.[a] The temperature must be slightly higher to move Q toward zero. Suppose we "guessed" that the temperature is 40.23°C.[b] Then the flash calculation gives $x_{EtOH} = 0.5173$, $y_{EtOH} = 0.3515$, $V/F = 0.1042$, $H^L(40.23°C) = -257,502$. The formula for H^V is similar to that for H^L but omits the ΔH_i^{vap} contribution and replaces x_i with y_i, so $H^V(40.23°C) = -212,088$. Following Eqn. 10.19, $Q = (1 - 0.1042)(-257,502) + 0.1042(-212,088) + 252,769 = 0.1$ J/mol. We may assume that 0.1 J/mol is sufficiently close to zero.

---

a. Note that the heats of vaporization must be recomputed at the new temperature.
b. Obviously, this was not our first guess. Alternatively, you could call the solver with the added constraint that $Q = 0$ while $\Sigma D_i = 0$ by changing T and V/F simultaneously.

10.5 EMISSIONS AND SAFETY

Hydrocarbon emission monitoring is an important aspect of environmentally conscious chemical manufacturing and processing. The United States Environmental Protection Agency (EPA) has published guidelines[8,9,10] on the calculations of emissions of **volatile organic compounds** (**VOC**s), and VOC emissions are monitored in the U.S. Most of the emission models apply the ideal gas law and Raoult's law and thus the calculation methods are easily applied. While many of the mixtures represented with these techniques are not accurately modeled for phase equilibria by Raoult's law, the method is suitable as a first approximation for emission calculations. This section explores emission calculations for batch processes. Batch processes are common in specialty chemical manufacture. In most cases, air or an inert gas such as nitrogen is present in the vapor space (also known as the **head space**). In some cases, the inert head space gas flows through the vessel, and is called a **purge** or **sweep** gas. These gases typically have negligible solubilities in the liquid phase and are thus considered **noncondensable.** There are several common types of unit operations encountered with VOC emissions, which will be covered individually.

Filling or Charging

During filling of a tank with a volatile component, gas is displaced from the head space. The displaced gas is assumed to be saturated with the volatile components as predicted with Raoult's law and the ideal gas law. Initially in the head space, $n_{head}^i = (P\underline{V}_{head}^i)/(RT)$ and after filling, $n_{head}^f = (P\underline{V}_{head}^f)/(RT)$, where the subscript *head* indicates the head space. The volume of liquid charged is equal to the volume change of the head space. The mole fractions of the VOC components are determined by Raoult's law, and the noncondensable gas makes up the balance of the head space. The moles of VOC emission from the tank are estimated by $y_i(n_{head}^i - n_{head}^f)$ for each VOC.

Purge Gas (Liquid VOC Present)

When a purge (sweep) gas flows through a vessel containing a liquid VOC, the effluent will contain VOC emissions. At the upper limit, the vessel effluent is assumed to be saturated with VOC as predicted by Raoult's law. For VOC component *i*,

$$\Delta n_i = n_{sweep} \cdot k_m (y_i/y_{nc}) \qquad\qquad 10.25$$

where $y_{nc} = 1 - \sum_i y_i$, where the sum is over VOCs only. The variable k_m is the saturation level,

and is set to 1 for the assumption of saturation and adjusted lower if justified when the purge gas is known to be unsaturated. The flow of noncondensables n_{sweep} can be related to a volumetric flow of purge gas using the ideal gas law,

$$n_{sweep} = \dot{n}t = \frac{P\dot{\underline{V}}_{sweep}}{RT}t \qquad\qquad 10.26$$

Purge Gas (No Liquid VOC Present)

Vessels need to be purged for changeover of reactants or before performing maintenance. After draining all liquid, VOC vapors remain in the vessel at the saturation level present before draining.

---

8. OAQPS, Control of Volatile Organic Compound Emissions from Batch Processes–Alternate Control Techniques Information Document, EPA-450/R-94-020, Research Triangle Park, NC 27711, February 1994.

9. OAQPS, Control of Volatile Organic Emissions from Manufacturing of Synthesized Pharmaceutical Products, EPA-450/2-78-029, December 1978.

10. U.S. E.P.A., Compilation of Air Pollution Emission Factors-Volume 1, (1993) EPA Publication AP-42.

The typical assumption upon purging is that the vessel is well mixed. A mole balance on the VOC gives $dn_i/dt = -y_i \dot{n}_{sweep}$; dividing by y_i the equation becomes $dn_i/y_i dt = -\dot{n}_{sweep}$. The left-hand side can subsequently be written $n_{tank} dy_i/y_i dt = (P\underline{V}_{tank})/(RT_{tank}) \cdot (dy_i)/(y_i dt)$, and the right-hand side can be written $(P\underline{\dot{V}}_{sweep})/(RT_{sweep})$. When the sweep gas and tank are at the same temperature, which is usually a valid case, the equation rearranges to $(dy_i)/y_i = (-\underline{\dot{V}}_{sweep}/\underline{V}_{tank})dt$, which integrates to

$$y_i^f = y_i^i \exp\left(\frac{-\underline{\dot{V}}_{sweep}}{\underline{V}_{tank}}t\right)$$

10.27

The emissions are calculated by

$$\Delta n_i = n_{tank}(y_i^i - y_i^f)$$

10.28

Heating

During a heating process, emissions arise because the vapors in the head space must expand as the temperature rises. Since vapor pressure increases rapidly with increasing temperature, VOC concentrations in the vapor phase increase also. Detailed calculations of emissions during heating are somewhat tedious, so an approximation is made; the emission of each VOC is based on the arithmetic average of the molar ratio of VOC to noncondensable gas at the beginning and the end of the heating multiplied by the total moles of noncondensable gas leaving the vessel. At the beginning of the heating, representing the VOC with subscript i and the noncondensables with subscript nc, the ratio of interest is $(n_i/n_{nc})^i = (y_i/y_{nc})^i = (y_i P/y_{nc}P)^i$, and at the end $(n_i/n_{nc})^f = (y_i/y_{nc})^f = (y_i P/y_{nc}P)^f$. The emission of VOC component i is calculated as

$$\Delta n_i = \frac{\Delta n_{nc}}{2}\left[\left(\frac{y_i}{y_{nc}}\right)^i + \left(\frac{y_i}{y_{nc}}\right)^f\right]$$

10.29

where $y_{nc} = 1 - \sum_i y_i$, and the sum is over VOCs only. The value of Δn_{nc} is given by

$$\Delta n_{nc} = \frac{\underline{V}_{head}}{R}\left[\left(\frac{\left(1-\sum_i y_i\right)P}{T}\right)^i - \left(\frac{\left(1-\sum_i y_i\right)P}{T}\right)^f\right] + \Delta n_{sweep}$$

10.30

where the summations are over VOCs only and Δn_{sweep} is the total moles of noncondensable that are swept (purged) through the vessel during heating and is set to zero when purging is not used. Eqns. 10.29 and 10.30 can overestimate the emissions substantially if the tank approaches the bubble point of the liquid because y_{nc} approaches zero, and then calculations are more accurately handled by a more tedious integration. The integration can be approximated by using the method presented here over small temperature steps and summing the results.

Depressurization

Three assumptions are made to model depressurizations: The pressure is decreased linearly over time; air leakage into the vessel is negligible; and the process is isothermal. The relationship is then the same as Eqn. 10.29, where Δn_{nc} is calculated by Eqn. 10.30 using $\Delta n_{sweep} = 0$.

Other Operations

Other operations involve condensers, reactors, vacuum vessels, solids drying, and tank farms. Condensers are commonly used for VOC recovery; however, the VOCs have a finite vapor pressure even at condenser temperatures and the emissions can be calculated by using Eqns. 10.25 and 10.26 at the condenser temperature. Reactors may convert or produce VOCs and, in a vented reactor, the emissions can be calculated by adapting one of the above techniques, keeping aware that generation of gas causes additional vapor displacement and possible temperature rise due to reaction. Vacuum units and solids-drying operations are also direct adaptations of the methods above. Tank farm calculations are more detailed and empirical. Tank emission calculations depend on factors such as the climate and the paint color of the tank. Fixed roof tanks must breathe as they warm during the day due to sunlight, and then cool during the night hours. There are also working losses due to routine filling of the tanks as covered above. Although heating and cooling in a tank with a static level can be treated by the methods presented above, when the levels are also changing due to usage, EPA publications are recommended for these more tedious calculation procedures.

Flash Point

The **flash point** is a property much different from that represented by the general flash or adiabatic flash discussed earlier. Fire requires fuel, an oxidizer (air in this case), and an ignition source. The flash point is the temperature above which a vapor mixture supports combustion when an ignition source is present. When liquids burn, fire occurs on a liquid surface; the vapors near the surface are burning, not the liquid itself. The flash point is important because it is the temperature at which the **Lower Flammability Limit** (**LFL**) concentration is reached at the liquid surface. A flash can also occur entirely in the vapor phase. When burning buildings explode in action movies, the movies are depicting the real condition of the vapors in the building reaching the flash point as plastics and other materials decompose. Fire fighters are very cautious entering buildings where a potential for such explosions exist.

The **Chemical Safety Board** (www.csb.gov) tracks accidents in the U.S. chemical industry. The CSB reports that workers continue to be careless with ignition sources near organic solvents. Earlier sections discussed the warming of a tank during the day. As a vented tank warms, emissions may reach the LFL as the day warms. Maintenance on metal tanks often involves grinding or welding, which introduces ignition sources and has resulted in numerous deaths and disfigurements as well as property damage. Raoult's law may be used to estimate the vapor phase concentrations, and the flash point temperature is approximately the temperature where $y \cdot (100\%) = \text{LFL}$. Technically, the LFL defined as a volume percent of combustible material, but it is the same as the mole fraction in an ideal gas. For a mixture, the flash point can be calculated approximately by the empirical relation

❶ The Chemical Safety Board is a federal agency charged with investigating chemical accidents. The goal is to learn from accidents and improve safety.

$$\sum_i \frac{y_i}{(LFL_i/100)} = 1 \qquad \text{flash point temperature condition for mixture} \qquad 10.31$$

Because combustion requires an oxidizer, each fuel also has an **Upper Flammability Limit (UFL)** above which there is not enough oxygen present to maintain combustion. Most accidents occur near the LFL, which is the motivation for more discussion of LFL. Both the LFL and the UFL are affected by inerts because of the effect on the ratio of oxidizer to fuel.

10.6 RELATING VLE TO DISTILLATION

We introduced some major points about the importance of distillation in Section 3.2. Roughly 80% of separations are done involving distillation and 70% of the capital cost of a chemical plant goes into distillation equipment, and thus the proper application of vapor-liquid equilibria and design are essential. Usually, one distillation column is required to separate any two components. To separate three components to high purity requires two columns. Obtaining four components to high purity requires three columns, and so forth. So a single reactor that requires two reactants and produces two products ($A + B \rightarrow C + D$) would probably require three distillation columns downstream if all the components are desired in high purity. Pharmaceutical and speciality chemical plants have more by-products than bulk chemical plants. This means that chemical engineers need to be fairly familiar with VLE, especially in the fine chemicals industry.

How Distillation Works

How does the laboratory experiment (Fig. 10.6(a)) relate to distillation in a chemical plant (Fig. 10.6(b) and (c))? We begin with an elementary introduction to the conceptual basis of distillation. We then follow up with more detailed descriptions of the basis for modeling the process.

Thermodynamics teaches us that the most volatile components are enriched in the vapor phase when a liquid solution is boiled. For example, suppose that you want to recover methanol (a potential transportation fuel or reactant for a fuel cell) from a mixture of 10% methanol (MeOH) and

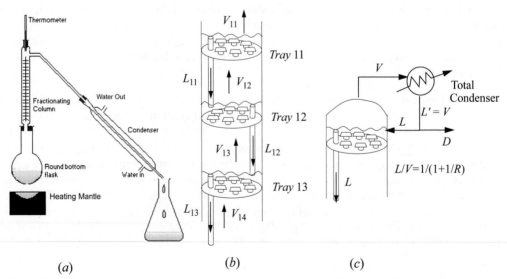

(a)   *(b)*   *(c)*

Figure 10.6 *Schematic diagrams of distillation columns. (a) A typical chemistry laboratory distillation apparatus; (b) close-up view of sieve trays showing the holes in the trays, downcomers, and liquid on each tray; (c) a partial condenser operates like a flash unit.*

90% water. The thermodynamics of the methanol + water system are summarized in Fig. 10.7. Fig. 10.7(a) shows that the solution bubble-point temperature is ~87°C. As it boils, the temperature remains constant, but the vapor composition leaving the vessel is ~40 mol% MeOH. Let us call this separation stage *a*. Forty percent is a big improvement over the initial 10%, but it is still mostly water. What can you do to make it more pure? Why not condense the vapor to a liquid and collect it in a separate pot to reboil it? After you have enough solution in that pot, take it to another boiler and perform stage *b*, then repeat for stages *c, d, e* as shown in Fig. 10.7(a). After ~5 stages, you could obtain 98% pure methanol. This is a simplification of multistage distillation at the conceptual level. (Note that most process simulation software numbers stages from the column top which is why we designated stages as letters rather than numbering from the bottom.) As outlined so far, it is inefficient and oversimplified because we considered the liquid phase composition to be invariant while the volatile component was boiling off. Separation textbooks provide the details on the mass balances.

The first law of thermodynamics tells us that energy is conserved. What if we could use the heat of condensing the vapor from stage *a* as the heat of boiling in stage *b*? That would be a big improvement. Furthermore, we can achieve this in minimal space if we use some clever plumbing. If we put the pot for stage *b* on top of stage *a*, and put little holes in the bottom of the pot, then the vapor boils through the holes faster than the liquid can weep back (Fig. 10.6(b)). The boiling point of the mixture on stage *b* is lower than the vapor temperature coming from below (*cf.* Fig. 10.7(a)). When the warmer vapor from stage *a* contacts the cooler liquid on stage *b*, it condenses. But the first law tells us that the heat of condensation must go somewhere. Where? It goes into boiling the liquid on stage *b*. Then we can do the same thing for stages *c, d, e*. This approach is called a tray distillation column and it is very common throughout the chemical industry. Roughly 70% of distillation columns are tray type.

We still have not addressed how tray distillation relates to the chemistry lab. The chemistry lab requires some gauze or glass wool in the glass tube above the boiling flask. Because of heat loss

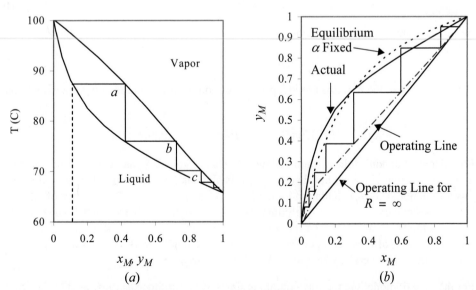

Figure 10.7 *Stage-wise separation of methanol (M) and water. (a) T-x,y diagram at 1 bar, showing stages. (b) McCabe-Thiele analysis based on assuming constant relative volatility,*
$$\alpha_{LH} = [(\alpha_{LH}^{T})(\alpha_{LH}^{B})]^{1/2} \text{ (dotted line) compared with experimental curve.}$$

through the condenser walls, some of the boiling liquid condenses, then trickles down the gauze and falls back into the boiling flask. This trickling liquid serves the same function as the liquid on the trays, but of course the contact and mass exchange is much less efficient. On the other hand, we can generate a lot of surface area and a lot of vapor-liquid contact by using a lot of gauze. In the chemical industry, columns based on this principle are called packed columns. Packed columns comprise the other 30% of distillation columns. There is another big difference between the chemical industry and chemistry lab. The vast majority of industrial columns run in continuous steady-state mode, not batch mode like the chemistry lab. This means that continuous feed enters on a stage that has a similar composition to the feed.

Most mixtures contain more than two components. It is common, however, to design the multi-component column based on the separation of two key components. Because boiling point (in the absence of azeotropes discussed later) increases with M_w, it is common to discuss separation based on *light* components (more volatile) moving up the column, and *heavy* components moving down. For preliminary column design a volatile **light key** (typically low M_w, thus *light*) and less volatile **heavy key** component are designated. Splitting two key components implicitly splits components lighter than the "light key" component from components heavier than the "heavy key" component. The **split** (S) is the fraction that exits with distillate. The light key (*LK*) component is the least volatile component with a split $S_{LK} > 0.5$. System components lighter than the light key must be even more volatile and exit as distillate. For example, consider a distillation of hexane, octane, decane, and dodecane. If we designate octane as the light key, then the hexane should also go out the top. The heavy key is the lightest component (most volatile) with a split $S_{HK} < 0.5$. In the example, we could select decane as the heavy key and thus dodecane would go out the bottom also. In a perfect world, the splits would be 100% for the light key and 0% for the heavy key, but that would require an infinitely tall distillation column. More typical splits are 99% and 1%.

The split fractions define the relevant mass balances in distillation. The thermodynamics relevant to distillation is implemented using the **relative volatility** in terms of the VLE *K*-ratios,

$$\alpha_{ij} = K_i/K_j \qquad\qquad 10.32$$

For the case of light and heavy key components,

$$\alpha_{LH} = K_{LK}/K_{HK} \qquad\qquad 10.33$$

For Raoult's law, the *K*-ratios are independent of composition, and thus is the relative volatility,

$$\alpha_{ij} = K_i/K_j = (P_i^{sat}/P)/(P_j^{sat}/P) = P_i^{sat}/P_j^{sat} \qquad \text{Raoult's law} \qquad 10.34$$

For systems that don't follow Raoult's law, the relative volatility may vary through the column owing to composition changes, but distillation is feasible as long as $\alpha_{LH} > 1$. (We will show the analysis for the nonideal α_{ij} calculation in Section 11.11 on page 442.) The presence of other components is of secondary concern for preliminary column design as long as $\alpha_{LH} > 1$, so shortcut column analysis treats LK and HK as if the mixture were binary. It may be possible to improve α_{LH} through the addition of other components (e.g., extractive distillation), but that merely reinforces the requirement of the overall system to the mandate that $\alpha_{LH} > 1$.

For the sake of modeling, the tray column is simplest to introduce as illustrated in Fig. 10.7(b). Fig. 10.7(b) focuses on the composition changes only, neglecting the temperature effects. Fig. 10.7(b) also shows the result of approximating that the relative volatility is constant. A larger α_{ij} results in a larger area under the *x-y* curve and an easier separation. If $\alpha_{ij} = 1$, the curve collapses on

Light key and heavy key components are used in preliminary column design.

Relative volatility.

the 45° diagonal. The diagram shows "steps" between the equilibrium line and the operating line. The equilibrium line represents the compositions at each tray as they leave. The operating line represents the compositions between trays. Moving up and down the column, the material balances are shown graphically by stepping back and forth as we relate the material balance "on stage" and "in-between stage." Comparing the curves for actual and constant α_{ij}, note that a similar number of stages is obtained. For the purposes of our model, constant relative volatility is a convenient approximation for the equilibrium curve as shown. But there is still a significant detail that has been omitted in our conceptual outline of distillation. Where did the liquid come from that is on the trays of the tray column?

Fig. 10.6(c) shows how the condenser on top of the column pours liquid back down to keep some on the trays. The part that we pour back down the column (L) is called **reflux.** The part that we recover as product is called distillate (D). The ratio of L/D is called the **reflux ratio** (R). The reflux ratio controls the amount of product recovered as distillate. If we actually want to recover some product (i.e., $D \neq 0$), then we must accept some value $R \neq \infty$. Finite values for R lead to the dashed-dot operating lines in Fig. 10.7(b). To understand this, consider that the 45° diagonal on Fig. 10.7(b) corresponds to $L/V = 1$. It turns out that taking distillate from the top of the column leads to slightly less separation on every stage, giving the dashed-dot lines of Fig. 10.7(b). Typical courses in mass transfer operations explain how to estimate the dashed-dot lines from values of R. The key point for now is that the value of R cannot be zero, or we will have no liquid on the trays, and it cannot be infinite or we will recover no product. We can go a little further and say that it must be closer to infinity for a distillation that has a relative volatility very near to unity, because the y-x curve in that case stays very close to the diagonal. Beyond that, we simply need to accept that somebody has analyzed this before and developed some equations for computing the minimum number of stages to achieve a desired separation (at infinite reflux), the minimum reflux ratio, and the actual number of stages. This is indeed the case, and the model equations are presented below.

A shortcut distillation calculation for the height of the column for constant relative volatility can be estimated from the Fenske equation,

$$N_{min} = \frac{\ln\left[\dfrac{S_{LK}(1 - S_{HK})}{S_{HK}(1 - S_{LK})}\right]}{\ln \alpha_{LH}^m} \quad \text{ideal solutions} \qquad 10.35$$

where N_{min} is the minimum number of theoretical trays at infinite reflux and α_{LH}^m is the geometric mean of the relative volatility calculated using the column top and bottom compositions, T and P. Typically, the number of actual trays is $N_{act} \sim 4N_{min}$, with the space between trays being 0.6m. So a column with 99 and 1% splits and a relative volatility of 3 would have $N_{min} = 8.4$ and a height of 20 m. With this background, the importance of the K-ratios and α_{LH} becomes clear. Note that if $\alpha_{LH} = 1$, then $N_{min} = \infty$. The relative volatility changes with composition for nonideal systems, and goes to 1 when an azeotrope exists. Then Eqn. 10.35 is not valid. We discuss such behavior in the following sections, and find significant motivation to understand the modeling of such systems in subsequent chapters.

10.7   NONIDEAL SYSTEMS

In Section 10.3 we introduced Raoult's law for mixtures where the components have very similar chemical functionality and molecular weight. We have seen how easy the ideal-solution calculations can be. However, Raoult's law is accurate for only a few of the systems you will encounter in practice. Examples of phase diagrams which deviate from Raoult's law are shown in Fig. 10.8 and Fig. 10.9. There are several features of these diagrams that introduce important concepts. First of all, the Raoult's law bubble lines are shown as dotted lines in the *P-x-y* diagrams to emphasize the deviations. Note again that the phase diagrams of each *P-x-y/T-x-y* pair can be qualitatively related by inverting one diagram of the pair.

In Fig. 10.8 the bubble line lies above the Raoult's law line, and these systems are said to have **positive deviations** from Raoult's law. Positive deviations occur when the components in the mixture would prefer to be near molecules of their own type rather than near molecules of the other component. Briefly, it is convenient to say that these components "dislike" each other. The 2-propanol + water system has vapor pressures that are close to each other relative to the deviation from ideality. As a result, the positive deviations are large enough to cause the pressure to reach a maximum (i.e,. $P^{max} > P_1^{sat} > P_2^{sat}$). The presence of a maximum (or minimum) causes the phase envelope to close at a composition known as the azeotropic composition. The nearness of the vapor pressures matters, because any deviation from ideality would give a maximum (or minimum), known as an **azeotrope,** if $P_1^{sat} = P_2^{sat}$. As a counterexample, the methanol + 3-pentanone system has significantly different vapor pressures for the components, and the deviations from ideality are not large enough to cause azeotrope formation. Recalling that the dew and bubble lines represent coexisting compositions at equilibrium, a maximum or minimum means that $x_i = y_i$ and relative volatility $\alpha_{ij} = 1$ at the azeotrope: $\alpha_{ij} > 1$ on one side of the azeotrope composition and $\alpha_{ij} < 1$ on the other side. This means that distillation ceases to provide separation at an azeotrope composition, and knowledge of azeotropes is critical for distillation design. When an azeotrope forms in a system with positive deviations, the azeotrope is a maximum on the *P-x-y* diagram and a minimum on the *T-x-y* diagram. To give a name to the type of azeotrope, the convention is to refer to azeotropes like that of 2-propanol + water as a **minimum boiling azeotrope,** referring to the boiling temperature reaching a minimum in composition. This can be confusing because the deviations from ideality are referred to as positive with respect to Raoult's Law on a *P-x-y* diagram. If you remember that "boiling" refers to boiling *temperature,* it may help you to reduce confusion. The azeotrope on a *P-x-y* diagram is a **maximum pressure azeotrope.** Since the vapor and liquid compositions are equivalent at the azeotrope, a flash drum or distillation column cannot separate a mixture at the azeotropic composition.

Azeotropes create challenges for chemical engineers. Azeotropic compositions for systems with either positive or negative deviations depend on temperature and pressure, however the dependencies are usually weak unless large pressure or temperature changes are made. For example, the ethanol + water system possesses an azeotrope that is widely known. This azeotrope causes a significant contribution to the high energy cost of bioethanol. Separating ethanol from the dilute fermentation product stream consumes about one-third of the energy content of the ethanol. Are there ways to separate ethanol more efficiently? Are there alternative fermentation products that can be produced without an azeotrope? Could fermentation (and copious amounts of water) be circumvented altogether and cellulose converted directly to chemical feedstocks similar to how heavy oils are cracked to ethylene? These are questions that a chemical engineering perspective brings to bear on these challenging problems. In the thermodynamics of phase equilibria, we are primarily concerned with distillation, liquid-liquid extraction, and azeotropes. *Ideal solutions do not form azeotropes and they do not form immiscible liquid phases. Thus, Raoult's law is incapable of representing these systems.*

In Fig. 10.9 the systems have **negative deviations** from Raoult's law because the bubble line lies below the Raoult's line. Similar azeotropic behavior is found in these systems if the vapor pressures are close to each other or the deviations are large. When an azeotrope forms in a system with negative deviations, the azeotrope is a minimum on the *P-x-y* diagram and a maximum on the *T-x-y*

---

❶ Positive deviations from Raoult's law. It is convenient to say that the components "dislike" each other.

❶ Relative volatility equals 1 for an azeotrope: $\alpha_{ij} > 1$ on one side of the azeotrope composition and $\alpha_{ij} < 1$ on the other side.

❶ Ideal solutions do not form azeotropes and they do not form immiscible liquid phases.

❶ Negative deviations from Raoult's law. It is convenient to say that the components "like" each other.

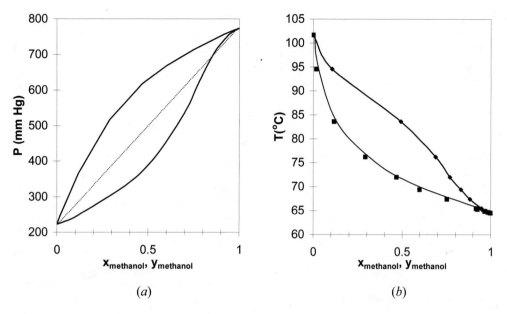

(a)

(b)

Figure 10.8 *(a), (b) Phase behavior of the methanol + 3-pentanone system. Left figure at 65°C. Right figure at 760 mm Hg. (T-x-y from Glukhareva, M.I., et al. 1976. Zh. Prikl. Khim. (Leningrad) 49:660, P-x-y calculated from fit of T-x-y.)*

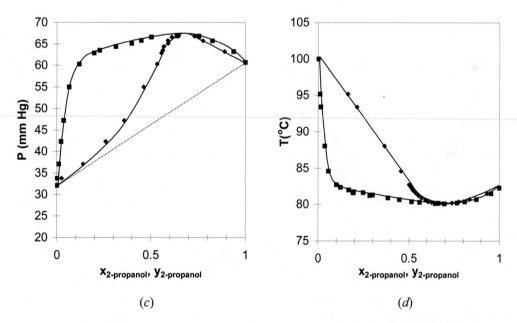

(c)

(d)

Figure 10.8 *(c), (d) Phase behavior of the 2-propanol + water system. Left figure at 30°C. Right figure at 760 mm Hg. (T-x-y from Wilson, A., Simons, E.L., 1952. Ind. Eng. Chem. 44:2214, P-x-y from Udovenko, V.V., and Mazanko. T.F. 1967. Zh. Fiz. Khim. 41:863.)*

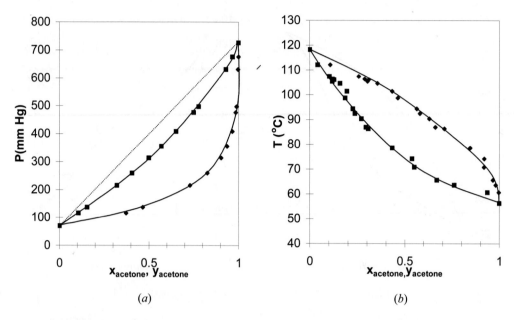

Figure 10.9 *(a), (b) Phase behavior of the acetone + acetic acid system. Left figure at 55°C. Right figure at 760 mm Hg. (T-x-y from York, R., Holmes, R.C. 1942. Ind. Eng. Chem. 34:345, P-x-y from Waradzin, W., Surovy, J., 1975. Chem. Zvesti 29:783.)*

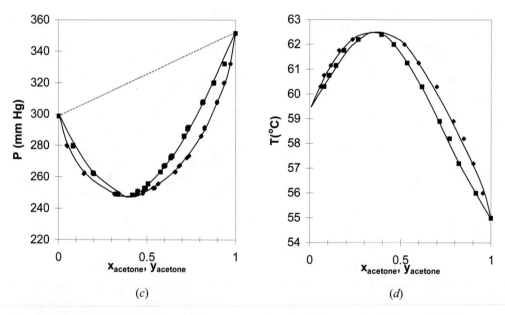

Figure 10.9 *(c); (d) Phase Behavior of the acetone + chloroform system. Left figure at 35.17°C. Right figure at 732 mm Hg. (T-x-y from Soday, F., Bennett, G.W., 1930. J.Chem. Educ. 7:1336, P-x-y from Zawidzki, V.J., 1900. Z. Phys. Chem. 35:129.)*

diagram. Therefore, this behavior is called a **maximum boiling azeotrope** or a **minimum pressure azeotrope.** From a chemical perspective, negative deviations indicate that the components "like" each other more than they like themselves. For example, mixing two acids may form an ideal solution, but mixing an acid with a base can give a negative deviation from ideality that feels warm to the touch. This is consistent with the negative sign on exothermic heats of reaction.

To better understand the reasons that mixtures deviate from Raoult's law, we need to explore the fundamental thermodynamic principles and the assumptions underlying Raoult's law. In doing so, we reveal the thermodynamic behavior that is necessary for Raoult's law, and more greatly appreciate the reasons for the occasional success and more frequent failure of the model. Raoult's law was developed empirically. To understand the true thermodynamic basis of Raoult's law, it is necessary to "evaluate" how thermodynamic properties depend on composition. In addition, we can show that the fugacity used in Chapter 9 for pure fluids extends to mixtures and the component fugacity is the starting point for phase equilibria in mixtures. Raoult's law involves specific assumptions about component fugacities for both the vapor and liquid phases. We develop models for deviations from Raoult's law in Chapters 11–13.

10.8   CONCEPTS FOR GENERALIZED PHASE EQUILIBRIA

Generalization of pure-component principles to multicomponent systems requires that we consider how the thermodynamic properties change with respect to changes in the amounts of individual components. For a pure fluid, the natural properties were simply a function of two state variables. In multicomponent mixtures, these energies and the entropy also depend on composition.

$$e.g., \ d\underline{U}(T, P, n_1, n_2, \ldots n_i) = \left(\frac{\partial \underline{U}}{\partial P}\right)_{T, n} dP + \left(\frac{\partial \underline{U}}{\partial T}\right)_{P, n} dT + \sum_i \left(\frac{\partial \underline{U}}{\partial n_i}\right)_{P, T, n_{j \neq i}} dn_i \qquad 10.36$$

$$e.g., \ d\underline{G}(T, P, n_1, n_2, \ldots n_i) = \left(\frac{\partial \underline{G}}{\partial P}\right)_{T, n} dP + \left(\frac{\partial \underline{G}}{\partial T}\right)_{P, n} dT + \sum_i \left(\frac{\partial \underline{G}}{\partial n_i}\right)_{P, T, n_{j \neq i}} dn_i \qquad 10.37$$

These equations extend the use of calculus from Chapter 6 to composition variables.

Note that these equations follow the mathematical rules developed in Chapter 6. Each term on the right-hand side consists of a partial derivative with respect to one variable, with all other variables held constant. The summation is simply a shorthand method to avoid writing a term for each component. The subscript $n_{j \neq i}$ means that the moles of all components except i are held constant. In other words, for a ternary system, $((\partial \underline{U})/(\partial n_1))_{P, T, n_{j \neq i}}$ means the partial derivative of $\underline{U}$ with respect to n_1 while holding P, T, n_2, and n_3 constant. For phase equilibria where P and T are manipulated, Eqn. 10.37 is more useful than Eqn. 10.36 because the Gibbs energy is a natural function of P and T. At constant moles and composition of material, the mixture must follow the same constraints as a pure fluid. That is, the state is dependent on only two state variables *if we keep the composition constant.*

$$\Rightarrow (\partial \underline{G}/\partial P)_{T,n} = \underline{V} \quad \text{and} \quad (\partial \underline{G}/\partial T)_{P,n} = -\underline{S};$$

These complicated-looking derivatives are really fundamental properties; therefore, we can rewrite Eqn. 10.37 as

$$dG = \underline{V}dP - \underline{S}dT + \sum_i \left(\frac{\partial \underline{G}}{\partial n_i}\right)_{P, T, n_{j \neq i}} dn_i \qquad 10.38$$

The quantity $(\partial \underline{G}/\partial n_i)_{P, T, n_{j \neq i}}$ tells us how the total Gibbs energy of the mixture changes with an infinitesimal change in the number of moles of species i, when the number of moles of all other species fixed, and at constant P and T. The quantity $(\partial \underline{G}/\partial n_i)_{P, T, n_{j \neq i}}$ will become very important in our later discussion. so we give it a name, called the **chemical potential,** and give it a symbol.

● Chemical potential.

$$\boxed{\mu_i \equiv (\partial \underline{G}/\partial n_i)_{P, T, n_{j \neq i}}} \qquad 10.39$$

We commonly write

$$\boxed{dG = \underline{V}dP - \underline{S}dT + \sum_i \mu_i dn_i} \qquad 10.40$$

Partial Molar Properties

Another name for the special derivative of Eqn. 10.39 is the **partial molar Gibbs energy.** We may generalize the form of the derivative and apply it to other properties. For any *extensive* thermodynamic property $\underline{M}$, we may write $(\partial \underline{M}/\partial n_i)_{T, P, n_{j \neq i}} \equiv \overline{M}_i$. Note that T, P, and $n_{j \neq i}$ are always held constant for a partial molar property. The overbar indicates a partial molar quantity, that is, for total volume $\underline{V}$, the quantity $(\partial \underline{V}/\partial n_i)_{T, P, n_{j \neq i}}$ is called the partial molar volume and given the symbol $\overline{V}_i$. Suppose we were considering a mixture of 500 red balls of size σ_R and 700 blue balls of size σ_B. How would the total volume change if we added one more red ball, keeping 700 blue balls? This is a finite difference example of the derivative called the partial molar volume of red balls. A special mathematical result of the differentiation is that we may write at constant temperature and pressure:

● Partial molar quantities provide a mathematical way to assign the overall mixture property according to composition expressed in moles or mole fractions.

$$\boxed{\underline{M} = \sum_i n_i \overline{M}_i \quad \text{or} \quad M = \sum_i x_i \overline{M}_i} \qquad 10.41$$

As a result, we may write

$$\underline{G} = \sum_i n_i \overline{G}_i = \sum_i n_i \mu_i \quad \text{or} \quad G = \sum_i x_i \overline{G}_i = \sum_i x_i \mu_i \qquad 10.42$$

We will return to these basic equations as we develop more relations for mixtures.

The chemical potential becomes an important term for each of the key properties, U, H, A, *and* G, because they are all related by an appropriate Legendre transform as shown in the following table. The constraints on the derivative are important. For example, $(\partial \underline{U}/\partial n_i)_{P,T,n_{j \neq i}}$ in Eqn. 10.36 is not equal to $(\partial \underline{U}/\partial n_i)_{\underline{S},\underline{V},n_{j \neq i}} = \mu_i$ in the first line of the table.

Closed	Open	Natural Variables
$dU = TdS - PdV$	$d\underline{U} = Td\underline{S} - Pd\underline{V} + \sum_i \mu_i dn_i$	S,V
$dH = TdS + VdP$	$d\underline{H} = Td\underline{S} + \underline{V}dP + \sum_i \mu_i dn_i$	S,P
$dA = SdT - PdV$	$d\underline{A} = -\underline{S}dT - Pd\underline{V} + \sum_i \mu_i dn_i$	T,V
$dG = SdT + VdP$	$d\underline{G} = -\underline{S}dT + \underline{V}dP + \sum_i \mu_i dn_i$	T,P

The open systems equations can be developed using the energy and entropy balances from the early chapters as shown in an online supplement.

Equilibrium Criteria

For equilibrium at constant T and P, the Gibbs energy is minimized and mathematically the minimum means $d\underline{G} = 0$ at equilibrium. Therefore, Eqn. 10.38 is equal to zero at a minimum, since dT and dP are zero, and for a closed system all dn_i are zero. Thus,

$$d\underline{G} = 0 \quad \text{at equilibrium, for constant } T \text{ and } P \qquad 10.43$$

This equation applies to whatever system we define. Suppose we define our system to consist of two components (e.g,. EtOH + H_2O) distributed between two phases (e.g., vapor and liquid), $d\underline{G} = d\underline{G}^L + d\underline{G}^V = 0$, and at constant T and P, the moles may redistribute between the two phases, by Eqn. 10.40 for both phases:

$$d\underline{G}^L + d\underline{G}^V = \mu_1^L dn_1^L + \mu_2^L dn_2^L + \mu_1^V dn_1^V + \mu_2^V dn_2^V = 0$$

But if component 1 leaves the liquid phase then it must enter the vapor phase (and similarly for component 2) because the overall system is closed.

$$\Rightarrow dn_1^L = -dn_1^V \quad \text{and} \quad dn_2^L = -dn_2^V$$

$$\Rightarrow (\mu_1^V - \mu_1^L)dn_1^V + (\mu_2^V - \mu_2^L)dn_2^V = 0$$

The only way to make this equal to zero in general is:

$$\boxed{\mu_1^V = \mu_1^L \quad \text{and} \quad \mu_2^V = \mu_2^L} \qquad 10.44$$

Setting the chemical potentials and T and P in each of the phases equal to each other provides a set of constraints (simultaneous equations) which may be solved for phase compositions provided we know the dependency of the chemical potentials on the phase compositions. Suppose the functions $\underline{G}^L(x,T,P)$ and $\underline{G}^V(y,T,P)$ are available for a binary system. Then

The chemical potential of each component must be the same in each phase at equilibrium.

$$\left(\frac{\partial \underline{G}^L}{\partial n_1}\right)_{T,P,n_2} = \left(\frac{\partial \underline{G}^V}{\partial n_1}\right)_{T,P,n_2} \; ; \; \left(\frac{\partial \underline{G}^L}{\partial n_2}\right)_{T,P,n_1} = \left(\frac{\partial \underline{G}^V}{\partial n_2}\right)_{T,P,n_1} \; ; \; x_2 = 1 - x_1; \; y_2 = 1 - y_1;$$

which gives four equations with four unknowns (x_1, x_2, y_1, y_2) that we can solve, in principle.[11] The first two equations are simply the equivalency of chemical potentials in the two phases.

Chemical Potential of a Pure Fluid

In Chapter 9, we showed the equilibrium constraint for a pure fluid is equality of the molar Gibbs energy for each of the phases (*cf.* Eqn. 9.3). How does this relate to Eqn. 10.44?

$$\mu_i \equiv (\partial (nG)/\partial n_i)_{P,T,n_{j \neq i}} = G(\partial n/\partial n_i)_{P,T,n_{j \neq i}} + n(\partial G/\partial n_i)_{P,T,n_{j \neq i}}$$

For a pure fluid, there is only one component, so $dn_i = dn$, and since $G(T,P)$ is intensive, then $n(\partial G/\partial n_i)_{T,P} = 0$. Also $(\partial n/\partial n_i)_{T,P} = 1$ by Eqn. 6.13. Therefore,

❗ The chemical potential of a pure fluid is simply the molar Gibbs energy.

$$\boxed{\mu_{i,pure} = G_i}$$

10.45

That is, the chemical potential of a pure fluid is simply the molar Gibbs energy. Pure components can be considered as a special case of the same general statement of the equilibrium constraint.

Component Fugacity

We introduced fugacity in Chapter 9. The chemical potential constraint is sufficient for solving any phase equilibrium problem, but the most popular engineering approach for actual computations makes use of the concept of fugacity. Fugacity is more "user-friendly" than "chemical potential" or "partial molar Gibbs energy." We have seen in Chapter 9 that the fugacity is the same as the pressure for an ideal gas, and that the fugacity is close to the vapor pressure for a liquid. Engineers use benchmarks such as these as guidelines for estimating fugacities and double-checking calculations. *Fuga-* comes from a Latin noun meaning flight, fleeing, or escape. The suffix *-ity* comes from a root meaning "character." Thus, fugacity was invented to mean flight-character, commonly called "escaping tendency." When phases are in equilibrium, the component moves ("escapes") from the phase where it has the higher fugacity to the phase where it has the lower fugacity until the fugacities are equal in both phases.[12]

❗ Fugacity is another way to express the chemical potential that is used more widely in engineering than chemical potential.

Let us generalize our pure component fugacity relations to apply to components in mixtures: At constant T, we defined $RTd\ln f \equiv dG$ (Eqn. 9.19) which can be generalized to define the fugacity of a component in a mixtures as

$$\boxed{RTd\ln \hat{f}_i \equiv d\mu_i}$$

10.46

---

11. Even though $\mu^L = \mu^V$ at equilibrium, the dependency of μ^V on composition will be quite different from the dependency of μ^L on composition because the molecules are arranged very differently.

12. The concept of fugacity becomes especially useful when we begin to discuss phase equilibrium in mixtures. In that case, it is conceivable that we could have some supercritical component dissolved in the liquid phase despite its high escaping tendency, (e.g. CO_2, in a carbonated beverage at 100°F). The possibility of a component that cannot be a liquid still dissolving in a liquid requires a very general concept of escaping tendency because the pure-component vapor pressure does not exist at those conditions. The definition of fugacity provides us with that general concept.

where $\hat{f}_i$ is the fugacity of component i in a mixture and μ_i is the chemical potential of the component. In the limit as the composition approaches purity, these properties become equal to the pure component values. A caret is used in the symbol for the fugacity of a component. The component fugacity is not a partial molar property, so the overbar cannot be used.

Equality of Fugacities as Equilibrium Criteria

The equality of chemical potentials at equilibrium can easily be reinterpreted in terms of fugacity in a manner analogous to our methods for pure components from Eqn. 10.44:

$$\boxed{\mu_i^V(T, P) = \mu_i^L(T, P)} \qquad 10.47$$

By integrating Eqn. 10.46 as a function of composition at fixed T from a state of pure i to a mixed state, we find

$$\mu_i^V - \mu_{i, pure} = RT\ln\frac{\hat{f}_i^V}{f_i} \qquad 10.48$$

where $\mu_{i, pure}$ and f_i are for the pure fluid at the system temperature. Writing an analogous expression for the liquid phase, and equating the chemical potentials using Eqn. 10.47, we find

$$\mu_i^V - \mu_i^L = RT\ln\left[\hat{f}_i^V / \hat{f}_i^L\right] = 0 \qquad 10.49$$

$$\boxed{\hat{f}_i^V = \hat{f}_i^L} \quad \text{at equilibrium.} \qquad 10.50$$

> *Note: Eqn. 10.47 or 10.50 becomes the starting point for all phase-equilibrium calculations. Therefore, we need to develop the capability to calculate chemical potentials or fugacities of components in vapor, liquid, and solid mixtures. Here we briefly introduce the framework for calculating the fugacities of components before we begin the direct calculations.*

Eqn 10.50 does not look much like Raoult's law, which was our motivation for exploring the thermodynamics of mixtures. To make this final connection between Raoult's law and Eqn. 10.50, we need to understand how energy and volume affect component fugacities.

10.9 MIXTURE PROPERTIES FOR IDEAL GASES

We have introduced the concepts of energy of mixing, enthalpy of mixing, and volume of mixing in Section 3.4. We can now relate the mixing behavior to the partial molar properties. The partial molar quantities for ideal gases must be the same as the pure component properties.

$$\overline{V}_i^{ig} = V_i, \ \ \overline{U}_i^{ig} = U_i, \ \overline{H}_i^{ig} = H_i \qquad 10.51$$

Entropy for an ideal-gas mixture is more complicated because, as shown in Chapter 4, even systems of fixed total energy have an entropy change associated with mixing due to the distinguishability of the components. The entropy of an ideal gas is calculated by the sum of the entropies of the components plus the entropy of mixing as given in Chapter 4:

Carets (^) are used to denote component properties in mixtures for f while without a caret the property is the pure component f. When working with μ, the meaning is inferred from the context of the situation.

$$\underline{S}^{ig} = \sum_i n_i S_i^{ig} + \Delta \underline{S}_{mix} = \sum_i n_i S_i^{ig} - R \sum_i n_i \ln y_i \quad \text{or} \quad S^{ig} = \sum_i y_i S_i^{ig} - R \sum_i y_i \ln y_i \quad \text{(ig) 10.52}$$

Therefore, the entropy of mixing is nonzero and positive:

$$\Delta S_{mix}^{ig} = S^{ig} - \sum_i y_i S_i^{ig} = -R \sum_i y_i \ln y_i > 0 \quad \text{(ig) 10.53}$$

and the partial molar entropy is

$$\bar{S}_i^{ig} = S_i^{ig} - R \ln y_i \quad \quad 10.54$$

The Gibbs energy and the fugacity will be at the core of phase equilibria calculations. The Gibbs energy of an ideal gas is obtained from the definition, $G \equiv H - TS$. Using H^{ig} an S^{ig} from above,

$$\underline{G}^{ig} \equiv \underline{H}^{ig} - T\underline{S}^{ig} = \sum_i n_i H_i^{ig} - T\left(\sum_i n_i S_i^{ig} - R \sum_i n_i \ln y_i\right) = \sum_i n_i G_i^{ig} + RT \sum_i n_i \ln y_i \quad \text{(ig) 10.55}$$

Therefore, the Gibbs energy of mixing is nonzero and negative:

$$\Delta G_{mix}^{ig} = G^{ig} - \sum_i y_i G_i^{ig} = RT \sum_i y_i \ln y_i < 0 \quad \text{(ig) 10.56}$$

The chemical potential of a component is given by Eqn. 10.39 and taking the derivative of Eqn. 10.55,

$$\mu_i^{ig} = (\partial \underline{G}^{ig}/\partial n_i)_{T, P, n_{j \neq i}} = G_i^{ig} + RT\left(\partial \left(\sum_i n_i \ln y_i\right)/\partial n_i\right)_{T, P, n_{j \neq i}} \quad \text{(ig) 10.57}$$

The derivative is most easily seen by expanding the logarithm before differentiation, $\ln y_i = \ln n_i - \ln n$. Then,

$$\left(\partial \left(\sum_i n_i \ln n_i - \ln n \sum_i n_i\right)/\partial n_i\right)_{T, P, n_{j \neq i}} = \ln n_i + 1 - 1 - \ln n = \ln y_i \quad \quad 10.58$$

Therefore, we find the **chemical potential of an ideal-gas component:**

$$\mu_i^{ig} = G_i^{ig} + RT \ln y_i \quad \text{(ig) 10.59}$$

❶ Chemical potential of an ideal-gas component.

By Eqn. 10.48, using Eqn. 10.59 and Eqn. 10.45,

$$\mu_i^{ig} - \mu_{i, pure}^{ig} = RT \ln \frac{\hat{f}_i^{ig}}{f_{i, pure}^{ig}} = RT \ln y_i \quad \text{or} \quad \hat{f}_i^{ig} = y_i f_{i, pure}^{ig} \quad \text{(ig) 10.60}$$

By Eqn. 9.22, $f^{ig}_{pure} = P$, resulting in the **fugacity of an ideal-gas component:**

$$\hat{f}^{ig}_i = y_i P$$

(ig) 10.61

> ❶ The fugacity of a component in an ideal-gas mixture is particularly simple; it is equal to the $y_i P$, the partial pressure!

Therefore, the fugacity of an ideal-gas component is simply its **partial pressure,** $y_i P$. This makes the ideal-gas fugacity easy to quantify rapidly for engineering purposes. One of the goals of the calculations that will be pursued in Chapter 15 is the quantification of the deviations of the fugacity from ideal-gas values quantified by the component fugacity coefficient.

10.10   MIXTURE PROPERTIES FOR IDEAL SOLUTIONS

Ideal solutions are similar to ideal-gas mixtures, but they to not follow the ideal-gas law. The internal energy and enthalpy for ideal solutions were introduced in Section 3.5. Since these properties are additive, the partial molar properties are equal to the pure component properties,

$$\overline{V}^{is}_i = V_i, \ \overline{U}^{is}_i = U_i, \ \overline{H}^{is}_i = H_i$$

10.62

As for the entropy change of mixing, the loss of order due to mixing is unavoidable, even for ideal solutions. During our consideration of the microscopic definition of entropy, we derived a general expression for the ideal entropy change of mixing, Eqn. 4.8.

$$\frac{\Delta S^{is}_{mix}}{R} = -\sum_i x_i \ln x_i$$

10.63

> ❶ The entropy of mixing is *nonzero* for an ideal solution.

Although we derived Eqn. 4.8 for mixing ideal gases, it also provides a reasonable approximation for mixing liquids of equal-sized molecules. The reason is that changes in entropy are related to the change in accessible volume. Even though a significant volume is occupied by the molecules themselves in a dense liquid, the void space in one liquid is very similar to the void space in another liquid if the molecules are similar in size and polarity. That means that the accessible volume for each component doubles when we mix equal parts of two equal-sized components. That is essentially the same situation that we had when mixing ideal gases. The partial molar entropy of mixing is

$$\overline{S}^{is}_i = S_i - R \ln x_i$$

10.64

Given the effect of mixing on these two properties, we can derive the effect on other thermodynamic properties, which has the same formula as that found for ideal gases:

$$\frac{\Delta G^{is}_{mix}}{RT} = \frac{\Delta H^{is}_{mix}}{RT} - \frac{\Delta S^{is}_{mix}}{R} = \sum_i x_i \ln x_i$$

10.65

The general relationship for ΔG_{mix} gives a relationship for fugacity. We can extend our definition of the enthalpy of mixing to the Gibbs energy of mixing,

$$\Delta G_{mix} \equiv G - \sum_i x_i G_i$$

10.66

But by using Eqns. 10.42 and 10.48,

$$\frac{\Delta G_{mix}}{RT} = \sum_i x_i \frac{(\mu_i - G_i)}{RT} = \sum_i x_i \ln\left(\frac{\hat{f}_i}{f_i}\right) \qquad 10.67$$

Thus, comparing Eqns. 10.67 and 10.65, for an ideal solution, $\dfrac{\Delta G_{mix}}{RT} = \sum_i x_i \ln x_i = \sum_i x_i \ln\left(\dfrac{\hat{f}_i}{f_i}\right)$.

By comparing the relations in the logarithms, we obtain the **Lewis-Randall rule** for ideal solutions:

$$\boxed{\hat{f}_i^{is}/f_i = x_i \quad \Rightarrow \quad \hat{f}_i^{is} = x_i f_i} \qquad 10.68$$

Lewis-Randall
rule for component
fugacity in an ideal
solution.

10.11   THE IDEAL SOLUTION APPROXIMATION AND RAOULT'S LAW

By our equilibrium constraint,

$$\hat{f}_i^V = \hat{f}_i^L \qquad 10.69$$

By our ideal solution approximation in both phases, the equilibrium criteria becomes

$$\boxed{y_i f_i^V = x_i f_i^L} \qquad 10.70$$

Now we need to substitute the expressions for f_i^V and f_i^L that we developed in Chapter 9. The fugacity of the pure vapor comes from Eqn. 9.25:

$$f_i^V = \varphi_i^V P \qquad 10.71$$

The fugacity of the liquid comes from Eqn. 9.39:

$$f_i^L = \varphi_i^{sat} P_i^{sat} \exp\left(\frac{V_i^L(P - P_i^{sat})}{RT}\right) \qquad 10.72$$

Combining Eqns. 10.70–10.72,

$$y_i \varphi_i^V P = x_i \varphi_i^{sat} P_i^{sat} \exp\left(\frac{V_i^L(P - P_i^{sat})}{RT}\right)$$

Writing in terms of the K_i ratio,

$$K_i = \frac{y_i}{x_i} = \frac{P_i^{sat}}{P}\left[\frac{\varphi_i^{sat}\exp[V_i^L(P - P_i^{sat})/RT]}{\varphi_i^V}\right]$$

Note: *at reasonably low pressures,*

$$\frac{\varphi_i^{sat}}{\varphi_i} \approx 1, \text{ and } \exp[V_i^L(P - P_i^{sat})/RT] \approx 1$$

resulting in Raoult's Law,

$$\boxed{K_i = \frac{P_i^{sat}}{P}} \quad \text{or} \quad \boxed{y_iP = x_iP_i^{sat}} \qquad\qquad 10.73 \quad \text{❶}_{\text{Raoult's law.}}$$

Raoult's law is only valid for molecularly similar components; ideal behavior is demonstrated in the Figs. 10.2 and 10.3 because of careful selection of molecularly similar binary pairs. If we mixed methanol from Fig. 10.2(a) and benzene from Fig. 10.3(c), the resultant system would be very non-ideal.

10.12   ACTIVITY COEFFICIENT AND FUGACITY COEFFICIENT APPROACHES

The discussion in Sections 10.8–10.11 sets the stage for the next few chapters. The principles discussed here form the basis for these chapters and a thorough understanding facilitates rapid understanding of the extensions. There are two main approaches to modeling nonideal fluids. They differ in the way that they treat the fugacities in the vapor and liquid phases. The first approach is to model deviations from ideal solution behavior in the liquid phase, using activity coefficients, which is covered in Chapters 11 through 14. When the vapors are nonideal, the vapor phase fugacities are modeled with an equation of state, an approach that usually requires a computer. As discussed for pure fluids, liquid phases may also be modeled with an equation of state by simply selecting the liquid root. The EOS approach is discussed in Chapter 15.

The presentation has been organized according to a hierarchical approach. First and foremost, it is necessary to recognize that deviations from Raoult's law can significantly alter the outlook on chemical processing. We introduce the activity approach for handling nonideal solutions in Chapter 11, and then extend to more complex models in Chapters 12 and 13. If you prefer a "one-size-fits-all" approach that can be applied to extremes of temperature, pressure, and component differences (e.g., CH_4 + eicosane at oil reservoir conditions) then you may want to study Chapter 15 sooner. The EOS approach must be handled carefully for hydrogen bonding components, however, as discussed in Chapter 19. We provide more coverage than can typically be incorporated into a single undergraduate course, so instructors need to be selective about which models to cover. Nevertheless, the principles from one approach typically extend to other approaches and by focusing on a solid understanding of the principles for the sections studied, the reader should be able to extend them when the need arises.

10.13   SUMMARY

The concepts in this chapter are relatively simple but far-reaching. A simple extension of the chain rule to multicomponent systems led to the equilibrium constraint for multicomponent multiphase equilibria. A simple application of the entropy of mixing derived in Chapter 4 led to the ideal solution model, Gibbs energy for a component in a mixture, and fugacity for a component in a mixture.

This simple solution model enabled us to demonstrate the computational procedures for dew points, bubble points, and flashes.

In the remaining chapters of Unit III we proceed in a manner that is extremely similar. We propose a solution model and apply the equilibrium constraint to derive an expression for the K-ratios. Then we follow exactly the same computational procedures as developed here. The primary difference is the increasing level of sophistication incorporated into each solution model. Thus, the chapters ahead focus increasingly on the detailed description of the molecular interactions and the impacts of assumptions on the accuracy of the resultant solution models.

In this context, this chapter represents one iteration in the typical sequence of "observe, predict, test, and evaluate." We observed that isothermal mixtures boil at higher pressures when more volatile components were present. We predicted that the bubble pressure varies linearly in mole fraction between the vapor pressures of the components (Raoult's law). We tested these predictions by checking the linearity of several mixtures. We evaluated the assumptions implicit in Raoult's law by extending the fundamental thermodynamic principles to mixtures. Successive iterations involve repeating this sequence, following up on clues from the previous evaluation.

Important Equations

We defined the K-ratio.

$$K_i \equiv \frac{y_i}{x_i}$$
10.3

Raoult's law is an important equation, but we have also seen that it has limited applicability.

$$K_i = \frac{P_i^{sat}}{P} \quad \text{or} \quad y_i P = x_i P_i^{sat}$$
10.6

We developed methods to calculate bubble, dew, and flash conditions summarized in Table 10.1 on page 373. We developed the relation for the relative volatility,

$$\alpha_{ij} = K_i / K_j$$
10.32

We discussed how the relative volatility varies dramatically with composition in an azeotropic system, approaching 1 at the azeotrope, and creating challenges. These variations undermine the utility of Raoult's law and the use of shortcut design equations for azeotropic systems.

For phase equilibria, including liquid-liquid equilibria, the fugacities of a component are equal,

$$\hat{f}_i^V = \hat{f}_i^L \quad \text{or, more generally,} \quad \hat{f}_i^\alpha = \hat{f}_i^\beta$$
10.50

where α and β indicate different phases, whether vapor, liquid(s), or solid. This is the starting point throughout Unit III. As we applied the generalized approach, we found

$$\hat{f}_i^{ig} = y_i P$$
10.61

and for an ideal solution,

$$\boxed{\hat{f}_i^{is}/f_i = x_i} \quad \Rightarrow \quad \boxed{\hat{f}_i^{is} = x_i f_i} \tag{10.68}$$

Two final equations that form the basis for future derivations are

$$\boxed{\mu_i \equiv (\partial \underline{G}/\partial n_i)_{P,\,T,\,n_{j \neq i}}} \text{ and } \boxed{RTd\ln\hat{f}_i \equiv d\mu_i} \qquad \text{10.39 and 10.46}$$

Through these equations, we can infer the fugacities of components in any solution, no matter how complex, by building a model for Gibbs energy. This is the subject of upcoming chapters. Models of the Gibbs energy may range from crudely empirical to sophisticated molecular analyses.

10.14   PRACTICE PROBLEMS

P10.1 The stream from a gas well consists of 90 mol% methane, 5 mol% ethane, 3 mol% propane, and 2 mol% n-butane. This stream is flashed isothermally at 233 K and 70 bar. Use the shortcut K-ratio method to estimate the L/F fraction and liquid and vapor compositions. (ANS. $L/F = 0.181$)

P10.2 An equimolar mixture of n-butane and n-hexane at pressure P is isothermally flashed at 373 K. The liquid-to-feed ratio is 0.35. Use the shortcut K-ratio method to estimate the pressure and liquid and vapor compositions. (ANS. $P = 0.533$ MPa, $x_{C6} = 0.78$)

P10.3 A mixture of 25 mol% n-pentane, 45 mol% n-hexane, and 30 mol% n-heptane is flashed isothermally at 365.9 K and 2 bar. Use the shortcut K-ratio method to estimate the L/F fraction and liquid and vapor compositions. (ANS. $L/F = 0.56$)

P10.4 A mixture containing 15 mol% ethane, 35 mol% propane, and 50 mole% n-butane is isothermally flashed at 9 bar and temperature T. The liquid-to-feed ratio is 0.35. Use the shortcut K-ratio method to estimate the pressure and liquid and vapor compositions. (ANS. 319.4 K, $x_{C4} = 0.74$)

10.15   HOMEWORK PROBLEMS

10.1 For a separations process it is necessary to determine the VLE compositions of a mixture of ethyl bromide and n-heptane at 30°C. At this temperature the vapor pressure of pure ethyl bromide is 0.7569 bar, and the vapor pressure of pure n-heptane is 0.0773 bar. Calculate the bubble pressure and the composition of the vapor in equilibrium with a liquid containing 47.23 mol% ethyl bromide assuming ideal solution behavior. Compare the calculated pressure to the experimental value of 0.4537 bar.

10.2 Benzene and ethanol (e) form azeotropic mixtures. Prepare a y-x and a P-x-y diagram for the benzene-ethanol system at 45°C assuming the mixture is ideal. Compare the results with the experimental data tabulated below of Brown and Smith, *Austral. J. Chem.* 264 (1954). (P in the data table is in bar.)

x_e	0	0.0374	0.0972	0.2183	0.3141	0.4150	0.5199	0.5284	0.6155	0.7087	0.9591	1.000
y_e	0	0.1965	0.2895	0.3370	0.3625	0.3842	0.4065	0.4101	0.4343	0.4751	0.8201	1.000
P	0.2939	0.3613	0.3953	0.4088	0.4124	0.4128	0.41	0.4093	0.4028	0.3891	0.2711	0.2321

10.3 The following mixture of hydrocarbons is obtained as one stream in a petroleum refinery on a mole basis: 5% ethane, 10% propane, 40% *n*-butane, 45% isobutane. Assuming the shortcut *K*-ratio model: (a) compute the bubble point of the mixture at 5 bar; (b) compute the dew point of the mixture at 5 bar; (c) find the amounts and compositions of the vapor and liquid phases that would result if this mixture were to be isothermally flash vaporized at 30 °C from a high pressure to 5 bar.

10.4 Consider a mixture of 50 mol% *n*-pentane and 50 mol% *n*-butane at 14 bar.

  (a) What is the dew-point temperature? What is the composition of the first drop of liquid?
  (b) At what temperature is the vapor completely condensed if the pressure is maintained at 14 bar? What is the composition of the last drop of vapor?

10.5 A 50 mol% mixture of propane(1) and *n*-butane(2) enters an isothermal flash drum at 37°C. If the flash drum is maintained at 0.6 MPa, what fraction of the feed exits as liquid? What are the compositions of the phases exiting the flash drum? Work the problem in the following two ways.

  (a) Use Raoult's law (use the Peng-Robinson equation to calculate pure component vapor pressures).
  (b) Assume ideal mixtures of vapor and liquid. (Use the Peng-Robinson equation to obtain f^{sat} for each component.)

10.6 A mixture of 55 mol% ethanol in n-propanol is at 0.2MPa and 80°C at 70 mol/s. The stream is fed to a adiabatic flash drum. Calculate the outlet stream flow rates, temperatures, and compositions at 0.05MPa.

  (a) Use the path of Fig. 2.6(a).
  (b) Use the path of Fig. 2.6(c).

10.7 An equimolar ternary mixture of acetone, *n*-butane, and ammonia at 1 MPa is to be flashed. List the known variables, unknown variables, and constraining equations to solve each of the cases below. Assume ideal solution thermodynamics and write the flash equations in terms of *K*-ratios, with the equations for calculating *K*-ratios written separately.

  (a) Bubble temperature
  (b) Dew temperature
  (c) Flash temperature at 25 mol% vapor.
  (d) Raised to midway between the bubble and dew temperature, then adiabatically flashed

10.8 Tank A is rapidly half-filled with volatile hydrocarbon. Tank B is 10 times as large and rapidly half-filled with the same hydrocarbon. Initially the gas space can be considered to be free of volatile organic and at the same pressure. The tanks are then closed. The tanks warm 20°C and the pressure in both tanks goes up. After warming, does one tank have a higher pressure than the other, or are the final pressures the same? Show your result with equations. Your answer should be general; it should not depend on numerical calculations.

❶ Do not confuse a flash point calculation with an isothermal flash calculation.

10.9 Above a solvent's flash point temperature, the vapor concentration in the headspace is sufficient that a spark will initiate combustion; therefore, extreme care must be exercised to avoid ignition sources. Calculate the vapor phase mole fraction for the following liquid solvents using flash points listed, which were obtained from the manufacturer's material safety data sheets (MSDS). The calculated vapor concentration is an estimate of the lower flammability limit (LFL). Assume that the headspace is an equilibrium mixture of air and solvent at 760 mmHg. The mole fraction of air dissolved in the liquid solvent is negligible for this calculation.

(a) Methane, −187.8°C   (d) Hexane, −21.7°C   (g) Toluene, 4.4°C
(b) Propane, −104.5°C   (e) Ethanol, 12.7°C   (h) *m*-xylene, 28.8°C
(c) Pentane, −48.9°C   (f) 2-butanone, −5.6°C   (i) Ethyl acetate, −4.5°C

10.10 Solvent vessels must be purged before maintenance personnel enter in order to ensure that: (1) sufficient oxygen is available for breathing; (2) vapor concentrations are below the flash point; (3) vapor concentrations are below the Occupational Safety and Health Administration (OSHA) limits if breathing apparatus is not to be used. Assuming that a 8 m^3 fixed-roof solvent tank has just been drained of all liquid and that the vapor phase is initially saturated at 22°C, estimate the length of purge necessary with 2 m^3/min of gas at 0.1 MPa and 22°C to reach the OSHA 8-hr exposure limit.[13]

(a) Hexane 500 ppm
(b) 1-butanol 100 ppm
(c) Chloroform 50 ppm
(d) Ethanol 1000 ppm
(e) Toluene 200 ppm

10.11 A pharmaceutical product is crystallized and washed with absolute ethanol. A 100 kg batch of product containing 10% ethanol by weight is to be dried to 0.1% ethanol by weight by passing 0.2 m^3/min of 50°C nitrogen through the dryer. Estimate the rate (mol/min) that ethanol is removed from the crystals, assuming that ethanol exerts the same vapor pressure as if it were pure liquid. Based on this assumption, estimate the residence time for the crystals in the dryer. The dryer operates at 0.1 MPa and the vapor pressure of the pharmaceutical is negligible.

10.12 Benzyl chloride is manufactured by the thermal or photochemical chlorination of toluene. The chlorination is usually carried out to no more than 50% toluene conversion to minimize the benzal chloride formed. Suppose reactor effluent emissions can be modeled ignoring by-products, and the effluent is 50 mol% toluene and 50 mol% benzyl chloride. Estimate the emission of toluene and benzyl chloride (moles of each) when an initially empty 4 m^3 holding tank is filled with the reactor effluent at 30°C and 0.1 MPa.

10.13 This problem explores emissions during heating of hexane(1) and toluene(2) in a tank with a fixed roof that is vented to the atmosphere through an open pipe in the roof. Atmospheric pressure is 760 mmHg. The tank volume is 2000 L, but the maximum operating liquid level is 1800 L. Determine the emissions of each VOC (in g) when the tank is heated.

(a) The liquid volume is 1800 L, $x_1 = 0.5$, $T^i = 10°C$, $\Delta T = 15°C$.
(b) The liquid volume is 1800 L, $x_1 = 0.5$, $T^i = 25°C$, $\Delta T = 15°C$.
(c) The liquid volume is 1500 L, $x_1 = 0.5$, $T^i = 25°C$, $\Delta T = 15°C$.
(d) The liquid volume is 1800 L, $x_1 = 1.0$, $T^i = 25°C$, $\Delta T = 15°C$.
(e) Explain why the ratio [(emission of toluene in part (b))/(emission of toluene in part (a))] is different from the corresponding ratio of hexane emissions.

❗ Do not confuse a flash point calculation with an isothermal flash calculation.

10.14 Use Raoult's law to estimate the flash point temperature for the following equimolar liquid mixtures in an air atmosphere at 750 mmHg total pressure:

---

13. OSHA may change these limits at any time.

(a) Pentane (LFL = 1.5%) and hexane (LFL = 1.2%)

(b) Methanol (LFL = 7.3%) and ethanol (LFL = 4.3%)

(c) Benzene (LFL = 1.3%) and toluene (LFL = 1.27%)

10.15 Go to www.csb.gov and watch a video assigned by your instructor. For the substance involved, look up the LFL. Use Raoult's law to estimate the flash point temperature and compare it with a literature value. For the scenario in the video, offer an explanation of how easy or difficult is was to reach the LFL under the conditions, and comment on the recommendations of the CSB.

AN INTRODUCTION TO ACTIVITY MODELS

11

First, do no harm.

Hippocrates

The subject of non-ideal solutions includes just about everything from aqueous acids to polymers to semiconductors. Not surprisingly, *there is no completely general model* for non-ideal solutions. But there are several popular approaches for specific situations like VLE of alcohols or LLE of organic solvents with water. We discuss typical models and briefly explain their forms and history. Moreover, the challenge of developing accurate descriptions of non-ideal solution behavior means that model development is still an active research area. The presentation here should provide enough background to understand the rationales behind new developments as well as the old.

Chapters 11 through 13 are concerned with correction factors to Raoult's law known as "activity coefficients." The difference is that Chapter 11 is concerned with purely empirical models of binary mixtures while Chapter 12 focuses on model equations that can be derived from the van der Waals equation of state. These models are not as complex as those in Chapter 13, but they convey the key concepts and serve as an introduction. The models in Chapter 13 reformulate the analysis in terms of the radial distribution function for mixtures. Students need to recognize that Raoult's law can lead to gross errors, failing to represent an azeotrope or liquid-liquid equilibrium (LLE), for example. With this in mind, we take a hierarchical approach. We begin with a simple illustration using the Margules 1-parameter model. This suffices to account crudely for mixtures of components that "like" or "dislike" each other. Later, we demonstrate that this simple model is related to a generalized empirical form called the Redlich-Kister expansion. To proceed beyond empirical fitting and begin understanding the molecular driving forces for nonideality, Chapter 12 returns to the van der Waals model and extends it to multicomponent mixtures. With molecular understanding, we can develop more accurate and general models, permitting us to formulate solutions that accomplish broad design goals, in addition to fitting and extrapolating specific data for specific systems.

Chapter Objectives: You Should Be Able to...

1. Compute VLE phase diagrams using modified Raoult's law. Perform bubble, dew, and flash calculations using modified Raoult's law.

2. Characterize adjustable parameters in activity models using experimental data.

3. Derive an expression for an activity coefficient given an arbitrary expression for the Gibbs excess energy.

4. Assess the degree of non-ideality of a given mixture based on the molecular properties of the components.

5. Comment critically on the merits and limitations of the following solution models: Henry's law, Margules models, and the Redlich-Kister expansion, including the ability to identify the most appropriate model for a given application.

6. Apply Henry's law to estimate fugacities of dilute components.

7. Explain osmotic pressure and compute its value and implications for biological systems.

11.1   MODIFIED RAOULT'S LAW AND EXCESS GIBBS ENERGY

In Section 10.7 we demonstrated that deviations from Raoult's law were manifested by changes in the bubble line and thus characterized positive and negative deviations. With a purely mathematical perspective for modeling the behavior, we could develop a "correction" to Raoult's law as illustrated in Fig 11.1.

Note that the bubble line for Raoult's law misses the shape completely. The bubble pressure formula for Raoult's law is linear in x_1 because $x_2 = 1 - x_1$ and the P^{sat} values are constants with respect to x_1. Suppose that we were to apply a "correction" to VLE K-ratio for each component and attribute the deviations to the liquid phase, where the molecules are closely packed and molecular geometry and intermolecular potentials are much more important than in the vapor phase. The resultant method provides

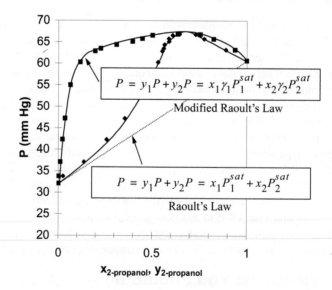

$$P = y_1 P + y_2 P = x_1 \gamma_1 P_1^{sat} + x_2 \gamma_2 P_2^{sat}$$

Modified Raoult's Law

$$P = y_1 P + y_2 P = x_1 P_1^{sat} + x_2 P_2^{sat}$$

Raoult's Law

Figure 11.1 *P-x-y diagram for isopropanol water at 30°C illustrating the rationale for activity coefficients using modified Raoult's law. Data from Udovenko, V.V., and Mazanko, T.F. 1967. Zh. Fiz. Khim. 41:1615.*

$$K_i \equiv \frac{y_i}{x_i} = \frac{\gamma_i P_i^{sat}}{P}$$ "modified Raoult's law"   11.1

The relation to the bubble line is not yet obvious. Cross-multiplying and summing gives,

$$P = y_1 P + y_2 P = x_1 \gamma_1 P_1^{sat} + x_2 \gamma_2 P_2^{sat}$$ 11.2

❶ Bubble pressure calculation.

Now we can begin to see how this approach adjusts the model to the bubble pressure. When a system has **positive deviations** from Raoult's law, the bubble line lies above the Raoult's law bubble line ($P = x_1 P_1^{sat} + x_2 P_2^{sat}$), therefore $\gamma_i > 1$. When a system has **negative deviations**, $\gamma_i < 1$. This correction factor γ_i, is referred to as the **activity coefficient.** Therefore, *P-x-y* data are related to the deviations of the activity coefficients from unity, or it may be helpful to consider the sign of $\ln \gamma_i$.[1] Look back at Fig. 11.1 and the figures in Section 10.7 and note that the deviations from Raoult's law disappear as pure compositions are approached. This means that the deviations depend on composition, and that the $\gamma_i(x)$ that we have introduced in Eqn. 11.1 must go to 1 as the solution becomes pure in the ith component.

As you might imagine, there are rules that we should follow to develop feasible functions. We discuss those theoretical aspects in the upcoming sections. The accepted method of modeling the system is to build a model for $\gamma_i(x)$ based in the **excess Gibbs energy.** The excess Gibbs energy is also discussed in upcoming sections, but briefly it is the Gibbs energy in "excess" of an ideal solution. It is generally positive when $\gamma_i > 1$ and negative when $\gamma_i < 1$. The model for excess Gibbs energy can be developed from a theoretical model of mixing behavior, and usually contains adjustable parameters to adjust the magnitude and skewness of the "excess" and thus fit the experiment.

Activity coefficients may be determined from experimental measurements by rearranging modified Raoult's law as implied by Eqn. 11.1:

$$\gamma_i = \frac{y_i P}{x_i P_i^{sat}}$$ 11.3

From the values of the activity coefficients, a value for excess Gibbs energy can be calculated:

$$\frac{G^E}{RT} = x_1 \ln \gamma_1 + x_2 \ln \gamma_2$$ 11.4

These values may be tabulated to provide G^E/RT vs. x_1 and the resultant curve can be regressed to fit a reasonable analytical expression for convenient interpolation at all compositions. Empirically, it is often preferable to fit the γ_i's directly. We demonstrate both approaches.

The three major stages of working with activity coefficients are shown in Fig. 11.2. In Stage I, the activity coefficients are determined at various compositions from the experiments. In Stage II, a model is selected and various techniques can be used to fit the model to the experimental data. Finally, in Stage III, the model is utilized to solve many different types of phase equilibria prob-

---

1. In polymer-solvent systems the activity coefficient of the solvent at high solvent concentrations is typically > 1, but the activity coefficient of the polymer is <<1. However, the overall system has positive deviations from the perspective of bubble pressure. We focus above on the behavior in systems of molecules of approximately the same size.

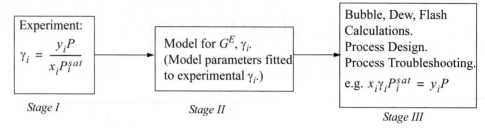

Figure 11.2 *Strategy for using excess Gibbs energy models for activity coefficients. Experiments are used to determine γ_i in Stage I. A model is selected and the model is fitted in Stage II. The model is utilized to extend and extrapolate the experimental results in Stage III.*

lems. Initially, we work some examples where Stage II is separate from Stage III. In advanced fitting, Stages II and III are combined. In addition to development of these skills, much of the chapter is devoted to demonstrating the theoretical foundations for the models so that you are familiar with some of the popular models available in process design software. We integrate examples of bubble, dew, and flash calculations throughout the discussions.

The One-Parameter Margules Equation

The simplest expression for the Gibbs excess function is the one-parameter Margules equation (also known as the two-suffix Margules equation). For a binary system,

$$\frac{G^E}{RT} = A_{12}x_1x_2 \qquad 11.5$$

Note: *The parameter A_{12} is a constant which is not associated with the other uses of the variable (equation of state parameters, Helmholtz energy, Antoine coefficients) in the text. The parameter A_{12} is typically used in discussions of the Margules equation, so we use it here.[2]*

The one-parameter Margules equation is symmetrical with composition. It has an extremum at $x_1 = 0.5$ in a binary system and becomes zero at purity of either component. The activity coefficients in a binary system for the one-parameter Margules equation are (derived later as Eqns. 11.29–11.31),

$$\ln\gamma_i = A_{12}(1-x_i)^2 \qquad 11.6$$

Let us fit the model to experimental data to demonstrate Stage I and Stage II procedures.

> ❶ The one-parameter Margules equation is the simplest excess Gibbs expression.

> ❶ Margules one-parameter model for a binary mixture.

> ▣ gammaModels/ Marg1P.m.

Example 11.1 Gibbs excess energy for system 2-propanol + water

Using data from the 2-propanol(1) + water(2) system presented in Fig. 10.8 calculate the excess Gibbs energy at $x_1 = 0.6369$ and fit the one-parameter Margules equation. Data from the original citation provide $T = 30°C$, $P_1^{sat} = 60.7$ mmHg, $P_2^{sat} = 32.1$ mmHg, and $y_1 = 0.6462$ when $x_1 = 0.6369$ at $P = 66.9$ mmHg.

Solution: The approach is to determine the activity coefficients and then relate them to the excess Gibbs energy. The Stage I step is

---

2. Another common characterization of the Margules one-parameter model is: $G^E = A_{12}x_1x_2$. This results in an explicit temperature dependence, $RT\ln\gamma_i = A_{12}(1-x_i)^2$. Use care when using parameters from various sources.

Example 11.1 Gibbs excess energy for system 2-propanol + water (Continued)

$$\gamma_1 = \frac{y_1 P}{x_1 P_1^{sat}} = \frac{0.6462 \cdot 66.9}{0.6369 \cdot 60.7} = 1.118$$

$$\gamma_2 = \frac{y_2 P}{x_2 P_2^{sat}} = \frac{0.3538 \cdot 66.9}{0.3631 \cdot 32.1} = 2.031$$

$$\frac{G^E}{RT} = x_1 \ln \gamma_1 + x_2 \ln \gamma_2 = 0.6369 \ln (1.118) + 0.3631 \ln (2.031) = 0.328 \qquad 11.7$$

If we were given more experimental data, we could repeat the calculation for each data point, thus creating a plot of G^E versus x_1 like the points shown in Fig. 11.3.

Then, we have been instructed to use the one-parameter Margules model for Stage II. Let us fit the model as given by Eqn. 11.5 and using the value from Eqn. 11.7.

$$\frac{G^E}{RT} = A_{12} x_1 x_2 = 0.328 \implies A_{12} = 0.328/[(0.6369)(0.3631)] = 1.42 \qquad 11.8$$

The curve of G^E versus x_1 is shown in Fig. 11.3 along with the two-parameter models to be discussed in Section 11.6 and Example 11.5.

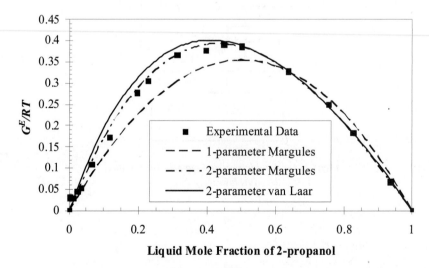

Figure 11.3 *Illustration of calculation of G^E from experiment and fitting of Margules models to a single point as discussed in Examples 11.1 and 11.5, for 2-propanol + water, with the experimental data points from Fig. 10.8 on page 395. Data are tabulated in Example 11.8. The van Laar model fit to a single point is explained in Section 12.2.*

Note from the example that the single point fit gives only an approximate representation of the excess Gibbs energy (because we compared to some additional data that were not included in the problem statement). However, let us proceed to see how this one data point can be leveraged to study the phase behavior.

11.2   CALCULATIONS USING ACTIVITY COEFFICIENTS

Once the activity coefficient model's parameters are known for a given system, the K-ratio can be calculated as a function of composition using Eqn. 11.1. For the one-parameter Margules equation, the activity coefficients are given by Eqn. 11.6. Then the bubble, dew, and flash routines can be executed from Table 10.1 on page 373. Because the activity coefficients depend on x_i, the algorithms where x_i is unknown require an initial guess to calculate a value for γ_i, and an iterative procedure to converge. Raoult's law is often used for the initial guess for x_i. Flow sheets for the methods are summarized in Appendix C, Section C.1. The method for **bubble pressure** does not require iteration because the activity coefficient depends on temperature and liquid composition and both are specified as inputs, as shown by Eqn. 11.2. This simple method is shown in Fig. 11.4.

Let us use the algorithm for bubble pressure to determine the pressure and vapor phase compositions predicted by the one-parameter Margules equation at new compositions based on the fit of G^E at the composition from Example 11.1. In fact, we can generate the entire diagram by repeating the bubble-pressure calculation across the composition range.

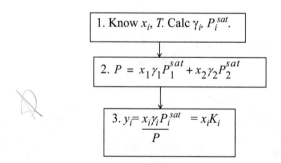

1. Know x_i, T. Calc γ_i, P_i^{sat}.

2. $P = x_1\gamma_1 P_1^{sat} + x_2\gamma_2 P_2^{sat}$

3. $y_i = \dfrac{x_i\gamma_i P_i^{sat}}{P} = x_i K_i$

Figure 11.4 *Bubble-pressure method for modified Raoult's law.*

Example 11.2 VLE predictions from the Margules equation

Use the fit of Example 11.1 to predict the P-x-y diagram for isopropanol + water at 30°C. The data used for Fig. 11.1 from Udovenko et al. for 2-propanol(1) + water(2) at 30°C show $x_1 = 0.1168$ and $y_1 = 0.5316$ at $P = 60.3$ mmHg.

Solution: This is a Stage III problem, since the first two stages have been completed earlier. Let us start by generating activity coefficients at the same composition where experimental data are provided, $x_1 = 0.1168$; we find

$$\ln \gamma_1 = A_{12} x_2^2 = 1.42(0.8832)^2 = 1.107, \Rightarrow \gamma_1 = 3.03$$

$$\ln \gamma_2 = A_{12} x_1^2 = 1.42(0.1168)^2 = 0.0194 \Rightarrow \gamma_2 = 1.02$$

Note that these activity coefficients differ substantially from those calculated in Example 11.1 because the liquid composition is different. We always recalculate the activity coefficients when new values of liquid composition are encountered.

Substituting into modified Raoult's law to perform a bubble-pressure calculation:

$$x_1\gamma_1 P_1^{sat} = (0.1168)(3.03)(60.7) = 21.48 \text{ mmHg} = y_1 P$$

$$x_2\gamma_2 P_2^{sat} = (0.8832)(1.02)(32.1) = 28.92 \text{ mmHg} = y_2 P$$

❶ Bubble-pressure calculation.

The total pressure is found by summing the partial pressures,

$$P = y_1 P + y_2 P = 50.4 \text{ mmHg}$$

We manipulate modified Raoult's law as shown in step 3 of Fig. 11.4:

$$y_1 = y_1 P/P = 21.48/50.4 = 0.426$$

Therefore, compared to the experimental data, the model underestimates the pressure and the vapor composition of y_1 is too low, but the use of one measurement and one parameter is a great improvement over Raoult's law. The estimation can be compared with the data by repeating the bubble-pressure calculation at selected x_i values across the composition range; the results are shown in Fig. 11.5. Recall that in Fig. 11.3 we noted that the excess Gibbs energy model using $A_{12} = 1.42$ fails to capture the skewness of the excess Gibbs energy curve. The deficiency is evident in the P-x-y diagram also. Fig. 11.5 includes a two-parameter fit that will be discussed later.

This example has demonstrated that a single experiment can be leveraged to generate an entire P-x-y diagram with a greatly improved representation of the system. There is an even better method to use a single experiment with a two-parameter model, but we can explain that later. Let us look at one more example using the one-parameter model, but let us integrate the fitting of the excess Gibbs energy (Stage II) simultaneously with the bubble-pressure calculation (Stage III).

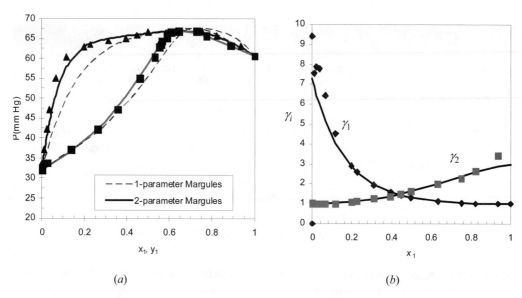

Figure 11.5 *(a) One-parameter and two-parameter Margules equation fitted to a single measurement in Examples 11.2 and 11.5 compared with the experimental data points from Fig. 10.8 on page 395. Data are tabulated in Example 11.8. (b) Activity coefficients predicted from the parameters fitted in Example 11.5 compared with points calculated from the data.*

Example 11.3 Gibbs excess characterization by matching the bubble point

The 2-propanol (1) + water (2) system is known to form an azeotrope at 760 mmHg and 80.37°C ($x_1 = 0.6854$). Estimate the Margules parameter by fitting the bubble pressure at this composition. Then compare your result to the Raoult's law approximation and to the data in Fig. 10.8(c) (at 30°C), where $P = 66.9$ mmHg at $x_1 = 0.6369$ as used in Example 11.1.

🛑 Bubble-pressure calculation.

Solution: The Antoine coefficients for 2-propanol and water are given in Appendix E. At $T = 80.37$°C, $P_1^{sat} = 694.0$ mmHg, and $P_2^{sat} = 359.9$ mmHg. We seek $P = 760$ mmHg. Let us use trial and error at the azeotropic composition to fit A_{12} to match the bubble pressure.

At $A_{12} = 1$, $\gamma_1 = \exp[1(1 - 0.6854)^2] = 1.104$; $\gamma_2 = \exp[1(1 - 0.3146)^2] = 1.600$; the bubble pressure is by Eqn. 11.2

$$P = 0.6854(694.)1.104 + 0.3146(359.9)1.600 = 706.3 \text{ mmHg}$$

Example 11.3 Gibbs excess characterization by matching the bubble point (Continued)

The pressure is too low. We need larger activity coefficients, so A_{12} must be increased. Typing the bubble-pressure formula into Excel or MATLAB (see file Ex11_03.m), we can adjust A_{12} until P = 760 mmHg.

Ex11_03.m.

at $A_{12} = 1.368$, $\gamma_1 = \exp[1.368(1 - 0.6854)^2] = 1.145$; $\gamma_2 = \exp[1.368(1 - 0.3146)^2] = 1.902$; the bubble pressure is

$$P = 0.6854(694.)1.145 + 0.3146(359.9)1.902 = 760.0 \text{ mmHg}$$

Now, for the second part of the problem, to apply this at $T = 30°C$, $P_1^{sat} = 58.28$ mmHg, $P_2^{sat} = 31.74$ mmHg. When $x_1 = 0.6369$ the ideal solution gives,

$$P = 0.6369(58.28) + 0.3631(31.74) = 48.64 \text{ mmHg}$$

At $A_{12} = 1.368$, $\gamma_1 = \exp[1.368(1 - 0.6369)^2] = 1.1976$; $\gamma_2 = \exp[1.368(1 - 0.3631)^2] = 1.7418$; the bubble pressure is

$$P = 0.6369(58.28)1.1976 + 0.3631(31.74)1.7418 = 64.53 \text{ mmHg}$$

Comparing, we see that the Raoult's Law approximation, $P = 48.6$ mmHg, deviates by 27% whereas the Margules model deviates by only 3.5%. Furthermore, the Margules model indicates an azeotrope because $64.5 > P_1^{sat} > P_2^{sat}$ means that there is a pressure maximum. Hence the Margules model "predicts" an azeotrope at this lower temperature, qualitatively consistent with Fig. 10.8(c), whereas the ideal solution model completely misses this important behavior.

x-y Plots

The "x-y" plot introduced in Fig. 10.4 can be prepared for the azeotropic system of Fig. 11.5 by plotting the pairs of y-x data/calculations at each pressure or temperature. Such a plot is shown in Fig. 11.6. The curve represents the two-parameter fit that is shown in Fig. 11.5. Note when an azeotrope exists that the y-x curve crosses the diagonal at the azeotropic composition.

Looking Ahead

Careful readers may notice that $A_{12} = 1.42$ from Example 11.1 and $A_{12} = 1.37$ from Example 11.3. The compositions were slightly different. We also noted in Example 11.1 when we peeked at additional data that the single parameter model was insufficient to represent the system all the way across the composition range, so this was also a factor in the difference. There is also a another possibility for fitting the activity model that we did not consider. After determining the activity coefficients from Eqn. 11.3, we could have used the values directly in the model Eqn. 11.6. This method was not used because the solution is overspecified with two equations and one A_{12} parameter value that would have been different for each. We could have calculated the two values and averaged them, but we chose instead to use the excess Gibbs energy or the bubble pressure directly—methods that used thermodynamic properties directly. We can see that improved models are desirable.

Clearly, the one-parameter Margules model has limitations, but it sets us on a path of continuing improvement that is fundamental to engineering: Observe, predict, test, evaluate, and improve. Observations for ideal solutions suggested a crude model in Raoult's law. When predictions with Raoult's law were tested for a broader range of mixtures, however, we observed deficiencies. Eval-

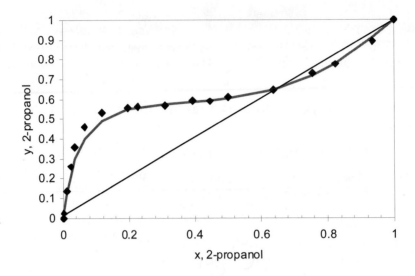

Figure 11.6 *Data and the two-parameter fit of Fig. 11.5 plotted as pairs of x and y. Both T-x-y and P-x-y data can be plotted in this way.*

uating the model, a correction factor was suggested that conformed to physical constraints like $\gamma_i(x_i) = 1$ when $x_i = 1$. Then a slightly more sophisticated model equation was suggested, the one-parameter Margules model. We followed through on several of the implications of this model (e.g., Eqns. 11.4 and 11.6) and arrived at new predictions. We tested those predictions and found improvements, but still deficiencies. Now we are ready to begin a new round of evaluation. Each successive round of evaluation requires deeper insight into the physical constraints, ultimately leading to careful consideration of the interactions at the molecular scale.

The activity coefficient models that we discuss in upcoming sections enable a broad range of engineering analyses. For example, we may wish to design a distillation column that operates at constant pressure and requires *T-x-y* data. However, the available VLE data may exist only as constant temperature *P-x-y* data. We may use the activity coefficient models to convert isothermal *P-x-y* data to isobaric *T-x-y* data, and vice versa. Furthermore, parameters from binary data can be combined and extended to multicomponent systems, even if no multicomponent data are available. Elementary techniques for fitting G^E (or activity coefficient) models are presented to fit single data points. Advanced techniques for fitting G^E or VLE data across the composition range are presented in specific examples and Section 11.9. In the next few pages, we fill in some of the theoretical development that we have skipped in our overview.

Preliminary Predictions Based on a Molecular Perspective

Ultimately, we would like to make predictions that go beyond fitting a single data point for a single binary mixture. We would like to design formulations to solve practical problems. For example, suppose somebody had sprayed graffiti on the *Mona Lisa*. Could we formulate a solvent that would remove the spray paint while leaving the original painting intact? What about an oil spill in the Gulf of Mexico? What kind of treatment could disperse it best? What kind of molecule could be added to break the azeotrope in ethanol + water? What formulation could promote the permeation of insulin through the walls of the small intestine? These may sound like very different problems, but they are all very similar to a thermodynamicist. To formulate a compatible solvent, we simply need to minimize the activity coefficient. For example, we should seek a solvent that has a low activity coefficient with polymethylmethacrylate (PMMA, a likely graffiti paint) and a high activity coefficient with linseed oil (the base of oil paint). We could imagine randomly testing many solvents, but then we might hope to observe patterns that would lead to predictions. These would be predictions of a higher order than simply extrapolating to a different temperature or composition, but they would enable us to contemplate the solutions to much bigger problems. You already possess sufficient molecular insight to begin this process. Elucidating that will simultaneously make these problems seem less daunting and help us on the way to more sophisticated model evaluation.

You know that acids and bases interact favorably. An obvious example would be mixing baking soda and vinegar which react. You could also mix acid into water. These interactions are "favorable" because they release energy, meaning they are exothermic. They release energy because their interaction together is stronger than their self-interactions with their own species. A subtler exothermic example is hydrogen bonding, familiar perhaps from discussions of DNA, where the molecules do not react, but form exothermic hydrogen bonds. Unlike a covalent bond, the hydrogen sits in a minimum energy position between the donor and acceptor sites. The proton of a hydroxyl (-OH) group is acidic while an amide or carbonyl group acts as a base. We can extend this concept and assign qualitative numerical values characterizing the acidity and basicity of many molecules as suggested by Kamlet et al.[3] These are the **acidity parameter,** α, and **basicity parameter,** β, values listed on the back flap. For example, this simple perspective suggests that chloroform ($\alpha > 0$) might make a good solvent for PMMA (a polymer with a molecular structure similar to methyl ethyl ketone, $\beta > 0$) because the α and β values should lead to favorable interactions. This is the perspective suggested by Fig. 11.7(a).

Hydrogen bonding may sound familiar, but there are subtleties that lead to complex behavior. These subtleties are largely related to the simultaneously acidic and basic behavior of hydroxyl species. We know that water contains -OH functionality, but its strong interaction with acids also indicates a basic character. It is both acidic and basic. The subtlety arises when we consider that its acidic and basic interactions link together when it exists as a pure fluid. Then a question arises about how water might interact with an "inert" molecule that is neither acidic nor basic, as illustrated in Fig. 11.7(b). Clearly, the water would squeeze the inert molecule out, so it could maximize its acid-base interactions. Referring to the back flap, $\alpha = \beta = 0$ for molecules like octane-hexadecane, and these components have molecular structures similar to oil. Thus, we see that this acid-base perspective correlates with the old guideline that oil and water do not mix. Oils are said to be "water-fearing," or **hydrophobic.** Finally, Fig. 11.7(c) illustrates what happens when two molecules have similar acidity and basicity, like methanol and ethanol. Then they can substitute for each other in the hydrogen bonding network and result in a solution that is nearly ideal. Molecules like alcohols are called **hydrophilic** ("water-liking").

This perspective is not a large leap from familiar concepts of acids, bases, and hydrogen bonding, but it does provide more insight than guidelines such as "like dissolves like" or "polarity leads to nonideality." Acids are not exactly "like" bases, but they do interact favorably. Methanol and ethanol are both polar, but they can form ideal solutions with each other.

We can go a step further by formulating numerical predictions using what we refer to as the **Margules acid-base (MAB) model.** The model provides first-order approximations. The model is:

> Margules acid-base (MAB) model.

$$A_{12} = (\alpha_2 - \alpha_1)(\beta_2 - \beta_1)(V_1 + V_2)/(4RT) \qquad 11.9$$

3. Kamlet, M. J., Abboud, J.-L.M., Abraham, M.H., Taft, R.W. 1983. *J. Org. Chem.* 48:2877.

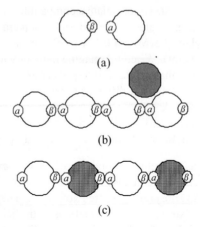

Figure 11.7 *Observations about complexation. (a) A mixture of acid with base suggests*
favorable interactions, as in acetone + chloroform. (b) Hydrogen bonding
leads to unfavorable interactions when one component associates strongly
and the other is inert, as in isooctane + water. (c) Hydrogen bonding
solutions can also be ideal solutions if both components have similar acidity
and basicity, as in methanol + ethanol.

where $V_i = M_{w,i}/\rho_i^L$ is the liquid molar volume at 298.15K in cm^3/mol. The MAB model is introduced here for pedagogical purposes. MAB is a simplification of SSCED[4] which is in turn a simplified adaptation of MOSCED[5], both of which are covered in Chapter 12. Typical values of V, α, and
β are presented in Table 11.1. For example, with chloroform + acetone at 60°C, this formula gives

$$A_{12} = (5.8 - 0) \cdot (0.12 - 11.14) \cdot (80.5 + 73.4)/[4(8.314)333] = -0.888 \qquad 11.10$$

Table 11.1 *Acidity (α) and Basicity (β) Parameters in $(J/cm^3)^{1/2}$ and*
Molar Volumes (cm^3/mol) for Various Substances as liquids at 298 K^a

Compound	α	β	V^L
Acetone	0	11.14	73.4
Benzene	0.63	2.24	89.7
Chloroform	5.80	0.12	80.5
Ethanol	12.58	13.29	58.2
n-Hexane	0	0	130.3
Isooctane	0	0	162.9
Isopropanol	9.23	11.86	76.8
Methanol	17.43	14.49	40.5
MEK	0	9.70	90.1
Water	50.13	15.06	18.0

a.   Additional parameters are on the back flap.

4. Elliott, J.R. 2010. *Chem. Eng. Ed.* 44(1):13–22.
5. Lazzaroni, M.J., Bush, D., Eckert, C.A., Frank, T.C., Gupta, S., Olson, J.D. 2005. *Ind. Eng. Chem. Res.* 44:4075.

Note how the order of subtraction results in a negative value for A_{12} when one of the components is acidic and the other is basic. If you switched the subscript assignments, then $\Delta\alpha$ would be negative and $\Delta\beta$ would be positive, but A_{12} would still be negative. This negative value makes the value of γ_i smaller, and that is basically what happens when hydrogen bonding is favorable. Something else happens when one compound forms hydrogen bonds but the other is inert. Taking isooctane(1) as representative of oil (or gasoline) and mixing it with water(2) at 25°C,

$$A_{12} = (50.13 - 0)(15.06 - 0)(18.0 + 162.9)/[4(8.314)298] = 12.33 \qquad 11.11$$

This large positive value results in $\gamma_1 > 7.5$ for the isooctane. We can use $\gamma_1 > 7.5$ to suggest a liquid phase split, as we should expect from the familiar guideline that oil and water do not mix. Furthermore, we can quantify the solubilities of the components in each other (aka. **mutual solubilities**) by noting that $x_i \approx 1/\gamma_i$ when $\gamma_i > 100$. Knowing the saturation limit of water contaminants can be useful in environmental applications. As a final example, note that we recover an ideal solution when both components hydrogen bond similarly, as in the case of ethanol + methanol at 70°C.

$$A_{12} = (17.43 - 12.58)(14.49 - 13.29)(40.5 + 58.2)/[4(8.314)343] = 0.05 \qquad 11.12$$

In this case, we see that hydrogen bonding by itself is not the cause of solution non-ideality. A *mismatch* of hydrogen bonding is required to create non-idealities.

Example 11.4 Predicting the Margules parameter with the MAB model

Predict the A_{12} value of the 2-propanol (1) + water (2) system using the MAB model at 30°C. Then compare your result to those of Examples 11.1 and 11.3.

Solution: From Eqn. 11.9, $A_{12} = (50.13 - 9.23)(15.06 - 11.86)(76.8 + 18.0)/[4(8.314)303] =$ 1.08. This compares to the value $A_{12}/RT = 1.42$ from Example 11.1 and $A_{12}/RT = 1.37$ from Example 11.3 at 30°C. The MAB model does not provide a precise prediction, but qualitatively indicates a positive deviation of the right magnitude.

With this perspective we can begin to contemplate formulations of very broad problems, but this is only the beginning. We will see in Section 12.5 that the MAB model overlooks an important contribution to the activity coefficient, even in the context of the relatively simple van der Waals perspective. In Section 13.1 we show limitations of the van der Waals perspective. Finally, Chapter 19 shows that accounting for hydrogen bonding as a chemical reaction results in a description of the Gibbs excess energy that is quite different from the perspectives in Chapters 11 to 13. All of these presentations focus primarily on relatively small molecules with single functionalities like alkyl, hydroxyl, or amide. Modern materials (including biomembranes and proteins) are composed of large molecules with deliberate arrangements of the functionalities resulting in self-assembly to perform remarkably diverse macroscopic functions.

11.3 DERIVING MODIFIED RAOULT'S LAW

In Chapter 10 we demonstrated that Raoult's law requires an ideal solution model for the vapor and liquid phases as well as conditions where the fugacity coefficients can be ignored. In Section 11.1 we have shown that a relatively simple function is able to capture a major correction to Raoult's law, but we have superficially made the connections to fundamental properties and we must now develop that understanding of how this function is related to component fugacity and Gibbs energy of the mixture.

To perform VLE calculations, we begin with the fundamental criterion $\hat{f}_i^V = \hat{f}_i^L$. The fugacity in non-ideal systems is modeled in terms of deviations from either the ideal gas model or the ideal solution model. The Venn diagram in Fig. 11.8 may be helpful in visualizing these relations. Ideal gas behavior is the simplest type of mixture behavior because the particles are non-interacting. This is shown in the center of Fig. 11.8. Clearly, ideal gas behavior is not followed by all mixtures, and therefore ideal gases are a subset of real mixtures. Strictly, ideal gas molecules cannot condense because they have no attractive forces; if fluids were ideal gases, there would be no liquids, and VLE would not occur. However, at low densities, gas molecules are frequently separated far enough that the effective intermolecular potentials are insignificant, and we can frequently model the gas phase *as if* it is an ideal gas. The fugacity of a component in an ideal-gas mixture is particularly simple; it is equal to y_iP, the partial pressure. The ideal gas model is acceptable for most small molecular weight vapors near atmospheric pressure;[6] and the primary failure of Raoult's law is due to the liquid phase where the molecules are closely packed. When the molecules are closely packed in a liquid phase the intermolecular potential is significant for the same molecules whose potential energies could be disregarded in the vapor phase. The size and shape of molecules and differences in chemical functionality at close distances often violate the assumptions of ideal solution behavior. We introduce the concept of the activity and activity coefficient, γ_i, to correct the ideal mixture model as summarized in Fig. 11.8.

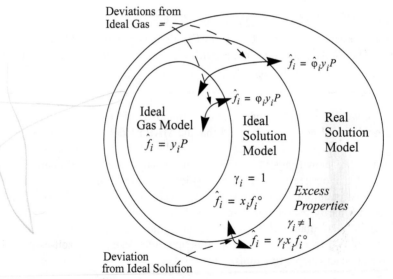

Figure 11.8 *Schematic of the relations between different fluid models. Ideal gases are a subset of ideal mixtures, which in turn are a subset of real mixtures. Departure functions (fugacity coefficients) characterize deviations from ideal-gas behavior, and excess properties (activity coefficients) characterize deviations from ideal-solution behavior.*

---

6. Carboxylic acids almost always have significant deviations from ideal gas behavior as we discuss in Chapters 16 and 19.

The activity coefficient is defined as a ratio of the component fugacity to the ideal solution fugacity at the same mole fraction:

$$\gamma_i \equiv \frac{\hat{f}_i}{x_i f_i^\circ}$$

11.13

❶ Activity coefficients are commonly used for highly non-ideal solutions.

A value of $\gamma_i = 1$ will denote an ideal solution; f_i° is the value of the fugacity at **standard state.** A standard state is slightly different from a reference state. The standard state is a specified temperature, pressure, and composition. The most common standard state in solution thermodynamics is the pure component at the same temperature pressure as the system. You can see that this is clearly different from a reference state that would stay fixed throughout a series of calculations. For this standard state, standard state fugacity is simply the pure component fugacity which we introduced in Chapter 9, f_i. In Fig. 11.8 we also introduce the **fugacity coefficient** $\hat{\varphi}_i$ for a component in a mixture to characterize the component fugacity relative to the ideal gas partial pressure. We defer most discussion of the component fugacity coefficients to Chapters 15 and 19.

Now, let us look rigorously at the development of modified Raoult's law. For the vapor phase, we begin with the rigorous expression from Fig. 11.8 including deviations from the ideal gas model, $\hat{f}_i^V = y_i \hat{\varphi}_i P$. For the liquid phase, we use an activity coefficient, γ_i, giving[7]

$$\hat{f}_i^L = x_i \gamma_i f_i^L$$

11.14

Typically the Poynting method (Section 9.8) is used to calculate the pure-component liquid phase fugacities, $f_i^L = \varphi_i^{sat} P_i^{sat} \exp\left(\dfrac{V_i^L(P - P_i^{sat})}{RT}\right)$. Combining these expressions,

$$y_i \hat{\varphi}_i P = x_i \gamma_i \varphi_i^{sat} P_i^{sat} \exp\left(\frac{V_i^L(P - P_i^{sat})}{RT}\right)$$

11.15

When used in this full form, Eqn. 11.15 is called the **gamma-phi** method. This may be written in terms of the K-ratio, $K_i = y_i/x_i$,

$$K_i \equiv \frac{y_i}{x_i} = \frac{\gamma_i P_i^{sat}}{P} \left[\frac{\varphi_i^{sat} \exp[V_i^L(P - P_i^{sat})/(RT)]}{\hat{\varphi}_i}\right]$$

11.16

At the low pressures of many chemical engineering processes the Poynting corrections and the ratios of $\varphi_i^{sat}/\hat{\varphi}_i$ for the components approach unity. Recalling $\hat{f}_i^L = \hat{f}_i^V$, we find

$$\boxed{K_i \equiv \frac{y_i}{x_i} = \frac{\gamma_i P_i^{sat}}{P}} \ ; \ \boxed{\hat{f}_i^L \approx x_i \gamma_i P_i^{sat}} \ ; \ \boxed{\hat{f}_i^V \approx y_i P}$$

11.17

❶ Modified Raoult's law.

We usually write

---

7. We ignore the pressure dependence of activity coefficients since most models ignore the effect, $RTd\ln\gamma_i = \bar{V}_i^E dP$ when the standard state pressure is the system pressure.

🛈 Modified Raoult's law.

$$\boxed{y_i P = x_i \gamma_i P_i^{sat}} \quad \text{or} \quad \boxed{K_i = \frac{\gamma_i^L P_i^{sat}}{P}}$$ 11.18

We can see that modified Raoult's law depends on the fugacity coefficient ratio being close to unity, not necessarily on the ideal gas law being exact. Next, let us demonstrate how the activity coefficient is related to the excess Gibbs energy.

11.4   EXCESS PROPERTIES

The deviation of a property from its ideal-solution value is called the **excess property.** For a generic property M, the excess property is given the symbol M^E, and M^E is the value of the property for the mixture relative to the property for an ideal mixture, $M^E = M - M^{is}$. Ideal solutions were discussed in Section 10.10. The molar volume of an ideal solution is just the weighted sum

🛈 Excess volume.

of the molar volumes of the components, $V^{is} = \sum_i x_i V_i$. The **excess volume** is then,

$$V^E = V - V^{is} = V - \sum_i x_i V_i$$ 11.19

Although the excess volumes *of liquids* are typically a very small percentage of the volume, the concepts of excess properties are easily grasped by first studying the excess volume and then exploring the more abstract quantities of excess enthalpy, entropy, or Gibbs energy.

The excess volume of the system 3-pentanone (1) + 1-chlorooctane (2) at 298.15 K has been measured by Lorenzana, et al.,[8] and is shown in Fig. 11.9. The molar volumes of the pure components are $V_1 = 106.44$ cm³/mol and $V_2 = 171.15$ cm³/mol. At the equimolar concentration, the excess volume is 0.204 cm³/mol. Therefore, the molar volume is $V = V^E + V^{is} = 0.204 + 0.5 \cdot 106.44 + 0.5 \cdot 171.15 = 139.00$ cm³/mol. The excess volume is only 0.15% of the total volume. The **partial molar excess volume** is calculated in a manner analogous to the **partial molar volume,** $\overline{V}_i^E = (\partial \underline{V}^E / \partial n_i)_{T, P, n_{j \neq i}}$. If an algebraic expression is available for the excess volume, it may be differentiated by this relation to yield formulas for the excess volumes. Graphically, the partial molar volumes at any point may be found by drawing the tangent line to the excess volume curve and reading the intercepts. At the composition shown at the tangent point in Fig. 11.9, the intercepts give $\overline{V}_1^E \approx 0.24$ cm³/mol and $\overline{V}_2^E \approx 0.17$ cm³/mol. The partial molar volumes depend on composition.

The excess enthalpy is very similar to the excess volume,

$$H^E = H - H^{is} = H - \sum_i x_i H_i$$ 11.20

8. Lorenzana, T., Franjo, C., Jiménez, E., Fernández, J., Paz-Andrade, M.I. 1994. *J. Chem. Eng. Data* 39:172.

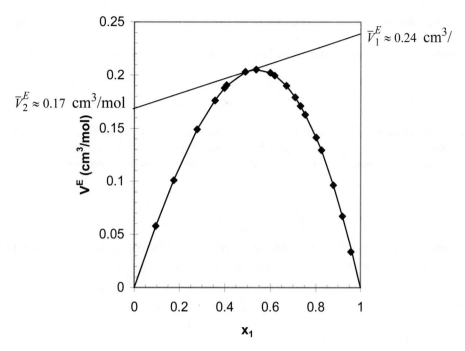

Figure 11.9 *Excess volume for the 3-pentanone (1) + 1-chlorooctane*
(2) system at 298.15 K.

A solution with $H^E > 0$ has an **endothermic heat of mixing,** and when $H^E < 0$, the heat of mixing is **exothermic.** In an adiabatic mixing process an endothermic mixing process will cool and an exothermic mixing process will heat.

In directly analogous fashion, the **excess Gibbs energy** can be defined as the difference between the Gibbs energy of the mixture and the Gibbs energy of an ideal solution, $G^E = G - G^{is}$. Then, instead of speaking of partial molar volumes, we speak of partial molar Gibbs energies. But you should recognize the **partial molar Gibbs energy** as the **chemical potential** as introduced in Section 10.8, and the significance of the chemical potential to phase equilibrium calculations should resonate strongly after reading Chapter 10. What remains is to rearrange the mathematics into the final relations to show how this property is related to the activity coefficients.

11.5 MODIFIED RAOULT'S LAW AND EXCESS GIBBS ENERGY

The excess Gibbs energy is

$$G^E \equiv G - G^{is}$$

$$= \left(G - \sum_i x_i G_i \right) - \left(G^{is} - \sum_i x_i G_i \right)$$

$$= \Delta G_{mix} - \Delta G^{is}_{mix} = \Delta G_{mix} - RT \sum_i x_i \ln(x_i)$$

❶ Excess Gibbs

11.21 energy.

where we have added and subtracted the sum of the component Gibbs energies in the second line, and used Eqns. 10.65 and 10.66 in the last line. Let us further examine $\Delta G_{mix} = \left(G - \sum_i x_i G_i \right)$.

Recall that $G = \sum_i x_i \mu_i$, (Eqn. 10.42). Previously in Eqn. 10.48 we expressed μ_i relative to a pure component value using a ratio of fugacities. Generalization of this equation will provide the link between Gibbs energies and component fugacities that will lead to the activity coefficient models. If we calculate the chemical potential relative to a **standard state**, we find by generalizing 10.48

$$\mu_i - \mu_i^o = RT \ln \frac{\hat{f}_i}{f_i^\circ} \qquad 11.22$$

This ratio of fugacities will appear often, and it is convenient to define the ratio as the **activity**, and it can be related to composition using the activity coefficient,

Activity.

$$a_i \equiv \frac{\hat{f}_i}{f_i^\circ} = x_i \gamma_i \qquad 11.23$$

Using a standard state at the T and P of the system (cf. page 425), $\mu_i^o = G_i$, $f_i^\circ = f_i$, we can develop an expression for ΔG_{mix},

$$\Delta G_{mix} = G - \sum_i x_i G_i = \sum_i x_i (\mu_i - G_i) = RT \sum_i x_i \ln \left(\frac{\hat{f}_i}{f_i} \right) \qquad 11.24$$

$$\Delta G_{mix} = RT \sum_i x_i \ln(a_i) = RT \sum_i x_i \ln(x_i \gamma_i) \qquad 11.25$$

Substituting into Eqn. 11.21:

$$G^E \equiv \Delta G_{mix} - RT \sum_i x_i \ln(x_i) = RT \sum_i x_i \ln(x_i \gamma_i) - RT \sum_i x_i \ln(x_i) = RT \sum_i x_i \ln(\gamma_i) \qquad 11.26$$

$$G^E = RT \sum_i x_i \ln(\gamma_i) \qquad 11.27$$

Excess Gibbs energy is zero for an ideal solution, and activity coefficients are unity.

Note that the activity coefficients and excess Gibbs energy are coupled—when the activity coefficients of all components are unity, the excess Gibbs energy goes to zero. The excess Gibbs energy is zero for an ideal solution.

Activity Coefficients as Derivatives

Activity coefficients are related to derivatives of the excess Gibbs energy, specifically the partial molar excess Gibbs energy. We have a very simple relation between partial molar quantities and molar quantities developed in Eqn. 10.41,

$$M = \sum_i x_i \overline{M}_i$$

Applying this relation to excess Gibbs energy,

$$G^E = \sum_i x_i \overline{G}_i^E$$

Comparing this with Eqn. 11.27, we see that

$$\left(\frac{\partial \underline{G}^E}{\partial n_i}\right)_{T,P,n_{j \neq i}} = \overline{G}_i^E = \mu_i^E = RT\ln\gamma_i \qquad\qquad 11.28$$

❶ Activity coefficients are related to the partial molar excess Gibbs energy.

where we have recognized that the partial molar excess Gibbs energy of a component is also the component excess chemical potential. So, for any expression of $G^E\,(T,P,x)$, we can derive γ's.

For the one-parameter Margules,

$$\frac{\underline{G}^E}{RT} = n(A_{12}x_1x_2) = (A_{12}n_2)\left(\frac{n_1}{n}\right) \qquad\qquad 11.29$$

Applying Eqn. 11.28 for $i = 1$, and using the product rule on $n_1/n = n_1(1/n)$,

$$\frac{1}{RT}\left(\frac{\partial \underline{G}^E}{\partial n_1}\right)_{T,P,n_2} = \ln\gamma_1 = An_2\left[\frac{1}{n} - \frac{n_1}{n^2}\right] = A\frac{n_2}{n}\left[1 - \frac{n_1}{n}\right] = Ax_2(1 - x_1) \qquad 11.30$$

$$\ln\gamma_1 = Ax_2^2 \;\;;\text{similarly } \ln\gamma_2 = Ax_1^2 \qquad\qquad 11.31$$

11.6 REDLICH-KISTER AND THE TWO-PARAMETER MARGULES MODELS

We noted in Example 11.1 a shortcoming in the one-parameter model's representation of the skewness of experimental excess Gibbs energy. In principle, adjusting both the magnitude and skewness of G^E is possible with a two-parameter model equation. The mathematical relations in Sections 11.3–11.5 liberate us to conjecture freely about forms of G^E that may fit any given set of VLE data. Based on any model, developing working equations for activity coefficients is a straightforward matter of taking derivatives. These are the considerations behind many empirical models like the Redlich-Kister and Margules two-parameter models.

Many models can be rewritten in the Redlich-Kister form given as[9]

$$\frac{G^E}{RT} = x_1x_2(B_{12} + C_{12}(x_1 - x_2) + D_{12}(x_1 - x_2)^2 + \ldots) \qquad\qquad 11.32$$

---

9. Redlich, O., Kister, T. 1948. *Ind. Eng. Chem.* 40:345–348.

Two-Parameter Margules Model

The two-parameter Margules model is a simplification of the Redlich-Kister,

$$\frac{G^E}{RT} = x_1 x_2 (A_{21} x_1 + A_{12} x_2) \qquad ^{10}$$

$$\text{11.33}$$

where we relate the constants to the Redlich-Kister via $A_{21} = B_{12} + C_{12}$, $A_{12} = B_{12} - C_{12}$, and $D_{12} = 0$. The constants A_{21} and A_{12} are fitted to experiment as we show below. Note that if $A_{21} = A_{12}$, the expression reduces to the one-parameter model. The expression for the activity coefficient of the first component can be derived as

$$\frac{G^E}{RT} = n_2 \frac{n_1}{n}(A_{21}(1-x_2) + A_{12}x_2) = n_2\left(\frac{n_1}{n}\right)\left(A_{21} + \left(\frac{n_2}{n}\right)(A_{12} - A_{21})\right) \qquad \text{11.34}$$

Applying Eqn. 11.28 for $i = 1$, and using the product rule on $n_1/n = n_1(1/n)$ and $n_2/n = n_2(1/n)$,

$$\frac{1}{RT}\left(\frac{\partial G^E}{\partial n_1}\right)_{T,P,n_2} = \ln \gamma_1 = n_2\left(A_{21} + \frac{n_2}{n}(A_{12} - A_{21})\right)\left[\frac{1}{n} - \frac{n_1}{n^2}\right] + n_2\left(\frac{n_1}{n}\right)\left(\frac{-n_2}{n^2}\right)(A_{12} - A_{21}) \quad \text{11.35}$$

$$\ln \gamma_1 = x_2^2 \left[(A_{21} + (1-x_1)(A_{12} - A_{21})) + (A_{21} - A_{12})x_1\right] \qquad \text{11.36}$$

Actcoeff.xlsx, sheet Margules; gammaModels/ Marg2P.m.

$$\ln \gamma_1 = x_2^2 \left[A_{12} + 2(A_{21} - A_{12})x_1\right] \text{; similarly } \ln \gamma_2 = x_1^2 \left[A_{21} + 2(A_{12} - A_{21})x_2\right] \qquad \text{11.37}$$

The two parameters can be fitted to a single VLE measurement using

$$A_{12} = \left(2 - \frac{1}{x_2}\right)\frac{\ln \gamma_1}{x_2} + \frac{2\ln \gamma_2}{x_1} \qquad A_{21} = \left(2 - \frac{1}{x_1}\right)\frac{\ln \gamma_2}{x_1} + \frac{2\ln \gamma_1}{x_2} \qquad \text{11.38}$$

where the activity coefficients are calculated from the VLE data. Care must be used before accepting the values from Eqn. 11.38 applied to a single measurement because experimental errors can occasionally result in questionable parameter values.

Example 11.5   Fitting one measurement with the two-parameter Margules equation

We mentioned following Example 11.2 that a single experiment could be used more effectively with the two-parameter model. Apply Eqn. 11.38 to the two activity coefficients values calculated in Example 11.1 and estimate the two parameters. This is an example of a Stage II calculation.

---

10. If RT is omitted from the excess Gibbs energy, then explicit temperature dependence of the γ_i results, and the "ln" terms of Eqn. 11.37 are preceded with RT. Use care when comparing parameters from various sources because both conventions are used.

Example 11.5 Fitting one measurement with the two-parameter Margules equation (Continued)

Solution: From Example 11.1, $x_1 = 0.6369$, $x_2 = 0.3631$, and $\gamma_1 = 1.118$, $\gamma_2 = 2.031$. From Eqn. 11.38,

$$A_{12} = \left(2 - \frac{1}{0.3631}\right)\frac{\ln 1.118}{0.3631} + \frac{2\ln 2.031}{0.6369} = 1.99$$

$$A_{21} = \left(2 - \frac{1}{0.6369}\right)\frac{\ln 2.031}{0.6369} + \frac{2\ln 1.118}{0.3631} = 1.09$$

The parameters from Example 11.5 provide the representation of G^E shown in Fig. 11.3. Using the concepts from earlier examples along with Eqn. 11.37 for the activity coefficients, bubble-pressure calculations across the composition range (Stage III calculations) result in the curve of Fig. 11.5 designated as the two-parameter model.

Example 11.6 Dew pressure using the two-parameter Margules equation

Use the parameters of Example 11.5 to predict the dew-point pressure and liquid composition for the 2-propanol(1) + water(2) system at $T = 30°C$, $y_1 = 0.4$, and compare with Fig. 11.5. Use the vapor pressures, $P_1^{sat} = 60.7$ mmHg, $P_2^{sat} = 32.1$ mmHg.

Solution: We will apply the procedure in Appendix C and refer to step numbers there.

Step 1. Refer to Chapter 10, $P = 1/(0.4/60.7 + 0.6/32.1) = 39.55$ mmHg,
$x_1 = 0.4(39.55)/60.7 = 0.26$. We skip Step 2 the first time.
Step 3. Using parameters from Example 11.5 in Eqn. 11.37,
$\gamma_1 = \exp(0.74^2(1.99 - 1.8(0.26)))=2.30$; $\gamma_2 = \exp(0.26^2(1.09 + 1.8(0.74)))=1.18$.
Step 4. $P = 1/(0.4/(2.3 \cdot 60.7) + 0.6/(1.18 \cdot 32.1)) = 53.46$ mmHg.
Note the jump in P compared to Step 1 for the first loop.
Step 5. $x_1 = 0.4(53.46)/(2.3 \cdot 60.7) = 0.153$. Continuing the loop:

γ_1	γ_2	P (mmHg)	x_1
3.421	1.063	51.26	0.0987
4.359	1.027	50.73	0.0767
4.849	1.016	50.62	0.0688

Continuing for several more iterations with four digits, $P = 50.63$ mmHg, and $x_1 = 0.0649$. The calculations agree favorably with Fig. 11.5. The dew calculations are consistent with a bubble calculation at $x_1 = 0.0649$.

Dew-pressure
calculation.

Because this is a long chapter, we summarize the relations between the activity coefficient models developed thus far in Table 11.2.

Table 11.2 *Summary of Empirical Activity Models and Simplifications Relative to the Redlich-Kister*

Model	G^E/RT	$\ln\gamma_1$	Simplification
Redlich-Kister	$x_1x_2[B_{12} + C_{12}(x_1-x_2) + D_{12}(x_1-x_2)^2 + \dots]$	*cf.*Practice problem P11.2	--
Margules two-parameter	$x_1x_2(A_{21}x_1 + A_{12}x_2)$	$x_2^2[A_{12} + 2(A_{21}-A_{12})x_1]$	$D_{12}=0$; $A_{21}=B_{12}+C_{12}$; $A_{21}=B_{12}-C_{12}$
Margules one-parameter	$x_1x_2A_{12}$	$x_2^2A_{12}$	$C_{12}=D_{12}=0$; $A_{12}=A_{12}=B_{12}$.
Ideal solution	0	0	$B_{12} = C_{12}=D_{12}=0$

11.7   ACTIVITY COEFFICIENTS AT SPECIAL COMPOSITIONS

Two parameter models provide sufficient flexibility with a balance of relative simplicity to provide successful VLE modeling. Determination of activity for each component permits two parameters to be fitted, and special compositions can be used.

Azeotropes

The location of an azeotrope is very important for distillation design because it represents a point at which further purification in a single distillation column is impossible. Look back at Fig. 11.1 on page 412. Looking at dilute isopropanol concentrations, note $x_{2\text{-propanol}} = 0.01 < y_{2\text{-propanol}}$, but near purity, $x_{2\text{-propanol}} = 0.99 > y_{2\text{-propanol}}$. The relative magnitudes have crossed and thus we expect $y_{2\text{-propanol}} = x_{2\text{-propanol}}$ (i.e., there is an azeotrope) somewhere in between. If the relative sizes are the same at both ends of the composition range, then we expect that an azeotrope does not exist.[11] Certainly, the best way to identify an azeotrope is to plot *T-x-y* or *P-x-y*, but a quick calculation at each end of the diagram is usually sufficient.

> A simple algorithm to decide if an azeotrope exists.

Note that the relative volatility introduced in Section 10.6 on page 390 also changes significantly in an azeotropic system. For the reasons above, $\alpha_{ij} > 1$ on one side of the azeotrope, $\alpha_{ij} = 1$ at the azeotrope, and $\alpha_{ij} < 1$ on the other side. Because shortcut distillation calculations fail at $\alpha_{ij} = 1$, they must not be used if ($\alpha_{ij} - 1$) changes sign between column ends. This means that screening systems for azeotropes such as using the algorithm above is important before blindly plugging numbers into shortcut distillation calculations.

> Relative volatility crosses 1 at an azeotrope.

We noted in Section 10.7 on page 393 that azeotropic behavior was dependent on the magnitude of deviations from ideality *and* the vapor pressure ratio. Look back at Fig. 11.1 on page 412 and recall that deviations from Raoult's law create the curve in the bubble line. When the pure component vapor pressures are nearly the same then a slight curve due to non-ideality can cause an azeotrope. The same size deviation in a system with widely different vapor pressure may not have an azeotrope. A plot of logP^{sat} versus $1/T$ with both components may show a point where the two curves cross when the heats of vaporization are different. This point is called a **Bancroft point**. Since the vapor pressures are exactly equal at the Bancroft point, any small non-ideality generates

> Any deviation from ideality will create an azeotrope at a Bancroft point.

---

11. Technically, this would not be true for a double azeotrope, but these are extremely rare.

an azeotrope. This might be avoidable if the system pressure is raised or lowered to circumvent the Bancroft point in the temperature range of a distillation column.

Many tables of known azeotropes are commonly available.[12] For systems with an azeotrope, the azeotropic pressure and composition provide a useful data point for fitting activity coefficient models because $x_1 = y_1$. Then $\gamma_1 = P/P_1^{sat}$; $\gamma_2 = P/P_2^{sat}$. Then the typical single point fitting formulas are used with the azeotrope composition to find the model parameters.

The azeotrope is a useful point to fit parameters.

Example 11.7 Azeotrope fitting with bubble-temperature calculations

Consider the benzene(1) + ethanol(2) system which exhibits an azeotrope at 760 mmHg and 68.24°C containing 44.8 mole% ethanol. Using the two-parameter Margules model, calculate the composition of the vapor in equilibrium with an equimolar liquid solution at 760 mmHg given the following Antoine constants:

$\log P_1^{sat} = 6.87987 - 1196.76/(T + 219.161)$
$\log P_2^{sat} = 8.1122 - 1592.86/(T + 226.18)$.

Solution: At $T = 68.24$°C, $P_1^{sat} = 519.7$ mmHg; $P_2^{sat} = 503.5$ mmHg, and the azeotrope composition is known, $x_1 = 0.552$; $x_2 = 0.448$. At this composition, the activity coefficients can be calculated.

$$\gamma_1 = (y_1 P)/(x_1 P_1^{sat}) = P/P^{sat} = 760/519.7 = 1.4624; \text{ likewise, } \gamma_2 = 760/503.5 = 1.5094$$

Positive deviations from Raoult's law, $\gamma_i > 1$.

Using Eqn. 11.38 with the composition and γ's just tabulated, $A_{12} = 1.2947$, $A_{21} = 1.8373$.
New activity coefficient values must be found at the composition, $x_1 = x_2 = 0.5$. Using Eqn. 11.37, $\gamma_1 = 1.583$; $\gamma_2 = 1.382$. The problem statement requires a bubble-temperature calculation. Using the method of Table 10.1 (a flow sheet is available in Appendix C, option (a); a MATLAB example is provided in Ex11_07.m),

Bubble-temperature calculation.

Ex11_07.m.

Guess $T = 60$°C $\Rightarrow P_1^{sat} = 391.63$ mmHg; $P_2^{sat} = 351.82$ mmHg. For this model, the activity coefficients do not change with temperature. The K-ratio depends on the activity coefficients:

$$y_i = x_i \gamma_i P_i^{sat}/P \Rightarrow y_1 = 0.408; y_2 = 0.320;$$

Checking the sum of y_i, $\sum_i y_i = 0.728 \Rightarrow T_{guess}$ is too low. Try a higher T.

After a few trials, at $T = 68.262$°C, $P_1^{sat} = 520.13$ mmHg; $P_2^{sat} = 504.1$ mmHg

$$y_i = x_i \gamma_i P_i^{sat}/P \Rightarrow y_1 = 0.542; y_2 = 0.458; \sum_i y_i = 1 \Rightarrow T_{guess} \text{ is } T_{azeotrope}.$$

Note: *The bubble temperatures at $x_1 = 0.55$ and 0.5 are almost the same. The T-x diagram is quite flat near an azeotrope. This has an important effect on temperature profiles in distillation columns.*

---

12. Gmehling, J. 1991. *Azeotropic Data.* Frankfort, Germany: DECHEMA Press; Weast, R.C. 2001. *Handbook of Chemistry and Physics.* Boca Raton, FL: CRC.

Purity and Infinite Dilution

A component is said to be infinitely dilute when only a trace is present. Thus, when a binary mixture is nearly pure in component 1, it is infinitely dilute in component 2. The activity coefficients take on special values at purity and infinite dilution.

$$\lim_{x_i \to 1} \gamma_i = 1 \quad \text{and} \quad \lim_{x_i \to 0} \gamma_i = \gamma_i^{\infty} \qquad\qquad 11.39$$

Find these limiting values in Fig. 11.5. As an example, consider the infinite dilution composition limits of Eqn. 11.37, $A_{12} = \ln \gamma_1^{\infty}$, $A_{21} = \ln \gamma_2^{\infty}$. Infinite dilution activity coefficients are sometimes available in the literature and can be useful for fitting if no data are available near the composition range of interest, but it should be recalled that extrapolations are less reliable than interpolations. In other words, one might experience significant errors in predictions of bubble pressures near equimolar compositions when basing parameters on infinite dilution activity coefficients. The same principle can be used with other activity coefficient models. Infinite dilution activity coefficients are especially important in applications requiring high purity. In those cases, several stages may be required in going from 99% to 99.999% purity.

11.8   PRELIMINARY INDICATIONS OF VLLE

Recall from Section 10.7 that azeotropes occur at $x=y$, where a maximum or minimum appears in all the plots. Also note that the bubble and dew lines do not cross, but they touch at the azeotrope composition. Occasionally when a P-x-y or T-x-y diagram is generated in a Stage III calculation, the diagram can look very odd. The two-parameter fit in Fig. 11.5 was generated using $A_{12} = 1.99$, $A_{21} = 1.09$ as fitted in Example 11.5. Suppose, due to a slight calculation error or programming typo, we generated a diagram using parameters $A_{12} = 2.99$, $A_{21} = 1.09$. The predicted phase diagram and y-x diagram would look like those shown in Fig. 11.10.

The behavior of the lines using these parameters actually predicts that two liquid phases exist. However, the diagram requires additional modification before coexisting compositions and the vapor-liquid-liquid equilibria (VLLE) can be read from the diagram. It is important to understand that the diagram has been generated assuming that only one liquid phase exists. Though we started the discussion by assuming that a parameter calculation error resulted in predictions, all systems that exhibit VLLE will have similarly odd diagrams when only one liquid phase is assumed to exist. This assumption is the default in common process simulators such as ASPEN Plus and ChemCAD because the calculations are faster when the simulator can avoid checking for two phases. When working with simulators, you should check the phase diagrams to see if liquid-liquid phase behavior exists and you should understand where to change the simulator settings to calculate liquid-liquid behavior when it exists. Within this chapter, you should be ready to recognize that such diagrams are indicative of two liquid phases. Also recall that a T-x-y diagram qualitatively resembles an inverted P-x-y, so peculiar loops appear on a T-x-y diagram if a similar situation exists. When models incorrectly predict VLLE behavior that we know to be incorrect, we need to check our calculations. We learn how to rigorously characterize VLLE phase diagrams and how to eliminate the loops in Chapter 14.

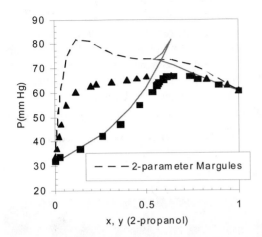
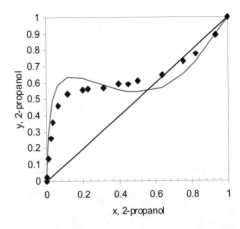

Figure 11.10 *Phase diagram calculations for the 2-propanol + water system at 30°C compared with data cited in Example 11.2. The parameters where selected as described in the text to illustrate how a numerical error can result in thermodynamically unstable loops. Note the dew line has a has a loop and the maximum in the bubble line is not at the azeotropic condition. Note in the y-x plot that the coexistence curve has maxima and minima. These calculated conditions are indicative of LLE as discussed in the text, though the experiments do not show LLE.*

11.9   FITTING ACTIVITY MODELS TO MULTIPLE DATA

Fitting of the Margules equations to limited data has been discussed in Examples 11.2 and 11.5. Fits to multiple points are preferred, which requires regression of the parameters to optimize the fit. In a few cases, the Gibbs excess function can be rearranged to form the basis for a linear regression. In general, a non-linear regression may be required. Modern computers facilitate either method.

Linear Fitting of the Margules Equation

Eqn. 11.33 can be linearized:.

$$\frac{G^E}{x_1 x_2 RT} = (A_{21} - A_{12})x_1 + A_{12}$$

11.40

Margules models can be linearized for fitting of parameters.

Therefore, plotting $G^E/(x_1 x_2 RT)$ versus x_1 gives a slope of $(A_{21} - A_{12})$ and an intercept of A_{12}. The value of A_{21} can also be determined by the value at $x_1 = 1$. Using the data for 2-propanol + water listed in Example 11.8 (ignoring the first mixture point) results in a slope = –0.9289, intercept = 2.001. Thus, $A_{12} = 2.001$, $A_{21} = 1.072$, slightly different from the single point fit of Example 11.5.

Nonlinear fitting techniques

Nonlinear parameter fitting is possible in Excel and MATLAB.

In general, parameters for excess Gibbs models are nonlinearly related to G^E or γ. Even in the cases of the Redlich-Kister and Margules equations, it may be more convenient to simply apply a nonlinear fitting procedure. The parameters can be fitted to the experimental data to optimize the fit to the experimental bubble pressure (*P-x-y* data) or bubble temperature (*T-x-y* data). In the case of fitting bubble-pressure data, the parameters can be used to generate a bubble-pressure calculation at each experimental x_i as demonstrated with the one-parameter Margules in Example 11.3. The Excel Solver tool or the MATLAB fminsearch or lsqnonlin can provide rapid fits. The spreadsheet Gammafit.xlsx or MATLAB m-file GammaFit.m permit nonlinear fitting of activity coefficient parameters for the Margules equation by fitting total pressure. Either can be easily modified to find parameters for any activity coefficient model. The strategy implemented here is to calculate the activity coefficients using assumed values for the parameters and generate a bubble-pressure calculation using Eqn. 11.2 for each of the experimental points. Then each bubble pressure is compared to the experimental bubble pressure, and $OBJ = \sum_{\text{all points}} (P_{\text{expt}} - P_{\text{calc}})^2$ is calculated. The optimizer is invoked to keep adjusting the parameters until OBJ is minimized.

Example 11.8 Fitting parameters using nonlinear least squares

Gammfit.xlsx, Gammafit.m.

Measurements for the 2-propanol + water system at 30°C have been published by Udovenko, et al. (1967).[a] Use the pressure and liquid composition to fit the two-parameter Margules equation to the bubble pressure. Plot the resultant P-x-y diagram.

Solution: In the experimental data, the researchers report experimental vapor pressures. It is best to use experimental values from the same publication to reduce the effect of systematic errors which may be present in the data due to impurities or calibration errors. The solution will be obtained by minimizing the sum of squares of error for the bubble pressures across the composition range.

MATLAB (condensed to show the major steps):

```
function GammaFit()
% statements omitted to load experiments into matrix 'Data'
x1 = Data(:,1); %data have been entered into columns of 'Data'
y1expt = Data(:,2); Pexpt = Data(:,3);
Ps1Calc = 60.7; Ps2Calc = 32.1; %experimental values used for Psat
x2 = 1-x1; % calculate x2
x  = [x1  x2];  % create a 2 column matrix of x1 & x2
A = [0 0];  % initial guess for A12 and A21
A = lsqnonlin(@calcError,A); %optimize, calling 'calcError' as needed
    function obj = calcError(A)
        A12 = A(1); %extract coeffs so eqns look like text
        A21 = A(2);
        Gamma1Calc = exp((x2.^2).*(A12 + 2* (A21 - A12).*x1));
        Gamma2Calc = exp((x1.^2).*(A21 + 2* (A12 - A21).*x2));
        Pcalc = (x1.*Gamma1Calc)*(Ps1Calc) + ...
                (x2.*Gamma2Calc)*(Ps2Calc);
        obj = Pcalc - Pexpt;
    end
```

The resultant parameters are $A_{12} = 2.173$, $A_{21} = 0.9429$. The distributed file includes statements to plot the final figure similar to that shown below. Note that fminsearch can be used if lsqnonlin is not available due to the toolboxes on your MATLAB installation. See the fit in Fig. 11.11.

Example 11.8 Fitting parameters using nonlinear least squares (Continued)

Excel: The spreadsheet "P-x-y fit P" in the workbook Gammafit.xlsx is used to fit the parameters as shown below. Antoine coefficients are entered in the table for the components shown at the top of the spreadsheet. The flag in the box in the center right determines whether experimental vapor pressures are used in the calculations or values calculated from the Antoine equation.

System Components		Parameters to adjust				Antoine Coefficents			Calculated P^{sat}(mm Hg)	Expt P^{sat}(mm Hg)	Selected P^{sat}(mm Hg)
		A_{12}	A_{21}	T(C)		A	B	C			
(1) 2-propanol					1	8.87829	2010.33	252.636	58.277622	60.7	60.7
(2) water		2.173055	0.942929	30	2	8.07131	1730.63	233.426	31.740167	32.1	32.1

| | | | | <---optional-----> | | | | | | | |

x_1	x_2	$\gamma_{1,calc}$	$\gamma_{2,calc}$	$y_{1,expt}$	$y_{2,expt}$	$y_{1,calc}$	$y_{2,calc}$	P_{expt}	P_{calc}	$(P_{error})^2$	
0	1	8.785079	1	0	1	0	1	32.1	32.1	0	Enter 1 to use Calculated P^{sat}
0.0015	0.9985	8.695982	1.000008	0.0254	0.9746	0.024107	0.975893	33.8	32.84386	0.9141954	Enter 0 to use Expt P^{sat}
0.0111	0.9889	8.152908	1.000416	0.1374	0.8626	0.147468	0.852532	37.1	37.25008	0.0225244	0
0.0231	0.9769	7.535179	1.001787	0.2603	0.7397	0.251681	0.748319	42.3	41.98014	0.102312	
0.0357	0.9643	6.95177	1.004234	0.3577	0.6423	0.326426	0.673574	47.2	46.14952	1.1035166	
0.0649	0.9351	5.815504	1.013755	0.4604	0.5396	0.42951	0.57049	55	53.33938	2.7576741	Objective Function
0.1168	0.8832	4.353254	1.043423	0.5316	0.4684	0.510602	0.489398	60.3	60.44532	0.0211192	14.26798435
0.197	0.803	2.970369	1.119928	0.5547	0.4453	0.551655	0.448345	62.9	64.38697	2.2110888	
0.2271	0.7729	2.62311	1.158008	0.5611	0.4389	0.557245	0.442755	63.5	64.88977	1.9314539	
0.312	0.688	1.945005	1.292474	0.5659	0.4341	0.563409	0.436591	64.4	65.3793	0.9590351	
0.3958	0.6042	1.549309	1.463142	0.5907	0.4093	0.567416	0.432584	65.1	65.59962	0.2496163	
0.4477	0.5523	1.386629	1.586209	0.589	0.411	0.572644	0.427356	65.8	65.80383	1.468E-05	
0.5009	0.4991	1.264065	1.724032	0.6098	0.3902	0.581846	0.418154	66.6	66.05434	0.2977452	
0.6369	0.3631	1.083192	2.106142	0.6462	0.3538	0.630433	0.369567	66.9	66.42417	0.2264146	
0.7542	0.2458	1.01937	2.411709	0.7296	0.2704	0.710348	0.289652	66.8	65.69551	1.2198966	
0.8245	0.1755	1.004463	2.546001	0.7752	0.2248	0.778018	0.221982	65.7	64.61353	1.1804165	
0.9363	0.0637	0.999471	2.622202	0.8892	0.1108	0.913749	0.086251	63.2	62.16513	1.0709608	
1	0	1	2.567492	1	0	1	0	60.7	60.7	0	

Experimental data for x_1 and P_{expt} are entered in columns A and I. Initial guesses for the constants A_{12} and A_{21} are entered in the labeled cells in the top table. Solver is then called to minimize the error in the objective function by adjusting the two parameters. Calculated pressures are determined by bubble-pressure calculations.

The results of the fit are shown by the plot on spreadsheet "P-x-y Plot." See the fit in Fig. 11.11.

Note that the system is the same used in Example 11.2 on page 417 and Example 11.5 on page 430. The fit in this example using all data is superior. The parameters are also slightly different from the linear fit discussed above because the objective function is different.

a. Udovenko, V.V., Mazanko, T.F. 1967. *Zh. Fiz. Khim.* 41:1615.

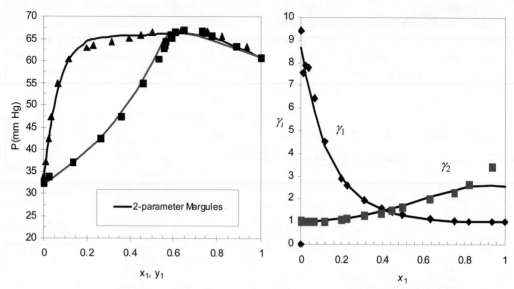

Figure 11.11 *Comparison of experimental data with regressed model as explained in Example 11.8.*

Alternative Objective Functions

An alternative choice of objective function for a given set of data usually results in a slightly different set of parameters. While the total pressure is often measured accurately, it may be desired to include vapor compositions in the objective function. For example, it is not uncommon for the pure component vapor pressures measured by investigators to differ from literature data; this is indicative of an impurity or a systematic error. One method of incorporating additional considerations into the fitting procedure is to use weighted objective functions, where recognition is made of probable errors in measurements. One of the most rigorous methods uses the maximum likelihood principles, which asserts that all measurements are subject to random errors and therefore have some uncertainty associated with them. Such techniques are discussed by Anderson, et al.,[13] and Prausnitz, et al.[14] The objective function for such an approach takes the form

$$\sum_{points} \left\{ \frac{(P_{true} - P_{expt})_i^2}{\sigma_P^2} + \frac{(T_{true} - T_{expt})_i^2}{\sigma_T^2} + \frac{(x_{1\,true} - x_{1\,expt})_i^2}{\sigma_x^2} + \frac{(y_{1\,true} - y_{1\,expt})_i^2}{\sigma_y^2} \right\} \qquad 11.41$$

where σ represents the variance for each type of measurement. The "true" values are calculated as part of the procedure. Typical values for variances are: $\sigma_P = 2$ mmHg, $\sigma_T = 0.2$ K, $\sigma_x = 0.005$, $\sigma_y = 0.01$, and therefore the weight of a measurement depends on the probable experimental error in the value.

13. Anderson, T.F., Abrams, D.S., Grens, E.A. 1978. *AICHE J.* 24:20.
14. Prausnitz, J., Anderson, T., Grens, E., Eckert, C., Hsieh, R., O'Connell, J. 1980. *Computer Calculations for Multicomponent Vapor-Liquid Equilibria,* Upper Saddle River, NJ: Prentice-Hall.

11.10   RELATIONS FOR PARTIAL MOLAR PROPERTIES

Gibbs-Duhem Equation

A useful expression known as the Gibbs-Duhem equation results when we analyze Eqn. 10.40 together with Eqn. 10.42. Consider the differential of Eqn. 10.42 using the product rule,

$$dG = \sum_i \mu_i dn_i + \sum_i n_i d\mu_i \qquad\qquad 11.42$$

Substituting for dG in Eqn. 10.40 results in

$$\sum_i \mu_i dn_i + \sum_i n_i d\mu_i = -SdT + VdP + \sum_i \mu_i dn_i \qquad\qquad 11.43$$

Simplifying, we obtain the Gibbs-Duhem equation,

$$\boxed{0 = -SdT + VdP - \sum_i n_i d\mu_i} \qquad\qquad 11.44$$

Therefore, we conclude at constant T and P:

$$\sum_i n_i d\mu_i = 0 \ \text{ at constant } T \text{ and } P \qquad\qquad 11.45$$

The relation is typically applied in the context of activity coefficients, as described below.

Gibbs-Duhem Relation for Activity Coefficients

To extend the Gibbs-Duhem equation to excess properties, the excess Gibbs energy can be manipulated in an manner analogous to the derivation above. Therefore,

$$0 = -S^E dT + V^E dP - \sum_i n_i d\mu_i^E \qquad\qquad 11.46$$

resulting in

$$\sum_i n_i d\mu_i^E = 0 \ \text{ at constant } T \text{ and } P \qquad\qquad 11.47$$

Inserting the relation between excess chemical potential and activity coefficients gives

$$\boxed{\sum_i n_i d\mu_i^E = RT \sum_i n_i d\ln\gamma_i = 0} \qquad\qquad 11.48$$

Technically it is not possible to vary composition for two coexisting phases in a binary without either T or P changing. However, experimental analysis of isothermal P-x-y data or isobaric T-x-y data shows that the $\underline{S}^E$ and $\underline{V}^E$ terms are almost always very small compared to the last term and Eqn. 11.48 is generally an excellent approximation. If we apply the expansion rule, Eqn. 6.17, using one of the mole fractions, and divide by n, in a binary

$$\boxed{x_1\left(\frac{\partial \ln \gamma_1}{\partial x_i}\right)_{T,P} + x_2\left(\frac{\partial \ln \gamma_2}{\partial x_i}\right)_{T,P} = 0} \quad \text{binary at fixed } T \text{ and } P \qquad 11.49$$

This equation means that the activity coefficients for a binary system, when plotted versus composition, must have slopes with opposite signs, and the slopes are related in magnitude by Eqn. 11.49. A further deduction is that if one of the activity coefficients in a binary system exhibits a maximum, the other must exhibit a minimum at the same composition. We find this relation useful in: 1) testing data for experimental errors (grossly inconsistent data); 2) generating the activity coefficients in a binary for a second component based on the behavior of the first component in experimental techniques where only one activity coefficient is measured; 3) for development of theories for the Gibbs energy of a mixture, since our model must follow this relation. The Gibbs-Duhem equation is also useful for checking thermodynamic consistency of data; however, the applications are subject to uncertainties themselves because the activity coefficient is itself derived from assumptions, (e.g., modified Raoult's law).[15] Fortunately, developers of activity coefficient models are generally careful to ensure the models satisfy the Gibbs-Duhem equation. An understanding of the restrictions of the Gibbs-Duhem equation is helpful when studying alternative standard states such as Henry's law or electrolyte models. A particularly useful application of the Gibbs-Duhem equation in a binary mixture is the use of the activity coefficient of one component to calculate the activity coefficient of the other component.[16] Often this can be done by fitting a model to the activity of the first component, but the Gibbs-Duhem equation provides a method that does not require the application of a particular mixture model with its associated assumptions.

Relations between Various Excess Properties

For the development of accurate process calculations, a thermodynamic model should accurately represent the temperature and pressure dependence of deviations from ideal solution behavior. The excess functions follow the same relations as the total functions, $H^E = U^E + PV^E$, $G^E = H^E - TS^E$, and $A^E = U^E - TS^E$. The derivative relations are also followed,

$$\left(\frac{\partial G^E}{\partial T}\right)_{P,x} = -S^E \, ; \quad \left(\frac{\partial \overline{G}_i^E}{\partial T}\right)_{P,x} = -\overline{S}_i^E \qquad 11.50$$

$$\left(\frac{\partial G^E}{\partial P}\right)_{T,x} = V^E \, ; \quad \left(\frac{\partial \overline{G}_i^E}{\partial P}\right)_{T,x} = \overline{V}_i^E \text{ for } P^\circ = P_{\text{system}} \qquad 11.51$$

resulting in

---

15. Van Ness, H.C., Byer, S.M., Gibbs, R.E. 1973. *AIChE J.* 19:238.

16. Examples of techniques that measure only one activity are osmotic pressure, partial pressure of solvent over a non-volatile polymer solution, the isopiestic method for measuring solvent activity in electrolyte systems, and electrochemical emf techniques in liquid metal solutions.

$$\left(\frac{\partial \ln \gamma_i}{\partial P}\right)_{T,x} = \frac{\bar{V}_i^E}{RT} \text{ for } P^\circ = P_{\text{system}} \qquad 11.52$$

The **Gibbs-Helmholtz** relation applies:

$$\left(\frac{\partial G^E/T}{\partial T}\right)_{P,x} = \frac{-H^E}{T^2} \ ; \ \left(\frac{\partial \bar{G}_i^E/T}{\partial T}\right)_{P,x} = \frac{-\bar{H}_i^E}{T^2} \qquad 11.53$$

Particularly useful is Eqn. 11.53 using the relation with activity coefficients:

$$\left(\frac{\partial \ln \gamma_i}{\partial T}\right)_{P,x} = \frac{-\bar{H}_i^E}{RT^2} \qquad 11.54$$

Therefore, excess enthalpy data from calorimetry may be used to check the temperature dependence of the activity coefficient models for thermodynamic consistency. Typically, activity coefficient parameters need to be temperature-dependent for representing data accurately, which implies an excess enthalpy. Likewise, any system with a heat of mixing will have temperature-dependent activity coefficients. A simple model modification is to replace the parameters with functions, for example, $A_{ij} = a_{ij} + b_{ij}/T$, where a_{ij} and b_{ij} are constants and T is in K. This sets the stage for computing heats of mixing of any activity model, as shown in Example 11.9.

Example 11.9   Heats of mixing with the Margules two-parameter model

Fitting the VLE of methanol + benzene[a] in the range of 308–328 K with the Margules two-parameter model and then fitting the parameters to $A_{ij} = a_{ij} + b_{ij}/T$ gives $A_{12} = 0.1671 + 714/T$ and $A_{21} = 2.3360 - 247/T$. Estimate the heat of mixing at 318 K and 50 mol% benzene.

Solution: The Margules two-parameter model is,

$$\frac{G^E}{RT} = x_1 x_2 (A_{21} x_1 + A_{12} x_2)$$

The relation between G^E and H^E is given by Eqn. 11.53. Noting the right side Margules parameters for the problem statement are simple functions of $(1/T)$, we can manipulate the derivative for this function. Since $d(1/T) = -T^{-2} dT$,

$$((\partial G^E/(RT))/(\partial T)) = -T^{-2}((\partial G^E/(RT))/(\partial(1/T))) = -H^E/(RT^2) \qquad 11.55$$

Thus,

$$\frac{H^E}{RT} = \frac{x_1 x_2}{T}\left(\frac{x_1 dA_{21}}{d(1/T)} + \frac{x_2 dA_{12}}{d(1/T)}\right) = x_1 x_2\left(x_1 \frac{714}{T} - x_2 \frac{247}{T}\right)$$

At 318 K and $x_1 = x_2 = 0.5$, $H^E = 8.314(0.5)(0.5(0.5\cdot714 - 0.5\cdot247)) = 485$ J/mol. Note that direct measurement of excess enthalpy is recommended when possible. Phase equilibria data must be very precise to provide an accurate enthalpy of mixing.

---

a. Gmehling J., et al., 1977-. *VLE Data Collection*. Frankfurt/Main: DECHEMA; Flushing, N.Y.: Distributed by Scholium International.

This example illustrates what it means for the "activity coefficient parameters to be temperature-dependent" and the manner of taking the derivative of G^E. Though we have calculated the excess Gibbs energy in the previous example, the parameters may also be used to calculate temperature dependence of activity coefficients. Activity coefficients are strong functions of composition but weak functions with respect to temperature. This becomes apparent as you study more systems.

Recalling that heat of mixing for an ideal solution is zero, we note that the heat of mixing and the excess enthalpy are one and the same. The heat capacity for liquid methanol is about 80 J/mol-K and about 130 J/mol-K for benzene. In an adiabatic mixing process, we would thus expect this equimolar mixture to be colder after mixing by roughly 5°C. (Note: The excess heat capacity, $C_p^E = C_p - C_p^{is}$, is the temperature derivative of the excess enthalpy; can you determine its value for this mixture?) In general, we could write that the enthalpy of any given composition is given by,

$$\frac{H}{RT} = \frac{x_1 H_1}{RT} + \frac{x_2 H_2}{RT} + x_1 x_2 \left(x_1 \frac{714}{T} - x_2 \frac{247}{T} \right) \qquad 11.56$$

In this way, we can represent the enthalpy of any stream to perform energy balances.

11.11   DISTILLATION AND RELATIVE VOLATILITY OF NONIDEAL SOLUTIONS

To illustrate the impact of activity coefficients on practical applications, it is helpful to revisit our discussion of distillation. The relative volatility of the light to heavy key, α_{LH}, is important to distillation, as discussed in Section 10.6. Since α_{LH} may not be constant over an entire distillation column, it is common to estimate the average value by the geometric mean of the bottom and top.

$$\alpha_{LH}^{m} = (\alpha_{LH}^{top} \alpha_{LH}^{bot})^{1/2} \qquad 11.57$$

Recalling the definition of α_{LH} from Eqn. 10.32, substituting Eqn. 11.18, and canceling pressures,

$$\alpha_{LH} = K_{LK}/K_{HK} = (\gamma_{LK} P_{LK}^{sat})/(\gamma_{HK} P_{HK}^{sat}) \qquad 11.58$$

Suppose in a binary mixture that we specify splits so that the top is $x_{LK}^{top} = 0.99$, and $x_{LK}^{bot} = 0.01$. Then recognizing that the activity coefficients go to unity near purity,

$$\alpha_{LH}^{top} = [(P_{LK}^{sat})/(\gamma_{HK}^{\infty} P_{HK}^{sat})]^{top} \qquad \alpha_{LH}^{bot} = [(\gamma_{LK}^{\infty} P_{LK}^{sat})/(P_{HK}^{sat})]^{bot} \qquad 11.59$$

Note: It is required that $\alpha_{LH} > 1$ at both ends of the column in order to avoid an azeotrope. In other words, Eqns. 10.35 and 11.57 CANNOT be applied unless $\alpha_{LH} > 1$ at both ends of the column.

Example 11.10   Suspecting an azeotrope

Make a preliminary estimate of whether we should suspect an azeotrope in the system benzene (B) + 2-propanol (I) at 80°C. Assume the MAB model. A convenient feature of Margules one-parameter models (including the MAB model) is that the infinite dilution activity coefficients are equal. (Note that "convenient" may not equate to "accurate.")

Example 11.10   Suspecting an azeotrope (Continued)

Solution: Note that this problem is isothermal rather than a distillation column design, but we can evaluate the relative volatility at either end of the composition range. Antoine.xlsx gives vapor pressures of $P_B^{sat} = 757$ mmHg and $P_I^{sat} = 683$ mmHg at 80°C, so benzene is the *LK*. For the MAB model,

$$A_{12} = (9.23 - 0.63)(11.86 - 2.24)(89.8 + 76.8)/[4(8.314)353] = 1.174;$$

$$\gamma_i^\infty = \exp(1.174) = 3.235$$

Using the component key assignments, $P_{LK}^{sat}/P_{HK}^{sat} = 757/683 = 1.108$. Therefore, at the end rich in *LK*, $\alpha_{LH} = (P_{LK}^{sat})/(\gamma_{HK}^\infty P_{HK}^{sat}) = 1.108/3.235 = 0.343$, and the end rich in *HK*, $\alpha_{LH} = (\gamma_{LK}^\infty P_{LK}^{sat})/(P_{HK}^{sat}) = 3.235 \cdot 1.108 = 3.58$. MAB predicts an azeotrope since $(\alpha_{LH} - 1)$ changes sign. The prediction should be validated with experimental data and/or more accurate models because of the approximations in the MAB model.

11.12   LEWIS-RANDALL RULE AND HENRY'S LAW

In this chapter, we have thus far introduced the standard state using the pure component properties at the state of the system (e.g., same *T, P,* and liquid state). What if the pure liquid substance does not exist at these conditions? For example, in liquid-phase hydrogenation reactions, H_2 is far above its critical temperature, yet exists in liquid solution at small concentrations. A pure standard state of liquid H_2 is impractical. Similarly, salts dissolve as ions in aqueous solution, but the ions cannot exist as pure liquids. A model for dilute liquid solutions would be convenient, particularly if it is possible to model the liquid as some type of ideal solution. To develop models for this behavior, we first consider the general compositional behavior of the component fugacities. Based on these observations, we introduce Henry's law to model the solution behavior relative to an ideal solution at dilute concentrations.

The Henry's Law Standard State

Consider the shape of the component-2 fugacity versus composition that results when Eqn. 11.13 is used along with an activity coefficient model developed in this chapter. If the liquid-phase model parameters provide positive deviations from Raoult's law, the shape of the component fugacity curve is represented by the curve in Fig. 11.12. In the figure, we follow the widely used convention for dilute binary solutions, where the solvent is designated as component-1, and the dilute solute is component-2. The use of a pure component property as a standard state creates an ideal solution line (Eqn. 10.68) known as the **Lewis-Randall rule** ideal solution line. Raoult's law is a special case of the Lewis-Randall ideal solution where we use the vapor pressure to approximate the standard state fugacity.[17] Thus, the activity coefficient models that we have developed previously are relative to the Lewis-Randall ideal solution. The activity coefficients follow the symmetrical normalization convention because $\lim_{x_i \to 1} \gamma_i = 1$ for all *i* (*cf.* Fig. 11.5). For a Lewis-Randall ideal solution, the activity coefficients approach one as the concentration approaches purity for that component, and the activity coefficients are usually farthest from unity at infinite dilution.

---

17. Do not be confused that we discuss Henry's "*law,*" Raoult's "*law,*" and the Lewis-Randall "*rule.*" The designations as "law" or "rule" are purely historical names, and are simply different perspectives on modeling the real solution.

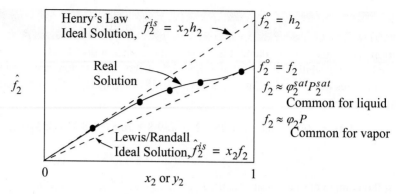

Figure 11.12 *Schematic representation of the fugacity of component 2 in a binary mixture.*

Consider that the fugacity curve in Fig. 11.12 is nearly linear at low concentrations. Thus, we could express the component fugacity as proportional to concentration using a tangent line near infinite dilution,

❶ Ideal Henry's law component.

$$\hat{f}_i^{is} = x_i h_i \qquad 11.60$$

which is the behavior of an ideal solution given by Henry's law.

The **Henry's law constant,** h_i, is usually determined experimentally, and depends on temperature, pressure, and solvent. The fact that it depends on solvent makes it very different from a pseudo-vapor pressure because a vapor pressure would be independent of solvent.

Looking at Fig. 11.12, note that Henry's law fails at high concentrations of a component unless an activity coefficient method is developed. Introducing a Henry's law activity coefficient to represent non-idealities,

❶ Henry's law for non-ideal solution.

$$\boxed{\hat{f}_i = x_i \gamma_i^* h_i} \qquad 11.61$$

where γ_i^* is the Henry's law activity coefficient. For the fugacity curve shown in Fig. 11.12 the Henry's law activity coefficient needs to be *less than one* at high concentrations, and the Henry's law activity coefficient goes to one at infinite dilution (compare relative to the Henry's law ideal solution line). In contrast, Fig. 11.12 shows that the activity coefficient relative to the Lewis-Randall ideal solution would be *greater than one*. This can be confusing if you have grown accustomed to activity coefficients less than one meaning that the components "like" each other. The component will have negative deviations from Henry's law and positive deviations from the Lewis-Randall rule.

Although the Henry's law activity coefficient goes to one at infinite dilution, it is inaccurate to designate the Henry's law standard state at that composition. In fact, the correct standard state designation is a hypothetical *pure* component fugacity (often not experimentally accessible) selected in a manner such that the infinite dilution activity coefficient goes to one. Applying the definition of activity, Eqn. 11.23, we see that $a_i = x_i \gamma_i^* = \hat{f}_i / f_i^\circ$. Comparing with Eqn. 11.61, we see that the standard state is $f_i^\circ = h_i$. So the important activity coefficient value is at infinite dilution, but the standard state composition is a hypothetical pure state. This perspective is especially useful for electrolyte solutions.

Relating γ_i for Henry's Law and Lewis-Randall Rule

Note in Fig. 11.12 that both Henry's approach and the Lewis-Randall approach must represent the same fugacity. Equating the two approaches,

$$\hat{f}_i = x_i \gamma_i^* h_i = x_i \gamma_i f_i \qquad\qquad 11.62$$

Taking the limit at infinite dilution where γ_2^* approaches one, and we see

$$\hat{f}_i = x_i h_i = x_i \gamma_i^\infty f_i \text{ at infinite dilution} \qquad\qquad 11.63$$

resulting in the relation between the Henry's and the Lewis-Randall fugacity and activity coefficient,

$$\boxed{h_i = \gamma_i^\infty f_i} \qquad\qquad 11.64$$

Relation between Henry's Law constant and Lewis-Randall fugacity.

In Fig. 11.13, look at the right side where $\ln\gamma_2$ approaches $\ln\gamma_2^\infty$ and $\ln\gamma_2^*$ approaches zero. The difference in the intercept at $x_2 = 0$ is $\ln(h_2/f_2)$. To model the Henry's law activity coefficient, the restriction that the activity coefficients must follow the Gibbs-Duhem Eqn. 11.49 remains; thus, the slope of the logarithm of the Henry's law activity coefficient must be the same as the slope of the logarithm of the Lewis-Randall activity coefficient—the shift is independent of composition. The shift is illustrated in Fig. 11.13. We may adapt any activity model developed for the Lewis-Randall rule to Henry's law by shifting the intercept values for the components modeled by Henry's law. Thus,

$$\boxed{\ln\gamma_i^* = \ln\gamma_i - \ln\gamma_i^\infty} \qquad\qquad 11.65$$

Formula to shift a Lewis-Randall activity model to a Henry's law activity model.

where any Lewis/Randall model can be used for γ_i and the same model is used for γ_i^∞. Usually the activity coefficient model is manipulated to obtain the infinite dilution activity coefficient

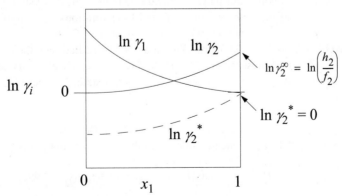

Figure 11.13 *Schematic illustration of the relation of the Henry's law activity coefficient compared to the Lewis-Randall rule activity coefficient.*

expressed in terms of the activity model parameters, and the difference is expressed analytically. For the one-parameter Margules equation, $\ln \gamma_i^{\infty} = A_{12}$. If applied to component 2 (dilute solute) of a binary, with component 1 (rich solvent) represented by the Lewis-Randall rule,

Unsymmetric activity coefficients for Henry's law based on the one-parameter Margules equation.

$$\ln \gamma_2^* = \ln \gamma_2 - \ln \gamma_2^{\infty} = A_{12}x_1^2 - A_{12} = A_{12}(x_1^2 - 1) \quad \ln \gamma_1 = A_{12}(x_2^2) \qquad 11.66$$

Readers should recognize that we have been careful to distinguish between the two standard states in the presentation here, including distinct symbols for the different activity coefficients. When one component is represented with Henry's law and the other represented by the Lewis-Randall rule, the overall model is described as using the **unsymmetrical normalization convention** for the activity coefficients.

Henry's Law on Molal Activity Scale

Eqn. 11.61 suggests that the units for the Henry's law constant should be pressure, but other conventions also exist. For example, a common way of presenting Henry's constants for gases is to express the liquid phase concentration in **molality** and provide a constant inverted relative to h_i. The result is

Molal activity coefficients and Henry's law.

$$\hat{f}_i = x_i \gamma_i^* h_i = m_i \gamma_i^{\square}/(K_H) \qquad 11.67$$

where the change in units for concentration requires a change in the activity coefficient and a change in units of the Henry's law constant.[18] Details on the molal activity coefficient $\gamma_i^{\square}$ are deferred until Chapter 18, but like γ_i^* it goes to one at infinite dilution. For a gas phase component, we have seen that $\hat{f}_i^V = \hat{f}_i^L$, and we may use the vapor phase fugacity in the Henry's law calculation. Many Henry's law constants in the NIST Chemistry WebBook follow the K_H convention for molality concentration units. The relation between molality and mole fraction in water is $x_i = m_i x_w 0.001 M_{w,w}$ as we derive later (cf. Eqn. 18.147) where $M_{w,w}$ is the molecular weight of water, and x_w is the mole fraction of water.

Example 11.11 Solubility of CO_2 by Henry's Law

Carbon dioxide solubility in water plays a critical role in biological physiology and environmental ocean chemistry, affects the accuracy of acid-base titrations in analytical chemistry, and makes many beverages fizzy. The Henry's law constant for CO_2 in water is listed on the NIST Chemistry WebBook[a] as $K_H = 0.035$ mol/kg-bar at 298.15 K. Estimate the mole fraction of CO_2 in water at 0.7 MPa total pressure and 298.15 K. Treat the vapor phase as an ideal gas and the liquid as an ideal solution with the Henry's law standard state. Neglect formation of ionic carbonate species.

Solution: $K_H y_{CO_2} P = m_{CO_2}$ where m_{CO_2} is mol/(kgH$_2$O). First find y_{CO_2} by using Raoult's law for water. Taking P_w^{sat} from the steam tables, Raoult's law for water gives

$$y_w = P_w^{sat}/P = 0.00317/0.7 = 0.0045 \Rightarrow y_{CO_2} = 0.9955.$$

---

18. The reason that the Henry's law constant appears to be inverted is because the solubilization is represented as an equilibrium constant for a "reaction" where the dissolved species is the "product" and the vapor phase species is the "reactant."

Example 11.11 Solubility of CO_2 by Henry's Law (Continued)

The solubility of CO_2 is thus,

$$K_H y_{CO_2} P = 0.035(\text{mol/kg-bar})(0.9955)(7 \text{ bar}) = 0.244 \text{ molal}$$

$$x_{CO_2} = m_{CO_2} x_w 0.001 M_{w,w} = m_{CO_2} 0.001 M_{w,w}(1 - x_{CO_2}) \Rightarrow$$

$$x_{CO_2} = \frac{m_{CO_2} 0.001 M_{w,w}}{1 + m_{CO_2} 0.001 M_{w,w}} = \frac{0.244(0.018)}{1 + 0.244(0.018)} = 0.0044$$

The ionic species ignored here in this binary system are sufficient to lower the pH, and though essential for comprehensive understanding, the concentrations are small relative to the molecular CO_2 modeled here. In physiology or ocean chemistry, many other salts are involved which make the equilibrium more complicated. Chapter 18 addresses several issues of ionization.

a. webbook.nist.gov

Dilute Solution Calculations Using Hypothetical Lewis-Randall Fugacities

Dissolved gas solubilities can be modeled by treating the liquid phase and vapor phase both with direct use of an equation of state (to be discussed in Chapter 15). However, Eqn. 11.63 suggests that we can model dilute solutions relative to the Lewis-Randall rule. Looking at Eqn. 11.63, you can appreciate why the Henry's law constant depends on solvent—the Lewis-Randall γ_i^∞ will be different for every solvent. The activity coefficient models we have developed can take γ_i^∞ into account. What we need is a manner to correlate the fugacity of hypothetical liquids above the critical point. A prevalent model for light gases in petrochemicals is the Grayson-Streed model (and the closely related Chao-Seader and Prausnitz-Shair models).[19]

Note that the shortcut vapor pressure equation yields finite numerical results for the Lewis-Randall fugacity even when $T > T_c$. Close to the critical temperature, it is sensible to simply extrapolate the shortcut equation. In that case, the shortcut result is more properly referred to as an estimate of f^L since the liquid vapor pressure is not rigorously defined at $T > T_c$. Common experience is that increasing temperature decreases gas solubility. At temperatures where $T > 2T_c$, however, a surprising thing happens. The *solubility* of the gas *increases* with increasing temperature. In other words, *fugacity* of the hypothetical gaseous component *decreases* with increasing temperature. This behavior is predicted by equations of state (Chapter 15), but for simple correlations the previously cited researchers developed correlations of the hypothetical Lewis-Randall fugacities that matched experiments together with simple mixture models (predominantly the regular solution models discussed in the next chapter). We introduce the concepts here using the MAB model.

Fig. 11.14 shows several generalized estimates for f^L as a function of reduced temperature. The Grayson-Streed estimates vary substantially depending on whether the general correlation is applied (GS-0) or specific correlations as for methane (CH_4) or hydrogen (H_2). The Grayson-Streed estimates have limits in their range of temperature that are reflected by their ranges in Fig. 11.14. The Prausnitz-Shair correlation gives a single curve for all compounds regardless of their acentric

19. Chao, K. C., Seader, J. D. 1961. *AIChE J.* 7:598; Grayson, H.G., Streed, C.W. Paper 20-PO7, 6th World Petroleum Conference, Frankfurt, June 1963; Prausnitz, J.M., Shair, F.H. 1961. *AIChE J.* 7:682.

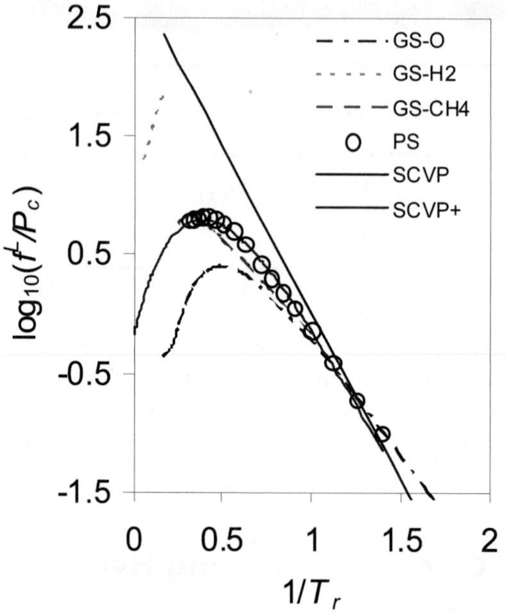

Figure 11.14 *Comparison of correlations for liquid fugacity at high temperature as described in the text. GS = Grayson/Streed, PS = Prausnitz/Shair, SCVP = shortcut vapor pressure equation., SCVP+ = extended shortcut vapor pressure equation. (ω = 0.21 for SCVP and SCVP+)*

factor. For our purposes, we would like to have a generalized correlation that improves qualitatively on the shortcut equation at high temperatures. This goal can be achieved with the following correlation for the Lewis-Randall fugacity:

$$\log_{10}(f^L / P_c) = 7(1 + \omega)(1 - T_c/T)/3 - 3\exp(-3T_c/T) \quad \text{SCVP+ equation} \qquad 11.68$$

> **❶** SCVP+ model for extrapolating Lewis-Randall fugacity above Tc.

This correlation is designed to match the shortcut vapor pressure (SCVP) equation at $T < T_c$. It provides a reasonable match of the Grayson-Streed estimates for CH_4 at $T > T_c$ and a fairly accurate match to the Prausnitz-Shair correlation when a value of $\omega = 0.21$ is applied to Eqn. 11.68. It also provides reasonable results for all temperatures. We refer to this as the SCVP+ equation.

Example 11.12   Henry's constant for CO_2 with the MAB/SCVP+ model

The solubility for CO_2 in water at 298 K and 7 bar can be estimated as $x_{CO_2} = 0.0044$. Treating the gas phase as an ideal gas and neglecting any aqueous ionic species, (a) fit γ_i^∞ using the Lewis-Randall rule and the SCVP+ equation for pure CO_2 and determine the one-parameter Margules parameter; (b) estimate A_{12} of the MAB model for CO_2 in water and γ_i^∞ and compare to part (a); (c) predict Henry's constant at 311 K using the MAB and the SCVP+ equation.

> ### Example 11.12   Henry's constant for CO_2 with the MAB/SCVP+ model (Continued)
>
> **Solution:**
>
> (a) First use the SCVP+ equation to predict the hypothetical liquid fugacity,
>
> $\log_{10}(f^L / 73.82) = 7(1 + 0.228)(1 - 304.2/298)/3 - 3\exp(-3 \cdot 304.2/298)$
>
> $\Rightarrow f_{CO_2}{}^L = 73.82 \cdot 10^{-0.200} = 47$ bar.
>
> (FYI: The Lewis-Randall standard state by the SCVP model would be 64 bar instead of 47 bar.) Referring to Example 11.11, the ideal gas vapor fugacity has been calculated there, and we can equate it with Henry's law and use the fugacity just calculated with the experimental x_{CO_2},
>
> $$y_{CO_2}P = 7 \cdot 0.9955 = x_{CO_2} \gamma_{CO_2}{}^\infty f_{CO_2}{}^L \quad \Rightarrow \gamma_i{}^\infty = 7 \cdot 0.9955/(0.0044 \cdot 47) = 34.$$
>
> $$\ln\gamma_{CO_2}{}^\infty = A_{12} = \ln(34) = 3.52$$
>
> (b) For MAB the default estimate is $A_{12} = (\alpha_2 - \alpha_1)(\beta_2 - \beta_1)(V_1 + V_2)/(4RT)$
>
> $$A_{12} = (1.87 - 50.13)(0 - 15.06)(44/1.18 + 18/1)/(4(8.314)298) = 4.05$$
>
> $$\gamma_i{}^\infty = \exp(4.05) = 57.4$$
>
> The MAB prediction for A_{12} is approximately $(100\%)(4.05 - 3.52)/3.52 = 15\%$ too high.
>
> (c) At 311 K, the fitted MAB model suggests that $A_{12} = 3.52(298/311) = 3.37 = \ln\gamma_{CO_2}{}^\infty$. So, $\gamma_{CO_2}{}^\infty = 29$.
>
> By Eqn. 11.68, $\log_{10}(f^L/73.82) = 7(1 + 0.228)(1 - 304.2/311)/3 - 3\exp(-3 \cdot 304.2/311) = -0.0968$
>
> By Eqn. 11.64, $h_{CO_2} = \gamma_{CO_2}{}^\infty f_{CO_2}{}^L = 29(10^{-0.0968}) = 23$ bar

Note that $\gamma_i{}^\infty > 10$ for CO_2 in H_2O with the SCVP+ model of Henry's law, suggesting that CO_2 and water are not very compatible. In fact, the $CO_2 + H_2O$ system does exhibit VLLE, affirming that this approach to Henry's law maintains consistency with the Lewis-Randall perspective. In a similar manner, other activity models, compounds, and conditions can be characterized.

11.13   OSMOTIC PRESSURE

Semi-permeable membranes exhibit the remarkable ability to sort molecules at the nanoscale. Semi-permeable membranes are used in reverse-osmosis water purification where water can permeate but salts cannot and in dialysis membranes where blood is purified. Cell walls and cell membranes in biological systems also have selective permeability to many species. Consider the membrane shown in Fig. 11.15(a) where pure W is on the left and a mixture of $W + C$ is on the right. (Often the solvent is water but in polymer chemistry organic solvents can be used.) The membrane is permeable to W but not to C. If the solutions are at the same pressure, P, then component W spontaneously flows from the left chamber (higher chemical potential because higher mole fraction) to the right chamber (lower chemical potential because lower mole fraction) in the condition of **osmosis.** If the pressure on the right side is increased, the degree of flow can be decreased. When the pressure has been increased by the **osmotic pressure, Π,** the sides achieve phase equilibrium and flow stops. If the pressure on the right side is increased by more than the osmotic pressure, a condition of **reverse osmosis** exists and component W flows from the right to the left. Reverse

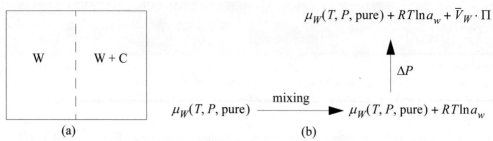

Figure 11.15 *(a) Illustration of a semipermeable membrane. The membrane is permeable to component W, but not component C. The label of W is convenient because water is a typical solvent. (b) Illustration of the path used in calculation of the chemical potential in the mixture at the osmotic pressure.*

osmosis is on the verge of becoming the largest scale chemical engineering unit operation in the world as populations grow and water becomes scarce.

At the pressure $(P + \Pi)$ on the right side, inward flow of W stops and the chemical potential is balanced. Let us create a convenient pathway to relate the chemical potential for the pure fluid at P to the mixture at $P + \Pi$. We can consider pressurizing the pure fluid and then mixing, or we can consider mixing the fluid and then pressurizing at fixed composition. Following historical derivation, it is common to use P as the standard state pressure for the mixing. The mixing process can be represented by the activity, $a_W = x_W \gamma_W$ (*cf.* Eqn. 11.23)

$$\mu_W(T, P, x_W) - \mu_W(T, P, \text{pure}) = RT\ln a_W \qquad 11.69$$

For W in a mixture, the pressure effect on chemical potential at constant T is $d\mu_W = \bar{V}_W dP$. Because the liquid is nearly incompressible, for the pressure step $\Delta\mu_W = \bar{V}_W \cdot ((P + \Pi) - P)$ and overall,

$$\mu_W(T, P + \Pi, x_W) = \mu_W(T, P, x_W) + \bar{V}_W \cdot ((P + \Pi) - P) = \mu_W(T, P, x_W) + \bar{V}_W \cdot \Pi \qquad 11.70$$

The calculation path is illustrated in Fig. 11.16(b). The initial state represents the left side of the membrane and the final state represents the right side. Equating the chemical potential expressions for the two sides of the membrane results in

$$\mu_W(T, P, \text{pure}) = \mu_W(T, P, \text{pure}) + RT\ln a_W + \bar{V}_W \cdot \Pi$$

Leading to the relation between osmotic pressure and activity of the permeable species,

$$\boxed{\Pi = -\frac{RT}{\bar{V}_W}\ln a_W} \qquad 11.71$$

> **ⓘ** Relation between osmotic pressure and activity of the permeable species.

The activity can be calculated from any activity coefficient model. Note that because the solution is very nearly pure W on a molar basis, we calculate activity relative to the Lewis-Randall rule for W, and it is common to replace the partial molar volume with the volume of pure W. A method known as the **McMillan-Mayer framework**[20] is used frequently in biology to express $\ln a_W$, writ-

---

20. Tester, J.W., Modell, M. 1997. *Thermodynamics and Its Applications*, 3rd ed. Upper Saddle River, NJ: Prentice-Hall, p. 469.

ing the logarithm of the activity as an expansion in terms of the *molar concentration* of the solute, [C]. Using the molar volume of W to normalize the expression per mole of W, the McMillan-Mayer framework results in

$$RT\ln a_W = -RTV_W([C] + B_2(T)[C]^2 + B_3(T)[C]^3 + \ldots) \qquad 11.72$$

where $B_2(T)$ and $B_3(T)$ are functions of temperature known as the **osmotic virial coefficients.** Combining the two expressions, eliminating the molar volume, and rewriting the expression using solute generic subscript i, C_i *the solute mass density* **(in units of grams/(volume of solution))**, and $M_{w,i}$ molecular weight, results in the form which is common in presenting data:

$$\boxed{\frac{\Pi}{RT} = \frac{C_i}{M_{w,i}}(1 + B_2(T)C_i + B_3(T)C_i^2 + \ldots)} \qquad 11.73$$

The osmotic virial coefficients are explicitly given temperature dependence, though they also depend on pH for biological molecules that change charge as a function of pH. Note that a plot of $\Pi/(RTC_i)$ will have an intercept related to the reciprocal of molecular weight and the plot can be used to determine molecular weight of solutes. Experimental data for osmotic pressure for the pig blood protein bovine serum albumin (BSA) in water at various pH values are shown in Fig. 11.16. The pH effects on charge are explained when we discuss electrolytes in Chapter 18.

Example 11.13 Osmotic pressure of BSA

Bovine serum albumin (BSA) has a molecular weight of 66399 g/mol. The osmotic pressure of an aqueous solution at 25°C and pH 5.4 is 74 mmHg when the concentration is 130 g/L and 260 mmHg at 234 g/L.[a] Using only these data, determine the second and third osmotic virial coefficients and estimate the pressure needed to concentrate a solution to 450 g/L across a membrane with pure water on the other side.

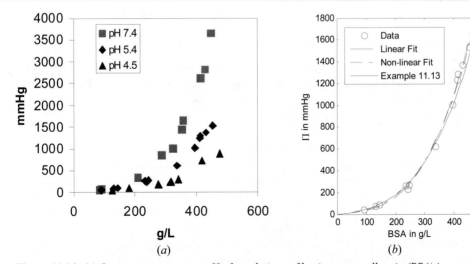

Figure 11.16 *(a) Osmotic pressure in mmHg for solutions of bovine serum albumin (BSA) in water at 25°C at different pH values. Data from Vilker, V.L.; Colton, C.K.; Smith, K.A. J. Colloid Int. Sci. 1981. 79:548. (b) Fits of osmotic pressure at pH 5.4 as explained in Example 11.13.*

Example 11.13 Osmotic pressure of BSA (Continued)

Solution: Since two points are given, let us linearize the equation for osmotic pressure to relate the coefficients to the slope and intercept. Defining a variable s to hold the rearranged variables,

$$\left(\frac{\Pi M_i}{RTC_i} - 1\right)/C_i \equiv s_i = B_2(T) + B_3(T)C_i \qquad 11.74$$

Converting the osmotic pressure to MPa, $\left(\dfrac{\Pi M_i}{RTC_i} - 1\right)/C_i$ is

$$s_i = \frac{\left(\dfrac{74\,\text{mmHg}}{\phantom{750\text{mmHg}}}\bigg|\dfrac{0.1\text{MPa}}{750\text{mmHg}}\bigg|\dfrac{\text{mol}\cdot\text{K}}{8.314\text{MPacm}^3}\bigg|\dfrac{1}{298.15\text{K}}\dfrac{66399\text{ g/mol}}{130\text{ g/L}}\dfrac{1000\text{cm}^3/\text{L}}{\phantom{130\text{ g/L}}}-1\right)}{130\,\text{g/L}} = 7.946\times10^{-3}\,\text{L/g}$$

Then at 234 g/L, $s_i = 1.269\times10^{-2}\,\text{L/g}$. The third coefficient is given by the slope of s_i versus C_i,

$$B_3(T) = \frac{(s_i)_{234} - (s_i)_{130}}{(C_i)_{234} - (C_i)_{130}} = \frac{1.269\times10^{-2} - 7.946\times10^{-3}}{234 - 130} = 4.562\times10^{-5}\,(\text{L/g})^2$$

The second coefficient is found using the third coefficient with either of the original data points. From the point at 130 g/L:

$$B_2(T) = s_i - B_3(T)C_i = 7.946\times10^{-3} - 4.562\times10^{-5}(130) = 2.015\times10^{-3}\,\text{L/g} \qquad 11.75$$

Now at 450 g/L,

$$\frac{\Pi}{RT} = \frac{C_i}{M_i}(1 + 2.015\times10^{-3}C_i + 4.562\times10^{-5}C_i^2)$$

$$= \frac{450}{66399}(1 + 2.015\times10^{-3}(450) + 4.562\times10^{-5}(450)^2) = 0.07548\,\text{mol/L}$$

$$\Pi = (0.07548\,\text{mol/L})(8.314)(298.15)(\text{L}/1000\text{cm}^3)(750\text{mmHg}/0.1\text{MPa}) = 1403\,\text{mmHg}$$

Therefore, we must apply a minimum estimated pressure of 1400 mmHg to concentrate the BSA to 450 g/L. The original paper cited gives a value of approximately 1500 mmHg. The estimate is within 10%. The prediction is sensitive to noise in the data points selected. A better method is to collect a few more data and regress a best fit.

Fig. 11.16(b) shows three fits of the data. For the "Linear Fit", the data are linearized following the procedure in this example, and then linear regression is used over all points. For the '"Non-Linear Fit", the error in the osmotic pressure prediction of Eqn. 11.73 is minimized using nonlinear regression. The "Example 11.13" curve uses the coefficients fitted in this example. The second osmotic coefficient for this data set is sensitive to the regression method. For the linear fit $[B_2\ B_3]$ = [1.93E-4 5.352E-5], for the nonlinear fit, [–3.57E-3 6.360E-5]. Careful analysis of the regression statistics shows that the uncertainty in the value of B_2 is larger than the value—the uncertainties for the 95% confidence limit of the nonlinear fit are ±[5.25E-3 1.25E-5].

a. Vilker, V.L., Colton, C.K., Smith, K.A. 1981. *J. Colloid Int. Sci.* 79:548. Note the original paper uses a molecular weight of 69000 g/mol.

Cell membranes are excellent examples of semipermeable membranes, especially when considering water permeability. One implication of this property is that altering the osmotic pressure in the cellular environment can make the cells "uncomfortable." Specifically, a higher salt concentration outside the cell might cause dehydration. On the other hand, zero salt concentration outside the cell might cause the cell to swell or rupture. This property extends to cell aggregates like skin, or the epithelium of the eye. For example, one requirement to minimize discomfort caused by eye drops is to make the solution **isotonic,** meaning that the osmotic pressure of water in the solution is the same as that of the reference cellular material, the eyes in this case.[21]

A common situation in pharmaceutical preparation is that the drug concentration is determined by the treatment protocol. However, the delivery solution should be isotonic with bodily fluids. Therefore, the solution must be supplemented with sodium chloride to make it isotonic. Fortunately, solute concentrations in cells are so low that B_2 and B_3 of Eqn. 11.73 can be neglected. This results in the interesting observation that osmotic pressure is independent of the nature of the compound as long as the molar concentration is the same. In other words, the concentrations of all constituents can simply be added up until the isotonic concentration is achieved. A property that follows this rule of adding up the constituents regardless of chemical nature is called a **colligative property,** of which osmotic pressure is an example (as long as the concentration is sufficiently low that B_2 and B_3 may be neglected).[22] As a point of reference, human blood is in the concentration range where colligative properties can be assumed and isotonic with any solution of 0.308 mol/L solute.

Example 11.14   Osmotic pressure and electroporation of E. coli

E. coli are bacteria commonly used to express desired proteins through genetic modification because they replicate and express whatever intracellular DNA they find. Introducing foreign DNA requires weakening the cell membrane by washing twice briefly (~10 min.) with pure water at 4°C, followed by a wash with 10wt% glycerol solution, centrifuging to isolate the cell pellet from the medium before washes. After the cells are rendered "electro-competent" through washing, all but 1 ml of the glycerol solution is removed and the aliquots are frozen for storage until the "electroporation" step (electrically shocking the cells) is conducted. What concentration of glycerol (wt%) is necessary to make a solution that is isotonic with human blood? Describe what happens to the water in the cells and the glycerol outside the cells when the medium is replaced with 10 wt% glycerol.

Solution: The molecular weight of glycerol can be found from the NIST Chemistry WebBook as 92.1. This means that a 0.308 mol/L solution has 0.308·92.1 g/L of glycerol. Assuming 1000g/L as the density (the same as water since the concentration is low), this gives a weight fraction of 0.308·92.1/1000 = 0.0284 = 2.84 wt%. Therefore, the 10 wt% is **hypertonic.** The activity of water is too low to be isotonic. The driving force is for water to come out of the cells, diluting the glycerol outside the cells. The cells will shrink and shrivel.

---

21. Connors, K.A. 2002. *Thermodynamics of Pharmaceutical Systems.* Hoboken, NJ: Wiley.

22. As you might imagine there are reasons why "normal saline" for medical IV has other salts dissolved. It is important to maintain balance of many specific electrolytes, and therefore we must be careful which species are delivered, though from a superficial level, any species can be used to make the solution isotonic.

11.14 SUMMARY

The strategies for problem solving remain much the same as the strategy set forth at the end of Chapter 10 and a review of that strategy is suggested. Use Table 10.1 on page 373 and the information in Sections 10.1–10.8 to identify known variables and the correct routine to use. Then apply the valid approximations.

The introduction of activity coefficients is new in this chapter. First, we showed that there are three different stages in working with activity coefficients: obtaining them from experiments; fitting a model to the experiments; and using the models to extrapolate to new compositions or different temperatures and pressures.

We provided several methods of fitting activity coefficient models to experiments, and we demonstrated bubble-pressure and bubble-temperature calculations. We presented the strategy for relating the non-idealities to the excess Gibbs energy. We hypothesized models to fit the correct shape of the excess Gibbs energy and we differentiated the models to obtain expressions for the activity coefficients. We related the nature of the non-idealities to the chemical structures in the mixture through the concepts of acidity and basicity.

We introduced the concept of activity. The foundation is laid here for relating the fugacity of a component in a mixture to its pure component fugacity. Subtle details pertain to the characterization of standard state, as discussed in the introductions to Henry's law and osmotic pressure and we superficially introduced the molality scale for Henry's law. We will refer back to this discussion in the context of electrolytes.

Recognize that the primary difference between this chapter and Chapter 10 is the γ_i used to calculate K_i. We also pointed out the need for a greater understanding of solution behavior to assist in developing theory-based activity models, which will lead us into the next two chapters.

Important Equations

The starting point for many phase equilibrium problems is Eqn. 11.13 on page 425:

$$\boxed{\hat{f}_i^L \equiv x_i \gamma_i f_i^\circ}$$

11.13

The various activity models alter the method of computing γ_i, but do not alter this basic equation. Eqn. 11.13 will appear in the simplified form for modified Raoult's law:

$$\boxed{\hat{f}_i^L \approx x_i \gamma_i P_i^{sat}}$$

11.14

$$\boxed{y_i P = x_i \gamma_i P_i^{sat}} \quad \text{or} \quad \boxed{K_i = \frac{\gamma_i^L P_i^{sat}}{P}}$$

11.18

Another significant equation can be summarized as the Redlich-Kister expansion (Eqn. 11.32 on page 429), in that this implicitly represents all the Margules models. When a G^E model is combined with Eqn. 11.28, the activity coefficients can be derived at any composition and substituted into modified Raoult's law to solve a wide variety of problems. The relations between G^E and activity coefficient are given by

$$G^E = RT \sum_i x_i \ln(\gamma_i)$$

$$\left(\frac{\partial G^E}{\partial n_i}\right)_{T,P,n_{j \neq i}} = \bar{G}_i^E = \mu_i^E = RT \ln \gamma_i \qquad \text{11.27 and 11.28}$$

The simplest binary phase equilibrium equation to keep in mind is the bubble pressure,

$$P = y_1 P + y_2 P = x_1 \gamma_1 P_1^{sat} + x_2 \gamma_2 P_2^{sat} \qquad \text{11.2}$$

Through this equation, it is very easy to compute the implications of non-ideality and assess qualitatively whether process complications like azeotropes or LLE are likely. A simple equation to guide your assessment is the MAB estimate of A_{12} in the Margules one-parameter model.

$$A_{12} = (\alpha_2 - \alpha_1)(\beta_2 - \beta_1)(V_1 + V_2)/(4RT) \qquad \text{11.9}$$

When considering distillation applications you must first check that $\alpha_{LH} > 1$ at top and bottom:

$$\alpha_{LH}{}^{top} = [(P_{LK}{}^{sat})/(\gamma_{HK}{}^{\infty} P_{HK}{}^{sat})]^{top} \qquad \alpha_{LH}{}^{bot} = [(\gamma_{LK}{}^{\infty} P_{LK}{}^{sat})/(P_{HK}{}^{sat})]^{bot} \qquad \text{11.59}$$

We also developed Henry's law,

$$\hat{f}_i = x_i \gamma_i^* h_i \qquad \text{11.61}$$

We showed how to relate Henry's law to the Lewis-Randall rule used for modified Raoult's law and how to predict the solubilities of supercritical gases in liquid solvents with the SCVP+ model.

11.15 PRACTICE PROBLEMS

P11.1 Ninov et al. (*J. Chem. Eng. Data,* 40:199, 1995) have shown that the system diethylamine(1) + chloroform(2) forms an azeotrope at 1 bar, 341.55 K and $x_1 = 0.4475$. Is this a maximum boiling or minimum boiling azeotrope? Determine the bubble temperature and vapor composition at $x_1 = 0.80$ and 1 bar. (ANS. 334 K, 0.92)

P11.2 Derive the expression for the activity coefficient of the Redlich-Kister expansion.

11.16 HOMEWORK PROBLEMS

11.1 The volume change on mixing for the liquid methyl formate(1) + liquid ethanol(2) system at 298.15 K may be approximately represented by J. Polack, Lu, B.C.-Y. 1972. *J. Chem Thermodynamics,* 4:469:

$$\Delta V_{mix} = 0.8 x_1 x_2 \text{ cm}^3/\text{mol}$$

(a) Using this correlation, and the data $V_1 = 67.28$ cm^3/mol, $V_2 = 58.68$ cm^3/mol, determine the molar volume of mixtures at $x_1 = 0, 0.2, 0.4, 0.6, 0.8, 1.0$ in cm^3/mol.

(b) Analytically differentiate the above expression and show that

$$\bar{V}_1^E = 0.8 x_2^2 \text{ and } \bar{V}_2^E = 0.8 x_1^2 \text{ cm}^3/\text{mol}$$

and plot these partial molar excess volumes as a function of x_1.

11.2 In vapor-liquid equilibria the relative volatility α_{ij} is defined by Eqn. 10.32.

(a) Provide a simple proof that the relative volatility is independent of liquid and vapor composition if a system follows Raoult's law.

(b) In approximation to a distillation calculation for a nonideal system, calculate the relative volatility α_{12} and α_{21} as a function of composition for the n-pentane(1) + acetone(2) system at 1 bar using experimental data in problem 11.11.

(c) In approximation to a distillation calculation for a non-ideal system, calculate the relative volatility α_{12} and α_{21} as a function of composition for the data provided in problem 10.2.

(d) Provide conclusions from your analysis.

11.3 After fitting the two-parameter Margules equation to the data below, generate a P-x-y diagram at 78.15°C.

Data at 78.15°C:
$P_1^{sat} = 1.006$ bar
$P_2^{sat} = 0.439$ bar
$\gamma_1^\infty = 1.6931;\ \ \gamma_2^\infty = 1.9523$

11.4 A stream containing equimolar methanol(1) + benzene(2) at 350 K and 1500 mmHg is to be adiabatically flashed to atmospheric pressure. The two-parameter Margules model is to be applied with $A_{12} = 1.85, A_{21} = 1.64$. Express all flash equations in terms of K_i values and K_i values in terms of Modified Raoult's law.

(a) List all the unknown variables that need to be determined to solve for the outlet.

(b) List all the equations that apply to determine the unknown variables.

11.5 In the system $A + B$, activity coefficients can be expressed by the one-parameter Margules equation with $A_{12} = 0.5$. The vapor pressures of A and B at 80°C are $P_A^{sat} = 900$ mmHg, $P_B^{sat} = 600$ mmHg. Is there an azeotrope in this system at 80°C, and if so, what is the azeotrope pressure and composition?

11.6 The system acetone(1) + methanol(2) is well represented by the one-parameter Margules equation using $A_{12} = 0.605$ at 50°C.

(a) What is the bubble pressure for an equimolar mixture at 30°C?

(b) What is the dew pressure for an equimolar mixture at 30°C?

(c) What is the bubble temperature for an equimolar mixture at 760 mmhg?

(d) What is the dew temperature for an equimolar mixture at 760 mmhg?

11.7 The excess Gibbs energy for a liquid mixture of n-hexane(1) + benzene(2) at 30°C is represented by $G^E = 1089\ x_1 x_2$ J/mol.

(a) What is the bubble pressure for an equimolar mixture at 30°C?

(b) What is the dew pressure for an equimolar mixture at 30°C?

(c) What is the bubble temperature for an equimolar mixture at 760 mmHg?

(d) What is the dew temperature for an equimolar mixture at 760 mmHg?

11.8 The liquid phase activity coefficients of the ethanol(1) + toluene(2) system at 55°C are given by the two-parameter Margules equation, where $A_{12} = 1.869$ and $A_{21} = 1.654$.

(a) Show that the pure saturation fugacity coefficient is approximately 1 for both components.

(b) Calculate the fugacity for each component in the liquid mixture at $x_1 = 0.0, 0.2, 0.4, 0.6,$ 0.8, and 1.0. Summarize your results in a table. Plot the fugacities for both components versus x_1. Label your curves. For each curve, indicate the regions that may be approximated by Henry's law and the ideal solution model.

(c) Using the results of part (b), estimate the total pressure above the liquid mixture at 55°C when a vapor phase coexists. Assume the gas phase is ideal for this calculation. Also estimate the vapor composition.

(d) Comment on the validity of the ideal gas assumption used in part (c).

11.9 (a) The acetone(1) + chloroform(2) system can be represented by the Margules two-parameter equation using $A_{12} = -1.149$, $A_{21} = -0.862$ at 35.17°C. Use bubble-pressure calculations to generate a P-x-y and y-x diagram and compare it with the selected values from the measurements of Zawidzki, *Z. Phys. Chem.*, 35, 129(1900).

(b) Compare the data to the predictions of the MAB model.

P(mmHg)	x_1	y_1
262.6	0.1953	0.1464
248.4	0.4188	0.4368
255.7	0.507	0.5640
272.2	0.6336	0.7271
290.5	0.7296	0.8273
320.1	0.8797	0.9377

11.10 (a) Fit the Margules two-parameter equation to the methanol(1) + benzene(2) system P-x-y data below at 90°C (Jost, W., Roek, H, Schroeder, W., Sieg, L., Wagner, H.G. 1957. *Z. Phys. Chem.* 10:133) by fitting to x_1=0.549. Plot the resultant fit together with the original data for both phases.

(b) Compare the data with the predictions of the MAB model.

x_1	y_1	P (mmHg)
0.117	0.502	1865
0.257	0.594	2113
0.376	0.618	2218
0.549	0.65	2273
0.707	0.689	2292
0.856	0.765	2208

11.11 (a) Fit the Margules two-parameter equation to the n-pentane(1) + acetone(2) system T-x-y data below at 1 bar (Lo et al. 1962. *J. Chem. Eng. Data* 7:32) by fitting to x_1=0.503. Plot the resultant fit together with the original data for both phases.

(b) Compare the data with the predictions of the MAB model.

x_1	0.021	0.134	0.292	0.503	0.728	0.953
y_1	0.108	0.475	0.614	0.678	0.739	0.906
$T\,(°C)$	49.15	39.58	34.35	33.35	31.93	33.89
P_1^{sat}	1.560	1.146	0.960	0.903	0.880	0.954
P_2^{sat}	0.803	0.551	0.453	0.421	0.410	0.445

11.12 For a particular binary system, data are available:

$$T = 45°C \quad P = 37 \text{ kPa} \quad x_1 = 0.398 \quad y_1 = 0.428$$

In addition, $P_1^{sat} = 27.78$ kPa and $P_2^{sat} = 29.82$ kPa. From these data,

(a) Fit the one-parameter Margules equation
(b) Fit the two-parameter Margules equation

11.13 The compositions of coexisting phases of ethanol(1) + toluene(2) at 55°C are $x_1 = 0.7186$, and $y_1 = 0.7431$ at $P = 307.81$ mmHg, as reported by Kretschmer and Wiebe, *J. Amer. Chem. Soc.*, 71, 1793(1949). Estimate the bubble pressure at 55°C and $x_1 = 0.1$, using

(a) The one-parameter Margules equation
(b) The two-parameter Margules equation

11.14 A vapor/liquid experiment for the carbon disulfide(1) + chloroform(2) system has provided the following data at 298 K: $P_1^{sat} = 46.85$ kPa, $P_2^{sat} = 27.3$ kPa, $x_1 = 0.2$, $y_1 = 0.363$, $P = 34.98$ kPa. Estimate the dew pressure at 298 K and $y_1 = 0.6$, using

(a) The one-parameter Margules equation
(b) The two-parameter Margules equation

11.15 The (1) + (2) system forms an azeotrope at $x_1 = 0.75$ and 80°C. At 80°C, $P_1^{sat} = 600$ mmHg, $P_2^{sat} = 900$ mmHg. The liquid phase can be modeled by the one-parameter Margules equation.

(a) Estimate the activity coefficient of component 1 at $x_1 = 0.75$ and 80°C. [Hint: The relative volatility (given in problem 11.2) is unity at the azeotropic condition.]
(b) Qualitatively sketch the P-x-y and T-x-y diagrams that you expect.

11.16 Ethanol(1) + benzene(2) form an azeotropic mixture. Compare the specified model to the experimental data of Brown and Smith cited in problem 10.2.

(a) Prepare a y-x and P-x-y diagram for the system at 45°C assuming the MAB model.
(b) Prepare a y-x and P-x-y diagram for the system at 45°C assuming the one-parameter Margules model and using the experimental pressure at $x_1 = 0.415$ to estimate A_{12}.
(c) Prepare a y-x and P-x-y diagram for the system at 45°C assuming the two-parameter model and using the experimental pressure at $x_1 = 0.415$ to estimate A_{12} and A_{21}.

11.17 The acetone + chloroform system exhibits an azeotrope at 64.7°C, 760 mmHg, and 20 wt% acetone.

(a) Use the MAB model to predict the T-x-y diagram at 1 bar.
(b) Use the Margules one-parameter model to estimate the T-x-y diagram at 1 bar.

11.18 For the Margules two-parameter model estimate the total pressure and composition of the vapor in equilibrium with a 20 mol% ethanol(1) solution in water(2) at 78.15°C using data at 78.15°C:

$$P_1^{sat} = 1.006 \text{ bar}; \quad P_2^{sat} = 0.439 \text{ bar}; \quad \gamma_1^\infty = 1.6931; \quad \gamma_2^\infty = 1.9523$$

11.19 Using the data from problem 11.18, fit the two-parameter Margules equation, and then generate a P-x-y diagram at 78.15°C.

11.20 A liquid mixture of 50 mol% chloroform(1) and 50% 1,4-dioxane(2) at 0.1013 MPa is metered into a flash drum through a valve. The mixture flashes into two phases inside the drum where the pressure and temperature are maintained at 24.95 kPa and 50°C. The compositions of the exiting phases are $x_1 = 0.36$ and $y_1 = 0.62$.

Your supervisor asks you to adjust the flash drum pressure so that the liquid phase is $x_1 = 0.4$ at 50°C. He doesn't provide any *VLE* data, and you are standing in the middle of the plant with only a calculator and pencil and paper, so you must estimate the new flash drum pressure. Fortunately, your supervisor has a phenomenal recall for numbers and tells you that the vapor pressures for the pure components at 50°C are $P_1^{sat} = 69.4$ kPa and $P_2^{sat} = 15.8$ kPa. What is your best estimate of the pressure adjustment that is necessary without using any additional information?

11.21 Suppose a vessel contains an equimolar mixture of chloroform(1) and triethylamine(2) at 25°C. The following data are available at 25°C:

	Chloroform(1)	Triethylamine(2)
MW	119.4	101.2
V^L (cm³/gmol)	80.19	139.
ΔH^{vap} (kJ/mol)	29.71	31.38
approx $\ln \gamma_i$	$-1.74x_2^2$	$-1.74x_1^2$
P^{sat} (mmHg)	193.4	67.3

(a) If the pressure in the vessel is 90 mmHg, is the mixture a liquid, a vapor, or both liquid and vapor? Justify your answer.
(b) Provide your best estimate of the volume of the vessel under these conditions. State your assumptions.

11.22 Ethanol(1) + benzene(2) form azeotropic mixtures.

(a) From the limited data below at 45°C, it is desired to estimate the constant A for the one-term Margules equation, $G^E/RT = Ax_1x_2$. Use all of the experimental data to give your best estimate.

x_1	0	0.3141	0.5199	1
y_1	0	0.3625	0.4065	1
P(bar)	0.2939	0.4124	0.4100	0.2321

(b) From your value, what are the bubble pressure and vapor compositions for a mixture with $x_1 = 0.8$?

11.23 An equimolar ternary mixture of acetone, n-butane, and ammonia at 1 MPa is to be flashed. List the known variables, unknown variables, and constraining equations to solve each of the cases below. Assume MAB solution thermodynamics with $G^E/RT=\Sigma\Sigma x_i x_j A_{ij}$ and write the flash equations in terms of K-ratios, with the equations for calculating K-ratios written separately. (Hint: Remember to include the activity coefficients and how to calculate them.)

(a) Bubble temperature
(b) Dew temperature
(c) Flash temperature at 25mol% vapor
(d) Raised to midway between the bubble and dew temperatures, then adiabatically flashed.

11.24 Fit the data from problem 11.11 to the following model by regression over all points, and compare with the experimental data on the same plot, using:

(a) One-parameter Margules equation
(b) Two-parameter Margules equation

11.25 Fit the specified model to the methanol(1) + benzene(2) system P-x-y data at 90°C by minimizing the sum of squares of the pressure residual. Plot the resultant fit together with the original data for both phases (data are in problem 11.10), using

(a) One-parameter Margules equation
(b) Two-parameter Margules equation

11.26 Fit the specified model to the methanol(1) + benzene(2) system T-x-y data at 760 mmHg by minimizing the sum of squares of the pressure residual. Plot the resultant fit together with the original data for both phases (Hudson, J.W., Van Winkle, M. 1969. *J. Chem. Eng. Data* 14:310), using

(a) One-parameter Margules equation
(b) Two-parameter Margules equation

x_1	y_1	$T(°C)$
0.026	0.267	70.67
0.05	0.371	66.44
0.088	0.457	62.87
0.164	0.526	60.2
0.333	0.559	58.64
0.549	0.595	58.02
0.699	0.633	58.1
0.782	0.665	58.47
0.898	0.76	59.9
0.973	0.907	62.71

11.27 VLE data for the system carbon tetrachloride(1) and 1,2-dichloroethane(2) are given below at 760 mmHg, as taken from the literature.[23]

x_1	y_1	$T(°C)$
0.040	0.141	81.59
0.091	0.185	80.39
0.097	0.202	80.27
0.185	0.310	78.73
0.370	0.473	76.62
0.506	0.557	75.78
0.880	0.831	75.71
0.900	0.848	75.86
0.923	0.875	75.95
0.960	0.907	76.20

(a) Fit the data to the one-parameter Margules equation.
(b) Fit the data to the two-parameter Margules equation.
(c) Plot the P-x-y diagram at 80°C, based on one of the fits from (a) or (b).

---

23. Yound, H.D., Nelson, O.A. 1932. *Ind. Eng. Chem. Anal. Ed.* 4:67.

11.28 When only one component of a binary mixture is volatile, the pressure over the mixture is determined entirely by the volatile component. The activity coefficient for the volatile component can be determined using modified Raoult's law and an activity coefficient model can be fitted. The model will satisfy the Gibbs-Duhem equation and thus an activity coefficient prediction can be made for the nonvolatile component. Consider a solution of sucrose and water. The sucrose is nonvolatile. The bubble pressures of water (1) + sucrose (2) solutions are tabulated below at three temperatures.

0°C		25°C		100°C	
x_1	$P(kPa)$	x_1	$P(kPa)$	x_1	$P(kPa)$
1	0.611	1	3.166	1.	101.33
0.9964	0.609	0.9982	3.16	0.995	100.78
0.9911	0.605	0.9928	3.14	0.9893	100.21
0.9823	0.599	0.9823	3.11	0.9858	99.81
0.9407	0.56	0.9653	3.03	0.9789	99.01
0.9251	0.542	0.9487	2.95	0.9737	98.35
0.9174	0.533	0.9328	2.87		
0.9025	0.516	0.9174	2.78		
0.8881	0.499				

(a) Fit the one-parameter Margules equation to the water data at the temperature(s) specified by your instructor. Report the values of A_{12}.

(b) Prepare a table of γ_1 values and plot of the experimental and fitted/predicted $\ln \gamma_1$ versus x_1 for water and sucrose over the range of experimental compositions for the temperature(s) specified by your instructor.

(c) Prepare a table of values and plot a curve for $\ln \gamma_2^*$ for the temperature(s) specified by your instructor.

(d) Prepare a table of values and a plot of osmotic pressure (in MPa) for the solution versus C_2 (g/L) at 25°C. The density at 25°C can be estimated using $\rho(g/mL) = 0.99721 + 0.3725w_2 + 0.16638w_2^2$ where w_2 is wt. fraction sucrose. Include a curve of the osmotic pressure expected for an ideal solution.

(e) Calculate the osmotic pressure (MPa) using the activity of water modeled with the one-parameter Margules equation at 25°C fitted in part (a). Add it to the plot in part (d).

(f) Calculate the second and third osmotic virial coefficients (for concentration units of g/L) at 25°C by fitting the calculations from part (d). Add the modeled osmotic pressure to the plot from (d).

(g) From the temperature dependence of the one-parameter Margules parameter fitted in (a), show that the parameter may be represented with $f(1/T(K))$. Then provide a model for the excess enthalpy and the parameter value(s) that represent the experimental data.

11.29 Red blood cells have a concentration of hemoglobin ($M_w \sim 68000$) at 0.02 M. The osmotic pressure a body temperature (37°C) is 0.83 MPa. Water can permeate the cells walls, but not hemoglobin.[24]

(a) Using only the second osmotic coefficient, determine the coefficient value (L/g), and determine the activity of water at the conditions given above.

(b) Calculate the ideal solution osmotic pressure at the conditions given above.

(c) Suppose we were to transfer red blood cells in a laboratory solution at 37°C (blood banks need to do this). We want the external glucose solution to match the red blood

---

24. Suggested by O'Connell, J.P. 2010. NSF BioEMB Workshop, San Jose, CA.

cell's internal osmotic pressure to avoid swelling or shrinking of the cells. If glucose has an osmotic pressure of 2 MPa at 0.7 M and 37°C, what glucose concentration (g/L) would match the internal osmotic pressure to keep the blood cells stable? What is molality of the resulting glucose solution? Comparing molalities, what can you infer about the solution non-idealities of the glucose solution compared to the hemoglobin solution?

11.30 Osmotic pressure of bovine serum albumin (BSA) has been measured at 298.15 K and various pH values by Vilker, V.L., Colton, C.K., Smith, K.A. 1981. *J. Colloid Int. Sci.* 79:548, as summarized in the table below. The investigators report the BSA molecular weight in their sample as 69,000.

pH = 7.4		pH = 5.4		pH = 4.5	
(g/L)	mmHg	(g/L)	mmHg	(g/L)	mmHg
84	48	91	41	126	47
91	59	130	74	182	93
211	332	144	90	278	182
211	334	234	260	317	228
289	844	240	229	318	244
325	996	245	269	343	284
325	996	338	618	418	716
354	1423	395	1005	475	889
357	1638	411	1230		
413	2620	414	1286		
428	2806	430	1370		
448	3640	454	1529		

(a) Regressing all data, determine the second and third osmotic virial coefficients for pH 7.4.

(b) Regressing all data, determine the second and third osmotic virial coefficients at pH 4.5.

11.31 Boric acid is a common supplement to make ophthalmic solutions isotonic. It is entirely undissociated at normal ophthalmic conditions.

(a) Estimate the concentration (wt%) of boric acid to prepare a solution that is isotonic with human blood.

(b) Estimate the concentration (wt%) of boric acid that should be added to a 0.025wt% solution of Claritin to make it isotonic. The molecular formula of Claritin is listed at ChemSpider.com as $C_{22}H_{23}ClN_2O_2$.

VAN DER WAALS ACTIVITY MODELS

Nothing is more practical than a good theory.[1]

Clausius

Empirical models like the Redlich-Kister expansion provide a significant improvement over the ideal solution approximation, but they lack the kind of connection with the molecular perspective that we have developed in Chapters 1 and 7. The empirical models of Chapter 11 are useful for determining the activity coefficients from a given expression for G^E, but they suggest little about the form that G^E should assume. As we develop models that incorporate physical insight more closely, we also obtain greater predictive power.

Section 12.1 introduces concepts common to many models based on the van der Waals approach including random mixing rules and the use of the "regular solution" assumption. Section 12.2 introduces the van Laar equation which has a meaningful functional form, but, like the Redlich-Kister and Margules equations, is fitted to experimental data. Section 12.3 introduces the Scatchard-Hildebrand theory, which represents historically the first widely-accepted predictive model, primarily useful for mixtures of nonpolar molecules. Though the original theory has limited direct use today, it is the basis of several other models, and solubility parameters developed therein are currently widely-used to characterize solvents. Section 12.4 develops the Flory equation and the Flory-Huggins model that combines the Flory representation of entropy of mixing with the Scatchard-Hildebrand theory for energy of mixing. The Flory-Huggins approach is widely used in polymer thermodynamics. Section 12.5 extends the Scatchard-Hildebrand theory by including acidity and basicity corrections in a manner that is very successful in estimating infinite dilution activity coefficients. That section also develops the SSCED and MAB models, which are pedagogical simplifications that apply the concepts with approximate methods. The final sections document the relation of the theories to the original van der Waals equation, and extend the models to multicomponent systems.

---

1. Quoted by Cor Peters on the occasion of his Area 1a lecturer award, 2010. AIChE National Meeting, Salt Lake City, UT.

Chapter Objectives: You Should Be Able to...

1. Compute VLE phase diagrams using modified Raoult's law with the van Laar, Scatchard-Hildebrand, SSCED, MOSCED, or Flory-Huggins models.

2. Compute the relative volatility of key components in a multicomponent mixture.

3. Explain the relationship between molecular properties like energy density, acidity, and basicity and macroscopic behavior like activity coefficients and azeotropes, enabling predictions and formulation design.

12.1   THE VAN DER WAALS PERSPECTIVE FOR MIXTURES

We have seen that the van der Waals EOS in Chapter 7 provides a simple basis for understanding the interplay between entropy, energy, repulsion, and attraction of pure fluids. Even the embellishments of the Peng-Robinson equation add little to the qualitative physical picture envisioned by van der Waals. Therefore, the van der Waals model provides a reasonable starting point for conceiving the physics of mixtures. The key quantities to be considered are van der Waals' a and b. If only we knew how to compute $a = a(x)$ and $b = b(x)$, then we could solve for Z for a mixture. Through Z, we can integrate our density-dependent formulas to obtain G, then differentiate with respect to composition to obtain partial molar Gibbs energies, chemical potentials, and component fugacities. This is the general strategy. The formulas for $a = a(x)$ and $b = b(x)$ are called **mixing rules**.

A Simple Model for Mixing Rules

Recognizing the significance of the Gibbs excess function, it should not be surprising that many researchers have studied its behavior and developed equations that can represent its various shapes. In essence, these efforts attempt to apply the same reasoning for mixtures that was so successful for pure fluids in the form of the van der Waals equation. The resultant expressions contain parameters that are intended to characterize the molecular interactions within the context of the theory. The utility of the theory is judged by how precisely the experimental data are correlated and by how accurately predictions can be made. Given that molecules in solution must actually interact according to some single set of laws of nature, one might wonder why there are so many different theories. The challenge with mixtures is that there are many different kinds of interactions occurring simultaneously, for example, disperse attractions, hydrogen bonding, size asymmetries, branching, rings, and various rotations and aspect ratios. As a result, many specific terms must be invoked to describe these many specific interactions. Because incorporation of all these effects makes the resultant model unwieldy, many researchers have made different approximations in their models. Each model has its proponents. It is difficult even to describe the various models without expressing personal prejudices. For practical applications, the perspective we adopt is that these equations usually fit the data, and that extrapolations beyond the experimental data must be performed at some risk.

❶ The composition dependence is introduced into an equation of state by *mixing rules* for the parameters. The basic equation form does not change.

The models considered in the remainder of this chapter are based on extension of the van der Waals equation of state (Eqn. 7.12) to the energy departure for mixtures. When we extend equations of state to mixtures, the basic form of the equation of state does not change. The fluid properties of the mixture are written in terms of the same equation of state parameters as for the pure fluids; however, equation of state parameters like a and b are functions of composition. The equations we use to incorporate compositional dependence into the mixture constants are termed **mixing rules.**

The parameter b represents the finite size of the molecules. For many mixtures of roughly equal-sized molecules, the dense volumes mix ideally. Therefore, it is reasonable to assume[2]

$$b = \sum_i x_i b_i \qquad\qquad 12.1$$

As for a, we must carefully consider how this term relates to the internal energy of mixing, because the Gibbs energy of mixing is closely related. The departure function can be quickly found using Eqn. 7.12 with the departure formula Eqn. 8.22, resulting in

$$U - U^{ig} = -a\rho = -a/V \qquad\qquad 12.2$$

The parameter a should represent the average attraction resulting from the many varied molecular interactions in the mixture.

In a binary mixture there are three types of interactions for molecules (1) and (2). First, a molecule can interact with itself (1+1 or 2+2 interactions), or it can interact with a molecule of the other type (a 1+2 interaction). Assuming a random fluid,[3] the probability of finding a (1) molecule is the fraction of (1) atoms, x_1. The **probability** of a 1+1 interaction is a **conditional probability.** A conditional probability is the probability of finding a second interacting molecule of a certain type given the first is a certain type. For independent events, a conditional probability is calculated by the product of the individual probabilities. Therefore, the probability of a 1+1 interaction is x_1^2. By similar arguments, the probability of a 2+2 interaction is x_2^2. The probability of a 1+2 interaction is $x_1 x_2$ and the probability of a 2+1 interaction is also $x_1 x_2$.[4] If the attractive interactions are characterized by a_{11}, a_{22}, and a_{12}, the mixing rule for a is given by:

$$a = x_1^2 a_{11} + 2x_1 x_2 a_{12} + x_2^2 a_{22} = \sum_i \sum_j x_i x_j a_{ij} \qquad\qquad 12.3$$

where the pure component a parameters are indicated by identical subscripts on a. In other words, a_{11} represents the contribution of 1+1 interactions, a_{22} represents 2+2 interactions, a_{12} represents the contribution of 1+2 interactions, and the mixing rule provides the mathematical method to sum up the contributions of the interactions. a_{12} is called the **cross coefficient,** indicating that it represents two-body interactions of unlike molecules. In the above sum, it is understood that a_{12} is equivalent to a_{21}. Note that a_{12} is *not* the a for the mixture. This type of mixing rule is called a **quadratic mixing rule,** because all cross-products of the compositions are included. It represents a fairly obvious approximation to the way mixing should be represented.

The quantity a_{12} plays a major role in the solution behavior. Predicting solution behavior is largely the same task as predicting a_{12}. This prediction takes the form of

$$a_{12} = (1 - k_{12})(a_{11} a_{22})^{\frac{1}{2}} \qquad\qquad 12.4$$

---

2. The variable x is customarily used as a generic composition variable for the mixing rule, whether applied to vapor or liquid roots.

3. The assumption of a random fluid is analyzed and evaluated in Section 13.7.

4. The actual probability for a 1+1 interaction is $\dfrac{N_1(N_1-1)}{N~(N-1)}$, but when N is large it is equal to x_1^2. Likewise, for a 2+2 the probability is $\dfrac{N_2(N_2-1)}{N~(N-1)}$. For a 1+2 it is $\dfrac{N_1~N_2}{N~(N-1)}$, which is effectively $x_1 x_2$.

where k_{12} is an adjustable parameter called the binary interaction parameter. The default is $k_{12} = 0$, giving a preliminary estimate of a_{12}. Going beyond this estimate requires contemplating whether we expect positive or negative deviations from Raoult's law. Eqns. 12.3 and 12.2 show that larger values of a_{12} result in more negative (exothermic) energies of mixing. More negative values of k_{12} lead to more negative energies of mixing, which is favorable to mixing.

We can relate a_{12} to the molecular properties by considering the square-well model. Fig. 1.2 shows the square-well model for a binary mixture. We are now in a better position to interpret its significance. When k_{12} is positive (case (c)), the depth of the attraction is shallower. The most positive value would be $k_{12} = 1$, in which case there would be no attraction at all. This is an easy way to remember that positive values of k_{12} lead to repulsion. When k_{12} is negative (case (b)), the attraction can be very deep. In fact, $k_{12} < -1$ is a possibility for very strong acid-base interactions. Recalling our discussion of the MAB model, we should begin to understand how to predict k_{12}. We further develop this understanding through the remaining discussion in this chapter.

Activity Models from the van der Waals Perspective

Several models derive from the van der Waals equation of state with the assumption of a constant packing fraction ($b\rho$). They are distinguished by different approximations of the terms in the resultant equations. For example, the entropic contributions to G^E are neglected by "regular" solution models (van Laar, Scatchard-Hildebrand, SSCED). The van Laar model is distinguished by treating the ratio of V_2/V_1 as an adjustable parameter, whereas Scatchard-Hildebrand and SSCED estimate that ratio from the liquid molar volume. SSCED differs from Scatchard-Hildebrand by accounting for the effects of hydrogen bonding (like the MAB model) as well as the influences of attractive dispersion forces ($a\rho$).

Many of these distinctions may seem superficial, but they are part of the historical development that forms the lexicon of activity models. Furthermore, they provide a convenient shorthand for referencing the various contributions to overall solution behavior. It is important to remember, however, that the assumed activity model in no way alters the procedures for computing VLE developed in Chapters 10 and 11. You should be able to adapt any solution model to solving VLE problems by simply substituting the appropriate expression for γ.

"Regular" Solutions

The Gibbs energy, $G^E = U^E + PV^E - TS^E$, is composed of three contributions. For liquids, the PV term is small to begin with, and PV^E is even smaller, so it makes sense to ignore the excess volume contribution and assume that the solution volume follows ideal mixing rules. Recalling how entropy is related to volume suggests the hypothesis that entropy might mix ideally also. These simple suppositions lead to the theories of van Laar or Scatchard and Hildebrand. We refer to "regular solution theories" when applying the assumptions of $V^E = S^E = 0$, leading to $G^E \approx U^E$.

The energetics of a of the mixture are given by Eqn. 12.3 and 12.4. For the volume, assuming zero excess volume, $V = \sum_i x_i V_i$ according to regular-solution theory. Combining into Eqn. 12.2,

$$(U - U^{ig}) = \frac{-\sum_i \sum_j x_i x_j a_{ij}}{\sum_i x_i V_i}$$

12.5

For the pure fluid, taking the limit as $x_i \to 1$,

$$(U - U^{ig})_i = \frac{-a_{ii}}{V_i} \Rightarrow (U - U^{ig})^{is} = -\sum_i (x_i a_{ii} / V_i) \qquad 12.6$$

For a binary mixture, subtracting the ideal solution result, $U^E = (U - U^{ig}) - (U - U^{ig})^{is}$ to get the excess energy gives,

$$U^E = x_1 \frac{a_{11}}{V_1} + x_2 \frac{a_{22}}{V_2} - \left(\frac{x_1^2 a_{11} + 2x_1 x_2 a_{12} + x_2^2 a_{22}}{x_1 V_1 + x_2 V_2} \right) \qquad 12.7$$

Collecting terms over a common denominator,

$$U^E = \frac{x_1 \frac{a_{11}}{V_1}(x_1 V_1 + x_2 V_2) + x_2 \frac{a_{22}}{V_2}(x_1 V_1 + x_2 V_2) - (x_1^2 a_{11} + 2x_1 x_2 a_{12} + x_2^2 a_{22})}{x_1 V_1 + x_2 V_2} \qquad 12.8$$

$$U^E = \frac{x_1^2 a_{11} + x_1 x_2 a_{11} \frac{V_2}{V_1} + x_2^2 a_{22} + x_1 x_2 a_{22} \frac{V_1}{V_2} - (x_1^2 a_{11} + 2x_1 x_2 a_{12} + x_2^2 a_{22})}{x_1 V_1 + x_2 V_2} \qquad 12.9$$

$$U^E = \frac{x_1 x_2 a_{11} \frac{V_2}{V_1} + x_1 x_2 a_{22} \frac{V_1}{V_2} - 2x_1 x_2 a_{12} \frac{V_2 V_1}{V_1 V_2}}{x_1 V_1 + x_2 V_2} \qquad 12.10$$

$$= \frac{x_1 x_2 V_1 V_2}{x_1 V_1 + x_2 V_2} \left(\frac{a_{11}}{V_1^2} + \frac{a_{22}}{V_2^2} - 2 \frac{a_{12}}{V_1 V_2} \right)$$

12.2 THE VAN LAAR MODEL

Johannes van Laar found that the parameters from the van der Waals equation of state were not accurate in predicting excess energy of mixing, and empirical fitting was required. He simplified the equation for the excess internal energy by arbitrarily defining a single symbol, "Q," to represent the final term in the equation:

J. van Laar was a student of Johannes van der Waals and Jacobus Henricus van't Hoff.

$$U^E = \frac{x_1 x_2 V_1 V_2}{x_1 V_1 + x_2 V_2} Q \quad \text{where} \quad Q \equiv \left(\frac{a_{11}}{V_1^2} + \frac{a_{22}}{V_2^2} - 2 \frac{a_{12}}{V_1 V_2} \right)$$

It would appear that this equation contains three parameters (V_1, V_2, and Q), but van Laar recognized that it could be rearranged such that only two adjustable parameters need to be determined.

$$A_{12} = \frac{QV_1}{RT}; \quad A_{21} = \frac{QV_2}{RT}; \quad \frac{A_{12}}{A_{21}} = \frac{V_1}{V_2} \qquad 12.11$$

with the final result:[5]

$$\boxed{\frac{G^E}{RT} = \frac{U^E}{RT} = \frac{A_{12}A_{21}x_1x_2}{(x_1A_{12} + x_2A_{21})}}$$

12.12

Differentiating Eqn. 12.12 gives expressions for the activity coefficients. To show this for γ_1,

$$\frac{G^E}{RT} = \frac{n_2n_1A_{12}A_{21}}{(n_1A_{12} + n_2A_{21})}$$

12.13

Applying Eqn. 11.28 for n_1 and differentiating the ratio using the product rule on $(n_2A_{12}A_{21})(n_1)(1/(n_1A_{12} + n_2A_{21}))$,

$$\frac{1}{RT}\left(\frac{\partial G^E}{\partial n_1}\right)_{T,P,n_2} = \frac{n_2A_{12}A_{21}}{(n_1A_{12} + n_2A_{21})} - \frac{n_2n_1A_{12}A_{21}}{(n_1A_{12} + n_2A_{21})^2} = \frac{x_2A_{12}A_{21}}{(x_1A_{12} + x_2A_{21})} - \frac{x_2x_1A_{12}^2A_{21}}{(x_1A_{12} + x_2A_{21})^2}$$

12.14

Obtaining a common denominator, rearranging, and applying symmetry for γ_2,

! Van Laar
model.

$$\boxed{\ln \gamma_1 = \frac{A_{12}}{\left[1 + \dfrac{A_{12}x_1}{A_{21}x_2}\right]^2}} \quad ; \quad \boxed{\ln \gamma_2 = \frac{A_{21}}{\left[1 + \dfrac{A_{21}x_2}{A_{12}x_1}\right]^2}}$$

12.15

> *Note: The parameters A_{12} and A_{21} for the van Laar and Margules equations have different values for the same data. Do not interchange them.*

When applied to binary systems, it is useful to note that these equations can be rearranged to obtain A_{12} and A_{21} from γ_1 and γ_2 given any one VLE point. This is the simple manner of estimating the parameters that we generally apply in this chapter. Similar to the two-parameter Margules, a single experimental point can be used as described in Section 11.6:

$$A_{12} = (\ln \gamma_1)\left[1 + \frac{x_2 \ln \gamma_2}{x_1 \ln \gamma_1}\right]^2 \qquad A_{21} = (\ln \gamma_2)\left[1 + \frac{x_1 \ln \gamma_1}{x_2 \ln \gamma_2}\right]^2$$

12.16

Care must be used before accepting the values of Eqns. 12.16 applied to a single experiment, because experimental errors can occasionally result in questionable parameter values. Eqn. 12.16 applied to the activity coefficients from Example 11.1 results in $A_{12} = 2.38$, $A_{21} = 1.15$, and G^E is plotted in Fig. 11.3. Methods of fitting the parameters in optimal fashion for many data are covered in Section 11.9.

---

5. Readers should be aware that sometimes Eqns 12.11 and 12.12 are written without the RT terms. With that parameterization, then the "ln" term in Eqn. 12.15 is multiplied by RT and the activity coefficients have explicit temperature dependence. Use care when using literature parameters.

> ### Example 12.1   Infinite dilution activity coefficients from the van Laar theory
>
> n-Propyl alcohol (1) forms an azeotrope with toluene (2) at $x_1 = 0.6$, 92.6°C, and 760 mmHg. Use the van Laar model to estimate the infinite dilution activity coefficients of these two species at this temperature.
>
> **Solution:** The vapor pressures using parameters from Antoine.xlsx are $P_1^{sat} = 637.86$ mmHg, $P_2^{sat} = 442.24$ mmHg.
>
> Applying the azeotropic data as explained in Section 11.7 gives: $\gamma_1 = P/P_1^{sat} = 760/637.86 = 1.191$; $\gamma_2 = P/P_2^{sat} = 1.719$. Eqn. 12.16 gives: $A_{12} = 1.643$; $A_{21} = 1.193$.
>
> Taking the limits of Eqn. 12.15 as the respective components approach zero composition results in $\ln \gamma_1^\infty = A_{12}$ and $\gamma_1^\infty = 5.17$; Similarly $\gamma_2^\infty = 3.30$.

Linear Fitting of the van Laar Model

Like the Margules models, the van Laar model can be linearized. Eqn. 12.12 can be rearranged:

$$\frac{x_1 x_2 RT}{G^E} = \left(\frac{1}{A_{21}} - \frac{1}{A_{12}}\right)x_1 + \frac{1}{A_{12}} \qquad 12.17$$

Therefore, if numerical values for the left-hand side are determined using G^E from experimental data as illustrated in Example 11.1 on page 414 and plotted versus x_1, the slope yields $(1/A_{21} - 1/A_{12})$, and the intercept yields $1/A_{12}$. The value of $1/A_{21}$ can also be determined by the value at $x_1 = 1$. Sometimes plots of the data are non-linear when fitting is attempted. This does not necessarily imply that the data are in error. It implies that an alternative model may fit the data better. By plotting the data in both the Margules and van Laar linearized forms, the better model can be identified as the one that is most linear. Fitting the data for 2-propanol + water presented in Example 11.8 (ignoring the first mixture point), results in an intercept = $1/A_{12} = 0.4993$, slope = 0.3486, resulting in $A_{12} = 2.00$, $A_{21} = 1.18$.

12.3   SCATCHARD-HILDEBRAND THEORY

Returning to Eqn. 12.10, G. Scatchard in Europe and Joel H. Hildebrand in the United States both made similar adjustments to match the van der Waals equation to experiment and provide a model capable of predictions for nonpolar fluids. They made an assumption that is equivalent to assuming $k_{12} = 0$ in Eqn. 12.4. Setting $a_{12} = \sqrt{a_{11}a_{22}}$, and collecting terms,

$$U^E = \frac{x_1 x_2 V_1 V_2}{x_1 V_1 + x_2 V_2}\left(\frac{a_{11}}{V_1^2} + \frac{a_{22}}{V_2^2} - 2\sqrt{\frac{a_{11}}{V_1^2}\frac{a_{22}}{V_2^2}}\right) = \frac{x_1 x_2 V_1 V_2}{x_1 V_1 + x_2 V_2}\left(\frac{\sqrt{a_{11}}}{V_1} - \frac{\sqrt{a_{22}}}{V_2}\right)^2 \qquad 12.18$$

J.H. Hildebrand is credited with suggesting that helium be mixed with breathing air in deep sea diving to minimize "the bends." He was awarded the ACS Priestly award in 1962. He continued to maintain an active professional life until he was 100.

Scatchard and Hildebrand recognized the unknown parameters in terms of volume fractions and disperse attraction energies that could be related to the pure component values. Defining a term called the "**solubility parameter**,"

$$U^E = \Phi_1 \Phi_2 (\delta_1 - \delta_2)^2 (x_1 V_1 + x_2 V_2) \qquad 12.19$$

where $\Phi_i \equiv x_i V_i / \sum_i x_i V_i$ is known as the "volume fraction" and 　　12.20

$\delta_i \equiv \sqrt{a_{ii}} / V_i$ is known as the "solubility parameter." 　　12.21

To estimate the value of δ_i, Scatchard and Hildebrand suggested that experimental data be used such that

$$\delta_i \equiv \sqrt{\frac{\Delta U_i^{vap}}{V_i}} = \sqrt{\frac{\Delta H_i^{vap} - RT}{V_i}} \qquad 12.22$$

(*Note the units on the "a" parameter from Eqn. 12.2, and from comparing Eqns. 12.21 and 12.22, and note the way V_i moves inside the root in Eqn. 12.22.*)

Table 12.1 *Solubility Parameters in $(J/cm^3)^{1/2}$ and Molar Volumes (cm^3/mol) for Various Substances as liquids at 298 K*

1-Olefins	δ	V^L	Napthenics	δ	V^L	Aromatics	δ	V^L
1-pentene	14.11	109	cyclopentane	17.80	93	benzene	18.82	88
1-hexene	15.14	124	cyclohexane	16.77	107	toluene	18.20	106
1,3 butadiene	14.52	86	Decalin	18.00	156	ethylbenzene	18.00	122
Amines	δ	V^L	**Ketones**	δ	V^L	styrene	19.02	114
ammonia	33.34	28	acetone	20.25	73	n-propylbenzene	17.59	139
methyl amine	22.91	46	2-butanone	19.02	89	anthracene	20.25	145
ethyl amine	20.45	65	2-pentanone	17.80	106	phenanthrene	20.05	186
pyridine	21.57	80	2-heptanone	17.39	139	naphthalene	20.25	125
n-Alkanes	δ	V^L	**Alcohols**	δ	V^L	**Ethers**	δ	V^L
n-pentane	14.32	114	water	47.86	18	dimethyl ether	18.00	68
n-hexane	14.93	130	methanol	29.66	40	diethyl ether	15.14	103
n-heptane	15.14	145	ethanol	25.57	58	dipropyl ether	15.95	136
n-octane	15.55	162	n-propanol	21.48	74	furan	19.23	72
n-nonane	15.95	177	n-butanol	27.82	91	THF	18.61	81
n-decane	16.16	194	n-hexanol	21.89	124			
			n-dodecanol	20.25	222			

In other words, δ_i is assumed to provide a standard measure of the "energy density" for each component. Because it represents the energy departure divided by volume, it is called the **cohesive**

energy density. As long as a standard reference condition is used, any convenient set of ΔU^{vap} and V_i may be applied. A convenient set of conditions that has become customary is the saturated liquid at 298 K. On this basis, we can tabulate a fair number of solubility parameters for ready reference, as shown in Table 12.1. Note that many similar compounds have similar values for their solubility parameters. Since similar solubility parameters yield small excess energies, solutions of similar components are predicted to be nearly ideal, as intuitively expected. By scanning the tables for the values of solubility parameters, we can quickly estimate whether the ideal solution model should be accurate or not. This approach gives a quantitative flavor to the old adage "like dissolves like."

Turning to the Gibbs energy, the regular solution assumptions give,

$$G^E = U^E = \Phi_1 \Phi_2 (\delta_1 - \delta_2)^2 (x_1 V_1 + x_2 V_2) \qquad 12.23$$

And the resultant activity coefficients are

$$\boxed{RT\ln\gamma_1 = V_1 \Phi_2^2 (\delta_1 - \delta_2)^2} \qquad 12.24$$

Scatchard-Hildebrand theory.

$$\boxed{RT\ln\gamma_2 = V_2 \Phi_1^2 (\delta_1 - \delta_2)^2} \qquad 12.25$$

Example 12.2 VLE predictions using the Scatchard-Hildebrand theory

Actcoeff.xlsx, worksheet REGULAR.

Benzene and cyclohexane are to be separated by distillation at 1 bar. Use the Scatchard-Hildebrand theory to predict whether an azeotrope should be expected for this mixture.

Solution: We will implement the algorithm to test for an azeotrope from Section 11.7 on page 432. Given x_B and P, we should perform bubble-temperature calculations.

Using parameters from Table 12.1, at $x_B = 0.99$, guess $T = 350$ K:
$$\Rightarrow \quad \Phi_B = 0.99(88)/[0.99(88) + 0.01(107)] = 0.9879$$
Calculating vapor pressures:
$$P_B^{sat} = 10^\wedge(6.87982 - 1196.76/(76.85 + 219.161)) = 686.9 \text{ mmHg}$$
$$P_C^{sat} = 10^\wedge(7.26475 - 1434.15/(76.85 + 246.721)) = 680 \text{ mmHg}$$

Bubble-temperature calculation.

Applying Eqns. 12.24 and 12.25:
$$\ln\gamma_B = 88(1 - 0.9879)^2(18.74 - 16.75)^2/(8.314 \cdot 350) = 1.8\text{E-}5 \Rightarrow \gamma_B = 1.00002$$
$$\ln\gamma_C = 107(0.98789)^2 (18.74 - 16.75)^2/(8.314 \cdot 350) = 0.1443 \Rightarrow \gamma_C = 1.1552$$

Calculating the pressure and vapor mole fractions:

$$\sum_i y_i P = \sum_i x_i \gamma_i P_i^{sat} = 0.99(1.00)(686.9) + 0.01(1.1552)(680.0) = 687.9 \text{ mmHg}$$

$$\Rightarrow y_B = 0.99(1.00)(686.9)/687 = 0.895, \ y_C = 0.01(1.1552)(680.0)/687 = 0.010$$

Since $\sum_i y_i < 1$, we must guess a higher temperature.

Guess $T = 354$ K $\Rightarrow P_B^{sat} = 777.7; P_C^{sat} = 770.2; \gamma_B = 1.00; \gamma_C = 1.1533$

Example 12.2 VLE predictions using the Scatchard-Hildebrand theory (Continued)

$$\sum_i y_i P = \sum_i x_i \gamma_i P_i^{sat} = 0.99(1.00)(777.7) + 0.01(1.1533)(770.2) = 778.8$$

$$\Rightarrow y_B = 1.013, y_C = 0.0113.$$

Interpolating between the first guesses:

$$T \approx 350 + 4(760 - 687)/(778.9 - 687) = 353.2$$

$$P_B^{sat} = 758.8; \; P_C^{sat} = 751.5; \; \gamma_B = 1.00; \; \gamma_C = 1.1537$$

$$\sum_i y_i P = 0.99(758.8)1.0 + 0.01(751.5)1.1537 = 760 \text{ mmHg}$$

converged with $\Rightarrow y_B = 0.9886 < x_B = 0.99$

At $x_B = 0.01$, guess $T = 353.5$ K $\Rightarrow \Phi_B = 0.01(88)/[0.01(88) + 0.99(107)] = 0.0082$

$$\ln \gamma_C = 107(0.0082)^2(18.74 - 16.75)^2/(8.314 \cdot 350) \approx 0 \Rightarrow \gamma_C = 1.00$$

$$\ln \gamma_B = 88(1 - 0.0082)^2(18.74 - 16.75)^2/(8.314 \cdot 350) = 0.1232 \Rightarrow \gamma_B = 1.126$$

$$\sum_i y_i P = \sum_i x_i \gamma_i P_i^{sat} = 0.01(1.126)(765.9) + 0.99(1.00)(758.4) = 759.5 \text{ mmHg}$$

converged with $\Rightarrow y_B = 0.011 > x_B = 0.01$

Therefore, $(y_B - x_B)$ changes sign between 0.01 and 0.99, so the system has an azeotrope.

When the Scatchard-Hildebrand solution theory is used, the $\{\delta_i\}$ and $\{V_i\}$ are available directly from pure component data, and in principle, there are no adjustable parameters. The theory is entirely predictive. The van Laar theory, on the other hand, treats both A_{12} and A_{21} as adjustable parameters. We can also obtain a compromise by assuming

$$a_{12} = \sqrt{a_{11} a_{22}} (1 - k_{12})$$

The activity coefficient expressions for binary solutions become:

$$\boxed{RT \ln \gamma_1 = V_1 \Phi_2^2 [(\delta_1 - \delta_2)^2 + 2k_{12}\delta_1\delta_2]} \; \boxed{RT \ln \gamma_2 = V_2 \Phi_1^2 [(\delta_1 - \delta_2)^2 + 2k_{12}\delta_1\delta_2]} \quad 12.26$$

In mixtures of compounds that deviate moderately from ideal-solution behavior, the Scatchard-Hildebrand solution theory with binary interaction parameters can be extremely helpful. The binary interaction parameter in those cases serves to adjust the magnitude of the excess Gibbs energy without addressing the skewness directly. Large deviations, however, are generally accompanied by non-ideal mixing in the volume and entropy. In those cases, the van Laar equations can often be useful in correlation, but the physical meaning behind the parameters is generally lost.

12.4 THE FLORY-HUGGINS MODEL

In deriving the entropy of mixing ideal gases in Eqn. 4.8 on page 138, we applied the notion that ideal gases are point masses and have no volume. We considered the entropy of mixing to be determined by the total volume of the mixture. When we consider the entropy of mixing liquids, however, we realize that the volume occupied by the molecules themselves is a significant part of the

---

Example 12.3 Deriving activity models involving volume fractions

The derivation of Scatchard-Hildebrand theory shows that volume fraction arises naturally as a characterization of composition, rather than mole fraction. This observation turns out to be true for many theories. Show that you can derive the relevant activity model from a Gibbs excess model involving volume fraction by deriving Eqn. 12.24 from Eqn. 12.28.

Solution: $\underline{G}^E = RT(\delta_2 - \delta_1)^2 (n_1 V_1 + n_2 V_2)\, \Phi_1\, \Phi_2$

Taking the derivative of the equation for G^E involves applying the chain rule to the three compositional factors: $(n_1 V_1 + n_2 V_2)$, Φ_1, and Φ_2.

The derivative of the first term is simply V_1 and,

$RT\ln\gamma_1 = \partial \underline{G}^E / \partial n_1 = RT(\delta_2 - \delta_1)^2 [\, V_1\, \Phi_1\, \Phi_2 + (n_1 V_1 + n_2 V_2)(\Phi_2\, \partial\Phi_1/\partial n_1 + \Phi_1\, \partial\Phi_2/\partial n_1)\,].$

It is helpful to maintain dimensional consistency in order to provide a quick check as we proceed. This can be achieved by multiplying and dividing by n, resulting in:

$RT\ln\gamma_1 = RT(\delta_2 - \delta_1)^2 [\, V_1\, \Phi_1\, \Phi_2 + (x_1 V_1 + x_2 V_2)(\Phi_2\, n\partial\Phi_1/\partial n_1 + \Phi_1\, n\partial\Phi_2/\partial n_1)\,]$

A key strategy in these derivations is to replace all compositional quantities with expressions depending only on $\{n_i\}$, not $\{x_i\}$, and not $\{\Phi_i\}$. For $\{\Phi_i\}$, this is easily achieved by multiplying the numerator and denominator by n, noting that $x_i = n_i/n$.

$\Phi_1 = x_1 V_1/(x_1 V_1 + x_2 V_2) = n_1 V_1/(n_1 V_1 + n_2 V_2); \quad \Phi_2 = x_2 V_2/(x_1 V_1 + x_2 V_2) = n_2 V_2/(n_1 V_1 + n_2 V_2).$

Taking the derivative involves product rule for Φ_1 and a simple reciprocal for Φ_2:

$\partial\Phi_1/\partial n_1 = V_1/(n_1 V_1 + n_2 V_2) - n_1 V_1^2/(n_1 V_1 + n_2 V_2)^2; \quad \partial\Phi_2/\partial n_1 = - n_2 V_2 V_1/(n_1 V_1 + n_2 V_2)^2.$

Multiplying by n and simplifying gives:

$$n\partial\Phi_1/\partial n_1 = V_1/(x_1 V_1 + x_2 V_2) - x_1 V_1^2/(x_1 V_1 + x_2 V_2)^2 = [V_1/(x_1 V_1 + x_2 V_2)](1 - \Phi_1) \qquad 12.27$$

$$n\partial\Phi_2/\partial n_1 = - n_2 V_2 V_1/(x_1 V_1 + x_2 V_2)^2 = -[V_1/(x_1 V_1 + x_2 V_2)](\Phi_2) \qquad 12.28$$

Substituting and noting that $(x_1 V_1 + x_2 V_2)$ cancels between numerator and denominator,

$$RT\ln\gamma_1 = RT(\delta_2 - \delta_1)^2 [\, V_1\, \Phi_1\, \Phi_2 + (\Phi_2 V_1(1 - \Phi_1) - \Phi_1 V_1(\Phi_2)]$$

The first and last terms in the brackets cancel. Also, for a binary mixture, $\Phi_2 = (1 - \Phi_1)$. So,

$$RT\ln\gamma_1 = RT(\delta_2 - \delta_1)^2 [V_1\Phi_2^2]$$

This procedure is extremely similar for all G^E models. In particular, Eqns. 12.27 and 12.28 can be easily adapted to any G^E model involving $\{\Phi_i\}$, if you first apply the chain rule thoughtfully.

---

total liquid volume. The volume occupied by one molecule is not accessible to the other molecules, and therefore, our assumptions regarding entropy may be inaccurate. One simple way of correcting for this effect is to subtract the volume occupied by the molecules from the total volume and treat the resultant "free volume" in the same way we treated ideal gas volume.

> **Free volume** is the difference between the volume of a fluid and the volume occupied by its molecules.

To use the concept, we assume that there is a fractional free volume, ϖ, globally applicable to all liquids and liquid mixtures. Let us further assume that the entropy change for a component is given by the change in free volume available to that component.

Example 12.4 Scatchard-Hildebrand versus van Laar theory for methanol + benzene

Fit the Scatchard-Hildebrand and van Laar models to the methanol + benzene azeotrope. Match the azeotropic pressure (and the composition in the case of the van Laar two-parameter model). The azeotrope appears at 58.3°C and $x_m = 0.614$. The vapor pressures at 58.3°C are 591.3 mmHg for methanol, 368.7 mmHg for benzene.

Solution: The van Laar parameters and binary interaction parameter are determined by matching the azeotropic pressure (and composition for the van Laar case) as described in previous examples. The resultant calculations are described in the worksheet REGULAR in the workbook Actcoeff.xlsx and the MATLAB file Ex12_04.m. Though not apparent from the figure below, the Scatchard-Hildebrand theory incorrectly predicts LLE until the binary interaction parameter is adjusted. See the supporting computer files. Fig. 12.1 illustrates the results of the fitting.

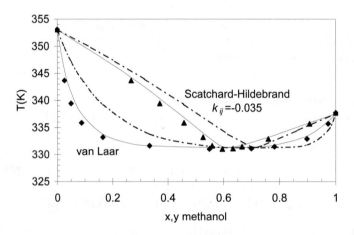

Figure 12.1 *T-x-y diagram for methanol and benzene for Example 12.4. The compositions are plotted in terms of mole fractions of methanol.*

The free volume available to any pure component is

$$\underline{V}_{f,i} = (n_i V_i)\varpi \qquad \qquad 12.29$$

If we assume that there is no volume change on mixing, the resultant free volume in the mixture is given by the same fraction, ϖ, and the mixture volume is

$$\underline{V}_{f,mixture} = (n_1 V_1 + n_2 V_2)\varpi \qquad \qquad 12.30$$

When two components mix, each component's entropy increases according to how much more space it has by an modification of Eqn. 4.6 using the free volume rather than the total volume:

$$\bar{S}_i - S_i = R \ln \frac{V_{f,mixture}}{V_{f,i}} = R \ln \frac{(n_1 V_1 + n_2 V_2)\varpi}{(n_i V_i)\varpi} = -R \ln \Phi_i \qquad 12.31$$

$$\Delta S_{mix} = S - \sum_i x_i S_i = \sum_i x_i (\bar{S}_i - S_i) = -R \sum_i x_i \ln \Phi_i \qquad 12.32$$

Note that Eqn. 12.32 reduces to the ideal solution result, Eqn. 10.63, when $V_1 = V_2$. The excess entropy is

$$S^E = -R \sum_i x_i \ln(\Phi_i / x_i) \qquad 12.33$$

This expression provides a simplistic representation of deviations of the entropy from ideal mixing. The entropy of mixing given by Eqn. 12.33 is frequently called the combinatorial entropy of mixing because it derives from the same combinations and permutations that we discussed in the case of particles in boxes. If entropy is the dominant factor in mixing, this formula can be used to find the excess Gibbs energy. When the excess enthalpy is zero, the mixture is called **athermal.**

$$\boxed{G^E = H^E - TS^E = RT(x_1 \ln(\Phi_1/x_1) + x_2 \ln(\Phi_2/x_2))} \qquad 12.34$$

Flory's equation.

It can also be combined with the Scatchard-Hildebrand solution theory[6] to derive the predictive theory of Blanks and Prausnitz[7] or the more common "Flory-Huggins" theory. These expressions are particularly important for solutions containing large molecules like polymers.

For a binary solution,

$$G^E = H^E - TS^E = RT(x_1 \ln(\Phi_1/x_1) + x_2 \ln(\Phi_2/x_2)) + \Phi_1 \Phi_2 (\delta_1 - \delta_2)^2 (x_1 V_1 + x_2 V_2) \quad 12.35$$

$$\ln \gamma_1 = \ln(\Phi_1/x_1) + (1 - \Phi_1/x_1) + \frac{V_1}{RT}\Phi_2^2 (\delta_1 - \delta_2)^2 \qquad 12.36$$

$$\ln \gamma_2 = \ln(\Phi_2/x_2) + (1 - \Phi_2/x_2) + \frac{V_2}{RT}\Phi_1^2 (\delta_1 - \delta_2)^2 \qquad 12.37$$

Frequently, for mixtures of polymer and solvent, the enthalpic term is fitted empirically to experimental data by adjusting the form of the equation to be the Flory-Huggins model,

$$\boxed{G^E = RT(x_1 \ln(\Phi_1/x_1) + x_2 \ln(\Phi_2/x_2)) + \Phi_1 \Phi_2 (x_1 + x_2 r)\chi RT} \qquad 12.38$$

Flory-Huggins model.

where component 1 is always the solvent, and component 2 is always the polymer. The variable $r = V_2/V_1$ denotes the ratio of volume of the polymer to the solvent. Similarly, $\chi \equiv V_1 (\delta_1 - \delta_2)^2/RT$. A solvent for which $\chi = 0$ is an athermal mixture. Plotting the result for S^E versus mole fraction for several size ratios, Fig. 12.2 shows that it is always positive, and it becomes larger and more

---

6. Huggins, M.L. 1941. *J. Phys. Chem.*, 9:440; and 1942. *Ann. N.Y. Acad. Sci.* 43:1.
7. Blanks, R.F., Prausnitz, J.M. 1964. *Ind. Eng. Chem. Fundam*, 3:1.

skewed as the size ratio increases. Thus, the size ratio has a large effect on the phase stability when the ratio is large.

Example 12.5 Polymer mixing

One of the major problems with recycling polymeric products is that different polymers do not form miscible solutions with one another; rather, they form highly non-ideal solutions. To illustrate, suppose 1g each of two different polymers (polymer A and polymer B) is heated to 127°C and mixed as a liquid. Estimate the activity coefficients of A and B using the Flory-Huggins model.

	MW	$V(cm^3/mol)$	$\delta(J/cm^3)^{1/2}$
A	10,000	1,540,000	19.2
B	12,000	1,680,000	19.4

Solution:

$$x_A = (1/10,000)/(1/10,000 + 1/12,000) = 0.546; \quad x_B = 0.454$$

$$\Phi_A = 0.546(1.54)/[0.546(1.54) + 0.454(1.68)] = 0.524; \quad \Phi_B = 0.476$$

$$\ln\gamma_A = \ln(0.5238/0.5455) + (1 - 0.5238/0.5455)$$
$$+ (1.54E6(19.4 - 19.2)^2(0.4762)^2)/(8.314(400))$$

$$= -0.0008 + 4.200 \Rightarrow \gamma_A = 66$$

$$\ln\gamma_P = \ln(0.4762/0.4545) + (1 - 0.4762/0.4545)$$
$$+ (1.68E6(19.4 - 19.2)^2(0.5238)^2)/(8.314(400))$$

$$= +0.0008 + 5.544 \Rightarrow \gamma_B = 258$$

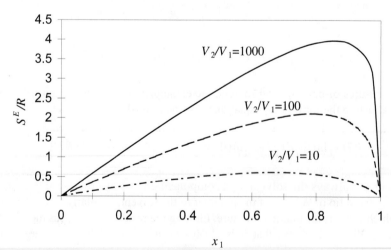

Figure 12.2 *Illustration of excess entropy according to Flory's equation for various pure component volume ratios.*

Several important implications can be interpreted from Example 12.5. It is often noted that the Flory-Huggins model is especially appropriate for polymer solution models. While the excess entropy is most significant for polymer-solvent mixtures, it is not so important for polymer-polymer mixtures. The key to polymer-polymer mixtures is noting that the activity coefficient is proportional to the exponential of the molar volume of the polymer. Therefore, even tiny differences in solubility parameter are amplified. Furthermore, the large activities computed for these components mean that the fugacities of these components would be greatly enhanced if intermingled at this composition. This means that they show a strong tendency to escape from each other. On the other hand, polymer compounds are too non-volatile to escape to the vapor phase. The only alternative is to escape into separate liquid phases. In other words, the liquids become immiscible. Computations of activity coefficients like those above play a major role in the liquid-liquid phase equilibrium calculations detailed in Chapter 14.

If you think creatively for a moment, you can imagine staggering possibilities for this amplification principle. To begin, we could synthesize molecules of just the right size to generate phase behavior that precisely measures the magnitude of the molecular interactions between, say, polyethylene and polypropylene. But suppose we would like to homogenize an immiscible blend of high molecular weight polymers. Then perhaps a polymer that was half of each type could help. Next comes a consideration that we generally avoid throughout this text. Would the *intra*molecular structure make a difference? In other words, it is possible to synthesize one (random) copolymer that alternates randomly between monomer types and another (block) copolymer that has a long section of one monomer type followed by another long section of a different monomer type. The properties resulting from these different intramolecular structures are very different. If the blocks are large enough, they can aggregate similar to phase separation. This is not exactly a phase separation, however, because the blocks might be part of the same molecule. Repeating this theme on a grand scale with 20 particular monomer types (amino acids) is called protein engineering. Modern science is just beginning to manipulate these kinds of interactions to synthesize self-assembled structures with specific design objectives. Exploring these possibilities would take us beyond the introductory level, however.

12.5   MOSCED AND SSCED THEORIES

The Scatchard-Hildebrand theory provides reasonable results for hydrocarbon mixtures, but the results can be highly unreliable if one of the components hydrogen bonds, especially if one of the components is water. The MOSCED, SSCED, and HSP models remedy this problem by accounting for hydrogen bonding as a separate contribution to the solubility parameter. MOSCED (pronounced moss-ked) stands for MOdified Separation of Cohesive Energy Density. SSCED (pronounced sked) stands for Simplified Separation of Cohesive Energy Density. Recall that the cohesive energy density is the term for δ^2. The HSP model (Hansen Solubility Parameters) is similar in concept, but does not distinguish between acidity and basicity.[8] Therefore, it cannot predict negative deviations from ideality in the manner of MOSCED or SSCED and is omitted from detailed discussion. It has been broadly applied, however, and provides impressive demonstrations of what can be achieved with these kinds of theories. The concept behind these models is that hydrogen bonding should be counted separately from the physical interactions envisioned by Scatchard and Hildebrand.

---

8. Hansen, C.M. 2007. *Hansen Solubility Parameters: A User's Handbook.* Boca Raton, FL: CRC Press, Inc.

The MOSCED Model

The MOSCED model is given by,[9]

$$\ln\gamma_2^\infty = \frac{V_2}{RT}\left[(\lambda_2 - \lambda_1)^2 + q_1^2 q_2^2 \frac{(\tau_2^T - \tau_1^T)^2}{\psi_1} + \frac{(\alpha_2^T - \alpha_1^T)(\beta_2^T - \beta_1^T)}{\xi_1}\right] + d_{12} \qquad 12.39$$

$$d_{12} = 1 - \left(\frac{V_2}{V_1}\right)^{aa} + aa\ln\left(\frac{V_2}{V_1}\right) \qquad 12.40$$

where λ_i is the dispersion factor (e.g., equal to δ for n-alkanes), τ_i is the polarity factor (e.g., 3.95 for benzene), q_i is a factor ranging from 0.9 to 1 (for our purposes, $q_i = 1$ for all i, cf. reference 9 for details) and α and β are acidity and basicity parameters. Values of λ and τ are given in Table 10.2. Values of α and β are given on the back flap. We recognize d_{12} as the Flory-Huggins contribution. aa, ψ_1 and ξ_1 are parameters characterizing solvent properties.

$$aa = 0.953 - 0.002314((\tau_2^T)^2 + \alpha_2^T \beta_2^T) \qquad 12.41$$

$$\alpha_i^T = \alpha_i\left(\frac{293}{T(K)}\right)^{0.8}; \beta_i^T = \beta_i\left(\frac{293}{T(K)}\right)^{0.8}; \tau_i^T = \tau_i\left(\frac{293}{T(K)}\right)^{0.4} \qquad 12.42$$

$$\psi_1 = POL + 0.002629\,\alpha_1^T \beta_1^T \qquad 12.43$$

$$\xi_1 = 0.68(POL - 1) + [3.24 - 2.4\exp(-0.002687(\alpha_1\beta_1)^{1.5})]^{(293/T)^2} \qquad 12.44$$

$$POL = 1 + 1.15q_1^4[1 - \exp(-0.002337(\tau_1^T)^3)] \qquad 12.45$$

Note that MOSCED is not intended to describe the entire solution behavior directly. Instead, it provides estimates for the infinite dilution activity coefficients. At that point, another activity coefficient model can be applied by fitting its parameters at infinite dilution to the MOSCED predictions.

The essential feature of this model is the explicit representation of acidity and basicity. Consider, for example, the acetone + chloroform system of Fig. 9.6(c). MOSCED model predicts a negative deviation from ideality for this system, as is the experimental behavior:

$$(\alpha_2 - \alpha_1)(\beta_2 - \beta_1) = (5.8 - 0)(0.12 - 11.14) = -63.9 \qquad 12.46$$

For 2-propanol + water, a positive deviation is properly indicated:

$$(\alpha_2 - \alpha_1)(\beta_2 - \beta_1) = (9.23 - 50.13)(11.86 - 15.06) = 131 \qquad 12.47$$

9. Lazzaroni, M.J., Bush, D., Eckert, C.A., Frank, T.C., Gupta, S., Olson, J.D. 2005. *Ind. Eng. Chem. Res.* 44:4075.

These measures of acidity and basicity provide useful insights into the chemical nature of compounds. It is notable that they can be characterized spectroscopically by mixing a range of compounds with standard bases and acids. To measure acidity, for example, one might individually mix pyridine with acetone, benzene, chloroform, and acetic acid. More acidic molecules would bind more strongly to the pyridine nitrogen and shift the ultraviolet absorption more strongly. So the magnitude of the shift would provide a relative measure of the acidity.[10] By verifying the trend with other standard bases, an average indicator could be developed for the acidity of each compound (and its variance). The spectroscopic measurement is independent from the VLE measurement, so the two observations strongly reinforce each other. This kind of chemical insight combined with spectroscopic evidence can be useful in a wide variety of settings. In catalyst design, for example, one might devise acidic adsorption sites to attract a reactant with high basicity, then measure spectroscopically whether the device was working. Students should look for creative opportunities to relate concepts like these across the curriculum.

The SSCED Model

Despite its attractions, the MOSCED model is relatively cumbersome. It has many terms, and in the end, another model (e.g., Redlich-Kister, van Laar, or a model from the next chapter) must be used to compute the phase behavior. As an pedagogical introduction to MOSCED, we would like to estimate the phase behavior with sufficient accuracy to predict whether an azeotrope or liquid-liquid separation may occur, but to make only approximate estimates with a single self-consistent theory. These motivations suggest a need for a simplified version of MOSCED.

Table 12.2 *Dispersion (λ) and Polarity (τ) Parameters in $(J/cm^3)^{1/2}$ liquids at 293 K*

1-Olefins	λ	τ	Napthenics	λ	τ	Aromatics	λ	τ
1-pentene	14.6	0.25	Cyclopentane	16.6	0	Benzene	16.7	3.95
1-hexene	15.2	0.23	Cyclohexane	16.7	0	Toluene	16.6	3.22
Amines	λ	τ	**Ketones**	λ	τ	Ethylbenzene	16.8	2.98
Aniline	16.5	9.41	Acetone	13.7	8.30	Naphthalene	17.8	4.53
Pyridine	16.4	6.13	2-butanone	14.7	6.64	Phenanthrene	18.5	5.31
			2-pentanone	15.1	5.49	Naphthalene	17.8	4.53
			2-heptanone	14.7	4.20			
n-Alkanes	λ	τ	**Alcohols**	λ	τ	**Ethers**	λ	τ
n-pentane	14.4	0.0	Water	10.6	10.5	Diethyl ether	14.0	2.79
n-hexane	14.9	0.0	Methanol	14.4	3.77	Dipropyl ether	15.2	2.00
n-heptane	15.2	0.0	Ethanol	14.4	2.53	MTBE	15.2	2.48
n-octane	15.4	0.0	*n*-propanol	14.0	1.95	THF	15.8	4.41
n-nonane	15.6	0.0	*n*-butanol	14.8	1.86			
n-decane	15.7	0.0	*n*-hexanol	15.0	1.27			
			n-octanol	15.1	1.31			

10. Kamlet, M.J., Abboud, J.-L.M., Abraham, M.H., Taft R.W. 1983. *J. Org. Chem.*, 48:2877. For solutes in liquids see Abraham, M.H., Andonian-Haftvan, J., Whiting, G.S., Leo, A., Taft, R.W. 1994. *J. Chem. Soc. Perkin Trans.* 2:1777.

Similar to MOSCED, the SSCED model retains the simple form of the Scatchard-Hildebrand model while correcting its gross misrepresentation of polar mixtures by taking advantage of the acidity and basicity measures of MOSCED. For a binary mixture, the SSCED model is

$$G^E = V\Phi_1\Phi_2[(\delta_2' - \delta_1')^2 + 2k_{12}\delta_2'\delta_1']$$

12.48

SSCED model.

$$RT\ln\gamma_k = V_k(1 - \Phi_k)^2[(\delta_2' - \delta_1')^2 + 2k_{12}\delta_2'\delta_1']$$

12.49

$$(\delta_i')^2 = \delta_i^2 - 2\alpha_i\beta_i$$

12.50

$$k_{12} = \frac{(\alpha_2 - \alpha_1)(\beta_2 - \beta_1)}{4\delta_2'\delta_1'}$$

12.51

Example 12.6 Predicting VLE with the SSCED model

Amines often function as bases that can moderate interactions with acidic compounds. In the case of triethylamine, however, the high hydrocarbon content competes with the basicity and it is difficult to intuitively assess how the solution ideality may turn out.

(a) Predict the bubble pressure and vapor composition of triethylamine (1) + ethanol (2) at 308 K and $x_1 = 0.59$ using the SSCED model.
(b) Compute the relative volatility, α_{LH}, at $x_1 = 0.01$ and 0.99, where triethylamine is the light key component. Is an azeotrope indicated? Copp and Everett (1953) report an azeotrope at $x_1 = 0.59$ and $P = 119$ mmHg.

SSCED and Antoine constants for triethylamine are:

M_w	ρ	δ	α	β	A	B	C	T_{Lo}°C	T_{Hi}°C
101	0.72	15.17	0.00	7.70	6.829	1206	216.8	-23	55

Solution:
(a) From the Antoine equation, $P_1^{sat} = 109$; $P_2^{sat} = 102$ mmHg (*cf.* Appendix E for ethanol);
$\delta_1' = 15.17 - 0 = 15.17$; $\delta_2' = (26.13^2 - 2\cdot12.58\cdot13.29)^{1/2} = 18.67$; $V_1 = 101/0.72 = 140$; $V_2 = 58.5$.

From Eqn. 12.51, $k_{12} = (12.58 - 0)(13.29 - 7.70)/(4\cdot15.17\cdot18.67) = 0.062$;
$[(\delta_2' - \delta_1')^2 + 2k_{12}\delta_1'\delta_2']/RT = [(18.67 - 15.17)^2 + 2\cdot0.062\cdot18.67\cdot15.17]/(8.314\cdot308) = 0.0185$.
At $x_1 = 0.59$, $\Phi_1 = 0.59\cdot140/(0.59\cdot140 + 0.41\cdot58.5) = 0.774$.

$\gamma_1 = \exp(140(1 - 0.774)^2 0.0185) = 1.141$; $\gamma_2 = \exp(58.5(0.774)^2 0.0185) = 1.916$;
$P = 0.59\cdot1.141\cdot109 + 0.41\cdot1.01\cdot102 = 154$; $y_1 = 0.59\cdot1.141\cdot109/154 = 0.477$.

(b) At $x_1 = 0.01$, $\Phi_1 = 0.01\cdot140/(0.01\cdot140 + 0.99\cdot58.5) = 0.024$;
$\gamma_1 = \exp(140(1 - 0.024)^2 0.0185) = 11.9$; $\gamma_2 = \exp(58.5(0.024)^2 0.0185) = 1.001$;
$P = 0.01\cdot11.9\cdot109 + 0.41\cdot1.001\cdot102 = 114$; $y_1 = 0.99\cdot1.001\cdot109/114 = 0.113$.
$\alpha_{LH}(0.01) = 11.9\cdot109/(1.001\cdot102) = 12.6$

Example 12.6 Predicting VLE with the SSCED model (Continued)

At $x_1 = 0.99$, $\Phi_1 = 0.99 \cdot 140/(0.99 \cdot 140 + 0.01 \cdot 58.5) = 0.996$;
$\gamma_1 = \exp(140(1 - 0.996)^2 0.0185) = 1.0001$; $\gamma_2 = \exp(58.5(0.996)^2 0.0185) = 2.93$;
$P = 0.99 \cdot 1.0001 \cdot 109 + 0.01 \cdot 2.93 \cdot 102 = 111$; $y_1 = 0.99 \cdot 1.0001 \cdot 109/111 = 0.973$.

$\alpha_{LH}(0.99) = 1.0001 \cdot 109/(2.93 \cdot 102) = 0.363$

Therefore, an azeotrope is suspected since $\alpha_{LH} - 1$ changes sign as discussed in Section 11.11. The system should be evaluated experimentally or with a literature search.

To follow up, Fig. 12.3 shows through comparison to experiment that the SSCED model over-estimates the nonideality of the solution, but the prediction of an azeotrope is valid. For a broader perspective, we can go beyond Example 12.6 and compare to the Scatchard-Hildebrand model, but the Scatchard-Hildebrand (ScHil) model is not even close. In fact, the Scatchard-Hildebrand model indicates VLLE where none exists. This is a common problem with Scatchard-Hildebrand theory in the presence of hydrogen bonding. It undermines the viability of the Scatchard-Hildebrand model for most applications, but the SSCED model retains its simplicity while providing a reasonable basis for conceiving formulations predictively. As a final note, the MAB model performs slightly better than the SSCED model for this mixture with $P = 148$ mmHg.

12.6 MOLECULAR PERSPECTIVE AND VLE PREDICTIONS

Can we just forget about some of these models? After all, the van der Waals perspective inherently accounts for the molecular properties through a and b. Why should we worry with so many variations? Is the SSCED model really so different from the Scatchard-Hildebrand model? What about the MAB model? The van Laar model? Which model is "best?" In every case, there is a factor mitigating against entirely eliminating any one of the models from consideration.

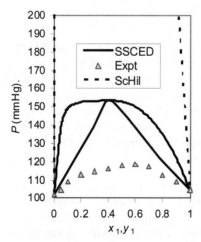

Figure 12.3 *VLE of triethylamine(1)+ethanol at 308 K. The basicity of the amine moderates the non-ideality of the solution by solvating the hydroxyl interaction. Data of Copp J.L., Everett D.H. 1953. Disc. Faraday Soc.15:174.*

To begin, it is possible to derive the MAB model as a special case of the SSCED model. When $V_1 = V_2$ and $\delta_1' = \delta_2'$, we obtain $\Phi_i = x_i$, and we obtain the Margules one-parameter model,

$$G^E = Vx_1x_2[2k_{12}\delta_2'\delta_1'] = \frac{(V_1 + V_2)}{2}x_1x_2\left[2\frac{\Delta\alpha\Delta\beta}{4(\delta_2'\delta_1')}\delta_2'\delta_1'\right] = A_{12}x_1x_2 \qquad 12.52$$

Technically, we could simply substitute V_1 or V_2 instead of $(V_1 + V_2)/2$, but writing it this way provides a small compensation for the observation that it is extremely unlikely that $V_1 = V_2$. So, if the MAB model is such an oversimplification, why not forget about it and just use the SSCED model? That would be a good argument, except that MAB is such a simple model. It does not require converting from mole fraction to volume fraction and you can anticipate the predicted sign on G^E without using a calculator.

On the other hand, the $\Delta\delta'$ term of SSCED is quite significant in some cases. For example, n-hexane + methylethylketone has a significant non-ideality that is overlooked by MAB. A similar argument could be made about polyethylene + polypropylene. Furthermore, the distinction between V_1 and V_2 in SSCED correctly indicates that it is much less favorable to squeeze a large molecule into a fluid of small molecules than vice versa, as in the n-butanol + water system.

Another argument could be made about the difference between the SSCED and Scatchard-Hildebrand models. Both models have the same skewness when k_{12} is fit to experimental data. The Scatchard-Hildebrand model takes precedence historically. So maybe we should forget the SSCED model. On the other hand, SSCED provides better *a priori* predictions of phase behavior when an associating component is involved. Eqn. 12.50 shows that the solubility parameter is unaltered if $\alpha = 0$ or $\beta = 0$, but it is substantially diminished for associating compounds like alcohols and water. With this change alone, the estimated nonideality is substantially diminished. This is a step in the right direction because overestimating the nonideality (Scatchard-Hildebrand) may cause more confusion than treating the solution as ideal (Fig. 12.4(a)). Remember to "First, do no harm." A similar observation is illustrated in Fig. 12.4(b). The peculiar lines of the Scatchard-Hildebrand (ScHil) model show what happens if a VLE model is applied when the activity model indicates VLLE. The experimental data indicate no LLE for either system, showing the qualitative deficiency of the Scatchard-Hildebrand model for associating mixtures, as well as its quantitative deficiency. The SSCED model, on the other hand, is qualitatively correct, and semi-quantitative in its predictions.

Another distinction between SSCED and the Scatchard-Hildebrand model is the guideline for k_{12} given by Eqn. 12.51. We can compute the VLE with this guideline and compare to the VLE at $k_{12} = 0$ to get a range of estimates that suggests the direction of the nonideality and its magnitude. Having multiple VLE estimates may seem like a bad idea, but computing a single crude prediction ignoring alternative estimates would be a much worse idea. Contradictory estimates should remind us of the value of using experimental data to correlate k_{12} whenever possible. The procedure for correlating k_{12} of SSCED based on experimental data is not different from the procedure for the Scatchard-Hildebrand model, nor from the procedure for the Margules or van Laar models.

One could also argue that the van Laar model provides the best fit of the VLE data, so it should be preferred. On the other hand, it offers no predictive capability, and we really would like to conceive designs for formulations. Furthermore, the van Laar model cannot be extended to multicomponent mixtures in a manner that is consistent with the van der Waals perspective, as discussed in the next section.

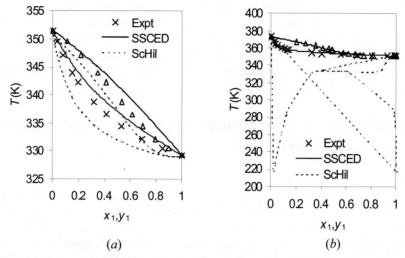

Figure 12.4 *Predictions of SSCED and Scatchard-Hildebrand theories at 1 bar. For Scatchard-Hilde-brand theory, $k_{ij} = 0$ is applied. For SSCED, Eqn. 12.51 applies. (a) Acetone+ethanol (data of Amer, H.H., Paxton, R.R. van Winkle, M. 1956. I&EC, 48:142); (b) ethanol+water (data of Bloom, C.H., Clump, C.W., Koeckert, A.H. 1961. Ind.Eng.Chem. 53:829).*

Finally, one might consider the MOSCED model to be the best model. First, it includes the Flory-Huggins correction for polymer-solvent interactions. Second, when combined with the van Laar or two-parameter Margules model, it provides predictive capability that is superior to the SSCED model. In a particular case study involving organic nitrates, the SSCED model gave roughly 10% deviation in predicted bubble pressure while the MOSCED model gave only 5%. Furthermore, the predictive insight of MOSCED is at least as good as that of SSCED because it uses the same values of α, β, δ, and V. On the other hand, it requires several more intermediate calculations than the SSCED model, including the translation of the infinite dilution activity coefficients into a different activity model and subsequent computation of the activity coefficient at the concentration of interest. The requirement of so many intermediate computations generally necessitates the use of a computer, in which case the methods of Chapter 13 are generally preferred. Overall, the argument in favor of MOSCED over SSCED is probably the best, however.

In summary, the "best" overall activity model should account for simple molecular properties in addition to providing a basis for fitting the data of a specific binary system, because our overall goal includes conceiving of formulations. Formulations generally involve more than two components and designing them requires simple intuitive insights like energy density (reflected in δ_i') and hydrogen bonding (reflected in α and β). A small molecule with a high energy density disfavors molecules with less energy density just as an associating compound squeezes out an inert one. The interplay between these types of molecular interactions is significant and it is best to contemplate both influences when predicting phase behavior. Through the SSCED model, several observations can be cited about this interplay: (1) Ignoring hydrogen bonding leads to overestimates of solution nonideality, as observed for the Scatchard-Hildebrand model; (2) accounting for hydrogen bonding reduces the differences in disperse interactions (i.e. $\Delta\delta_i' < \Delta\delta_i$); (3) competition between the effects of complexation and energy density can make it difficult to predict whether deviations from Raoult's law are positive or negative; and (4) large nonidealities result when both association and energy density indicate unfavorable mixing, typified by the hydrophobic effect.

In this context the SSCED and MOSCED models would appear to provide the best overall models within the van der Waals perspective because they both account for the interplay between energy density and hydrogen bonding. The trade-off between them is one of precision versus simplicity. Resolving this trade-off requires the context of a particular application. If you are always working with similar solvents and species, the greater detail of the MOSCED model may provide opportunities for refining your predictions.

12.7   MULTICOMPONENT EXTENSIONS OF VAN DER WAALS' MODELS

Most systems encountered in chemical processes and formulations are multicomponent. If the application requires bypassing an azeotrope, a third component (called an **entrainer**) might be added. If a biomembrane is to be penetrated by a pharmaceutical treatment, the formulation must at least account for water, the pharmaceutical, the biomembrane, and any additive. The output of a simple reaction like A + B → C would mean that at least three components must be separated if conversion was less than 100%. These examples and more lead to the conclusion that a multicomponent solution model is a necessity.

Unfortunately, the van Laar model makes the assumption that V_2/V_1 can be treated as an adjustable parameter. While this is fine for a binary mixture, it becomes problematic for a ternary mixture because knowing V_2/V_1 and V_3/V_2 means that V_3/V_1 must be implied. The normal procedure for the van Laar model would fit the binaries for 2-1, 3-2, and 3-1 independently, however. The likelihood of achieving a consistent value of V_3/V_1 from such a fit is practically zero.

On the other hand, the simple relation of the Scatchard-Hildebrand theory to the van der Waals equation permits a simple extension to multicomponent systems. The derivation of this extension is given by practice problem 12.4. The result is

$$G^E = \Sigma\, x_k V_k (\delta_k - <\delta>)^2 + V<<k_{mm}>>] \qquad\qquad 12.53$$

$$RT\ln\gamma_k = V_k[(\delta_k - <\delta>)^2 + 2<k_{km}>\delta_k - <<k_{mm}>>] \qquad\qquad 12.54$$

where $<\delta> = \Sigma\Phi_j\delta_j$; $<<k_{mm}>> = \Sigma\Phi_i\delta_i<k_{im}>$; and $<k_{im}> = \Sigma\Phi_j\delta_j<k_{ij}>$.

The SSCED model is similar to the Scatchard-Hildebrand model by design. This similarity guarantees that the algebra extending to multicomponent mixtures is identical, giving

$$RT\ln\gamma_k = V_k[(\delta_k' - <\delta'>)^2 + 2<k_{km}>\delta_k' - <<k_{mm}>>] \qquad\qquad 12.55$$

where $<\delta'> = \Sigma\Phi_j\delta_j'$; $<<k_{mm}>> = \Sigma\Phi_i\delta_i'<k_{im}>$; and $<k_{im}> = \Sigma\Phi_j\delta_j'<k_{ij}>$.

At first sight, the notation may seem confusing. The following example clarifies the necessary summations.

Example 12.7 Multicomponent VLE using the SSCED model

A partial condenser is generally indicated when distilling a stream that has a light gas impurity, even if the gas is dilute. A total condenser is impractical because of the "noncondensable" gas. Of course, we know that it must condense at some sufficiently low temperature, but it is wasteful to cool the stream that far or reflux the light gas component. A partial condenser condenses just as much liquid as necessary to keep the column functioning then sends the remainder downstream as a vapor distillate product. Downstream, a second partial condenser will provide liquid and the remaining gas can be purged to a flare tower.

Fermentation of corn to form acetone (A), n-butanol (B), and ethanol (E) in water (W) followed by a drying step results in the stream below to be distilled.[a] Dissolved carbon dioxide (CO_2) is a prevalent by-product of the fermentation process. Use the SSCED model and assume 2 bar pressure with splits of 99% ethanol and 2.8% water. Assume that acetone moderates the ethanol+water interaction to achieve this tops composition without an azeotrope interfering. You may estimate vapor pressures using the shortcut equation.

Comp.	CO_2	A	E	B	W
kmol/h	0.7	13	5	15	42
V^L	37.3	73.4	58.5	91.8	18
δ	14.6	19.6	26.1	23.4	47.9
α	1.87	0	12.58	8.44	50.13
β	0	11.01	13.29	11.11	15.06
δ'	14.6	19.6	18.67	18.96	27.94

where δ' has been computed from δ in accordance with Eqn. 12.50.

(a) Estimate the flow rate and composition of the distillate based on the key component splits.
(b) Estimate the dew temperature of the distillate stream as a preliminary estimate of the operating temperature in the partial condenser.

Solution: We will indicate the vapor outlet of the partial condenser stream as D.

(a) E will be the light key, and W the heavy key. Everything lighter than the light key is assumed to go out the top, and everything heavier than the heavy key is assumed to go out the bottom. We estimate the boiling temperatures (K) at 2 bar in the first row. Technically, CO_2 does not have a boiling temperature at 2 bar. Nevertheless, we apply the shortcut model here (at 351 K, where CO_2 is supercritical) merely as an estimate, as suggested in the problem statement. Components are sorted in decreasing volatility to show that a split between E and W sends CO_2 and A completely to distillate and B completely to the bottoms. The third row shows the consequent distillate (D) flows and fourth row shows the mole fractions.

Example 12.7 Multicomponent VLE using the SSCED model (Continued)

Comp.	CO$_2$	A	E	W	B
T_b (K)	197	350	371	392	414
D(kmol/h)	0.7	13	4.95	0.42	0
y_i^D	0.037	0.682	0.260	0.022	0

The tops composition of ethanol to water on a solvent-free basis is 4.95/(4.95+0.42) = 92%, which exceeds the azeotropic composition only slightly.

Dew -
temperature
calculation.

(b) For the **dew-temperature** calculation, we use option (b) from Appendix C and refer to the step numbers. As an initial estimate, $T \sim \Sigma \, y_i^D T_b = 351$ K and assuming Raoult's law (step 1), $K_i^{(0)} = P_i^{sat}/P$ gives $x_i^{(0)} = y_i^D/K_i^{(0)}$:

Comp.	CO$_2$	A	E	W	B	
y_i^D	0.037	0.682	0.260	0.022	0	
P_i^{sat}/P	89	1.01	0.51	0.25	0.13	(See note in part (a) about CO$_2$)
$x_i^{(0)}$	0.00	0.672	0.513	0.088	0	$\Sigma x_i^{(0)} = 1.273$ (step 1)
$x_i^{(0')}$	0.00	0.528	0.403	0.069	0	(step 1)

In the last row, we divide $x_i^{(0)}$ by $\Sigma x_i^{(0)}$ (the process is known as **normalization**), such that the mole fractions $x_i^{(0')}$ sum to one. To prepare for step 3, the SSCED model gives the values below for k_{ij}. For example, the value for ethanol + water is

$$k_{EW} = (12.58 - 50.13)(13.29 - 15.06)/(4(18.67)27.94) = 0.0319$$

Similarly, $k_{AE} = 0.0184$; $k_{AW} = 0.0896$.

Finally, we are ready to apply the SSCED model. We drop butanol and CO$_2$ from further calculations because their liquid compositions are zero. $x_i^{(1)}$ is the result of the first iteration in which the γ_i are applied. $x_i^{(2)}$ results from varying the temperature while assuming the γ_i are constant (even though the composition is changing slightly too). We are approximating to force $x_T = 1$ to get a good T guess (step 5) for the next iteration.

Comp.	CO$_2$	A	E	W	B	
$x_i^{(0')}$	0.00	0.528	0.403	0.069	0	
$\Phi_i^{(0')}$	0.00	0.6094	0.3710	0.0196	0	$<\delta'> = 19.4$
$<k_{im}>$	NA	0.177	0.239	1.294	NA	$<<k_{mm}>> = 4.478$
γ_i	NA	1.065	1.106	2.374	NA	(step 3)
$\gamma_i P_i^{sat}/P$	NA	1.070	0.566	0.605	NA	(step 3)
$x_i^{(1)}$	0.00	0.637	0.459	0.036	0	(step 3) $\Sigma x_i^{(1)} = 1.133$ (step 4)
$x_i^{(1')}$	0.00	0.563	0.405	0.032	0	$\Sigma x_i^{(1')} = 1.000$ (step 4)
$\gamma_i P_i^{sat}/P$	NA	1.208	0.645	0.684	NA	$T=354.88$K (step 5 ignoring new γ_i)
$x_i^{(2)}$	0.00	0.565	0.403	0.032	0	$\Sigma x_i^{(2)} = 1.000$

where γ_A, γ_E, γ_W, $<\delta'>$, $<k_{im}>$, and $<<k_{mm}>>$ are computed by Eqn. 12.55. For example,

---

Example 12.7   Multicomponent VLE using the SSCED model (Continued)

$<\delta'>$ = 0.6094(19.6) + 0.3710(18.67) + 0.0196(27.94)0.0319 = 19.4

$<k_{Em}>$ = 0.6094(19.6)0.0184 + 0.3710(18.67)**0.0** + 0.0196(27.94)0.0319 = 0.239

$<<k_{mm}>>$ = 0.6094(19.6)0.177 + 0.3710(18.67)0.239 + 0.0196(27.94)1.294 = 4.478

γ_E = exp(58.5((18.67 − 19.4)^2 + 2(0.239)18.67 − 4.478)/(8.314·351)) = 1.106

In going from $x_i^{(1')}$ to $x_i^{(2)}$, we have varied the temperature assuming that the γ_i are unchanged. This must be validated with one final iteration at T = 354.88 K.

Comp.	CO$_2$	A	E	W	B	
$x_i^{(2)}$	0.00	0.565	0.403	0.032	0	(step 3)
$\Phi_i^{(2)}$	0.00	0.6378	0.3535	0.0087	0	$<\delta'>$ = 19.4
$<k_{im}>$	NA	0.143	0.239	1.334	NA	$<<k_{mm}>>$ = 3.696
γ_i	NA	1.051	1.120	2.413	NA	(step 3)
$\gamma_i P_i^{sat}/P$	NA	1.192	0.653	0.695	NA	(step 3)
$x_i^{(3)}$	0.00	0.572	0.397	0.032	0	(step 3) $\Sigma x_i^{(3)}$ = 1.001 ~ 1.0 (step 7)

Note:

1. The liquid composition of less volatile components is enhanced in a partial condenser.

2. For this particular condenser, $K_W > K_E$ means there is a limit to splitting W from E.

a.   This problem is based on the AIChE 2009 National Student Design Competition, Richard L. Long Coordinator, AIChE New York (2008).

Noting that MAB is a special case of the SSCED model, the expression for the multicomponent SSCED model suggests a similar form for the MAB model.

$$G^E = V<<k_{mm}>> = \Sigma\Sigma \, x_i x_i A_{ij} \qquad \text{12.56}$$

The relation between the Margules form and the Redlich-Kister form suggests a similar relation.

$$G^E = \Sigma\Sigma \, x_i x_i B_{ij} + \Sigma\Sigma \, x_i x_i C_{ij}(x_i - x_j) + \Sigma\Sigma \, x_i x_i D_{ij}(x_i - x_j)^2 + ... \qquad \text{12.57}$$

In this manner, we can envision extending many activity models to multicomponent applications.

The power to design formulations that solve practical problems generally involves multicomponent systems. Design requires synthesis of many fundamental principles into a working toolbox as well as creative thinking. If we want to disperse an oil spill, we should consider the fate of the additive as well as the oil. After the oil is dispersed, where will the additive go? What is its toxicity? If we want to circumvent an azeotrope, where will the entrainer go? Should it be more volatile or less volatile than the key components? How should its molecular properties relate to those of the key components? What about the volatility of a solvent to remove paint? What about the solvent's water solubility? The multicomponent SSCED perspective empowers us to think creatively about these kinds of problems.

Example 12.8 Entrainer selection for gasohol production

The ethanol + water system has a well-known azeotrope at 89.4 mol% ethanol and 353 K, frustrating efforts to distill fuel grade ethanol cheaply and easily. Industrially, the azeotrope is broken using adsorption. Another strategy is to add a third component (an "entrainer") that reduces the activity coefficient of the water. In this way, the relative volatility of ethanol to water can remain greater than one. Noting that activity coefficients pertain to the liquid phase, our **entrainer** should stay in the liquid phase. In other words, it should be less volatile than either key component. We can envision pouring it into the top of the distillation column and letting it trickle down. This process is called **extractive distillation.** We would like to use as little entrainer as possible and it should not form another azeotrope with water. One suggested entrainer is 2-pyrrolidone, for which the key properties are given below. You can assume the shortcut VP model for 2-pyrrolidone and SSCED predictions for k_{ij} interactions with 2-pyrrolidone. How much 2-pyrrolidone would be needed to keep the relative volatility greater than 1.1 at 99mol% ethanol? Can you suggest any other prospective entrainers based on their molecular structure?

Molecular properties for 2-pyrrolidone are

T_c	P_c	ω	M_W	ρ	δ	α	β
802.0	6.17	0.432	85.1	1.11	29.21	2.39	27.59

Solution: Noting that azeotropes are sensitive to vapor pressure, we should use the Antoine constants from Appendix E for ethanol and water. The vapor pressure of the entrainer is less important because it must be substantially lower than that of water to prevent another azeotrope. The SSCED predicted value of $k_{12} = 0.0319$ fails to reproduce the experimentally observed azeotrope, but a value of $k_{12} = 0.058$ gives an azeotrope with $x_E = 0.894$ and 353 K. Checking the relative volatility at $x_E = 0.99$ and 353 K we find $\alpha_{LH} = 0.94$. We should add entrainer until $\alpha_{LH} = 1.1$.

We can assume $T = 353$ K for now, so the vapor pressure ratio stays constant. We can check our result at the bubble temperature of the final formulated composition. Recalling the definition of relative volatility from Section 10.6 and substituting modified Raoult's law,

$$\alpha_{LH} = K_L/K_H = (P_L^{sat}/P_H^{sat})\,(\gamma_L/\gamma_H) \qquad\qquad 12.58$$

From the Antoine equation, $P_L^{sat}/P_H^{sat} = 2.03$. Therefore, we seek $\gamma_L/\gamma_H > 1.1/2.03 = 0.542$.

Taking a basis of 1 mole of ethanol (and 0.0101 water), increasing the ratio of 2-pyrrolidone (P) to ethanol to 0.05 gives a final composition of {0.9433, 0.0095, 0.0472} for {x_E, x_W, x_P}. This gives a value of $\alpha_{LH} = 2.03(1.001/2.075) = 1.11$. Setting the pressure to 0.101 MPa and finding the bubble temperature gives $T = 352.6$, so the temperature is altered very little.

Another candidate for entrainer might be ethylene glycol. From its molecular structure, it is similar to half ethanol and half water. Following the same procedure, we find that a final composition of {0.8532, 0.0086, 0.1382} is required to achieve the same α_{LH}. This is roughly a factor of 3 for entrainer on a mole basis or a factor of 2 on a weight basis. 2-pyrrolidone is predicted to be superior because of the large value for its basicity, larger than water's. This causes the k_{ij} parameters with 2-pyrrolidone to be negative, so less is needed. Glycol is less effective.

12.8   FLORY-HUGGINS AND VAN DER WAALS THEORIES

We have shown that the contribution to the excess internal energy in the Flory-Huggins theory is identical to that in the Scatchard-Hildebrand theory. We derived the Scatchard-Hildebrand theory from the excess internal energy function of the van der Waals equation on page 468 and 12.3 on page 471. Therefore, any potential difference between the Flory-Huggins theory and the van der Waals equation must pertain to the entropy. Reviewing briefly, the van der Waals equation of state gives

$$\frac{G(T,P) - G^{ig}(T,P)}{RT} = -\ln(1 - b\rho) - \frac{a\rho}{RT} + Z - 1 - \ln(Z) \qquad 12.59$$

$$\frac{G^E}{RT} = -\ln(1 - b\rho) - \frac{a}{RTV} + Z - 1 + \sum_i x_i \ln(1 - b_i \rho_i) \qquad 12.60$$

$$+ \sum_i x_i \frac{a_{ii}}{RTV_i} - \sum_i x_i Z_i + 1 - \sum_i x_i \ln(Z/Z_i)$$

Recall that the van der Waals equation gives $U - U^{ig} = \frac{a}{V}$.

Therefore, $U^E = U - \sum_i x_i U_i = -\frac{a}{V} + \sum_i x_i \frac{a_{ii}}{V_i}$. Comparing $\left(-\frac{a}{V} + \sum_i x_i \frac{a_{ii}}{V_i} \right)$ to the result for regular solutions, we see that,

$$U^E = \Phi_1 \Phi_2 (\delta_1 - \delta_2)^2 (x_1 V_1 + x_2 V_2)$$

which is the same. We may also note that $Z - \sum_i x_i Z_i$ is a very small number because: 1) These are liquid compressibility factors, so all Z's are small numerically; and 2) the excess volume is usually a small percentage of the total volume, $V \approx \sum_i x_i V_i$. Thus, we may neglect $Z - \sum_i x_i Z_i$.

Turning to the differences between the entropy terms, the van der Waals equation gives

$$TS^E = H^E - G^E = U^E + RT\left(Z - \sum_i x_i Z_i \right) - G^E \qquad 12.61$$

$$\frac{S^E}{R} = \sum_i x_i \ln\left(\frac{1 - b\rho}{1 - b_i \rho_i} \right) + \sum_i x_i \ln\left(\frac{PV/(RT)}{PV_i/(RT)} \right) \qquad 12.62$$

Note: $(1 - b_i \rho_i) = (V_i - b_i)/V_i \equiv (V_i \varpi)/V_i = \varpi$. This means that $b_i \rho_i = b\rho$ for all i.

If we assume that ϖ is a universal constant for all fluids, including the mixture, then

$$\frac{S^E}{R} = \sum_i x_i \ln\left(\frac{PV/(RT)}{PV_i/(RT)}\right) = -\sum_i x_i \ln\left(\frac{V_i}{V}\right) = \sum_i x_i \ln\left(\frac{x_i V_i}{V} \frac{1}{x_i}\right) = -\sum_i x_i \ln\left(\frac{\Phi_i}{x_i}\right) \qquad 12.63$$

This expression is identical to Flory's equation (and note the importance of the ln(Z) term as the second term on the right-hand side of Eqn. 12.62, which derived from the ideal gas reference state). Therefore, the only difference between van der Waals' and Flory's theories is the assumption that ϖ is a universal constant. This is equivalent to saying that the packing fraction ($b\rho$) is a constant (the packing fraction is one minus the void fraction). *In other words, the Flory-Huggins theory is simply the van der Waals theory with the assumption that $b\rho$ = constant.* The difference between the Flory-Huggins theory and the Scatchard-Hildebrand theory is accounting for mixing at constant pressure instead of mixing at constant packing fraction. This is related to the argument about free volume being larger for larger molecules because fitting a polymer in the same volume as a solvent must lead to a deviation from the ideal gas law at some degree of polymerization. Therefore, the V in PV/RT must be proportional to the volume of the molecule.

The suggestion that $b\rho$ = *constant* is actually quite consistent with another observation that should seem more familiar. That is, the mass density of a polyatomic species is only weakly dependent on its molecular weight. For example, the mass density of decane is 0.73 g/cm^3 and the density of *n*-hexadecane is 0.77 g/cm^3. Since the molar density decreases inversely as molecular weight increases but the *b*-parameter increases proportionally as molecular weight increases, a constant value for the mass density implies a constant value for $b\rho$. When you consider that the mass density for almost all hydrocarbons, alcohols, amides, amines, and their polymers lies between 0.7 and 1.3 g/cm^3, you begin to get an idea of how broadly applicable this approximation is.

Nevertheless, there are some obvious limitations to the assumption of a constant packing fraction. A little calculation would make it clear that the ϖ for liquid propane at $T_r = 0.99$ is significantly larger that ϖ for toluene at $T_r = 0.619$. Thus, a mixture of propane and toluene at 366 K would not be very accurately represented by the Flory-Huggins theory. Note that deviations of ϖ from each other are related to differences in the compressibilities of the components. Thus, it is common to refer to the Flory-Huggins theory as an "incompressible" theory and to develop alternative theories to represent "compressible" polymer mixtures. Not surprisingly, these alternative theories closely resemble the van der Waals equation (with a slightly modified temperature dependence of the *a* parameter). This observation lends added significance to Rayleigh's statement: "I am more than ever an admirer of van der Waals."

12.9   SUMMARY

This has been a somewhat theoretical chapter. We have gone through iterations of observation, prediction, testing, and evaluation with several theories (e.g., van Laar, Scatchard-Hildebrand, Flory-Huggins, SSCED, and MOSCED). With each iteration, we have achieved increasing precision and insight. Sulfuric acid and water may react very favorably toward each other ($G^E \ll 0$), while 2-propanol and water have enhanced escaping tendencies because the energy required for forcing them to mix is counteracting the entropic driving force ($G^E > 0$). These facts are due in large part to the interaction energies characterized by molecular quantities like the solubility parameter, acidity, and basicity. We showed that two major effects influence the mixing: energy density and complexation. This molecular perspective provides clarification of the familiar guideline that "like dissolves like." With this perspective, we begin to acquire the capability to intuitively reason about chemical formulations. We illustrated the value of this reasoning with several practical examples. You should be able to distinguish favorable molecular interactions from unfavorable.

Carrying forward the molecular perspective, we can characterize these tendencies in terms of the cross-interaction energy (a_{12}). If the cross-interaction energy is weaker than the geometric mean ($k_{12} > 0$), then each component prefers its own company. If the cross-interaction energy is stronger than the geometric mean (i.e., $k_{12} < 0$), then the components are strongly attracted, releasing energy as they fall into the well of their mutual attraction. The energy to break the favorable interactions must be added to separate such a mixture and this may show up as a maximum boiling azeotrope. Conversely, mixtures with $k_{12} > 0$ tend to exhibit minimum boiling azeotropes, or VLLE if $\gamma_i > 10$.

Molecular insight can help a lot when conceiving of formulations, but these conceptions must ultimately be tested and validated experimentally.

Important Equations

This chapter has built the connection between the molecular perspective offered by the van der Waals model and semi-empirical estimates of activity coefficients for binary and multicomponent mixtures. The key equations in this extension are

$$a = x_1^2 a_{11} + 2x_1 x_2 a_{12} + x_2^2 a_{22} = \sum_i \sum_j x_i x_j a_{ij} \qquad 12.3$$

$$a_{12} = (1 - k_{12})(a_{11} a_{22})^{\frac{1}{2}} \qquad 12.4$$

where k_{12} is an adjustable parameter called the binary interaction parameter.

This extension results in several closely related activity models of varying complexity. The choice of the "best" model depends on the application and the individual's comfort level with a particular degree of complexity. As a summary, the SSCED model provides a reasonable compromise.

$$\boxed{RT\ln\gamma_k = V_k(1 - \Phi_k)^2[(\delta_2' - \delta_1')^2 + 2k_{12}\delta_2'\delta_1']} \qquad 12.49$$

$$\boxed{(\delta_i')^2 = \delta_i^2 - 2\alpha_i\beta_i} \qquad 12.50$$

$$\boxed{k_{12} = \frac{(\alpha_2 - \alpha_1)(\beta_2 - \beta_1)}{4\delta_2'\delta_1'}} \qquad 12.51$$

With this equation, the important roles of energy density ($\delta^2 \equiv a/V^2$), molecular size (V_k), and hydrogen bonding are all evident. Understanding these roles enables you to go beyond fitting data and make predictions about the behavior to be expected when various chemicals are combined.[11] With practice, these predictions evolve to provide intuitive insight into formulations that achieve specific engineering objectives.

---

11. When predicting non-ideality with Scatchard-Hildebrand, SSCED, or MOSCED theories, it is important to recognize that the volume of the molecule affects the non-ideality. It is "more nonideal" to squeeze a large molecule into a dense fluid of small ones than vice versa. For an unambiguous assessment of nonidealiy, evaluate γ^∞ of the largest molecule.

12.10 PRACTICE PROBLEMS

P12.1 Acrolein(1) + water(2) exhibits an atmospheric (1 bar) azeotrope at 97.4 wt% acrolein and 52.4°C.

(a) Determine the values of A_{ij} for the van Laar equation that match this bubble-point pressure at the same liquid and vapor compositions and temperature. (ANS. 1.91, 2.42)

(You may use the shortcut vapor pressure equation for acrolein: $T_c = 506$ K; $P_c = 51.6$ bar; and $\omega = 0.330$; MW = 56.)

(b) Tabulate P at 325.55 K and $x = \{0.1, 0.3, 0.5\}$ via the van Laar equation using the A_{12} and A_{21} determined above. (ANS. P(bar) = {0.653, 0.998, 1.03})

P12.2 The system α-epichlorohydrin(1) + n-propanol(2) exhibits an azeotrope at 760 mmHg and 96°C containing 16 mol% epichlorohydrin. Use the van Laar theory to estimate the composition of the vapor in equilibrium with a 90 mol% epichlorohydrin liquid solution at 96°C. (α-epichlorohydrin has the formula C_3H_5ClO, and IUPAC name 1-chloro-2,3-epoxypropane. Its vapor pressure can be approximated by: $\log_{10} P^{sat} = 8.0270 - 2007/T$, where P^{sat} is in mmHg and T is in Kelvin. You can use the shortcut vapor pressure equation for n-propanol.) (ANS. 0.72, 0.63 bar)

P12.3 The following free energy model has been suggested for a particularly unusual binary liquid-liquid mixture. Derive the expression for the activity coefficient of component 1,

$$\frac{\Delta \underline{G}^E}{nRT} = (\Phi_1 \Phi_2)^{0.75} V \frac{(\delta_1 - \delta_2)^2}{RT}$$

where, $\Phi_i = (x_i V_i)/V$ and $V = \sum_k x_k V_k$.

(ANS. $\dfrac{V_1(\delta_1 - \delta_2)^2}{RT} \Phi_2^{0.75}[0.25\Phi_1^{0.75} + 0.75\Phi_1^{0.25}\Phi_2]$)

P12.4 The Scatchard-Hildebrand model can be extended to multicomponent mixtures in the following manner. Setting $a_{ij} = (a_{ii}a_{jj})^{\frac{1}{2}}$ and $a_{ii}/V_i = V_i \delta^2$ Eqn. 12.7 can be rewritten as

$$U^E = \sum_i x_i V_i \delta_i^2 - \left(\frac{\sum \sum x_i V_i \delta_i x_j V_j \delta_j}{\sum x_k V_k} \right) \qquad 12.64$$

Recognizing that the quadratic term is separable and simplifying the square-root of the square:

$$U^E = \Sigma x_i V_i \delta_i^2 - V{<}\delta{>}^2 \qquad 12.65$$

where $<\delta> \equiv \Sigma x_i V_i \delta_i / V \equiv \Sigma \Phi_i \delta_i$ and $V \equiv \Sigma x_i V_i$.

This result can be made even simpler by adding and subtracting $V{<}\delta{>}^2$ and rearranging to obtain:

$$U^E = \Sigma x_i V_i \delta_i^2 - V{<}\delta{>}^2 = \Sigma x_i V_i \delta_i^2 - 2V{<}\delta{>}(\Sigma x_i V_i \delta_i / V) + {<}\delta{>}^2 \Sigma x_i V_i \qquad 12.66$$

where we have substituted the definition of $<\delta>$ in the second term and the definition of V in the third.

Collecting all terms in a common summation, we obtain:

$$U^E = \Sigma x_i V_i(\delta_i^2 - 2\delta_i<\delta> + <\delta>^2) => G^E = U^E = \Sigma x_i V_i(\delta_i - <\delta>)^2 \qquad 12.67$$

This is a remarkably simple result. Derive an equally simple expression for the activity coefficient of a component in a multicomponent mixture. (Hint: It is easier to start with Eqn. 12.65.) (ANS. Eqn. 12.54)

12.11   HOMEWORK PROBLEMS

12.1 The compositions of coexisting phases of ethanol(1) + toluene(2) at 55°C are $x_1 = 0.7186$, and $y_1 = 0.7431$ at $P = 307.81$ mmHg, as reported by Kretschmer and Wiebe, 1949. *J. Amer. Chem. Soc.,* 71:1793. Estimate the bubble pressure at 55°C and $x_1 = 0.1$, using

(a) The Scatchard-Hildebrand model with $k_{12} = 0$
(b) The SSCED model with a default value of k_{12}
(c) The SSCED model with k_{12} matched to the data
(d) The van Laar equation

12.2 A vapor/liquid experiment for the carbon disulfide(1) + chloroform(2) system has provided the following data at 298 K: $P_1^{sat} = 46.85$ kPa, $P_2^{sat} = 27.3$ kPa, $x_1 = 0.2$, $y_1 = 0.363$, and $P = 34.98$ kPa. Estimate the dew pressure at 298 K and $y_1 = 0.6$, using

(a) The Scatchard-Hildebrand model with $k_{12} = 0$
(b) The SSCED model with a default value of k_{12}
(c) The SSCED model with k_{12} matched to the data
(d) The van Laar equation

12.3 The (1) + (2) system forms an azeotrope at $x_1 = 0.75$ and 80°C. At 80°C, $P_1^{sat} = 600$ mmHg, $P_2^{sat} = 900$ mmHg. The liquid phase can be modeled by the van Laar model with $V_2 = 1.3V_1$.

(a) Estimate the activity coefficient of component 1 at $x_1 = 0.75$ and 80°C. [Hint: The relative volatility (given in Eqn. 10.32) is unity at the azeotropic condition.]
(b) Qualitatively sketch the *P-x-y* and *T-x-y* diagrams that you expect.

12.4 Ethanol(1) + benzene(2) form azeotropic mixtures. Compare the specified model to the experimental data of Brown and Smith cited in problem 10.2.

(a) Prepare a *y-x* and *P-x-y* diagram for the system at 45°C assuming the van Laar model and using the experimental pressure at $x_1 = 0.415$ to estimate A_{12} and A_{21}.
(b) Prepare a *y-x* and *P-x-y* diagram for the system at 45°C with the predictions of the Scatchard-Hildebrand theory with $k_{12} = 0$.
(c) Prepare a *y-x* and *P-x-y* diagram for the system at 45°C assuming the SSCED model and using the standard guideline to estimate k_{12}.
(d) Prepare a *y-x* and *P-x-y* diagram for the system at 45°C assuming the SSCED model and using the experimental pressure at $x_1 = 0.415$ to estimate k_{12}.

12.5 The CRC Handbook lists the azeotrope for the acetone + chloroform system as 64.7°C and 20 wt% acetone.

(a) Use the van Laar model to estimate the *T-x-y* diagram at 1 bar.

(b) Use the SSCED model to estimate the *T-x-y* diagram at 1 bar with predicted k_{12}.

(c) What value of V_2/V_1 is implied by the van Laar parameters?

12.6 Using the van Laar model and the data from problem 11.3, estimate the total pressure and composition of the vapor in equilibrium with a 20 mol% ethanol(1) solution in water(2) at 78.15°C.

12.7 A liquid mixture of 50 mol% chloroform(1) and 50% 1,4-dioxane(2) at 0.1013 MPa is metered into a flash drum through a valve. The mixture flashes into two phases inside the drum where the pressure and temperature are maintained at 24.95 kPa and 50°C. The compositions of the exiting phases are $x_1 = 0.36$ and $y_1 = 0.62$.

Your supervisor asks you to adjust the flash drum pressure so that the liquid phase is $x_1 = 0.4$ at 50°C. He doesn't provide any *VLE* data, and you are standing in the middle of the plant with only a calculator and pencil and paper, so you must estimate the new flash drum pressure. Fortunately, your supervisor has a phenomenal recall for numbers and tells you that the vapor pressures for the pure components at 50°C are $P_1^{sat} = 69.4$ kPa and $P_2^{sat} = 15.8$ kPa. What is your best estimate of the pressure adjustment that is necessary without using any additional information?

12.8 Fit the data from problem 11.11 to the following model by regression over all points, and compare with the experimental data on the same plot, using

(a) The Scatchard-Hildebrand model with $k_{12} = 0$

(b) The SSCED model with a default value of k_{12}

(c) The SSCED model with k_{12} matched to the data

(d) The van Laar equation

(e) Plot the *P-x-y* diagram at 80°C, based on the fits specified by your instructor.

12.9 Fit the data from problem 11.10 to the following model by regression over all points, and compare with the experimental data on the same plot, using

(a) – (d) as in problem 12.8.

(e) Plot the *T-x-y* diagram at 1 bar, based on the fits specified by your instructor.

12.10 Fit the data from problem 11.26 to the following model by regression over all points, and compare with the experimental data on the same plot, using

(a) – (e) as in problem 12.8.

12.11 Fit the data from problem 11.27 to the following model by regression over all points, and compare with the experimental data on the same plot, using

(a) – (e) as in problem 12.8.

12.12 Starting from the excess Gibbs energy formula for Flory's equation, derive the formula for the activity coefficient of component 1 in a binary mixture.

12.13 Crime scene investigators have determined that an acrylic spray paint (polymethylmethacrylate, PMMA) was used to deface the *Mona Lisa*. Leonardo used linseed oil. We would like a solvent that interacts more strongly with acrylic than with linseed oil. Based on their chemical structures, we can approximate the SSCED parameters of linseed oil as

n-hexadecane and acrylic paint as methylethylketone. Do you recommend $CHCl_3$, toluene, or acetone as the solvent? Explain.

12.14 R410a is a replacement for R22 in air conditioners and heat pumps. Air conditioners require a different refrigerant because they operate in a different temperature range. R410a avoids the problems with the ozone layer caused by chlorofluorocarbons, but its longevity may be limited because it has a relatively high global warming factor (1725 times the effect of CO_2). Roughly, it is a 50wt% mixture of difluoromethane (D) and pentafluoroethane (P) (i.e., 70mol% D). Kobayashi and Nishiumi (1998) report a pressure of 1.098 MPa at 283.05K.[12] You may assume the SCVP equation

Compound	Tc(K)	Pc(MPa)	ω	Cp^{ig}/R	M_W	ρ_{298}	$\delta(J/cc)^{1/2}$	$\alpha(J/cc)^{1/2}$	$\beta(J/cc)^{1/2}$
Difluoromethane	351.6	5.83	0.273	5.2	52.0	1.21	20.67	5.00	3.00
Pentafluoroethane	342.0	3.44	0.259	11.4	120.0	1.49	14.67	?	0.00

(a) Assuming $k_{ij} = 0$ for the binary interaction parameter of the SSCED equation, predict whether an azeotrope should be expected in this system at 283.05 K. Tabulate the relative volatilities at $x_D = 0.01$ and 0.99.

(b) Solve for the value of k_{12} that matches the reported pressure.

(c) What acidity value for pentafluoroethane matches the value of k_{ij} determined in part (b)?

12.15 As part of a biorefining effort, butanediols are being produced by fermentation. The problem is that the isomers are all mixed up. Furthermore, 1,3-propanediol comprises roughly 30mol% of the mixture on a dry basis (i.e., water has been removed). The problem is to assess the prospects for azeotrope formation and avoidance. The following steps should shed some light on the problem.

(a) Plot $\log_{10}(P^{sat})$ versus $1000/T(K)$ in the range of $T^\circ C = [100,400]$ for 1,3-propanediol, 1,3-butanediol, and 1,4-butanediol on the same axes. Are there Bancroft points?

(b) Compile a table of γ_i^∞ for each component in each solvent based on the SSCED model. Which combinations show the greatest tendency to form azeotropes?

---

12. 1998. *Fluid Phase Equil.* 144:191.

LOCAL COMPOSITION ACTIVITY MODELS

I have constructed three thousand different theories in connection with the electric light... Yet, in only two cases did my experiments prove the truth of my theory.

Thomas A. Edison

It is evident from Chapter 12 that describing solution nonideality with van der Waals' mixing rules is imprecise. With a single parameter like k_{12} or A_{12}, we can match the magnitude of the excess energy, but not the skewness. The Margules two-parameter and van Laar models address this problem in an ad hoc fashion, but there is no physical basis for extending these to multiple components. A rational basis for extending the analysis of mixtures should seek a legitimate explanation for the source of varied skewness. One such explanation is that molecules do not distribute themselves randomly. Instead, they may tend to form clusters. You might witness this kind of distribution at a dinner party, where the children are in one area, the men are discussing sports, and the ladies are discussing anything but sports. There is sufficient mixing among the clusters that you cannot consider it a phase separation. Some children and ladies like sports. Some men do not. Nevertheless, the distribution is not entirely random. In other words, the local composition around a child at a dinner party, for example, may differ from the bulk composition. We examine this prospect graphically in Example 13.2. Local composition models recognize this possibility and account for the local composition enhancement in terms of two parameters, just enough to characterize both the magnitude and the skewness of the excess energy. Careful analysis of the molecular scale energies in terms of the local compositions facilitates straightforward extension to multicomponent mixtures.

Chapter Objectives: You Should Be Able to...

1. Characterize adjustable parameters in local composition activity models using experimental data.

2. Comment critically on the merits and limitations of the following activity models: Wilson, UNIQUAC, UNIFAC, and NRTL, including the ability to identify the most appropriate model for a given mixture.

A Preliminary Glimpse of UNIFAC — A Predictive Method

This chapter is densely packed with theoretical details. In the interest of "telling you what we are going to tell you," it is useful to see how the final equations are applied before being concerned with their derivations. One popular activity coefficient model is the predictive model of UNIFAC. It is the closest thing to a universally applicable predictive model that we currently have, so it makes sense to get right to the point and introduce the rudiments of implementing this model at an early stage. It is a rather complicated model and deriving it must await several other derivations. Nevertheless, the availability of a computer program for applying the method makes it possible to apply it as a "black box" at this stage, and the utility of the model should inspire us to learn more about it. Detailed calculations are illustrated in the UNIFAC spreadsheet in the file Actcoeff.xlsx introduced in Chapter 11, and in the MATLAB file Ex13_01.m.

Example 13.1 VLE prediction using UNIFAC activity coefficients

See Actcoeff.xlsx - worksheets UNIFAC, unifacVLE/ MATLAB: Ex13_01.m

The 2-propanol (also known as isopropyl alcohol or *IPA*) + water (*W*) system is known to form an azeotrope at atmospheric pressure and 80.37°C ($x_W = 0.3146$).[a] Use UNIFAC to estimate the conditions of the azeotrope at 760 mmHg. Is UNIFAC accurate for this mixture?

Solution: The Antoine coefficients for IPA and water are given in Appendix E. To begin, the VLE can be computed at the true experimental conditions to see if it looks like there is an azeotrope nearby. This system is presented in Figs. 10.8 and 10.9, so we know what the answer looks like. We will use the principles developed in Section 11.7 to determine if an azeotrope exists. Note that the azeotrope condition is at a maximum or minimum on the diagrams.

The pressure is given as 760 mmHg. We develop a detailed description of the UNIFAC model later, but you need to know a little bit about it just to run the program. The UNIFAC model is based on structural and energetic information for the functional groups that comprise the molecules in the mixture. The UNIFAC model estimates the activity coefficients using the groups by calculating size, shape, and energy parameters based on the number and types of groups in the molecules.[b] The structures of the molecules for this problem and the UNIFAC groups are:

$$\underset{\underset{H}{|}}{\overset{\overset{OH}{|}}{H_3C-C-CH_3}} \qquad \underset{H \qquad H}{\overset{O}{\diagdown}}$$

	Water	IPA
	H2O - 1	CH3 - 2
		CH - 1
		OH - 1

The UNIFAC model can be operated from either the Excel spreadsheet or the MATLAB routine. To operate the spreadsheet program, simply type the temperature of interest (80.37) and the number of functional groups of each type in the appropriate column for each component. In MATLAB, the groups are entered into the cell matrix compArray. Group names available in MATLAB can be determined by using load 'unifacAij.mat'.

Bubble-temperature calculation.

Example 13.1   VLE prediction using UNIFAC activity coefficients (Continued)

Enter the mole fractions (e.g., $[x_W = 0.3146, x_{IPA} = 0.6854]$). The activity coefficients are $\Rightarrow \gamma_w = 2.1108$; $\gamma_{IPA} = 1.0885$. According to modified Raoult's law, $P = y_1 P + y_2 P = x_1 \gamma_1 P_1^{sat} + x_2 \gamma_2 P_2^{sat}$. By entering the proper Antoine coefficients, the pressure is computed using this formula, and we can keep guessing temperatures (which changes γs and P^{sat}s) until the pressure equals 760 mmHg (or we can apply the Excel solver tool or MATLAB fzero $\Rightarrow T = 80.47°C$):

T	P_{IPA}^{sat}	P_W^{sat}	$P = \sum_i x_i \gamma_i P_i^{sat}$	y_W
80.37	695	360	757	0.3158
82.50	760	395	829	0.3164
80.47	697	361	760	0.3158

The vapor phase mole fractions are calculated using $y_1 = (y_1 P)/P = (x_1 \gamma_1 P_1^{sat})/P$ and an analogous expression for the second component (or by using $y_2 = 1 - y_1$). Since the vapor and liquid compositions are not equal ($x_w = 0.3146 \neq 0.3158 = y_w$), we did not find the azeotrope. We must try several values of x to find the azeotrope composition.

Try $x_w = 0.3177 \Rightarrow \gamma_W = 2.1035$; $\gamma_{IPA} = 1.0904$:

T	P_{IPA}^{sat}	P_W^{sat}	$\sum_i x_i \gamma_i P_i^{sat}$	y_W
80.47	697	361	760	0.3177

Since $x_w = 0.3177 = y_w$, this is the composition of the azeotrope estimated by UNIFAC. UNIFAC seems to be fairly accurate for this mixture at the azeotrope. Also note that T versus x is fairly flat near an azeotrope; this is why it was unnecessary to modify the guess for bubble temperature at the new composition. This is generally true, and important in the distillation of azeotropic systems.

a.   Perry, R.H., Chilton, C.H. 1973. *Chemical Engineers' Handbook,* 5 ed, New York NY: McGraw-Hill, Chapter 13.
b.   The functional groups for a given molecule are often determined automatically by process design software. Several examples of group assignments are given in Table 13.2 on page 512.

13.1   LOCAL COMPOSITION THEORY

Now that we see the capabilities of the predictions, we have motivation to understand the model. One of the major assumptions of van der Waals mixing was that the mixture interactions were independent of each other such that quadratic mixing rules would provide reasonable approximations as shown in Eqn. 12.3 on page 467. But in some cases, like radically different strengths of attraction, the mixture interaction can be strongly coupled to the mixture composition. That is, for instance, the cross parameter could be a function of composition: $a_{12} = a_{12}(x)$. One way of treating this prospect is to recognize the possibility that the **local compositions** in the mixture might deviate strongly from the bulk compositions. As an example, consider a lattice consisting primarily of type A atoms but with two B atoms right beside each other. Suppose all these atoms were the same size and that the coordination number was 10. Then the local compositions around a B atom are $x_{AB} = 9/10$ and $x_{BB} = 1/10$ (notation of subscripts is $AB \Rightarrow$ "A around B"). Specific interactions such as hydrogen bonding and polarity might lead to such effects, and thus, the basis of the hypothesis is that energetic differences lead to the nonrandomness that causes the quadratic mixing rules to break down. To develop the theory, we first introduce nomenclature to identify the local compositions summarized in Table 13.1.

Table 13.1 *Nomenclature for Local Composition Variables*

Composition around a "1" Molecule	Composition around a "2" Molecule
x_{21} – mole fraction of "2's" around "1"	x_{12} – mole fraction of "1's" around "2"
x_{11} – mole fraction of "1's" around "1"	x_{22} – mole fraction of "2's" around "2"
local mole balance, $x_{11} + x_{21} = 1$	local mole balance, $x_{22} + x_{12} = 1$

We assume that the local compositions are given by some weighting factor, Ω_{ij}, relative to the overall compositions.

$$\frac{x_{21}}{x_{11}} = \frac{x_2}{x_1}\Omega_{21}$$

13.1

$$\frac{x_{12}}{x_{22}} = \frac{x_1}{x_2}\Omega_{12}$$

13.2

Therefore, if $\Omega_{12} = \Omega_{21} = 1$, the solution is random. Before introducing the functions that describe the weighting factors, let us discuss how the factors may be used.

Local Compositions around "1" Molecules

Let us begin by considering compositions around "1" molecules. We would like to write the local mole fractions x_{21} and x_{11} in terms of the overall mole fractions, x_1 and x_2. Using the local mole balance,

$$x_{11} + x_{21} = 1$$

13.3

Rearranging Eqn. 13.1,

$$x_{21} = x_{11}\frac{x_2}{x_1}\Omega_{21}$$

13.4

Substituting Eqn. 13.4 into Eqn. 13.3,

$$x_{11}\left(1 + \frac{x_2}{x_1}\Omega_{21}\right) = 1$$

13.5

Rearranging,

$$x_{11} = \frac{x_1}{x_1 + x_2\Omega_{21}}$$

13.6

Substituting Eqn. 13.6 into Eqn. 13.4,

$$x_{21} = \frac{x_2 \Omega_{21}}{x_1 + x_2 \Omega_{21}} \qquad\qquad 13.7$$

Local Compositions around "2" Molecules

Similar derivations for molecules of type "2" results in

$$x_{22} = \frac{x_2}{x_1 \Omega_{12} + x_2} \qquad x_{12} = \frac{x_1 \Omega_{12}}{x_1 \Omega_{12} + x_2} \qquad\qquad 13.8$$

Example 13.2 Local compositions in a two-dimensional lattice

The following lattice contains x's, o's, and void spaces. The coordination number of each cell is 8. Estimate the local composition (x_{xo}) and the parameter Ω_{xo} based on rows and columns away from the edges.

		O		O		X		O	
	X		O	X			X		
	X		X	X		O	X	O	
O		X		O			X		
						O			X
	O		X		O			X	
	X			O		O		X	
O	X				X		X		O

Solution: There are 9 o's and 13 x's that are located away from the edges. The number of x's and o's around each o are as follows:

o#	1	2	3	4	5	6	7	8	9	
#x's	3	3	3	2	1	1	0	2	2	= 17
#o's	2	0	0	0	1	0	3	1	1	= 8

$x_{xo} = 17/25 = 0.68; \; x_o = 9/22; \; \Omega_{xo} = \;\; 17/8 \cdot 9/13 \;\; = 1.47 \quad = 1.47$

Note: Fluids do not really behave as though their atoms were located on lattice sites, but there are many theories based on the supposition that lattices represent reasonable approximations. In this text, we have elected to omit detailed treatment of lattice theory on the basis that it is too approximate to provide an appreciation for the complete problem and too complicated to justify treating it as a simple theory. This is a judgment call and interested students may wish to learn more about lattice theory. Sandler presents a brief introduction to the theory which may be acceptable for readers at this level.[a]

a.   Sandler, S.I. 1989. *Chemical Entineering Thermodynamic,* 2nd ed. Hoboken NJ: Wiley.

Local Composition and Gibbs Energy of Mixing

We need to relate the local compositions to the excess Gibbs energy. The perspective of representing all fluids by the square-well potential lends itself naturally to the local composition concept. Then the intermolecular energy is given simply by the local composition times the well-depth for that interaction. We simply ignore all but the nearest neighbors because they are outside the square-well. In equation form, the energy equation for mixtures can be reformulated in terms of local compositions. The local mole fraction can be related to the bulk mole fraction by defining a quantity Ω_{ij} as follows:

$$\frac{x_{ij}}{x_{jj}} \equiv \frac{x_i}{x_j}\Omega_{ij} \qquad 13.9$$

The next step in the derivation requires scaling up from the molecular-scale local composition to the macroscopic energy in the mixture. The rigorous procedure for taking this step requires integration of the molecular distributions times the molecular interaction energies, analogous to the procedure for pure fluids as applied in Section 7.11. This rigorous development is presented below in Section 13.7. On the other hand, it is possible to simply present the result of that derivation for the time being. This permits a more rapid exploration of the practical implications of local composition theory. The form of the equation is not so difficult to understand from an intuitive perspective, however. The energy departure is simply a multiplication of the local composition (x_{ij}) by the local interaction energy (ε_{ij}). The departure properties are calculated based on a general model known as the **two-fluid theory.**[1] According to the two-fluid theory, any intensive departure function in a binary is given by

$$(M - M^{ig}) = x_1(M - M^{ig})^{(1)} + x_2(M - M^{ig})^{(2)} \qquad 13.10$$

Where the local composition environment of the type 1 molecules determines $(M - M^{ig})^{(1)}$, and the local composition environment of the type 2 molecules determines $(M - M^{ig})^{(2)}$. Note that $(M - M^{ig})^{(i)}$ is composition-dependent and is equal to the pure component value only when the local composition is pure i.

Using the concept of a square-well model and thus counting only nearest neighbors, noting that $\varepsilon_{12} = \varepsilon_{21}$, and recalling that the local mole fractions must sum to unity, we have for a binary mixture

$$U - U^{ig} = \frac{N_A}{2}[x_1 N_{c,1}(x_{11}\varepsilon_{11} + x_{21}\varepsilon_{21}) + x_2 N_{c,2}(x_{12}\varepsilon_{12} + x_{22}\varepsilon_{22})] \qquad 13.11$$

where $N_{c,j}$ is the coordination number (total number of atoms in the neighborhood of the jth species), and where we can identify

$$(U - U^{ig})^{(1)} = \frac{N_A}{2}N_{c,1}(x_{11}\varepsilon_{11} + x_{21}\varepsilon_{21}) \text{ and } (U - U^{ig})^{(2)} = \frac{N_A}{2}N_{c,2}(x_{12}\varepsilon_{12} + x_{22}\varepsilon_{22}) \quad 13.12$$

When x_1 approaches unity, x_2 goes to zero, and from Eqn. 13.1 x_{21} goes to zero, and x_{11} goes to one. The limits applied to Eqn. 13.11 result in $(U - U^{ig})_{pure1} = (N_A/2)N_{c,1}\varepsilon_{11}$. Similarly, when x_2 approaches unity, $(U - U^{ig})_{pure2} = (N_A/2)N_{c,2}\varepsilon_{22}$. For an ideal solution,

$$(U - U^{ig})^{is} = x_1(U - U^{ig})_{pure1} + x_2(U - U^{ig})_{pure2} = \frac{N_A}{2}[x_1 N_{c,1}\varepsilon_{11} + x_2 N_{c,2}\varepsilon_{22}] \qquad 13.13$$

1. Scott, R.L. 1956. *Annu. Rev. Phys.Chem* 7:43.

The excess energy is obtained by subtracting Eqn. 13.13 from Eqn. 13.11,

$$U^E = U - U^{is} = \frac{N_A}{2}[x_1 N_{c,1}((x_{11}\varepsilon_{11} + x_{21}\varepsilon_{21}) - \varepsilon_{11}) + x_2 N_{c,2}((x_{12}\varepsilon_{12} + x_{22}\varepsilon_{22}) - \varepsilon_{22})] \quad 13.14$$

Collecting terms with the same energy variables, and using the local mole balance from Table 13.1 on page 502, $(x_{11}-1)\varepsilon_{11} = -x_{21}\varepsilon_{11}$, and $(x_{22}-1)\varepsilon_{22} = -x_{12}\varepsilon_{22}$, resulting in

$$U^E = \frac{N_A}{2}[x_1 x_{21} N_{c,1}(\varepsilon_{21} - \varepsilon_{11}) + x_2 x_{12} N_{c,2}(\varepsilon_{12} - \varepsilon_{22})] \quad 13.15$$

Substituting Eqn. 13.7 and Eqn. 13.8,

$$U^E = \frac{N_A}{2}\left[\frac{x_1 x_2 \Omega_{21} N_{c,1}(\varepsilon_{21} - \varepsilon_{11})}{x_1 + x_2 \Omega_{21}} + \frac{x_2 x_1 \Omega_{12} N_{c,2}(\varepsilon_{12} - \varepsilon_{22})}{x_1 \Omega_{12} + x_2}\right] \quad 13.16$$

At this point, the traditional local composition theories deviate from regular solution theory in a way that really has nothing to do with local compositions. Instead, the next step focuses on one of the subtleties of classical thermodynamics. Example 6.7 shows that the derivative of Helmholtz energy is related to internal energy. Therefore, we can integrate energy to find the change in Helmholtz energy,

$$\int_\infty^T d\left(\frac{A^E}{RT}\right) = \frac{A^E}{RT} - \frac{A^E}{RT}\bigg|_\infty = -\int_\infty^T \frac{U^E}{RT^2}dT \quad 13.17$$

where $A^E/(RT)\big|_\infty$ is the infinite temperature limit at the given liquid density, independent of temperature but possibly dependent on composition or density. We need to insert Eqn. 13.16 into Eqn. 13.17 and integrate. We need to have some algebraic expression for the dependence of Ω_{ij} on temperature, which is what distinguishes the local composition theories from each other.

13.2 WILSON'S EQUATION

Wilson[2] made a bold assumption regarding the temperature dependence of Ω_{ij}. Wilson's original parameter used in the literature is Λ_{ji}, but it is related to Ω_{ij} in a very direct way. Wilson *assumes*[3]

$$\Omega_{ij} = \Lambda_{ji} = \frac{V_i}{V_j}\exp\left(\frac{-N_A N_{c,j}(\varepsilon_{ij} - \varepsilon_{jj})}{2RT}\right) = \frac{V_i}{V_j}\exp\left(\frac{-A_{ji}}{RT}\right) \quad 13.18$$

(note: $\Lambda_{ii} = \Lambda_{jj} = 1$, and $A_{ij} \neq A_{ji}$ even though $\varepsilon_{ij} = \varepsilon_{ji}$), and integration with respect to T becomes very simple. Assuming $N_{c,j} = 2$ for all j at all ρ,

2. Wilson, G.M. 1964. *J. Am. Chem. Soc.* 86:127.

3. Advanced readers may note that our definition of local compositions differs slightly from Wilson's. Wilson's original derivation combined the two-fluid theory of local compositions with an ad hoc "one-fluid" Flory equation. The same result can be derived more consistently using a two-fluid theory. The difference is that the local compositions are dependent on size as well as energies as defined by Eqns. 13.1, 13.2, and 13.18. This gives $x_{ij}/x_{jj} = (\Phi_i/\Phi_j)\exp(-A_{ji}/(RT))$ where the original was $x_{ij}/x_{jj} = (x_i/x_j)\exp(-A_{ji}/(RT))$.

$$\frac{A^E}{RT} = -x_1 \ln(\Phi_1 + \Phi_2 \exp(-A_{12}/RT)) - x_2 \ln(\Phi_1 \exp(-A_{21}/RT) + \Phi_2) + \left.\frac{A^E}{RT}\right|_\infty \qquad 13.19$$

A convenient simplifying assumption before proceeding further is that $G^E \sim A^E$. This corresponds to neglecting the excess volume of mixing relative to the other contributions and is quite acceptable for liquids. The customary way of interpreting G^E/RT is to separate it into an energetic part known as the **residual contribution**, $(G^E/RT)^{RES}$, that vanishes at infinite temperature or when $\varepsilon_{12} - \varepsilon_{22} = 0$ and $\varepsilon_{21} - \varepsilon_{11} = 0$, and a size/shape part known as the **combinatorial contribution**, $(G^E/RT)^{COMB}$, that represents the infinite temperature limit at the liquid density. For Wilson's equation, the first two terms vanish at high T, so

$$(G^E/RT)^{RES} = -x_1 \ln(\Phi_1 + \Phi_2 \exp(-A_{12}/RT)) - x_2 \ln(\Phi_1 \exp(-A_{21}/RT) + \Phi_2) \qquad 13.20$$

For the combinatorial contribution, Wilson used Flory's equation,

$$G^E/(RT)\big|_\infty = (G^E/RT)^{COMB} = x_1 \ln(\Phi_1/x_1) + x_2 \ln(\Phi_2/x_2) \qquad 13.21$$

It should be noted that the assumption of the temperature dependence of Eqn. 13.18 has been made for convenience, but there is some justification for it, as we show in Section 13.7. Wilson's equation becomes

$$\frac{G^E}{RT} = -x_1 \ln\left(\Phi_1 + \Phi_2 \exp\left(\frac{-A_{12}}{RT}\right)\right) - x_2 \ln\left(\Phi_1 \exp\left(\frac{-A_{21}}{RT}\right) + \Phi_2\right) + x_1 \ln\frac{\Phi_1}{x_1} + x_2 \ln\frac{\Phi_2}{x_2} \qquad 13.22$$

Algebraic rearrangement of Wilson's equation results in the form that is usually cited,

$$\boxed{\frac{G^E}{RT} = -x_1 \ln(x_1 + x_2\Lambda_{12}) - x_2 \ln(x_1\Lambda_{21} + x_2)} \qquad 13.23$$

For a binary system, the activity coefficients from the Wilson equation are:

$$\ln\gamma_1 = 1 - \ln(x_1\Lambda_{11} + x_2\Lambda_{12}) - \frac{x_1\Lambda_{11}}{x_1\Lambda_{11} + x_2\Lambda_{12}} - \frac{x_2\Lambda_{21}}{x_1\Lambda_{21} + x_2\Lambda_{22}}$$

$$\ln\gamma_2 = 1 - \ln(x_1\Lambda_{21} + x_2\Lambda_{22}) - \frac{x_1\Lambda_{12}}{x_1\Lambda_{11} + x_2\Lambda_{12}} - \frac{x_2\Lambda_{22}}{x_1\Lambda_{21} + x_2\Lambda_{22}}$$

Noting that $\Lambda_{11} = \Lambda_{22} = 1$, and looking back at Eqn. 13.6, we can also see that for the first equation

$$1 - \frac{x_1}{x_1 + x_2\Lambda_{12}} = x_{21} = \frac{x_2\Lambda_{12}}{x_1 + x_2\Lambda_{12}}.$$ We can rearrange this expression to obtain the slightly more compact relation:

Wilson's equation.

$$\boxed{\ln\gamma_1 = -\ln(x_1 + x_2\Lambda_{12}) + x_2\left(\frac{\Lambda_{12}}{x_1 + x_2\Lambda_{12}} - \frac{\Lambda_{21}}{x_1\Lambda_{21} + x_2}\right)} \qquad 13.24$$

Similar rearrangement of the second expression gives:

$$\ln \gamma_2 = -\ln(x_1 \Lambda_{21} + x_2) - x_1 \left(\frac{\Lambda_{12}}{x_1 + x_2 \Lambda_{12}} - \frac{\Lambda_{21}}{x_1 \Lambda_{21} + x_2} \right)$$

13.25

where $\quad \Lambda_{12} = \dfrac{V_2}{V_1} \exp\left(\dfrac{-A_{12}}{RT} \right) \quad$ and $\quad \Lambda_{21} = \dfrac{V_1}{V_2} \exp\left(\dfrac{-A_{21}}{RT} \right)$

13.26

> **Parameters for the Wilson equation, A_{ji}. Note that the literature values often include energy units. Use the correct R!**

One limitation of Wilson's equation is that it is unable to model liquid-liquid equilibria, but it is reasonably accurate for modeling the liquid phase when correlating the vapor liquid equilibria.

> **Wilson's equation is incapable of representing LLE.**

Extending Eqn. 13.23 to a multicomponent solution,

$$\frac{G^E}{RT} = -\sum_j x_j \ln\left(\sum_i x_i \Lambda_{ji} \right) \Rightarrow \frac{G^E}{RT} = -\frac{1}{n}\sum_j n_j \ln\left(\sum_i \frac{n_i \Lambda_{ji}}{n} \right) \quad \text{where } \Lambda_{ii} = 1 \qquad 13.27$$

$$\frac{G^E}{RT} = -\left[\sum_j n_j \ln\left(\sum_i n_i \Lambda_{ji} \right) - n\ln(n) \right] \qquad 13.28$$

To determine activity coefficients, the excess Gibbs energy is differentiated. Differentiating the last term,

$$\left(\frac{\partial(n\ln n)}{\partial n_k} \right)_{T,P,n_{i,j \neq k}} = \ln n + 1$$

and letting "sum" stand for the summation of Eqn. 13.28, and combining,

$$\left(\frac{\partial(\text{sum})}{\partial n_k} \right)_{T,P,n_{i,j \neq k}} = -\ln\left(\sum_i n_i \Lambda_{ki} \right) - \sum_j \left(\frac{n_j \Lambda_{jk}}{\sum_i n_i \Lambda_{ji}} \right)$$

combining,

$$\ln \gamma_k = 1 - \ln\left(\sum_i x_i \Lambda_{ki} \right) - \sum_j \frac{x_j \Lambda_{jk}}{\sum_i x_i \Lambda_{ji}} \qquad 13.29$$

Example 13.3 Application of Wilson's equation to VLE

For the binary system n-pentanol(1) + n-hexane(2), the Wilson equation constants are $A_{12} = 1718$ cal/mol, $A_{21} = 166.6$ cal/mol. Assuming the vapor phase to be an ideal gas, determine the composition of the vapor in equilibrium with a liquid containing 20 mole% n-pentanol at 30°C. Also calculate the equilibrium pressure. $P_1^{sat} = 3.23$ mmHg; $P_2^{sat} = 187.1$ mmHg.

Solution: From CRC,

$$\rho_1 = 0.8144 \text{ g/ml } (1\text{mole}/88\text{g}) \Rightarrow V_1 = 108 \text{ cm}^3/\text{mole}$$
$$\rho_2 = 0.6603 \text{ g/ml } (1\text{mole}/86\text{g}) \Rightarrow V_2 = 130 \text{ cm}^3/\text{mole}$$

Note: ρ_1 and ρ_2 are functions of T but $\rho_1/\rho_2 \approx$ const. $\Rightarrow V_2/V_1 = 1.205$ assumed at all T.

Utilizing Eqn. 13.26,

$$\Lambda_{12} = 1.205 \exp(-1718/1.987/303) = 0.070; \quad \Lambda_{21} = 1/1.205 \exp(-166.6/1.987/303) = 0.629$$

$$\frac{\Lambda_{12}}{x_1 + x_2\Lambda_{12}} - \frac{\Lambda_{21}}{x_1\Lambda_{21} + x_2} = \frac{0.070}{0.2 + 0.8 \cdot 0.070} - \frac{0.629}{0.2 \cdot 0.629 + 0.8} = -0.4062$$

$$\ln \gamma_1 = 1.0408 \Rightarrow \gamma_1 = 2.822; \qquad \ln \gamma_2 = 0.1584 \Rightarrow \gamma_2 = 1.172$$

$$P = (y_1 + y_2)P = x_1\gamma_1 P_1^{sat} + x_2\gamma_2 P_2^{sat} = 0.2 \cdot 2.822 \cdot 3.23 + 0.8 \cdot 1.172 \cdot 187.1 = 177.2 \text{ mmHg}$$
$$y_1 = x_1\gamma_1 P^{sat}/P = 0.2 \cdot 2.822 \cdot 3.23/177.2 = 0.0103$$

 Bubble-pressure calculation.

13.3 NRTL

The NRTL model[4] (short for Non-Random Two Liquid) equates U^E from Eqn. 13.16 directly to G^E, ignoring the proper thermodynamic integration. At the same time, it introduces a third binary parameter that generates an extremely flexible functional form for fitting activity coefficients.

See Actcoeff.xlsx, worksheet NRTL MATLAB: nrtl.m

$$G^E = U^E = \frac{N_A}{2}\left[\frac{x_1 x_2 \Omega_{21} N_{c,1}(\varepsilon_{21} - \varepsilon_{11})}{x_1 + x_2\Omega_{21}} + \frac{x_2 x_1 \Omega_{12} N_{c,2}(\varepsilon_{12} - \varepsilon_{22})}{x_1\Omega_{12} + x_2}\right] \qquad 13.30$$

$$N_{c,1} = N_{c,2} = 2; \quad \tau_{ij} = \frac{N_A N_{c,j}(\varepsilon_{ij} - \varepsilon_{jj})}{2RT} = \frac{(g_{ij} - g_{jj})}{RT} = \frac{\Delta g_{ij}}{RT} \qquad 13.31$$

$$\Omega_{ij} = G_{ij} = \exp\left(\frac{-\alpha_{ij}N_A N_{c,j}(\varepsilon_{ij} - \varepsilon_{jj})}{2RT}\right) = \exp(-\alpha_{ij}\tau_{ij}); \quad \tau_{ii} = 0; \quad G_{ii} = 1; \quad g_{ij} = g_{ji} \quad 13.32$$

For a binary mixture, the activity equations become

---

4. Renon, H., Prausnitz, J.M. 1969. *Ind. Eng. Proc. Des. Dev.* 8:413.

$$\frac{G^E}{RT} = x_1 x_2 \left[\frac{\tau_{12} G_{12}}{x_1 G_{12} + x_2} + \frac{\tau_{21} G_{21}}{x_1 + x_2 G_{21}} \right]$$

13.33

For a binary mixture, the activity equations become

$$\ln \gamma_1 = x_2^2 \left[\frac{\tau_{12} G_{12}}{(x_1 G_{12} + x_2)^2} + \tau_{21} \left(\frac{G_{21}}{x_1 + x_2 G_{21}} \right)^2 \right]$$

13.34

$$\ln \gamma_2 = x_1^2 \left[\frac{\tau_{21} G_{21}}{(x_1 + x_2 G_{21})^2} + \tau_{12} \left(\frac{G_{12}}{x_1 G_{12} + x_2} \right)^2 \right]$$

13.35

$$G_{ij} = \exp(-\alpha_{ij} \tau_{ij}); \quad \tau_{ij} = \frac{\Delta g_{ij}}{RT}$$

13.36

> **!** Literature NRTL parameter values, Δg_{ij}, typically have units of energy. Use the correct value of R!

When $\alpha_{ij} = 0$, the binary model simplifies to the one-parameter Margules model,

$$\ln \gamma_1 = x_2^2 [\tau_{12} + \tau_{21}]; \quad \ln \gamma_2 = x_1^2 [\tau_{21} + \tau_{12}]; \quad \text{for } \alpha_{ij} = 0$$

13.37

The NRTL model is not very appealing from a theoretical perspective, but its flexibility has led to a broad range of applications including combinations with electrolyte models. As a practical matter, a value of $\alpha_{ij} = 0.3$ is taken as a default and the equation works much like the Wilson equation, except that it is able to model LLE. The parameter α_{ij} is adjusted for additional flexibility when necessary, such as when modeling LLE where the value is commonly increased. The multicomponent form of NRTL is

$$\ln \gamma_i = \frac{\sum_j x_j \tau_{ji} G_{ji}}{\sum_k x_k G_{ki}} + \sum_j \frac{x_i G_{ij}}{\sum_k x_k G_{kj}} \left(\tau_{ij} - \frac{\sum_m x_m \tau_{mj} G_{mj}}{\sum_k x_k G_{kj}} \right)$$

13.38

> **!** Multicomponent NRTL.

13.4 UNIQUAC

UNIQUAC[5] (short for UNIversal QUAsi Chemical model) builds on the work of Wilson by making three primary refinements. First, the temperature dependence of Ω_{ij} is modified to depend on surface areas rather than volumes, based on the hypothesis that the interaction energies that determine local compositions are dependent on the relative surface areas of the molecules. If the parameter q_i is proportional to the surface area of molecule i,

$$\Omega_{ij} = \frac{q_i}{q_j} \exp\left(\frac{-N_A z(\varepsilon_{ij} - \varepsilon_{jj})}{2RT} \right) = \frac{q_i}{q_j} \exp\left(\frac{-a_{ij}}{T} \right) = \frac{q_i}{q_j} \tau_{ij} = \frac{q_i}{q_j} \exp\left(\frac{-a_{ij}}{T} \right)$$

13.39

5. Abrams, E.S., Prausnitz, J.M. 1975. *AIChE J.* 21:116.

where $z = 10$. The intermediate parameter $\tau_{ij} = \exp(-a_{ij}/T)$ is used for compact notation where $\tau_{ii} = \tau_{jj} = 1$, $a_{ii} = a_{jj} = 0$.[6] In addition, when the energy equation 13.16 is written with $N_{c,j} = zq_j = 10q_j$ for all j at all ρ, the different sizes and shapes of the molecules are implicitly taken into account. Qualitatively, the number of molecules that can contact a central molecule increases as the size of the molecule increases. Using surface fractions attempts to recognize the branching and overlap that can occur between segments in a polyatomic molecule. The inner core of these segments is not accessible, only the surface is accessible for energetic interactions. Therefore, the model of the energy is proposed to be proportional to surface area. Unfortunately, it is not straightforward to construct a more rigorous argument in favor of surface fractions from the energy equation itself. Inserting Eqns. 13.39 and 13.16 into Eqn. 13.17, the excess Helmholtz for a binary solution becomes

$$A^E/(RT) = -x_1 q_1 \ln(\theta_1 + \theta_2 \tau_{21}) - x_2 q_2 \ln(\theta_1 \tau_{12} + \theta_2) + A^E/(RT)\big|_\infty \qquad 13.40$$

where θ_i is the surface area fraction, and $\theta_i = x_i q_i/(x_1 q_1 + x_2 q_2)$ for a binary. Analogous to Wilson's equation, G^E is calculated as A^E, a good approximation. The first two terms represent $(G^E/RT)^{RES}$,

$$(G^E/(RT))^{RES} = -x_1 q_1 \ln(\theta_1 + \theta_2 \tau_{21}) - x_2 q_2 \ln(\theta_1 \tau_{12} + \theta_2) \qquad 13.41$$

that can be compared with Eqn. 13.20 and the final term $A^E/(RT)\big|_\infty$ represents $(G^E/RT)^{COMB}$. The $(G^E/RT)^{COMB}$ term is attributed to the entropy of mixing hard chains, and an approximate expression for this contribution is applied by Maurer and Prausnitz.[7] This representation of the entropy of mixing traces its roots back to the work of Staverman[8] and Guggenheim[9] and was discussed more recently by Lichtenthaler et al.[10] It is very similar to the Flory term, but it corrects for the fact that large molecules are not always large balls, but sometimes long "strings." By noting that the ratio of surface area to volume for a sphere is different from that for a string, Guggenheim's form (the form actually applied in UNIFAC and UNIQUAC) provides a simple but general correction, giving an indication of the degree of branching and nonsphericity. Nevertheless, the Staverman-Guggenheim term represents a relatively small correction to Flory's term. As shown in Fig. 13.1, the extra correction of including the "surface to volume" parameter serves to decrease the excess entropy to some value between zero and the Flory-Huggins estimate. The combinatorial part of UNIQUAC ($A^E/(RT)\big|_\infty$) for a binary system takes a form that can be compared with 13.21

$$\left(\frac{G^E}{RT}\right)^{COMB} = \left(x_1 \ln\frac{\Phi_1}{x_1} + x_2 \ln\frac{\Phi_2}{x_2}\right) - 5\left[q_1 x_1 \ln\left(\frac{\Phi_1}{\theta_1}\right) + q_2 x_2 \ln\left(\frac{\Phi_2}{\theta_2}\right)\right]. \qquad 13.42$$

The Guggenheim form of the excess entropy is based on the molecular volume fractions, Φ_j, and the surface fractions, θ_j. Instead of using macroscopic property data to calculate volume fractions and surface fractions they are based on relative molecule volumes, r, and relative molecule surface areas, q, for each type of molecule.

6. Note that these assumptions create local compositions of the form $x_{ij}/x_{jj} = (\theta_i/\theta_j)\exp(-a_{ij}/T)$. Compare this with the form of Wilson's equation (footnote page 505). Note that the use of the subscripts for the local composition energetic parameters τ and a are switched for the UNIQUAC relative to the Wilson equation Λ and A.

7. Maurer, G., Prausnitz, J.M. 1978. *Fluid Phase Equil.* 2:91.

8. Staverman, A.J. 1950. *Recl. Trav. Chem. Pays Bas.* 69:163.

9. Guggenheim, E.A. *Mixtures,* 1952. Oxford, England: Oxford University Press.

10. Lichtenthaler, R.N., Abrams, D.S., Prausnitz, J.M. 1973. J.M. *Can. J. Chem.* 51:3071.

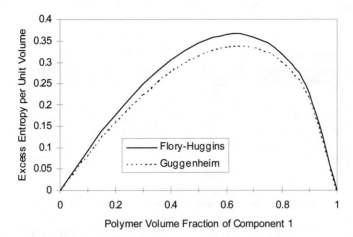

Figure 13.1 *Excess entropy according to the Flory-Huggins equation versus Guggenheim's equation at $V_2/V_1 = 1695$ for a polymer solvent mixture.*

$$\Phi_j \equiv \frac{x_j r_j}{\sum\limits_i x_i r_i} \qquad\qquad \theta_j \equiv \frac{x_j q_j}{\sum\limits_i x_i q_i} \qquad\qquad 13.43$$

r and q are the relative volume and relative surface area of a molecule.

The relative molecular parameters r and q may be calculated from group size and surface area parameters using the concept of **group contributions.** The size/area parameters are ratios to the equivalent size/area for the -CH$_2$- group in a long chain alkane.[11] The group parameters are added in the same manner as the UNIFAC method discussed in the next section and as given in Table 13.2 on page 512, except the UNIQUAC r and q values for alcohols are typically not calculated by group contributions (see the footnote to Table. 13.2). In the table, the uppercase R_k parameter is for the group volume, and the uppercase Q_k parameter is for group surface area. From these values, the molecular size (r_j) and shape (q_j) parameters may be calculated by multiplying the group parameter by the number of times each group appears in the molecule, and summing over all the groups in the molecule,

R_k and Q_k are the relative volume and relative surface area of a functional group.

$$r_j = \sum_k v_k^{(j)} R_k \; ; \qquad\qquad q_j = \sum_k v_k^{(j)} Q_k \qquad\qquad 13.44$$

where $v_k^{(j)}$ is the number of groups of the kth type in molecule j. The subdivision of the molecule into groups is sometimes not obvious because there may appear to be more than one way to subdivide, but the conventions have been set forth in examples in the table and these conventions should be followed. The large number of possible functional groups is divided into **main groups** and further subdivided into structurally similar **subgroups.** Usually the functional groups include a nearest neighbor atom as part of the group. The group parameters are calculated from the **van der Waals volume** and van der Waals surface area. Note that the **van der Waals volume** and **van der Waals**

Care is necessary when subdividing a molecule into functional groups.

---

11. The group parameters are based on the relative van der Waals volume and surface area of sites, $R_k = V_k$ (cm^3/mol)/ 15.15, $Q_k = A_k$(cm^2/mol)/2.5E9; see Abrams, Prausnitz, 1975. *AIChE J.* 21:116. Even though the reducing parameters are based on -CH2-, the values of R_k and Q_k are nonunity. See the reference for details. The unity value for -OH volume is a coincidence.

area are *not* calculated from the van der Waals EOS. They are inferred from x-ray and other property data.[12].

Table 13.2 *Selected Group Parameters for the UNIFAC and UNIQUAC Equations*[a]

Main Group	Sub-group	R(rel.vol.)	Q(rel.area)	Example
CH2	CH3	0.9011	0.8480	
	CH2	0.6744	0.5400	*n*-hexane: 4 CH2 + 2 CH3
	CH	0.4469	0.2280	Isobutane: 1CH + 3 CH3
	C	0.2195	0	Neopentane: 1C + 4 CH3
C=C	CH2=CH	1.3454	1.1760	1-hexene: 1 CH2=CH + 3 CH2 + 1 CH3
	CH=CH	1.1167	0.8670	2-hexene: 1 CH=CH + 2 CH2 + 2 CH3
	CH2=C	1.1173	0.9880	
	CH=C	0.8886	0.6760	
	C=C	0.6605	0.4850	
ACH	ACH	0.5313	0.4000	Benzene: 6 ACH
	AC	0.3652	0.1200	Benzoic acid: 5 ACH + 1 AC + 1 COOH
ACCH2	ACCH3	1.2663	0.9680	Toluene: 5 ACH + 1 ACCH3
	ACCH2	1.0396	0.6600	Ethylbenzene: 5 ACH + 1 ACCH2 + 1 CH2
	ACCH	0.8121	0.3480	
OH[b]	OH	1.0000	1.2000	*n*-propanol: 1 OH + 1 CH3 + 2 CH2
CH3OH	CH3OH	1.4311	1.4320	Methanol is an independent group
water	H2O	0.9200	1.4000	Water is an independent group
furfural	furfural	3.1680	2.484	Furfural is an independent group
DOH	(CH2OH)2	2.4088	2.2480	Ethylene glycol is an independent group
ACOH	ACOH	0.8952	0.6800	Phenol: 1 ACOH + 5 ACH
CH2CO	CH3CO	1.6724	1.4880	Dimethylketone: 1 CH3CO + 1 CH3 Methylethylketone: 1 CH3CO + 1 CH2 + 1 CH3
	CH2CO	1.4457	1.1800	Diethylketone: 1 CH2CO + 2 CH3 + 1 CH2
CHO	CHO	0.9980	0.9480	Acetaldehyde: 1 CHO+1 CH3
CCOO	CH3COO	1.9031	1.7280	Methyl acetate: 1 CH3COO + 1 CH3
	CH2COO	1.6764	1.4200	Methyl propanate: 1 CH2COO + 2 CH3
COOH	COOH	1.3013	1.2240	Benzoic acid: 5 ACH + 1 AC + 1 COOH

a. AC in the table means aromatic carbon. The main groups serve as categories for similar subgroups as explained in the UNIFAC section.

b. Alcohols are usually treated in UNIQUAC without using the group contribution method. Accepted UNIQUAC values for the set of alcohols [MeOH, EtOH, 1-PrOH, 2-PrOH, 1-BuOH] are r = [1.4311, 2.1055, 2.7799, 2.7791, 3.4543], q = [1.4320, 1.9720, 2.5120, 2.5080, 3.0520]. See Gmehling, J., Oken, U. 1977-*Vapor-Liquid Equilibrium Data Collection*. Frankfort, Germany: DECHEMA.

Example 13.4 Combinatorial contribution to the activity coefficient

In polymer solutions, it is not uncommon for the solubility parameter of the polymer to nearly equal the solubility parameter of the solvent, but the mixture is still nonideal. To illustrate, consider the case when 1 g of benzene is added to 1 g of pentastyrene to form a solution. Estimate the activity coefficient of the benzene (B) in the pentastyrene (PS) if $\delta_{ps} = \delta_B = 9.2$ and V_{ps} and V_B are estimated using group contributions. Use the Flory activity model and group contributions of UNIQUAC/UNIFAC to estimate volume fractions.

12. Bondi, A. 1968. *Physical Properties of Molecular Crystals, Liquids and Glasses*. Hoboken NJ: Wiley.

Example 13.4 Combinatorial contribution to the activity coefficient (Continued)

Solution: Since $\delta_{ps} = \delta_B = 9.2$, we can ignore the residual contribution. Therefore,

$$\ln \gamma_B = \ln(\Phi_B/x_B) + (1 - \Phi_B/x_B) \qquad 13.45$$

Benzene is composed of 6(ACH) groups @ 0.5313 R-units per group $\Rightarrow V_B \propto 3.1878$. Pentastyrene is composed of 25(ACH) + 1(ACCH2) + 4(ACCH) + 4(CH2) + 1(CH3) $\Rightarrow V_{ps} \propto 21.17$:

$$MW_B = 78.114 \text{ and } MW_{ps} = 520.76 \Rightarrow x_B = 0.8696 \qquad 13.46$$

$$\Phi_B = 0.8696(3.1878)/[0.8696(3.1878) + 0.1304(21.17)] = 0.5010 \qquad 13.47$$

Note: The volume fraction is close to the weight fraction because they are so structurally similar.

$$\ln \gamma_B = \ln(0.5010/0.8696) + (1 - 0.5010/0.8696) = -0.1275 \quad \Rightarrow \quad \gamma_B = 0.8803 \qquad 13.48$$

Flory's model (no energetic contribution) predicts that the partial pressure of benzene in the vapor, $y_B P = x_B \gamma_B P_B^{sat}$, would be about 12% *less* than the ideal solution model.

The parameters to characterize the volume and surface area fractions have already been tabulated, so no more adjustable parameters are really introduced by writing it this way. The only real problem is that including all these group contributions into the formulas makes hand calculations extremely tedious. Fortunately, computers and spreadsheets make this task much simpler. As such, we can apply the UNIQUAC method almost as easily as the van Laar method.

For a binary mixture, the activity equations become

$$\begin{aligned} \ln \gamma_1 &= \ln(\Phi_1/x_1) + (1 - \Phi_1/x_1) - 5q_1[\ln(\Phi_1/\theta_1) + (1 - \Phi_1/\theta_1)] \\ &+ q_1[1 - \ln(\theta_1 + \theta_2 \tau_{21}) - \theta_1/(\theta_1 + \theta_2 \tau_{21}) - \theta_2 \tau_{12}/(\theta_1 \tau_{12} + \theta_2)] \end{aligned} \qquad 13.49$$

See Actcoeff.xlsx, worksheets uniquac, uniquac5 MATLAB: uniquac.m.

$$\begin{aligned} \ln \gamma_2 &= \ln(\Phi_2/x_2) + (1 - \Phi_2/x_2) - 5q_2[\ln(\Phi_2/\theta_2) + (1 - \Phi_2/\theta_2)] \\ &+ q_2[1 - \ln(\theta_1 \tau_{12} + \theta_2) - \theta_1 \tau_{21}/(\theta_1 + \theta_2 \tau_{21}) - \theta_2/(\theta_1 \tau_{12} + \theta_2)] \end{aligned} \qquad 13.50$$

$$\tau_{ij} = \exp\left(\frac{-a_{ij}}{T}\right) \qquad 13.51$$

Literature UNIQUAC parameter values, a_{ij}, typically are in K.

Like the Wilson equation, the UNIQUAC equation requires that two adjustable parameters be characterized from experimental data for each binary system. The inclusion of the excess entropy in UNIQUAC by Abrams et al. (1975) is more correct theoretically, but Wilson's equation can be as accurate as the UNIQUAC method for many binary vapor-liquid systems, and much simpler to apply by hand. UNIQUAC supersedes the Wilson equation for describing liquid-liquid systems, however, because the Wilson equation is incapable of representing liquid-liquid equilibria as long as the Λ_{ij} parameters are held positive (as implied by their definition as exponentials, and noting that exponentials cannot take on negative values).

Extending Eqn. 13.40 to a multicomponent solution, the UNIQUAC equation becomes

$$\frac{G^E}{RT} = \sum_j x_j \ln(\Phi_j/x_j) - 5\sum_j q_j x_j \ln(\Phi_j/\theta_j) - \sum_j q_j x_j \ln\left(\sum_i \theta_i \tau_{ij}\right) \qquad 13.52$$

Note that the leading sum is simply Flory's equation. The first two terms are the combinatorial contribution and the last is the residual contribution. The parts can be individually differentiated to find the contributions to the activity coefficients,

$$\ln \gamma_k = \ln \gamma_k^{COMB} + \ln \gamma_k^{RES} \qquad 13.53$$

Multicomponent UNIQUAC.

$$\ln \gamma_k^{COMB} = \ln(\Phi_k/x_k) + [1 - \Phi_k/x_k] - 5q_k[\ln(\Phi_k/\theta_k) + (1 - \Phi_k/\theta_k)] \qquad 13.54$$

$$\ln \gamma_k^{RES} = q_k\left[1 - \ln\sum_i \theta_i \tau_{ik} - \sum_j (\theta_j \tau_{kj})\middle/\left(\sum_i \theta_i \tau_{ij}\right)\right] \qquad 13.55$$

13.5 UNIFAC

UNIFAC[13] (short for UNIversal Functional Activity Coefficient model) is an extension of UNIQUAC with no user-adjustable parameters to fit to experimental data. Instead, all of the adjustable parameters have been characterized by the developers of the model based on group contributions that correlate the data in a very large database. The assumptions regarding coordination numbers, and so forth, are similar to the assumptions in UNIQUAC. The same strategy is applied:

$$\boxed{\ln \gamma_k = \ln \gamma_k^{COMB} + \ln \gamma_k^{RES}}$$

The combinatorial term is therefore identical and given by Eqn. 13.54. The major difference between UNIFAC and UNIQUAC is that, for the residual term, UNIFAC considers interaction energies between *functional groups* (rather than the whole molecule). Interactions of functional groups are added to predict relative interaction energies of molecules. Examples are shown in Table 13.2. Each of the subgroups has a characteristic size and surface area; however, the energetic interactions *are considered to be the same for all subgroups with a particular main group*. Thus, representative interaction energies (a_{ij}) are tabulated for only the main functional groups, and it is implied that all subgroups will use the same energetic parameters. An illustrative sample of values for these interactions is given in Table 13.3. Full implementations of the UNIFAC method with large numbers of functional groups are typically available in chemical engineering process design software. A subset of the parameters is provided on the UNIFAC spreadsheet in the Actcoeff.xlsx spreadsheet included with the text. Knowing the values of these interaction energies permits estimation of the properties for a really impressive number of chemical solutions. The limitation is that we are *not always entirely sure of the accuracy* of these predictions.

Although UNIFAC is closely related to UNIQUAC, keep in mind that there is no direct extension to a correlative equation like UNIQUAC. If you want to fit experimental data that might be on hand, you cannot do it within the defined framework of UNIFAC; UNIQUAC or NTRL is the preferred choice when adjustable parameters are desired. Although it is tedious to estimate the a_{ij}

13. Fredenslund, Aa., Jones, R.L.; Prausnitz, J.M. 1975. *AIChE J.* 21:1086.

parameters of UNIQUAC or NRTL from UNIFAC, some implementations of chemical engineering process design software have included facilities for estimating UNIQUAC or NRTL parameters from UNIFAC. This approach can be useful for estimating interactions for a few binary pairs in a multicomponent mixture when most of the binary pairs are known from experimental data specific to those binary interactions.

Table 13.3 *Selected VLE Interaction Energies a_{ij} for the UNIFAC Equation in Units of Kelvin*

Main Group, i	CH2 $j=1$	ACH $j=3$	ACCH2 $j=4$	OH $j=5$	CH3OH $j=6$	water $j=7$	ACOH $j=8$	CH2CO $j=9$	CHO $j=10$	COOH $j=20$
1,CH2	---	61.13	76.5	986.5	697.2	1318	1333	476.4	677	663.5
3,ACH	−11.12	---	167	636.1	637.3	903.8	1329	25.77	347.3	537.4
4,ACCH2	−69.7	−146.8	---	803.2	603.3 .	5695	884.9	−52.1	586.8	872.3
5,OH	156.4	89.6	25.82	---	−137.1	353.5	−259.7	84	−203.6	199
6,CH3OH	16.51	−50	−44.5	249.1	---	−181	−101.7	23.39	306.4	−202.0
7,water	300	362.3	377.6	−229.1	289.6	---	324.5	−195.4	−116.0	−14.09
8,ACOH	275.8	25.34	244.2	−451.6	−265.2	−601.8	---	−356.1	−271.1	408.9
9,CH2CO	26.76	140.1	365.8	164.5	108.7	472.5	−133.1	---	−37.36	669.4
10,CHO	505.7	23.39	106.0	529	−340.2	480.8	−155.6	128	---	497.5
20,COOH	315.3	62.32	89.86	−151	339.8	−66.17	−11.00	−297.8	−165.5	---

The basic approach to understanding UNIFAC is the generalization of the methods for calculating the residual activity coefficient. Imagine the interactions of a CH_3 group in a mixture of isopropanol (1) and component (2). The isopropanol consists of $2(CH_3) + 1(CH) + 1(OH)$. Therefore, in the mixture, a CH_3 will encounter CH_3, CH, OH groups, and the groups of component (2), and the interaction energies depend on the number of each type of group available in the solution. Therefore, the interaction energy of CH_3 groups can be calculated relative to a hypothetical solution of 100% CH_3 groups. The mixture can be approximated as a solution of groups (SOG)[14] (rather than a solution of molecules), and the interaction energies can be integrated with respect to temperature to arrive at chemical potential in a manner similar to the development of Eqn. 13.40.

Therefore, it is possible to calculate $\dfrac{\mu_{CH_3}^{SOG} - \mu_{CH_3}^{o}}{RT} = \ln\Gamma_{CH_3}$ where $\mu_{CH_3}^{o}$ is the chemical potential in a hypothetical solution of 100% CH_3 groups and Γ_{CH_3} is the activity coefficient of CH_3 in the solution of groups. The chemical potential of CH_3 groups in pure isopropanol (1), given by $\mu_{CH_3}^{(1)}$, will differ from $\mu_{CH_3}^{o}$ because even in pure isopropanol CH_3 will encounter a mixture of CH_3, CH, and OH groups in the ratio that they appear in pure isopropanol, and therefore the activity coefficient of CH_3 groups in pure isopropanol, $\Gamma_{CH_3}^{(1)}$, is not unity, where the superscript (1) indicates pure component (1). The difference that is desired is the effect of mixing the CH_3 groups in isopropanol with component (2), relative to pure isopropanol,

---

14. A model exists in the literature that is called ASOG, which is different from the UNIFAC approach, but also uses functional groups.

$$\frac{\mu_{CH_3}^{SOG} - \mu_{CH_3}^{(1)}}{RT} = \frac{\mu_{CH_3}^{SOG} - \mu_{CH_3}^{o}}{RT} - \frac{\mu_{CH_3}^{(1)} - \mu_{CH_3}^{o}}{RT} = \ln\Gamma_{CH_3} - \ln\Gamma_{CH_3}^{(1)}$$

Fig. 13.2 provides an illustration of the differences that we seek to calculate, with water as a component (2).

If the chemical potential of a molecule consists of the sum of interactions of the groups,

$$\mu_1 = 2\mu_{CH_3}^{SOG} + \mu_{CH}^{SOG} + \mu_{OH}^{SOG} \quad \text{and} \quad \mu_1^o = 2\mu_{CH_3}^{(1)} + \mu_{CH}^{(1)} + \mu_{OH}^{(1)}$$

Therefore, we arrive at the important result that is utilized in UNIFAC,

$$\ln\gamma_1^{RES} = \frac{\mu_1 - \mu_1^o}{RT} = \sum_m v_m^{(1)}[\ln\Gamma_m - \ln\Gamma_m^{(1)}] \qquad 13.56$$

where the sum is over all function groups in molecule (1) and $v_m^{(1)}$ is the number of occurrences of group m in the molecule. The activity coefficient formula for any other molecular component can be found by substituting for (1) in Eqn. 13.56. Note that Γ_m is calculated in a solution of groups for all molecules in the mixture, whereas $\Gamma_m^{(1)}$ is calculated in the solution of groups for just component (1). Note that we use uppercase letters to represent the group property analog of the molecular properties, with the following exceptions: Uppercase τ looks too much like T, so we substitute Ψ, and the a_{ij} for UNIFAC is understood to be a group property even though the same symbol is represented by a molecular property in UNIQUAC. The relations are shown in Table 13.4.

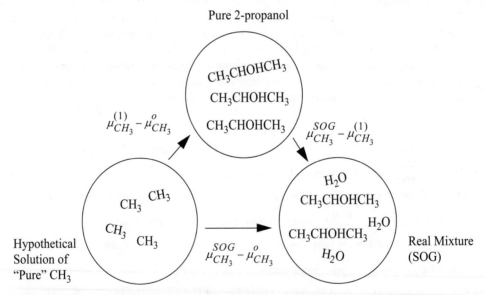

Figure 13.2 *Illustration relating the chemical potential of CH_3 groups in pure 2-propanol, a real solution of groups where water is component (2), and a hypothetical solution of CH_3 groups. The number of groups sketched in each circle is arbitrary and chosen to illustrate the types of groups present. The chemical potential change that we seek is $\mu_{CH_3}^{SOG} - \mu_{CH_3}^{(1)}$. We calculate this difference by taking the difference between the other two paths.*

Table 13.4 *Comparison of Group Variables and Molecular Variables for UNIFAC*

	Group Variable	Molecular Variable
Volume	R	r
Surface area	Q	q
Activity coefficient	Γ	γ
Surface fraction	Θ	θ
Energy variable	Ψ_{ij}	τ_{ij}
Energy parameter	a_{ij}	a_{ij}
Mole fraction	X	x

$\ln\Gamma_m$ is calculated by generalizing the UNIQUAC expression for $\ln\gamma_m^{RES}$. Generalizing Eqn. 13.55 and supporting equations,

$$\ln\Gamma_m = Q_m\left[1 - \ln\sum_i \Theta_i\Psi_{im} - \sum_j \frac{\Theta_j\Psi_{mj}}{\sum_i \Theta_i\Psi_{ij}}\right] \qquad 13.57$$

UNIFAC. See Actcoeff.xlsx, worksheets UNIFAC and UNIFACLLE, and MATLAB UnifacCaller.m.

$$\Theta_j = (\text{surface area fraction of group } j) \equiv \frac{X_jQ_j}{\sum_i X_iQ_i} \qquad 13.58$$

$$\Psi_{mj} = \exp\left(\frac{-a_{mj}}{T}\right) \text{ (not user-adjusted)} \qquad 13.59$$

$$X_j = \frac{\displaystyle\sum_{molecules\ i} v_j^{(i)}x_i}{\displaystyle\sum_{molecules\ i}\ \sum_{groups\ k} v_k^{(i)}x_i} \qquad 13.60$$

where $v_k^{(i)}$ is the number of groups of type k in molecule i. Fortunately, the spreadsheet and programs provided with the textbook save us from doing the tedious calculations for UNIFAC, although an understanding of the principles is important.

Example 13.5 Calculation of group mole fractions

Calculate the group mole fraction for CH_3 in a mixture of 60 mole% 2-propanol, 40 mole% water.

Solution: The two molecules are illustrated in Example 13.1 on page 500 and the group assignments are tabulated there. On a basis of 10 moles of solution, there are six moles of 2-propanol, and four moles of H_2O. The table below summarizes the totals of the functional groups.

Group	Moles	X_j
CH3	12	0.429
CH	6	0.214
OH	6	0.214
H2O	4	0.143
Σ	28	

The mole fraction of CH_3 groups is then $X_{CH_3} = 12/28 = 0.429$. The mole fractions of the other groups are found analogously and are also summarized in the table. The results are consistent with Eqn. 13.60 which is more easily programmed,

$$X_{CH_3} = \frac{2(0.6) + 0(0.4)}{(2(0.6) + 1(0.6) + 1(0.6)) + (1(0.4))} = 0.429$$

Example 13.6 Detailed calculations of activity coefficients via UNIFAC

Actcoeff.xlsx, UNIFAC, and UnifacVLE/ unifacCaller.m.

Let's return to the example for the IPA + water system mentioned in Example 13.1. Compute the surface fractions, volume fractions, group interactions, and summations that go into the activity coefficients for this system at its azeotropic conditions. The isopropyl alcohol (*IPA*) + water (*W*) system is known to form an azeotrope at atmospheric pressure and 80.37°C ($x_W = 0.3146$)[a].

Applying Eqn. 13.57,

$$\ln\Gamma_{CH_3}^{(1)} = 0.848\left\{1 - \ln[0.543 + 0.073 + 0.384(0.642)] - \frac{0.543}{0.543 + 0.073 + 0.384(0.642)}\right.$$

$$\left. - \frac{0.073}{0.543 + 0.073 + 0.384(0.642)} - \frac{0.384(0.061)}{0.543(0.061) + 0.073(0.061) + 0.384}\right\} = 0.3205$$

Solution (this calculation can be followed interactively in the UNIFAC spreadsheet):

The molecular size and surface area parameters are found by applying Eqn. 13.44. Isopropanol has 2 CH_3, 1 OH, and 1 CH group. The group parameters are taken from Table 13.2.

Example 13.6 Detailed calculations of activity coefficients via UNIFAC (Continued)

For IPA: $r = 2 \cdot 0.9011 + 0.4469 + 1.0 = 3.2491$; $q = 2 \cdot 0.8480 + 0.2280 + 1.2 = 3.124$
For water: $r = 0.920$; $q = 1.40$

At $x_W = 0.3146$, $\Phi_W = 0.1150$, and $\theta_W = 0.1706$ using the same combinatorial contribution as UNIQUAC, Eqn. 13.54,
$$\ln\gamma_W^{COMB} = \ln(\Phi_W/x_W) + (1 - \Phi_W/x_W) - 5q_W[\ln(\Phi_W/\theta_W) + (1 - \Phi_W/\theta_W)] = 0.10724$$
$$\ln\gamma_I^{COMB} = \ln(\Phi_I/x_I) + (1 - \Phi_I/x_I) - 5q_I[\ln(\Phi_I/\theta_I) + (1 - \Phi_I/\theta_I)] = -0.00204$$

Note that these combinatorial contributions are computed on the basis of the total molecule. This is because the space-filling properties are the same whether we consider the functional groups or the whole molecules.

For the residual term, we break the solution into a solution of groups. Then we compute the contribution to the activity coefficients arising from each of those groups. We have four functional groups altogether (2CH3, CH, OH, H2O). We will illustrate the concepts by calculating $\ln\Gamma_{CH_3}^{(1)}$ and simply tabulate the results for the remainder of the calculations since they are analogous.

First, let us tabulate the energetic parameters we will need. We can summarize the calculations in tabular form as follows:

$\Psi_{\text{row-col}}$	CH3	CH	OH	H2O
CH3	1	1	0.061	0.024
CH	1	1	0.061	0.024
OH	0.642	0.642	1	0.368
H2O	0.428	0.428	1.912	1

For pure isopropanol, we tabulate the mole fractions of functional groups, and calculate the surface fractions:

j	$v_j^{(1)}$	X_j	Q_j	Θ_j
CH_3	2	0.5	0.848	0.543
CH	1	0.25	0.228	0.073
OH	1	0.25	1.2	0.384
			$\sum_j X_j Q_j = 0.781$	

The same type of calculations can be repeated for the other functional groups. The calculation of $\ln\Gamma_{H_2O}^{(2)}$ is not necessary, since the whole water molecule is considered a functional group.

Performing the calculations in the mixture, the mole fractions, X_j, need to be recalculated to reflect the compositions of groups in the overall mixture. Table 13.5 summarizes the calculations.

Example 13.6 Detailed calculations of activity coefficients via UNIFAC (Continued)

The pure component values of $\ln\Gamma_j^{(i)}$ can be easily verified on the spreadsheet after unhiding the columns with the intermediate calculations or in MATLAB by removing the appropriate ";". Entering values of 0 and 1 for the respective molecular species mole fractions causes the values of $\ln\Gamma_j^{(i)}$ to be calculated. (Note that values will appear on the spreadsheet computed for infinite dilution activity coefficients of the groups which do not exist in the pure component limits, but these are not applicable to our calculation so we can ignore them.) Subtracting the appropriate pure component limits gives the final row in Table 13.5. All that remains is to combine the group contributions to form the molecules, and to add the residual part to the combinatorial part.

Table 13.5 *Summary of Calculations for Mixture of Isopropanol and Water at 80.37 °C and $x_w = 0.3146$*

	CH3 $j=1$	CH $j=2$	OH $j=3$	H2O $j=4$
Q_j	0.848	0.228	1.200	1.400
Θ_j	0.4503	0.0605	0.3186	0.1706
$\sum_i \Theta_i \Psi_{ij}$	0.7885	0.7885	0.6762	0.3000
$\ln\Gamma_j$	0.4641	0.1248	0.3538	0.6398
$\ln\Gamma_j^{(2)}$	Not Applicable	NA	NA	0
$\ln\Gamma_j^{(1)}$	0.3205	0.0862	0.5927	NA
$\ln\Gamma_j - \ln\Gamma_j^{(i)}$	0.1336	0.0386	−0.2388	0.6398

$\ln\gamma_I = \ln\gamma_I^{RES} + \ln\gamma_I^{COMB} = [2(0.1336) + 0.0386 - 0.2388] - 0.00204 = 0.0848; \quad \gamma_I = 1.0885$
$\ln\gamma_W = \ln\gamma_W^{RES} + \ln\gamma_W^{COMB} = [0.6398] + 0.1072 = 0.7470; \quad \gamma_W = 2.1108$

a. Perry, R.H., Chilton, C.H. 1973. *Chemical Engineers' Handbook,* 5 ed. New York, NY: McGraw-Hill, Chapter 13.

13.6 COSMO-RS METHODS

In principle, all electronic and molecular interactions are described by quantum mechanics, so you may wonder why we have not considered computing mixture properties from this fundamental approach. In practice, two considerations limit the feasibility of this approach. First, quantum mechanical computations tend to be time consuming. Precise computations can require days for a single molecule the size of naphthalene and the computation time increases as N^7, where N is the number of atoms in the molecule. Second, the intermolecular interaction energy, which affects the mixing properties, is at least an order of magnitude smaller than the intramolecular energy. Therefore, high precision is required to compute the intermolecular interactions directly. Circumventing these limitations has been the focus of the "COSMO-RS" approach.

COSMO-RS is an abbreviation for "COnductor-like Screening MOdel for Real Solvents." It refers to a method of performing quantum mechanical calculations as if the simulated molecule were in a conductive bath rather than a vacuum. The method was developed originally by Klamt and coworkers[15] as an extension of previous work on a continuum solvation model (CSM).[16] The implementation of Klamt and coworkers is marketed commercially as COSMO*therm* and updated continually. A free educational version with graphical user interface is available that includes capability for about 350 compounds. The implementation of Klamt and coworkers is currently based on the TURBOMOLE package for quantum mechanical simulations, and a few other packages also provide consistent results. Later, Lin and Sandler developed an implementation based on the DMOL^3 simulation package,[17] which is included in the Accelrys Materials Studio. We refer to this implementation as COSMO-RS/SAC. It is available as a free download including capability for about 1500 compounds from a web site maintained by Y.A. Liu at Va. Tech.[18] The method includes several empirical parameters that have been characterized by fitting experimental data. The specific parameters depend on the quantum simulation method. In the example below, we have applied the SAC parameters.

The key idea of the COSMO-RS approach has been to focus on the polarization of the surface surrounding a molecule. The significance of the surface polarization can be likened to the acidity and basicity of the SSCED model. If the surrounding surface is positively charged, then the molecular surface must be basic; if the surrounding surface is negatively charged, then the molecular surface must be acidic. If we imagine coloring these acidic interactions as blue and the basic interactions as red, we could represent the molecular surfaces in the manner of the pictures on the cover of the textbook. The overall surface energy, including positive, negative, and neutral influences, can be correlated with activity coefficients and other properties like the heat of vaporization.

To calculate the surface polarization an approximate quantum mechanical method can be applied, known as density functional theory (DFT). DFT computations typically require only a few hours for a molecule like naphthalene, and the increase in computation time scales as N^3. If we integrate over the interactions between the surfaces of molecules, we can imagine how results similar to SSCED could be achieved. An additional advantage of COSMO-RS is that local composition effects are implicit in the integration of the local polarization interactions over all orientations between the two molecules. Furthermore, acidity and basicity are not limited to a single characteristic value per molecule, but are characterized by a range of polarizations over the entire surface of each molecule.

To implement the method, the molecular surface charges are calculated using DFT. The observed polarization is referred to as a σ-profile, where σ refers to the surface charge density (Coulombs/Å^2). Typical σ-profiles are illustrated in Fig. 13.3. The $p(\sigma)$ represents the amount of area per σ-interval (Å^2/(Coulombs/Å^2)) plotted against the surface charge density (Coulombs/Å^2). In other words, $p(\sigma)$ is proportional to a count of how many segments possess a given amount of surface polarization. It is analogous to the number of occurrences of a group in UNIFAC. The curve is normalized such that an integral of the σ-profile gives the total surface area of the molecule. The effective area per segment is divided out near the end of the calculation when computing the activity coefficient. The area under the curve in a particular region shows the total area of the molecule with the charge density. Fig. 13.3(a) shows how ethanol is a smaller molecule than octanol based on the total area under the curve.

15. Klamt, A. 1995. *J. Phys. Chem.* 99:2224. Klamt, A., Jonas, V., Buerger, T., Lohrenz, J.C.W. 1998. *J. Phys. Chem.* 102:5074. Klamt, A., Eckert, F. 2000. *Fluid Phase Equil.* 172:43.
16. Klamt, A., Schurmann, G.J. 1993. *J. Chem. Soc. Perkin Trans.* 2:799.
17. Lin, S.T., Sandler, S.I. 2002. *Ind. Eng. Chem. Res.* 41:899. DMOL^3 was the original basis of Klamt's work and is still supported.
18. www.design.che.vt.edu/VT-Databases.html. Mullins, et al. 2006. *Ind. Eng. Chem. Res.* 45:3973.

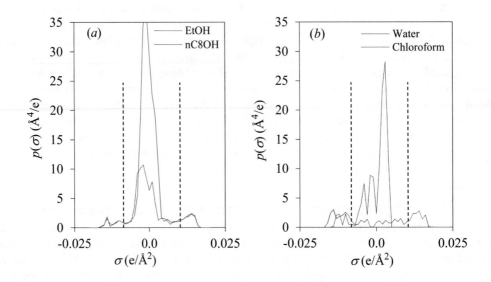

Figure 13.3 *Samples of σ-profiles for application to the COSMO-RS method, $A_{eff}n_i(\sigma)$.*
Dashed vertical lines show the threshold values for hydrogen bonding.

Since the charge density, σ, ranges over negative and positive values, all values less than –0.0084 are considered to contribute to acidity. A similar consideration applies to the β contribution except that all $\sigma > 0.0084$ contribute to basicity. In Fig. 13.3(a), note that the two alcohols have extremely similar contributions of total area for both acidity and basicity. Fig. 13.3(b) shows that water is a very small molecule (small area under the curve) with extensive hydrogen bonding (both acidic and basic outside the dispersion bounds) while chloroform is a relatively large molecule with strong acidity and no basicity. All of these behaviors make sense qualitatively, so what remains is the translation into a quantitative method.

Computing activity coefficients for a binary solution is similar to the UNIFAC method if you can imagine transforming the earlier summation over groups to a summation over discretized polarization segments of the σ-profile. Note that use of the term "segment" does not refer to a geometric segment, but to an "interval" of surface polarization. A particular amount of surface polarization may occur at various places over the surface of the molecule, but all would be added together to get $p(\sigma)$. It may be helpful to think of this quantity in mathematical terms rather than as a physical entity. Typically, the σ-profile is discretized into 51 values ranging from –0.025 to 0.025 (Coulombs/Å^2).[19] Therefore, we can refer to σ_k where $k = [1,51]$. To simplify the computations, the integration $\int p(\sigma)d\sigma \sim \Delta\sigma \Sigma p(\sigma_k)$ is represented as a sum, $\Sigma p(k)$. The larger size of a particular molecule is reflected in the $\Sigma p(k)$ in a comparable manner to the molecular volume in the Scatchard-Hildebrand or SSCED models. Discretized segments on molecule i interact with segments on all molecules, including the ith molecule itself. In a mixture, $p_i(k)$ designates $p(k)$ for the ith component. The activity coefficient contribution for a segment k (Γ_k) in a solution of polarization segments is analogous to Eqn. 13.57, but with a significantly different functional form.

---

19. Each discretized value of σ represents the center of the bin for that range of σ.

$$n\Gamma_k = -\ln \sum_j \Theta(j)\Gamma_j \Psi_{jl} \qquad\qquad 13.61$$

The surface area fraction of polarization segments in a binary mixture of molecules type $1 + 2$ is

$$\Theta(k) = \frac{x_1 p_1(k) + x_2 p_2(k)}{x_1 q_1 + x_2 q_2} \qquad\qquad 13.62$$

$$\Psi_{jk} = \exp\!\left(\frac{-a_{jk}}{T}\right) \qquad\qquad 13.63$$

where a_{jk} is defined below and the molecular surface area is

$$q_i = \Sigma\, p_i(k) \qquad\qquad 13.64$$

A practical difference between Eqns. 13.57 and 13.61 is that the activity coefficient contribution of a given σ-interval depends on itself through the summation over all interactions in Eqn. 13.61. Conceptually, the influence of a segment on its neighbors alters its own behavior as those neighbors respond to the local activity. This necessitates iteration, initiated with $\Gamma_k = 1$ for all k.

A fundamental difference between Eqns. 13.57 and 13.61 is that the formulas are derived entirely differently. The matrix of a_{jk} is related to balancing electrostatic charges, as given by:

$$a_{jk} = [\, 8233(\,\sigma(j) + \sigma(k)\,)^2 + 85580\,\alpha(j,k)\,\beta(j,k)\,]/0.001987 \qquad\qquad 13.65$$

where $\alpha(j,k)$ and $\beta(j,k)$ characterize the hydrogen bonding between the jth and k^{th} segments, similar to the α and β parameters of the SSCED model. We assume that hydrogen bonding occurs regardless of which σ value (j or k) surpasses the threshold value because the energetic reward is sufficient for them to find each other regardless of where the segments are. Mathematically, this becomes,

$$\beta(j,k) = \max(\,[\max\{\sigma(j),\sigma(k)\}-0.0084],\,0\,\}; \;\; \alpha(j,k) = \min\{\,[\min\{\sigma(j),\sigma(k)+0.0084],\,0\,\} \qquad 13.66$$

Similar to UNIFAC, this approach leads to a nonunity value for the activity coefficient of a pure fluid. So,

$$\ln\gamma_i = \ln\gamma_i^{COMB} + \Sigma_k\, p_i(k)(\ln\Gamma_k - \ln\Gamma_k^{(i)})/A_{eff} \qquad\qquad 13.67$$

where $\ln\gamma_i^{COMB}$ is computed using the same formula as for UNIFAC or UNIQUAC; $A_{eff} = 7.5$ is the normalization for the area, and $\Gamma_k^{(i)}$ is the segment activity in pure component i, computed by applying Eqn. 13.62 and so forth, with the appropriate composition.

Although we can conceive of COSMO-RS as being similar to UNIFAC, it is really much more general. The UNIFAC method requires experimental data to characterize the a_{ik} matrix. COSMO-RS, on the other hand, computes this matrix based on σ-profiles that have been computed with no experimental data except the empirical constants in Eqn. 13.65. Usually, more precise results are obtained with UNIFAC if all the groups have been characterized, but the COSMO-RS approach can be used to supplement the UNIFAC method when no experimental data exist.

Example 13.7 Calculation of activity coefficients using COSMO-RS/SAC

σ-profiles for methanol and acetone are listed below. (a) Use these to compute the activity coefficient at $x_M = 0.425$ and 55°C assuming the SAC values of the COSMO-RS parameters as given by Lin and Sandler. (b) Compute the activity coefficients over the entire range of compositions and compare to the fit of UNIQUAC to experimental data and the SSCED model.

Table of σ-profiles. Note that $\sigma(k) = [-0.025, 0.025]$, a total of 51 values. Cells omitted if all zeros.

k	7	8	9	10	11	12	13	14	15	16
$p_1(k)$	0	0	0	0.598	1.081	1.304	0.607	0.584	0.644	1.318
$p_2(k)$	0	0	0	0	0	0	0	0	0	0
k	17	18	19	20	21	22	23	24	25	26
$p_1(k)$	1.160	0.876	0.428	1.833	3.576	6.431	6.897	5.273	7.463	5.094
$p_2(k)$	0	0	3.548	11.256	10.867	9.795	8.820	8.261	13.617	6.570
k	27	28	29	30	31	32	33	34	35	36
$p_1(k)$	1.582	2.271	1.181	1.282	1.762	1.460	0.279	1.024	0.810	2.056
$p_2(k)$	4.665	1.551	1.622	1.050	0.154	0.846	0.812	1.879	1.541	3.309
k	37	38	39	40	41	42	43	44	45	46
$p_1(k)$	0.822	1.436	3.112	1.806	1.803	0.048	0	0	0	0
$p_2(k)$	4.498	3.771	2.594	1.620	0	0	0	0	0	0

Solution:
(a) Applying Eqn. 13.64 gives $q_1 = 67.9$ and $q_2 = 102.6$. The mixture's segmental area fraction, $\Theta(k)$, is zero for $k = 1$ to 9. $\Theta(10) = (0.425(0.598) + (0.575)0)/(0.425(67.9) + (0.575)102.6) = 2.89(10^{-3})$ and so forth for $\Theta(11)$ to $\Theta(42)$.

For $j = k = 1$, $\sigma(j) = \sigma(k) = -0.025$. So, $\max(\sigma(j), \sigma(k)) = -0.025$, and $-0.025 - 0.0084 = -0.0334$, but $\max(0, -0.0334) = 0$.

So $\alpha(1, 1) = 0$, and even though $\beta(1, 1) = -0.0166$, the product $\alpha(1, 1)\beta(1, 1) = 0$ and hydrogen bonding contributes zero for $i = k = 1$. For the dispersion term, however, $a_{1,1} = [8233(\sigma(1) + \sigma(1))^2]/0.001987 = 10359$ and $\Psi(1,1) = \exp(-10359/328.15) = 1.951(10^{-14})$.

The first term for which hydrogen bonding is nonzero is $i = 1$, $k = 35$. Then, $\alpha(1, 35) = \max(\sigma(j), \sigma(k)) - 0.0084 = 0.0006$ and $\beta(1, 35) = -0.0166$.
The total is $a_{1,35} = [8233(-0.025 + 0.009)^2 + 85580(0.0006)(-0.0166)]/0.001987 = -428.98$ and $\Psi(1, 35) = \exp(428.98/328.15) = 3.696$. Note how $p(k)$ contributes to Θ but not Ψ.

Example 13.7 Calculation of activity coefficients using COSMO-RS/SAC (Continued)

These two cases ($i = k = 1$) and ($i = 1$, $k = 35$) illustrate important behaviors. Briefly, segment pairs involving the disperse interactions between two like-charged segments are disfavored, as indicated by the small value of Ψ. Even when the polarizations are exactly opposite and sum to zero, the best that can happen for the dispersion term is $\Psi = 1$. All other values of polarization lead to $\Psi<1$ for the dispersion term (because the contribution is squared), reflecting an unfavorable interaction. This is vaguely reminiscent of the $(\Delta\delta)^2$ term that applies to dispersion interactions in the SSCED model. The second case shows that acid-base interactions are opposite in sign to the dispersion interactions, and generally are sufficient to provide $\Psi > 1$, indicating favorable interactions. Again, this is reminiscent of hydrogen bonding in the SSCED model.

Completing the solution for part (a) involves summing all the terms, both for the mixture and for the pure fluid. After the first iteration of Eqn. 13.61 for Γ_k, the first 10 values are roughly 0.5. Other sample values are $\Gamma_{20} = 1.561$, $\Gamma_{30} = 1.203$, $\Gamma_{40} = 0.736$, and $\Gamma_{50} = 0.500$. After the last iteration, these values become $\Gamma_1 = 7.864(10^{-5})$, $\Gamma_{10} = 0.0279$, $\Gamma_{20} = 1.478$, $\Gamma_{30} = 1.107$, $\Gamma_{40} = 2.522$, $\Gamma_{50} = 3.312(10^{-3})$.

Similarly, after the first iteration for pure fluids: $\Gamma_k^{(1)} = 0.5$ for $k = 1$ to 10, ..., $\Gamma_{20}^{(1)} = 1.559$, $\Gamma_{30}^{(1)} = 1.267$, ... $\Gamma_1^{(2)} = 0.501$, $\Gamma_{10}^{(2)} = 0.563$, $\Gamma_{20}^{(2)} = 1.562$, and $\Gamma_{30}^{(2)} = 1.176$.

After the final iteration, $\Gamma_1^{(1)} = 7.05(10^{-5})$, $\Gamma_{10}^{(1)} = 2.931(10^{-2})$, $\Gamma_{20}^{(1)} = 1.546$, $\Gamma_{30}^{(1)} = 1.204$, ... $\Gamma_1^{(2)} = 3.322(10^{-4})$, $\Gamma_{10}^{(2)} = 3.748(10^{-3})$, $\Gamma_{20}^{(2)} = 1.453$, $\Gamma_{30}^{(2)} = 1.063$, ...

Summing the $\Gamma_k - \Gamma_k^{(i)}$ gives $\ln\gamma_1 = 0.0957 - 0.0353 = 0.0605$ and $\ln\gamma_2 = 0.0746 - 0.0145 = 0.0601$, where the negative terms are the Guggenheim-Staverman contribution using q_i and $r_1 = 48.8$ and $r_2 = 86.4$. The values of r_1 and r_2 come from the (VT) database of Mullins et al.
(b) The σ-profiles for methanol and acetone are shown below, along with a comparison to the activity coefficients from the SSCED model and the UNIQUAC model.[a]

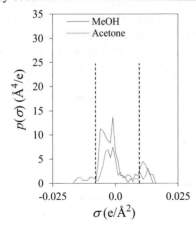
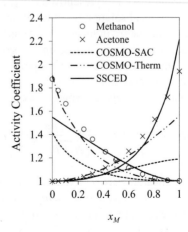

> ### Example 13.7 Calculation of activity coefficients using COSMO-RS/SAC (Continued)
>
> Regarding the σ-profiles, it is apparent from the high positive polarization (proton acceptor) of the acetone that hydrogen bonding should play a significant role. The right figure shows the comparison of the COSMO-RS model with a UNIQUAC fit (considered the benchmark in this case). The COSMO-RS/SAC model and the SSCED model are almost scaled mirror images, and neither is precisely correct.[b] The COSMO-RS/SAC model underestimates the nonideality of this mixture.

a. The UNIQUAC model (right figure) was fit to the data of Marinichev A.N., Susarev M.P. 1965. *Zhur. Prikl. Khim*, 38(2):371.

b. These components are available in the educational version of COSMOtherm. Estimates are more accurate with COSMOtherm, but the mirror-image effect is consistent.

13.7 THE MOLECULAR BASIS OF SOLUTION MODELS

As discussed during the development of quadratic mixing rules, there comes a point at which the assumption of random mixing cannot completely explain the nonidealities of the solution. Local compositions are examples of nonrandomness. The popularity of local composition models like Wilson's equation or UNIFAC means that we need to develop some appreciation of the strength of the underlying theory and its limitations. Similar to the situation for the random mixing models, there are limitations to the local composition models. At this time, however, we are not exactly sure what all the limitations are. This is still a question for active research. Nevertheless, we can provide an understanding of the assumptions in these models, because the assumptions are the sources of errors that impose limitations.

Extending the Energy Equation to Mixtures

We begin the discussion with the energy equation, not the pressure equation as we did for pure fluids. This is because we are presently concerned with the Gibbs energy of mixing and its excess change relative to ideal solution behavior. It turns out that the excess Gibbs energy is dominated by the excess internal energy in most cases. In other words, the entropy of mixing is given with reasonable accuracy by the mixing rule on "b" given in Eqn. 12.1. Therefore, we focus our attention on extending the energy equation to mixtures. This requires revisiting our development of the energy equation for pure fluids and applying the same principles to extend it to mixtures. With two small modifications, the energy equation we developed for pure fluids becomes:

$$\frac{U_1 - U_1^{ig}}{N_1 kT} = \frac{N_1}{2V} \int_{}^{u_{11}} \frac{u_{11}}{kT} g_{11} 4\pi r^2 dr \qquad 13.68$$

The small modifications are: (1) We have put the equation on an atomic basis instead of a molar basis by noting $n_i N_A = N_i$ and $Nk = nR$; (2) We recognize that these are the contributions to the energy departure that arise from atoms of type "1." For the pure fluid, it so happens that we only have atoms of type "1."

In developing the energy equation for pure fluids, we recognized that the average internal energy departure per atom [i.e., $(U_1 - U_1^{ig})/N_1$] was equal to the energy per pair per unit volume times the local density in that volume integrated over the total volume. To make the extension to a mixture, we must simply recognize that there are now atoms of "type 2" around those atoms of "type 1." To illustrate, consider a parking lot full of blue cars and green cars. If one parking lot had only green cars, then the average energy per green car would involve the average number of green cars at each distance around one green car times the energy associated with green cars being that distance from each other. If you pack them too close, you will have to work hard to pack them, and so forth. Now consider the next parking lot, where blue cars are mixed with green cars. The average energy per *green* car will now involve contributions from green-green interactions and blue-green interactions. In equation form, this becomes

$$\frac{U_g - U_g^{ig}}{N_g kT} = \frac{N_g}{2\underline{V}}\int \frac{u_{gg}}{kT} g_{gg} 4\pi r^2 dr + \frac{N_b}{2\underline{V}}\int \frac{u_{bg}}{kT} g_{bg} 4\pi r^2 dr \qquad 13.69$$

We can check this equation by noting that it approaches the pure green car equation when all the blue cars leave the parking lot (i.e., as $N_b \to 0$). We may next write the average energy per blue car by symmetry and the total energy by addition.

$$\frac{U_b - U_b^{ig}}{N_b kT} = \frac{N_b}{2\underline{V}}\int \frac{u_{bb}}{kT} g_{bb} 4\pi r^2 dr + \frac{N_g}{2\underline{V}}\int \frac{u_{bg}}{kT} g_{bg} 4\pi r^2 dr \qquad 13.70$$

$$\frac{U - U^{ig}}{NkT} = \frac{N_g}{N}\left(\frac{U_g - U_g^{ig}}{N_g kT}\right) + \frac{N_b}{N}\left(\frac{U_b - U_b^{ig}}{N_b kT}\right)$$

$$\frac{U - U^{ig}}{NkT} = \frac{N_g}{N}\left(\frac{N_g}{2\underline{V}}\int \frac{u_{gg}}{kT} g_{gg} 4\pi r^2 dr + \frac{N_b}{2\underline{V}}\int \frac{u_{bg}}{kT} g_{bg} 4\pi r^2 dr\right)$$
$$+ \frac{N_b}{N}\left(\frac{N_b}{2\underline{V}}\int \frac{u_{bb}}{kT} g_{bb} 4\pi r^2 dr + \frac{N_g}{2\underline{V}}\int \frac{u_{bg}}{kT} g_{bg} 4\pi r^2 dr\right)$$

Finally, making the substitutions in terms of the mole fractions, recognizing that $u_{bg} = u_{gb}$ and $g_{bg} = g_{gb}$, and converting back to a molar basis, we see that for multicomponent mixtures,

$$\boxed{\frac{U - U^{ig}}{RT} = \frac{N_A \rho}{2}\sum_i \sum_j x_i x_j \int \frac{N_A u_{ij}}{RT} g_{ij} 4\pi r^2 dr} \qquad 13.71$$

🛑 The energy equation for mixtures.

Comparing to the van der Waals equation for pure fluids,

$$a_{ii} = -\frac{1}{2}\int N_A^2 u_{ii} g_{ii} 4\pi r^2 dr \qquad 13.72$$

$$\Rightarrow \frac{U - U^{ig}}{RT} = -\frac{\rho}{RT}\sum_i \sum_j x_i x_j a_{ij} \Rightarrow a = \sum_i \sum_j x_i x_j a_{ij} \qquad 13.73$$

where $a_{ij} = -\frac{1}{2}\int N_A^2 u_{ij} g_{ij} 4\pi r^2 dr$ and it is understood that $a_{ij} = a_{ji}$.

In this form we can recognize that the radial distribution functions may be dependent on composition as well as temperature and density. Therefore, assuming the quadratic mixing rule simply neglects the composition dependence of the a parameter, as well as the temperature and density dependence. We found in Unit II that the assumptions about temperature and density in the van der Waals equation were flawed and that is what motivated the Peng-Robinson equation. Similarly, neglecting the composition dependence of the radial distribution functions leads to some limitations that give rise to local composition theory.

Local Compositions in Terms of Radial Distribution Functions

Recalling the energy equation for mixtures, Eqn. 13.71 multiplied by RT,

$$U - U^{ig} = \frac{N_A \rho}{2} \sum_i \sum_j x_i x_j \int N_A u_{ij} g_{ij} 4\pi r^2 dr \qquad 13.74$$

We may now define the local compositions in terms of the radial distribution functions.

$$x_{ij} = \frac{N_i \sigma_{ij}^3}{VN_{c,j}} \int_0^{R_{ij}} g_{ij} 4\pi r_{ij}^2 dr_{ij} = \frac{x_i n N_A \sigma_{ij}^3}{VN_{c,j}} \int_0^{R_{ij}} g_{ij} 4\pi r_{ij}^2 dr_{ij} \qquad 13.75$$

where $\quad r_{ij} = r / \sigma_{ij}$

$\qquad R_{ij} =$ "neighborhood"

$\qquad N_{c,i} =$ total # of atoms around sites of type "i," that is, the coordination number.

Rearrangement gives the molecular definition of the local composition parameter Ω_{ij},

$$\frac{x_{ij}}{x_{jj}} = \frac{N_{c,j} N_i \sigma_{ij}^3 \int g_{ij} 4\pi r_{ij}^2 dr_{ij}}{N_{c,j} N_j \sigma_{jj}^3 \int g_{jj} 4\pi r_{jj}^2 dr_{jj}} \equiv \frac{x_i}{x_j} \Omega_{ij} \qquad 13.76$$

and we note the similarity between the integral in the energy equation and the integral in the definition of local composition.

For a square-well fluid, $\varepsilon_{ij} =$ constant, so we can factor it out of the integral,

$$\int u_{ij} g_{ij} 4\pi r^2 dr = \varepsilon_{ij} \int g_{ij} 4\pi r^2 dr \qquad 13.77$$

$$U - U^{ig} = \frac{1}{2} \sum_i \sum_j x_j \frac{x_i N_A \sigma_{ij}^3}{V} n N_A \varepsilon_{ij} \int g_{ij} 4\pi r^2_{ij} dr_{ij} \qquad 13.78$$

Substituting $N_{c,j}$, and x_{ij} into the energy equation for mixtures (multiply Eqn. 13.75 by $N_{c,j}$ and substitute into Eqn. 13.78),

$$U - U^{ig} = \frac{1}{2}\sum_j x_j N_{c,j} \sum_i x_{ij} N_A \varepsilon_{ij} \qquad 13.79$$

This is the equation previously applied as the starting point for development of the Gibbs excess energy model from a local composition perspective. The rest of the derivation proceeds as before.

Assumptions in Local Composition Models

In the previous section, we discussed some of the currently popular expressions for activity coefficients. We listed the assumptions involved in developing the expressions but we did not take time to discuss those assumptions. Instead, we directly applied the expressions as a practical necessity and moved on. In this section, we recall those assumptions and attempt to put them in perspective. After developing this perspective, we conclude with a word of caution; the reliability of the predictions depends largely on the accuracy of the assumptions.

The local composition theory, upon which UNIQUAC and others are based, has the general intent of correcting regular solution theory for asymmetries in solution behavior due to fluid structure near a central species. Relaxing the assumption that $S^E = 0$ also leads to the necessity of considering the entropy of mixing, and differences between the UNIQUAC model and Wilson's model are primarily due to differences in treatment of S^E. In review, four assumptions are shared by the local composition theories when considered with respect to spheres:

1. The average energy of an i-j interaction is independent of temperature, density, and other species present.

2. $(A - A^{is}) = (G - G^{is})$.

3. The "coordination number" of a specie in a mixture is the same as that of the pure species.

4. The temperature-dependent part of the energy of mixing is given by
   $\Omega_{ij} = (\sigma_{ij}/\sigma_{jj})^3 \exp[z(\varepsilon_{ij} - \varepsilon_{jj})/2kT]$ where z is a "coordination number."

Wilson's equation makes the following assumptions:

5w. $N_{c,j} = z = 2$ for all j at all densities.

6w. $(A^E/(RT))^{COMB} = \sum_j x_j \ln(\Phi_j/x_j)$.

7w. $(\sigma_{ij}/\sigma_{jj})^3 = V_i/V_j$ for all i, j, and $\dfrac{A_{ji}}{RT} = \dfrac{z(\varepsilon_{ij} - \varepsilon_{jj})}{2kT}$, $\Lambda_{ji} = \Omega_{ij}$.[20]

UNIFAC(QUAC) makes the following assumptions:

5u. $z = 10$, $N_{c,j} = 10 \cdot \sum_k v_k^{(j)} Q_k$.

6u. $(A^E/(RT))^{COMB} = \sum_j \left[x_j \ln(\Phi_j/x_j) + \frac{1}{2}N_{c,j}x_j \ln(\theta_j/\Phi_j) \right]$.

7u. $(\sigma_{ij}/\sigma_{jj})^3 = N_{c,i}/N_{c,j} = q_i/q_j$ for all i, j, $\dfrac{a_{ij}}{T} = \dfrac{z(\varepsilon_{ij} - \varepsilon_{jj})}{2kT}$, $\Omega_{ij} = \dfrac{q_i}{q_j}\tau_{ij}$.

---

20. The order of subscripts for the local compositions in Wilson's equation is nonintuitive; see Eqns. 13.1 and 13.2.

Assumption 1 involves factoring some average energy out of the energy integral such that the local composition integral is obtained. As noted, this assumption would be correct for a square-well potential, so we can probably trust that it would be reasonable for other similar potentials like the Lennard-Jones. The doubt which arises, however, involves the application of this approximation to highly nonideal mixtures. The square-well and Lennard-Jones potentials rarely give rise to highly nonideal mixtures when realistic values for their parameters are chosen. There is very little evidence to judge whether the ε_{ij} factored out in this way is really independent of temperature and density for nonideal mixtures. In fact, in Chapter 1, we showed that dipole interactions *are* temperature dependent.

Assumption 2 has to do with neglecting $\ln(Z/Z^{is})$. For liquids, this may seem dangerous until one realizes that it amounts to neglecting $\ln(1 + \rho^E/\rho) \approx \rho^E/\rho$. Relative to the density of a liquid, the excess density is generally small (but easy to measure with a high degree of accuracy) and this assumption is acceptable.

Assumption 3 has to do with convenience. If the coordination number of each species was assumed to change with mixing, the theory could become very complicated. That is not a very good physical reason, of course. Physically speaking, this assumption could become quite poor if the sizes of the molecules (or segments in the case of UNIFAC) were very different.

Assumption 4 is the primary assumption behind all of the current local composition approaches, but it is not required by the concept of local compositions. It is simply computationally expedient in the equations that develop. The crucial aspect of the assumption is the simple form of the temperature dependence of Ω_{ij}. The main motivation for this assumption appears to be obtaining an expression which can be integrated analytically. But how accurate is this assumption on a physical basis? Moreover, how can we determine the physical behavior for the behavior of Ω_{ij} in an unequivocal manner? Merely fitting experimental data for the Gibbs excess function is equivocal because some set of adjustable parameters will provide a good fit even if the model has no physical basis. An alternative available to us that was not available to van der Waals is to apply computer simulation of square-well mixtures over a specific range of densities and temperatures and test the validity of Wilson's approximation directly through the simulated local compositions. This approach was undertaken by Sandler and Lee.[21]

Sandler and Lee have developed a correlation for what amounts to Ω_{ij} of a square-well potential.

$$\Omega_{ij} = (\sigma_{ij}/\sigma_{jj})^3 \exp[N_A(\varepsilon_{ij} - \varepsilon_{jj})/(2RT)] \left\{ \frac{\sqrt{2} + \rho\sigma_{jj}^3[\exp(\varepsilon_{jj}/(kT)) - 1]}{\sqrt{2} + \rho\sigma_{ij}^3[\exp(\varepsilon_{ij}/(kT)) - 1]} \right\} \qquad 13.80$$

This expression reproduces the local compositions of a substantial set of molecular simulation data which Sandler and coworkers have generated. We can therefore use this expression along with the molecular simulation data as a guide to the accuracy of Wilson's assumption.

The most important consideration is the temperature dependence of this parameter because it is the integration with respect to temperature that allows us to get from energy to free energy. As for density, it could be argued that the density of all liquids is roughly the same, so it is not unreasonable to pick a specific density and just study the temperature effects. Suppose

$$(\sigma_{22}/\sigma_{11})^3 = 1, \quad \rho\sigma^3 = 0.7, \quad \text{and} \quad \varepsilon_{22} = 2\varepsilon_{11} \qquad 13.81$$

---

21. Sandler, S.I., Lee, K-H. 1986. *Fluid Phase Equil.* 30:135.

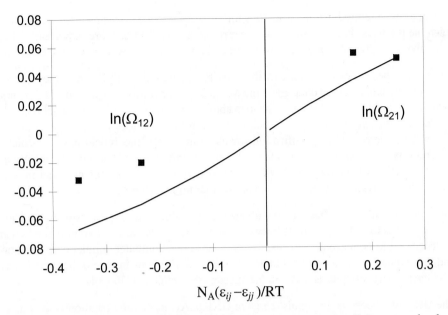

Figure 13.4 *Temperature-dependence of local composition parameters. Points are molecular simulation data and the curve is the correlation of Sandler and Lee.[21] The approximate linearity of the plot lends support to the assumption applied in integrating the internal energy to obtain the free energy.*

$$\ln\Omega_{ij} = \frac{(\varepsilon_{ij} - \varepsilon_{jj})}{2kT} + \ln\left[\frac{0.714 + 0.7\exp(\varepsilon_{jj}/(kT))}{0.714 + 0.7\exp(\varepsilon_{ij}/(kT))}\right] \qquad 13.82$$

If the natural logarithm term on the right-hand side in the above equation is small or gives rise to a contribution which is linear in $1/kT$, then Wilson's assumption is basically correct and his definition of ε_{ij} would just be a little different from Eqn. 13.18. If $\ln\Omega_{ij}$ versus $1/kT$ is not linear, however, then Wilson's assumption is very questionable. Fig. 13.4 shows $\ln\Omega_{ij}$ is fairly linear over certain ranges of temperature. This suggests that the primary assumption of Wilson and UNIFAC(QUAC) is not unreasonable. Does this mean that the problem of nonideal solutions is solved? Maybe, but maybe not. Unfortunately, we must look closely at the range of temperatures that are applicable. This range is limited by the tendency for the mixture to phase-separate. That is, dropping the temperature at constant overall density eventually places the conditions inside the binodal curve. For the composition and density listed above, the binodal occurs when $N_A(\varepsilon_{12} - \varepsilon_{12})/RT < -0.4$. This corresponds to a maximum value of A_{12} in Wilson's equation on the order of ~300 cal/mol at temperatures near 400 K. To use larger values for A_{12} would be unsupported, but larger values are often used, as illustrated in Example 13.3.

Another indication of the potential for error with the local composition approach is given by experimental data for excess enthalpies of mixing. Relations from classical thermodynamics make it possible to estimate the enthalpy of mixing by taking the derivative of the Gibbs energy of mixing with respect to temperature. Larsen et al., have developed a modified form of UNIFAC to address this problem.[22] Not surprisingly, the modification involves the introduction of a substantial number of additional adjustable parameters. Even though the modified form does improve the

---

22. Larsen, B., Rasmussen, P.S., Fredenslund, Aa. 1987. *Ind. Eng. Chem. Res.* 26:2274.

accuracy of all the thermodynamic properties for a large number of systems there are many systems for which the predicted heats of mixing are in error by 100%–700%. More importantly, there is no way of knowing in advance when the predictions will be in error or when they will be accurate.

So why do these approximations fit the activity coefficient data? Because they have enough adjustable parameters to fit the data. Even the Margules one-parameter equation is good enough for that in many cases, but we suffer few delusions about its physical accuracy. In conclusion, we must say that local composition theory has much to recommend it. It does fit a great wealth of experimental data and there is some justification for its form via the theoretical physics which can be applied. But it is often extrapolated too far and that can lead to miscalculations by unwary users. In the end, we must never underestimate the value of experimental data for nonideal mixtures and apply the currently available theory with a careful and mildly critical view.

Assumptions 5 through 7 have to do with the entropy of mixing. The inclusion of the Staverman-type modifications to address the differences between surface fraction and volume fraction are generally recognized to be reasonable based on polymer lattice computer simulations. This modification and the estimate of molecular volumes by group contributions instead of liquid molar volumes comprise the primary differences between the Wilson and UNIQUAC models.

The UNIFAC theory is distinguished from UNIQUAC in that the solution is assumed to be a mixture of functional groups, not molecules. The UNIQUAC theory is then applied to each type of group interaction. The values for the group interactions are then regressed from a data base that includes phase equilibrium data for many, many systems. In one sense, the UNIFAC method is more like a massive regression than a truly predictive method. Thus, it lies somewhere between the purely correlative method of fitting van Laar constants and the purely predictive method of the Scatchard-Hildebrand theory (or any equation of state with $k_{ij} = 0$). Like any regressed equation, it can be unreliable if extrapolated far beyond the originally applied data. If you are ever in the position of designing truly novel chemical systems, you should be especially sensitive to the need for specific experimental data.

13.8   SUMMARY

The theories developed in this chapter are based on the local composition concept. Similar to models developed in the previous chapter, accurate representation of highly nonideal solutions requires the introduction of at least two adjustable parameters. These adjustable parameters permit us to compensate for our ignorance in a systematic fashion. By determining reasonable values for the parameters from experimental data, we can interpolate between several measurements, and in some cases extrapolate to systems where we have no measurements. Learning how to determine reasonable values for the parameters and apply the final equations is an important part of this chapter. We also introduce the UNIFAC model, which is useful for predicting behavior when no experimental measurements are available. Similar to UNIFAC, COSMO-RS methods are predictive, but are based on quantum mechanical simulations that can be applied when experimental data are entirely lacking for a particular functional group. UNIFAC would require at least a small quantity of experimental data to characterize the group contributions.

We should also note, however, that using an equation of state is similarly simplified by using a computer, so the basic motivation for developing solution models specific to liquids is simultaneously undermined by requiring computers for implementation. From this perspective, what we should be doing is analyzing the mixing rules and models of interaction energies in equations of state if we intend to use a computer anyway. We return to this point in Unit IV, when we discuss hydrogen-bonding equations of state for nonideal solutions.

13.9   IMPORTANT EQUATIONS

The UNIFAC method receives broad application throughout thermodynamic modeling. In fact, occasional applications of UNIFAC may be *too* broad in the sense that experimental data specific to a particular binary system are ignored and UNIFAC predictions are not validated with actual measurements. A literature search should be conducted for experimental data pertaining to every molecular interaction in a mixture and compared to the UNIFAC predictions. If significant deviations are observed, then UNIQUAC (or Wilson or NRTL) should be applied as the general activity model with system-specific parameters whenever possible and parameters inferred from UNIFAC predictions only in cases where no experimental data are available. In that context, an important equation for this chapter is best characterized by the UNIQUAC model.

$$\frac{G^E}{RT} = \sum_j x_j \ln(\Phi_j/x_j) - 5\sum_j q_j x_j \ln(\Phi_j/\theta_j) - \sum_j q_j x_j \ln\left(\sum_i \theta_i \tau_{ij}\right) \qquad 13.52$$

Fundamentally, the energy equation for mixtures summarizes all local composition models:

$$\frac{U - U^{ig}}{RT} = \frac{N_A \rho}{2}\sum_i \sum_j x_i x_j \int \frac{N_A u_{ij}}{RT} g_{ij} 4\pi r^2 dr \qquad 13.71$$

13.10   PRACTICE PROBLEMS

P13.1 The following lattice contains x's, o's, and void spaces. The coordination number of each cell is 8. Estimate the local composition (X_{xo}) and the parameter Ω_{ox} based on rows and columns away from the edges. (ANS. 0.68,1.47)

		o		o		x		o	
	x			o	x			x	
	x		x	x		o	x	o	
o		x		o			x		
						o			x
	o		x		o			x	
	x		o		o		x		
o	x				x		x		o

13.11   HOMEWORK PROBLEMS

13.1 Show that Wilson's equation reduces to Flory's equation when $A_{ij} = A_{ji} = 0$. Further, show that it reduces to an ideal solution if the energy parameters are zero, *and* the molecules are the same size.

13.2 The actone(1) + chloroform(2) system has an azeotrope at $x_1 = 0.38$, 248 mmHg, and 35.17°C. Fit the Wilson equation, and predict the *P-x-y* diagram.

13.3 Model the behavior of ethanol(1) + toluene(2) at 55°C using the UNIQUAC equation and the parameters $r_1 = 2.1055$, $r_2 = 3.9228$, $q_1 = 1.972$, $q_2 = 2.968$, $a_{12} = -76.1573$ K, and $a_{21} = 438.005$ K.

13.4 The UNIFAC and UNIQUAC equations use surface fraction and volume fractions. This problem explores the differences.

(a) Calculate the surface area and volume for a cylinder of diameter $d = 1.0$ and length $L = 5$ where the units are arbitrary. Calculate the surface area for a sphere of the same volume. Which object has a higher surface area to volume ratio?

(b) Calculate the volume fractions and surface area fractions for an equimolar mixture of the cylinders and spheres from part (a). Use subscript *s* to denote spheres and subscript *c* to denote cylinders.

(c) For this equimolar mixture, calculate the local composition ratios x_{cs}/x_{ss} and x_{sc}/x_{cc} for the UNIQUAC equation if the energy variables τ_{cs} and τ_{sc} are unity. For the equimolar mixture, substitute the values of volume fraction and surface fraction into the expression for UNIQUAC activity coefficients, and simplify as much as possible, leaving the *q*'s as unknowns.

(d) Consider *n*-pentane and 2,2-dimethyl propane (also known as neopentane). Calculate the UNIQUAC *r* and *q* values for each molecule using group contribution methods. Compare the results with part (a). [Hint: You might want to think about the -C-C-C- bond angles.]

13.5 Consider a mixture of isobutene(1) + butane(2). Consider a portion of the calculations that would need to be performed by UNIFAC or UNIQUAC.

$$
\begin{array}{cc}
\begin{array}{c}
CH_3 \\
| \\
CH_2 = C - CH_3
\end{array}
&
CH_3CH_2CH_2CH_3 \\
(1) & (2)
\end{array}
$$

(a) Calculate the surface area and volume parameters for each molecule.

(b) Provide reasoning to identify which component has a larger liquid molar volume. Which compound has a larger surface area?

(c) Calculate the volume fractions for an equimolar mixture.

13.6 Solve problem 10.14 using UNIFAC to model the liquid phase.

13.7 The flash point of liquid mixtures is discussed in Section 10.5. For the following mixtures, estimate the flash point temperature at 760 mmHg for the following components and their equimolar mixtures using UNIFAC:

(a) methanol (LFL = 7.3%) + 2-butanone (LFL = 1.8%)
(b) ethanol (LFL = 4.3%) + 2-butanone (LFL = 1.8%).

13.8 Use the UNIFAC model to predict the VLE behavior of the n-pentane(1) + acetone(2) system at 1 bar and compare to the experimental data in problem 11.11.

13.9 According to Gmehling et al. (1994),[23] the system acetone + water shows azeotropes at: (1) 2793 mmHg, 95.1 mol% acetone, and 100°C; and (2) 5155 mmHg, 88.4 mol% acetone and 124°C. What azeotropic pressures and compositions does UNIFAC indicate at 100°C and 124°C? Othmer et al. (1946) (cf. Gmehling[24]) have studied this system at 2570 mmHg. Prepare T-x-y or P-x-y plots comparing the UNIFAC predictions to the experimental data.

13.10 Consider the experimental data of Brown and Smith (1954) cited in problem 10.2. Prepare a P-x-y plot and a plot of experimental activity coefficients vs. composition. Then use UNIFAC to predict the activity coefficients across the composition range and add the calculations to the plots.

13.11 Flash separations are fundamental to any process separation train. A full steady-state process simulation consists largely of many consecutive flash calculations. Use UNIFAC to determine the temperature at which 20 mol% will be vaporized at 760 mmHg of an equimolar mixture liquid feed of n-pentane and acetone.

13.12 A preliminary evaluation of a new process concept has produced a waste stream of the composition given below. It is desired to reduce the waste stream to 10% of its original mass while recovering essentially pure water from the other stream. Since the solution is very dilute, we can use a simple equation known as Henry's law to represent the system. According to Henry's law, $f_i^L = h_i x_i = x_i \gamma_i^\infty P_i^{sat}$. Use UNIFAC to estimate the Henry's law constants when UNIFAC parameters are available. Use the Scatchard-Hildebrand theory when UNIFAC parameters are not available. Estimate the relative volatilities (relative to water) of each component. Relative volatilities are defined in problem 11.2.

Compositions in mg/liter are:

Methanol	H2S	Methyl Mercaptan	Dimethyl Sulfoxide	Dimethyl Disulfide
5100	30	50	50	100

13.13 Derive the form of the excess enthalpy predicted by Wilson's equation assuming that A_{ij}'s and ratios of molar volumes are temperature-independent.

13.14 Orbey and Sandler (1995. *Ind. Eng. Chem. Res.* 34:4351.) have proposed a correction term to be added to the excess Gibbs energy of mixing given by UNIQUAC. To a reasonable degree of accuracy the new term can be written

23. Gmehling, et al., 1994. *Azeotropic Data,* NY: VCH.
24. Gmehling, J., Onken, V., Arlt, W. 1977. *Vapor-Liquid Equilibrium Data Collection*, Frankfurt, Germany: DECHEMA.

$$\frac{G^E_{HB}}{RT} = \sum_i x_i \left[-2\ln(1 + a_i F) + \frac{a_i F}{1 + a_i F} - C^{pure}_i \right]$$

where

$$F \equiv \sum_j \frac{x_j a_j}{1 + a_j F}$$

Derive an expression for the correction to the activity coefficient. [Hint: Do you remember how to differentiate implicitly?]

13.15 The energy equation for mixtures can be written for polymers in the form:

$$\frac{U - U^{ig}}{RT} = \frac{\rho}{2} \sum_i \sum_j x_i x_j N_{d,i} N_{d,j} \int \frac{N_A u_{ij}}{RT} g_{ij} 4\pi r^2 dr$$

By analogy to the development of the Scatchard-Hildebrand theory, this can be rearranged to:

$$\frac{U - U^{ig}}{RT} = \frac{-\rho}{2RT} \sum_i \sum_j x_i x_j N_{d,i} N_{d,j} N_A \varepsilon_{ij} \sigma^3_{ij} a^*_{ij}$$

where

$N_{d,i}$ = degree of polymerization for the ith component

ρ = molar density

x_i = mole fraction of the ith component

N_A = Avogadro's number

U = molar internal energy.

$a_{ii}^* = 3 + 2/Nd_i$

$a_{ij}^* = (a_{ii}^* \, a_{jj}^*)^{1/2}$

$\sigma^3_{ij} = (\sigma^3_i + \sigma^3_i)/2$

$\varepsilon_{ij} = (\varepsilon_{ii} \, \varepsilon_{jj})^{1/2}$

Derive an expression for $\ln\gamma_1$ for the activity coefficient model presented above.

13.16 Use the UNIFAC model to predict P-x-y data at 90°C and x_1, = {0, 0.1, 0.3, 0.5, 0.7, 0.9, 1.0} for propanoic acid + water. Fit the UNIQUAC model to the predicted P-x-y data and report your UNIQUAC a_{12} and a_{21} parameters in kJ/mole.

13.17 (a) Rearrange Eqn. 13.22 to obtain Eqn. 13.23.

(b) Use Eqns. 13.16 and 13.18 in Eqn. 13.17 and perform the integration to obtain Eqn. 13.19.

(c) Use Eqns. 13.16 and 13.31 in Eqn. 13.17 and perform the integration to obtain Eqn. 13.40.

13.18 Fit the data from problem 11.11 to the following model by regression over all points, and compare with the experimental data on the same plot, using the

   (a) Wilson equation
   (b) NRTL equation
   (c) UNIQUAC equation

13.19 Work problem 11.25 using the

   (a) Wilson equation
   (b) NRTL equation
   (c) UNIQUAC equation

13.20 Work problem 11.26 using the

   (a) Wilson equation
   (b) NRTL equation
   (c) UNIQUAC equation

13.21 Using the data from problem 11.27, fit the specified model equation and then plot the P-x-y diagram at 80°C using the

   (a) Wilson equation
   (b) NRTL equation
   (c) UNIQUAC equation

LIQUID-LIQUID AND SOLID-LIQUID PHASE EQUILIBRIA

In the field of observation, chance favors the prepared mind.

Pasteur

The large magnitudes of the activity coefficients in the polymer mixing example should suggest an interesting possibility. Perhaps the escaping tendency for each of the polymers in the mixture is so high that they would prefer to escape the mixture to something besides the vapor phase. In other words, the components might separate into two distinct liquid phases. This can present quite a problem for blending plastics and recycling them because they do not stay blended. The next question is: How can we tell if a liquid mixture is stable as a single liquid phase? Also, crystallization is used for many pharmaceuticals and industrial products. An understanding of solid solubility is important for designing their separation.

Chapter Objectives: You Should Be Able to...

1. Compute LLE and VLLE phase behavior, including the ability to identify the onset of liquid instability.

2. Compute SLE phase behavior.

3. Predict the partitioning of a solute between two fluid phases (e.g., *n*-octanol and water).

4. Construct and interpret triangular phase diagrams.

14.1 THE ONSET OF LIQUID-LIQUID INSTABILITY

Our common experience tells us that oil and water do not mix completely, even though both are liquids. If we consider equilibria between the two liquid phases, we can label one phase α and the other β. For such a system we can quickly show that the equilibrium compositions are given by

$$\hat{f}_i^{\,\alpha} = \gamma_i^\alpha x_i^\alpha P_i^{sat} = \gamma_i^\beta x_i^\beta P_i^{sat} = \hat{f}_i^{\,\beta}$$
$$\gamma_i^\alpha x_i^\alpha = \gamma_i^\beta x_i^\beta$$

14.1

where superscripts identify the liquid phase. The coexisting compositions are known as **mutual solubilities.** Note that we have assumed an activity coefficient approach here even though we could formulate an entirely analogous treatment by an equation of state approach. There is also the possibility that three phases can coexist, two liquids and a vapor, which is illustrated below and is known as vapor-liquid-liquid equilibria, or VLLE. In this case we have an additional fugacity relation for the gas phase, where we assume in the figure that the ideal gas law is valid for the vapor phase. An example of how such a system could be solved is given below. The phase equilibria can be solved by starting with whichever two phases we know the most about, and filling in the details for the third phase.

$$\hat{f}_i^{\,\alpha} = \hat{f}_i^{\,\beta} = \hat{f}_i^{\,V} = \gamma_i^\alpha x_i^\alpha P_i^{sat} = \gamma_i^\beta x_i^\beta P_i^{sat} = y_i P$$

(ig) 14.2

Example 14.1 Simple vapor-liquid-liquid equilibrium (VLLE) calculations

At 25°C, a binary system containing components 1 and 2 is in a state of three-phase LLVE. Analysis of the two equilibrium liquid phases (α and β) yields the following compositions:

$$x_2^\alpha = 0.05; \qquad x_1^\beta = 0.01$$

Vapor pressures for the two pure components at 25°C are $P_1^{sat} = 0.7$ bar and $P_2^{sat} = 0.1$ bar.

Making reasonable assumptions, determine good estimates for the following.

   (a) The activity coefficients γ_1 and γ_2 (use Lewis-Randall standard states).
   (b) The equilibrium pressure.
   (c) The equilibrium vapor composition.

Solution:

$$y_i P = x_i \, \gamma_i \, P_i^{sat}$$

Assume $\gamma_1^\alpha \approx 1$, $\gamma_2^\beta \approx 1$ because these are practically pure in the specified phases.

$$y_1 P = \gamma_1^\alpha \, x_1^\alpha \, P_1^{sat} = 0.95(0.7); \quad y_2 P = \gamma_2^\beta \, x_2^\beta P_2^{sat} = 0.99(0.1)$$

Example 14.1 Simple vapor-liquid-liquid equilibrium (VLLE) calculations (Continued)

$$P = \sum_i y_i P = 0.764 \text{ bar}, \qquad y_1 = 0.95(0.7)/0.764 = 0.8704, \qquad y_2 = 0.1296$$

$$\gamma_2^\alpha = y_2 P/0.05(0.1) = 19.8; \quad \gamma_1^\beta = y_1 P/0.01(0.7) = 95$$

We also can determine activity coefficients from a theory to determine infinitely dilute concentrations. When a phase is nearly pure, the activity coefficient is nearly one at the same time x_i is nearly one.

Example 14.2 LLE predictions using Flory-Huggins theory: Polymer mixing

One of the major problems with recycling polymeric products is that different polymers do not form miscible solutions with each other, but form highly nonideal solutions. To illustrate, suppose 1 g each of two different polymers (polymer A and polymer B) is heated to 127°C and mixed as a liquid. Estimate the mutual solubilities of A and B using the Flory-Huggins equation. Predict the energy of mixing using the Scatchard-Hildebrand theory. Polymer data:

	MW	$V \, (\text{cm}^3/\text{mol})$	$\delta \, (\text{J/cm}^3)^{\frac{1}{2}}$
A	10,000	1,540,000	19.2
B	12,000	1,680,000	19.4

Solution: This is the same mixture that we considered as an equal-weight-fraction mixture in Example 12.5. Based on that calculation, we know that the solution is highly nonideal. We must now iterate on the guessed solubilities until the implied activity coefficients are consistent. Let's start by guessing that the two polymer phases are virtually pure and infinitely dilute in the other component.

$$\lim_{x_A^\alpha \to 0} \Phi_A^\alpha = 0 \text{ and } \lim_{x_B^\beta \to 0} \Phi_B^\beta = 0$$

$$\lim_{x_i \to 0} \frac{\Phi_i}{x_i} = \lim_{x_i \to 0} \frac{x_i V_i}{x_i V_i + x_j V_j} \cdot \frac{1}{x_i} = \frac{V_i}{V_j}$$

Using Eqns. 12.36 and 12.37,

$$\ln\gamma_A^\alpha = \ln(0.91) + (1 - 0.91) + 1.54\text{E6}(19.4 - 19.2)^2 \, (1.0)^2 \, /[8.314(400)] = 18.5$$
$$\ln\gamma_B^\beta = \ln(1.09) + (1 - 1.09) + 1.68\text{E6}(19.4 - 19.2)^2 \, (1.0)^2 \, /[8.314(400)] = 20.2$$

Since $\gamma_i^\alpha x_i^\alpha = \gamma_i^\beta x_i^\beta$, then $\gamma_B^\beta x_B^\beta = 1$ and $\gamma_A^\alpha x_A^\alpha = 1$.

$$\Rightarrow \gamma_A^\alpha = 1.1\text{E8} \Rightarrow x_A^\alpha = 9.1\text{E-9}$$
$$\Rightarrow \gamma_B^\beta = 5.9\text{E8} \Rightarrow x_B^\beta = 1.7\text{E-9}$$

Good guess. The polymers are totally immiscible. No further iterations are needed.

14.2   STABILITY AND EXCESS GIBBS ENERGY

Expressions for activity coefficients are the same for LLE as they are for VLE. The difference is that multiple liquid compositions can give the same activities or total pressure at a given temperature. This behavior is implied in Fig. 11.10, where we commented that the calculated lines indicate LLE. The time has come to analyze why this happens and how to properly represent the phase behavior of a system with two liquid phases.

Keep in mind that nature dictates phase stability by minimizing the Gibbs energy when T and P are fixed. Gibbs energies of mixtures using the one-parameter Margules equation for three values of A_{12} are plotted in Fig. 14.1 (the curves). In this plot, the important quantity is the Gibbs energy of the mixture G. In Fig. 14.1, the endpoints represent the values of $G_i = H_i - TS_i$ for the pure components, where the references states have been arbitrarily chosen, and only component 2 is at its reference state. If we imagine computing the Gibbs energy of two separate beakers containing the two separate components on the basis of one mole of total fluid, we see that this **overall molar Gibbs energy** is simply a molar average of the two component Gibbs energies along line 1 in the plot. There is no contribution from the entropy of mixing, because they are not mixing; they are in separate beakers. Substituting $(1-x_1)$ for x_2 in this molar average formula shows that this formula is simply a linear relation in terms of x_1. Deviations from this line are the changes due to the mixing process. Note how the shape of the Gibbs energy changes with the value of A_{12}. Since $\Delta G_{mix} = G - \sum_i x_i G_i$, then rearranging,

$$G = \Delta G_{mix} + \sum_i x_i G_i = G^E + \Delta G_{mix}^{is} + \sum_i x_i G_i \qquad 14.3$$

where the last term represents the sum of component enthalpies represented by the upper straight line. The increasing excess Gibbs energy of mixing (larger A_{12}) ultimately causes a "w-shaped" curve to form. The individual contributions of Eqn. 14.3 to the Gibbs energy are shown in Fig. 14.2.

The procedure of summing together Gibbs energies along a line also applies to any tie line. As shown in Fig. 14.1, $A_{12} = 1$ leads to a situation where connecting any two compositions by a straight line gives a value of G at the overall composition that is higher than the G along the curve. Line 2 on the plot shows this for a mixture with an overall composition $z_1 = 0.4$ and assuming that two phases are formed $x_1^\alpha \approx 0.17$ (point a) and $x_1^\beta \approx 0.74$ (point b). The overall Gibbs energy would be given by point c on the tie line, $G/RT = -0.15$. However, if the mixture stays as one phase, along the mixture line the Gibbs energy would be $G/RT = -0.23$ (point c'), a lower value, which means this is more stable. (Note that c' must be at the overall composition along the line, $z_1 = 0.4$.) Since the curve for $A_{12} = 1$ is concave up, a straight line between two points always gives a higher value for two phases.

This situation is quite different when we consider the case where $A_{12} = 3$. Along this curve, the mixture at $z_1 = 0.5$ would have $G \approx 0.31$, (point d), if it remained as a single phase. However, if the solution splits into two phases, line 3 can be drawn between the compositions of the phases (one point on either side of d) as shown by points e ($x_1^\alpha \approx 0.07$) and f ($x_1^\beta \approx 0.93$), and the overall energy is given by the intersection of this line with the overall composition $z_1 = 0.5$ as shown by point d'. The lowest energy is obtained when the line is tangent to the humps as shown in the figure, where $G/RT \approx 0.18$, (point d'). Then, by splitting into two phases, the system clearly has a lower value for G/RT. Any other line that is drawn would force point d' to have a higher Gibbs energy than this point. (Try it.) Considering these points at different values of A_{12} indicates that there is no

A system will split into two phases if it results in a lower Gibbs energy.

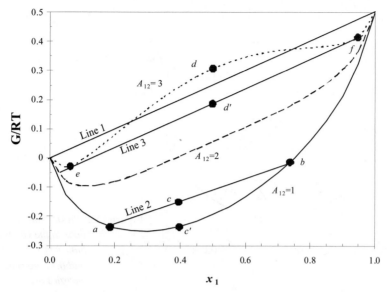

Figure 14.1 *Illustration of the Gibbs energy of a mixture represented by the Margules one-parameter equation.*

phase split unless there is a hump in G/RT that makes it concave down. Note that A_{12} must be positive to create this curvature. This means that the activity coefficients must be greater than one, and the system must also have positive deviations from Raoult's law for VLE.

One more point needs to be made before working some examples. Note that the line construction seems similar to what was done for VLE in a flash calculation at the beginning of Chapter 10. In fact, this is completely analogous mathematically to the flash in those diagrams, and the lever rule applies. The ratios of the phases can be found in a similar fashion. The only difference is that two liquid phases are formed upon a liquid-liquid flash rather than a vapor and liquid flash. For the example in the figure above, the fraction of the overall mixture that is the α phase (left side of diagram) is given by $\overline{fd'}/\overline{fe} = (0.93 - 0.5)/(0.93 - 0.07) = 0.5$, so the mixture of this example splits into equal portions of the two phases. Note that the compositions for points e and f are the same for *any overall composition between the two points*. So a different overall composition in a binary mixture shifts the relative amounts of the two phases, but not their composition. (This simplification does not hold for more components; the Gibbs phase rule says $F = C - P + 2$.)

14.3 BINARY LLE BY GRAPHING THE GIBBS ENERGY OF MIXING

Fig. 14.2 shows the contributions to the Gibbs energy of a mixture for $A_{12} = 3$ of Fig 14.1. The pure component Gibbs energies do not contribute to the curvature in the Gibbs energy of a mixture, and therefore are not needed for LLE calculations—we need just ΔG_{mix}. In principle, all that is required to make predictions of LLE partitioning is some method of calculating activity coefficients. In this section we use specific models (MAB and UNIFAC) to demonstrate calculation of LLE using ΔG_{mix}. The plotting/tangent line method can be extended to any activity coefficient model. This method is often the easiest method to use for binary solutions, though we show that it is subject to uncertainties from drawing/reading the tangent line.

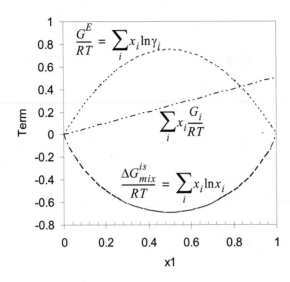

Figure 14.2 *Illustration of the contributions to the Gibbs energy of a binary mixture when $A_{12} = 3$ and the pure component Gibbs energies are as in Fig. 14.1.*

The MAB and UNIFAC models are convenient for demonstrating the calculations, but there is a certain danger in applying too much confidence in such predictions. LLE is more sensitive to the accuracy of the activity coefficients than VLE. Furthermore, the empirical nature of UNIFAC means that the same parameter set, $\{a_{mn}\}$, is not generally accurate for both VLE and LLE, so a different predictive parameter set is used. As for the sensitivity problem, the best advice is not to take any predictions too seriously. They can be used as a guide to assess miscibility in a way that is slightly better than looking at solubility parameter tables, but should never be considered as a substitute for experimental data. With these cautions in mind, it is useful to show how LLE can be predicted using UNIFAC and MAB. We have provided the LLE parameters on the spreadsheet UNIFAC(LLE) within Actcoeff.xlsx, and within Matlab/gammaModels/unifacLLE.

⚠ UNIFAC parameters for LLE differ from those for VLE.

Example 14.3 LLE predictions by graphing

💻 Actcoeff.xlsx– UNIFAC(LLE); MATLAB Ex14_03.m.

Arce et al.[a] give the compositions for the tie lines in the system water(1) + propanoic acid(2) + methylethylketone (MEK)(3) at 298 K and 1 bar. As limiting conditions, the mutual solubilities of water + MEK ($1CH_3 + 1CH_3CO + 1CH_2$) binary are also listed as $x_1^\alpha = 0.342$, $x_1^\beta = 0.922$.

(a) Use MAB to roughly estimate the water + MEK binary mutual solubilities to ± 5 mole%.
(b) Use UNIFAC to roughly estimate the water + MEK binary mutual solubilities to ± 5 mole%.

Solution:
(a) $A_{12} = (50.13 - 0)(15.06 - 9.70)(90.1 + 18.0)/(4 \cdot 8.314 \cdot 298) = 2.931$, virtually the same as the parameter used above.
Adding $G^E/RT = A_{12}x_1x_2$ and $\Delta G_{mix}^{is}/(RT) = \sum_i x_i \ln x_i$ gives $\Delta G_{mix}/(RT)$. Using the drawing tool shows $x_1^\alpha = 0.93$ and $x_1^\beta = 0.07$
(b) Selecting the appropriate groups from the UNIFAC menu, then copying the values of the activity coefficient, we can develop Figs. 14.3 and 14.4 using increments of $x_w = 0.05$. In MAT-LAB we can set up a vector $x_1 = 0:0.05:1$, and then insert a loop into unifacCallerLLE.m. Noting $G^E/(RT) = \sum_i x_i \ln \gamma_i$ and programming the formula for $\Delta G_{mix}/(RT)$.

Example 14.3 LLE predictions by graphing (Continued)

Using the line drawing tool we obtain tangents at $x_1^\alpha = 0.35$ and $x_1^\beta = 0.94$.

These are sufficiently precise for the problem statement as given above. Note how the MAB model results in symmetric estimates of the compositions, a serious deficiency for LLE, and UNIFAC happens to be fairly close.

---

a.   1995. *J. Chem. Eng. Data* 40:225.

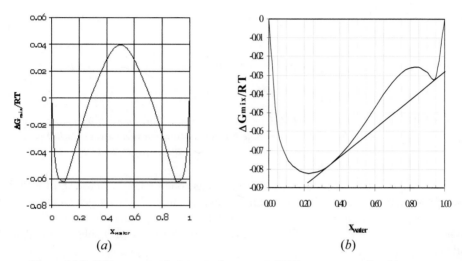

Figure 14.3 *Gibbs energy of mixing in the water + MEK system as predicted by*
(a) MAB and (b) UNIFAC.

14.4   LLE USING ACTIVITIES

Usually we require higher precision than obtained by graphing the Gibbs energy. Furthermore, we may encounter multicomponent mixtures, for which the extension of the above method is not straightforward. We can develop an entirely general method for computing the phase partitioning given relative activities in Eqn 14.1. In Fig. 14.4 are plotted the activities for the water + MEK system of Example 14.3. The extrema in the activity plot are characteristic of LLE. The vertical lines indicate the compositions where the activities are equal in each phase. The horizontal lines indicate the activity values. This analysis is a graphical solution to Eqn. 14.1. We need a method to search for this condition numerically. Rearranging Eqn. 14.1 we have,

$$K_i = \frac{x_i^\alpha}{x_i^\beta} = \frac{\gamma_i^\beta}{\gamma_i^\alpha} \qquad\qquad 14.4$$

Note that the K-ratios calculated using the ratio of mole fractions should be identical to the value calculated using the ratio of activity coefficients at the stable LLE condition. This form is

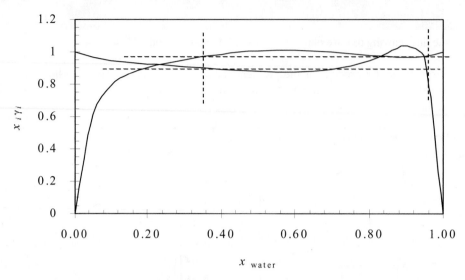

Figure 14.4 *Activities of water and MEK as a function of mole fraction water as predicted by UNIFAC. The activity versus mole fraction plots will have a maximum when LLE exists. The dashed lines show the compositions where the activities of components are equal in both phases simultaneously.*

entirely analogous to the *K*-ratios in VLE. To find this condition, we can use an LLE flash calculation. We can iterate on the system by assuming values for mole fractions, generating activity coefficients at those *x* values to get *K*-ratios and then generating new values for mole fractions from the *K*-ratios. If the loop is constructed properly, it will create a successive substitution algorithm that will converge.

We can develop such a procedure by noting for a binary mixture that x_i^α must sum to unity. We can calculate the concentrations using the compositions x_1^β in the other phase, and use the *K*-ratio to generate a new guess of composition. The principle balance equation is $x_1^\beta K_1 + (1 - x_1^\beta)K_2 = 1$ which leads to

Iterative flash procedure for binary LLE.

$$x_1^\beta = \frac{1 - K_2}{K_1 - K_2}; \text{ and } x_1^\alpha = x_1^\beta K_1 \qquad\qquad 14.5$$

The method is initialized by assuming the two phases are virtually immiscible with an infinitely dilute trace of the other component. The method is as follows.

1. Assume that phase β is nearly pure 1, $x_1^\alpha = 1/\gamma_1^{\alpha,\infty}$, and α is nearly pure 2, $x_2^\beta = 1/\gamma_2^{\beta,\infty}$. These represent initialization of the iteration procedure. The procedure is most stable with an initial guess of mutual solubility outside the two-phase region.

2. Calculate $K_{i,old} = \gamma_i^\beta/\gamma_i^\alpha$ where the γ_i's are evaluated at the initial compositions.

3. Calculate $x_{1,new}^\beta = (1 - K_{2,old})/(K_{1,old} - K_{2,old})$, $x_{2,new}^\beta = 1 - x_{1,new}^\beta$.

4. Calculate $x_{i,new}^\alpha = K_{i,old}x_{i,new}$.

5. Determine $\gamma_{i,new}$ values for each liquid phase from the $x_{i,new}$ values.

6. Calculate $K_{i, new} = \gamma_i^\beta / \gamma_i^\alpha$.

7. Replace all $x_{i,old}$ and $K_{i,old}$ values with the corresponding new values.

8. Loop to step 3 until calculations converge. The calculations converge slowly.

A similar method for ternary systems is explored in a homework problem. Note that the Rachford-Rice flash method given for VLE in Section 10.3 can be adapted and provides an even more robust solution, but it is not as easy to implement in Excel without a macro. The method is provided in Matlab/Chap14/LLEflash.m.

LLEflash.m implements the Rachford-Rice flash method.

Let us apply the binary algorithm above to the water and MEK system studied in the previous example.

Example 14.4 The binary LLE algorithm using MAB and SSCED models

Compute the mutual solubilities of water and MEK at 298 K and compare to the experimental data of Example 14.3 assuming the following models: (a) MAB (b) SSCED.

Solution:
(a) From Example 14.3, $A_{12} = 2.931$. The symmetry of the MAB model gives $x_1^\alpha = x_2^\beta = 1/\exp(2.931) = 0.05335$. Computing γ_i's at these compositions, $K_W = 1.0084/13.83 = 0.0729$; $K_{MEK} = 13.83/1.0084 = 13.72$. Then Eqn. 14.5 gives $x_{1new}^\alpha = 0.93205$; $x_{1new}^\beta = 0.06795$ for the first iteration. Unfortunately, the LLE calculations converge more slowly than VLE flash calculations. The calculations may drift a couple mole percent in compositions after they are changing at step sizes in the tenths of mole percents, so patience is required in converging the calculations. Section 14.9 provides details on setting up a macro or circular calculation. The table below summarizes the initial iterations. This same model is used above and the results are the same, but numerically known to better precision than the graphical method.

Iteration	x_1^α	x_1^β	$K_{1,new}$	$K_{2,new}$
Initialize	0.9466	0.0534	0.0729	13.72
1	0.9321	0.0680	0.0794	12.59
2	0.9264	0.0736	0.0821	12.18
Converged	0.9225	0.0775	0.0840	11.91

(b) The SSCED model gives:
$k_{12} = (50.13 - 0)(15.06 - 9.70)/(4 \cdot 27.94 \cdot 18.88) = 0.1274.$

From $\ln\gamma_1^\infty = 18[(27.94 - 18.88)^2 + 2(0.1274)27.94(18.88)]/(8.314 \cdot 298) = 1.573$, $x_1^\alpha = 1/\exp(1.573) = 0.2072.$

Applying the same formulas to MEK: $x_2^\beta = 1/2626 = 0.9997.$

The table below shows the improved predictions from SSCED relative to MAB. Note how the molecular size difference is reflected by the much greater activity of trying to squeeze the large molecule among the small ones. This reflects a significantly improved insight for SSCED relative to the MAB model.

Example 14.4 The binary LLE algorithm using MAB and SSCED models (Continued)

Iterating further on x_1^α through Eqn. 14.5 gives $x_1^\alpha = 0.2509$.

Iteration	x_1^α	x_1^β	$K_{1,new}$	$K_{2,new}$
Initialize	0.9997	0.2072	0.2073	2623
1	0.9997	0.2413	0.2414	2515
2	0.9997	0.2487	0.2488	2493
Converged	0.9997	0.2509	0.2510	2486

A similar approach to this example could be applied to solve for the ternary problem of partitioning of the propanoic acid between water and MEK, starting with the above result and infinite dilution of the propanoic acid. By steadily increasing the propanoic acid and performing flash calculations each time, the impact of the propanoic acid on the water + MEK partitioning could be studied. Can you guess whether the propanoic acid causes relatively more MEK to dissolve into the water phase or vice versa? The answer is explored later in a homework problem.

14.5 VLLE WITH IMMISCIBLE COMPONENTS

A special case of VLLE is obtained when one of the liquid-phase components is almost entirely insoluble in other components, and all other components are essentially insoluble in it, as occurs with many hydrocarbons with water. When a mixture forms two liquid phases, the mole fractions sum to unity in each of the phases. When a vapor phase coexists, it is a mixture of all components. The bubble pressure can be calculated using Eqn. 14.2, where the liquid phase fugacities are used to calculate the vapor phase partial pressures. For example, water and n-pentane are extremely immiscible. Applying the strategy used in Example 14.1, each liquid phase is essentially pure, resulting in the bubble pressure $P = P_{water}^{sat} + P_{pentane}^{sat}$ which is greater than the bubble point of either component. Rather than considering the bubble pressure, think about the bubble temperature. Since each component contributes to the partial pressure, the bubble temperature of a mixture of two immiscible liquids is lower than the bubble temperature of either pure component at a specified pressure. This phenomenon can be use to permit boiling at a lower temperature as shown in the next example.

Example 14.5 Steam distillation

Consider a steam distillation with the vapor leaving the top of the fractionating column and entering the condenser at 0.1 MPa with the following analysis:

	y_i	T_c	P_c	ω
n-C8	0.20	568.8	2.486	0.396
C12 fraction	0.40	660.0	2.000	0.540
H_2O	0.40	(use steam tables)		

Example 14.5 Steam distillation (Continued)

Assuming no pressure drop in the condenser and that the water and hydrocarbons are completely immiscible, calculate the maximum temperature which ensures complete condensation at 0.1 MPa. Use the shortcut K-ratio method for the hydrocarbons.

Solution: Apply the following notation to designate the phases:

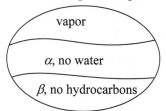

The temperature that we seek is a bubble temperature of the liquid phases. The hydrocarbons and the water are essentially immiscible. We may approximate the hydrocarbon liquid phase, α, as an ideal solution of C8 and C12 with no water present. Therefore, two liquid phases will form: one of pure H_2O and the other a mixture of 1/3 n-C8 + 2/3 C12 fraction. We may apply Raoult's law with $x_w = 1$ for water in the β phase. The vapor mixture is a single phase, however, and must conform to: $1 = \sum_i y_i$.

	Phase	x_i	VLE K_i (353 K)	y_i	VLE K_i (368 K)	y_i
n-C8	α	0.333	0.254	0.084	0.415	0.138
C12 fraction	α	0.667	0.015	0.010	0.028	0.019
H_2O	β	1.0	0.474	0.474	0.846	0.846
			$\Sigma y_i =$	0.568	$\Sigma y_i =$	1.0027

So the bubble temperature with water present is ~95°C. Note that the bubble temperature is below the bubble temperature of pure water. What would it be without water?

	x_i	VLE K_i (400K)	y_i	VLE K_i (440K)	y_i
n-C8	0.333	1.049	0.349	2.727	0.908
C12 fraction	0.667	0.092	0.061	0.319	0.213
		$\Sigma y_i =$	0.410	$\Sigma y_i =$	1.121

Then, interpolating $T \approx 400 + (1 - 0.41)/(1.121 - 0.41) \cdot 40 = 433$ K. Thus, we reduced the bubble temperature by 65°C in the steam distillation.

14.6 BINARY PHASE DIAGRAMS

Liquid-liquid mutual solubilities in partially miscible systems change with temperature at a given pressure. Whether the solubilities increase or decrease can be due to a number of factors including hydrogen bonding. When one species H-bonds and the other does not, then as the temperature is raised and hydrogen bonds are broken, the fluids become more "similar," and the LLE region decreases in size. The fluids can become miscible before the boiling point as shown in Fig. 19.12 on page 804 for methanol + cyclohexane. The liquid-liquid envelope is the dome in the figure. The temperature where the liquids become totally miscible is known as the **upper critical solution temperature** (UCST). Pressure affects the VLE curve, but has virtually no effect on the liquid phases or LLE. Thus, as the pressure is lowered, the VLE shifts to lower temperatures, and the VLE curve will overlap the LLE curve, giving a diagram similar to Fig. 16.2B on page 616, insets (e) and (f). For other systems that differ more greatly in vapor pressure, an azeotrope may not appear in the VLE and the diagram will appear as in Fig. 16.2A on page 616, insets (b) and (c).

When VLE is predicted by process simulators, the common default settings check for only one liquid phase. The result can be the odd diagrams as in Fig. 11.10 on page 435. How do we resolve such diagrams? Consider the *T-x-y* diagram in Fig. 14.5(a) for ethyl acetate + water generated using UNIQUAC parameters in the ASPEN Plus database. By default, ASPEN Plus models VLE; the dotted line is vapor and the dashed line is liquid. The solid line and dots have been added manually. Note the L^β is in equilibrium with the left *V* branch all the way from 373 K to the two left-most dots. Likewise the L^α is in equilibrium with the right *V* branch all the way from the pure ethyl acetate boiling point to the two rightmost dots. The horizontal line is the VLLE condition because both *L* branches are simultaneously in equilibrium with the center *V* dot, and thus they are also in equilibrium with each other. The liquid compositions are at the outside dots and the vapor composition in this case is given by the center dot. This state is known as a **heteroazeotrope** because two liquid phases co-exist at the azeotrope condition. The odd loops below this VLLE temperature are not equilibrium states and are discarded. Fig. 14.5(b) is a comparison of experimental data with the diagram generated by specifying that the process simulator check for VLLE, though the dotted LLE lines are added manually to show the general type of expected behavior. Note how the lower loops

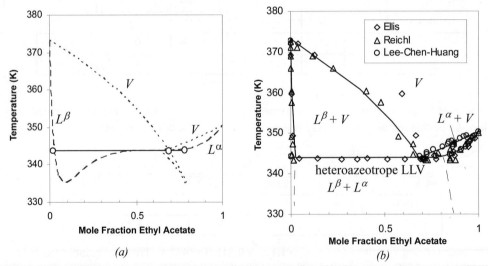

Figure 14.5 *(a) VLE predictions of ethyl acetate + water as predicted by literature parameters in ASPEN using UNIQUAC. The horizontal line and the dots have been added manually. (b) The correct phase behavior after specifying to check for VLLE. Note that the liquid-liquid envelope is sketched by hand based on general behavior that may be expected, not predicted, but the two liquid compositions at the bubble temperature are the ends of the horizontal VLLE line. Data from Ellis, S. R. M.; Garbett, R. D. 1960. Ind. Eng. Chem. 52:385-388; Reichl, A.; et al, 1998. Fluid Phase Equil. 153:113–134; Lee, L.-S.; et al. 1996. J. Chem. Eng. Japan 96:427–438.*

are eliminated. The horizontal line method is easy to apply manually to any diagrams that you may generate.

14.7   PLOTTING TERNARY LLE DATA

Graphical representation of ternary LLE data is important for design of separation processes. For ternary systems, triangular coordinates simultaneously represent all three mole fractions, or alternatively, all three weight fractions. Triangular coordinates are shown in Fig. 14.6(a), with a few grid lines displayed. The fraction of component A is represented by lines parallel to the BC axis: Along $\overline{de}$, the composition is $x_A = 0.25$; along $\overline{ab}$, the composition is $x_A = 0.5$. The fraction of B is represented by lines parallel to the AC axis; along $\overline{ac}$, the composition is $x_B = 0.5$; along $\overline{fg}$, the composition is $x_B = 0.25$. The fraction of C is along lines parallel to the AB axis; along $\overline{bc}$, $x_C = 0.5$. Combining these conventions, at point h, $x_A = 0.25$, $x_B = 0.25$, $x_C = 0.5$. An example of plotted LLE phase behavior is shown on triangular coordinates in Fig. 14.6(b). The compositions of coexisting α and β phases are plotted and connected with tie lines. The lever rule can be applied. For example, in Fig. 14.6(b), a feed of overall composition d will split into two phases: (moles β)/(moles α) = $\overline{de}/\overline{cd}$, and (moles β)/(moles feed) = $\overline{de}/\overline{ce}$.

To specify the composition of an arbitrary phase α in a ternary system, only two mole fractions are required. Since the mole fractions must sum to unity, if x_A^{α} and x_C^{α} are known, then the third mole fraction can be found, $x_B^{\alpha} = 1 - x_A^{\alpha} - x_C^{\alpha}$. The subscripts may be interchanged, and the same constraints hold for phase β. Therefore, cartesian coordinates can be used to plot two mole (or

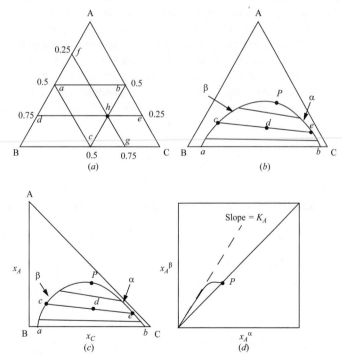

Figure 14.6 *Illustrations of graphical representation of ternary data on triangular diagrams. (a) Illustration of grid lines on an equilateral triangle; (b) illustration of LLE on an equilateral triangle; (c) illustration of LLE on a right triangle; (d) illustration of tie-line data on a right triangle.*

weight) fractions of each phase. Fig. 14.6(c) presents the data from Fig. 14.6(b) on cartesian coordinates. The tie line and lever rule concepts also apply on this diagram. Tie line data can be plotted on cartesian coordinates as in Fig. 14.6(d). This representation of tie line data permits improved accuracy when interpolating between experimental tie lines. The slope of the tie line data, as presented in Fig. 14.6(d), is frequently linear near the origin; the slope of the line in this region is x_A^β/x_A^α, which is the distribution coefficient, K_i. For the LLE phase behavior shown in Fig. 14.6, the $B + C$ miscibility is increased by the addition of A. Along the phase boundary, the tie lines become shorter upon approach to point P, the **Plait point.** Note that at the Plait point, all distribution coefficients are one.

Other Examples of LLE Behavior

The ternary LLE of Fig. 14.6 is one of several common types. In this example, $A + C$ is totally miscible (there is no immiscibility on the AC axis), as is $A + B$. When two of the three pairs of components are immiscible, the type of Fig. 14.7(a) can result, and when all three pairs have immiscibility regions, then type of Fig. 14.7(b) can result. In Fig. 14.7(b), the center region is LLL; since the T and P are fixed, any overall composition in the center triangle will split into the three phases of compositions a, b, and c, because $F = C - P + 2 = 3 - 3 + 2 = 2$.

14.8   CRITICAL POINTS IN BINARY LIQUID MIXTURES

Referring back to Fig. 14.1, we may wish to find the combination of x_1 and A_{12} where the system just begins to phase-split. This is known as a liquid-liquid critical point. If it is the highest T where two phases exist (called the UCST, upper critical solution temperature), then we must seek x_1 where the concavity is equal to zero at only one composition (if it was less than zero anywhere, then we would just have a regular phase split, not a critical point—refer back to the curve for $A_{12} = 3$ in Fig. 14.1). If the concavity is equal to zero at x_1 and greater everywhere else, then this must represent a minimum in the concavity as well a point where it equals zero. These two conditions provide two equations for the two unknowns, x_1 and T, involved in determining the critical point. Recalling that concavity is given by the second derivative:

$$\frac{\partial^2 G}{\partial x_1^2} = 0; \qquad \frac{\partial^3 G}{\partial x_1^3} = 0 \qquad\qquad 14.6$$

A generalization of these concepts to multicomponent mixtures gives

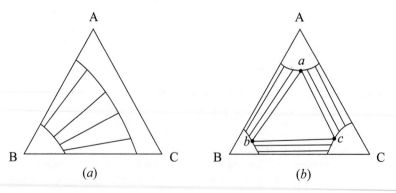

(a)   *(b)*

Figure 14.7 *Illustration of other types of LLE behavior.*

$$\frac{\partial^2 (G/(RT))}{\partial n_1^2} = 0; \qquad \frac{\partial^3 (G/(RT))}{\partial n_1^3} = 0 \qquad\qquad 14.7$$

By analogy to the phase diagram for pure fluids, the locus where the second derivative equals zero represents the boundary of instability and is called the liquid-liquid **spinodal.**

Example 14.6 Liquid-liquid critical point of the Margules one-parameter model

Based on Fig. 14.1 and the discussion of concavity above, it looks like the value of $A_{12} = 2$ may be close to the critical point. Use Eqns. 14.6 and 14.7 to determine the exact value of the critical parameter.

Solution: Multiplying through by n and recognizing that we have previously performed the initial part of this derivation (see Eqns. 11.29–11.31) gives

$$\underline{G}/(RT) = n_1 \ln x_1 + n_2 \ln(1 - x_1) + A_{12} n_2 x_1$$

$$\frac{\partial (\underline{G}/(RT))}{\partial n_1} = \frac{\partial (\underline{G}^{is}/(RT))}{\partial n_1} + \frac{\partial (\underline{G}^{E}/(RT))}{\partial n_1} = \ln x_1 + \ln \gamma_1 = \ln x_1 \gamma_1 = \ln x_1 + A_{12} x_2^2 \qquad 14.8$$

Since this is a binary solution, there is a simple finite relationship between the derivative with respect to mole number and the derivative with respect to mole fraction, leading expeditiously to the expected conclusion:

$$\frac{\partial^2 (\underline{G}/(RT))}{\partial n_1^2} = \frac{\partial^2 (\underline{G}/(RT))}{\partial x_1^2}\left(\frac{\partial x_1}{\partial n_1}\right)^2 = 0 \Rightarrow \frac{\partial^2 (\underline{G}/(RT))}{\partial x_1^2} = 0 = \frac{1}{x_1} - 2(1 - x_1)A_{12}$$

$$\frac{\partial^3 (\underline{G}/(RT))}{\partial n_1^3} = \frac{\partial^3 (\underline{G}/(RT))}{\partial x_1^3}\left(\frac{\partial x_1}{\partial n_1}\right)^3 = 0 \Rightarrow \frac{\partial^3 (\underline{G}/(RT))}{\partial x_1^3} = 0 = \frac{-1}{x_1^2} + 2A_{12} \qquad 14.9$$

$$2A_{12} = \frac{1}{x_1^2} \Rightarrow 1 - \frac{1 - x_1}{x_1} \Rightarrow x_{1,c} = \frac{1}{2}$$

$$\boxed{\frac{1}{x_{1,c}} = 0.5; \quad A_{12,c} = 2} \text{ one-parameter Margules model} \qquad 14.10$$

From this example, we can draw a useful conclusion regarding the magnitude of activity coefficients that leads to immiscibility. Based on the Margules model, LLE should occur whenever $A_{12} > 2$, and a useful engineering rule of thumb for immiscibility is when the activity coefficients exceed the value of 7.4,

$$\boxed{\gamma_i^\infty = \exp(A_{12}) = \exp(2) = 7.4} \text{ limit for Margules one-parameter miscibility} \qquad 14.11$$

With the capability to determine immiscibility, it is possible to convey some background on one of the most significant scientific challenges currently occupying modern thermodynamicists. This is the problem of phase separation in polymer solutions and polymer blends. The same forces driving these phase separations lead to the extremely important problem of the collapse transition of a single polymer chain in a solvent or mixture. A special kind of collapse describes the protein folding problem. Imagine a bowl of spaghetti formed from a single noodle. After stretching the noodle until it is completely straight, release and watch it collapse into the bowl again. If the noodle was really a protein, it would collapse into exactly the same hooks and crooks as it was before being stretched. The folding is driven by hydrophobic effects, hydrogen bonding, and intramolecular interactions like helix formation. A complete understanding of this phenomenon would greatly facilitate drug design.

Limiting this introductory presentation to liquid-liquid equilibria, the phase partitioning of polymer mixtures is somewhat simpler in that we care only about collections of chains rather than the details of individual chains. Polymer solutions are classified differently from polymer blends in a manner that is superfluous to our mathematical analysis: **polymer solutions** are blends where the degree of polymerization of one of the components is unity (the small one is known as the solvent). With this minimal background, we can phrase the following problem:

Example 14.7 Liquid-liquid critical point of the Flory-Huggins model

Determine the critical value of the Flory-Huggins χ parameter considering the degrees of polymerization of each component.

Solution: Note that we have already solved this problem for the special case where the two components are identical in size. Then the excess entropy is zero, the volume fractions are equal to the mole fractions, and the Margules one-parameter model is recovered with A_{12} having nearly the same meaning as the Flory-Huggins χ parameter. To consider the problem of including the degree of polymerization, N_d, we must define the parameter with respect to a standard unit of volume. N_d is the number of monomer repeat units in the polymer. In the presentation below (and most other presentations of the same material), the volume of a *monomer* of component 1 is assigned as this standard volume ($\chi' = V_{std} \cdot [\delta_1 - \delta_2]^2 / RT$; $V_{std} = V_1/N_{d,1}$). Note $\chi' = \chi / N_{d,1}$, and that we are introducing temperature dependence into χ'. Recalling the formula for the activity coefficient with this notational adaptation, the starting point (Eqn. 12.36) for this derivation becomes:

$$\frac{\partial (G/(RT))}{\partial n_1} = \frac{\partial (G^{is}/(RT))}{\partial n_1} + \frac{\partial (G^E/(RT))}{\partial n_1} = \ln x_1 \gamma_1$$

$$= \ln \Phi_1 + (1 - \Phi_1/x_1) + N_{d,1} \chi' \Phi_2^2$$

The next step is greatly simplified if we recognize a simple relationship that is very similar to the formula for computing the number average molecular weight from the weight fractions of each component. The analogous formula for the volume can be rearranged in terms of the volume ratio $r = V_2/V_1$ (where 1 is solvent) as follows:

Example 14.7 Liquid-liquid critical point of the Flory-Huggins model (Continued)

$$V = \left[\frac{\Phi_1}{V_1} + \frac{\Phi_2}{V_2}\right]^{-1} \Rightarrow \frac{\Phi_1}{x_1 V_1} = \frac{1}{V} = \frac{\Phi_1}{V_1} + \frac{\Phi_2}{V_2}$$

$$\Rightarrow \frac{\Phi_1}{x_1} = \Phi_1 + \Phi_2(1/r) \Rightarrow 1 - \frac{\Phi_1}{x_1} = \Phi_2(1 - 1/r)$$

Since this is a binary solution, there is a simple finite relationship between the derivative with respect to mole number and the derivative with respect to *volume* fraction, leading expeditiously to the *general* conclusion (note $d\Phi_1 = -d\Phi_2$):

$$\frac{\partial^2(G/RT)}{\partial n_1^2} = \frac{\partial^2(G/RT)}{\partial \Phi_1^2}\left(\frac{\partial \Phi_1}{\partial n_1}\right)^2 = 0 \Rightarrow \frac{\partial^2(G/(RT))}{\partial \Phi_1^2} = 0$$

$$= \frac{1}{\Phi_1} - (1 - 1/r) - 2(1 - \Phi_1)Nd_1\chi' \qquad 14.12$$

$$\frac{\partial^3(G/(RT))}{\partial n_1^3} = \frac{\partial^3(G/(RT))}{\partial \Phi_1^3}\left(\frac{\partial \Phi_1}{\partial n_1}\right)^3 = 0 \Rightarrow \frac{\partial^3(G/(RT))}{\partial \Phi_1^3} = 0 = \frac{-1}{\Phi_1^2} + 2N_{d,1}\chi'$$

which leads to two important results:

$$2N_{d,1}\chi' = \frac{1}{\Phi_1^2} \Rightarrow 0 = \frac{1}{\Phi_1} - (1 - 1/r) - \frac{(1 - \Phi_1)}{\Phi_1^2} \Rightarrow \Phi_{1,c} = \frac{1}{1 + \sqrt{1/r}} = \frac{\sqrt{V_2}}{\sqrt{V_1} + \sqrt{V_2}}$$

$$\chi'_c = \frac{1}{2N_{d,1}}(1 + \sqrt{1/r})^2 = \frac{V_1}{2N_{d,1}}\left(\frac{1}{\sqrt{V_1}} + \frac{1}{\sqrt{V_2}}\right)^2 \Rightarrow \frac{(\delta_1 - \delta_2)^2}{RT_c} = \frac{1}{2}\left(\frac{1}{\sqrt{V_1}} + \frac{1}{\sqrt{V_2}}\right)^2 \qquad 14.13$$

These results suggest that critical concentration decreases to zero with increasing polymer size but the critical temperature approaches a finite limit that is related to the solvent size.

There are a number of significant conclusions that may be drawn from the above example. First, for polymer solutions where (1) is the solvent (i.e., $V_1 \ll V_2$), the critical composition of polymer (2) approaches zero, and the critical temperature is a finite value that depends on the solvent and polymer molar volumes and solubility parameters. Furthermore, the critical temperatures for all polymers in a given solvent should be given by a universal curve with respect to molecular weight when reduced by the solubility parameter difference, although these predictions are only semi-quantitative due to the approximate nature of the Scatchard-Hildebrand theory. For blends where $V_1 \sim V_2$, the critical temperature should be proportional to the molecular weight. We have applied several approximations in deriving these formulas, however. Therefore, significant efforts are currently underway to determine whether the formulas presented above are sufficiently accurate to describe the complex phase behavior often observed in polymer solutions and blends. Hanging in the balance is the ability to tailor-make polymer solutions and blends with many commercial advantages, because the ability to manipulate phase behavior successfully often relies on operating within the very sensitive critical region and knowing how to maneuver appropriately.

14.9   NUMERICAL PROCEDURES FOR BINARY, TERNARY LLE

Numerical procedures using Excel and MATLAB are provided in online supplements. The Excel procedure extends Actcoeff.xlsx and explains details on setting up the macro or circular reference for binary or ternary mixtures. The MATLAB Rachford-Rice procedure can be extended more easily to multicomponent mixtures.

14.10   SOLID-LIQUID EQUILIBRIA

Solid-liquid equilibria (SLE) calculations begin just as VLE and LLE calculations, by equating fugacities. From Eqn. 11.13, $\hat{f}_i^L = x_i \gamma_i f_i^o$. The next step is to equate $\hat{f}_i^L = \hat{f}_i^S$ and derive an equation to solve for temperature or composition depending on the problem statement. We have deliberately avoided substituting $f_i^o = P_i^{sat}$, however, as we did for VLE and LLE. This is because pure components below their melting temperature do not exist as liquids, so the vapor pressure is not relevant. This peculiarity necessitates a derivation to characterize a hypothetical liquid state. The treatment of the solid phase is usually much simpler, however, because most solids exist in a practically pure state. The purity of crystallized solid phases is one of the reasons for the prevalence of crystallization and filtration in the pharmaceutical industry.

Pressure Effects

For SLE, as for LLE, pressure changes usually have very small effects on the equilibria unless the pressure changes are large (10 to 100 MPa), because the enthalpies and entropies of *condensed* phases are only weakly pressure-dependent. Since $dG = RT\, d \ln f = dH - T\, dS = V\, dP$ for a pressure change at constant T, the Gibbs energy and fugacity exhibit only small changes with pressure when the enthalpy and entropy exhibit small changes. (Recall that the Poynting correction factor is usually very near one.) In a mixture of liquids, the analysis must be done with partial molar enthalpies and partial molar entropies; however, these properties also depend only weakly on pressure. In the following subsections, we calculate properties at the triple-point, or other low pressures, and use the results at atmospheric pressure without pressure correction due to the weak pressure dependence.

> ❶ Pressure effects on SLE are usually neglected.

SLE in a Single Component System

To begin our discussion, we will consider a single component. At 1 bar, water freezes at 0°C. Ice exists below this temperature and liquid exists above. From the principles of thermodynamics, the Gibbs energy is minimized at constant pressure and temperature; therefore, above 0°C, the Gibbs energy of liquid water must be lower than the Gibbs energy of solid water. In order to express this concept quantitatively, we must consider how the Gibbs energies of each of these phases changes with temperature.

The effect of temperature on the Gibbs energy of any phase may be determined most easily at constant pressure. We may write $dG = -S\, dT + V\, dP$, and recognize using the concepts of Chapter 6

$$(\partial G/\partial T)_P = -S$$

The temperature dependence of Gibbs energy is then dependent on the entropy of the phase. Entropy is a positive quantity; therefore, the Gibbs energy of any phase must decrease with increasing temperature. However, the entropy of a liquid phase is greater than the entropy of a solid phase; thus, the Gibbs energy of a liquid phase decreases more rapidly as the temperature increases. Since

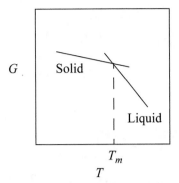

Figure 14.8 *Illustration of Gibbs energies for pure SLE.*

the Gibbs energies of the phases are equal at the freezing temperature, the Gibbs energy of the liquid will lie below the Gibbs energy of the solid above the freezing temperature (see Fig. 14.8).[1] Below the freezing temperature, we can still extrapolate the Gibbs energy of the liquid, but we must refer to this as a hypothetical liquid because it does not exist as a stable phase. The portion of the solid Gibbs energy curve above the melting temperature represents the Gibbs energies of a hypothetical solid, and a melting process will occur spontaneously at constant temperature and pressure because the ΔG^{fus} for the process is negative. Alternatively, an equilibrium solid will not be formed above the melting temperature because ΔG^{fus} is positive. A discussion for behavior below the melting temperature is not presented, although the ideas are similar. Melting of a solid below the freezing temperature requires a $\Delta G^{fus} > 0$ at constant temperature and pressure. The mixing process balances this positive Gibbs energy change to create mixtures below the normal melting temperature.

> ❶ $\Delta G^{fus} > 0$ below the normal melting temperature.

The Calculation Pathway for Mixtures

We know that solids dissolve in liquid mixtures well below their normal melting temperatures. Sugar and salt both dissolve in water at room temperature, although the pure solid melting temperatures are far above room temperature. Also, salt is spread on the highways in the winter to lower the temperature at which solid ice forms. We may choose to address several problems such as: 1) How much sugar may be dissolved in water before the solubility limit is reached? 2) In a water/salt solution, at what temperature will a solid form, and will the crystals be water or salt? (Salt is introduced as a practical example, although rigorous treatment of the problem involves electrolyte thermodynamics.) 3) How may a solvent or solvent mixture be modified to regulate crystallization? In order to deal with these concepts mathematically, we use state properties to calculate thermodynamic changes along convenient pathways that involve hypothetical steps.

Let us consider a practical example of dissolving naphthalene (2) in *n*-hexane (1) at 298 K. Since the normal melting temperature for pure naphthalene is 353.3 K, how can we explain the phenomenon that the naphthalene dissolves in hexane? First, recall that the naphthalene will dissolve in the *n*-hexane if the total Gibbs energy of the system (*n*-hexane and naphthalene) decreases upon dissolution. Thus, more and more solid may be added to the liquid solution until any further addition causes the total Gibbs energy to increase rather than decrease. This method of calculating equi-

---

1. To begin the calculations, we must specify a reference state. In any thermodynamic analysis, we must have only one reference state for each chemical species; for our example here, the reference state for water must be the same whether the water is solid or liquid. See Section 9.12 on page 361 and footnote therein.

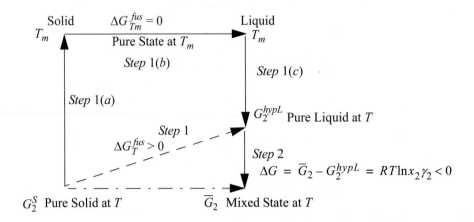

Figure 14.9 *Illustration of the two-step process for calculating solubility of solids in liquids. Overall,* $G_2^S = \overline{G}_2$. *Note that the Gibbs energy goes up in Step 1 to create liquid, below the normal melting* T_m, *but the Gibbs energy goes down when the liquid is mixed.*

librium is fairly tedious to apply, and a preferred method is used that gives identical results: The solubility limit is reached when the chemical potential of the naphthalene in the liquid is the same as the chemical potential of the pure solid naphthalene. Therefore, we solve the problem by equating the chemical potentials for naphthalene in the liquid and solid phases.

The equilibrium can be written as $\mu_2 = \mu_2^S$ or, recognizing the notational definitions $\mu_2 = \overline{G}_2$ and $\mu_2^S = G_2^S$, therefore,

$$\overline{G}_2 - G_2^S = 0 \qquad\qquad 14.14$$

Next, envision the hypothetical pathway shown in Fig. 14.9 to calculate the chemical potential difference given by Eqn. 14.14, which consists of two primary steps.

Step 1. Naphthalene is melted to form a hypothetical liquid at 298 K. The Gibbs energy change for this step is positive as discussed above. The Gibbs energy change is:

$$\Delta G_T^{fus} = G_2^{hypL} - G_2^S \qquad\qquad 14.15$$

where the superscript *hypL* indicates a hypothetical liquid.

Step 2. The hypothetical liquid naphthalene is mixed with liquid *n*-hexane. If the solution is nonideal, the Gibbs energy change for component 2 is

$$\overline{G}_2 - G_2^{hypL} = RT\ln(x_2\gamma_2) \qquad\qquad 14.16$$

The Gibbs energy change for this step is always negative if mixing occurs spontaneously, and must be large enough to cancel the Gibbs energy change from step 1.

Then clearly, from Fig. 14.9 and Eqns. 14.14 through 14.16,

$$\overline{G}_2 - G_2^S = (\overline{G}_2 - G_2^{hypL}) + (G_2^{hypL} - G_2^S) = RT\ln(x_2\gamma_2) + \Delta G_T^{fus}$$

Solubility is determined by a balance between the positive ΔG^{fus} and the negative Gibbs energy effect of mixing.

or

$$\Delta G_T^{fus} = -RT\ln(x_2\gamma_2) \qquad \text{14.17}$$

where T is 298 K for our example. Relations for the activity coefficients in the right-hand side of the equation have been developed in previous chapters. In the next subsection, the calculation of ΔG_T^{fus} is explained.

Formation of a Hypothetical Liquid

The Gibbs energy change for step 1 is most easily calculated using the entropies and enthalpies. For an isothermal process, we can write:

$$dG = dH - T\,dS$$

Continuing with the example of dissolving naphthalene at 298 K, the Gibbs energy change for melting naphthalene at 298 K can be calculated from the enthalpy and entropy of fusion,

$$\Delta G_T^{fus} = \Delta H_T^{fus} - T\Delta S_T^{fus} \qquad \text{14.18}$$

where $T = 298$ K. ΔH_{298}^{fus} can be calculated by determining the enthalpies of the liquid and solid phases relative to the normal melting temperature, where the heat of fusion, $\Delta H_{353.3}^{fus} = 18.8$ kJ/mol. The enthalpy of solid at 353.3 K relative to solid at 298 K (step 1(a) of Fig. 14.9) is

$$H_{353.3}^{S} - H_{298}^{S} = \int_{298}^{353.3} C_P^S dT$$

The enthalpy of the liquid at 298 K relative to liquid at 353.3 K (step 1(c) of Fig. 14.9) is:

$$H_{298}^{hypL} - H_{353.3}^{L} = \int_{353.3}^{298} C_P^L dT$$

Thus, the ΔH_{298}^{fus} for melting is

$$\Delta H_{298}^{fus} = H_{298}^{hypL} - H_{298}^{S} = (H_{298}^{hypL} - H_{353.3}^{L}) + (H_{353.3}^{L} - H_{353.3}^{S}) + (H_{353.3}^{S} - H_{298}^{S})$$

$$= \Delta H_{353.3}^{fus} + \int_{353.3}^{298} (C_P^L - C_P^S)dT$$

which we can generalize to

$$\Delta H_T^{fus} = H_T^{hypL} - H_T^{S} = \Delta H_{T_m}^{fus} + \int_{T_m}^{T} (C_P^L - C_P^S)dT \qquad \text{14.19}$$

A similar derivation for the entropy gives

$$\Delta S_{298}^{fus} = \Delta S_{353.3}^{fus} + \int_{353.3}^{298} \frac{(C_P^L - C_P^S)}{T}dT$$

which we can generalize to

$$\Delta S_T^{fus} = \Delta S_{T_m}^{fus} + \int_{T_m}^{T} \frac{(C_P^L - C_P^S)}{T} dT$$

14.20

In addition to these relationships, at the normal melting temperature, since $\Delta G_{T_m}^{fus} = 0$,

$$\Delta S_{T_m}^{fus} = \frac{\Delta H_{T_m}^{fus}}{T_m}$$

14.21

where $T_m = 353.3$ K. Combining the results and neglecting the integrals (which are nonzero, but essentially cancel each other), results in the **Schröder-van Laar** equation for component 2:

$$\Delta G_T^{fus} = \Delta H_{T_m}^{fus} - T\Delta S_{T_m}^{fus} = \Delta H_{T_{m,2}}^{fus}\left(1 - \frac{T}{T_{m,2}}\right)$$

14.22

where for our example, $T = 298.15$ and $T_m = 353.3$. This is the ΔG_T^{fus} that we desire for Eqn. 14.17.

Criteria for Equilibrium

In general, combining Eqn. 14.22 with Eqn. 14.17, we arrive at the equation for the solubility of component 2,

Solubility equation for crystalline solids.

$$\ln(x_2\gamma_2) = \frac{-\Delta H_2^{fus}}{R}\left(\frac{1}{T} - \frac{1}{T_{m,2}}\right) = \frac{-\Delta S_2^{fus}}{R}\left(\frac{T_{m,2}}{T} - 1\right)$$

14.23

where heat of fusion is at the normal melting temperature of 2, and heat capacity integrals are neglected.

Example 14.8 Variation of solid solubility with temperature

Estimate the solubility of naphthalene in *n*-hexane for the range $T = [298, 350K]$ using the SSCED model. Plot $\log_{10}(x_N)$ versus $1000/T$.

Solution: From Appendix E, $T_{m,2} = 353.3$ K and $\Delta H^{fus} = 18,800$J/mol.

We can begin at 298 K, assuming an ideal solution. Then

$$x_N = \exp[(-18800/8.314)\cdot(1/298 - 1/353.3)] = 0.305.$$

Starting with $x_N = 0.305$ as an initial guess,

$$\Phi_N = 0.305\cdot130.6/(0.695\cdot130.3 + 0.305\cdot130.6) = 0.306.$$

Noting that $\delta_1' = 14.93$, $\delta_2' = 19.19$ and $k_{12} = 0.0052 \Rightarrow \gamma_2 = 1.693$.

Example 14.8 Variation of solid solubility with temperature (Continued)

$x_N = 0.305/1.693 = 0.1802$. Iterating on x_N to achieve consistency, $x_N = 0.135$. Repeating this procedure at other temperatures gives Fig. 14.10.

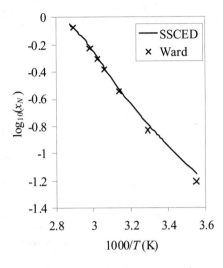

Figure 14.10 *Freezing curve for the system n-hexane(1) +*
naphthalene(2). Experimental data of H.L. Ward,
1926. J. Phys. Chem., 30:1316.

Hexane also dissolves in a hexane–naphthalene solution below its melting temperature. The general relationship for solving SLE can be written as:

$$\ln(x_i \gamma_i) = \frac{-\Delta H_i^{fus}}{R}\left(\frac{1}{T} - \frac{1}{T_{m,i}}\right)$$

14.24

where the heat of fusion is for the pure i^{th} component at its normal melting temperature, $T_{m,i}$. Note that Eqns. 14.23 and 14.24 may be used to determine crystallization temperatures at specified compositions.

Example 14.9 Eutectic behavior of chloronitrobenzenes

Fig. 14.11 illustrates application to the system *o*-chloronitrobenzene (1) + *p*-chlorornitrobenzene (2).[a] The compounds are chemically similar; thus, the liquid phase may be assumed to be ideal, and the activity coefficients may be set to 1. The two branches represent calculations performed from Eqns. 14.23 and 14.24, each giving one-half the diagram. The curves are hypothetical below the point of intersection. This temperature at the point of intersection of the two curves is called the **eutectic temperature,** and the composition is the **eutectic composition.**

Eutectic.

Example 14.9 Eutectic behavior of chloronitrobenzenes (Continued)

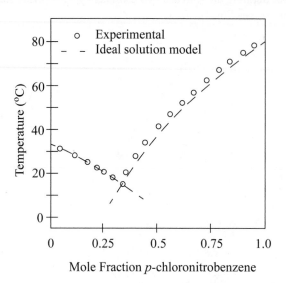

Figure 14.11 *Freezing curves for the system o-chloronitrobenzene(1) + p-chloronitrobenzne(2).*

a. Holleman, A.F. 1900. *Rec. trav. chim.,* 19:101; Kohman, G.T. 1925. *J. Phys. Chem.,* 25:1048; *cf.* Timmermans, J. 1936. *Les Solutions concentrées*, Paris.

Example 14.10 Eutectic behavior of benzene + phenol

In most systems, an activity coefficient model must be included. Fig. 14.12 shows an example where the ideal solution model is not a good approximation, and the activity coefficients are modelled with the UNIFAC activity coefficient model. To solve for solubility given a temperature, the following procedure may be used (taking component 2, for example):

1. Assume the $\gamma_2 = 1$.

2. Solve Eqn. 14.23 for x_2.

3. At this value of x_2, determine γ_2 from the activity model.

4. Return to step 2, including the value of γ_2 in Eqn. 14.23, iterating to converge.

This is a relatively stable iteration, and Excel Solver can iterate on activity by adjusting composition, bypassing the successive substitution method above.

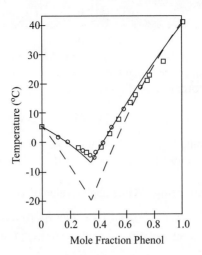

Figure 14.12 *Freezing curves for the system benzene(1) + phenol(2). Solid*
line, UNIFAC prediction; dashed line, ideal solution prediction;
squares, Tsakalotos, D., Guye, P. 1910. J. Chim. Phys. 8:340;
circles, Hatcher, W., Skirrow, F. J. Am. Chem. Soc., 1917.
39:1939. Based on figure of Gmehling, J., Anderson, T., Praus-
nitz, 1978. J. Ind. Eng. Chem Fundam. 17:269.

A common procedure when crystallizing products of a pharmaceutical process is to add an
antisolvent. The SSCED model is especially convenient for this kind of application because: (1)
The concentrations are generally small for crystalline products left in solution; and (2) the infinite
dilution activity coefficient of the precipitate involves a simple average of the solvent and antisolvent properties. With these concepts in mind, it is straightforward to tailor a solvent to achieve a
target composition at a given temperature.

Example 14.11 Precipitation by adding antisolvent

Ephedrine is a commonly used stimulant, appetite suppressant, and decongestant, related to
pseudoephedrine. It can be extracted from the Chinese herb, ma huang. Ephedrine is to be crystallized from ethanol at 278 K by adding water as an antisolvent.

(a) Estimate the mole fraction of water needed to reduce the concentration of ephedrine in
solution to 0.1mol% using the SSCED model.

(b) Yalkowsky and Valvani (1980) have suggested that ΔS^{fus} = 56.5 J/mol-K for rigid molecules.[a] Evaluate this relation in comparison to the estimated value of ΔH^{fus} = 25kJ/mol.

Additional data for ephedrine are M_w = 165.2; T_m = 313K; and ΔH^{fus} = 25kJ/mol.

Comp.	$\alpha(J/cm^3)^{0.5}$	$\beta(J/cm^3)^{0.5}$	$\delta(J/cm^3)^{0.5}$	$V^L(cm^3/mol)$
Ethanol(1)	12.58	13.29	18.68	58.3
Water(2)	50.13	15.06	27.94	18.0
Ephedrine(3)	7.70	12.60	16.36	172.3

Example 14.11 Precipitation by adding antisolvent (Continued)

Solution:
(a) From Eqn. 14.25, $T_{m,i} = 313$ K and $\Delta H^{fus} = 25,000$ J/mol. We can begin by solving for the target value of the activity coefficient, noting that the concentration of drug is practically infinitely dilute. Then using $x_i = 1E-3$ to approximate infinite dilution,

$$\gamma_i^{\infty} = \exp[(-25000/8.314)\cdot(1/278 - 1/313)]/0.001 = 245 \text{ is what we seek to find.}$$

The solution requires iteration using Eqn. 12.55. As the mole fraction of water is increased, the activity coefficient of ephedrine increases because water is the antisolvent.

Since we have worked the problem before, "Guessing" a value of $x_1 = 0.2102$,

$$\Phi_1 = 0.2102(58.3)/(0.2102\cdot58.3 + 0.7898\cdot18.0) = 0.4627$$
$$<\delta'> = 0.4627(18.68) + 0.5373(27.94) = 23.66$$

From Eqn. 12.51,

$$k_{12} = 0.0318; \quad k_{13} = 0.0028; \quad k_{23} = 0.0571;$$
$$=> <k_{3m}> = 0.4627(18.68)0.0028 + 0.5373(27.94)0.0571 = 0.8808;$$

$$\text{Similarly, } <k_{1m}> = 0.4779; \quad <k_{2m}> = 0.2752;$$
$$<<k_{mm}>> = 0.4627(18.68)0.4779 + 0.5373(27.94)0.2752 = 8.262.$$

By Eqn. 12.55, $RT \ln\gamma_3 = 172.3((16.36 - 23.66)^2 + 2(16.36)0.8808 - 8.262) => \gamma_3 = 245$. So the solution should be 79 mol% water. Good "guess!"

(b) With $\Delta S^{fus} = 56.5$ and $T_m = 313$K, $\Delta H^{fus} = 56.5(313) = 17,700$ J/mol, 29% lower than 25,000. The rule does not appear to apply to this compound.

---

a. Connors, K.A. 2002. *Thermodynamics of Pharmaceutical Systems: An Introduction for Students of Pharmacy*, Hoboken, NJ: Wiley, p. 129.

A special feature of Example 14.11 is the way it shows how to tailor a solvent to achieve a particular environment for a target solute. A similar approach could be applied to compatibilizing a liquid solvent to avoid LLE. For example, how much methanol should be added to isooctane to reduce the activity coefficient of water below a value of 7.4? This is the calculation behind "dry gas," used to dissolve water from gas tanks.

SLE with Solid Mixtures

So far, we have only covered phase behavior in systems where the solids are completely immiscible in each other. Fig. 14.13 illustrates a case where the solids form solid solutions, and Fig. 14.14 illustrates behavior where compounds are formed in the solid complexes. Also, the case of wax precipitation from petroleum results in a range of n-C_{20} to n-C_{35} straight chain alkanes being mixed in the solid phase. Paraffin wax that you can purchase in the grocery store is primarily composed of n-C_{20} to n-C_{35} straight chain alkanes. For the case of liquid and solid mixtures in equilibrium, the derivation of the equilibrium relationship can be modified by adding a step for "unmixing" of solid solutions to the schematic of Fig. 14.9 on page 558. This step is analogous to a reversal of step 2 of the diagram except involving solid solutions and pure solids rather than liquid solutions and pure liquids. For each component in the mixtures:

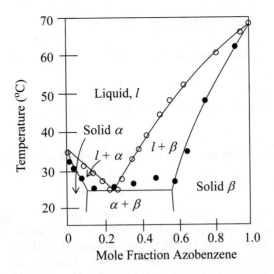

Figure 14.13 *Freezing curves for the Azoxybenzene(1) + azobenzene(2) system illustrating a system with solid-solid solubility. Based on Hildebrand, J.H., Scott, R.L., Solubility of Nonelectrolytes, New York, NY: Dover, 1964.*

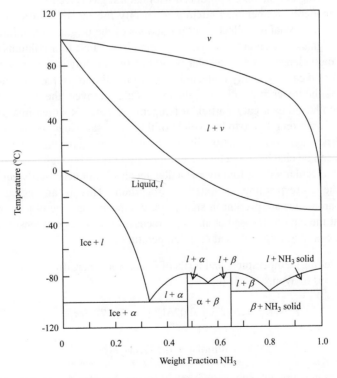

Figure 14.14 *Solid-liquid and vapor-liquid behavior for the ammonia(1) + water(2) system at 1.013 bar. NH_3 and H_2O form two crystals in the stoichiometries: (α) $NH_3 \cdot H_2O$; (β) $2NH_3 \cdot H_2O$. (Based on Landolt-Börnstein, 1960. II/2a:377.)*

$$-RT\ln(x_i^S \gamma_i^S) + \Delta G_i^{fus} + RT\ln(x_i^L \gamma_i^L) = 0 \qquad\qquad 14.25$$

Thus, we can recognize an SLE K-ratio on the left-hand side, $K_{SLE} = (x_i^L \gamma_i^L)/(x_i^S \gamma_i^S)$,

$$RT\ln\frac{(x_i^L \gamma_i^L)}{(x_i^S \gamma_i^S)} = -\Delta G_i^{fus} \quad \text{or} \quad \boxed{\ln K_{SLE} = \frac{-\Delta H_i^{fus}}{R}\left(\frac{1}{T} - \frac{1}{T_{m,i}}\right)} \qquad 14.26$$

and we can recognize Eqn. 14.24 as a simplification where for a pure solid, $x_i^S \gamma_i^S = 1$.

Petroleum Wax Precipitation

An especially difficult problem in the recovery of natural gas is the clogging of pipes caused by small amounts of wax that accumulate over time. In the Gulf of Mexico, natural gas at the bottom of the well can be 250 bars and 100°C, but it must be reduced to 100 bars to be permitted in the pipeline, and the sea floor can drop to 5°C. The reduction in pressure and temperature results in a loss of carrying power and the small amounts of heavy liquid hydrocarbons can condense, eventually coating the walls with viscous liquid. After the liquid has formed, further cooling can cause solid wax to deposit on the walls of the pipe. These deposits cause constrictions and larger pressure drops that lead to more deposits, and so forth.

Naturally occurring petroleum co-produced with natural gas is a complex mixture of hundreds of individual components. Rather than attempt to specify the identity and composition of every component, it is conventional to collect several fractions of the original according to the ranges of their molecular weights. Hansen et al.[2] provide the data in the first four columns of Table 14.1 for the composition, mass density, and molecular weight of several fractions from a typical petroleum stream. This kind of data is typically collected by distilling the initial sample and collecting fractions over time. As the lower molecular weight species are removed, the boiling temperature rises and the distillate collected over each particular temperature range is stored in a separate container. The weight of each fraction relative to the weight of the initial sample gives the composition of that species fraction. The mass of each fraction divided by its volume gives the density. And the average molecular weight can be characterized by gas chromatography or through correlations with viscosity. Note that the molecular weight for any particular species is not necessarily equal to the molecular weight for the corresponding saturated hydrocarbon. This is an indication that olefins, naphthenics, and aromatics are present in significant compositions. The objective of this example problem is to treat the data of Hansen et al. as characteristic of a gas condensate and compute the fractions of the stream that form solids at each temperature.

The fusion (melting) temperatures and heats of fusion for n-paraffins can be calculated according to the correlations of Won.[3]

$$T_i^{fus}(K) = 374.5 + 0.02617M_i - 20172/M_i \qquad\qquad 14.27$$

$$\Delta H_i^{fus}(\text{cal/mol}) = 0.1426M_i T_i^{fus} \qquad\qquad 14.28$$

Noting that each species fraction can contain many species besides the n-paraffins that are responsible for practically all wax formation, it is necessary to estimate the portion of each species frac-

---

2. Hansen, J.H., Fredenslund, Aa., Pedersen, K.S., Ronningsen, H.P. 1988. *AIChE J.* 34:1937.
3. Won, K.W. 1986. *Fluid Phase Equil.* 30:265. See also Won, K.W. 1989. *Fluid Phase Equil.* 53:377.

tion that can form a wax. Since the densities of *n*-paraffins are well known, it is convenient to use the difference between the observed density and the density of an *n*-paraffin of the same molecular weight as a measure of the *n*-paraffin content. The correlations for estimating the percentage of wax-forming components in the feed (z_i^W) are taken from Pedersen.[4]

$$z_i^W = z_i^{tot}\left[1 - \left(0.8824 + \frac{0.5353\,M_i}{1000}\right)\left(\frac{\rho_i - \rho_i^P}{\rho_i^P}\right)^{0.1144}\right]$$ 14.29

$$\rho_i^P(\text{g/cm}^3) = 0.3915 + 0.0675\ln(M_i)$$ 14.30

where z_i^{tot} is the species overall mole fraction in the initial sample.

z_i^W is the portion of that fraction which is wax-forming (i.e., *n*-paraffin).

Example 14.12 Wax precipitation

Use the data from the first four columns of Table 14.1 and correlations for wax to estimate the solid wax phase amount and the composition of the solid as a function of temperature. Use your estimates to predict the temperature at which wax begins to precipitate. Hansen et al. give the experimental value as 304 K.

Solution: This problem is basically a multicomponent variation of the binary solid-liquid equilibrium problems discussed above. The main difference is that the solid phase is not pure. We can adapt the algorithm as follows.

Assuming ideal solution behavior for both the solid and liquid phases, we define $K_i = x_i^L / x_i^S$, and as before, we assume the difference in heat capacities between liquid and solid is negligible relative to the heat of fusion,

$$K_i = \exp\left[\frac{-\Delta H_i^{fus}}{RT}\left(1 - \frac{T}{T_i^{fus}}\right)\right]$$ 14.31

which is independent of the compositions of the liquid and solid phases because of the ideal solution assumptions. The solid solution mole fraction is given by $x_i^S = x_i^L / K_i$. Compare this method to the vapor-liquid calculations using the shortcut *K*-ratio in Chapter 9. This is a liquid-solid freezing temperature analog to the vapor-liquid dew-temperature procedure. The liquid mole fractions are given by the z_i^W values in the table below. All that remains is to guess values of *T*, which changes all K_i until $\sum_i x_i^S = 1.0$. Hand calculations would be easy with a couple of components, but spreadsheets are recommended for a multicomponent mixture. Using Solver for spreadsheet Wax.xlsx distributed with the textbook software gives *T* = 320.7 K. Intermediate results are tabulated in Table 14.1. The *T* is slightly higher than the experimental value, but reasonably accurate considering the complex nature of the petroleum fractions and their variabilities from one geographic location to another.

---

4. Pedersen, K.S. 1995. *SPE Prod. and Fac.* Feb:46.

Table 14.1 *Summary of Data for Wax Fractions and Calculations of the Precipitate Composition as Calculated by Example 14.12*

Species	Wt%	M_i	$\rho(g/cm^3)$	$\rho^P(g/cm^3)$	z_i^{tot}	z_i^W	T^{fus}	ΔH^{fus} (J/mol)	K_i^{320}	x_i^S
<c4	0.031	29	0.416	0.619	0.003	0.000	---	---		
c5	0.855	71	0.632	0.679	0.031	0.000	92	3904		
c6	0.377	82	0.695	0.689	0.012	0.000	131	6386		
c7	2.371	91	0.751	0.696	0.068	0.021	155	8419	28.98	0.0007
c8	2.285	103	0.778	0.704	0.058	0.016	181	11134	24.75	0.0006
c9	2.539	116	0.793	0.712	0.057	0.015	204	14080	20.81	0.0007
c10	2.479	132	0.798	0.721	0.049	0.013	225	17714	16.78	0.0008
c11	1.916	147	0.803	0.728	0.034	0.009	241	21128	13.67	0.0006
c12	2.352	163	0.817	0.735	0.038	0.009	255	24777	10.95	0.0008
c13	2.091	175	0.836	0.740	0.031	0.007	264	27519	9.26	0.0008
c14	3.677	190	0.843	0.746	0.051	0.011	273	30952	7.49	0.0015
c15	3.722	205	0.849	0.751	0.047	0.010	281	34393	6.04	0.0017
c16	2.034	215	0.853	0.754	0.025	0.005	286	36691	5.22	0.0010
c17	4.135	237	0.844	0.761	0.046	0.010	296	41757	3.78	0.0026
c18	3.772	251	0.846	0.764	0.039	0.008	301	44989	3.07	0.0027
c19	3.407	262	0.857	0.767	0.034	0.007	304	47532	2.60	0.0026
c20	2.781	268	0.868	0.769	0.027	0.005	306	48921	2.38	0.0021
c21	3.292	284	0.862	0.773	0.030	0.006	311	52631	1.86	0.0031
c22	3.14	299	0.863	0.776	0.027	0.005	315	56116	1.48	0.0035
c23	3.445	315	0.963	0.780	0.029	0.003	319	59841	1.15	0.0027
c24	3.254	330	0.865	0.783	0.026	0.005	322	63340	0.91	0.0052
c25	2.975	342	0.867	0.785	0.023	0.004	324	66144	0.75	0.0054
c26	3.038	352	0.869	0.787	0.023	0.004	326	68485	0.64	0.0061
c27	2.085	371	0.873	0.791	0.015	0.002	330	72941	0.47	0.0052
c28	2.74	385	0.877	0.793	0.019	0.003	332	76231	0.37	0.0079
c29	3.178	399	0.881	0.796	0.021	0.003	334	79527	0.29	0.0107
>c30	31.12	578	0.905	0.821	0.141	0.011	355	122213	0.01	0.9308

Solid-Liquid Equilibria Summary

Phase equilibrium involving solids is an extension of previous modeling concepts. Like liquid-liquid equilibria, the condensed phase fugacities are quite insensitive to pressure, so the partition coefficients are simply functions of temperature. The main difference is that the heats of fusion are used to relate the component fugacities in their various states of matter. In multicomponent mixtures, the solid-liquid procedures for calculation are analogs of the vapor-liquid procedures, where the partition coefficients are calculated in a different manner.

14.11 SUMMARY

The presentations of LLE and SLE have been brief, but have opened a broad new frontier of phase behavior analysis: multiphase equilibrium. You might find it incredible to see how many phases and peculiar behaviors can be observed with just "oil," water, and special third components known as surfactants. (Soap is an example of a surfactant.) The short introduction here is a branching point that barely scratches the surface of such oleic and aqueous systems.[5] Such molecules are more complicated than we can represent with the simple models here because of the way that they organize to make films and micelles. Far down this road you may begin to understand the forces that hold cell membranes and living organisms together. At the more practical engineering level, you should be able to perform preliminary designs of liquid extraction and crystallization equipment with little more thermodynamical background than has been presented here.

This kind of breadth is possible with such short introductions because the fundamentals have been laid out previously. The key equation in both LLE and SLE is familiar from VLE (Eqn. 11.13):

$$\hat{f}_i = \gamma_i x_i f_i^\circ$$ 11.13

The liquid phases split because the Gibbs excess energy becomes so large that the stability limit is exceeded. In other words, the fugacities are so high that the components must escape each other, even if their volatilities are too low for VLE. An entropic penalty must be paid, but a highly unfavorable energy of interaction may more than compensate. A convenient guideline is,

Suspect LLE if $\gamma_i^\infty > 7.4$, in which case $x_i \approx 1/\gamma_i^\infty$ is a good initial guess.[6]

With this guideline, LLE computation is a simple coincidence that may occur occasionally during VLE computations. To paraphrase Pasteur, its observation presents no problem if we know how to recognize it. The procedure for calculating Gibbs energy is the same as for VLE. The calculation of the phase distributions requires a flash calculation in terms of liquid-liquid K-ratios. But the flash algorithm has been discussed before and liquid-liquid flash calculations are hardly different from vapor-liquid flash calculations. The introduction to the prospect of LLE has reminded us of the need to check for stability, however. We first encountered the concept of stability in contemplating the critical points of pure fluids. With the prospect of multiple phases, we begin to realize the need to explore critical phase behavior in a systematic fashion. We have treated that problem for binary and ternary mixtures but have not generalized to multicomponent systems. The generalization of critical phase behavior requires a fair amount of calculus and matrix manipulation,[7] but leads to a much deeper understanding of stability behavior than we have provided here. As a practi-

5. Hiemenz, P.C. 1986. *Principles of Colloid and Surface Chemistry,* 2nd ed. New York, NY: Marcel Dekker, NY.
6. ... and reasonable estimate when $\gamma_i^\infty > 100$
7. Tester, J.W., Modell, M. 1997. *Thermodynamics and Its Applications,* 3rd ed. Upper Saddle River, NJ: Prentice-Hall.

cal matter, activity coefficient models must be used carefully because the precision of the models is usually inferior to that of VLE predictions for the same systems. In particular, the temperature dependence of LLE is usually not predicted well. Frequently, the miscibility increases with temperature more quickly than predicted by the activity coefficients even with more sophisticated models like UNIQUAC or UNIFAC. Temperature-dependent parameters can improve the fit, but have little theoretical basis, making extrapolations more tenuous than for VLE.

Important Equations

For LLE, the key equation comes from setting the two expressions for f_i^L equal and canceling f_i^o,

$$\gamma_i^\alpha x_i^\alpha = \gamma_i^\beta x_i^\beta \tag{14.1}$$

The iteration for LLE is a little tricky because it relies on the γ's being large but they get smaller as the iteration proceeds. If you guess a composition too close to equimolar, you might miss the LLE.

For SLE, the key equation is (Eqn. 14.23),

$$\ln(x_i \gamma_i) = (-\Delta H^{fus}/R)(1/T - 1/T_m) \tag{14.24}$$

If $\gamma_i \gg 1$ then $T \to T_m$. (It is nonsense if the computations yield a value of $T > T_m$.) Otherwise, the solubility may be significant at $T < T_m$. This observation suggests changing the solvent (by adding anti-solvent to increase γ_i), in addition to simply cooling. This is a common technique in pharmaceutical crystallization, among other applications. SLE computation requires iteration if $\gamma_i \neq 1$, because the liquid composition must be guessed in order to estimate γ_i. But iterative problems like this are familiar from VLE experience and present no problem.

14.12 PRACTICE PROBLEMS

P14.1 It has been suggested that the phase diagram of the hexane + furfural system can be adequately represented by the Margules one-parameter equation, where $\ln \gamma_i = x_j^2 \cdot 800/T$ (K). Estimate the liquid-liquid mutual solubilities of each component in each liquid phase at 298 K. (ANS. ~10% each, by symmetry)

P14.2 Suppose the solubility of water in ethyl benzene was measured by Karl-Fisher analysis to be 1 mol%. Use UNIFAC to estimate the solubility of ethylbenzene in the water phase. (ANS. 0.003wt% or 30ppmw)

P14.3 According to Perry's Handbook, the system water + isobutanol forms an atmospheric pressure azeotrope at 67.14 mol% water and 89.92°C. Based on these data, we can estimate the van Laar coefficients to be $A_{12} = 1.566$; $A_{21} = 3.833$ at 273 K. Estimate the liquid-liquid mutual solubilities of each component in each liquid phase at 273 K. (ANS. 0.33,0.97)

P14.4 Use the SSCED model to predict the solubility of iodine in carbon tetrachloride at 298 K. Iodine's melting point is 387 K and $H^{fus} = 15.5$ kJ/mol.

14.13 HOMEWORK PROBLEMS

14.1 Suppose the (1) + (2) system exhibits liquid-liquid immiscibility. Suppose we are at a state where $G_1/RT = 0.1$ and $G_2/RT = 0.3$. The Gibbs energy of mixing quantifies the Gibbs energy of the mixture relative to the Gibbs energies of the pure components. Suppose the excess Gibbs energy for the (1) + (2) mixture is given by:

$$G^E/RT = 2.5\, x_1\, x_2$$

(a) Combine this with the Gibbs energy for ideal mixing to calculate the Gibbs energy of mixing across the composition range and plot the results against x_1 to illustrate that the system exhibits immiscibility.

(b) Draw a tangent to the humps to illustrate that the system is one phase at compositions greater than $z_1 = 0.854$ and less than $z_1 = 0.145$, but will split into two phases with compositions at any intermediate overall composition. Most systems with liquid-liquid immiscibility must be modeled with a more complex formula for excess Gibbs energy. The humps on the diagram are usually off-center, as in Fig. 14.3 on page 545 in the text. The simple model used for the calculations here results in the symmetrical diagram.

(c) When a mixture splits into two phases, the over-all fractions (of total moles) of the two phases are found by the lever rule along the composition coordinate. Suppose 0.6 mol of (1) and 0.4 mol of (2) are mixed. Use the lever rule to calculate the total number of moles which would be found in each phase of the actual system. Designate the (1)-rich phase as the β phase.

(d) What is the value of the hypothetical Gibbs energy, (expressed as G/RT), of a mixture of 0.6 mol of (1) and 0.4 mol of (2) if the mixture were to remain as one phase? Calculate the Gibbs energy of the total system considering the phase split into two phases, and show that the Gibbs energy is less than the Gibbs energy of the single-phase system.

14.2 Assume solvents A and B are virtually insoluble in each other. Component C is soluble in both.

(a) Use the Scatchard-Hildebrand theory to estimate the distribution coefficient at low concentrations of C given as (mole fraction C in A)/(mole fraction C in B).

(b) If the phase containing A is 0.1 mol% C, estimate the composition of the phase containing B.

(c) If an extractor was designed and constructed, is the distribution coefficient favorable for extraction from B into A? Data:

	Volume (cm^3/mole)	$\delta\,(cal/cm^3)^{1/2}$
A	50	5.8
B	250	10.4
C	100	9.8

14.3 A new drug is to have the formula para-CH_3CH_2-(C_6H_4)-CH_2CH_2COOH, where (C_6H_4) designates a phenyl ring. A useful method for assessing the extent of partitioning between the bloodstream and body fat is to determine the infinite dilution partitioning coefficient for the drug between water and n-octanol. Use UNIFAC to make this determination. The body temperature is 37°C. Will the new drug stay in the bloodstream or move into fatty body parts?

14.4 Use the Scatchard-Hildebrand theory to generate figures of activity as a function of composition and ΔG_{mix} as a function of composition for neopentane and dichloromethane at 0°C. Determine the compositions of the two phases in equilibrium. Data:

	V (cm³/mole)	δ (cal/cm³)$^{1/2}$
Neopentane	122	6.2
Dichloromethane	64	9.7

14.5 The bubble point of a liquid mixture of n-butanol and water containing 4 mol% butanol is 92.7°C at 1 bar. At 92.7°C the vapor pressure of pure water is 0.784 bar and that of pure n-butanol is 0.427 bar. Assuming the activity coefficient of water in the 4% butanol solution is near unity, estimate the composition of the vapor and activity coefficient of butanol that gives the correct bubble pressure and compare to the values estimated by UNIFAC.

14.6 Schulte et al.[8] discuss a linear solvation energy relationship (LSER) method for the partitioning of 41 environmentally important compounds between hexane + water phases at 25°C. The LSER method is based on the idea that contributions to the Gibbs excess energy (and to the logarithm of the partition coefficient) from effects like van der Waals forces and hydrogen bonding are independent of each other. Therefore, these contributions can be added up as separate linear contributions. We can test this hypothesis by plotting partition data for several compounds based on experimental data and LSER. We can also test the predictive capabilities of alternative theories by plotting their results with different curves. Table 14.2 presents the required parameters for the LSER method for several compounds. These parameters are to be substituted into the equation:

$$\log K_{H/W} = 0.404 + 5.382 \, v_i - 1.786 \, \pi_i + 0.856 \, \delta_i^S - 4.644 \, \beta_i^S - 3.078 \, \alpha_i^S$$

where v = volume parameter, π = polarity parameter, δ^S = polarizability parameter, β^S = hydrogen bond acceptor parameter, and α^S = hydrogen bond donor parameter. Compute the log partition coefficients for the following compounds by the LSER method and plot them against the experimental values listed in Table 14.2. Include predictions using the following methods. (Hint: compounds in the environment usually exist at ppm concentrations.)

(a) the MAB model.
(b) the SSCED model.
(c) the UNIFAC model.

Table 14.2 *LSER Parameters and Experimental Hexane + Water Partition Coefficients for Several Compounds*

Compd	logK	v	π	δ^S	β^S	α^S	δ	α	β
Phenol	−0.96	0.536	0.72	0.52	0.33	0.61	24.63	25.14	5.35
o-cresol	−0.12	0.637	0.68	1.0	0.41	0.54	22.87	27.15	2.17
2,4-dimethyl phenol	0.36	0.738	0.64	1.0	0.42	0.54	22.46	21.6	4.6
Benzaldehyde	0.36	0.606	0.92	1.39	0.44	0	21.60	0	8.80
p-chlorobenzaldehyde	1.60	0.698	0.92	1.39	0.40	0	20.34	0	7.65

8. 1998. *J. Chem. Eng. Data* 43:72.

Table 14.2 *LSER Parameters and Experimental Hexane + Water Partition Coefficients for Several Compounds (Continued)*

Compd	log$K$	ν	π	δ^S	β^S	α^S	δ	α	β
Benzene	2.06	0.491	0.59	0.68	0.10	0	18.74	0.63	2.24
Toluene	2.75	0.592	0.55	1.0	0.11	0	18.33	0.57	2.23

14.7 Predict the compositions of the coexisting liquid phases for the system methanol (1) + cyclohexane (2) at 298 K. Let α be the methanol-rich phase.

    (a) Use the MAB model.
    (b) Use the SSCED model.
    (c) Use the UNIFAC model.

14.8 Predict the compositions of the coexisting liquid phases for the system methanol (1) + cyclohexane (2) at 285.15 K and 310.15 K. Let α be the methanol-rich phase. Compare quickly with the data from Fig. 19.12 on page 804 and comment on the accuracy of the results. (Include a printout of your results including converged compositions and activity coefficients of both phases at one of the temperatures.)

    (a) Use the MAB model.
    (b) Use the SSCED model.
    (c) Use the UNIFAC model.

14.9 Benzene and water are virtually immiscible. What is the bubble pressure of an overall mixture that is 50 mol% of each at 75°C?

14.10 Water + hexane and water + benzene are immiscible pairs.

    (a) The binary system water + benzene boils at 69.4°C and 760 mmHg. What is the activity coefficient of benzene in water if the solubility at this point is $x_B = 1.6E{-}4$, using only this information and the Antoine coefficients?
    (b) What is the vapor composition at the bubble pressure at room temperature (292 K) for a ternary mixture consisting of 1 mole overall of each component if the organic layer is assumed to be an ideal solution?
    (c) What is the vapor composition at the bubble pressure at room temperature (292 K) for a ternary mixture consisting of 1 mole overall of each component if the activity coefficients of the organic layer are predicted by UNIFAC?

The following problems concern LLE in ternary systems. Experimental data for the systems are listed in Tables 14.3 and 14.4.

Table 14.3 *Water(1) + Methylethylketone(MEK)(2) + Acetic Acid(3) System at T = 299.85 K[a]*

α phase mol%			β phase mol%		
1	2	3	1	2	3
92.689	7.311	0.	36.383	63.617	0.
91.644	8.049	0.307	38.601	60.547	0.851
90.839	8.623	0.538	40.531	57.986	1.482
90.681	8.733	0.586	41.835	56.287	1.878

Table 14.3 *Water(1) + Methylethylketone(MEK)(2) + Acetic Acid(3) System at T = 299.85 K[a] (Continued)*

α phase mol%			β phase mol%		
1	2	3	1	2	3
89.325	9.717	0.958	44.866	52.640	2.494
88.631	10.228	1.140	49.069	47.994	2.937
88.084	10.669	1.247	52.180	44.769	3.051

a. Skrzec, A.E., Murphy, N.F. 1954. *Ind. Eng. Chem.* 46:2245.

Table 14.4 *1-Butanol(1) + water(2) + methanol(3) at 288.15 K[a] as reported by Mueller*

α phase mol%			β phase mol%		
1	2	3	1	2	3
2.115	95.071	2.813	45.254	51.598	3.148
2.319	92.876	4.804	41.183	53.459	5.358
2.548	91.304	6.148	35.997	56.276	7.727
2.966	89.460	7.574	30.372	59.851	9.777
4.043	86.874	9.083	23.296	65.170	11.534
5.171	85.094	9.736	18.482	69.704	11.814

a. Mueller, A.J., Pugsley, L.I., Ferguson, J.B. 1931. *J. Phys. Chem.* 35:1314.

14.11 Consider the system water(1) + MEK(2) at 299.85 K. The solubilities measured by Skrzec, A.E., Murphy, N.F., 1954. *Ind. Eng. Chem.*, 46:2245, are $x_1^\alpha = 0.927$ and $x_1^\beta = 0.364$. For a binary system, the LLE iteration procedure has been outlined in Example 14.4. Apply the procedure to determine the mutual solubilities predicted by UNIQUAC. The mixture parameters are $r = [0.92, 3.2479]$, $q = [1.40, 2.876]$, $a_{12} = -2.0882$ K, and $a_{21} = 345.53$ K. Let phase α be the water-rich phase.

14.12 For a binary system, iterations can be performed by finding a new value of $x_{1,new}^\alpha$ from only the K-ratios as shown in Eqn. 14.5. For a ternary system, we need at least one composition. Derive the iteration formula, using $x_{3,old}^\alpha$ as the specified composition

$$x_{1,new}^\alpha = \frac{1 - x_{3,old}^\alpha(K_{3,old} - K_{2,old}) - K_{2,old}}{K_{1,old} - K_{2,old}}.$$

14.13 Consider the system water(1) + methylethylketone(MEK)(2) + acetic acid(AA)(3) at 299.85 K. For a ternary LLE system, estimate tie lines at $x_3^\alpha = 0.005$, 0.01, 0.02, using UNIQUAC, where the parameter values are $r = [0.92, 3.2479, 2.2024]$, $q = [1.40, 2.876, 2.072]$, and the a values (in K) are $a_{12} = -2.0882$, $a_{21} = 345.53$, $a_{13} = 254.15$, $a_{31} = -301.02$, $a_{23} = -254.13$, $a_{32} = -4.5537$. Let α be the water-rich phase. Plot the results on rectangular coordinates, using x_1 as the abscissa and x_3 as the ordinate. Connect the tie lines on the plot. Add the experimental tie lines (Table 14.3) to the same plot using different symbols.

14.14 One mole of a stream containing pentane, acetone, methanol, and water in proportions $z =$ 0.75, 0.13, 0.11, 0.01, respectively, is to be mixed and decanted with 1 mol of pure water at 25°C. Estimate the partition coefficients for each of the components and the proportion of lower-phase/Feed where "Feed" includes both the pentane-rich and pure water streams with the specified model.

(a) The MAB model.
(b) The SSCED model.
(c) The UNIFAC model.

14.15 Calculate the LLE in the system 1-butanol(1) + water(2) + methanol(3) at 288.15 K, using UNIQUAC with the following parameters: $r =$ [3.4543, 0.92, 1.4311]; $q =$ [3.052, 1.4, 1.432]; and the a values (in K) are $a_{13} = 355.54$; $a_{31} = -164.09$; $a_{12} = -82.688$; $a_{21} = 443.56$; $a_{32} = -85.451$; $a_{23} = -321.92$; and compare graphically with the data from Table 14.4.

14.16 Solve problem 14.13, except use the specified model.

(a) The MAB model.
(b) The SSCED model.
(c) The UNIFAC model.

14.17 Consider the system water(1) + methylethylketone(MEK)(2) + propanoic acid(PA)(3). Use UNIFAC to predict the compositions for the coexisting phases at $x_3{}^\alpha = 0.01, 0.05,$ and 0.10 at 298.15 K. Let α be the water-rich phase. Plot the results on rectangular coordinates by using x_1 as the abscissa and x_3 as the ordinate. Connect the tie lines on the plot.

14.18 Solve problem 14.15, except use the specified model.

(a) The MAB model.
(b) The SSCED model.
(c) The UNIFAC model.

14.19 Derive the formulas for the spinodal curves of the Flory-Huggins model and plot the spinodals (T versus Φ^α, Φ^β) of several polystyrenes in cyclohexane using UNIFAC parameters to estimate the volume of polystyrene relative to ethylbenzene and taking the experimental values of solubility parameters and molar volumes for the species when $N_d = 1$. Take the degrees of polymerization of polystyrene to be 100, 200, 500, 1000. Plot the estimated reciprocal critical temperatures versus $N_d{}^{-1/2}$. Mark the infinite molecular weight critical temperature on both plots with a big **X**.

Solid-Liquid Behavior

14.20 In the treatment of solid-liquid equilibria, the effects of pressure on melting points are neglected.

(a) Draw a schematic of the Gibbs energy of liquid and solid phases versus pressure at constant temperature for a compound for which the molar volume of the solid is less than the molar volume of the liquid. Plot both curves on the same figure, and indicate the melting pressure. Most chemicals follow this type of behavior.
(b) For water, the molar volume of the solid is greater than the molar volume of the liquid. Sketch the Gibbs energy of liquid and solid phases as a function of pressure at constant temperature for this type of behavior. Plot both curves on the same figure, and indicate the melting pressure.

(c) Calculate the hypothetical Gibbs energy for melting solid naphthalene at 5 bar and the normal melting temperature, 80.2°C. You may assume that the liquid and solid are incompressible. Be sure to clearly specify the path you use for your calculation. $V^L = 133$ cm^3/gmole, $V^S = 124.8$ cm^3/gmole.

(d) Calculate the hypothetical Gibbs energy for melting solid naphthalene at 1 atm and 78°C. Compare the magnitude with the results of part (c) to verify that the pressure effects are small relative to temperature effects.

14.21 Generate a solid-liquid equilibrium T-x diagram for naphthalene(1) + biphenyl(2) assuming ideal solutions. What are the predicted eutectic temperature and composition? The experimental eutectic point is 39.4°C and x biphenyl = 0.555 (Lee, H.H., Warner, J.C., 1935. *J. Amer. Chem. Soc.* 57:318).

14.22 At 25°C, the solubility of naphthalene in *n*-hexane is 11.9 mol%. The liquid phase is nonideal. Use the simple solution model $G^E/RT = Ax_1x_2$ to predict the solubility at 10°C. (The experimental solubility at 10°C is 6.5mol%, Sunier, A. 1930. *J. Phys. Chem.* 34:2582).

14.23 Phenanthrene and anthracene are structurally very similar. Would you expect them to have similar solubilities in benzene at 25°C? Provide a quantitative answer, and an explanation.

14.24 Predict the solubility (in mole fraction) of phenol at the cited conditions using the specified model. (i) Use the MAB model. (ii) Use the SSCED model. (iii) Use the UNIFAC model.

(a) Solubility in *n*-heptane at 25°C.
(b) Solubility in ethanol at 25°C.
(c) Solubility in a 50/50 mole ratio of heptane and ethanol at 25°C.

14.25 A 50 wt% (22.5 mol%) solution of ethylene glycol + water freezes at about 240 K.

(a) What freezing temperature would be predicted by assuming that ethylene glycol and water form an ideal solution? The freezing occurs by formation of water crystals.
(b) Does your calculation indicate that the system has positive or negative deviations from Raoult's law? Why?

14.26 Determine the ideal solubility of naphthalene in any solvent at 40°C. Then predict the solubility and compare with the experimental solubility (shown in parentheses) for the specified solvent and specified model. (i) Use the MAB model. (ii) Use the SSCED model. (iii) Use the UNIFAC model.

(a) Methanol (4.4)
(b) Ethanol (7.3)
(c) 1-propanol (9.4)
(d) 2-propanol (7.6)
(e) 1-butanol (11.6)
(f) *n*-hexane (22.2)
(g) Cyclohexanol (22.5)
(h) Acetic acid (11.7)
(i) Acetone (37.8)
(j) Chloroform (47.3)

14.27 Determine the ideal solubility of anthracene in any solvent at 20°C. Then predict the solubility and compare with the experimental solubility (shown in parentheses) for the specified solvent and specified model. (i) Use the MAB model. (ii) Use the SSCED model. (iii) Use the UNIFAC model.

    (a) Acetone (0.31)
    (b) Chloroform (0.94)
    (c) Ethanol (0.05)
    (d) Methanol (0.02)

14.28 Determine the ideal solubility of phenanthrene in any solvent at 20°C. Then predict the solubility and compare with the experimental solubility (shown in parentheses) for the specified solvent and specified model. (i) Use the MAB model. (ii) Use the SSCED model. (iii) Use the UNIFAC model.

    (a) Acetone (14.5)
    (b) Chloroform (23.8)
    (c) Ethanol (1.23)
    (d) Acetic acid (1.92)
    (e) Methanol (0.64)

14.29 Determine the solubility curve for naphthalene in the specified solvent, and compare with the literature data:

    (a) Acetic acid[9]
    (b) *n*-hexane[9]
    (c) Cyclohexanol[10]
    (d) Acetone[9]
    (e) Chloroform[11]
    (f) Methanol[9]
    (g) *n*-butanol[9]
    (h) Ethanol[12]
    (i) *n*-propanol[12]
    (j) 2-propanol[12]

14.30 The gas condensate from a new gas well in Prudhoe Bay, Alaska has the following weight% of C5, C10, C15, C20, C25, C30, C35, C40, C45, C50, and >C50, respectively: 1, 4, 7, 10, 12, 12, 12, 12, 8, 8, and 14. Estimate the temperature at which wax may begin to precipitate from this liquid.

14.31 Generate an SLE phase diagram for phenol(1) + cyclohexane(2).

    (a) Assume an ideal solution.
    (b) Use MAB to model liquid phase nonidealities.
    (c) Use SSCED to model liquid phase nonidealities.
    (d) Use UNIFAC to model liquid phase nonidealities.
    (e) Make a comment about how the solubility of phenol in cyclohexane differs from the solubility in benzene at the same temperature. (See Fig. 14.12 on page 563.)

---

9. Ward, H. 1926. *J. Phys. Chem.* 30:1316.
10. Weissenberger, G. 1927. *Z. Agnew. Chem.* 40:776.
11. Hildebrand, J. 1920. *J. Am. Chem. Soc.* 42:2180.
12. Sunier, A. 1930. *J. Phys. Chem.* 34:2582.

14.32 Create a flow sheet analog to VLE or LLE calculations to find the melting temperature and liquid phase composition for a given solid mixture composition for the following.

(a) Ideal solutions of solid and liquid
(b) Nonideal solutions of solid and liquid

14.33 Create a flow sheet analog to VLE or LLE calculations to find the freezing temperature and solid composition for a given liquid composition when the liquids and solids form a non-ideal solution.

14.34 Create a flowsheet analog to VLE or LLE flash calculations to find the coexisting liquid and solid compositions that exist for a liquid-solid mixture of specified overall composition that is between the conditions of first freezing and first melting.

14.35 Salicylic acid is similar in structure to aspirin. Shalmashi et al.[13] have measured the data in Table 14.5.

(a) Find the value of α for salicylic acid in water that best correlates the data, assuming $\beta = 0$.
(b) Predict the solubility of the acid in ethanol.
(c) Plot $\log(x_{acid})$ versus $1000/T$ including correlated and measured values.
(d) Plot all the calculated versus experimental values. This is known as a parity plot.

Table 14.5 *Solubility of Salicylic Acid in Various Solvents. Compositions in Weight%*

Solvent\T(°C)	25	35	45	55	65	75
Water	0.22	0.35	0.54	0.74	1.16	1.71
Ethanol	32.54	38.51	42.74	47.79	52.52	56.98
Ethyl Acetate	0.30	0.49	0.77	1.26	1.80	2.50
Carbon Tetrachloride	20.08	23.58	26.92	30.10	33.47	36.92
p-Xylene	0.62	1.10	1.75	2.70	4.01	5.78

14.36 Sometimes we would like to enhance the solubility of a drug by adding a cosolvent, instead of adding antisolvent to precipitate. Making optimal use of the data in the previous problem, estimate the amount of ethanol that should be added to water to prepare an aqueous solution of salicylic acid with concentration of 10 wt% at 25°C.

14.37 Yalkowsky and Rubino (1985)[14] have observed roughly linear behavior for logarithmic solubility in mixed solvents when plotted as volume fraction of the solvent/cosolvent. That is, $\log(x_i) = \Phi_1'\log(x_{i,1}) + \Phi_2'\log(x_{i,2})$ where $x_{i,j}$ is the solubility in the jth pure solvent and Φ_i' is the volume fraction on a solute-free basis; for example, $\Phi_E' = \Phi_E/(\Phi_E + \Phi_W)$. Use the SSCED model to make your best estimates of the solubility of salicylic acid at 318 K in a range of mixtures with ethanol/water mole fraction ratios of 3/1, 1/1, and 1/3. Plot your results as $\log_{10}(x_{acid})$ versus Φ_E'. Include the experimental results and the guideline of Yalkowsky and Rubino. Comment on whether the SSCED model is consistent with the observation of Yalkowsky and Rubino.

13. Shalmashi A., Eliassi, A. 2008. *J. Chem. Eng. Data* 53:199-200.
14. Connors, K.A, 2002. *Thermodynamics of Pharmaceutical Systems: An Introduction for Students of Pharmacy*, Hoboken, NJ: Wiley. p. 131.

CHAPTER 15

PHASE EQUILIBRIA IN MIXTURES BY AN EQUATION OF STATE

The whole is simpler than the sum of its parts.

J.W. Gibbs

Suppose it was required to estimate the vapor-liquid *K*-ratio of methane in a mixture at room temperature. For an initial guess, we might assume it follows ideal-solution behavior. It is a relatively simple molecule (e.g., no polar moments, no hydrogen bonding). But we cannot use Raoult's law because the required temperature is well above the critical temperature. We could use Henry's law, or the SCVP+ model (Section. 11.12), but the assumption of low concentrations may be inappropriate at very high pressures. The equation of state method discussed here is an attractive alternative.

We begin this chapter with a review of the mixing rules introduced in Section 12.1. Then we show how the mixing rule leads to the fugacities and *K*-ratios needed for VLE calculations. We then provide algorithms and illustrate how VLE calculations are programmed using an equation of state (EOS). Finally we provide some insight into how critical behavior in mixtures differs from critical behavior in pure components, and that some "counterintuitive" behavior can exist, such as quality that decreases when pressure is increased.

Chapter Objectives: You Should Be Able to...

1. Compute VLE phase diagrams using an EOS.

2. Characterize adjustable parameters in EOS models using experimental data.

3. Derive an expression for a fugacity coefficient given an arbitrary EOS and mixing rules.

4. Comment critically on the merits and limitations of the PREOS relative to the activity models of Chapters 11–13, including the ability to suggest ways that the PREOS can be systematically improved.

15.1 MIXING RULES FOR EQUATIONS OF STATE

Virial Equation of State

The virial equation was introduced for pure fluids in Section 7.4. Previously, we have also given a strategy for relating parameters to composition in Section 12.1. If we extend this mixing rule to the virial equation,

$$B = \sum_i \sum_j x_i x_j B_{ij}$$

15.1

which for a binary mixture becomes

$$B = y_1^2 B_{11} + 2 y_1 y_2 B_{12} + y_2^2 B_{22}$$

15.2

Similar to our previous discussion, it is understood that B_{12} is equivalent to B_{21}. *The cross coefficient B_{12} is not the virial coefficient for the mixture.*

❶ *Combining rules are used to quantify the parameters that represent unlike molecule interactions.*

To obtain the cross coefficient, B_{12}, we must create a **combining rule** to propose how the cross coefficient depends on the properties of the pure components 1 and 2. For the virial coefficient, the relationship between the pair potential and the virial coefficient was given in Section 7.11. However, a less rigorous method is often used in engineering applications. Rather, combining rules are created to use the **corresponding state** correlations developed for pure components in terms of T_{c12} and P_{c12}. The combining rules used to determine the values of the cross coefficient critical properties are:

$$T_{c12} = (T_{c1} T_{c2})^{1/2} (1 - k'_{12})$$

15.3

❶ *Binary interaction parameters are used to adjust the combining rule to better fit experimental data, if available.*

The parameter k'_{12} is an adjustable parameter (called the **binary interaction parameter**) to force the combining rules to more accurately represent the cross coefficients found by experiment.[1] However, in the absence of experimental data, it is customary to set $k'_{12} = 0$.

$$V_{c12} = \left(\frac{V_{c1}^{1/3} + V_{c2}^{1/3}}{2} \right)^3$$

15.4

$$Z_{c12} = \frac{1}{2} (Z_{c1} + Z_{c2})$$

15.5

and

$$\omega_{12} = \frac{1}{2} (\omega_1 + \omega_2)$$

15.6

The first three of these combining rules lead to:

$$P_{c12} = Z_{c12} R T_{c12} / V_{c12}$$

15.7

---

1. Reid, R., Prausnitz, J.M., Poling, B. 1987. *The Properties of Gases and Liquids.* 4th ed. New York, NY: McGraw-Hill, p. 133.

Then, T_{c12}, P_{c12}, and ω_{12} are used in the virial coefficient correlation presented in Chapter 7 to obtain B_{12} (Eqns. 7.6–7.10) which is subsequently incorporated into the equation for the mixture. If Z_c (or V_c) is not available, it may be estimated using $Z_c = 0.291 - 0.08\omega$. The virial equation for a binary mixture is implemented on the spreadsheet Virialmx.xlsx furnished with the text.

Virialmx.xlsx.

Example 15.1   The virial equation for vapor mixtures

Calculate the molar volume for a 60 mole% mixture of neopentane(1) in CO_2(2) at 310 K and 0.2 MPa.

Solution: The conditions are entered in the spreadsheet Virialmx.xlsx, with the following results:

```
X Microsoft Excel - Virialmx.xls                                    _ □ X
 File  Edit  View  Insert  Format  Tools  Data  Window  Help        _ 8 X
      A      B         C         D        E        F       G       H      I     J
  2      Virial Equation for a Mixture
  3
  4               T (K)              310
  5               P (MPa)            0.2
  6               kij                0
  7
  8      Compound   Tc (K)   Pc (MPa)    ω    Vc (cm³/mol)  Zc      Tr      Pr    criteria
  9      Neopentane  433.8    3.199    0.196    303.28    0.269   0.7146  0.0625  0.00117
 10      CO2(2)      304.2    7.382    0.228     93.87    0.274   1.0191  0.0271  0.32117
 11      (1)-(2)     363.27   4.59     0.212    178.62    0.2715  0.8534  0.0436
 12
 13
 14      Compound    Bᵁ        Bᴵ      BPc/RTc  Bij(cm³/mol)      Bij matrix
 15      Neopentane -0.6394321 -0.5663837 -0.7504433 -846.06     -846.06  -330.50
 16      CO2(2)     -0.3264383 -0.0198832 -0.3309717 -113.39     -330.50  -113.39
 17      (1)-(2)    -0.460867  -0.195774  -0.5023711 -330.50
 |◄ ◄ ► ►|\ Introduction \ Virial Mix /        |◄|              ►|
```

The original spreadsheet is modified slightly for this solution. Cells J9 and J10 are programmed with a rearranged form of Eqn. 7.10, $T_r - 0.686 - 0.439P_r$, and if these cells are positive, then the virial equation is suitable. The critical volume is calculated from T_c, P_c, and Z_c. Cells F15–F17 list the virial coefficients for neopentane, CO_2, and the cross coefficient, respectively.

The virial coefficient for the mixture is given by Eqn. 15.1,
$$B = 0.6^2 \cdot (-846.06) + 2(0.6)(0.4)(-330.5) + 0.4^2 \cdot (-113.39) = -481.36 \text{ cm}^3/\text{mol}$$
$$V = RT/P + B = 8.314 \cdot 310/0.2 - 481.36 = 12{,}405 \text{ cm}^3/\text{mol}$$
The volumetric behavior of the mixture depends on composition. The mixture volume differs from an ideal solution, $V^{is} = \sum_i x_i V_i$. The difference $V - V^{is}$ is called the excess volume, V^E.

The molar volume of pure neopentane is
$$V = RT/P + B = 8.314 \cdot 310/0.2 - 846.1 = 12{,}041 \text{ cm}^3/\text{mol}$$

> **Example 15.1 The virial equation for vapor mixtures (Continued)**
>
> The molar volume of pure CO_2 is
> $$V = RT/P + B = 8.314 \cdot 310/0.2 - 113.4 = 12{,}773 \text{ cm}^3/\text{mol}$$
> The molar volume of an ideal solution of a 60 mole% neopentane mixture is
> $$V^{is} = 0.6(12{,}041) + 0.4(12{,}773) = 12{,}334 \text{ cm}^3/\text{mol}$$
> and the excess volume is
> $$V^E = 12{,}405 - 12{,}334 = 71.2 \text{ cm}^3/\text{mol}.$$
> The molar volume and excess volume can be determined across the composition range by changing y's in the formulas.

Cubic Equations of State

The customary mixing rules for cubic equations of state have been introduced in Section 12.1:

$$a = \sum_i \sum_j x_i x_j a_{ij} \quad \text{and} \quad b = \sum_i x_i b_i \qquad 15.8$$

Note the mathematical similarity of the mixing rule for a with the mixing rule used for the virial coefficient. All of the compositional dependence of the equation of state is incorporated into the two relations. A combining rule is not necessary for the b term, however the a term does require a combining rule. The customary combining rule is

$$a_{ij} = (1 - k_{ij})(a_{ii} a_{jj})^{1/2} \qquad 15.9$$

where k_{ij} is referred to as a **binary interaction parameter.** This is similar to the form of the geometric mean rule for critical temperatures used for virial coefficients. The adjustable parameter k_{ij} is used to adjust the combining rule to fit experimental data more closely. Technically, this just transfers our ignorance into the adjustable parameter k_{ij}. Values for k_{ij} for various binary combinations are tabulated in the literature.[2]

In the absence of experimental data or literature values for k_{ij}, we may make a first-order approximation by letting $k_{ij} = 0$. This approximation serves our purpose nicely, because the equation of state approach then requires no more information than the ideal solution approach (T_c, P_c, ω, T, P, x, y), but it offers the possibility of more realistic representation of the phase diagram because of the more fundamental molecular basis. We can demonstrate this improved accuracy by considering some examples.

❶ Review of the concepts from Section 10.8 may help put the approaches in context. Keep in mind that the objective is still to perform bubble, dew and flash calculations, but after relaxing the ideal solution assumption.

15.2 FUGACITY AND CHEMICAL POTENTIAL FROM AN EOS

We begin with a reminder that for phase equilibria calculations, that the fugacities of components are needed. The tool that we need for VLE calculations is the K-ratio and an expression for the component fugacity. In Section 10.9 we demonstrated that the component fugacity for an ideal gas component is equal to the partial pressure. In this chapter we develop a method of "correcting" the partial pressure to provide the fugacity. As a variation of the Venn diagram presented in Fig. 11.8,

---

2. Reid, R., Prausnitz, J.M., Poling, B. 1987. *The Properties of Gases and Liquids*. 4th ed. New York, NY: McGraw-Hill, p. 83.

we present the schematic shown in Fig. 15.1. Because the equation of state is capable of representing liquid phases by using the smaller root, we show both vapor and liquid phases.

The method of deriving the fugacity is an extension of Eqn. 10.39. If we compare the chemical potential in the real mixture to the chemical potential for an ideal gas, we see that the difference is given by the component derivative of the Gibbs departure.

$$\frac{\mu_i(T,P) - \mu_i^{ig}(T,P)}{RT} = \left(\frac{\partial \underline{G}/RT}{\partial n_i}\right)_{T,P,n_{j \neq i}} - \left(\frac{\partial \underline{G}^{ig}/RT}{\partial n_i}\right)_{T,P,n_{j \neq i}} = \left(\frac{\partial (\underline{G} - \underline{G}^{ig})/RT}{\partial n_i}\right)_{T,P,n_{j \neq i}}$$

15.10

We have seen the Gibbs departure in Eqns. 9.23 and 9.31. For the virial equation, we have

$$\frac{G - G^{ig}}{RT} = \frac{BP}{RT}$$

15.11

where we recognize that the virial coefficient depends on composition via Eqn. 15.2. By differentiation of this expression, we obtain the chemical potential. We can calculate the component fugacity if we use Eqn. 11.22 and replace the standard state with the ideal gas mixture state. Since the component fugacity in the ideal gas state is the partial pressure, the fugacity coefficient becomes

$$\boxed{\ln\left(\frac{\hat{f}_i}{y_i P}\right) = \frac{(\mu_i - \mu_i^{ig})}{RT} = \left(\frac{\partial (\underline{G} - \underline{G}^{ig})/(RT)}{\partial n_i}\right)_{T,P,n_{j \neq i}}}$$

15.12

General form of fugacity coefficient in a mixture useful for EOSs of the form $Z(T,P)$.

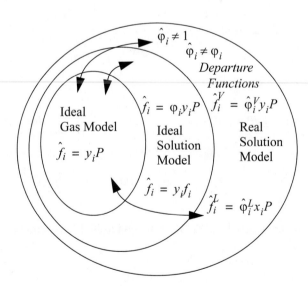

Figure 15.1 *Schematic showing the equation of state approach to modeling fugacities of components. Departure function (fugacity coefficient) methods are used for both the vapor and liquid phases. Superscripts are used to distinguish the fugacity coefficients of each phase. Liquid-phase compositions are conventionally denoted by x_i and vapor-phase compositions by y_i.*

We define the ratio of the component fugacity to the partial pressure (ideal gas component fugacity) as the component fugacity coefficient.

Component fugacity coefficient.

$$\boxed{\hat{\varphi}_i \equiv \frac{\hat{f}_i}{y_i P}} \quad \text{or} \quad \boxed{\hat{f}_i = y_i \hat{\varphi}_i P}$$

15.13

Differentiation of the Gibbs departure leads to the component fugacity coefficients for a binary,

Fugacity coefficient for virial equation of state.

$$\boxed{\ln \hat{\varphi}_1 = [2y_1 B_{11} + 2y_2 B_{12} - B]P/RT} \quad \boxed{\ln \hat{\varphi}_2 = [2y_1 B_{12} + 2y_2 B_{22} - B]P/RT}$$

15.14

which will be shown in more detail later. The fugacity coefficient of a component in a mixture may be directly determined at a given T and P by evaluating the virial coefficients at the temperature, then using this equation to calculate the fugacity coefficient.

Differentiation of the Gibbs departure function is difficult for a pressure-explicit equation of state like the Peng-Robinson equation of state. The difficulty arises because the Gibbs departure function is given in terms of volume and temperature rather than pressure (Eqns. 8.36 and 9.33), and differentiation at constant pressure as required by Eqn. 15.12 is difficult. As in the case of pure fluids, classical thermodynamics provides the means to solve this problem. Instead of differentiating the Gibbs departure function, we differentiate the Helmholtz departure function. Recalling,

$$dG = \underline{V}dP - \underline{S}dT + \sum_i (\partial \underline{G}/\partial n_i)_{T, P, n_{j \neq i}} dn_i$$

and noting,

$$d\underline{A} = -Pd\underline{V} - \underline{S}dT + \sum_i (\partial \underline{A}/\partial n_i)_{T, \underline{V}, n_{j \neq i}} dn_i$$

we also use $A = G - PV$, or $dA = dG - d(PV)$:

$$d\underline{A} = d\underline{G} - Pd\underline{V} - \underline{V}dP = \underline{V}dP - \underline{S}dT + \sum_i (\partial \underline{G}/\partial n_i)_{T, P, n_{j \neq i}} dn_i - Pd\underline{V} - \underline{V}dP$$

$$\Rightarrow -Pd\underline{V} - \underline{S}dT + \sum_i (\partial \underline{A}/\partial n_i)_{T, \underline{V}, n_{j \neq i}} dn_i = -Pd\underline{V} - \underline{S}dT + \sum_i (\partial \underline{G}/\partial n_i)_{T, P, n_{j \neq i}} dn_i$$

Equating coefficients of dn_i we see an alternative method to find the chemical potential,

$$(\partial \underline{A}/\partial n_i)_{T, \underline{V}, n_{j \neq i}} = (\partial \underline{G}/\partial n_i)_{T, P, n_{j \neq i}} = \mu_i(T,P)$$

15.15

Note that T, P, and $\underline{V}$ identify the same conditions for the real fluid. Therefore, when we evaluate the departure, the ideal gas state must be corrected from V^{ig} to V,

$$\frac{\mu_i(T,P) - \mu_i^{ig}(T,P)}{RT} = \left(\frac{\partial A/RT}{\partial n_i}\right)_{T,\underline{V},n_{j\neq i}} - \left(\frac{\partial A^{ig}/RT}{\partial n_i}\right)_{T,\underline{V}^{ig},n_{j\neq i}}$$

$$= \left(\frac{\partial (A - A^{ig})_{TV}/RT}{\partial n_i}\right)_{T,\underline{V},n_{j\neq i}} - \ln Z \qquad 15.16$$

where the notation $(A - A^{ig})_{TV}$ denotes a departure function at the same T, V, which is the integral of Eqn. 8.27. The last term, $\ln Z$, represents the correction of the ideal gas Helmholtz energy from V to V^{ig}. Careful inspection of the true form on the integral leading to $\ln Z$ should convince you that differentiation does not change this term, and only the integral for the departure in Eqn. 15.16 must be differentiated.

Therefore, the fugacity coefficient is calculated using

> **!** General form for fugacity coefficient for a pressure explicit equation of state such as the Peng-Robinson, that is, $Z(T,\rho)$.

$$\boxed{\ln(\hat{\varphi}_i) = \frac{(\mu_i - \mu_i^{ig})}{RT} = \left(\frac{\partial (A - A^{ig})_{TV}/(RT)}{\partial n_i}\right)_{T,\underline{V},n_{j\neq i}} - \ln Z} \qquad 15.17$$

To apply this, consider the Peng-Robinson equation as an example.

$$\frac{(A - A^{ig})_{TV}}{(RT)} = -\ln(1 - B/Z) - \frac{A}{B\sqrt{8}}\ln\left(\frac{Z + (1 + \sqrt{2})B}{Z + (1 - \sqrt{2})B}\right)$$

$$= -\ln(1 - b\rho) - \frac{a}{bRT\sqrt{8}}\ln\left(\frac{1 + (1 + \sqrt{2})b\rho}{1 + (1 - \sqrt{2})b\rho}\right)$$

By extending the method of reducing the equation of state parameters developed in Eqns. 7.21 and 7.21, $A^V = \sum_i \sum_j y_i y_j A_{ij}$ and $B^V = \sum_i y_i B_i$, where $A_{ij} = \sqrt{A_{ii}A_{jj}}(1 - k_{ij})$. Then, differentiation as we will show in Example 15.5 on page 592, yields for a binary system

> **!** Fugacity coefficient for the Peng-Robinson equation of state in dimensionless form.

$$\boxed{\begin{aligned}\ln\left(\frac{\hat{f}_i^V}{y_i P}\right) &= \frac{B_i}{B^V}(Z^V - 1) - \ln(Z^V - B^V) - \\ &\quad \frac{A^V}{B^V\sqrt{8}}\left(\frac{2(y_1 A_{i1} + y_2 A_{i2})}{A^V} - \frac{B_i}{B^V}\right)\ln\left(\frac{Z^V + (1 + \sqrt{2})B^V}{Z^V + (1 - \sqrt{2})B^V}\right)\end{aligned}} \qquad 15.18$$

As we saw in the case of equations of state for pure fluids, there is no fundamental reason to distinguish between the vapor and liquid phases except by the magnitude of Z. The equation of state approach encompasses both liquids and vapors very simply. We replace the vapor phase mole fractions with liquid phase mole fractions in all formulas including those for A and B, resulting in

$$\ln\left(\frac{\hat{f}_i^L}{x_i P}\right) = \frac{B_i}{B^L}(Z^L - 1) - \ln(Z^L - B^L) -$$

$$\frac{A^L}{B^L\sqrt{8}}\left(\frac{2(x_1 A_{i1} + x_2 A_{i2})}{A^L} - \frac{B_i}{B^L}\right)\ln\left(\frac{Z^L + (1 + \sqrt{2})B^L}{Z^L + (1 - \sqrt{2})B^L}\right)$$

15.19

Recalling that $\hat{f}_i^V = \hat{f}_i^L$ at equilibrium, we write the equality and rearrange to find the expression for the K-ratio used to solve VLE problems.

$$y_i \hat{\varphi}_i^V P = x_i \hat{\varphi}_i^L P \quad \text{or} \quad K_i = \hat{\varphi}_i^L / \hat{\varphi}_i^V$$

15.20

❗ Eqn. 15.20 provides the primary equations for VLE via equations of state. Different equations of state provide different formulas for $\hat{\varphi}_i$.

Given K_i for all i, it is straightforward to solve VLE problems using the same procedures as for ideal solutions.

Note: Eqns. 15.20 provide the primary equations for VLE via equations of state. These equations are implemented by iteration procedures summarized in Appendix C. Only the bubble method will be presented in the chapter in detail. Although cubic equations can represent both vapor and liquid phases, note that the virial equation cannot be used for liquid phases.

Bubble-Pressure Method

For a bubble-pressure calculation, the T and all x_i are known as shown in Table 10.1 on page 373. Like the simple calculation performed in the preceding chapter, the criterion for convergence is

$\sum_i y_i = 1$ which needs to be expressed in terms of variables for the current method. Rearranging

Eqn. 15.20, this sum becomes $\sum_i \frac{x_i \hat{\varphi}_i^L P}{\hat{\varphi}_i^V P} = \sum_i x_i K_i = 1$. Unlike the activity model calculations,

we cannot explicitly solve for pressure because all $\hat{\varphi}_i^L$ and $\hat{\varphi}_i^V$ depend on pressure. Additionally,

all $\hat{\varphi}_i^V$ depend on composition of the vapor phase, which is not exactly known until the problem is solved. Typically, we use Raoult's law with the shortcut vapor pressure equation for the first guesses of y_i and P. From these values, we determine all K_i and check the sum of y values. If the sum is greater than one, the pressure guess is increased, if less than one, the pressure guess is decreased. A complete flowchart and example will be discussed in Section 15.4, but for now, let us explore the methods for calculating the fugacity coefficients.

As we observed for pure fluids, it is important to select the proper root when applying an equation of state. Considering the Workbook Prfug.xlsx, for the one-root region, we should select that row for the fugacity coefficients. For the three-root region, we should choose the root with the lowest mixture fugacity. At low pressure and near room temperature, systems are usually in the three-root region for both liquid and vapor compositions, but that may change as we approach the critical region. The number of roots depends on composition as well as T and P. For example, it often occurs that one root occurs using the vapor composition when one component is supercritical in equilibrium with a liquid phase. This means we need to select among at least four possibilities for each phase when computing the K-values: largest Z root for vapor composition, smallest Z root with liquid composition, single root with vapor composition, single root with liquid composition. If we compute K-values with all four ratios, only one of the possibilities provides meaningful results and these are the ones to apply in the next iteration.

Example 15.2 K-values from the Peng-Robinson equation

The bubble-point pressure of an equimolar nitrogen (1) + methane (2) system is to be calculated by the Peng-Robinson equation and compared to the shortcut K-ratio estimate at 100 K. The shortcut K-ratio estimate will be used as an initial guess: $P = 0.4119$ MPa, $y_{N_2} = 0.958$. Apply the formulas for the fugacity coefficients to obtain an estimate of the K-values for nitrogen and methane and evaluate the sum of the vapor mole fractions based on this initial guess.

Prfug.xlsx may be helpful in following this example.

Solution:

	T_c(K)	P_c(MPa)	ω
N_2	126.1	3.394	0.040
CH_4	190.6	4.604	0.011

The spreadsheet Prfug.xlsx may be used to follow the calculations. The K-values using the vapor root with vapor composition and liquid root with liquid composition are valid throughout the iterations of this example. From the shortcut calculation, $P = 0.4119$ MPa at 100 K. Applying Eqns. 7.21 and 7.22 for the pure component parameters:

$$\text{For } N_2: A_{11} = 0.09686; \; B_1 = 0.011906;$$
$$\text{For } CH_4: A_{22} = 0.18242; \; B_2 = 0.013266$$
$$\text{By the square-root combining rule Eqn. 15.9: } A_{12} = 0.13293$$

Based on the vapor composition of the shortcut estimate at $y_1 = 0.958$, the mixing rule gives $A^V = 0.099913$; $B^V = 0.01196$; Solving the cubic for the vapor root at this composition gives $Z^V = 0.9059$.

$$\ln \frac{\hat{f}_i^V}{y_i P} = \frac{B_i}{B^V}(Z^V - 1) - \ln(Z^V - B^V)$$

$$- \frac{A^V}{B^V \sqrt{8}} \ln\left(\frac{Z^V + (1 + \sqrt{2})B^V}{Z^V + (1 - \sqrt{2})B^V}\right)\left(\frac{2(y_1 A_{i1} + y_2 A_{i2})}{A^V}\right) - \frac{B_i}{B^V}$$

Then

$$\ln \hat{\varphi}_1^V = \frac{0.011906}{0.01196}(0.9059 - 1) - \ln(0.9059 - 0.01196) -$$

$$\frac{0.099913}{0.01196 \cdot 2.8284} \ln\left(\frac{0.9059 + 2.414 \cdot 0.01196}{0.9059 - 0.4142 \cdot 0.01196}\right)$$

$$\left(\frac{2(0.958 \cdot 0.09686 + 0.042 \cdot 0.13293)}{0.099913} - \frac{0.011906}{0.01196}\right)$$

Example 15.2 K-values from the Peng-Robinson equation (Continued)

$$= -0.08756, \ \hat{\varphi}_1^V = 0.9162$$

Many of the terms are the same for the methane in the mixture:

$$\ln \hat{\varphi}_2^V = \frac{0.013266}{0.01196}(0.9059 - 1) - \ln(0.9059 - 0.01196) -$$

$$\frac{0.099913}{0.01196 \cdot 2.8284} \ln\left(\left(\frac{0.9059 + 2.414 \cdot 0.01196}{0.9059 - 0.4142 \cdot 0.01196}\right) \cdot \right.$$

$$\left.\left(\frac{2(0.958 \cdot 0.13293 + 0.042 \cdot 0.18242)}{0.099913} - \frac{0.013266}{0.01196}\right)\right)$$

$$= -0.16571, \ \hat{\varphi}_2^V = 0.8473$$

To save some tedious calculations, the liquid formulas have already been applied at $x = \{0.5, 0.5\}$ to obtain: $\hat{\varphi}_1^L = 1.791$; $\hat{\varphi}_2^L = 0.0937$. Determining the K values,

$$K_1 = \frac{\hat{\varphi}_1^L}{\hat{\varphi}_1^V} = \frac{1.791}{0.9162} = 1.955, \quad K_2 = \frac{\hat{\varphi}_2^L}{\hat{\varphi}_2^V} = \frac{0.0937}{0.8473} = 0.1106$$

$$y_1 = 0.5 \cdot 1.955 = 0.978; y_2 = 0.5 \cdot 0.1106 = 0.055; \sum_i y_i = 1.033$$

A higher guess for P would be appropriate for the next iteration in order to make the K-values smaller. $\hat{\varphi}_1^L$ and $\hat{\varphi}_2^L$ would need to be evaluated at the new pressure. The calculations are obviously tedious. K_i calculations are possible in Excel by first copying the "Fugacities" sheet on Prfug.xlsx, using one sheet for liquid and the other for vapor, and then referencing cells on one of the sheets to calculate the K_i. We provide an example of such an arrangement in Prmix.xlsx. More details on the entire procedure will follow in Section 15.4.

15.3   DIFFERENTIATION OF MIXING RULES

Since the compositional dependence is within the mixing rule, if we understand how to differentiate the general mixing rules, then we can easily apply them to the models that use them.

Since a compositional derivative is necessary to obtain the partial molar quantities, and the compositions are present in summation terms, we must understand the procedures for differentiation of the sums. Since *all of the compositional dependence* is embedded in these terms, if we understand how these terms are handled, we can then apply the results to *any* equation of state. Only three types of sums appear in most forms of equations of state, which have been introduced above. The first type of derivative we will encounter is of the form

$$\left(\frac{\partial nb}{\partial n_k}\right)_{T, \underline{V}, n_{j \neq k}}$$

15.21

where $b = \sum_i y_i b_i$. For a binary $nb = n_1 b_1 + n_2 b_2$, and k will be encountered once in the sum, whether $k = 1$ or $k = 2$, thus:

$$\left(\frac{\partial nb}{\partial n_1}\right)_{T, \underline{V}, n_2} = b_1 \qquad \left(\frac{\partial nb}{\partial n_2}\right)_{T, \underline{V}, n_1} = b_2 \qquad \text{15.22}$$

and the general result is

$$\left(\frac{\partial nb}{\partial n_k}\right)_{T, \underline{V}, n_{j \neq k}} = b_k \qquad \text{15.23}$$

The second type of derivative which we will encounter is of the form

$$\left(\frac{\partial n^2 a}{\partial n_k}\right)_{T, \underline{V}, n_{j \neq k}} \qquad \text{15.24}$$

$n^2 a$ may be written as $n^2 \sum_i \sum_j x_i x_j a_{ij}$. For a binary mixture, $n_1^2 a_{11} + 2 n_1 n_2 a_{12} + n_2^2 a_{22}$. Taking the appropriate derivative,

$$\left(\frac{\partial n^2 a}{\partial n_1}\right)_{T, \underline{V}, n_2} = 2 n_1 a_{11} + 2 n_2 a_{12} \quad \text{and} \quad \left(\frac{\partial n^2 a}{\partial n_2}\right)_{T, \underline{V}, n_1} = 2 n_1 a_{12} + 2 n_2 a_{22} \qquad \text{15.25}$$

The general result is

$$\left(\frac{\partial n^2 a}{\partial n_k}\right)_{T, \underline{V}, n_{j \neq k}} = 2 \sum_j n_j a_{jk} \qquad \text{15.26}$$

For the virial equation, we need to differentiate a function that will look like:

$$\left(\frac{\partial nB}{\partial n_k}\right)_{T, n_{j \neq k}} = \left(\frac{\partial \left(\frac{1}{n}\right) \left(\sum_i \sum_j n_i n_j B_{ij}\right)}{\partial n_k}\right)_{T, P, n_{j \neq k}} \qquad \text{15.27}$$

Differentiation by the product rule gives

$$\frac{1}{n} \left(\frac{\partial \left(\sum_i \sum_j n_i n_j B_{ij}\right)}{\partial n_k}\right)_{T, P, n_{j \neq k}} - \frac{\sum_i \sum_j n_i n_j B_{ij}}{n^2} \qquad \text{15.28}$$

The double sum in the derivative is $n^2 B$ which we have evaluated in equivalent form in Eqn. 15.25. The second term is just B given by Eqn. 15.1. Therefore, we have for a binary mixture

$$\left(\frac{\partial nB}{\partial n_1}\right)_{T,P,n_2} = \left(\frac{\partial\left(\frac{1}{n}\right)\left(\sum_i\sum_j n_i n_j B_{ij}\right)}{\partial n_1}\right)_{T,P,n_2} = 2y_1 B_{11} + 2y_2 B_{12} - B \quad \text{and}$$

$$\left(\frac{\partial nB}{\partial n_2}\right)_{T,P,n_1} = \left(\frac{\partial\left(\frac{1}{n}\right)\left(\sum_i\sum_j n_i n_j B_{ij}\right)}{\partial n_2}\right)_{T,P,n_1} = 2y_1 B_{12} + 2y_2 B_{22} - B \qquad 15.29$$

The general result is

$$\left(\frac{\partial\left(\frac{1}{n}\right)\left(\sum_i\sum_j n_i n_j B_{ij}\right)}{\partial n_k}\right)_{T,P,n_{j\ne k}} = 2\sum_j y_j B_{jk} - B \qquad 15.30$$

Example 15.3 Fugacity coefficient from the virial equation

For moderate deviations from the ideal-gas law, a common method is to use the virial equation given by:

$$Z = 1 + BP/RT$$

where $B = \sum_i\sum_j y_i y_j B_{ij}$. Develop an expression for the fugacity coefficient.

Solution: For the virial equation, we have the result of Eqn. 9.30, $\dfrac{G - G^{ig}}{RT} = \ln\varphi = \dfrac{BP}{RT}$

Applying Eqn. 15.12

$$\left(\frac{\partial(G - G^{ig})/RT}{\partial n_k}\right)_{T,P,n_{j\ne k}} = \frac{P}{RT}\left(\frac{\partial\left[n\sum_i\sum_j y_i y_j B_{ij}\right]}{\partial n_k}\right)_{T,P,n_{j\ne k}}$$

the argument we need to differentiate looks like $n\sum_i\sum_j y_i y_j B_{ij} = \dfrac{1}{n}\sum_i\sum_j n_i n_j B_{ij}$.

Example 15.3 Fugacity coefficient from the virial equation (Continued)

Differentiation has been performed in Eqn. 15.29, which we can generalize as

$$\ln\frac{\hat{f}_k}{y_k P} = \left(2\left(\sum_j y_j B_{jk}\right) - B\right)\frac{P}{RT} \qquad 15.31$$

which has been shown earlier for a binary in Eqn. 15.14.

Example 15.4 Fugacity coefficient from the van der Waals equation

Van der Waals' equation of state provides a simple but fairly accurate representation of key equation of state concepts for mixtures. The main manipulations developed for this equation are the same for other equations of state but the algebra is a little simpler. Recalling van der Waals' equation from Chapter 6,

$$Z = \frac{1}{1-b\rho} - \frac{a\rho}{RT} = 1 + \frac{b\rho}{1-b\rho} - \frac{a\rho}{RT}$$

where $a = \sum_i \sum_j y_i y_j a_{ij}$ and $b = \sum_i y_i b_i$. Develop an expression for the fugacity coefficient.

Solution: We need to apply Eqn. 15.17. For the departure, we apply Eqn. 8.27 because the differentiation indicated above is performed at constant volume, not constant pressure.

$$\frac{(\underline{A} - \underline{A}^{ig})_{TV}}{RT} = \int_0^{b\rho} (Z-1)\frac{d(b\rho)}{b\rho} = \int_0^{b\rho} \left(\frac{b\rho}{1-b\rho} - \frac{a}{bRT}b\rho\right)\frac{d(b\rho)}{b\rho} = -\ln(1-b\rho) - \frac{a}{bRT}b\rho$$

$$\frac{(\underline{A} - \underline{A}^{ig})_{TV}}{RT} = -n\ln(1-b\rho) - \frac{an^2}{\underline{V}RT}$$

Apply Eqn. 15.17, but instead of differentiating directly, use the chain rule, Eqn. 6.16.

$$\left(\frac{\partial(term)}{\partial n_k}\right)_{T,\underline{V},n_{k\neq i}} = \left(\frac{\partial(term)}{\partial(b\rho)}\right)_{T,\underline{V},n_{k\neq i}}\left(\frac{\partial(b\rho)}{\partial n_k}\right)_{T,\underline{V},n_{k\neq i}} \quad \text{or}$$

$$\left(\frac{\partial(term)}{\partial n_k}\right)_{T,\underline{V},n_{k\neq i}} = \left(\frac{\partial(term)}{\partial(n^2 a)}\right)_{T,\underline{V},n_{k\neq i}}\left(\frac{\partial(n^2 a)}{\partial n_k}\right)_{T,\underline{V},n_{k\neq i}}$$

$$\left(\frac{\partial(\underline{A} - \underline{A}^{ig})_{TV}/RT}{\partial n_k}\right)_{T,\underline{V},n_{k\neq i}} = -\ln(1-b\rho) + \frac{n}{1-b\rho}\left(\frac{\partial b\rho}{\partial n_k}\right)_{T,\underline{V},n_{k\neq i}} - \frac{1}{\underline{V}RT}\left(\frac{\partial n^2 a}{\partial n_k}\right)_{T,\underline{V},n_{k\neq i}}$$

Example 15.4 Fugacity coefficient from the van der Waals equation (Continued)

$$bp = \frac{nb}{\underline{V}} \Rightarrow \left(\frac{\partial nb/\underline{V}}{\partial n_k}\right)_{T,\underline{V},n_{i \ne k}} = \frac{b_k}{\underline{V}}$$

$$\ln(\hat{\varphi}_k) = -\ln(1 - bp) + \frac{b_k p}{1 - bp} - \frac{2\sum_j n_j a_{kj}}{\underline{V}RT} - \ln Z$$

15.32

$$= -\ln(1 - bp) + \frac{b_k p}{1 - bp} - \frac{2p\sum_j x_j a_{kj}}{RT} - \ln Z$$

$$bp \equiv \frac{B}{Z}; \quad \frac{a}{bRT} \equiv \frac{A}{B}; \quad \frac{a_{jk}}{a} \equiv \frac{A_{jk}}{A}; \quad \frac{b_k}{b} \equiv \frac{B_k}{B}$$

$$\boxed{\ln(\hat{\varphi}_k) = -\ln(Z - B) + \frac{B_k}{Z - B} - \frac{2\sum_j x_j A_{kj}}{Z}}$$

15.33

Example 15.5 Fugacity coefficient from the Peng-Robinson equation

The Peng-Robinson equation is given by

$$Z = \frac{1}{1 - bp} - \frac{ap}{RT}\frac{1}{(1 + 2bp - b^2 p^2)}$$

where $a = \sum_i \sum_j y_i y_j a_{ij}$ and $b = \sum_i y_i b_i$. Develop an expression for the fugacity coefficient.

Solution: We need to apply Eqn. 15.17. From integration for the pure fluid,

$$\frac{(A - A^{ig})_{TV}}{RT} = -\ln(1 - bp) - \frac{a}{bRT\sqrt{8}}\ln\left(\frac{1 + (1 + \sqrt{2})bp}{1 + (1 - \sqrt{2})bp}\right)$$

$$\frac{(\underline{A} - \underline{A}^{ig})_{TV}}{RT} = -n\ln(1 - bp) - \frac{an^2}{nbRT\sqrt{8}}\{\ln[1 + (1 + \sqrt{2})bp] - \ln[1 + (1 - \sqrt{2})bp]\}$$

Example 15.5   Fugacity coefficient from the Peng-Robinson equation (Continued)

The next steps look intimidating. Basically, they apply the same procedure for differentiation as the last example.

$$\left(\frac{\partial(\underline{A}-\underline{A}^{ig})_{TV}/RT}{\partial n_k}\right)_{T,V,n_{k\neq i}} = -\ln(1-b\rho)+\frac{n}{1-b\rho}\left(\frac{\partial b\rho}{\partial n_k}\right)$$

$$-\frac{an^2}{nbRT\sqrt{8}}\left\{\frac{(1+\sqrt{2})\left(\frac{\partial b\rho}{\partial n_k}\right)}{1+(1+\sqrt{2})b\rho}-\frac{(1-\sqrt{2})\left(\frac{\partial b\rho}{\partial n_k}\right)}{1+(1-\sqrt{2})b\rho}\right\}$$

$$-\left(\frac{\left(\frac{\partial an^2}{\partial n_k}\right)}{nbRT\sqrt{8}}-\frac{an^2\left(\frac{\partial nb}{\partial n_k}\right)}{RT\sqrt{8}\,(nb)^2}\right)\ln\left[\frac{1+(1+\sqrt{2})b\rho}{1+(1-\sqrt{2})b\rho}\right]$$

$$\ln(\hat\varphi_k) = -\ln(1-b\rho)-\ln Z+\frac{b_k\rho}{1-b\rho}-\frac{ab_k\rho}{bRT\sqrt{8}}\left\{\frac{(1+\sqrt{2})}{1+(1+\sqrt{2})b\rho}-\frac{(1-\sqrt{2})}{1+(1-\sqrt{2})b\rho}\right\}$$

$$-\frac{a}{bRT\sqrt{8}}\left[\frac{2\sum_j x_j a_{jk}}{a}-\frac{b_k}{b}\right]\ln\left[\frac{1+(1+\sqrt{2})b\rho}{1+(1-\sqrt{2})b\rho}\right]$$

Note a simplification that is not obvious:

$$\frac{b_k\rho}{1-b\rho}-\frac{ab_k\rho}{bRT\sqrt{8}}\left\{\frac{(1+\sqrt{2})}{1+(1+\sqrt{2})b\rho}-\frac{(1-\sqrt{2})}{1+(1-\sqrt{2})b\rho}\right\}=$$

$$\frac{b_k}{b}\left[\frac{b\rho}{1-b\rho}-\frac{ab\rho}{bRT\sqrt{8}}\left\{\frac{(1+\sqrt{2})}{1+(1+\sqrt{2})b\rho}-\frac{(1-\sqrt{2})}{1+(1-\sqrt{2})b\rho}\right\}\right]=\frac{b_k}{b}[Z-1]$$

$$\ln(\hat\varphi_k) = -\ln(1-b\rho)-\ln Z+\frac{b_k}{b}[Z-1]-\frac{a}{bRT\sqrt{8}}\left[\frac{2\sum_j x_j a_{jk}}{a}-\frac{b_k}{b}\right]\ln\left[\frac{1+(1+\sqrt{2})b\rho}{1+(1-\sqrt{2})b\rho}\right]$$

Substituting the following definitions,

$$b\rho\equiv\frac{B}{Z};\qquad \frac{a}{bRT}\equiv\frac{A}{B};\qquad \frac{a_{jk}}{a}\equiv\frac{A_{jk}}{A};\qquad \frac{b_k}{b}\equiv\frac{B_k}{B}$$

Example 15.5   Fugacity coefficient from the Peng-Robinson equation (Continued)

$$\ln(\hat{\varphi}_k) = -\ln(Z - B) + \frac{B_k}{B}\{Z - 1\} - \frac{A}{B\sqrt{8}}\left[\frac{2\sum_j x_j A_{jk}}{A} - \frac{B_k}{B}\right]\ln\left[\frac{Z + (1 + \sqrt{2})B}{Z + (1 - \sqrt{2})B}\right] \qquad 15.34$$

which has been shown in Eqns. 15.18–15.19 for a binary.

15.4   VLE CALCULATIONS BY AN EQUATION OF STATE

❶ The engineering objective is to use equations of state for bubble, dew, and flash calculations.

❶ Flow sheets for bubble temperature, dew, and flash routines are in Appendix C.

▣ Bubble-pressure calculations are enabled with the spreadsheet Prmix.xlsx.

At the end of Section 15.2, the bubble-pressure method was briefly introduced to show how the fugacity coefficients are incorporated into a VLE calculation, without concentrating on the details of the iterations. Section 15.3 offered derivations of formulas for the fugacity coefficients that were presented without proof at the beginning of the chapter. Now, it is time to turn to the applied engineering objective: calculation of phase equilibria. Refer again to Table 10.1 on page 373, that lists the types of routines that are needed and the convergence criteria. Note that Table 10.1 is independent of the model used for calculating VLE. As an example of the iteration procedure for cubic equations of state, the bubble-pressure flow sheet is presented in Fig. 15.2. The flow sheet puts detail to the procedure discussed superficially in Example 15.2 and immediately preceding the example. Flow sheets for bubble temperature, dew, and flash routines are available in Appendix C. As with ideal solutions, the bubble-pressure routine is the easiest to apply, so we cover it in detail in the following examples. Iterative phase equilibrium calculations can be tedious and difficult to automate. We can facilitate the calculations to some extent by combining two copies of the PrFug spreadsheet into a single workbook, which we call Prmix.xlsx. The four possible K-value representations are included for convenient selection, as described in Example 15.2. This workbook forms only a starting basis with an emphasis on clearly showing the fundamental steps.

Example 15.6   Bubble-point pressure from the Peng-Robinson equation

Use the Peng-Robinson equation ($k_{ij} = 0$) to determine the bubble-point pressure of an equimolar solution of nitrogen (1) + methane (2) at 100 K.

Solution: The calculations proceed by first calculating the short-cut K-ratio as in Example 15.2

on page 587. The ideal-solution (^{is}) bubble pressure was $P \approx \sum_i x_i P_i^{sat} \approx 0.4119$ bars;

$y_{N2}^{is} = 0.958$. In fact, the K-values for the first iteration have already been determined in that example, in great detail. The values from that example are $K_1 = 1.955$, $K_2 = 0.1106$. The new estimates of vapor mole fractions are obtained by multiplying $x_i K_i$. In Example 15.2, the sum

was found to be $\sum_i y_i = 1.033$. These calculations are summarized in the first column of Table

10.1.

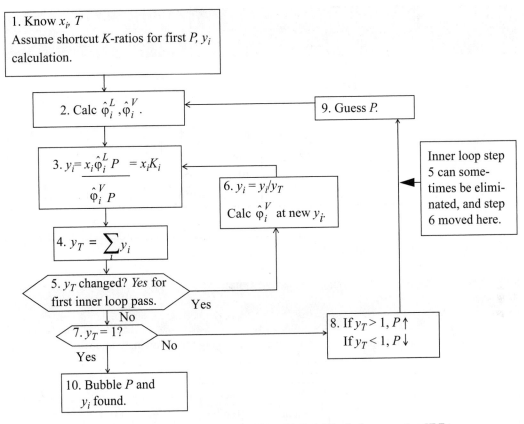

Figure 15.2 *Bubble-pressure flow sheet for the equation of state method of representing VLE. Other routines are given in Appendix C.*

Example 15.6 Bubble-point pressure from the Peng-Robinson equation (Continued)

Noting that these sum to a number greater than unity, we must choose a greater value of pressure for the next iteration. Before we can start the next iteration, however, we must develop new estimates of the vapor mole fractions; the ones we have do not make sense because they sum to more than unity. These new estimates can be obtained simply by dividing the given vapor mole fractions by the number to which they sum. This process is known as **normalization** of the mole fractions. For example, to start the second iteration, $y_1 = 0.978/1.033 = 0.947$. After repeating the process for the other component, the mole fractions will sum to unity. Since the result for the second iteration is less than one, the pressure guess is too high.

Normalization of mole fractions.

The third iteration consists of applying the interpolation rule to obtain the estimate of pressure and use of the normalization procedure to obtain the estimates of vapor mole fractions. $P = 0.4119 + (1 - 1.033)/(0.956 - 1.033) \cdot (0.45 - 0.4119) = 0.428$ MPa. Since the estimated vapor mole fractions after the third iteration sum very nearly to unity, we may conclude the calculations here. This is the bubble pressure. Note how quickly the estimate for y_1 converges to the final estimate of 0.945.

Example 15.6 Bubble-point pressure from the Peng-Robinson equation (Continued)

Comp	x_i	$P = 0.4119,$ $y_1 = 0.958$		$P = 0.45,$ $y_1 = 0.947$		$P = 0.428,$ $y_1 = 0.946$	
		K_i	y_i	K_i	y_i	K_i	y_i
N_2	0.5	1.957	0.978	1.808	0.904	1.890	0.945
CH_4	0.5	0.1106	0.055	0.1031	0.052	0.1073	0.054
			$\Sigma y_i = 1.033$		$\Sigma y_i = 0.956$		$\Sigma y_i = 0.999$

Example 15.7 Isothermal flash using the Peng-Robinson equation

A distillation column is to produce overhead products having the following compositions:

Component	z_i
Propane	0.23
Isobutane	0.67
n-Butane	0.10

Suppose a partial condenser is operating at 320 K and 8 bars. What fraction of liquid would be condensed according to the Peng-Robinson equation, assuming all binary interaction parameters are zero ($k_{ij} = 0$)?

Solution: This is an isothermal flash calculation. Refer back to the same problem (Example 10.1 on page 382) for an initial guess based on the shortcut K-ratio equation. $V/F = 0.25 \Rightarrow \{x_i\} = \{0.1829, 0.7053, 0.1117\}$ and $\{y_i\} = \{0.3713, 0.5642, 0.0648\}$. Substituting these composition estimates for the vapor and liquid compositions into the routine for estimating K-values (*cf.* Example 15.2 on page 587), we can obtain the estimates for K-values given below:

	$T_c(K)$	$P_c(bar)$	ω	z_i	K_i
C_3	369.8	42.49	0.152	0.23	1.729
i-C_4	408.1	36.48	0.177	0.67	0.832
n-C_4	425.2	37.97	0.193	0.10	0.640

Example 15.7 Isothermal flash using the Peng-Robinson equation (Continued)

The computations for the flash calculation are basically analogous to those in Example 10.1, except that K_i values are calculated from Eqn. 15.20. A detailed flow sheet is presented in Appendix C. For this example, the K-values are not modified until the iteration on V/F converges. After convergence on V/F, the vapor and liquid mole fractions are recomputed using Eqns. 10.15 and 10.16, followed by recomputed estimates for the K-values. If the new estimates for K-values are equal to the old estimates for K-values, then the overall iteration has converged. If not, then the new estimates for K-values are substituted for the old values, and the next iteration proceeds just like the last. This method of iteratively solving for the vector of K-values is known in numerical analysis as the "successive substitution" method.

		$V/F = 0.25$	$V/F = 0.10$	$V/F = 0.132$		
z_i	K_i	D_i	D_i	D_i	x_i	y_i
0.23	1.729	−0.142	−0.1563	−0.1529	0.2098	0.3627
0.67	0.832	0.118	0.1145	0.1151	0.6852	0.5701
0.10	0.640	0.040	0.0373	0.0378	0.1050	0.0672
		$\Sigma D_i = 0.016$	$\Sigma D_i = -0.004$	$\Sigma D_i = 0.0000$	1.000	1.000

Using these x's and y's for guesses we find $K = 1.7276, 0.8318$, and 0.6407, respectively. These K-values are similar to those estimated at the compositions derived from the ideal-solution approximation, and will yield a similar V/F. Therefore, we conclude that this iteration has converged (a general criterion is that the average % change in the K-values from one iteration to the next is less than 10^{-4}). Comparison to the shortcut K-ratio approximation shows small but significant deviations—$V/F = 0.13$ for Peng-Robinson versus 0.25 for the shortcut K-ratio method.

Based on this example, we may conclude that the shortcut K-Ratio approximation provides a reasonable first approximation at these conditions. Note, however, that none of the components is supercritical and all the components are saturated hydrocarbons.

$K_i^{Peng-Rob}$	$K_i^{shortcut}$
1.727	2.03
0.832	0.80
0.641	0.58

It is tempting to expand further on Prmix.xlsx to facilitate greater automation and simple-minded application. An online supplement provides a very preliminary step in this direction through the use of macro's. Ultimately, however, this literature comprises specialized research that is beyond our introductory scope. In general, the analysis requires detailed consideration of phase stability and criticality. References cited in Chapter 16 describe works by Michelsen and Mollerup, Eubank, and Tang that can help to create more reliable algorithms. It is a useful exercise to customize your workbooks to increase your confidence in achieving reliable solutions, but do not spend excessive time trying to program every possibility. Chapter 16 and the references cited there are recommended for advanced programming.

Example 15.8 Phase diagram for azeotropic methanol + benzene

Methanol and benzene form an azeotrope. For methanol + benzene the azeotrope occurs at 61.4 mole% methanol and 58°C at atmospheric pressure (1.01325 bars). Additional data for this system are available in the *Chemical Engineers' Handbook*. Use the Peng-Robinson equation with $k_{ij} = 0$ (see Eqn. 15.9) to estimate the phase diagram for this system and compare it to the experimental data on a *T-x-y* diagram. Determine a better estimate for k_{ij} by iterating on the value until the bubble point pressure matches the experimental value (1.013 bar) at the azeotropic composition and temperature. Plot these results on the *T-x-y* diagram as well. Note that it is impossible to match both the azeotropic composition and pressure with the Peng-Robinson equation because of the limitations of the single parameter, k_{ij}.

The experimental data for this system are as follows:

x_m	0.000	0.026	0.050	0.088	0.164	0.333	0.549	0.699	0.782	0.898	0.973	1.000
y_m	0.000	0.267	0.371	0.457	0.526	0.559	0.595	0.633	0.665	0.760	0.907	1.000
$T(K)$	353.25	343.82	339.59	336.02	333.35	331.79	331.17	331.25	331.62	333.05	335.85	337.85

Solution: Solving this problem is computationally intensive, but still approachable with Prmix.xlsx. The strategy is to manually set a guessed k_{ij} and then perform a bubble pressure calculation at the azeotrope temperature (331.15 K) and composition, $x_m = 0.614$. The program will give a calculated pressure and vapor phase composition. The vapor-phase composition may not match the liquid-phase composition because the azeotrope is not perfectly predicted; however, we continue to manually change k_{ij}, and repeat the bubble pressure calculation until we match the experimental pressure of 1.013 bar. The following values are obtained for the bubble pressure at the experimental azeotropic composition and temperature with various values of k_{ij}.

k_{ij}	0.0	0.1	0.076	0.084
P(bar)	0.75	1.06	0.9869	1.011

The resultant k_{ij} is used to perform bubble-temperature calculations across the composition range, resulting in Fig. 15.3. Note that we might find a way to fit the data more accurately than the method given here, but any improvements would be small relative to estimating $k_{ij} = 0$. We see that the fit is not as good as we would like for process design calculations. This solution is so nonideal that a more flexible model of the thermodynamics is necessary. Note that the binary interaction parameter alters the magnitude of the bubble-pressure curve very effectively but hardly affects the skewness at all. Since this mixture is far from the critical region, a two-parameter activity model like van Laar or UNIQUAC would be recommended as shown in Fig. 12.1. The Peng-Robinson model with van der Waals' mixing rule comes closest to the Scatchard-Hildebrand activity model. We observed that the Scatchard-Hildebrand model performed poorly for hydrogen bonding mixtures. Two approaches are common when precision is needed in the critical region for hydrogen bonding mixtures. Either a multiparameter activity model can be adapted as a basis for an advanced mixing rule, or hydrogen bonding can be treated explicitly. References to these are approaches are presented in the chapter summary.

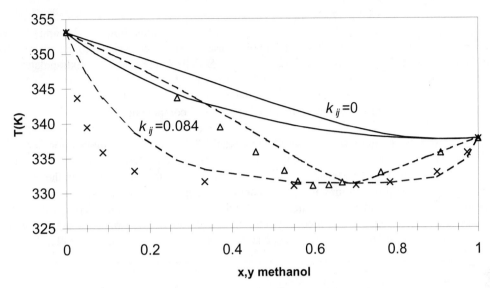

Figure 15.3 *T-x-y diagram for the azeotropic system methanol + benzene. Curves show the predictions of the Peng-Robinson equation ($k_{ij} = 0$) and correlation ($k_{ij} = 0.084$) based on fitting a single data point at the azeotrope; x's and triangles represent liquid and vapor phases, respectively.*

Example 15.9 Phase diagram for nitrogen + methane

Prmix.xlsx.

Use the Peng-Robinson equation ($k_{ij} = 0$) to determine the phase diagram of nitrogen + methane at 150 K. Plot P versus x, y and compare the results to the results from the shortcut K-ratio equations.

Solution: First, the shortcut K-ratio method gives the dotted phase diagram in Fig. 15.4. Applying the bubble-pressure procedure with the program Prmix.xlsx, we calculate the solid line in Fig. 15.4. For the Peng-Robinson method we assume K-values from the previous solution as the initial guess to get the solutions near $x_{N_2} = 0.685$. The program Prmix.xlsx assumes this automatically, but we must also be careful to make small changes in the liquid composition as we approach the critical region.

Fig. 15.4 was generated by entering liquid nitrogen compositions of 0.10, 0.20, 0.40, 0.60, 0.61, 0.62..., 0.68, and 0.685. This procedure of starting in a region where a simple approximation is reliable and systematically moving to more difficult regions using previous results is often necessary and should become a familiar trick in your accumulated expertise on phase equilibria in mixtures. We apply a similar approach in estimating liquid-liquid equilibria.

Comparing the two approximations numerically and graphically, it is clear that the shortcut approximation is significantly less accurate than the Peng-Robinson equation at high concentrations of the supercritical component. This happens because the mixture possesses a critical point, above which separate liquid and vapor roots are impossible, analogous to the situation for pure fluids. Since the mixing rules are in terms of a and b instead of T_c and P_c, the equation of state is generating effective values for A_c and B_c of the mixture.

Example 15.9 Phase diagram for nitrogen + methane (Continued)

Instead of depending simply on T and P as they did for pure fluids, however, A_c and B_c also depend on composition. The mixture critical point varies from one component to the other as the composition changes. Since the shortcut (and also SCVP+) approximation extrapolates the vapor pressure curve to obtain an effective vapor pressure of the supercritical component, that approximation does not reflect the presence of the mixture critical point and this leads to significant errors as the mixture becomes rich in the supercritical component.

The mixture critical point also leads to computational difficulties. If the composition is excessively rich in the supercritical component, the equation of state calculations may obtain the same solution for the vapor root as for the liquid root and, since the fugacities are equal, the program will terminate. The program may indicate accurate convergence in this case due to some slight inaccuracies that are unavoidable in the critical region. Or the program may diverge. It is often up to the competent engineer to recognize the difference between accurate convergence and a spurious answer. Plotting the phase envelope is an excellent way to stay out of trouble.

🛑 The shortcut *K*-ratio method provides an initial estimate when a supercritical component is at low liquid-phase compositions, but incorrectly predicts VLE at high liquid-phase concentrations of the supercritical component.

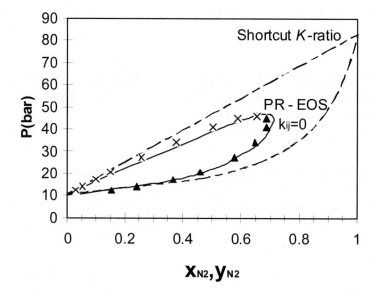

Figure 15.4 *High pressure P-x-y diagram for the nitrogen + methane system comparing the shortcut K-ratio approximation and the Peng-Robinson equation at 150 K. The data points represent experimental results. Both theories are entirely predictive since the Peng-Robinson equation assumes that $k_{ij} = 0$.*

Example 15.10 Ethane + heptane phase envelopes

Use the Peng-Robinson equation ($k_{ij} = 0$) to determine the phase envelope of ethane + n-heptane at compositions of $x_{C7} = [0, 0.1, 0.2, 0.3, 0.5, 0.7, 0.9, 1.0]$. Plot P versus T for each composition by performing bubble-pressure calculations to their terminal point and dew-temperature calculations until the temperature begins to decrease significantly and the pressure approaches its maximum. If necessary, close the phase envelope by starting at the last dew-temperature state and performing dew-pressure calculations until the temperature and pressure approach the terminus of the bubble-point curve. For each composition, mark the points where the bubble and dew curves meet with X's. These X's designate the "mixture critical points." Connect the X's with a dashed curve. The dashed curve is known as the **critical locus** of the mixture.

Prmix.xlsx facilitates bubble, dew, and flash calculations.

❶ Mixture critical points.

Solution: Note that these phase envelopes are similar to the one from the previous problem, except that we are changing the temperature instead of the composition along each curve. They are more tedious in that both dew and bubble calculations must be performed to generate each curve. The lines of constant composition are sometimes called **isopleths**. The results of the calculations are illustrated in Fig. 15.5. The results at mole fractions of 0 and 1.0 are indicated by dash-dot curves to distinguish them as the vapor pressure curves. Phase equilibria on the P-T plot occurs at the conditions where a bubble line of one composition intersects a dew line of a different composition.

❶ Isopleths.

Some practical considerations for high pressure processing can be inferred from the diagram. Consider what happens when starting at 90 bars and ~445 K and dropping the pressure on a 30 mole% C7 mixture at constant temperature. Similar situations could arise with flow of natural gas through a small pipe during natural gas recovery. As the pressure drops, the dew-point curve is crossed and liquid begins to condense. Based on intuition developed from experiences at lower pressure, one might expect that dropping the pressure should result in more vapor-like behavior, not condensation. On the other hand, dropping the pressure reduces the density and solvent power of the ethane-rich mixture. This phenomenon is known as **retrograde condensation**. It occurs near the critical locus when the operating temperature is less than the maximum temperature of the phase envelope. Since this maximum temperature is different from the mixture critical temperature, it needs a distinctive name. The name applied is the "critical condensation temperature" or **cricondentherm**. Similarly, the maximum pressure on the phase envelope is known as the **cricondenbar**. Note that an analogous type of phase transition can occur near the critical locus when the pressure is just above the critical locus and the temperature is changed.

❶ Retrograde condensation.

Example 15.10 Ethane + heptane phase envelopes (Continued)

To extend the analysis, imagine what happens in a natural gas stream composed primarily of methane but also containing small amounts of components as heavy as C80. A retrograde condensation region exists where the heavy components begin to precipitate, as discussed in Example 14.12 on page 567. But a different possibility also exists because the melting temperature of the heavy components may often exceed the operating temperature, and the precipitate that forms might be a solid that could stick to the walls of the pipe. This in turn generates a larger constriction which generates a larger pressure drop during flow, right in the vicinity of the deposit. In other words, this deposition process may tend to promote itself until the flow is substantially inhibited. Wax deposition is a significant problem in the oil and natural gas industry and requires considerable engineering expertise because it often occurs away from critical points, as well as in the near-critical regions of this discussion. A wide variety of solubility behavior can occur, as we show in Chapter 16.

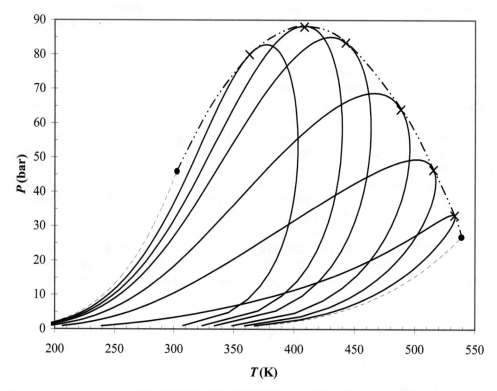

Figure 15.5 *High pressure phase envelopes for the ethane + heptane system comparing the effects of composition according to the Peng-Robinson equation. The theory is entirely predictive because the Peng-Robinson equation has been applied with $k_{ij} = 0$. X's mark the mixture critical points and the dashed line indicates the critical locus. The first and last curves represent the vapor pressures for the pure components.*

15.5   STRATEGIES FOR APPLYING VLE ROUTINES

For some problems, such as generation of a phase diagram, the examples in this chapter can be followed directly. Often, however, it takes some thought to decide which type of VLE routine is appropriate to apply in a given situation. Do not discount this step in the problem solution strategy. Part of the objective of the homework problems is to increase understanding of phase behavior by encouraging thought about which routine to use. Since the equation of state routines are complicated enough to require a computer, they are also solved relatively rapidly, *once the VLE routine has been identified*. Use Table 10.1 on page 373 and the information in Section 10.1 to identify the correct procedure to apply before turning to computational techniques. Use the problem statement to identify whether the liquid, vapor, or overall mole fractions are known. Study the problem statement to see whether the P and T are known. Often this information together with Table 10.1 will determine the routines to apply. Sometimes more than one approach is satisfactory. Sometimes the ideal-solution approximation may be applied, and a review of Section 10.8 may be helpful. Before using software that accompanies the text, be sure to read the appropriate section of Appendix A, and the instructions or readme.txt files that accompany the program.

15.6   SUMMARY

The essence of the equation of state approach to mixtures is that the equation of state for mixtures is the same as the equation of state for pure fluids. The expressions for Z, A, and U are exactly the same. The only difference is that the parameters (e.g., a and b of the Peng-Robinson equation) are dependent on the composition. That should come as no large surprise when you consider that these parameters must transform from pure component to pure component in some continuous fashion as the composition changes. What may be surprising is the wealth of behaviors that can be inferred from some fairly simple rules for modeling this transformation. Everything from azeotropes to retrograde condensation, and even liquid-liquid separation can be represented with qualitative accuracy based on this simple extension of the equation of state.

So, are we done with phase behavior modeling? Unfortunately, the keyword in the preceding paragraph is "qualitative." Equations of state are sufficiently accurate for most applications involving hydrocarbons, gases, and to some extent, ethers, esters, and ketones. For many oil and gas wells, it may suffice to treat the hydrocarbon-rich phases in this way and treat water separately. But any applications involving strongly hydrogen-bonding species tend to require greater accuracy than currently attainable from equations like the Peng-Robinson EOS. For example, if methanol is used as a hydrate inhibitor in a gas well, its partitioning may require a more sophisticated treatment. One idea is to adapt multiparameter activity models like the UNIQUAC model as the basis for mixing rules. This is the approach of the Wong-Sandler model.[3] Another approach is to analyze hydrogen bonding as simultaneous reaction and phase equilibria, as discussed in Chapter 19.

Important Equations

Once again the mixing rules play an important role in defining the thermodynamics. Since the Peng-Robinson mixing rules have the same form as the van der Waals mixing rules, including a single binary interaction parameter, k_{ij}, the Peng-Robinson model cannot match the skewness of the Gibbs excess curve, only the magnitude. Outside the critical region, you might as well use an activity model. The advantage of the Peng-Robinson model is that it provides a holistic framework that

---

3. Wong, D.S.H.; Sandler, S.I. 1992. *AIChE J.* 38:671.

applies seamlessly to vapor, liquid, and critical region. Noting how activity models artificially designate different methods for different phases, it is gratifying to see that such conceptual simplicity is feasible. The key equation for establishing this feasibility is deriving the fugacity coefficient for a pressure-explicit equation of state:

$$\ln(\hat{\varphi}_i) = \frac{(\mu_i - \mu_i^{ig})}{RT} = \left(\frac{\partial(\underline{A} - \underline{A}^{ig})_{TV}/(RT)}{\partial n_i}\right)_{T, \underline{V}, n_{j \neq i}} - \ln Z \qquad 15.17$$

Given this equation, it is straightforward to derive fugacity coefficients, and K-ratios, for any equation of state or mixing rule. Two related equations that often appear are Eqns. 15.23 and 15.26.

$$\left(\frac{\partial nb}{\partial n_k}\right)_{T, \underline{V}, n_{j \neq k}} = b_k \qquad 15.23$$

$$\left(\frac{\partial n^2 a}{\partial n_k}\right)_{T, \underline{V}, n_{j \neq k}} = 2\sum_j n_j a_{jk} \qquad 15.26$$

Look for ways to rearrange the equations before differentiating such that these terms appear and then differentiation becomes much simpler, often reducing to simple substitution. Finally, the EOS method melds with every other phase equilibrium computational procedure when the expression is derived for the partition coefficient, as given by a slight variation on Eqn. 15.20.

$$y_i \hat{\varphi}_i^U P = x_i \hat{\varphi}_i^L P \quad \text{or} \quad K_i = \hat{\varphi}_i^L / \hat{\varphi}_i^U \qquad 15.20$$

Here we have generalized Eqn. 15.20 slightly by recognizing that the upper phase could be vapor or it could be the upper phase of LLE. The beauty of the EOS perspective is that the fluid phase model is the same for liquid or vapor; only the proper root must be selected.

15.7 PRACTICE PROBLEMS

P15.1 Repeat all the practice problems from Chapter 10, this time applying the Peng-Robinson equation.

P15.2 Acrolein (C_3H_4O) + water exhibits an atmospheric (1 bar) azeotrope at 97.4 wt% acrolein and 52.4°C. For acrolein: $T_c = 506$ K; $P_c = 51.6$ bar; and $\omega = 0.330$; $MW = 56$.

(a) Determine the value of k_{ij} for the Peng-Robinson equation that matches this bubble pressure at the same liquid composition and temperature. (ANS. 0.015)

(b) Tabulate P, y at 326.55 K and $x = \{0.57, 0.9, 0.95, 0.974\}$ via the Peng-Robinson equation using the k_{ij} determined above.
(ANS. (1.33, 0.575), (1.16, 0.736), (1.06, 0.841), (1.0, 0.860))

P15.3 Laugier and Richon (*J. Chem. Eng. Data*, 40:153, 1995) report the following data for the H_2S + benzene system at 323 K and 2.010 MPa: $x_1 = 0.626$; $y_1 = 0.986$.

(a) Quickly estimate the vapor-liquid K-value of H_2S at 298 K and 100 bar. (ANS. 0.21)

(b) Use the data to estimate the k_{ij} value, then estimate the error in the vapor phase mole fraction of H_2S. (ANS. 0.011, 0.1%)

P15.4 The system ethyl acetate + methanol forms an azeotrope at 27.8 mol% EA and 62.1°C. For ethyl acetate, $T_c = 523.2$ K; $P_c = 38.3$ bar; and $\omega = 0.362$.

(a) What is the estimate of the bubble-point pressure from the Peng-Robinson equation of state at this composition and temperature when it is assumed that $k_{ij} = 0$? (ANS. 0.98 bars)
(b) What value of k_{ij} gives a bubble-point pressure of 1 bar at this temperature and composition? (ANS. 0.0054)
(c) What is the composition of the azeotrope and value of the bubble-point pressure at the azeotrope estimated by the Peng-Robinson equation when the value of k_{ij} from part (b) is used to describe the mixture? (ANS. $x_{EA} = 0.226$)

P15.5 (a) Assuming zero for the binary interaction parameter ($k_{ij} = 0$) of the Peng-Robinson equation, predict whether an azeotrope should be expected in the system CO_2 + ethylene at 222 K. Estimate the bubble-point pressure for an equimolar mixture of these components. (ANS. No, 8.7 bar)
(b) Assuming a value for the binary interaction parameter ($k_{ij} = 0.11$) of the Peng-Robinson equation, predict whether an azeotrope should be expected in the system CO_2 + ethylene at 222 K. Estimate the bubble-point pressure for an equimolar mixture of these components. (ANS. Yes, 11.3 bar)

P15.6 (a) Assuming zero for the binary interaction parameter ($k_{ij} = 0$) of the Peng-Robinson equation, estimate the bubble pressure and vapor composition of the pentane + acetone system at $x_p = 0.728$, 31.9°C. (ANS. 0.78 bars, $y_1 = 0.83$)
(b) Use the experimental liquid composition and bubble condition of the pentane + acetone system at $x_P = 0.728$, $T = 31.9$°C, $P = 1$ bar to estimate the binary interaction parameter (k_{ij}) of the Peng-Robinson equation, and then calculate the bubble pressure of a 13.4 mol% pentane liquid solution at 39.6°C. (ANS. 0.117, 1.12 bar)

P15.7 Calculate the dew-point pressure and corresponding liquid composition of a mixture of 30 mol% carbon dioxide, 30% methane, 20% propane, and 20% ethane at 298 K using

(a) The shortcut K-ratios (ANS. 32 bar)
(b) The Peng-Robinson equation with $k_{ij} = 0$ (ANS. 44 bar)

P15.8 The equation of state below has been suggested for a new equation of state. Derive the expression for the fugacity coefficient of a component.

$$Z = 1 + 4cb\rho/(1 - b\rho)$$

where $\quad b = \sum_i x_i b_i$

$$c = \sum_i \sum_j x_i x_j c_{ij}$$

$$c_{ij} = \sqrt{c_{ii} c_{jj}}$$

(ANS. $\ln \hat{\varphi}_k = 4 \left(c - 2 \sum_i x_i c_{ik} \right) \ln(1 - b\rho) + \dfrac{b_k}{b}(Z - 1) - \ln Z$)

15.8 HOMEWORK PROBLEMS

15.1 Using Fig. 15.5 on page 602, without performing additional calculations, sketch the *P-x-y* diagram at 400 K showing the two-phase region. Make the sketch semi-quantitative to show the values where the phase envelope touches the axes of your diagram. Label the bubble and dew lines. Also indicate the approximate value of the maximum pressure.

15.2 Consider two gases that follow the virial equation. Show that an ideal mixture of the two gases follows the relation $B = y_1 B_{11} + y_2 B_{22}$.

15.3 Consider phase equilibria modeled with $y_i \hat{\varphi}_i^V P = x_i \hat{\varphi}_i^L P$. When might $\hat{\varphi}_i$ be replaced by φ_i for each phase? When might $\hat{\varphi}_i = 1$ be used for each phase? Discuss the appropriateness of using the virial equation for mixtures to solve phase behavior using the expression $y_i \hat{\varphi}_i^V P = x_i \hat{\varphi}_i^L P$.

15.4 Calculate the molar volume of a binary mixture containing 30 mol% nitrogen(1) and 70 mol% *n*-butane(2) at 188°C and 6.9 MPa by the following methods.

(a) Assume the mixture to be an ideal gas.
(b) Assume the mixture to be an ideal solution with the volumes of the pure gases given by

$$Z = 1 + \frac{BP}{RT}$$

and the virial coefficients given below.

(c) Use second virial coefficients predicted by the generalized correlation for *B*.
(d) Use the following values for the second virial coefficients.

Data:

$B_{11} = 14 \qquad B_{22} = -265 \qquad B_{12} = -9.5 \qquad$ (Units are cm³/gmole)

(e) Use the Peng-Robinson equation.

15.5 For the same mixture and experimental conditions as problem 15.4, calculate the fugacity of each component in the mixture, $\hat{f}_i$. Use methods (a) – (e).

15.6 A vapor mixture of CO_2 (1) and *i*-butane (2) exists at 120°C and 2.5 MPa. Calculate the fugacity of CO_2 in this mixture across the composition range using

(a) The virial equation for mixtures
(b) The Peng-Robinson equation
(c) The virial equation for the pure components and an ideal mixture model.

15.7 Use the virial equation to consider a mixture of propane and *n*-butane at 515 K at pressures between 0.1 and 4.5 MPa. Verify that the virial coefficient method is valid by using Eqn. 7.10.

(a) Prepare a plot of fugacity coefficient for each component as a function of composition at pressures of 0.1 MPa, 2 MPa, and 4.5 MPa.

(b) How would the fugacity coefficient for each component depend on composition if the mixture were assumed to be ideal, and what value(s) would it have for each of the pressures in part (a)? How valid might the ideal-solution model be for each of these conditions?

(c) The excess volume is defined as $V^E = V - \sum_i x_i V_i$, where V is the molar volume of

the mixture, and V_i is the pure component molar volume at the same T and P. Plot the prediction of excess volume of the mixture at each of the pressures from part (a). How does the excess volume depend on pressure?
(d) Under which of the pressures in part (a) might the ideal gas law be valid?

15.8 Consider a mixture of nitrogen(1) + n-butane(2) for each of the options: (*i*) 395 K and 2 MPa; (*ii*) 460 K and 3.4 MPa; (*iii*) 360 K and 1 MPa.

(a) Calculate the fugacity coefficients for each of the components in the mixture using the virial coefficient correlation. Make a table for your results at $y_1 = 0.0, 0.2, 0.4, 0.6, 0.8, 1.0$. Plot the results on a graph. On the same graph, plot the curves that would be used for the mixture fugacity coefficients if an ideal mixture model were assumed. Label the curves.
(b) Calculate the fugacity of each component in the mixture as predicted by the virial equation, an ideal-mixture model, and the ideal-gas model. Prepare a table for each component, and list the three predicted fugacities in three columns for easy comparison. Calculate the values at $y_1 = 0.0, 0.2, 0.4, 0.6, 0.8, 1.0$.

15.9 The virial equation $Z = 1 + BP/RT$ may be used to calculate fugacities of components in mixtures. Suppose $B = y_1 B_{11} + y_2 B_{22}$. (This simple form makes calculations easier. Eqn. 15.1 gives the correct form.) Use this simplified expression and the correct form to calculate the respective fugacity coefficient formulas for component 1 in a binary mixture.

15.10 The Lewis-Randall rule is usually valid for components of high concentration in gas mixtures. Consider a mixture of 90% ethane and 10% propane at 125°C and 170 bar. Estimate $\hat{f}_i$ for ethane.

15.11 One of the easiest ways to begin to explore fugacities in nonideal solutions is to model solubilities of crystalline solids dissolved in high pressure gases. In this case, the crystalline solids remain as a pure phase in equilibrium with a vapor mixture, and the fugacity of the "solid" component must be the same in the crystalline phase as in the vapor phase. Consider biphenyl dissolved in carbon dioxide, using $k_{ij} = 0.100$. The molar volume of crystalline biphenyl is 156 cm³/mol.

(a) Calculate the fugacity (in MPa) of pure crystalline biphenyl at 310 K and 330 K and 0.1, 1, 10, 15, and 20 MPa.
(b) Calculate and plot the biphenyl solubility for the isotherm over the pressure range. Compare the solubility to the ideal gas solubility of biphenyl where the Poynting correction is included, but the gas phase nonidealities are ignored.

15.12 Repeat problem 15.11, except consider naphthalene dissolved in carbon dioxide, using $k_{ij} = 0.109$. The molar volume of crystalline naphthalene is 123 cm³/mol.

15.13 A vessel initially containing propane at 30°C is connected to a nitrogen cylinder, and the pressure is isothermally increased to 2.07 MPa. What is the mole fraction of propane in the vapor phase? You may assume that the solubility of N_2 in propane is small enough that the liquid phase may be considered pure propane. Calculate using the following data at 30°C.

	C_3	N_2
V^L (cm³/gmole)	75.6	
P^{sat} (MPa)	1.065	
B (cm³/mole)	−380	−4.0
B_{12} (cm³/mole)		−70

15.14 A 50-mol% mixture of propane(1) + n-butane(2) enters a flash drum at 37°C. If the flash drum is maintained at 0.6 MPa, what fraction of the feed exits as a liquid? What are the compositions of the phases exiting the flash drum? Work the problem the following two ways.
(a) Use Raoult's law.
(b) Assume ideal mixtures of vapor and liquid (K_i is independent of composition).

Data: $B_{11} = -369.5$ cm³/mol $B_{22} = -665.1$ cm³/mol $B_{12} = -486.9$ cm³/mol
   $P_1^{sat} = 1.269$ MPa $P_2^{sat} = 0.343$ MPa

15.15 A mixture containing 5 mol% ethane, 57 mol% propane, and 38 mol% n-butane is to be processed in a natural gas plant. Estimate the bubble-point temperature, the coexisting vapor compositions, and K-ratios for this mixture at all pressures above 1 bar at which two phases exist. Set $k_{ij} = 0$. Plot ln P versus $1/T$ for your results. What does this plot look like? Plot log K_i versus $1/T$. What values do the K_i approach?

15.16 Vapor-liquid equilibria are usually expressed in terms of K factors in petroleum technology. Use the Peng-Robinson equation to estimate the values for methane and benzene in the benzene + methane system with equimolar feed at 300 K and a total pressure of 30 bar and compare to the estimates based on the shortcut K-ratio method.

15.17 Benzene and ethanol form azeotropic mixtures. Prepare a y-x and a P-x-y diagram for the benzene + ethanol system at 45°C assuming the Peng-Robinson model and using the experimental pressure at $x_E = 0.415$ to estimate k_{12}. Compare the results with the experimental data of Brown and Smith cited in problem 10.2.

15.18 A storage tank is known to contain the following mixture at 45°C and 15 bar on a mole basis: 31% ethane, 34% propane, 21% n-butane, 14% i-butane. What is the composition of the coexisting vapor and liquid phases, and what fraction of the molar contents of the tank is liquid?

15.19 The CRC Handbook lists the atmospheric pressure azeotrope for ethanol + methylethylketone at 74.8°C and 34 wt% ethanol. Estimate the value of the Peng-Robinson k_{12} for this system.

15.20 The CRC Handbook lists the atmospheric pressure azeotrope for methanol + toluene at 63.7°C and 72 wt% methanol. Estimate the value of the Peng-Robinson k_{12} for this system.

15.21 Use the Peng-Robinson equation for the ethane/heptane system.

(a) Calculate the P-x-y diagram at 283 K and 373 K. Use $k_{12} = 0$. Plot the results.

(b) Based on a comparison of your diagrams with what would be predicted by Raoult's law at 283 K, does this system have positive or negative deviations from Raoult's law?

15.22 One mol of *n*-butane and one mol of *n*-pentane are charged into a container. The container is heated to 90°C where the pressure reads 7 bar. Determine the quantities and compositions of the phases in the container.

15.23 Consider a mixture of 50 mol% *n*-pentane and 50 mol% *n*-butane at 15 bar.

(a) What is the dew temperature? What is the composition of the first drop of liquid?
(b) At what temperature is the vapor completely condensed if the pressure is maintained at 15 bar? What is the composition of the last drop of vapor?

15.24 LPG gas is a fuel source used in areas without natural gas lines. Assume that LPG may be modeled as a mixture of propane and *n*-butane. Since the pressure of the LPG tank varies with temperature, there are safety and practical operating conditions that must be met. Suppose the desired maximum pressure is 0.7 MPa, and the lower limit on desired operation is 0.2 MPa. Assume that the maximum summertime tank temperature is 50°C, and that the minimum wintertime temperature is −10°C. [Hint: On a mass basis, the mass of vapor within the tank is negligible relative to the mass of liquid after the tank is filled.]

(a) What is the upper limit (mole fraction) of propane for summertime propane content?
(b) What is the lowest wintertime pressure for this composition from part (a)?
(c) What is the lower limit (mole fraction) of propane for wintertime propane content?
(d) What is the highest summertime pressure for this composition from part (b)?

15.25 The k_{ij} for the pentane + acetone system has been fitted to a single point in problem P15.6. Generate a *P-x-y* diagram at 312.75 K.

15.26 The synthesis of methylamine, dimethylamine, and trimethylamine from methanol and ammonia results in a separation train involving excess ammonia and converted amines. Use the Peng-Robinson equation with $k_{ij} = 0$ to predict whether methylamine + dimethylamine, methylamine + trimethylamine, or dimethylamine + trimethylamine would form an azeotrope at 2 bar. Would the azeotropic behavior identified above be altered by raising the pressure to 20 bar? Locate experimental data relating to these systems in the library. How do your predictions compare to the experimental data?

Compound	T_c (K)	P_c (MPa)	ω
Methylamine	430.0	7.43	0.292
Dimethylamine	437.7	5.31	0.302
Trimethylamine	433.3	4.09	0.205

15.27 For the gas/solvent systems below, we refer to the "gas" as the low molecular weight component. Experimental solubilities of light gases in liquid hydrocarbons are tabulated below. The partial pressure of the light gas is 1.013 bar partial pressure. Do the following for each assigned system.

(a) Estimate the partial pressure of the liquid hydrocarbon by calculating the pure component vapor pressure via the Peng-Robinson equation, and by subsequently applying Raoult's law for that component.
(b) Estimate the total pressure and vapor composition using the results of step (a).

(c) Use the Peng-Robinson equation with $k_{ij} = 0$ to calculate the vapor and liquid compositions that result in 1.013 bar partial pressure of the light gas and compare the pressure and gas phase composition with steps (a) and (b).

(d) Henry's law asserts that $\hat{f}_i = h_i x_i$, when x_i is near zero, and h_i is the Henry's law constant. Calculate the Henry's law constant from the calculations of part (c).

(e) Calculate the solubility expected at 2 bar partial pressure of light gas by using Henry's law as well as by the Peng-Robinson equation and comment on the results.

	Gas	Liquid	T (°C)	x_{gas}	Source
(*i*)	Methane	Cyclohexane	25	2.83 E-3	a
(*ii*)	Methane	Carbon tetrachloride	25	2.86 E-3	a
(*iii*)	Methane	Benzene	25	2.07 E-3	a
(*iv*)	Methane	n-hexane	25	3.15 E-3	b
	Methane	n-hexane	25	4.24 E-3	c

a. Hildebrand, J.H., Scott, R.L. 1950. *The Solubility of Nonelectrolytes.* 3rd ed. New York: Reinhold. Table 4, p. 243.

b. McDaniel, A.S. 1911. *J. Phys. Chem,* 15:587.

c. Guerry Jr., D. 1944. Thesis, Vanderbilt Univ.

15.28 Estimate the solubility of carbon dioxide in toluene at 25°C and 1 bar of CO_2 partial pressure using the Peng-Robinson equation with a zero binary interaction parameter. The techniques of problem 15.27 may be helpful.

15.29 Oxygen dissolved in liquid solvents may present problems during use of the solvents.

(a) Using the Peng-Robinson equation and the techniques introduced in problem 15.27, estimate the solubility of oxygen in n-hexane at an oxygen partial pressure of 0.21 bar.

(b) From the above results, estimate the Henry's law constant.

15.30 Estimate the solubility of ethylene in n-octane at 1 bar partial pressure of ethylene and 25°C. The techniques of problem 15.27 may be helpful. Does the system follow Henry's law up to an ethylene partial pressure of 3 bar at this temperature? Provide the vapor compositions and total pressures for the above states.

15.31 Henry's law asserts that $\hat{f}_i = h_i x_i$, when x_i is near zero, and h_i is the Henry's law constant. Gases at high reduced temperatures can exhibit peculiar trends in their Henry's law constants. Use the Peng-Robinson equation to predict the Henry's law constant for hydrogen in decalin at $T = [300\ K, 600\ K]$. Plot the results as a function of temperature and compare to the prediction from the shortcut prediction. Describe in words the behavior that you observe.

15.32 A gas mixture follows the equation of state

$$\frac{PV}{RT} = 1 + \left(b - \frac{a}{T}\right)\frac{P}{RT}$$

where b is the size parameter, $b = \sum_i x_i b_i$, and a is the energetic parameter,

$a = \sum_i \sum_j x_i x_j a_{ij}$. Derive the formula for the partial molar enthalpy for component 1 in a binary mixture, where the reference state for both components is the ideal gas state of T_R, P_R, and the pure component parameters are temperature-independent.

15.33 The procedure for calculation of the residual enthalpy for a pure gas is shown in Example 8.5 on page 316. Now consider the residual enthalpy for a binary gas mixture. For this calculation, it is necessary to determine da/dT for the mixture.

(a) Write the form of this derivative for a binary mixture in terms of da_1/dT and da_2/dT based on the conventional quadratic mixing rule and geometric mean combining rule with a nonzero k_{ij}.

(b) Provide the expression for the residual enthalpy for a binary mixture that follows the Peng-Robinson equation.

(c) A mixture of 50 mol% CO_2 and 50 mol% N_2 enters a valve at 7 MPa and 40°C. It exits the valve at 0.1013 MPa. Explain how you would determine whether CO_2 precipitates, and if so, whether it would be a liquid or solid.

15.34 A gaseous mixture of 30 mol% CO_2 and 70 mol% CH_4 enters a valve at 70 bar and 40°C and exits at 5.3 bar. Does any CO_2 condense? Assume that the mixture follows the virial equation. Assume that any liquid that forms is pure CO_2. The vapor pressure of CO_2 may be estimated by the shortcut vapor pressure equation. CO_2 sublimes at 0.1013 MPa and −78.8°C, although freezing is less likely.

15.35 The vapor-liquid equilibria for the system acetic acid(1) + acetone(2) needs to be characterized in order to simulate an acetic anhydride production process. Experimental data for this system at 760 mmHg have been reported by Othmer (1943)[4] as summarized below. Use the data at the equimolar composition to determine a value for the binary interaction parameter of the Peng-Robinson equation. Based on the value you determine for the binary interaction parameter, determine the percent errors in the Peng-Robinson prediction for this system at a mole fraction of $x_{(1)} = 0.3$.

T(C)	103.8	93.1	85.8	79.7	74.6	70.2	66.1	62.6	59.2
$x_{(1)}$	0.9	0.8	0.7	0.6	0.5	0.4	0.3	0.2	0.1

15.36 A mixture of methane and ethylene exists as a single gas phase in a spherical tank (10 m^3) on the grounds of a refinery. The mixture is at 298 K and 1 MPa. It is a spring day, and the atmospheric temperature is also 298 K. The mole fraction of ethylene is 20 mol%. Your supervisor wants to draw off gas quickly from the bottom of the tank until the pressure is 0.5 MPa. However, being astute, you suggest that depressurization will cause the temperature to fall, and might cause condensation.

(a) Provide a method to calculate the change in temperature with respect to moles removed or tank pressure valid up until condensation starts. Assume the depressurization is adiabatic and reversible. Provide relations to find answers, and ensure that enough equations are provided to calculate numerical values for all variables, but you do not need to calculate a final number.

(b) Would the answer in part (a) provide an upper or lower limit to the expected temperature?

4. Gmehling, J., Onken, U., Arlt, W. 1977– *Vapor-Liquid Equilibrium Data Collection*, Frankfurt: DECHEMA.

(c) Outline how you could find the P, T, n of the tank where condensation starts. Provide relations to find answers, and ensure that enough equations are provided to calculate numerical values for all variables, but you do not need to calculate a final number.

15.37 (a) At 298 K, butane follows the equation of state: $P(V - b) = RT$ at moderate pressures, where b is a function of temperature. Calculate the fugacity for butane at a temperature of 298 K and a pressure of 1 MPa. At this temperature, $b = -732$ cm^3/mol.

(b) Pentane follows the same equation of state with $b = -1195$ cm^3/mol at 298 K. In a mixture, b follows the mixing rule: $b = x_1^2 b_1 + 2x_1x_2b_{12} + x_2^2 b_2$ where $b_{12} = -928$ cm^3/mol. Calculate the fugacity of butane in a 20 mol% concentration in pentane at 298 K and 1 MPa, assuming the mixture is an ideal solution.

15.38 The Soave equation of state is:

$$Z = \frac{1}{1 - b\rho} - \frac{a\rho}{RT} \frac{1}{(1 + b\rho)}$$

where the mixing and combining rules are given by Eqns. 15.8 and 15.9. Develop an expression for the fugacity coefficient and compare it to the expression given by Soave (1972. *Chem. Eng. Sci.* 27:1197).

15.39 The following equation of state has been proposed for hard-sphere mixtures:

$$\frac{A - A^{ig}}{RT} = \frac{2}{(1 - \eta_P)^2}$$

where $\eta_P = \sum_i x_i b_i / V$

Derive an expression for the fugacity coefficient.

15.40 The equation of state below has been suggested. Derive the expression for the fugacity coefficient.

$Z = 1 + 4c\rho/(1-b\rho)$

where $b = \sum_i x_i b_i$; $c = \sum_i \sum_j x_i x_j c_{ij}$; $c_{ij} = $ constant

15.41 The following free energy model has been suggested as part of a new equation of state for mixtures. Derive the expression for the fugacity coefficient of component 1.

$$\frac{A(T, V) - A^{ig}(T, V)}{RT} = -\frac{B^2}{C} \ln\left(1 + \frac{C\rho}{B}\right)$$

where $B = \sum_i \sum_j x_i x_j B_{ij}$; $C = \sum_i \sum_j \sum_k x_i x_j x_k C_{ijk}$

CHAPTER

ADVANCED PHASE DIAGRAMS

16

Whenever new fields of technology are developed, they will involve atoms and molecules. Those will have to be manipulated on a large scale, and that will mean that chemical engineering will be involved—inevitably.

Isaac Asimov (1988)

This chapter discusses topics related to phase behavior at a higher level of conception. We have seen that VLE, LLE, and SLE can occur for various isolated systems under specific conditions. We can even anticipate conditions when one type of phase behavior may dominate. For example, a large value of k_{12} is likely to lead to LLE. But the LLE must subside if the temperature is higher than the liquid-liquid critical temperature. Furthermore, VLE must subside if the temperature is higher than the vapor-liquid critical temperature. Is the vapor-liquid critical temperature always higher than the liquid-liquid critical temperature? Do the particular molecular properties (size, energy density, complexation) matter or do all molecules conform to a single "universal curve" of some sort? Noting that phase behavior is a function of many variables, how can we conceive of the behavior in a way that permits intuitive reasoning? Are there diagrams that might permit us to envision how to achieve certain practical goals? Or perhaps to prove that some goals are unattainable?

These are the questions that motivate the study of advanced phase diagrams. Through this study, we find that the molecular properties do matter, but six distinct diagrammatic types are sufficient to classify the types of phase behavior that have been observed to date. As an example of how diagrams can be applied, we illustrate the procedure for residue curve analysis, illustrating how processes like azeotropic and extractive distillation can be conceived and planned.

16.1 PHASE BEHAVIOR SECTIONS OF 3D OBJECTS

Several types of phase behavior may occur in binary systems.[1] In earlier chapters, we explored phase behavior by examining *P-x-y* or *T-x-y* diagrams. In this section we demonstrate how these

---

1. Much of this section has been published in Lira, C.T. 1996. "Thermodynamics of Supercritical Fluids with Respect to Lipid-Containing Systems," in *Supercritical Fluid Technology in Oil and Lipid Chemistry,* King, J.W., List, G.R., eds., Champaign, IL: AOCS Press. Reproduced with permission.

phase diagrams are related to the three-dimensional *P-T-x-y* diagrams. The *P-x-y* and *T-x-y* diagrams are two-dimensional cross sections of the three-dimensional phase envelope, and by studying the phase envelope, the progressions of changing shapes of the two-dimensional cross sections can be more easily grasped. The relations between the three-dimensional phase envelope and the cross sections are shown in Figs. 16.1A–16.1C for three different systems that all fall under the classification as Type I systems. Type I is the simplest class of phase behavior because there is no LLE behavior. Note that nonazeotrope as well as minimum boiling and maximum boiling homogeneous azeotropes are in this class. In each of the three-dimensional diagrams, short dashes are used to denote the pure component vapor pressures which terminate at the critical points denoted by *A* and *B*. Horizontal cross sections of the three-dimensional phase envelopes are shown with long dashes and are *T-x-y* diagrams. Vertical cross sections of the three-dimensional phase envelopes are shown with solid lines, and are *P-x-y* diagrams. (A summary of special notation used in this section appears at the end of the section.) There is a one-to-one correspondence of the cross sections in the three-dimensional plots to the phase envelopes plotted on the *P-x-y* and *T-x-y* cross sections. The solid line running from the critical point of *A* to the critical point of *B* is the **locus of critical points** of the mixture where the vapor and liquid become identical. Note the branches of the two-dimensional cross sections do not span the composition range when the critical locus is intersected. Each lobe on the cross sections will have a critical point. As discussed in Example 15.10 on page 601, the critical points are frequently not at the maximum temperature or pressure of the phase envelope. Refer back to phase diagrams from Chapters 10–15 to see how the phase diagrams in those chapters relate to the diagrams shown here.

❶ A summary of special notation appears at the end of this section.

❶ *P-x-y* and *T-x-y* diagrams are cross sections of 3*D* phase envelopes.

P-x-y and *T-x-y* diagrams for systems with LLE are shown in Figs. 16.2A and 16.2B. The liquid-liquid behavior occurs in the region to the left of the **critical endpoint *U*,** in the *U*-shaped region above the vapor-liquid envelope. Three-phase *llv* occurs on the surface marked with tie lines. The intersection of this surface with *P-x-y* and *T-x-y* diagrams results in the *llv* tie lines on the cross-section diagrams. At the upper critical endpoint, *U*, a vapor and liquid phase become identical, denoted with the notation *l–l=v*. When the vapor pressures of the components are significantly different from each other, the system may not have an azeotrope as shown in Fig. 16.2A(a–c). When the vapor pressures are closer to each other, azeotropes and heteroazeotropes will form. Fig. 16.2B(d–f) shows a system with heteroazeotropic behavior below the temperature of the upper critical endpoint, T_U, and azeotropic behavior above T_U.

Perspective

Before advancing further into the phase behavior classifications, some justification for such study is offered. High pressure can be used to create dense fluids that are useful for processing. For example, high pressure gases are employed industrially for petroleum fractionations in the oil industry and for hops and spice extractions and coffee decaffeination in the food industry. Dense gases are also under study for fractionation of specialty vegetable and fish oil components, as reaction media for polymerizations and other chemical synthesis and separations. High pressure processing and supercritical fluid extraction rely on control of solubility through manipulation of temperature and pressure. Solubility behaviors follow clear patterns which depend on similarities/differences in the thermodynamic and structural properties of the solute and the solvent. This section serves as an overview of phase behavior and systematic trends in phase behavior.

Natural materials such as foods and oils are multicomponent mixtures. Polymers typically contain a molecular weight distribution. Frequently, these types of mixtures are not well identified or characterized. Solubilities and extractabilities for these mixtures are currently difficult to predict quantitatively; however, significant knowledge regarding solubility trends may be obtained by

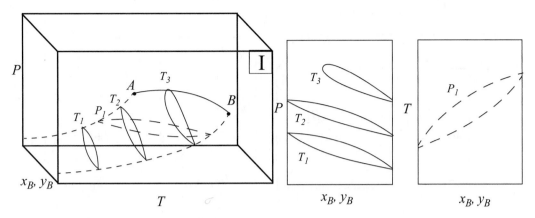

Figure 16.1A *Illustration of a system which does not form an azeotrope. The two-dimensional envelopes correspond to cross sections shown on the three-dimensional diagram. The P-x-y diagrams are vertical cross sections, and the T-x-y is a horizontal cross section.*

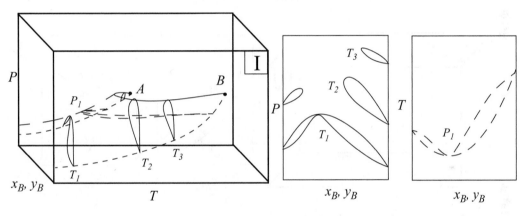

Figure 16.1B *Illustration of a system which forms a minimum boiling azeotrope due to positive deviations from Raoult's law. The two-dimensional envelopes correspond to cross sections shown on the three-dimensional diagram.*

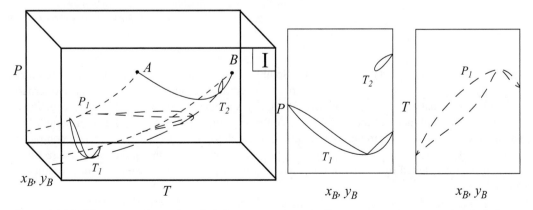

Figure 16.1C *Illustration of a system which forms a maximum boiling azeotrope due to negative deviations from Raoult's law. The two-dimensional envelopes correspond to cross sections shown on the three-dimensional diagram.*

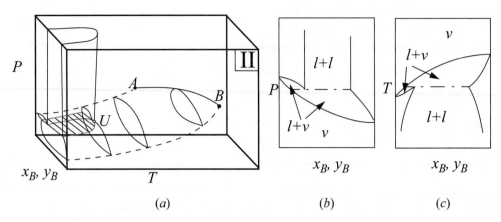

Figure 16.2A *(a) Type II system where the vapor pressures are significantly different; (b) P-x-y at a temperature below T_U; (c) T-x-y at a pressure below P_U.*

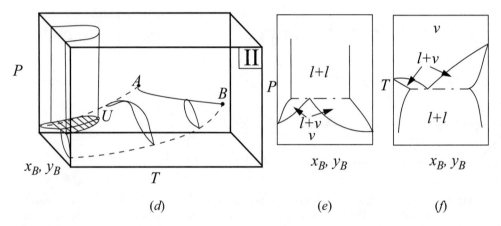

Figure 16.2B *(d) Type II system where the vapor pressures are relatively similar; (e) P-x-y at a temperature below T_U; (f) T-x-y at a pressure below P_U. It is also possible for an azeotrope to lie to the right or left of the liquid-liquid region, rather than the heteroazeotropic behavior that is shown.*

studying simpler binary and ternary systems. Solubility represents a saturation condition; therefore, solubility is represented as a boundary on a phase diagram. Systematic study of binary and ternary systems shows that the phase boundaries of ternary systems are intermediate to the constituent binary systems, and many of the same trends continue in multicomponent systems,[2] although fundamental exploration of the trends is an ongoing research topic. Process simulators and computers continue to simplify the calculation of phase equilibria; however, the interpretation of the resultant phase behavior is aided by a general understanding of the classes of phase behavior presented in this section.

2. Luks, K.D. 1986. *Fluid Phase Equil.* 29:209–24.

16.2 CLASSIFICATION OF BINARY PHASE BEHAVIOR

Since 1970 there have been several reviews and classifications of phase behavior[2,3,4,5,6,7,8,9,10,11]. The types are usually summarized by the projection of their phase boundaries onto two-dimensional pressure-temperature diagrams. Type I and II phase behavior have already been discussed, and they are shown by the upper two plots in Fig. 16.3. Note that azeotropic behavior is a subset of the major classes of behavior and it is not shown explicitly on the projections in Fig. 16.3.

> There are six major types of phase behavior.

For this discussion, the convention of van Konynenburg and Scott[10] is followed for classification of phase behavior. Following a trend that should be familiar by now, van Konynenburg and Scott explored the implications of the van der Waals equation in pursuit of answers to questions like those raised in the introduction. There are six major types of phase behavior shown in the plots which use special notation denoted in the figures and at the end of this section. The types of phase behavior are indicated with the Roman numerals in the upper left of each plot, and the distinctions between the types are due to the location of critical points and critical loci. Types I–V are exhibited by the van der Waals equation at various proportions of size and energy density. Type VI phase behavior appears to be exclusive to aqueous systems.[11]

The Gibbs phase rule is helpful in interpreting the *P-T* projections. In a pure system, if two phases coexist, one degree of freedom is available; therefore, two-phase coexistence appears as a line on a *P-T* projection. At the triple point, solid, liquid, and gas coexist, and no degrees of freedom are available; therefore, the condition is a point on the *P-T* projection. At critical points, all intensive properties of two phases become identical, so the number of degrees of freedom is reduced, and this also appears as a point for pure systems. To avoid misuse of the **phase rule,** only intensive variables are used for the degrees of freedom. Also, the intensive variables must be varied over a finite range; for example, two fluid phases are impossible above the critical temperature of a pure substance. More detailed discussions of the correct use of the degrees of freedom are available.[12]

> The Gibbs phase rule is helpful in interpreting the *P-T* projections.

In a binary system, liquid-vapor phase behavior may occur with two degrees of freedom. Liquid-vapor critical behavior occurs with one degree of freedom. Therefore, liquid-vapor behavior appears within a region on the *P-T* projection, and critical behavior occurs along a line known as the critical locus. In Fig. 16.3, pure component *l-v* lines are indicated by the dashed lines. Pure component critical points are indicated by solid circles. Invariant critical points in the binary are indicated by open circles and critical lines are indicated by solid lines. Critical lines are hashed to indicate the side on which two phases coexist. Since the critical locus is a projection on the *P-T* dia-

3. Jangkamolkulchai, A., Lam, D.H., Luks K.D. 1989. *Fluid Phase Equil.* 50:175–187.

4. Estrera, S.S., Arbuckle, M.M., Luks, K.D. 1987. *Fluid Phase Equil.* 35:291–307.

5. McHugh, M.A., Krukonis, V.J. 1986. *Supercritical Fluid Extraction: principles and practice*, Stoneham, MA: Butterworths.

6. Miller, M.M., Luks, K.D. 1989. *Fluid Phase Equil.* 44:295–304.

7. Schneider, G.M. 1991. *Pure and Applied Chem.* 63:1313–1326. Also published as Schneider, G.M. 1991. *J. Chem. Therm.* 23:301–26; Schneider, G.M. 1970. *Adv. Chem. Phys.* 1:1–42.

8. White, G.L., Lira, C.T. 1992. *Fluid Phase Equil.* 78:269; White, G.L., Lira C.T. 1989. in *Supercritical Fluid Science and Technology.* Johnston, K.P., Penninger, J.M.L. eds., Washington, DC: American Chemical Society, pp. 111–120.

9. Schneider, G.M. 1978. *Angew. Chem. Int. Ed. Engl.* 17:716–27.

10. van Konynenburg, P.H., Scott, R.L. 1980. *Phil. Trans. Roy. Soc. London, Ser. A* 298(1442):495–540.

11. Streett, W.B. 1983. in *Chemical Engineering at Supercritical Fluid Conditions.* Paulaitis, M.E., Penninger, J.M.L., Gray, R.D., Davidson, P. eds., Ann Arbor, MI: Ann Arbor Science, pp. 3–30.

12. Modell, M., Reid R. 1983. *Thermodynamics and Its Applications,* 2ed. Upper Saddle River, NJ: Prentice-Hall, pp. 259–264; Denbigh, K. 1981. *The Principles of Chemical Equilibria,* 4ed. Cambridge, UK: Cambridge University Press, pp. 188–190.

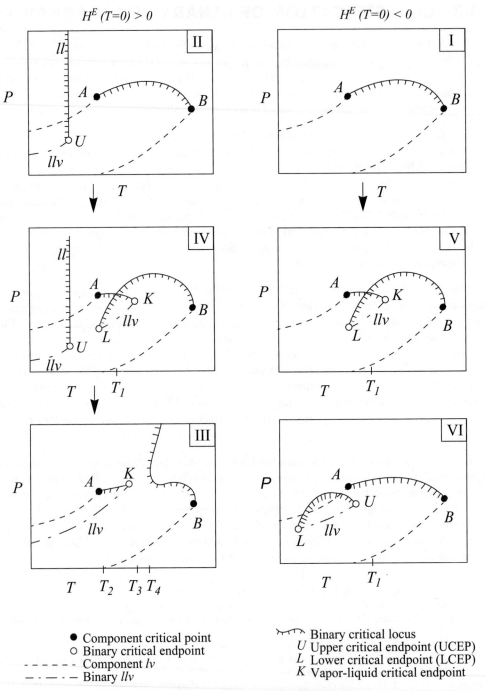

Figure 16.3 *Progression of binary phase behavior with increasing molecular asymmetry according to van Konynenburg and Scott.[10] Arrows denote progressions of phase behavior expected by theory. Experimental progressions frequently differ.*

gram, the diagram and phase rule indicate that two phases coexist on one side of the curve *over a finite composition range*. The diagram does not imply that two phases will coexist at *all* compositions below the critical locus. Also, in some areas, two critical curves are superimposed over the same temperature-pressure range (e.g., Types IV and V); however, the overlaps are due to the projection onto the two-dimensional diagram—the two overlapping regions of critical behavior occur over different composition ranges. The critical locus represents conditions where two phases become identical but this does not necessarily imply that only one phase exists above the critical locus because: 1) the two phases which become identical at the critical locus may coexist at temperatures and/or pressures slightly above the critical locus as nonidentical phases (see Example 15.10 on page 601; and 2) if three phases exist below the critical line, two phases will coexist above the critical line. Also, the absence of a critical line on a region of a *P-T* trace does not imply that only one phase is present; it simply means that no phases become identical within the range of the diagram.

Phase Behavior in the Presence of Solids

Figs. 16.1A–16.5 illustrate only fluid phase behavior. Solid-liquid-vapor coexistence can interfere with the fluid phase behavior when the triple point temperature of the higher molecular weight (*heavier*) component approaches the range of experiments. (The general trend is for the melting point to increase with molecular weight.) The superposition of solidus lines on these diagrams will be discussed later. For all the phase diagrams sketched here, the light component *A*, has the lower critical temperature and appears at the left in Fig. 16.3 and in the rear of Fig. 16.4.

Molecular Asymmetry

The type of phase behavior depends on the molecular asymmetry of the mixture. **Molecular asymmetry** is a term used to describe size differences (molecular weight) for functionally similar molecules or polarity or functional differences for molecules of similar molecular weight. As the molecular asymmetry of the system increases, the critical points of the species generally move farther from each other on the *P-T* traces. With increasing disparity of the critical points, all phase behavior spans increasingly larger areas of *P-T* space.

❶ The type of phase behavior depends on the molecular asymmetry of the mixture.

Type I and Type II phase behavior occurs in systems where the molecular asymmetry is relatively small. When the molecular asymmetry is greater, the immiscibility regions become larger, as illustrated by the three-dimensional diagrams for Types III and IV in Fig. 16.4. Note that the region below *U* in Type IV can look the same as the region below *U* in Type II. Fig. 16.5 shows some isothermal sections where the temperatures used for the plots are denoted in Figs. 16.3 and 16.4. Note that the phase behavior in Figs. 16.5(a) and 16.5(b) differ only in the existence of a *ll* critical point of the first diagram. The effect of temperature on the phase behavior of a Type III system is interesting because the narrow neck region can pinch off as the temperature is increased, resulting in the cross sections of Figs. 16.5(c) and 16.5(d).

Molecular asymmetry was characterized by van Konynenberg and Scott by the variables along the axes of Fig. 16.6(a). We can refer to Fig. 16.6(a) as a master phase map because it shows where to find certain types of behavior, but it does not provide a system-specific diagram. These must have been extremely painstaking computations in 1970. Fortunately, modern programming makes it feasible to compute phase diagrams with relative ease. A particular example is the GPEC project being developed by Cismondi and coworkers.[13] We can gain insight into the process required to develop a phase map by running the (free) GPEC program for mixtures of *n*-alkanes with N_2, CH_4,

13. Cismondi, M., Michelsen, M. 2007. *J. Supercrit. Fluids* 39:287. gpec.efn.uncor.edu. accessed 11/2011.

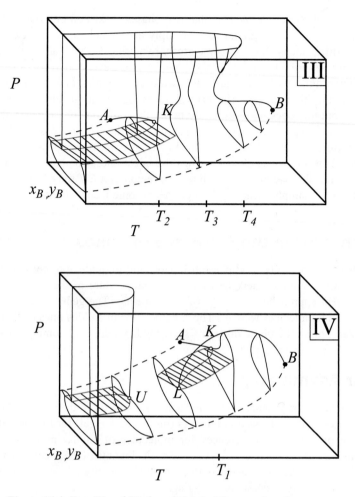

Figure 16.4 *Type III and IV phase behavior illustrated on three-dimensional diagrams. Symbols are the same as Fig. 16.3. The labeled temperatures are the same as Fig. 16.3 and are used for plotting cross sections in other figures.*

C_2H_6, CO_2, CH_3OH, and H_2O using the Peng-Robinson equation. To generate a map, each point in Fig. 16.6(b) represents a computation for a different mixture. X represents type II, Δ represents type III, and $\square$ represents type IV. Consider N_2 + n-alkanes. The phase diagram for N_2 + methane is type I, N_2 + ethane is type IV, but N_2 + propane and all higher molecular weight n-alkanes are Type III, based on the calculations. This is expected based on the progressions in Fig. 16.3 (Also, see Assimilation of Experimental Data on Homologous Series on page 629.) Type III behavior may seem somewhat surprising. It means that the vapor-liquid critical locus never quite merges from one pure component to the other. For the N_2 + propane system, the propane-rich critical temperature locus extends below N_2's critical temperature, but the pressure is higher and it is impossible to squeeze enough N_2 into the propane such that the critical loci connect. Instead, the N_2-propane phase-split widens with increasing pressure into something resembling $l + l$ (Fig. 16.5(c)). The driving force in this case is the high compressibility near the critical region, and the difference in energy density between N_2 and propane. Type IV for N_2 + ethane is peculiar; it exhibits a liquid-liquid phase split, even though neither component is polar.

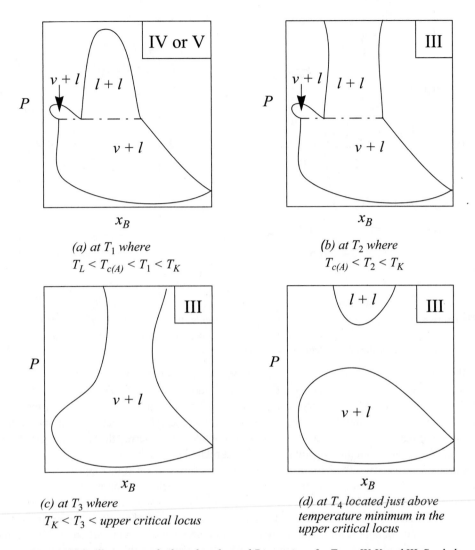

Figure 16.5 *Illustration of selected isothermal P-x sections for Types IV, V, and III. Symbols and labeled temperatures are the same as Figs. 16.3 and 16.4.*

The axes in Fig. 16.6 are defined such that differences in size, solvation, and energy density are characterized.

$$\zeta \equiv (a_{11}/b_{11}^2 - a_{22}/b_{22}^2)/(a_{11}/b_{11}^2 + a_{22}/b_{22}^2) \qquad 16.1$$

$$\Lambda \equiv (a_{11}/b_{11}^2 - 2(a_{12}/b_1 b_2) + a_{22}/b_{22}^2)/(a_{11}/b_{11}^2 + a_{22}/b_{22}^2) \qquad 16.2$$

$$\xi \equiv (b_2 - b_1)/(b_2 + b_1) \qquad 16.3$$

where a and b are the van der Waals parameters. Recall from Eqn. 12.21 that $a/b^2 \sim \delta^2$.

Several notes should be added to put Fig. 16.6(b) into context. Perhaps the most important observation is that the boundaries between types of behavior are not perfectly distinct. Types IV

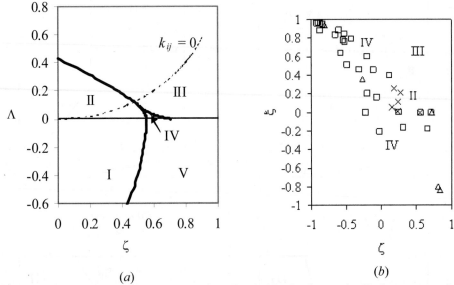

(a)   (b)

Figure 16.6   *Global phase maps (a) for equal sized molecules for the van der Waals equation, after van Konynenberg and Scott. The dashed line corresponds to $k_{12} = 0$. (b) Variable sized molecules with the Peng-Robinson equation at $k_{12} = 0$.*

and III may overlap occasionally depending on the specific components in the mixture. Nevertheless, some general trends are discernible. For example, the range of Type IV behavior appears to be much larger when the molecules vary in size than one might infer from Fig. 16.6(a). Furthermore, it should be noted that, for the Peng-Robinson equation, the ξ–ζ relation increases up and to the left along a homologous series; so the number of components in a series that overlap the Type IV region may be greater than initially anticipated. On a related note, when the molecular weight of the larger molecule exceeds ~500 the mapped points lie in a tiny portion at the upper left corner of the figure. Thus, a vast amount of polymer solution experience lies in a region where Types II, III, and IV are barely distinguishable based on the current analysis, and small variations in k_{12} would drastically alter the phase behavior (as one might expect in the presence of alcohols, for example). First, Types I and V do not appear although many phase diagrams have been classified as Types I and V experimentally. Similar to van Konynenburg and Scott, we attribute this to interference from solid phase boundaries in the experimental systems (as may become more obvious when studying below), hypothesizing that a *ll* region must exist at $T \rightarrow 0$ when the geometric mean is applied. Hence Type II systems may be classified experimentally as Type I if the *ll* region lies below the solid phase boundary, and similarly for Type IV and V systems.

Phase Stability and Critical Phase Behavior Computations

We have touched on the subject of phase stability previously, but we have not delved deeply into it because it can be a complicated subject if treated in detail. Referring to the critical points of pure fluids, we developed Eqns. 7.27 based on a simple analysis of the inflection behavior of isotherms. A similar analysis led to the identification of spinodal conditions and formulation of the equal area rule as a computational method for vapor pressure in the form of Eqn. 9.49 on page 358. Conceptually, stability analysis is always similar to the perspective developed for pure fluids, but the derivatives become more complicated when the dimensionality of the problem expands to include multiple components.

A hint of the development for mixtures was given in the LLE ($l + l$) analysis leading to Eqn. 14.7. Once again, we considered the inflection behavior with respect to a single variable (x_1) to infer the necessary conditions for LLE, and, coincidentally, the critical point. If you consider that pressure is the first derivative of the Helmholtz energy, you may be struck by the similarity of Eqns. 7.27 and 14.7. At constant volume, both involve second and third derivatives of "free energy" with respect to mole number.

The multicomponent analysis has been discussed in depth by many authors. The state of the art as of this writing is well represented by the work of Michelsen and Mollerup.[14] Briefly, the extension requires a generalization of the tangent line evident in the LLE ($l + l$) development to become a tangent plane. This leads to a need for many derivatives with respect to density, mole number, and temperature along with tests for the stability of phases detected. This approach forms the basis of the GPEC program by Cismondi and coworkers.

An alternative formulation has been presented by Eubank and coworkers as an extension of the equal area rule (EAR) for pure fluids.[15] As in the case of pure fluids, stability checking is an inherent part of the EAR method because the integrand is easily checked for extrema before performing further analysis. The formulation of EAR for binary mixtures is particularly simple since it involves only a single variable, x_1.

$$\underline{G} = (n_1 + n_2)G(x_1) \tag{16.4}$$

$$\mu_1 = \partial \underline{G} / \partial n_1 = G(x_1) + (dG/dx_1)(ndx_1/dn_1) = G(x_1) + (dG/dx_1)(x_2) \tag{16.5}$$

Note that the right-hand steps are only possible for binary mixtures. From Eqn. 10.42, $G = x_1 \mu_1 + x_2 \mu_2$, so

$$\mu_1 = x_1 \mu_1 + x_2 \mu_2 + x_2(dG/dx_1) => \mu_1(1 - x_1) = x_2 \mu_2 + x_2(dG/dx_1) \tag{16.6}$$

Setting $(1 - x_1) = x_2$ and rearranging,

$$(dG/dx_1) = \mu_1 - \mu_2 \tag{16.7}$$

This equation facilitates computing (dG/dx_1) using a typical program for chemical potentials. Since $\mu_i^\alpha = \mu_i^\beta$,

$$(dG/dx_1)^\alpha = (dG/dx_1)^\beta \tag{16.8}$$

To recognize the EAR form, we can multiply by $(x_1^\alpha - x_1^\beta)$ and expand to obtain

$$(x_1^\alpha - x_1^\beta)(dG/dx_1)^\alpha = (x_1^\alpha - x_1^\beta)(\mu_1^\alpha - \mu_2^\alpha) = x_1^\alpha(\mu_1^\alpha - \mu_2^\alpha) - x_1^\beta(\mu_1^\beta - \mu_2^\beta) \tag{16.9}$$

$$x_1^\alpha(\mu_1 - \mu_2) - x_1^\beta(\mu_1 - \mu_2) = x_1^\alpha\mu_1 - x_1^\alpha\mu_2 - x_1^\beta\mu_1 + x_1^\beta\mu_2 + \mu_2 - \mu_2 \tag{16.10}$$

Combining the $\mu_2 - x_1^\alpha\mu_2 = x_2^\alpha\mu_2$. A similar combination gives $x_2^\beta\mu_2$. So,

$$x_1^\alpha \mu_1 + x_2^\alpha\mu_2 - x_1^\beta \mu_1 - x_2^\beta\mu_2 = G^\alpha - G^\beta \tag{16.11}$$

Returning to Eqn. 16.9, substituting and rearranging, we have

14. Michelsen, M., Mollerup, J.M. 2004. *Thermodynamic Models: Fundamentals and Computational Aspects*, Denmark: Tie-Line Publications, www.tie-tech.net, (accessed 11/2011).
15. Eubank, P.T., Elhassan, A.E., Barrufet, M.A., Whiting, W.B. 1992. *Ind. Eng. Chem. Res.* 31:942.

$$\left(\frac{dG}{dx_1}\right)^\alpha = \left(\frac{dG}{dx_1}\right)^\beta = \frac{G^\alpha - G^\beta}{x_1^\alpha - x_1^\beta} = \frac{1}{x_1^\alpha - x_1^\beta}\int_\beta^\alpha\left(\frac{dG}{dx_1}\right)dx_1 \qquad 16.12$$

Eqn. 16.12 is analogous to Eqn 9.49 and implementation follows the analogous procedure: Find the spinodals, guess the value of $(dG/dx_1)^\alpha$, solve for x_2^α and x_1^β, compute the next guess for $(dG/dx_1)^\alpha$, and repeat to convergence. One practical note for polymer systems would be to search for spinodals based on volume fraction since $dG/d\Phi_1 = 0$ when $dG/dx_1 = 0$, and using volume fraction spreads the concentration range over more reasonable values for polymer systems. Advantages of the EAR approach include (1) relatively few derivative properties are required; (2) a stability test is an inherent part of the procedure; and (3) convergence is stable, even near the critical point.

Experimental Studies of Homologous Series

One way to understand the progression of phase behavior is to review literature measurements which classify the phase behaviors. Classifications are based on experimental studies and may require revision if additional phase transitions are found to occur.[16,17] Types I or II occur when the molecules are fairly similar in structure or critical properties. Van Konynenburg and Scott found the van der Waals equation predicted the progression II $\Rightarrow$ V $\Rightarrow$ III for increasing molecular asymmetry in systems with an endothermic low-temperature heat of mixing, and they found the progression I $\Rightarrow$ V for increasing asymmetry in systems with an exothermic low-temperature heat of mixing. Most nonpolar systems are expected to have endothermic heats of mixing (e.g., ethane-butane, benzene-hexane) and might be expected to be Type II by theory; however, they are classified as based on experimental measurements, and most are Type I. Type II phase behavior includes a liquid-liquid immiscibility below the critical temperature of both components. The liquid-liquid behavior is often relatively insensitive to pressure when the liquids are incompressible far below the critical temperature. In the liquid-liquid region, as the temperature is increased, the liquids become increasingly miscible in each other until they become identical at the UCEP. The UCEP occurs at extremely low temperatures for small endothermic heats of mixing, and is frequently not found experimentally due to lack of experimental exploration at low temperatures or due to freezing of the liquids before they become immiscible. While a significant number of liquid-liquid experiments have been performed to locate liquid-liquid UCEPs at atmospheric pressure,[18] phase behavior characterization near the critical point of the lighter component is necessary to permit classification as Type II or IV. A summary of some homologous series is provided below to explore the progressions of phase behavior. To study the series, the asymmetry of the systems is systematically increased by varying the molecular weight of the heavier component by one functional group at a time, and observing the trends in the location of critical points and the class of phase behavior.

Ethane/n-Alkane Series Trends

Studies of the ethane family are summarized by Peters, et al.,[16] and Miller and Luks.[6,] Type I behavior exists up through n-heptadecane because a UCEP has not been reported, possibly because the components freeze before they become immiscible. Beginning with n-octadecane, a liquid-liquid-vapor region develops near the ethane critical point characteristic of Type V. Once again, a UCEP is not found experimentally. This phase behavior continues through n-tricosane. With n-tetracosane and n-pentacosane a modification of Type V or III occurs where the s_Bl_2v line interferes with the fluid behavior as shown in Fig. 16.7(a). The three phase s_Bl_2v line extends from the triple

16. Peters, C.J., Lichtenthaler, R.N. de Swaan Arons, J. 1986. *Fluid Phase Equil.* 29:495–504.
17. Enick, R., Holder, G.D., Morsi, B.I. 1985. *Fluid Phase Equil.* 22:209–24.
18. Hildebrand, J.H., Prausnitz, J.M., Scott, R.L. 1970. *Regular and Related Solutions.* Van Nostrand.

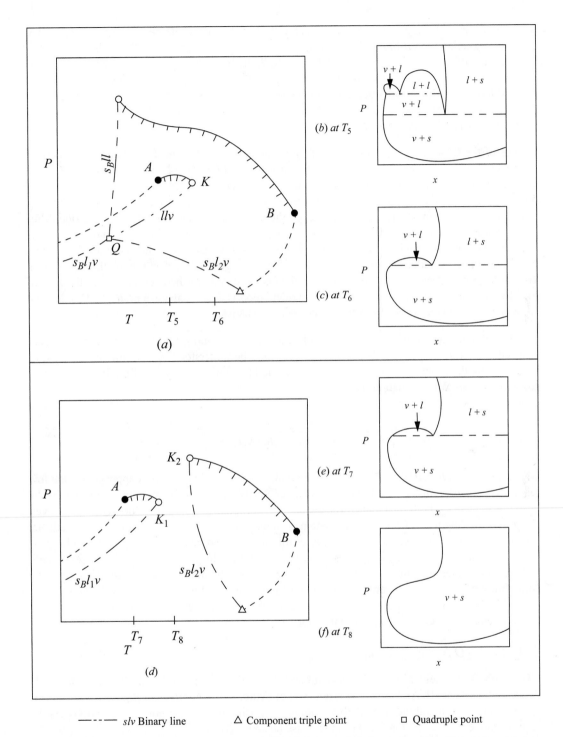

---···--- *slv* Binary line   △ Component triple point   □ Quadruple point

Figure 16.7 *Pressure-temperature projections of phase behavior when the triple point of B is in the vicinity of the critical point of A. Symbols are the same as Fig. 16.3, except as noted.*

point of the heavier component and intersects the *llv* line at a quadruple point Q where four phases coexist. Beginning with *n*-hexacosane, the $s_B l_2 v$ line moves to temperatures above the K point and the Fig. 16.7(d) phase behavior begins.[16] In Fig. 16.7, the line labeled $s_B l_2 v$ denotes the conditions where the solid melts over some composition ranges.[8] Schneider reports that squalane, a branched C_{30}, exhibits Type III,[9] which does not fit the pattern of *n*-alkanes at the same carbon number.

CO₂/n-Alkane Series Trends

Studies of the CO_2 family are summarized by Schneider,[7,9] Miller and Luks,[6] and Enick et al.[17] Pure CO_2 freezes at a relatively high reduced temperature ($T_r = 0.712$) while ethane freezes at a comparatively low reduced temperature ($T_r = 0.295$). Therefore, it may initially seem less likely to observe liquid-liquid immiscibility in CO_2 systems because the systems might freeze before the liquid-liquid critical point, U, is reached; however, the opposite is found, and although the liquid-liquid region is not seen for the lightest alkanes, *n*-heptane clearly shows Type II behavior.[6] Type II behavior continues through *n*-dodecane. Enick, et al.,[17] show that *n*-tridecane is Type IV, and that beginning with *n*-tetradecane, the type of Fig. 16.7(a) emerges. Schneider[9] reports the type of Fig. 16.7(a) for *n*-hexadecane. The type of Fig. 16.7(a) is exhibited at *n*-heneicosane[19] and the type of Fig. 16.7(d) appears with *n*-docosane.[6,20] Schneider[9] reports that squalane, a branched C_{30}, is Type III, and as with ethane, varies from the pattern with *n*-alkanes.

The homologous series of *n*-alkanes with CO_2 exhibits liquid-liquid behavior at much higher reduced temperatures compared to ethane.[21] This may be understood by considering the heat of mixing using the propane system as an example. The low temperature heat of mixing is estimated from the van der Waals equation by

$$\lim_{T \to 0} H^E = \left(\frac{a_1}{b_1^2} - \frac{2a_{12}}{b_1 b_2} + \frac{a_2}{b_2^2} \right) \frac{x_1 x_2 b_1 b_2}{x_1 b_1 + x_2 b_2} \qquad 16.13$$

where the *a*'s and *b*'s are the van der Waals' parameters. Using Eqn. 16.13 to approximate the low temperature heat of mixing and the normal van der Waals' geometric mean combining rule for the cross parameter a_{12}, both the ethane/propane and CO_2/propane systems are endothermic; however, the heat of mixing in the CO_2/propane system is roughly 100 times the heat of mixing for ethane/propane. Therefore, the excess Gibbs energy of mixing will also be considerably larger and liquid-liquid behavior is expected to occur to higher temperatures. In general, the *n*-alkane series with CO_2 shows greater asymmetry than the same ethane series, as might be expected. Therefore, the solubilities are lower at a given temperature and pressure, and the progression of behavior from Type II to the type of Fig. 16.7(d) occurs at lower carbon numbers, even though the critical temperatures of ethane and CO_2 are approximately the same.

CO₂ and N₂O Series Trends

CO_2/alcohols: Schneider[9] reports 2-hexanol and 2-octanol as Type II and 2,5-hexanediol and 1-dodecanol[7] as Type III. As expected, the addition of hydroxyls on component B increases the asymmetry of the system and the same carbon number by raising the critical point of B, and lowers the carbon numbers for transitions between the phase behavior types.

19. Fall, D.J., Fall, J.L., Luks, K.D. 1985. *J. Chem. Eng. Data* 30:82–88.

20. Fall D.J., Luks, K.D. 1984. *J. Chem. Eng. Data* 29:413–417.

21.   For comparative purposes the reduced temperature is calculated based on the critical temperature of ethane or CO_2 for the respective systems.

CO$_2$/aromatics and ethane/aromatics: Both CO$_2$ and ethane have been studied with n-alkylbenzenes through C$_{21}$ and C$_{20}$, respectively.[6] The over-all trends in behavior are very similar to the comparisons made in the n-alkane series. Polycyclic aromatics have also been studied, and are typically the type of Fig. 16.7(d).[8]

N$_2$O/n-alkanes: Nitrous oxide has been studied with the n-alkanes, and the phase behavior is very similar to the ethane series[3] in both the carbon number where phase behavior changes, and the critical endpoint temperature values.

CO$_2$/triglycerides/fatty acids: Many common triglycerides and fatty acids are solids at 30°C, and thus exhibit the type of Fig. 16.7(a) or Fig. 16.7(d) behavior in binaries with CO$_2$. Compounds which are normally liquids will probably be Type III, IV, or V. Relatively few studies of model systems are available and in all studies only solubilities are reported.[22,23,24,25] Bamberger, et al.[25] did determine that lauric and myristic acid, trilaurin, and trimyristin melted at the average pressure of their experiments, but the experiments were insufficient to characterize the phase behavior types as Fig. 16.7(a) or (d). Tripalmitin is reported to remain a solid although the triple point is only 5°C greater than trimyristin. Also, most of the mixtures involving tripalmitin were reported to remain solid, which would not be expected if the solids formed an eutectic mixture.[8] The other publications have not provided any characterization of the melting behavior of the solids. Unfortunately, disagreements of up to an order of magnitude exist in a few of the solubility measurements. The disagreements have been attributed in part to purity of materials.[23,24] Nilsson, et al.[24] report solubility data for mono- and di-glycerides. One interesting conclusion from the analysis of Czubryt, et al.[26] is that stearic acid appears to be dimerized in the CO$_2$ phase. Phase behavior of oil mixtures will be discussed in the section regarding solubility behavior.

Propane/Triglycerides/Fatty Acids Series Trends

Hixson, et al. published most of the available information on these systems in the 1940's,[27,28,29] and recently, more complete information has been published on the propane/tripalmitin system.[30] Propane refining of oils was practiced industrially as the Solexol process and descriptions are available.[31] Similar to the case of stearic acid in CO$_2$ discussed above, the acids appear to dimerize in the fluid phase. This hypothesis is supported by the correlation of phase behavior with the effective molecular weight, which is double the molecular weight of an acid and equal to the molecular weight of an ester or triglyceride. (See the trend for the LCEP in Fig. 16.8). Binary mixtures of propane with lighter molecular weight acids and esters is Type I (a UCEP has not been located), while the triglycerides and heavier acids and esters exhibit Type IV. For lauric acid (effective M.W. = 400.6) and for myristic acid (effective M.W. = 456.7) a lower critical end point (LCEP) has not

Critical points vary systematically as molecular asymmetry is changed.

22. King, M.B., Bott, T.R., Barr, M.J., Mahmud, R.S. 1987. *Sep. Sci. Techn.* 22:1103–20; Brunetti, L., Daghetta, A. Fedeli, E., Kikic, I., Zanderighi, L. 1989. *J. Am. Oil Chem. Soc.* 66:209–17; Chrastil, J. 1982. *J. Phys. Chem.* 86:3016–21.

23. Goncalves, M, Vasconcelos, A.M.P., Gomes de Azevedo, E.J.S., Chaves das Neves, H.J., Nunes da Ponte, M. 1991. *J. Am. Oil Chem. Soc.* 68:474–80.

24. Nilsson, W.B., Gauglitz, E.J., Hudson, J.K. 1991. *J. Am. Oil Chem. Soc.* 68:87–91.

25. Bamberger, T., Erickson, J.C., Cooney, C.L., Kumar, S.K. 1988. *J. Chem. Eng. Data* 33:23.

26. Czubryt, J.J., Myers, M.N., Giddings J.C. 1970. *J. Phys. Chem.* 74:4260–66.

27. Hixson, A.W., Bockelmann, J.B. 1942. *Trans. Am. Inst. Chem. Eng.* 38:891–930; Drew, D.A., Hixson, A.N. 1944. *Trans. Am. Inst. Chem. Eng.* 40:675–694; Hixson, A.N., Miller, R. 1940. U.S. Pat. 2,219,652; 1944. U.S. Pat. 2,344,089; 1945. U.S. Pat. 2,388,412.

28. Hixson, A.W., Hixson, A.N. 1941. *Trans. Am. Inst. Chem. Eng.* 37:927–957.

29. Bogash, R., Hixson, A.N. 1949. *Chem. Eng. Progress* 45:597–601.

30. Coorens, H.G.A., Peters, C.J., de Swaan Arons, J. 1988. *Fluid Phase Equil.* 40:135–151.

31. Passino, H.J. 1949. *Ind. Eng. Chem.* 41:280–287, Dickinson, N.L., Meyers, J.M. 1952. *J. Am. Oil Chem. Soc.* 29:235–39.

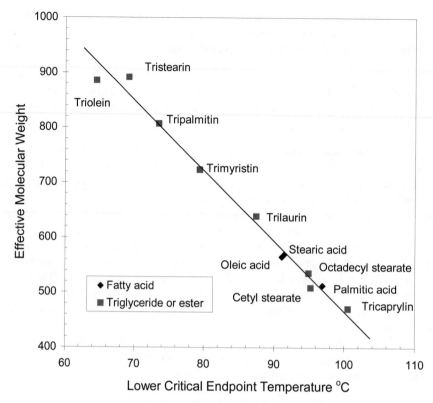

Figure 16.8 *Trend in the lower critical endpoint (LCEP) for components in binary mixtures with propane (Type IV systems).*

been located at propane concentrations between 25 and 95 wt%.[29] Tricaprylin (M.W. = 470.7) and higher molecular weight compounds exhibit Type IV. The correlation in Fig. 16.8 is based on only saturated and mono-unsaturated compounds. Conjugation of unsaturation may have a dramatic effect as exhibited by abietic acid which does not show an LCEP above 30°C.[28]

The Solexol process is based on refining of the oils using the *llv* region in Type IV or V systems between the LCEP and the vapor-liquid critical endpoint, *K*. The phases of interest are the two liquid phases, and Hixson and coworkers did not measure the vapor phase solubilities. As might be expected, the compounds with the lowest effective molecular weight have the greatest solubility in the propane-rich liquid, and for a given chain length solubility increases with decreasing *effective* molecular weight: triglyceride - fatty acid - ester. Hixson's studies concentrate on the measurement of *l-l* solubilities on the three-phase *llv* line in binaries and extension of the measurements to model ternary systems. Hixson's publications address the incorporation of staged countercurrent flow, the importance of density differences for counter-current flow, and internal column recycle using temperature gradients which have been discussed recently in applications with CO_2 extractions.[32,33] These engineering concepts remain important in industrial applications of high pressure technology. In the Solexol process, virtually all the feed oil leaves the top of the first tower dissolved in

32. Stahl, E., Quirin, K.-W., Gerard, D. 1988. *Dense Gases for Extraction and Refining,* New York, NY: Springer-Verlag.

33. Eisenbach, W. 1984. *Ber. Bunsenges. Phys. Chem.* 88:882; Nilsson, W.B., E.J. Gauglitz, Jr., J.K. Hudson, V.F. Stout, J. Spinelli 1988. *J. Am. Oil Chem. Soc.* 65:109–117; Nilsson, W.B., Gauglitz, E.J., Hudson, J.K. 1989. *J. Am. Oil Chem. Soc.* 66:1596–1600.

propane, and the stream is then fractionated by molecular weight and saturation. The majority of color bodies leave the bottom of the first column. The conjugated unsaturated components are in general less soluble than the saturated or mono-unsaturated components. One of the distinct capabilities of the Solexol process was the concentration of vitamin A but, as alternative methods for obtaining vitamin A became available, the process was abandoned.

Assimilation of Experimental Data on Homologous Series

Schneider[7] suggests that Types II, IV, and III follow a trend which can be understood if the phase behaviors are considered as a result of effects that are superimposed on each other and can be best visualized on Fig. 16.3. When a homologous series of systems is studied, with only the heavier compound varying, the first system to exhibit Type III behavior has a strong maximum and minimum in the critical line extending from the heavier component critical point. If this strong minimum exists for less dissimilar systems, it is expected to intersect the *llv* line in two places, giving Type IV behavior. These trends also appear consistent with the calculations of Chai[34] using the Peng-Robinson equation. Another trend is obvious which supports this theory. As the dissimilarity of components increases, the UCEP temperature at *U* increases which tends to decrease the difference in the temperatures of *U* and *L* of Type IV systems until they merge and Type III results (Fig. 16.9). It is also possible for the phase behavior to progress directly from Type II to Type III.[10]

❶ There is a systematic progression from Type II → IV → III.

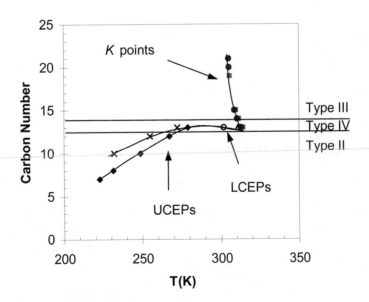

Figure 16.9 *Illustration of trends in critical endpoint temperatures as the molecular asymmetry of systems increases for carbon dioxide + n-alkyl benzenes (diamonds, triangle, filled circles) and carbon dioxide + n-paraffins (x's, o, filled squares) as compiled by Miller and Luks.*[11] *In both series, the systems are Type II below C13, Type IV for C13, then Type III above C13.*

34. Chai, C.-P., 1981. "Phase Equilibrium Behavior for Carbon-Dioxide and Heavy Hydrocarbons," Ph.D. dissertation, University of Delaware.

Critical Phase Behavior Summary

Phase behavior is determined by the degree of molecular asymmetry. A range of phase behavior exists, and a natural progression of phase behaviors occurs as the asymmetry of the system is increased. This progression can be visualized with two-dimensional master phase maps that represent sections through a three-dimensional space characterized by asymmetry in size, solvation, and energy density. These asymmetries can explain five of the six types of phase diagram observed in nature. The sixth type of diagram seems to be peculiar to aqueous systems and presumably may require explicit accounting for hydrogen bonding. Trends in critical phase behavior suggest that organic acids dimerize in nonpolar solvents.

P-T projections are specific to the type of phase behavior and summarize conditions where various processes are feasible or infeasible.

List of Special Symbols for This Section

1–$2 = 3$	phases 1,2,3 coexist, and phases 2 and 3 are identical at a critical state.
$1 = 2$–3	phases 1,2,3 coexist, and phases 1 and 2 are identical at a critical state.
A	component with lower critical temperature.
B	component with higher critical temperature.
K	a K point—a critical endpoint where the phases are $l - l = v$.
l	liquid phase.
LCEP	a lower critical endpoint where the phases are $l = l - v$.
Q	a Q point, or quadruple point, which has four phases in equilibria; herein, the four phases are $sllv$.
s	solid phase.
$UCEP$	an upper critical endpoint where the phases are $l - l = v$.
v	vapor phase.

16.3   RESIDUE CURVES

Using residue curves can save many design hours.

Distillation is among the most highly developed and reliable separation techniques for the chemical industry, so it is usually among the first separation techniques considered during process design. Design of multicomponent distillation columns can be complicated by azeotropes and heteroazeotropes. Shortcut distillation techniques are useful for distillation column screening, but azeotrope systems require modified shortcut equations and relative volatilities.[35] Design of multicomponent steady-state distillation columns is usually performed by process simulators; nevertheless, many design hours are saved if the phase behavior is explored using residue curves before working with the detailed column calculations. In this section, residue curves for ternary systems with a single feed and one distillate product and one bottoms product will be covered.[36] Residue curve analysis for ternary systems, under these conditions, involves the following concepts that will be described below.

1. The steady-state column feed, distillate, and bottoms compositions are co-linear on ternary diagrams in accordance with the lever rule.

35. Frank, T.C. 1997. *Chem. Eng. Prog.*, 93(4):52–63.

36. Interested readers may find a review article helpful. See Fien, G.-J.A.F., Liu, Y.A. 1994. *Ind. Eng. Chem. Res.* 33:2505. The topic is also available in chemical engineering textbooks intended for separations courses.

2. Two combined streams may be represented by a mixing point co-linear with the feed streams at a point between the streams given by the lever rule.

3. Binary and ternary azeotropes are connected on ternary diagrams by boundary lines called **separatrices.** (A single boundary is called a **separatrix.**)

4. Separatrices divide the ternary residue curve map diagram into regions.

5. As a reliable rule, the distillate, feed, and bottoms compositions will all be in the same residue curve map region.[37]

6. Residue curves begin and end in a single region—they do not cross separatrices.

Therefore, only certain composition regions in a ternary system describe possible products for a given feed, and the residue curves provide the guide to attainable compositions.

> ❶ Residue curves can provide a guide to attainable compositions.

Residue curves represent the trace of liquid-phase (the residue) composition during a single-stage constant-pressure batch distillation. Single-stage batch distillation calculations are not complex, and will be introduced below using the VLE K-ratios that have been the topics of earlier chapters. As the batch is distilled, the boiling point of the residue will increase until the composition reaches either a maximum boiling azeotrope or the pure composition of the highest boiling component. The residue curves will **terminate** at these compositions. The residue compositions will move away from any minimum boiling azeotropes and/or the pure composition of the lowest boiling component. The residue curves **originate** at these compositions. In the diagrams in this section, we use the letter o to represent the low boiling point residue curves' origin and the letter t to represent the high boiling point residue curves' terminus.

The Bow-Tie Approximation

Let us begin analysis by considering residue curves for a ternary system without azeotropes, propane + butane + pentane. As this ternary mixture boils, bubble-point calculations show that propane and butane are the most prevalent vapor species. As the vapor is removed, the residue becomes increasingly rich in pentane. Residue curves calculated with the Peng-Robinson equation are shown in Fig. 16.10(a). Starting from any initial composition, the curves on the diagram can be used to follow the trace of the residue composition. The residue from any ternary composition in this system will become increasingly rich in pentane. The residue curve map has a single region, and the residue curves originate at pure propane (the lowest boiling composition in the region) and terminate at pentane (the highest boiling point in the region). Residue curves are helpful in the design of steady-state continuous flow distillation columns because the accessible compositions of distillate and bottoms products from a steady-state distillation column can be found by studying the region of the residue curve map that contains the feed. *The distillate of a one-feed steady-state column will tend to approach the origin of the residue curves, and bottoms of the column will tend to approach the terminus of the residue curves within the restrictions summarized in points (1) through (6) above.*

Recognize the importance of point 1, that the feed, distillate, and bottoms for a ternary system must be colinear when plotted. Consider the implications for the propane + butane + pentane system shown in Fig. 16.10(a). For a feed, F, one limit of operation is for the bottoms product of a distillation column B_1 to be highest boiling point of the region (the terminus); the distillate (top product), D_1, must also be colinear on the line through B_1 and F, located to satisfy the lever rule material balance, $D_1/F + B_1/F = 1$, and $D_1/B_1 = (\overline{FB_1})/(\overline{FD_1})$. The other limit for operating the

---

37. This guideline may be overcome when the region boundary is highly curved and the feed is near the boundary on the concave side; however, such application is rarely economical.

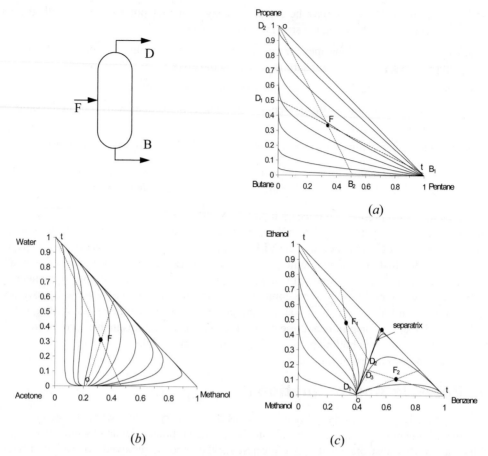

Figure 16.10 *(a) Residue curves for the system propane + n-butane + n-pentane as calculated with the Peng-Robinson equation at 1 bar; (b) methanol + acetone + water as modeled with the UNIQUAC equation at 1 bar; (c) methanol + ethanol + benzene as modeled with the UNIQUAC equation at 1 bar. Calculations performed using Aspen Plus ver. 9.2.*

column is to obtain the lowest boiling point in the region (the origin) as distillate, D_2, pure propane. In this case, B_2 must lie on a line extending through D_2 and F, similarly satisfying the lever rule. *The first approximation of reachable compositions is given by the bow-tie shaped region D_1-D_2-B_2-B_1.*[38] The distillate and bottoms compositions are not required to be on the boundary of the region as we have chosen to show in the figures, but the colinearity and lever rule must be followed. Note that it is not possible to simultaneously obtain pure propane and pure pentane in a single-feed, two-product (distillate and bottoms) column from feed F. Note that products near pure butane are unattainable, so design can focus on other alternatives. The bow-tie shaped region is an approximation that will be adequate for the screening introduction intended here.

> ❗ The bow-tie method provides a first approximation for attainable compositions.

Azeotropes

When one azeotrope exists in a ternary system, the residue curve map will still be a single region, but the azeotrope will be an origin or a terminus for the residue curves. Maximum boiling azeo-

---

38. The products of a two-feed column may lie outside the bow-tie region drawn from the overall feed, a principle exploited in extractive distillation. See Wahnschafft, O.M., Westerberg, A.W. 1993. *Ind. Eng. Chem. Res.*, 32:1108.

tropes may become the terminus of some residue curves, and minimum boiling azeotropes may become the origin of some residue curves. For example, in the methanol + acetone + water system shown in Fig. 16.10(b), the methanol + acetone system forms a minimum boiling azeotrope that becomes an origin for the residue curves because it is the lowest boiling point in the region. The distillate product for any steady-state column with a feed in the region will tend toward this composition. The highest boiling point in the region is the terminus of the residue curves at the composition of pure water, and the steady-state column bottoms will tend toward this composition. The bow-tie approximation of attainable compositions is shown for a feed F, and note that a distillate composition of pure methanol or pure acetone is not attainable because the residue curves' origin is the azeotrope. The lever rule concepts apply as in nonazeotropic systems.

When two or more azeotropes exist, the residue curve map will often be divided into two or more regions of residue curves. The residue curve maps show the distillation regions of attainable compositions, and the location of the separatrices. For example, residue curves for methanol + ethanol + benzene are shown in Fig. 16.10(c). In the left region, ethanol is the high boiling point (residue curve terminus) and the methanol + benzene azeotrope is the low boiling point (residue curve origin). In the right region, the methanol + benzene azeotrope is also the residue curve origin (low boiling point) and benzene is the residue curve terminus (high boiling point). The ethanol + benzene azeotrope is at an intermediate temperature between the origin and terminus of each region, and therefore, is neither an origin nor a terminus, and is called a saddle point. Two different arbitrary feed compositions are plotted on the diagram, and the corresponding bow-tie shaped regions of approximate accessible compositions are shown. Since the design rule is that a single column cannot cross a separatrix, the practical boundary for distillate compositions from the feed F_1 is approximated by the region F_1-D_1-D_2. For a feed of composition F_2, the practical region of attainable distillate compositions is approximated by the region F_1-D_1-D_3. Although an example with a maximum boiling azeotrope is not shown, systems with these azeotropes can be screened using the same lever rule techniques; however, the bottoms of the column will be affected by the azeotropes.

Heteroazeotropes — Systems with LLE

Ternary systems exhibiting LLE form minimum-boiling heteroazeotropes and can often be separated in a system of two columns as shown in Fig. 16.11. The LLE behavior often spans a separatrix on the residue curve map. This means that an overhead stream, D_M, can be condensed in a decanter, and one of the liquid phases will be on the same side of the separatrix as D_1 and can be returned into that column (left in the figure). The other decanter liquid phase will be on the other side of the separatrix and can be used as a feed to another column (right in the figure). An example of this procedure is given by the separation of ethanol + water using benzene. In this case, benzene is intentionally added to break the azeotrope and permit water to be recovered from one column and ethanol from the other column. The system involves an ethanol + benzene minimum boiling azeotrope, an ethanol + water minimum boiling azeotrope, and a benzene + water minimum boiling heteroazeotrope. The system has three separatrices. The ternary azeotrope, o, is the lowest boiling point in the system and it is the origin of the residue curves for all three regions. The left column operates in the right region of the residue curve map, and the right column operates in the upper left region. Care is taken to avoid having F_2 fall in the lower left region of the residue curve map; such a feed would result in a bottoms of benzene rather than ethanol. Illustrative material balance lines are provided on the diagram, and the LLE curve is superimposed on the residue curve map to clearly show how the tie lines span the separatrices. For this example, the residue curve origin can be moved farther into the LLE region by increasing the system pressure.

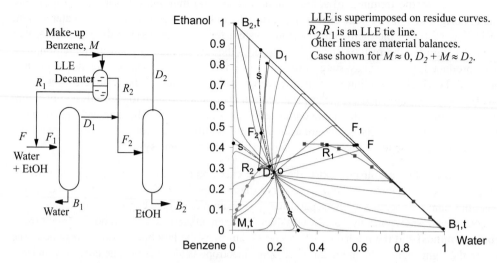

Figure 16.11 *Illustration of the column configuration, residue curves, and LLE behavior for the separation of ethanol and water using benzene. Residue curves calculated using Aspen Plus ver. 9.2 with UNIQUAC at 760 mmHg, LLE data at 25°C from Chang, Y.-I., Moultron, R.W. 1953. Ind. Eng.Chem. 45:2350. LLE tie lines are not plotted to avoid clutter, but tie-line data are represented by pairs of points along the binodal line.*

Generating Residue Curves

Residue curves are generated by a bubble-temperature algorithm for a simple single-stage batch distillation without reflux. If the total moles vaporized from a multicomponent mixture is dn^L, the moles of component i leaving are calculated by the composition of the vapor phase, $dn_i = y_i dn^L$. For component i in the liquid phase, $dn_i = d(x_i n^L) = n^L dx_i + x_i dn^L$. Equating the vapor and liquid expressions for dn_i,

$$y_i \, dn^L = x_i \, dn^L + n^L dx_i \qquad 16.14$$

Rearranging,

$$(y_i - x_i) dn^L/n^L = dx_i \qquad 16.15$$

which may be written

$$(K_i - 1) x_i d(\ln[n^L]) = dx_i \qquad 16.16$$

where K_i values can be calculated by any appropriate bubble-temperature method from Chapters 10–14 for the system at composition x_i. To generate residue curves in an N-component mixture, differential values of $d(\ln[n^L])$ may be chosen arbitrarily to generate differential values of dx_i. Only $N-1$ values of dx_i are required since

$$\sum_{i=1}^{N-1} dx_i + dx_N = 0 \qquad 16.17$$

MATLAB residue.m and residue.xlxm are available to perform calculations.

From an arbitrarily selected initial composition, the trace of x_i values yields the residue curve. Special care should be taken to include liquid-liquid behavior, if present in the system, requiring a three-phase bubble-temperature calculation as illustrated in Chapter 14. Table 16.1 presents the first few residue curve calculations for the n-pentane(1) + n-hexane(2) + n-heptane(3) system at 0.1013 MPa via Eqn. 16.16 using the shortcut K-ratio method to calculate the K_i values. The initial composition for the residue curve is arbitrarily selected as $x_1 = 0.1$, $x_2 = 0.6$ and $x_3 = 0.3$ denoted in cells with double ruling in Table 16.1. The step size for Eqn. 16.16 is arbitrarily selected as $d(\ln[n^L]) = -0.15$ to generate the dx_i values listed in the table. Eqn. 16.16 is used directly as a finite difference formula to provide a quick estimate as $x_i^{new} = x_i^{old} + dx_i^{old}$ to move down the table from the initial point. The residue curve is generated towards the increasing n-pentane liquid compositions by

Table 16.1 *Example Calculation of Residue Curves for n-butane(1) + n-hexane(2) + n-heptane(3) as Predicted by the Shortcut K-ratio Method*

x_1	x_2	x_3	$T(°C)$	P_1^{sat} (MPa)	P_2^{sat} (MPa)	P_3^{sat} (MPa)	K_1	K_2	dx_1	dx_2
0.156	0.596	0.248	65.02	0.2477	0.0902	0.0360	2.446	0.890	−0.034	9.8E−3
0.126	0.601	0.273	67.09	0.2622	0.0961	0.0386	2.588	0.949	−0.030	4.6E−3
0.1	0.6	0.3	69.07	0.2766	0.1021	0.0412	2.731	1.008	−0.026	−7.3E−4
0.074	0.600	0.327	71.17	0.2926	0.1088	0.0442	2.888	1.074	−0.021	−6.7E−3
0.053	0.593	0.354	73.09	0.3078	0.1152	0.0471	3.039	1.137	−0.016	−0.012

moving up the table from the initial point and using $x_i^{new} = x_i^{old} - dx_i^{old}$. More accurate finite difference methods can be employed for important applications. Residue curves are calculated in other composition ranges by selecting other initial compositions. The location of the separatrices is obvious after generating enough residue curves. Residue curve calculations are offered by some process simulation software due to the importance of screening separations in the chemical industry. More detailed discussions are available.[39]

Residue Curve Summary

Residue curves provide useful tools for designing separation schemes. Although a single feed column with one overhead and one bottoms product cannot cross a **separatrix,** a separation scheme can be constructed using multiple columns to achieve most separations. The residue curve maps are useful in screening additives for selection of the most promising systems. Residue curve calculations for homogeneous liquids are straightforward, and an algorithm has been presented.

39. Barnicki, S.D., Siirola, J.J. 1997. "Separations Process Synthesis," and Doherty, M.F., Knapp, J.P. 1997. "Distillation, Azeotropic and Extractive," in *Kirk-Othmer Encyclopedia of Chemical Technology,* 4th ed., New York, NY: Wiley.; also Van Dongen, D.B., Doherty, M.F. 1985. *Ind. Eng. Chem. Fundam.* 24:454.

16.4 PRACTICE PROBLEMS

P16.1 Consider the methanol(1) + water(2) + acetone(3) system with a feed shown in Figure 16.10(b). Rate each of the following products as impossible or possible, and explain.

	$D(x_1, x_2)$	$B(x_1, x_2)$
(a)	(1,0)	(0,0.45)
(b)	(0.4,0)	(0,1)
(c)	(0.4, 0)	(0.1, 0.9)
(d)	(0.3, 0.7)	(0.35,0)

(ANS. impossible; impossible; possible; impossible.)

16.5 HOMEWORK PROBLEMS

Phase Behavior

16.1 A binary mixture obeys a simple one-term equation for excess Gibbs energy, $G^E = Ax_1x_2$, where A is a function of temperature: $A = 2930 + 5.02E5/T(K)$ J/mol.

(a) Does this system exhibit partial immiscibility? If so, over what temperature range?
(b) Suppose component 1 has a normal boiling temperature of 310 K, and component 2 has a normal boiling temperature of 345 K. The enthalpies of vaporization are equal, both being 4475 cal/mol. Sketch a qualitatively correct T-x-y diagram including all LLE,VLE behavior at 1 bar.

16.2 As the research scientist of a company where phase equilibria is under study, you are approached by a laboratory technician with the most recent (and incomplete) high pressure phase equilibria results. The technician expresses concern that the data have not been collected correctly, or that some of the samples may have been mislabeled for some of the runs, or there is a problem with the automated equipment, because the data appear different from what he has seen before. The most recent results are summarized on the two P-x-y diagrams over the same P range shown below at T_1 and T_2, where $T_1 < T_2$. The circles and triangles give coexisting phase compositions at each pressure. What is your response?

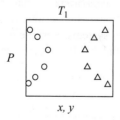
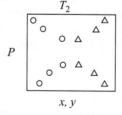

16.3 Using the value $k_{ij} = 0.084$ fitted to the system methanol(1) + benzene(2), use the Peng-Robinson equation to plot P-x-y behavior of the system at the following temperatures: 350 K, 500 K, 520 K. Plot the T-x-y behavior at the following pressures: 0.15 MPa, 2 MPa, 5.5 MPa. Sketch the P-T projection of the system.

16.4 Use UNIFAC to model the methanol(1) + methylcyclohexane(2) system. Provide a *P-x-y* diagram at 60°C and 90°C, and a *T-x-y* at 0.1013 MPa.

16.5 Use UNIFAC to model the methanol(1) + hexane(2) system. Provide a *P-x-y* diagram at 50°C and 25°C, and a *T-x-y* at 0.1013 MPa.

16.6 Model the system CO_2(1) + tetralin(2) using $k_{ij} = 0.10$. Generate a *P-x-y* diagram at 20°C, 45°C, and a *T-x-y* at 12 MPa and 22 MPa. What type of phase behavior does the system exhibit?[40]

16.7 The system ethanol(1) + bromoethane(2) forms an azeotrope containing 93.2 mol% bromoethane and having a minimum boiling point of 37.0°C at 760 mmHg. The following vapor pressure data are available:

P (mmHg)	$T^{sat}(°C)$	
	Ethanol	Bromoethane
100	34.9	---
400	63.5	21.0
760	78.4	38.4
1520	97.5	60.2
3800	---	95.0

(a) Use all of the available data to determine coefficients for equation $\log_{10} P^{sat} = A + B/T$ by performing a linear regression to find the coefficients A and B. Be sure to use temperature in Kelvin.
(b) Use the azeotropic point to determine the parameters for the van Laar equation.
(c) Generate a *P-x-y* diagram at 100°C. Tabulate the data used to plot the diagram.
(d) Generate a *T-x-y* diagram at 760 mmHg. Assume that the activity coefficients are independent of temperature over the range of temperature involved. (This is the type of calculation necessary for generating an *x-y* diagram for a distillation at 760 mmHg.)
(e) Experimentally, the system is found to have a homogeneous maximum pressure azeotrope (Smyth, C.P., Engel, E.W. 1929. *J. Am. Chem. Soc.,* 51:2660). How does this compare with your findings?

16.8 Benzaldehyde is known in the flavor industry as bitter almond oil. It has a cherry or almond essence. It may be possible to recover it using CO_2 for a portion of the processing. Explore the phase behavior of the CO_2(1) + benzaldehyde(2) system, using the Peng-Robinson equation to categorize the system among the types discussed in this chapter. Experimental data are not available, so use an interaction coefficient $k_{ij} = 0.1$, which has been fitted to the CO_2-benzene system. Calculate *P-x-y* diagrams at 295 K and 323 K. Also determine *T-x-y* diagrams at 34.5 bar and 206.9 bar. Note that 34.5 bar is below the critical pressure of both substances. If a bubble calculation fails, try a dew, and vice versa.

---

40. Experimental data are available in Leder, F., Irani, C.A. 1975. *J. Chem. Eng. Data* 20:323.

Residue Curves

16.9 For the systems specified below, obtain residue curve maps and plot the range of possible distillate and bottoms compositions for the given feeds. UNIQUAC parameters are provided.

(a) methanol(1) + ethanol(2) + ethyl acetate(3), $z_1 = 0.25$, $z_2 = 0.25$, $z_3 = 0.5$,
$r = [1.4311, 2.1055, 3.4786]$, $q = [1.432, 1.972, 3.116]$,
a parameters (in K); $a_{12} = -91.226$, $a_{21} = 124.485$, $a_{13} = -74.2746$, $a_{31} = 389.285$,
$a_{23} = -31.5531$, $a_{32} = 183.154.0$
Note: Aspen 2006 gives poor VLE using LLE parameters for this system. Use the provided parameters.

(b) Solve part (a), but use $z_1 = 0.4$, $z_2 = 0.4$, $z_3 = 0.2$.

(c) methanol(1) + 2-propanol(2) + water(3), $z_1 = 0.2$, $z_2 = 0.6$, $z_3 = 0.2$,
$r = [1.4311, 2.7791, 0.92]$, $q = [1.432, 2.508, 1.4]$,
a parameters (in K); $a_{12} = 39.717$, $a_{21} = -26.6935$, $a_{13} = -112.697$, $a_{31} = 176.077$,
$a_{23} = 151.061$, $a_{32} = 55.1272$.

(d) Solve part (c), but use $z_1 = 0.6$, $z_2 = 0.2$, $z_3 = 0.2$.

(e) methanol(1) + ethanol(2) + benzene(3), $z_1 = 0.25$, $z_2 = 0.25$, $z_3 = 0.5$,
$r = [1.4311, 2.1055, 3.1878]$, $q = [1.432, 1.972, 2.4]$,
a parameters (in K); $a_{12} = -91.2264$, $a_{21} = 124.485$, $a_{13} = -27.2253$, $a_{31} = 559.486$,
$a_{23} = -42.6567$, $a_{32} = 337.739$.

(f) Solve part (e), but use $z_1 = 0.4$, $z_2 = 0.4$, $z_3 = 0.2$.

16.10 Consider an equimolar mixture of methanol(1) + water(2) + acetone(3). Demonstrate that it is not possible to obtain pure acetone as distillate from a single-feed column by generating a residue curve map and applying the bow-tie approximation.[41] What compositions are attainable from this feed? Data: $r = [1.4311, 0.92, 2.5735]$, $q = [1.432, 1.4, 2.3360]$, a parameters (in K); $a_{12} = -112.697$, $a_{13} = -50.9396$, $a_{21} = 176.077$, $a_{23} = 29.1681$, $a_{31} = 218.872$, $a_{32} = 228.990$.

---

41. It is possible to obtain acetone distillate from a two-feed column if an equimolar acetone(1) + methanol(2) is fed near the center of the column and water is fed near the top of the column, however such separations are not predicted by the bow-tie approximation. A separation of this type is called an extractive distillation.

REACTION EQUILIBRIA

Problems involving reactions are affected by equilibrium limitations as well as their kinetics. Briefly, the Gibbs energy related to the "partitioning" of atoms between species must be minimized in a fashion analogous to the way that it was minimized for phase partitioning. For example, the formation of methanol from carbon monoxide and hydrogen is favorable based on the energy from the bond formation, but limited by the reduction in entropy as the carbon monoxide and hydrogen are squeezed into a single molecule. Thus, even though the heat of reaction may be favorable, a balance is struck between the loss in entropy and the favorable energy as the methanol is formed. At 100 bars and 510 K, this balance occurs at about 60% conversion of stoichiometric $CO + H_2$ to CH_3OH. This unit introduces the principles for determining the effect of temperature and pressure on equilibrium conversion. A few overly zealous practitioners of the past have erroneously assumed that conversion of reactants is determined solely by the rate of reaction, and have attempted to increase conversion beyond the equilibrium limit by increasing reactor size. For example, Le Châtelier (1888) wrote:

> It is known that in the blast furnace the reduction of iron oxide is produced by carbon monoxide, according to the reaction: $Fe_2O_3 + 3CO = 2Fe + 3CO_2$, but the gas leaving the chimney contains a considerable proportion of carbon monoxide, which thus carries away an important quantity of unutilized heat. Because this incomplete reaction was thought to be due to an insufficiently prolonged contact between carbon monoxide and the iron ore, the dimensions of the furnaces have been increased. In England they have been made as high as thirty meters. But the proportion of carbon monoxide escaping has not diminished, thus demonstrating, by an experiment costing several hundred thousand francs, that the reduction of iron oxide by carbon monoxide is a limited reaction. Acquaintance with the laws of chemical equilibrium would have permitted the same conclusion to be reached more rapidly and far more economically.

Let us hope that this is one bit of history that you will not repeat.

639

17

REACTION EQUILIBRIA

We must first speak a little concerning contact or mutual touching, action, passion and reaction.

Daniel Sennert (1660)

Another important aspect of the thermodynamics of multicomponent systems is the rearrangement of atoms within and between molecules, known as chemical reaction. Equilibrium thermodynamic considerations tell us the direction and extent to which a reaction will go. As with phase equilibria, the constraint of minimum Gibbs energy dictates the equilibrium results at a fixed T and P.

We begin this chapter by noting that the material from Section 3.6 is important for this chapter and you may wish to read that section again. There are several steps to understand before the equilibrium conversion is calculated, and some steps may seem very theoretical. We begin in Section 17.1 by relating the reaction coordinate to the minimum in Gibbs energy at equilibrium. Then in Section 17.3 we introduce the standard state Gibbs energy of reaction using the Gibbs energies of formation. Next, we relate the Gibbs energies of the components in the reacting system to the chemical potentials and finally develop the equilibrium constant in terms of the ideal gas law and begin to calculate equilibrium conversions. However, the standard state Gibbs energy used to calculate the equilibrium constant depends on temperature, and thus the equilibrium "constant" also changes with temperature which is discussed in Section 17.7. We then proceed with more advanced topics such as energy balances, use of the Gibbs minimization method, and multiple phases in reaction equilibrium.

Chapter Objectives: You Should Be Able to...

1. Solve for the equilibrium reaction coordinate values and the equilibrium mole fractions for a given K_{aT} and P for single and multiple reactions.

2. Understand the influences of pressure, nonstoichiometric feed, and inerts on reaction equilbrium.

3. Calculate $\Delta G^{o}_{298.15}$ and $\Delta H^{o}_{298.15}$ for a given reaction.

4. Calculate ΔG^{o}_{T} and K_{aT} using the van't Hoff equation.

5. Set up the energy balance for a given feed and equilibrium conversion, testing for closure or solving for heat transfer.

6. Incorporate solid species and liquid components into equilibrium calculations.

7. Understand the Gibbs minimization method for calculating reaction equilibrium.

17.1   INTRODUCTION

You have probably performed some reaction equilibrium computations before, usually in high school or freshman chemistry. This chapter shows how the "activities" (partial pressures for ideal gases) of products divided by reactants can be related to a quantity, K_a, that does not depend on pressure or composition, and despite its dependence on temperature, it is called the **equilibrium constant.** Developing the relationship between activities utilizes the concept of minimizing Gibbs energy and rearranging the basic relation. By study of the derivation we learn how to generalize reaction equilibrium analysis to multiple reactions and simultaneous reaction and phase equilibria.

We begin the chapter with an example to provide an overview of some of the methods developed in the chapter. We have selected an introductory reaction where all species are approximated as ideal gases. For ideal gases, we show in upcoming sections that the relation between equilibrium constraint and partial pressure is written

$$K_a = \prod (y_i P)^{\nu_i}$$

(ig) 17.1

where the symbol $\prod$ designates a product (analogous to the symbol Σ representing the summation sign), $y_i P$ is the partial pressure (always expressed in bar) of the ith component, and ν_i is the stoichiometric coefficient discussed in Section 3.6. Since stoichiometric coefficients are negative for reactants, the product symbol results in a ratio of products over reactants. The solution primarily requires a mass balance relating the partial pressure to the reaction coordinate (also discussed in Section 3.6). The major steps to solving an equilibrium problem are as follows.

1. Ascertain how many phases are present and the method to be used for the equilibrium calculations. Our initial examples will use only a gas phase and determine equilibrium compositions using an equilibrium constant method. Later we will show how to use liquid and solid phases and how to use the Gibbs energy directly.

2. Use standard state properties to obtain the value of the equilibrium constant at the reaction temperature, or for the Gibbs minimization method find the Gibbs energies of the species. Usually this consists of two substeps:

   (a) Perform a calculation using the standard state Gibbs energies at a reference temperature and pressure.

   (b) Correct the temperature (and pressure for Gibbs method) to the reaction conditions.

3. Perform a material balance on the reactant and product species and relate the composition to the equilibrium constant or standard state properties from steps (1) and (2).

4. Solve for the equilibrium compositions.

The steps are made clearer by a series of examples. Steps (1) and (2) are lengthy, and the applications are easier to see by first studying steps (3) and (4) as we show in the next example. This example will help to provide motivation for understanding how to use the standard state Gibbs energies in steps (1) and (2).

Example 17.1   Computing the reaction coordinate

CO and H_2 are fed to a reactor in a ratio of 2:1 at 500 K and 20 bar, where the equilibrium constant is $K_a = 0.00581$. (We will illustrate how to calculate K_a in Section 17.7.)

$$CO_{(g)} + 2H_{2(g)} \rightleftarrows CH_3OH_{(g)}$$

Compute the equilibrium conversion of CO.

Solution: In the expression for K_a we insert each $y_i P$ with the appropriate exponent and then insert the numerical value of pressure:

$$K_a = (y_{CH_3OH}P)^1(y_{CO}P)^{-1}(y_{H_2}P)^{-2} \qquad 17.2$$

$$0.00581 = \frac{y_{CH_3OH}P}{y_{CO}P(y_{H_2}P)^2} \text{ but } P = 20 \text{ bar} \Rightarrow \frac{y_{CH_3OH}}{y_{CO}y_{H_2}^2} = 0.00581P^2 = 2.32$$

To relate the composition to the mass balance, we select a basis and use the reaction coordinate. **Basis:** 2 mole CO fed. Note the excess CO at the feed conditions. The **reaction coordinate** and method of selecting a basis have already been introduced in Section 3.6. The stoichiometry table becomes

$$
\begin{aligned}
n_{CO} &= 2 - \xi \\
n_{H_2} &= 1 - 2\xi \\
n_{CH_3OH} &= \xi \\
\hline
n_T &= 3 - 2\xi
\end{aligned}
$$

Note that all n values must stay positive, constraining the range for a physically acceptable solution to be $0 \leq \xi \leq 0.5$. The mole fractions can be written in terms of ξ using the stoichiometry table.

$$y_{CO} = \frac{2-\xi}{3-2\xi} \quad y_{H_2} = \frac{1-2\xi}{3-2\xi} ; \quad y_{CH_3OH} = \frac{\xi}{3-2\xi} \qquad 17.3$$

Substituting the mole fractions into the equilibrium constant expression,

$$\frac{\dfrac{\xi}{(3-2\xi)}}{\dfrac{(2-\xi)(1-2\xi)^2}{(3-2\xi)^3}} = 2.32 \quad \Rightarrow \quad \frac{\xi(3-2\xi)^2}{(2-\xi)(1-2\xi)^2} = 2.32 \qquad 17.4$$

$$\text{Rearranging} \Rightarrow F(\xi) \equiv 2.32 \cdot (2-\xi)(1-2\xi)^2 - \xi(3-2\xi)^2 = 0 \qquad 17.5$$

Example 17.1 Computing the reaction coordinate (Continued)

A trial-and-error solution is much more robust by using the difference of Eqn. 17.5 rather than the ratio of Eqn. 17.4. We solve by trial and error and substitute to get ξ recalling $0 \le \xi \le 0.5$. A summary of guesses:

ξ	$F(\xi)$
0.10	2.04
0.25	−0.548
0.20	0.151
0.2099	0.0006

At reaction equilibrium for the given feed conditions, equilibrium is represented by $\xi = 0.21$. Now Eqn. 17.3 may be used to find the y's. The conversion of CO is $0.21/2\cdot100\% = 10.5\%$; conversion of H_2 is $2(0.21)/1\cdot100\% = 42\%$. Note the conversion is species-dependent with nonstoichiometric feed. Conversion can be increased further by increasing the pressure further, or by changing T where K_a is larger, provided a catalyst is available and kinetics are adequate at that T.

This example demonstrates the method to use K_a to calculate the reaction coordinate. Readers should note that the value of K_a is fixed at a given temperature, but the equilibrium value of ξ may vary for different feed conditions and often pressure for gas phase reactions as we will show in other examples. To relate the equilibrium conditions to reaction engineering textbooks, we note that most reaction engineering textbooks use conversion rather than reaction coordinate to track reaction progress. By convention, **conversion** is tracked for the limiting species (the species used up first at the value of ξ closest to zero in the direction of the reaction). A relation is shown in the footnote of Section 3.6.

Several other concepts are important for a general understanding of calculating reaction equilibria. First, we must understand: fundamental relations between the Gibbs energies, activities, and the equilibrium constant (Sections 17.2–17.4); simplifications that are applied for ideal gases and the effect of pressure and inerts (Section 17.5); and calculations of the temperature dependence of the equilibrium constants (Sections 17.7–17.9). Later sections illustrate the adaptation of the fundamental equations to broader applications like multiple reactions with simultaneous phase equilibria.

17.2 REACTION EQUILIBRIUM CONSTRAINT

Several sub-steps are involved in the procedure outlined in Section 17.1 steps (1) and (2) to find the equilibrium constant. In this section, we derive the equilibrium constraint, and then show how the thermodynamic properties are used to simplify to Eqn. 17.1. At reaction equilibria, the total Gibbs energy is minimized. If the composition of a system is changing, the change in the Gibbs energy is given by:

$$d\underline{G} = -\underline{S}dT + \underline{V}dP + \sum_i \mu_i dn_i \quad (\text{Recall } \mu_i \equiv (\partial\underline{G}/\partial n_i)_{T,P,n_{i\ne j}}) \qquad 17.6$$

The fact that species are being created or consumed by a reaction does not alter this equation. At constant temperature and pressure, the first two terms on the right-hand side drop out:[1]

$$d\underline{G} = \sum_i \mu_i dn_i \qquad\qquad 17.7$$

Substituting the definition of reaction coordinate from Eqn. 3.39,

$$d\underline{G} = \sum_i \mu_i v_i d\xi \qquad\qquad 17.8$$

Because $\underline{G}$ is minimized at equilibrium at fixed T and P, the derivative with respect to reaction coordinate is zero:

$$d\underline{G}/d\xi = \sum_i \mu_i v_i = 0 \qquad\qquad 17.9$$

Now there is one unknown, ξ, in terms of which we can determine the changes in moles for all of the components. We make a further manipulation before we apply the equilibrium constraint. In phase equilibria, we found fugacity to be a convenient property to use because it simplified to the partial pressure for a component in an ideal gas mixture. We can rewrite Eqn. 17.9 in terms of fugacities. We recall our definition of fugacity $dG = RT\,d\ln f$. Integrating from the standard state to the mixture state of interest (*cf.* generalizing Eqn. 10.48),

$$\int_{G^o}^{\mu} dG_i = RT \int_{f^o}^{\hat{f}} d\ln f_i \Rightarrow \mu_i - G_i^o = RT\ln\left[\frac{\hat{f}_i}{f_i^o}\right] \Rightarrow \mu_i = G^o + RT\ln\left[\frac{\hat{f}_i}{f_i^o}\right] \qquad 17.10$$

where G_i^o is the standard state Gibbs energy of species i and f_i^o is the standard state fugacity. A standard state is introduced for liquids in Section 11.3, and now we generalize the approach. The standard state is at the reaction temperature, but a specified composition (often pure) and pressure P^o. Substitution of Eqn. 17.10 into Eqn. 17.9,

$$0 = \sum_i v_i\left(G_i^o + RT\ln\left[\frac{\hat{f}_i}{f_i^o}\right]\right) \text{ or } 0 = \left(\sum_i v_i G_i^o\right)/(RT) + \sum_i v_i \ln\left[\frac{\hat{f}_i}{f_i^o}\right] \qquad 17.11$$

or

---

1. An alternative method of arriving at the same result is to write at fixed T, P, $\underline{G} = \sum_i \mu_i n_i$. The derivative is mathematically found by using the product rule for differentiation resulting in $d\underline{G} = \sum_i \mu_i dn_i + \sum_i n_i d\mu_i$ and the last sum is zero by the Gibbs-Duhem equation.

$$0 = \left(\sum_i \nu_i G_i^\circ\right)/(RT) + \sum_i \ln\left[\frac{\hat{f}_i}{f_i^\circ}\right]^{\nu_i} \quad \text{(equilibrium constraint)} \qquad 17.12$$

We will need to calculate both summations appearing in Eqn. 17.12, and then combine the results. Qualitatively, this equation indicates how atoms should be arranged within molecules to minimize Gibbs energy. The connection between Eqn. 17.12 and Eqn. 17.1 will become obvious as we move through the next sections. Let us work on the two summations separately. We will show that the second summation is related to the product of partial pressures for gas phase species which we define as the equilibrium constant. The first summation will relate to the negative numerical value of the equilibrium constant because the two terms of Eqn. 17.12 are equal and opposite.

17.3   THE EQUILIBRIUM CONSTANT

We now focus on the second summation of Eqn. 17.12. The ratio appearing in the logarithm is known as the activity, (*cf.* Eqns. 11.23 for a liquid, but now in a general sense):

❗ activity

$$a_i \equiv \frac{\hat{f}_i}{f_i^\circ} \qquad 17.13$$

The numerator $\hat{f}_i$ represents a mixture property that changes with composition. We have developed methods to calculate $\hat{f}_i$ in Eqns. 10.61 (ideal gases), 10.68 (ideal solutions), 11.14 (real solution using γ_i), Eqn. 15.13 (real gases using $\hat{\varphi}_i$). The denominator represents the component at a specific standard state, which includes specification of a fixed composition (which can be pure or a mixture state).

The second sum of Eqn. 17.12 can be manipulated after inserting the activity notation,

$$\sum_i \ln\left[\frac{\hat{f}_i}{f_i^\circ}\right]^{\nu_i} = \sum_i \ln(a_i)^{\nu_i} = \ln\prod_i a_i^{\nu_i} \qquad 17.14$$

In a reacting mixture $\hat{f}_i$ and/or a_i will change as the reacting composition moves toward equilibrium. However, at equilibrium, the product term of activities is extremely important. We define the product term at equilibrium as the **equilibrium constant** K_a with the a subscript to denote that activity is used:

❗ General equilibrium constraint.

$$K_a = \prod_i a_i^{\nu_i} = \prod_i \left[\frac{\hat{f}_i}{f_i^\circ}\right]^{\nu_i} \qquad 17.15$$

Combining the definition of the equilibrium constant with Eqn. 17.12, the first summation can be used to find the value of the constant:

$$0 = \left(\sum_i v_i G_i^\circ\right)/(RT) + \ln K_a \ \text{ or } \ \exp\left(-\left(\sum_i v_i G_i^\circ\right)/(RT)\right) = K_a \qquad 17.16$$

Note that use of the term *constant* can be misleading because it depends on temperature. It is constant with respect to feed composition and changing mole numbers of reacting species as we will show below. We use a subscript a on the equilibrium constant to stress that it depends on activities. As we show later, there are other approximations for the equilibrium constant, and subscripts are used to differentiate between different conventions.

The Equilibrium Constant for Ideal Gases

The activity is a general property defined by Eqn. 17.13. We have seen it applied to liquids in Section 11.5. For ideal gases, the numerator of the activity is $\hat{f}_i^{ig} = y_i P$. We complete the formula for activity by selecting the standard state. For gaseous reacting species, the convention is to use a standard state of the pure gas at $P^\circ = 1$ bar. For an ideal gas, $f_i^\circ = P^\circ$ (Eqn. 9.29). Thus, $f_i^\circ = 1$ bar. The fugacity ratio (activity) is dimensionless provided that we always express the partial pressure in bar. The second sum of Eqn. 17.12 for ideal gases simplifies to

$$K_a = \prod_i a_i^{v_i} = \prod_i \left[\frac{\hat{f}_i}{f_i^\circ}\right]^{v_i} = \prod_i (y_i P)^{v_i} \qquad \text{(ig) 17.17}$$

where the first two equalities are general, but the last is restricted to ideal gases. We will later reevaluate the fugacity ratio for nonideal gases, liquids, and solids. Now let us examine the first sum of Eqn. 17.12 which will give us the value of K_a.

17.4   THE STANDARD STATE GIBBS ENERGY OF REACTION

The first term on the right side of Eqns. 17.12 and 17.16, $\displaystyle\sum_i v_i G_i^\circ$, is called the **standard state Gibbs energy of reaction** at the temperature of the reaction, which we will denote ΔG_T°. The standard state Gibbs energy of reaction is analogous to the standard state heat of reaction introduced in Section 3.6. The standard state Gibbs energy for reaction can be calculated using **Gibbs energies of formation**.

$$\Delta G_T^o \equiv \sum_i v_i G_i^o = \sum_{\text{products}} |v_i| \Delta G_{f,i}^o - \sum_{\text{reactants}} |v_i| \Delta G_{f,i}^o \qquad 17.18$$

As an example, for $CH_{4(g)} + H_2O_{(g)} \rightarrow CO_{(g)} + 3H_{2(g)}$

$$\Delta G_T = \sum_i v_i G_i^o = \Delta G_{f(CO_{(g)})}^o + 3\Delta G_{f(H_{2(g)})}^o - \Delta G_{f(CH_{4(g)})}^o - \Delta G_{f(H_2O_{(g)})}^o$$

It may be helpful to think of the sum as representing a path via **Hess's law** where the reactants are "unformed" to the elements and then "formed" into the products. The signs of the formation Gibbs energies of the products are positive and the signs for the reactants are negative. Thus,

❶ Standard state Gibbs energy of reaction.

$$\Delta G_T^o = \sum_i v_i \Delta G_{f,\,i}^o$$

17.19

   The Gibbs energies of formation are typically tabulated at 298.15 K and 1 bar, and special calculations must be performed to calculate ΔG_T^o at other temperatures—the calculations will be covered in Section 17.7. Like the enthalpy of formation, the Gibbs energy of formation is taken as zero for elements that naturally exist as molecules at 298.15 K and 1 bar, and the same cautions about the state of aggregation apply. Gibbs energies of formation are tabulated for many compounds in Appendix E at 298.15 K and 1 bar. Note that for water, the difference between $\Delta G_{f\,298.15(g)}^o$ and $\Delta G_{f\,298.15(l)}^o$ is the Gibbs energy of vaporization at 298 K. The difference is nonzero because liquid is more stable. (Which phase will have a lower Gibbs energy of formation at 298.15 K and 1 bar?)

   The standard state Gibbs energy of reaction is related to the equilibrium constant through Eqn. 17.16,

❶ Relation between Gibbs energy of reaction and the equilibrium constant.

$$\exp(-\Delta G_T^o/(RT)) = K_a$$

17.20

Once the value of the equilibrium constant is known, equilibrium compositions can be determined, as shown in Example 17.1. The next example illustrates calculation of the Gibbs energy of reaction and the equilibrium constant.

Example 17.2   Calculation of standard state Gibbs energy of reaction

Butadiene is prepared by the gas phase catalytic dehydrogenation of 1-butene:

$$C_4H_{8(g)} \underset{\leftarrow}{\rightarrow} C_4H_{6(g)} + H_{2(g)}$$

Calculate the standard state Gibbs energy of reaction and the equilibrium constant at 298.15 K.

Solution: We find values tabulated for the standard state enthalpies of formation and standard state Gibbs energy of formation at 298.15 K.

Compound	$\Delta H_{f,\,298.15}^o$ (J/mole)	$\Delta G_{f,\,298.15}^o$ (J/mole)
1-butene (g)	−540	70,240
1,3-butadiene (g)	109,240	149,730
Hydrogen (g)	0	0

$$\Delta G_{298.15}^o = 149{,}730 + 0 - 70{,}240 = 79{,}490 \text{ J/mole}$$

**Example 17.2 Calculation of standard state Gibbs energy of reaction
(Continued)**

The equilibrium constant is determined from Eqn. 17.16;

$$\exp\left(\frac{-\Delta G^o_{298.15}}{RT}\right) = \exp\left(\frac{-79,490}{8.314(298.15)}\right) = 1.18 \times 10^{-14} = K_{a,\,298.15}$$

This reaction is not favorable at room temperature because the equilibrium constant is small.

Composition and Pressure Independence of K_a

The use of standard states for calculating ΔG^o_T has important implications on the composition and pressure independence of K_a. The standard state is at a fixed pressure, P^o. Thus, ΔG^o_T is independent of pressure. The standard states are also at fixed composition (often pure), and thus ΔG^o_T is independent of equilibrium composition. Looking at Eqn. 17.20, we conclude that because ΔG^b_T is independent of equilibrium composition and pressure, K_a is independent of equilibrium composition and pressure. One important point is that the state of aggregation in the standard state *is* important and the values of ΔG^o_T and K_a do depend on the state of aggregation in the standard state; this point will be clarified in later sections.

> ❶ The equilibrium constant K_a is independent of composition and pressure.

We have now demonstrated steps 2(a), 3, and 4 for the procedure given in Section 17.1. The concepts have been demonstrated, but we must correct the temperature before doing calculations at temperatures other than 298.15 K. The butadiene reaction of Example 17.2 becomes more favorable with a larger K_a at higher temperatures. We will discuss some important aspects of the effects of pressure and inerts and also discuss reaction spontaneity before showing the calculation of temperature corrections.

17.5 EFFECTS OF PRESSURE, INERTS, AND FEED RATIOS

At a given temperature, equilibrium values of the reaction coordinate are affected by pressure, inerts, and feed ratios. The principle that changing the quantities affects equilibrium conversions is known as **Le Châtelier's principle** in honor of Henry Louis Le Châtelier who first characterized the phenomenon. An understanding of Le Châtelier's principle is important for operating industrial reactions. Two important modifications led to significant hydrogen conversions in Example 17.1 even though the equilibrium constant was small—use of pressure and nonstoichiometric feed.

> Henry Louis Le Châtelier (1850–1936) was a French chemist. He was elected to the French Académie des Sciences and the Royal Swedish Academy of Sciences in 1907.

Pressure Effects

Pressure has little effect on the activities of condensed species (e.g., the Poynting correction is typically small) and thus it has a primary significance only for reactions with gas phase components. Pressure has important effects when both 1) gas species are involved in reactions and 2) the stoichiometric numbers of gas species are different for reactants and products. When the stoichiometric moles of gas species are the same for reactants and products, P has no effect by the ideal gas approximation, and for nonideal gases only indirect effects due to fugacity coefficients.

The equilibrium constants for ideal gases can be written

$$\left(\prod y_i^{\nu_i}\right)P^{\Sigma\nu_i} = K_a \qquad \qquad \text{(ig) 17.21}$$

This form makes the pressure effect more obvious. As mentioned above, when the stoichiometric number of gas moles is the same for products and reactants, $\Sigma\nu_i = 0$ and the pressure effect drops out. When the stoichiometric numbers of vapor reactant moles is greater than the stoichiometric numbers of product vapor moles, an increase in pressure will drive the reaction to higher conversions, $\Sigma\nu_i < 0$. When the stoichiometric gas mole ratios are reversed, a decrease in pressure will help drive the reaction to higher conversions, $\Sigma\nu_i > 0$. In Example 17.1 the pressure of 20 bar was important to yield significant conversions. It can be helpful to consider that qualitatively the pressure "squeezes" the reaction towards the side with fewer gas moles. As an exercise, determine the reaction coordinate for the same feed when the pressure is 1 bar.

Inerts

A component that does not participate in a reaction is called **inert.** Inert gas components often have an indirect, but important effect on the equilibrium reaction coordinate when gas phase species are present. Inerts change the overall mole fractions and thus mitigate the pressure effects. When $\Sigma\nu_i > 0$, adding an inert will increase conversion at a fixed total pressure. However, when $\Sigma\nu_i < 0$, the mitigation of the pressure effect is undesirable and inerts should be avoided. Qualitatively, the presence of an inert decreases the "squeezing" effect mentioned above.

Nonstoichiometric Feed

Conversions of specific reactants are influenced using nonstoichiometric feed. In Example 17.1, excess CO was fed to the reactor; conversions were 42% (H_2), and 10.5% (CO). Generally, an excess of one reactant will tend to increase conversion of the other reactant. The effect can be seen qualitatively using Eqn. 17.2. For a given K_a, at a certain value of y_{CH_3OH}, a higher value of y_{CO} results in a lower value of y_{H_2}. When using stoichiometric feed (CO:H_2 = 1:2) in Example 17.1, the equilibrium conversions of H_2 and CO are equal (40.6%). Example 17.1 includes both excess CO and a pressure effect. The excess CO in Example 17.1 is high enough to mitigate the beneficial pressure effect in a manner similar to an inert gas. The feed ratio giving highest H_2 conversion for the specified conditions uses less excess CO, (CO:H_2 = 1:1), which results in conversions of 45.2% (H_2) and 22.6% (CO). Use of nonstoichiometric feed is common in industrial reactions because in some cases it helps avoid side reactions in addition to effects on equilibrium.

Example 17.3 Butadiene production in the presence of inerts

Consider again the butadiene reaction of Example 17.2 on page 648. Butadiene is prepared by the gas phase catalytic dehydrogenation of 1-butene, at 900 K and 1 bar.

$$C_4H_{8(g)} \overset{\rightarrow}{\leftarrow} C_4H_{6(g)} + H_{2(g)}$$

(a) In order to suppress side reactions, the butene is diluted with steam before it passes into the reactor. Estimate the conversion of 1-butene for a feed consisting of 10 moles of steam per mole of 1-butene.

(b) Find the conversion if the inerts were absent and side reactions are ignored.

(c) Find the total pressure that would be required to obtain the same conversion as in (a) if no inerts were present.

Example 17.3 Butadiene production in the presence of inerts (Continued)

In the earlier example, we determined the value at 298.15 K for ΔG_f°. Now we need a value at 900 K. The next section explains how the value at 900 K may be obtained. For now, use the following data for ΔG_f° (kJ/mole) at 900 K and 1 bar:

Component	$\Delta G_{f,\,900}^\circ$ (kJ/mole)
1,3 Butadiene	243.474
1-Butene	232.854
Hydrogen	0

Solution:

$$\Delta G_{900}^o = 243.474 - 232.854 = 10.62 \text{ kJ/mole}$$

$$K_a = \exp(-\Delta G_{900}^o / RT) = 0.242$$

(a) Basis of 1 mole 1-butene feed. Set up reaction coordinate, using I to indicate inerts,.

$$
\begin{array}{ll}
n_{C_4H_8} & 1 - \xi \\[4pt]
n_{C_4H_6} & \xi \\[4pt]
n_{H_2} & \xi \\[4pt]
n_I & 10 \\
\hline
n_T & 11 + \xi
\end{array}
$$

$$0.242 = \left(\frac{\xi}{11 + \xi}\right)^2 P \Big/ \left(\frac{1 - \xi}{11 + \xi}\right)$$

The physical range of the solution is $0 \le \xi \le 1$. $P = 1$ bar $\Rightarrow 1.242\xi^2 + 2.42\,\xi - 2.662 = 0 \Rightarrow \xi = 0.784$. For the basis of 1 mol 1-butene feed, the conversion is 78.4%.

(b) $n_I = 0$ and the basis of feed is the same and $0 \le \xi \le 1$. The total number of moles is $n_T = 1 + \xi$; $1.242\xi^2 - 0.242 = 0$; $\xi = 0.44$, so conversion decreases to 44% without inert.

(c) Rearranging the equilibrium expression for pressure,
$P^{-1} = \xi^2 / [0.242 \cdot (1 - \xi) \cdot (1 + \xi)], \ 0 \le \xi \le 1$.
Inserting a reaction coordinate of $\xi = 0.784$ gives $P = 0.152$ bar. So the reaction would need to run at a much lower pressure without the inerts to achieve the same conversion. In other words, inerts serve to dilute the fugacities of the products and suppress the reverse reaction since there are more moles of product than reactant.

17.6   DETERMINING THE SPONTANEITY OF REACTIONS

In our preliminary examples, we have assumed rather idealized cases where none of the products are present in the inlet. However, in some cases, products may be present and then the reaction direction may not be as we anticipate. We can look at the reaction thermodynamics in a slightly different way to determine the direction of the reaction under given compositions, T and P. Starting from Eqn. 17.10, we may add by weighting with the stoichiometric numbers, resulting in

$$\sum_i v_i \mu_i = \sum_i v_i G_i^\circ + RT \sum_i v_i \ln\left[\frac{\hat{f}_i}{f_i^\circ}\right] \qquad 17.22$$

The term $\sum_i v_i \mu_i$ on the left side is called the **Gibbs energy of reaction** and is given the symbol ΔG_T. Note that this is a different term than the **standard state Gibbs energy of reaction** (the second term) that uses the superscript $^\circ$. Thus, we can write,

$$\Delta G_T = \Delta G_T^\circ + RT \sum_i v_i \ln\left[\frac{\hat{f}_i}{f_i^\circ}\right] = \Delta G_T^\circ + RT \ln \prod_i \left[\frac{\hat{f}_i}{f_i^\circ}\right]^{v_i} \quad \text{(general relation)} \qquad 17.23$$

> ❶ The propensity for a reaction to go forward or backward under actual conditions is determined by ΔG_T, not ΔG_T°.

A reaction with $\Delta G_T^\circ < 0$ is called **exergonic** and results in $K_a > 1$, and a reaction with $\Delta G_T^\circ > 0$ is called **endergonic,** resulting in $K_a < 1$. This provides an indication of whether the equilibrium favors products or reactants, but does not mean that reactions with small values of K_a cannot be conducted industrially. For example, Example 17.1 involved a small K_a (thus endergonic with $\Delta G_T^\circ > 0$), yet the conversion of H_2 was 42%. The propensity for the reaction to go forward or backward depends instead on the Gibbs energy of reaction ΔG_T at the concentrations represented by the fugacity ratios. If the conditions provide $\Delta G_T < 0$, then the Gibbs energy is lowered when the reaction proceeds in the forward direction. If we evaluate conditions and $\Delta G_T > 0$, then the reaction goes in the reverse direction than what we have written. In either case, the concentrations adjust until the system reaches the equilibrium condition, $\Delta G_T = 0$, and then Eqn. 17.12 applies. In summary, the direction a reaction proceeds is determined by ΔG_T, not by ΔG_T°. Note when evaluating the fugacity term for determining spontaneity that the actual conditions are used, not the equilibrium conditions. At the feed conditions of Example 17.1, $y_{CH_3OH} = 0$, which ensures $\Delta G_T < 0$ at the feed conditions even though $\Delta G_T^\circ > 0$.

17.7   TEMPERATURE DEPENDENCE OF K_a

Always remember that ΔG_T° depends on the standard state, which changes with temperature. Comparing Examples 17.2 and 17.3, $\Delta G_T^\circ = 79{,}490$ J/mole at 298 K ($K_a = $ 1E-14), but decreases to $\Delta G_T^\circ = 10{,}620$ J/mol at 900 K ($K_a = 0.242$). In order to calculate ΔG_T°, it may seem that we need to know ΔG_f° for each compound at all temperatures. Fortunately this is not necessary because the ΔG_T° can be determined from the Gibbs energy for the reaction at a certain **reference temperature** (usually 298.15 K) together with the enthalpy for the reaction and the heat capacities of the species.

Suppose we have a table of standard energies of formation at 298.15 K but we would like the value for ΔG_T^o at some other temperature. We can account for temperature effects by applying classical thermodynamics to the change in Gibbs energy with respect to temperature using the **Gibbs-Helmholtz relation,**

$$\frac{\partial(\Delta G/RT)}{\partial T} = \frac{1}{RT}\left(\frac{\partial \Delta G}{\partial T}\right)_P - \frac{\Delta G}{RT^2} = -\frac{\Delta S}{RT} - \left(\frac{\Delta H}{RT^2} - \frac{\Delta S}{RT}\right) = \frac{-\Delta H}{RT^2} \qquad 17.24$$

Jacobus Henricus van't Hoff (1852–1911) was awarded the Nobel Prize in chemistry in 1901.

which results in the **van't Hoff equation:**

$$\boxed{\frac{\partial(\Delta G_T^\circ/RT)}{\partial T} = \frac{\partial(-\ln K_a)}{\partial T} = \frac{-\Delta H_T^\circ}{RT^2}} \qquad 17.25$$

❗ van't Hoff equation.

$$\boxed{\frac{\Delta G_T^\circ}{RT} = -\int_{T_R}^T \frac{\Delta H_T^\circ}{RT^2}dT + \frac{\Delta G_R^\circ}{RT_R} = -\ln K_a = -\int_{T_R}^T \frac{\Delta H_T^\circ}{RT^2}dT - \ln K_{aR}} \qquad 17.26$$

For accurate calculations, we must recognize that the heat of reaction depends on temperature. We have developed the standard heat of reaction in Section 3.6 and discussed the temperature dependence there. We show later that an assumption of a temperature-independent heat of reaction results in a short-cut approximation that is often close to the full calculation. We first show the full calculation. Substituting into the van't Hoff equation (Eqn. 17.26) and integrating again,

$$\frac{\Delta G^\circ}{RT} = \frac{J}{R}\left(\frac{1}{T} - \frac{1}{T_R}\right) - \frac{\Delta a}{R}\ln\frac{T}{T_R} - \frac{\Delta b}{2R}(T - T_R) - \frac{\Delta c}{6R}(T^2 - T_R^2) - \frac{\Delta d}{12R}(T^3 - T_R^3) + \frac{\Delta G_R^\circ}{RT_R} \qquad 17.27$$

where we previously described finding J in Eqn. 3.46 on page 113. If desired, all values at T_R can be lumped together in a constant, I.

$$\boxed{\frac{\Delta G^\circ}{RT} = -\ln K_a = \frac{J}{RT} - \frac{\Delta a}{R}\ln T - \frac{\Delta bT}{2R} - \frac{\Delta cT^2}{6R} - \frac{\Delta dT^3}{12R} + I} \qquad 17.28$$

The constant I may be evaluated from a knowledge of $\Delta G^o{}_{298}$ by plugging in $T = 298.15$ on the right-hand side as illustrated below.

Example 17.4 Equilibrium constant as a function of temperature

The heat capacities of ethanol, ethylene, and water can be expressed as $C_P = a + bT + cT^2 + dT^3$ where values for a, b, c, and d are given below along with standard energies of formation. Calculate the equilibrium constant [$\exp(-\Delta G_T^\circ/RT)$] for the vapor phase hydration of ethylene at 145°C and 320°C.

$$C_2H_{4(g)} + H_2O_{(g)} \ \overset{\rightarrow}{\leftarrow} \ C_2H_5OH_{(g)}$$

Example 17.4   Equilibrium constant as a function of temperature (Continued)

	$\Delta H^o_{f,\,298}$ kJ/mol	$\Delta G^o_{f,\,298}$ kJ/mol	a	b	c	d
Ethylene	52.51	68.43	3.806	1.566E-01	−8.348E-05	1.755E-08
Water	−241.835	−228.614	32.24	1.924E-03	1.055E-05	−3.596E-09
Ethanol	−234.95	−167.73	9.014	2.141E-01	−8.390E-05	1.373E-09

The workbook Kcalc.xlsx or MATLAB Kcalc.m are helpful in doing these calculations.

Solution:

$$C_2H_4 \;+ H_2O \;\; \overset{\rightarrow}{\leftarrow} \;\; C_2H_5OH$$

$$\nu_i \qquad -1 \qquad -1 \qquad\qquad +1$$

Taking 298.15 K as the reference temperature,

$\Delta H^\circ_R = \Delta H^\circ_{298.15} = -234.95 - [52.51 + (-241.835)] = -45{,}625$ J/mole
$\Delta G^\circ_R = \Delta G^\circ_{298.15} = -167.73 - [68.43 + (-228.614)] = -7546$ J/mole

The variable J may be found with Eqn. 3.46 on page 113 at 298.15 K.

$$\Delta H^\circ_{298.15} = -45{,}625 = J + \Delta a T + (\Delta b/2)\cdot T^2 + (\Delta c/3)\cdot T^3 + (\Delta d/4) T^4 = J + (9.014 - 3.806 -$$

$$32.24)\, T + [(0.2141 - 0.1566 - 0.0019)/2]\, T^2 + [(-8.39 + 8.348 - 1.055)(1E\text{-}5)/3]\, T^3 + [(1.373$$

$$-17.55 + 3.596)(1E\text{-}9)/4]\, T^4$$

$$= J - 27.032\, T + 0.02779\, T^2 - (3.657E\text{-}6) T^3 - (3.145E\text{-}9) T^4$$

Plugging in $T = 298.15$ K, and solving for J, $J = -39.914$ kJ/mole. Using this result in Eqn. 17.28 at 298.15 K will yield the variable I.

$$\Delta G_T^\circ/RT = -39{,}914/(8.314\cdot T) + 27.032/8.314 \ln T - [(5.558E\text{-}2)/(2\cdot 8.314)]\, T$$

$$+[(1.097E\text{-}5)/(6\cdot 8.314)]\, T^2 + [(1.258E\text{-}8)/(12\cdot 8.314)]\, T^3 + I$$

Plugging in ΔG_R° at 298.15K, $\Delta G_T^\circ/RT = -7546/8.314/298.15 = 3.0442$. Plugging in for T on the right-hand side results in $I = -4.494$.

The resultant formula to calculate ΔG_T° at any temperature is

$$\Delta G_T^\circ = -39{,}914 + 27.032\, T \ln T - 0.0278\, T^2 + (1.828E\text{-}6) T^3 + (1.048E\text{-}9) T^4 - 37.363\, T$$

at 145°C $\Delta G_T^\circ = 7997$ J/mol $\Rightarrow K_a = 0.1002$;

at 320°C $\Delta G_T^\circ = 31{,}045$ J/mol $\Rightarrow K_a = 0.00185$

17.8   SHORTCUT ESTIMATION OF TEMPERATURE EFFECTS

Recall Eqn. 17.25, which we refer to as the *general* van't Hoff equation:

$$\frac{\partial(\Delta G_T^\circ/RT)}{\partial T} = -\frac{\Delta H_T^\circ}{RT^2} = -\frac{\partial \ln(K_a)}{\partial T}$$

We can make rapid estimates of the equilibrium constant when we make the approximation that ΔH_T° is independent of temperature. That is, suppose $\Delta C_P = \Delta a = \Delta b = \Delta c = \Delta d = 0$, which means the sensible heat effects for the reactants and products are the same. This is most closely approximated when all species are about the same molecular size and the same state of aggregation. With this approximation, $J = \Delta H_R^\circ$ in Eqn. 17.27, or we can integrate Eqn. 17.25 directly to obtain what we refer to as the **shortcut van't Hoff equation:**

$$\ln\left(\frac{K_a}{K_{aR}}\right) = \frac{\Delta G_R^{\,\circ}}{RT_R} - \frac{\Delta G_T^{\,\circ}}{RT} = \frac{-\Delta H_R^{\,\circ}}{R}\left(\frac{1}{T} - \frac{1}{T_R}\right) \qquad 17.29$$

❶ Shortcut van't Hoff equation.

This equation enables rapid screening for the effects of temperature and the detailed van't Hoff can be used as a follow-up calculation. As a particular observation, we take special note from the above equation that exothermic reactions ($\Delta H_T < 0$) lead to K_a decreasing as temperature increases, and endothermic reactions ($\Delta H_T > 0$) lead to K_a increasing as temperature increases. This means that *equilibrium* conversion (for a specified feed) decreases with increasing temperature for exothermic reactions, and increases for endothermic reactions. This effect is illustrated in Fig.17.1. The emphasis is placed on *equilibrium* because *reaction rates increase* with temperature. Industrial application of exothermic reactions are almost always run at elevated temperatures even though the equilibrium constant decreases at high temperature. For all reactions, the reaction rate will approach zero as equilibrium is approached. The benefit of faster kinetics typically outweighs the smaller equilibrium constant for exothermic reactions when economics are considered. There connection between equilibrium constants and kinetic rates that approach zero is explained in Section 17.15.

The approximate results of the shortcut van't Hoff equation should be followed with the detailed van't Hoff for critical applications. To improve shortcut estimates, the detailed van't Hoff

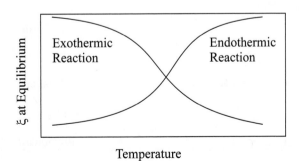

Figure 17.1 *Qualitative behavior of equilibrium conversion for exothermic and endothermic reactions.*

can be used at T_{near} within 100 K of the temperatures of interest to calculate $\Delta G_{Tnear}°$ and $\Delta H_{Tnear}°$. Then the values at T_{near} can be used as the reference values in Eqn. 17.29.

Example 17.5 Application of the shortcut van't Hoff equation

Apply the shortcut approximation to the vapor phase hydration of ethylene. This reaction has been studied in the previous example, and the Gibbs energy of reaction and heat of reaction can be obtained from that example.

Solution:

$$K_{a,298.15} = \exp(7546/8.314/298) = 21.03$$

$$\ln\left(\frac{K_a}{21.03}\right) = \frac{45,625}{8.314}\left(\frac{1}{T} - \frac{1}{298.15}\right)$$

$$K_a = 0.106 \text{ at } 145°C; \; K_a = 0.0022 \text{ at } 320°C$$

The results are very similar to the answer obtained by the general van't Hoff equation in Example 17.4.

17.9   VISUALIZING MULTIPLE EQUILIBRIUM CONSTANTS

Plots of equilibrium constants provide a rapid method to visualize the gross trends and orders of magnitude. Fig. 17.2 illustrates how several reactions can be illustrated in a single graph. The equilibrium constants are calculated with the full temperature dependence. Note that the plots are nearly linear as would be approximated by the short-cut van't Hoff. Exothermic reactions have a positive slope and endothermic reactions have a negative slope. When dealing frequently with a set of reactions, such a graph can serve as a "road map" where the optimal temperature window of operation maximizes desired products while minimizing by-products and potential coupling of reactions.

The reactions of Fig. 17.2 are typically involved in many high-profile applications including: combustion, chemical-vapor infiltration, reforming, coking during reforming, space station gas management, electrolysis, and the hydrogen economy. Several of these reactions have common names which are listed below.

Reaction	Name	
$H_2O_{(g)} + C_{(s)} \rightleftarrows CO_{(g)} + H_{2(g)}$	Syngas	17.30
$CO_{2(g)} + 4H_{2(g)} \rightleftarrows CH_{4(g)} + 2H_2O_{(g)}$	Sabatier	17.31
$CO_{(g)} + H_2O_{(g)} \rightleftarrows CO_{2(g)} + H_{2(g)}$	Water-gas shift	17.32
$CO_{2(g)} + C_{(s)} \rightleftarrows 2CO_{(g)}$	Boudouard	17.33
$0.5O_{2(g)} + C_{(s)} \rightleftarrows CO_{(g)}$	Partial oxidation	17.34

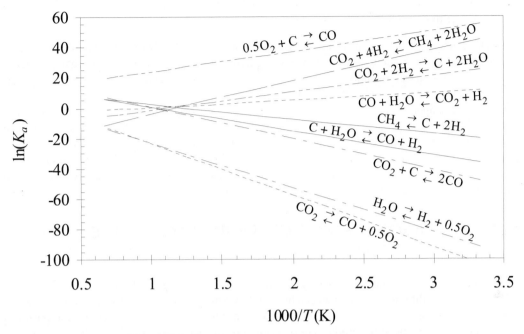

Figure 17.2 *Graphical analysis of competing reactions.*

Reaction	Name	
$CH_{4(g)} \rightleftarrows C_{(s)} + 2H_{2(g)}$	Methane pyrolysis	17.35
$CO_{2(g)} + 2H_{2(g)} \rightleftarrows C_{(s)} + 2H_2O_{(g)}$	Bosch	17.36
$H_2O_{(g)} \rightleftarrows H_{2(g)} + 0.5O_{2(g)}$	Water decomposition	17.37
$CO_{2(g)} \rightleftarrows CO_{(g)} + 0.5O_{2(g)}$	CO oxidation	17.38

To illustrate interpretation of the graph, consider an application of the above reactions to material management in space station gas management, where an objective is to remove CO_2 and provide O_2. There are many ways that the reactions could be combined. On a space station sunlight is relatively abundant. Therefore, high temperatures and solar cells are available, but food must be imported. Note that the Sabatier reaction would convert waste CO_2 to fuel but requires H_2. Fig. 17.2 shows that the equilibrium constant is favorable below 900 K. The Bosch reaction also favors products at temperatures below 900 K. The Bosch reaction produces graphitic carbon, which can be collected in dense form and conveniently disposed. The hydrogen required for the Bosch reaction could be generated by water decomposition, which could be achieved with electrolysis or pyrolysis, with the benefit of co-producing oxygen for respiration. A small extrapolation of Fig. 17.2 shows that water pyrolysis is favorable above 2300 K.[2] Coupling the Sabatier reaction with methane pyrolysis has been suggested. Methane pyrolysis is favorable above 900 K. This would produce hydrogen for other use. Hydrogen production could also be achieved by the syngas reaction, if graphitic carbon was available. H_2 could be enhanced and CO removed by the water-gas shift. Catalysts can selectively alter the kinetics to minimize undesired products, although they cannot alter

---

2. By comparison, CO pyrolysis is less favored above 700 K, owing to its smaller heat of reaction.

the equilibrium constraints. Nevertheless, all combinations are constrained by material balances, which dictate the overall reactions.

This kind of reaction network analysis is typical of many applications. For example, some simple economic considerations show why producing hydrogen by steam reforming of methane (natural gas) is the preferred method compared to electrolysis. The energetic cost of water electrolysis raises serious doubts about electrolysis feasibility on Earth. With abundant electrical energy, it might be more appropriate to operate electric vehicles. It is not practical to articulate all the ways that this kind of network analysis can be applied to modern problems, but these illustrations should suggest the manner of proceeding for many such analyses. Noting that energies of reaction are an implicit part of the analysis, a tremendous wealth of information is implied by a single graph like Fig. 17.2. Later, in Section 17.11, we demonstrate how combining an unfavorable reaction with a strongly favorable reaction can help to drive the unfavorable reaction.

17.10   SOLVING EQUILIBRIA FOR MULTIPLE REACTIONS

When the equilibrium state in a reacting system depends on two or more simultaneous chemical reactions, the equilibrium composition can be found by a direct extension of the methods developed for single reactions. Each reaction will have its own reaction coordinate in which the compositions can be expressed. Some of the products of one reaction may act as reactants in another reaction, but the amount of that substance can still be written in terms of the extents of the reactions. Eventually, the material balances lead us to a system of N nonlinear equations in terms of N unknowns. We illustrate a solution by hand and then demonstrate how numerical solvers can be used.

Example 17.6 Simultaneous reactions that can be solved by hand

We can occasionally come across multiple reactions which can be solved without a computer. These are generally limited to textbook problems, but provide a starting point and test case for applying the general approach. Consider the two series/parallel gas phase reactions:

$$A + B \rightleftharpoons C + D \qquad K_{a1} = 2.667$$
$$A + C \rightleftharpoons 2E \qquad K_{a2} = 3.200$$

The reactions are considered series reactions because C is a product of the first reaction, but a reactant in the second. They are parallel because A is a reactant in both reactions. The pressure in the reactor is 10 bar, and the feed consists of 2 moles of A and 1 mole of B. Calculate the composition of the reaction mixture if equilibrium is reached with respect to both reactions.

Solution: The material balance gives:

Example 17.6 Simultaneous reactions that can be solved by hand (Continued)

$$n_A = 2 - \xi_1 - \xi_2$$
$$n_B = 1 - \xi_1$$
$$n_C = \quad \xi_1 - \xi_2$$
$$n_D = \quad \xi_1$$
$$n_E = \qquad 2\xi_2$$

$$\overline{\quad n_T = 3 \quad}$$

Note that for a physical solution, $0 \le \xi_1 \le 1$, $0 \le \xi_2 \le \xi_1$ to ensure that all mole numbers are positive. This reaction network is independent of P because $\Sigma v_i = 0$. The equilibrium constants are

$$\frac{y_C y_D P^2}{y_A y_B P^2} = 2.667 \; ; \text{ and } \quad \frac{y_E^2 P^2}{y_A y_C P^2} = 3.200 \; ; \text{ or in terms of reaction coordinates,}$$

$$\frac{(\xi_1 - \xi_2)\xi_1}{(2 - \xi_1 - \xi_2)(1 - \xi_1)} = 2.667 \; ; \quad \frac{4\xi_2^2}{(2 - \xi_1 - \xi_2)(\xi_1 - \xi_2)} = 3.2$$

Solving the first equation for ξ_1 using the quadratic equation,

$$\xi_1 = 2.4 - 1.1\xi_2 - \sqrt{(2.4 - 1.1\xi_2)^2 - 1.6(2 - \xi_2)} \qquad\qquad 17.39$$

Similarly, for the second reaction,

$$\xi_2 = -4 + \sqrt{16 + 4\xi_1(2 - \xi_1)} \qquad\qquad 17.40$$

We may now solve by trial and error. The procedure is: 1) guess ξ_1; 2) solve Eqn. 17.40 for ξ_2; 3) solve Eqn. 17.39 for ξ_1^{new}; 4) if $\xi_1^{new} \ne \xi_1$, go to step 1. The iterations are summarized below.

MATLAB
Ex17_06.m.

ξ_1	ξ_2	ξ_1^{new}
1.0000	0.4721	0.8355
0.8355	0.4600	0.8342
0.8342	0.4598	0.83415

Further iteration results in no further significant change.

These equations were amenable to the quadratic formula, but in general equilibrium criteria can be more complicated. Fortunately, standard programs available that are formulated to solve numerically multiple nonlinear systems of equations, so we can concentrate on applying the program to thermodynamics instead of developing the numerical analysis. Many software packages like Mathematica, Mathcad, MATLAB, and even Excel offer the capability to solve nonlinear systems of equations. Excel provides an especially convenient basis for illustrating the methods presented here.

Example 17.7 Solving multireaction equilibria with Excel

Methanol has a lower vapor pressure than gasoline. That can make it difficult to start a car fueled by pure methanol. One potential solution is to convert some of the methanol to methyl ether *in situ* during the start-up phase of the process (i.e., automobile). At a given temperature, 1 mole of MeOH is fed to a reactor at atmospheric pressure. It is assumed that only the two reactions given below take place. Compute the extents of the two simultaneous reactions over a range of temperatures from 200°C to 300°C. Also include the equilibrium mole fractions of the various species.

$$CH_3OH_{(g)} \;\rightleftharpoons\; CO_{(g)} + 2H_{2(g)} \qquad (1)$$

$$2CH_3OH_{(g)} \;\rightleftharpoons\; CH_3OCH_{3(g)} + H_2O_{(g)} \qquad (2)$$

Solution: A worksheet used for this solution is available in the workbook Rxns.xlsx.

Data for reaction (1) have been tabulated by Reactions Ltd.[a]—at 473.15 K, $\Delta H_T = 96{,}865$ J/mol and $\ln K_{a1,473} = 3.8205$. Over the temperature range of interest we can apply the shortcut van't Hoff equation assuming constant heat of reaction using the data at 200°C as a reference.

$$\ln K_{a1} = -\frac{96865}{8.314}\left(\frac{1}{T} - \frac{1}{473.15}\right) + 3.8205$$

Data for reaction (2) can be obtained from Appendix E for MeOH and water. For DME, the values are from Reid et al. (1987).[b]

Component		ΔH_f° kJ/mole	ΔG_f° kJ/mole
MeOH	−2	−200.94	−162.24
H₂O	1	−241.835	−228.614
MeOMe	1	−184.2	−113.0
		−24.155	−17.134

The shortcut van't Hoff equation for this reaction gives:

$$\text{at } 298 \text{ K}, \; \ln K_{a2,\,298.15} = \frac{17134}{8.314(298)} = 6.9156$$

$$\ln K_{a2} = \frac{24,155}{8.314}\left(\frac{1}{T} - \frac{1}{298.15}\right) + 6.9156$$

Example 17.7 Solving multireaction equilibria with Excel (Continued)

Material balances:

Specie	Initial	Final
1 MeOH	1	$1 - \xi_1 - 2\xi_2$
2 CO	0	ξ_1
3 H_2	0	$2\xi_1$
4 MeOMe	0	ξ_2
5 H_2O	0	ξ_2
Total	1	$1 + 2\xi_1$

Writing equations for reaction coordinates for reaction 1:

$$\frac{4\xi_1^3}{(1 - \xi_1 - 2\xi_2)(1 + 2\xi_1)^2} = K_{a1} \Rightarrow 4\xi_1^3 - K_{a1} \cdot (1 - \xi_1 - 2\xi_2) \cdot (1 + 2\xi_1)^2 = 0 = \text{"err1"}$$

and for reaction 2:

$$\frac{\xi_2^2}{(1 - \xi_1 - 2\xi_2)^2} = K_{a2} \Rightarrow \xi_2^2 - K_{a2} \cdot (1 - \xi_1 - 2\xi_2)^2 = 0 = \text{"err2"}$$

These two equations are solved simultaneously for ξ_1 and ξ_2. We have rearranged the objective functions to eliminate the ratios of ξ functions and use differences instead because the Excel Solver is much more robust with this mathematical form. The solution is implemented in the worksheet DUALRXN in Rxns.xlsx or Matlab Ex17_07.m. In the example here (see Fig. 17.3), the ΔC_P for both reactions is neglected. The equations derived above are entered directly into the cells, and the Solver tool is called.[c] You will need to designate one of the reaction equations as the target cell, the value of which is set to zero. The other reaction equation should be designated as a constraint (also set to zero). The cells with the reaction coordinates are the variables to be changed to obtain a solution. Under "options," you may want to specify the "conjugate" method, since that generally seems to converge more robustly for the reacting systems typically encountered. Generally, the Solver tool will require a reasonably accurate initial guess to keep it from converging on absurd results (e.g., $y_i < 0$). The initial guess can be easily developed by varying the values in the reaction-extent cells until the target cells move in the right direction. It sounds difficult, but the given worksheet will get you started, then you can experiment with initial guesses and experience how good your initial guesses need to be.

Rxns.xlsx, Worksheet DUALRXN of MATLAB Ex17_07.m.

a. These data are slightly different from values calculated using tabulated properties from Appendix E, but such variations are common in thermochemical data. The equilibrium compositions are about the same if the example is reworked using data from Appendix E.

b. Reid, R., Prausnitz, J.M., Poling, B. 1987. *The Properties of Gases and Liquids*, 4th ed. New York: McGraw-Hill.

c. See the online supplement for an introduction to Solver.

Sample solution of two simultaneous reactions:

$$CH_3OH = CO + 2H_2$$

$$2CH_3OH = CH_3OCH_3 + H_2O$$

(Details of input equations described in text by Elliott and Lira)

T(K)	473	493	513	533	553	573
K_{a1}	45.272	122.971	308.986	724.512	1597.281	3332.341
K_{a2}	27.4786	21.4179	17.0214	13.7627	11.3002	9.4069
ξ_1	0.9048	0.9651	0.9870	0.9951	0.9979	0.9991
ξ_2	0.0435	0.0158	0.0058	0.0022	0.0009	0.0004
y_1	0.0030	0.0012	0.0005	0.0002	0.0001	0.0000
y_2	0.3220	0.3294	0.3319	0.3328	0.3331	0.3332
y_3	0.6441	0.6587	0.6637	0.6656	0.6662	0.6665
y_4	0.0155	0.0054	0.0020	0.0007	0.0003	0.0001
y_5	0.0155	0.0054	0.0020	0.0007	0.0003	0.0001

Objective Functions

err1	0.0000	0.0000	0.0000	0.0000	0.0000	0.0000
err2	0.0000	0.0000	0.0000	0.0000	0.0000	0.0000

Figure 17.3 *Worksheet DUALRXN from workbook Rxns.xlsx for Example 17.7 showing converged answers at several temperatures.*

17.11 DRIVING REACTIONS BY CHEMICAL COUPLING

Frequently, one may encounter a reaction that is not favored by K_a, and manipulation of temperature or pressure or feed composition provides only limited benefit for the desired conversion. In these cases, it may be possible to couple the reaction to another, more favorable, reaction to drive the overall production forward. Biological systems use coupling extensively. The building of sugars and biological tissue from CO_2 and water is thermodynamically unfavorable. Carbon is fully oxidized and it must be reduced to create carbohydrates, and the reactions are endergonic at room temperature. These reactions are achieved by coupling an unfavorable carbon reduction with a strongly exergonic reaction.

To illustrate the principles of chemical coupling with a simple set of reactions, let us consider the production of butadiene from butene dehydrogenation at 900 K. We have investigated this reaction in Example 17.3 where we showed that the reaction is endergonic: $K_a = 0.242$ is small. The example showed that conversion is improved by diluting with steam. Consider instead if CO_2 is fed to the reactor and a catalyst is provided for the water-gas shift reaction (c.f. Eqn. 17.32). The CO_2 could then react with H_2 product of the dehydrogenation, *inducing* higher conversion for the dehydrogenation. The hydrogen product is removed by Le Châtelier's principle and the dehydrogenation reaction is pulled forward.

Example 17.8 Chemical coupling to induce conversion

Example 17.3(a) considered use of steam as a diluent where the conversion was found to be 78% using 10 moles of steam as diluent and only 44% without the diluent. Consider the conversion by inducing higher conversion by replacing the 10 mole steam with 10 mole CO_2 which adds the water-gas shift reaction. For the water-gas shift written as $CO_{2(g)} + H_{2(g)} \rightleftarrows CO_{(g)} + H_2O_{(g)}$, $K_{a2} = 0.441$ at 900 K. What is the conversion of 1-butene at 900 K and 1 bar?

Solution: The butadiene reaction has been written in Example 17.3(a) and ξ_1 will be used for that 1-butene reaction and ξ_2 will be used for the water-gas shift reaction. The stoichiometry table is,

$$n_{C_4H_8} \qquad 1 - \xi_1$$

$$n_{C_4H_6} \qquad \xi_1$$

$$n_{H_2} \qquad \xi_1 - \xi_2$$

$$n_{CO_2} \qquad 10 - \xi_2$$

$$n_{CO} \qquad \xi_2$$

$$n_{H_2O} \qquad \xi_2$$

$$\overline{n_T \qquad 11 + \xi_1}$$

$$K_{a1} = 0.242 = \left(\frac{\xi_1}{11 + \xi_1}\right)\left(\frac{\xi_1 - \xi_2}{11 + \xi_1}\right)P \Big/ \left(\frac{1 - \xi_1}{11 + \xi_1}\right) \text{ or}$$

$$0.242(1 - \xi_1)(11 + \xi_1) - (\xi_1)(\xi_1 - \xi_2) = 0 \tag{17.41}$$

$$K_{a2} = 0.441 = \frac{\xi_2^2}{(10 - \xi_1)(\xi_1 - \xi_2)} \text{ or } 0.441(10 - \xi_1)(\xi_1 - \xi_2) - \xi_2^2 = 0 \tag{17.42}$$

Physical limits for the reaction coordinates are $0 \le \xi_1 \le 1$ and $0 \le \xi_2 \le \xi_1$. Solving Eqns. 17.41 and 17.42 simultaneously, we find $\xi_1 = 0.949$ and $\xi_2 = 0.792$. Reviewing previous examples, the conversion at 1 bar was only 44% without an inert, increased to 78% with an inert, and increased to 95% using CO_2 to induce conversion by reaction coupling. Note that even though the water-gas shift equilibrium constant is not very large, it makes a significant difference in the conversion of 1-butene. Whether this is implemented depends on the feasibility of economically separating the products.

Chemical coupling can be classified in three ways: (1) induction, where a second reaction "pulls" a desired reaction by removing a product as in Example 17.8; (2) pumping, where the second reaction creates additional reactant for the desired reaction to "pump"; or (3) complex, where both induction and pumping are operative.[3] An example of complex coupling starts with the Sabatier reaction of Eqn. 17.31 with the goal of converting CO_2:

$$CO_{2(g)} + 4H_{2(g)} \overset{\rightarrow}{\leftarrow} CH_{4(g)} + 2H_2O_{(g)} \qquad\qquad K_{a,900K} = 0.3101 \qquad\qquad 17.43$$

Consier the possibility that equilibrium conversion may be inproved by the methane pyrolysis reaction of Eqn. 17.35,

$$CH_{4(g)} \overset{\rightarrow}{\leftarrow} C_{(s)} + 2H_{2(g)} \qquad\qquad K_{a,900K} = 3.546 \qquad\qquad 17.44$$

The $C_{(s)}$ will not be present in the gas phase, and $K_a = (y_{H_2}P)^2/(y_{CH_4}P)$ applying the principles that we will discuss in Section 17.14. Suppose we consider feeding H_2:CO_2 in a 2:1 ratio which is less than the stoichiometric H_2. However, Eqn. 17.44 will produce H_2 to pump Eqn. 17.43, and will remove some CH_4 to induce Eqn 17.43. Under the specified feed conditions at 900K and 2 bar, equibrium conversion of CO_2 is 28.1% if no pyrolyis occurs. If we promote pyrolysis in the same reactor, and add the second reaction, CO_2 conversion will increase to 34.9%. This illustrates the possibilities of coupling reactions, but the subtle effect may not justify the coupled system. Biology uses coupling extensively as illustrated by homework problem 18.15 which is delayed to introduce the concepts for biological standard states in Chapter 18.

17.12 ENERGY BALANCES FOR REACTIONS

We have previously introduced the energy balance in Section 3.6 and also discussed adiabatic reactors. In this section we consider that there may be a there is a maximum possible value of ξ (outlet conversion) due to chemical equilibrium. Equilibrium may affect both adiabatic and nonadiabatic reactors, but we cover adiabatic reactors, and the extension to nonadiabatic should be obvious with the inclusion of the heat term.

Adiabatic Reactors

The energy balance for a steady-state adiabatic flow reactor is given in Eqn. 3.53 on page 118. The variables T^{out} and ξ from the energy balance also appear in the equilibrium constraint that will govern maximum conversion. Earlier, in Chapter 3, we considered the reaction coordinate to be specified. However, in a reaction-limited adiabatic reactor, we must solve the energy balance together with the equilibrium constraint to simultaneously determine the maximum conversion and adiabatic outlet temperature. Using the energy balance from Eqn. 3.53, do the following.

1. Write the energy balance, Eqn 3.53. Calculate the enthalpy of the inlet components at T^{in}.

2. Guess the outlet temperature, T^{out}. Calculate the enthalpy of the outlet components at T^{out}.

3. Determine $\dot{\xi}$ at T^{out} using the chemical equilibrium constant constraint.

4. Calculate $\dot{\xi}\Delta H_R^o$ for this conversion.

5. Check the energy balance for closure.

6. If the energy balance does not close, go to step 2.

As you might expect, this type of calculation lends itself to numerical solution, such as the Solver in Excel.

---

3. O'Connell, J.M., Fernandez, E., Komives, C. July, 2010. NSF BioEMB Workshop on Thermodynamics, San Jose, CA.

Example 17.9 Adiabatic reaction in an ammonia reactor

Estimate the outlet temperature and equilibrium mole fraction of ammonia synthesized from a stoichiometric ratio of N_2 and H_2 fed at 400 K and reacted at 100 bar. How would these change if the pressure was 200 bar?

$$1/2\ N_{2(g)} + 3/2\ H_{2(g)} \ \rightleftarrows \ NH_{3(g)}$$

Solution: For a rough estimate we will use the shortcut approximation of temperature effects. Furthermore, we will assume $K_\varphi \approx 1$. (Is this a good approximation or not?[a]) Therefore we obtain,

$$P\,K_a = y_{NH3} / [(y_{N2})\,(y_{H2})^3]^{\frac{1}{2}}$$

Basis: Stoichiometric ratio in feed.

$$\dot{n}_{N_2}^{out} = \frac{1}{2} - \frac{1}{2}\dot{\xi} = \frac{1}{2}(1 - \dot{\xi}); \ \dot{n}_{H_2}^{out} = \frac{3}{2} - \frac{3}{2}\dot{\xi} = \frac{3}{2}(1 - \dot{\xi}); \ \dot{n}_{NH_3}^{out} = \dot{\xi}; \ \dot{n}_T^{out} = 2 - \dot{\xi} \quad 17.46$$

For the purposes of the example, the shortcut van't Hoff equation will be used to iterate on the adiabatic reactor temperature. However, the full van't Hoff method will be used to obtain ΔG_{Tnear}° and ΔH_{Tnear}° at an estimated nearby temperature $T_{near} = 600K$ as suggested in Section 17.8. Then the shortcut van't Hoff equation will be used over a limited temperature range for less error. The energy balance will also use ΔH_{Tnear}° ; we will create an energy balance path through $T_{near} = 600K$ rather than 298.15K. We will compare the approximate answer with the full van't Hoff method at the end of the example.

For ammonia, $\Delta G_{f,\,298.15}^{\circ} = -16,401.3$ J/mol, $\Delta H_{f,\,298.15}^{\circ} = -45,940$ J/mol. Since the reactants are in the pure state, the respective reactant formation values are zero, and therefore the formation values for ammonia represent the standard state values for the reaction. Inserting the formation values along with the heat capacities into the detailed van't Hoff equation—one of the K_a calculators highlighted in the margin note to Example 17.4 on page 653 is used—at an assumed temperature of 600 K, the values obtained are $\Delta H_{600}^{\circ} = -51,413$ J/mol and $K_{a,600} = 0.0417659$. Then the shortcut van't Hoff in the vicinity will be

$$\ln\left(\frac{K_{a,\,T}}{0.0417659}\right) = \frac{51,413}{8.314}\left(\frac{1}{T} - \frac{1}{600}\right) \qquad 17.47$$

From an assumed value of T, this equation will provide the equilibrium constant. Some manipulation is necessary to obtain the material balance from $K_{a,T}$. Plugging the mole fraction expressions into Eqn. 17.17, and collecting the fractions 1/2 and 3/2,

$$PK_{a,\,T}\left(\frac{3^3}{2^4}\right)^{1/2} = \frac{\dot{\xi}(2-\dot{\xi})}{(1-\dot{\xi})^2} \Rightarrow 2\dot{\xi} - \dot{\xi}^2 = \frac{\sqrt{27}}{4}PK_{a,\,T}(1-\dot{\xi})^2$$

$$\text{defining } M = \frac{\sqrt{27}}{4}PK_{a,\,T} \Rightarrow (M+1)\dot{\xi}^2 - 2(M+1)\dot{\xi} + M = 0 \Rightarrow \dot{\xi}^2 - 2\dot{\xi} + \left(\frac{M}{M+1}\right) = 0$$

Example 17.9 Adiabatic reaction in an ammonia reactor (Continued)

Applying the quadratic formula,

$$\dot{\xi} = \frac{2 \pm \sqrt{4 - 4M/(1+M)}}{2} = 1 - \sqrt{1 - M/(1+M)} \qquad 17.48$$

The strategy will be to guess T, and calculate $K_{a,T}$, M, and $\dot{\xi}$. $\dot{\xi}$ will be used in Eqn. 17.46 to perform the material balance. The material balance will be combined with the energy balance using the Heat of Reaction method (*cf.* Example 3.6), until the energy balance closes as represented by:

$$F(T) = \sum_{components} \dot{n}_i^{in} \int_{T_R}^{T^{in}} C_{P,i} dT - \sum_{components} \dot{n}_i^{out} \int_{T_R}^{T^{out}} C_{P,i} dT - \dot{\xi} \Delta H_R^o = 0 \qquad 17.49$$

Heat capacity integrals and the energy balance have been entered in the workbook Rxns.xlsx. At the initial guess of 600 K, the $F(T)$ of Eqn. 17.49 is 19.4 kJ. A converged result is found at 699 K shown in Fig. 17.4 and the $\dot{\xi} = 0.33$, conversion of feed is 33%. At 200 bar, the answer is 739 K, and conversion is 38%.

Workbook
Rxns.xlsx,
worksheet
RxnAdia-shortcut
or MATLAB
Ex17_09.m.

The detailed van't Hoff is available in the same workbook and results in 698 K and 33% conversion at 100 bar, and 737 K and 37% conversion at 200 bar.

Adiabatic Synthesis of Ammonia   Protected without a password

Feed Temperature (K)	400
Outlet Temperature(K)	699.07
P(bar)	100
T_R(K)	600 K
Standard State Hrxn(T_R)	-51413 J/mol
K_a (T_R)	0.04177
$\ln[K_a$ (T)]	-4.63623
K_a at reaction T	0.00969

	$\Delta H°_{f,298}$	$\Delta G°_{f,298}$	Heat Capacity Constants			
	(kJ/mol)	(kJ/mol)	a	b	c	d
H2	0	0	2.71E+01	9.27E-03	-1.38E-05	7.65E-09
N2	0	0	3.12E+01	-1.36E-02	2.68E-05	-1.17E-08
NH3	-45.94	-16.4013	2.73E+01	2.38E-02	1.71E-05	-1.19E-08
Δ	-45.94	-16.4013				
M			1.259311 ξ		0.3347085	

	Inlet			Outlet		
	moles	H(J/mol)	totals	moles	H(J/mol)	totals
H2	1.5	-5854.46	-8781.7	0.99794	2915.96	2909.94
N2	0.5	-5927.19	-2963.59	0.33265	3016.21	1003.33
NH3	0	-8401.78	0	0.33471	4630.33	1549.81
Total			-11745.3			5463.08

Balance($\Sigma H^{in} n^{in} - \Sigma H^{out} n^{out} - \xi \Delta H$)=	6.665E-07 J

NOTE: The inlet moles cannot be changed without recalculating a formula for ξ
Use solver to set value of Balance to zero by adjusting Feed Temperature, Outlet Temperature, or P.

Figure 17.4 *Display from Rxns.xlsx showing a converged answer.*

a. We can evaluate this assumption by calculating the reduced temperatures at the end of our calculation and estimating the virial coefficients, then fugacity coefficients.

Graphical Visualization of the Energy Balance

The energy balance is presented in Fig. 3.6 on page 119. The difference here is that the appropriate curve from Fig. 3.6 is superimposed on the plot and the outlet conversion and outlet temperature are determined by the intersection of the energy balance line and the equilibrium line. Fig. 17.5 illustrates an exothermic reaction. In the event that the reaction does not reach equilibrium because

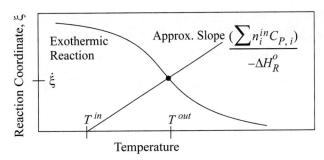

Figure 17.5 *Approximate energy balance for an exothermic reaction. The dot simultaneously represents the equilibrium outlet conversion and reaction coordinate value at the adiabatic outlet temperature. The plot for an endothermic reaction will be a mirror image of this figure as explained in the text.*

of kinetic limitations, the reaction coordinate must be located along the energy balance line below the equilibrium value. For the case of the ammonia reaction, the equilibrium constraint curve could be generated by inserting various temperatures in Eqn. 17.47 and then determining the reaction coordinate from Eqn. 17.48. The energy balance is plotted using Eqn. 3.55. The dot in the figure represents the point where the energy balance and equilibrium constraint are both satisfied. Note that an endothermic reaction will have an energy balance with a negative slope, and the equilibrium line will change shape as shown in Fig. 3.6, making the plot for an endothermic reaction a mirror image of Fig. 17.5 reflected across a vertical line at T^{in}.

17.13   LIQUID COMPONENTS IN REACTIONS

When a liquid component is involved in a reaction, the fugacity ratio for activity in Eqn. 17.15 is typically expressed using activity coefficients. Thus,

$$\frac{\hat{f}_i}{f_i^\circ} = a_i = \frac{x_i \gamma_i \varphi_i^{sat} P_i^{sat} \exp\left(\frac{V^L(P - P^{sat})}{RT}\right)}{\varphi_i^{sat} P_i^{sat} \exp\left(\frac{V^L(1 - P^{sat})}{RT}\right)} = x_i \gamma_i \exp\left(\frac{V^L(P - 1)}{RT}\right) \approx x_i \gamma_i \qquad 17.50$$

where P is expressed in bar, and the Poynting correction is often negligible, as shown. Another important change in working with liquid components is that in determining K_a liquid phase values are used for $\Delta G^o_{f,i}$, not the ideal gas values. Frequently these values are not available in the literature, so it is common to express equilibrium in terms of temperature-dependent correlations for K_a as described in Section 17.18.

Equilibrium constants calculated using liquid phase species are different from the equilibrium constants for the same reactants and products in the gas phase. Consider that $\hat{f}_i/f_i^\circ \approx y_i P$ for the vapor phase. If a vapor phase reaction is simultaneously in phase equilibrium with a liquid phase with the same components, by modified Raoult's law we could replace $y_i P = x_i \gamma_i P_i^{sat}$. However, for the liquid equilibrium constant we should use only the activity (Eqn. 17.50) without the vapor pressure. Thus, we conclude that the equilibrium constants for the liquid phase reaction must be different from the equilibrium constant for the same vapor phase reaction, and also that the standard state Gibbs energy change for the same reaction must be different.

Example 17.10 Oligomerization of lactic acid

Lactic acid is a bio-derived chemical intermediate produced in dilute solution by fermentation. Lactic acid is an α-hydroxy carboxylic acid. As an aqueous solution of lactic acid is concentrated by boiling off water, the carboxylic acid on one molecule reacts with a hydroxyl on another forming a dimer and releasing water. Denoting the "monomer" as L_1 and a dimer as L_2,

$$2L_1 \rightleftarrows L_2 + H_2O \qquad\qquad 17.51$$

The dimer has a hydroxyl and carboxylic acid that can react further to form trimer L_3,

$$L_2 + L_1 \rightleftarrows L_3 + H_2O \qquad\qquad 17.52$$

As more water is removed, the chain length grows, forming oligomers. The oligomerization can be represented by a recurring reaction for chain formation. Each liquid phase reaction that adds a lactic acid molecule can be modeled with a universal temperature-independent value of $K_a = 0.2023$ and the solutions may be considered ideal.[a] Commercial lactic acid solutions are sold based on the wt% of equivalent lactic acid monomer. So 100 g of 50 wt% solution would be composed of 50 g of lactic acid monomer and 50 g of water that react to form an equilibrium distribution of oligomers. The importance of including modeling of higher oligomers increases as the concentration of lactic acid increases.

(a) Determine the mole fractions and wt% of species in a 50 wt% lactic acid solution in water where the distribution is approximated by only reaction 17.51.
(b) Repeat the calculations for an 80 wt% lactic acid solution in water where both reactions are necessary to approximate the distribution.

MATLAB
Ex17_10.m may be helpful in calculations for this example.

Solution:
(a) Basis: 100 g total, 50 g of L_1 = (50 g)/(90.08 g/mol) = 0.555 mol initially, 50 g = (50 g)/(18.02 g/mol) = 2.775 mol water initially, and 3.330 mol total. The equilibrium relation is $K_a = 0.2023 = x_{L_2}x_{H_2O}/(x_{L_1})^2$. Since the total number of moles does not change with reaction, it cancels out of the ratio, and we can write $0.2023 = n_{L_2}n_{H_2O}/(n_{L_1})^2$. Introducing reaction coordinate,

$$0.2023 = \xi(2.775 + \xi)/(0.555-2\xi)^2 \qquad\qquad 17.53$$

Solving, we find, $\xi = 0.0193$, $x_{L_1} = (0.555 - 2(0.0193))/3.33 = 0.155$, $x_{L_2} = 0.0193/3.33 = 0.006$, $x_{H_2O} = (2.775 + 0.0193)/3.33 = 0.839$. Note that although the mole fraction of L_2 seems small, converting to wt%, the water content is $(2.775 + 0.0193)(18.02 \text{ g/mol})/(100 \text{ g})\cdot 100\% = 50.4$ wt%, L_1 is $(0.555 - 2(0.0193))(90.08 \text{ g/mol})/(100 \text{ g})\cdot 100\% = 46.5$ wt%, and L_2 is $0.0193(162.14 \text{ g/mol})/(100 \text{ g})\cdot 100\% = 3.1$ wt%.

(b) Basis: 100 g total, 80 g of L_1 = (80 g)/(90.08 g/mol) = 0.888 mol initially, 20 g = (20 g)/(18.02 g/mol) = 1.110 mol water initially, and 1.998 mol total. Moles are conserved in both reactions. The equilibrium relations are $K_{a1} = 0.2023 = x_{L_2}x_{H_2O}/(x_{L_1})^2$, $K_{a2} = 0.2023 = x_{L_3}x_{H_2O}/(x_{L_2}x_{L_1})$. Introducing the mole numbers and reaction coordinates,

$$0.2023 = (\xi_1 - \xi_2)(1.110 + \xi_1 + \xi_2)/(0.888 - 2(\xi_1) - \xi_2)^2 \qquad\qquad 17.54$$

$$0.2023 = \xi_2(1.110 + \xi_1 + \xi_2)/((\xi_1 - \xi_2)(0.888 - 2(\xi_1) - \xi_2)) \qquad\qquad 17.55$$

Example 17.10 Oligomerization of lactic acid (Continued)

Solving simultaneously, $\xi_1 = 0.0907$, $\xi_2 = 0.009$, $x_{L_1} = (0.888 - 2(0.0907) - 0.009)/1.998 = 0.349$, $x_{L_2} = (0.0907 - 0.009)/1.998 = 0.041$, $x_{L_3} = 0.009/1.998 = 0.0045$, $x_{H_2O} = (1.110 + 0.0907 + 0.009)/1.998 = 0.6055$. The weight fractions are: $(1.110 + 0.0907 + 0.009)(18.02 \text{ g/mol})/(100 \text{ g}) \cdot 100\% = 21.8$ wt% water, $(0.888 - 2(0.0907) - 0.009)(90.08 \text{ g/mol})/(100 \text{ g}) \cdot 100\% = 62.8$ wt% L_1, and $(0.0907 - 0.009)(162.14 \text{ g/mol})/(100 \text{ g}) \cdot 100\% = 13.2$ wt% L_2, $0.009(234.2 \text{ g/mol})/(100 \text{ g}) \cdot 100\% = 2.1$ wt% L_3.

a. Vu, D. T., Kolah, A.K., Asthana, N.S., Peereboom, L., Lira, C.T., Miller, D.J. 2005. "Oligomer distribution in concentrated lactic acid solutions." *Fluid Phase Equil.* 236:125–135.

If a vapor state coexists with a liquid phase during a reaction, the phase equilibria and reaction equilibria are coupled. Reactions need to be considered in only one of the two phases, and the equilibrium compositions will be consistent with compositions that would have been determined by the same reaction equilibria in the other phase. Similarly, some reaction equilibria constants may be known for only one or the other of the phases, and the equilibria can be solved by using reaction equilibria for whichever phase is most convenient. Simultaneous reaction and phase equilibria can be extremely useful for driving reactions in preferred directions as we illustrate in Example 17.15.

17.14   SOLID COMPONENTS IN REACTIONS

When a solid component is involved in a reaction, the fugacity ratio for activity in Eqn. 17.15 is typically expressed using activity coefficients. For a solid solution,

$$\frac{\hat{f}_i}{f_i^\circ} = a_i = \frac{x_i \gamma_i \varphi_i^{sat} P_i^{sat} \exp\left(\dfrac{V^S(P - P^{sat})}{RT}\right)}{\varphi_i^{sat} P_i^{sat} \exp\left(\dfrac{V^S(1 - P^{sat})}{RT}\right)} = x_i \gamma_i \exp\left(\frac{V^L(P - 1)}{RT}\right) \approx x_i \gamma_i \quad \text{solid solution} \quad 17.56$$

where P is expressed in bar, P^{sat} represents the solid sublimation pressure, and the Poynting correction is often negligible. Commonly, multiple solids exist as physical mixtures of pure crystals as discussed in Section 14.10 on page 556. When the solids are immiscible, Eqn. 17.56 simplifies to

$$\frac{f_i}{f_i^\circ} = a_i = \frac{\varphi_i^{sat} P_i^{sat} \exp\left(\dfrac{V^S(P - P^{sat})}{RT}\right)}{\varphi_i^{sat} P_i^{sat} \exp\left(\dfrac{V^S(1 - P^{sat})}{RT}\right)} = \exp\left(\frac{V^L(P - 1)}{RT}\right) \approx 1 \quad \text{pure solid} \quad 17.57$$

Similar to working with liquids, solid phase data are used for $\Delta G_{f,i}^o$, not the ideal gas values. When these values are not available in the literature, it is common to express equilibrium in terms of temperature-dependent correlations for K_a as described in Section 17.18.

Consider the reaction:

$$CO_{(g)} + H_{2(g)} \; \underset{\leftarrow}{\overset{\rightarrow}{}} \; C_{(s)} + H_2O_{(g)} \qquad\qquad 17.58$$

The carbon formed in this reaction comes out as coke, a solid which is virtually pure carbon and separate from the gas phase. What is the activity of this carbon? Since it is pure, $a_C = 1$. Would its presence in excess ever tend to push the reaction in the reverse direction? Since the activity of solid carbon is always 1 it cannot influence the extent of this reaction. How can we express these observations quantitatively? Eqn. 17.15 becomes

$$K_a = \exp\left[\frac{-\Delta G_T^\circ}{RT}\right] = \frac{\left(\hat{f}_{H2O}/f_{H2O}^\circ\right)\cdot 1.0}{\left(\hat{f}_{H2}/f_{H2}^\circ\right)\left(\hat{f}_{CO}/f_{CO}^\circ\right)} = \frac{y_{H2O}}{y_{H2}y_{CO}}\frac{1}{P} \qquad 17.59$$

To compute ΔG°_T as a function of temperature, we apply the usual van't Hoff procedure. This means that $C_{P,c}$ can be treated just like C_P of the gaseous species.

Example 17.11 Thermal decomposition of methane

A 2-liter constant-volume pressure vessel is evacuated and then filled with 0.10 moles of methane, after which the temperature of the vessel and its contents is raised to 1273 K. At this temperature the equilibrium pressure is measured to be 7.02 bar. Assuming that methane dissociates according to the reaction $CH_{4(g)} \rightleftarrows C_{(s)} + 2H_{2(g)}$, compute K_a for this reaction at 1273 K from the experimental data.

Solution:

$$K_a = \frac{\left(\hat{f}_{H_2}/f_{H_2}^\circ\right)^2\cdot 1.0}{\left(\hat{f}_{CH_4}/f_{CH_4}^\circ\right)} = \frac{y_{H_2}^2}{y_{CH_4}}P$$

We can calculate the mole fractions of H_2 and CH_4 as follows. Since the temperature is high, the total number of moles finally in the vessel can be determined from the ideal gas law (assuming that the solid carbon has negligible volume): $n = P\underline{V}/RT = 0.702\cdot2000/(8.314)(1273) = 0.1327$. Now assume that ξ moles of CH_4 reacted. Then we have the following total mass balance: $n_T = 0.10 + \xi$. Therefore, $\xi = 0.0327$ and

$$K_a = \frac{\left(\dfrac{2\xi}{0.1327}\right)^2 P}{\left(\dfrac{(0.1 - \xi)}{0.1327}\right)} = \frac{0.493^2}{0.507}7.02 = 3.37$$

Note that the equilibrium constant indicates that significant decomposition will occur (the reaction is exergonic, $K_a > 1$) and that graphite forms. Such behavior is known as "coking" and is common during industrial catalysis. Industrial application of catalysis often includes consideration of "regeneration'" of the catalyst by burning off the coke and using the heat of combustion elsewhere in the chemical plant.

17.15   RATE PERSPECTIVES IN REACTION EQUILIBRIA

We have avoided discussing rate effects until now with the rationale that most coverage for reaction kinetics will occur in a course focused on reactor design. Nevertheless, there is overlap between the topics of reaction equilibria and reaction rates that can serve as a bridge between the two subjects. In all equilibrium phenomena, it is important to recognize that the balance achieved is dynamic, not static. For example, the molecules at the interface between a vapor and liquid are not stationary; they are perpetually exchanging between the vapor and liquid. Application of thermodynamics helps us understand the conditions where the balance occurs. Similarly, under conditions of chemical reaction equilibrium, the species are continuously interconverting with equal rates for the forward and reverse reactions.

From a thermodynamic perspective, the true driving force for chemical reaction is the activity. When the activities are balanced as given by Eqn. 17.15 the reaction reaches equilibrium and the forward are reverse rates are equal. The activities are directly proportional to the concentrations for liquids and vapors. Thus, it is common to use concentrations instead of activities for simple kinetic models. Consider the vapor-phase reaction,

$$2A_{(g)} + B_{(g)} \underset{\leftarrow}{\rightarrow} C_{(g)} + D_{(g)} \qquad 17.60$$

For example, if two components, A and B, react to form C and D, then the rate of accomplishing the reaction must depend on the probability of the two components colliding with each other. This probability decreases as the concentration of one of the components diminishes. By convention the rates are typically written for the stoichiometrically limiting component. Also, they are typically written for the rate of *formation* per volume of reacting mixture.[4] If A is the limiting reactant, the rate of the formation of A due to reaction of A with B would then be,

$$-r_{A,f} = k_f [A]^2 [B] \qquad 17.61$$

where the minus sign acknowledges that A is disappearing rather than forming, and the subscript f indicates reaction in the forward direction and k_f is known as the forward rate constant. When the exponents on the rate equation match the stoichiometric coefficients, the reaction is called an **elementary reaction.** When a reaction is equilibrium-limited, it is considered kinetically reversible. Recognizing that A is formed by reaction of C and D, the reverse reaction rate for formation of A is

$$r_{A,r} = k_r [C][D] \qquad 17.62$$

The net rate of formation of A must be zero at reaction equilibrium,

$$r_{A,f} + r_{A,r} = -k_f [A]^2 [B] + k_r [C][D] = 0 \qquad 17.63$$

Recognizing that concentration of a gas phase component is related to partial pressure, $[A] = y_A P / RT$, and similarly for other components. Rearranging Eqn. 17.63 and inserting the partial pressure results in

$$\frac{[C][D]}{[A]^2[B]} = \frac{k_f}{k_r} = \frac{(y_C P)(y_D P)}{(y_A P)^2 (y_B P)} (RT)^1 = K_a (RT)^{-\Sigma \nu_i} \qquad 17.64$$

---

4. Fogler, H.S. 1999. *Elements of Chemical Reaction Engineering*, 3rd ed. Upper Saddle River, NJ: Prentice-Hall, pp. 340ff.

where in this case, $\Sigma \nu_i = -1$, but the general expression is written to help readers remember the general relation for gas phase reactions. Note the manner in which the forward and reverse reaction rate constants are related to the equilibrium constant. This means that if the forward rate constant is measured in an experiment when the product concentrations are low, then the reverse rate constant can be determined from the equilibrium constant. Note that similar relations can be written for liquid-phase elementary reactions.

Certainly many reactions have rate expression more complicated than the elementary reactions discussed here. For example, enzyme catalyzed reactions often involve a binding step that is not represented by the simple statistical concept of the elementary reaction. Many reactions involve intermediate species that must be included in the mechanism and kinetic rate law. Understanding more complex rate laws is an important skill covered in reaction engineering courses. Our intention here was to show the relation for elementary reactions and to communicate the concept of forward and reverse rates approaching each other at reaction equilibrium. Note that this means that if the reaction approaches equilibrium in the forward direction, the overall rate of disappearance of A will slow and become zero. At slow rates, the reactor volume must be large to achieve meaningful change in reaction coordinate, so it is rarely economical to run commercial reactors all the way to equilibrium. However, the calculation of the equilibrium condition is important for any reactor design in order to know the limiting conversion, and usually avoid the conditions! Often, the equilibrium constant is used to calculate the reverse rate constant from the forward rate constant as discussed above.

17.16   ENTROPY GENERATION VIA REACTIONS

When introducing entropy and reversibility in Section 4.11 on page 175, we made a general statement that spontaneous reactions generate entropy. Then, in Section 4.12 on page 177 we derived relations between availability and entropy generation. In that section, we treated a single nonreactive stream. For a reaction in a steady-state open system, Eqn. 4.54 becomes

$$\sum_{\text{outlets}} (B^{out})dn^{out} - \sum_{\text{inlets}} (B^{in})dn^{in} = d\underline{W}_S - T_o d\underline{S}_{gen} \qquad 17.65$$

$B^{out} = H^{out} - T_o S^{out}$ and $B^{in} = H^{in} - T_o S^{in}$ involve H and S evaluated at the respective T^{out} and T^{in}. Enthalpies of mixed streams were first introduced in Chapter 3 and entropies for mixed streams in Chapter 4.[5] Models for departures functions and excess properties in Chapters 11–15 can be added to improve the mixture property values. The concepts for proper choice of a reference state for properties is important as discussed in Chapter 3 in the section Energy Balances for Reactions on page 113. The flow terms in Eqn. 17.65 are not the same as the Gibbs energy, but the availability will decrease with a spontaneous reaction. Therefore, both sides of the equation will be negative, and shaft work will not be obtained (a normal situation in an industrial reactor), and then entropy is generated. Note that entropy generation can be decreased for a spontaneous reaction only if work is produced. For an electrochemical redox reaction (Chapter 18), the process can produce some electrical work, which is one reason that fuel cells are of much current interest. For a closed-system process, the analysis is similar to the open-system process. The relation is seen most readily if $T_o = T$ (and additional work can be obtained using a heat engine between T and T_o), and at constant pressure, the closed system balance (Eqn. 4.57) becomes

---

5. The entropy of formation can be obtained by inserting formation values into $S = (H - G)/T$.

$$dG = dW_S - TdS_{gen}, \text{ constant } P \text{ closed system, } T_o = T \qquad 17.66$$

We conclude that production of nonexpansion/contraction work equal to changes in the Gibbs energy is necessary to eliminate entropy generation. Note that it is possible to relate the total work from a reaction to Helmholtz energy using Eqn. 4.58.

17.17 GIBBS MINIMIZATION

A remarkably simple technique can be applied to solve for the equilibrium compositions of species. It is most effective when only a gas phase is present. This technique recognizes the simplicity of the fundamental problem of minimizing the Gibbs energy at equilibrium. By expressing the total Gibbs energy of the mixture in terms of its ideal solution components, we can simply request that the value of the Gibbs energy be minimized. The Gibbs energy of the mixture is calculated by Eqn. 10.42 and the needed chemical potential (partial molar Gibbs energy) is given by Eqn. 10.59:

$$G = \sum n_i \overline{G}_i \quad \text{and} \quad \frac{\overline{G}_i}{RT} = \frac{G_i}{RT} + \left(\frac{\overline{G}_i - G_i}{RT}\right) = \frac{G_i}{RT} + \ln y_i \qquad \text{(ig) } 17.67$$

where the last equality assumes all components are ideal gases. If we take the reference state as the elements in their natural form at the standard state, then, at the standard state pressure, $G_i/(RT) = (\Delta G_f^o)/(RT)$. However, frequently the reactions are not at standard state pressure. The pressure effect on Gibbs energy is given by Eqn. 9.17. When the pressure effect is added,

$$\frac{G_i}{RT} = \frac{\Delta G_{f,i}^o}{RT} + \ln \frac{P}{P^o} \qquad \text{(ig) } 17.68$$

Combining Eqns. 17.67 and 17.68 results in

$$\frac{G}{RT} = \sum_i n_i \left(\frac{\Delta G_{f,i}^o}{RT}\right) + \sum_i n_i \ln y_i P \qquad \text{(ig) } 17.69$$

To find equilibrium compositions, we just need to minimize Eqn. 17.69 by varying the mole numbers n_i of each component while simultaneously satisfying the atom balance. Note that the mole fractions in the equation will also change as the mole numbers are varied. We do not need to explicitly write out the reactions. This method assumes that equilibrium is reached by whatever system of reactions is necessary. Most process simulators provide Gibbs minimization as a process unit.

Example 17.12 Butadiene by Gibbs minimization

Review Example 17.3(a) where steam is used to enhance conversion for 1-butene dehydrogenation. Gibbs energies of formation at 900 K for the hydrocarbons are summarized in that example.

Example 17.12 Butadiene by Gibbs minimization (Continued)

The Gibbs energy of formation for water at 900 K is −198.204 kJ/mol. Vary conversion by selecting values of the reaction coordinate, calculating the Gibbs energy by Eqn. 17.69, and plotting the total Gibbs energy as a function of reaction coordinate. Demonstrate that Gibbs energy is minimized. Compare the equilibrium composition with that found in Example 17.3(a).

Solution: The initial moles of feed are 1 mol of 1-butene and 10 moles of steam. As an example calculation, select $\xi = 0.1$. Then the material balance provides, $n_{C_4H_8} = 0.9$, $n_{C_4H_6} = n_{H_2} = 0.1$, $n_{H_2O} = 10$. The mole fractions are $y_{C_4H_8} = 0.9/(0.9 + 2(0.1) + 10) = 0.08108$, $y_{C_4H_6} = y_{H_2} = 0.0090$, $y_{H_2O} = 0.9009$, and inserting the quantities into Eqn. 17.69, gives (inserting components in the order given above)

$$\frac{G}{RT} = \sum_i n_i \left(\frac{\Delta G^o_{f,i}}{RT} \right) + \sum_i n_i \ln y_i P$$

$$= 0.9 \frac{(232854)}{8.3145(900)} + 0.1 \frac{(243474)}{8.3145(900)} + 0.1(0) + 10 \frac{(-198204)}{8.3145(900)} +$$

$$0.9 \ln(0.08108 \cdot 1) + 0.1 \ln(0.009 \cdot 1) + 0.1 \ln(0.009 \cdot 1) + 10 \ln(0.9009 \cdot 1) = -237.858$$

Repeating the calculation at various extents of reaction results in the following plot:

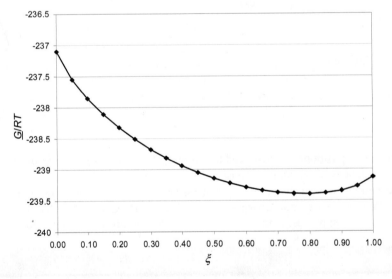

Careful analysis would show that the minimum is at $\xi = 0.784$ as in the earlier example. Note the changes in values are a small percentage of the absolute values of the total Gibbs energy, a numerical observation that is important in setting up convergence in Excel.

The previous example is somewhat contrived because the reaction was specified and the reaction coordinate was varied. The method is applicable without specifying reactions as long as the atom material balances are satisfied when selecting mole numbers.

Example 17.13 Direct minimization of the Gibbs energy with Excel

Apply the Gibbs minimization method to the problem of steam cracking of ethane at 1000 K and 1 bar where the ratio of steam to ethane in the feed is 4:1. Determine the distribution of C_1 and C_2 products, neglecting the possible formation of aldehydes, carboxylic acids, and higher hydrocarbons.

Solution: The solution is obtained using the worksheet GIBBSMIN contained in the workbook Rxns.xlsx (see Fig. 17.6).

One fundamental problem and one practical problem remain to be faced; there are several constraints that must be respected during the minimization process. These are the atom balances. We must keep in mind that we are not destroying matter, only rearranging it. So the number of carbon atoms, say, must be the same at the beginning and end of the process. Atom balance constraints must be written for every atom present. The atom balances are given straightforwardly by the stoichiometry of the species. For this example the balances are as follows.

O-balance: $n_O^{feed} - 2n_{CO_2} - n_{CO} - 2n_{O_2} - n_{H_2O} = 0$

H-balance: $n_H^{feed} - 4n_{CH_4} - 4n_{C_2H_4} - 2n_{C_2H_2} - 6n_{C_2H_6} - 2n_{H_2} - 2n_{H_2O} = 0$

C-balance: $n_C^{feed} - n_{CH_4} - 2n_{C_2H_4} - 2n_{C_2H_2} - n_{CO_2} - n_{CO} - 2n_{C_2H_6} = 0$

The practical problem that remains is that the numerical solver often attempts to substitute negative values for the prospective species. This problem is easily treated by solving for the $\log(n_i)$ during the iterations and determining the values of n_i after the solution is obtained. Large negative values for the $\log(n_i)$s cause no difficulty. They simply mean that the concentrations of those species are small.

In order to apply Gibbs minimization, the Gibbs energy of formation is required for each component at the reaction temperature. This preliminary calculation is the same type of calculation as performed in Example 17.4 on page 653, but is not shown here. For example, the Gibbs energy of methane is simply the Gibbs energy of the formation reaction $C_{(s)} + 2H_{2(g)} \rightleftarrows CH_{4(g)}$ at 1000 K. The values are embedded in the calculation of G_i shown in the worksheet in Fig. 17.6.

The primary product of this particular process is hydrogen. Fracturing hydrocarbons is a common problem in the petrochemical industry. This kind of process provides the raw materials for many downstream processes. The extension of this method to other reactions is straightforward. Some examples of interest would include several systems with environmental applications: carbon monoxide and NO_x from a catalytic converter, or by-products from catalytic destruction of chlorinated hydrocarbons. Evaluating the equilibrium possibilities by this method is so easy that it should be a required preliminary calculation for any gas phase process reaction study.

Workbook, Rxns.xlsx, Worksheet GIBBSMIN, or MATLAB Gibbsmin.m.

Values for $\Delta G_i^\circ/RT$ are determined by using Kcalc.xls at 1000 K prior to using this spreadsheet.

	$\Delta G_i/RT$ 1000 K	Moles Fed	log(n_i)	n_i	y_i	$n_i(G_i/RT+\ln y_i)$
CH$_4$	2.456		-1.207132929	0.062067903	0.007	-0.15560
C$_2$H$_4$	14.27		-6.988574695	1.02666E-07	0.000	0.00000
C$_2$H$_2$	20.427		-9.430100874	3.71449E-10	0.000	0.00000
CO$_2$	-47.612		-0.258598372	0.55131731	0.062	-27.78131
CO	-24.089		0.141955136	1.386612582	0.156	-35.97632
O$_2$	0		-20.11203877	7.72612E-21	0.000	0.00000
H$_2$	0		0.729578484	5.365108819	0.604	-2.70090
H$_2$O	-23.178	4	0.179193408	1.510752799	0.170	-37.69136
C$_2$H$_6$	13.33	1	-6.000000002	1E-06	0.000	0.00000
Total				8.875859886		-104.30549

Balances				
	O-bal	4	4	
	H-bal	14	14	
	C-bal	2	2	

Figure 17.6 *Worksheet GIBBSMIN from the workbook Rxns.xlsx for Example 17.13.*

Example 17.14 Pressure effects for Gibbs energy minimization

Apply the Gibbs energy minimization method to the methanol synthesis reaction using stoichiometric feed at 50 bar and 500 K. The reaction has been discussed in Example 17.1 on page 643, and in Section 17.5 on page 649.

Solution: It is convenient to first find $\Delta G^o_{f,1\,bar}$ and then find G_i/RT for each species, and then apply these values in the Gibbs minimization.

Compound	$\Delta G_{f,1bar}$(kJ/mole)	G_i/RT	
Methanol	−134.04	−134.04/8.314E-3/500 + ln(50)	=−28.332
CO	−155.38	−155.38/8.314E-3/500 + ln(50)	=−33.466
H$_2$	0	ln(50)	= 3.9120

For a basis of 1 mole CO: $n_{C,feed} = 1 = n_{CO} + n_{MeOH}$; $n_{H,feed} = 4 = 2n_{H2} + 4n_{MeOH}$; $n_{O,feed} = n_{CO} + n_{MeOH}$; the C-balance is redundant with the O-balance so only one of these should be included as a constraint to improve convergence. Minimizing the Gibbs energy gives $y_{MeOH} = 0.42$ in agreement with the other method in Section 17.5 on page 649.

Note: The objective function changes weakly with mole numbers near the minimum, so tighten the convergence criteria or re-run the Solver after the first convergence. Convergence is sensitive to the initial guess. An initial guess which works is log(n_i) = −0.1 for all i.

17.18   REACTION MODELING WITH LIMITED DATA

Frequently chemical engineers must model chemical reactions without information on the Gibbs energy of formation. Although it can be estimated, often a better method is to conduct experiments to find equilibrium conditions and calculate K_a directly from compositions. Experiments can be performed over a temperature range, resulting in an empirical representation of Eqn. 17.28. In fact, methods like this are used to experimentally determine Gibbs energies of formation. The data can be fitted to determine constants I and J. If the data are scattered, it is better to use the simplified van't Hoff Eqn. 17.29 in the form $\ln K_a = a + b/T$, where a and b are fitted to the experimental data, and b is an indirect measurement of $-\Delta H_R°/R$. Thus, equilibrium data for new compounds or for liquid or solid reactions (discussed next) are often presented in the literature in the form of an equation for K_a instead of providing Gibbs energies of formation.

17.19   SIMULTANEOUS REACTION AND VLE

It is not difficult to imagine situations in which reactions take place in the presence of multiple phases. Absorption of CO_2 into NaOH solution involves a reaction of the CO_2 as it dissolves to form sodium bicarbonate and sodium carbonate. Hydrogen bonding in "pure" fluids implies reaction and phase equilibrium at saturation conditions. The production of methyl t-butyl ether (MTBE) as an oxygenated fuel additive is an interesting process in which catalyst is placed on the trays of a distillation column. As catalysts are developed which are active at progressively lower temperatures, multiphase reactions should become even more common. In low-temperature methanol synthesis (~240°C), it can be advantageous to add a liquid phase to absorb the heat of reaction, as described in the example below. Biological pathways are also in development for many products, which frequently involve multiple phases.

The thermodynamic analysis of this seemingly complex kind of process is actually very similar to the analysis of multireaction equilibria. The extent of formation of a second phase is analogous to a reaction coordinate. The easiest way to illustrate the formulation of the problem is to consider an example. Suppose the methanol synthesis reaction was carried out at 75 bars and 240°C such that the gas phase mole fractions were 0.25 CO, 0.25 MeOH, and 0.50 H_2. Based on a stoichiometric feed composition, the conversion would then be 50%. Now suppose this gas phase was placed in contact with a liquid phase with a nonvolatile solvent. The K-ratios at these conditions are about 10 for CO and H_2, and about 1.0 for MeOH. What would be the composition in the liquid phase and the extent of conversion if only liquid was removed? The composition of the liquid would be 0.025 CO and 0.25 MeOH. Therefore, the extent of conversion would be 0.25/0.275 = 91%. Thus, the addition of a liquid phase greatly enhances conversion of this process. The example below elaborates on these findings in a much more formal manner.

Example 17.15   The solvent methanol process

In a process being considered for methanol synthesis, a heavy liquid phase is added directly to the reactor to absorb the heat of reaction. The liquid is then circulated through an external heat exchanger. Usually, the catalyst is slurried in the liquid phase. An alternative to be considered is putting the catalyst in a fixed bed and adding just enough liquid so that a fairly small amount of vapor is left at the end of the reaction. Supposing naphthalene was used as the heavy liquid phase, use the Peng-Robinson equation to obtain approximate vapor-liquid K-value expressions of the form

Example 17.15 The solvent methanol process (Continued)

$$K_i = \frac{a_i 10^{[b_1(1-1/Tr)]}}{P}$$

for each component at a temperature of 200–250°C and pressures from 50–100 bar.[a] In the worksheet computations, you may assume the K-value of naphthalene to be negligible.

Solution: A worksheet used for this solution is available in the workbook Rxns.xlsx.

Computing the K-value would normally require calling the Peng-Robinson equation during every flash and reaction iteration. This approximate correlation enables you to use Excel to perform the calculations since it is independent of any external programming requirements. The correlation should be suitably accurate if you "guess" compositions for developing the correlation that are reasonably close to the compositions at the outlet of the reactor. We suggest a guess for feed composition of {0.02, 0.10, 0.02, 0.035, 0.005, 0.82} for {CO, H_2, CO_2, methanol, water, and naphthalene}.

As an example of a way to develop a synthetic data base, perform flash calculations at 75 bars and temperatures of {200, 210, 220, 230, 240, 250} and the suggested feed composition. Tabulate the K-values for each component and plot them logarithmically with reciprocal temperature on the abscissa. Select a set of points, then select "add trendline" from the Chart menu. Select the options for a logarithmic fit, and displaying the equation on the chart. The coefficients of the equation give the a and b for the local "shortcut" correlation. (This step simplifies the implementation in Excel, but would be unnecessary if you were writing a dedicated program with access to a Peng-Robinson subroutine.)

Designating the solvent as component 6, the vapor-liquid K-values can be estimated as follows.
$K_1 = (9.9/P) \cdot 10 ^\wedge (2.49(1-133/T))$
$K_2 = (970/P)$
$K_3 = (23/P) \cdot 10 ^\wedge (2.87(1-304/T))$
$K_4 = (61/P) \cdot 10 ^\wedge (3.64(1-513/T))$
$K_5 = (410/P) \cdot 10 ^\wedge (3.14(1-647/T))$
$K_6 \approx 0$

Solve for the simultaneous reaction and phase equilibria at 240°C and $P = 100$ bars considering the following two reactions:

$$CO + 2H_2 \; \underset{\leftarrow}{\overset{\rightarrow}{}} \; CH_3OH \quad (1)$$

$$CO_2 + 3H_2 \; \underset{\leftarrow}{\overset{\rightarrow}{}} \; CH_3OH + H_2O \quad (2)$$

Example 17.15 The solvent methanol process (Continued)

Add moles of the heavy liquid until 9 moles of liquid is obtained for every mole of vapor *output*. The gases are fed in proportions 2:7:1 $CO:H_2:CO_2$.

Applying the shortcut van't Hoff equation (calculated at 503 K using $\ln K_a$ and $\Delta H°_R$, all reacting species gases),

$$\ln K_{a1} = 11746/T - 28.951$$

$$\ln K_{a2} = 6940/T - 24.206$$

Stoichiometry

Comp #		n_i^i	ν_{1i}	ν_{2i}	n_i^f
1	CO	2	−1	0	$2 - \xi_1$
2	H_2	7	−2	−3	$7 - 2\xi_1 - 3\xi_2$
3	CO_2	1	0	−1	$1 - \xi_2$
4	MeOH	0	1	1	$\xi_1 + \xi_2$
5	H_2O	0	0	1	ξ_2
					$10 - 2\xi_1 - 2\xi_2$

Imagine performing a flash at each new extent of conversion:

$$n_T = n_{T0} - 2\xi_1 - 2\xi_2$$

$$z_i = (n_{0i} + \nu_{1i}\xi_1 + \nu_{2i}\xi_2)/n_T$$

$$y_i = \frac{z_i K_i}{K_i + \dfrac{L}{F}(1 - K_i)}$$

Writing objective functions:

$$F(1) = err1 = (P^2 K_{a1} y_1 y_2^2 - y_4)$$

$$F(2) = err2 = (P^2 K_{a2} y_3 y_2^3 - y_4 y_5)$$

$$F(3) = 1 - \sum_i y_i$$

This worksheet is called SMPRXN. An example of the output from a feed of 2,7,1,0,0 mole each of CO, H_2, CO_2, CH_3OH, H_2O is shown in Fig. 17.7.

Rxns.xlsx, Worksheet SMPRXN.

Example 17.15 The solvent methanol process (Continued)

The method of solving this problem is extremely similar to the DUALRXN problem. The only significant addition is an extra constraint equation which specifies that the vapor mole fractions must sum to unity. Note that ξ_1 is greater than unity. This is because we have 2 moles of CO in the feed, so 1.3 moles converted is about 65%.

P(bar)	50					
T(K)	513	533	553	573	593	613
Ka_1	0.002347712	0.000994295	0.0004481	0.0002135	0.00010694	5.6037E-05
Ka_2	2.30525E-05	1.38758E-05	8.66454E-06	5.59125E-06	3.71625E-06	2.53674E-06
K_1	13.839	14.633	15.410	16.170	16.912	17.637
K_2	19.400	19.400	19.400	19.400	19.400	19.400
K_3	6.792	7.867	9.016	10.235	11.520	12.867
K_4	1.220	1.671	2.237	2.934	3.779	4.788
K_5	1.241	1.747	2.399	3.223	4.245	5.491
K_6	0.000	0.000	0.000	0.000	0.000	0.000
ξ_1	1.845	1.604	1.185	0.694	0.302	0.058
ξ_2	0.361	0.253	0.198	0.179	0.180	0.191
L/F	0.900	0.900	0.900	0.900	0.900	0.900
moles solv	16.368	23.680	32.671	41.780	48.658	52.826
y_1	0.043	0.082	0.129	0.168	0.192	0.206
y_2	0.692	0.691	0.691	0.693	0.693	0.692
y_3	0.125	0.116	0.101	0.087	0.080	0.076
y_4	0.120	0.097	0.069	0.043	0.025	0.014
y_5	0.020	0.014	0.010	0.009	0.010	0.012
err1	0.000	0.000	0.000	0.000	0.000	0.000
err2	0.000	0.000	0.000	0.000	0.000	0.000
$\Sigma(y_i)$	1.000	1.000	1.000	1.000	1.000	1.000
z_1	0.007	0.013	0.020	0.026	0.029	0.031
z_2	0.101	0.101	0.101	0.101	0.102	0.101
z_3	0.029	0.025	0.020	0.016	0.014	0.013
z_4	0.101	0.062	0.035	0.017	0.008	0.004
z_5	0.016	0.008	0.005	0.004	0.003	0.003

Figure 17.7 *Worksheet SMPRXN from workbook Rxns.xlsx for Example 17.15 at several temperatures.*

a. Note: The symbols *a* and *b* are simply regression coefficients, not the equation of state parameters *a* and *b*.

Example 17.16 NO$_2$ absorption[a]

The strength of concentrated acid which can be produced is limited by the back pressure of NO$_2$ over the acid leaving the absorbers. The overall reaction, obtained by adding reactions (a) and (b), is shown as (c). Here we assume that N$_2$O$_4$ is equivalent to 2NO$_2$.

$$2NO_2 + H_2O \; \overset{\rightarrow}{\leftarrow} \; HNO_3 + HNO_2 \quad (a)$$

$$3HNO_2 \; \overset{\rightarrow}{\leftarrow} \; HNO_3 + 2NO + H_2O \quad (b)$$

$$3NO_{2(g)} + H_2O_{(l)} \; \overset{\rightarrow}{\leftarrow} \; 2HNO_{3(l)} + NO_{(g)} \quad (c)$$

The gas entering the bottom plate of a nitric acid absorber contains 0.1 mole of NO per mole of mixture and 0.25 mole of NO$_2$ per mole mixture. The entering gas also contains 0.3 bar partial pressure of oxygen, in addition to inert gas. The total pressure is 1 bar. The acid made by the absorption operation contains 50% by weight of HNO$_3$, and the operation is isothermal at 86°F. Estimate the composition of the gas entering the second plate and the strength of the gas leaving the second plate.

Solution: (Basis: 1 mole gaseous feed)
Assume $y_w = y_{HNO_3} = 0$ and $x_{NO_2} = x_{NO} = 0$.

For liquid:

	In	Δ	Out
H$_2$O	W	$-\xi$	$W - \xi$
HNO$_3$	0	2ξ	2ξ
Total			$W + \xi$

For vapor:

	In	Δ	Out
NO$_2$	0.25	-3ξ	$0.25 - 3\xi$
NO	0.10	ξ	$0.10 + \xi$
O$_2$	0.30	0	0.30
I	0.35	0	0.35
Total			$1 - 2\xi$

Example 17.16 NO_2 absorption[a] (Continued)

$$K_a = \frac{\left(\hat{f}_{NO}/f_{NO}^{\circ}\right)\left(\hat{f}_{HNO_3}/f_{HNO_3}^{\circ}\right)^2}{\left(\hat{f}_{H_2O}/f_{H_2O}^{\circ}\right)\left(\hat{f}_{NO_2}/f_{NO_2}^{\circ}\right)^3} = \frac{y_{NO}(x_{HNO_3}\gamma_{HNO_3})^2}{y_{NO_2}^3(x_{H_2O}\gamma_{H_2O})} \frac{1}{P^2}$$

We can determine the mole fractions from the weight fractions:

$$x_{HNO_3} = (0.5/63)/[(0.5/63) + (0.5/18)] = 0.222$$

Noting from the *CRC Handbook*[b] the vapor pressure of HNO_3 is 64.6 mmHg, we can estimate the activity coefficients of HNO_3 and water from the *x-y* data in *The Chemical Engineers' Handbook*.[c]

$$\gamma_{HNO_3} = \frac{y_{HNO_3}P}{x_{HNO_3}P_{HNO_3}^{sat}} = \frac{0.39 \text{ mmHg}}{0.222 \cdot 64.6 \text{ mmHg}} = 0.027$$

$$\gamma_{H_2O} = \frac{10.7 \text{ mmHg}}{0.778 \cdot 23.8 \text{ mmHg}} = 0.5785$$

Gibbs energies of formation are available in Appendix E for all but nitrogen dioxide, and Reid et al.[d] give the standard Gibbs energy of formation as 52 kJ/mol and the standard heat of formation as 33.87 kJ/mol. Performing a shortcut calculation using Kcalc.xlsx, the equilibrium constant at 303.15 K is $K_a = 0.0054$.

At $P = 1$ bar:

$$K_a = \frac{y_{NO}(0.222 \cdot 0.027)^2}{y_{NO_2}^3(0.778 \cdot 0.5785)} = 0.0054 = \frac{y_{NO}}{y_{NO_2}^3} \cdot 0.00007983$$

Substituting for the reaction coordinate:

$$\frac{y_{NO}}{y_{NO_2}^3} = \frac{0.054}{0.00007983} = 67.65 = \left(\frac{0.1 + \xi}{1 - 2\xi}\right) \Big/ \left(\frac{0.25 - 3\xi}{1 - 2\xi}\right)^3$$

Solving the cubic equation, $\xi = 0.0431$.

$x_{HNO_3} = (2\xi/(W + \xi) = 2 \cdot 0.0431/(W + 0.0431) = 0.222$ tells us that

$W = -0.0431 + 2 \cdot 0.0431/0.222 = 0.345$ moles

So the composition of gas entering the second stage is $y_{NO} = 0.157$; $y_{NO_2} = 0.132$; $y_{O_2} = 0.328$; $y_I = 0.383$. Computations for further stages would be similar.

---

a. S. Lee, Personal Communication, 1993.

b. Weast, R.C. 1979. *CRC Handbook of Chemistry and Physics*, 60th ed, Boca Raton, FL: CRC Press, p. D-224.

c. Perry, R.H., Chilton, C.H. 1986. *The Chemical Engineers' Handbook,* 6th ed. New York, NY: McGraw-Hill, p. 3–70.

d. Reid, R.C., Prausnitz, J.M., Poling, B. 1987. *The Properties of Gases & Liquids,* 4th ed, New York, NY: McGraw-Hill.

17.20 SUMMARY

We have greatly enlarged the scope of our coverage of engineering thermodynamics with very little extension of the conceptual machinery. All we really did conceptually was recall that the Gibbs energy should be minimized. The provision that atoms can be moved from one chemical species to another with commensurate changes in energy and entropy simply means that reference states must be assigned to elemental standard states instead of standard states based on pure components, such that the free energies of all the components can be compared. In this sense we begin to comprehend in a new light the broad range of applications mentioned in Einstein's quote at the end of Chapter 1. Instead of conceptual challenges, reaction equilibria focus primarily on the computational aspects of setting up and solving the problems. Notably, equation solvers can provide multidimensional capability. These are tools that can be adapted to many problems, even those beyond the scope of thermodynamics. You should familiarize yourself with such tools and build the expertise that will permit you to enhance your productivity.

Important Equations

The shortcut van't Hoff equation provides a rapid method for screening the effect of temperature on the equilibrium constant. For best results, the temperature range should be limited as suggested in Section 17.8.

$$\ln\left(\frac{K_a}{K_{aR}}\right) = \frac{\Delta G_R^{\circ}}{RT_R} - \frac{\Delta G_T^{\circ}}{RT} = \frac{-\Delta H_R^{\circ}}{R}\left(\frac{1}{T} - \frac{1}{T_R}\right) \qquad 17.29$$

When combined with the equilibrium constraint Eqn. 17.15, reaction conversions can be estimated for many common scenarios.

$$K_a = \prod_i a_i^{\nu_i} = \prod_i \left[\frac{\hat{f}_i}{f_i^{\circ}}\right]^{\nu_i} \qquad 17.15$$

The most common scenario is for ideal gases, because most industrial reactions are conducted at high temperature (to accelerate rates) and low pressure (to minimize cost).

$$K_a = \prod_i (y_i P)^{\nu_i} \qquad \text{(ig) } 17.17$$

For liquid phase components, we showed that the activity should be written as $a_i = x_i \gamma_i$. For solid species, we showed that the activity is unity when the solid is pure. The relevant Gibbs energy of formation must be used when liquids or solids appear in the equilibrium constant expression.

For simultaneous reaction and phase equilibria, we showed that the reaction could be treated in any one of the phases. Since the phase equilibrium constraint asserts equality of fugacity between phases, conceiving of the reaction in any particular phase is inconsequential, providing that the activities and Gibbs energies of formation use the same standard state.

Test Yourself

1. Reproduce the values given for the Gibbs energy of formation at 900 K for the component in Example 17.3. What values are given by the shortcut van't Hoff equation?

2. Explain why a plot of $\ln K$ versus $1/T$ exhibits slight curvature.

3. When T is raised, the equilibrium constant is found to increase. Is the reaction endothermic or exothermic?

4. Explain why formation of a pure solid product does not inhibit a gas-solid reaction from going forward.

5. A reaction occurs with all components present in liquid and gas phases. Compositions in both phases are measured and the activities are calculated. The equilibrium constant value calculated using vapor phase activities is different from the equilibrium constant calculated using liquid phase activities. Is something wrong?

6. Describe the behavior of the reaction rate as equilibrium is approached. Explain why most industrial reactors are not run near equilibrium. Explain why calculation of an equilibrium constant is a good idea whenever designing a reactor.

17.21   PRACTICE PROBLEMS

P17.1 An equimolar mixture of H_2 and CO is obtained by the reaction of steam with coal. The product mixture is known as "water-gas." To enhance the H_2 content, steam is mixed with water-gas and passed over a catalyst at 550°C and 1 bar so as to convert CO to CO_2 by the reaction:

$$H_2O + CO \quad \rightleftarrows \quad H_2 + CO_2$$

Any unreacted H_2O is subsequently condensed and the CO_2 is subsequently absorbed to give a final product that is mostly H_2. This operation is called the water-gas shift reaction. Compute the equilibrium compositions at 550°C based on an equimolar feed of H_2, CO, and H_2O.

Data for 550°C:

	ΔG_f° (kJ/gmol)	ΔH_f° (kJ/gmol)
H_2O	−202.25	−246.60
CO	−184.47	−110.83
CO_2	−395.56	−294.26

P17.2 One method for the production of hydrogen cyanide is by the gas-phase nitrogenation of acetylene according to the reaction: $N_2 + C_2H_2 \rightleftarrows 2HCN$. The feed to a reactor in which the above reaction takes place contains gaseous N_2 and C_2H_2 in their stoichiometric proportions. The reaction temperature is controlled at 300°C. Estimate the product composition if the reactor pressure is: (a) 1 bar; (b) 200 bar. At 300°C, $\Delta G_T^\circ = 30.08$ kJ/mole.

P17.3 Butadiene can be prepared by the gas-phase catalytic dehydrogenation of 1-butene: $C_4H_8 \rightleftarrows C_4H_6 + H_2$. In order to suppress side reactions, the butene is diluted with steam before it passes into the reactor.

(a) Estimate the temperature at which the reactor must be operated in order to convert 30% of the 1-butene to 1,3-butadiene at a reactor pressure of 2 bar from a feed consisting of 12 mol of steam per mole of 1-butene.

(b) If the initial mixture consists of 50 mol% steam and 50 mol% 1-butene, how will the required temperature be affected?

ΔG_f° (kJ/gmol)	600 K	700 K	800 K	900 K
1,3-Butadiene	195.73	211.71	227.94	244.35
1-Butene	150.92	178.78	206.89	235.35
ΔG°_T	44.81	32.93	21.05	9.00

P17.4 Ethylene oxide is an important organic intermediate in the chemical industry. The standard Gibbs energy change at 298 K for the reaction $C_2H_4 + \frac{1}{2}O_2 \rightleftarrows C_2H_4O$ is −79.79 kJ/mole. This large negative value of ΔG°_T indicates that equilibrium is far to the right at 298 K. However, the direct oxidation of ethylene must be promoted by a catalyst selective to this reaction to prevent the complete combustion of ethylene to carbon dioxide and water. Even with such a catalyst, it is thought that the reaction will have to be carried out at a temperature of about 550 K in order to obtain a reasonable reaction rate. Since the reaction is exothermic, an increase in temperature will have an adverse effect on the equilibrium. Is the reaction feasible (from an equilibrium standpoint) at 550 K, assuming that a suitable catalyst selective for this reaction is available? For ethylene oxide, $\Delta H^f_{298} = -52.63$ kJ/mol. Heat capacity equations (in J/mole-K) for the temperature range involved may be approximated by $C_{P,C_2H_4O} = 6.57 + 0.1389\,T(K)$; $C_{P,C_2H_4} = 15.40 + 0.0937\,T(K)$; $C_{P,O_2} = 26.65 + 0.0084\,T(K)$.

P17.5 The water-gas shift reaction is to be carried out at a specified temperature and pressure employing a feed containing only carbon monoxide and steam. Show that the maximum equilibrium mole fraction of hydrogen in the product stream results when the feed contains CO and H_2O in their stoichiometric proportions. Assume ideal gas behavior.

P17.6 Assuming ideal gas behavior, estimate the equilibrium composition at 400 K and 1 bar of a reactive gaseous mixture containing the three isomers of pentane. Standard formation data at 400 K are

	ΔG_f° (kJ/mol) (400 K)
n-pentane	40.17
Isopentane	34.31
Neopentane	37.61

P17.7 One method for the manufacture of synthesis gas depends on the vapor-phase catalytic reaction of methane with steam according to the equation $CH_4 + H_2O \rightleftarrows CO + H_2$. The only other reaction which ordinarily occurs to an appreciable extent is the water-gas shift reaction. Gibbs energies and enthalpies for the problem are tabulated below in kJ/mol.

	ΔH°_f (600 K)	ΔH°_f (1300 K)	ΔG°_f (600 K)	ΔG°_f (1300 K)
CH_4	−83.22	−91.71	−22.97	52.30
H_2O	−244.72	−249.45	−214.01	−175.81
CO	−110.16	−113.85	−164.68	−226.94
CO_2	−393.80	−395.22	−395.14	−396.14

Compute the equilibrium compositions based on a 1:1 feed ratio at 600 K and 1300 K and 1 bar and 100 bars.

P17.8 Is there any danger that solid carbon will form at 550°C and 1 bar by the reaction $2CO \rightleftarrows C_s + CO_2$? (ANS. Yes)

P17.9 Calculate the equilibrium percent conversion of ethylene oxide to ethylene glycol at 298 K and 1 bar if the initial molar ratio of ethylene oxide to water is 3.0.

$$C_2H_4O_{(g)} + H_2O_{(l)} \rightleftarrows (CH_2OH)_2 \text{ (1 M aq solution)} \quad \Delta G_T^{\circ} = -7824 \text{ J/mole}$$

To simplify the calculations, assume that the gas phase is an ideal gas mixture, that $\gamma_w = 1.0$, and that the shortcut K value is applicable for ethylene oxide and ethylene glycol.

P17.10 Acetic acid vapor dimerizes according to $2A_1 \rightleftarrows A_2$. Assume that no higher-order associations occur. Supposing that a value for K_a is available, and that the monomers and dimers behave as an ideal gas, derive an expression for y_{A_1} in terms of P and K_a. Then develop an expression for PV/n_oRT in terms of y_{A_1}, where n_o is the superficial number of moles neglecting dimerization. Hint: Write n_o/n_T in terms of y_{A_1} where $n_T = n_1 + n_2$.

17.22   HOMEWORK PROBLEMS

17.1 For their homework assignment three students, Julie, John, and Jacob, were working on the formation of ammonia. The feed is a stoichiometric ratio of nitrogen and hydrogen at a particular T and P.

Julie, who thought in round numbers of product, wrote:

$$\frac{1}{2}N_2 + \frac{3}{2}H_2 \rightleftarrows NH_3$$

John, who thought in round numbers of nitrogen, wrote:

$$N_2 + 3H_2 \rightleftarrows 2NH_3$$

Jacob, who thought in round numbers of hydrogen, wrote:

$$\frac{1}{3}N_2 + H_2 \rightleftarrows \frac{2}{3}NH_3$$

(a) How will John's and Jacob's standard state Gibbs energy of reactions compare to Julie's?
(b) How will John's and Jacob's equilibrium constants be related to Julie's?
(c) How will John's and Jacob's equilibrium compositions be related to Julie's?
(d) How will John's and Jacob's reaction coordinate values be related to Julie's?

17.2 The simple statement of the Le Châtelier principle leads one to expect that if the concentration of a reactant were increased, the reaction would proceed so as to consume the added reactant. Nevertheless, consider the gas-phase reaction, $N_2 + 3H_2 \rightleftarrows 2NH_3$ equilibrated with excess N_2 such that N_2's equilibrium mole fraction is 0.55. Does adding more N_2 to the equilibrated mixture result in more NH_3? Why?

17.3 The production of NO by the direct oxidation of nitrogen occurs naturally in internal combustion engines. This reaction is also used to produce nitric oxide commercially in electric arcs in the Berkeland-Eyde process. If air is used as the feed, compute the equilibrium conversion of oxygen at 1 bar total pressure over the temperature range of 1300–1500°C. Air contains 21 mol% oxygen and 79% N_2.

17.4 The following reaction reaches equilibrium at the specified conditions.

$$C_6H_5CH = CH_{2(g)} + H_{2(g)} \rightleftharpoons C_6H_5C_2H_{5(g)}$$

The system initially contains 3 mol H_2 for each mole of styrene. Assume ideal gases. For styrene, $\Delta G^\circ_{f,298} = 213.18$ kJ/mol, $\Delta H^\circ_{f,298} = 147.36$ kJ/mol.

(a) What is K_a at 600°C?
(b) What are the equilibrium mole fractions at 600°C and 1 bar?
(c) What are the equilibrium mole fractions at 600°C and 2 bar?

17.5 For the cracking reaction,

$$C_3H_{8(g)} \rightleftharpoons C_2H_{4(g)} + CH_{4(g)}$$

the equilibrium conversion is negligible at room temperature but becomes appreciable at temperatures above 500 K. For a pressure of 1 bar, neglecting any side reactions, determine:

(a) The temperature where the conversion is 75%. [Hint: conversion = amount reacted/ amount fed. Relate ξ to the conversion.]
(b) The fractional conversion which would be obtained at 600 K if the feed to a reactor is 50 mol% propane and 50 mol% nitrogen (inert). (Consider the reaction to proceed to equilibrium.)

17.6 Ethanol can be manufactured by the vapor phase hydration of ethylene according to the reaction: $C_2H_4 + H_2O \rightleftharpoons C_2H_5OH$. The feed to a reactor in which the above reaction takes place is a gas mixture containing 25 mol% ethylene and 75 mol% steam.

(a) What is the value of the equilibrium constant, K_a, at 125°C and 1 bar?
(b) Provide an expression to relate K_a to ξ. Solve for ξ.

17.7 Ethylene is a valuable feedstock for many chemical processes. In future years, when petroleum is not as readily available, ethylene may be produced by dehydration of ethanol. Ethanol may be readily obtained by fermentation of biomass.

(a) What percentage of a pure ethanol feed stream will react at 150°C and 1 bar if equilibrium conversion is achieved?
(b) If the feed stream is 50 mol% ethanol and 50 mol% N_2, what is the maximum conversion of ethanol at 150°C and 1 bar?

17.8 The catalyzed methanol synthesis reaction, $CO_{(g)} + 2H_{2(g)} \rightleftharpoons CH_3OH_{(g)}$, is to be conducted by introducing equimolar feed at 200°C. What are the mole fractions and the temperature at the outlet if the system is adiabatic at 10 bar and the catalyst provides equilibrium conversion without any competing reactions?

17.9 A gas stream composed of 15 mol% SO_2, 20 mol% O_2, and 65 mol% N_2 enters a catalytic reactor operating and forms SO_3 at 480°C and 2 bar.

(a) Determine the equilibrium conversion of SO_2.
(b) Determine the heat transfer required per mole of reactor feed entering at 295 K and 2 bar.

17.10 The feed gas to a methanol synthesis reactor is composed of 75 mol% H_2, 12 mol% CO, 8 mol% CO_2, and 5 mol% N_2. The system comes to equilibrium at 550 K and 100 bar with respect to the following reactions:

$$2H_{2(g)} + CO_{(g)} \rightleftharpoons CH_3OH_{(g)}$$

$$H_{2(g)} + CO_{2(g)} \rightleftharpoons CO_{(g)} + H_2O_{(g)}$$

Assuming ideal gases, derive the equations that would be solved simultaneously for ξ_1, ξ_2 where 1 refers to the first reaction listed. Provide numerical values for the equilibrium constants. Determine ξ_1 and ξ_2 ignoring any other reactions.

17.11 The 10/25/93 issue of *Chemical and Engineering News* suggests that the thermodynamic equilibrium in the isomerization of *n*-butene (CH3CH=CHCH3, a mix of cis and trans isomers) is reached at a temperature of 350°C using a zeolite catalyst. The products are isobutene and 1-butene (CH_2=CHCH$_2$CH$_3$). The isobutene is the desired product, for further reaction to MTBE. Determine the equilibrium composition of this product stream at 1 bar.

17.12 Acrylic acid is produced from propylene by the following gas phase reaction:

$$C_3H_6 + \frac{3}{2}O_2 \rightleftharpoons CH_2CHCOOH + H_2O$$

A significant side reaction is the formation of acetic acid:

$$C_3H_6 + \frac{5}{3}O_2 \rightleftharpoons CH_3COOH + CO_2 + H_2O$$

The reactions are carried out at 310°C and 4 bar pressure using a catalyst and air as an oxidant. Steam is added in the ratio 8:1 steam to propylene to stabilize the heat of reaction. If 50% excess air is used (sufficient air so that 50% more oxygen is present than is needed for all the propylene to react by the first reaction), calculate the equilibrium composition of the reactor effluent.

17.13 (a) As part of a methanol synthesis process similar to problem 17.10, one side reaction that can have an especially unfavorable impact on the catalyst is coke formation. As a first approximation of whether coke (carbon) formation would be significant, estimate the equilibrium extent of coke formation based solely on the reaction: $CO + H_2 \rightleftharpoons C_{(s)} + H_2O$. Conditions for the reaction are 600 K and 100 bar.
(b) Is coke formation by the reaction from part (a) expected at the conditions cited in problem 17.10?

17.14 Hydrogen gas can be produced by the following reactions between propane and steam in the presence of a nickel catalyst:

$$C_3H_8 + 3H_2O \rightleftharpoons 3CO + 7H_2$$

$$CO + H_2O \rightleftharpoons CO_2 + H_2$$

Neglecting any other competing reactions:

(a) Compute the equilibrium constants at 700 K and 750 K.
(b) What is the equilibrium composition of the product gas if the inlet to the catalytic reactor is propane and steam in a 1:5 molar ratio at each of the temperatures and 1 bar?

17.15 Write and balance the chemical reaction of carbon monoxide forming solid carbon and carbon dioxide vapor. Determine the equilibrium constant at 700 K and 750 K. Will solid carbon form at the conditions of problem 17.14?

17.16 Catalytic converters on automobiles are designed to minimize the NO and CO emissions derived from the engine exhaust. They generally operate between 400°C and 600°C at 1 bar of pressure. K.C. Taylor (1993. *Cat. Rev. Sci. Eng.* 35:457.) gives the following compositions (in ppm, molar basis) for typical exhaust from the engine:

NO	CO	O_2	CO_2	N_2	H_2	H_2O	hydrocarbons(~propane)
1050	6800	5100	135000	724000	2300	125000	750

The additional products of the effluent stream include NO_2, N_2O, N_2O_4, and NH_3. Estimate the compositions of all species at each temperature {400°C, 500°C, 600°C} and plot the ratio of NH_3/CO as a function of temperature. (Note: Use the options of the Solver software to set the precision of the results as high as possible.)

17.17 Styrene can be hydrogenated to ethyl benzene at moderate conditions in both the liquid and the gas phases. Calculate the equilibrium compositions in the vapor and liquid phases of hydrogen, styrene, and ethyl benzene at each of the following conditions:

(a) 3 bar pressure and 298 K, with a starting mole ratio of hydrogen to styrene of 2:1
(b) 3 bar pressure and 423 K, with a starting mole ratio of hydrogen to styrene of 2:1
(c) 3 bar pressure and 600 K, with a starting mole ratio of hydrogen to styrene of 2:1

17.18 Habenicht et al. (1995. *Ind. Eng. Chem. Res.*, 34:3784) report on the reaction of t-butyl alcohol (TBA) and ethanol (EtOH) to form ethyltertiary-butyl ether (ETBE). The reaction is conducted at 170°C. A typical feed stream composition (in mole fraction) is:

TBA	EtOH	H_2O
0.027	0.832	0.141

Isobutene is the only significant by-product. Assuming that equilibrium is reached in the outlet stream, estimate the minimum pressure at which the reaction must be conducted in order to maintain everything in the liquid phase. Do isobutene or ETBE exceed their liquid solubility limits at the outlet conditions?

17.19 Limestone ($CaCO_3$) decomposes upon heating to yield quicklime (CaO) and carbon dioxide. At what temperature does limestone exert a decomposition pressure of 1 bar?

17.20 Two-tenths of a gram of $CaCO_{3(s)}$ is placed in a 100 cm^3 pressure vessel. The vessel is evacuated of all vapor at 298 K and sealed. The reaction $CaCO_{3(s)} \underset{\leftarrow}{\rightarrow} CaO_{(s)} + CO_{2(g)}$ occurs as the temperature is raised. At what temperature will the conversion of $CaCO_3$ be 50%, and what will the pressure be?

17.21 One suggestion for sequestering CO_2 is to synthesize carbonate polymers. Polycarbonate is well known for its strength and transparency. To gauge the feasibility of this approach, consider the synthesis of dimethyl carbonate (DMC) from methanol and CO_2 at 350 K.

(a) Write a balanced stoichiometric equation for this reaction. Highlight any by-products.
(b) Estimate the K_a value for the reaction. What pressure is required to achieve $P \cdot K_a = 0.1$?
(c) Methanol and DMC are both liquids at this temperature. Explain how to estimate their partial pressures at a given extent of conversion. You may neglect the CO_2 in the liquid for this question.

17.22 Ethyl acetate is to be produced by a liquid phase reaction.

(a) Use the shortcut van't Hoff equation to calculate the expected conversion of HOAc for equimolar feeds of EtOH and HOAc in a batch reactor at 80°C.

(b) Repeat part (a) with a 3:1 ratio of EtOH to HOAc at 80°C.

	EtOH	HOAc	EtOAc	H2O
$\Delta H^o_{f(l)}$ (kJ/mol)	-277.69	-484.5	-480.	-285.83
$\Delta G^o_{f(l)}$ (kJ/mol)	-174.78	-389.9	-332.2	-237.129

17.23 Hamilton, et al.,[6] have studied the binding of DNA chromosomes to proteins WT1 and EGR1. WT1 uses a zinc binding site to suppress a certain type of kidney tumor. EGR1 binds to regulate cell proliferation. There may be an important regulatory link between the two proteins.[7]

	WT1	EGR1
ln$K$ at 4°C	19.6	20.4
ln$K$ at 11°C	20	20
ln$K$ at 37°C	20.9	19

(a) Use the binding data above to determine the standard state Gibbs energies of reaction at each temperature, the heat of reaction over the temperature range, and the entropy of reaction over the temperature range.

(b) The binding constants are about the same size. Evaluate the relative magnitudes of the enthalpy and entropy of formation for each binding type. Does the thermodynamic analysis imply that the binding types are similar or different?

17.24 The enthalpy of reaction for many biological reactions and surfactants is strongly tempera-ture-dependent, but instead of using full heat capacities, the temperature dependence can be characterized by differences in heat capacities $\Delta C_P = C_{P,prod} - C_{P,react}$, typically assumed to be independent of temperature.[7]

(a) The heat of reaction is represented by $\Delta H = \Delta H_R + \Delta C_P(T - T_R)$. Show that

$$\Delta G_T = \Delta G_R\left(\frac{T}{T_R}\right) + \Delta C_P\left[(T - T_R) - T\ln\left(\frac{T}{T_R}\right)\right] - [T - T_R]\frac{\Delta H_R}{T_R}$$

(b) Provide an equation consistent with part (a) for the temperature dependence of the entropy of reaction.

17.25 Lysozyme (MW = 14.313 kDa) undergoes a phase transition from a native folded (N) to unfolded state (U) that can be considered a reversible reaction, $N_{(aq)} \overset{\rightarrow}{\leftarrow} U_{(aq)}$, where the subscript ($aq$) indicates that the protein is in an aqueous solution. At high temperature the protein is in state U and at low temperature, it is in state N. The melting temperature (T_m) is where the concentrations are equal. The melting temperature can be changed by pH, ionic strength (added salt), denaturant (solvents), or pressure. The results of problem 17.24 apply. The protein unfolding has been studied by Cooper, et al.[8] The heat to unfold 1 mg/

6. Hamilton, T.B., Borel, F., Romaniuk, P.J. 1998. *Biochemistry*, 37:2051–2058.
7. Suggested by O'Connell, J.P. July 2010. BioEMB Workshop, San Jose, CA.
8. Cooper, A.; Eyles, S.J.; Radford, S.E.; Dobson, C.M. 1992. *J. Mol. Biol.*, 225:939–943.

mL of protein in 2 mL of pH = 4 solution at T_m = 78°C is 0.0755 J, ΔC_P = 6.3 kJ/mol-K. The ΔV = –40 mL/mol.[7]

(a) What is the Gibbs energy change for the unfolding at the melting temperature? (Hint: The answer requires conceptual thought, not many calculations.)

(b) Show that at the melting temperature at pH = 4, ΔH = 540 kJ/mol and ΔS = 1.54 kJ/mol.

(c) Evaluate the contributions to ΔG at 23°C and pH = 4. Is the process driven by enthalpy or entropy under these conditions?

(d) Show that the relation for $[N]/[U]$ is $\ln\dfrac{[U]}{[N]} = \dfrac{-\Delta C_P}{R}\left[\left(1 - \dfrac{T_m}{T}\right) - \ln\left(\dfrac{T}{T_m}\right)\right] - \left[1 - \dfrac{T_m}{T}\right]\dfrac{\Delta S_{T_m}}{R}$

where from problem 17.24, $T_R = T_m$. For a solution of overall concentration 1 mg/mL, plot $[U]$ as a function of temperature for $20 \le T \le 110$ °C, and provide a tabular summary of $[U]$ at 60°C and 90°C.

(e) When pH is varied, the net charges on the protein change (as will be explained in Chapter 18), resulting in different T_m and ΔH. In the range $20 \le T_m \le 78$ °C, the relation is $\Delta H(\text{kJ/mol}) = 48.6 + 6.3\, T_m(°C)$. Derive the relation between ΔS and T_m.

(f) At pH = 1, T_m = 43°C. Provide a plot of plot $[U]$ as a function of temperature for $20 \le T \le 110$ °C, and provide a tabular summary of $[U]$ at 25°C and 55°C.

(g) (advanced) Lysozyme has four disulfide bonds that are important in the folding behavior. The authors created a mutant protein with only three disulfide bonds. The relation in (e) still holds. At pH = 2.5, T_m = 23°C for the mutant. Evaluate the contributions of enthalpy and entropy to the folding.

(h) (advanced) Provide an interpretation for the sign of ΔV. At pH = 4, how much pressure is needed to lower the melting temperature 5°C to 73°C?[9]

17.26 Surfactants clump together to form organized structures called micelles that can be spheres, rods, and so forth. The formation of the clump can be modeled as a "chemical" reaction, though there are no chemical bonds formed or broken. When surfactants are in solution below the critical micelle concentration, CMC, the surfactant molecules are almost all "free" in solution. At the CMC, micelles start to form. Above the CMC, the amount of free surfactant is almost constant in solution and as more surfactant is added to the solution, more micelles form. It is conventional to provide the property changes of surfactants per mol of surfactant molecules, not per mole of micelle.

(a) Given the data below for nonylglucoside (NG) *demicellization* in water, calculate the *micellization* value, ΔC_{Pmic}, as a function of temperature. Is the heat capacity of a micelle greater than or less than the heat capacity of the molecules that make up the micelle?

(b) Calculate ΔS_{mic} and for the surfactant as a function of temperature. Is the *overall* solution (including water) more ordered or less ordered after micellization?

NG data:

T (K)	$\Delta H°_{demic}$ (kJ/mol)	$\Delta G°_{demic}$ (kJ/mol)	ΔC_{Pmic} (kJ/mol-K)
285	-13.53	20.03	
293	-9.00	20.78	
303	-3.50	21.60	
313	0.07	22.37	

9. Different values of ΔV are given by Li, T.M., Hook. J.W., Drickamer, H.G., Weber, G. 1976. *Biochem*, 15:5571–5580. Samarasinghe, S.D., Campbell, D.M., Jonas, A., Jonas, J. 1992. *Biochem* 31:7773–7778.

T (K)	ΔH^0_{demic} (kJ/mol)	ΔG^0_{demic} (kJ/mol)	ΔC_{Pmic} (kJ/mol-K)
323	4.30	23.05	
333	8.25	23.62	

17.27 Micelle formation in surfactants is described in problem 17.26. Solve the problem using the data for sodium docecyl sulfate, SDS in water.

SDS data:

T (K)	ΔH^0_{demic} (kJ/mol)	ΔG^0_{demic} (kJ/mol)	$\Delta C_{P,demic}$ (kJ/mol-K)
283	-5.74	36.18	
293	-1.29	37.64	
303	3.53	38.64	
313	7.92	39.80	
323	11.78	40.76	
333	15.46	41.63	

17.28 For nonylglucoside, NG, thermodynamic data for demicellization in water are presented in problem 17.26. Model the micelle reaction as $nS \rightleftarrows M_n$ where S is free surfactant and M_n is a micelle. Treat the solution as an ideal solution. Vary the total concentration of NG from 0 up to 20 mmol/L. Water is 55.5 mol/L. Calculate mole fractions and molar concentrations (mmol/L) of the free surfactant (in mM) and the micelles (in µM) of NG at T=285 K, where n = 58. Plot curves for the concentrations. Identify the approximate CMC in mmol/L. Hint to simplify calculations: The mol/L added is very small relative water molarity (55.5 mol/L). The density can be assumed to be constant. Also, the thermodynamic data are per mol of surfactant, thus model the reaction written as $S \rightleftarrows (1/n)M_n$.

CHAPTER

ELECTROLYTE SOLUTIONS

18

Water, water, every where,
Nor any drop to drink.

S.T. Coleridge, "The Rime of the Ancient Mariner"

Electrolyte solutions are as common as seawater. They exist in every biological organism, in underground reservoirs, and in numerous industrial processes. Thermodynamic models of electrolyte solutions can range from extremely simple to extremely complex. At the simple end of the spectrum are solubility product constants, familiar from introductory chemistry courses. At the complex end of the spectrum, we recognize that concentrated solutions can lead to ion-ion interactions that influence the activity coefficients, altering the simultaneous reaction and phase equilibria that pervade the entire subject. Our goal in this chapter is to introduce the vocabulary and the manner of setting up typical problems involving electrolytes.

Chapter Objectives: You Should Be Able to...

1. Compute the equilibrium concentrations of ionic species in "ideal" electrolyte solutions.

2. Compute the effect of salts on boiling point, freezing point, osmotic pressure, and acid/base dissociation.

3. Understand and articulate the reason for charges on biological molecules and the isoelectric point.

4. Quantitatively describe the "salting in" and "salting out" effects and the common ion effect for solubility.

5. Estimate the true and apparent mole fractions of gaseous solutes in electrolyte solutions.

6. Apply the extended Debye-Hückel model for activity coefficients in dilute solutions.

18.1 INTRODUCTION TO ELECTROLYTE SOLUTIONS

Briefly, an electrolyte is a substance that dissociates into charged species in a liquid phase. The behavior can occur in solution, or in the case of ionic liquids used as nonvolatile solvents, occurs in the pure state. Electrolytes exist in biological and industrial systems and are thus important to our everyday life.

Sodium chloride almost totally dissociates into sodium and chloride ions in dilute aqueous solutions near room temperature. Due to this strong dissociation, sodium chloride is described in these circumstances with a **strong electrolyte** model by assuming that the dissociation is complete. On the other hand, acetic acid, the dominant ingredient in household vinegar, partially dissociates into hydrogen ions and acetate ions in dilute aqueous solution and is modeled as a **weak electrolyte.** A key caution is that the literature tends to describe compounds *as* strong or weak electrolytes. However, the extent of dissociation depends on the environment. For example, Fig. 18.1 shows the distribution of species in sulfuric acid as a function of concentration measured experimentally[1] and modeled.[2] Both hydrogens dissociate only in very dilute solution (which is difficult to discern from the figure). The second dissociation is suppressed at moderate and high concentrations. The first dissociation disappears at high concentrations and sulfuric acid exhibits limited dissociation when pure. Experimental information on the degree of dissociation is important as well as thermodynamic models to represent the behavior to characterize the strong or weak dissociation under the conditions of interest.

The extremely polar nature of water and the capability of water to adjust its partial charges by adjusting the H-O bond distance make it capable of **hydrating** or **solvating** the ions. The water surrounding the ion is known as the **water of hydration** or **solvation** and the ions are described as **solvated.** Fundamental studies show that three to five water molecules hydrate ions in dilute to moderately concentrated solutions. The number is inexact due to the difficulty of the experiments and the fact that the hydration shell is not forming stoichiometric bonds, so the whole hydrated ion is constantly undergoing exchange resulting in various packing effects and various sizes. The key point is to understand that ions must be solvated in solution, and the degree of electrolyte dissocia-

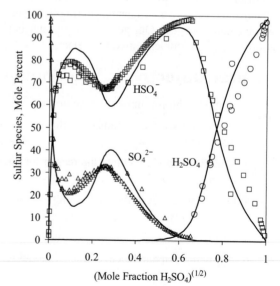

Figure 18.1 *Speciation of sulfuric acid in aqueous solutions as measured by experiments[1] and modeled by OLI Systems, Inc.[2] Note the square root scale to emphasize the dilute region.*

---

1. Experimental data from Young, T.F., Maranville, L.F., Smith, H.M. 1959. *The Structure of Electrolyte Solutions*, p. 35; Clegg, S.L., Rard, J.A., Pitzer, K.S. 1994. *J. Chem. Soc., Faraday Trans.* 90:1875; Clegg, S.L., Brimblecombe, P. 1995. *J. Chem. Eng. Data* 40:43; Walrafen, G.E., Yang, W.-H., Chu, Y.C., Hokmabadi, M.S. 2000. *J. Solution Chem.* 29:905.
2. Wang, P.; Anderko, A.; Springer, R.D.; Young, R.D. 2006. *J. Mol. Liq,* 125:37–44. Another model using a symmetrical convention is available, Que, H; Song, Y.; Chen, C.-C. J. 2011. *Chem Eng. Data* 56:963–977.

tion depends on the ability of a solvent to dissolve ions and the competing driving force for the electrolyte to stay undissociated. Solvents must have a high dielectric constant to be capable of solvating ions. Sodium chloride has an infinitesimal solubility in hexane because the ions are not solvated effectively.

Electrolytes can also have important effects in inhomogeneous fluids like surfaces and colloids. For example, the ions may align next to a surface of opposite charge causing a phenomenon known as an electric double layer. Ions can also strongly affect surfactants in micelles, emulsions, and microemulsions. We restrict our attention here to bulk, homogeneous systems, which are affected in relatively straightforward ways by reaction and dilution.

18.2 COLLIGATIVE PROPERTIES

Boiling points, freezing points and osmotic pressure are sometimes termed **colligative properties.** The adjective "colligative" describes phenomena that are dependent on molar concentration and ignore the solution nonidealities. This of course is an approximation. We understand from previous chapters discussing nonelectrolytes that solution nonidealities can be important. However, in the case of dissolved solids such as sodium chloride, which have negligible vapor pressure under common conditions, the partial pressure of water/solvent determines the boiling point because the salt does not contribute measurably to the vapor composition. Likewise, the osmotic pressure and freezing points are dominated by concentration effects. We know that sodium chloride is commonly used to melt snow and ice on sidewalks and roads in cold climates. The salt works because the freezing temperature of the solution is lower than that of pure water. It is more effective on a molar basis than a compound that does not dissociate because it decreases the mole fraction of water to a greater extent, thus decreasing the chemical potential of water. François-Marie Raoult wrote about electrolytes and their melting point depression compared to molecular solutes: "These facts show that, contrary to what I thought until now, the general law of freezing does not apply to the salts dissolved in water. They tend to show that it applies to radicals (ions) constituting the salts, almost as if these radicals (ions) were simply mixtures in dissolutions."[3] To be fair, however, the degree of melting depression is not always the most critical factor when selecting a system for melting depression or boiling point elevation. Automobile radiators commonly use ethylene glycol as an antifreeze. On a molar basis, the ethylene glycol is less effective than a dissociating salt in lowering the freezing point and increasing the boiling point. Nevertheless, the liquid additive is used instead of salt for practical reasons like avoiding solid precipitation or corrosion (especially for NaCl).

Raoult's study of electrolyte dissociation led to further developments by Jacobius van't Hoff and Wilhelm Ostwald.

Example 18.1 Freezing point depression

Compare NaCl (used on icy roads), ethylene glycol (used in car radiators), and glucose (used by hibernating frogs) as alternatives for freezing point depression. Consider 5 g of each for 0.1 L (5.55 mol) of water and then compare 0.1 mol of each in 0.1 L of water. For the molar basis, compare the masses used and the effectiveness. Assume NaCl totally dissociates, and use an ideal solution approximation.

---

3. van't Hoff, J.H., 1902. "Raoult Memorial Lecture." *J. Chem. Soc., Trans.* 81: 969–981. doi:10.1039/CT9028100969

Example 18.1 Freezing point depression (Continued)

Solution: The melting point is calculated with Eqn. 14.24. To calculate mole fractions, the molecular weights are NaCl 58.44, ethylene glycol (EG) 62.07, and glucose 180.16. For 5 g of each, the molar amounts are 0.0855 mol, 0.0805 mol, and 0.027 mol, respectively. The mole fractions of water in the solutions (recall that NaCl forms two moles of ions!) are $x_{H_2O} = 5.55/(5.55 + 2 \cdot 0.0855) = 0.970$, $x_{H_2O} = 5.55/(5.55 + 0.0805) = 0.986$, $x_{H_2O} = 0.995$, respectively, and

$$\ln(x_{H_2O}) = -\frac{6009.5}{(8.314)(273.15)}\left(\frac{273.15}{T} - 1\right)$$

$$\Rightarrow T = \frac{273.15}{1 - (8.314)(273.15)\ln(x_{H_2O})/6009.5} \qquad 18.1$$

The freezing points for 5 g of each are 270.0 K, 271.7 K, 272.7 K for depressions of 3.2°C, 1.5°C, and 0.5°C respectively. NaCl is more effective than an equivalent mass of EG. Frogs must generate a very concentrated solution of glucose to keep from freezing while hibernating (though concentrated glucose also forms a metastable subcooled liquid easily). For 0.1 mol of each, the mol fractions are $x_{H_2O} = 0.965$, $x_{H_2O} = 0.982$, $x_{H_2O} = 0.982$, with freezing points of 269.6 K, 271.3 K, 271.3 K for depressions of 3.6°C, 1.9°C, and 1.9°C, respectively. There is no difference between the last two solutes because they do not dissociate. The masses needed for 0.1 mol are 5.8 g, 6.2 g, 18.0 g. On a mass basis, NaCl is more effective than glucose even though only one-third as much is used. For EG and glucose, 0.1 mol of each gives the same melting depression, but the mass of glycol is about one-third because the molecular weight is smaller.

Osmotic pressure was discussed in nonelectrolytes in Section 11.13 on page 449. For electrolyte systems a primary difference is that the dissociation of strong electrolytes creates a larger effect on osmotic pressure at the same molar concentration. Look back at Eqn. 11.71. The osmotic pressure is related to the logarithm of the activity of water. When a monovalent electrolyte dissociates, it doubles the effect on the activity relative to an undissociated molecular species (with the ideal solution approximation). Sodium chloride (MW = 58.4 g/mol) and propanol (MW = 60.1 g/mol) have similar molecular weights, but when dissolved in water, to achieve a given osmotic pressure, only about half as much mass of salt is required at low concentrations.

Example 18.2 Example of osmotic pressure

Consider the solutes from Example 18.1 assuming complete dissociation of NaCl and ideal solutions. (a) Compare the osmotic pressure for 0.1 mol of each in 0.1 L of water at 298.15 K. (b) What concentration of NaCl (wt%) is isotonic with human blood?

Solution: (a) The mole fractions have been calculated in Example 18.1 as 0.965, 0.982, and 0.982. The osmotic pressure is given by Eqn. 11.71. The osmotic pressure for an ideal solution is

$$\Pi = \frac{-RT}{V_W}\ln(x_W) = -\frac{(8.314)(298.15)}{18.07}\ln(x_W) \qquad 18.2$$

Example 18.2 Example of osmotic pressure (Continued)

Inserting the mole fractions of each, the osmotic pressures are 4.89 MPa, 2.5 MPa, 2.5 MPa. In a reverse osmosis system, a solution of NaCl requires much more pressure to purify than a solution of a nonelectrolyte with the same apparent concentration.

(b) Isotonicity with human blood is defined in Section 11.13 on page 449 as having a concentration that is 0.308 mol/L of solute. Since two ions are obtained for each NaCl that dissociates, this corresponds to 0.154 mol/L of NaCl, or 8.99 g/L. Assuming the concentration is sufficiently low, a dilute aqueous solution corresponds to a density of 1000 g/L. Therefore, the weight fraction is 9/1000 = 0.009 or 0.9wt%. This is commonly known as "physiological saline" or just "saline."

Vapor-liquid equilibria is also affected by electrolytes. In many cases the electrolyte can be considered to be nonvolatile, such as with sodium chloride. Below we consider the equilibrium condition where salt is only in the liquid phase. In actual application, some salt may be entrained in aerosol droplets as is well known in ocean-side communities where corrosion from salty aerosols is common, but this is not an equilibrium phenomenon. On the other hand, many electrolytes are volatile, such as HCl and acetic acid, so the following analysis will not apply in exactly the same manner.

Example 18.3 Example of boiling point elevation

Consider the solutes from Example 18.1. Compare the bubble points for 0.1 mol of each in 0.1 L of water at 1.013 bar. Consider complete dissociation of NaCl and ideal solutions. Ignore volatility of EG.

Solution: This will be a bubble-temperature calculation. Because the solutes are nonvolatile (ignoring volatility of EG), $y_{H2O} = 1$. The bubble-pressure condition is

$$y_{H_2O} P = 760 \text{ mmHg} = x_{H_2O} P^s_{H2O} = x_{H_2O} 10^{\wedge}(8.07131 - 1730.63/(T + 233.426) \qquad 18.3$$

Using the Antoine equation for water and the mole fractions from Example 18.1, the bubble temperatures are found by using an iterative solver to be 101°C, 100.5°C, and 100.5°C, respectively. Again, the salt has a larger effect due to its dissociation.

Typically, the analysis of electrolyte dissociation is treated in the reaction equilibrium framework termed **speciation,** modeling the dissociation into ionic species as a chemical reaction. When multiple phases exist, simultaneous reaction and phase equilibria must be solved. The general term **electrolyte** refers to a species that dissociates into ions in solution.

18.3 SPECIATION AND THE DISSOCIATION CONSTANT

The term **speciation** refers to a cataloging of the species that exist in solution. The species are characterized by writing dissociation reactions that identify the species and material balance constraints that exist in solution. To introduce the concepts of speciation, consider the dissociation of water:

$$H_2O_{(l)} \rightleftarrows H^+_{(aq)} + OH^-_{(aq)} \qquad\qquad 18.4$$

Thus, in pure water, the species in solution are H_2O, H^+, and OH^-. From introductory chemistry courses, we are familiar with this reaction and the equilibrium constant at 25°C:

$$K_{a,\,298} = [H^+][OH^-] = 10^{-14} \qquad\qquad 18.5$$

Another important concept in speciation is that the solution must satisfy a **charge balance**, and the net charge in solution must be zero. While the charge balance is trivial for this example, it becomes important in setting up the mathematical solutions to the material balances.

Eqn. 18.5 contains some implicit assumptions that are rarely explained in introductory chemistry books. From our discussion in Chapter 17, we know that a rigorous calculation should use activities for a chemical reaction:

$$K_{a,\,298} = \exp(-\Delta G^o_{298}/RT) = \frac{a_{H^+}\,a_{OH^-}}{a_{H_2O}} \qquad\qquad 18.6$$

Comparing the last two equations, we recognize that Eqn. 18.6 can be simplified to Eqn. 18.5 if the activities of ions are replaced with concentrations, and if the activity of undissociated water is unity. Certainly the approximations of Eqn. 18.5 are valid under common conditions or the introductory textbooks would have been in error. But what is meant by the activity of ions, and why does the activity of water not appear in Eqn. 18.5? To understand the simplifications, we must understand the various concentration conventions. The subtleties are important because the values of the equilibrium constants that characterize the reactions are coupled to the concentration and activity scales.

To partially clarify the relation between Eqns. 18.5 and 18.6, it is necessary to recall that activity is dimensionless but molarity has dimensions of mol/L. This issue is subtly handled by using molal concentrations and defining the standard state for electrolytes to be an ideal solution at $m^o = 1$ molal. Therefore, Eqn. 18.6 can be successively converted using $K_{a,\,298} = (m_{H^+}\gamma^{\square}_{H^+} m_{OH^-}\gamma^{\square}_{OH^-})/(m^o_{H^+} m^o_{OH^-} a_{H_2O}) \approx m_{H^+} m_{OH^-} \approx [H^+][OH^-]$ where the m^o states are equal to 1 molal, the molal activity coefficients are ignored, and the molar concentration is used as an approximation to the molality. Water disappears from the relation because the water standard state is taken as purity and a neutral solution is virtually pure water.[4] Supporting discussion is provided in Sections 18.13–18.15 and summarized in Section 18.24.

Measuring Speciation

Quantitative speciation is important in development of a proper thermodynamic model. Various techniques are used, including absorption, NMR and Raman spectroscopies, conductance, emf, solubilities, and rates of reaction. Most techniques require estimation of the activity coefficients to extrapolate to infinite dilution where the K_a is calculated from ideal solution approximations. Because modeling extrapolation is always necessary to "measure" the constants, considerable scatter among literature values is common.

---

4. A similar "trick" was applied when tabulating standard state fugacities at 1 bar in Eqn. 17.17 on page 647.

18.4   CONCENTRATION SCALES AND STANDARD STATES

To discuss the concentration of an electrolyte, some terminology conventions are important for clarity. For example, when 0.01 mole of sodium chloride is added to water and diluted to 1 liter at 25°C, the solution results in 0.02 mol/L of ions because it acts as a strong electrolyte. However, we need a method to communicate the solubility of an electrolyte **as an entity** or on a **superficial basis** or **apparent basis** or **nominal basis**. These terms are used interchangeably in the literature; they refer to the totality of electrolytes in solution as if they were molecular, without dissociation. **Formal concentrations** (mol/L) are used in chemistry to refer to the molar concentration based on the chemical *formula* of the substance. The terms "undissociated" and "un-ionized" are different from "apparent." For species that partially dissociate, the sum of the undissociated and dissociated species comprises the apparent composition. For example, acetic acid does not completely dissociate in aqueous solution, and to communicate most clearly, the terms "nominal," "superficial," "as an entity," and "apparent" are used to refer to the total amount in solution.[5] VLE of a weak electrolyte like acetic acid occurs between the undissociated species in the vapor and the undissociated species in the liquid (as we will show later), but a "apparent" perspective would simply consider that there is acetic acid in the vapor and liquid and the apparent activity coefficient of acetic acid is different from what it would be without dissociation. The true distribution of species should be determined by reaction equilibria and Le Châtelier's principle to characterize the NH_4OH that forms and the NH_4^+ and OH^- speciation. Although the apparent perspective may seem oversimplified, the apparent concentration is very important in engineering calculations because it is often the most accessible measure of concentration when multiple species are present. In this text, we strive to consistently refer to "apparent concentrations," but we may use the equivalent term "superficial concentrations."

> ❶ The apparent basis refers to the hypothetical situation as a molecular species.

The NaCl solution of the preceding paragraph would be described as a 0.01 M apparent concentration of NaCl. On the other hand, the 0.01 M apparent solution of NaCl would have **true** concentrations of 0.01 M Na^+ and 0.01 M Cl^-. The convention of terminology is to describe modeled concentrations as "true" concentrations even though the modeled true concentrations often vary for different models for weak electrolytes. [Apparently, some "true" concentrations are more true than others ;-).] In a good model, true concentrations from experiments are represented accurately. Note how the apparent concentration is the only quantity on which everybody can agree.

Concentration Scales For Electrolytes

Throughout introductory chemistry texts, the convention is to express the concentrations in Eqn. 18.5 using molarity. For introductory courses, frequently all the calculations are at room temperature (25 °C), and thus temperature effects are disregarded. For biological systems, the electrolytes are almost always very near room temperature and the density of the solutions is almost always constant, so the convention is to use molarity. However, this choice of concentration as a composition scale has a disadvantage when the temperature changes in industrial processes because temperature affects density, which affects molar concentrations. Thus, an increase in temperature decreases molar concentrations of all species, even when the solution composition does not change. Therefore, for fundamental calculations as a function of temperature, alternative concentration scales are required.

---

5. For example, the apparent moles would be computed from the mass of acetic acid divided by the molecular weight of an undissociated monomer of acetic acid. We discuss the dimerization of acetic acid in Chapter 19.

In the case of electrolytes for industrial reactions, it is often easiest to perform calculations based on the number of solute (i.e., electrolyte) molecules relative to the number of solvent (e.g., water) molecules. This can be done with mole fraction or molality. Molality has a disadvantage that the quantity diverges to infinity at high concentration. Nevertheless, it is the dominant convention in the older electrolyte literature for nonideal solution behavior and dissociation constants. Molality approaches molarity at low concentrations near room temperature and thus is a natural extension from the use of molarity. Also, molality has a convenient magnitude at common concentrations. **Molality** is the moles of species existing in 1 kg of solvent (e.g., water) molecules. If water is the solvent, the number of moles per kg of solvent is 55.509. We use the notation "m" for molality and "M" for molarity. For example, a one **molal** aqueous solution of sodium chloride is prepared by adding 1 mole of NaCl (58.44 g) to 1 kg of water. The **molarity** of that solution is slightly less than 1 M since the total volume after addition is greater than 1 liter. Molality and molarity are subtly but distinctly different. The ratio of solute to solvent molecules in all 1 molal solutions will be the same for all solutes in a given solvent. To clarify, a comparison of molality, molarity, and mole fractions for NaCl solutions is provided in Table 18.1. The molality and molarity approach each other at low aqueous concentration near room temperature. For dilute solutions, the molarity and molality can be interchanged as a good approximation. The mole fraction scale, typically with a Henry's law standard state, is preferred when working with concentrated solutions.[6] Molality has a disadvantage that it goes to infinity when the water concentration goes to zero. Formulas for interconversions between concentration scales are summarized in Section 18.24

> **We use the notation "m" for molality and "M" for molarity.**

Table 18.1 *Apparent Mole Fraction, Molality, Density and Molarity for Aqueous Sodium Chloride Solutions[a]*

wt%	Mol Frac x_{NaCl}	Molality m	Density (kg/L) 0°C	Density (kg/L) 25°C	Density (kg/L) 100°C	Molarity (M) 0°C	Molarity (M) 25°C	Molarity (M) 100°C
1	0.003104	0.1728	1.00747	1.00409	0.9651	0.1724	0.1718	0.1651
2	0.006252	0.3492	1.01509	1.01112	0.9719	0.3474	0.3460	0.3326
8	0.026105	1.4879	1.06121	1.05412	1.0134	1.4526	1.4429	1.3872
16	0.055459	3.2592	1.12419	1.11401	1.0713	3.0777	3.0498	2.9329
26	0.097722	6.0119	1.20709	1.19443	1.1492	5.3701	5.3138	5.1125

a. Densities are from Washburn, E.W., ed., 1926–1930. *International Critical Tables*, National Research Council, vol. 3, p. 79.

Standard State for Electrolyte Systems

When molality or molarity is used for concentration, the standard state is selected such that the activity coefficients of the electrolyte species, including the undissociated species, go to unity at infinite dilution. This is similar to the Henry's law standard state discussed in Section 11.12.[7] An unsymmetric convention (*cf.* Section 11.12) is used in that a Lewis-Randall standard state is applied for the solvent (typically water) and the activity coefficient of the solvent goes to unity when the solvent is pure.

To understand how the conventions are useful, look back at Table 18.1. Note that the mole fraction of water is approximately $x_w = 0.9$ when the solution is 3 m in NaCl (remember NaCl dissociates!). So the use of an activity coefficient of unity for water is reasonable for approximate cal-

6. The mole fraction scale is least ambiguous, and more consistent with previous chapters. The electrolyte literature for dilute solutions historically uses molality or molarity, and thus we follow those conventions. Molarity has the disadvantage of requiring a density calculation which often must be based on a model. Process simulators typically use the mole fraction scale.

7. There are subtle distinctions between this and the Henry's law standard state, as detailed in Sections 18.13 and 18.24.

culations. The use of an activity coefficient of unity for the electrolytes is exact only at infinite dilution, but can be used as a rough approximation for introductory calculations. We take that approach initially, using ideal solution calculations and the infinite dilution standard state for electrolytes and the Lewis-Randall standard state for the solvent, and later introduce the more rigorous calculation using activity coefficients.

A caution is that various conventions of molarity, molality, and mole fractions exist in the literature.[8] Therefore, when using equilibrium constants, the reader must be careful to understand the conventions used to tabulate the values, and must carefully convert the constants if a different scale/convention is to be used for calculations. The convention can sometimes be inferred if the constant is given with units. Interconversion of constants is discussed in Section 18.25.

Solubility and degree of dissociation depend on the solvent. The majority of published equilibrium constants are for aqueous systems. However, many solvents, including amines, pyridines, ammonia, alcohols, esters, and carboxylic acids, are also capable of solvating ions to varying extents. Because the environment of the ion is critical in determination of the extent of solvation, care should be used when modeling dissociation in nonaqueous solvents.

18.5   THE DEFINITION OF pH

The pH of a solution is defined to be

$$pH \equiv -\log_{10}(a_{H^+})$$

(18.7)

where the activity is expressed on the molal scale. Methods for calculating the activity are covered later. Commonly as an approximation, the concentration (mol/L) is substituted, and the "p" notation stands for the negative of the common logarithm:

$$pH \approx -\log_{10}[H^+]$$

(18.8)

We use this common approximation for introductory examples. The primary use of pH is to characterize a solution as acidic or basic.

Relation of H and OH in Water

Recall at room temperature[9]

$$K_{a,w} = \frac{a_{H^+} a_{OH^-}}{a_{H_2O}} = 10^{-14}$$

(18.9)

The "p" notation is extended to other ions and equilibrium constants. When working with dilute solutions, this means that $-\log(a_{H^+} a_{OH^-}) = -\log(a_{H^+}) - \log(a_{OH^-}) = -\log K_{a,w}$, or

$$pH + pOH = pK_{a,w}$$

(18.10)

---

8. A common notation in the older literature is to use *f* for the rational (Henry's law) activity coefficient, which is very confusing with the use of *f* as fugacity.
9. A more rigorous value is $10^{-13.995}$, but the value of 10^{-14} will be used for casual calculations.

At room temperature, $pK_{a,w} = 14$. Thus, when the solution is acidic, the concentration of $[OH^-] < 10^{-7}$ mol/L, and when the solution is basic, the concentration of $[H^+] < 10^{-7}$ mol/L. Often one of these concentrations can be neglected when taking sums or differences when it is small relative to other terms.

Importance of Solvent

In water, the pH typically varies in the range 0–14. However, because of the definition, strong acids at high concentrations can have negative pH. In ammonia, however, the pH varies from 0 to 32. Therefore, the environment is important in determining the range of pH.

18.6   THERMODYNAMIC NETWORK FOR ELECTROLYTE EQUILIBRIA

Equilibrium Constants

Suppose an electrolyte has a chemical formula C_2A where C^+ is a monovalent cation and A^{2-} is a bivalent anion. Succinic acid, H_2Succ, is an example of an electrolyte with this formula. An equilibrium network can be created as shown in Fig. 18.2. An electrolyte with an arbitrary C_2A composition is shown, with the expectation that readers can properly generalize the network for a specific electrolyte. If completely general notation was used, the figure notation would be unwieldy. While consideration of simultaneous equilibrium between solid, liquid, and vapor phases is rarely necessary, certain subsets of the general network are very common. For example, when an aqueous solution of succinic acid is saturated with solid H_2Succ, the aqueous solution contains H_2Succ, $HSucc^{2-}$, and H^+ and the prevalence of a particular species depends on the pH, but the H_2Succ in the vapor is probably negligible.

Fig. 18.2 emphasizes that the undissociated electrolyte species is the key connection to vapor-liquid equilibria, and also can be used for solid-liquid equilibria. It is proven in Section 18.23 that the apparent chemical potential of C_2A in the liquid phase μ_i^L, is equal to the chemical potential of the dissolved undissociated species $\mu_{i(aq)}$:

$$\mu_i^L = \mu_{i(aq)} \qquad\qquad 18.11$$

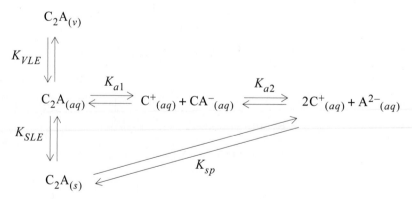

Figure 18.2 *Reaction network for an example electrolyte where C is a monovalent cation and A is a divalent cation. The equilibrium constants are related as explained in the text.*

This may seem odd in the case of a strong electrolyte where the amount of undissociated species is infinitesimal, but it is in fact rigorous. From an engineering perspective this apparent chemical potential is important because it is used to relate the apparent chemical potential to the component chemical potentials in the phases where the electrolyte is undissociated.

Note that the reaction for the solubility product constant

$$C_2A_{(s)} \overset{K_{sp}}{\underset{\leftarrow}{\rightarrow}} 2C^+{}_{(aq)} + A^{2-}{}_{(aq)} \qquad\qquad 18.12$$

can be obtained by summing the reactions for melting along with the two dissociation reactions. Since the Gibbs energies are related to the logarithms of the equilibrium constants, we may write

$$K_{sp} = K_{SLE}K_{a1}K_{a2} \qquad\qquad 18.13$$

Thus, the solubility of electrolytes is frequently written using the solubility product constant, K_{sp} instead of representing all the intermediate dissociation behavior. While this is rigorous thermodynamically using activities, an approximate solution is often used with concentrations rather than activities.

Whether a given electrolyte acts as a strong or weak electrolyte is sometimes unknown at the time a model is required for process design. In other cases, the electrolytes are assumed to act as strong electrolytes for convenience. A confusing aspect of the literature is that often the reader must infer whether a strong electrolyte model is applied depending on whether the dissociation constants are used. When a weak electrolyte is modeled using a strong electrolyte model, the adjustable activity model parameters can be forced to fit, but often the parameters are unusual. For example,, when activity coefficients are determined for a weak electrolyte assuming that it is strong, the activity coefficients are very small. This is because the assumed "true" concentration of ions is much larger than the actual concentration, so the activity coefficients must be smaller than expected to match the experimental activities.

18.7   PERSPECTIVES ON SPECIATION

An important consideration about speciation is the dissociation reaction stoichiometry of reactions that form H^+. The concept of hydration was discussed in Section 18.1. Positive ions usually require water of hydration, and do not float freely in solution as implied by Eqn. 18.4. For example, the species $H_9O_4^+$ is spectroscopically identifiable even at the normal boiling point of water.[10] However, the generally accepted method of writing the reaction is given by Eqn. 18.4 with the understanding that it is actually hydrated. Nevertheless, omitting the water of hydration in the reactions can lead to unexpected results from calculations in mixed solvents. When water is the dominant solvent, there is sufficient water to hydrate the ions. However, at lower concentrations of water, Eqn. 18.4 has no requirement for water of hydration to be present. A more realistic method of writing reactions that form H^+ is to write them including at least one water, for example,

$$2H_2O_{(l)} \overset{\rightarrow}{\leftarrow} H_3O^+{}_{(aq)} + OH^-{}_{(aq)} \qquad\qquad 18.14$$

---

10. Butler, J.N. 1998. *Ionic Equilibria*. New York: Wiley, p. 8.

where $H_3O^+_{(aq)}$ is known as the **hydronium** ion. For acetic acid, the ionization forming hydronium can be written

$$AcOH_{(aq)} + H_2O_{(l)} \rightleftarrows H_3O^+_{(aq)} + AcO^-_{(aq)} \qquad 18.15$$

In this manner, at least some water of hydration must be present for ionization to occur. When the dissociation reactions are written this way, it has no effect on the equilibrium constants from the literature which are measured when the activity of water is essentially one. On the other hand, when Gibbs energies are used, the Gibbs energy and enthalpy of formation of the hydronium includes the corresponding energy and enthalpy of formation for water, resulting in the same Gibbs energies and enthalpies of reaction for Eqns. 18.14 and 18.4, and the same equilibrium behavior when water is the dominant component. When the speciation is calculated in a mixture with small concentrations of water where the activity of water deviates from unity, Eqns. 18.4 and 18.14 lead to different results. For example, the dissociation of H_2SO_4 requires water, as seen in Fig. 18.1, but a dissociation in terms of H^+ would result a dissociation in pure sulfuric acid. The requirement for a solvent in the dissociation reaction is more realistic in aqueous systems.[11] Despite the importance of the hydration water in calculations, the convention in the literature is to write the reactions using only the H^+ ion, and we follow that convention here where we work with dilute aqueous solutions.

The nature of hydration changes with concentration. As the ion concentrations increases, the ions are often paired with counter-ions and are called **ion pairs.** Ion pairs often contain water between them. Under other circumstances water is excluded from within the pair and the pair is hydrated. The ion pair phenomenon is used in ion pair chromatography (IPC) to influence the retention time using ions that are bulky or interact strongly with the solid stationary phase.

Charge Balance

When ions are dissolved in solution, an extra constraint of charge balance is needed, which is often called the condition of **electroneutrality.** The net charge on a solution must be zero. Electroneutrality can be expressed using moles, molarity, or molality. This condition provides an important constraint that is used in all calculations, supplementing the component balances and equilibrium constraints. Note that the charge balance always uses concentrations, not activities.

Approximate Calculations

We begin quantitative discussion by using concentrations instead of activities. We expect that students can make the transition to activities as their skill level develops. Further, as we show, the calculations using concentrations are frequently the first step to a more rigorous solution including activities. To avoid clumsy notation, we write many equilibrium constant relations using concentrations instead of activities. When a solution is dilute, we also use concentration in place of molality without further description. Near the end of the chapter we provide a complete example using activities.

18.8   ACIDS AND BASES

The terms **strong** and **weak** are used when referring to acids and bases in a manner similar to salts. The terms do not imply anything directly about pH. Rather, like other electrolytes, they refer to a

---

11. As a practical note, water is important to include in aqueous systems, but ionic liquids and some other species (HF) may exist dissociated or partially dissociated when pure.

compound's degree of dissociation. When the dissociation constant is extremely large the acid/base is considered strong; when dissociation is incomplete the acid/base is considered weak.

Strong Acids/Bases and the Leveling Effect

The magnitude of the dissociation constant (or the associated Gibbs energy of dissociation) determines whether an acid/base is strong or weak. **An acid is a proton source** and reacts with the solvent (usually water) to create an increased activity of $[H^+]$ or $[H_3O^+]$; for example, AcOH in Eqn. 18.15 is a proton donor and an acid. HCl is a strong acid and dissociates totally at common concentrations to form H_3O^+ (denoted as H^+) and Cl^-. **A base acts as a proton sink,** reacting with water to withdraw a proton and increase the hydroxide activity, as shown by the reaction of the weak base ammonia with water,

$$NH_3 + H_2O \rightleftarrows NH_4^+ + OH^- \qquad\qquad 18.16$$

Hydroxides are also common strong bases, such as NaOH or $Ca(OH)_2$. These strong bases are a direct source of hydroxide because when they dissociate the $[OH^-]$ must be high to balance the positive cation charges in solution.

The strength of an acid or base is determined by the solvent, which governs the degree of dissociation/reaction. Table 18.2 illustrates acid/base strength. Strong acids are to the upper left. Strong bases are to the lower right. Weak acids are in the left column below H_3O^+. Weak bases are in the right column above OH^-. The reaction of H_3O^+ with water may initially look like little has happened: $H_3O^+ + H_2O \rightleftarrows H_2O + H_3O^+$. All acids in the left column above H_3O^+ are equally strong when they are dilute in water because the protons released from the acid immediately react with H_2O to give hydronium ions. Thus, the strongest acid in water is H_3O^+ and any stronger acids are **leveled** to be the same strength. In an analogous way, OH^- is the strongest base that can exist in water. Any other proton sinks added to solution from stronger bases react immediately with water to give OH^-. (Often these reactions are violent and superbases must be handled with care near water.) Strong bases are leveled by water and do not produce a stronger base solution than an equivalent concentration of NaOH. Sometimes organic chemists use strong acids or strong bases in nonaqueous solvents to overcome the leveling effect.

Each acid in the left column in the table has a **conjugate base** in the right column. Similarly, each base in the right column has a **conjugate acid** in the left column. Strong acids have weak conjugate bases, and strong bases have weak conjugate acids.

Strong Acids

Consider the behavior of HCl in solution at an apparent molarity of C_A. The behavior is characterized by the dissociation of HCl (essentially complete) and the dissociation of water. Further, a charge balance must exist in solution, yielding the following conditions

$$[Cl^-] = C_A \text{ material balance for complete dissociation} \qquad\qquad 18.17$$

$$K_{a,w} = [H^+][OH^-] = 10^{-14} \quad \text{equilibrium} \qquad\qquad 18.18$$

$$[H^+] = [Cl^-] + [OH^-] \quad \text{charge balance} \qquad\qquad 18.19$$

The next step is important and we use it throughout our calculations. Though trivial here, later we find it important to combine the material balance and charge balance to arrive at a balance known

Table 18.2 *Reference Table for Relative Acid and Base Strengths at 25°C Based on*
$pK_{a,water} = -13.995.$ $pK_b = 13.995 - pK_a$

	Species	Acid	$pK_{a,A}$	Base	$pK_{a,B}$	
Strong Acids	Perchloric acid	$HClO_4$	~ −7	ClO_4^-	~21	**Increasing Base Strength**
	Hydrogen chloride	HCl	~ −3	Cl^-	~17	
	Sulfuric acid	H_2SO_4	~ −3	HSO_4^-	~17	
	Nitric acid	HNO_3	−1	NO_3^-	~15	
	Hydronium ion	H_3O^+	0	H_2O	13.995	
	Sulfurous acid	H_2SO_3	1.857	HSO_3^-	12.138	
	Bisulfate	HSO_4^-	1.987	SO_4^{-2}	12.008	
	Phosphoric acid	H_3PO_4	2.148	$H_2PO_4^-$	11.847	
	Hydrofluoric acid	HF	3.17	F^-	10.825	
	Acetic acid	CH_3COOH	4.756	CH_3COO^-	9.239	
	Total dissolved CO_2[a]	$CO_{2(aq)} + H_2CO_3$	6.351	HCO_3^-	7.644	
	Hydrogen sulfide	H_2S	7.02	HS^-	6.98	
	Dihydrogen phosphate	$H_2PO_4^-$	7.198	HPO_4^{-2}	6.797	
	Bisulfite ion	HSO_3^-	7.172	SO_3^{-2}	6.823	
	Hypochlorous acid	$HOCl$	7.53	OCl^-	6.47	
	Hydrogen cyanide	HCN	9.21	CN^-	4.79	
	Boric acid	H_3BO_4	9.237	$B(OH)_4^-$	4.758	
	Ammonium ion	NH_4^+	9.245	NH_3	4.750	
	Bicarbonate	HCO_3^-	10.329	CO_3^{-2}	3.666	
	Hydrogen phosphate	HPO_4^{-2}	12.375	$PO4^{-3}$	1.620	
Increasing Acid Strength	Water [b]	H_2O	13.995	OH^-	0	**Strong Bases**
	Bisulfide	HS^-	~14	S^{-2}	~0	
	Ammonia	NH_3	~23	$NH2^-$	~ −9	
	Hydroxide ion	OH^-	~24	O^{-2}	~ −10	

a. By common use. Based on $([CO_{2(aq)}] + [H_2CO_3])K_a = [HCO_3^-][H^+]$. $[H_2CO_3]$
    $\approx 0.002[CO_{2(aq)}]$ at 298.15 K. Thus, $(1.002[CO_{2(aq)}])K_a = [HCO_3^-][H^+]$ at
    298.15 K.

b. If water is acting as a solute rather than a solvent, as it must if the acid strength
    of H_2O is being compared with that of other very weak acids, then $pK_{a,A} \sim 16$
    should be used. See 1990. *J. Chem. Ed.* 67(5):386–388.

as the **proton condition.** The principle is to insert the known constants from the problem statement, which often cancel to leave the intermediate and smaller concentrations. In this case no terms drop out, but the proton condition becomes

$$[H^+] = C_A + [OH^-] \quad \text{proton condition} \qquad\qquad 18.20$$

Think about the size of the $[H^+]$ and $[OH^-]$ and how they are coupled to $K_{a,w}$. Suppose that $C_A = 10^{-2}$ mol/L. Then, by the coupling of the equilibrium and charge balance, because HCl is an acid, we expect $[OH^-] < 10^{-7}$ mol/L by Eqn. 18.10, thus $[OH^-] << [Cl^-] = 10^{-2}$ mol/L. The charge balance involves terms that are *very* different in size. Now, when $C_A > 10^{-6.5}$, it is a good approximation to ignore $[OH^-]$ and we can calculate pH = pC_A. At low acid concentrations below about $10^{-6.5}$, we must insert the dissociation constant for water into the charge balance, to obtain $[H^+] = C_A + K_{a,w}/[H^+]$ and solve the resultant quadratic equation for $[H^+]$. The behavior of strong acids is plotted in Fig. 18.3. *Note that a good approximation for the strong acid curve on this log-log plot is a diagonal line with slope −1 running from (0,0) through (7, −7). It is an excellent approximation from (0,0) through (6.5,−6.5).*

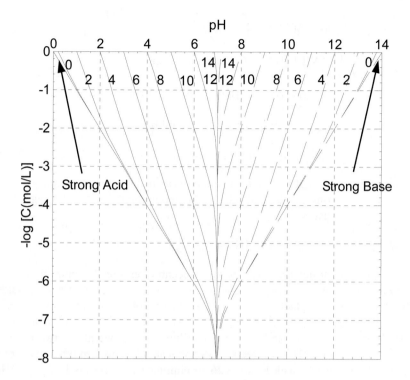

Figure 18.3 *Flood diagram showing the behavior of strong monoprotic acids, weak monoprotic acids, salts of weak monoprotic bases, and strong monovalent bases.*

Strong Bases

Consider a strong monovalent base, such as $[OH^-]$. In a solution of apparent molarity C_B of NaOH the constraints are

$$[Na^+] = C_B \quad \text{material balance for complete dissociation} \qquad 18.21$$

$$K_{a,w} = [H^+][OH^-] = 10^{-14} \quad \text{equilibrium} \qquad 18.22$$

$$[H^+] + [Na^+] = [OH^-] \quad \text{charge balance} \qquad 18.23$$

Combining the material balance and charge balance results in the proton condition

$$[H^+] + C_B = [OH^-] \quad \text{proton condition} \qquad 18.24$$

When $C_B > 10^{-6.5}$, it will be a good approximation to ignore $[H^+]$ and we can calculate pH = 14 − pC_B. At low concentrations, the quadratic equation becomes $[H^+] + C_B = K_{a,w}/[H^+]$. As evident in Fig. 18.3, *a good approximation for the strong base curve on this log-log plot is a diagonal line with slope 1 running from (7,−7) through (14, 0).*

Weak Monoprotic Acids

Consider a weak monoprotic acid such as acetic acid, $pK_{a,A} = 4.7$. We denote as the undissociated acid as HA. The weak acid requires an additional dissociation reaction because it is incomplete.

The material balance is written including the acid HA and **conjugate base** A⁻, because the acid is only partially dissociated. The acid equilibrium constant is required to quantify the dissociation.

$$[A^-] + [HA] = C_A \quad \text{material balance} \tag{18.25}$$

$$K_{a,A} = [H^+][A^-]/[HA] \quad \text{equilibrium} \tag{18.26}$$

$$K_{a,w} = [H^+][OH^-] = 10^{-14} \quad \text{equilibrium} \tag{18.27}$$

$$[H^+] = [A^-] + [OH^-] \quad \text{charge balance} \tag{18.28}$$

The rigorous solution to pH at a given concentration can be obtained from Eqn. 18.26 by eliminating [HA] and [A⁻] in terms of [H⁺] and [OH⁻] and then eliminating [OH⁻] using $[OH^-] = K_w/[H^+]$, resulting in a cubic equation,

$$[H^+]^3 + K_{a,A}[H^+]^2 - (C_A K_{a,A} + K_{a,w})[H^+] - K_{a,A}K_{a,w} = 0 \quad \text{(Hint: limited utility)} \tag{18.29}$$

Solutions of the cubic for various $pK_{a,A}$ values and concentrations are shown in the solid lines on the left side of Fig. 18.3. The graph can serve as a useful guide for monoprotic weak acids, but is only applicable for the charge balance of Eqn. 18.28 when no other ions are present. Rather than solving the cubic, or becoming too reliant on the graph, real problems usually involve other ions, and thus the Flood diagram and cubic have limited utility. Instead, we utilize two equations that relate [A⁻] and [HA] to C_A and [H⁺] and are always applicable, even when other ions are present. Combining the material balance with Eqn. 18.26 to eliminate [A⁻] results in $K_{a,A} = [H^+](C_A - [HA])/[HA]$ which becomes

$$\boxed{[HA] = C_A[H^+]/([H^+] + K_{a,A})} \quad \text{useful for undissociated acid} \tag{18.30}$$

Combining the material balance with Eqn. 18.26 to eliminate [HA] results in $K_{a,A} = [H^+][A^-]/(C_A - [A^-])$ which becomes

$$\boxed{[A^-] = K_{a,A}C_A/([H^+] + K_{a,A})} \quad \text{useful for conjugate base} \tag{18.31}$$

Note that the denominators of the last two equations are the same and we need to know the pH to solve for the concentrations. Example 18.4 shows how these equations are useful for problem solving.

Weak Monoprotic Bases

Consider a weak monovalent base such as acetate ion, denoted as A⁻, $pK_{a,B} = 9.3$, which might be added to solution as sodium acetate. The sodium acetate dissociates completely, making it a strong electrolyte, but the acetate equilibrates with water to form undissociated acid (HA) and hydroxide, $A^- + H_2O \rightleftarrows HA + OH^-$, so sodium acetate is called a weak base. The material balance is written for the cation in this case as well as the two forms of A, and the governing equations are:

$$[Na^+] = [HA] + [A^-] = C_B \quad \text{material balance} \tag{18.32}$$

$$K_{a,B} = [HA][OH^-]/[A^-] \quad \text{equilibrium} \tag{18.33}$$

$$K_{a,w} = [H^+][OH^-] = 10^{-14} \quad \text{equilibrium} \qquad 18.34$$

$$[H^+] + [Na^+] = [A^-] + [OH^-] \quad \text{charge balance} \qquad 18.35$$

In an analogous way to treating weak acids, the material balance and charge balance are combined to eliminate [HA] and [A$^-$] in terms of [H$^+$] and [OH$^-$]. Then the dissociation constant of water is used to eliminate [OH$^-$], resulting in

$$K_{a,B}[H^+]^3 + (C_B K_{a,B} + K_{a,w})[H^+]^2 - K_{a,B}K_{a,w}[H^+] - K_{a,w}^2 = 0 \quad \text{limited use} \qquad 18.36$$

The same equation results from a weak neutral base with capacity for one ion, such as ammonia. Note, however, for ammonia that the base NH$_3$ is neutral and the **conjugate acid** NH$_4^+$ is charged, but when the charge balance is modified and the same method is used, Eqn. 18.36 results. As with a weak acid, the pH values at various base concentrations have been plotted in Fig. 18.3. As with weak acids, the charge balance that leads to the cubic equation does not hold when other ions are present. However, we can solve problems using Eqns. 18.30 and 18.31. Thus, we develop a single method of solution regardless of whether acid or base is added to water. Note that

$$\boxed{pK_{a,A} + pK_{a,B} = pK_{a,w} \quad \text{or} \quad K_{a,A}K_{a,B} = K_{a,w}} \qquad 18.37$$

As an example, consider the fluconazole shown in Fig. 18.4. This is a drug used for treating fungal infections. Fluconazole is a base that is protonated in water, depending on pH:

$$\text{Fluconazole} + H_2O \rightleftarrows \text{Fluconazole}^+ + OH^- \qquad 18.38$$

By lowering the pH, the [OH$^-$] is lowered, driving the reaction to the right. At high pH values, the reaction shifts to the left, as we show in Example 18.4. The equilibrium shift affects solubility as we consider in Example 18.8 on page 726.

Example 18.4 Dissociation of fluconazole

Fluconazole is a drug used for treating fungal infections. Behavior of drugs at various pH conditions is important because the stomach system is at low pH, but the intestinal system has a higher pH. Thus, models for the dissociation and solubility are desirable. Fluconazole equilibrium written as Eqn 18.38 can be modeled with the expression

$$\ln(K_a) = -1.28 - 8000/T \qquad 18.39$$

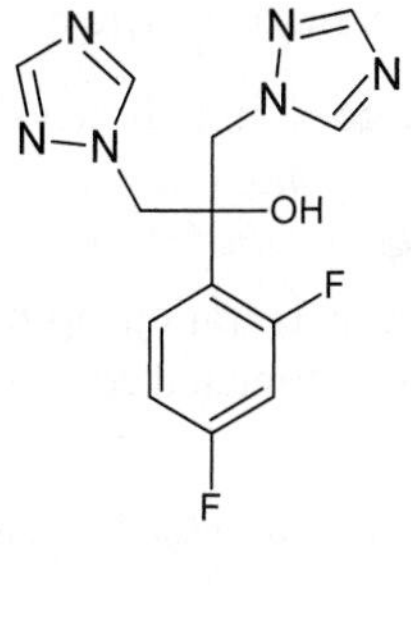

Figure 18.4 *Structure of Fluconazole.*

Example 18.4 Dissociation of fluconazole (Continued)

where fluconazole and its ion and the hydroxyl are on the molality scale and water is on the Lewis-Randall scale. Determine the percentage of fluconazole dissociated at pH 7 and pH 1.5 when the apparent amount of fluconazole in aqueous solution is 1.5E-3m. The molecular weight of fluconazole is 306.27. Assume ideal solutions.

Solution: First consider the chemical reaction to identify the acid and base. In this case, the fluconazole is the base and the fluconazole$^+$ is the acid. Therefore Eqn. 18.39 represents $K_{a,B}$. As with this example, it is common in the literature that acids and bases are not explicitly identified, and recognition is an important step in the solution. We can rewrite the reaction in the acid form as,

$$\text{Fluconazole}^+ + H_2O \rightleftarrows \text{Fluconazole} + H_3O^+$$

(We adopt this approach of writing reactions in the acid form as a standard method, as further implemented in Section 18.9 below.) For Eqn. 18.39, the equilibrium constant at 298.15 K is $K_{a,B} = 6.181\text{E-13}$, or $K_{a,A} = 10^{-14}/6.181\text{E-13} = 0.01617$, or $pK_{a,A} = 1.79$. Therefore, the fluconazole is predominately protonated below pH = 1.8 and largely neutral above. The calculations should bear out this rule of thumb.

At pH = 7 we have from Eqn. 18.30,

$[\text{Fluc}^+] = C_{\text{fluc}}[H^+] / ([H^+] + K_{a,A}) = 1.5\text{E-3m}(10^{-7})/(10^{-7} + 0.01617) = 9.27\text{E-9m}$. Therefore, $[\text{Fluc}] = 1.5\text{E-3m}$, and the fraction protonated is $9.27\text{E-9}/(1.5\text{E-3})\cdot100\% = 0\%$ (trace).

At pH = 1.5, the system is near the $pK_{a,A}$ and both terms are important in the denominator, $[\text{Fluc}^+] = C_{\text{fluc}}[H^+] / ([H^+] + K_{a,A}) = 1.5\text{E-3m}(10^{-1.5})/(10^{-1.5} + 0.01617) = 9.9\text{E-4m}$. Thus, $[\text{Fluc}] = 1.5\text{E-3m} - 9.9\text{E-4m} = 5.1\text{E-4m}$. The concentration of protonated species is higher, as expected. The fraction protonated is $9.9\text{E-4}/(1.5\text{E-3})\cdot100\% = 66\%$.

Perspectives on Calculations

Calculations for electrolyte systems can be challenging to converge because the concentrations of important species vary by several orders of magnitude. Each pH unit is an order of magnitude; thus, at pH 2 compared to pH 7 the $[H^+]$ is five orders of magnitude larger. Calculations using Excel Solver are insensitive to the latter condition if the Solver tolerance is set to 1E-3! Also, issues may occur if iterating on values near zero because the concentration of 1E-3 can easily jump to a negative value on the next iteration if not constrained. Iterations can also converge slowly for some situations.[12] Because charges are involved, the net charge in a solution must be zero; a charge balance is required when iterating on concentrations. However, a charge balance often involves adding terms that are different by orders of magnitude. There are several general recommendations.

❗ The $pK_{a,A}$ represents the pH condition where the reaction coordinate will be 50% between acid and conjugate base, and thus their concentrations will be equal.

1. Develop a good initial guess using techniques that we discuss below. *The $pK_{a,A}$ represents the pH condition where the reaction coordinate will be 50% between acid and conjugate base and thus their concentrations will be equal.*

2. Constrain concentrations to be positive or use logarithms of concentrations for iterations when using automated equation solvers.

---

12. Butler, James N. 1998. *Ionic Equilibria*. New York: Wiley. pp. 20–35.

3. Set the convergence criteria in Solver or optimization routine to an extremely small number (1E-30) and the number of iterations to a high number.

4. Check results to ensure convergence, especially if the specified number of iterations is reached.

How does one generate a good first guess when the concentrations differ by orders of magnitude? The best way to generate a good first guess is to use a Sillèn diagram. Sillèn diagrams, originally developed by Swedish chemist Lars Gunnar Sillèn in the 1950s, are quick to sketch. The next section discusses how to construct and use a Sillèn diagram. Often the results from the Sillèn are sufficiently accurate for routine practical applications or as first estimates for more detailed calculations using activity coefficients.

Other important aspects of the initial examples include the standard states, concentration units, and composition independence of K. The typical convention used for standard states of charged species is similar to Henry's law, but subtly different. Without belaboring the details until later in the chapter, the activity coefficients for the charged species are unity at infinite dilution, and we will disregard the activity coefficients for the introductory examples. The corrections are typically small when the concentrations of ions (measured by the **ionic strength**) is low, less than approximately 10^{-2} m. We will use this approximation at even higher concentrations to develop problem-solving strategies. The reader should be cautioned that the activity coefficients for charged species can become large rapidly and can be very large, but typically above 5m. A good understanding of the standard state for water and uncharged solute species is also important. The standard states for water and uncharged solutes are different from that used for charged species. Uncharged solutes, such as molecular acids and bases, like acetic acid or ammonia, are treated with Henry's law. This shares a similarity to the treatment of charged species, because the activity coefficients are unity. Water, on the other hand, is treated relative to the Lewis-Randall standard state of pure water. Because the water concentration on a mole fraction basis is nearly one, it will be sufficient to approximate the activity coefficient of water as one and the activity of water will be approximately one. In summary, we extend the concepts of using unsymmetric standard states introduced in Chapter 11.

> ❗ Ions are treated with a molal standard state. Aqueous molecular solutes are treated with Henry's law. Water is treated with the Lewis-Randall rule.

Another important approximation is that we use molar concentration rather than molality to work the examples early in the chapter. This follows the conventions used in introductory chemistry, and, as shown in Table 18.1, can be a good approximation at low concentrations. Technically, the units should be molality for the electrolyte species, and certainly the examples can be reworked with those units, but use of molality requires more unit conversions with little pedagogical advantage. The later examples in the chapter use molality to demonstrate the more rigorous approach.

Temperature, Pressure, and Composition Effects on K

Initially, the discussions and examples provide values for equilibrium constants. Commonly in introductory chemistry texts values are provided for 298.15 K, though the designation is often omitted in those texts. The values of K_a change with temperature as with any reactive system.

In Chapter 17, the equilibrium constant did not depend on pressure. This is not the case for electrolytes when the typical electrolyte standard state is used for ions, though it is common to neglect it as a first approximation. When an ion is dissolved in water at infinite dilution, the hydrogen bonding is disrupted resulting in a pressure-dependence for the infinite-dilution standard state.

We do not develop the details further in this text, but readers should consult advanced texts or handbooks when working at high pressures.[13]

There are two conventions to correct for solution nonidealities. Extending the concepts of Chapter 17, the most rigorous method of including solution nonidealities is to use activity coefficients as we show later. However, another method used in the literature is to determine the dependence of $K = K_a / K_\gamma$ on ionic strength of the solution, and then proceed with calculations using K rather than K_a. This is possible because to the first approximation the activity coefficients depend on charge and ionic strength and are independent of species (as we show in Section 18.15). Therefore, when working with values from the literature, some care is necessary to discern if the authors' values for K are at the standard states or if they are corrected for ionic strength. When the dependence of ionic strength is included and K is substituted for K_a, the methods developed below can be used directly by replacing K_a with K in the calculations.

18.9   SILLÈN DIAGRAM SOLUTION METHOD

Monoprotic Acids and Bases

Seven main steps are necessary to solve electrolyte problems using a Sillèn diagram (*cf.* Fig. 18.5), which is similar to a Flood diagram. We summarize the steps and then work an example for sodium acetate. Skim the procedure initially, and then follow closely with Example 18.5.

1. Create a coordinate system like the Flood diagram. (A template is available on the textbook web site.) Draw straight lines for the strong acid and strong base lines. The detail of the taper at pH = 7 should be ignored, and cross the lines. Note that the sum of the two lines is always –14 on the log scale and represents the ion product for water. Label these lines [H$^+$] (left) and [OH$^-$] (right).

    Some rules of thumb are helpful for plotting on common logarithmic coordinates. Note that when [B] = 2[A], the ordinate of [B] on a $\log_{10}$ scale is 0.3 units higher than [A]. Likewise, when [B] = 0.5[A], then the [B] ordinate will be 0.3 units lower. A factor of 5 is 0.7 units. And of course a factor of 10 is one unit. For convenience the pairs of (linear factor, log10 translation) are (2, 0.3), (3, 0.47), (4, 0.6), (5, 0.7), (6, 0.78), (7, 0.85), (8, 0.9), (9, 0.95), (10, 1).

2. Write the material balance for the dissociating species to relate the apparent species to the species in solution; for example, Eqn. 18.32.

3. Write the equilibria relations using the dissociation constants for weak acids or bases. If the acid/base is strong it will completely dissociate, and thus the relation is not needed. *Always write the reactions in the acid form* (even if bases are involved); for example, Eqn. 18.26 for acetate or acetic acid. Write the dissociation reaction for water. Using the acid form provides a consistent solution strategy, but is not theoretically required.

---

13. A widely-accepted method for calculating infinite dilution properties uses the Helgeson-Kirkham-Flowers (HKF) EOS. This equation of state is specifically developed for the infinite dilution properties as a function of temperature and pressure. The EOS is extremely detailed, as explained in four papers: (1) Helgeson, H.C., Kirkham, D.H. 1974. *Am. J. Sci.* 274:1089–1198; (2) Helgeson, H.C., Kirkham, D.H. 1974. Am. J. Sci. 274:1199–1261; (3) Helgeson, H.C., Kirkham, D.H. 1976. Am. J. Sci. 276:97–240; (4) Helgeson, H.C., Kirkham, D.H., Flowers, G.C. 1981. Am. J. Sci. 281:1249–1516. A database of parameters is available at www.predcent.org in the SUPCRT and OBIGT (includes some extensions) databases. A windows interface is provided to use with some of the data, though the tables can be used directly for calculations at the reference state of 298.15 K and 1 bar. Application of the EOS is illustrated in Wang, P., Anderko, A., Young, R.D. 2002. Fluid Phase. Equil. 203:141–176 and references 26–30 therein include parameters.

4. Write the electroneutrality constraint.

5. Sketch Eqns. 18.30 and 18.31 without calculations on the diagram using these steps. (See the example.) The steps are: **(a) create a system point** at C_A (or C_B) and $pK_{a,A}$; the procedure always uses $pK_{a,A}$, even for bases; **(b) create an acid/base intersection point** at $(pK_{a,A}, \log C_A - 0.3)$. (The value of 0.3 represents a decrease of 50% in the concentration, which is where the acid and base concentrations will match.); **(c) sketch diagonal lines** with slope +1 and –1 (parallel to the H+ and OH– lines) *below* C_A that project through the system point but extend downwards starting about $\log C_A - 1$; **(d) draw horizontal lines** on either side of the system point leaving a gap of approximately 1 pH unit on either side of the system point and label the line on the left (low pH) as the acid and the line to the right (high pH) as the base; **(e) connect the sloping lines with the horizontal lines** with smooth curves that pass through the acid/base intersection point.

6. Decide which concentrations are largest and which are least significant. Let C_i be the apparent concentration. The goal is to simplify the balances and provide a good guess for true concentrations. This is almost always done by converting the charge balance to a proton condition by inserting the mass balance to eliminate terms that are largest and leave smaller terms that are more similar in magnitude. Use the diagram as a guide to decide which concentrations are insignificant in the pH range expected. The goal is to use the proton condition to identify the intersection of the positive and negative charges of the proton condition. Unless some of the diagonal curves are very close to each other this will be easy. There can be various proton conditions that are equally valid when many ions are present at similar concentrations. Hints: Remember that each unit on the log scale is an order of magnitude. Acids by themselves result in pH < 7; bases alone result in pH > 7; salts of a strong acid and weak base (e.g., $NH_4^+Cl^-$) are acidic; salts of a weak acid and strong base (e.g., NaOAc) are basic.

7. Check the result. The results can be checked by iterating on charge balance pH by inserting Eqns. 18.30 and 18.31 or the analogs.

Example 18.5 Sillèn diagram for HOAc and NaOAc

Sodium acetate, NaOAc, is dissolved in water at an apparent concentration of $C_B = 10^{-2}$ mol/L. Construct a Sillèn diagram and estimate the pH. For acetic acid, $pK_{a,A} = 4.76$ at room temperature in dilute solutions.

Solution:

Here we replace the generic A^- with OAc^- to denote acetate. The Sillèn diagram is presented in Fig. 18.5. The approximate solution (thick lines) is shown below superimposed on the exact equations (thin lines).

Step 1: The lines for $[H^+]$ and $[OH^-]$ have been drawn and labeled in the figure.

$$\text{Step 2:} \quad [Na^+] = [HOAc] + [OAc^-] = C_B \quad \text{material balance} \qquad 18.40$$

$$\text{Step 3: } K_{a,A} = [OAc^-][H^+]/[HOAc] \quad \text{equilibrium} \qquad 18.41$$

$$K_{a,w} = [H^+][OH^-] = 10^{-14} \quad \text{equilibrium} \qquad 18.42$$

$$\text{Step 4: } [H^+] + [Na^+] = [OAc^-] + [OH^-] \quad \text{charge balance} \qquad 18.43$$

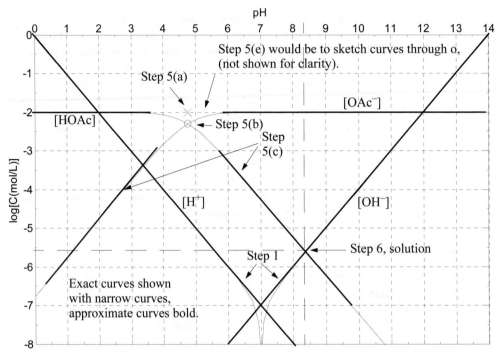

Figure 18.5 *Sillèn plot for Example 18.5, acetate with an overall concentration 0.01 M.*

Example 18.5 Sillèn diagram for HOAc and NaOAc (Continued)

Step 5: See the diagram labels denoting steps 5(a) and 5(b). Referring to the procedure above indicates the system point (**x**) should be at C_B and $pK_{a,A}$. On the diagram below, use a straight edge to verify that the lines for step 5(c) extrapolate through the **x** from step 5(a). Note that the curves for step 5(e) are not shaded for clarity, but it is obvious that the curves could be easily drawn through o. Recall that step 5 plots Eqns. 18.30 and 18.31 without calculations.

Step 6: Develop the proton condition. This step is very important and can be the most confusing. It is best understood by using equations together with the diagram. Since we have dissolved the salt of a weak acid and strong base, we expect the pH to be above 7. Looking at the diagram in this range, $[OAc^-] \gg [HOAc]$ and we will be unable to reliably calculate $[HOAc] = C_B - [OAc^-]$ with the material balance because the last two terms are nearly equal. Let us use the material balance to eliminate the large terms $[Na^+]$ and $[OAc^-]$ from the charge balance. Note that we can replace $[Na^+]$ with C_B and $[OAc^-] = C_B - [HOAc]$. This causes all the largest terms to drop from the charge balance, giving

$$[H^+] + [HOAc] = [OH^-] \quad \text{proton condition} \qquad 18.44$$

> ### Example 18.5 Sillèn diagram for HOAc and NaOAc (Continued)
>
> Eqns. 18.40–18.43 are now all condensed to using Eqn. 18.44 with the graph, looking for where the proton condition is satisfied. Looking at the lines on the graph where pH > 7, it is obvious that [HOAc] is almost three orders of magnitude larger than [H$^+$] above pH = 5. Thus, the left side of the proton condition becomes [HOAc] + [H$^+$] = [HOAc] + … where … denotes a very small number. The proton condition becomes [HOAc] + … = [OH$^-$], and the solution is given for practical purposes by the intersection of the [HOAc] curve with the [OH$^-$] curve as shown in the diagram. The approximate concentrations are
>
> $$pH = 8.4, pOH = 14 - 8.4 = 5.6, [HOAc] = 10^{-5.6}, [Na^+] = [OAc^-] = 10^{-2}$$
>
> **Step 7:** The proton condition is in terms of Eqns. 18.27 and 18.30, and avoiding taking differences, [OH$^-$] = 10^{-14}/[H$^+$] = [H$^+$] + [HOAc] = [H$^+$] + C_A[H$^+$] / ([H$^+$] + $K_{a,A}$)
> Rearranging for successive substitution on [H$^+$], and inserting the initial guess of pH = 8.4, iterate on the highest power of [H$^+$],
> $$[H^+]^2 = 10^{-14}/(1 + C_A/([H^+] + K_{a,A})) = 10^{-8.4} + (10^{-2})(10^{-8.4})/(10^{-8.4} + 10^{-4.76})$$
> $$[H^+] = \sqrt{10^{-14}/(1 + C_A/([H^+] + K_{a,A}))} = \sqrt{10^{-14}/(1 + 10^{-2}/(10^{-8.4} + 10^{-4.76}))}$$
> $$= 4.165E\text{-}9$$
> pH = 8.38. Plugging this back in results in no further changes. Recall that if successive substitution results in divergence rather than convergence, that the equation needs to be rearranged. See Appendix A, Section A.4.

This example has demonstrated that a relatively complex problem can be solved rapidly with a quick sketch. A key simplification used in this introductory example was that [H$^+$] << [HOAc]. Do not generalize this approximation. If the same problem is repeated with $C_B = 10^{-4}$ mol/L, this translates the lines for [HOAc] and [OAc$^-$] downwards two orders of magnitude and the lines for [H$^+$] and [HOAc] become very close. The approximation [H$^+$] << [HOAc] is not valid then. Instead, the left side of the proton condition, [H$^+$] + [HOAc], can be calculated at a selected point from the two lines. The [H$^+$] + [HOAc] sum becomes a line through this point with the same slope as the individual lines. The answer is where this "summed" line crosses [OH$^-$].

A remarkable feature of this solution technique is that the solution to the four simultaneous equations did not require sophisticated algebra or a cubic equation. As an exercise, consider a solution of acetic acid, $C_A = 10^{-2}$ mol/L. The diagram is the same as used above. However, the material balance and proton condition are different. Use the diagram to show that the pH = 3.35 approximately.

Polyprotic Acids and Bases

The phosphoric system (H_3PO_4, $H_2PO_4^-$, HPO_4^{2-}, PO_4^{3-}) and the CO_2 (CO_2, HCO_3^-, CO_3^{2-}) systems are important for both biology and environmental applications. Succinic acid, a dicarboxylic acid produced by fermentation, is expected to become more widely produced via fermentation in future years, typically as a salt. Amino acids, the building blocks for proteins, combine a basic amine and a carboxylic acid on the same molecule. Let us begin by considering the nonvolatile phosphate system.

Phosphate System

The equilibria can be written (using all acid equilibrium constants, but without the A subscript for convenience),

$$\frac{[H_2PO_4^-]}{[H_3PO_4]} = \frac{K_{a1}}{[H^+]} \qquad\qquad 18.45$$

$$\frac{[HPO_4^{2-}]}{[H_2PO_4^-]} = \frac{K_{a2}}{[H^+]} \quad \text{or} \quad \frac{[HPO_4^{2-}]}{[H_3PO_4]} = \frac{K_{a1}K_{a2}}{[H^+]^2} \qquad\qquad 18.46$$

$$\frac{[PO_4^{3-}]}{[HPO_4^{2-}]} = \frac{K_{a3}}{[H^+]} \quad \text{or} \quad \frac{[PO_4^{3-}]}{[H_3PO_4]} = \frac{K_{a1}K_{a2}K_{a3}}{[H^+]^3} \qquad\qquad 18.47$$

The material balance on phosphorous is

$$C = [H_3PO_4] + [H_2PO_4^-] + [HPO_4^{2-}] + [PO_4^{3-}] \qquad\qquad 18.48$$

Defining a variable α_i to denote the fraction of each species relative to the total phosphate concentration where the subscript denotes the number of protons,

$$\alpha_3 = \frac{[H_3PO_4]}{C}, \; \alpha_2 = \frac{[H_2PO_4^-]}{C}, \; \alpha_1 = \frac{[HPO_4^{2-}]}{C}, \; \alpha_0 = \frac{[PO_4^{3-}]}{C} \qquad 18.49$$

Dividing the material balance by $[H_3PO_4]$, we find the reciprocal of α_3,

$$\frac{1}{\alpha_3} = \frac{C}{[H_3PO_4]} = 1 + \frac{[H_2PO_4^-]}{[H_3PO_4]} + \frac{[HPO_4^{2-}]}{[H_3PO_4]} + \frac{[PO_4^{3-}]}{[H_3PO_4]} = 1 + \frac{K_{a1}}{[H^+]} + \frac{K_{a1}K_{a2}}{[H^+]^2} + \frac{K_{a1}K_{a2}K_{a3}}{[H^+]^3}$$

$$18.50$$

Inverting and simplifying,

$$\alpha_3 = \frac{[H_3PO_4]}{C} = \frac{[H^+]^3}{[H^+]^3 + K_{a1}[H^+]^2 + K_{a1}K_{a2}[H^+] + K_{a1}K_{a2}K_{a3}} \qquad 18.51$$

Then α_2 as the fraction of acid with three protons is

$$\alpha_2 = \frac{[H_2PO_4^-]}{C} = \frac{[H_3PO_4][H_2PO_4^-]}{C} \frac{}{[H_3PO_4]} = \alpha_3 \frac{K_{a1}}{[H^+]}$$

$$\alpha_2 = \frac{[H_2PO_4^-]}{C} = \frac{K_{a1}[H^+]^2}{[H^+]^3 + K_{a1}[H^+]^2 + K_{a1}K_{a2}[H^+] + K_{a1}K_{a2}K_{a3}} \qquad 18.52$$

Recognizing the recurring relation between the fractions, the arguments for one and no protons are

$$\alpha_1 = \frac{[HPO_4^{2-}]}{C} = \frac{K_{a1}K_{a2}[H^+]}{[H^+]^3 + K_{a1}[H^+]^2 + K_{a1}K_{a2}[H^+] + K_{a1}K_{a2}K_{a3}} \qquad 18.53$$

$$\alpha_0 = \frac{[PO_4^{3-}]}{C} = \frac{K_{a1}K_{a2}K_{a3}}{[H^+]^3 + K_{a1}[H^+]^2 + K_{a1}K_{a2}[H^+] + K_{a1}K_{a2}K_{a3}} \qquad 18.54$$

The Sillèn diagram for the phosphate system is slightly more complicated than a monoprotic system, but can still be quickly drawn by hand. The concentration of each species i can rigorously be calculated at each pH by $\alpha_i C$ where α_i is calculated from Eqns. 18.51–18.54. The exact concentrations are shown in Fig. 18.6 for $C = 10^{-2}$ mol/L. The upper part of the Sillèn diagram is analogous to the monoprotic diagram. The difference is that the slopes change to $+2$ or -2 when the species concentration crosses the $pK_{a,A}$ for a neighboring dissociation. The curve passes through a y coordinate approximately 0.3 "$\log_{10}$ units" below the point where the upper extrapolated line crosses the neighboring $pK_{a,A}$. Though not apparent from this diagram, at very low concentrations when the species concentration crosses additional $pK_{a,A}$ values, the slope increments (decrements) again by one.

Example 18.6 Phosphate salt and strong acid

A solution of NaH_2PO_4 and HCl is prepared such that the total phosphorous concentration is 1E-2 M and the total Cl concentration is 5E-3M. Calculate the pH and concentrations of species present.

Solution: Begin a problem with multiple ions with the material balance. The material balances on the sodium, chloride, and phosphate are:

$$1 \times 10^{-2}M = [H_3PO_4] + [H_2PO_4^-] + [HPO_4^{2-}] + [PO_4^{3-}] = [Na^+] \qquad 18.55$$

$$[Cl^-] = 5 \times 10^{-3}M \qquad 18.56$$

$$[H^+] + [Na^+] = [H_2PO_4^-] + 2[HPO_4^{2-}] + 3[PO_4^{3-}] + [OH^-] + [Cl^-] \text{ charge balance} \quad 18.57$$

Note the coefficients on the ions in the charge balance. The Sillèn diagram for the phosphate system is shown below. It may seem daunting that so many species are present, but when you look at the Sillèn diagram, notice that only two phosphate species at a time are important at any pH range. This occurs because the $pK_{a,A}$ values are well separated. The curves are drawn with the exact relations, but can be quickly sketched. Practice the sketch using the rules given above and compare with Fig. 18.6.

Steps 1-5 of the procedure have already been executed.

Step 6. The proton condition is developed by eliminating $[Na^+]$ and $[Cl^-]$ using the material balances since they are both known constants. The material balance for phosphate is also inserted, resulting in

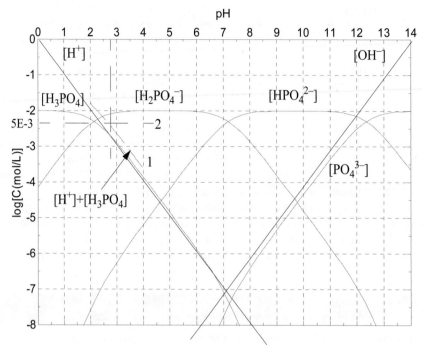

Figure 18.6 *Phosphoric system at C = 10^{-2} mol/L discussed in Example 18.6.*

Example 18.6 Phosphate salt and strong acid (Continued)

$$[H^+] + 10^{-2} = 10^{-2} - [H_3PO_4] + [HPO_4^{2-}] + 2[PO_4^{3-}] + [OH^-] + 5 \cdot 10^{-3} \qquad 18.58$$

which becomes

$$[H^+] + [H_3PO_4] = [HPO_4^{2-}] + 2[PO_4^{3-}] + [OH^-] + 5 \cdot 10^{-3} \text{ proton condition} \qquad 18.59$$

Understanding where to find the solution requires some thought and reasoning rather than a direct numerical manipulation. Both terms on the left side of the proton condition are almost equal at 2.5 < pH < 7. The values are added on the short dashed line marked "1" (since they are virtually equal in most of the range, the sum is double, or about 0.3 units higher on the log$_{10}$ scale). Note that [H$_2$PO$_4^-$] does not appear in the proton condition. On the right-hand side, the term 5E-3 dominates at pH < 6. Solutions at high pH are impossible because the decreasing left-hand side is too small to balance the value of 5E-3 plus increasing concentrations of the negative phosphate and hydroxide ions in the proton condition. Therefore, the solution must be a low pH where the concentration of negative phosphate and hydroxide ions in the proton condition are small. The solution occurs where [H$^+$] + [H$_3$PO$_4$] = 5E-3 (the line marked "2"), and pH = 2.6. The approximate concentrations from the diagram are [H$^+$] = [H$_3$PO$_4$] = 2.5E-3. Eqn. 18.45 simplifies to [H$_2$PO$_4^-$] = K_{a1} = 10$^{-2.15}$, [HPO$_4^{2-}$] = 10$^{-6.5}$, [Na$^+$] = 1E-2, [Cl$^-$] = 5E-3.

| **Example 18.6 Phosphate salt and strong acid (Continued)** |

Step 7. The detailed calculations are often tedious. Inserting Eqn. 18.51 into the proton condition, where the first three terms on the right side are negligible,

$$[H^+] + \frac{C[H^+]^3}{[H^+]^3 + K_{a1}[H^+]^2 + K_{a1}K_{a2}[H^+] + K_{a1}K_{a2}K_{a3}} = 5E\text{-}3 + \dots \qquad 18.60$$

$$[H^+]\left(1 + \frac{C[H^+]^2}{[H^+]^3 + K_{a1}[H^+]^2 + K_{a1}K_{a2}[H^+] + K_{a1}K_{a2}K_{a3}}\right) = \qquad 18.61$$

$$[H^+]\left(1 + \frac{C/[H^+]}{1 + K_{a1}/[H^+] + (K_{a1}K_{a2})/[H^+]^2 + (K_{a1}K_{a2}K_{a3})/[H^+]^3}\right) = 5E\text{-}3$$

Inserting the initial guess,

$$[H^+] = (5E\text{-}3)/\left[1 + \frac{C/[H^+]}{1 + K_{a1}/[H^+] + (K_{a1}K_{a2})/[H^+]^2 + (K_{a1}K_{a2}K_{a3})/[H^+]^3}\right] \qquad 18.62$$

$$[H^+] = (5E\text{-}3)/\left[1 + \frac{10^{-2}/[H^+]}{1 + 10^{-2.148}/[H^+] + \dots}\right]$$

$$= (5E\text{-}3)/\left[1 + \frac{10^{-2}/10^{-2.6}}{1 + 10^{-2.148}/10^{-2.6} + \dots}\right] = (5E\text{-}3)/\left[1 + \frac{3.981}{1 + 2.8314 + \dots}\right] \qquad 18.63$$

$$(5E\text{-}3)/[2.039] = 2.45E\text{-}3 = 10^{-2.611}$$

Repeating the iteration results in $[H^+] = 2.44E\text{-}3$, pH = 2.613. Note how close we were with the graphical value of pH = 2.6.

Amino Acids

Amino acids are the fundamental building blocks from which all **proteins** are built. DNA encodes the formulas used to assemble 22 standard amino acids into the multitudes of proteins. Proteins with specific catalytic functions are called **enzymes.** Amino acids include at least one carboxylic acid group and one amine group. The 20 amino acids summarized in Fig. 18.7 are encoded in the universal genetic code. Together with selenocysteine and pyrrolysine which are encoded in special situations, the 22 amino acids link together to provide the functionalities required for biological life by use of various side chains. When biological machinery assembles amino acids into proteins, a carboxylic acid from one amino acid is covalently bonded to the amine on the next amino acid. One end of any protein backbone is an amine and the other end is a carboxylic acid. Note that some side chains in Fig. 18.7 include acidic and basic side chains. These acidic and basic side chains lead to charges on proteins, which change as a function of pH. Since biological systems usually have buffered pH near 7, which is above the $pK_{a,A}$ for the carboxylic acids, those groups are in the conjugate base form, leading to negative charges on the side chains. Similarly, basic groups below the $pK_{a,A}$ are protonated. For example, serum albumin, a globular (round) blood protein, has negative charges on the surface at physiological pH. At physiological pH, it has an intrinsic charge of -17 and binds

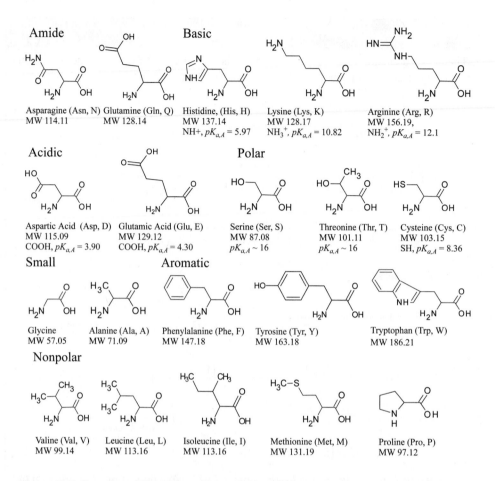

Figure 18.7 *Summary of 20 amino acids encoded by the universal genetic code. The amine and carboxylic groups on the bottom of each molecule are where the amino acid is linked into the biomolecule. The acidic and basic side chains are shown uncharged.*

6 monovalent anions giving a total net charge of –23. At low pH, the intrinsic charge becomes positive and the protein binds more anions. Hence the charge and binding change with pH.[14] Serum albumin is critical for controlling the intravascular hydrostatic pressure by regulating the osmotic pressure. About half of the osmotic pressure is controlled by Donnan equilibria, as described on page 725. Other proteins have a larger number of basic side groups than acidic groups. The basic side chains are protonated at neutral pH, contributing a positive charge.

Glycine

Glycine is the simplest amino acid. The side chain is simply a hydrogen; that is, there is no side chain. For glycine, $pK_{a,A1} = 2.35$ for the carboxylic acid and $pK_{a,A2} = 9.78$ for the amine. Thus, the amino acids combine the concepts of acids and bases with the concepts of a polyprotic system. The species at various pH values and nomenclature are shown in Fig. 18.8. Note that the species at neutral pH has a net charge of zero, but includes a positive and negative charge. Such a molecule with

14. Fogh-Andersen, N., Bjerrum, P.J., Siggaard-Andersen O. 1993. "Ionic Binding, Net Charge, and Donnan Effect of Human Serum Albumin as a Function of pH." *Clin. Chem.* 39:48–52.

glycinium   glycine   glycinate
H_2Gly^+ HGly Gly^-
$pH < 2.35$ $2.35 < pH < 9.78$ $pH > 9.78$

$$^+H_3NCH_2COOH \rightleftarrows \, ^+H_3NCH_2COO^- \rightleftarrows H_2NCH_2COO^-$$

$$pK_{a,A} = 2.35 \qquad\qquad pK_{a,A} = 9.78$$

Figure 18.8 *Dominant species for glycine at various pH levels. $pK_{a,A} = 2.35$ for the carboxylic acid and $pK_{a,A} = 9.78$ for NH_3^+. The top line shows the nomenclature and the second line shows abbreviations.*

both positive and negative charges is called a **zwitterion** (German for "hybrid ion"). This combination of positive and negative charges results in large dipole moments for biological molecules at neutral pH. The relevant equilibria are (written using acid dissociation constants, but omitting the A subscript):

$$K_{a1} = \frac{[H^+][HGly]}{[H_2Gly^+]} \quad \text{or} \quad \frac{[HGly]}{[H_2Gly^+]} = \frac{K_{a1}}{[H^+]} \qquad 18.64$$

$$K_{a2} = \frac{[H^+][Gly^-]}{[HGly]} \quad \text{or} \quad \frac{[Gly^-]}{[HGly]} = \frac{K_{a2}}{[H^+]} \quad \text{or} \quad \frac{[Gly^-]}{[H_2Gly^+]} = \frac{K_{a1}K_{a2}}{[H^+]^2} \qquad 18.65$$

Example 18.7 Distribution of species in glycine solution

(a) Calculate the pH of a 0.1 M solution of glycine.
(b) What is the distribution of species for glycine at a physiological pH of 7.4?

Solution: A Sillèn diagram for glycine (Fig. 18.9) is sketched by the standard procedures.
(a) The relevant equilibria are given in Eqns. 18.64 and 18.65. The material balance is

$$C = [H_2Gly^+] + [HGly] + [Gly^-] = 0.1M \qquad 18.66$$

The charge balance is

$$[H^+] + [H_2Gly^+] = [Gly^-] + [OH^-] \qquad 18.67$$

The pH is expected to be near neutral because the glycine added is neither an acid nor a base, though it has both functionalities. Look at the charge balance and the Sillèn plot of concentrations near neutral pH. On the left-hand side of the charge balance, the concentration of $[H^+]$ is about an order of magnitude smaller than $[H_2Gly^+]$ making the total positive charge concentration about $1.1[H_2Gly^+]$. On the right side, $[Gly^-]$ is over three orders of magnitude larger than $[OH^-]$. Thus, the charge balance is effectively

$$1.1[H_2Gly^+] = [Gly^-] + \ldots \qquad 18.68$$

Example 18.7 Distribution of species in glycine solution (Continued)

The answer is found at pH = 6.1. $[OH^-] = 10^{-(14-6.1)} = 10^{-7.9}$ M. The glycine species concentrations are quickly read from the graph, $[HGly] = 10^{-2}$ M, $[Gly^-] = 10^{-4.65}$ M, $[H_2Gly^+] = 10^{-4.75}$ M. Note the glycine is almost totally in the zwitterion form and the charged forms are about 3.5 orders of magnitude smaller. The final verification of the concentrations is left as a homework problem. Note that if the concentration of glycine was lower, $[H^+]$ would become more important in the charge balance.

(b) First, $[OH^-] = 10^{-(14-7.4)} = 10^{-6.6}$ M. The glycine species concentrations are quickly read from the graph, $[HGly] = 10^{-2}$ M, $[Gly^-] = 10^{-3.4}$ M, $[H_2Gly^+] = 10^{-6.05}$ M.

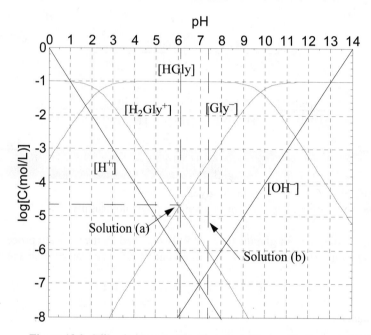

Figure 18.9 *Sillèn diagram for 0.1 M glycine discussed in Example 18.7.*

Summary for Use of Sillèn Plots

Throughout this section, we have demonstrated the use of Sillèn plots for weak acids and bases, and polyprotic systems including amino acids. We have demonstrated that the equilibrium relations plotted on the diagram are not directly dependent on the other ions present in the system. The material balance is often combined with the charge balance to yield a proton condition. The proton condition focuses on the intermediate concentrations, by canceling out the overwhelmingly large concentrations, enhancing precision. The proton condition is used along with the plot to determine an approximate solution. In cases where the curves for the dominant species with the same charge sign in the proton condition are close to each other, the curve values are added together and replotted to find the solution graphically (*cf.* Example 18.6).

For polyprotic systems we have demonstrated that the distribution can easily be calculated by the Sillén method (*cf.* Example 18.6). For the polyprotic systems in the examples, the pK_a values were well separated and each species has a pH where it is the dominant species and the species concentration is virtually equal to the overall concentration, C. For example, in the case of glycine, the intermediate species HGly is virtually equal to C in the range of $4 < \text{pH} < 8$ (*cf.* Fig. 18.9). In some polyprotic systems, the $pK_{a,A}$ values are close together—for example, in succinic acid, where the species are $H_2\text{Succ}$, HSucc^-, and Succ^{2-}. In these cases, the plot maxima in the intermediate species concentration (HSucc^- in the case of succinic acid) will not extend all the way up to C. When the diagram is prepared, the regions with sloping lines should be drawn first. Then, if the maxima are not obvious, the full equations for the distribution can be developed as illustrated in Example 18.6 and then plotted for the intermediate species.

18.10   APPLICATIONS

To introduce the concepts of electrolytes, we started in Section 18.2 with examples of freezing point depression, osmotic pressure, and boiling point elevation. Here we consider other applications where some of the subtler effects of charges are important.

Buffers

When salts and acids that share a common ion are present in solution, the solution is buffered. A buffered solution is resistant to changes in pH, and such behavior is critical in biology. For example, blood is buffered to be at pH 7.4 with carbonates, and slight deviations can cause severe illness. The buffering capacity is dependent on the concentrations of the acid and salt. For a given overall concentration, the buffering capacity is best understood relative to a titration curve. The buffering capacity (for a change in either pH direction) is greatest when the acid is "half" titrated. The fundamental explanation for the buffering phenomenon is because the titration curve is steepest when the acid is half titrated. For a monoprotic acid, this occurs when the acid and salt concentrations are equal, and the buffered pH = $pK_{a,A}$ unless the buffer is very dilute such that the acid/base lines are close to the $[H^+]$ or $[OH^-]$ lines on a Sillèn diagram. Consider a weak monoprotic acid HA (species A) and the sodium salt NaA, (species B). A specific case would be a mixture of acetic acid and sodium acetate in water. The equations are an extension of Eqns. 18.25–18.28:

$$[HA] + [A^-] = C_A + C_B \text{ and } [Na^+] = C_B \quad \text{material balances} \qquad 18.69$$

$$K_{a,A} = [H^+][A^-]/[HA] \quad \text{equilibrium} \qquad 18.70$$

$$K_w = [H^+][OH^-] = 10^{-14} \quad \text{equilibrium} \qquad 18.71$$

$$[Na^+] + [H^+] = [A^-] + [OH^-] \quad \text{charge balance} \qquad 18.72$$

The sodium material balance is inserted into the charge balance, and solving for $[A^-]$,

$$[A^-] = C_B + [H^+] - [OH^-] \qquad 18.73$$

If this is substituted for $[A^-]$ in the acid balance,

$$[HA] = C_A - [H^+] + [OH^-] \qquad 18.74$$

Substituting into the equilibrium Eqn. 18.70,

$$[H^+] = K_{a,A} \frac{C_A - [H^+] + [OH^-]}{C_B + [H^+] - [OH^-]} \qquad 18.75$$

When the $[H^+]$ and $[OH^-]$ are much smaller than C_A and C_B the right-hand side simplifies, resulting in the **Henderson-Hasselbalch** equation:

$$[H^+] \approx K_{a,A} \frac{C_A}{C_B} \text{ or } pH \approx pK_{a,A} - \log\frac{C_A}{C_B} = pK_{a,A} + \log\frac{C_B}{C_A} \qquad 18.76$$

The Sillèn graph method can certainly be used, but the Henderson-Hasselbalch equation is convenient under proper conditions. When C_A and C_B are near $[H^+]$ or $[OH^-]$, the equation can lead to absurd results. A better approximation is

$$pH \approx pK_{a,A} - \log\frac{C_A - [H^+]}{C_B + [H^+]} \text{ (acidic } pK_{a,A}) \text{ or } pH \approx pK_{a,A} - \log\frac{C_A + [OH^-]}{C_B - [OH^-]} \text{ (basic } pK_{a,A})18.77$$

The equations can be used iteratively from an initial assumed value of $[H^+]$ or $[OH^-]$.

Isoelectric Point and Ionic Strength

Proteins and biomolecules frequently have charged surfaces at neutral pH due to the carboxylic acid and amine side chains. Basic amines are protonated at neutral pH values and acidic carboxylic acids are deprotonated. When the biomolecules have a net charge, they repel each other, and are thus more soluble, enabling them to provide important biological functions by remaining soluble. Two important phenomena exhibited by charged molecules are the change in solubility as the pH is changed, and the dependence of solubility on ionic strength (salt concentration).

First of all, as the pH is varied, the charges on the side chains change. The pH value at which the biomolecule has no net charge is called the **isoelectric point.** Solubility is usually smallest at the isoelectric point because the lack of net charges permits the large macromolecules to approach each other and the large cooperative physical forces cause them to precipitate. Therefore, solubility typically increases rapidly on either side of the isoelectric point. The isoelectric point is often characterized by the **pI** which is the isoelectric point pH. Solubility of a milk protein β-lactoglobulin is shown in Fig. 18.10.

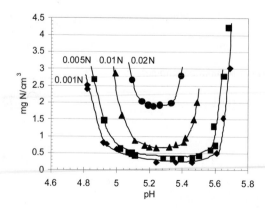

Figure 18.10 *Salting in and illustration of minimum solubility at the isoelectric point for milk protein* β*-lactoglobulin as a function of pH and ionic strength. Gronwall, A., 1942. C.R. Trav. Lab. Carlsberg, Ser. Chim. 24:185-200.*

The solubility of biomolecules increases at low ionic strength when the ionic strength (ion concentration) increases. This effect is called **salting in** and occurs because the ions in solution screen the surface charges. This reduces attractions between positive and negative charges, even near the isoelectric point. There are usually positive and negative charges when the *net* charge is zero, (*cf.* amino acid and zwitterion section above) which lead to net attraction. However, at high ionic strength (high salt concentrations), the opposite effect is seen and increasing salt concentrations result in decreasing biomolecule solubility, known as **salting out.** Salting out occurs because the ionic strength is so high that it screens the repulsive forces that would normally prevent precipitation. Thus, the solubility increases with salt concentration at low salt loading, but decreases with salt concentration at high salt loading, causing a maximum in solubility at intermediate salt concentrations. Observe in Fig. 18.10 the salting in behavior and the minimum solubility near the isoelectric point at all ionic strengths.

Donnan Equilibria

Membranes can have interesting effects when they are impermeable to certain ions or charged species. DNA is a polyanion and requires cations to balance its negative charge. For example, consider a membrane impermeable to DNA shown in Fig. 18.11 where an arbitrary DNA of charge $-z$ is shown in the presence of KCl. The chemical potential of the apparent KCl must be the same on both sides of the membrane, resulting in

$$\left(a_{K^+}a_{Cl^-}\right)_\alpha = \left(a_{K^+}a_{Cl^-}\right)_\beta \qquad 18.78$$

Electroneutrality requires

$$[K^+]_\alpha = [Cl^-]_\alpha \quad \text{and} \quad [K^+]_\beta = z[DNA^{z^-}]_\beta + [Cl^-]_\beta \qquad 18.79$$

Using concentrations to approximate activities and combining the charge and equilibrium relations results in

$$[K^+]_\alpha = [K^+]_\beta\left(1 - \frac{z[DNA^{z^-}]}{[K^+]_\beta}\right)^{1/2} \qquad 18.80$$

It is clear that the concentration of potassium on the β side is larger due to the minus sign in the parentheses. Similarly, manipulation for chlorine shows that the concentration on the α side is larger, due to the plus sign in the parentheses:

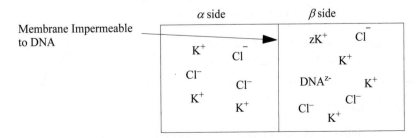

Figure 18.11 *Illustration of Donnan Equilibria for DNA. DNA cannot cross the membrane. A larger concentration of ions will exist on the β side, creating higher pressure on the β side due to osmotic pressure.*

$$[\text{Cl}^-]_\alpha = [\text{Cl}^-]_\beta\left(1 + \frac{z[\text{DNA}^{z-}]}{[\text{Cl}^-]_\beta}\right)^{1/2} \qquad\qquad 18.81$$

What is the relative magnitude of the terms in parentheses in Eqns. 18.80 and 18.81? Substitute Eqns. 18.80 and 18.81 into the left side of Eqn. 18.78, and it becomes obvious that the product of the two terms in parentheses is 1. What may not be immediately clear is that mathematically there are more dissolved species on the β side, creating a **higher osmotic pressure on the β side.** Hemoglobin in red blood cells contributes to a Donnan effect because it is confined to the cells. Another interesting effect is that counter anions such as chloride, bicarbonate, and hydroxyl can pass through the membrane and impact the pH of blood because they are bases.

Solubility and K$_{sp}$

Dissociation of species can have a dramatic effect on solubility in water. Consider the behavior of fluconazole in Example 18.4. How might the dissociation affect the solubility? Since the solubility is dependent on the activity of the un-ionized species, the solubility goes up appreciably as the equilibrium shifts to the protonated form below the $pK_{a,A}$.

Example 18.8 Dissociation and solubility of fluconazole

In Example 18.4 the dissociation of fluconazole (fluc) was considered. The solubility can be modeled using (on the molality scale)

$$K_{SLE} = a_{\text{fluc}(aq)} \quad \text{and} \quad \ln(K_{SLE}) = 8.474 - 3721.9/T \qquad\qquad 18.82$$

Determine the solubility of fluconazole at pH 7 and pH 1.5 and the distribution of species in solution at 298.15K. Assume ideal solutions.

Solution: This involves two simultaneous equilibria, dissociation and solubility. Note that the acid form of the reaction equilibrium constant is

$$K_{a,A} = [\text{fluc}][\text{H}^+]/([\text{fluc}^+]a_{\text{H}_2\text{O}})$$

Since the pH is specified, the solution is so dilute (thus $a_{\text{H}_2\text{O}} = 1$), and the $K_{a,A}$ is constant, the ratio $[\text{fluc}]/[\text{fluc}^+]$ is constant at a given pH. From the earlier example, at pH 7, the reciprocal is $[\text{fluc}^+]/[\text{fluc}] = 0$, and at pH 1.5, $[\text{fluc}^+]/[\text{fluc}] = 9.9/5.1 = 1.94$.

At 298.15 K, $K_{SLE} = 0.018$. Using the ideal solution approximation, $[\text{fluc}] = 0.018$ m. This is independent of pH. At pH 7, virtually no $[\text{fluc}^+]$ is present and thus the solubility is 0.018 mol/L, or using the molecular weight, 0.018 mol/L(306.27 g/mol) = 5.5 g/L.

At pH 1.5, the $[\text{fluc}] = 0.018$ m and $[\text{fluc}^+] = 1.94[\text{fluc}]$, thus the total solubility is 2.94(0.018) = 0.0529 mol/L, or 0.0529(306.27) = 16.2 g/L. The pH makes a large difference in the solubility!

Common Ion Effect

When compounds in solution share a common ion and one of the compounds is near or at the solubility limit, addition of the other species can induce precipitation of the first. For example, consider a solution of water saturated with KCl. If a small amount of NaCl is added to the solution, additional KCl precipitates because the equilibrium is disrupted.

$$KCl_{(s)} \underset{\leftarrow}{\overset{\rightarrow}{}} K^+_{(aq)} + Cl^-_{(aq)} \quad K_{sp} = a_{K^+} a_{Cl^-} \qquad 18.83$$

Because the activity of Cl is increased when NaCl is added, the activity of K must be decreased to match K_{sp}, thus additional KCl precipitates. Likewise, KCl affects the solubility of NaCl near its solubility limit.

18.11   REDOX REACTIONS

Chemical reactions involving electron transfers are fundamental steps in diverse applications ranging from biological systems to corrosion, batteries, and fuel cells. Electron losses are called oxidations and reactions involving gains in electrons are called reductions. A simple acronym to remember the conventions is OILRIG; oxidation is loss, reduction is gain. To balance electrons, an oxidation reaction is always coupled with a reduction reaction, and the combined reactions are termed **redox reactions.**

Redox reactions can be conducted in any phase. Even combustion of methane is a redox reaction where carbon is oxidized into CO_2 and O_2 is reduced into water. Though "oxidation" sounds like it is limited to reaction with oxygen, the process is much more general, relating to the loss of electrons. The gas phase combustion of methane does not produce useful electrons. However, biological oxidation carries out the oxidation of glucose in a series of smaller steps, capturing the electrons and by coupling the favorable oxidation to otherwise unfavorable reactions. In a battery, the redox reaction is enabled by electron flow through an external circuit and a **voltage** is generated. By permitting the spontaneous redox reaction through an external circuit, we obtain electrical power.

Li-ion batteries power the majority of portable devices today. They are constructed of electrodes of CoO_2 and graphite (represented here as C_6 to stress the aromatic ring structure and the way each ring can host one cation). Li^+ and e^- are shuttled back and forth during discharge and charge. During battery discharge, an oxidation of graphite is occurring at the **anode,** $LiC_6 \underset{\leftarrow}{\overset{\rightarrow}{}} C_6 + Li^+ + e^-$ (remember that *oxidation is loss*, so electrons are generated at the anode), and reduction of Co from Co^{+4} to Co^{+3} at the **cathode,** $CoO_2 + Li^+ + e^- \underset{\leftarrow}{\overset{\rightarrow}{}} LiCoO_2$. The overall discharge reaction is $LiC_6 + CoO_2 \underset{\leftarrow}{\overset{\rightarrow}{}} C_6 + LiCoO_2$. The electrons are carried through the external circuit while the Li^+ are not capable of flowing through the circuit, but travel through an electrolyte solution between the electrodes in the opposite direction as the electrons. The combination of the electrodes and electrolyte is called an electrochemical **cell** or **battery.** The Li-ion cell produces about 3.9V. To charge the cell, an opposing voltage > 3.9V must be applied. The life of the cell depends upon chemical reversibility, side reactions, and physical changes that may affect transport or the ability of the Li^+ to be hosted by the graphite sheets or the CoO_2.

Half-Reactions

Redox reactions can be decoupled into the corresponding reduction and oxidation processes as we have indicated with the battery example above. For example, look at the overall reaction for the battery above and note that the reaction does not show the electrons! From the overall battery reaction, how do we determine the two half-reactions, how many electrons are transferred, the voltage, and the direction of electron flow?

As with other thermodynamic tables a reference is used to establish a relative scale. Because reduction reactions are always coupled to oxidation reactions, measurement of absolute potentials is not possible. The reference for redox reactions is to measure/tabulate reduction reactions relative to a standard H_2 electrode. In the **standard H_2 electrode,** an acid solution is used to establish $a_{H^+} = 1$ (pH = 0) at 298.15K and $H_{2(g)}$ is bubbled through the solution at 1 bar partial pressure;[15] the reaction $2H^+ + 2e^- \rightleftarrows H_{2(g)}$ is conducted on a submersed platinum electrode. This reaction is a reduction of H^+. The direction of the reaction depends on whether the reduction potential for the other electrode is greater than or less than that of hydrogen. The potential for hydrogen ion reduction is arbitrarily set to zero at the standard conditions cited above, permitting determination of other reduction potentials. The convention established by the International Union of Pure and Applied Chemistry, (IUPAC) is to always tabulate the potential for *reduction* reactions.[16] The oxidation potential is the negative of the reduction potential. Standard reduction potentials determined at 298.15 K and molal activity of unity (usually near 1 m) for a variety of reactions are tabulated in Appendix E.[17] The half-reaction with the higher *reduction* potential undergoes reduction at standard conditions in the electrochemical reaction, regardless of the number of electrons in the half-reaction. The other reaction proceeds in the direction of oxidation at standard conditions. The relation between the overall cell voltage and the Gibbs energy of reaction at equilibrium is

$$\Delta G° = -n_e F E° = -RT \ln K_a \qquad 18.84$$

Michael Faraday (1791–1867) established the concept of an electromagnetic field, and also popularized the terms "anode," "cathode," "electrode," "ion."

where n_e is the number of electrons transferred in the balanced pair of redox reactions, F is **Faraday's constant** 96,485 J/V, and $E° = E°_{red} - E°_{ox}$ is the difference in *reduction* potentials of both half-reactions (or you may wish to think of adding the reduction potential and the oxidation potential). *Note that the potential is measured per electron, and thus the potentials can be combined without balancing.* However, the number of balanced electrons *is* necessary to relate the potential to the equilibrium constant, Gibbs energy, or nonstandard conditions. As discussed in Section 18.6, Gibbs energies for steps are additive for an overall reaction; the principle can be applied to the half-reactions, $\Delta G° = -n_e F E°_{red} + n_e F E°_{ox}$, where $E°_{red}$ and $E°_{ox}$ both represent *reduction* potentials. A redox reaction at equilibrium is not useful because there are no driving forces. Instead, the direction of the reaction at nonstandard conditions is dependent on ΔG. The principles of Chapter 17 also apply, even under nonstandard conditions, and the reaction goes forward for $\Delta G < 0$, and backward for $\Delta G > 0$. We write

$$\Delta G = \Delta G° + RT \ln \prod a_i^{v_i} \quad \text{or} \quad n_e F E = n_e F E° - RT \ln \prod a_i^{v_i} \qquad 18.85$$

Walther Nernst (1864–1941) is credited with development of the third law of thermodynamics, for which he was awarded the 1920 Nobel Prize in chemistry.

This is often presented as the **Nernst equation.** Also, since reactions frequently occur near 298.15 K, and log is more convenient than ln for quick calculations, the equation is frequently written with values inserted,

$$E = E° - \frac{RT}{n_e F} \ln \prod a_i^{v_i} = E° - \frac{0.05916}{n_e} \log \prod a_i^{v_i} \qquad \text{Nernst equation} \qquad 18.86$$

where the last equation is limited to 298.15 K. The voltage of a cell is determined by E, not directly by $E°$.

---

15. Conventionally, the reference state is shifted to pH = 7 for biological reactions as will be discussed in Section 18.12.
16. Readers should be careful when using other resources to note if the IUPAC convention is followed.
17. The standard potentials are sometimes called "redox" potentials, which can be confusing as to whether they refer to reduction or oxidation. Here we follow the IUPAC convention and always tabulate the "reduction" potentials.

Balancing Redox Reactions

Suppose that you want to design a battery and evaluate the voltage that would be generated. The conventional presentation of thermochemical information is organized in tables of half-cell reactions. After determining the two half-cell reactions and finding them in tables, the two half-cell reactions are balanced and then combined to determine the overall cell voltage. The steps to identify the half-reactions and reaction direction are: 1) identify the oxidized and reduced species using **oxidation states** (the procedure to determine oxidation states is summarized in Table 18.3)[18]; 2) break the reactions into two half-reactions; 3) for each half-reaction, balance the number of atoms for the species oxidized or reduced; 4) balance the change in oxidation state by adding the correct number of electrons to one side of each half-reaction; 5) balance oxygen by adding H_2O to one side of each half-reaction; 6) balance the hydrogen by adding H^+ to the appropriate side (the total charge on both sides of the reactions should now be the same.); 7) look up the reduction potential for each reaction and write the reaction with the smaller reduction potential as the oxidation reaction; 8) multiply the reactions by the smallest integers such that when they are added the electrons cancel; and 9) determine the Gibbs energy and equilibrium condition from Eqn. 18.85 or 18.86 for nonstandard conditions. Note that if the reaction is under basic conditions, it may be more appropriate to work with the basic form. Because OH^- and H_2O both involve H and O, it is easiest to balance with H^+ and then use the water dissociation reaction to convert the reaction as shown in Example 18.9.

Steps to determine standard cell potential and direction of reaction.

Table 18.3 *Procedure to Determine Oxidation States*

1. The oxidation state of an element is 0, e.g. H_2, $Zn_{(s)}$. 2. The oxidation state of H is +1 except in metal hydrides, where it is –1, (e.g. NaH). 3. When oxygen is in a molecule, the oxidation state is –2 except when bonded to another oxygen, as in hydrogen peroxide, H_2O_2, where it is –1.	4. The alkali metals almost always have an oxidation state of +1. The alkali earth metals almost always have a value of +2. 5. Halogens usually have a value of –1, except when bonded to another halogen, then the most electronegative is given –1 and the other is determined by overall charge. 6. The net charge on a molecule equals the sum of oxidation states to determine oxidation states of other atoms.

Including Nonredox Reactions

Frequently, it is necessary to combine redox reactions with reactions that do not involve oxidation and reduction, such as a dissociation or solubility, or use the dissociation reaction of water to convert the acid form (reactions using H^+) to a basic form (reactions using OH^-). This is rigorously done using the Gibbs energy of formation. To combine the reactions: (1) write all the individual reactions balanced so that they add to give the overall balanced reaction as explained above and include the desired nonredox reaction; (2) determine the Gibbs energies for the constituent balanced reactions; (3) add the reactions and Gibbs energies together and then divide the overall Gibbs energy by F and n_e using the Nernst equation to find $E°$. The concepts for combining reactions are shown in Example 18.9 where acidic half-cell reactions are transformed to basic reactions.

---

18. Oxidation *states* are distinct from oxidation *numbers*. They are usually, but not always, the same in complex molecules.

Example 18.9 Alkaline dry-cell battery

Consumer portable electronics are commonly powered by 'alkaline' dry-cell batteries. These cells use an alkaline paste instead of an aqueous solution. The moisture content is low to minimize leakage, and the alkaline solution is used instead of acid because the degradation of the electrodes is slower in alkali compared to acid. The relevant species are $Zn_{(s)}$, $ZnO_{(s)}$, γ-$MnO_{2(s)}$, and α-$MnOOH_{(s)}$. A new battery has $Zn_{(s)}$ and γ-$MnO_{2(s)}$ electrodes.

(a) Determine the balanced reactions for H^+ and then transform them to use OH^-. Then provide the balanced overall reaction. (b) Determine the voltage generated by the cell when $[OH^-] = 1$ m and $[OH^-] = 1.1$ m, and the Gibbs energy of reaction.

Solution: The oxidation states of Zn are 0 for $Zn_{(s)}$ and +2 for $ZnO_{(s)}$; of Mn are +4 for $MnO_{2(s)}$ and +3 for α-$MnOOH_{(s)}$. Since the initial electrode is $Zn_{(s)}$ and γ-$MnO_{2(s)}$, Zn is being oxidized (losing electrons) and Mn is being reduced during battery use.

(a) For Mn, the half-cell reduction reaction is found to be γ-$MnO_{2(s)} + H^+ + e^- \rightleftarrows \alpha$-$MnOOH_{(s)}$, through the following procedure. Start with the Mn species ($MnO2$ and $MnOOH$) on each side of the reaction (more reduced on the right). The reduction requires one electron to go from +4 to +3, so one electron is added to the left. At this point, the O is already balanced, and one H^+ is added to the left to balance hydrogen. The total charge is 0 on each side of the reaction. To convert to the base form, we add $H_2O \rightleftarrows H^+ + OH^-$, giving γ-$MnO_{2(s)} + H_2O + e^- \rightleftarrows \alpha$-$MnOOH_{(s)} + OH^-$ and the total charge is -1 on each side of the reaction.

For the other electrode, the half-cell reduction reaction is found to be $ZnO_{(s)} + 2H^+ + 2e^- \rightleftarrows Zn_{(s)} + H_2O$ through the following procedure. After writing the Zn species on each side (more reduced on the right), we note that the reaction requires two electrons and add them to the left, water is added on the right side to balance oxygen, then $2H^+$ are added to the left side to balance H. The total charge is 0 on each side. To convert to the base form, we add $2H_2O \rightleftarrows 2H^+ + 2OH^-$, giving $ZnO_{(s)} + H_2O + 2e^- \rightleftarrows Zn_{(s)} + 2OH^-$.

For the overall reaction, to balance electrons, two Mn must be reduced for each Zn oxidized. Combining, $Zn_{(s)} + 2\gamma$-$MnO_{2(s)} + H_2O \rightleftarrows ZnO_{(s)} + 2\alpha$-$MnOOH_{(s)}$.

(b) The voltage is found by taking the difference in reduction potentials found in Appendix E. The standard potential is found by the differences in *reduction* potentials, $E^\circ = 0.3 - (-1.26) = 1.56$ V. The potential under operating conditions is given by

$$E = E^\circ - \frac{0.05916}{2}\log\frac{a_{ZnO}a_{MnOOH}}{a_{Zn}a_{MnO_2}a_{H_2O}}$$

Since all the species except for H_2O are solids, they exist in the pure state as a first approximation. (In actual practice the $MnOOH$ forms a solid solution with MnO_2, but we ignore the effect here.) The activity of water is near 1 in the paste and $[OH^-]$ does not appear, and thus it has no effect on the equilibrium voltage. Therefore, the battery should give a constant 1.56 V throughout its life.

Example 18.9 Alkaline dry-cell battery (Continued)
Note that we are neglecting transport effects and the solid solution behavior. Thus, the actual voltage drops as the battery dies owing in part to these effects. The Gibbs energy of reaction is $\Delta G = -n_e F E = -2(96485)1.56 = -301$ kJ/mol, a spontaneous reaction when the circuit is closed.

Fuel Cells

Fuel cells offer many potential advantages for energy usage. They are similar to a battery in that they involve oxidation at the anode and reduction at the cathode. Like the battery discussed above, they also involve transport of molecular cations between the electrodes. The primary difference is that the oxidizing and reducing species are considered to be "fuels" that either flow past the electrodes, or are fuels that can be replenished.

Fuel cell technology is in a state of rapid change. Typical issues revolve around the economical choice of fuels and the longevity of fuel cell devices. Nevertheless, the promise of converting chemical energy into electrical energy without the limitations of the Carnot cycle is a significant motivation. A biological fuel cell is considered in Example 18.11. The status of this topic is addressed in an online supplement with particular emphasis on the thermodynamic aspects of this technology.

18.12   BIOLOGICAL REACTIONS

Oxidation States and Degree of Reduction

Oxidation states, introduced in Section 18.11, provide an important balance condition for any chemical process, but particularly for biochemical reactions and fermentations. Recall that glucose oxidation to CO_2 and H_2O is an important energy-generating reaction in eurakyrotic cells to permit synthesis reactions. The oxidation of glucose or other foodstuffs provides electrons for reducing other species. For biological reaction networks, an electron balance can provide critical analysis of feasible products. CO_2, H_2O, N_2, and O_2 in any mixture cannot sustain biological life in the absence of other energy inputs. Therefore, such a mixture constitutes a useful reference point for a scale known as the **degree of reduction.**[19] The degree of reduction provides a means to compare the overall electrons in a substance and the energy that can be gained by metabolically converting them to a mixture of CO_2 and H_2O. Combustion of a carbon-containing substance with the generic formula $C_f H_a O_b$ follows the balance,

$$C_f H_a O_b + r O_2 \rightleftarrows f CO_2 + \frac{a}{2} H_2O \qquad \text{18.87}$$

It is easy to show using stoichiometry that $r = f + a/4 - b/2$. The oxidation state of oxygen in O_2 is 0, and in products is -2. Thus, four electrons are transferred to oxygen atoms for each mole of O_2. The moles of electrons transferred to oxygen from the carbon compound are thus $4r = 4f + a - 2b$, where the degree of reduction multipliers are $+4$ for C, $+1$ for H, and -2 for O. Nitrogen, sulphur, and phosphorous are often supplied to fermentations, and the reduction multipliers are

---

19. Roels, J.A. 1987. "Thermodynamics of Growth." in *Basic Biotechnology*, Bu'lock, J., Kristiansen, B., eds., Academic Press, pp. 57–74.

selected such that reduction numbers are zero in the supply.[20] When the N supply is ammonia, the multiplier for N is given a value of –3. For H_2SO_4 as the source, the multiplier for S is +6, and for phosphoric acid as a source, the multiplier for P is +5. To apply an electron balance, the reduction multipliers are used, not the oxidation states. Consider reaction of acetaldehyde. For acetaldyhde (C_2H_4O) the reduction calculation is $2(+4) + 4(+1) + 1(–2) = 10$; for O_2 the calculation is $r \cdot 2 \cdot (–2) = (5/2) \cdot 2 \cdot (–2) = –10$; for a net of 0 on the left-side. For CO_2, $1(+4) + 2(–2) = 0$, for H_2O, $2(+2) + 1(–2) = 0$, and the right-side is also 0. Though each side is not always zero, the two sides will balance.

For carbon-containing compounds the degree of reduction, γ_{red}, is often expressed per mole of carbon, (known as a basis of **C-moles**). For $C_fH_aO_bN_cS_dP_e$,

$$\gamma_{red} = (4f + a - 2b - 3c + 6d + 5e)/f \qquad \text{carbon-containing} \qquad 18.88$$

Thus for acetaldehyde above, $\gamma_{red} = 10/2 = 5$ per C-mole. Glucose ($C_6H_{12}O_6$), has a degree of reduction of $(4(6) + 12 - 2(6))/6 = 4$ per C-mole. Hexane (C_6H_{14}) is a more highly reduced species, with a degree of reduction of $(4(6) + 14)/6 = 6.33$. For molecules containing multiple atoms of carbon, the degree of reduction can expressed in terms of moles or C-moles. For example, 180 g of glucose ($M_w = 180$), can be described as 1 mole of $C_6H_{12}O_6$, or as six C-moles of CH_2O, ($M_w = 30$). An average elemental formula for cell mass is $CH_{1.8}O_{0.5}N_{0.2}$, with a degree of reduction of 4.2 per C-mole, slightly higher than glucose. For compounds not containing carbon, the degree of reduction is expressed per mole of that compound. For the compound $H_aO_bN_cS_dP_e$,

$$\gamma_{red} = (a - 2b - 3c + 6d + 5e) \qquad \text{not carbon-containing} \qquad 18.89$$

A fermentation can be represented with a pseudo-reaction, balancing inputs and outputs. For example, on the basis of one C-mole of substrate $CH_aO_bN_cS_d$,

$$CH_aO_bN_cS_d + Y_o(O_2) + Y_n(NH_3) + Y_{aux}(CH_eO_fN_gS_h) \rightarrow$$

$$Y_{biomass}(CH_iO_jN_kS_l) + Y_{product}(CH_mO_nN_pS_q) + Y_{CO_2}(CO_2) + Y_w(H_2O) + Y_S(H_2SO_4)$$

where the Y values on the left are for the nutrients and on the right are for the products and by-products. The number of moles for each species is the value of the corresponding coefficient Y.

The number of electrons must balance for reactants and products using the degree of reduction relative scale. An electron balance is a useful method for performing mass balances on fermentation processes. The number of electrons in a feed or product is simply $\sum_i (\text{C-moles or } Y)_i \gamma_i$. Thus, you can see that if you envision a biological process converting a mole of glucose to a mole of hexane, the fermentation needs an additional source of electrons to perform the reduction and the required C-mols of the substance supplying the electrons can be calculated. Some fermentations "fix" CO_2, such as the succinic acid ($C_4H_6O_4$) fermentation. CO_2 and water have a degree of reduction of 0, and succinic acid has a degree of reduction of 3.5, so 3.5 electrons must be furnished from some other source for each C-mole of succinic acid produced (14 electrons per mole of succinic acid). Roels[19] has presented a simple approximate correlation between the degree of reduction and availability and degree of reduction and heat of combustion. Applying the availability concepts from Section 4.12 (with different notation), Roels analyzes heat production and irreversibilities in aerobic and anaerobic fermentations. Grethlein, et al. used electron balances to determine CO and

20. For special nitrogen sources see Roels, J.A. 1983. *Energetics and Kinetics in Biotechnology,* New York: Elsevier Biomedical Press, p. 40.

CO_2 utilization in syngas (mixtures of H_2, CO, CO_2) fermentations to produce methanol.[21] Shuler and Kargi show that the combination of the electron balance and elemental balances together can be used to determine the fraction of product, biomass by-product, and fraction lost to CO_2.[22]

Binding Polynomials

To treat driving forces for reactions such as electron transfers and chemical equilibria, transformed Gibbs energies of formation are used along with their related apparent equilibrium constants. To perform the transformation, we use a binding polynomial when a species can exist in several bound states. Here we discuss binding polynomials that are helpful for relating the apparent molar concentration to the concentrations of individual species.

When we discussed H_3PO_4 in Section 18.9, we developed a recurring relation for the dissociation in Eqn. 18.50. In that section, we considered H_3PO_4 to be the "parent" molecule that lost successive hydrogens with each dissociation. However, an alternative perspective is to consider PO_4^{3-} to be a binding receptor for H^+ "ligands." If we consider the addition of H^+ to be successive binding reactions, the first binding constant is the reciprocal of the last dissociation constant, $K_{a1}^{bind} = 1/K_{a3}$, and other binding/dissociations can be similarly related. From this perspective, the binding receptor, PO_4^{3-} is the species of interest. A total balance on the species from this perspective replaces Eqn. 18.50 with the equivalent relation (left as a homework problem),

$$C = [PO_4^{3-}]\left(1 + \frac{[H^+]}{K_{a3}} + \frac{[H^+]^2}{K_{a3}K_{a2}} + \frac{[H^+]^3}{K_{a3}K_{a2}K_{a1}}\right)$$

$$= [PO_4^{3-}](1 + K_{a1}^{bind}[H^+] + K_{a1}^{bind}K_{a2}^{bind}[H^+]^2 + K_{a1}^{bind}K_{a2}^{bind}K_{a3}^{bind}[H^+]^3)$$

18.90

Either of the arguments in parentheses is called a **binding polynomial,** P_{bind}. Many successive binding events can be represented by this recursion pattern with either the dissociation constants or the binding constants. The concept illustrated here for three protons as ligands can be generalized to other binding receptors and ligands. Each term in the binding polynomial is proportional to the concentration of a bound species, and the sum represents all possibilities. The fraction of the binding receptor in a given state is (in terms of the binding reaction constant),

$$\frac{K_i[x]^i}{P_{bind}} \quad 0 \le i \le t$$

18.91

where we have generalized to an arbitrary ligand concentration $[x]$, t is the maximum number of ligands, and we have generalized the bonding constant, $(K_0 = 1)$. For our phosphoric acid example, $K_2 = K_{a1}^{bind}K_{a2}^{bind} = 1/(K_{a3}K_{a2})$ (recall for our example that the first binding is related to the third dissociation). In the analogy here, the H^+ serves as the ligand. A key quantity in comparing models of binding to experiments is known as the average number of ligands bound per receptor as a function of ligand concentration. Using binding constants, the average ligands per receptor are calculated by the sum of the number of ligands multiplied by the fraction given by Eqn. 18.91,

21. Grethlein, A.J., Worden, R.M., Jain, M.K., Datta, R. 1990. *Appl. Biochem. Biotechnol.* 24/25:875.
22. Shuler, M.L.; Kargi, F. 2006. *Bioprocess Engineering: Basic Concepts,* 2nd ed. Upper Saddle River, NJ: Prentice-Hall, pp. 209–216.

$$\langle i \rangle = \frac{(0(1) + 1K_1[H^+] + 2K_2[H^+]^2 + 3K_3[H^+]^3)}{P_{bind}}$$

$$= P_{bind}^{-1} \sum_{i=0}^{t} iK_i[x]^i = \frac{[x]}{P_{bind}} \frac{dP_{bind}}{d[x]} = \frac{d\ln P_{bind}}{d\ln[x]}$$

18.92

An equivalent expression can be obtained using P_{bind} in terms of the dissociation constants from Eqn. 18.90.[23] If the binding constants (or dissociation constants, which are the reciprocal of each) are known, then the average binding number can be found as a function of ligand concentration. Note that the average binding number does not depend on the receptor concentration. The binding polynomials are used in transforming the individual Gibbs energies to the apparent Gibbs energy of formation for a family of receptors as we show later.

Energy Carriers in Biological Systems

When introducing biological reactions in Section 3.7, we mentioned the use of carbohydrates, fats, and proteins as food sources. We know that sugars are not "burned" using a single step in the human body. The human body could not survive the adiabatic temperatures of a single-step oxidation. However, biological systems oxidize sugars to CO_2 and water. The reactions that disassemble these foods or other energy storage molecules are termed **catabolic reactions.** Reactions that build new structures are termed **anabolic reactions.** Biological systems have a clever way of carrying out the energy transformations. The body carries out the oxidation in small steps, using enzymes to pair endergonic steps with highly exergonic steps. Biological systems transfer and store energy by either 1) forming and breaking bonds; 2) performing redox reactions using electron carriers. A main carrier of energy captured by forming bonds is a molecule called **adenosine triphosphate,** or **ATP** as shown in Fig. 18.12. In the process of glycolysis, two ATP molecules are used to transfer

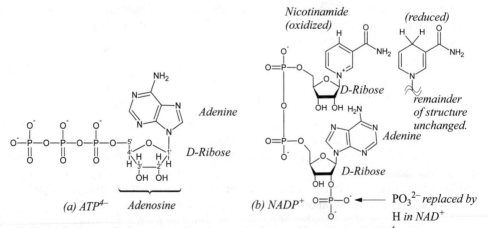

Figure 18.12 *(a) Structure of adenosine triphosphate in the form stable at high pH, ATP^{4-}. Adenosine diphosphate (ADP) has two phosphates; adenosine monophosphate (AMP) has one; adenosine has none. (b)Nicotinamide adenine dinucleotide phosphate, $NADP^+$. Nicotinamide also shown reduced, as in NADPH or NADH.*

---

23. The binding polynomial is an analogy to a partition function in statistical mechanics. It represents a normalization constant given by the sum of the proportionality function for the probability of each binding event. The average binding number is analogous to the expectation value.

two phosphates to glucose (a 6-carbon sugar), modifying it to facilitate subsequent isomerizations and production of two molecules of glyceraldehyde 3-phosphate. Then, the two aldehydes are oxidized to a phosphorylated carboxylic acid in coupled reactions, reducing **nicotinamide adenine dinucleotide,**[24] **NAD⁺** to **NADH** (see Fig. 18.12), storing electrons for other reactions. The two ATP molecules which earlier became ADP are regenerated. In the final step of glycolysis, pyruvate is produced, along with transforming two additional molecules of ADP to ATP, storing additional energy. CO_2 is produced in subsequent reactions when the pyruvate is decarboxylated. Another biological electron carrier is **flavin adenine dinucleotide, FAD** (oxidized), which is reduced to **FADH$_2$** in other reactions. Similar to the NAD reduction that occurs on the nicotinamide, the riboflavin moiety is reduced, in this case with two hydrogens. The biological networks are quite complex, and this text is not intended to serve as an introduction to biological networks. The goal of the next few subsections is to explain the methods used for calculation of thermodynamic driving forces and equilibrium constants in biochemical reactions.

Adenosine triphosphate (ATP), diphosphate (ADP) and monophosphate (AMP) shown in Fig. 18.12 are primary carriers of energy in eukaryotic biological systems, and the distribution of phosphate species is important to represent. ATP can bind up to five protons at low pH (the four on the phosphates plus one on the NH_2). As the pH is lowered, the average number of bindings will undergo a continuous increase until all "receptor" sites are filled at low pH. The individual species are denoted ATP^{4-}, $HATP^{3-}$, H_2ATP^{2-}, H_3ATP^-, H_4ATP, and H_5ATP^+. ATP can also bind other cations, one of principle importance being Mg^{2+}, and the relevant species are $MgATP^{2-}$, $MgHATP^-$, and Mg_2ATP. Similar to the situation discussed above with phosphoric acid, each ATP dissociation has a known dissociation constant. The notation [ATP] will be used to represent the apparent concentration of ATP in all forms. The distribution of phosphate species is also important as discussed here because phosphates are transferred to/from molecules during many of the biological cycles. The pH and pMg are natural variables for determining the distribution of species using P_{bind} and Eqn. 18.91.

Biological Standard State and Apparent Equilibrium Constants

Biologists work on the molar concentration scale with standard state properties at 298.15 K, 1 bar and a standard concentration of 1 M, except water is kept on the Lewis-Randall scale, analogous to the molal treatment. The 1 M standard state is awkward in biological systems because the standard state of [H⁺] is a 1 M solution with a pH near 0. Gibbs energy changes based on such a standard state requires a large correction to physiological pH. However, it is convenient to transform the Gibbs energy such that pH and/or pMg may be held constant. Until this point in the text, we have utilized Gibbs energy for analyzing chemical and phase equilibria because it is minimized at constant T and P. If a system were at constant T and V, then the Helmholtz energy would be the correct property minimized, and if at constant S and V, then U would be minimized (note the relation between the natural variables and the minimized property). Biological systems are pH buffered, and when a biological reaction occurs, it occurs at a constant T, P, and pH, and often other ion concentrations are constant, such as Mg. The convention is to transform the Gibbs energy calculations to a pH (and ion concentration) of interest and use a potential that measures the driving forces at constant pH. The process of transformation is special for H⁺ "receptors" such as ATP^{4-}, PO_4^{3-}, and other species such as ADP and AMP which lead to a distribution of species. The collection of a

---

24. Note that the actual charge on NAD varies depending on protonation of the NH_2 and phosphates. The usual convention is to indicate the oxidized form as NAD⁺, indicating the charge on nitrogen, not the overall charge. The reduced form is conventionally represented as NADH, again, not necessarily reflecting the overall charge on the molecule.

given receptor populated with various numbers of ligands are known as a family of **pseudoisomers**. The transformed properties are denoted with ′. In a similar way, binding of Mg^{2+} is important for ATP, ADP, and AMP. An additional transformation can be made to provide Gibbs energies when **pMg** = $-\log[Mg^{2+}]$ is held constant, denoted using the same ′.

The transformed Gibbs energy has some interesting effects on the way that reactions are written and balanced. In this section, we present two main concepts: 1) balancing of reactions in the transformed framework; and 2) relationships between the apparent equilibrium constant and the equilibrium or nonequilibrium concentrations. In this section, we focus on applications where the apparent equilibrium constants are known or determined from apparent equilibrium concentrations. Details on the steps to calculate the apparent equilibrium constants from Gibbs energies of formation are provided in Section 18.17.

When the Gibbs energy is transformed for H, the pH of the solution is considered to be buffered and the surrounding solution is then a sink/reservoir for H^+ ions. This means that when we write isolated chemical reactions, H is not conserved because the surrounding solution is a sink/reservoir. Therefore, a single reaction in this environment does not cause the pH to go up or down. When we write chemical reactions, we write them without balancing H or H^+. Because we ignore a cation, H^+, we also ignore the charge balance for chemical reactions. Analogous arguments apply if the transformation is done for Mg^{2+}, and we ignore the balance on Mg.

Another convention of biological thermodynamics uses the apparent concentrations/Gibbs energies for families of pseudoisomers instead of tracking the individual species. This applies to species like phosphate and ATP. Looking at Eqn. 18.90, it is obvious that the distribution of phosphate species is completely determined by the buffered pH. Similar arguments apply to ATP, ADP, or other H^+ receptors except that Mg^{2+} is simultaneously considered. The approach is to write equilibrium constants that use the apparent concentrations of H^+ and Mg^{2+} receptors, and absorb the calculations of the distribution and electrolyte nonidealities into Gibbs energies of formation and the equilibrium constants. Details on the mathematics and thermodynamics are explained in Section 18.17, but the details are not important for applications. For applications, the important principle is to recognize that the Gibbs energies of formation and equilibrium constants change significantly with pH, pMg, ionic strength, and temperature. *The quantities must be available or calculated at the specific conditions before the equilibrium calculations are performed.* However, once they are available, they can be applied with easy hand calculations to determine driving forces or equilibrium conditions. The biological molar standard states and the transformations result in

$$\Delta G' = \Delta G'^\circ + RT\ln\prod_i [i]^{v_i} \qquad \Delta G'^\circ = -RT\ln K_c' \qquad \text{18.93}$$

Consider the hydrolysis reaction of ATP to release a phosphate and produce ADP and phosphate,

$$ATP + H_2O \rightleftarrows ADP + H_3PO_4 \qquad \text{or} \qquad ATP + H_2O \rightleftarrows ADP + P_i \qquad \text{18.94}$$

where the left-hand notation writes phosphate as phosphoric acid, and the right-hand notation writes phosphate as a generic P_i. Since ATP, ADP, and phosphoric acid are all distributions of receptor pseudoisomers, the right-side notation is more common in biological publications. The equilibrium constant, at a specified T, pH, and pMg would be written,

$$K_c' = \frac{[\text{ADP}][\text{P}_i]}{[\text{ATP}]} = \exp\left(\frac{\Delta G'^\circ_{f,\text{ADP}} + \Delta G'^\circ_{f,\text{P}_i} - \Delta G'^\circ_{f,\text{ATP}} - \Delta G'^\circ_{f,\text{H}_2\text{O}}}{RT}\right) \qquad 18.95$$

Since the biological standard state uses molar concentrations, the transformed equilibrium constant does also. Note that water is included in the Gibbs energy calculation, but not in the equilibrium constant because the standard state for water is purity and the solution is nearly pure, even though it is transformed.

Example 18.10   ATP hydrolysis

(a) Calculate the transformed standard state Gibbs energy of reaction and equilibrium constant K_c' for hydrolysis of ATP at $\text{pH}_c = 7$, $\text{pMg} = 3$, 298.15 K, ionic strength, $I = 0.25$ m, where the following data apply.

(b) Show whether the reaction is endergonic or exergonic at the above conditions when the apparent concentrations are[a] [ATP] = 0.00185 M, [ADP] = 0.0014 M, [P$_i$] = 0.001 M. If the reaction is exergonic, at what concentration of ADP does it reach equilibrium if the concentration of phosphate and ATP are constant?

Gibbs energies of formation at $\text{pH}_c = 7$, $\text{pMg} = 3$, 298.15K, $I = 0.25$ mol/kg

	ATP	H_2O	ADP	P_i
$\Delta G'^\circ_{f,i}$ (kJ/mol)	−2298	−156	−1426	−1060

Solution:

(a) First, note that the Gibbs energy of water is different from the value in Appendix E because of the transformation. The transformed standard state Gibbs energy of reaction is
$-1426 - 1060 + 2298 + 156 = -32$ kJ/mol. The equilibrium constant will be

$$\exp\left(\frac{-\Delta G'^\circ_{f,i}}{RT}\right) = K_c' = \exp\left(\frac{32000}{8.314(298.15)}\right) = 4.04 \times 10^5 \ .$$

(b)   The   propensity   for   reaction   at   the   given   concentrations   is

$$\Delta G'_{f,i} = \Delta G'^\circ_{f,i} + RT\ln\frac{[\text{ADP}][\text{P}_i]}{[\text{ATP}]} = -32000 + 8.314(298.15)\ln\frac{[0.0014][0.001]}{0.00185} = -49.8\text{kJ/mol}$$

The reaction is even more strongly exergonic than the standard state. Equilibrium occurs when
$[\text{ADP}] = K_c'[\text{ATP}]/[\text{P}_i] = 5.3 \times 10^8 (0.00185)/0.001 = 9.8 \times 10^8 \text{M}$. Of course, such a high concentration never happens, so the reaction is always favorable at reasonable concentrations. Instead of hydrolyzing ATP and "losing the energy," the phosphate is transferred to glucose in a coupled reaction, the subject of a homework problem.

---

a.   In the human body, [ATP]/[ADP] ~ 10. Alberts, B.; Bray, D.; Hopkin, K.; Johnson, A.; Lewis, J.; Raff J.; Roberts, K.; Walter, P. *Essential Cell Biology*, 3rd ed., New York: NY, Garland Science, (2010), pg. 465.

Example 18.11   Biological fuel cell

A biological fuel cell is a portable electrical source that can be
refueled. Electrical current is generated by a biological redox
couple. In an ideal fuel cell, the enzymes would be immobilized
on the electrodes and maintain the same activity as if free. In the
conceptualized fuel cell on the right, glucose is to be oxidized to
gluconolactone in the right cell, catalyzed by immobilized glucose oxidase. Oxygen is excluded from the right cell to avoid
loss of electrons by bulk oxidation. The left cell is saturated
with air, and a reduction of O_2 to H_2O_2 catalyzed by immobilized laccase is envisioned. Electrons are to flow through the

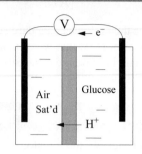

external circuit and H^+ is to flow through the membrane. Each side of the cell is buffered to
$pH_c = 7$, $I = 0.25$ M at $T = 298.15$ K. Suppose the concentrations on the right side are [glucose]
$= 0.1$ M, [gluconolactone] $= 0.05$ M, and on the left side $[H_2O_2] = 0.05$ M. Determine the transformed standard state half-cell potentials and the voltage expected from the cell under stated
concentrations. The standard state Gibbs energies of relevant species are shown below at the
stated conditions.

	$H_2O_{2(aq)}$	Glucose	Gluconolactone
$\Delta G'^{\circ}_{f,\,i}$ (kJ/mol)	−52	−427	−496

Solution: Note that two hydrogen ions are generated by the oxidation and two are consumed by
the reduction of O_2. The two half-cell reactions are glucose $\rightleftarrows$ gluconolactone $+ 2H^+ + 2e^-$,
and $O_{2(g)} + 2H^+ + 2e^- \rightleftarrows H_2O_{2(aq)}$. Note that we could use $O_{2(aq)}$ in the reaction, but that
would require an extra calculation using Henry's law. Since the solution is saturated, we may use
the partial pressure in the gas phase where the standard state Gibbs energy is 0. The standard
state *reduction* potential for the glucose reaction is $\Delta G'^{\circ} = -427 + 496 = 69$ kJ/mol, thus $E'^{\circ} =
-\Delta G'^{\circ}/n_eF = -69000/2/96485 = -0.357$ V.

For the oxygen reaction, the standard state *reduction* potential is $\Delta G'^{\circ} = -52 = -52$ kJ/mol (the
Gibbs energy of formation for $O_{2(g)}$ is 0), thus $E'^{\circ} = -\Delta G'^{\circ}/n_eF = 52,000/2/96485 = 0.269$ V.
The potential expected from a standard state cell would be $E'^{\circ} = 0.269 + 0.357 = 0.626$ V, which is
favorable. Let us evaluate E under the proposed conditions. Using the Nernst equation,

$$E = E^{\circ} - \frac{0.05916}{2}\log\frac{[\text{gluconolactone}][H_2O_2]}{[\text{glucose}](y_{O_2}P)} = 0.626 - \frac{0.0592}{2}\log\frac{(0.05)(0.05)}{(0.1)(0.21)} = 0.653 \text{ V}.$$

Thus, the cell is favorable. Note that other factors are important before the cell can be implemented, such as the rate that electrons can be produced, which requires preserving the activity
(turnover number) of the enzymes. Immobilized enzymes have much slower kinetics compared
with free enzymes. Many of the challenges have been summarized by Calabrese Barton, et. al.[a]

a.   Calabrese Barton, S., Gallaway, J., Atanassov, P. 2004. *Chem. Rev.* 104:4867–4886.

18.13   NONIDEAL ELECTROLYTE SOLUTIONS: BACKGROUND

To this point in the chapter, we have considered solutions to be ideal or absorbed the nonidealities into the Gibbs energies or equilibrium constants. The representation of nonidealities is important for applications where the concentrations are above approximately 0.01 m. The literature through the 1970s has been largely developed by chemists and a variety of notations and models are used in the literature. More recently, chemical engineers have become actively involved in model development and applications. The remainder of this chapter limits the discussion to the extended Debye-Hückel model and its use as a starting point for more sophisticated models. There are common underlying themes in most of the literature. The use of a standard state in which the dilute activity coefficient goes to unity is common for electrolytes and often for molecular species, though the Lewis-Randall scale is used almost always for water. Due to the prevalent use of molality in literature, coverage of the chemical potential and activity coefficients on that scale is necessary.

As with nonelectrolytes, the chemical potential is the primary property that determines phase equilibria and is independent of the scale used to characterize the value. On the molality scale, the chemical potential of the electrolyte is written in a manner analogous to the Lewis-Randall rule on the mole fraction scale,

$$\mu_i = \mu_i^{\square} + RT\ln(m_i\gamma_i^{\square})$$ 18.96

where μ_i° is replaced with $\mu_i^{\square}$ and $m_i\gamma_i^{\square}$ is the activity of the component.[25] The details of $\mu_i^{\square}$ are subtle as elaborated in Section 18.24. A key consideration is that molality has dimensions of mol/kg that are inconsistent with the (dimensionless) activity and activity coefficient. By defining the standard state at 1 molal, we are implicitly writing $\mu_i = \mu_i^{\square} + RT\ln(m_i\gamma_i^{\square}/m_o)$, but $m_o = 1$ mol/kg. We defer the details because most problems are solved by incorporation of $\mu_i^{\square}$ into the Gibbs energy of formation that is commonly tabulated. In other words, the infinite dilution chemical potential is subsumed into the Gibbs energy of formation and reflected in the computation of ΔG_{rxn} that produces K_a. Then,

$$a_i = f_i/f_i^{\square} = \exp((\mu_i - \mu_i^{\square})/(RT)) = m_i\gamma_i^{\square}/m_o$$ 18.97

As long as $\gamma_i^{\square} \approx \gamma_i^{\square\infty} = 1$, $a_i = m_i$. This perspective highlights why the concept is unnecessary to solve problems using K_a at low concentrations as we have done early in the chapter. Because it is common to keep the Lewis-Randall standard state for water, reaction Eqn. 18.14 is written,

$$(\mu_{H_3O^+}^{\square} + (\mu_{OH^-})^{\square} - \mu_{H_2O}^o)/(RT) = -\ln K_a = -\ln(m_{H_3O^+}\gamma_{H_3O^+}^{\square}m_{OH^-}\gamma_{OH^-}^{\square}/(x_{H_2O}\gamma_{H_2O}))$$ 18.98

Thus, the activity coefficients and Gibbs energies are related to the equilibrium constant that we have used earlier. Since a solution of pure H^+ or H_3O^+ or OH^- cannot exist, we cannot easily use a pure state for either unless it is a hypothetical pure state. However, we can measure the behavior of ions in extremely dilute solutions by various means including spectroscopy and electrochemical cells. Even though measurements are taken at finite concentrations, the behavior can be extrapolated to infinite dilution. To illustrate, suppose the activity of H^+ can be measured and plotted, as

---

25. Various symbols are used in the literature for the molal standard state and molal activity coefficient and there is not a standard convention. We use the square ($\square$) to designate the 1 molal standard state and use it throughout for clarity. It is quite common in the electrolyte literature that the special symbol be omitted on the activity coefficient, obfuscating the implemented standard state.

shown in Fig. 18.13. Then extrapolating the measurements to infinite dilution yields an ideal solution that is similar to Henry's law. Similar to Henry's law, the activity coefficient goes to unity at infinite dilution. At very dilute concentrations, the molality can usually be approximated by the molarity as we discussed above.

When first working with electrolytes, this standard state can be very confusing. Because the activity coefficient goes to unity at infinite dilution, a common misconception is that the standard state composition is infinitely dilute. However, the standard state composition is 1 molal with a slope taken such that the infinite dilution activity coefficient is unity. An ideal solution on the basis of the standard state follows the dashed line as shown in Fig. 18.13 and extrapolates to higher compositions. Any composition along the dashed line could be taken as the standard state and the activity coefficient would still be unity, but choosing 1 molal as the standard state permits us to (deceptively, but conveniently) drop m_o numerically from many equations. Referring back to nonelectrolyte systems, the Henry's law standard state applies a similar concept, but extrapolates the ideal solution line to the fugacity of the hypothetical pure fluid. Like the discussion here, the slope of the Henry's law ideal solution is selected such that the infinite dilution value also goes to unity. The difference is that the Henry's law line is based on fugacity as the y-axis, whereas the electrolyte standard state is based on activity. Like Henry's law and the Lewis-Randall rule, the activity coefficient quantifies the deviation from the ideal solution line. Readers should refer to Section 18.24 for further clarification of the relations between Henry's law, the Lewis-Randall rule, and molality/molarity.

18.14 OVERVIEW OF MODEL DEVELOPMENT

The relation between chemical potentials of the ions and the molal activity coefficients is analogous to nonelectrolytes:

$$\mu_+ = \mu_+^{\square} + RT\ln\gamma_+^{\square}m_+ \qquad\qquad 18.99$$

$$\mu_- = \mu_-^{\square} + RT\ln\gamma_-^{\square}m_- \qquad\qquad 18.100$$

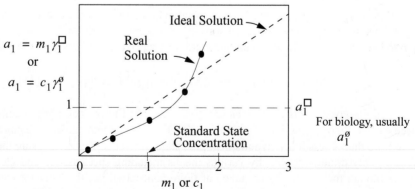

Figure 18.13 *Schematic representation of the activity of an ion at low concentration in a binary mixture. The standard states of ions for industrial calculations are typically 1 molal. For biological systems, they are usually 1 molar. The molality and molarity scales are used alternatively, not simultaneously. When using molal concentration, use molal activity. When using molar concentration, use the molar activity scale.*

The activity coefficients of the ions are determined from the chemical potentials. The details of model development are beyond the intentions of the overview provided here. Briefly, the Gibbs energy of the solution is equal to the work performed in placing an ion in the solution of other ions. The work is determined from solving Poisson's equation for electrostatics using various approximations for the charge density,

$$\frac{1}{r^2}\frac{\partial}{\partial r}\left[r^2\frac{\partial \Phi}{\partial r}\right] = \frac{-\rho_\pm(r)}{\varepsilon}$$

18.101

where r is the radial position, Φ is the electric potential, $\rho_\pm(r)$ is the charge distribution as a function of radial distance, $\varepsilon = \varepsilon_o D$, where ε_o is the permittivity of a vacuum, and D is the dielectric constant. Different models result from different approximations to the charge distribution and the screening. The **Extended Debye-Hückel** model discussed next is one example of a solution.[26] An excellent overview of the approximations and methods of solving the equation is available[27] but is beyond the scope intended here. Briefly, once again, the Debye-Hückel approximation results from assuming that the charge distribution follows the low density radial distribution function form

$$g(r) \sim \exp(-u_{Coul}/kT)$$

The solvent is considered to be a continuum during the calculation represented by the dielectric constant. Various mathematical approximations are made to develop the solution, and different approximations lead to slightly different approximate formulations used in the literature. The Extended Debye-Hückel model is limited to dilute concentrations, generally below ionic strengths of 0.1 molal, and significant errors result from using the model outside this range. The excess Gibbs energy for the Extended Debye-Hückel theory is[28]

$$\frac{G^E}{RT} = \frac{-4x_s M_{w,s} A_\gamma}{1000(Ba)^3}\left[\ln(1 + Ba\sqrt{I}) - Ba\sqrt{I} + \frac{(Ba)^2 I}{2}\right]$$

18.102

$$\text{where } A_\gamma = \frac{N_A^2}{8\pi}\left(\frac{e^2}{\varepsilon_o \varepsilon_r RT}\right)^{3/2}\frac{(2000\rho_s)^{1/2}}{2.303} = \frac{1.8249E6(\rho_s)^{1/2}}{(\varepsilon_r T)^{3/2}}$$

18.103

$$\text{and } A_\gamma = 0.510(\text{kg/mol})^{1/2} \text{ for water at } 25°C$$

18.104

$$\text{ionic strength } I = 0.5\sum_{\text{ions}} m_i z_i^2$$

18.105

where $e = 1.60218E{-}19$ C, $\varepsilon_o = 8.85419E{-}12$ C^2 N^{-1} m^{-2} is the permittivity of vacuum, ε_r is the dielectric constant or relative permittivity of the solvent, R is the gas constant in J/mol-K, T is the temperature in K, ρ_s is the density of the solvent in g/cm^3, z_i is the valence of the Coulombic charge on each ion type, and I is the ionic strength which characterizes the overall charges in the charge distribution. The parameter A_γ is *not a Helmholtz energy*. Dielectric constants for water as a function

26. Zemaitis, J.F., Clark, D.M., Rafal, M., Scrivner, N.C. 1986. *Handbook of Aqueous Electrolytes.* New York: AIChE-DIPPR. p. 595.

27. Tester, J., Modell, M. 1999. *Thermodynamics and Its Applications,* Upper Saddle River, NJ: Prentice-Hall.

28. In addition to the mathematical approximations, other common approximations are included in Eqn. 18.102. The ionic strength should technically be written in terms of ion concentration rather than molality which results in the solution density rather than solvent density in Eqn. 18.103 which makes little significance for dilute solutions.

of temperature are provided in Appendix E. The parameter a represents the average distance of closest approach, which is larger than the ion size due to water hydration which is always present. The term $1/(BI^{1/2})$ is an approximate distance known as the screening, shielding, or Debye length. It represents the screening of the coulombic potential due to the presence of other ions. A common assumption is $Ba = 1$ (kg/mol)$^{1/2}$, though in biological systems, $Ba = 1.6$ (kg/mol)$^{1/2}$.

18.15   THE EXTENDED DEBYE-HÜCKEL ACTIVITY MODEL

Activity Coefficients for Ions

The activity coefficients for ions are obtained by differentiating the excess Gibbs energy of Eqn. 18.102.[29] The resultant formula is

$$\log_{10}\gamma_i^{\square} = \frac{-z_i^2 A_\gamma \sqrt{I}}{1 + Ba\sqrt{I}} \quad \text{up to } I = 0.1 \text{ m} \qquad 18.106$$

where the constants are defined in Eqns. 18.103–18.105. The ionic strength is calculated based on the actual ion concentrations, which means that for weak electrolytes calculation of I must be repeated when the concentration of ions changes during iterations on concentration. The model predicts activity coefficients that are unity at infinite dilution of ions and decrease to a finite limit at high concentration. Experimentally, activity coefficients usually pass through a minimum at concentrations above 0.1 m, which is not captured by the model. Note that all species with the same charge will have the same activity coefficient values at a given ionic strength. More sophisticated models are available for higher concentrations as we discuss in Section 18.20.

Activity and Osmotic Coefficient for Water

The solvent activity coefficients from the Extended Debye-Hückel theory can be obtained by differentiation of the model for excess Gibbs energy. The result is

$$\log_{10}\gamma_s = \frac{2A_\gamma M_{w,s}}{1000(Ba)^3}\left\{1 + Ba\sqrt{I} - \frac{1}{1 + Ba\sqrt{I}} - 2\ln(1 + Ba\sqrt{I})\right\} \qquad 18.107$$

Recall that the activity of the solvent is expressed on the Lewis-Randall standard state. The mole fractions are typically near unity, and thus many significant digits are required to characterize activity coefficients of solvent. Commonly the activity of the solvent is expressed in terms of the **"practical" osmotic coefficient**, Φ,

$$\Phi = \frac{-1000}{M_{w,s}\displaystyle\sum_{electrolytes} m_i}\ln a_s \qquad 18.108$$

---

29. The activity coefficient for Eqn. 18.106 is rigorously γ^* rather than $\gamma^{\square}$. These differ by a factor of x_{H_2O} (see Eqn. 18.163). However, when ion molality is 0.1 m, $x_{H_2O} = 0.998$. At higher concentrations the difference is more important. At ion molality of 6 m, $x_{H_2O} = 0.9$. For more discussion of this detail see Lee, L. 2008. *Molecular Thermodynamics of Electrolyte Solutions*. Hackensack, NJ: World Scientific Publishing, and Robinson, R.A.; Stokes, R.H. 1959. *Electrolyte Solutions*, 2nd Ed., Butterworths. p. 229.

In literature, activities of ions are often measured indirectly by measuring or controlling the partial pressure (isopiestic method) of water above the solution and then reporting the osmotic coefficient. The results are very sensitive to whether complete dissociation is assumed for activity calculation and in the summation in the denominator. Readers must pay careful attention to the assumptions applied in the experimental interpretation. The osmotic pressure can be converted to the ion activity using the Gibbs-Duhem equation to obtain the mean ionic activity coefficient described in Section 18.19. The osmotic coefficient approaches 1 at infinite dilution of ions. The osmotic coefficient is related to the **osmotic pressure,**

$$\Pi = \frac{RT}{V_s} \frac{M_{w,s} \sum_{electrolytes} m_i}{1000} \Phi \qquad 18.109$$

Nonidealities for Nonelectrolyte Solutes

When nonelectrolytes exist in solution with electrolytes, such as with acetic acid, the undissociated acetic acid is typically treated with a molal standard state with the corresponding unity infinite dilution activity coefficient. To fit experimental data, the activity coefficient can be represented as $\ln\gamma_u^\square = bI$ where b is a constant fitted empirically. While this does not satisfy the Gibbs-Duhem equation, it is a common model.

18.16   GIBBS ENERGIES FOR ELECTROLYTES

In Chapter 17, we determined K_a from the Gibbs energy of formation. However, we also noted that occasionally results are summarized in terms of a temperature-dependent K_a. Treatment of the model using K_a requires less thermodynamic information, and it is quite common for electrolytes. Tabulating in terms of Gibbs energies requires consistency with a large database that is tedious to maintain. Such data bases are available for common ions only.

The Gibbs energy of reaction can be represented by

$$\Delta G = \sum_i v_i \mu_i^\circ + RT \sum_i v_i \ln[a_i] = \Delta G_T^o + RT \sum_i v_i \ln[a_i] = -RT\ln K_a + RT \sum_i v_i \ln[a_i] \quad 18.110$$

Writing a completely general notation is difficult because different standard states are often used for different components. For example, we have already discussed using $\mu_+^\square$ for cations and similar notation for anions. *Writing a general sum is clumsy because of the different standard states used for components, so we leave the generic superscript ° and expect that readers apply the appropriate standard states.*

Two key steps in understanding the tables for Gibbs energies are to consider the dissociation constant of water, and the selection of zero as the Gibbs energy of formation for H^+. These steps and choices become clearer in upcoming descriptions. In Chapter 17, we introduced the use of Gibbs energy of the reaction as

$$\frac{\Delta G_T}{RT} = \frac{\Delta G_T^\circ}{RT} + \ln\left[\prod_{i=1}^{NC}[a_i]^{\nu_i}\right] \qquad 18.111$$

Consider again the dissociation constant for water, as shown in Eqn. 18.4. The equilibrium constant for this reaction at 298.15 K is well known as $K_a = 10^{-14}$. The Gibbs energy of the reaction is thus:

$$\Delta G_{298}^o = -RT\ln K_a = -RT\ln(10^{-14}) = -8.314(298.15)\ln(10^{-14}) = 79.908 \text{ kJ/mol} \quad 18.112$$

Now consider that the Gibbs energy of the reaction can be calculated by the Gibbs energies of formation at 298.15 K using the value from Eqn 18.112:

$$\begin{aligned}\Delta G_{298}^o &= \Delta G_{f,298}^\square(H^+) + \Delta G_{f,298}^\square(OH^-) - \Delta G_{f,298}^o(H_2O_{(l)}) \\ &= \Delta G_{f,298}^\square(H^+) + \Delta G_{f,298}^\square(OH^-) + 237.18 = 79.908 \text{ kJ/mol}\end{aligned} \qquad 18.113$$

Note that, like the chemical potentials, writing a general notation for ΔG_{298}^o is slightly imprecise and we use the default $^\circ$ superscript, expecting readers to insert the ion standard state for ions. Note that the standard state Gibbs energy for pure water at 298.15 K has been inserted from the tables for nonelectrolytes. This detail makes an important connection with the standard tables for all other molecular components, and thus the values from the usual tables can be applied when compounds appear in reactions with electrolytes, as long as this convention is used.

Looking at Eqn. 18.113, two values are unknown, both $\Delta G_{f,298}^\square(H^+)$ and $\Delta G_{f,298}^\square(OH^-)$. The dilemma is resolved with the arbitrary choice at 298.15 K and 1 bar:

$$\boxed{\Delta G_{f,298,\,1bar}^\square(H^+) = 0} \qquad 18.114$$

This convention then determines the value

$$\Delta G_{298}^\square(OH^-) = -157.3 \text{ kJ/mol} \qquad 18.115$$

With these values, the remainder of the tables can be developed. Other acid reactions involving H^+ can then be characterized based on the degree of dissociation and the above standard selection of $\Delta G_{f,298}^\square(H^+)$. For example, the value of $\Delta G_{298}^\square(Cl^-)$ can be determined by the dissociation behavior of HCl. Once the Gibbs energy of Cl^- can be determined the dissociation of NaCl will lead to the Gibbs energy of Na^+. The remainder of the tables are developed using similar calculations.

Strategy for Using Gibbs Energies

There is important perspective regarding tabulation of standard state Gibbs energies. Have you considered how scientists created the tables for $\Delta G_{f,i}^\square$? Scientists used experimental equilibrium concentration measurements with models to calculate a_i, and then inserted them into the equilibrium relation:

$$\frac{\Delta G_T}{RT} = 0 = \frac{\Delta G_T^\circ}{RT} + \ln\left[\prod_{i=1}^{NC}[a_i]^{\nu_i}\right] \quad \text{equilibrium} \qquad 18.116$$

Experiments were performed where $\Delta G^o_{f,i}$ (or $\Delta G^{\square}_{f,i}$) was known for all but one of the species. Then the value of $\Delta G^o_{f,i}$ for the species was determined from the experiment by difference. Calculations from multiple investigators using different reactions refined the values that we use from the tables today. When we solve an applied problem, we are using the equation in the opposite direction: looking up $\Delta G^o_{f,i}$ (and $\Delta G^{\square}_{f,i}$) and using models of a_i to determine concentrations. Calculations are reliable as long as a_i's are calculated using methods consistent with the standard state $\Delta G^o_{f,i}$ and $\Delta G^{\square}_{f,i}$.

The steps to solving a problem usually involve reverting the procedures used to develop the tables. Tables are used to calculate ΔG°_{298} (or $\Delta G^{\square}_{f,i}$) from standard state. Temperature and pressure corrections are applied to determine $\Delta G_{T,P}^{\circ} = -RT\ln K$ and then selecting methods to calculate a_i consistent with the standard state. Concentrations are thus determined from the activity, often assuming ideal solutions. Frequently, the Gibbs energies of individual species are not calculated, or are not available. Rather, scientists report the values of K_a as used in Example 18.4. A reliable database of Gibbs energies is available as documented in footnote 13 of this chapter.

18.17   TRANSFORMED BIOLOGICAL GIBBS ENERGIES AND APPARENT EQUILIBRIUM CONSTANTS

The transformed Gibbs energies in Section 18.12 are a convenient method to handle biological reactions but the details were not discussed earlier. The transformation of Gibbs energy to a field of buffered pH is analogous to the other Legendre transforms used previously. To obtain Gibbs energy from internal energy, starting from $dU = TdS - PdV$, we introduced $G = U - TS + PV$, resulting in a potential where T and P are the natural variables, $dG = -SdT + VdP$, and G is minimized when the natural variables are constrained. By introducing $G' = U - TS + PV - N_H\mu_{H^+}$ we arrive at a potential that is a natural function of pH. The variable N_H is the number of hydrogens in the molecule. To obtain a transform that is also a natural function of pMg $= -\log[Mg2+]$, we use $G' = U - TS + PV - N_H\mu_{H^+} - N_{Mg}\mu_{Mg^{2+}}$. Standard-state transformed Gibbs energies of formation are developed for species, and they are used analogously to the untransformed Gibbs energies to calculate the Gibbs energy for reactions and the transformed equilibrium constant.

Also, to correct for solution nonidealities, the extended Debye-Hückel model is added, and a convention is to use $Ba = 1.6$ in the model. A further difference from previous models is that molar concentrations are used for the equilibrium constants, though it makes little difference numerically because solutions of biological molecules are typically dilute on a molar basis. The effect of nonidealities is typically calculated using the molal form of the Debye-Hückel model, using the overall solution molal ionic strength since the model does not differentiate between charged species.

Gibbs Energy Transformations for Species with a Single Form

For a species containing hydrogen or Mg,

$$\Delta G'^{\circ}_{f,T,i}(I, pH, pMg) = \Delta G^{\circ}_{f,T,i}(I) - N_{H,i}\{\Delta G^{\circ}_{f,T,H^+}(I) - RTpH_c\ln(10)\}$$
$$- N_{Mg,i}\{\Delta G^{\circ}_{f,T,Mg^{2+}}(I) - RTpMg\ln(10)\}$$

18.117

> Transformed Gibbs energy of formation at a specified pH and I. Workbook Gprime-Calc.xlsx or MATLAB GprimeCalc.m are helpful.

The Gibbs energy of formation appearing on the right side for $j = i$, H^+, and Mg^{2+} is

$$\Delta G^{\circ}_{f,T,j}(I) = \Delta G^{\circ}_{f,T,j}(I=0) - RT\ln(10)z_j^2 A_\gamma(\sqrt{I}/(1 + Ba\sqrt{I}))$$

18.118

where $Ba = 1.6$ (kg/mol)$^{1/2}$, $pH_c = -\log[H^+]$, $pMg = -\log[Mg^{2+}]$, and I is measured in molality. By assuming that the heat of formation is independent of temperature in the small temperature range where biological reactions occur, the short-cut van't Hoff equation can be applied before inserting the Gibbs energy of formation into

$$\Delta G^\circ_{f,j}(I=0) = \left(\frac{T}{298.15}\right)\Delta G^\circ_{f,\,298.15,\,j}(I=0) + \left(1 - \frac{T}{298.15}\right)\Delta H^\circ_{f,\,298.15,\,j}(I=0) \qquad 18.119$$

The equations are presented in the reverse order compared to how they are used. Standard state values are inserted into Eqn. 18.119 for $j = i$, H^+, and Mg^{2+}. The results are inserted into Eqn. 18.118 for each, and then each of those results is inserted into Eqn. 18.117. Conversion of the Gibbs energy of formation is easy using the Excel workbook GprimeCalc.xlsx or MATLAB m-file GprimeCalc.m. Note that the solution nonidealities are a minor correction compared to the transformations on H^+ and Mg^{2+}. The ′ notation is used quite widely for this transformed Gibbs energy at any specified pH, though in some literature the transformation is restricted to pH 7. The context of applications must be studied to discern if Mg^{2+} is included.

Enthalpy Transformations for Species with a Single Form

The enthalpy is obtained via the Gibbs-Helmholtz relation, where the heat of formation at $I = 0$, pH = 0, and pMg = 0 is independent of temperature as a first approximation. The Debye-Hückel temperature dependence introduces a correction. The heat of formation is

Transformed enthalpy of formation at a specified pH and I. Workbook Gprime-Calc.xlsx or MATLAB GprimeCalc.m are helpful.

$$\Delta H'^\circ_{f,T,i}(I, pH, pMg) = \Delta H^\circ_{f,T,i}(I) - N_{H,i}\Delta H^\circ_{f,T,H^+}(I) - N_{Mg,i}\Delta H^\circ_{f,T,Mg^{2+}}(I) \qquad 18.120$$

and for each enthalpy of formation on the right side, $j = i$, H^+, and Mg^{2+}:

$$\Delta H^\circ_{f,T,j}(I) = \Delta H^\circ_{f,T,j}(I=0) + RT^2\ln(10)z_j^2(dA_\gamma/(dT))(\sqrt{I}/(1 + Ba\sqrt{I})) \qquad 18.121$$

Gibbs Energy Transformations for Pseudoisomers

For families of receptors with different numbers of ligands, the transformed Gibbs energy of formation depends on the distribution of species, which changes with T, pH_c, and pMg, but the dependence is easily represented by the binding polynomial. The Gibbs energy of formation for the apparent species takes a very simple modification relative to the Gibbs energy of formation of the completely bare receptor,

$$\Delta G'^\circ_{f,T,i}(I, pH_c, pMg) = \Delta G'^\circ_{f,T,i(1)}(I, pH_c, pMg) - RT\ln P_{bind} \qquad 18.122$$

where $G'^\circ_{f,T,i(1)}(I, pH_c, pMg)$ on the right is for the completely bare receptor as determined by Eqns. 18.117–18.119. The binding polynomial can most easily be expressed in terms of the transformed Gibbs energies of formation of each pseudoisomer relative to the most bare receptor,

$$P_{bind} = \left(1 + \exp\left(\frac{\Delta G'^\circ_{f,i(1)} - \Delta G'^\circ_{f,i(2)}}{RT}\right) + \exp\left(\frac{\Delta G'^\circ_{f,i(1)} - \Delta G'^\circ_{f,i(3)}}{RT}\right) + ...\right) \qquad 18.123$$

where all Gibbs energies are for aqueous solutions using Eqns. 18.117 - 18.119 at a specified pH_c, pMg, and ionic strength, but the designations have been omitted for brevity. Alternative forms of

Eqn. 18.123 are sometimes written, but the given form is recommended to avoid exponentials of very large or very small numbers because the differences in Gibbs energies of formation are much smaller than the values.[30,31] Note that the exponential terms in Eqn. 18.123 are equivalent to the products of the equilibrium constants as shown in Eqn. 18.90.

The pH_c, pMg, and nonideal solution effects are already incorporated into the reactions, and thus the binding polynomial in terms of the transformed equilibrium constants for phosphate would give

$$P_{bind} = \left(1 + \frac{1}{K'_{a3}} + \frac{1}{K'_{a3}K'_{a2}} + \frac{1}{K'_{a3}K'_{a2}K'_{a1}}\right)$$

$$= (1 + K'^{bind}_{a1} + K'^{bind}_{a1}K'^{bind}_{a2} + K'^{bind}_{a1}K'^{bind}_{a2}K'^{bind}_{a3}) \qquad 18.124$$

The fraction of receptor in each pseudoisomer form j in family i, can be calculated by two different methods. P_{bind} can be used directly, or we can use the transformed Gibbs energy of formation for the pseudoisomer with the transformed Gibbs energy of pseudoisomer form j,

$$r_j = \frac{1}{P_{bind}} \exp\left(\frac{\Delta G'^{\circ}_{f, i(1)} - \Delta G'^{\circ}_{f, i(j)}}{RT}\right) = \exp\left(\frac{\Delta G'^{\circ}_{f, T, i} - \Delta G'^{\circ}_{f, i(j)}}{(RT)}\right) \qquad 18.125$$

where all Gibbs energies are at the specified T, pH, and pMg. The argument of the first exp( ) is the same as used in 18.123, and the values of the first exp( ) are the same as the values for each term in Eqn. 18.124. The transformed enthalpy for the apparent species is most easily calculated by using the individual r_i values where each pseudoisomer values is transformed by Eqn. 18.120,

$$\boxed{\Delta H'^{\circ}_{f, T, i}(I, pH, pMg) = \sum_j r_j \Delta H'^{\circ}_{f, T, i(j)}(I, pH, pMg)} \qquad 18.126$$

This section has been intended as an introduction to biological thermodynamics. Readers interested in more depth will find more details in the work of Alberty[31,32,33] or Goldberg.[34] While the transformation of the standard state potentials shifts the standard state values of Gibbs energies, such that, for a reaction, $\Delta G'^{\circ}_{T, i}(I, pH_c, pMg) \neq \Delta G^{\circ}_{T, i}(I, pH_c, pMg)$, the actual driving force for a reaction at nonequilibrium conditions is the same,[35] thus, $\Delta G'_{T, i}(I, pH_c, pMg) = \Delta G_{T, i}(I, pH_c, pMg)$ as in Eqn 18.93 and in the analogous untransformed relation. Also, this means that $K'_c \neq K_c$, but the reaction is at equilibrium at the same true concentrations regardless of the transformation. This relation again emphasizes that the true driving force for the reaction is not represented by the standard state values.

30. Alberty, R.A. 2001. *J. Phys. Chem. B* 105:7865–7870.

31. Alberty, R.A. 2003. *Thermodynamics of Biochemical Reactions*. Hoboken, NJ: Wiley, p. 69. Note that updated values for ATP, ADP, AMP, and transformations with Mg are explained in ref. 32.

32. Alberty, R.A. 2003. *J. Phys. Chem. B* 107:12324–12330.

33. Alberty, R.A. 2006. *Biochemical Thermodynamics: Applications of Mathematica*. Hoboken, NJ: Wiley.

34. Goldberg R.N., Tewari, Y.B., Bhat, T.N. 2004. *Bioinformatics* 20(16):2874-2877, http://xpdb.nist.gov/enzyme_thermodynamics/ accessed 11/7/2011.

35. Iotti, S.; Sabatini, A.; Vacca, A. 2010. *J. Phys Chem B*. 114:1985–1993.

Example 18.12 Gibbs energy of formation for ATP

The Gibbs energies of formation of ATP species are available in Appendix E. Using the available Gibbs energies, calculate the apparent Gibbs energy of formation for ATP at $T = 298.15$ K, $pH_c = 7$, $pMg_c = 3$, $I = 0.25$ mol/kg, the pK', and the percentage of each species present.

Solution: The Gibbs energies for the molal standard state are $ATP^{4-} = -2768.1$, $HATP^{3-} = -2811.48$, $H_2ATP^{2-} = -2838.18$, $MgATP^{2-} = -3258.68$, $MgHATP^- = -3287.5$, and $Mg_2ATP = -3729.33$ kJ/mol. The charges are -4, -3, -2, -2, -1, and 0, respectively. The number of H's are 12, 13, 14, 12, 13, and 12, respectively. Inserting the values into GprimeCalc.xlsx, at the stated conditions, the transformed Gibbs energies of the species are $ATP^{4-} = -2291.9$, $HATP^{3-} = -2288.8$, $H_2ATP^{2-} = -2270.7$, $MgATP^{2-} = -2297.1$, $MgHATP^- = -2282.7$, and $Mg_2ATP = -2288.8$ kJ/mol. The rest of the problem must be solved with hand calculations.

The third dissociation reaction is $HATP^{3-} \rightleftarrows ATP^{4-}$,

$$\frac{1}{K'_{a3}} = \exp\left(\frac{\Delta G'^\circ}{RT}\right) = \exp\left(\frac{\Delta G'^\circ_{f(1)} - \Delta G'^\circ_{f(2)}}{RT}\right) = \exp\left(\frac{-2291.9 + 2288.8}{0.008314(298.15)}\right) = 0.2863$$

The product of the first and second is given by $H_2ATP^{2-} \rightleftarrows ATP^{4-}$,

$$\frac{1}{K'_{a3}K'_{a2}} = \exp\left(\frac{\Delta G'^\circ}{RT}\right) = \exp\left(\frac{\Delta G'^\circ_{f(1)} - \Delta G'^\circ_{f(3)}}{RT}\right) = \exp\left(\frac{-2291.9 + 2270.7}{0.008314(298.15)}\right) = 1.9E-4$$

The remaining terms for $H_nATP^{(-4+n)}$ in the binding polynomial are not important at pH = 7 because they will be even smaller than the last term calculated since the pH is far above the pK_a (review Example 18.6 on page 718 for an analogy with phosphoric acid). For the species involving Mg, defining K'_{aMg} for $MgATP^{2-} \rightleftarrows ATP^{4-}$,

$$\frac{1}{K'_{aMg}} = \exp\left(\frac{\Delta G'^\circ_{f(1)} - \Delta G'^\circ_{f(MgATP)}}{RT}\right) = \exp\left(\frac{-2291.9 + 2297.1}{0.008314(298.15)}\right) = 8.1480$$

Defining K'_{aMgH} for $MgHATP^- \rightleftarrows ATP^{4-}$,

$$\frac{1}{K'_{aMgH}} = \exp\left(\frac{\Delta G'^\circ_{f(1)} - \Delta G'^\circ_{f(MgHATP)}}{RT}\right) = \exp\left(\frac{-2291.9 + 2282.7}{0.008314(298.15)}\right) = 0.0244$$

defining K'_{aMg2} for $Mg_2ATP \rightleftarrows ATP^{4-}$,

$$\frac{1}{K'_{aMg2}} = \exp\left(\frac{\Delta G'^\circ_{f(1)} - \Delta G'^\circ_{f(Mg2ATP)}}{RT}\right) = \exp\left(\frac{-2291.9 + 2288.8}{0.008314(298.15)}\right) = 0.2863$$

The binding polynomial is $P_{bind} = 1 + 0.2863 + 2E-4 + 8.148 + 0.0244 + 0.2863 = 9.7452$.

The fraction of each pseudoisomer is given by Eqn. 18.91, where C is the total ATP concentration in all forms:

$$\frac{[ATP^{4-}]}{C} = \frac{1}{P_{bind}} = \frac{1}{9.7452} = 0.10 \qquad \frac{[HATP^{3-}]}{C} = \frac{[MgATP]}{C} = \frac{0.2863}{9.7452} = 0.03$$

Example 18.12 Gibbs energy of formation for ATP (Continued)

$$\frac{[\text{MgATP}^{2-}]}{C} = \frac{8.148}{9.7452} = 0.84,$$

The other species make up the remainder and are insignificant at pH = 7.

The Gibbs energy of formation for the apparent species is

$$\Delta G'^{\circ}_{f, \text{ATP}} = \Delta G'^{\circ}_{f, \text{ATP}(1)} - RT\ln P_{\text{bind}} = -2291.9 - 0.008314(298.15)\ln 9.7452$$

$$= -2297.5 \text{ kJ/mol}$$

Note this is the value used in Example 18.10.

18.18   COUPLED MULTIREACTION AND PHASE EQUILIBRIA

Many texts are available to facilitate more advanced study. The following example uses thermochemical data from the OBIGT database documented in footnote 13 of this chapter. The example here is implement using the extended Debye-Hückel model and ignoring pressure corrections.

Chlorination of Water

Chlorination is one method of water treatment for drinking. When Cl_2 dissolves in pure water, it undergoes reaction with water to simultaneously form the strong acid HCl and the weak acid hypochlorous acid (HClO). The reaction is sensitive to pH, and at low pH it is shifted toward molecular Cl_2. At high pH, the reaction shifts to hypochlorous acid which is an oxidizer as well as a weak acid. Chlorine bleach is prepared by stabilizing the hypochlorous acid at high pH by reacting Cl_2 with a solution of NaOH.

Consider the situation with pure water. There are three reactions to be taken into account:

$$Cl_{2(aq)} + H_2O \; \rightleftarrows \; H^+ + Cl^- + HClO_{(aq)} \qquad K_{a1} = \frac{a_{H^+}a_{Cl^-}a_{HClO_{(aq)}}}{a_{Cl_{2(aq)}}a_{H_2O}} \qquad \text{18.127}$$

$$HClO_{(aq)} \; \rightleftarrows \; H^+ + ClO^-_{(aq)} \qquad K_{a2} = \frac{a_{H^+}a_{ClO^-}}{a_{HClO_{(aq)}}} \qquad \text{18.128}$$

$$H_2O \; \rightleftarrows \; H^+ + OH^- \qquad K_w = \frac{a_{H^+}a_{OH^-}}{a_{H_2O}} \qquad \text{18.129}$$

We can also write the liquid-vapor equilibria as "reactions," noting the convention in the electrolyte literature is that the liquid phase is always the "product." This is the convention used for Henry's law constants (*cf.* Section 11.12) in Eqns. 18.130 and 18.131:

$$H_2O_{(v)} \; \rightleftarrows \; H_2O_{(l)} \qquad K = \frac{a_{H_2O}}{y_{H_2O}P} = \frac{1}{P^{sat}_{H_2O}} \qquad \text{18.130}$$

$$\text{Cl}_{2(v)} \rightleftharpoons \text{Cl}_{2(aq)} \quad K_{H(\text{Cl}_2)} = \frac{a_{\text{Cl}_2}}{y_{\text{Cl}_2} P} \qquad \text{18.131}$$

Example 18.13 Chlorine + water electrolyte solutions

Determine the concentration and species present when chlorine is in equilibrium with water at 298.15 K and 0.8 atm. Develop an approximate solution and then use extended Debye-Hückel with $Ba = 1$ $(\text{kg/mol})^{1/2}$. Thermodynamic properties from the OBIGT documented in footnote 13 of this chapter are tabulated in Table 18.4.

Table 18.4 *Thermochemical Data for the Species*

Species	ΔG^o_f(kJ/mol)	ΔH^o_f(kJ/mol)
$H_2O_{(l)}$	−237.21	−285.83
$H_2O_{(v)}$	−228.61	−241.84
H^+	0	0
OH^-	−157.30	−230.02
$Cl_{2(aq)}$	6.95	−23.39
$Cl_{2(v)}$	0	0
$HClO_{(aq)}$	−79.91	−120.92
Cl^-	−131.29	−167.08
ClO^-	−36.82	−107.11

Solution: We first work the problem assuming ideal solutions. This provides an approximate answer. Then we may use the activity coefficients to refine the answer. Using the Gibbs energies of formation, the equilibrium constants are: $pK_{a1} = 3.339$, $pK_{a2} = 7.549$, and $K_{H(\text{Cl}2)} = 0.0606$, where K_H is Henry's constant for Cl_2.

Since chlorine forms the strong acid HCl and the weak hypochlorous acid when dissolving in pure water, we expect pH < 7. Note that the weak hypochlorous acid should be almost totally protonated below pH = pK − 1 = 6.5. Since a strong acid HCl is being formed, this seems very likely. Let us proceed with that assumption. This enables us to disregard the dissociation of Eqn. 18.128 as a first approximation.

The three reaction equilibria are summarized in Eqns. 18.127–18.129. The charge balance is

$$[H^+] = [Cl^-] + [ClO^-] + [OH^-] = [Cl^-] + \ldots \qquad \text{18.132}$$

where $[ClO^-]$ is ignored because the dissociation of hypochlorous is small when the pH is small and $[OH^-]$ is ignored when pH is small. Thus, the equilibria of Eqn. 18.127 can be approximated as

$$[HClO_{(aq)}] \approx \frac{K_{a1}[Cl_{2(aq)}]a_{H_2O}}{[H^+]([H^+] - \ldots)} \approx \frac{K_{a1}[Cl_{2(aq)}]}{[H^+]^2} \qquad \text{18.133}$$

Example 18.13 Chlorine + water electrolyte solutions (Continued)

Approximate Solution:

The partial pressure for water can be estimated by first assuming that the water is almost pure. This approximation can be refined later if we find significant concentrations of chlorine species. We also use molar concentrations to approximate molalities. Using Raoult's law for water, $y_{H_2O}P = P_{H_2O}^{sat}$. From the steam tables, $y_{H_2O}P = P_{H_2O}^{sat} = 0.0317$ bar $= 0.0313$ atm, and $y_{H2O} = 0.0313/0.8 = 0.039$. Then $y_{Cl_2}P = 0.8 - 0.0313 = 0.7687$ atm, $y_{Cl_2} = 1 - 0.039 = 0.961$. Using Henry's law coefficient (K_H) for Cl_2 at 298.15 K, the concentration of $Cl_{2(aq)}$ is (independent of pH):

$$[Cl_{2(aq)}] = K_H(y_{Cl_2}P) = (0.0606)(0.7687) = 0.0484 \text{ mol/L} \qquad 18.134$$

The concentration $[HOCl_{(aq)}] = 10^{-3.339}(0.0484)/[H^+]^2$ (Eqn. 18.133) at small values of pH is plotted in a Sillèn diagram. The weak acid dissociation of hypochlorous acid $[OCl^-]$ is to be calculated from Eqn. 18.31 using the concentration of $[HOCl]$ as a function of pH. As expected, the dissociation is small at low pH.

The weak acid curve in Fig. 18.14 is much different from curves in previous examples because, in this case, the overall concentration of weak acid is changing rapidly with pH. Now consider the material balance associated with Eqns. 18.127 and 18.128. Since Eqn. 18.128 does not occur to a significant extent, to a good approximation by the stoichiometry of Eqn. 18.127 $[H^+] = [Cl^-] = [HOCl]$. This occurs at the intersection shown by the dotted lines. The approximate solution is pH = 1.55, $[H^+] = [Cl^-] = [HOCl] = 10^{-1.55} = 0.0282$ mol/L. Note on the diagram that $[OCl-] = 10^{-7.5} = 3.2E-8$. Now, we can use these as initial guesses in a more rigorous answer.

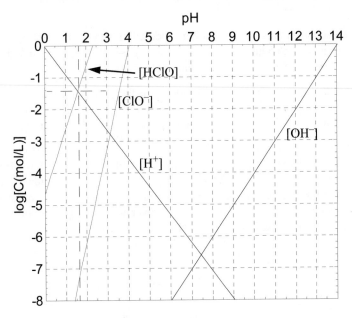

Figure 18.14 *Determination of equilibria for the chlorine system.*

Example 18.13   Chlorine + water electrolyte solutions (Continued)

Calculation with Activity Coefficients:

Thermodynamic properties for the components are tabulated below and in the spreadsheet: CL2H2O.xlsx. Note that the data tabulated below include values for Cl_2 and H_2O in both the vapor and aqueous phases. The Gibbs energies are used to calculate the VLE distribution coefficients as a "reaction."

To solve the nine equations (five equilibria, three atom balances, and one charge balance) simultaneously, we must identify nine unknowns. The nine unknowns selected here are the species listed in Table 18.4: the liquid moles of H_2O, Cl_2, $HClO$, H^+, Cl^-, ClO^-, and OH^-, and the vapor moles of $H_2O_{(v)}$ and $Cl_{2(v)}$. The basis is 1 liter of liquid water ($n^i_{H_2O} = 55.51$ moles) and $n^i_{Cl_2} = 0.9$ moles initially.

The detailed calculation are handled as a reactive flash. Three atom balances must be satisfied–H, O, Cl, along with the charge balance. The atom balances and charge balance are shown in Table 18.5 for the basis of 1 liter of liquid water and 0.9 moles of Cl_2. The compositions for iteration are summarized in Table 18.6.

Table 18.5 *Atom Balance and Electroneutrality Constraints*

	Initial	Final	Final–Initial
H-bal	111.02	111.02	0.00E+00
Cl-bal	1.8	1.8	0.00E+00
O-bal	55.51	55.51	0.00E+00
e-bal	0	0	0.00E+00

The results from the approximate ideal solution calculation above are used as initial guesses. Excel Solver is used to adjust the moles of each species (in the second and sixth columns of Table 18.6) until all equations are simultaneously satisfied.

Note from Table 18.6 that the $\gamma^\square$ for all the ionic species is the same. This occurs because the Debye-Hückel model is too simple to make distinctions as long as all species have the same valence. The activity coefficients for Cl_2 and $HClO$ are assumed to be unity.

The calculations summarized in Table 18.6 show that the chlorine solubility is enhanced beyond what might be predicted from the Henry's law constant alone due to formation of $HClO$ and Cl^- in solution. The Cl in $HClO$ and Cl^- together is about two-thirds of the Cl atoms in $Cl_{2(aq)}$ at $P = 0.8$ atm. Open the spreadsheet Solver to see how the constraints were implemented. Compare with the approximate answer to see that the approximate answer is pretty close. For example, [HClO] = 0.0312 versus 0.0282.

Example 18.13   Chlorine + water electrolyte solutions (Continued)

Table 18.6 *Electrolyte Component Mole Numbers and Activities at the Converged Composition*

Species	Liquid Moles	Molality	$\gamma^{\square}$	Species	Vapor Moles	y
H_2O	55.44564	----	1.000060649	H_2O	0.03322	3.88E-02
$Cl_{2(aq)}$	4.65E-02	0.046591564	1	Cl_2	0.82233	9.61E-01
HClO	3.11E-02	0.031168318	1	*sum*	0.85555	1.00
H^+	3.11E-02	0.031168399	0.838440619			
OH^-	4.55E-13	4.55732E-13	0.838440619	$y_{Cl2}P = 0.77$		
ClO^-	4.01E-08	4.01834E-08	0.838440619			
Cl^-	3.11E-02	0.031168359	0.838440619			
tot moles	55.5856					

Intermediate Calculations	$I = 0.03117$
	$x_{H2O} = 0.99748$ $pH = -\log_{10}((0.0312)(0.838))$
	$a_{H2O} = 0.9975$ $= 1.58$

18.19   MEAN IONIC ACTIVITY COEFFICIENTS

Mean ionic activity coefficients are often used for electrolytes modeling in the literature. The mean ionic activity coefficients provide an alternative method to express the activity of the apparent electrolyte species. This section provides the background to relate those activity coefficients to the ion activity coefficients. The chemical potential of the apparent electrolyte species is the same as the undissociated electrolyte species as shown in Supplement Section 18.23, however only the undissociated chemical potential is used here to keep the equations shorter.

$$\mu_u = v_+\mu_+ + v_-\mu_- \qquad\qquad 18.135$$

If we insert Eqns. 18.99 and 18.100,

$$\mu_u = v_+(\mu_+^{\square} + RT\ln\gamma_+ m_+) + v_-(\mu_-^{\square} + RT\ln\gamma_- m_-) = (v_+\mu_+^{\square} + v_-\mu_-^{\square}) + RT\ln[(m_+^{v_+}m_-^{v_-})(\gamma_+^{v_+}\gamma_-^{v_-})] \qquad 18.136$$

Looking at the last terms in parentheses, we can define,

$$\mu_{\pm}^{\square} \equiv (v_+\mu_+^{\square} + v_-\mu_-^{\square}) \qquad\qquad 18.137$$

The molality and activity coefficients are averaged by taking the geometric mean. In the literature, these are commonly called the mean, and the clarification as the geometric mean is commonly omitted. The **mean molality** is

$$m_{\pm} \equiv (m_+^{v_+}m_-^{v_-})^{1/v} \qquad\qquad 18.138$$

where $\nu \equiv (\nu_+ + \nu_-)$. The **mean ionic activity coefficient** is found from the geometric mean of the ion activity coefficients:

$$\gamma_\pm \equiv (\gamma_+^{\nu_+}\gamma_-^{\nu_-})^{1/\nu} \qquad 18.139$$

This results in the following relation for the chemical potential of the undissociated species

$$\mu_u = \mu_\pm^\square + \nu\ln(m_\pm\gamma_\pm) \text{ and } a_\pm = \gamma_\pm^\nu m_\pm^\nu \text{ and } a_u = a_\pm^\nu \qquad 18.140$$

Note that in the formula for chemical potential, the variable ν appears before the ln term, unlike the similar expression for ions or nonelectrolyte species. Therefore, *the mean ionic activity is not the same as the activity of the undissociated (or apparent) activity.* There is no new information in these equations. They are simply an alternative method of expressing the activity coefficients, molalities, and chemical potentials. The mean molal activity coefficient is

$$\log_{10}\gamma_\pm = \frac{-A_\gamma|z_+z_-|\sqrt{I}}{1 + Ba\sqrt{I}} \qquad 18.141$$

The final equation requires a proof that $\nu_+z_+^2 + \nu_-z_-^2$ is related to $|z_+z_-|$. To see this, start with the charge balance, $\nu_+z_+ + \nu_-z_- = 0$. Obtain one equation by multiplying by z_+ and another by multiplying by z_-. Add them and rearrange:

$$\nu_+z_+^2 + \nu_-z_-^2 = -(\nu_+ + \nu_-)z_+z_- = -\nu(z_+z_-) = \nu|z_+z_-| \qquad 18.142$$

The extended Debye-Hückel model is valid for 1-1 electrolytes up to about 0.1 molal, and to lower concentrations for species with multiple charges: The extended Debye-Hückel is compared with experimental mean ionic activity coefficients for NaCl in Fig. 18.15. Note that the experimental

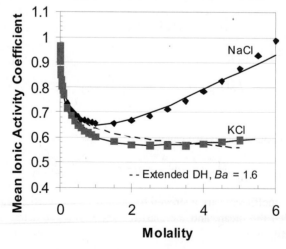

Figure 18.15 *Mean ionic activity coefficient for NaCl, KCl and those predicted by the extended Debye-Hückel (DH) model at 298K. Dashed line is the extended Debye-Hückel, Solid lines are the unsymmetric eNRTL using default parameters in ASPEN Plus ver. 7.1. Experimental activity coefficients are from Hamer, W.J.; Wu, Y.-C., 1972. J. Phys. Chem. Ref. Data, 1:1047.*

activity coefficients for NaCl are nearly 1 near 6 m. This happens to be about the solubility limit at room temperature, but the solution is quite nonideal at lower concentrations.

The osmotic pressure can be manipulated using the Gibbs-Duhem equation to obtain the mean ionic activity coefficient. The derivation is beyond the scope intended here, but the equation is

$$\ln \gamma_{\pm} = \Phi - 1 + \int_0^m (\Phi - 1) \frac{dm^{1/2}}{m^{1/2}} \qquad 18.143$$

The use of the square root is a necessary mathematical manipulation. The integral may be done numerically. One difficulty is that the experimental data must extend to low concentrations.

18.20 EXTENDING ACTIVITY CALCULATIONS TO HIGH CONCENTRATIONS

This chapter has served as an introduction to electrolyte models, and the extended Debye-Hückel leaves much to be desired in its limitations to concentrations lower than 0.1 m. However, the model has been used as an introduction, and those who work with electrolytes can find more models in the literature. In the older literature, the model was primarily improved by making modifications to the Debye-Hückel approximations. For example, Bromley and Davies add to the activity coefficient a term CI, where C is a parameter and I is ionic strength. Fig. 18.15 illustrates that the parameter C must be system-dependent to represent the data. One suggestion is that the ionic strength modifies the dielectric constant of the medium. Others propose that the ions begin to interact with each other in a way that the Debye-Hückel model cannot capture. Molecular simulations are relatively complicated in the presence of long-range electrostatic interactions, delaying the conclusive resolution of such arguments. Since the mid-1980s significant success has been achieved by combining various versions of the Debye-Hückel model with activity models such as NRTL or UNIQUAC.[36] The Debye-Hückel model is considered to represent the "long range" electrostatic interactions, and the conventional activity models are considered to represent the "short range" physical interactions. Often, the short-range model parameters are lumped to minimize the number of parameters to be adjusted. Plotted in Fig. 18.15 are the activity coefficients calculated with ASPEN Plus using the unsymmetric electrolyte-NRTL (eNRTL) model. The ASPEN electrolyte wizard was used to set up the dissociations and pull parameters from the database. Owing to the importance of electrolytes in industrial processes and corrosion management, and the complexities of correct modeling, companies such as OLI Systems, Inc., specialize in electrolyte modeling.

18.21 SUMMARY

This chapter began with a review of acid-base behavior to stress the importance of pH on equilibrium. Compounds are in the acid form below the $pK_{a,A}$ and in the base form above. Techniques including Sillèn's graphical method were provided to determine solution pH values and species distributions at various concentrations. We explained the origin of charges on biological molecules and why the charges change with pH, as well as the concept of zwitterions. Applications such as solubility, osmotic coefficients, and isoelectric point were developed.

---

36. The electrolyte NRTL model, available in the ASPEN Plus simulator, is a common example. Originally developed as an unsymmetric model, a symmetric version is now available. Song, Y.; Chen, C.-C. 2009. "Symmetric Electrolyte Nonrandom Two-Liquid Activity Coefficient Model." *Ind. Eng. Chem. Res.* 48:7788–7797.

Concepts of redox reactions were developed, relating the voltage to the Gibbs energy of reaction. Procedures were given to determine oxidation states, degree of reduction for molecules, and voltages in cells. The concept of oxidation and reduction in biological systems was introduced in the context of a biological fuel cell.

Binding polynomials were introduced as a method of representing simultaneous equilibria of families of pseudoisomers. The concepts of transformed Gibbs energies were introduced for biological systems buffered in pH or pMg.

Then solution nonidealities were introduced using the extended Debye-Hückel model. We finished the chapter by providing an example calculation of ATP distribution in nonideal solutions, and an example that couples phase equilibria with electrolyte equilibria. The later example demonstrates that for dilute solutions the graphical technique and simple arguments comparing the pH with the $pK_{a,A}$ values provides rapid estimates that are valuable for converging to more precise values. Because pH and species concentrations vary over many orders of magnitude, the approximate methods are important to use first, and in many cases they are adequate for approximate engineering work. Ultimately, activity coefficient modeling is important for accurate calculations. The chapter concludes with some supplemental sections that are extremely valuable for conversions of the units used in the literature.

Important Equations

The most important equations of this chapter are the material balance, reaction equilibria, and electroneutrality relations. Unfortunately, these are different for every electrolyte system, so there is not much point in listing them the way we usually do in the chapter summary. A key step in using Sillèn's method is to use the acid form of dissociations for weak electrolytes,

$$[HA] = C_A[H^+]/([H^+] + K_{a,A})$$ useful for undissociated acid   18.30

$$[A^-] = K_{a,A}C_A/([H^+] + K_{a,A})$$ useful for conjugate base   18.31

A new equation is the electroneutrality constraint. It is not a surprising equation, but it can cause difficulty because some species may be present in very small quantities that make very big differences–pH, for instance. Solving for these small quantities often requires rearranging the equations into a proton condition to avoid the precision problems that come with adding small numbers to large numbers.

In Section 18.11, the Nernst equation is important to relate voltage to standard potentials and actual concentrations, and the number of electrons transferred:

$$E = E° - \frac{RT}{n_e F}\ln\prod a_i^{v_i} = E° - \frac{0.05916}{n_e}\log\prod a_i^{v_i}$$ Nernst equation   18.86

In Section 18.12 the Gibbs energy was transformed to use apparent concentrations. All the pH, pMg, and solution nonidealities were transferred to the Gibbs energies of reaction and the equilibrium constants:

$$\Delta G' \;=\; \Delta G'^{\circ} + RT\ln \prod_i [i]^{v_i} \qquad \Delta G'^{\circ} \;=\; -RT\ln K_c'$$

18.93

Besides the primary equations, the extended Debye-Hückel equation (Eqn. 18.107) is introduced in this chapter to account for nonideal behavior of ionic species. It is best limited to concentrations of 0.1 molal or less, but it conveys the concept that electrolyte solutions may deviate from ideality just as nonelectrolytes do. Going beyond Eqn. 18.107 would generally involve developing expertise beyond the introductory level.

Although the notation and reference states are obscure and frustrating, the implications are impressive. Salting in and salting out, protonation versus pH, osmosis, buffering, and leveling are just a few examples of implications that play significant roles in commonly encountered chemical systems, especially biological systems and corrosive environments. All the new jargon may seem overwhelming at first, but it can be assimilated if you only remember the three P's: practice, practice, practice.

18.22   SUPPLEMENT 1: INTERCONVERSION OF CONCENTRATION SCALES

Throughout this chapter, subscript s indicates solvent, and $M_{w,i}$ represents molecular wt in (g/mol).

$$\text{Molality: } m_i \;=\; n_i/(0.001 n_s M_{w,s}) \text{ mol/(kg solvent)}$$

18.144

$$\text{Concentration: } n_i/\underline{V} \text{ mol/(liter solution)}$$

18.145

Volume related to mass density (ρ in g/L):

$$\underline{V} \;=\; \left(n_s M_{w,s} + \sum_{\text{electrolytes}} n_i M_{w,i} \right) / \rho \quad \text{(liter solution)}$$

18.146

relation of mole fraction to molality:

$$x_i \;=\; \frac{n_i}{\sum\limits_i n_i} \;=\; \frac{n_i}{\sum\limits_i n_i} \left(\frac{n_s M_{w,s}}{n_s M_{w,s}} \right) \;=\; \frac{n_i}{(n_s M_{w,s})} \frac{n_s}{\sum\limits_i n_i} M_{w,s} \;=\; 0.001 m_i x_s M_{w,s}$$

18.147

The relation for molality leads to another commonly used substitution for x_s,

$$x_s \;=\; 1 - \sum_{\text{electrolytes}} x_i \;=\; 1 - \sum_{\text{electrolytes}} 0.001 m_i x_s M_{w,s} \;=\; \frac{1}{1 + 0.001 M_{w,s} \sum\limits_{\text{electrolytes}} m_i}$$

18.148

18.23   SUPPLEMENT 2: RELATION OF APPARENT CHEMICAL POTENTIAL TO SPECIES POTENTIALS

To understand the origin of the models for the mean activity coefficient, some discussion of the chemical potentials is necessary. Many of the concepts are extensions of the methods used for reaction engineering. For example, consider the general case of an electrolyte dissociating in solvent. When an electrolyte dissociates, the material balance gives the molality of positive and negative charges in solution,

$$n_{pos} = v_+(n_i - n_u) \qquad\qquad 18.149$$

$$n_{neg} = v_-(n_i - n_u) \qquad\qquad 18.150$$

where the subscript u represents the amount of electrolyte that is un-ionized. The change in Gibbs energy at fixed temperature and pressure is given by the changes in composition and the chemical potentials,

$$d\underline{G} = \mu_u dn_u + v_+\mu_+(dn_i - dn_u) + v_-\mu_-(dn_i - dn_u) + \mu_w dn_w \qquad 18.151$$

which rearranges to

$$d\underline{G} = (\mu_u - v_+\mu_+ - v_-\mu_-)dn_u + (v_+\mu_+ + v_-\mu_-)dn_i + \mu_w dn_w \qquad 18.152$$

The quantities μ_+ and μ_- are analogous to the partial molar Gibbs energies of other components and the subscript w represents water or solvent. Since the positive and negative charges cannot change independently in a physical mixture as required by the rigorous definition of the partial molar quantity where all but one species is constrained when the derivative is evaluated, they are nonphysical, but they can be calculated by theory. Recall that the quantity dn_i relates to changes in the apparent composition. As a solution of fixed apparent concentration equilibrates (e.g., $dn_i = 0$), the un-ionized concentration changes until Gibbs energy is minimized and $d\underline{G} = 0$. Thus, we conclude that at the equilibrium ionization the first term in parentheses is zero. Thus,

$$\mu_u = v_+\mu_+ + v_-\mu_- \qquad\qquad 18.153$$

This relationship between the "products" and "reactants" of the ionization is analogous to the relations developed for molecular reacting systems. This equality can be inserted into the second term, at equilibrium ionization resulting in

$$d\underline{G} = \mu_u dn_i + \mu_w dn_w \qquad\qquad 18.154$$

On an apparent basis, we also must satisfy

$$d\underline{G} = \mu_i dn_i + \mu_w dn_w \qquad\qquad 18.155$$

where μ_i is the apparent chemical potential. *By comparison of the last two equations we conclude that the apparent chemical potential must equal the chemical potential of the un-ionized species,* which also must equal the weighted sum of the chemical potentials of the ions:

$$\mu_i = \mu_u = v_+\mu_+ + v_-\mu_-$$ 18.156

Thus, the approach for developing a model for the apparent chemical potential and apparent activity coefficient is based on developing models for the ions and then using the weighted sum.

18.24 SUPPLEMENT 3: STANDARD STATES

An important principle of the following discussion is that the chemical potential should be a property of the state of the system. All models should result in an identical value for the chemical potential at the same state. The standard state provides a convenient reference condition, but is slightly different from a reference state because it is at the same temperature as the system.[37]

The typical convention for nonelectrolytes uses mole fractions and the **Lewis-Randall standard state** μ_i^o:

$$\mu_i = \mu_i^o + RT\ln(x_i\gamma_i)$$ 18.157

An alternative convention is related to the **Henry's law standard state** μ_i^* and the activity coefficient is known as the **rational activity coefficient.**[38]

$$\mu_i = \mu_i^* + RT\ln(x_i\gamma_i^*)$$ 18.158

The activity coefficients on the two scales are related.

$$\gamma_i^* = \gamma_i/\gamma_i^\infty$$ 18.159

Inserting the activity coefficient relation into Eqn. 18.158 results in

$$\mu_i^* = \mu_i^o + RT\ln\gamma_i^\infty$$ 18.160

Consider Eqn. 18.158. Inserting Eqn. 18.147 to replace the mole fraction,

$$\mu_i = \mu_i^* + RT\ln(0.001m_ix_sM_{w,s}\gamma_i^*)$$ 18.161

Because molality is not dimensionless, but activity is dimensionless, we must introduce some manipulations. We wish to introduce a molal activity coefficient, $\gamma_i^\square$, to use with molal concentrations. The convention is to set the standard state as a hypothetical ideal solution, $\gamma_i^\square = 1$, at unit molality, $m_o = 1$ mol/(kg solvent). Introducing the standard state concentration (twice):

$$\mu_i = \mu_i^* + RT\ln(0.001 M_{w,s}m_o) + RT\ln(m_ix_s\gamma_i^*/m_o)$$ 18.162

The unit value of m_o is traditionally omitted and thus "transparent." We define the molal activity coefficient:

---

37. personal communication with Professor Kaj Thomsen of Department of Chemical and Biochemical Engineering, Technical University of Denmark is acknowledged for development of material in this section.
38. "Rational" means based on mole fraction.

$$\boxed{\gamma_i^{\square} \equiv x_s \gamma_i^* = x_s \gamma_i / \gamma_i^{\infty}}$$

18.163

Note that Eqn. 18.147 can be reinserted into Eqn. 18.163 to eliminate x_s, if desired. Inserting Eqn. 18.163 into Eqn. 18.162,

$$\mu_i = \mu_i^* + RT\ln(0.001 M_{w,s} m_o) + RT\ln(m_i \gamma_i^{\square}/m_o)$$

18.164

We can see that the standard state reference potential must be given by the first two terms on the right-hand side of the equation:

$$\mu_i^{\square} = \mu_i^* + RT\ln(0.001 M_{w,s} m_o) = \mu_i^o + RT\ln(0.001 \gamma_i^{\infty} M_w m_o)$$

18.165

Substituting Eqn. 18.165 into Eqn. 18.164 results in the **molal standard state and molal activity coefficient:**

$$\boxed{\mu_i = \mu_i^{\square} + RT\ln\left(\frac{m_i \gamma_i^{\square}}{m_o}\right)}$$

18.166

The activities corresponding to the standard states are thus

$$a_i = x_i \gamma_i \quad \text{Lewis-Randall activity}$$

18.167

$$a_i^* = x_i \gamma_i^* \quad \text{Henry's law (rational) activity}$$

18.168

$$\boxed{a_i^{\square} = \gamma_i^{\square}(m_i/m_o) = \gamma_i^{\square} m_i} \quad \text{molal activity } (m_o = 1\text{m})$$

18.169

where the *activities from the different scales are not equal* at a given concentration because of the difference in standard states. Combining Eqns. 18.162 and 18.147,

$$m_i \gamma_i^{\square} = m_o x_i \gamma_i^*(1000/M_{w,s}) \quad (m_o = 1\text{ m})$$

18.170

Finally, we note that the value of m_o is dropped from all the final expressions in application, based on the assertion that its value is 1 molal by the definition of the standard state. This is the basis for the equations presented in Sections 18.4 and 18.13.

18.25   SUPPLEMENT 4: CONVERSION OF EQUILIBRIUM CONSTANTS

Equilibrium constants in electrolyte literature are often presented on the molal scale. For clarity in this section, we will use $K_{a,m}$ to denote the molality equilibrium constant and K_a to denote the rational (Henry's law scale) using mole fractions. Recall that the solvent (usually water) is on the Lewis-Randall scale. Using the molality scale for the electrolytes,

$$K_{a,m} = (x_s \gamma_s)^{\nu_s} \prod_{i \neq s} (m_i \gamma_i^{\square})^{\nu_i}$$

18.171

On the rational (Henry's law) mole fraction scale, we have

$$K_a = (x_s \gamma_s)^{\nu_s} \prod_{i \neq s} (x_i \gamma_i^*)^{\nu_i}$$

18.172

To convert, consider the ln of each equation and take the difference. Inserting Eqn. 18.170,

$$\ln K_{a,m} = \ln K_a + \sum_{i \neq s} \nu_i \ln \left(\frac{m_i \gamma_i^\square}{x_i \gamma_i^*} \right) = \ln K_a + \ln \left(\frac{1000}{M_{w,s}} \right) \sum_{i \neq s} \nu_i$$

18.173

For a 1-1 electrolyte such as NaCl, $\sum_{i \neq s} \nu_i = 1$, thus $\ln K_{a,m} = \ln K_a + \ln \left(\dfrac{1000}{M_{w,s}} \right)$. If a K_a is

desired on the Lewis-Randall scale, similar conversions can be done using infinite dilution activity coefficients.

18.26   PRACTICE PROBLEMS

P18.1 (a) Compute the freezing point depression for an aqueous solutions that is 3 wt% NaCl.
  (b) Compute the boiling point elevation for an aqueous solutions that is 3 wt% NaCl.
  (c) Compute the osmotic pressure for an aqueous solutions that is 3 wt% NaCl.

18.27   HOMEWORK PROBLEMS

18.1 Calcium chloride is used occasionally as an alternative to sodium chloride for de-icing walkways. It is rumored to maintain puddles even a day or so after all evidence of sodium chloride has disappeared.

  (a) Compute the freezing point depression for aqueous solutions that are 5 wt% $CaCl_2$ and NaCl.
  (b) Compute the boiling point elevation for aqueous solutions that are 5 wt% $CaCl_2$ and NaCl at 760 mmHg.
  (c) Compute the osmotic pressure for aqueous solutions that are 5 wt% $CaCl_2$ and NaCl at 25 °C.

18.2 Ammonia is a weak base, as indicated by the $pK_{a,A}$ and $pK_{a,B}$ values in Table 18.2. Determine the percentage of NH_3 dissociated at pH 7 and pH 1.5 when the apparent amount of NH_3 in aqueous solution is 0.15 m. Assume ideal solutions.

18.3 Sodium fluoride, NaF, is dissolved in water at an apparent concentration of $C_B = 10^{-3}$ mol/ L. Construct a Sillèn diagram and estimate the pH. Refer to the $pK_{a,A}$ and $pK_{a,B}$ values in Table 18.2.

18.4 A solution of $NaHCO_3$ and HCl is prepared such that the total carbon concentration is 1E-3 M and the total Cl concentration is 2E-3 M. Calculate the pH and concentrations of species present. Assume that the pressure is sufficiently that any evolved CO_2 remains in solution. Estimate the partial pressure of the CO_2 by
  (a) using Henry's Law.
  (b) assuming the MAB model.

18.5 Plot the "apparent molality" of Cl_2 in solution against the partial pressure of Cl_2. The apparent molality is the sum of all Cl species in solution (Cl_2 counts twice) divided by 2 (to put it on a Cl_2 basis). Compare your plot to the experimental data of Whitney and Vivian (1941).[39]

18.6 Model a soft drink as a solution of water with CO_2 dissolved at 298.15 K. In this way we ignore the sugar, flavor, and color. The Henry's law constant for CO_2 at 298.15 K is 0.035 (mol/kg-bar).

   (a) What pH and composition exist when the vapor phase is 3.5 bar absolute at room temperature ignoring O_2 or N_2 present? This approximates conditions in the unopened container.
   (b) After the soft drink is opened, and the liquid equilibrates with atmosphere, what pH and composition exist when the CO_2 vapor mole fraction is y_{CO_2} = 0.0003 (the normal ambient value) and the pressure is 1 bar?

18.7 Sodium bicarbonate, $NaHCO_3$, commonly known as baking soda, is dissolved in water at 10^{-2} m at 298.15 K. Assume ideal solutions.

   (a) Determine the pH and the dominant species concentrations. For this part of the problem, ignore the potential loss of CO_2 escaping from the solution as vapor.
   (b) Now evaluate whether CO_2 may have a propensity to come out of solution at the conditions determined in (a) at 1 bar total pressure. y_{CO_2} = 0.0003 is the normal ambient value.

18.8 Sodium carbonate is mixed into a solution of acetic acid and the container is rapidly closed before the container components react. The amount of sodium carbonate is such that the total sodium concentration is 0.05 m and the total acetate concentration is also 0.05 m. When the mixture equilibrates, the partial pressure of CO_2 over the solution is measured to be 0.5 bar. Determine the pH and concentrations of the acetate species. The Henry's law constant for CO_2 at 298.15 K is 0.035 (mol/kg-bar).

18.9 Thermodynamic data for Gibbs energy of formation is shown below (kJ/mol for molal standard states) at 298.15 K. A saturated solution of NaCl is approximately an ideal solution.

KCl(s)	NaCl(s)	Cl^-(aq)	K^+(aq)	Na^+(aq)
−408.92	−384.12	−131.29	−282.46	−261.88

   (a) Use the Gibbs energy of formation to determine the solubility of NaCl in molality at 298.15 K. Treat the equilibrium between the solids and ions as a K_{sp}.
   (b) Determine the NaCl solubility (molal) when the concentration of KCl is 1 m.
   (c) Prove that the solution in part (b) is not saturated with KCl. (An ideal solution is actually a poor approximation for a saturated solution of KCl, but provide the proof based on an ideal solution.)

18.10 Suppose 0.1 mol of CO_2 were mixed with 0.9 mol of Cl_2 and 1 liter of water. What would be the concentrations of the aqueous species and the mole fractions in the vapor phase at 0.8 atm in that case?

---

39. 1941. *Ind. Eng. Chem.* 33:741.

18.11 Corrosion resistant alloys (such as nickel alloys and stainless steels) can be susceptible to crevice corrosion in solutions where no corrosion is observed in the bulk solution[40]. For example, nickel base alloys are immune to corrosion in seawater; however, in areas where two pieces of this alloy are joined (typically by a flange and an o-ring) crevice corrosion may be observed. This phenomenon occurs as the result of two conditions, restricted mass transport and water hydrolysis which both act to make the solution inside a crevice more aggressive. Water hydrolysis occurs when metal cations react with water to form acid (H^+):

$$xM^{z+} + yH_2O \rightleftarrows (M_x(OH)_y^{(xy-y)+})_{(s)} + yH^+ \qquad 18.174$$

In a bulk solution diffusion, convection and migration transport the acid away from the surface and no damaging effects are observed. However, the restricted mass transport inside a crevice results in accumulation of metal ions under the crevice former and acidification. As a result, the alloy can be exposed to a very aggressive environment. The pH inside the crevice can be calculated from knowledge of empirically determined concentration quotients (Q_{xy}) where:

$$Q_{xy} = \frac{[H^+]^y}{[M^{z+}]^x} \qquad 18.175$$

You will note that Q_{xy} is similar to K_a in form; however, here we use concentrations and not activities. Q_{xy} is another way to express $K = K_a/K_\gamma$ as given in then end of Section 18.8. As a practical example of this phenomenon, consider austenitic stainless steels which are generally composed of Fe, Ni, and Cr. Corrosion results in the formation of metal cations in solution, the most aggressive cation for stainless steel being Cr^{+3}.

(a) Given that $\log(Q_{13}) = -4.6$ for Cr^{+3} at a temperature and ionic strength of interest, write the hydrolysis reaction (Eqn. 18.174). Then, solve the corresponding concentration quotient (Eqn. 18.175) to obtain a relation between pH and the concentration of $[Cr^{+3}]$.
(b) Make a table of the crevice concentrations that result in pH = {6, 4, 2}.
(c) Explain why the concentration of $M_x(OH)_y^{(xy-y)+}$ does not appear in Eqn. 18.175.
(d) For Fe^{2+}, $\log Q_{11} = -9.5$, and for Ni^{2+}, $\log Q_{11} = -10.5$ and the hydrolyzed ion is soluble for each. Repeat (a) for these ions but relate pH to the ratio of ion concentrations.

18.12 Ruthenium (Ru) is a strong oxidation catalyst for organic compounds typically in the form $RuO_{4(aq)}$ represented as $H_2RuO_{5(aq)}$, but it is a stoichiometric catalyst because it is reduced during the oxidation of the organic species. Ru species also undergo redox reactions with water and dissolved oxygen. Assume ideal solutions.

(a) Show that $RuO_{4(s)}$ is not stable in contact with an air saturated solution of water and that it will revert to $RuO_{2(s)}$, independent of pH. (Hint: Water will not appear explicitly in the final reaction.)
(b) Show that the solubility of $H_2RuO_{5(aq)}$ is independent of pH in an aqueous solution saturated with air at 1 bar in contact with $RuO_{2(s)}$. Calculate the concentration of $H_2RuO_{5(aq)}$ that would exist at equilibrium.
(c) Consider the equilibrium of $RuO_4^-{}_{(aq)}$ and $H_2RuO_{5(aq)}$ in the presence of $RuO_{2(s)}$. Determine the coefficients for the equation $\log[H_2RuO_5]^a/[RuO_4^-] = b + c(pH)$.
(d) Consider the equilibrium of $RuO_4^-{}_{(aq)}$ and $RuO_4^{2-}{}_{(aq)}$ in the presence of $RuO_{2(s)}$. Determine the coefficients for the equation $\log[RuO_4^{2-}]^a/[RuO_4^-] = b + c(pH)$.

---

40. Problem suggested by R. S. Lillard, University of Akron.

18.13 (a) Rank the following molecules in order of increasing oxidation of carbon and give the oxidation state of C for each: CO2, -CHO(aldehyde), -COOH(carboxylic acid), -CO-(ketone), -CH2OH(alcohol), -CH2-, -CH3, CH4.

(b) Rank the following C5 molecules in order of decreasing degree of reduction, pentane, valeric acid, 2-pentanone, propyl-ethyl ether, 1-pentanol, 2-methyl butanol, and 1-pentanal.

18.14 The human body processes ethanol by oxidizing it to acetaldehyde via the NAD_{ox}/NAD_{red} dehydrogenase redox reaction. The reaction is

$$NAD_{ox} + ethanol \rightleftarrows NAD_{red} + acetaldehyde$$

The values for properties in the order the species appear in the reaction are

$$\Delta G'^{\circ}_{f, 310.15}(pH_c=7.4, I=0.25) = \{1163.9, 91.45, 1231.0, 43.2\}\,kJ/mol$$
$$\Delta H'^{\circ}_{f, 310.15}(pH_c=7.4, I=0.25) = \{-11.9, -291.2, -42.9, -214.1\}\,kJ/mol$$

(a) Determine the magnitude of the heat of reaction under the stated conditions.

(b) Calculate the equilibrium constant. If we assume the ratio of the two forms of NAD are near unity, what is implied about the ratio of acetaldehyde:ethanol? What is the importance of sign of the Gibbs energy for the subsequent oxidation of acetaldehyde to acetic acid?[41]

18.15 The first step in biological glycolysis (the catabolic reaction for glucose consumption) involves addition of a phosphate to create glucose 6-phosphate^{2-}. If the reaction were to occur in aqueous solution "chemically" (as compared to "biochemically'"), it would be written

$$glucose_{(aq)} + P_{i(aq)} \rightleftarrows glucose\ 6\text{-}phosphate^{2-}_{(aq)} + H_2O_{(l)} \qquad 18.176$$

However, in a biological solution, ATP and ADP are carriers of phosphate. Another reaction in solution is the hydrolysis of ATP to ADP:

$$ATP + H_2O_{(l)} \rightleftarrows ADP + P_{i(aq)} \qquad 18.177$$

The transformed values of the Gibbs energies of formation and enthalpy of formation (kJ/mol) at 298.15K, pH_c 7.0, I 0.25m, pMg 3.0 are tabulated below along with the physiological concentrations.

Species	Glucose	P_i	Gluc 6	H_2O	ATP	ADP
$\Delta G'^{\circ}_f$	−427	−1060	−1319	−156	−2298	−1426
$\Delta H'^{\circ}_f$	−1267	−1299	−2279	−286	−3605	−2622
C (M)	0.005	0.001	0.000083	56	0.00185	0.00014

(a) Evaluate the standard state Gibbs energy and enthalpy for Eqn. 18.176 and $\Delta G'$ under the actual concentrations.

(b) Repeat part (a) for Eqn. 18.177.

---

41. Acetic acid can be metabolized to CO_2 and water. Ethylene glycol is toxic because the oxidation products are glycolic acid and oxalic acid. Isopropanol is toxic because it produces acetone. Methanol is toxic because it is metabolized to formic acid.

(c) Biological glycolysis works by coupling the two reactions. Write the overall reaction and evaluate $\Delta G'$ under physiological conditions. The reactions can be added because the P_i and H_2O concentrations are constant, independent of these two reactions.

18.16 When we discussed H_3PO_4 in Section 18.9, we developed a recurring relation for the dissociation in Eqn. 18.50. Later we gave with verbal argument a binding polynomial in Eqn. 18.90.

(a) Write the series of binding reactions for PO_4^{3-} and derive the Eqn. 18.90.
(b) Create a plot of $<i>$ vs. pH for PO_4^{3-}.

18.17 Write a binding polynomial for CO_2 in aqueous systems and determine the transformed standard state Gibbs energy of total CO_2 at $pH_c = 7$, $I = 0.25$ m, and $T = 25$ °C. Give the distribution of aqueous species at these conditions.

18.18 (a) Write the binding polynomial for ATP at 298.15K in terms of binding constants in the absence of Mg for application between $3 < pH_c < 14$. Assume ideal solutions. Hint: Use untransformed equilibrium constants calculated from the Gibbs energies of formation, and ignore the species that don't have Gibbs energies tabulated in the appendix.
(b) Convert the binding constants to dissociation constants and give the $pK_{a,A}$ for each dissociation constant.
(c) Using the binding constants, calculate and plot the $<i>$ vs. pH for ATP between $3 < pH_c < 14$. Mark the $<i>$ at the $pK_{a,A}$ values determined in (b).
(d) Give the fraction of each species at a $pH_c = 7.6$.

18.19 Repeat problem 18.18, but use ADP.

18.20 Beginning with the untransformed Gibbs energies of formation, document the intermediate calculations for the value of apparent Gibbs energy of formation of ADP at the conditions of Example 18.10, using the extended Debye-Hückel activity coefficient model and transformed Gibbs energies. Also calculate the distribution of each species.

18.21 Repeat problem 18.20, but use H_3PO_4.

18.22 At $pH_c = 7$, $I = 0.25$ m, $T = 25$ °C, beginning with the untransformed Gibbs energies of formation, document the intermediate calculations for the value of apparent Gibbs energy of formation of CO_2 at the conditions using the extended Debye-Hückel activity coefficient model and transformed Gibbs energies. Also calculate the distribution of each species.

CHAPTER

19

MOLECULAR ASSOCIATION AND SOLVATION

The satisfaction and good fortune of the scientist lie not in the peace of possessing knowledge, but in the toil of continually increasing it.

Max Planck

When specific chemical forces act between molecules, there is a possibility of complex formation. Chapter 17 dealt with systems where the interactions were strong enough to create new molecules. The interactions of complexes are much weaker but affect solution thermodynamics in a way that is fundamentally different from van der Waals interactions. The solution thermodynamics of complex formation is best represented as simultaneous phase and reaction equilibrium. The complexes usually cannot be isolated, but their existence is certain from measurements such as spectroscopic studies. Hydrogen bonding is an example of this type of behavior, as well as Lewis acid/base interactions. When complexation occurs between molecules that are all from the same component, the phenomenon is called **association.** For example, acetic acid dimerizes in pure solutions. When complexation occurs between molecules that are from different components, the phenomenon is called **solvation.** Complexation can occur in familiar ways, like hydrogen bonding, or in less familiar settings like charge transfer complexes. Mathematically, these phenomena can be treated in the same manner, which we will refer to as the chemical contribution to the Helmholtz energy, A^{chem}. By taking the limit of the resultant formulas as the complexation energy approaches infinity, we can also derive a contribution of the Helmholtz energy that characterizes the formation of dimers and chains, A^{chain}. Until now, we have described nonspherical contributions with empirical corrections, like $\kappa(\omega)$ of the Peng-Robinson equation. Chemical theory provides a fundamental description of how nonspherical molecules have nonzero acentric factors. Chapter 7 demonstrates how contributions to Z are attributable, and the contributions to the Helmholtz departure of Eqn. 8.27 follow,

$$(A - A^{ig})_{TV} = A^{rep} + A^{att} + A^{chem} + A^{chain}$$ 19.1

$$Z - Z^{ig} = Z - 1 = Z^{rep} + Z^{att} + Z^{chem} + Z^{chain}$$ 19.2

These equations comprise a new perspective on how equations of state should be formulated. Rather than limiting molecular interactions to a spherically symmetric perspective like the square-well model, chemical theory accounts for orientationally specific interactions. For example, the proton acceptors in water do not interact with the proton donors in the same way that they interact with other proton acceptors. Clearly, the orientations of water molecules must matter. Chemical theory provides a rigorous and practical means to recognize this.

Imagine for a moment how complex the analysis of complexation could be. For example, alcohols have proton acceptors and donors that can bond in chains of dimers, trimers, ... Each oligomer might need a specific activity coefficient and an independent reaction equilibrium constant. Then similar specifications could be necessary for all solvation interactions. Fortunately, the reaction equilibrium constants tend to follow predictable patterns, which we need to understand. Furthermore, the oligomer activities can be related through the equilibrium constraints. In the end, the model equations are only slightly more complex than the Peng-Robinson equation.

Furthermore, the mathematical formalism of chain formation can be extended by simply extrapolating the complexation energy to infinity. This provides a self-consistent theory of polymeric molecules. In previous discussions, we assumed that the shape factor in a model like ESD was somewhat arbitrarily imposed. In the present discussion, the shape factor is derived as a natural outcome of chain formation. This opens the door to a discussion of intramolecular interactions, whereas previous discussions have focused on intermolecular interactions. To get an idea of what lies beyond that door, note that proteins, DNA, and RNA are polymers with very specific intramolecular arrangements. Factor in the complexity of pH and other polyelectrolyte interactions, and you can understand why this text really is merely an introduction.

Chapter Objectives: You Should Be Able to...

1. Explain the relations between reaction and phase equilibria that yield the model equations for the PC-SAFT and ESD models.

2. Solve the model equations for the mole fraction of monomer, dimer, and so on, and the resultant fugacity expressions to obtain K-values.

3. Critically assess the intramolecular characterization implicit in PC-SAFT and ESD models and suggest deficiencies that should be addressed in future research.

19.1   INTRODUCING THE CHEMICAL CONTRIBUTION

Evidence of complexation is often subtle and could be overlooked or ignored in many situations, but its effects are irrefutable in other situations, so a question arises about ignoring facts because they are inconvenient. For example, dimerization of carboxylic acids is observable in the saturated vapor compressibility factor of small acids, but less so for acids of 10 carbons or more. On the other hand, the trend in LCEP discussed in Chapter 16 shows that dimerization is present even in long-chain fatty acids. Another commonly cited example of association is HF vapor, important in petroleum refining and the manufacture of refrigerants, modeled as $(HF)_n$ with n predominately 2 or 6. Evidence of this stoichiometry is found in vapor density data. Alcohols are also common substances in the chemical process industry that exhibit association. Spectroscopic data are the best source for characterizing complex formation and further information is available.[1] On the other hand, infrared spectra of the hydroxyl stretch, for example, can be difficult to interpret because they include a broad band of stretches indicative of the various oligomers that form in chains. Solid-liquid phase boundaries are indicative of strong complex formation, and may be used to infer complex

---

1. Prausnitz, J. M., Lichtenthaler, R. N., Azevedo, E. G. 1986. *Molecular Thermodynamics of Fluid-Phase Equilibria,* 2nd ed. Upper Saddle River, NJ: Prentice-Hall, Chapters 4 and 7.

stoichiometries. For example, the phase diagram for NH_3 + water in Fig. 14.14 shows two complexes that form in the solid phase, and might be expected to appear in fluid phases in addition to ionic species. Acetone and chloroform show a 1:1 compound in the solid phase.[2] Altogether, the chemical evidence and the results of molecular simulations are converging on the outlook that complexation is an effect that can and should be included in any systematic theory of molecular interactions and phase behavior.

Hydrogen Bonding

Hydrogen bonding is the most common phenomenon leading to association or solvation. As an example, consider the saturated vapor phase compressibility factor of acetic acid. Table 19.1 shows that, even though the pressure is very low for the saturated vapor at low temperature, the vapor compressibility factor of acetic acid deviates considerably from unity, when we would expect the ideal gas law to hold and $Z = 1$. For comparison, benzene under the same conditions is very nearly an ideal gas.

Table 19.1 *P-V-T Evidence of Association from the Compressibility Factors of Saturated Vapors*

Acetic Acid			Benzene		
T	P^{sat}(mmHg)	Z^V	T	P^{sat}(mmHg)	Z^V
293	11.8	0.507	290	65.78	0.994
323	56.20	0.540	320	241.52	0.989
373	416.50	0.586	370	1240.60	0.962

In the case of acetic acid, the small compressibility factor is due to the dimerization of the carboxylic acid. Even though the pressure is very low, the compressibility factor approaches 0.5 because the number of molecules in the vapor is actually half the apparent amount, owing to the dimerization.[3] Note that the compressibility factor increases as the temperature is increased. Why should the conversion to dimer be more complete at low temperature? Because association is an exothermic reaction and the van't Hoff relation clearly dictates that conversion of exothermic reactions decreases with increasing temperature. Another implication of associating fluids relative to nonassociating fluids is that higher temperatures are needed to make the associating network break into a gas, and this means that T_c for an associating component will be significantly higher than a nonassociating component of similar structure (e.g., H_2O versus CH_4). The chemical association of acetic acid during dimerization is illustrated in Fig. 19.1. Note that the structure forms two hydrogen bonds simultaneously which makes the dimerization quite strong. Note also that a property of

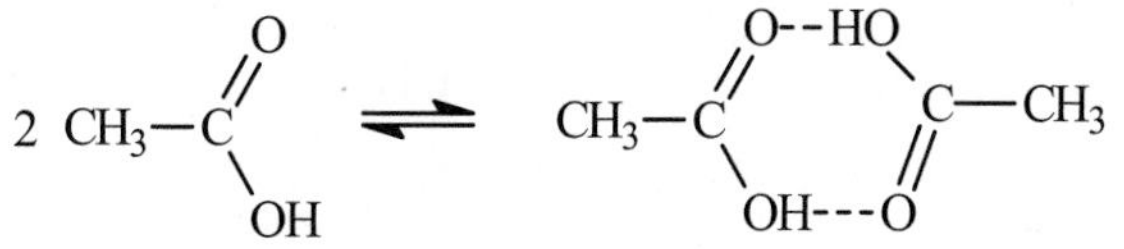

Figure 19.1 *Schematic of association in acetic acid.*

---

2. Campbell, A. N., Kartzmark, E. M. 1960. *Can. J. Chem.* 38:652.

3. The apparent number of moles is given by the species mass divided by the molecular weight of the monomer. On an apparent basis, we ignore the effect of association to dimers, trimers, and so on.

hydrogen bonding is that the O-H-O bond angle is nearly linear. Although the ring has eight atoms, the carboxylic acid structure is close to a six-sided ring, not an eight-sided ring.

Hydrogen bonding is also common between different species. For example, the hydroxyl hydrogen in phenol is fairly acidic while trimethyl amine is basic. The trimethyl amine would form a solvation complex in this case, even though it is incapable of association. Hydrogen bonding in solvating systems can enhance miscibility. For example, water and triethylamine are miscible in all proportions below 18°C (and above 0°C where water freezes). However, above 18–19°C, the increased thermal energy breaks the hydrogen bonds resulting in an immiscibility region which increases with increasing temperature[4] until the hydrogen bonding energy becomes too weak to influence the phase behavior. To put this into perspective, water miscibility with triethylmethane (aka 3-ethylpentane) is practically negligible at all conditions.

Usually, hydrogen bonding refers to protons bonding when they are attached to O, N, or F. In rare instances, protons bound to carbons can form a complex and it is common to refer to this as hydrogen bonding as well.[1] The hydrogen on a highly chlorinated molecule can form a complex with a Lewis base, such as a carbonyl, ether, or amine group. Well-known examples are chloroform + acetone, chloroform + diethyl ether, and acetylene + acetone, as depicted in Fig. 19.2. The chloroform + acetone in homework problem 11.15 should be reconsidered in light of complex formation.

Charge-Transfer Complexes

Solvation can also occur in the absence of "hydrogen bonding," per se. Similar kinds of interactions can be called "charge-transfer complexes," in which case one component is electron-rich (loosely bound electron) and another is strongly electron-attracting (low energy vacant orbital). For example, common electron donors are benzene rings with donating groups like $-OH$, $-OCH_3$, $-N(CH_3)_2$. Common electron acceptors are compounds with several nitro groups (1,3,5-trinitrobenzene), quinones, and compounds with several CN groups (tetracyanoethane, 2,3-dicyano-1,4-benzoquinone). For example, nitrobenzene and mesitylene form a complex when they are mixed as shown in Fig. 19.3. Tetrahydrofuran (THF) and toluene form a charge transfer complex when mixed together. When a charge-transfer complex forms, the energy typically corresponds to about 5 kJ/mole, and the strength is the same order of magnitude as a hydrogen bond. Chemically, these situations are all a bit different, but the mathematics is the same.

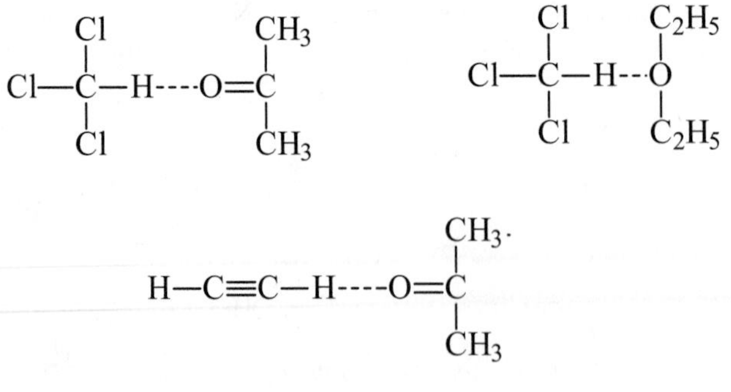

Figure 19.2 *Schematic of solvation in several pairs where association of the pure components is negligible.*

4. Hales, B. J., Bertrand, G. L., Hepler, L.G. 1966. *J. Phys. Chem.* 70:3970.

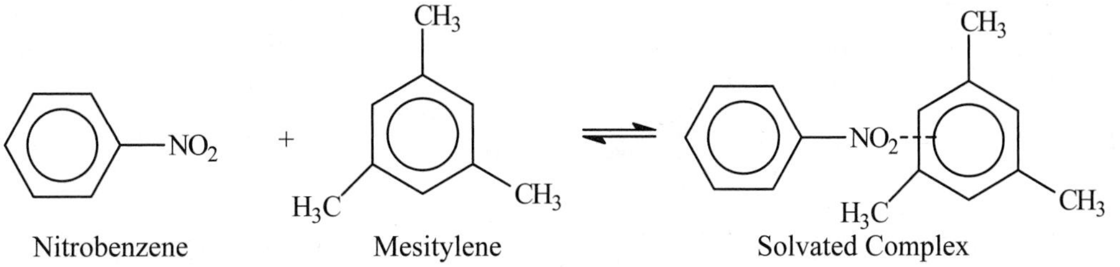

Figure 19.3 *Schematic of charge-transfer complexation.*

We discussed acidity and basicity in connection with the MOSCED and SSCED models in Chapter 11. Recall that a measure of the relative acidity/basicity is available from spectroscopic measurements tabulated in the Kamlet-Taft acidity/basicity parameters.[5] The Kamlet-Taft parameters are determined by comparing spectroscopic behavior of probe molecules in the solvents of interest. A small table of parameters has been included on the back flap.[6] These acidity and basicity parameters were designed for the SCED perspective, but they can provide guidelines for the energies of hydrogen bonding and complex formation too. The primary difference is that the SCED perspective does not alter the underlying solution model in a fundamental sense. To clarify, the similarity of the SSCED model to the Scatchard-Hildebrand means that the same solution model would be obtained with proper choices of k_{ij}. The k_{ij} values of the SSCED model are more predictable, but they do not alter the skewness of the Gibbs excess energy any more than the k_{ij} values of the Scatchard-Hildebrand model. Recognizing hydrogen bonding and complexation as mild reactions does alter the skewness of the Gibbs excess energy. In this sense, complexation theory provides an alternative to local composition theory. Whereas local composition theory correlates qualitatively with molecular simulation results, complexation theory correlates quantitatively. This means that a systematic step can be taken in connecting the molecular interactions with their macroscopic behavior.

Preliminary Considerations of Stoichiometry and Notation

In a generalized binary solution,[7] the ith complex can be represented by the general form $A_{a_i}B_{b_i}$, where the values of a_i and b_i are integers which will depend on the particular system (note that for an associated specie, either a_i or b_i is zero, but that won't affect the proof). The integers a_i and b_i are the stoichiometric coefficients:

$$a_i A + b_i B = A_{a_i}B_{b_i} \qquad\qquad 19.3$$

For example, a hypothetical system is shown in Fig. 19.4 that exhibits both association and solvation where the components A and B are added to the solution in quantities n_A and n_B.

The concentrations $x_A = \dfrac{n_A}{n_A + n_B}$ and $x_B = \dfrac{n_B}{n_A + n_B}$ are the mole fractions that are experimentally important for macroscopic characterization and are the conventional mole fractions. Since

---

5. Kamlet, M. J., Abboud, J.-L. M., Abraham, M. H., Taft, R. W. 1983. *J. Org. Chem.* 48:2877. For solutes in liquids see Abraham, M. H., Andonian-Haftvan, J., Whiting, G. S., Leo, A., Taft, R. W., 1994. *J. Chem. Soc. Perkin Trans.* 2:1777.

6. Lazzaroni, M.J., Bush, D., Eckert, C.A., Frank, T.C., Gupta, S., Olson, J.D. 2005. *Ind. Eng. Chem. Res.* 44:4075.

7. We restrict the discussion to binaries simply because the notation becomes unwieldy in multicomponent solutions, and the reader should recognize that the concepts and proofs extend to additional components.

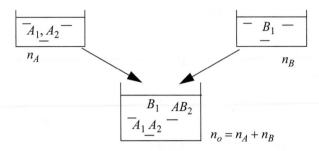

Figure 19.4 *Illustration of mixing of species A with species B to form a solution with both association and solvation. The labels in the containers indicate the true species present, not the
relative concentrations. The species have been chosen for illustrative purposes.*

these mole fractions do not represent the true species in solution, they are also called the *apparent*
mole fractions. The mole fractions in the actual solution are called the *true* mole fractions, and also

are denoted by *x*'s; for example, $x_{A_1} = \dfrac{n_{A_1}}{n_{A_1} + n_{A_2} + n_{B_1} + n_{AB_2}}$, $x_{B_1} = \dfrac{n_{B_1}}{n_{A_1} + n_{A_2} + n_{B_1} + n_{AB_2}}$.

Note that x_A (the apparent mole fraction) is not the same as x_{A_1} (the *A* monomer mole fraction).

 **The subscript
M is used for
monomer.**

Because the monomer mole fraction will end up being so important in later proofs, we give the special subscript *M* to help distinguish this as the monomer in the true mixture. Therefore, $x_{A_1} \equiv x_{AM}$,
and $x_{B_1} \equiv x_{BM}$. The true species that are present in solution will usually be inferred from fitting
experimental data, and the true mole fractions are usually *modeled* quantities rather than experimental quantities since they are subject to the assumptions of the model. The nomenclature is nevertheless the established convention in the literature. Note that the implication that the mixture of
this discussion is a *liquid* mixture is not restrictive. The same balances and notation will be used to
refer to vapor mixtures, but *y*'s will be substituted for the *x*'s.

19.2 EQUILIBRIUM CRITERIA

Mole Balance

One may wonder how quantification of the phenomena can be approached in a generalized fashion,
but the criteria are presented clearly by Prigogine and Defay (1954) whose proof we reproduce here
with modified notation. The first balances that must be satisfied are the material balances. For a
binary solution created from n_A moles of *A*, and n_B moles of *B*, where n_i is the moles of each specie
formed by Eqn. 19.3,

$$n_A = \sum_i a_i n_i \quad n_B = \sum_i b_i n_i \qquad\qquad 19.4$$

where the summations are over all *i* true species found in the solution. For example, in the binary
solution of Fig. 19.4, Eqn. 19.4 becomes

$$n_A = 1n_{AM} + 0n_{BM} + 2n_{A_2} + 1n_{AB_2}$$

$$n_B = 0n_{AM} + 1n_{BM} + 0n_{A_2} + 2n_{AB_2}$$

Chemical Potential Criteria

The chemical potentials of the true species are designated by μ_{AM}, μ_{BM}, $\mu_{A_{a_i}B_{b_i}}$ and so forth, and the apparent chemical potentials by μ_A and μ_B. Applying the principles of chemical equilibria to binary Eqn.19.3, we find

$$\mu_{A_{a_i}B_{b_i}} = a_i\mu_{AM} + b_i\mu_{BM} \qquad\qquad 19.5$$

If the total differential of G (Eqn. 17.7) is evaluated at constant T and P, allowing the species to come to equilibria,

$$dG = 0 = \sum_i \mu_{A_{a_i}B_{b_i}} dn_{A_{a_i}B_{b_i}}$$

we find by incorporating Eqn. 19.5 and the differential of Eqn. 19.4

$$0 = \mu_{AM}\sum_i a_i dn_{A_{a_i}B_{b_i}} + \mu_{BM}\sum_i b_i dn_{A_{a_i}B_{b_i}}$$

$$0 = \mu_{AM}dn_A + \mu_{BM}dn_B \qquad\qquad 19.6$$

On the other hand, for any binary solution at constant P, T, according to the apparent components

$$d\underline{G} = 0 = -\underline{S}dT + \underline{V}dP + \sum \mu_i dn_i = \mu_A dn_A + \mu_B dn_B \qquad\qquad 19.7$$

By comparing 19.6 and 19.7 we conclude

$$\boxed{\mu_A = \mu_{AM} \qquad \mu_B = \mu_{BM}} \qquad\qquad 19.8$$

Therefore, the apparent (conventional) chemical potential is quantified by a model that calculates the chemical potential of the true monomer species. It should be noted that this proof is independent of the number or stoichiometry of species that are formed in solution.

Fugacity Criteria

The chemical potential criteria may be extended to fugacity. For the apparent chemical potential,

$$\mu_A = \mu_A^o + RT\ln\frac{\hat{f}_A}{f_A^o} \qquad\qquad 19.9$$

and for the monomer,

$$\mu_{AM} = \mu_{AM}^o + RT\ln\frac{\hat{f}_{AM}}{\hat{f}_{AM}^o} \qquad\qquad 19.10$$

Note that for a species that associates, the standard state for the monomer is a mixture state since, even when A is pure, there is a mixture of true associated species. Applying Eqn. 19.8,

$$\mu_A^o + RT\ln\frac{\hat{f}_A}{f_A^o} = \mu_{AM}^o + RT\ln\frac{\hat{f}_{AM}}{\hat{f}_{AM}^o} \qquad 19.11$$

where the standard state is pure component A.

In the event that component A does not associate, the true solution is completely a monomer when A is pure, and

$$\mu_A^o = \mu_{AM}^o \qquad f_A^o = \hat{f}_{AM}^o = f_{AM}^o \qquad 19.12$$

which leads to

$$\boxed{\hat{f}_A = \hat{f}_{AM}} \qquad 19.13$$

The situation when A associates is slightly more complex. Recognizing that the apparent state neglects complexation, the change in chemical potential of monomer μ_{AM}^o is

$$\int_{\text{pure monomer}}^{\text{pure component }A} d\mu_{AM}^o = RT\int_{\text{pure monomer}}^{\text{pure component }A} d\ln\hat{f}_{AM}^o$$

where the lower limit is the hypothetical standard state of pure monomer, and the upper limit is the monomer standard state that actually exists in a pure associating solution of A. Integrating both sides, and recognizing the lower limit of each integral is the same as the apparent standard state

$$\mu_{AM}^o - \mu_A^o = RT\ln\frac{\hat{f}_{AM}^o}{f_A^o}$$

Combining with Eqn. 19.11 again results in Eqn. 19.13. Note that a parallel proof would show

$$\boxed{\hat{f}_B = \hat{f}_{BM}} \qquad 19.14$$

Activity Criteria

Returning to Eqn. 19.9, it can be rewritten in terms of the apparent activity and activity coefficient.

$$\mu_A = \mu_A^\circ + RT\ln\frac{x_A\gamma_A f_A^\circ}{f_A^\circ} = \mu_A^\circ + RT\ln x_A\gamma_A \qquad 19.15$$

Defining an activity coefficient, α_{AM}, of the true monomer species, the chemical potential is

$$\mu_{AM} = \mu_{AM}^\circ + RT\ln\frac{x_{AM}\alpha_{AM}\hat{f}_{AM}^\circ}{f_{AM}^\circ} = \mu_{AM}^\circ + RT\ln x_{AM}\alpha_{AM} \qquad 19.16$$

Using Eqn. 19.8 to equate Eqns. 19.15 and 19.16,

$$x_A \gamma_A = x_{AM} \alpha_{AM} \exp\left(\frac{\mu^\circ_{AM} - \mu^\circ_A}{RT}\right) \qquad 19.17$$

where the exponential term is a constant at a given temperature. The symmetrical convention of apparent activity requires $\lim_{x_A \to 1} x_A \gamma_A = 1$. For a nonassociating species, the exponential term of Eqn. 19.17 is unity by Eqn. 19.12, and thus $\lim_{x_A \to 1} \alpha_{AM} = 1$, $\lim_{x_A \to 1} x_{AM} = 1$. For an associating species $\lim_{x_A \to 1} x_{AM} = x^\circ_{AM}$ which is the true mole fraction of monomer in pure A, which is not unity. Therefore, the exponential term is simply the reciprocal of the limiting value of the monomer activity $\exp\left(\frac{-(\mu^\circ_{AM} - \mu^\circ_A)}{RT}\right) = \lim_{x_A \to 1} x_{AM} \alpha_{AM} = (x^\circ_{AM} \alpha^\circ_{AM})$. As such, we write

$$x_A \gamma_A = \frac{x_{AM} \alpha_{AM}}{(x^\circ_{AM} \alpha^\circ_{AM})} \qquad 19.18$$

where $x^\circ_{AM} \alpha^\circ_{AM} = 1$ for a nonassociating component. A parallel proof would show that

$$x_B \gamma_B = \frac{x_{BM} \alpha_{BM}}{(x^\circ_{BM} \alpha^\circ_{BM})} \qquad 19.19$$

These equations show how the activity coefficient could be less than one even for an "ideal" solution. For example, acetone and chloroform might form an ideal solution in the sense that $\alpha_{AM} = 1 = \alpha_{BM}$ at all concentrations. Complexation would result in $x_{AM} < x^\circ_{AM}$ when B was present, such that $\gamma_A < 1$. We explore this prospect extensively in Section 19.4.

19.3 BALANCE EQUATIONS FOR BINARY SYSTEMS

The balance equations to be solved take the same form for both vapors and liquids. The liquid equations will be shown, and the reader should recognize the vapor equations by analogy. First, the true mole fractions must sum to unity:

$$\sum_i x_i = 1 \qquad 19.20$$

In a binary system, a balance equation can be written for either component to match the apparent mole fraction:

$$x_A = \frac{\text{moles of all A in compounds}}{\text{moles of all A and B in compounds}} = \frac{\sum_i a_i n_i}{\sum_i (a_i + b_i) n_i}$$

Dividing numerator and denominator by the true number of moles, n_T,

$$x_A = \frac{\sum_i a_i x_i}{\sum_i (a_i + b_i) x_i} \qquad\qquad 19.21$$

Rearranging Eqn. 19.21 to facilitate implementation of the balance, multiply by the denominator,

$$\sum_i x_A(a_i + b_i)x_i - \sum_i a_i x_i = 0$$

and collecting the true mole fraction results in a form of Eqn. 19.21 that is easier to implement:

$$\boxed{\sum_i (b_i x_A - a_i x_B) x_i = 0} \qquad\qquad 19.22$$

Eqns. 19.20 and 19.22 are not yet ready to implement because all of the true mole fractions are unknown and only two equations have been developed. In the next section, we show that the true mole fractions can be written in terms of an equilibrium constant and the monomer mole fractions, which will provide sufficient information once the equilibrium constants are known.

19.4   IDEAL CHEMICAL THEORY FOR BINARY SYSTEMS

The simplest method of modeling complex behavior is to neglect the nonidealities by modeling a vapor phase as an ideal gas mixture including the complexes (true fugacity coefficients equal to 1), and to model a liquid phase as an ideal solution containing complexes (true activity coefficients equal to 1). This approach is called Ideal Chemical Theory and can be used to calculate the non-ideal apparent fugacity or activity coefficients. Two brief observations at the outset help to put the chemical perspective into context. First, in reference to systems that can only solvate (not associate), the observed activity coefficients of A must be less than one because $x_{AM}/x^{\circ}_{AM} < x_A$ when B is present. Second, for systems in which component A associates and B can neither associate nor solvate, the observed activity coefficients for A must be greater than one because $x_{AM}/x^{\circ}_{AM} > x_A$ when B is present owing to the interference of B with A from dilution. If you combine these two observations for mixtures that can solvate and associate, you can see how an entire range of activity coefficients may be obtained.

Modeling complex formation with ideal chemical theory, Eqn. 19.3 can be expressed in terms of an equilibrium constant:

$$x_i = K_{a,i} x_{AM}^{a_i} x_{BM}^{b_i} \qquad\qquad 19.23$$

Plugging into Eqns. 19.20 and 19.22, the equations to be solved are obtained:

$$\boxed{x_{AM} + x_{BM} + \sum_i K_{a,i} x_{AM}^{a_i} x_{BM}^{b_i} - 1 = 0} \qquad\qquad 19.24$$

$$x_A x_{BM} - x_B x_{AM} + \sum_i (b_i x_A - a_i x_B) K_{a,i} x_{AM}^{a_i} x_{BM}^{b_i} = 0 \qquad\qquad 19.25$$

Once the K_i are known, then x_{AM} and x_{BM} can be determined by solving these two equations at a specified apparent concentration. Subsequently, all true mole fractions, x_i (Eqn. 19.23) and the apparent mole fractions γ_A, γ_B (Eqns. 19.18 and 19.19), can be calculated. If γ_A, γ_B are known from experiment, and the complex stoichiometry is known, K_i values can be adjusted to fit the data using optimization methods. A spreadsheet is provided for solving for the true species for given values of K_i in the programs Ichemt.xlsx and Ichemt.m.

For ideal chemical theory applied to the vapor phase, the x_i are replaced with y_i and Eqn. 19.3 is expressed as

$$y_i = K_{a,i} P^{(a_i + b_i - 1)} y_{AM}^{a_i} y_{BM}^{b_i} \qquad\qquad 19.26$$

Eqns. 19.20 and 19.22 then become

$$y_{AM} + y_{BM} + \sum_i K_{a,i} P^{(a_i + b_i - 1)} y_{AM}^{a_i} y_{BM}^{b_i} - 1 = 0 \qquad\qquad 19.27$$

$$y_A y_{BM} - y_B y_{AM} + \sum_i (b_i y_A - a_i y_B) K_{a,i} P^{(a_i + b_i - 1)} y_{AM}^{a_i} y_{BM}^{b_i} = 0 \qquad\qquad 19.28$$

These equations are ideal gas equations from the perspective of the true solution. As with the liquid-phase calculation, if the K_i values are known, y_{AM} and y_{BM} can be determined.

Example 19.1 Compressibility factors in associating/solvating systems

Derive a formula to relate the true mole fractions to the compressibility factor of a vapor phase where the true species follow the ideal gas law.

Solution: A vessel of volume $\underline{V}$ holds n_o apparent moles. However, experimentally, in the same total volume, there would be a smaller number of true moles n_T. Applying the ideal gas law,

$$P\underline{V} = n_T RT \qquad \frac{P\underline{V}}{n_T RT} = 1 \qquad\qquad 19.29$$

Experimentally, we wish to work in terms of the apparent number of moles,

$$Z = \frac{P\underline{V}}{n_o RT} = \frac{P\underline{V}}{n_T RT} \cdot \frac{n_T}{n_o} = \frac{n_T}{n_o} \qquad\qquad \text{(ig)}$$

Example 19.1 Compressibility factors in associating/solvating systems (Continued)

Note that this equation is labelled as an ideal gas equation because the true species follow the ideal gas law, even though from the perspective of the apparent species, the ideal gas law will not be followed. From the total mole balances, $n_T = \sum_i n_i$, and $n_o = \sum_i (a_i + b_i)n_i$; therefore,

$$Z = \frac{\sum_i n_i}{\sum_i (a_i + b_i)n_i}$$

Dividing numerator and denominator by n_T,

$$Z = \frac{\sum_i y_i}{\sum_i (a_i + b_i)y_i} = \frac{1}{\sum_i (a_i + b_i)y_i} \qquad 19.30$$

Therefore, once the true mole fractions have been determined, the compressibility factor can be calculated. Determining the true mole fractions requires solving the reaction equilibria, as discussed in the next example.

Example 19.2 Dimerization of carboxylic acids

P-V-T measurements of acetic and propionic acid vapors are available.[a] The equilibrium constants for acetic and propionic acids at 40°C are 375 bar^{-1} and 600 bar^{-1} respectively. At a pressure of 0.01 bar, determine the true mole fractions, the compressibility factor, and the fugacity coefficients.

Solution: Beginning with Eqn. 19.27, letting A be the acid of interest,

$$y_{AM} + K_{A_2}Py_{AM}^2 - 1 = 0$$

Eqn. 19.28 is not required since the system is a single component. This simple equation can be solved with the quadratic formula,

$$y_{AM} = \frac{-1 + \sqrt{1 + 4K_{A_2}P}}{2K_{A_2}P}$$

At $P = 0.01$ bar, for acetic acid, $y_{AM} = 0.4$, $y_{A_2} = 0.6$, and even at this low pressure, Eqn. 19.30 gives $Z = 0.625$. For the fugacity coefficient, starting with Eqn. 19.13,

$$y_A \hat{\varphi}_A P = y_{AM}\hat{\varphi}_{AM}P$$

Example 19.2 Dimerization of carboxylic acids (Continued)

Since the system is pure, $y_A = 1$ and the fugacity coefficient on the left-hand side will be for a pure species. Since the model uses ideal chemical theory, $\hat{\varphi}_{AM} = 1$. Therefore

$$\hat{\varphi}_A = y_{AM} = 0.4$$

The same procedure can be repeated for propionic acid, however it will be even more nonideal. The answers are: $y_{AM} = 0.333$, $Z = 0.6$, $\hat{\varphi}_A = 0.333$.

---

a. McDougall, F. H., 1936. *J. Amer.Chem.Soc.* 58:2585; 1941. 63:3420.

Example 19.3 Activity coefficients in a solvated system

1,4-dioxane (component B) is a cyclic 6-member ring, $C_4H_8O_2$, with oxygens in the 1 and 4 positions. When mixed with chloroform (component A) the oxygens provide solvation sites for the hydrogen on chloroform. Since there are two sites on 1,4-dioxane, two complexes are possible, AB and A_2B. McGlashan and Rastogi[a] have studied this system and report the liquid phase can be modeled with ideal chemical theory using $K_{AB} = 1.11$, $K_{A_2B} = 1.24$ at 50°C. Calculate the true mole fractions and activity coefficients across the composition range.

Solution: We will use the program Ichemt.xlsx to solve Eqns. 19.24 and 19.25. For the A_2B compound, $a_i = 2$, $b_i = 1$, $K_i = 1.24$. For the compositions, we enter increments of 0.05 for the apparent compositions. Near the endpoints, we enter $x_A = 0.001$ and $x_A = 0.999$ to avoid numerical underflows and overflows. The activity coefficients are easily calculated using $\gamma_A = z_A/x_A$, $\gamma_B = z_B/x_B$ since neither component exhibits association. The results are shown in Fig. 19.5. Note that the solvation causes negative deviation from Raoult's law. Also note the relation between the complex stoichiometry and the maxima in the complex concentration. Can you rationalize why the infinite dilution activity coefficient of 1,4-dioxane is smaller than the infinite dilution activity coefficient of chloroform?

IdChemTheory.xlsx
IdChemTheory.m

---

a. McGlashan, M. L., Rastogi, R. P. 1958. *Trans. Faraday Soc.* 54:496.

19.5 CHEMICAL-PHYSICAL THEORY

The assumptions of ideal chemical theory are known to be oversimplifications for many systems and physical interactions must be included. For a liquid phase, the activity coefficients of the true species can be reintroduced. Then

$$x_i = \frac{K_{a,i}(x_{AM}\alpha_{AM})^{a_i}(x_{BM}\alpha_{BM})^{b_i}}{\alpha_i} \qquad \text{19.31}$$

Utilizing this result with Eqns. 19.20 and 19.22, the following equations are obtained:

$$x_{AM} + x_{BM} + \sum_i \frac{K_{a,i}(x_{AM}\alpha_{AM})^{a_i}(x_{BM}\alpha_{BM})^{b_i}}{\alpha_i} - 1 = 0 \qquad \text{19.32}$$

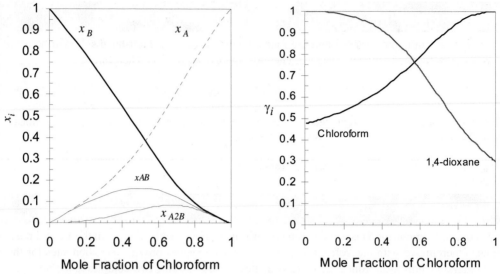

Figure 19.5 *Ideal chemical theory applied to the chloroform + 1,4-dioxane system as calculated in Example 19.3.*

$$x_A x_{BM} - x_B x_{AM} + \sum_i (b_i x_A - a_i x_B) \frac{K_{a,i} (x_{AM} \alpha_{AM})^{a_i} (x_{BM} \alpha_{BM})^{b_i}}{\alpha_i} = 0 \qquad 19.33$$

Since most activity coefficient models require two parameters per pair of molecules, the number of parameters becomes large. In addition, any parameters for the complex must be estimated or fit to experiment since the complex cannot be isolated. Solution of the equations is more challenging because the true activity coefficients must be updated with each iteration on x_{AM} and x_{BM}.

For chemical-physical theory applied to the vapor phase,

$$y_i = \frac{K_{a,i} P^{(a_i + b_i - 1)} (y_{AM} \hat{\varphi}_{AM})^{a_i} (y_{BM} \hat{\varphi}_{BM})^{b_i}}{\hat{\varphi}_i} \qquad 19.34$$

Eqns. 19.20 and 19.22 then become

$$y_{AM} + y_{BM} + \sum_i \frac{K_{a,i} P^{(a_i + b_i - 1)} (y_{AM} \hat{\varphi}_{AM})^{a_i} (y_{BM} \hat{\varphi}_{BM})^{b_i}}{\hat{\varphi}_i} - 1 = 0 \qquad 19.35$$

$$y_A y_{BM} - y_B y_{AM} + \sum_i (b_i y_A - a_i y_B) \frac{K_{a,i} P^{(a_i + b_i - 1)} (y_{AM} \hat{\varphi}_{AM})^{a_i} (y_{BM} \hat{\varphi}_{BM})^{b_i}}{\hat{\varphi}_i} = 0 \qquad 19.36$$

The physical properties of the complex must also be modeled with this approach, and the same challenges for solving the equations are present as discussed above for chemical-physical theory of liquid phases.

An interesting study has been performed by Harris[8] for acetylene in *n*-hexane, butyrolactone, and *n*-methyl pyrrolidone at 25°C. In this study, a simplified van Laar model was used to model the physical deviations, which resulted in one physical parameter. Naturally, the acetylene + *n*-hexane does not exhibit solvation, but the other binaries do, with the pyrrolidone showing the strongest complexation. Further, the *n*-hexane system has positive deviations from Raoult's law across the composition range, the pyrrolidone shows negative deviations, and the lactone shows both positive and negative deviations. All three systems are accurately modeled using two parameters each—one chemical parameter and one physical parameter.

Another approach to the chemical-physical theory is to use the Flory-Huggins theory for the physical contributions. This is the approach of Coleman and Painter in modeling polymer solutions. The Coleman-Painter model leads to complications in the extension to ternary mixtures, however, owing to several details in their perspective on chemical networks.[9]

Multicomponent chemical-physical theory can be achieved most elegantly with Wertheim's theory which we will discuss in the next section.[10] Wertheim's theory characterizes chemical interactions from the perspective of the acceptor or donor sites instead of the species. This simplifies to the counting of nonbonded sites, especially for multicomponent systems, and the nonbonded sites can be related to the monomer fraction, which suffices to define the solution thermodynamics. Wertheim's theory requires a complementary physical theory for the nonchemical attractive and repulsive interactions. Briefly, chemical interaction is short-ranged, so variations in bonding are affected by the frequency of species coming into contact. Repulsive interactions dominate the frequency of contact (specifically, $g(\sigma)$). We can estimate $g(\sigma)$ for spherical molecules with the Carnahan-Starling equation. For nonspherical molecules, we can imagine that they are composed of spherical segments. Then the role of the attractive contribution is like that of a spherical molecule, to provide a disperse field of attractive energy that acts between spherical entities and reduces the pressure. This leads to a remarkably compact and self-consistent model of chemical-physical equilibria.

Before we begin our discussions of Wertheim's theory, let us mention an additional approach to chemical-physical theory is provided by Heidemann and Prausnitz.[11] They showed that reasonable assumptions about the van der Waals parameters of monomers, dimers, trimers, and so on leads to a closed form solution for the compressibility factor and fugacity coefficient. Similar to Wertheim's theory for pure fluids, the Heidemann-Prausnitz method provides a complete chemical-physical theory, describing all variations with density, temperature, composition, and chemistry. However, similar to the Coleman-Painter theory, this method has complications in the extension to multicomponent mixtures. Suresh and Elliott[12] showed that the Heidemann-Prausnitz method is equivalent to Wertheim's theory subject to certain assumptions about the change in heat capacity due to reaction. In the interest of covering the most general method, we focus now on Wertheim's theory, but we introduce concepts using the Heidemann-Prausnitz perspective as a simple way of illustrating several of the more striking results derived from Wertheim's theory. This is necessary because the rigorous proofs of Wertheim's theory of Wertheim's original publications go beyond the introductory scope envisioned here.

8. Harris, H. G., Prausnitz, J. M. 1969. *Ind. Eng. Chem. Fundam.* 8:180.

9. Coleman, M. M., Painter, P. C., Graf, J. F. 1995. *Specific Interactions and the Miscibility of Polymer Blends.* Boca Raton, FL: CRC Press.

10. Wertheim, M. S. 1984. *J. Stat. Phys.* 35:19.

11. Heidemann, R. A., Prausnitz, J. M. 1976. *Proc. Nat. Acad. Sci.* 73:1773.

12. Suresh, S.J., Elliott, J.R. 1992. *Ind. Eng. Chem. Res.* 31:2783.

19.6   WERTHEIM'S THEORY FOR COMPLEX MIXTURES

The general approach is exactly what you would expect: Write all the reaction and phase equilibrium constraints and then solve the nonlinear system of equations. Making this approach into a practical alternative to, say, the Peng-Robinson model requires several clever observations, approximations, and rearrangements, however. Wertheim's theory is based on the contribution to the Helmholtz energy. In the end, A^{chem} is recognizable as a distinct contribution with a firm foundation in experimental observation and molecular simulation that adds just one intermediate (but robust) step in solving for the density given temperature and pressure.

Wertheim's theory has the same objective as this chapter: to develop a theory for the chemical contribution to the Helmholtz energy, A^{chem}, and consequently Z^{chem}, through the derivative relations in Chapter 6. Because the volume derivative of A results in P, the volume (or density) derivative of A can be used to calculate the contribution of chemical interactions to Z. Wertheim's theory refers to the concepts of monomers and dimers discussed previously, but develops a self-contained and self-consistent model based on a given equation of state for the nonchemical contributions. In Chapter 7, we demonstrated that an equation of state expression for Z could be written in terms of the repulsive Z^{rep} and attractive Z^{att} contributions. (i.e., A^{rep} and A^{att}, aka the "physical" contributions). Wertheim's original development uses sophisticated statistical mechanics beyond the scope of this introduction. Nevertheless, we can understand his results in terms of contributions to the reaction and equilibrium equations. Whenever we arrive at a set of terms that seems difficult to simplify, we can apply Wertheim's result as a "clever guess," and show how this result leads to a self-consistent interpretation for specific physical contributions (e.g., the van der Waals model). Once we have an expression for A^{chem} expressed in terms of ρ_0 and T, it can be added to the physical contributions and the equation of state can be applied like any other equation of state.

Because the notation is complicated, we develop this section using pure component dimers, then chains, and later generalize the results. Wertheim's theory is applied to equations of state, so we use the notation x to represent mole fractions in both the vapor and liquid phases and the state of aggregation will be determined by the size of Z. Also, we omit the "A" from x_{AM} when there is only one component. The starting point for Wertheim's theory is to rearrange the analysis in terms of the true numbers of bonding *sites* in the fluid. The extent of association is then characterized in terms of the fraction of bonding sites *not* bonded, X. The relation for a species with one dimer-forming site is:

$$X = n_M / n_0 \qquad \text{(for one-site molecules)} \qquad 19.37$$

This "fraction of bonding sites not bonded" is closely related to the fraction monomer, x_M. The relevant mass balances are discussed below. To understand Wertheim's theory, you must understand what is meant by a hydrogen bonding site. A key element of Wertheim's perspective is to characterize the bonding sites as small, off-center "blisters" of attractive energy. This gives orientational specificity because the sites can only bond if the angle from the left repulsive site to the bonding sites to the right repulsive site is close to $180°$; any other orientation would be inconsequential. Furthermore, the smallness of the blister relative to the repulsive site means that three sites cannot bond simultaneously because it would require the third repulsive to overlap with the two that were already bound as shown in Fig. 19.6. This captures the short-range nature, orientational specificity, and steric hindrance that we recognize in hydrogen bonding, and complexation in general.

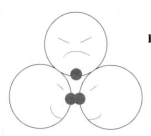

Figure 19.6 *Wertheim's perspective on bonding one-site molecules. The shaded portions represent the blisters. The lower two molecules are happily bonded, but the upper molecule can't join the same bond.*

Dimer Formation (one-site molecules)

We begin with association to form a dimer. For the formation of a dimer (denoted with subscript D), Eqn. 19.4 and the sum of true mole fractions can be combined and rearranged to relate the apparent moles n_0 and total true moles n_T to the fraction of unbonded sites X:

$$n_0 = n_M + 2n_D \text{ and } n_T = n_M + n_D \Rightarrow n_0 - n_T = n_D = (n_0 - n_M)/2 = n_0(1 - X)/2 \qquad 19.38$$

Dividing Eqn. 19.38 by n_0 gives $1 - n_T/n_0 = (1 - X)/2$, where X is the fraction of bonding sites not bonded:

$$n_T/n_0 = (1 + X)/2 \qquad 19.39$$

$$x_M = n_M/n_T = (X \cdot n_0)/n_T = X(n_0/n_T) = 2X/(1 + X) \qquad 19.40$$

Noting that the solution must satisfy the mass balance, Eqn. 19.20, $x_M + x_D = 1$. Together with the reaction equilibrium (law of mass action, Eqn. 19.34) we may write,

$$x_D = 1 - x_M = x_M^2 K_a(P/(P°))(\hat{\varphi}_M^2/\hat{\varphi}_D) \qquad 19.41$$

where the standard state $P°$ is included for clarity. Substituting Eqn. 19.40 gives $1 - 2X/(1 + X) = 4X^2 K_a(P/(P°))(\hat{\varphi}_M^2/\hat{\varphi}_D)/(1 + X)^2$, which can be rearranged to

$$1 - X = 2X^2 K_a(P/(P°))(\hat{\varphi}_M^2/\hat{\varphi}_D)(n_0/n_T) = 2X^2 \Delta \qquad 19.42$$

Defining

$$\boxed{\Delta = K_a(P/(P°))(\hat{\varphi}_M^2/\hat{\varphi}_D)(n_0/n_T)} \qquad 19.43$$

may seem odd at first glance, but it is one of the major simplifications derived from Wertheim's theory. Often in the literature Δ_{lit} is defined such that $\rho\Delta_{lit} = \Delta$. We outline Wertheim's analysis below, but a key step is that this conglomeration of symbols can be simplified to

$$\boxed{\Delta^C = \rho g(\sigma) K^C [\exp(\varepsilon^C/(kT)) - 1]} \qquad 19.44$$

where $g(\sigma)$ is the radial distribution function at contact distance σ, ε^C is the bond energy of the complex, and K^C is the bonding volume related to the size of the "blisters."[13] Eqn. 19.42 shows that X can be solved directly from the density and temperature since Δ is a function of ρ and T. Since the density and temperature must be specified in applying the physical contributions of the equation of

---

13. The 'C' denotes a 'C-type' dimerization of identical molecules. Later we use 'AD' to denote acceptor/donor association.

state anyway, the interjection of this contribution does not complicate matters in the way that solving simultaneous phase and reaction equilibria does. Imagine how cumbersome this model might become if all complexation required iterative solution. The principles are the same, but the feasibility is radically altered.

Aside from the advanced statistical mechanical analysis of Wertheim's paper, we can appreciate the phenomenology of his analysis in two ways. First, we can recognize Δ as an equilibrium constant of a reversible reaction, the ratio of forward and reverse rates. The forward reaction is proportional to the probability of the sites finding each other. This probability is zero if the density is zero, and it is enhanced by $g(\sigma)$. The reverse reaction is inhibited by the strength of association. The stronger the bonding energy, the slower the dissociation.[14] Second, we can apply the van der Waals model with some simple assumptions. If we assume that $b_D = 2b_M$, $a_{DD} = 4a_{MM}$, and $a_{DM} = 2a_{MM}$, we obtain a similar result, in the manner of Heidemann and Prausnitz as shown in Fig. 19.7. Applying these assumptions to the fugacity coefficients of the van der Waals equation and evaluating $\ln(\hat{\varphi}_M^2 / \hat{\varphi}_D)$ of Eqn. 19.42 by adapting 15.32 to monomer and dimer, we obtain,

$$\ln(\hat{\varphi}_D) - 2\ln(\hat{\varphi}_M) = -\ln(1 - \eta_P) + \frac{b_D \rho_T}{(1 - \eta_P)} - \frac{2\rho_T \sum_i x_i a_{iD}}{RT} - \ln\left(\frac{P\underline{V}}{n_T RT}\right)$$

$$-2\left(-\ln(1 - \eta_P) + \frac{b_M \rho_T}{(1 - \eta_P)} - \frac{2\rho_T \sum_i x_i a_{iM}}{RT} - \ln\left(\frac{P\underline{V}}{n_T RT}\right)\right) = \ln(1 - \eta_P) + \ln\left(\frac{P\underline{V}}{n_T RT}\right)$$

19.45

Inserting into 19.43 gives,

$$\Delta \equiv K_a \frac{P n_0 \hat{\varphi}_M^2}{P^\circ n_T \hat{\varphi}_D} = \frac{K_a}{(1 - \eta_P)} \frac{P n_T RT n_0}{P^\circ} \frac{P \underline{V}}{n_T} = \frac{\rho RT K_a}{P^\circ (1 - \eta_P)}$$

19.46

Comparing with 19.44 the vdW EOS corresponds roughly to $g(\sigma) = 1/(1 - \eta_P)$. Also, the equilibrium constant, K_a, can be referenced to the critical temperature and written as,

$$\ln\left(\frac{K_a}{K_{ac}}\right) = \ln(T_r)^{\left(\frac{\Delta Cp}{R}\right)} + \left(\frac{\Delta H_{Tc}^o}{RT_c} - \frac{\Delta Cp}{R}\right)\left(1 - \frac{1}{T_r}\right)$$

19.47

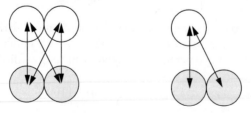

Figure 19.7 *Illustration that $a_{DD} = 4\,a_{MM}$ and $a_{DM} = 2\,a_{MM}$ are reasonable by adding the number of pair interactions.*

---

14. You might wonder what happens when the bond energy approaches that of a covalent bond. See the section on Wertheim's theory of polymerization.

The "best" expression for $\Delta Cp/R$ is debatable. Experimental measurements are unlikely to provide sufficient precision to resolve the debate. From a practical perspective, we would prefer a compact expression for K_a. From a theoretical perspective, Wertheim's analysis is the most sophisticated. Suresh and Elliott[15] used T_c as a reference temperature, and showed that Wertheim's analysis is consistent with the assumption that,

$$\frac{\Delta Cp}{R} = \frac{\ln\left[\exp\!\left(\dfrac{\varepsilon^{C}}{kT}\right) - 1\right] + \left(\dfrac{\varepsilon^{C}}{kT_c}\right)\!\left(1 - \dfrac{1}{T_r}\right) - \ln(T_r)}{\ln(T_r) - \left(1 - \dfrac{1}{T_r}\right)}$$ 19.48

Substitution shows the resulting relation between K_a and a portion of Eqn. 19.44,

$$(K_a RT)/(P^\circ) = K^{C}[\exp(\varepsilon^{C}/(kT)) - 1]$$ 19.49

where $(K_{ac} RT_c)/(P^\circ) = K^{C}$ and $-\Delta H_{Tc}/R = \varepsilon^{C}/k$. The superscript "$C$" denotes a C-type association but the relations can be extended to an AD-type (though an arbitrary reference temperature must be selected for AD interactions).[16] Substituting this expression for K_a in 19.46 and $g(\sigma)$ for $1/(1 - \eta_P)$ results in Eqn. 19.44, eliminating the need to solve iteratively for the density, monomer and dimer concentrations, and fugacity coefficient ratio. An additional step is required to transform the extent of reaction (implicit in X) into a remarkably simple thermodynamic contribution, A^{chem}.

Low Density

The next objective is to evaluate the impact on Helmholtz energy, A^{chem}, the change in Helmholtz energy due to bonding. As discussed in the introduction, A^{chem} is the chemical contribution as it pertains to an equation of state. Let's begin by rewriting the Gibbs energy for a single bond at low density, noting that $P\underline{V} = n_T RT$ then.[17]

$$\underline{G}^{chem} = \underline{A}^{chem} + (P\underline{V})^{chem} \approx \underline{A}^{chem} + RT\,(n)^{chem} = \underline{A}^{chem} + RT(n_T - n_0)$$ 19.50

Or, on a apparent molar basis, (dividing by n_0),

$$G^{chem} = A^{chem} + RT(n_T/n_0 - 1)$$ 19.51

In rearranging, note that $G^{chem} = \Delta\mu^{chem} = RT\ln(\hat{f}_M/f_0)$, where $\hat{f}_M$ is the monomer fugacity with association fully recognized and f_0 is the apparent fugacity based on zero association.

$$\mu = \mu_M \;\Rightarrow\; \hat{f}_M \approx y_M P = n_M RT/\underline{V}$$ 19.52

Further noting that

$$f_0 \approx n_0 RT/\underline{V} \;\Rightarrow\; \Delta\mu^{chem} = RT\ln(n_M/n_0) = G^{chem}$$ 19.53

Substituting Eqns. 19.39 and 19.53 into Eqn. 19.51, and recalling $X = n_M/n_0$ for dimers, we have

---

15. They also compared to $\Delta Cp/R = -1$, an alternative that yields a similarly compact expression for K_a. Suresh, S.J., Elliott, J.R. 1992. *Ind. Eng. Chem. Res.* 31:2783.
16. Wertheim used the symbol "*assoc*" instead of "C" when treating dimerization of a single component. His theory has been widely adapted to all forms of complexation, however, and for acceptor-donor bonding, we later use "AD."
17. This introductory derivation is adapted from a presentation by W.G. Chapman at EquiFase, October 13–19, 2002.

$$A^{chem}/RT = \ln(X) + (1 - X)/2 \quad \text{(pure dimerizing fluid)} \qquad 19.54$$

This turns out to be a very powerful equation.

All Densities

The remarkable aspect of Eqn. 19.54 is that it is accurate for all densities and extents of association, although it has been derived here only for binary association at low density. In fact, the significance of Wertheim's work is that he provides a rigorous statistical mechanical derivation of this identity at all conditions. Once again, we can support this result phenomenologically through the van der Waals model. Adapting Eqn. 19.38, with the mixing rule of Eqn. 15.8,

$$\rho_T b = (n_T/\underline{V})(x_M b_M + x_D b_D) = (n_T/\underline{V}) b_M (n_M + 2n_D)/n_T = b_M \rho_0 = \eta_P \qquad 19.55$$

This equation shows that there is no overlap of repulsive sites when a hydrogen bond occurs, so the volume occupied by molecules is the same regardless of association. Similarly,

$$\rho_T \Sigma x_j a_{Mj} = \rho_T (x_M a_{MM} + x_D a_{MD}) = \rho_T (x_M a_{MM} + x_D 2a_{MM}) = a_{MM} \rho_0 \qquad 19.56$$

We can express the fugacity of the fluid in two ways, noting that $\hat{f}_M = f$ where $\hat{f}_M$ is defined by the monomer fugacity in the true mixture, that is, $\ln(\hat{\varphi}_M) = \ln(\hat{f}_M /x_M P)$, and f is defined by $\ln(f/P) = (G - G^{ig})/RT$ for the "pure" fluid based on the apparent perspective. The expression in terms of φ_M relates to the true species and uses the fugacity expression for a component in a mixture. The expression in terms of f/P emphasizes that we are still discussing a single pure component. Using 15.32, (note by 19.55 $b_M = b\rho_T/\rho_0 = bn_T/n_0$).

$$\ln(\hat{\varphi}_M) = -\ln(1 - b\rho_T) + \frac{\left(\dfrac{n_T}{n_0}\right) b\rho_T}{(1 - b\rho_T)} - \frac{2\rho_T \sum x_j a_{Mj}}{RT} - \ln\left(\frac{P\underline{V}}{n_T RT}\right) = \ln\left(\frac{\hat{f}_M}{x_M P}\right) \qquad 19.57$$

By Eqns. 19.55 to 19.56,

$$\ln\left(\frac{\hat{f}_M}{P}\right) = \ln(x_M) - \ln(1 - \eta_P) + \frac{\left(\dfrac{n_T}{n_0}\right) \eta_P}{(1 - \eta_P)} - \frac{2a_{MM}\rho_0}{RT} - \ln\left(\frac{P\underline{V}}{n_T RT}\right) \qquad 19.58$$

We modify 15.32 to recognize A^{chem} and Z^{chem} contributions based on the apparent perspective,

$$\ln\left(\frac{f}{P}\right) = -\ln(1 - \eta_P) - \frac{\rho_0 a_{MM}}{RT} + \frac{A^{chem}}{n_0 RT} + \frac{\eta_P}{(1 - \eta_P)} - \frac{\rho_0 a_{MM}}{RT} + Z^{chem} - \ln\left(\frac{P\underline{V}}{n_0 RT}\right) \qquad 19.59$$

Equating 19.58 to 19.59, we can immediately cancel terms of $\ln(1 - \eta_P)$ and $\rho_0 a_{MM}$. Also noting that $\ln(P\underline{V}/n_0 RT) - \ln(P\underline{V}/n_T RT) = \ln(n_T/n_0)$,

$$\ln(x_M) + \left(\frac{n_T}{n_0} - 1\right)\frac{\eta_P}{(1 - \eta_P)} + \ln\left(\frac{n_T}{n_0}\right) = \frac{A^{chem}}{n_0 RT} + Z^{chem} \qquad 19.60$$

noting that $x_M \cdot n_T/n_0 = X$, and from Eqn. 19.39 $(n_T/n_0 = (1+X)/2)$, we obtain,

$$\ln(X) + \left(\frac{1+X}{2} - 1\right)\frac{\eta_P}{(1 - \eta_P)} - Z^{chem} = \frac{A^{chem}}{n_0 RT} = \ln(X) - \left(\frac{1-X}{2}\right)\frac{\eta_P}{(1 - \eta_P)} - Z^{chem} \qquad 19.61$$

Recall that $Z/\eta_P = d(A/RT)/d\eta_P$. Taking Eqn. 19.54 as a trial solution and checking that it is consistent with Eqn. 19.61,

$$Z^{chem} = \eta_P \frac{d}{d\eta_P}\left(\frac{A^{chem}}{n_0 RT}\right) = \left(\frac{1}{X} - \frac{1}{2}\right)\eta_P \frac{\partial X}{\partial \eta_P} = \left(\frac{2-X}{2}\right)\frac{\eta_P}{X}\frac{\partial X}{\partial \eta_P} \qquad 19.62$$

We can evaluate $\partial X/\partial \eta_P$ through Eqn. 19.42 by differentiating implicitly.

$$-\frac{\partial X}{\partial \eta_P} = 4X\Delta\frac{\partial X}{\partial \eta_P} + 2X^2 \frac{\partial \Delta}{\partial \eta_P} \Rightarrow (1 + 4X\Delta)\frac{\partial X}{\partial \eta_P} = -2X^2\frac{\partial \Delta}{\partial \eta_P} \qquad 19.63$$

Multiplying and dividing the left side by X and replacing $2X^2\Delta$ with $1 - X$, then multiplying and dividing the right side by Δ and replacing $2X^2\Delta$ with $1 - X$, multiplying by η_P we obtain Z^{chem} by Eqn. 19.61,

$$Z^{chem} = \frac{(2-X)}{2}\frac{\eta_P}{X}\frac{\partial X}{\partial \eta_P} = \frac{-(1-X)}{2}\frac{\eta_P}{\Delta}\frac{\partial \Delta}{\partial \eta_P} \qquad 19.64$$

Recalling the definition of Δ from Eqn. 19.44,

$$\Delta_{vdW} = \frac{\eta_P}{(1-\eta_P)}\frac{K^C}{b}(\exp(\beta\varepsilon^C)-1) \Rightarrow \frac{\eta_P \partial \Delta_{vdW}}{\Delta_{vdW}\partial \eta_P} = \frac{\partial \ln \Delta_{vdW}}{\partial \ln \eta_P} = \frac{1}{(1-\eta_P)} \qquad 19.65$$

Substituting into 19.64 gives

$$Z^{chem} = -(1-X)/[2(1-\eta_P)] \quad \text{(pure dimerizing fluid)}$$

Substituting Z^{chem} into 19.61 gives

$$\frac{A^{chem}}{n_0 RT} = \ln(X) - \frac{\eta_P}{(1-\eta_P)}\left(\frac{1-X}{2}\right) + \left(\frac{1}{1-\eta_P}\right)\left(\frac{1-X}{2}\right) = \ln(X) + \frac{(1-\eta_P)}{(1-\eta_P)}\left(\frac{1-X}{2}\right) \qquad 19.66$$

Hence we have recovered Eqn. 19.54 and verified our trial solution using the van der Waals model, without the assumption of low density. The online chapter notes include a demonstration that Eqn. 19.54 is also recovered with the ESD model. Altogether, we can thoroughly appreciate the results of Wertheim's analysis, even if the rigors of Wertheim's statistical mechanics exceed our current scope. We can derive the framework of the simultaneous reaction and phase equilibria and see the crucial terms requiring simplification. At that point, Wertheim's "clever guesses" provide a tremendous simplification of an immensely complex problem, all the more remarkable when recognizing that they are thoroughly grounded in a rigorous fundamental analysis.

Given Eqn. 19.54 for the free energy and Eqn. 19.42 to solve for X, the model is essentially solved. $Z = 1 + Z^{rep} + Z^{att} + Z^{chem}$ can be solved at a given T and P by iterating on ρ. The algorithm to solve for apparent density is illustrated in Fig. 19.8. Then the Helmholtz energy departure can be differentiated to give the component fugacity. Relative to the binary VLE calculations, this pure component result might not seem so impressive. Nevertheless, the upcoming sections show that the extension to chain association in mixtures will involve only slightly more computations.

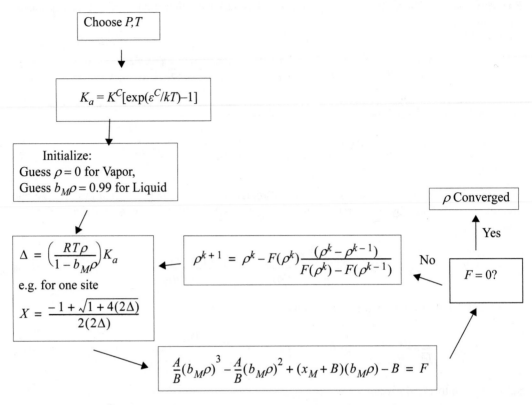

Figure 19.8 *Flow sheet for calculating density by the van der Waals associating fluid model.*

Example 19.4 The chemical contribution to the equation of state

Assuming 2Δ is about 1000 at 300 K and $\rho = 1.04$ g/cm^3, estimate A^{chem}/RT, Z^{chem}, and x_M of liquid acetic acid. You may assume that $b_M = 23$ cm^3/mol and the van der Waals model for Z.

Solution: Referring to Eqn. 19.66, $A^{chem}/RT = \ln(X) + (1 - X)/2$ and $Z^{chem} = -(1 - X)/[2(1 - \eta_P)]$

$$X = [-1 + \sqrt{1 + 4(2\Delta)}\,]/(2(2\Delta)) = 0.03113 \Rightarrow A^{chem}/RT = -2.985$$

$$\eta_P = 23\cdot 1.04/60 = 0.3987 \Rightarrow Z^{chem} = -0.806$$

$$\text{Referring to Eqn. 19.39, } x_M = 2X/(1 + X) = 0.06037$$

Comparing x_M to X, the true solution is 94% dimer and 97% of the acid molecules exist in the dimer form. It is also interesting that $Z^{chem} < -0.5$ for this liquid phase. The amount of dimerization would decrease at lower density, and for the gas phase it would be significantly lower, with Z^{chem} smaller in magnitude.

Chain Association

To extend the analysis from dimer formation to model chain formation, the primary adjustment is to assign two sites per molecule, consistent with one proton acceptor (A) and one proton donor (D), as we might expect for an alcohol. We can easily count the number of acceptors and donors in such linear chains by noting that one unbonded acceptor is left in each bonded chain, referring to Fig. 19.9. The equations for donors are entirely symmetrical and are omitted for simplicity. Note that n^A (the mole number of acceptors *not* bonded) is something quite different from n_A (the mole number of an "A-mer"). The extent of association is then characterized in terms of the fraction of acceptor sites *not* bonded, X_j^A. To see the relationship, consider the mass balances We obtain,

$$n^A = \Sigma\, n_j = n_T \qquad\qquad 19.67$$

But the total number of acceptors is given by noting that there are "j" total acceptors per j-mer,

$$n_0^{\ A} = \Sigma\, jn_j = n_0 \qquad\qquad 19.68$$

Note that n_0 refers to the same apparent number of moles discussed previously. The fraction of unbonded sites is a ratio of 19.67 and 19.68:

$$\boxed{X^A \ = \ n^A / n_0^A \ = \ n_T / n_0} \qquad \text{(pure chain-forming fluid)} \qquad 19.69$$

There is a further simplification that results from treating the bonding sites instead of the bonding molecules. The fraction of sites bonded can be perceived as a simple product of the bonding probabilities. First, note that the fraction of monomers bonded, $x^{AD} = n^{AD}/n_0$, and the fraction of monomers not bonded, X^A, must sum to unity.

$$X^A + x^{AD} = 1 \qquad\qquad 19.70$$

x^{AD} is the fraction of acceptors that are bonded, regardless of whether they are bonded in monomers, dimers, trimers, ... In principle, the second term is an infinite sum. From an acceptor site perspective, however, we assume that the thermodynamic change from the unbonded state to the bonded state is the same, regardless of the degree of polymerization for that i-mer. That is, adding one more monomer to the end of a chain has the same equilibrium constant regardless of the chain length. Chemically, we have

$$A_{i-1} + A = A_i \text{ and } \Delta^{AD}_{\ i} = \Delta^{AD}_{\ i-1} \text{ for all } i \qquad\qquad 19.71$$

That transition can be represented by

$$x^{AD} = 1 - X^A = X^A \cdot X^D \cdot \Delta^{AD} \qquad\qquad 19.72$$

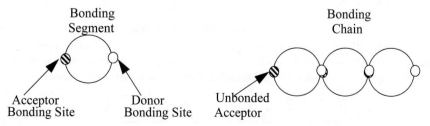

Figure 19.9 *Wertheim's theory of chain association in a two-site model.*

where X^D is the fraction of unbonded donors and $\Delta^{AD} = \rho g(\sigma) \, K^{AD}[\exp(\varepsilon^{AD}/kT) - 1]$ adapts Eqn. 19.44 to AD interactions. The term on the left is the fraction of acceptors that are bonded, and the term on the right expresses the observation that acceptors and donors must be unbonded in order to be available for bonding. By noting that one donor bonds for every acceptor, we see that $X^A = X^D$. Then we write Eqn. 19.72 in terms of X^A, $(X^A)^2$, and Δ^{AD} and we obtain $\Delta^{AD}(X^A)^2 + X^A - 1 = 0$ or

$$X^A = \frac{-1 + \sqrt{1 + 4\Delta^{AD}}}{2\Delta^{AD}} \quad \text{(pure chain-forming fluid)} \qquad 19.73$$

The extension of the Helmholtz energy to chain formation simply applies the same formula developed for dimer formation. This formula accounts for the change in entropy and energy each time a bond is formed. Whether the bond is formed as part of a dimer or part of a chain, the reduction in entropy by forming a bond is the same. So is the energy released by the bond formation. In terms of acceptors and donors for a pure chain-forming fluid, Eqn. 19.54 becomes,

$$A^{chem}/RT = \ln(X^A) + (1 - X^A)/2 + \ln(X^D) + (1 - X^D)/2 = 2\ln(X^A) + (1 - X^A) \quad \text{(pure chains)} \quad 19.74$$

Eqns. 19.73, 19.74, and 19.44 characterize the chemical contribution for molecules like alcohols. Given a temperature and density, Eqn. 19.44 gives Δ^{AD}, then Eqn. 19.73 gives X^A, then Eqn. 19.74 gives A^{chem}, then Eqn. 19.1 gives $(A - A^{ig})$. Altogether, just one extra step (Eqn. 19.73) is required to compute the Helmholtz energy relative to the van der Waals model, but the rigor of the chemical perspective is greatly enhanced. In the MOSCED model, for example, the contribution involving $(\alpha_i - \alpha_j)(\beta_i - \beta_j)$ is entirely empirical. It is contrived to give the right sign when mixing acids and bases, but there is no basis for it in theory, and it does not alter the skewness of the excess Gibbs energy. On the other hand, Wertheim's theory is based on a rigorous derivation relating the molecular scale bonding volume and energy to the macroscopic properties, and fundamentally altering the behavior of the Gibbs energy.

Extension to Mixtures

Analyzing the impact of the chemical contribution on excess Gibbs energy requires extension of Wertheim's theory to mixtures. The beauty of Wertheim's perspective is that the extension of the reaction equilibrium relation (Eqn. 19.72) is entirely straightforward. One donor must bond for each bonded acceptor, whether the molecules are mixed or pure. The only issue is which molecule possesses the acceptor and which possesses the donor, but that is a notational detail. Furthermore, the fraction of bonded acceptors of type i must be in equilibrium with the all unbonded donors which can interact, including those on other types of molecules. We simply need to sum all the transition probabilities and the extension becomes:

$$\boxed{1 - X_i^A = X_i^A \sum_j x_j N_{d,j} X_j^D \Delta_{ij}^{AD}} \quad \text{or for solving use} \quad \boxed{X_i^A = 1 \bigg/ \left(1 + \sum_j x_j N_{d,j} X_j^D \Delta_{ij}^{AD}\right)} \qquad 19.75$$

where the sum is over donor sites, x_j is the apparent molecule mole fraction of the site host. $N_{d,j}$ is the number of identical donor sites of type j on the host, like the hydroxyl sites in polyvinyl alcohol. Analogous equations are written for each donor site, where the sum is over each interacting acceptor site and we replace Δ_{ij}^{AD} with Δ_{ij}^{DA}. Extending Eqn. 19.44 to 'AD type' association,

$$\boxed{\Delta_{ij}^{AD} = \rho g_{ij}(\sigma_{ij}) K_{ij}^{AD}[\exp(\varepsilon_{ij}^{AD}/(kT)) - 1]} \qquad 19.76$$

The ordering of the subscripts and superscripts in Eqn. 19.76 provides the notational detail that permits accounting for which bonding site is the acceptor and which is the donor. For example, by writing $\varepsilon_{ij}{}^{AD}$, the acceptor index is i and the donor is j. When the donor is i and the acceptor j, we write $\varepsilon_{ij}{}^{DA}$. To clarify, this distinction might be important in a mixture of alcohols and amines. An amine is usually a weak proton donor (indicated by its low acidity, α) but a strong proton acceptor (indicated by its basicity, β), whereas a typical alcohol has roughly equal acidity and basicity ($\alpha = \beta$). Numbering the acceptor and donor on the alcohol with subscript 1, and those on the amine with subscript 2, if we suppose that $\beta_{\text{alcohol}} \sim \beta_{\text{amine}}$, then $\varepsilon_{11}{}^{AD} \sim \varepsilon_{12}{}^{AD} > \varepsilon_{12}{}^{DA} \sim \varepsilon_{22}{}^{DA}$. This would mean strong solvation for the amines in alcohols, and negative deviations from Raoult's law, as observed experimentally. Eqn. 19.75 states that the bonding probability for an acceptor increases when there are donors on other molecules, and it decreases proportional to the mole fraction when the donor species are diluted by nonassociating species. The precise extent of chemical interaction is controlled by $\Delta_{ij}{}^{AD}$, which could range from zero (for alkanes in water, for example) to a substantial quantity (when mixing carboxylic acids, for example). The extension of A^{chem} to any number of bonding sites or components becomes recalling that the sums are over sites,

$$
\frac{A^{chem}}{RT} = \sum_{i \in A} x_i N_{d,\,i} [\ln X_i^A + (1 - X_i^A)/2] + \sum_{i \in D} x_i N_{d,\,i} [\ln X_i^D + (1 - X_i^D)/2]
$$
$$
+ \sum_{i \in C} x_i N_{d,\,i} [\ln X_i^C + (1 - X_i^C)/2]
$$

19.77

Briefly, this equation indicates that the change in Helmholtz energy due to bonding is the same regardless of how those bonds are formed. In other words, the reduction in entropy due to bond formation is universal when the packing fraction is unchanged. We know that entropy is the primary contribution because energy does not appear explicitly in Eqn. 19.77. Bonding energy affects A^{chem} implicitly through Δ, because a larger energy gives a larger value of Δ and a smaller value of X_i^B. We have included X^C here to represent dimerization (e.g., carboxylic acid bonding) as something distinct from AD interaction. A's can only bond with D's, and this leads to chain formation. On the other hand, C's can only bond with C's, confining these sites to dimerization. Although we wrote the equation for these three types of bonds, there is really no limit and Eqn. 19.77 can be extended straightforwardly to many situations. In the description here, X^C would not affect X^A because C's can only interact with C's. If a carboxylic acid (e.g., acetic acid) is to interact with an alcohol (e.g., water), A's or D's would need to be included as part of the carboxylic acid segment, in addition to the C's, as illustrated in Fig. 19.10.[18]

Eqns. 19.75–19.77 provide a powerful and versatile complement to our treatment of phase equilibria. In Chapters 10–12, we might have alluded to hydrogen bonding, for example, as a reason why oil and water do not mix, but our models did not truly recognize it as bonding. The van der Waals models and local composition theories treat attractive energy as spherically symmetric, like the square-well potential. But complexation is stereospecific and this alters the description of the

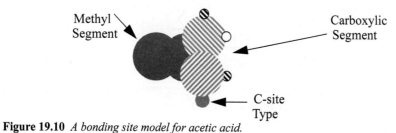

Methyl Segment

Carboxylic Segment

C-site Type

Figure 19.10 *A bonding site model for acetic acid.*

---

18. An alternative perspective would allow C to interact with A, C, or D. This would not change Eqn 19.77, but would require terms like Δ^{CA} and Δ^{CD}. *cf.* Muro-Suñé N, et al. 2008. *Ind. Eng. Chem. Res.* 47:5660–5668.

Helmholtz energy. The Helmholtz energy of hydrogen bonding is as different from that of the van der Waals model as Eqn. 19.77 is from $a\rho/RT$.

19.7   MASS BALANCES FOR CHAIN ASSOCIATION

The thermodynamics and phase behavior are sufficiently described by Eqns. 19.75 and 19.77, but you may be curious about the true mole fractions of the species. Furthermore, it is interesting to see how this "fraction of acceptor sites not bonded" is closely related to the fraction monomer, x_M. This turns out to be a bit subtle, and it should not distract you from the primary issue of phase behavior. If you are interested, we can use material balances to obtain two simple relations between the true number of moles in the solution, n_T, and the apparent number of moles that we would expect if there was no association,[19] n_0. Note that n_0 is the number of moles one would compute based on dividing the mass of solution by the molecular weight of a monomer as taught in introductory chemistry. For example, in 100 cm^3 of water one would estimate

$$n_0 = 100 \text{ cm}^3 \cdot 1.0 \text{ g/cm}^3/(18 \text{ g/mole}) = 5.556 \text{ moles}$$

But how many moles of H_2O monomer do you think truly exist in that beaker of water? We will return to this question shortly. Note that each i-mer contains "i" monomers, such that the contribution to the apparent number of moles is $i \cdot n_i$. Note also that the true mole fractions, x_i, are given by n_i/n_T, but it may not look so simple at first.

For molecules with one donor and one acceptor site, the fraction of unbonded acceptors is the joint probability that the acceptors and donors are both unbonded, and since by mass balance $X^A = X^D$, then $n_M/n_0 = X^A X^D = (X^A)^2 = (n_M/n_T)(n_T/n_0) = x_M(n_T/n_0)$. We can write the equilibrium relation for i-mer formation using 19.43 as $x_i/(x_M x_{i-1}) = ((\hat{\phi}_M \hat{\phi}_{i-1})/(\hat{\phi}_i P°))PK_{a,i} = (\Delta)(n_T/n_0)$. Combining the probability and equilibrium relations, $x_i = x_{i-1}(X^A)^2\Delta$. A recursive relation results, $x_i = x_M((X^A)^2\Delta)^{i-1}$. Now consider the balances

$$n_0 = \Sigma \, i \cdot n_i = n_T \Sigma \, i \cdot x_i \qquad\qquad 19.78$$

Substituting the recursive relation,

$$n_0 = n_T \Sigma \, i \cdot x_M((X^A)^2\Delta)^{i-1} = x_M n_T [1 + 2((X^A)^2\Delta) + 3((X^A)^2\Delta)^2 + 4((X^A)^2\Delta)^3 + \ldots]$$

This series is a common converging series. We find from a math handbook that

$$n_T x_M [1 + 2((X^A)^2\Delta) + 3((X^A)^2\Delta)^2 + 4((X^A)^2\Delta)^3 + \ldots] = n_T x_M [1/(1 - (X^A)^2\Delta)^2]$$

$$\boxed{\frac{n_0}{n_T} = \frac{x_M}{(1 - (X^A)^2\Delta)^2} = \sum_i i \cdot x_i} \qquad\qquad 19.79$$

Since the mole fractions must sum to unity, we can write a second balance, for x_i,

$$\boxed{1 = \sum_i x_i = x_M[1 + ((X^A)^2\Delta) + ((X^A)^2\Delta)^2 + ((X^A)^2\Delta)^3 + \ldots]} \qquad\qquad 19.80$$

---

19. Here we choose to use subscript 0 to clearly distinguish the notation for apparent moles, even though it would be the quantity normally reported from a macroscopic experiment.

and again recognizing the series,

$$1 = x_M \left[1 + ((X^A)^2 \Delta) + ((X^A)^2 \Delta)^2 + ((X^A)^2 \Delta)^3 + \ldots \right] = x_M \left[1/(1 - (X^A)^2 \Delta) \right]$$

$$\boxed{x_M = (1 - (X^A)^2 \Delta)} \qquad\qquad 19.81$$

Substituting x_M for $(1 - (X^A)^2 \Delta)$ in Eqn. 19.79 results in,

$$n_0 / n_T = x_M / x_M^2 = 1/x_M$$

Noting Eqn. 19.69

$$\boxed{\frac{n_T}{n_0} = X^A = x_M} \; ; \text{and} \; \frac{n_M}{n_0} = (X^A)^2 \text{ (pure chains)} \qquad\qquad 19.82$$

This equation makes clear that the properties of the mixture are closely related to the properties of the monomer.

Example 19.5   Molecules of H_2O in a 100 ml beaker

Modeling water as chains, assuming Δ is about 100 at room temperature and $\rho = 1$ g/cm^3, estimate the moles of H_2O monomer in a 100 ml beaker of liquid water.

Solution: Note that the problem statement requests moles of H_2O, not $(H_2O)_2$ or $(H_2O)_3$, and so on, so we are interested in the true number of H_2O monomer moles. We know $n_0 = 5.556$ by applying the monomer molecular weight, but the number of monomer moles $n_M = x_M n_T$ will be significantly less. Proceeding, using Eqn. 19.73,

$$X^A = [-1 + \sqrt{1 + 4\Delta}\,]/2\Delta = 0.095 = n^A/n_0 = n_T/n_0 = x_M = 0.095 = n_T/n_0$$

$$\Rightarrow n_M = 0.095 \cdot (n_T/n_0) \cdot n_0 = 0.095^2 \cdot n_0 = 0.05 \text{ moles}$$

Therefore, the true number of moles is 100 times less than the apparent number of moles.

19.8   THE CHEMICAL CONTRIBUTION TO THE FUGACITY COEFFICIENT AND COMPRESSIBILITY FACTOR

The solution to phase equilibrium problems can be achieved in the manner of Chapter 15 (Eqn 15.20), where Eqns. 19.1 and 19.2 describe the enhanced equation of state. Eqns. 19.75–19.77 completely characterize the temperature, density, and composition dependence of the chemical contribution to Helmholtz energy. The Z^{chem} contribution is implied, but requires differentiation as in $RT \cdot Z^{chem} = -V(\partial A^{chem}/\partial V)_T$. Similarly, the fugacity coefficient is implicitly determined through differentiation. Nevertheless, the differentiation can be complicated relative to the fugacity coefficient of the van der Waals model. The summation of Eqn. 19.77 means that terms like $\partial X_i/\partial n_j$ contribute and Eqn. 19.75 implies a nonlinear system of equations that must be solved to determine these contributions. For example, consider a mixture of three alcohols, with $X_i^A = X_i^D$ and $N_{d,i} = 1$ for all i. Numbering the acceptors and donors with the same subscript as their host, Eqn. 19.75 implies that

$$- \partial X_1^A / \partial n_1 = X_i^A \left[x_1 X_1^A \Delta_{11}^{AD} \partial X_1^A / \partial n_1 + x_2 X_2^A \Delta_{12}^{AD} \partial X_1^A / \partial n_2 + x_1 X_3^A \Delta_{13}^{AD} \partial X_1^A / \partial n_3 \right] \qquad 19.83$$

The only way to fully determine all $\partial X_i / \partial n_j$ is to apply Eqn. 19.75 eight more times to obtain nine equations for the nine unknown values implied by $\partial X_i / \partial n_j$. Once again, Wertheim's theory seems to become impractical.

Fortunately, this particular nonlinear system of equations possesses subtle but advantageous properties. Briefly, there are many symmetries in the calculus that lead to surprising simplifications when cleverly manipulated. Michelsen and Hendriks showed that $\underline{A}^{chem}/(RT)$ can be rewritten as the stationary point of a generalized function $\underline{Q}$ where $\underline{A}^{chem}/(RT)$ is minimized and $\underline{Q}$ is maximized.[20] A result of the chemical equilibria is that $\partial \underline{Q} / \partial X_i^B = 0$, for all i and B, where B is an acceptor or donor. The beauty of the generalized function is that derivatives with respect to X_i^B can be separated from derivatives with respect to $\underline{V}$ or n_j. Applying the expansion rule to $\underline{Q}(T, \underline{V}, n, X)$, and using $(\partial \underline{Q}/(\partial X_i^B))_{T, \underline{V}, n_{j \neq k}} = 0$ at the stationary point,

$$\left(\frac{\partial \underline{Q}_{sp}}{\partial \underline{V}} \right)_{T, n} = \left(\frac{\partial \underline{Q}}{\partial \underline{V}} \right)_{T, n, X} + \sum_{i \in B} \left(\frac{\partial \underline{Q}}{\partial X_i^B} \right)_{T, n, \underline{V}} \left(\frac{\partial X_i^B}{\partial \underline{V}} \right) = \left(\frac{\partial \underline{Q}}{\partial \underline{V}} \right)_{T, n, X} + \sum_{i \in B} (0) \left(\frac{\partial X_i^B}{\partial \underline{V}} \right) = \left(\frac{\partial \underline{Q}}{\partial \underline{V}} \right)_{T, n, X}$$

$$19.84$$

Similarly,

$$\left(\frac{\partial \underline{Q}_{sp}}{\partial n_k} \right)_{T, \underline{V}, n_{j \neq k}} = \left(\frac{\partial \underline{Q}}{\partial n_k} \right)_{T, \underline{V}, X, n_{j \neq k}} + \sum_{i \in B} \left(\frac{\partial \underline{Q}}{\partial X_i^B} \right)_{T, \underline{V}, n_{j \neq k}} \left(\frac{\partial X_i^B}{\partial n_k} \right) = \left(\frac{\partial \underline{Q}}{\partial n_k} \right)_{T, \underline{V}, X, n_{j \neq k}} \qquad 19.85$$

The generalized function, $\underline{Q}$, is intentionally created by shifting $\underline{A}^{chem}/(RT)$ a manner such that derivatives with respect to X_i^B will cancel and that $\underline{Q} = \underline{A}^{chem}/(RT)$ at the stationary point. Ignoring C-type sites of Eqn. 19.77 for this discussion, summing over sites using x_i as in Eqn. 19.75

$$Q(T, \underline{V}, n, X) = \sum x_i N_{d,i} [\ln(X_i^A) + (1 - X_i^A)] + \sum x_i N_{d,i} [\ln(X_i^D) + (1 - X_i^D)] - h/2 \qquad 19.86$$

Each site summation has been shifted by $\sum x_i N_{d,i} (1 - X_i^B)/2$. The term h is constructed to cancel this shift at the stationary point (equilibrium) by using the right hand side of Eqn. 19.75 for each x_i.

$$h = \sum_{i \in A} \sum_{j \in D} x_i x_j N_{d,i} N_{d,j} X_i^A X_j^D \Delta_{ij}^{AD} + \sum_{i \in D} \sum_{j \in A} x_i x_j N_{d,i} N_{d,j} X_j^A X_i^D \Delta_{ij}^{DA}$$

$$\underline{h} = \sum_{i \in A} \sum_{j \in D} n_i n_j N_{d,i} N_{d,j} X_i^A X_j^D (\Delta_{ij}^{AD}/n) + \sum_{i \in D} \sum_{j \in A} n_i n_j N_{d,i} N_{d,j} X_j^A X_i^D (\Delta_{ij}^{DA}/n) \qquad 19.87$$

where all x and n are for apparent moles. We can write the extensive expression of Eqn. 19.86, $\underline{Q}$

$$\underline{Q}(T, \underline{V}, n, X) = \sum_{i \in A} n_i N_{d,i} [\ln(X_i^A) + (1 - X_i^A)] + \sum_{i \in D} n_i N_{d,i} [\ln(X_i^D) + (1 - X_i^D)] - \underline{h}/2 \qquad 19.88$$

The balance Eqn. 19.75 should not be inserted except at the stationary point. Even though the right and left side of this equation were used to create $\underline{Q}$, the equality holds only at the stationary point. Because $\underline{Q}_{sp} = \underline{A}^{chem}/(RT)$ at the stationary point compositions determined by Eqn. 19.75, we can take advantage of Eqns. 19.84 and 19.86 to obtain,

---

20. Michelsen, M.L., Hendriks, E.M. 2001. *Fluid Phase Equil.* 180:165.

$$RT \cdot Z^{chem} = -V\partial(A^{chem}/\partial V) = \eta_P RT(\partial Q/\partial \eta_P)_\mathbf{X} = -0.5RT\eta_P(\partial h/\partial \eta_P)_\mathbf{X}$$

Note that Q and $\underline{Q}$ are not manipulated before differentiation, and remember that only h is density dependent. Differentiating Eqn. 19.87,

$$-2Z^{chem} = \eta_P(\partial h/\partial \eta_P)_\mathbf{X} = \sum_{i \in A}\sum_{j \in D} x_i x_j N_{d,i} N_{d,j} X_i^A X_j^D (\eta_P \partial \Delta_{ij}{}^{AD}/\partial \eta_P)$$
$$+ \sum_{i \in D}\sum_{j \in A} x_i x_j N_{d,i} N_{d,j} X_i^D X_j^A (\eta_P \partial \Delta_{ij}{}^{DA}/\partial \eta_P) \qquad 19.89$$

Condensing notation for 19.77, $\Delta = \rho g(\sigma) K^{AD\dagger}$ where $K^{AD\dagger} = K^{AD}[\exp(\varepsilon/(kT)) - 1]$. To simply Eqn. 19.89, consider the derivative,

$$\eta_P \frac{\partial \Delta_{ij}}{\partial \eta_P} = \frac{\eta_P K_{ij}^{AD\dagger}}{b} \frac{\partial(\eta_P g(\sigma))}{\partial \eta_P} = \rho K_{ij}^{AD\dagger}\left(g(r) + \eta_P \frac{\partial g(\sigma)}{\partial \eta_P}\right) = \Delta_{ij}\left(1 + \frac{\partial \ln g(\sigma)}{\partial \ln \eta_P}\right)$$

Inserting into 19.89 and recognizing that we can insert 19.75 at the stationary point,

$$\boxed{Z^{chem} = -0.5\left(1 + \frac{\partial \ln g(\sigma)}{\partial \ln \eta_P}\right)\left(\sum_{i \in A} x_i N_{d,i}(1 - X_i^A) + \sum_{i \in D} x_i N_{d,i}(1 - X_i^D)\right)} \qquad 19.90$$

$$\ln(\hat{\varphi}_k^{chem}) = (\partial \underline{Q}/\partial n_k)_\mathbf{X} = \sum N_{d,k}[\ln(X_k^A) + (1 - X_k^A)] + \sum N_{d,k}[\ln(X_k^D) + (1 - X_k^D)] - 0.5(\partial \underline{h}/\partial n_k)_\mathbf{X} \quad 19.91$$

$$(\partial \underline{h}/\partial n_k)_X = 2\sum x_j(N_{d,k} N_{d,j} X_k^A X_j^D \Delta_{kj}{}^{AD} + N_{d,k} N_{d,j} X_k^D X_j^A \Delta_{kj}{}^{DA})$$
$$+ \sum\sum x_i x_j N_{d,i} N_{d,j}[X_i^A X_j^D n^2 \partial(\Delta_{ij}{}^{AD}/n)/\partial n_k + X_i^D X_j^A n^2 \partial(\Delta_{ij}{}^{DA}/n)/\partial n_k] \qquad 19.92$$

where all x and n are for apparent moles. Cancellation of terms by Eqn. 19.75 results in

$$\ln(\hat{\varphi}_k^{chem}) = \sum N_{d,k}\ln(X_k^A) + \sum N_{d,k}\ln(X_k^D) - 0.5\sum\sum x_i x_j N_{d,i} N_{d,j}[X_i^A X_j^D n^2 \partial(\Delta_{ij}{}^{AD}/n)/\partial n_k +$$
$$X_i^D X_j^A n^2 \partial(\Delta_{ij}{}^{DA}/n)/\partial n_k] \qquad 19.93$$

The derivative can be simplified

$$\frac{n^2 \partial(\Delta_{ij}/n)}{\partial n_k} = \frac{n^2 K_{ij}^{AD\dagger}}{\underline{V}} \frac{\partial g(\sigma)}{\partial n_k} = \frac{n\rho K_{ij}^{AD\dagger} g(\sigma)}{g(\sigma)} \frac{\partial g(\sigma)}{\partial n_k} = (\Delta_{ij})n\frac{\partial \ln g(\sigma)}{\partial n_k}$$

Inserting into 19.93 and recognizing $h_{sp} = \sum_{i \in A} x_i N_{d,i}(1 - X_i^A) + \sum_{i \in D} x_i N_{d,i}(1 - X_i^D)$ can replace the quadratic sum,

$$\boxed{\begin{aligned}\ln(\hat{\varphi}_k^{chem}) &= \sum_{i \text{ on } k}(N_{d,i}\ln(X_i^A) + N_{d,i}\ln(X_i^D)) - \\ &0.5\left(n\frac{\partial \ln g(r)}{\partial n_k}\right)\left(\sum_{i \in A} x_i N_{d,i}(1 - X_i^A) + \sum_{i \in D} x_i N_{d,i}(1 - X_i^D)\right)\end{aligned}} \qquad 19.94$$

The computational complexity of Eqns. 19.75–19.77 is reduced for the case with one acceptor and one donor per molecule assuming $\Delta_{(\text{molec } i)(\text{molec } j)}^{AD} = (\Delta_{(\text{molec } i)(\text{molec } i)}^{AD}\Delta_{(\text{molec } j)(\text{molec } j)}^{AD})^{1/2}$, which we refer to as the square root combining rule (SRCR). The SRCR is suitable for Δ_{ij} of alcohols + aldehydes + water, but not for alcohols + amines. In general, Eqns. 19.75–19.77 require an iterative solution, as illustrated in Example 19.7. An initial guess for any j,B adapts the SRCR method[21]

$$\frac{1}{X_j^B} \approx 1 + \sqrt{\Delta_{(\text{molec } j)(\text{molec } j)}^{AD}} \sum_{\text{molec } i} \{(x_i\sqrt{\Delta_{(\text{molec } i)(\text{molec } i)}^{AD}})/[1 + \sqrt{\Delta_{(\text{molec } i)(\text{molec } i)}^{AD}}]\} \quad 19.95$$

When the SRCR rule is not valid (e.g. alcohols + amines), Eqn. 19.95 can be adapted by replacing $\sqrt{\Delta_{(\text{molec } i)(\text{molec } i)}^{AD}\Delta_{(\text{molec } j)(\text{molec } j)}^{AD}}$ in the numerator with $\Delta_{(\text{molec } i)(\text{molec } j)}^{AD}$.

This concludes the theoretical development for the chemical contributions to phase equilibrium. Eqns. 19.75–19.77 and 19.90 and 19.94 permit solution of Eqns. 19.1 and 19.2 for mixtures as well as pure fluids and computation of the fugacity coefficients to perform any phase equilibrium determination. Wertheim's theory of solution thermodynamics is more challenging than that of van der Waals or local compositions, but it replaces the empirical conjectures of those models with rigorous analysis that has been verified with molecular simulations. The perspective offered by Wertheim's theory will be extended to nonspherical molecules in the following section.

Example 19.6 Complex fugacity for the van der Waals model

A sample calculation with a specific reference equation of state will clarify application for trimethylamine(t) + methanol(m). With A_t on component t, and A_m and D_m on component m, let $K_{tm}^{AD} = K_{mm}^{AD} = 0.72$ cm³/mol and $\varepsilon_{tm}^{AD} = \varepsilon_{mm}^{AD} = 20$ kJ/mol, $b_t = 27.5$ and $b_m = 20.4$ cm³/mol. For the associating van der Waals equation, assuming chains form,
(a) Derive Z^{chem} and $\ln(\hat{\varphi}_k^{chem})$ adapting the definition of Δ from Eqn. 19.44.
(b) Evaluate the expressions at $x_t = 0.5$, $\rho = 0.0141$ mol/cm³, and $T = 300$K.
Solution: (a) For Z^{chem}, we need

$$1 + \frac{\partial \ln g(\sigma)}{\partial \ln \eta_P} = 1 + \frac{\eta_P}{g(\sigma)}\frac{(\partial(1-\eta_P)^{-1})}{\partial \eta_P} = 1 + \frac{\eta_P}{g(r)(1-\eta_P)^2} \frac{1}{} = \frac{1}{(1-\eta_P)} \quad 19.96$$

$$Z^{chem} = \frac{-0.5}{(1-\eta_P)}\left(\sum_{i \in A} x_i N_{d,i}(1-X_i^A) + \sum_{i \in D} x_i N_{d,i}(1-X_i^D)\right) \quad 19.97$$

For $(\hat{\varphi}_k^{chem})$ we need

$$n\frac{\partial \ln g(\sigma)}{\partial n_k} = \frac{n}{g(\sigma)}\frac{\partial(1-\eta_P)^{-1}}{\partial n_k} = \frac{n}{g(\sigma)}\frac{(\partial \eta_P)/(\partial n_k)}{(1-\eta_P)^2} = \frac{b_k \rho}{(1-\eta_P)} \quad 19.98$$

$$\ln(\hat{\varphi}_k^{chem}) = \sum_{i \text{ on } k}(N_{d,i}\ln(X_i^A) + N_{d,i}\ln(X_i^D))$$
$$-\frac{0.5 b_k \rho}{(1-\eta_P)}\left(\sum_{i \in A} x_i N_{d,i}(1-X_i^A) + \sum_{i \in D} x_i N_{d,i}(1-X_i^D)\right) \quad 19.99$$

---

21. Elliott, J.R. 1996. *Ind. Eng. Chem. Res.* 35:1624. To relate Elliott's variable F to the h used here, $F^2 = h/2 = [\sum_i x_i X_j^A (\Delta_{jj}^{DA})^{1/2}]^2$.

Example 19.6 Complex fugacity for the van der Waals model (Continued)

b) Because only one acceptor/donor value exists for the specified interactions, $\Delta_{ij}^{AD} = \rho K^{AD}(\exp(\beta \varepsilon_{ij}^{AD}) - 1)/(1 - \eta_P) = \Delta$.

$$b = 0.5 \cdot 27.5 + 0.5 \cdot 20.4 = 24.0; \quad \eta_P = 0.0141 \cdot 24.0 = 0.338 \qquad \text{19.100}$$

$$\Delta_{tm}^{AD} = \Delta_{mm}^{AD} = 0.338/(1 - 0.338) \cdot (0.72/24.0) \cdot (\exp(20000/8.314/300) - 1) = 46.4 \qquad \text{19.101}$$

To solve Eqns. 19.92 for X_t^A, X_m^A, and X_m^D,

$$1 - X_t^A = 0.5 X_t^A X_m^D \Delta \qquad \text{19.102}$$

$$1 - X_m^D = 0.5 X_t^A X_m^D \Delta + 0.5 X_m^A X_m^D \Delta \qquad \text{19.103}$$

$$1 - X_m^A = 0.5 X_m^A X_m^D \Delta \qquad \text{19.104}$$

This gives three equations. Note that $X_t^A = X_m^A$ for this case. We use these in Eqn. 19.103 to obtain a quadratic equation in terms of X_m^D. Usually, we would need to iterate to solve for X_i^B.

$$X_t^A = X_m^A = 1/(1 + 0.5 X_m^D \Delta) \qquad \text{19.105}$$

$$X_m^D = 1/(1 + 0.5 X_t^A \Delta + 0.5 X_m^A \Delta) = 1/(1 + X_m^A \Delta) = 1/(1 + \Delta/(1 + 0.5 X_m^D \Delta)) \qquad \text{19.106}$$

$$X_m^D = (1 + 0.5 X_m^D \Delta)/(1 + (1 + 0.5 X_m^D)\Delta)$$

$$X_m^D + X_m^D \Delta + 0.5(X_m^D)^2 \Delta = 1 + 0.5 X_m^D \Delta$$

$$X_m^D = [-(1 + \Delta/2) + \sqrt{(1 + \Delta/2)^2 + 2\Delta}\,]/\Delta = 0.0397 \qquad \text{19.107}$$

$$X_m^A = X_t^A = 1/(1 + 0.5 \cdot 0.0397 \cdot 46.4) = 0.520 \qquad \text{19.108}$$

This shows that D_m is almost completely bonded. Calculating the sum in Z^{chem} of 19.97,

$$h_{sp} = 0.5(1 - 0.52) + 0.5(1 - 0.52) + 0.5(1 - 0.0397)] = 0.960$$

$$Z^{chem} = -0.5 h_{sp}/(1 - \eta_P) = -0.480/(1 - 0.338) = -0.725$$

By Eqn. 19.99, recognizing Z^{chem} within the last term of each

$$\ln(\hat{\varphi}_t^{chem}) = \ln(X_t^A) - 0.5 h_{sp} b_t \rho/(1 - \eta_P) = -0.6539 - 0.725(27.5)(0.0141) = -0.935$$

$$\ln(\hat{\varphi}_m^{chem}) = \ln(X_m^A) + \ln(X_m^D) - 0.5 h_{sp} b_m \rho/(1 - \eta_P)$$
$$= -0.6539 - 3.226 - 0.725(20.4)(0.0141) = -4.09$$

There are several points of interest in this result. The acceptors in this mixture outnumber donors by two to one. Therefore, it is impossible that $X_i^A < 0.5$, and, in fact, $X_m^D \sim 2 \cdot (X_t^A - 0.5)$ because the lack of donor saturation is reflected twice, in X_t^A and X_m^A.   The compressibility factor is depressed in a simple way that sums over all donors and acceptors, but the fugacity is depressed more for the alcohol than for the amine. There are two ways for the alcohol to interact, but only one for the amine, so the depression of the fugacity is much greater. On the other hand, the fugacity of the alcohol is depressed less in the mixture than in the pure fluid because relatively fewer acceptors are bonded ($\ln(\hat{\varphi}_m^{chem}) = -6.105$ at $x_m = 1$). So the mixture activity for the alcohol is enhanced by less hydrogen bonding relative to the pure component, while the activity of the amine is depressed by more hydrogen bonding at all compositions relative to the pure component.

Example 19.7 More complex fugacity for the van der Waals model

Evaluate the expressions for Z^{chem} and $\ln(\hat{\varphi}_k^{chem})$ of trimethylamine(t) + methanol(m) at $x_t = 0.4$, $\rho = 0.0141$ mol/cm^3, and $T = 300$ K. With A_t on component t, and A_m and D_m on component m, let $K_{tm}^{AD} = K_{mm}^{AD} = 0.72$ cm^3/mol and $1.25\varepsilon_{tm}^{AD} = \varepsilon_{mm}^{AD} = 20$ kJ/mol, $b_t = 27.5$, and $b_m = 20.4$ cm^3/mol.

Solution: The difference between this example and the previous is that $\varepsilon_{tm}^{AD} \neq \varepsilon_{mm}^{AD}$, indicating that the amine + alcohol is slightly weaker than the alcohol + alcohol association. Because of this lack of symmetry, an iterative solution for X_i^B is required.

Substituting the mole fractions and solving for Δ's,

$b = 0.4 \cdot 27.5 + 0.6 \cdot 20.4 = 23.4$; $\eta_P = 0.0141 \cdot 23.4 = 0.328$. This is slightly less than Eqn 19.100.

$$\Delta_{mm}^{AD} = \rho K_{mm}^{AD}(\exp(\beta\varepsilon_{mm}^{AD}) - 1)/(1 - \eta_P) = 45.8; \quad \Delta_{tm}^{AD} = 9.21; \Delta_{tt}^{DA} = \Delta_{tm}^{DA} = 0.$$

$$1 - X_t^A = 0.6X_t^A X_m^D \Delta_{tm}^{AD}; \quad 1 - X_m^D = 0.4X_m^D X_t^A \Delta_{tm}^{AD} + 0.6X_m^D X_m^A \Delta_{mm}^{AD};$$

$$1 - X_m^A = 0.6X_m^A X_m^D \Delta_{mm}^{AD}; \quad \text{rearranging all three:}$$

$$X_t^A = 1/(1 + 0.6X_m^D \Delta_{tm}^{AD}); X_m^A = 1/(1 + 0.6X_m^D \Delta_{mm}^{AD});$$

$$X_m^D = 1/(1 + 0.4X_t^A \Delta_{tm}^{AD} + 0.6X_m^A \Delta_{tm}^{AD});$$

Unlike the previous example, an explicit solution is not found. The previous example was contrived to achieve an exact solution, but this is rarely possible. Normally, we must iterate to achieve a numerical solution. It is convenient to guess X_m^D, then compute X_t^A and X_m^A, then use successive substitution to converge all X_i^B. Adapting Eqn. 19.95 for the non-SRCR case,

$$1/X_m^D \approx 1 + 0.4(9.2)/(1 + \sqrt{0}) + 0.6(45.8)/(1 + \sqrt{45.8}) = 8.218 \Rightarrow X_m^D = 0.122;$$

$$X_t^A = 0.597; X_m^A = 0.230; X_m^D = 0.105;$$

Eleven more iterations gives $X_t^A = 0.677$; $X_m^A = 0.296$; $X_m^D = 0.0864$. The large number of iterations is necessary because this particular mixture deviates substantially from the SRCR.

$$X_t^A = 1/(1 + 0.6 \cdot 0.0864 \cdot 9.21) = 0.677$$
$$X_m^A = 1/(1 + 0.6 \cdot 0.0864 \cdot 45.8) = 0.296$$
$$X_m^D = 1/(1 + 0.4 \cdot 0.677 \cdot 9.21 + 0.6 \cdot 0.296 \cdot 45.8) = 0.086$$

Calculating the sum in Z^{chem} of 19.97,

$$h_{sp} = 0.4(1 - 0.677) + 0.6(1 - 0.296) + 0.6(1 - 0.086)] = 1.100$$
$$Z^{chem} = -0.5h_{sp}/(1 - \eta_P) = -0.55/(1 - 0.328) = -0.818$$
$$\ln(\hat{\varphi}_t^{chem}) = \ln(X_t^A) - 0.5h_{sp}b_t\rho/(1 - \eta_P) = -0.390 - 0.55(27.5)(0.0141)/(0.672) = -0.707$$
$$\ln(\hat{\varphi}_m^{chem}) = \ln(X_m^A) + \ln(X_m^D) - 0.5h_{sp}b_m\rho/(1 - \eta_P) = -3.91$$

These results show that a 20% decrease in ε_{tm}^{AD} compared to Example 19.6 gives a 80% decrease in Δ_{tm}^{AD}. That is fairly sensitive. This change in Δ_{tm}^{AD} is primarily responsible for the increase in X_t^A from 0.52 to 0.68 and the decrease of X_m^A from 0.520 to 0.30.

19.9 WERTHEIM'S THEORY OF POLYMERIZATION

Now that we have an accounting for the thermodynamics of bond formation, it is natural to wonder what happens to the thermodynamics as the bond energy approaches infinity. This would be a natural limit for covalent bond formation. Having a theoretical basis for nonspherical molecules would be a big step forward, considering that all theories discussed until now have been based on spherical molecules. Of course, we added correction terms like $\alpha(T, \omega)$ to the Peng-Robinson model, but this was done with no theoretical basis. Wertheim's theory provides an opportunity to develop meaningful guidelines for shape effects.

The key step is to find the contribution to the equation of state from forming a bond in the limit of infinite bond energy. The result for dimerization, Eqn. 19.54, is convenient to illustrate the key points. At first glance, the limit may not seem obvious, because the X term in A^{chem} must approach zero and the log term would then be undefined. This issue can be resolved by substituting, $1 - X = X^2 \Delta$. We use A^{bond} to denote the covalent nature of the bonds and take the limit of no monomer.

$$\frac{A^{bond}}{n_o RT} = \ln(X) - \left(\frac{1-X}{2}\right) = 0.5 \ln\left(\frac{1-X}{\Delta}\right) - \left(\frac{1-X}{2}\right) \rightarrow \lim_{X \to 0} \frac{A^{bond}}{n_o RT} = -0.5 \ln(\Delta) - 0.5 \qquad 19.109$$

Eqn. 19.109 is helpful when $\Delta \to \infty$ because Z^{bond} can be obtained by differentiation of A^{bond}. Referring to Eqns. 19.44 and 19.49 and taking the derivative,

$$\frac{P^{bond}V}{n_T RT} = \frac{n_o \rho}{n_T} \frac{\partial}{\partial \rho}\left(\frac{A^{bond}}{n_o RT}\right) = \frac{-2\rho}{2\Delta} \frac{\partial}{\partial \rho}(\rho g(\sigma) K_a) = -\left[1 + \frac{\rho \partial g(\sigma)}{g \partial \rho}\right] \equiv Z^{bond} \qquad 19.110$$

From a model for $g(\sigma)$, the bonding contribution to the EOS results. For example, if $g(\sigma)$ is given by the van der Waals model,

$$Z^{bond}(\text{vdW, dimer}) = -\left[1 + \frac{\eta_P}{g_{vdW}} \frac{\partial}{\partial \eta_P}(g_{vdW})\right] = -\left[1 + \frac{\eta_P}{(1 - \eta_P)}\right] = \frac{-1}{(1 - \eta_P)} \qquad 19.111$$

Generalizing this result to a polymer chain with m segments, there are $(m-1)$ bonds per chain. For example, continuing with the vdW model,

$$Z^{polybond}(\text{vdW, }m\text{-mer}) = -(m - 1)\left[1 + \frac{\eta_P}{(1 - \eta_P)}\right] = \frac{-(m - 1)}{(1 - \eta_P)} \qquad 19.112$$

This is essentially Wertheim's theory of polymerization, although Wertheim specifically treated the case resulting in a mixture with a range of molecular weights and average degree of polymerization of $<m>$.[22]

19.10 STATISTICAL ASSOCIATING FLUID THEORY (THE SAFT MODEL)

Shortly after Wertheim's work appeared, Chapman et al. formulated an equation of state that incorporated the bonding contribution and complexation as well as the disperse repulsive and attractive terms. Their perspective was to treat any solution in the conventional way as a fluid of independent spheres, then to add the bonding contribution required to assemble the spheres into chains. Then the equation of state becomes

$$Z = mZ^{HS} + (m - 1)Z^{bond} + mZ^{att} + Z^{chem} \qquad 19.113$$

---

22. Wertheim, M.S. 1986. *J. Stat. Phys.* 42:459.

Adding and subtracting $(1 - m)$ to isolate the ideal gas limit,

$$Z = mZ^{HS} + (1 - m) + (m - 1) + (m - 1)Z^{bond} + mZ^{att} + Z^{chem}$$

$$Z = 1 + m(Z^{HS} - 1) + (m - 1)(1 + Z^{bond}) + mZ^{att} + Z^{chem}$$

$$Z = 1 + m(Z^{HS} - 1) - (m - 1)\rho\partial\ln g/\partial\rho + mZ^{att} + Z^{chem} = 1 + m(Z^{HS} - 1) + Z^{chain} + mZ^{att} + Z^{chem} \quad 19.114$$

Recognizing the significance of Wertheim's statistical mechanical theory for associating (and solvating) systems, Chapman et al. named their model SAFT. In principle, any equation of state can be applied for the dispersion interactions, but Chapman et al. adopted the Carnahan-Starling model for the hard-sphere systems, including the Mansoori-Carnahan-Starling-Leland (MCSL) model for hard-sphere mixtures.[23] That choice has remained consistent in most variations of the SAFT model, but several alternatives have been adopted to describe the attractive dispersion interactions, Z^{att}. The original version suggested using second order perturbation contributions of the Lennard-Jones fluid for Z^{att}.[24] Huang and Radosz adopted a 20-parameter equation of state for argon (HR-SAFT).[25] More recently, Gross and Sadowski took a slightly different approach.[26] They treated the hard-sphere and chain contributions in the usual manner of SAFT, but treated Z^{att} by a second order perturbation theory that takes the tangent-sphere-chain as the reference fluid, instead of the tangent spheres themselves. They refer to their method as Perturbed Chain SAFT (PC-SAFT). In the conventional SAFT approach, Z^{att} would be a universal curve, but PC-SAFT shows a mild variation in this quantity with chain length. We focus our discussion on PC-SAFT for the most part.

Example 19.8 The SAFT model

Chapman et al. (1990)[a] suggested that second order perturbation theory could be applied for the segment term of the SAFT model, with the hard-sphere contribution described by the Carnahan-Starling (CS) equation and the A^{att} given by:

$$A^{att}/RT = A_1\beta\varepsilon + A_2(\beta\varepsilon)^2$$
$$A_1 = -11.61\eta_P - 8.28\eta_P^2 - 5.24\eta_P^3 + 34.21\eta_P^4$$
$$A_2 = -25.76\eta_P + 181.87\eta_P^2 - 547.17\eta_P^3 + 529.00\eta_P^4$$

Express this model as an equation of state for alcohols, including Z^{chain}; that is $Z = Z(m, \eta_P, \beta\varepsilon)$.
Solution: The CS equation is given by $Z^{HS} - 1 = 4\eta_P(1 - \eta_P/2)/(1 - \eta_P)^3$. This corresponds to $g(\sigma) = (1 - \eta_P/2)/(1 - \eta_P)^3$ and, by Eqn. 19.44, $\Delta = \rho(1 - \eta_P/2)K^{AD}(\exp(\beta\varepsilon_{ij}^{AD}) - 1)/(1 - \eta_P)^3$. Then, $\eta_P\partial\ln g/\partial\eta_P = \eta_P\{-0.5/(1-\eta_P)^3 + 3(1-\eta_P/2)/(1-\eta_P)^4\}/g = \eta_P[-0.5/(1 - \eta_P/2) + 3/(1 - \eta_P)]$
Rearranging,
$$Z^{chain} = -(m - 1)(5\eta_P - 2\eta_P^2)/[(2 - \eta_P)(1 - \eta_P)]$$

$Z^{chem} = -0.5h\,\partial\ln\Delta/\partial\ln\eta_P = -(1 - X^A)\,\partial\ln\Delta/\partial\ln\eta_P = -(1 - X^A)\{1 + (5\eta_P - 2\eta_P^2)/[(2 - \eta_P)(1 - \eta_P)]\}$; and $Z^{att} = Z_1\beta\varepsilon + Z_2(\beta\varepsilon)^2$, where $Z_1 = -11.61\eta_P - 16.56\eta_P^2 - 15.72\eta_P^3 + 136.84\eta_P^4$ and $Z_2 = -25.76\eta_P + 363.74\eta_P^2 - 1641.51\eta_P^3 + 2116.00\eta_P^4$.

Putting it all together,

---

23. Mansoori, G.A., Carnahan, N.F., Starling, K.E., Leland, T.W. 1971. *J. Chem. Phys.* 54:1523.
24. Chapman, W.G., Gubbins, K.E., Jackson, G., Radosz, M. 1990. *Ind. Eng. Chem. Res.* 29:1709.
25. Huang, S.H., Radosz, M. 1990. *Ind. Eng. Chem. Res.* 29:2284.
26. Gross, J., Sadowski, G. 2001. *Ind. Eng. Chem. Res.* 40:1244.

Example 19.8 The SAFT model (Continued)

$$Z(m,\eta_P,T_r)= 1 + m[4\eta_P(1 - \eta_P/2)/(1-\eta_P)^3 + Z_1\beta\varepsilon + Z_2(\beta\varepsilon)^2]$$

$$- (m - 1)(5\eta_P - 2\eta_P^2)/[(2 -\eta_P)(1-\eta_P)] - (1 - X^A)\{1+(5\eta_P - 2\eta_P^2)/[(2 -\eta_P)(1 - \eta_P)]\}$$

where $X^A = [-1 + (1 + 4\Delta)^{1/2}]/(2\Delta)$, $\Delta= \eta_P(1 - \eta_P/2)(K^{AD}/b)(\exp(H\beta\varepsilon) - 1)/(1 - \eta_P)^3$, $H = \varepsilon_{ij}^{AD}/\varepsilon$. Since all the terms can be computed based on $m, \eta_P, \beta\varepsilon$, the equation of state is complete.

a. Chapman, W.G., Gubbins, K.E., Jackson, G., Radosz, M. 1990. *Ind. Eng. Chem. Res.* 29:1709.

The tangent-sphere-chain that lays the foundation of all SAFT models is well defined and relatively simple to treat by molecular simulation. This makes it possible to evaluate the accuracy of Wertheim's theory for the hard chain reference system. With only slightly more effort we can also evaluate the accuracy for a reference fluid of fused sphere chains with 110° bond angles, as in *n*-alkane chains. As shown in Fig. 19.11, the comparison is quite favorable for tangent sphere chains, showing that Wertheim's theory and the related SAFT models have a solid theoretical foundation that is validated by molecular simulation.

The PC-SAFT model has the same form as Example 19.8 except for A^{att}.

$$(A - A^{ig})/RT = mA^{HS} + A^{chain} + mA^{att} + A^{chem} \qquad 19.115$$

$$A^{att} = -12\eta_P I_1\beta\varepsilon - 6\eta_P(\beta\varepsilon)^2 I_2/D_2 \qquad 19.116$$

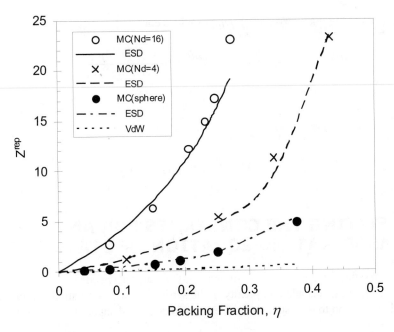

Figure 19.11 *Comparison of molecular simulations, the van der Waals equation, and the ESD equation of state for Z^{rep}, where $Z^{rep} = m(Z^{HS}- 1)+Z^{chain}$. N_d is the number of spheres in a chain.*

$$I_n = \sum_{j=0}^{6} a_{j,n}(m)\eta_P^j \qquad (19.117)$$

$$a_{j,n}(m) = a_{0j,n} + a_{1j,n}\frac{(m-1)}{m} + a_{2j,n}\frac{(m-1)(m-2)}{m} \qquad (19.118)$$

$$D_2 = \frac{8\eta_P - 4\eta_P^2}{(1-\eta_P)^4} - \frac{(m-1)}{m}\frac{20\eta_P - 27\eta_P^2 + 12\eta_P^3 - 2\eta_P^4}{[(1-\eta_P)(2-\eta_P)]^2} + \frac{1}{m} \qquad (19.119)$$

Eqns. 19.117 and 19.118 include 42 coefficients listed in the original reference.

You might wonder whether there is a simpler form of the SAFT model that is more sophisticated than the van der Waals model, but not as complicated as the PC-SAFT model. Such a model would be convenient for illustrating the key advantages of an association model without losing the simplicity of a cubic model like the PR model. One alternative is simply to add the association contribution of Example 15.8 to the PR model. This is the basis of the CPA model of Kontogeorgis et al.[27] This is a feasible model and it has been applied in many practical settings, but it is not entirely faithful to the Wertheim perspective in that it uses $g(\sigma)$ from one model and Z^{HS} from another, while ignoring Z^{chain} completely. Another alternative is to reconsider the ESD model in light of the SAFT analysis. Then we can rewrite the ESD model as a "simplified SAFT" model:[28]

$$Z = 1 + q(Z^{HS} - 1 + Z^{att}) - (q-1)\frac{1.9\eta_P}{(1-1.9\eta_P)} - \frac{(1-X^A)}{(1-1.9\eta_P)} \qquad (19.120)$$

In this form, recognizing that $g(\sigma)=1/(1-1.9\eta_P)$ provides consistency as a SAFT model. Then,

$$Z^{HS} = 1 + \frac{4\eta_P}{(1-1.9\eta_P)}; \quad Z^{att} = \frac{(-9.5)Y\eta_P}{1+1.7745Y\eta_P} \qquad (19.121)$$

$$Z^{chain} = -(q-1)\frac{1.9\eta_P}{(1-1.9\eta_P)} \qquad (19.122)$$

$$Z^{chem} = \frac{-(1-X^A)}{(1-1.9\eta_P)} \quad \text{(pure chains)} \qquad (19.123)$$

19.11   FITTING THE CONSTANTS FOR AN ASSOCIATING EQUATION OF STATE

To this point in the discussion, we have assumed that the constants needed for a fluid are available. However, association models add complexity in the sense that two association parameters must be characterized in addition to the usual size (b), energy (a or ε), and shape (k, m, q, or c). One simple

---

27. Kontogeorgis, G.M., Michelsen, M.M., Folas, G.K., Derawi, S., von Solms, N., Stenby, E.H. 2006. *Ind. Eng. Chem. Res.*, 45:4855.

28. See problem 7.19, $q = 1 + 1.90476(c - 1)$. Algebraically, $4\cdot0.90476/1.90476=1.900$. The coefficient 1.90476 was inferred originally in a very different way, but it is entirely consistent with Wertheim's theory when $g(\sigma)=1/(1-1.9\eta_P)$.

approach is to assign standardized values to the bonding volume and energy. For example, alcohols can be assigned an energy of 17 kJ/mol. Aldehydes, amides, amines, and nitriles can be assigned an energy of 5.2 kJ/mol. Given the bonding energy and volume, three parameters remain to be determined in a manner equivalent to three parameter corresponding states.

The simplest case is when the association energy is zero. Then the critical method can be applied in the usual way. For the ESD model, this is especially simple, because it is cubic. The approach of setting $(Z-Z_c)^3 = 0$ can be applied over a range of values of c from 1 to infinity. For each value of c, the acentric factor can be computed once the critical point has been determined. Then the value of c can be regressed as a function of acentric factor, and values of b and ε can be correlated as functions of c. This results in the following correlations,

$$c = 1.0 + 3.535\omega + 0.533\omega^2 \tag{19.124}$$

$$Z_c = (1 + 0.115/c^{1/2} - 0.186/c + 0.217/c^{3/2} - 0.173/c^2)/3 \tag{19.125}$$

$$b = \frac{RT_c Z_c^3 [-(1.9k_1 Z_c + 3a) + \sqrt{(1.9k_1 Z_c + 3a) + 4a(4c - 1.9)(9.5q - k_1)/Z_c}]}{P_c \qquad 2a} \tag{19.126}$$

$$Y_c = \left(\frac{RT_c}{bP_c}\right)^2 \frac{Z_c^3}{a} \tag{19.127}$$

$$\frac{\varepsilon}{k} = T_c \ln(Y_c + 1.0617) \tag{19.128}$$

where $a = 1.9(9.5q - k_1) + 4ck_1$ and $k_1 = 1.7745$.

An interesting implication of this result is that $Z_c \rightarrow {}^1/_3$ in the infinite chain limit. Lue et al. showed that this is a general result for all SAFT models, despite the experimental observation that Z_c appears to approach zero for long chains.[29] They attributed this deficiency to inaccurate characterization of the intramolecular interactions by SAFT models at low density and high temperature.

At a slightly higher level of complexity, the bonding energy and volume can be treated as adjustable parameters and regressed to minimize deviations in vapor pressure and density. This is the predominate method for most SAFT models. In fact, the critical point method has been systematically avoided for SAFT models other than the ESD model. The regression method requires extensive pure component data. Unfortunately, sufficient data exist for relatively few compounds to regress optimal values or even critical values, and those regressions have already been performed and the results are available. Therefore, the important problem is to characterize the constants when data are few or nonexistent.

Emami et al. have formulated a convenient method that requires little or no experimental data.[30] Their method has been developed for the ESD, HR-SAFT, and PC-SAFT models. The method refers to standard literature correlations for ΔH^{vap} and $\rho^{liq}_{298.15}$ and provides UNIFAC group contribution correlations for the shape factors. This method is facilitated by spreadsheets that are available in Chapter 19 supplements on the textbook's web site.

29. Lue, L., Friend, D.G., Elliott, J.R. 2000. *Mol. Phys.* 98:1473.
30. Emami, F.S., Vahid, A., Elliott, J.R., Feyzi, F. 2008. *Ind. Eng. Chem. Res.*, 47:8401–8411.

Implementations of ESD, HR-SAFT, and PC-SAFT are available from the various authors. A convenient set of implementations that also provides the capability to generate global phase diagrams is available from Cismondi et al.[31]

19.12 SUMMARY

A simple way of remembering the qualitative conclusions of this analysis can be derived by considering the behavior of the fugacity coefficient. One can easily demonstrate that the fugacity coefficient of the monomeric species is insensitive to the extent of association if it is expressed on the basis of the true number of moles in the associated mixture. But all of our phase equilibrium algorithms are based on the fugacity divided by the apparent mole fraction; for example, the flash algorithm is the same for any equation of state. The relation between the two fugacity coefficients is given by

$$\hat{\varphi}_i = \frac{\hat{f}_i}{x_i P} = \frac{\hat{f}_i}{x_M P} \frac{x_M}{x_i} = \frac{\hat{\varphi}_M x_M}{x_i}$$

19.129

This means that we must simply multiply the fugacity coefficient from the usual equation of state expression by the ratio of true to apparent mole fraction. Since this ratio is always less than one, we see that the effect of association is to suppress the effective fugacity of the associating species.

For mixtures, elevation of the monomer mole fraction by breaking the association network accounts for VLE quite accurately. Fig. 19.12 illustrates the benefit of a chemical physical model

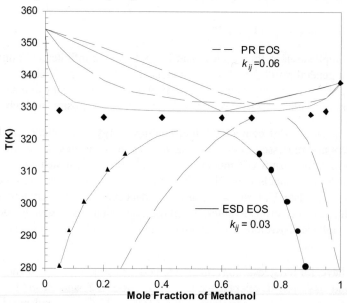

Figure 19.12 *T-x-y diagram for the system methanol + cyclohexane. Data from Soerensen, J. M.; Arlt, W. Liquid-Liquid Equilibrium Data Collection; DECHEMA: Frankfurt/Main, 1979 Vol. V, Part 1.*

---

31. www.gpec.plapiqui.edu.ar

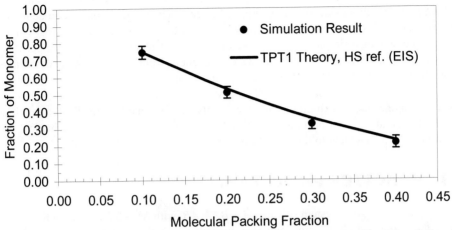

Figure 19.13 *DMD-B simulation of hard dumbbell methanol with reduced bond length l/σ = 0.4, at T = 300 K and $N_A \varepsilon_{HB}/R = 2013$ K. TPT1 theory is an adaptation of Wertheim's theory.*

relative to a purely physical equation like the Peng-Robinson equation. The figure depicting the methanol + cyclohexane system shows the improved accuracy in representing simultaneous LLE and VLE when hydrogen bonding is recognized. Notice the change in the skewness of the curves when hydrogen bonding is applied. The hydrogen bonding model is accomplishing this change in skewness as a clear and understandable explanation of the physics. By contrast, the van Laar model in Chapter 11 altered the skewness by adjusting constants that ignore the physics. We would expect that the stronger physical basis would provide greater capability for extrapolations to multicomponent mixtures. Unfortunately, remarkably few multicomponent studies have been performed to date. Hence, there is no single recommended method for treating nonideal multicomponent solutions at this time.

From a theoretical perspective, however, we may still feel uncomfortable with having made several sweeping assumptions with little justification besides their making the equations easier to solve. This may not seem like much of an improvement over local composition theory. On the other hand, the assumptions could be reasonably accurate; they simply need to be tested. As in the case of local composition theory, molecular simulations provide an effective method of testing the assumptions implicit in the development of a theory. Fig. 19.13 shows a comparison to molecular simulation results and to Wertheim's theory.[32] It can be seen that the above assumptions lead to reasonably accurate agreement with the molecular simulations and therefore they represent at least a self-consistent theory of molecular interactions.

This is not to say that chemical theory completely solves all problems. Local composition effects are real and should be incorporated into the mixing rules. Evidence supporting this step can be found in the anomalous behavior of the methane + hexane system. If such local composition effects are so prominent for nonassociating solutions, they should be accounted for at all times. As an example of other problems, the association network of water seems to be different enough from that of alcohols that a more sophisticated model will be necessary to represent difficult solutions like hydrocarbons + water to the high degree of accuracy (ppm) required by organizations like the Environmental Protection Agency. Furthermore, the solvation between different species can be

32. Liu, J-X., Elliott, J. R., 1996. *Ind. Eng. Chem. Res.* 35:1234.

extremely complicated and require substantially more investigation to develop reliable engineering models. Finally, it is well known that "nonadditive" effects play a significant role in aqueous and alcoholic solutions.[33] That is, the energy of network formation changes in a way that cannot be understood based only on a simple potential model for a single water molecule. These peculiarities may seem esoteric, but they are key obstacles which prevent us from revealing many of the mysteries of biomolecular solutions. Other areas of application such as polymer solutions involving association, as in nylon, can also be imagined. These are the areas which remain to be explored. The methods for engaging in this exploration predominantly involve mathematically formalizing our treatment of the radial distribution function through applications of statistical mechanics. At this point, we leave this engagement to the "satisfaction and good fortune" of the reader.

19.13 PRACTICE PROBLEMS

P19.1 (a) A gas-phase A+B system solvates $A + B \rightleftarrows AB$ with $K_a = 0.5$ at 298.15 K. Calculate the compressibility factor, apparent fugacity coefficients, and the true vapor phase mole fractions in a mixture at 298.15 K and 2 bar when the apparent concentration is $y_A = 0.45$ using ideal chemical theory.

(b) A liquid-phase A+B system solvates $A + B \rightleftarrows AB$ with $K_a = 0.7$ at 298.15 K. Calculate the true liquid-phase mole fractions in a mixture at 298.15 K and 1 bar when the apparent concentration is $x_A = 0.45$ using ideal chemical theory.

(c) A gas-phase A+B system associates $2A \rightleftarrows A_2$ with $K_a = 0.5$ at 298.15 K. Calculate the compressibility factor, apparent fugacity coefficients, and the true vapor phase mole fractions in a mixture at 298.15 K and 2 bar when the apparent concentration is $y_A = 0.45$ using ideal chemical theory.

19.14 HOMEWORK PROBLEMS

19.1 Consider a dilute isothermal mixing process of acetic acid(1) in benzene(2). For the dilute region (say, up to 5 mol% acid), draw schematically curves for the following:

$\bar{S}^E_1$ versus x_1; $\bar{H}^E_1$ versus x_1; $\bar{G}^E_1$ versus x_1.

Briefly justify your schematic graphs with suitable explanations. Take standard states as the pure substances.

19.2 Acetic acid dimerizes in the vapor phase. Show that the fugacity of the dimer is proportional to the square of the fugacity of the monomer.

19.3 By assuming that the equilibrium constant for each successive hydrogen bond is equal in the generalized association approach developed in this chapter, what assumptions are being made about the Gibbs energy, enthalpy, and entropy for each successive hydrogen bond?

19.4 The value of the excess Gibbs energy at 298 K for an equimolar chloroform(1) + triethylamine(2) system is $G^E = -0.91$ kJ/mol. Assuming only a 1-1 compound is formed, model the excess Gibbs energy with ideal chemical theory, and plot the P-x-y diagram.

19.5 Suppose that, due to hydrogen bonding, the system $A + B$ forms a 1-1 complex in the vapor phase when mixed. Neither pure species self-associates in the vapor phase. The equilibrium constant for the solvation is $K_{AB} = 0.8$ bar^{-1} at 80°C. At 80°C, a mixture with a appar-

---

33. Hait et al., 1993. *Ind. Eng. Chem. Res.* 32:2905.

ent (bulk) mole fraction of $y_A = 0.5$ is all vapor at 0.78 bar. Calculate the fugacity coefficient of A in the vapor phase using ideal chemical theory at this composition, temperature, and pressure. Use hand calculations.

19.6 At 143.5°C, the vapor pressure of acetic acid is 2.026 bar. The dimerization constant for acetic acid vapor at this temperature is 1.028 bar^{-1}. The molar liquid volume of acetic acid at this temperature is 57.2 cm^3/mol. Calculate the fugacity of pure acetic acid at 143.5°C and 10 bar. Use hand calculations.

19.7 An $A + B$ mixture exhibits solvation in the liquid phase, which is to be represented using ideal chemical theory. Because of a Lewis acid/base interaction, the system is expected to form a 1-1 compound.

   (a) Which one of the following sets of true mole fractions are correct for the system using an equilibrium constant of 3.2 to represent the complex formation at an apparent composition $x_A = 0.4$?

	X_{AM}	X_{BM}	X_{AB}
Set I	0.2096	0.4731	0.3173
Set II	0.2646	0.3983	0.3372

   (b) Based on your answer for part (a), what are the apparent activity coefficients of A and B?

19.8 Water and acetic acid do not form an azeotrope at 760 mmHg. The normal boiling point of acetic acid is 118.5°C. Therefore, at 118.5°C and 760 mmHg, the mixture will exhibit only vapor behavior across the composition range. The following equilibrium constants have been fitted to represent the vapor-phase behavior:[34]

	Dimer, $-\log_{10}K$, where K in $(\text{mmHg})^{-1}$	Trimer, $-\log_{10}K$, where K in $(\text{mmHg})^{-2}$
Acetic acid	10.108−3018/T(K)	18.63−4960/T(K)
Water	6.881−808.2/T(K)	
Acetic acid/water complex	Same as water	

   (a) Let compound A be acetic acid and B be water. Calculate the true mole fractions of all the species from $y_A = 0.05$ to $y_A = 0.95$. At what apparent mole fraction does each specie show a maximum true mole fraction? What is the relation of this apparent mole fraction with the compound's stoichiometry?

   (b) Plot the fugacity coefficient of acetic acid and water as a function of acetic acid mole fraction. What is the physical interpretation of the rapid change of the acetic acid fugacity coefficient in the dilute region, if the water fugacity coefficient doesn't show such a dramatic trend in its dilute region?

19.9 (a) The molar Gibbs energy of mixing (per mole of superficial solution) for a liquid binary system

$$\Delta G_{mix}/RT = x_A \ln(x_A \gamma_A) + x_B \ln(x_B \gamma_B) \qquad 19.130$$

   expressed extensively, this becomes

---

34. Tsonopoulos, C., Prausnitz, J. M. 1970. *Chem Eng. J.* 1:273.

$$\Delta G_{mix}/RT = n_A\ln(x_A\gamma_A) + n_B\ln(x_B\gamma_B) \qquad 19.131$$

Introduce the concepts of chemical theory into Eqn. 19.131 to prove that the Gibbs energy of mixing is equivalently given by the sums over true species,

$$\frac{\Delta G_{mix}}{RT} = \sum_{true\ i} n_i\ln((x_i\alpha_i)/((x_A^o\alpha_A^o)^{a_i}(x_B^o\alpha_B^o)^{b_i})) - \sum_{true\ i} n_i\ln(K_i) \qquad 19.132$$

where K_i is unity for the monomers. Hint: $n_A = \Sigma a_i n_i$.

(b) Show that on a molar basis for an ideal chemical theory solution that has only solvation, per mole of *true* solution, the equation reduces to

$$\frac{\Delta G_{mix}}{RT} = \sum_{true\ i} x_i\ln(x_i) - \sum_{true\ i} x_i\ln(K_i) \qquad 19.133$$

and provide a physical interpretation relating the Gibbs energy of formation to K.

(c) Considering a system where A associates, show that the Gibbs energy of mixing by ideal chemical theory is per mole of *true* solution given by

$$\frac{\Delta G_{mix}}{RT} = \sum_{true\ i} x_i\ln(x_i/(x_A^o)^{a_i}) - \sum_{true\ i} x_i\ln(K_i) \qquad 19.134$$

Below are tabulated calculations for ideal chemical theory for an $A + B$ system where A forms dimers with $K=140$. Use Eqns. 19.134 and 19.130 to tabulate the respective Gibbs energies of mixing over RT. Then tabulate n_T/n_0 (the number of true moles divided by the number of apparent moles) and multiply Eqn. 19.134 by this number and compare with Eqn. 19.130.

Apparent x_A	x_{Am}	x_{Bm}	x_{A2}
1.00E-02	4.46E-03	0.99276	2.78E-03
0.1	1.76E-02	0.93903	4.34E-02
0.2	2.63E-02	0.87728	9.65E-02
0.3	3.35E-02	0.80975	0.15679
0.4	4.01E-02	0.73505	0.22488
0.5	4.65E-02	0.65118	0.30235
0.6	5.28E-02	0.55633	0.39083
0.7	5.93E-02	0.44785	0.49282
0.8	6.61E-02	0.32233	0.61157
0.9	7.33E-02	0.17516	0.75157
0.99	8.02E-02	1.90E-02	0.90078
1	0.081019	0	0.918981

19.10 Furnish a proof that the concentration of true species i is maximum at composition $x_A^* = a_i/(a_i + b_i)$, $x_B^* = b_i/(a_i + b_i)$ where a_i and b_i are given in Eqn. 15.1. [Hint: The Gibbs-Duhem equation is useful for relating derivatives of activity.]

19.11 Show that the result for Z^{chem} is obtained by taking the appropriate derivative of A^{chem}.

19.12 Use the ESD equation[35] to model the monomer, dimer, and trimer in the vapor and liquid phases of saturated water at 373 K, 473 K, and 573 K. How does the monomer fraction of saturated vapor change with respect to temperature? How does monomer fraction of saturated liquid change?

19.13 Derive the equations for determining the critical point of the ESD equation[35] based on ε_{HB} and K^{AD} being zero by noting that $dF/dZ = 0$ and $d^2F/dZ^2 = 0$, where $F = Z^3 + a_2Z^2 + a_1Z + a_0$ when hydrogen bonding is negligible.

19.14 Plot P against V at 600 K for water with the ESD equation using the characterization of Suresh and Elliott (1992).[35] Apply the equal area rule and determine the vapor pressure at that temperature. Raise the temperature until the areas equal zero and compare this temperature to the true value of 647.3 K. For simplifications, see also Elliott (1996).[21]

19.15 Apply the ESD equation[35] to the methanol + benzene system and compare to the data in Perry's Handbook based on matching the bubble pressure at the azeotropic point. Prepare a *T-x-y* diagram and determine whether the ESD equation indicates a liquid-liquid phase split for any temperatures above 250 K. Perform the same analysis for the Peng-Robinson EOS. Do you see any differences? Compare to Fig. 19.12. (Hint: check chethermo.net for ESD.)

19.16 Use the ESD equation[35] to estimate the mutual LLE solubilities of methanol and *n*-hexane at 285.15 K, 295.15 K, and 310.15 K. Use the value of $k_{ij} = 0.03$ as fitted to a similar system in Fig. 19.12 on page 804. (Hint: check chethermo.net for ESD program.)

19.17 The hydrogen halides are unusual. For example, here are the critical properties of various hydrogen halides:

	MW	T_c(K)	P_c(bar)	Z_c	ω
HF	20.00	461.0	64.88	0.12	0.372
HCl	36.46	324.6	83.07	0.249	0.120
HBr	80.91	363.2	85.50	0.283	0.063

Experimental data for vapor pressure and apparent molecular weight of HF vapor are:

T(K)	P^{sat}	M_w^{sat}	M_w at 1 bar
227.3	0.0519	92.8	117.6
243.9	0.1265	85.0	112.5
277.8	0.5780	69.8	85.4
303	1.4353	58.4	43.0
322.6	2.6178	50.3	21.8

These apparent molecular weights have been found by measuring the mass density of the vapor and comparing with an ideal gas of molecular weight 20. Assuming that HF forms only monomers and hexamers, use the ESD EOS with $c = q = 1$ for both monomer and hexamer to fit the vapor density data as accurately as possible in the least squares sense and estimate the corresponding value of Z_c.

19.18 (a) Compute the values of K_c^C, a/bRT_c, x_{Mc}, and $b\rho_c$ for methanol and ethanol according to the van der Waals hydrogen bonding equation of state, with $Z_c=0.375(Z_c^{expt}/Z_c^{homo})$ where Z_c^{homo} is the Z_c of the homologous n-alkane (e.g. ethane for methanol).

---

35. Suresh, S.J., Elliott, J.R. 1992. *Ind. Eng. Chem. Res.* 31:2783-2794.

(b) Assuming an enthalpy of hydrogen bonding of 24 kJ/mole, calculate the acentric factors for methanol and ethanol according to the vdw-HB EOS.

19.19 Derive the association model for the Peng-Robinson model, with $Z_c = 0.3074(Z_c^{expt}/Z_c^{homo})$ where Z_c^{homo} is the Z_c of the homologous n-alkane (e.g. ethane for methanol). Extend the homomorph concept by applying $\omega^{PR} = \omega^{homo}$, where ω^{homo} is the acentric factor for the nonassociating homomorph and ω^{PR} is the acentric factor substituting for the associating compound into the Peng-Robinson expression for a.

(a) For methanol, determine the values of K_a', b, a/bRT_c, x_{Mc} that match the critical point.
(b) Determine the vapor pressure at $T_r = 0.7$ for methanol assuming a hydrogen bonding energy of 15 kJ/mole, and compare to the experimental value. Infer the acentric factor and compare to the experimental value.
(c) Plot log P_r^{sat} versus T_r^{-1} for the Peng-Robinson EOS and the Peng-Robinson hydrogen bonding EOS, and experiment.

19.20 Acetic acid has a much stronger tendency to dimerize than any alcohol. Therefore, it is not reasonable to assume that $K_{a2} = K_{a3} = \ldots$ for acetic acid. The assumption is reasonable for $K_{a3} = K_{a4} = \ldots$, however. We can supplement the theory by adding a single additional equation for the dimerization reaction with an effective equilibrium constant equal to the ratio of the true K_{a2} divided by the linear association value. Assume that the linear (A-D) association is negligible for the saturated vapor at ~300–350 K.

(a) Determine the value of K_{a2} that matches the saturated vapor compressibility factor in that range. Recommend whether $N_A \varepsilon_{HB}/R = 4000$ K or 10000 K fits dimerization best.
(b) Determine the values of K_c^{AD}, b, and ε/k that match the homologous Z_c for the ESD equation with $c = 1$.

19.21 Extend the ESD equation[35] to compounds with more than one bonding segment. Consider ethylene glycol as a compound with both an associating head and tail. Extend the mixture analysis to treat this case with two bonding segments ($N_d = 2$).

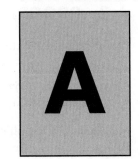

APPENDIX
A

SUMMARY OF COMPUTER PROGRAMS

Several programs are furnished with the text to help you learn the material and to assist in repetitive and/or complex calculations. Programs are available for Microsoft Excel and MATLAB. There is significant duplication of capabilities among the two platforms. The software is updated periodically. Visit the web site listed in the front flap for the latest version of both the software and this appendix.

A.1 PROGRAMS FOR PURE COMPONENT PROPERTIES

Matlab/3D/**PVT.m**—3-D PVT for water fluid phases; **PHT.m**—3-D PHT for water fluid phases.

Matlab/Props/**Props.dat**—Critical properties, acentric factor, heat capacity constants, and Kamlet-Taft parameters for a variety of substances. Use 'load Props.dat' to load to workspace. **Props-Browse.m**–A quick summary of the compound name and ID number. Many programs load Props.dat.

Matlab/Psat/**AntoineGet.m**—Returns Antoine constants from database using ID numbers; **Antoine-TableBrowse.m**—Use for a quick overview of ID numbers and temperature limits; **Antoine-Table.mat**—The database, use `load AntoineTable.mat` to load to workspace for viewing; **Tsat.m**—Utility program for calculating the saturation temperature. Many programs load Antoine-Table.mat.

Excel/**Antoine.xlsx**—A tabulation of Antoine coefficients for a variety of common chemicals.

Excel/**Props.xlsx**—Critical properties, acentric factor, heat capacity constants, and Kamlet-Taft parameters for a variety of substances. Use to copy/paste values into other workbooks.

Excel/**Steam.xls**—Steam property calculator. Same formulation implemented in Harvey, A. P., Peskin, A. P., Klein, S. A., NIST/ASME Steam Properties, Version 2-1, NIST Standard Reference Data Program, December 1997. Another good resource for steam properties for Excel and MATLAB is Xsteam available at www.x-eng.com/ (accessed 11/2011).

Excel/**Preos.xls**—An Excel workbook for calculating properties using the Peng-Robinson equation. Also calculates departure functions and thermodynamic properties.

811

Matlab/Chap07/**Preos.m**—Pure component *P-V-T* calculations via the Peng-Robinson equation of state. Loads data from Matlab/Props/Props.dat.

Matlab/Chap08-09/**PreosPropsMenu.m**—A GUI interface for calculating *U, H, S,* and *f* using the Peng-Robinson equation of state. Loads data from Matlab/Props/Props.dat.

A.2 PROGRAMS FOR MIXTURE PHASE EQUILIBRIA

Matlab/**ExamplePlot.m**—Plot with multiple data sets, markers, lines, labels, and legend.

Matlab/TextExamples/**Ex10_02.m**—Raoult's law adiabatic flash; **Ex11_03.m**—Fitting 1-parameter Margules; **Ex11-07.m**—Bubble *T* for MRL; **Gammafit.m**—Fitting of activity coefficients, **Ex12_04.m**—Scatchard-Hildebrand; **Ex13_01.m**—UNIFAC-VLE; **Ex14_03.m**—UNIFAC-LLE

Matlab/Chap10/**RaoultTxy.m**—*T-x-y* diagram using Raoult's Law; **Flshr.m**—Isothermal VLE flash using Raoult's law.

Matlab/Chap11-13/—See the gammaModels and TextExamples folders for appropriate routines.

Matlab/gammaModels/**Marg1P.m**—One-parameter Margules; **Marg2P.m**—Two-parameter Margules; **uniquac.m**—UNIQUAC equation; **vanLaar.m**—Van Laar equation; **nrtl.m**—NRTL equation; unifacVLE/**unifacCaller.m**—Example of function calls for VLE UNIFAC; unifacLLE/**unifacCallerLL.m**—Example of function calls for LLE UNIFAC.

Matlab/Chap14/**LLEflash.m**—Flash using UNIFAC or UNIQUAC.

Matlab/Chap15/**Prfug.m**—Peng-Robinson fugacities at a given state; **Prmix.m**—VLE using Peng-Robinson.

Matlab/Chap16/**Residue.m** – Residue curve calculator programmed to use UNIQUAC.

Excel/**Flshr.xls**—Two-phase isothermal flash using Raoult's law. Currently set for a binary system using the short-cut Psat equation.

Excel/**Actcoeff.xls**—Activity coefficients as a function of composition. These spreadsheets may be modified to calculate excess Gibbs energy, fugacities, and *P-x-y* diagrams.

1. MARGULES—A spreadsheet to use with the MARGULES activity coefficient model.

2. REGULAR—A spreadsheet to calculate VLE for methanol + benzene using the van Laar and Scatchard-Hildebrand models.

3. NRTL, NRTL5—Binary and multicomponent (up to five components) NRTL activity coefficient models.

4. UNIQUAC, UNIQUAC5—Binary and multicomponent (up to five components) UNIQUAC activity coefficient model. The multicomponent model can be used for LLE.

5. UNIFAC(VLE)—A spreadsheet to use with the UNIFAC activity coefficient model for up to five components to model VLE.

6. UNIFAC(LLE)—Two spreadsheets to use with the UNIFAC activity coefficient model for up to five components to model LLE.

Excel/**Gammafit.xls**—Fitting activity coefficient parameters.

Excel/**Prfug.xlsx**—Component fugacities via the Peng-Robinson equation for up to three components. Useful for understanding mixing rules, and for manually following iterative steps for phase equilibria calculations.

Excel/**Prmix.xlsx**—Peng Robinson phase equilibria.

Excel/**Residue.xlsm**—Spreadsheet with macro for calculating residue curves for homogeneous systems using UNIQUAC for up to three components.

Excel/**Virialmx.xls**—Spreadsheet that calculates the second virial coefficient for a binary mixture using the critical temperature, pressure, and volume.

Excel/**Wax.xls**—A spreadsheet to calculate wax solubilities.

Excel/**Ichemt.xls**—A chemical theory spreadsheet for calculating the true mole fractions at a given superficial mole fraction. The equilibrium constants need to be known before using the program.

A.3   REACTION EQUILIBRIA

Matlab/TextExamples/**Ex17_09.m**—Shortcut adiabatic reaction.

Matlab/Chap17-19/**Kcalc.m**—Equilibrium constant calculator for detailed van't Hoff; **RxnAdia.m**—Adiabatic reaction with full van't Hoff; **Gibbsmin.m**—Gibbs minimization; **GprimeCalc.m**—Transformed properties for biological systems; **IChemTheory.m**—Ideal chemical theory.

Excel/**Kcalc.xls** – Workbook to calculate equilibrium constants as a function of temperature.

Excel/**Rxns.xls** – Workbook with spreadsheets used for multiple reaction equilibria in the text.

1. RXNADIA-shortcut—Shortcut van't Hoff for ammonia example.
2. RXNADIA—Detailed van't Hoff for ammonia example.
3. DUALRXN—Simultaneous reactions.
4. GIBBSMIN—Gibbs minimization.
5. SMPRXN—Simultaneous reactions with phase equilibria.

Excel/**CL2H2O.xlsx**—Worked electrolyte example.

Excel/**GprimeCalc.xlsx**—Transformed Gibbs energies for biological systems.

Excel/**IdChemTheory.xlsx**—Ideal chemical theory.

A.4   NOTES ON EXCEL SPREADSHEETS

*.XLSX Files

These workbooks are a starting point for homework problems. You may need to modify the existing spreadsheet to work a homework problem. The *.XLSX files are provided in "document protected" format so that inadvertent modification will not occur. Only the unlocked cells which appear blue on the screen may be modified without turning the document protection off. (To change the document protection, select "Protection...." from the "Tools" menu. No passwords are used on the distributed spreadsheets. To change the protection status of individual cells, choose "Cell Protection" from the "Format" menu when document protection is turned off.)

While the tabular format of Excel is a benefit, the disadvantage is that it is very difficult to figure out how a pre-programmed sheet works and/or to debug. To find out the interdependencies of cells, first unprotected the sheet, then use the "auditing" tools.

Online Supplements

An online supplement covers topics including naming of variables, importing text files, plotting multiple data sets on the same axes, setting up successive substitution, use of Solver, and array operations. The textbook web site provides links for tutorial videos for the software.

A.5   NOTES ON MATLAB

MATLAB content is arranged in folders to maximize computing flexibility and power. Files in the folders gammaModels, Props, and Psat should not be moved because other programs call routines in those folders.

When MATLAB first starts, the current directory that holds the code of interest must be selected in the "current directory" drop-down box. Some of the code for this text is organized into subfolders relative to that directory; if a routine cannot be found, this may be the reason. Use the "path" command to see the full search path. Use the "addpath" command or the "genpath" nested into the addpath as shown in the Example files. For example, `addpath(../Psat)` will permit calling any functions in the Psat folder at the same folder level as any routine that is run in that session. The statement `addpath(genpath(../gammaModels))` will add a path to include any routines in the gammaModels or any subfolder of gammaModels for routines that are run at the same folder level as gammaModels. Use additional "../" to move up additional folder levels. Absolute paths can be used in code, but can cause problems if .m files are shared and the recipient does not put the routines in exactly the same folder. The addpath statement must be run once each session unless savepath is used. The convention we use for the text is to embed the addpath statements into the routines that need to access functions in different folders, and we use relative paths.

The display of MATLAB windows can be controlled through the "Desktop" menu; use the menu if a command window, workspace, or other window is not visible. MATLAB files for code for this text are unprotected; create a backup if you need to make changes. An overview of important commands and methods is furnished as an online supplement.

(), [], and {} have special meanings. Do not interchange them or you may get unexpected results. () are used for indicating arithmetic precedence. [] are used to indicate vectors and matrices. {} are used for cell arrays.

*.m and *.dat Files and File Naming

The MATLAB code is contained in .m files which may be scripts or functions. Scripts return the output to the workspace, while functions offer the capability to be called by other functions and return the results to the calling routine. For good programming practice, file names should match the function names, and each should be unique. At the time of this writing, MATLAB does not work well with spaces in file names, and such use leads to confusing error messages.

Online Supplements

Note that a MATLAB supplemental quick reference is available as an online supplement. An additional online supplement covers topics such as importing of text files, plotting, solving routines, minimization and optimizers, and matrix operations.

Note that "element-by-element" processing is powerful. For example, if *A, B,* and *C* are vectors with Antoine coefficients for multiple components, `Psat = 10^(A - B./(T+C))` will produce a vector of vapor pressures at *T*. See the online MATLAB supplement for more details.

Matlab/SuppExamples contains files for the examples in the online supplement.

A.6   DISCLAIMER

The programs provided with this text are for educational use only. They are provided AS IS, without any warranty. They must not be sold under any circumstances.

MATHEMATICS

B.1 IMPORTANT RELATIONS

Algebra

Some functions like logarithms and exponentials appear so often in thermodynamics that it makes sense to summarize some of them here. Also, integrations and differentiations are frequently performed, so a few important formulae are presented.

$$\ln(ab) = \ln a + \ln b \qquad \text{B.1}$$

$$\ln\left(\frac{a}{b}\right) = \ln a - \ln b \qquad \text{B.2}$$

$$\ln a^y = y \ln a \qquad \text{B.3}$$

$$\exp(a + b) = e^a e^b \qquad \text{B.4}$$

$$\ln N! \approx N \ln N - N \text{ for large } N \text{ (Stirling's approximation)} \qquad \text{B.5}$$

The quadratic formula provides roots to the equation $ax^2 + bx + c = 0$:

$$x = \frac{-b \pm \sqrt{b^2 - 4ac}}{2a} \qquad \text{B.6}$$

Cubic equations are discussed in Section B.2.

Beginning in Chapter 10, summation notation is used extensively. Many of the formulas are easily programmed using matrices and linear algebra. A matrix is a rectangular representation of the elements of an array. The elements of an array are identified by subscripts. For example:

$$X = [0.4, 0.6], \; x_1 = 0.4, x_2 = 0.6$$

For a multidimensional array, the first element subscript represents the row and the second element subscript identifies the column, for example:

$$A = \begin{bmatrix} 1 & 3 \\ 2 & 4 \end{bmatrix} = \begin{bmatrix} a_{11} & a_{12} \\ a_{21} & a_{22} \end{bmatrix} \quad a_{11} = 1, a_{21} = 2, a_{12} = 3, a_{22} = 4.$$

The transpose of a matrix is obtained by exchanging a_{ij} with a_{ji}. In shorthand notation, the transpose is represented by a superscript T. The number of rows and columns interchange after the transpose operation:

$$A = \begin{bmatrix} 1 & 4 & 7 \\ 2 & 5 & 8 \\ 3 & 6 & 9 \end{bmatrix}; \quad A^T = \begin{bmatrix} 1 & 2 & 3 \\ 4 & 5 & 6 \\ 7 & 8 & 9 \end{bmatrix},$$

$$X = \begin{bmatrix} 0.4 & 0.6 \end{bmatrix}; \quad X^T = \begin{bmatrix} 0.4 \\ 0.6 \end{bmatrix}$$

$$B = \begin{bmatrix} 1 & 4 \\ 2 & 5 \\ 3 & 6 \end{bmatrix}; \quad B^T = \begin{bmatrix} 1 & 2 & 3 \\ 4 & 5 & 6 \end{bmatrix}$$

Matrices can be multiplied. The product of `array1` and `array2` becomes `array3` and:

(a) The number of columns in `array1` must equal the number of rows in `array2`.

(b) `array3` has the same number of rows as `array1` and the same number of columns as `array2`.

(c) Element ij of `array3` is obtained by multiplying the elements in the ith row of `array1` by the elements in the jth column of `array2` and summing the products.

For one-dimensional arrays,

$$XB = \begin{bmatrix} x_1 & x_2 \end{bmatrix} \begin{bmatrix} b_1 \\ b_2 \end{bmatrix} = \sum_{i=1}^{2} x_i b_i \qquad\qquad \text{B.7}$$

which is the linear mixing rule, Eqn. 15.8. Suppose $X = \begin{bmatrix} x_1 & x_2 \end{bmatrix}$ and $Ps = \begin{bmatrix} P_1^{sat} & P_2^{sat} \end{bmatrix}$, then

$$XPs^T = \begin{bmatrix} x_1 & x_2 \end{bmatrix} \begin{bmatrix} P_1^{sat} \\ P_2^{sat} \end{bmatrix} = \sum_{i=1}^{2} x_i P_i^{sat} \qquad\qquad \text{B.8}$$

which is a bubble formula for Raoult's Law. A set of compositions can be stored in a matrix. For example, using two sets of compositions,

$$Pbub = XPs^T = \begin{bmatrix} x_{11} & x_{12} \\ x_{21} & x_{22} \end{bmatrix} \begin{bmatrix} P_1^{sat} \\ P_2^{sat} \end{bmatrix} = \begin{bmatrix} Pbub_1 \\ Pbub_2 \end{bmatrix} \tag{B.9}$$

where the first subscript in designation for X indicates the data set, and the subscripts on $Pbub$ indicate the corresponding bubble pressure for that composition.

An example of a one- and two-dimensional array is:

$$YA = \begin{bmatrix} y_1 & y_2 \end{bmatrix} \begin{bmatrix} a_{11} & a_{12} \\ a_{21} & a_{22} \end{bmatrix} = \begin{bmatrix} y_1 a_{11} + y_2 a_{21} & y_1 a_{12} + y_2 a_{22} \end{bmatrix} = \begin{bmatrix} \sum_{i=1}^{2} y_i a_{i1} & \sum_{i=1}^{2} y_i a_{i2} \end{bmatrix}$$

Multiplying the result by Y^T,

$$YAY^T = \begin{bmatrix} y_1 a_{11} + y_2 a_{21} & y_1 a_{12} + y_2 a_{22} \end{bmatrix} \begin{bmatrix} y_1 \\ y_2 \end{bmatrix} = y_1(y_1 a_{11} + y_2 a_{21}) + y_2(y_1 a_{12} + y_2 a_{22})$$

when $a_{ij} = a_{ji}$, we may write

$$YAY^T = y_1^2 a_{11} + 2y_1 y_2 a_{12} + y_2^2 a_{22} = \sum_{i=1}^{2}\sum_{j=1}^{2} y_j y_i a_{ij} \tag{B.10}$$

which is the quadratic mixing rule, Eqn. 15.8. For an example using two multidimensional arrays:

$$AB = \begin{bmatrix} a_{11} & a_{12} \\ a_{21} & a_{22} \end{bmatrix} \begin{bmatrix} b_{11} & b_{12} & b_{13} \\ b_{21} & b_{22} & b_{23} \end{bmatrix} = \begin{bmatrix} c_{11} & c_{12} & c_{13} \\ c_{21} & c_{22} & c_{23} \end{bmatrix} = C$$

$$= \begin{bmatrix} (a_{11}b_{11} + a_{12}b_{21}) & (a_{11}b_{12} + a_{12}b_{22}) & (a_{11}b_{13} + a_{12}b_{23}) \\ (a_{21}b_{11} + a_{22}b_{21}) & (a_{21}b_{12} + a_{22}b_{22}) & (a_{21}b_{13} + a_{22}b_{23}) \end{bmatrix}$$

MATLAB offers element-by-element algebra for vectors and matrices. If two matrices have the same dimensions, they may be dot multiplied. For example:

$$X.*GAM = \begin{bmatrix} x_{11} & x_{12} \\ x_{21} & x_{22} \end{bmatrix} .* \begin{bmatrix} \gamma_{11} & \gamma_{12} \\ \gamma_{21} & \gamma_{22} \end{bmatrix} = \begin{bmatrix} x_{11}\gamma_{11} & x_{12}\gamma_{12} \\ x_{21}\gamma_{21} & x_{22}\gamma_{22} \end{bmatrix} \tag{B.11}$$

Therefore, modified Raoult's law can be programmed for bubble-pressure calculations as

$$X.*GAM*(Ps^T) = \begin{bmatrix} x_{11} & x_{12} \\ x_{21} & x_{22} \end{bmatrix} .* \begin{bmatrix} \gamma_{11} & \gamma_{12} \\ \gamma_{21} & \gamma_{22} \end{bmatrix} \begin{bmatrix} P_1^{sat} \\ P_2^{sat} \end{bmatrix} = \begin{bmatrix} x_{11}\gamma_{11} & x_{12}\gamma_{12} \\ x_{21}\gamma_{21} & x_{22}\gamma_{22} \end{bmatrix} \begin{bmatrix} P_1^{sat} \\ P_2^{sat} \end{bmatrix} = \begin{bmatrix} Pbub_1 \\ Pbub_2 \end{bmatrix} \tag{B.12}$$

For an overview of programming arrays in Excel, see Appendix A.

Calculus

Differentiation:

$$\frac{d[\exp(f(x))]}{dx} = \exp(f(x))\frac{d[f(x)]}{dx}$$ B.13

$$\frac{d(\ln[f(x)])}{dx} = \frac{1}{f(x)}\frac{d[f(x)]}{dx}$$ B.14

General differentiation of composite functions:

(Product rule)

$$\frac{d[f(x) \cdot g(x) \cdot h(x)]}{dx} = f(x) \cdot h(x)\frac{d[g(x)]}{dx} + g(x) \cdot h(x)\frac{d[f(x)]}{dx} + f(x) \cdot g(x)\frac{d[h(x)]}{dx}$$ B.15

$$\frac{d\left[\dfrac{g(x)}{h(x)}\right]}{dx} = \frac{h(x)\dfrac{d[g(x)]}{dx} - g(x)\dfrac{d[h(x)]}{dx}}{h(x)^2}$$ B.16

$$\frac{d\left[g(x) \cdot \dfrac{1}{h(x)}\right]}{dx} = \frac{d[g(x)]}{dx} \cdot \frac{1}{h(x)} + g(x) \cdot \frac{d\left[\dfrac{1}{h(x)}\right]}{dx}$$ B.17

$$d[f(x, y, z)] = \left(\frac{\partial f}{\partial x}\right)_{y, z} dx + \left(\frac{\partial f}{\partial y}\right)_{x, z} dy + \left(\frac{\partial f}{\partial z}\right)_{x, y} dz$$ B.18

$$\frac{d[f(x)^x]}{dx} = f(x)^x\left\{\ln[f(x)] + \frac{d\{\ln[f(x)]\}}{d\{\ln[x]\}}\right\}$$ B.19

Integration:

$$\int\frac{1}{x}dx = \ln x$$ B.20

$$\int\frac{1}{ax + b}dx = \frac{1}{a}\ln(ax + b)$$ B.21

$$\int\frac{x\,dx}{ax + b} = \frac{x}{a} - \frac{b}{a^2}\ln(ax + b)$$ B.22

$$\int\frac{x^2\,dx}{ax + b} = \frac{(ax + b)^2}{2a^3} - \frac{2b(ax + b)}{a^3} + \frac{b^2}{a^3}\ln(ax + b)$$ B.23

$$\int \frac{x^3 dx}{ax+b} = \frac{(ax+b)^3}{3a^4} - \frac{3b(ax+b)^2}{2a^4} + \frac{3b^2(ax+b)}{a^4} - \frac{b^3}{a^4}\ln(ax+b) \qquad \text{B.24}$$

$$\int \frac{dx}{x(ax+b)} = \frac{1}{b}\ln\left(\frac{x}{ax+b}\right) \qquad \text{B.25}$$

$$\int \frac{dx}{(ax+b)^2} = \frac{-1}{a(ax+b)} \qquad \text{B.26}$$

$$\int \frac{dx}{x^2(ax+b)} = -\frac{1}{bx} - \frac{a}{b^2}\ln\left(\frac{x}{ax+b}\right) \qquad \text{B.27}$$

$$\int \frac{x dx}{(ax+b)^2} = \frac{b}{a^2(ax+b)} + \frac{1}{a^2}\ln(ax+b) \qquad \text{B.28}$$

$$\int \frac{x^2 dx}{(ax+b)^2} = \frac{(ax+b)}{a^3} - \frac{b^2}{a^3(ax+b)} - \frac{2b}{a^3}\ln(ax+b) \qquad \text{B.29}$$

$$\int \frac{x^3 dx}{(ax+b)^2} = \frac{(ax+b)^2}{2a^4} - \frac{3b(ax+b)}{a^4} + \frac{b^3}{a^4(ax+b)} + \frac{3b^2}{a^4}\ln(ax+b) \qquad \text{B.30}$$

$$\int \frac{x dx}{(ax+b)^3} = \frac{-1}{a^2(ax+b)} + \frac{b}{2a^2(ax+b)^2} \qquad \text{B.31}$$

$$\int \frac{x^2 dx}{(ax+b)^3} = \frac{2b}{a^3(ax+b)} - \frac{b^2}{2a^3(ax+b)^2} + \frac{1}{a^3}\ln(ax+b) \qquad \text{B.32}$$

$$\int \frac{x^3 dx}{(ax+b)^3} = \frac{x}{a^3} - \frac{3b^2}{a^4(ax+b)} + \frac{b^3}{2a^4(ax+b)^2} - \frac{3b}{a^4}\ln(ax+b) \qquad \text{B.33}$$

$$\int \frac{dx}{a+bx+cx^2} = \frac{1}{\sqrt{-q}}\ln\frac{(2cx+b-\sqrt{-q})}{(2cx+b+\sqrt{-q})} \qquad \text{where } q \equiv 4ac - b^2 \text{ for } q < 0.$$

$$= \frac{2}{\sqrt{q}}\arctan\frac{2cx+b}{\sqrt{q}} \text{ for } q > 0 \qquad \text{B.34}$$

Integration by parts $\int u\,dv = uv - \int v\,du$.

Numerical integration by trapezoidal rule:

$$\int_{X_0}^{X_n} f(x)\,dx = \Delta x \sum_{i=0}^{n} f(x_i) - \Delta x\left[\frac{(f(x_o) - f(x_n))}{2}\right] \qquad \text{B.35}$$

where Δx is a constant step size between discrete values of $f(x)$.

See also Chapter 6 for additional mathematical relationships.

B.2 SOLUTIONS TO CUBIC EQUATIONS

A cubic equation of state may be solved by trial and error or analytically,

$$Z^3 + a_2 Z^2 + a_1 Z + a_0 = 0 \qquad \text{B.36}$$

where the a_2, a_1, and a_0 are constants for the purposes of solving the cubic.

Iterative Method

The Newton-Raphson method uses an initial guess along with the derivative value to rapidly converge on the solution. This discussion focuses on an example solution for the following equation:

$$Z^3 + a_2 Z^2 + a_1 Z + a_0 = F \qquad \text{B.37}$$

We seek the value of Z where $F = 0$. Suppose we have made an initial guess Z_{old} which gives a value F_{old}, as shown in the upper-left graph in Fig. 7.6 on page 265. We are seeking a value of Z that results in $F = 0$. If F_{old} is the current value, and if we use the derivative of F as a linear approximation of the function behavior, then $0 = m \cdot Z_{new} + b$, (where the slope m can be calculated analytically from Eqn. B.37 as $dF/dZ = (3Z^2 - 2(1 - B)Z + (A - 3B^2 - 2B))$. Since the current point is on the same line, we may also write $F_{old} = m \cdot Z_{old} + b$. Taking the difference we get $0 - F_{old} = m \cdot (Z_{new} - Z_{old}) + (b - b)$ or rearranging, $-F_{old}/m + Z_{old} = Z_{new}$. Since $m = (dF/dZ)$, we have $Z_{new} = Z_{old} - F/(dF/dZ)$. The procedure can be repeated until the answer is obtained. A summary of steps is:

1. Guess $Z_{old} = 1$ or $Z_{old} = 0$ and compute $F_{old}(Z_{old})$.
2. Compute dF/dZ.
3. Compute $Z_{new} = Z_{old} - F/(dF/dZ)$.
4. If $|\Delta Z/Z_{now}| < 1.E - 5$, print the value of Z_{now} and stop.
5. Compute $F_{new}(Z_{new})$ and use this as F_{old}. Return to step 3 until step 4 terminates.

Note that an initial guess of $Z = 0$ converges on the smallest real root. An initial guess of $Z = 1$ almost always converges on the largest real root. At very high reduced pressures, an initial guess greater than one is sometimes required since the compressibility factor can exceed one (see Fig. 7.4 on page 257).

Analytical Method

Below are summaries of two methods for solving analytically. These techniques are implemented for the Peng-Robinson equation in the spreadsheet Preos.xlsx. Eqn. B.36 can be reduced to the form

$$x^3 + px + q = 0 \qquad \text{B.38}$$

by substituting for Z the value

$$Z = x - \frac{a_2}{3} \qquad \text{B.39}$$

The values of p and q for Eqn. B.38 will then be

$$p = \frac{1}{3}(3a_1 - a_2^2) \text{ and } q = \frac{1}{27}(2a_2^3 - 9a_2a_1 + 27a_0) \qquad \text{B.40}$$

If a_2, a_1, and a_0 are real (which they are for an EOS), then defining

$$R \equiv \frac{q^2}{4} + \frac{p^3}{27} \qquad \text{B.41}$$

results in one real root and two conjugate roots if $R > 0$;

results in three real roots, of which two are equal if $R = 0$;

results in three real and unequal roots if $R < 0$.

Solution Method I: Algebraic Solution

Let

$$P = \sqrt[3]{\frac{-q}{2} + \sqrt{R}} \text{ and } Q = \sqrt[3]{\frac{-q}{2} - \sqrt{R}} \qquad \text{B.42}$$

The values of x are given by using Eqns. B.40, B.41, and B.42:

$$x = P + Q, \quad \frac{-(P+Q)}{2} + \frac{P-Q}{2}\sqrt{-3}, \quad \frac{-(P+Q)}{2} - \frac{P-Q}{2}\sqrt{-3} \qquad \text{B.43}$$

Values of Z are then found with Eqn. B.39.

Solution Method II: Trigonometric Solution

Let $x = m \cos \theta$; then

$$x^3 + px + q = m^3 \cos^3\theta + pm\cos\theta + q = 4\cos^3\theta - 3\cos\theta - \cos(3\theta) = 0 \qquad \text{B.44}$$

therefore,

$$\frac{4}{m^3} = \frac{-3}{pm} = \frac{-\cos(3\theta)}{q} \qquad \text{B.45}$$

which leads to

$$m = 2\sqrt{\frac{-p}{3}}, \quad \cos(3\theta) = \frac{3q}{pm} \qquad \text{B.46}$$

therefore,

$$\theta_1 = \frac{1}{3}\cos^{-1}\left(\frac{3q}{pm}\right) \qquad \text{B.47}$$

where θ_1 is in radians. By the functionality of the cosine function, two other solutions will be

$$\theta_1 + \frac{2\pi}{3} \text{ and } \theta_1 + \frac{4\pi}{3} \tag{B.48}$$

The values of x are given by using Eqns. B.40 and B.48:

$$x = m\cos\theta_1, \ m\cos\left(\theta_1 + \frac{2\pi}{3}\right), \ m\cos\left(\theta_1 + \frac{4\pi}{3}\right) \tag{B.49}$$

Values of Z are found with Eqn. B.39.

Sorting Roots

Meaningful roots for the Peng-Robinson and many other common EOSs are of the form

$$P = P^{rep} + P^{att}$$

where $P^{rep} = \dfrac{RT}{V-b}$. To ensure that $P^{rep} > 0$, we must have $Z > B$, where $B = bP/(RT)$.

When three real, positive roots exist, the meaningful roots must satisfy $\kappa_T = -\dfrac{1}{V}\left(\dfrac{\partial V}{\partial P}\right)_T \geq 0$, that is, the isothermal compressibility must be positive. The value of κ_T may also be used to determine whether a root is vapor or liquid in cases where only one root is found by identifying the phase with the larger isothermal compressibility as the vapor phase. κ_T is always greater for a vapor root than for a liquid root except at the critical point.[1]

Determination of Equation of State Constants from the Critical Point

Determination of equation of state constants from the critical conditions has been the prevalent method of characterizing fluids. When a cubic equation of state is fit to the critical point, the parameters may be determined in a couple of ways. First, we can evaluate the derivatives,

$$\left(\frac{\partial P}{\partial \rho}\right)_T = 0 \qquad \text{and} \qquad \left(\frac{\partial^2 P}{\partial \rho^2}\right)_T = 0 \tag{B.50}$$

applied at T_c. By differentiating the equation of state, the resultant equations can be simultaneously solved to find two equation of state constants. A simpler approach using significantly less calculus and algebra is to write out $(Z - Z_c)^3$, which goes to zero at the critical point.

$$Z^3 - (3Z_c)Z^2 + (3Z_c^2)Z - (Z_c^3) = 0 \tag{B.51}$$

If we compare Eqn. B.51 with Eqn. B.36 at the critical point, we find

$$-a_2 = 3Z_c; \quad a_1 = 3Z_c^2; \quad -a_0 = Z_c^3 \tag{B.52}$$

For the van der Waals equation, comparing Eqns. B.51 and B.52 with Example 7.7 on page 271,

---

1. Poling, B. E., Grens II, E. A., Prausnitz, J. M. 1981. *Ind. Eng. Chem. Process Des. Dev.* 20:127–130.

$$1 + B_c = 3Z_c^{EOS} \tag{B.53}$$

$$A_c = 3(Z_c^{EOS})^2 \tag{B.54}$$

$$A_c B_c = (Z_c^{EOS})^3 \tag{B.55}$$

where the superscript *EOS* has been added to explicitly show that this is the value predicted by the equation of state. Plugging Eqn. B.54 into Eqn. B.55, we find

$$B_c = \frac{Z_c^{EOS}}{3} \tag{B.56}$$

Plugging into Eqn. B.53,

$$\boxed{Z_c^{EOS} = \frac{3}{8}} \tag{B.57}$$

Thus, the van der Waals equation predicts a universal value of $Z_c = 0.375$. Plugging this into Eqn. B.56, we find

$$\boxed{b = \frac{RT_c}{8P_c}} \tag{B.58}$$

and into Eqn. B.54,

$$\boxed{a = \frac{27}{64}\frac{(RT_c)^2}{P_c}} \tag{B.59}$$

B.3   THE DIRAC DELTA FUNCTION

Understanding many of the terms appearing in thermodynamic functions like the repulsive contribution to the equation of state requires an understanding of a somewhat peculiar mathematical function known as the Dirac delta function. This function has the property of "filtering" the value of a function being integrated and focusing attention on the function value at a particular value of the independent variable. That is, over the interval $x_1 < x_0 < x_2$,

$$F(x_0) = \int_{x_1}^{x_2} F(x)\delta(x - x_0)dx \tag{B.60}$$

where δ is the Dirac delta function defined by

$$\delta(x - x_0) = \begin{cases} 0 \text{ for } x < x_0 \\ \infty \text{ for } x = x_0 \\ 0 \text{ for } x > x_0 \end{cases}$$

B.61

and shown in the figure below. The following discussion should clarify the function.

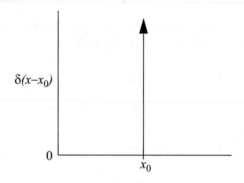

The Dirac delta function

Suppose for the moment that we approximate the delta function as a square pulse function, such that,

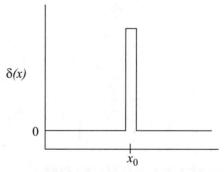

Approximation of the Dirac delta function as a square pulse function.

$$\delta(x) = \begin{cases} 0 \text{ for } x < (x_0 - 0.05) \\ 10 \text{ for } (x_0 - 0.05) < x < (x_0 + 0.05) \\ 0 \text{ for } x > (x_0 + 0.05) \end{cases}$$

B.62

Then when we evaluate the integral,

$$I = \int_0^\infty F(x)\delta(x)dx$$

B.63

a reasonable approximation would be that $F(x) = F(x_0)$ over the relatively short interval where δ holds a nonzero value. Since $F(x_0)$ is approximately constant over the short interval, it may be factored out of the integral, giving,

$$I = \int_0^\infty F(x)\delta(x)dx = F(x_0)\int_0^\infty \delta(x)dx$$

$$= F(x_0)\left(\int_0^{(x_0-0.05)} 0\,dx + \int_{(x_0-0.05)}^{(x_0+0.05)} 10\,dx + \int_{(x_0+0.05)}^\infty 0\,dx \right)$$

B.64

$$= F(x_0)\cdot 10[(x_0+0.05)-(x_0-0.05)] = F(x_0)\cdot 10(0.1) = F(x_0)$$

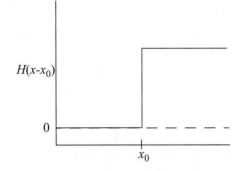

Heaviside step function.

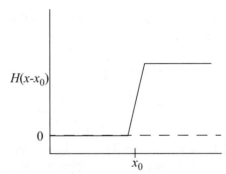

Approximation to the Heaviside step function whose derivative is the square pulse function explained above.

The next step is obvious. Let δ be zero except for an interval of ±0.005 around x_0. Within the interval, set $\delta = 100$. Then our approximation of $F(x) = F(x_0)$ throughout the interval is even more reasonable. Proceeding in this fashion leads to an interval that is differentially deviating from x_0 with δ approaching infinity.

This seems like a crazy thing to do, since we have now defined the function value to be itself times an integral that must be unity. But note that our ultimate goal is to leave the function inside and we will show how this helps us momentarily. Looking at this final result, we see that the function δ has filtered out the value of $F(x)$ at a single value of x.

Despite its peculiarity, the δ function is a well-defined function and therefore integrable. To understand the importance of the δ function, consider taking the derivative of a step function, $H(x - x_0)$, shown below. The derivative of $H(x - x_0)$ is positively infinite at x_0 but zero on either side of x_0. It may help you to imagine slanting the corners of the step function and plotting the derivative first and then making the corners systematically more square. If you plot this derivative function, you will see that it is the square pulse function and becomes the Dirac delta function in the limit of the Heaviside step function. *Therefore, the Dirac delta function is the derivative of a Heaviside step function. And the integral of the Dirac delta function is equal to the Heaviside step function.* It is really this last observation that permits us to derive important results for the integrals of discontinuous functions.

Example B.1 The hard-sphere equation of state

A significant application of the Dirac delta function is suggested by the pressure equation developed for hard-sphere fluids. The discussion in Unit II indicates that, at low density, the radial distribution function (rdf) is given by a simple Boltzmann weighting probability:

$$g(r) \sim exp(-u/kT)$$

Note that for hard-sphere fluids, this function is discontinuous and behaves as the Heaviside step function at $r = \sigma$.

The pressure equation is:

$$\frac{P}{\rho RT} = 1 - \frac{N_A \rho}{6kT} \int_0^\infty r \frac{\partial u}{\partial r} g(r) 4\pi r^2 dr \qquad \text{B.65}$$

If we were to substitute the hard-sphere potential and a discontinuous rdf like the low-density one into the pressure equation, we would have a combination of discontinuities that we could not resolve. Note what happens if we postulate a function $y(r)$ such that

$$y(r) = exp(u/kT) \cdot g(r) \qquad \text{B.66}$$

Clearly, $y(r)$ would be continuous at low density and approaches a value of one. Furthermore, it turns out that $y(r)$ can be rigorously proven to be continuous for all densities and all potentials.[1] Apply these insights to develop the equation of state for a hard-sphere fluid in terms of its rdf value at $r = \sigma$.

Solution: Substituting $g(r) = y(r)exp(-u/kT)$ and recognizing

$$\frac{\partial exp\left(-\dfrac{u}{kT}\right)}{\partial r} = -\frac{1}{kT} exp\left(-\frac{u}{kT}\right) \frac{\partial u}{\partial r}, \text{ then} \qquad \text{B.67}$$

$$Z = \frac{P}{\rho RT} = 1 - \frac{N_A \rho}{6kT} \int_0^\infty \frac{\partial u}{\partial r} y(r) exp\left(-\frac{u}{kT}\right) 4\pi r^3 dr = 1 + \frac{N_A \rho}{6} \int_0^\infty \frac{\partial exp\left(-\dfrac{u}{kT}\right)}{\partial r} y(r) 4\pi r^3 dr$$

$$\text{B.68}$$

If you plot the function $exp(-u/kT)$ versus r for the hard-sphere potential, you will see that it is a Heaviside step function. This means that its derivative is a Dirac delta. So the pressure equation becomes, using $y^{HS}(r)$ to emphasize the constraint of the hard sphere function,

$$Z = \frac{P^{HS}}{\rho RT} = 1 + \frac{N_A \rho}{6} \int_0^\infty \delta(r - \sigma) y^{HS}(r) 4\pi r^3 dr \qquad \text{B.69}$$

Applying Eqn. B.60,

Example B.1   The hard-sphere equation of state (Continued)

$$Z = \frac{P^{HS}}{\rho RT} = 1 + \frac{4\pi\sigma^3 N_A \rho}{6} y^{HS}(\sigma)$$

(B.70)

Although it may seem that we don't know $y(\sigma)$, recall that it is a continuous function. Therefore the value of $y(\sigma)$ will be approximated by $y(\sigma^+)$. Further, recognizing that $u^{HS}(\sigma^+) = 0$, therefore $\exp(u^{HS}(\sigma^+)/kT) = 1$ in Eqn. B.66, and therefore $y^{HS}(\sigma^+) = g^{HS}(\sigma^+)$.

$$Z = \frac{P^{HS}}{\rho RT} = 1 + \frac{4\pi\sigma^3 N_A \rho}{6} g^{HS}(\sigma^+)$$

(B.71)

Defining $\eta_P = \frac{1}{6}\pi\sigma^3 N_A \rho$

$$\boxed{Z = \frac{P^{HS}}{\rho RT} = 1 + 4\eta_P g^{HS}(\sigma^+)}$$

(B.72)

The relationship between Z^{HS} and $g^{HS}(\sigma^+)$ is used often. It provides the functional form for building Z^{rep} in the equation of state. It also provides a basis for converting our approximate equation of state for Z^{rep} back into a quantitative estimate of $g^{HS}(\sigma)$. Recognize that Eqn. B.72 is a virial equation for hard spheres, $Z = 1 + B\rho$. At low density, $g^{HS}(\sigma^+) = \exp(-u^{HS}(\sigma^+)/kT) = 1$, resulting in Boltzmann's value for the second virial coefficient for hard spheres.

Example B.2   The square-well equation of state

The preceding example illustrates the use of all the tools necessary to develop any equation of state, but one more illustration may help to clarify the way that the tools can be used in combination to derive a wide variety of results. The problem statement is as follows: Develop a formula for deriving the equation of state for any fluid described by the square-well potential ($\lambda = 1.5$), given an estimate for the radial distribution function (rdf). Apply this formula to obtain the equation of state from the following approximation for the rdf,

$$g(r) \approx \begin{cases} 0 \text{ for } r < \sigma \\ \left(1 + \dfrac{b\rho\varepsilon}{kT}\dfrac{1}{x^4}\right) \text{ for } r \geq \sigma \end{cases}$$

where $x = r/\sigma$, and $b = \pi N_A \sigma^3/6$. Before looking at the solution, think for yourself. Can you conceive of how to solve the problem without any help?

Solution: The trick is to realize that the exponential of the square-well potential is composed of two step functions, each of a different height. The step up is of height, $\exp(\varepsilon/kT)$ whereas the step down is of height, $\exp(\varepsilon/kT) -1$.

$$\exp(-u/kT) = \exp(\varepsilon/kT) \, H(r - \sigma) - [\exp(\varepsilon/kT) -1] \cdot H(r - \lambda\sigma)$$

Example B.2 The square-well equation of state (Continued)

Taking the derivative of the Heaviside function gives the Dirac Delta in two places:

$$Z = 1 + \frac{N_A \rho}{6} \int \{\delta(r - \sigma)y(r)\exp(\varepsilon/kT) - \delta(r - \lambda\sigma)y(r)[\exp(\varepsilon/kT) - 1]\}4\pi r^2 dr$$

$$= 1 + \frac{4\pi N_A \rho \sigma^3}{6}\{y(\sigma)\exp(\varepsilon/kT) - \lambda^3 y(\lambda\sigma)[\exp(\varepsilon/kT) - 1]\}$$

Noting that $y(r) = g(r)\exp(u/kT)$ and that $\exp(u/kT)$ is best evaluated inside the well:

$$Z = 1 + 4b\rho\{g(\sigma^+) - \lambda^3[1 - \exp(-\varepsilon/kT)]g(\lambda\sigma^-)\} \qquad \text{B.73}$$

This is valid for the square-well fluid with any g(r).

For the above expression for $\lambda = 1.5$: $g(\sigma^+) = 1 + b\rho\varepsilon/kT$ and $g(\lambda\sigma^-) = g(1.5\sigma^-) = 1 + 0.198\ b\rho\varepsilon/kT$

$$Z = 1 + 4b\rho\{1 + b\rho\varepsilon/kT - 1.5^3[1 - \exp(-\varepsilon/kT)](1 + 0.198\ b\rho\varepsilon/kT)\} \qquad \text{B.74}$$

At the given conditions: $Z = 1 + 4(0.2)\{1 + 0.2 - 2.1333(1.0396)\} = 0.1858$

It would be straightforward at this point to develop the entire phase diagram for this new equation of state. The result would yield expressions for ε/k and b in terms of T_c and P_c. The value for the acentric factor would be a fixed value since there is no third parameter to affect it. Its numerical value could be determined in the same way that the value was determined for the van der Waals equation in Chapter 8.

At first glance, one may wonder why we have expended so much effort to represent this problem in terms of the rdf when we must approximate it anyway. It may seem fruitless to have translated our ignorance of the equation of state into ignorance of the radial distribution function. It turns out that the thermodynamic properties are fairly insensitive to details of the rdf. If you doubt this, reflect on what van der Waals achieved by effectively assuming that $g(r) = 1$ for all temperatures and densities. Furthermore, if you refer to the discussion of local composition theory in Chapter 13, you will see that the fundamental basis for virtually all of the activity coefficient models currently in use is: $g_{ij}(r)/g_{jj}(r) = \exp[(\varepsilon_{ij} - \varepsilon_{jj})/kT]$. As a slightly oversimplified summary of modern research in equilibrium thermodynamics, one could say that it is a search for better approximations of the radial distribution function. The thermodynamics of hydrogen bonding can be related to the rdf as discussed in Chapter 19. The thermodynamics of polymers and folding of proteins simply require generalization in terms of an intramolecular rdf known as the conformation. Electrolyte and solution thermodynamics are given very directly in terms of the rdf as discussed in Chapter 13. Adsorption and slit thermodynamics can be expressed in terms of the rdf between the fluid molecules and the adsorbent surface. Clearly, even rough approximations can be very useful when developed at the level of the rdf and carried to their logical conclusion. It may not be worth the effort for every chemical engineer, but it should not be beyond the grasp of many engineering students who sincerely want to understand the method behind the madness of fugacity estimation.

STRATEGIES FOR SOLVING VLE PROBLEMS

In earlier chapters, we have discussed applications of VLE using simplified procedures such as bubble calculations using modified Raoult's law. This appendix summarizes flow sheets for modified Raoult's law and cubic equations of state.

- Section C.1 focuses on modified Raoult's Law and offers a summary of flow sheets that quickly converge. Though you may be able to figure out other ways to converge calculations, it is often best to use strategies that have been well tested.

- Section C.2 discusses equations of state. Software permits solution of VLE using cubic equations without knowing the details of how the iterations are performed, and so a strategy is presented here also. There are many strategies throughout the literature, and the reader should be aware that other strategies are also successful.

- Section C.3 covers the gamma-phi method for nonideal gases with activity coefficients. These calculations require more sophistication, and are summarized online.

For each approach, there are five flowcharts presented—bubble *P*, bubble *T*, dew *P*, dew *T*, and isothermal flash. Specific routines may also be written for *VLLE* or other multiphase applications, which are not summarized here.

C.1   MODIFIED RAOULT'S LAW METHODS

The equation that must be solved is: $y_i P = x_i \gamma_i P_i^{sat}$

Bubble *P*

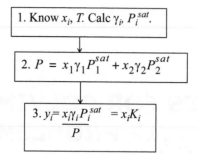

1. Know x_i, T. Calc γ_i, P_i^{sat}.

2. $P = x_1 \gamma_1 P_1^{sat} + x_2 \gamma_2 P_2^{sat}$

3. $y_i = \dfrac{x_i \gamma_i P_i^{sat}}{P} = x_i K_i$

Bubble *T*

(Choose one flow sheet.)

Option (*a*)

1. Know x_i, P
Guess T (e.g., Eqn. 10.22).

2. Calc γ_i, P_i^{sat}. ← 7. Guess T

3. $y_i = \dfrac{x_i \gamma_i P_i^{sat}}{P} = x_i K_i$

4. $y_T = \sum_i y_i$

5. $y_T = 1$?   6. If $y_T > 1$, $T\downarrow$
    Yes   No   If $y_T < 1$, $T\uparrow$

8. Bubble T and y_i found

Option (*b*)

1. Know x_i, P. Guess T (e.g., Eqn. 10.22).

2. $P_{calc} = \sum_i x_i \gamma_i P_i^{sat}$

Adjust T until $P_{calc} = P$

3. Bubble T found,
$y_i = \dfrac{x_i \gamma_i P_i^{sat}}{P} = x_i K_i$

Dew *P*

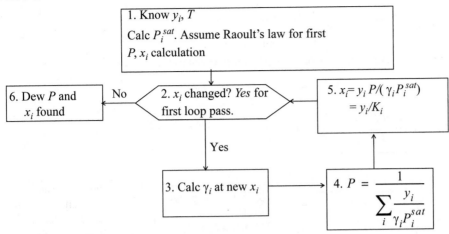

1. Know y_i, T
Calc P_i^{sat}. Assume Raoult's law for first P, x_i calculation

6. Dew P and x_i found ← No — **2. x_i changed? *Yes* for first loop pass.** ← **5. $x_i = y_i P/(\gamma_i P_i^{sat})$ $= y_i/K_i$**

Yes ↓

3. Calc γ_i at new x_i → **4. $P = \dfrac{1}{\displaystyle\sum_i \dfrac{y_i}{\gamma_i P_i^{sat}}}$**

Dew *T*

(Choose one flow sheet.)

Option (*a*)

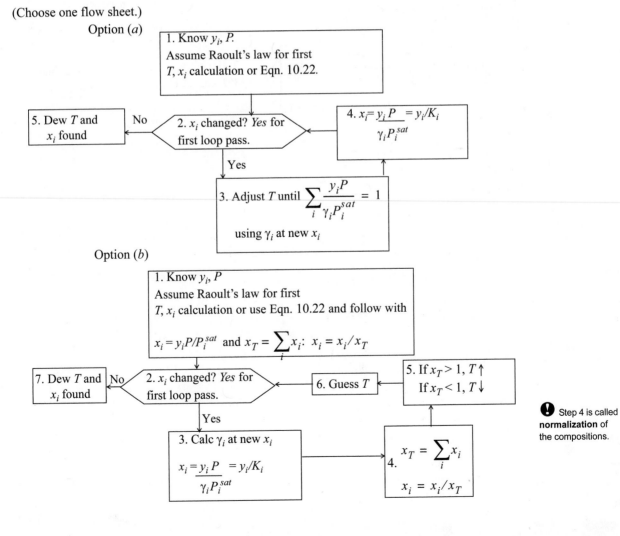

1. Know y_i, P.
Assume Raoult's law for first T, x_i calculation or Eqn. 10.22.

5. Dew T and x_i found ← No — **2. x_i changed? *Yes* for first loop pass.** ← **4. $x_i = \dfrac{y_i P}{\gamma_i P_i^{sat}} = y_i/K_i$**

Yes ↓

3. Adjust T until $\displaystyle\sum_i \dfrac{y_i P}{\gamma_i P_i^{sat}} = 1$ using γ_i at new x_i

Option (*b*)

1. Know y_i, P
Assume Raoult's law for first T, x_i calculation or use Eqn. 10.22 and follow with
$x_i = y_i P/P_i^{sat}$ and $x_T = \displaystyle\sum_i x_i$: $x_i = x_i/x_T$

7. Dew T and x_i found ← No — **2. x_i changed? *Yes* for first loop pass.** ← **6. Guess T** ← **5. If $x_T > 1$, $T\uparrow$ If $x_T < 1$, $T\downarrow$**

Yes ↓

3. Calc γ_i at new x_i
$x_i = \dfrac{y_i P}{\gamma_i P_i^{sat}} = y_i/K_i$
→ **4. $x_T = \displaystyle\sum_i x_i$ $x_i = x_i/x_T$**

🛈 Step 4 is called **normalization** of the compositions.

Isothermal Flash

1. Know z_i, P, T
Apply shortcut K-ratio method for first K_i calculation.
Skip step 2 first time, set x_i, y_i to force outer loop below to execute at least once.

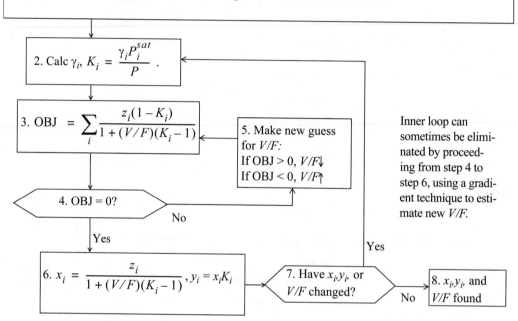

2. Calc γ_i, $K_i = \dfrac{\gamma_i P_i^{sat}}{P}$.

3. OBJ $= \displaystyle\sum_i \dfrac{z_i(1 - K_i)}{1 + (V/F)(K_i - 1)}$

4. OBJ $= 0$?

5. Make new guess for V/F:
If OBJ > 0, $V/F\downarrow$
If OBJ < 0, $V/F\uparrow$

Inner loop can sometimes be eliminated by proceeding from step 4 to step 6, using a gradient technique to estimate new V/F.

No

Yes

6. $x_i = \dfrac{z_i}{1 + (V/F)(K_i - 1)}$, $y_i = x_i K_i$

7. Have x_i, y_i, or V/F changed?

Yes

No

8. x_i, y_i, and V/F found

C.2   EOS METHODS

The equation that must be solved is: $y_i \hat{\varphi}_i^V P = x_i \hat{\varphi}_i^L P$.

Bubble *P*

(The bubble pressure flow sheet is presented in Section 15.4)

Bubble *T*

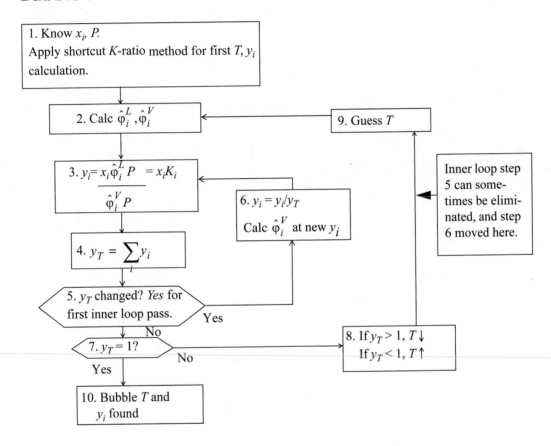

Dew *P*

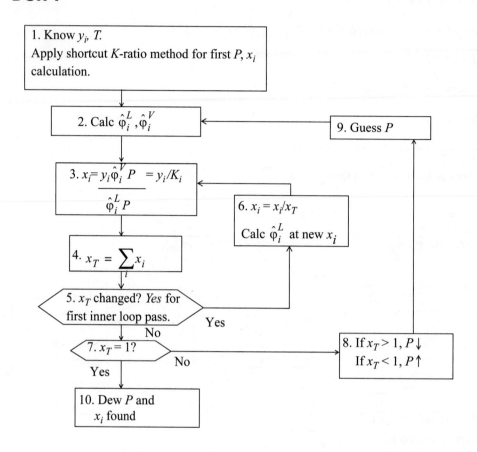

Dew *T*

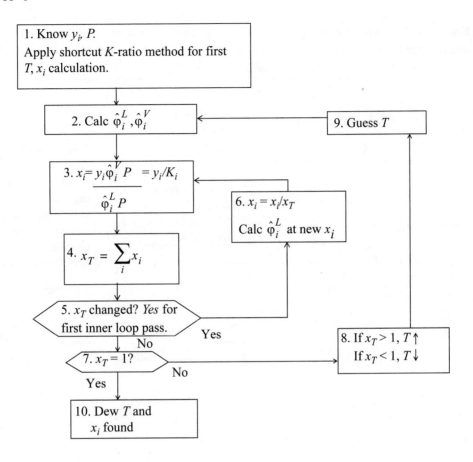

Isothermal Flash

1. Know z_i, P, T. Perform bubble and dew to be sure system will flash at conditions given. Apply shortcut K-ratio method for first K_i calculation. Skip step 2 first time, set $x_i = y_i = z_i$ to force outer loop below to execute at least once.

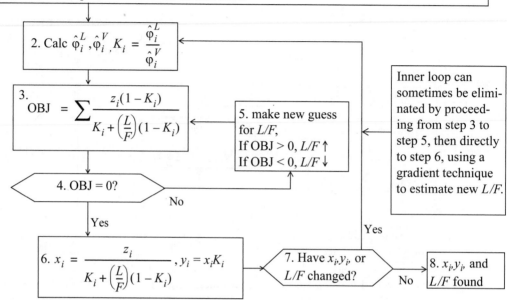

2. Calc $\hat{\varphi}_i^L$, $\hat{\varphi}_i^V$, $K_i = \dfrac{\hat{\varphi}_i^L}{\hat{\varphi}_i^V}$

3. $\text{OBJ} = \sum \dfrac{z_i(1 - K_i)}{K_i + \left(\dfrac{L}{F}\right)(1 - K_i)}$

4. OBJ = 0?

5. make new guess for L/F,
If OBJ > 0, $L/F \uparrow$
If OBJ < 0, $L/F \downarrow$

Inner loop can sometimes be eliminated by proceeding from step 3 to step 5, then directly to step 6, using a gradient technique to estimate new L/F.

6. $x_i = \dfrac{z_i}{K_i + \left(\dfrac{L}{F}\right)(1 - K_i)}$, $y_i = x_i K_i$

7. Have x_i, y_i, or L/F changed?

8. x_i, y_i, and L/F found

C.3 ACTIVITY COEFFICIENT (GAMMA-PHI) METHODS

The equation that must be solved is: $y_i \hat{\varphi}_i^V P = x_i \gamma_i \varphi_i^{sat} P_i^{sat} \exp(V_i^L (P - P_i^{sat})/RT)$.

These flow sheets are available online as a supplement to Unit III.

APPENDIX D

MODELS FOR PROCESS SIMULATORS

D.1 OVERVIEW

There are so many thermodynamic models commonly used in chemical process simulators that it would be overwhelming to cover all of them in great detail. This is why the discussion in the text focuses on a few representative models. Nevertheless, students interested in process engineering will often face the need to choose the most appropriate thermodynamic model, and the most appropriate model may not be one of those that we have covered in detail. Fortunately, the differences between many of the thermodynamic models and the ones that we have studied are generally quite small. In this appendix, we review some of most common thermodynamic models and put them into context with others that we have studied. This should help students to feel a bit more comfortable wading through the wealth of models from which to choose.

Students interested in becoming process simulation experts will be interested in reading the recent articles reviewing the selection of thermodynamic models. Schad[1] and Carlson[2] provide some significant examples and cites several relevant articles. A common thread throughout these articles is the emphasis on accurate application of thermodynamic principles. It is interesting to see the large number of examples in which practical engineering applications were so deeply affected by the fundamentals of thermodynamics.

D.2 EQUATIONS OF STATE

We have covered the Peng-Robinson and virial equations in fair detail, but there are many others. Some that we have mentioned but not treated in detail are the Redlich-Kwong equation (homework problem 7.9),[3] the Lee-Kesler equation (Eqn. 7.11 on page 260),[4] and a popular form of its extension to mixtures, the Lee-Kesler-Plocker equation,[5] the Soave equation (Eqn. 7.65 on page 294),[6]

---

1. Schad, R.C. 1998. *Chem. Eng. Prog.* 94(1).:21–27.
2. Carlson, E.C., 1996. *Chem. Eng. Prog.* Oct.:35.
3. Redlich, O, Kwong, J.N.S. 1949. *Chem. Rev.* 44:233.
4. Lee, B.I., Kesler, M.G. 1975. *AIChE J.* 21:510.
5. Plocker, U., Knapp, H., Prausnitz, J.M. 1978. *Ind. Eng.Chem.Proc.Des.Dev.* 17:324.

also known as the Soave-Redlich-Kwong or SRK equation), the ESD equation,[7] and the SAFT equation.[8] A slight variation on the Soave equation is the API equation;[9] it changes only the value of κ as a function of acentric factor in order to obtain a slight improvement in the predicted vapor pressures of hydrocarbons. A specific implementation of the virial equation useful for associating systems is the Hayden-O'Connell method.[10] The Soave equation, Peng-Robinson equation, Lee-Kesler-Plocker equation, and API equation are all very similar in their predictions of VLE behavior of hydrocarbon mixtures. They are accurate to within ~5% in correlations of bubble-point pressures of hydrocarbons and gases (CO, CO_2, N_2, O_2, H_2S) and about ~15% for predictions based on estimated binary interaction parameters. The Lee-Kesler-Plocker equation can be slightly more accurate for enthalpy and liquid density for some hydrocarbon mixtures, but the advantage is generally slight with regard to enthalpy and there are better alternatives to equations of state for liquid densities if you want accurate values. The cubic equations have some convergence advantages for VLE near critical points and their relative simplicity makes them more popular choices for adaptations of semi-empirical mixing rules to tune in an accurate fit to the thermodynamics of a specific system of interest. The best choice among these is generally the one for which the binary interaction parameters have been determined with the greatest reliability. (Accurate reproduction of the most experimental data at the conditions of your specific interest wins.)

The primary role of equations of state is that they can predict thermodynamic properties at any conditions of temperature and pressure, including the critical region. The disadvantage is that they tend to be inaccurate for strongly hydrogen-bonding mixtures. This disadvantage is diminishing in importance with the development of hydrogen-bonding equations of state (like the SAFT and ESD equations), but it is not clear at this time whether these newer equations of state will displace any of the long-standing cubic equations of state with their semi-empirical modifications.

D.3   SOLUTION MODELS

We have covered many solution models in fair detail: the Margules equation, the Redlich-Kister expansion, the van Laar equation, the Scatchard-Hildebrand theory (the most common implementation of regular solution theory), the Flory-Huggins equation, the Wilson equation, NRTL, UNIQUAC, and UNIFAC. Once again, the best choice will most often depend on the availability of binary interaction parameters which are relevant to the specific conditions of interest.

The primary role of solution models is to provide semi-empirical models which have a greater degree of flexibility than equation of state models, owing to the greater number of adjustable parameters and their judicious choice such that both magnitude and skewness of the free energy curves can be accurately tuned.

D.4   HYBRID MODELS

Another set of models that have been developed relatively recently can be referred to as "hybrid" models in the sense that they combine equation of state models with solution models. The two most prevalent of these are the Modified Huron-Vidal (MHV) method and the Wong-Sandler mixing rules.[11] The basic idea is to apply a solution model at high density or pressure to characterize the

6. Soave, G. 1972. *Chem. Eng. Sci.* 27:1197.

7. Elliott Jr., J. R., Suresh, S.J.; Donohue, M.D. 1990. *Ind. Eng. Chem. Res.* 29:1476.

8. Chapman, W.G., Gubbins, K.E., Jackson, G., Radosz, M. 1990. *Ind. Eng. Chem. Res.* 29:1709.

9. Graboski, M.S., Daubert, T.E. 1979. *Ind. Eng.Chem.Proc.Des.Dev.* 17:448.

10. Hayden, J.G., O'Connell, J.P. 1975. *Ind. Eng.Chem.Proc.Des.Dev.* 14:3.

mixing rules of the equation of state and then interpolate from this result to the virial equation at low density. These methods tend to compete with the hydrogen bonding models in the sense that they enhance accuracy for nonideal solutions at high temperatures and pressures. They are more empirical, but they tend to leverage the well-developed solution models (like UNIFAC) more directly. They also tend to be more efficient computationally than the hydrogen bonding equations of state.

D.5   RECOMMENDED DECISION TREE

When faced with choosing a thermodynamic model, it is helpful to at least have a logical procedure for deciding which model to try first. A decision tree is included in Fig. D.1. For nonpolar fluids, an equation of state may suffice. For polar fluids, a fitted activity coefficient model is preferred, possibly in combination with the Hayden-O'Connell method or in combination with some other equation of state for the vapor phase (like the Peng-Robinson equation). This approach can often provide satisfactory predictions as long as the pressures are 10 bar or less. Predictions by this approach should be checked against literature data to the greatest extent possible. If there are no experimental data for one of the binary systems in this event, then UNIFAC can be used to generate "pseudo-data" that can be used to predict the Gibbs excess energy for that binary, and these pseudo-data can be used to regress UNIQUAC or NRTL parameters if desired (homework problem 13.16). Above 10 bars, the choices are not so obvious. The most obvious method to try if you are satisfied with the correlations below 10 bars is to apply the MHV or Wong-Sandler approach. If you need to predict phase behavior over a broad range of conditions based on few data in a narrow range of conditions, a hydrogen bonding equation might provide more reliable leverage in light of its clearer connection with the physical chemistry in the solution. If you are dealing with compounds which dissociate electrolytically or associate strongly and specifically in solution, then it will probably be necessary to apply a simultaneous reaction and phase equilibrium approach. These kinds of systems are common in gas strippers for compounds like CO_2, H_2S, and amines. For these systems, it is especially important to check your correlations against experimental data near the conditions of your specific interest.

---

11. Wong, D. S. H., Sandler, S. I. 1992. *AICHE J.* 38:671.

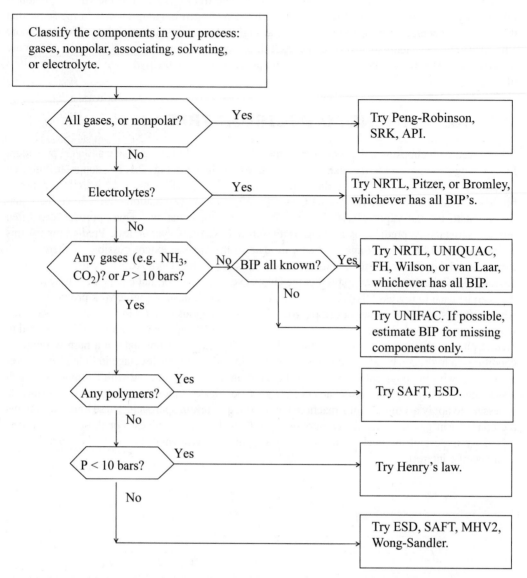

Figure D.1 *Flowchart to select the best thermodynamic model. The abbreviation BIP is used to mean binary interaction parameters.*

APPENDIX

THEMODYNAMIC PROPERTIES

E.1 THERMOCHEMICAL DATA

The heat capacity coefficients are used in the equation:

$$C_P = A + BT + CT^2 + DT^3$$

where T is in K and C_P in J/mol-K The heat capacities tabulated on the back flap are only suitable for quick order of magnitude calculations. The full form of the heat capacity should be used when possible.

Note that the value of the heat capacity at room temperature is not given by the first coefficient A when other coefficients are present. The polynomial should not be truncated.

		$\Delta H_{f,298.15}$	$\Delta G_{f,298.15}$	Heat Capacity Constants			
		kJ/mol	kJ/mol	A	B	C	D
Solids: Standard state is pure solid at 298.15 K, 1 bar.							
Carbon (graphite) (298-1300 K)		0	0	-3.958	5.586E-02	-4.548E-05	1.517E-08
Calcium carbonate, $CaCO_3$		-1206.92	-1128.79				
Calcium oxide, CaO		-635.09	-604.03				
Glucose		-1273.3	-908.1				
Liquids, over the temperature range from 273.15 to 373.15 K[a]							
Acetone				135.6	-0.1770	2.837E-4	6.89E-7
Ammonia				188.1	8.377E-01	1.602E-03	
Aniline				131.5	2.414E-01	-1.314E-04	
Benzene				-6.211	5.650E-01	-3.141E-04	
1,3-Butadiene				188.8	-7.313E-01	1.711E-03	
Carbon tetrachloride				175.9	-4.014E-01	8.409E-04	
Chlorobenzene				93.77	2.732E-01	-2.652E-04	
Chloroform				159.8	-3.566E-01	6.902E-04	
Cyclohexane				-75.23	1.175	-1.344E-03	

843

		$\Delta H_{f,298.15}$	$\Delta G_{f,298.15}$	Heat Capacity Constants			
		kJ/mol	kJ/mol	A	B	C	D
	Ethanol			281.6	-1.435	2.903E-03	
	Ethylene oxide			174.9	-7.184E-01	1.432E-03	
	Methanol			111.7	-4.264E-01	1.090E-03	
	Lactic acid	-682.96[b]	-521.04[b]				
	n-Propanol			3.463E+02	-1.749E+00	3.552E-03	
	Sulfur trioxide			-2.436E+01	1.140E+00	-7.045E-04	
	Toluene			1.258E+02	5.645E-02	1.359E-04	
	Triolean	-2161[b]	17,010[b]				
	Water	-285.83	-237.214	7.243E+01	1.039E-02	-1.497E-06	

Ideal gases: Standard State: Ideal Gas 298.15 K and 1 bar. [c]
USE PRESSURE IN BAR WHEN USING THESE PROPERTIES

Paraffins

1	Methane	-74.8936	-50.45	19.25	0.05213	1.197E-05	-1.132E-08
2	Ethane	-83.82	-31.86	5.409	0.1781	-6.938E-05	8.713E-09
3	Propane	-104.68	-24.29	-4.224	0.3063	-1.586E-04	3.215E-08
4	n-Butane	-125.79	-16.57	9.487	0.3313	-1.108E-04	-2.822E-09
5	Isobutane	-134.99	-20.8781	-1.39	0.3847	-1.846E-04	2.895E-08
7	n-Pentane	-146.76	-8.65	-3.626	0.4873	-2.580E-04	5.305E-08
8	Isopentane			-9.525	0.5066	-2.729E-04	5.723E-08
9	Neopentane (2,2-Dimethylpropane)			-16.59	0.5552	-3.306E-04	7.633E-08
11	n-Hexane	-166.92	-0.16736	-4.413	0.528	-3.119E-04	6.494E-08
17	n-Heptane	-187.8	8.2	-5.146	0.6762	-3.651E-04	7.658E-08
27	n-Octane	-208.75	16.4013	-6.096	0.7712	-4.195E-04	8.855E-08
41	Isooctane(2,2,4-Trimethylpentane)	-224.01	13.6817	-7.461	0.7779	-4.287E-4	9.173E-8
46	n-Nonane	-228.74	24.97	-8.374	0.8729	-4.823E-04	1.031E-07
56	n-Decane	-249.46	33.18	-7.913	0.9609	-5.288E-04	1.131E-07
64	n-Dodecane			-9.328	1.149	-6.347E-04	1.359E-07
66	n-Tetradecane			-10.98	1.338	-7.423E-04	1.598E-07
68	n-Hexadecane	-374.17	82.15	-13.02	1.529	-8.537E-04	1.850E-07

Naphthenes

104	Cyclopentane	-78.4	38.8693	-53.62	0.5426	-3.031E-04	6.485E-08
105	Methylcyclopentane	-106.023	35.7732	-50.11	0.6381	-3.642E-04	8.014E-08
137	Cyclohexane	-123.3	31.7566	-54.54	0.6113	-2.523E-04	1.321E-08
138	Methylcyclohexane	-154.766	27.2797	-61.92	0.7842	-4.438E-04	9.366E-08
153	cis-Decalin	-168.95	85.8138				
154	trans-Decalin	-182.297	73.4292				

Olefins and Acetylene

201	Ethylene	52.51	68.43	3.806	0.1566	-8.348E-05	1.755E-08
202	Propylene	19.71	62.14	3.71	0.2345	-1.160E-04	2.205E-08
204	1-Butene	-0.54	70.24	-2.994	0.3532	-1.990E-04	4.463E-08
205	cis-2-Butene	-7.4	65.5				
206	trans-2-Butene	-11	63.4				
207	Isobutene	-17.1	57	16.05	0.2804	-1.091E-04	9.098E-09
209	1-Pentene			-0.134	0.4329	-2.317E-04	4.681E-08
216	1-Hexene	-41.67	87.4456				
260	1-Decene	-124.69	119.83				
270	Cyclohexene	-4.6	106.859				
303	1,3-Butadiene	109.24	149.73	-1.687	0.3419	-2.340E-04	6.335E-08
309	2-Methyl-1,3-butadiene (Isoprene)			-3.412	0.4585	-3.337E-04	1.000E-07
401	Acetylene	226.731	209.2	26.82	0.07578	-5.007E-05	1.412E-08
	Styrene			-28.25	0.6159	-4.023E-4	9.935E-8

Aromatics

501	Benzene	82.88	129.75	-33.92	0.4739	-3.017E-04	7.130E-08
502	Toluene	50.17	122.29	-24.35	0.5125	-2.765E-04	4.911E-08
504	Ethylbenzene	29.92	130.73	-43.1	0.7072	-4.811E-04	1.301E-07

		$\Delta H_{f,298.15}$	$\Delta G_{f,298.15}$	Heat Capacity Constants			
		kJ/mol	kJ/mol	A	B	C	D
505	1,2-Dimethyl benzene	19	122.22	-15.85	0.5962	-3.443E-04	7.528E-08
506	1,3-Dimethyl benzene	17.24	119	-29.17	0.6297	-3.747E-04	8.478E-08
507	1,4-Dimethyl benzene	17.95	121.26	-1.509	0.6042	-3.374E-04	6.820E-08
510	Isopropylbenzene (Cumene)	3.93	137.15	-33.936	0.7842	-5.087E-04	1.291E-07
558	Biphenyl	182.42	281.08	-97.07	1.106	-8.855E-04	2.790E-07
601	Phenylethene	147.36	213.802				
701	Naphthalene	150.959	223.593	-68.8	0.8499	-6.506E-04	1.981E-07
702	1-Methylnaphthalene			-64.82	0.9387	-6.942E-04	2.016E-07
706	1,2,3,4-Tetrahydro-naphthalene (Tetralin)	24.2	167.1				
803	Indene	163.28	233.97				
805	Phenanthrene	206.9	308.1				
Oxygenated Hydrocarbons							
1001	Formaldehyde	-117.152	-112.968				
1002	Acetaldehyde	-166.021	-133.302				
1051	Acetone	-215.7	-151.2	6.301	0.2606	-1.253E-04	2.038E-08
1052	2-Butanone	-239	-151.9	10.94	0.3559	-1.900E-04	3.920E-08
1101	Methanol	-200.94	-162.24	21.15	0.07092	2.587E-05	-2.852E-08
1102	Ethanol	-234.95	-167.73	9.014	0.2141	-8.390E-05	1.373E-09
1103	Propanol	-255.2	-161.795	2.47	0.3325	-1.855E-04	4.296E-08
1104	2-Propanol	-272.295	-173.385	32.43	0.1885	6.406E-05	-9.261E-08
1105	Butanol	-274.6	-150.666	3.266	0.418	-2.242E-04	4.685E-08
1107	Isobutanol			-7.708	0.4689	-2.884E-04	7.231E-08
1114	1-Hexanol	-316.5	-135.562				
1181	Phenol	-96.3993	-32.8988				
1181	Glucose	-1273.3	-908.40	-54.64	1821	1.084E-3	3.775E-7
1201	Ethylene glycol	-392.878	-304.47				
1211	Propylene glycol	-425.429	-304.483				
1251	Acetic acid	-434.425	-376.685	4.84	0.2549	-1.753E-04	4.949E-08
1256	Butyric acid	-475	-355				
1281	Benzoic acid	-294.1	-210.413				
1289	Terephthalic acid	-663.331	-597.179				
1312	Methyl acetate	-408.8	-321.5				
1313	Ethyl acetate	-444.5	-327.4				
1381	Dimethyl terephthalate	-643.583	-473.646				
1479	Tetrahydrofuran (THF)			19.1	0.5162	-4.132E-04	1.454E-07
1402	Diethyl ether	-252	-122	21.42	0.3359	-1.035E-04	-9.357E-09
1403	Isopropyl ether	-318.821	-121.88				
	Ethylene oxide			-7.519	0.2222	-1.256E-04	2.592E-08
Halocarbons							
1501	Carbon tetrachloride	-95.8136	-60.6261	40.72	0.2049	-2.270E-04	8.843E-08
1502	Methyl chloride	-80.7512	-62.8855	13.88	0.1014	-3.889E-05	2.567E-09
1503	Ethyl chloride	-112.26	-59.9986				
1511	Dichloromethane	-95.3952	-68.8687				
1521	Chloroform (CHCl3)	-103.345	-68.5339	24	0.1893	-1.841E-04	6.657E-08
1562	Chlorobenzene	51.09	98.29	-33.89	0.5631	-4.522E-04	1.426E-07
1591	Vinylidene chloride (1,1-C2H2Cl2)	2.38	25.39				
1601	Freon-12 (CCl2F2)	-493.294	-453.964	31.6	0.1782	-1.509E-04	4.342E-08
	Freon-22 (CClF2)			17.3	0.1618	-1.170E-04	3.058E-08
	Freon-11 (CCl3F)			40.98	0.1668	-1.416E-04	4.146E-08
	Freon-113 (C2Cl3F3)			61.14	0.2874	-2.420E-04	6.904E-08
Miscellaneous Gases							
899	Nitrous oxide	82.05	103.638	21.62	0.07281	-5.778E-05	1.830E-08
901	Oxygen	0	0	28.11	-3.7E-06	1.746E-05	-1.065E-08
902	Hydrogen (equilibrium)	0	0	27.14	0.009274	-1.381E-05	7.645E-09

		$\Delta H_{f,298.15}$	$\Delta G_{f,298.15}$	Heat Capacity Constants			
		kJ/mol	kJ/mol	A	B	C	D
905	Nitrogen	0	0	31.15	-0.01357	2.680E-05	-1.168E-08
908	Carbon monoxide	-110.53	-137.16	30.87	-0.01285	2.789E-05	-1.272E-08
909	Carbon dioxide	-393.51	-394.38	19.8	0.07344	-5.602E-05	1.715E-08
910	Sulfur dioxide	-296.81	-300.14	23.85	0.06699	-4.961E-05	1.328E-08
911	Sulfur trioxide	-396.0	-371.3	19.21	0.1374	-1.176E-04	3.700E-08
912	Nitric oxide	90.25	86.58	29.35	-0.00094	9.747E-06	-4.187E-09
913	Helium-4	0	0	20.8			
914	Argon	0	0	20.8			
917	Fluorine	0	0				
918	Chlorine	0	0	26.93	0.03384	-3.869E-05	1.547E-08
919	Neon	0	0	20.8			
920	Krypton	0	0	20.8			
922	Bromine	0	0	33.86	0.01125	-1.192E-05	4.534E-09
959	Xenon	0	0	20.8			
Nitrogen, Phophorous, and Sulfur Gases, etc.							
	Hydrazine (N2H4)			9.768	0.1895	-1.657E-04	6.025E-08
	Hydrogen cyanide			21.86	0.06062	-4.961E-05	1.815E-08
1701	Methylamine	-23	32.0913				
1704	Ethylamine	-48.6599	37.2794				
1772	Acetonitrile			20.48	0.1196	-4.492E-05	3.203E-09
1801	Methanethiol	-22.6	-9.5				
1802	Ethanethiol	-46	-4.5				
1820	Dimethyl sulfide	-37.2	7.4				
1877	Urea	245.81	-152.716				
1901	Sulfuric acid	-740.568	-653.469				
1902	Phosphoric acid	-1250.22	-1106.42				
1903	Nitric Acid	-133.863	-73.4459				
1904	Hydrogen chloride	-92.3074	-95.2864	30.67	-0.0072	1.246E-05	-3.898E-09
1911	Ammonia	-45.94	-16.4013	27.31	0.02383	1.707E-05	-1.185E-08
1912	Sodium hydroxide	-197.757	-200.46				
1921	Water(g)	-241.835	-228.614	32.24	0.001924	1.055E-05	-3.596E-09
1922	Hydrogen sulfide	-20.63	-33.284	31.94	0.001436	2.432E-05	-1.176E-08
1938	Carbon disulfide	116.943	66.9063	27.44	0.08127	-7.666E-05	2.673E-08

a. Smith, J.M., Van Ness, H.C., Abbott, M.M. 1996. *Introduction to Chemical Engineering Thermodynamics,* 5th edition. New York: McGraw-Hill, pg 639, and Perry's Handbook.

b. Predicted.

c. Reid, R.C., Prausnitz, J.M., Poling, B.E. 1987. *The Properties of Gases and Liquids*, 4th ed. New York: McGraw-Hill, p. 656–732.

E.2   LATENT HEATS

	$T_m(°C)$	ΔH^{fus} kJ/mol	$T^{sat}(°C)$ at 1.01325 bar	ΔH^{vap} kJ/mol
Acetic acid	16.6	12.09	118.2	23.7
Acetone	−95.	5.69	56.	30.2
Anthracene	216.5	28.86		
Benzene	5.53	9.837	80.1	30.765
Biphenyl	69.2	18.58		
n-Butane	−138.3	4.661	−0.6	22.305
Cyclohexane	6.7	2.677		
Ethanol	−114.6	5.021	78.5	38.58
n-Hexane	−95.32	13.03	68.74	28.85
Naphthalene	80.2	18.80		
Phenanthrene	99.2	16.46		
Phenol	40.9	11.43		
Water	0.00	6.0095	100.0	40.656

E.3 ANTOINE CONSTANTS

The following constants are for the equation

$$\log_{10} P^{sat} = A - \frac{B}{T + C}$$

where P^{sat} is in mmHg, and T is in Celsius. Additional Antoine constants are tabulated in Antoine.xls.

	A	B	C	T range (°C)	Source
Acetic acid	8.02100	1936.01	258.451	18–118	a
Acetic acid	8.26735	2258.22	300.97	118–227	a
Acetone	7.63130	1566.69	273.419	57–205	a
Acetone	7.11714	1210.595	229.664	−13–55	a
Acrolein (2-propenal)	8.62876	2158.49	323.36	2.5–52	b
Benzene	6.87987	1196.76	219.161	8–80	a
Benzyl chloride	7.59716	1961.47	236.511	22–180	b
Biphenyl (solid)	13.5354	4993.37	296.072	20–40	c
1-Butanol	7.81028	1522.56	191.95	30–70	d
1-Butanol	7.75328	1506.07	191.593	70–120	d
2-Butanone	7.28066	1434.201	246.499	−6.5–80	b
Chloroform	6.95465	1170.966	226.232	−10–60	a
Ethanol	8.11220	1592.864	226.184	20–93	a
Hexane	6.91058	1189.64	226.28	−30–170	a
1-Propanol	8.37895	1788.02	227.438	−15–98	a
2-Propanol	8.87829	2010.33	252.636	−26–83	a
Methanol	8.08097	1582.271	239.726	15–84	a
Naphthalene (solid)	8.62233	2165.72	198.284	20–40	e
Pentane	6.87632	1075.78	233.205	−50–58	a
3-Pentanone	7.23064	1477.021	237.517	36–102	a
Toluene	6.95087	1342.31	219.187	−27–111	a
Water	8.07131	1730.63	233.426	1–100	a

a. Gmehling, J., 1977-. *Vapor-liquid Equilibrium Data Collection*, Frankfort, Germany: DECHEMA.
b. Fit to data from Stull, D.R. in *Perry's Chemical Engineering Handbook*, 5th ed., McGraw-Hill, pp. 3-46 to 3-62.
c. Timmermans, J., 1950. *Physio-Chemical Constants of Pure Organic Compounds*, New York: Elsevier.
d. Fit to data from *Handbook of Chemistry and Physics*, 56th ed, R.C. Weast, ed., CRC Press, 1974–75, pp. D191–D210.
e. Fit to data of Ambrose, D., Lawerenson, I.J., Sprake, C.H.S. 1975. *J. Chem. Therm.* 7:1173.

E.4 HENRY'S CONSTANT WITH WATER AS SOLVENT

Selected from the compilation of Sander. $K_H(T) = K°_H \exp(d(\ln(K_H))/d(1/T) ((1/T) - 1/(298.15 \text{ K})))$

Compound	$K°_H$(mol/kg-bar)	$d\ln(K_H)/d(1/T)$ (K)	Ref.[a]
O_2	1.30E-03	1500	Lide and Fredrickse (1995)[b]
H_2	7.80E-04	500	Lide and Fredrickse (1995)
NH_3	61	4200	Clegg and Brimblecombe (1989)[c]
N_2	6.10E-04	1300	Kavanaugh and Trussle (1980)[d]

Compound	$K°_H$ (mol/kg-bar)	$d\ln(K_H)/d(1/T)$ (K)	Ref.[a]
Cl_2	9.50E-02	2100	Lide and Fredrickse (1995)
Br_2	7.60E-01	4100	Dean (1992)[e]
H_2S	1.00E-01	2000	Lide and Fredrickse (1995)
He	3.80E-04	92	Wilhelm et al (1977)
Ne	4.50E-04	450	Wilhelm et al (1977)
Ar	1.40E-04	1500	Wilhelm et al (1977)
Hg	9.30E-02		Clegg&Brimblecombe(1989)
CH_4	1.40E-03	1600	Lide and Fredrickse (1995)
C_2H_6	1.90E-03	2300	Lide and Fredrickse (1995)
C_3H_8	1.50E-03	2700	Lide and Fredrickse (1995)
C_6H_{14}	1.00E-03	7500	Ashworth (1988)[f]
CyC_6H_{12}	5.60E-03	3200	Ashworth (1988)
$1-C_6H_{12}$	2.40E-03		MacKay and Shiu (1981)[g]
Benzene	1.90E-01	4300	Kavanaugh and Trussle (1980)
MTBE	1.60E+00	7700	Robbins et al. (1993)[h]
CO	9.90E-04	1300	Lide and Fredrickse (1995)
CO_2	3.50E-02	2400	Lide and Fredrickse (1995)
CH_3F	5.90E-02	2200	Wilhelm et al (1977)[i]
CHF_3	1.30E-02	3200	Wilhelm et al (1977)
CF_4	2.10E-04	1800	Wilhelm et al (1977)
CH_3Cl	1.00E-01	2800	Wilhelm et al (1977)
$CHCl_3$	2.50E-01	4600	Kavanaugh and Trussle (1980)
CCl_4	3.30E-02	4000	Ashworth (1988)

a. Selected from Sander, R. 1999. "Compilation of Henry's Law Constants for Inorganic and Organic Species of Potential Importance in Environmental Chemistry." www.mpch-mainz.mpg.de/~sander/res/henry.html (accessed 11/2011). The units in the original work are mol/(L-atm). However, note that the conversion to mol/(kg-bar) makes no numerical difference in the values truncated to the two significant digits.

b. Lide, D.R. Frederikse, H.P.R., eds. 1995. *CRC Handbook of Chemistry and Physics,* 76th Edition. Boca Raton, FL: CRC Press.

c. Brimblecombe, P. Clegg, S.L. 1988. "The solubility and behaviour of acid gases in the marine aerosol." *J. Atmos. Chem.* 7:1–18.

d. Kavanaugh, M.C. Trussell, R.R. 1980. "Design of aeration towers to strip volatile contaminants from drinking water." *J. Am. Water Works Assoc* 72:684–692.

e. Dean, J. A. 1992. *Lange's Handbook of Chemistry.* New York, NY: McGraw-Hill.

f. Ashworth, R. A., Howe, G. B., Mullins, M. E., Rogers, T. N. 1988. "Air-water partitioning coefficients of organics in dilute aqueous solutions." *J. Hazard. Mater.* 18:25–36.

g. Mackay, D., Shiu, W. Y. 1981. "A critical review of Henry's law constants for chemicals of environmental interest." *J. Phys. Chem. Ref. Data* 10:1175–1199.

h. Robbins, G. A., Wang, S., Stuart., J.D., 1993. "Using the headspace method to determine Henry's law constants." *Anal. Chem.* 65:3113–3118.

i. Wilhelm, E., Battino, R., Wilcock, R.J. "Low-pressure solubility of gases in liquid water." *Chem. Rev.* 77:219–262.

E.5 DIELECTRIC CONSTANT FOR WATER

T °C	D[a]	T °C	D
0	87.90	50	69.88

$T\,°C$	D^a	$T\,°C$	D
5	85.90	55	68.30
10	83.95	60	66.76
15	82.04	65	65.25
18	80.93	70	63.78
20	80.18	75	62.34
25	78.36	80	60.93
30	76.58	85	59.55
35	74.85	90	58.20
38	73.83	95	56.88
40	73.15	100	55.58
45	71.5		

a. Values from NSRDS-NBS 24, W.J. Hamer

E.6 DISSOCIATION CONSTANTS OF POLYPROTIC ACIDS

The constants are at 25°C and $I = 0$.

Diprotic Acid[a]	$pK_{a,A1}$	$pK_{a,A2}$
Carbon dioxide + H_2O[b]	6.351	10.329
H_2S	7.02	14.0
Oxalic acid	1.27	4.266
d-Tartaric acid	3.04	4.37
Succinic acid	4.207	5.636
Glutaric acid	4.34±0.01	5.43±0.02
Adipic acid	4.42±0.01	5.52±0.01
o-Phthalic acid	2.950	5.408

a. Data selected from Harris, D.C. 2007. *Quantitative Chemical Analysis.* 7th ed., New York NY:
    W.H. Freeman, and Smith, W., Martell, A.E., 1976–1989. *Critical Stability Constants.* 6 volumes.
    New York, NY: Plenum.
b. See footnote comment, Table 18.2 on page 706.

Triprotic: o-phosphoric acid, 2.148, 7.198, 12.375.
Tetraprotic: citric acid, 3.128, 4.761, 6.396, ~16?(hydroxyl)

E.7 STANDARD REDUCTION POTENTIALS

All species are at 298.15 K and 1 bar aqueous unless otherwise noted. Gases are at 1 bar partial pressure and pressure is measured in bar. The molal standard state is used for soluble species, with a molal activity coefficient of one at infinite dilution. For water the standard state composition is pure water. The temperature coefficients permit calculation of $E°_T = E°_{298.15} + (dE°/dT)(T - 298.15)$. This small subset is representative.

Reaction[a]	$E°$ (volts)	$dE°/dT$ (mV/K)
Aluminum		
$Al^{3+} + 3e^- \rightleftarrows Al_{(s)}$	−1.677	0.533
$AlCl^{2+} + 3e^- \rightleftarrows Al_{(s)} + Cl^-$	−1.802	
$Al(OH)^{4-} + 3e^- \rightleftarrows Al_{(s)} + 4OH^-$	−2.328	-1.13

Reaction[a]	E° (volts)	dE°/dT (mV/K)
Bromine		
$BrO_4^- + 2H^+ + 2e^- \; \rightleftarrows \; BrO_3^- + H_2O$	1.745	−0.511
$HOBr + H^+ + e^- \; \rightleftarrows \; \frac{1}{2}Br_{2(l)} + H_2O$	1.584	−0.75
$BrO_3^- + 6H^+ + 5e^- \; \rightleftarrows \; \frac{1}{2}Br_{2(l)} + 3H_2O$	1.513	−0.419
$Br_{2(aq)} + 2e^- \; \rightleftarrows \; 2Br^-$	1.098	−0.499
$Br_{2(l)} + 2e^- \; \rightleftarrows \; 2Br^-$	1.078	−0.611
$Br_3^- + 2e^- \; \rightleftarrows \; 3Br^-$	1.062	−0.512
$BrO^- + H_2O + 2e^- \; \rightleftarrows \; Br^- + 2OH^-$	0.766	−0.94
$BrO_3^- + 3H_2O + 6e^- \; \rightleftarrows \; Br^- + 6OH^-$	0.613	−1.287
Carbon		
$C_2H_{2(g)} + 2H^+ + 2e^- \; \rightleftarrows \; C_2H_{4(g)}$	0.731	
$CH_3OH + 2H^+ + 2e^- \; \rightleftarrows \; CH_{4(g)} + H_2O$	0.583	−0.039
Dehydroascorbic acid $+ 2H^+ + 2e^- \; \rightleftarrows \;$ Ascorbic acid $+ H_2O$	0.390	
$H_2CO + 2H^+ + 2e^- \; \rightleftarrows \; CH_3OH$	0.237	
$HCO_2H + 2H^+ + 2e^- \; \rightleftarrows \; H_2CO + H_2O$	−0.029	−0.63
$CO_{2(g)} + 2H^+ + 2e^- \; \rightleftarrows \; CO_{(g)} + H_2O$	−0.1038	−0.3977
$CO_{2(g)} + 2H^+ + 2e^- \; \rightleftarrows \; HCO_2H$	−0.114	−0.94
Chlorine		
$HClO_2 + 2H^+ + 2e^- \; \rightleftarrows \; HOCl + H_2O$	1.674	0.55
$HClO + H^+ + e^- \; \rightleftarrows \; \frac{1}{2}Cl_{2(g)} + H_2O$	1.630	−0.27
$ClO_3^- + 6H^+ + 5e^- \; \rightleftarrows \; \frac{1}{2}Cl_{2(l)} + 3H_2O$	1.458	−0.347
$Cl_{2(aq)} + 2e^- \; \rightleftarrows \; 2Cl^-$	1.396	−0.72
$Cl_{2(g)} + 2e^- \; \rightleftarrows \; 2Cl^-$	1.3604	−1.248
$ClO_4^- + 2H^+ + 2e^- \; \rightleftarrows \; ClO_3^- + H_2O$	1.226	−0.416
$ClO_3^- + 3H^+ + 2e^- \; \rightleftarrows \; HClO_2 + H_2O$	1.157	−0.180
$ClO_3^- + 2H^+ + e^- \; \rightleftarrows \; ClO_2 + H_2O$	1.13	0.074
$ClO_2 + e^- \; \rightleftarrows \; ClO_2^-$	1.068	−1.335
Chromium		
$Cr_2O_7^{2-} + 14H^+ + 6e^- \; \rightleftarrows \; 2Cr^{3+} + 7H_2O$	1.36	−1.32
$CrO_4^{2-} + 4H_2O + 3e^- \; \rightleftarrows \; Cr(OH)_{3(s,\,hydrated)} + 5OH^-$	−0.12	−1.62
$Cr^{3+} + e^- \; \rightleftarrows \; Cr^{2+}$	−0.42	1.4
$Cr^{3+} + 3e^- \; \rightleftarrows \; Cr_{(s)}$	−0.74	0.44
$Cr^{2+} + 2e^- \; \rightleftarrows \; Cr_{(s)}$	−0.89	−0.04

Reaction[a]	$E°$ (volts)	$dE°/dT$ (mV/K)
Copper		
$Cu^+ + e^- \rightleftarrows Cu_{(s)}$	0.518	−0.754
$Cu^{2+} + 2e^- \rightleftarrows Cu_{(s)}$	0.339	0.011
$Cu^{2+} + e^- \rightleftarrows Cu^+$	0.161	0.776
$Cu(OH)_{2(s)} + 2e^- \rightleftarrows Cu_{(s)} + 2OH^-$	−0.222	
Hydrogen		
$2H^+ + 2e^- \rightleftarrows H_{2(g)}$	0.000	0
$H_2O + e^- \rightleftarrows \frac{1}{2}H_{2(g)} + OH^-$	−0.8280	−0.8360
Iron		
$FeO_4^{2-} + 3H_2O + 3e^- \rightleftarrows FeOOH_{(s)} + 5OH^-$	0.80	−1.59
$Fe^{3+} + e^- \rightleftarrows Fe^{2+}$	0.771	1.175
$FeOOH_{(s)} + 3H^+ + e^- \rightleftarrows Fe^{2+} + 2H_2O$	0.74	−1.05
$Fe(glutamate)^{3+} + e^- \rightleftarrows Fe(glutamate)^{2+}$	0.240	
$FeOH^+ + H^+ + 2e^- \rightleftarrows Fe_{(s)} + H_2O$	−0.16	0.07
$Fe^{2+} + e^- \rightleftarrows Fe_{(s)}$	−0.44	0.07
$FeCO_{3(s)} + 2e^- \rightleftarrows Fe_{(s)} + CO_3^{2-}$	−0.756	−1.293
Lead		
$PbO_{2(s)} + 4H^+ + 2e^- \rightleftarrows Pb^{2+} + 2H_2O$	1.458	−0.253
$3PbO_{2(s)} + 2H_2O + 4e^- \rightleftarrows Pb_3O_{4(s)} + 4OH^-$	0.269	−1.136
$Pb_3O_{4(s)} + H_2O + 2e^- \rightleftarrows 3PbO_{(s,red)} + 2OH^-$	0.224	−1.211
$Pb_3O_{4(s)} + H_2O + 2e^- \rightleftarrows 3PbO_{(s,yellow)} + 2OH^-$	0.207	−1.177
$Pb^{2+} + 2e^- \rightleftarrows Pb_{(s)}$	−0.126	−0.395
Manganese		
$MnO_4^- + 4H^+ + 3e^- \rightleftarrows \beta\text{-}MnO_{2(s)} + 2H_2O$	1.692	−0.671
$Mn^{3+} + e^- \rightleftarrows Mn^{2+}$	1.56	1.8
$MnO_4^- + 8H^+ + 5e^- \rightleftarrows Mn^{2+} + 2H_2O$	1.507	−0.646
$Mn_2O_{3(s)} + 6H^+ + 2e^- \rightleftarrows 2Mn^{2+} + 3H_2O$	1.485	−0.926
$\beta\text{-}MnO_{2(s)} + 4H^+ + 2e^- \rightleftarrows Mn^{2+} + 2H_2O$	1.23	−0.609
$MnO_4^- + e^- \rightleftarrows MnO_4^{2-}$	0.56	−2.05
$\gamma\text{-}MnO_{2(s)} + H_2O + e^- \rightleftarrows \alpha\text{-}MnOOH_{(s)} + OH^-$	0.30	
$3Mn_2O_{3(s)} + H_2O + 2e^- \rightleftarrows 2Mn_3O_{4(s)} + 2OH^-$	0.002	−1.256
$Mn_3O_{4(s)} + 4H_2O + 2e^- \rightleftarrows 3Mn(OH)_{2(s)} + 2OH^-$	−0.352	−1.61
$Mn^{2+} + 2e^- \rightleftarrows Mn_{(s)}$	−1.182	−1.129
$Mn(OH)_{2(s)} + 2e^- \rightleftarrows Mn_{(s)}$	−1.565	−1.10

Reaction[a]	E° (volts)	dE°/dT (mV/K)
Oxygen		
$O_{3(g)} + 2H^+ + 2e^- \;\rightleftharpoons\; O_{2(g)} + H_2O$	2.075	−0.489
$H_2O_{2(g)} + 2H^+ + 2e^- \;\rightleftharpoons\; 2H_2O$	1.763	−0.698
$\tfrac{1}{2}O_{2(g)} + 2H^+ + 2e^- \;\rightleftharpoons\; H_2O$	1.2291	−0.8456
$O_{2(g)} + 2H^+ + 2e^- \;\rightleftharpoons\; H_2O_2$	0.695	−0.993
Ruthenium		
$RuO_{4(s)} + H_2O \;\rightleftharpoons\; H_2RuO_5 \qquad \log[H_2RuO_5] = -0.88$		
$RuO_4^{2-} + 4H^+ + 2e^- \;\rightleftharpoons\; RuO_{2(s)} + 2H_2O$	2.005	
$RuO_4^- + 4H^+ + 3e^- \;\rightleftharpoons\; RuO_{2(s)} + 2H_2O$	1.533	
$H_2RuO_5 + 4H^+ + 4e^- \;\rightleftharpoons\; RuO_{2(s)} + 3H_2O$	1.40	
$RuO_{4(s)} + 4H^+ + 4e^- \;\rightleftharpoons\; RuO_{2(s)} + 2H_2O$	1.387	
$H_2RuO_5 + e^- \;\rightleftharpoons\; RuO_4^- + H_2O$	1.001	
$RuO_4^- + e^- \;\rightleftharpoons\; RuO_4^{2-}$	0.589	
Zinc		
$ZnOH^+ + H^+ + 2e^- \;\rightleftharpoons\; Zn_{(s)} + H_2O$	−0.497	0.03
$Zn^{2+} + 2e^- \;\rightleftharpoons\; Zn_{(s)}$	−0.762	0.119
$Zn(OH)_3^- + 2e^- \;\rightleftharpoons\; Zn_{(s)} + 3OH^-$	−1.183	
$Zn(OH)_4^{2-} + 2e^- \;\rightleftharpoons\; Zn_{(s)} + 4OH^-$	−1.199	
$Zn(OH)_{2(s)} + 2e^- \;\rightleftharpoons\; Zn_{(s)} + 2OH^-$	−1.249	
$ZnO_{(s)} + H_2O + 2e^- \;\rightleftharpoons\; Zn_{(s)} + 2OH^-$	−1.260	

a. Most values from Harris, D.C. 2007. *Quantitative Chemical Analysis.* 7th ed., New York NY: W.H. Freeman. A good source for more values is Bratsch, S.G. 1989. *J. Chem. Ref. Data* 18:1, available at www.nist.gov/data/PDFfiles/ jpcrd355.pdf (Oct 2011), and Bard, A.J., Parsons, R., Jordan, J. 1985. *Standard Potentials in Aqueous Solution*, New York: Marcel Dekker.

E.8   BIOCHEMICAL DATA

Standard state for soluble species is an ideal solution at 1 M except for water, which is relative to the Lewis-Randall standard state. The data are for the untransformed Gibbs energies and enthalpies.

Name[a]	$\Delta G^o_{f,\,298.15}$ (kJ/mol)	$\Delta H^o_{f,\,298.15}$ (kJ/mol)	z_i	N_H	N_{Mg}
NADox	0	0	−1	26	0
NADred	22.65	−31.94	−2	27	0
ATP^{4-}	−2768.1	−3619.21	−4	12	0
ATP^{3-}	−2811.48	−3612.91	−3	13	0
ATP^{2-}	−2838.18	−3627.91	−2	14	0
$MgATP^{2-}$	−3258.68	−4063.31	−2	12	1
$MgHATP^{1-}$	−3287.5	−4063.01	−1	13	1
Mg_2ATP	−3729.33	−4519.51	0	12	2

Name[a]	$\Delta G^o_{f, 298.15}$ (kJ/mol)	$\Delta H^o_{f, 298.15}$ (kJ/mol)	z_i	N_H	N_{Mg}
ADP^{3-}	−1906.13	−2626.54	−3	12	0
$HADP^{2-}$	−1947.1	−2620.94	−2	13	0
H_2ADP^-	−1971.98	−2638.54	−1	14	0
$MgADP^-$	−2387.97	−3074.54	−1	12	1
$MgHADP$	−2416.67	−3075.44	0	13	1
AMP^{2-}	−1040.45	−1635.37	−2	12	0
$HAMP^-$	−1078.86	−1629.97	−1	13	0
H_2AMP	−1101.63	−1648.07	0	14	0
$MgAMP$	−1511.68	−2091.07	0	12	1
$HPO4^{2-}$	−1096.1	−1299	−2	1	0
$H_2PO_4^-$	−1137.3	−1302.6	−1	2	0
$MgHPO_4$	−1566.87	−1753.8	0	1	1
H^+	0	0	1	1	0
Mg^{2+}	−455.3	−467	2	0	1
$O_{2(g)}$	0	0	0	0	0
$CO_{2(g)}$	−394.36	−393.5	0	0	0
CO_3^{2-}	−527.81	−677.14	−2	0	0
HCO_3^-	−586.77	−691.99	−1	1	0
H_2CO_3	−623.11	−699.63	0	2	0
Acetaldehyde	−139	−212.23	0	4	0
Ethanol	−181.64	−288.3	0	6	0
Formate	−351	−425.55	−1	1	0
Glucose	−915.9	−1262.19	0	12	0
Glucose 6-phoshate^{2-}	−1763.94	−2276.44	−2	11	0
Glucose 6-phoshate$^-$	−1800.59	−2274.64	−1	12	0
Gluconolactone	−903.5		0	10	0
H_2O	−237.19	−285.83	0	2	0
H_2O_2	−134.03	−191.17	0	2	0
HPO_4^{2-}	−1096.1	−1299	−2	1	0
$H_2PO_4^-$	−1137.3	−1302.6	−1	2	0
Pyruvate$^-$	−472.27	−596.22	−1	3	0

a. Alberty, R.A. "Mathematical Functions for Thermodynamic Properties of Biochemical Reactants." (7/2005), Mathmatica notebook. BasicBioChemData3.nb, Wolfram Library Archive, http://library.wolfram.com/info-center/MathSource/5704/ Accessed 11/2011.

E.9   PROPERTIES OF WATER[1]

I. Saturation Temperature

T (°C)	P (MPa)	V^L m³/kg	V^V m³/kg	U^L kJ/kg	ΔU^{vap} kJ/kg	U^V kJ/kg	H^L kJ/kg	ΔH^{vap} kJ/kg	H^V kJ/kg	S^L kJ/kg-K	ΔS^{vap} kJ/kg-K	S^V kJ/kg-K
0.01	0.000612	0.001000	205.9912	0.00	2374.92	2374.92	0.00	2500.92	2500.92	0.0000	9.1555	9.1555
5	0.000873	0.001000	147.0113	21.02	2360.76	2381.78	21.02	2489.04	2510.06	0.0763	8.9485	9.0248
10	0.001228	0.001000	106.3032	42.02	2346.63	2388.65	42.02	2477.19	2519.21	0.1511	8.7487	8.8998
15	0.001706	0.001001	77.8755	62.98	2332.51	2395.49	62.98	2465.35	2528.33	0.2245	8.5558	8.7803
20	0.002339	0.001002	57.7567	83.91	2318.41	2402.32	83.91	2453.52	2537.43	0.2965	8.3695	8.6660
25	0.003170	0.001003	43.3373	104.83	2304.30	2409.13	104.83	2441.68	2546.51	0.3672	8.1894	8.5566
30	0.004247	0.001004	32.8783	125.73	2290.18	2415.91	125.73	2429.82	2555.55	0.4368	8.0152	8.4520
35	0.005629	0.001006	25.2053	146.63	2276.04	2422.67	146.63	2417.92	2564.55	0.5051	7.8466	8.3517
40	0.007385	0.001008	19.5151	167.53	2261.86	2429.39	167.53	2405.98	2573.51	0.5724	7.6831	8.2555
45	0.009595	0.001010	15.2521	188.43	2247.65	2436.08	188.43	2394.00	2582.43	0.6386	7.5247	8.1633
50	0.012400	0.001012	12.0269	209.33	2233.40	2442.73	209.34	2381.95	2591.29	0.7038	7.3710	8.0748
55	0.015800	0.001015	9.5643	230.24	2219.10	2449.34	230.26	2369.83	2600.09	0.7680	7.2218	7.9898
60	0.019900	0.001017	7.6672	251.16	2204.74	2455.90	251.18	2357.65	2608.83	0.8313	7.0768	7.9081
65	0.025000	0.001020	6.1935	272.09	2190.32	2462.41	272.12	2345.38	2617.50	0.8937	6.9359	7.8296
70	0.031200	0.001023	5.0395	293.03	2175.83	2468.86	293.07	2333.03	2626.10	0.9551	6.7989	7.7540
75	0.038600	0.001026	4.1289	313.99	2161.25	2475.24	314.03	2320.57	2634.60	1.0158	6.6654	7.6812
80	0.047400	0.001029	3.4052	334.96	2146.60	2481.56	335.01	2308.01	2643.02	1.0756	6.5355	7.6111
85	0.057900	0.001032	2.8258	355.95	2131.86	2487.81	356.01	2295.32	2651.33	1.1346	6.4088	7.5434
90	0.070200	0.001036	2.3591	376.97	2117.00	2493.97	377.04	2282.49	2659.53	1.1929	6.2852	7.4781
95	0.084600	0.001040	1.9806	398.00	2102.04	2500.04	398.09	2269.52	2667.61	1.2504	6.1647	7.4151
100	0.101400	0.001043	1.6718	419.06	2086.96	2506.02	419.17	2256.40	2675.57	1.3072	6.0469	7.3541
105	0.120900	0.001047	1.4184	440.15	2071.75	2511.90	440.27	2243.12	2683.39	1.3633	5.9319	7.2952
110	0.143400	0.001052	1.2093	461.26	2056.41	2517.67	461.42	2229.64	2691.06	1.4188	5.8193	7.2381
115	0.169200	0.001056	1.0358	482.41	2040.92	2523.33	482.59	2215.99	2698.58	1.4737	5.7091	7.1828
120	0.198700	0.001060	0.8912	503.60	2025.26	2528.86	503.81	2202.12	2705.93	1.5279	5.6012	7.1291
125	0.232200	0.001065	0.7700	524.83	2009.44	2534.27	525.07	2188.03	2713.10	1.5816	5.4954	7.0770
130	0.270300	0.001070	0.6680	546.09	1993.44	2539.53	546.38	2173.70	2720.08	1.6346	5.3918	7.0264
135	0.313200	0.001075	0.5817	567.41	1977.24	2544.65	567.74	2159.13	2726.87	1.6872	5.2900	6.9772
140	0.361500	0.001080	0.5085	588.77	1960.85	2549.62	589.16	2144.28	2733.44	1.7392	5.1901	6.9293
145	0.415700	0.001085	0.4460	610.19	1944.23	2554.42	610.64	2129.16	2739.80	1.7907	5.0919	6.8826
150	0.476200	0.001091	0.3925	631.66	1927.39	2559.05	632.18	2113.75	2745.93	1.8418	4.9953	6.8371
155	0.543500	0.001096	0.3465	653.19	1910.32	2563.51	653.79	2098.02	2751.81	1.8924	4.9002	6.7926
160	0.618200	0.001102	0.3068	674.79	1892.99	2567.78	675.47	2081.97	2757.44	1.9426	4.8065	6.7491
165	0.700900	0.001108	0.2724	696.46	1875.39	2571.85	697.24	2065.57	2762.81	1.9923	4.7143	6.7066
170	0.792200	0.001114	0.2426	718.20	1857.53	2575.73	719.08	2048.82	2767.90	2.0417	4.6233	6.6650
175	0.892600	0.001121	0.2166	740.02	1839.37	2579.39	741.02	2031.69	2772.71	2.0906	4.5335	6.6241
180	1.002800	0.001127	0.1938	761.92	1820.91	2582.83	763.05	2014.16	2777.21	2.1392	4.4448	6.5840
185	1.123500	0.001134	0.1739	783.91	1802.13	2586.04	785.19	1996.22	2781.41	2.1875	4.3572	6.5447

1. Harvey, A. P., Peskin, A. P., Klein, S. A., December 1997. NIST/ASME Steam Properties, Version 2.1, NIST Standard Reference Data Program.

T (°C)	P (MPa)	V^L m³/kg	V^V m³/kg	U^L kJ/kg	ΔU^{vap} kJ/kg	U^V kJ/kg	H^L kJ/kg	ΔH^{vap} kJ/kg	H^V kJ/kg	S^L kJ/kg-K	ΔS^{vap} kJ/kg-K	S^V kJ/kg-K
190	1.25520	0.001141	0.1564	806.00	1783.01	2589.01	807.43	1977.85	2785.28	2.2355	4.2704	6.5059
195	1.39880	0.001149	0.1409	828.18	1763.56	2591.74	829.79	1959.03	2788.82	2.2832	4.1846	6.4678
200	1.55490	0.001157	0.1272	850.47	1743.73	2594.20	852.27	1939.74	2792.01	2.3305	4.0997	6.4302
205	1.72430	0.001164	0.1151	872.87	1723.53	2596.40	874.88	1919.95	2794.83	2.3777	4.0153	6.3930
210	1.90770	0.001173	0.1043	895.39	1702.92	2598.31	897.63	1899.64	2797.27	2.4245	3.9318	6.3563
215	2.10580	0.001181	0.0947	918.04	1681.90	2599.94	920.53	1878.79	2799.32	2.4712	3.8488	6.3200
220	2.31960	0.001190	0.0861	940.82	1660.43	2601.25	943.58	1857.37	2800.95	2.5177	3.7663	6.2840
225	2.54970	0.001199	0.0784	963.74	1638.50	2602.24	966.80	1835.35	2802.15	2.5640	3.6843	6.2483
230	2.79710	0.001209	0.0715	986.81	1616.09	2602.90	990.19	1812.71	2802.90	2.6101	3.6027	6.2128
235	3.06250	0.001219	0.0653	1010.04	1593.16	2603.20	1013.77	1789.40	2803.17	2.6561	3.5214	6.1775
240	3.34690	0.001229	0.0597	1033.44	1569.69	2603.13	1037.55	1765.41	2802.96	2.7020	3.4403	6.1423
245	3.65120	0.001240	0.0547	1057.02	1545.65	2602.67	1061.55	1740.67	2802.22	2.7478	3.3594	6.1072
250	3.97620	0.001252	0.0501	1080.79	1521.00	2601.79	1085.77	1715.16	2800.93	2.7935	3.2786	6.0721
255	4.32290	0.001264	0.0459	1104.77	1495.72	2600.49	1110.23	1688.84	2799.07	2.8392	3.1977	6.0369
260	4.69230	0.001276	0.0422	1128.97	1469.75	2598.72	1134.96	1661.64	2796.60	2.8849	3.1167	6.0016
265	5.08530	0.001289	0.0387	1153.41	1443.04	2596.45	1159.96	1633.53	2793.49	2.9307	3.0354	5.9661
270	5.50300	0.001303	0.0356	1178.10	1415.57	2593.67	1185.27	1604.42	2789.69	2.9765	2.9539	5.9304
275	5.94640	0.001318	0.0328	1203.07	1387.26	2590.33	1210.90	1574.27	2785.17	3.0224	2.8720	5.8944
280	6.41660	0.001333	0.0302	1228.33	1358.06	2586.39	1236.88	1542.99	2779.87	3.0685	2.7894	5.8579
285	6.91470	0.001349	0.0278	1253.92	1327.89	2581.81	1263.25	1510.48	2773.73	3.1147	2.7062	5.8209
290	7.44180	0.001366	0.0256	1279.86	1296.67	2576.53	1290.03	1476.67	2766.70	3.1612	2.6222	5.7834
295	7.99910	0.001385	0.0235	1306.19	1264.30	2570.49	1317.27	1441.43	2758.70	3.2080	2.5371	5.7451
300	8.58790	0.001404	0.0217	1332.95	1230.67	2563.62	1345.01	1404.63	2749.64	3.2552	2.4507	5.7059
305	9.20940	0.001425	0.0199	1360.18	1195.67	2555.85	1373.30	1366.13	2739.43	3.3028	2.3629	5.6657
310	9.86510	0.001448	0.0183	1387.93	1159.14	2547.07	1402.22	1325.73	2727.95	3.3510	2.2734	5.6244
315	10.55620	0.001472	0.0169	1416.28	1120.89	2537.17	1431.83	1283.22	2715.05	3.3998	2.1818	5.5816
320	11.28430	0.001499	0.0155	1445.31	1080.70	2526.01	1462.22	1238.37	2700.59	3.4494	2.0878	5.5372
325	12.05100	0.001528	0.0142	1475.11	1038.30	2513.41	1493.52	1190.81	2684.33	3.5000	1.9908	5.4908
330	12.85810	0.001561	0.0130	1505.80	993.35	2499.15	1525.87	1140.16	2666.03	3.5518	1.8904	5.4422
335	13.70730	0.001597	0.0118	1537.56	945.40	2482.96	1559.45	1085.90	2645.35	3.6050	1.7856	5.3906
340	14.60070	0.001638	0.0108	1570.62	893.82	2464.44	1594.53	1027.32	2621.85	3.6601	1.6755	5.3356
345	15.54060	0.001685	0.0098	1605.30	837.79	2443.09	1631.48	963.42	2594.90	3.7176	1.5586	5.2762
350	16.52940	0.001740	0.0088	1642.13	776.01	2418.14	1670.89	892.75	2563.64	3.7784	1.4326	5.2110
355	17.57010	0.001808	0.0079	1681.96	706.44	2388.40	1713.72	812.93	2526.65	3.8439	1.2941	5.1380
360	18.66600	0.001895	0.0069	1726.28	625.50	2351.78	1761.66	719.83	2481.49	3.9167	1.1369	5.0536
365	19.82140	0.002017	0.0060	1777.79	526.00	2303.79	1817.77	605.18	2422.95	4.0014	0.9483	4.9497
370	21.04360	0.002215	0.0050	1844.07	386.19	2230.26	1890.69	443.83	2334.52	4.1112	0.6900	4.8012
373.95	22.06400	0.003106	0.0031	2015.73	0.00	2015.73	2084.26	0.00	2084.26	4.4070	0.0000	4.4070

II. Saturation Pressure

T (°C)	P (MPa)	V^L m³/kg	V^V m³/kg	U^L kJ/kg	ΔU^{vap} kJ/kg	U^V kJ/kg	H^L kJ/kg	ΔH^{vap} kJ/kg	H^V kJ/kg	S^L kJ/kg-K	ΔS^{vap} kJ/kg-K	S^V kJ/kg-K
6.97	0.001	0.001000	129.1780	29.30	2355.19	2384.49	29.30	2484.37	2513.67	0.1059	8.8690	8.9749
17.50	0.002	0.001001	66.9869	73.43	2325.47	2398.90	73.43	2459.45	2532.88	0.2606	8.4620	8.7226
24.08	0.003	0.001003	45.6532	100.98	2306.90	2407.88	100.98	2443.86	2544.84	0.3543	8.2221	8.5764
28.96	0.004	0.001004	34.7911	121.38	2293.12	2414.50	121.39	2432.28	2553.67	0.4224	8.0510	8.4734
32.87	0.005	0.001005	28.1853	137.74	2282.06	2419.80	137.75	2422.98	2560.73	0.4762	7.9176	8.3938
36.16	0.006	0.001006	23.7334	151.47	2272.76	2424.23	151.48	2415.15	2566.63	0.5208	7.8082	8.3290
39.00	0.007	0.001008	20.5245	163.34	2264.71	2428.05	163.35	2408.37	2571.72	0.5590	7.7155	8.2745
41.51	0.008	0.001008	18.0989	173.83	2257.58	2431.41	173.84	2402.37	2576.21	0.5925	7.6348	8.2273
43.76	0.009	0.001009	16.1992	183.24	2251.19	2434.43	183.25	2396.97	2580.22	0.6223	7.5635	8.1858
45.81	0.01	0.001010	14.6701	191.80	2245.36	2437.16	191.81	2392.05	2583.86	0.6492	7.4996	8.1488
60.06	0.02	0.001017	7.6480	251.40	2204.58	2455.98	251.42	2357.52	2608.94	0.8320	7.0752	7.9072
69.10	0.03	0.001022	5.2284	289.24	2178.46	2467.70	289.27	2335.28	2624.55	0.9441	6.8234	7.7675
75.86	0.04	0.001026	3.9930	317.58	2158.75	2476.33	317.62	2318.43	2636.05	1.0261	6.6429	7.6690
81.32	0.05	0.001030	3.2400	340.49	2142.72	2483.21	340.54	2304.68	2645.22	1.0912	6.5018	7.5930
85.93	0.06	0.001033	2.7317	359.85	2129.10	2488.95	359.91	2292.95	2652.86	1.1455	6.3856	7.5311
89.93	0.07	0.001036	2.3648	376.68	2117.20	2493.88	376.75	2282.67	2659.42	1.1921	6.2869	7.4790
93.49	0.08	0.001039	2.0871	391.63	2106.58	2498.21	391.71	2273.47	2665.18	1.2330	6.2009	7.4339
96.69	0.09	0.001041	1.8694	405.10	2096.97	2502.07	405.20	2265.11	2670.31	1.2696	6.1247	7.3943
99.61	0.1	0.001043	1.6939	417.40	2088.15	2505.55	417.50	2257.45	2674.95	1.3028	6.0561	7.3589
120.21	0.2	0.001061	0.8857	504.49	2024.60	2529.09	504.70	2201.53	2706.23	1.5302	5.5967	7.1269
133.52	0.3	0.001073	0.6058	561.11	1982.04	2543.15	561.43	2163.45	2724.88	1.6717	5.3199	6.9916
143.61	0.4	0.001084	0.4624	604.22	1948.88	2553.10	604.66	2133.39	2738.05	1.7765	5.1190	6.8955
151.83	0.5	0.001093	0.3748	639.54	1921.17	2560.71	640.09	2108.02	2748.11	1.8604	4.9603	6.8207
158.83	0.6	0.001101	0.3156	669.72	1897.07	2566.79	670.38	2085.76	2756.14	1.9308	4.8285	6.7593
164.95	0.7	0.001108	0.2728	696.23	1875.58	2571.81	697.00	2065.75	2762.75	1.9918	4.7153	6.7071
170.41	0.8	0.001115	0.2403	719.97	1856.06	2576.03	720.86	2047.44	2768.30	2.0457	4.6159	6.6616
175.35	0.9	0.001121	0.2149	741.55	1838.09	2579.64	742.56	2030.47	2773.03	2.0941	4.5272	6.6213
179.88	1	0.001127	0.1944	761.39	1821.36	2582.75	762.52	2014.59	2777.11	2.1381	4.4469	6.5850
187.96	1.2	0.001139	0.1633	796.96	1790.87	2587.83	798.33	1985.41	2783.74	2.2159	4.3058	6.5217
195.04	1.4	0.001149	0.1408	828.36	1763.40	2591.76	829.97	1958.88	2788.85	2.2835	4.1840	6.4675
201.37	1.6	0.001159	0.1237	856.60	1738.23	2594.83	858.46	1934.36	2792.82	2.3435	4.0764	6.4199
207.11	1.8	0.001168	0.1104	882.37	1714.87	2597.24	884.47	1911.44	2795.91	2.3975	3.9800	6.3775
212.38	2	0.001177	0.0996	906.15	1692.97	2599.12	908.50	1889.79	2798.29	2.4468	3.8922	6.3390
223.95	2.5	0.001197	0.0799	958.91	1643.15	2602.06	961.91	1840.02	2801.93	2.5543	3.7015	6.2558
233.85	3	0.001217	0.0667	1004.69	1598.47	2603.16	1008.34	1794.81	2803.15	2.6456	3.5400	6.1856
242.56	3.5	0.001235	0.0571	1045.47	1557.47	2602.94	1049.80	1752.84	2802.64	2.7254	3.3989	6.1243
250.35	4	0.001253	0.0498	1082.48	1519.24	2601.72	1087.49	1713.33	2800.82	2.7968	3.2728	6.0696
257.44	4.5	0.001270	0.0441	1116.53	1483.15	2599.68	1122.25	1675.70	2797.95	2.8615	3.1582	6.0197
263.94	5	0.001286	0.0394	1148.21	1448.77	2596.98	1154.64	1639.57	2794.21	2.9210	3.0527	5.9737
275.59	6	0.001319	0.0324	1206.01	1383.89	2589.90	1213.92	1570.67	2784.59	3.0278	2.8623	5.8901
285.83	7	0.001352	0.0274	1258.20	1322.78	2580.98	1267.66	1504.97	2772.63	3.1224	2.6924	5.8148
295.01	8	0.001385	0.0235	1306.23	1264.25	2570.48	1317.31	1441.37	2758.68	3.2081	2.5369	5.7450

T (°C)	P (MPa)	V^L m³/kg	V^V m³/kg	U^L kJ/kg	ΔU^{vap} kJ/kg	U^V kJ/kg	H^L kJ/kg	ΔH^{vap} kJ/kg	H^V kJ/kg	S^L kJ/kg-K	ΔS^{vap} kJ/kg-K	S^V kJ/kg-K
303.35	9	0.001418	0.0205	1351.11	1207.42	2558.53	1363.87	1379.07	2742.94	3.2870	2.3921	5.6791
311.00	10	0.001453	0.0180	1393.54	1151.65	2545.19	1408.06	1317.43	2725.49	3.3607	2.2553	5.6160
327.81	12.5	0.001546	0.0135	1492.26	1013.35	2505.61	1511.58	1162.73	2674.31	3.5290	1.9348	5.4638
342.16	15	0.001657	0.0103	1585.35	870.27	2455.62	1610.20	1000.50	2610.70	3.6846	1.6260	5.3106
354.67	17.5	0.001803	0.0079	1679.22	711.32	2390.54	1710.77	818.53	2529.30	3.8394	1.3037	5.1431
365.75	20	0.002040	0.0059	1786.41	508.63	2295.04	1827.21	585.14	2412.35	4.0156	0.9159	4.9315
373.95	22.06400	0.003106	0.0031	2015.73	0.00	2015.73	2084.26	0.00	2084.26	4.4070	0.0000	4.4070

III. Superheated Steam

P = 0.01 MPa (45.8)

T(°C)	V(m³/kg)	U(kJ/kg)	H(kJ/kg)	S(kJ/kg-K)
45.8	14.6701	2437.2	2583.9	8.1488
50	14.9139	2443.3	2592.4	8.1755
100	17.1964	2515.5	2687.5	8.4489
150	19.5132	2587.9	2783.0	8.6892
200	21.8256	2661.3	2879.6	8.9049
250	24.1361	2736.1	2977.4	9.1015
300	26.4456	2812.3	3076.7	9.2827
350	28.7545	2890.0	3177.5	9.4513
400	31.0631	2969.3	3279.9	9.6094
450	33.3714	3050.3	3384.0	9.7584
500	35.6796	3132.9	3489.7	9.8998
550	37.9876	3217.2	3597.1	10.0344
600	40.2956	3303.3	3706.3	10.1631
650	42.6035	3391.2	3817.2	10.2866
700	44.9113	3480.8	3929.9	10.4055
750	47.2191	3572.2	4044.4	10.5202
800	49.5269	3665.3	4160.6	10.6311
850	51.8347	3760.3	4278.6	10.7386
900	54.1424	3856.9	4398.3	10.8429
950	56.4501	3955.2	4519.7	10.9442
1000	58.7578	4055.2	4642.8	11.0428
1050	61.0655	4156.8	4767.5	11.1389
1100	63.3732	4260.0	4893.7	11.2325
1150	65.6808	4364.7	5021.5	11.3239
1200	67.9885	4470.9	5150.7	11.4132
1250	70.2961	4578.4	5281.4	11.5004
1300	72.6038	4687.4	5413.4	11.5857

P = 0.05 MPa (81.3)

T(°C)	V(m³/kg)	U(kJ/kg)	H(kJ/kg)	S(kJ/kg-K)
81.3	3.2400	2483.2	2645.2	7.5930
100	3.4187	2511.5	2682.4	7.6953
150	3.8897	2585.7	2780.2	7.9413
200	4.3562	2660.0	2877.8	8.1592
250	4.8206	2735.1	2976.1	8.3568
300	5.2840	2811.6	3075.8	8.5386
350	5.7469	2889.4	3176.8	8.7076
400	6.2094	2968.9	3279.3	8.8659
450	6.6717	3049.9	3383.5	9.0151
500	7.1338	3132.6	3489.3	9.1566
550	7.5957	3217.0	3596.8	9.2913
600	8.0576	3303.1	3706.0	9.4201
650	8.5195	3391.0	3816.9	9.5436
700	8.9812	3480.6	3929.7	9.6625
750	9.4430	3572.0	4044.2	9.7773
800	9.9047	3665.2	4160.4	9.8882
850	10.3663	3760.1	4278.5	9.9957
900	10.8280	3856.8	4398.2	10.1000
950	11.2896	3955.1	4519.6	10.2014
1000	11.7513	4055.1	4642.7	10.3000
1050	12.2129	4156.8	4767.4	10.3960
1100	12.6745	4259.9	4893.7	10.4897
1150	13.1361	4364.6	5021.4	10.5811
1200	13.5977	4470.8	5150.7	10.6703
1250	14.0592	4578.4	5281.3	10.7576
1300	14.5208	4687.3	5413.3	10.8428

P = 0.10 MPa (99.6)

T(°C)	V(m³/kg)	U(kJ/kg)	H(kJ/kg)	S(kJ/kg-K)
99.6	1.6939	2505.6	2675.0	7.3588
100	1.6959	2506.2	2675.8	7.3610
150	1.9367	2582.9	2776.6	7.6148
200	2.1724	2658.2	2875.5	7.8356
250	2.4062	2733.9	2974.5	8.0346
300	2.6388	2810.6	3074.5	8.2172
350	2.8710	2888.7	3178.6	8.3866
400	3.1027	2968.3	3278.6	8.5452
450	3.3342	3049.4	3382.8	8.6946
500	3.5655	3132.2	3488.7	8.8361
550	3.7968	3216.6	3596.3	8.9709
600	4.0279	3302.8	3705.6	9.0998
650	4.2590	3390.7	3816.6	9.2234
700	4.4900	3480.4	3929.4	9.3424
750	4.7209	3571.8	4043.9	9.4572
800	4.9519	3665.0	4160.2	9.5681
850	5.1828	3760.0	4278.2	9.6757
900	5.4137	3856.6	4398.0	9.7800
950	5.6446	3955.0	4519.5	9.8813
1000	5.8754	4055.0	4642.6	9.9800
1050	6.1063	4156.6	4767.3	10.0761
1100	6.3371	4259.8	4893.5	10.1697
1150	6.5680	4364.5	5021.3	10.2611
1200	6.7988	4470.7	5150.6	10.3504
1250	7.0296	4578.3	5281.2	10.4376
1300	7.2604	4687.2	5413.2	10.5229

P = 0.20MPa (120.3)

T(°C)	V(m³/kg)	U(kJ/kg)	H(kJ/kg)	S(kJ/kg-K)
120.3	0.8857	2529.1	2706.2	7.1269
150	0.9599	2577.1	2769.1	7.2810
200	1.0805	2654.6	2870.7	7.5081
250	1.1989	2731.4	2971.2	7.7100
300	1.3162	2808.8	3072.1	7.8941
350	1.4330	2887.3	3173.9	8.0644
400	1.5493	2967.1	3277.0	8.2236
450	1.6655	3048.5	3381.6	8.3734
500	1.7814	3131.4	3487.7	8.5152
550	1.8973	3215.9	3595.4	8.6502
600	2.0130	3302.2	3704.8	8.7792
650	2.1287	3390.2	3815.9	8.9030
700	2.2443	3479.9	3928.8	9.0220
750	2.3599	3571.4	4043.4	9.1369
800	2.4755	3664.7	4159.8	9.2479
850	2.5910	3759.6	4277.8	9.3555
900	2.7066	3856.3	4397.6	9.4598
950	2.8221	3954.7	4519.1	9.5612
1000	2.9375	4054.8	4642.3	9.6599
1050	3.0530	4156.4	4767.0	9.7560
1100	3.1685	4259.6	4893.3	9.8497
1150	3.2839	4364.3	5021.1	9.9411
1200	3.3994	4470.5	5150.4	10.0304
1250	3.5148	4578.1	5281.1	10.1176
1300	3.6302	4687.0	5413.1	10.2029

P = 0.30MPa (133.5)

T(°C)	V(m³/kg)	U(kJ/kg)	H(kJ/kg)	S(kJ/kg-K)
133.5	0.6058	2543.2	2724.9	6.9916
150	0.6340	2571.0	2761.2	7.0791
200	0.7164	2651.0	2865.9	7.3131
250	0.7964	2728.9	2967.9	7.5180
300	0.8753	2807.0	3069.6	7.7037
350	0.9536	2885.9	3172.0	7.8750
400	1.0315	2966.0	3275.5	8.0347
450	1.1092	3047.5	3380.3	8.1849
500	1.1867	3130.6	3486.6	8.3271
550	1.2641	3215.3	3594.5	8.4623
600	1.3414	3301.6	3704.0	8.5914
650	1.4186	3389.7	3815.3	8.7153
700	1.4958	3479.5	3928.2	8.8344
750	1.5729	3571.0	4042.9	8.9494
800	1.6500	3664.3	4159.3	9.0604
850	1.7271	3759.3	4277.4	9.1680
900	1.8042	3856.0	4397.3	9.2724
950	1.8812	3954.4	4518.8	9.3739
1000	1.9582	4054.5	4642.0	9.4726
1050	2.0352	4156.2	4766.7	9.5687
1100	2.1122	4259.4	4893.1	9.6624
1150	2.1892	4364.1	5020.9	9.7538
1200	2.2662	4470.3	5150.2	9.8431
1250	2.3432	4577.9	5280.9	9.9303
1300	2.4202	4686.9	5412.9	10.0156

P = 0.40MPa (143.6)

T(°C)	V(m³/kg)	U(kJ/kg)	H(kJ/kg)	S(kJ/kg-K)
143.6	0.4624	2553.1	2738.1	6.8955
150	0.4709	2564.4	2752.8	6.9306
200	0.5343	2647.2	2860.9	7.1723
250	0.5952	2726.4	2964.5	7.3804
300	0.6549	2805.1	3067.1	7.5677
350	0.7140	2884.4	3170.0	7.7399
400	0.7726	2964.9	3273.9	7.9002
450	0.8311	3046.6	3379.0	8.0508
500	0.8894	3129.8	3485.5	8.1933
550	0.9475	3214.6	3593.6	8.3287
600	1.0056	3301.0	3703.2	8.4580
650	1.0636	3389.1	3814.6	8.5820
700	1.1215	3479.0	3927.6	8.7012
750	1.1794	3570.6	4042.4	8.8162
800	1.2373	3663.9	4158.8	8.9273
850	1.2951	3759.0	4277.0	9.0350
900	1.3530	3855.7	4396.9	9.1394
950	1.4108	3954.2	4518.5	9.2409
1000	1.4686	4054.3	4641.7	9.3396
1050	1.5264	4155.9	4766.5	9.4357
1100	1.5841	4259.2	4892.8	9.5295
1150	1.6419	4363.9	5020.7	9.6209
1200	1.6997	4470.1	5150.0	9.7102
1250	1.7574	4577.8	5280.7	9.7975
1300	1.8152	4686.7	5412.8	9.8828

P = 0.50MPa (151.8)

T(°C)	V(m³/kg)	U(kJ/kg)	H(kJ/kg)	S(kJ/kg-K)
151.8	0.3748	2560.7	2748.1	6.8207
200	0.4250	2643.3	2855.8	7.0610
250	0.4744	2723.8	2961.0	7.2724
300	0.5226	2803.2	3064.6	7.4614
350	0.5702	2883.0	3168.1	7.6346
400	0.6173	2963.7	3272.3	7.7955
450	0.6642	3045.6	3377.7	7.9465
500	0.7109	3129.0	3484.5	8.0892
550	0.7576	3213.9	3592.7	8.2249
600	0.8041	3300.4	3702.5	8.3543
650	0.8505	3388.6	3813.9	8.4784
700	0.8970	3478.5	3927.0	8.5977
750	0.9433	3570.2	4041.8	8.7128
800	0.9897	3663.6	4158.4	8.8240
850	1.0360	3758.6	4276.6	8.9317

P = 0.60MPa (158.8)

T(°C)	V(m³/kg)	U(kJ/kg)	H(kJ/kg)	S(kJ/kg-K)
158.8	0.3156	2566.8	2756.1	6.7593
200	0.3521	2639.3	2850.6	6.9683
250	0.3939	2721.2	2957.6	7.1832
300	0.4344	2801.4	3062.0	7.3740
350	0.4743	2881.6	3166.1	7.5481
400	0.5137	2962.5	3270.8	7.7097
450	0.5530	3044.7	3376.5	7.8611
500	0.5920	3128.2	3483.4	8.0041
550	0.6309	3213.2	3591.8	8.1399
600	0.6698	3299.8	3701.7	8.2695
650	0.7085	3388.1	3813.2	8.3937
700	0.7472	3478.1	3926.4	8.5131
750	0.7859	3569.8	4041.3	8.6283
800	0.8246	3663.2	4157.9	8.7395
850	0.8632	3758.3	4276.2	8.8472

P = 0.80MPa (170.4)

T(°C)	V(m³/kg)	U(kJ/kg)	H(kJ/kg)	S(kJ/kg-K)
170.4	0.2403	2576.0	2768.3	6.6616
200	0.2609	2631.0	2839.7	6.8176
250	0.2932	2715.9	2950.4	7.0401
300	0.3242	2797.5	3056.9	7.2345
350	0.3544	2878.6	3162.2	7.4106
400	0.3843	2960.2	3267.6	7.5734
450	0.4139	3042.8	3373.9	7.7257
500	0.4433	3126.6	3481.3	7.8692
550	0.4726	3211.9	3590.0	8.0054
600	0.5019	3298.7	3700.1	8.1354
650	0.5310	3387.1	3811.9	8.2598
700	0.5601	3477.2	3925.3	8.3794
750	0.5892	3569.0	4040.3	8.4947
800	0.6182	3662.4	4157.0	8.6061
850	0.6472	3757.6	4275.4	8.7139

Continuation tables (T from 900 to 1300 °C):

T(°C)	V(m³/kg)	U(kJ/kg)	H(kJ/kg)	S(kJ/kg-K)
900	1.0823	3855.4	4396.6	9.0362
950	1.1285	3953.9	4518.2	9.1377
1000	1.1748	4054.0	4641.4	9.2364
1050	1.2210	4155.7	4766.2	9.3326
1100	1.2673	4259.0	4892.6	9.4263
1150	1.3135	4363.7	5020.5	9.5178
1200	1.3597	4470.0	5149.8	9.6071
1250	1.4059	4577.6	5280.5	9.6944
1300	1.4521	4686.6	5412.6	9.7797

T(°C)	V(m³/kg)	U(kJ/kg)	H(kJ/kg)	S(kJ/kg-K)
900	0.9018	3855.1	4396.2	8.9518
950	0.9404	3953.6	4517.8	9.0533
1000	0.9789	4053.7	4641.1	9.1521
1050	1.0175	4155.5	4766.0	9.2482
1100	1.0560	4258.7	4892.4	9.3420
1150	1.0946	4363.5	5020.3	9.4335
1200	1.1331	4469.8	5149.6	9.5228
1250	1.1716	4577.4	5280.4	9.6101
1300	1.2101	4686.4	5412.5	9.6954

T(°C)	V(m³/kg)	U(kJ/kg)	H(kJ/kg)	S(kJ/kg-K)
900	0.6762	3854.5	4395.5	8.8185
950	0.7052	3953.1	4517.2	8.9201
1000	0.7341	4053.2	4640.5	9.0189
1050	0.7630	4155.0	4765.4	9.1151
1100	0.7920	4258.3	4891.9	9.2089
1150	0.8209	4363.1	5019.8	9.3004
1200	0.8498	4469.4	5149.2	9.3898
1250	0.8787	4577.1	5280.0	9.4771
1300	0.9076	4686.1	5412.2	9.5625

P = 1.00MPa (179.9)

T(°C)	V(m³/kg)	U(kJ/kg)	H(kJ/kg)	S(kJ/kg-K)
179.9	0.1944	2582.8	2777.1	6.5850
200	0.2060	2622.2	2828.3	6.6955
250	0.2327	2710.4	2943.1	6.9265
300	0.2580	2793.6	3051.6	7.1246
350	0.2825	2875.7	3158.2	7.3029
400	0.3066	2957.9	3264.5	7.4669
450	0.3304	3040.9	3371.3	7.6200
500	0.3541	3125.0	3479.1	7.7641
550	0.3777	3210.5	3588.1	7.9008
600	0.4011	3297.5	3698.6	8.0310
650	0.4245	3386.0	3810.5	8.1557
700	0.4478	3476.2	3924.1	8.2755
750	0.4711	3568.1	4039.3	8.3909
800	0.4944	3661.7	4156.1	8.5024
850	0.5176	3757.0	4274.6	8.6103
900	0.5408	3853.9	4394.8	8.7150
950	0.5640	3952.5	4516.5	8.8166
1000	0.5872	4052.7	4639.9	8.9155
1050	0.6104	4154.5	4764.9	9.0118
1100	0.6335	4257.9	4891.4	9.1056
1150	0.6567	4362.7	5019.4	9.1972
1200	0.6798	4469.0	5148.9	9.2866
1250	0.7030	4576.7	5279.7	9.3739
1300	0.7261	4685.8	5411.9	9.4593

P = 1.20MPa (188.0)

T(°C)	V(m³/kg)	U(kJ/kg)	H(kJ/kg)	S(kJ/kg-K)
188.0	0.1633	2587.8	2783.7	6.5217
200	0.1693	2612.9	2816.1	6.5909
250	0.1924	2704.7	2935.6	6.8313
300	0.2139	2789.7	3046.3	7.0335
350	0.2346	2872.7	3154.2	7.2139
400	0.2548	2955.5	3261.3	7.3793
450	0.2748	3038.9	3368.7	7.5332
500	0.2946	3123.4	3476.7	7.6779
550	0.3143	3209.1	3586.3	7.8150
600	0.3339	3296.3	3697.0	7.9455
650	0.3535	3385.0	3809.2	8.0704
700	0.3730	3475.5	3922.9	8.1904
750	0.3924	3567.3	4038.2	8.3060
800	0.4118	3661.0	4155.2	8.4176
850	0.4312	3756.3	4273.8	8.5256
900	0.4506	3853.3	4394.0	8.6303
950	0.4699	3952.0	4515.9	8.7320
1000	0.4893	4052.2	4639.4	8.8310
1050	0.5086	4154.1	4764.4	8.9273
1100	0.5279	4257.5	4891.0	9.0212
1150	0.5472	4362.3	5019.0	9.1128
1200	0.5665	4468.7	5148.5	9.2022
1250	0.5858	4576.4	5279.3	9.2895
1300	0.6051	4685.4	5411.5	9.3749

P = 1.40MPa (195.0)

T(°C)	V(m³/kg)	U(kJ/kg)	H(kJ/kg)	S(kJ/kg-K)
195.0	0.1408	2591.8	2788.9	6.4675
200	0.1430	2602.7	2803.0	6.4975
250	0.1636	2698.9	2927.9	6.7488
300	0.1823	2785.7	3040.9	6.9552
350	0.2003	2869.7	3150.1	7.1379
400	0.2178	2953.1	3258.1	7.3046
450	0.2351	3037.0	3366.1	7.4594
500	0.2522	3121.8	3474.8	7.6047
550	0.2691	3207.7	3584.5	7.7422
600	0.2860	3295.1	3695.4	7.8730
650	0.3028	3384.0	3807.8	7.9982
700	0.3195	3474.4	3921.7	8.1183
750	0.3362	3566.5	4037.2	8.2340
800	0.3529	3660.2	4154.3	8.3457
850	0.3695	3755.6	4273.0	8.4538
900	0.3861	3852.7	4393.3	8.5587
950	0.4027	3951.4	4515.2	8.6604
1000	0.4193	4051.7	4638.8	8.7594
1050	0.4359	4153.6	4763.9	8.8558
1100	0.4525	4257.0	4890.5	8.9497
1150	0.4690	4361.9	5018.6	9.0413
1200	0.4856	4468.3	5148.1	9.1308
1250	0.5021	4576.0	5279.0	9.2182
1300	0.5187	4685.1	5411.2	9.3036

P = 1.60MPa (201.4)

T(°C)	V(m³/kg)	U(kJ/kg)	H(kJ/kg)	S(kJ/kg-K)
201.4	0.1237	2594.8	2792.8	6.4199
250	0.1419	2692.9	2919.9	6.6753
300	0.1587	2781.6	3035.4	6.8863
350	0.1746	2866.6	3146.0	7.0713
400	0.1901	2950.7	3254.9	7.2394
450	0.2053	3035.0	3363.5	7.3950
500	0.2203	3120.1	3472.6	7.5409
550	0.2352	3206.3	3582.6	7.6788

P = 1.80MPa (207.1)

T(°C)	V(m³/kg)	U(kJ/kg)	H(kJ/kg)	S(kJ/kg-K)
207.1	0.1104	2597.2	2795.9	6.3775
250	0.1250	2686.7	2911.7	6.6087
300	0.1402	2777.4	3029.9	6.8246
350	0.1546	2863.6	3141.8	7.0120
400	0.1685	2948.3	3251.6	7.1814
450	0.1821	3033.1	3360.9	7.3380
500	0.1955	3118.5	3470.4	7.4845
550	0.2088	3205.0	3580.8	7.6228

P = 2.00MPa (212.4)

T(°C)	V(m³/kg)	U(kJ/kg)	H(kJ/kg)	S(kJ/kg-K)
212.4	0.0996	2599.1	2798.3	6.3390
250	0.1115	2680.2	2903.2	6.5475
300	0.1255	2773.2	3024.2	6.7684
350	0.1386	2860.5	3137.7	6.9583
400	0.1512	2945.9	3248.3	7.1292
450	0.1635	3031.1	3358.2	7.2866
500	0.1757	3116.9	3468.2	7.4337
550	0.1877	3203.6	3579.0	7.5725

(continued)

T(°C)	V(m³/kg)	U(kJ/kg)	H(kJ/kg)	S(kJ/kg-K)
600	0.2500	3293.9	3693.9	7.8100
650	0.2647	3382.9	3806.5	7.9354
700	0.2794	3473.5	3920.5	8.0557
750	0.2940	3565.7	4036.1	8.1716
800	0.3087	3659.5	4153.3	8.2834
850	0.3232	3755.0	4272.2	8.3916
900	0.3378	3852.1	4392.6	8.4965
950	0.3523	3950.9	4514.6	8.5984
1000	0.3669	4051.2	4638.2	8.6974
1050	0.3814	4153.1	4763.4	8.7938
1100	0.3959	4256.6	4890.0	8.8878
1150	0.4104	4361.5	5018.2	8.9794
1200	0.4249	4467.9	5147.7	9.0689
1250	0.4394	4575.7	5278.7	9.1563
1300	0.4538	4684.8	5410.9	9.2417

P = 2.50MPa (224.0)

T(°C)	V(m³/kg)	U(kJ/kg)	H(kJ/kg)	S(kJ/kg-K)
224.0	0.0799	2602.1	2801.9	6.2558
250	0.0871	2663.3	2880.1	6.4107
300	0.0989	2762.2	3009.6	6.6459
350	0.1098	2852.5	3127.0	6.8424
400	0.1201	2939.8	3240.1	7.0170
450	0.1302	3026.2	3351.6	7.1767
500	0.1400	3112.8	3462.7	7.3254
550	0.1497	3200.1	3574.3	7.4653
600	0.1593	3288.5	3686.8	7.5979
650	0.1689	3378.2	3800.4	7.7243
700	0.1783	3469.3	3915.2	7.8455
750	0.1878	3562.0	4031.5	7.9620
800	0.1972	3656.2	4149.2	8.0743
850	0.2066	3752.0	4268.5	8.1830
900	0.2160	3849.4	4389.3	8.2882
950	0.2253	3948.4	4511.7	8.3904
1000	0.2347	4048.9	4635.6	8.4896
1050	0.2440	4151.0	4761.0	8.5863
1100	0.2533	4254.7	4887.9	8.6804
1150	0.2626	4359.7	5016.2	8.7722
1200	0.2719	4466.2	5146.0	8.8618
1250	0.2812	4574.1	5277.1	8.9493
1300	0.2905	4683.3	5409.5	9.0349

(continued)

T(°C)	V(m³/kg)	U(kJ/kg)	H(kJ/kg)	S(kJ/kg-K)
600	0.2220	3292.7	3692.3	7.7543
650	0.2351	3381.9	3805.1	7.8799
700	0.2482	3472.6	3919.4	8.0004
750	0.2613	3564.9	4035.1	8.1164
800	0.2743	3658.8	4152.4	8.2284
850	0.2872	3754.3	4271.3	8.3367
900	0.3002	3851.5	4391.9	8.4416
950	0.3131	3950.3	4514.0	8.5435
1000	0.3261	4050.7	4637.6	8.6426
1050	0.3390	4152.7	4762.8	8.7391
1100	0.3519	4256.2	4889.5	8.8331
1150	0.3648	4361.1	5017.7	8.9248
1200	0.3777	4467.5	5147.3	9.0143
1250	0.3905	4575.3	5278.3	9.1017
1300	0.4034	4684.5	5410.6	9.1872

P = 3.00MPa (233.9)

T(°C)	V(m³/kg)	U(kJ/kg)	H(kJ/kg)	S(kJ/kg-K)
233.9	0.0667	2603.2	2803.2	6.1856
250	0.0706	2644.7	2856.5	6.2893
300	0.0812	2750.8	2994.3	6.5412
350	0.0906	2844.4	3116.1	6.7449
400	0.0994	2933.5	3231.7	6.9234
450	0.1079	3021.2	3344.8	7.0856
500	0.1162	3108.6	3457.2	7.2359
550	0.1244	3196.6	3569.7	7.3768
600	0.1324	3285.5	3682.8	7.5103
650	0.1405	3375.6	3796.9	7.6373
700	0.1484	3467.0	3912.2	7.7590
750	0.1563	3559.9	4028.9	7.8758
800	0.1642	3654.3	4146.9	7.9885
850	0.1720	3750.3	4266.5	8.0973
900	0.1799	3847.9	4387.5	8.2028
950	0.1877	3947.0	4510.1	8.3051
1000	0.1955	4047.7	4634.1	8.4045
1050	0.2033	4149.9	4759.7	8.5012
1100	0.2111	4253.6	4886.7	8.5955
1150	0.2188	4358.7	5015.2	8.6874
1200	0.2266	4465.3	5145.0	8.7770
1250	0.2343	4573.3	5276.2	8.8646
1300	0.2421	4682.5	5408.8	8.9502

(continued)

T(°C)	V(m³/kg)	U(kJ/kg)	H(kJ/kg)	S(kJ/kg-K)
600	0.1996	3291.5	3690.7	7.7043
650	0.2115	3380.8	3803.8	7.8302
700	0.2233	3471.6	3918.2	7.9509
750	0.2350	3564.0	4034.1	8.0670
800	0.2467	3658.0	4151.5	8.1790
850	0.2584	3753.6	4270.5	8.2874
900	0.2701	3850.9	4391.1	8.3925
950	0.2818	3949.8	4513.3	8.4945
1000	0.2934	4050.2	4637.0	8.5936
1050	0.3051	4152.2	4762.3	8.6901
1100	0.3167	4255.7	4889.1	8.7842
1150	0.3283	4360.7	5017.3	8.8759
1200	0.3399	4467.2	5147.0	8.9654
1250	0.3515	4575.0	5278.0	9.0529
1300	0.3631	4684.1	5410.3	9.1384

P = 3.50MPa (242.6)

T(°C)	V(m³/kg)	U(kJ/kg)	H(kJ/kg)	S(kJ/kg-K)
242.6	0.0571	2602.9	2802.6	6.1243
250	0.0588	2624.0	2829.7	6.1764
300	0.0685	2738.8	2978.4	6.4484
350	0.0768	2836.0	3104.8	6.6601
400	0.0846	2927.2	3223.2	6.8427
450	0.0920	3016.1	3338.0	7.0074
500	0.0992	3104.5	3451.6	7.1593
550	0.1063	3193.1	3565.0	7.3014
600	0.1133	3282.5	3678.9	7.4356
650	0.1202	3372.9	3793.5	7.5633
700	0.1270	3464.7	3909.3	7.6854
750	0.1338	3557.8	4026.3	7.8027
800	0.1406	3652.5	4144.6	7.9156
850	0.1474	3748.6	4264.4	8.0247
900	0.1541	3846.4	4385.7	8.1303
950	0.1608	3945.6	4508.4	8.2328
1000	0.1675	4046.4	4632.7	8.3324
1050	0.1742	4148.7	4758.4	8.4292
1100	0.1809	4252.5	4885.6	8.5235
1150	0.1875	4357.7	5014.1	8.6155
1200	0.1942	4464.4	5144.1	8.7053
1250	0.2009	4572.4	5275.4	8.7929
1300	0.2075	4681.7	5408.0	8.8785

P = 4.00 MPa (250.4)

T(°C)	V(m³/kg)	U(kJ/kg)	H(kJ/kg)	S(kJ/kg-K)
250.4	0.0498	2601.7	2800.8	6.0696
300	0.0589	2726.2	2961.7	6.3639
350	0.0665	2827.4	3093.3	6.5843
400	0.0734	2920.7	3214.5	6.7714
450	0.0800	3011.0	3331.2	6.9386
500	0.0864	3100.3	3446.0	7.0922
550	0.0927	3189.5	3560.3	7.2355
600	0.0989	3279.4	3674.9	7.3705
650	0.1049	3370.3	3790.1	7.4988
700	0.1110	3462.4	3906.3	7.6214
750	0.1170	3555.8	4023.6	7.7390
800	0.1229	3650.6	4142.3	7.8523
850	0.1289	3747.0	4262.4	7.9616
900	0.1348	3844.8	4383.9	8.0674
950	0.1406	3944.2	4506.8	8.1701
1000	0.1465	4045.1	4631.2	8.2697
1050	0.1524	4147.5	4757.1	8.3667
1100	0.1582	4251.4	4884.4	8.4611
1150	0.1641	4356.7	5013.1	8.5532
1200	0.1699	4463.5	5143.1	8.6430
1250	0.1757	4571.5	5274.5	8.7307
1300	0.1816	4680.9	5407.2	8.8164

P = 4.50 MPa (257.4)

T(°C)	V(m³/kg)	U(kJ/kg)	H(kJ/kg)	S(kJ/kg-K)
257.4	0.0441	2599.7	2798.0	6.0197
300	0.0453	2713.0	2944.2	6.2854
350	0.0514	2818.6	3081.5	6.5153
400	0.0584	2914.2	3205.6	6.7070
450	0.0648	3005.8	3324.2	6.8770
500	0.0708	3096.0	3440.4	7.0323
550	0.0765	3186.0	3555.6	7.1767
600	0.0821	3276.4	3670.9	7.3127
650	0.0877	3367.7	3786.6	7.4416
700	0.0931	3460.0	3903.3	7.5646
750	0.0985	3553.7	4021.0	7.6826
800	0.1038	3648.8	4140.0	7.7962
850	0.1092	3745.3	4260.3	7.9057
900	0.1145	3843.3	4382.1	8.0118
950	0.1197	3942.8	4505.2	8.1146
1000	0.1250	4043.9	4629.8	8.2144
1050	0.1302	4146.4	4755.8	8.3115
1100	0.1354	4250.4	4883.2	8.4060
1150	0.1406	4355.8	5012.0	8.4981
1200	0.1458	4462.5	5142.2	8.5880
1250	0.1510	4570.7	5273.7	8.6758
1300	0.1562	4680.1	5406.4	8.7615
	0.1614			

P = 5.00 MPa (263.9)

T(°C)	V(m³/kg)	U(kJ/kg)	H(kJ/kg)	S(kJ/kg-K)
263.9	0.0394	2597.0	2794.2	5.9737
300	0.0453	2699.0	2925.7	6.2110
350	0.0520	2809.5	3069.3	6.4516
400	0.0578	2907.5	3196.7	6.6483
450	0.0633	3000.6	3317.2	6.8210
500	0.0686	3091.7	3434.7	6.9781
550	0.0737	3182.4	3550.9	7.1237
600	0.0787	3273.3	3666.8	7.2605
650	0.0836	3365.0	3783.2	7.3901
700	0.0885	3457.7	3900.3	7.5136
750	0.0934	3551.6	4018.4	7.6320
800	0.0982	3646.9	4137.7	7.7458
850	0.1029	3743.6	4258.3	7.8556
900	0.1077	3841.8	4380.2	7.9618
950	0.1124	3941.5	4503.6	8.0648
1000	0.1171	4042.6	4628.3	8.1648
1050	0.1219	4145.2	4754.5	8.2620
1100	0.1266	4249.3	4882.0	8.3566
1150	0.1312	4354.8	5011.0	8.4488
1200	0.1359	4461.6	5141.2	8.5388
1250	0.1406	4569.8	5272.8	8.6266
1300	0.1453	4679.3	5405.7	8.7124

P = 6.00 MPa (275.6)

T(°C)	V(m³/kg)	U(kJ/kg)	H(kJ/kg)	S(kJ/kg-K)
275.6	0.0324	2589.9	2784.6	5.8901
300	0.0362	2668.4	2885.5	6.0703
350	0.0423	2790.4	3043.9	6.3357
400	0.0474	2893.7	3178.2	6.5432
450	0.0522	2989.9	3302.9	6.7219
500	0.0567	3083.1	3423.1	6.8826
550	0.0610	3175.2	3541.3	7.0307
600	0.0653	3267.2	3658.7	7.1693
650	0.0694	3359.6	3776.2	7.3001
700	0.0735	3453.0	3894.3	7.4246
750	0.0776	3547.5	4013.2	7.5438
800	0.0816	3643.2	4133.1	7.6582
850	0.0857	3740.3	4254.2	7.7685
900	0.0896	3838.8	4376.6	7.8751
950	0.0936	3938.7	4500.3	7.9784
1000	0.0976	4040.1	4625.4	8.0786
1050	0.1015	4142.9	4751.9	8.1760
1100	0.1054	4247.1	4879.7	8.2709
1150	0.1093	4352.8	5008.9	8.3632
1200	0.1133	4459.8	5139.3	8.4534
1250	0.1172	4568.1	5271.1	8.5413
1300	0.1211	4677.7	5404.1	8.6272

P = 7.00 MPa (285.8)

T(°C)	V(m³/kg)	U(kJ/kg)	H(kJ/kg)	S(kJ/kg-K)
285.8	0.0274	2581.0	2772.6	5.8148
300	0.0295	2633.5	2839.9	5.9337
350	0.0353	2770.1	3016.9	6.2304
400	0.0400	2879.5	3159.2	6.4502
450	0.0442	2979.0	3288.3	6.6353
500	0.0482	3074.3	3411.4	6.8000
550	0.0520	3167.9	3531.6	6.9506
600	0.0557	3260.9	3650.6	7.0910
650	0.0593	3354.3	3769.3	7.2231
700	0.0629	3448.3	3888.2	7.3486
750	0.0664	3543.3	4007.9	7.4685
800	0.0699	3639.5	4128.4	7.5836
850	0.0733	3736.9	4250.1	7.6944
900	0.0768	3835.7	4373.0	7.8014
950	0.0802	3935.9	4497.1	7.9050
1000	0.0836	4037.5	4622.5	8.0055
1050	0.0870	4140.5	4749.3	8.1031
1100	0.0903	4245.0	4877.3	8.1981
1150	0.0937	4350.8	5006.7	8.2907
1200	0.0971	4457.9	5137.4	8.3810
1250	0.1004	4566.4	5269.4	8.4690
1300	0.1038	4676.1	5402.6	8.5551

P = 8.00 MPa (295.0)

T(°C)	V(m³/kg)	U(kJ/kg)	H(kJ/kg)	S(kJ/kg-K)
295.0	0.0235	2570.5	2758.7	5.7450
300	0.0243	2592.3	2786.5	5.7937
350	0.0300	2748.3	2988.1	6.1321
400	0.0343	2864.6	3139.4	6.3658
450	0.0382	2967.8	3273.3	6.5579
500	0.0418	3065.4	3399.5	6.7266
550	0.0452	3160.5	3521.8	6.8799
600	0.0485	3254.7	3642.4	7.0221
650	0.0517	3348.9	3762.3	7.1556
700	0.0548	3443.6	3882.2	7.2821
750	0.0579	3539.1	4002.6	7.4028
800	0.0610	3635.7	4123.8	7.5184
850	0.0641	3733.5	4246.0	7.6297
900	0.0671	3832.6	4369.3	7.7371
950	0.0701	3933.1	4493.8	7.8411
1000	0.0731	4035.0	4619.6	7.9419
1050	0.0761	4138.2	4746.7	8.0397
1100	0.0790	4242.8	4875.0	8.1350
1150	0.0820	4348.8	5004.6	8.2277
1200	0.0849	4456.1	5135.5	8.3181
1250	0.0879	4564.6	5267.7	8.4063
1300	0.0908	4674.5	5401.0	8.4924

P = 9.00 MPa (303.4)

T(°C)	V(m³/kg)	U(kJ/kg)	H(kJ/kg)	S(kJ/kg-K)
303.4	0.0205	2558.5	2742.9	5.6791
350	0.0258	2724.9	2957.3	6.0380
400	0.0300	2849.2	3118.8	6.2876
450	0.0335	2956.3	3258.0	6.4872
500	0.0368	3056.3	3387.4	6.6603
550	0.0399	3153.0	3512.0	6.8164
600	0.0429	3248.4	3634.1	6.9605
650	0.0458	3343.4	3755.2	7.0953
700	0.0486	3438.8	3876.1	7.2229
750	0.0514	3534.9	3997.1	7.3443
800	0.0541	3632.0	4119.1	7.4606
850	0.0569	3730.2	4241.9	7.5724
900	0.0596	3829.6	4365.7	7.6802
950	0.0622	3930.3	4490.6	7.7844
1000	0.0649	4032.4	4616.7	7.8855
1050	0.0676	4135.9	4744.0	7.9836
1100	0.0702	4240.6	4872.7	8.0790
1150	0.0729	4346.8	5002.5	8.1719
1200	0.0755	4454.2	5133.6	8.2625
1250	0.0781	4562.9	5266.0	8.3508
1300	0.0807	4672.9	5399.5	8.4370

P = 10.00 MPa (311.0)

T(°C)	V(m³/kg)	U(kJ/kg)	H(kJ/kg)	S(kJ/kg-K)
311.0	0.0180	2545.2	2725.5	5.6160
350	0.0224	2699.6	2924.0	5.9459
400	0.0264	2833.1	3097.4	6.2141
450	0.0298	2944.5	3242.3	6.4219
500	0.0328	3047.0	3375.1	6.5995
550	0.0357	3145.4	3502.0	6.7585
600	0.0384	3242.0	3625.8	6.9045
650	0.0410	3337.9	3748.1	7.0408
700	0.0436	3434.0	3870.0	7.1693
750	0.0461	3530.7	3992.0	7.2916
800	0.0486	3628.2	4114.5	7.4085
850	0.0511	3726.8	4237.8	7.5207
900	0.0535	3826.5	4362.0	7.6290
950	0.0560	3927.5	4487.3	7.7335
1000	0.0584	4029.9	4613.8	7.8349
1050	0.0608	4133.5	4741.4	7.9332
1100	0.0632	4238.5	4870.3	8.0288
1150	0.0656	4344.8	5000.4	8.1219
1200	0.0679	4452.3	5131.7	8.2126
1250	0.0703	4561.2	5264.2	8.3010
1300	0.0727	4671.3	5397.9	8.3874

P = 12.50 MPa (327.8)

T(°C)	V(m³/kg)	U(kJ/kg)	H(kJ/kg)	S(kJ/kg-K)
327.8	0.0135	2505.61	2674.31	5.4638
350	0.0161	2624.8	2826.6	5.7130
400	0.0200	2789.6	3040.0	6.0433
450	0.0230	2913.7	3201.4	6.2749
500	0.0256	3023.2	3343.6	6.4650
550	0.0280	3126.1	3476.5	6.6317
600	0.0303	3225.8	3604.6	6.7828
650	0.0325	3324.1	3730.2	6.9227
700	0.0346	3422.0	3854.6	7.0539
750	0.0367	3520.1	3978.6	7.1782
800	0.0387	3618.7	4102.8	7.2967
850	0.0407	3718.3	4227.5	7.4102
900	0.0427	3818.9	4352.9	7.5194
950	0.0447	3920.6	4479.2	7.6249
1000	0.0466	4023.5	4606.5	7.7269
1050	0.0486	4127.7	4734.9	7.8258
1100	0.0505	4233.1	4864.5	7.9219
1150	0.0524	4339.8	4995.1	8.0154
1200	0.0543	4447.7	5127.0	8.1065
1250	0.0562	4556.6	5260.0	8.1952
1300	0.0581	4667.3	5394.1	8.2819

P = 15.00 MPa (342.2)

T(°C)	V(m³/kg)	U(kJ/kg)	H(kJ/kg)	S(kJ/kg-K)
342.2	0.0103	2455.6	2610.7	5.3106
350	0.0115	2520.9	2693.1	5.4437
400	0.0157	2740.6	2975.7	5.8819
450	0.0185	2880.7	3157.9	6.1434
500	0.0208	2998.4	3310.8	6.3480
550	0.0229	3106.2	3450.4	6.5230
600	0.0249	3209.3	3583.1	6.6796
650	0.0268	3310.1	3712.1	6.8233
700	0.0286	3409.8	3839.1	6.9572
750	0.0304	3509.4	3965.2	7.0836
800	0.0321	3609.2	4091.1	7.2037
850	0.0338	3709.8	4217.1	7.3185
900	0.0355	3811.2	4343.7	7.4288
950	0.0372	3913.6	4471.0	7.5350
1000	0.0388	4017.1	4599.2	7.6378
1050	0.0404	4121.8	4728.4	7.7373
1100	0.0421	4227.7	4858.6	7.8339
1150	0.0437	4334.8	4989.9	7.9278
1200	0.0453	4443.1	5122.3	8.0192
1250	0.0469	4552.6	5255.7	8.1083
1300	0.0485	4663.2	5390.3	8.1952

P = 17.50 MPa (354.7)

T(°C)	V(m³/kg)	U(kJ/kg)	H(kJ/kg)	S(kJ/kg-K)
354.7	0.0079	2390.5	2529.3	5.1431
400	0.0125	2684.3	2902.4	5.7211
450	0.0152	2845.4	3111.4	6.0212
500	0.0174	2972.4	3276.7	6.2424
550	0.0193	3085.8	3423.6	6.4266
600	0.0211	3192.5	3561.3	6.5890
650	0.0227	3295.8	3693.8	6.7366
700	0.0243	3397.5	3823.5	6.8734
750	0.0259	3498.6	3951.7	7.0019
800	0.0274	3599.7	4079.3	7.1236
850	0.0289	3701.2	4206.8	7.2398
900	0.0303	3803.4	4334.5	7.3511
950	0.0318	3906.6	4462.9	7.4582
1000	0.0332	4010.7	4592.0	7.5616
1050	0.0346	4115.9	4721.9	7.6617
1100	0.0360	4222.3	4852.8	7.7588
1150	0.0374	4329.8	4984.6	7.8531
1200	0.0388	4438.4	5117.5	7.9449
1250	0.0402	4548.3	5251.5	8.0343
1300	0.0416	4659.2	5386.4	8.1215

P = 20.00 MPa (365.8)

T(°C)	V(m³/kg)	U(kJ/kg)	H(kJ/kg)	S(kJ/kg-K)
365.8	0.0059	2295.0	2412.4	4.9315
400	0.0100	2617.9	2816.9	5.5525
450	0.0127	2807.2	3061.7	5.9043
500	0.0148	2945.3	3241.2	6.1446
550	0.0166	3064.7	3396.1	6.3389
600	0.0182	3175.3	3539.0	6.5075
650	0.0197	3281.4	3675.3	6.6593
700	0.0211	3385.1	3807.8	6.7990
750	0.0225	3487.7	3938.1	6.9297
800	0.0239	3590.1	4067.5	7.0531
850	0.0252	3692.6	4196.4	7.1705
900	0.0265	3795.7	4325.4	7.2829
950	0.0278	3899.5	4454.7	7.3909
1000	0.0290	4004.3	4584.7	7.4950
1050	0.0303	4110.0	4715.4	7.5957
1100	0.0315	4216.9	4846.9	7.6933
1150	0.0327	4324.8	4979.4	7.7880
1200	0.0340	4433.8	5112.8	7.8802
1250	0.0352	4544.0	5247.2	7.9699
1300	0.0364	4655.2	5382.6	8.0574

P = 25.00MPa

T(°C)	V(m³/kg)	U(kJ/kg)	H(kJ/kg)	S(kJ/kg-K)
400	0.0060	2428.5	2578.7	5.1400
450	0.0092	2721.2	2950.6	5.6759
500	0.0111	2887.3	3165.9	5.9642
550	0.0127	3020.8	3339.2	6.1816
600	0.0141	3140.0	3493.5	6.3637
650	0.0154	3251.9	3637.7	6.5242
700	0.0166	3359.9	3776.0	6.6702
750	0.0178	3465.8	3910.9	6.8054
800	0.0189	3570.7	4043.8	6.9322
850	0.0200	3675.4	4175.6	7.0523
900	0.0211	3780.2	4307.1	7.1668
950	0.0221	3885.5	4438.5	7.2765
1000	0.0232	3991.5	4570.2	7.3820
1050	0.0242	4098.3	4702.5	7.4839
1100	0.0252	4206.0	4835.4	7.5825
1150	0.0262	4314.8	4969.0	7.6781
1200	0.0272	4424.6	5103.5	7.7710
1250	0.0281	4535.4	5238.8	7.8613
1300	0.0291	4647.2	5375.1	7.9493

P = 30.00MPa

T(°C)	V(m³/kg)	U(kJ/kg)	H(kJ/kg)	S(kJ/kg-K)
400	0.0028	2071.9	2156.2	4.4808
450	0.0067	2618.9	2821.0	5.4421
500	0.0087	2824.0	3084.7	5.7956
550	0.0102	2974.5	3279.7	6.0402
600	0.0114	3103.4	3446.7	6.2373
650	0.0126	3221.7	3599.4	6.4074
700	0.0137	3334.3	3743.9	6.5598
750	0.0147	3443.6	3883.4	6.6997
800	0.0156	3551.2	4020.0	6.8300
850	0.0166	3658.0	4154.9	6.9529
900	0.0175	3764.6	4288.8	7.0695
950	0.0184	3871.4	4422.3	7.1810
1000	0.0192	3978.6	4555.8	7.2880
1050	0.0201	4086.5	4689.6	7.3910
1100	0.0210	4195.2	4823.8	7.4906
1150	0.0218	4304.8	4958.7	7.5871
1200	0.0226	4415.3	5094.2	7.6807
1250	0.0235	4526.8	5230.5	7.7716
1300	0.0243	4639.2	5367.6	7.8602

P = 35.00MPa

T(°C)	V(m³/kg)	U(kJ/kg)	H(kJ/kg)	S(kJ/kg-K)
400	0.0021	1914.8	1988.5	4.2142
450	0.0050	2497.5	2671.0	5.1945
500	0.0069	2755.3	2997.9	5.6331
550	0.0083	2925.8	3218.0	5.9092
600	0.0095	3065.6	3398.9	6.1228
650	0.0106	3190.9	3560.7	6.3030
700	0.0115	3308.3	3711.6	6.4622
750	0.0124	3421.2	3855.9	6.6069
800	0.0133	3531.5	3996.3	6.7409
850	0.0141	3640.5	4134.2	6.8665
900	0.0149	3748.9	4270.6	6.9853
950	0.0157	3857.2	4406.2	7.0985
1000	0.0165	3965.8	4541.5	7.2069
1050	0.0172	4074.8	4676.8	7.3112
1100	0.0179	4184.4	4812.4	7.4118
1150	0.0187	4294.8	4948.4	7.5091
1200	0.0194	4406.1	5085.0	7.6034
1250	0.0201	4518.2	5222.2	7.6950
1300	0.0208	4631.2	5360.1	7.7841

P = 40.00MPa

T(°C)	V(m³/kg)	U(kJ/kg)	H(kJ/kg)	S(kJ/kg-K)
400	0.0019	1854.9	1931.4	4.1145
450	0.0037	2364.2	2511.8	4.9449
500	0.0056	2681.6	2906.5	5.4744
550	0.0070	2875.0	3154.4	5.7857
600	0.0081	3026.8	3350.4	6.0170
650	0.0091	3159.5	3521.6	6.2078
700	0.0099	3282.0	3679.1	6.3740
750	0.0107	3398.6	3828.4	6.5236
800	0.0115	3511.8	3972.6	6.6612
850	0.0123	3623.1	4113.6	6.7896
900	0.0130	3733.3	4252.5	6.9106
950	0.0137	3843.1	4390.2	7.0256
1000	0.0144	3952.9	4527.3	7.1355
1050	0.0150	4063.0	4664.2	7.2409
1100	0.0157	4173.7	4801.1	7.3425
1150	0.0163	4284.9	4938.3	7.4406
1200	0.0170	4396.5	5075.9	7.5357
1250	0.0176	4509.7	5214.1	7.6279
1300	0.0182	4623.3	5352.8	7.7175

P = 50.00MPa

T(°C)	V(m³/kg)	U(kJ/kg)	H(kJ/kg)	S(kJ/kg-K)
400	0.0017	1787.8	1874.4	4.0029
450	0.0025	2160.3	2284.7	4.5896
500	0.0039	2528.1	2722.6	5.1762
550	0.0051	2769.5	3025.3	5.5563
600	0.0061	2947.1	3252.5	5.8245
650	0.0070	3095.6	3443.4	6.0373
700	0.0077	3228.7	3614.6	6.2178
750	0.0084	3353.1	3773.9	6.3775
800	0.0091	3472.2	3925.8	6.5225
850	0.0097	3588.0	4072.9	6.6565
900	0.0103	3702.0	4216.8	6.7819
950	0.0109	3814.9	4358.7	6.9004
1000	0.0114	3927.3	4499.4	7.0131
1050	0.0120	4039.7	4639.3	7.1209
1100	0.0125	4152.2	4778.9	7.2244
1150	0.0131	4265.1	4918.4	7.3242
1200	0.0136	4378.6	5058.1	7.4207
1250	0.0141	4492.7	5198.1	7.5141
1300	0.0146	4607.4	5338.4	7.6048

P = 60.00MPa

T(°C)	V(m³/kg)	U(kJ/kg)	H(kJ/kg)	S(kJ/kg-K)
400	0.0016	1745.2	1843.2	3.9317
450	0.0021	2055.1	2180.2	4.4140
500	0.0030	2393.2	2570.3	4.9356
550	0.0040	2664.5	2901.9	5.3517
600	0.0048	2866.8	3156.8	5.6527
650	0.0056	3031.3	3366.7	5.8867
700	0.0063	3175.4	3551.3	6.0814
750	0.0069	3307.6	3720.5	6.2510
800	0.0075	3432.6	3880.0	6.4033
850	0.0080	3553.2	4033.1	6.5428
900	0.0085	3670.9	4182.0	6.6725
950	0.0090	3786.9	4328.1	6.7944
1000	0.0095	3901.9	4472.2	6.9099
1050	0.0100	4016.5	4615.1	7.0200
1100	0.0104	4130.9	4757.3	7.1255
1150	0.0109	4245.5	4899.1	7.2269
1200	0.0113	4360.4	5040.8	7.3248
1250	0.0118	4475.8	5182.5	7.4194
1300	0.0122	4591.8	5324.5	7.5111

IV. Compressed Liquid

P = 5 MPa

T(°C)	V(m³/kg)	U(kJ/kg)	H(kJ/kg)	S(kJ/kg-K)
0	0.000998	0.0	5.0	0.0001
20	0.001000	83.6	88.6	0.2954
40	0.001006	166.9	172.0	0.5705
60	0.001015	250.3	255.4	0.8287
80	0.001027	333.8	339.0	1.0723
100	0.001041	417.6	422.9	1.3034
120	0.001058	501.9	507.2	1.5236
140	0.001077	586.8	592.2	1.7344
160	0.001099	672.5	678.0	1.9374
180	0.001124	759.5	765.1	2.1338
200	0.001153	847.9	853.7	2.3251
220	0.001187	938.4	944.3	2.5127
240	0.001227	1031.6	1037.7	2.6983
260	0.001275	1128.5	1134.9	2.8841

P = 10 MPa

T(°C)	V(m³/kg)	U(kJ/kg)	H(kJ/kg)	S(kJ/kg-K)
0	0.000995	0.1	10.1	0.0003
20	0.000997	83.3	93.3	0.2943
40	0.001003	166.3	176.4	0.5685
60	0.001013	249.4	259.6	0.8260
80	0.001024	332.7	342.9	1.0691
100	0.001038	416.2	426.6	1.2996
120	0.001055	500.2	510.7	1.5191
140	0.001074	584.7	595.5	1.7293
160	0.001095	670.1	681.0	1.9315
180	0.001120	756.5	767.7	2.1271
200	0.001148	844.3	855.8	2.3174
220	0.001181	934.0	945.8	2.5037
240	0.001219	1026.1	1038.3	2.6876
260	0.001265	1121.6	1134.3	2.8710
280	0.001323	1221.8	1235.0	3.0565
300	0.001398	1329.4	1343.3	3.2488

P = 15 MPa

T(°C)	V(m³/kg)	U(kJ/kg)	H(kJ/kg)	S(kJ/kg-K)
0	0.000993	0.2	15.1	0.0004
20	0.000995	83.0	97.9	0.2932
40	0.001001	165.7	180.8	0.5666
60	0.001011	248.6	263.7	0.8234
80	0.001022	331.6	346.9	1.0659
100	0.001036	414.8	430.4	1.2958
120	0.001052	498.5	514.3	1.5148
140	0.001071	582.7	598.7	1.7243
160	0.001092	667.6	684.0	1.9259
180	0.001116	753.6	770.3	2.1206
200	0.001144	840.8	858.0	2.3100
220	0.001175	929.8	947.4	2.4951
240	0.001212	1021.0	1039.2	2.6774
260	0.001256	1115.1	1134.0	2.8586
280	0.001310	1213.4	1233.0	3.0409
300	0.001378	1317.6	1338.3	3.2279
320	0.001473	1431.9	1454.0	3.4263
340	0.001631	1567.9	1592.4	3.6555

P = 20 MPa

T(°C)	V(m³/kg)	U(kJ/kg)	H(kJ/kg)	S(kJ/kg-K)
0	0.000990	0.2	20.0	0.0005
20	0.000993	82.7	102.6	0.2921
40	0.000999	165.2	185.2	0.5646
60	0.001008	247.8	267.9	0.8208
80	0.001020	330.5	350.9	1.0627
100	0.001034	413.5	434.2	1.2920
120	0.001050	496.8	517.8	1.5105
140	0.001068	580.7	602.1	1.7194
160	0.001089	665.3	687.0	1.9203
180	0.001112	750.8	773.0	2.1143
200	0.001139	837.5	860.3	2.3027
220	0.001170	925.8	949.2	2.4867
240	0.001205	1016.1	1040.2	2.6676
260	0.001247	1109.0	1134.0	2.8469
280	0.001298	1205.5	1231.5	3.0265
300	0.001361	1307.1	1334.4	3.2091
320	0.001445	1416.6	1445.5	3.3996
340	0.001569	1540.2	1571.6	3.6086
360	0.001825	1703.6	1740.1	3.8787

P = 50 MPa

T(°C)	V(m³/kg)	U(kJ/kg)	H(kJ/kg)	S(kJ/kg-K)
0	0.000977	0.3	49.1	-0.0010
20	0.000980	80.9	130.0	0.2845
40	0.000987	161.9	211.3	0.5528
60	0.000996	243.1	292.9	0.8055
80	0.001007	324.4	374.8	1.0442
100	0.001020	405.9	456.9	1.2705
120	0.001035	487.7	539.4	1.4859
140	0.001052	569.8	622.4	1.6916
160	0.001070	652.3	705.8	1.8889
180	0.001091	735.5	790.1	2.0790
200	0.001115	819.4	875.2	2.2628
220	0.001141	904.4	961.4	2.4414
240	0.001171	990.6	1049.1	2.6156
260	0.001204	1078.2	1138.4	2.7864
280	0.001243	1167.7	1229.9	2.9547
300	0.001288	1259.6	1324.0	3.1218
320	0.001341	1354.3	1421.4	3.2888
340	0.001405	1452.9	1523.1	3.4575
360	0.001485	1556.5	1630.7	3.6301
380	0.001588	1667.1	1746.5	3.8101

P = 100.0 MPa

T(°C)	V(m³/kg)	U(kJ/kg)	H(kJ/kg)	S(kJ/kg-K)
0	0.000957	-0.3	95.4	-0.0085
20	0.000962	78.0	174.2	0.2699
40	0.000969	157.0	253.9	0.5328
60	0.000978	236.2	334.0	0.7809
80	0.000988	315.6	414.5	1.0153
100	0.001000	395.1	495.1	1.2375
120	0.001014	474.6	576.0	1.4487
140	0.001028	554.4	657.2	1.6501
160	0.001045	634.3	738.8	1.8429
180	0.001063	714.5	820.8	2.0280
200	0.001083	795.1	903.4	2.2064
220	0.001104	876.3	986.7	2.3788
240	0.001128	958.0	1070.8	2.5459
260	0.001154	1040.3	1155.8	2.7084
280	0.001183	1123.5	1241.8	2.8669
300	0.001215	1207.6	1329.1	3.0219
320	0.001250	1292.8	1417.8	3.1740
340	0.001290	1379.1	1508.2	3.3238
360	0.001335	1466.8	1600.3	3.4717
380	0.001385	1556.0	1694.5	3.6182

E.10 PRESSURE-ENTHALPY DIAGRAM FOR METHANE

(Source: NIST, Thermophysics Division, Boulder, CO, USA, used with permission.)

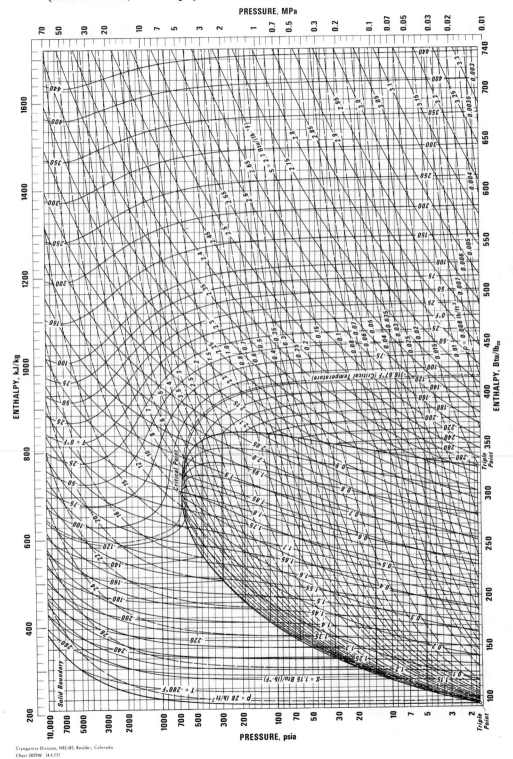

Cryogenics Division, NBS-IBS, Boulder, Colorado
Chart 2029W (4-1-77)

E.11   PRESSURE-ENTHALPY DIAGRAM FOR PROPANE

(Source: NIST, Thermophysics Division, Boulder, CO, USA, used with permission.)

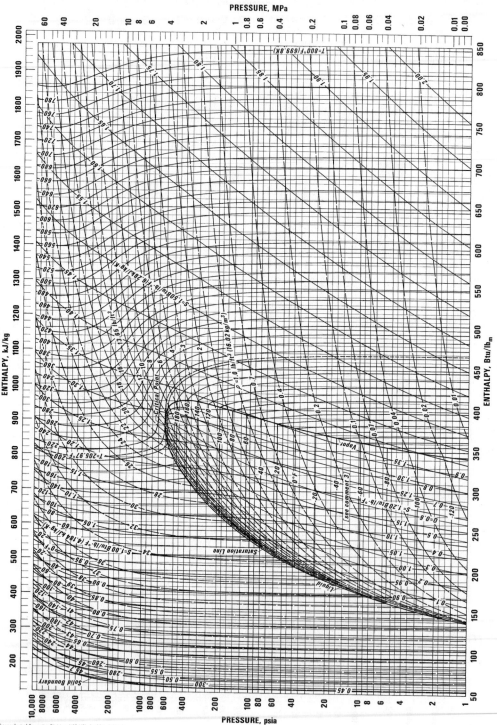

E.12 PRESSURE-ENTHALPY DIAGRAM FOR R134A (1,1,1,2-TETRAFLUOROETHANE)

(Source: NIST, Thermophysics Division, Boulder, CO, USA, used with permission.)

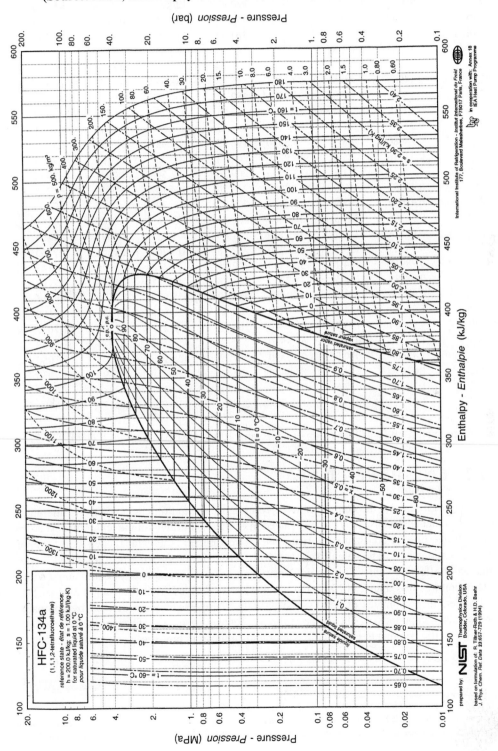

Properties of Saturated HFC-134a.

T	P	ρ^L	ρ^V	H^L	H^V	S^L	S^V
K	MPa	kg/m^3	kg/m^3	kJ/kg	kJ/kg	kJ/kg-K	kJ/kg-K
240	0.07248	1397.7	3.8367	156.78	378.33	0.8320	1.7552
244	0.08784	1385.8	4.5965	161.87	380.85	0.8530	1.7505
248	0.10568	1373.8	5.4707	166.99	383.35	0.8738	1.7462
252	0.12627	1361.7	6.4715	172.14	385.84	0.8943	1.7423
256	0.14989	1349.5	7.6117	177.33	388.31	0.9147	1.7388
260	0.17684	1337.0	8.9051	182.55	390.75	0.9348	1.7356
264	0.20742	1324.4	10.3660	187.81	393.17	0.9548	1.7327
268	0.24197	1311.6	12.0110	193.11	395.56	0.9747	1.7301
272	0.28080	1298.5	13.8570	198.45	397.93	0.9943	1.7277
276	0.32426	1285.3	15.9230	203.84	400.25	1.0139	1.7255
280	0.37271	1271.7	18.2270	209.26	402.54	1.0332	1.7235
284	0.42651	1258.0	20.7940	214.74	404.79	1.0525	1.7217
288	0.48603	1243.9	23.6450	220.27	406.99	1.0717	1.7200
292	0.55165	1229.5	26.8080	225.85	409.14	1.0907	1.7184
296	0.62378	1214.7	30.3130	231.49	411.23	1.1097	1.7169
300	0.70282	1199.6	34.1920	237.18	413.26	1.1286	1.7155
304	0.78918	1184.1	38.4830	242.95	415.22	1.1475	1.7142
308	0.88330	1168.1	43.2280	248.78	417.11	1.1663	1.7128
312	0.98560	1151.5	48.4750	254.69	418.92	1.1850	1.7114
316	1.09650	1134.5	54.2820	260.68	420.63	1.2038	1.7100
320	1.21660	1116.7	60.7140	266.76	422.25	1.2226	1.7085
324	1.34620	1098.3	67.8510	272.94	423.74	1.2414	1.7068
328	1.48600	1079.0	75.7890	279.23	425.10	1.2603	1.7050
332	1.63640	1058.8	84.6440	285.63	426.31	1.2793	1.7030
336	1.79810	1037.5	94.5630	292.18	427.34	1.2984	1.7007
340	1.97150	1015.0	105.7300	298.88	428.17	1.3177	1.6980

Abstracted from R. Tillner-Roth; H. D. Baehr, 1994. *J. Phys. Chem. Ref. Data,* 23:657.

E.13   END PAPER DATA

Themochemical data on the end paper were compiled from multiple sources over time and have become entrenched because they are referenced in examples throughout the text. The best estimates of these properties are constantly evolving and we recommend the values listed here for approximate calculations only. The default source for property data was Reid et al.[2] Italicized entries are estimated or effective values. For liquid density and solubility parameter of compounds with boiling temperatures less than 298.15 K, these values were derived from the compilation of Daubert and Danner.[3] For estimates of acidity and basicity, these values were derived by comparing to homologous compounds in the compilation of Lazzaroni et al.[4] The spreadsheet props.xlsx includes specific comments for specific compounds indicating how these values were estimated.

2. Reid, R.C., Prausnitz, J.M., Poling, B.E. 1987. *The Properties of Gases and Liquids, 4th ed.,* McGraw-Hill.

3. Daubert, T.E., Danner, R.P. 1989. *Physical and Thermodynamic Properties of Pure Chemicals:Data Compilation,* Hemisphere.

4. Lazzaroni, M.J., Bush, D., Eckert, C.A., Frank, T.C., Gupta, S., Olson, J.D. 2005. *Ind. Eng. Chem. Res.* 44:4075.

INDEX

PROPERTIES OF SELECTED COMPOUNDS

*Heat capacities are for **ideal gas at 298K** and should be used for **order of magnitude calculations** only. ρ is for liquid at 298.15K. See appendices for temperature-dependent formulas and constants.*

ID	Compound	T_c (K)	P_c (MPa)	ω	ρ g/cm³	MW	$C_P{}^{ig}/R$	δ (J/cm³)^½	α (J/cm³)^½	β (J/cm³)^½
Aliphatics										
1	METHANE	190.6	4.604	0.011	*0.29*	16	4.30	11.7	0	0
2	ETHANE	305.4	4.880	0.099	*0.43*	30	6.31	13.5	0	0
3	PROPANE	369.8	4.249	0.152	*0.58*	44	8.85	13.1	0	0
4	*n*-BUTANE	425.2	3.797	0.193	*0.60*	58	11.89	13.5	0	0
5	ISOBUTANE	408.1	3.648	0.177	*0.55*	58	11.70	12.5	0	0
7	*n*-PENTANE	469.7	3.369	0.249	0.62	72	14.45	14.3	0	0
8	ISOPENTANE	460.4	3.381	0.228	0.62	72	14.28	13.9	0	0
9	NEOPENTANE	433.8	3.199	0.196	0.60	72	14.62	13.1	0	0
11	*n*-HEXANE	507.4	3.012	0.305	0.66	86	17.21	14.9	0	0
17	*n*-HEPTANE	540.3	2.736	0.349	0.68	100	19.95	15.3	0	0
27	*n*-OCTANE	568.8	2.486	0.396	0.70	114	22.70	15.5	0	0
27	ISOOCTANE	544.0	2.570	0.303	0.70	114	22.50	14.1	0	0
46	*n*-NONANE	595.7	2.306	0.437	0.71	128	25.45	15.6	0	0
56	*n*-DECANE	618.5	2.123	0.484	0.73	142	28.22	15.7	0	0
64	*n*-DODECANE	658.2	1.824	0.575	0.75	170	33.71	15.9	0	0
66	*n*-TETRADECANE	696.9	1.438	0.570	0.76	198	39.22	16.1	0	0
68	*n*-HEXADECANE	720.6	1.419	0.747	0.77	226	44.54	16.2	0	0
Naphthenes										
104	CYCLOPENTANE	511.8	4.502	0.194	0.74	70	9.97	16.5	0	0
105	METHYLCYCLOPENTANE	532.8	3.785	0.230	0.74	84	13.21	16.1	0	0
137	CYCLOHEXANE	553.5	4.075	0.215	0.77	84	12.74	16.8	0	0
138	METHYLCYCLOHEXANE	572.2	3.471	0.235	0.77	98	16.25	16.1	0	0
153	DECALIN(cis)	703.6	3.200	0.279	0.89	138	20.04	17.6	0	0
Olefins and Acetylene										
201	ETHYLENE	282.4	5.032	0.085	*0.43*	28	5.26	13.5	0	0.40
202	PROPYLENE	364.8	4.613	0.142	*0.61*	42	7.69	13.2	0	0.40
207	1-BUTENE	419.6	4.020	0.187	*0.63*	56	10.31	13.7	0	0.40
204	ISOBUTENE	417.9	3.999	0.189	*0.59*	56	10.72	13.7	0	0.40
209	1-PENTENE	464.8	3.529	0.233	0.63	70	13.17	14.5	0	0.24
401	ACETYLENE	308.3	6.139	0.187	*0.50*	28	5.32	*18.68*	0.40	0.40
303	1,3-BUTADIENE	425.4	4.330	0.193	*0.65*	54	9.56	15.6	0	0.70
309	ISOPRENE	484	3.850	0.158	0.68	68	12.78	15.3	0	0.70
Aromatics										
501	BENZENE	562.2	4.898	0.211	0.87	78	9.82	18.7	0.63	2.24
502	TOLUENE	591.8	4.109	0.264	0.86	92	12.49	18.3	0.57	2.23
504	ETHYLBENZENE	617.2	3.609	0.304	0.86	106	15.44	18.0	0.23	1.83
505	*o*-XYLENE	630.4	3.734	0.313	0.88	106	16.03	18.4	0.10	1.80
506	*m*-XYLENE	617.1	3.541	0.326	0.86	106	15.35	18.1	0.19	1.84
507	*p*-XYLENE	616.3	3.511	0.326	0.86	106	15.26	17.9	0.27	1.87
510	CUMENE	631.2	3.209	0.338	0.86	121	18.25	17.4	0.20	2.57
558	BIPHENYL	789.3	3.847	0.366	0.99	154	19.52	19.3	0.50	4.00
563	DIPHENYLMETHANE	768	2.920	0.461	1.00	168	21.87	19.6	0.50	4.00
701	NAPHTHALENE	748.4	4.051	0.302	*0.98*	128	16.03	19.5	0.86	6.87
702	METHYLNAPHTHALENE	772	3.650	0.292	1.02	142	19.08	20.1	0.77	6.13
706	TETRALIN	720.2	3.300	0.286	0.97	132	18.63	19.3	0.60	4.82